XV International Conference "Topical Problems of Architecture, Civil Engineering, Energy Efficiency and Ecology – 2016"

MATEC Web of Conferences Volume 73 (2016)

Tyumen, Russia
27 - 29 April 2016

Part 1 of 2

Editors:

Tatiana Maltseva

ISBN: 978-1-5108-2871-1

Printed from e-media with permission by:

Curran Associates, Inc.
57 Morehouse Lane
Red Hook, NY 12571

Some format issues inherent in the e-media version may also appear in this print version.

Printed by Curran Associates, Inc. (2016)

For additional information, please contact EDP Sciences – Web of Conferences at the address below.

EDP Sciences – Web of Conferences
17, Avenue du Hoggar
Parc d'Activité de Courtabœuf
BP 112
F-91944 Les Ulis Cedex A
France

Phone: +33 (0) 1 69 18 75 75
Fax: +33 (0) 1 69 28 84 91

contact-edps@webofconferences.org

Additional copies of this publication are available from:

Curran Associates, Inc.
57 Morehouse Lane
Red Hook, NY 12571 USA
Phone: 845-758-0400
Fax: 845-758-2633
Email: curran@proceedings.com
Web: www.proceedings.com

TABLE OF CONTENTS

PART 1

PART 2

ISSN (electronic edition): 2261-236X

XV International Conference

«Topical Problems of Architecture, Civil engineering, Energy efficiency and Ecology - 2016»

TPACEE 2016

The TPACEE 2016 conference is organized on behalf of Tyumen state university of architecture and civil engineering, the leading centre of Western Siberia, Russia, involved into the development construction and civil engineering technologies.

TPACEE conferences are held annually and bring together the professionals working on the issues of energy-efficient construction, ecology and the Arctic development. The first conference was held in Tyumen in 2000. Since then, the venue remains traditionally unchanged.

Popularity for TPACEE events shows the need for regular special events devoted to energy efficient and environmentally friendly civil engineering solutions, including the construction in the Arctic and permafrost.

Date: April 27-29, 2016

Venue: Tyumen state university of architecture and civil engineering, Tyumen, Russia.

We would like to express our gratitude to Editorial Board of MATEC Web of Conferences and personally Isaline Augusto for your noble effort.

Conference Scientific Editor:
Tatyana Maltseva
Tyumen state university of architecture and civil engineering, Russia

Conference Chair:
Miroslav Premrov
University of Maribor, Slovenia

edp sciences

MATEC Web of Conferences

Statement of Peer review

In submitting conference proceedings to MATEC Web of Conferences, I certify to the Publisher that I adhere to the **Policy on Publishing Integrity** of the journal in order to safeguard good scientific practice in publishing.

1. All articles have been subjected to peer review administered by the proceedings editors.
2. Reviews have been conducted by expert referees, who have been requested to provide unbiased and constructive comments aimed, whenever possible, at improving the work.
3. Proceedings editors have taken all reasonable steps to ensure the quality of the materials they publish and their decision to accept or reject a paper for publication has been based only on the merits of the work and the relevance to the journal.

Title, date and place of the conference

XV International Conference
«Topical Problems of Architecture, Civil engineering, Energy efficiency and Ecology - 2016»
April 27-29, 2016

Tyumen State University of Architecture and Civil Engineering,
ul. Lunacharskogo, 2, Tyumen, 625001, Russia

Proceedings editor(s): Tatiana Maltseva

Date and editor's signature

Progressive Collapse of High-Rise Buildings from Fire

Valerii Pershakov[1], *Andrii* Bieliatynskyi[1], *Ivanna* Popovych[1,*], *Katerina* Lysnytska[1] and *Vladimir* Krasheninnikov[2]

[1]National Aviation University, 03058 Kosmonavta Komarova, 1, Kyiv, Ukraine
[2]Peter the Great St. Petersburg Polytechnic University, 195251 Politekhnicheskaya st., 29, Saint-Petersburg, Russia

Abstract. Considers ensuring the stability of structures of high-rise buildings against progressive collapse due to fire, proposed measures to ensure the stability of high-rise buildings due to progressive collapse. The analysis of large fires in high-rise buildings with progressive collapse and review of the literature on the issue of progressive collapse. The analysis of the Ukrainian normative documents on progressive collapse resistance.

1 Introduction

The worldwide and also Ukraine problem of ensuring the stability of structures of high-rise buildings against progressive collapse as a result of fire is becoming more urgent because, leads to very serious consequences. Wear and tear of fixed assets of the country, increasing the rate and density of construction in urban areas, an increase in recent years, the number of terrorist acts (bombings, arson, etc.) - all this creates the potential value of progressive collapse.

Since the world's terrorist attack in the United States on September 11 examines projects for the construction of high-rise buildings are able to withstand explosions and fires. The questions of limiting the height of high-rise buildings, increasing the number of routes for evacuation, tightening security zone requirements. Advances maintenance of elevators performance requirements during a time of emergency situations, including fires.

2 Literature review

Designing of residential buildings is carried out taking into account the factors that caused emergencies, including fire. The structural system of residential buildings must be designed so as to ensure its overall stability during emergency Norm local destructive loads on the individual load-bearing structures, at least for the time required for evacuation of people (the explosions of various types, fires, falling heavy objects, impact of heavy vehicles, etc.) [1 - 3].

* Corresponding author : popovich_ivanna@mail.ru

As a result of fires in high-rise buildings with the appropriate conclusions, and the reaction of the authorities of different countries, their effects have been amended in the regulations [1-6], due to the fact that the fires in high-rise buildings, more traumatic and lead to more damage than normal fires buildings. A fire in a building higher than 25 floors, is 3-4 times more victims than the 9-16-storey building [7 - 9].

So there are special requirements for fire resistance of structures, elevators for firefighters to escape routes to the backup power supply [10 - 27]. To estimate the necessary measures to limit the spread of fire through the building should be a closer look at the fires that have influenced the formation of the regulatory requirements for fire resistance of structures and putting out fires in high rise buildings.

For example, a fire in the 62-storey building in Los Angeles in 1988, despite the fact that virtually the entire area of the building was equipped with sprinklers, the water in case of fire has been given. Only through the optimum fire protection of load-bearing elements, steel skyscraper construction withstood a three-hour exposure to a flame. On fire 64 firefighters were involved calculation, which is half of the city fire forces [28].

Fire-fighting system has not worked well, and a fire in 2004 in the 56-storey skyscraper in Caracas government. The building was completely equipped with sprinkler, but it was defective. 2 hours after the start of the fire, the fire spread to the roof, covering the floors with the upper 34 to 56 floors, and lasted more than 17 hours [29].

A fire in a 37-storey high-rise building in Chicago in 2003. The fire started on the 12th floor. The building was not protected by sprinkler system and fire spread beyond the room, because the walls of the corridor were not implemented on the entire height of the floor. Hot flue gases also got into the ventilation system of the corridor. As the doors of floors in the stairwell was in order to protect the security locked from the stairs, the people who came into the stairwell, they could not get away with it. As a result, six people were killed there, locked up 12 floors. The subsequent fire analysis showed that if the doors are not blocked by the stairs, people would not have died. Also, if the building was equipped with protective materials, flue gases would not be so hot and would not receive such an extension [29].

In Odessa, in 2015 he lit paneling on the upper floors of the unfinished buildings of the complex "Arcadia". the fire quickly spread to the first from the upper floors. Burned paneling. The fire was extinguished by two tall ladders 30 and 50 meters. But the problem was that the sleeves did not have enough pressure, and does not provide internal fire extinguishing system in the building (Fig. 1a). [30]. Another complicating factor in extinguishing fires in high-rise buildings, and not only in them, is not sufficiently fast arrival of fire units. This is not always due to the late arrival of the transport infrastructure problems.

In case of fire in the 106-meter 32-storey "Vidzor Building" office building in Madrid in 2005, the staff tried to put out the fire yourself, causing firefighters arrived at the scene just over two hours. The fire spread from the 1 st to the 32 th floor.

The building was provided by the central reinforced concrete core and steel outer frame, which is a result of a fire struck the 6 upper floors (Fig. 1b), subsequently decided to demolish the building was [31]. This fire was the main discussion example destruction of bearing structures of the building up to the tragedy of the World Trade Center (WTC, World Trade Centre) September 11, 2001 in New York, the United States, which killed 2,451 people, and damage amounted to $ 33.4 billion [32 - 37].

The collapse of the WTC buildings - the first case of complete destruction of the high-rise building. The report National Institute of Standards and Technology (USA) [39] according to the analysis emphasized the importance of adequate fire protection. The main idea of the recommendations was binding regulatory requirements for the design basis threat. The recommendations, which were presented at the NIST report, are reflected in the annex to the US national standards.

After the tragedy of September 11, many standards have been revised on the basis of this report. As previously occurred fires in high-rise buildings do not lead to the complete destruction of the buildings, some of the rules in different countries allow the evacuation plan with only the fire floor and adjacent floors above and below. Who is considered the evacuation, including, if possible, the whole building [39].

Fire Spread height of the building, as opposed to the spread of fire in the horizontal direction due to the need to overcome the fire building slabs, each of which is fire-resistant. Thus in the ceiling, in contrast to the walls and partitions, no doorways and windows. Protecting designs of staircases, lift shafts and ducts crossing the building vertically, also performed fireproof. It is clear that the division of the building into fire compartments in the vertical is in this case purely nominal.

The spread of fire on high-rise buildings with the most serious consequences, as a rule, takes place on the facades. A typical example is a fire in a government 32-storey skyscraper "Transport Tower" in the Kazakh capital Astana, 30 May 2006 (Figure 1, a)). Flames destroyed facing about 15 floors on one side of the building and all the 32 floor to another. The fire was accompanied by the dispersion of shards of glass façade [40].

a)  b) 

Fig. 1. a) Fire in skyscraper «Transport-Tower», Astana; b) Hotel Mandarin Oriental after fire, Beijing

In case of fire in the 159-meter 44-story Mandarin Oriental hotel building in Beijing February 9, 2009 (Figure 1, b)), the fire quickly spread down the facade. The fire lasted for about 5 hours. During this time, approximately 80% of building burnt. In fire fighting involved about 600 firefighters. Due to the severe fire in the city fell ashy snow, which happens on volcanic eruption. The fire is called the fireworks, the remains of which firefighters discovered on the roof of the building. According to various estimates prior to the fire, 65 people were - the hotel staff. 5-star hotel with 241 rooms did not have time to put into operation. The total investment in its construction amounted to $ 700 million [41].

 DOI: 10.1051/matecconf/20167301001

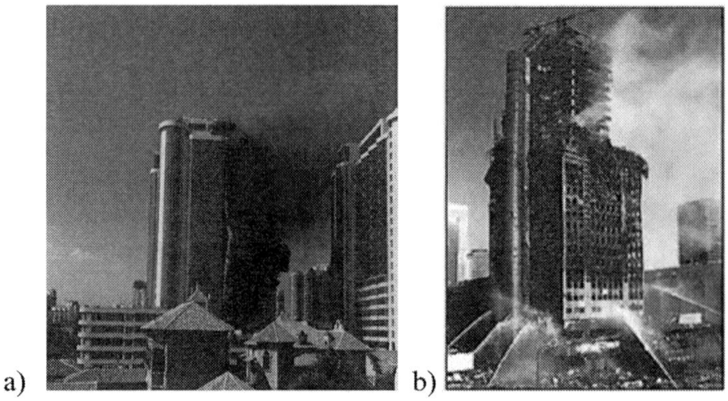

Fig. 2. a) Burning high-rise 24-storey building in Odessa; b) The fire in the 32-storey skyscraper, Madrid

In conventional buildings, in the absence of horizontal barriers in the facade, the flames from the window opening 15-20 minutes from the start of a fire in a room can spread up the balconies, loggias, window frames, igniting flammable building elements and furnishings in floor space located above .

In high-rise buildings are widely used ventilated facade systems. In some types of facade used combustible materials and as a supporting framework used aluminum profiles, which considerably increases the class of constructive fire danger buildings. The use of easily flammable insulation (thermal insulation) can lead to the rapid spread of the fire and the formation of highly toxic combustion products, as evidenced by the above example. The use of such materials should be limited to the development of high-rise buildings requirements.

There are two principles of valuation fire compartments: they allowed the area and functional characteristics [5, 6, 8, 9, 20, 22, 31, 33 - 37].

However, in the fire compartment can be placed several functional processes, different in purpose and fire hazard. Fire resistance is an international fire-technical characteristic, which regulated building codes and regulations, and describes the ability of structures to resist the action of fire. Measuring fire resistance parameters standardized international norms and standards in different countries, based on the standard ISO 834-11:2014 [42] which is designed to cover a range of thicknesses of the fire protection material, a range of steel sections characterized by their section factors, a range of design temperatures, and a range of valid fire resistance classification periods. In this regard, the fire resistance is a basic element of the whole system of fire protection of buildings, as is the decisive parameter for selecting other security features.

3 Discussion

Since the phenomenon of progressive collapse is probabilistic in nature, completely eliminate the risk of its occurrence is impossible, but with prior calculation and circuit simulation collapse due to fire damage can be minimized in the local destruction and to provide the necessary time to evacuate people in the building.

Resistance to progressive collapse due to fire means that in case of fire in high-rise buildings allowed local caving individual vertical bearing elements within a floor or areas overlap one floor, but the initial collapse should not cause the destruction of structures that are transmitted load before seen elements damaged emergency impact [1].

As a local collapse, understand:

- vertical structures one floor house a limited area of around 80 m² (diameter 10 m):

- two walls intersect, in areas of their place of intersection of the nearest hole in each wall or to the next vertical cross section of the wall the other direction;

- separate columns from the surrounding areas of walls located on one floor in the area of local caving;

- areas overlap one floor to collapse the local area.

In all cases, the total cross-sectional area remote vertical elements, located on a plot of 80 m^2, should not exceed: for concrete elements 0,9 m^2 for fiber concretes elements 0,7 m^2 for tough fixtures 15% [1].

The most dangerous fire to the upper floors, water pump not enough power, not enough fire ladder length.

Multifunctional buildings height over 16 floors should have a special degree of fire resistance. According to the fire resistance of buildings over 16 floors height, special requirements towards their increase [2], for example:

- constructions of these buildings should be made of non-combustible materials.

- minimum fire resistance of structures multipurpose buildings increased storey with a special degree of fire resistance shall be not less than: load-bearing walls - REI 180 fire walls - REI 180 columns - R180, wall stairwells - REI 180 items beams (beams, girders, frames, trusses) - R 180 180 -REI fire floor, elevator shafts envelope - REI 90 fire pits envelope elevators - REI 120 communication envelope mines - REI 60, etc.

The average fire load of these buildings do not exceed 50 kg / m2 (in terms of wood).

For high buildings over 100 meters above limit fire main structures is recommended to increase by 1 hour.

This requirement seems excessive, leading to a rise in construction, difficulties in the design of ultra-high buildings.

Effectively the requirement to use during building construction designs that with standard fire resistance testing should resist temperatures up to 1100 ° C for 4 hours. This corresponds to the fire load in the room in terms of wood, 150 kg/m2, exceeding actual and permissible fire load value in areas of tall buildings in 2 times. For example, two 110-storey office buildings of the World Trade Center in New York City fire load in the office during the events of September 11, 2001 was 40 kg / m2 equivalent timber and according to regulations of Ukraine, allowed fire loads in areas of tall buildings do not exceed 50 kg/m2. Because of this, it is possible to limit norms for maximum fire resistance of structures tall buildings the size of 180 minutes (3 hours).

The special nature of the fire danger high-rise buildings is determined by:

- presence of conditions conducive to the emergence of a fire;

- possibility occupancy in the building;

- height of the building, exceeding the possibilities of using to save people mechanical stairs that are in the fire service garrisons;

- possibility of partial or complete destruction by fire of individual elements of the building, some parts of the building or the entire building;

- widely adopted in high-rise building flame, smoke, toxic substances on the premises, corridors and technical communications, and through gaps in building constructions;

- blocking lifts and failure of control elevators;

- lack or insufficiency of resources and time to rescue people in the middle of the building;

- lack of clear rules in the regulations on the assessment of fire danger analyzed objects.

Fire protection systems height of buildings above 16 floors and includes 15 security features that are intended can be grouped into the following components [2]:

Block 1. Measures to ensure the sustainability of buildings or parts thereof against progressive collapse:

• providing fire resistance of structures and buildings;

• ensuring explosion building, which can be explosive premises.

It should be noted the particular importance of this unit measures of fire protection system for high-rise buildings, as it provides a "primary safety" object in the form of reserve stability of the system, its ability to resist over time not only the impact of the fire, but others, including combined influences. If the object has insufficient stability of the entire system to ensure the safety of people and buildings in general becomes useless.

At high temperatures that occur during fires reduced strength overlap and they can collapse. The collapse could happen after the fire, because after lowering the temperature toughness ceiling is not restored. Cover also collapses due to accumulation of water on them, which poured fire.

So, for backup strength and stability of buildings against progressive collapse, its constructive scheme should include reserve items. These elements can be diaphragm stiffness, columns, autryherni floors, which in case of fire can perform additional functions redistribution of loads.

High buildings should be calculated using software systems that allow you to take into account the physical and geometric nonlinearity follows [7]:

1. A calculation of all circuits in physically nonlinear statement on permanent and temporary load that are emergency communications. The resulting stress strain state is starting to calculate the load of elements removed.

2. Calculation of the additional load of items that are removed conducted in physically and geometrically nonlinear statement. Loads of removing items correspond to the obtained them at the preliminary stage of calculation and increased by a factor of 1.2 dynamism. Test of strength elements remaining performed excluding buckling. The calculation also need to perform physically and geometrically nonlinear statement. If it appears that some elements of the model are not satisfying strength (destroyed), the calculation goes the same way for the next step without such elements.

3. Every floor high-rise buildings should be calculated on the weight of perception overlap areas top floor (permanent and long loads with dynamic coefficient = 1.5) over an area of 80 m2.

4. The calculation will be completed or localized progressive damage or complete destruction of the settlement scheme.

This simulation of force majeure can detect additional, backup, load-bearing capabilities are primarily associated with the effect of adaptability. For example, floor-designed for large troughs starts as membrane and objectives Designer ensure such work [3].

Block 2. Measures to limit the spread of fires in high-rise buildings:

• unit fire barriers in the middle of the building;

• placement of fire breaks between buildings.

Block 3. Measures to ensure timely and unimpeded evacuation and rescue them at HB in high-rise buildings. The main provisions of the regulations on the safety of people at fires in buildings are the requirements provided in case of fire:

• the possibility of evacuation of people, regardless of age and physical condition before the threat to their life and health;

• access personnel fire departments and fire fighting to feed the fire, and interventions to save lives and property.

The solution to this problem in the design of new and renovated buildings implemented the following measures:

• measures for timely evacuation and protection against smoky;

• fire alarm systems and evacuation management;

• measures to save lives and limit the fire risk of materials, structures and buildings.

Block 4. Active Protection System buildings from fire:

• fire alarm system and fire fighting;

• stronghold of the fire;
• Central device remote control system of fire protection building.

4 Conclusions

Thus, conclude that resistance to progressive collapse in fires serves:

- margin basic building structures that should provide most profitable way: rational design and planning solutions building, structural measures to ensure the integrity of the construction, use of materials and design solutions that ensure development in structural elements and their compounds plastic deformation, designing technical floors in a spatial system, etc.
- use of 4 units of fire protection systems.
- preservation of fire resistance of the main load-bearing structures with the free development of fire;
- exception progressive collapse with loss of fire resistance of individual bearing structures.
- division into fire compartments tall building height does not solve the problem of limiting the spread of fire and height of the building can be replaced by a zoning technical systems. Efforts should be focused on limiting the spread of fire on the facade.

References

1. DBN V.2.2-15-2005
2. DBN V.1.1.-7-2002
3. DBN V.1.2-2: 2006
4. NFPA 5000, Building Construction and Safety Code
5. International Building Code 2009, International Code Council, Inc., USA (2009)
6. PD7974-6: The application of fire safety engineering principles to fire safety design of buildings, Human factors: life safety strategies—occupant evacuation, behaviour and condition (British Standards Institution, London, 2004)
7. V.M. Pershakov, M.S. Barabash, A.O. Bielyatynskyi, K.M. Lysnytska, *Problems countering progressive collapse of structures of buildings* (NAU, Kyiv, 2015)
8. B. Stafford Smith, A. Coull, *Tall building structures: analysis and design* (John Wiley & Sons. Inc., USA, 1991)
9. U.S. Fire Administration, Highrise Fires, Topical fire research series, **2(18)** (2002).
10. C. H. Chen, Y. F. Zhu, Y. Yao, Y. Huang, X. Long, Journal of Constructional Steel Research. **122**, 238-250 (2016)
11. B.A. Izzuddin, A.G. Vlassis, A.Y. Elghazouli, D.A. Nethercot, Engineering Structures, **30(5)**, 1308–1318 (2008)
12. A.G.Vlassis, B.A.Izzudin, A.Y.Elghazouli, D.A.Nethercot, Engineering Structures, **30(5)**, 1424–1438 (2008)
13. M.A Qian-li, Huang Ting-lin, Procedia Engineering, **11**, 302–307 (2011)
14. P. Kolarkar, M. Mahendran, Fire Saf. J., **53**, 85–104 (2012)
15. M. Lazarevska, M., Cvetkovska, M., Knezevic, V. Murgul, N. Vatin, Applied Mechanics and Materials, **627**, 276-282 (2014)
16. N. Vatin, A. Sinelnikov, M. Garifullin, D. Trubina, AMM, **1037–1041**, 633–634 (2014)
17. G. Craighead, Butterworth-Heinemann, **3**, 39-43 (2009)
18. M. Jocovic, B. Melovic, N. Vatin, V. Murgul, Applied Mechanics and Materials, **678**, 644-647 (2014)

19. L. Huanga, T. Chen, H. Yuan, Transportation Research Procedia, **2**, 518 – 523 (2014)
20. V.A. Macovei, Journal Emergencies. Industrial and environmental safety, **12**, 28-36 (2015)
21. J. Jianga , L. Chenb, S. Jiangc , G.-Q. Lic, Case Studies in Fire Safety, **4**, 28–38 (2015)
22. Ahmad, S.A. Hassan, A. Ripin, M. W. Ali, S. Haron, Fire Safety Journal, **58**, 160–169 (2013)
23. O. A. Mohamed, Structural Engineering, **3**, 11-18 (2015)
24. S. Elkoly, B.E. Ariss, Engineering Failure Analysis, **40**, 33-47 (2014)
25. J. Li, H. Hao, Engineering Structures, **52**, 101-113 (2013)
26. K. Mela, T. Tiainen, M. Heinisuo, Advanced Engineering Informatics, **26(4)**, 716-726 (2012)
27. M. Salminen, M. Heinisuo, Journal of Constructional Steel Research, 97, 105-113 (2014)
28. R. W. Archibald, J. J. Medby, B. Rosen, J. Schachter, *Security and Safety in Los Angeles High-Rise Buildings After 9/11*, (RAND, USA, 2002)
29. Information on: http://911research.wtc7.net/wtc/analysis/compare/fires.html
30. Information on: http://odpublic.net/news/2015/08/29/v-arkadii-gorit-vysotka/
31. R. Pope, New Steel Construction, **14(3)**, 26–8 (2006)
32. G. Roger Morse, Fire Engineering, **10**, 45-53 (2002)
33. V.M. Roitman, Global Security, **9**, 30-35 (2006)
34. M. Gravit, A. Vaititckii, M. Imasheva, D. Nigmatullina, A. Shpakova, Classification of fire-technical characteristic of roofing materials in European and Russian regulation documents, MATEC Web of Conferences **53**, 01031 (2016)
35. A. Krivtcov, M. Gravit, S. Zimin, O. Nedryshkin, V. Pershakov, Calculation of limits of fire resistance for structures with fire retardant coating, MATEC Web of Conferences **53**, 01032 (2016)
36. L. Svatovskaya, A. Sychova, M. Sychov, V. Okrepilov, MATEC Web of Conferences, **53**, Article Number 01023 (2016)
37. M. Gravit, V. Gumenyuk, M. Sychov, O. Nedryshkin, Estimation of the pores dimensions of intumescent coatings for increase the fire resistance of building structures, Procedia Engineering, **117**, 119 – 125 (2015)
38. N.C. McConnell, K.E. Boyce, J. Shields, E.R. Galea, R.C. Day, L.M. Hulse, Fire Safety Journal. **45**, 1-34 (2010)
39. *Federal Building and Fire Safety Investigation of the Word Trade Center Disaster: Final Report of the National Construction Safety Team on the Collapses of the World Trade Center Towers*. NIST NCSTAR 1, (National Institute of Standards and Technology, Gaithersburg, MD, 2005)
40. Information on: http://www.newsru.com/world/30may2006/tower.html
41. Information on: http://911research.wtc7.net/wtc/analysis/compare/fires.html
42. ISO 834-11:2014. Fire resistance tests -- Elements of building construction - Part 11: Specific requirements for the assessment of fire protection to structural steel elements.

MATEC Web of Conferences **73**, 01002 (2016) DOI: 10.1051/matecconf/20167301002
TPACEE-2016

Distributed System of Dams for Flood Protection of Urban Land to Ensure Sustainable Development

Mikhail Fedorov[1], *Vladimir* Badenko[1,*], *Vladimir* Maslikov[1] and *Alexander* Chusov[1]

[1]St. Petersburg State Polytechnical University, 195251, Politekhnicheskaya, 29, Saint-Petersburg, Russia

Abstract. The article considers innovative materials and structures for multipurpose flood protection self-regulated dams. Russian and foreign experiences in flood consequence, as well as conceptions for protection of urban land in the context of climate change are analyzed. Method for flood protection is proposed and is proved by creation of distributed system of self-regulated flood-control distributed system of dams in a river basin.

1 Introduction

Flood protection is one of the main development problems of modern society, related to environment. In conditions of evidencing trend to global climate change the goal becomes even more important, as well as a concern rate of stream flow.

From the economic point of view flooding was always considered a major natural disaster. Observed climate change was led to sharp increase frequency of heavy flooding across the World, which is accompanied by increasing of economic losses and a number of injured persons. Rates of growth of economic losses are comparable with increasing of gross output across the globe. Showery flooding causes particularly drastic consequences. Primary hazard consists in their abruptness and sharp increasing amount of stream runoff, that is in a great deal exceed volume river flow of spring tide.

In last decades there were several heavy flooding in Europe which connected with long-duration of heavy rain. From the period 1950 to 2006, 12 flooding took place with large human losses. Frequently flooding covered entire regions, thus, 2002 was recorded by number of countries, which territories were simultaneously flooded (Austria, Czech Republic, France, Germany, Hungary, Rumania). Moreover, flooding, that covered territories of several countries, happened in other periods: Netherlands, Belgium, France and Germany (1993 and 1995), Czech Republic, Poland and Germany (1997), France, Italy and Poland (2010), Germany and Italy (2012). Approximately one fifth part of European cities with population more than 100 000 citizens per city are exposed to flash flood, generated by extremes precipitations [1-6].

* Corresponding author: vbadenko@gmail.com

The issues of rain floods are permanent exist in Russia, and in consequence of climate change it is expected to further aggravation [7]. Extreme flood, generated by heavy rain, were observed in the north Caucasus, at the coast of the Black sea, in the Russian Far East, etc. In the regions with monsoon climate, that is typical for Russian's Far East, flash rain flood covers great territories. Flood event in 2013, in consequence of heavy rains, continued approximately about two months inundated territories of the Far East and northern-east China [8]. Going forward in the middle of the 21 Century precipitation extremeness in summer period in mountainous area of the Caucasus, in Siberia and in the Far East is able to increase, therefore frequency and value of rainfall flood will grow up [9].

Nowadays, flood protection of territory is carried out according risk-based flood management as a system approach that assesses and compares the structural and non-structural (management and administration) ways to pursue the optimal ameliorating effects [4, 10]. Common flood protection measures are orientated on existing situation and does not take into account its uncertainty in the nearest future and further such as urbanization development, climate change and land utilization (deforestation, wetland reduction), etc. Thus, cost-expensive actions occur inefficient. In this respect a conception of flood mitigation is changing from protective measures to risk management. In particular, in EU a Directive was adopted [11] by flood risk management, which connected with European Water Framework Directive on management of river basin water sources [12]. Flood risk management in dynamical conditions of climate change is started with river basin spatial planning of water, agricultural and forestry, settlement, etc., and then assessment of flood risk area and corresponding damages is made. Based on the evidence found suitable protection systems from flooding in river basins are developed.

In Russia, as in other countries, regulation of maximum runoff by great reservoirs [13] is widely used and effective method for flood mitigation. Corresponding dams are located on main stream, and can protect areas located downstream. The objective of this paper is to develop a method for design of distributed system of self-regulated dams for flood protection of urban land to ensure sustainable development.

2 Materials and methods

In recent times, there is a tendency to build on the mainstream low-pressure multipurpose dam with reservoir (DWR) (with minimal operational level and small flood control capacity) for the purpose of reduction land inundation. In conditions decreasing of adjustability extreme of water flow, additional required flood-prevention reservoir is possible to place on local inflows. In this case large areas can be protected in flood plain, upstream and local inflows, which percent can put together up to 50 % on the part of total inundated area in the river basin, due to spatial distribution of flood protection regulating reservoir. This approach accords with modern strategy of flood risk decreasing by management of flood discharge for all river basins by distributed system of DWRs with minimal impact on the environment [14, 15].

For specific river basins zoning of economic development of territories and flooding areas are carried out, allocation scheme of flood-control DWRs is developed, whose quantity, type and parameters are chosen depending on physical and climatic, social and economic, environmental conditions, requirements of environmental protection, etc. For this purpose there is necessity to development of mathematical models, describing natural and economic situation, operating mode of flood-control DWRs, providing a flood risk assessment and planning activities on risk reduction.

For example, a system of non-energy flood-control DWRs is created in a river basin, locating in riverhead of a base stream and local inflows (fig. 1).

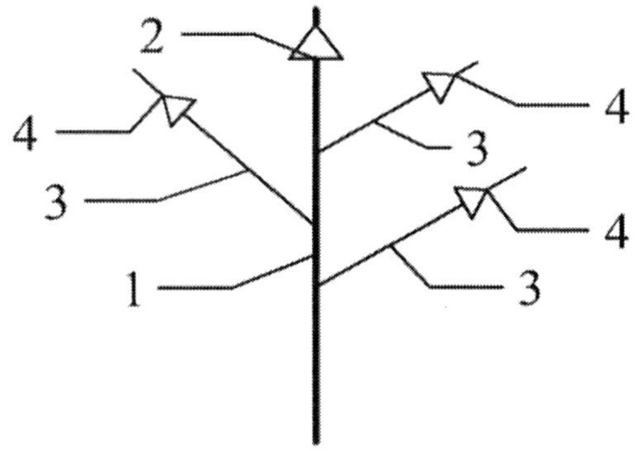

Fig. 1. The flood control waterworks in a river basin. 1 - the main watercourse, 2 - reservoir in the upper reaches of the river, 3 - side confluents, 4 - flood-control waterworks.

Previously executed researches showed reasonability of creating reservoirs on local inflows by way of temporary filling self-regulatory storages [16]. A DWR culvert is performed with uncontrolled spillway without the use of flow regulating valve, which increase reliability of operation and reduce the cost of maintenance. During accumulation of flood discharge this type of reservoir bed is flooded in a short-time, and from its self-discharge to the next flood keeps dry in natural state.

Flood-control DWR can be performed in form of combination structure of regulating culvert. In this DWR the passing discharge is carried out on two stages: downstream is through bottom discharge, which works in free-flow mode, and upstream is through overflow spillway (Fig. 2).

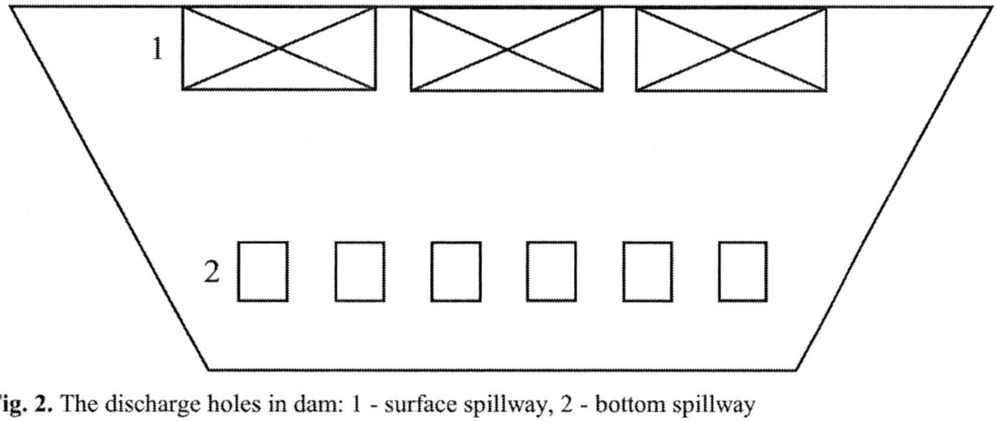

Fig. 2. The discharge holes in dam: 1 - surface spillway, 2 - bottom spillway

For selected scheme of flood management on watershed algorithms were developed and mathematical models of operational mode of flood-control DWRs in local inflows were implemented in software, which allowing to vary by parameters of discharge structures for determination of regulated flow in tail-water, watermarks in headrace, water volume at accumulation of flood control discharge, areas and duration of flooding lands [17-19].

The calculation of flood discharge volume, accumulated in reservoir at the moment t, is carried out by balance method:

$$V(t)=V(t-\Delta t) + [Q1(t) - [Q2(t) + Q3(t) + Q4(t)]]\cdot\Delta t, \qquad (1)$$

where: V(t) – volume of water in a reservoir at time t, m3; Q1(t) – water flow entering a reservoir, m3/s; Q2(t) – water discharge to a downstream, m3/s; Q3(t) – loss of water by evaporation from surface of a reservoir, m3/s; Q4(t) – seepage water discharge, m3/s; t – time, s. Δt – time step,s.

For DWRs in local inflows is provided a realization of main environmental requirements. Preservation of flood inundation processes is dammed escapade discharge in tail-water of DWR that should comply of flood probability of 10 %; conservation of nature ecosystems by their temporary flooding is the maximum water level in headrace of DWR does not exceed of acceptable level.

3 Results and discussion

As an example an operational mode of flood-control DWR was considered [15]. The dam with 10 bottoms and 3 open spillways was analyzed in one of local inflow of river basin of the river Selemdzha in the Far East of Russia, where the most frequently catastrophic floods in Russia are usually registered.

In figure 3 calculation results of water discharge in tail-water DWR is shown. Peak water discharge 1 % of probability is decreased approximately on one third, that is reach to rate of probability discharge 10 %, which is allowed to preserve natural flood and inundation processes, but at the time, avoided catastrophic flooding.

These results were obtained with special software developed. The food hydrograph analized is one of typically observed in this point on the river.

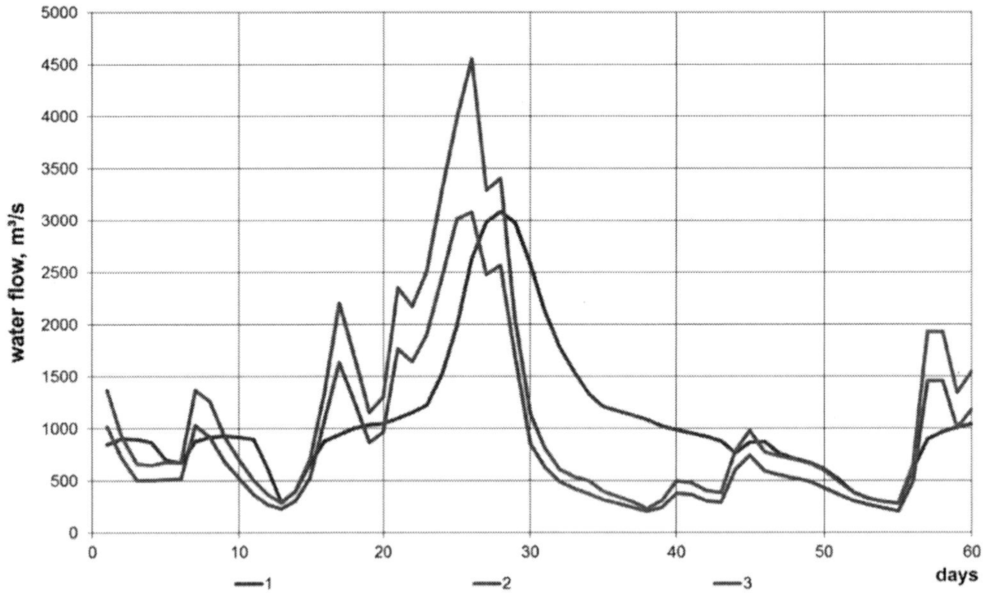

Fig. 3. Changes of water flow in the downstream of DWS
1 – the regulated discharge to the downstream, 2 – the natural flow with 1% probability,
3 – the natural flow with 10% probability.

In Figure 4 calculated changes of area and time of flooding of landscapes in the river basin obtained using software developed are shown.

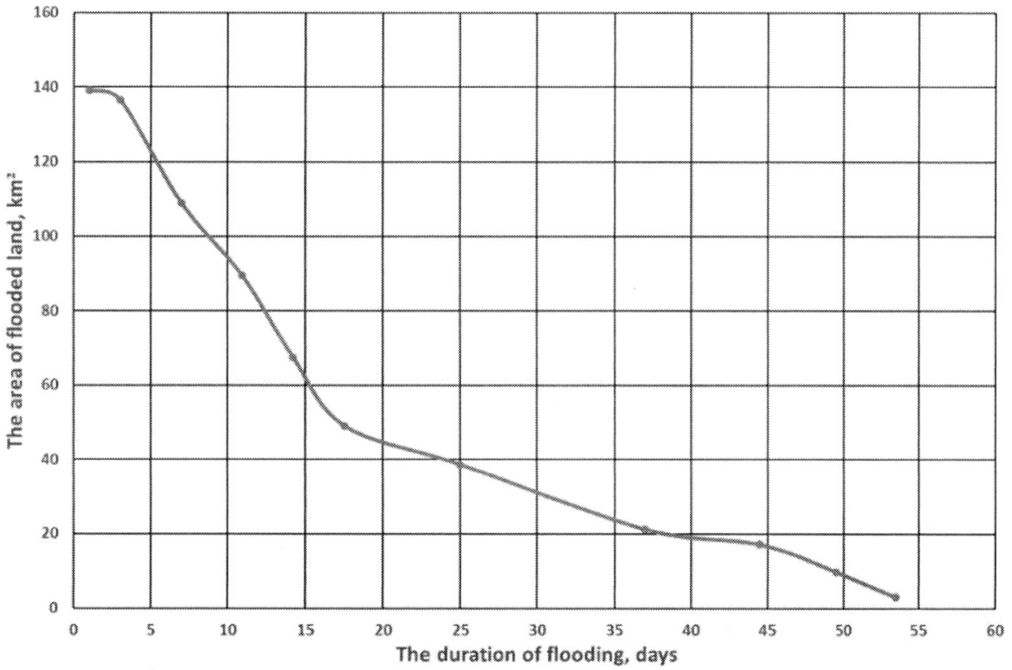

Fig. 4. The area and duration of land flooding

Conservation assessment of ecosystem biodiversity was made on temporary flooded lands with use of non-decrease criteria of biodiversity, as well as relation natural and man-made ecosystems [20]. It is shown that DWR is corresponded to requirements of environmental safety. It should be noted, that areas of flooding in this example was defined based on paper maps.

At the present time, methodologies for modelling with use of GIS are developed [21-24]. In Figure 5 a part of satellite image of river basin in cross-section of flood-control DWR and flood hazard area corresponding to water plane of flood passing with different probability, designed with use of digital elevation model in GIS environment, is shown.

Using GIS-modeling allows considering variants of regulation for flood discharges using different DWR parameters with detail analysis of potential flooding areas in upstream and downstream of DWRs taking into account both social-ecological-economical aspects and climate changes. This approach can solve problems with population growth, urbanisation and climate change which represent significant pressures on sustainable urban development, requiring flood events managers to consider a wider array of management options that account for economic, social and environmental factors using modeling in GIS environment [25-31].

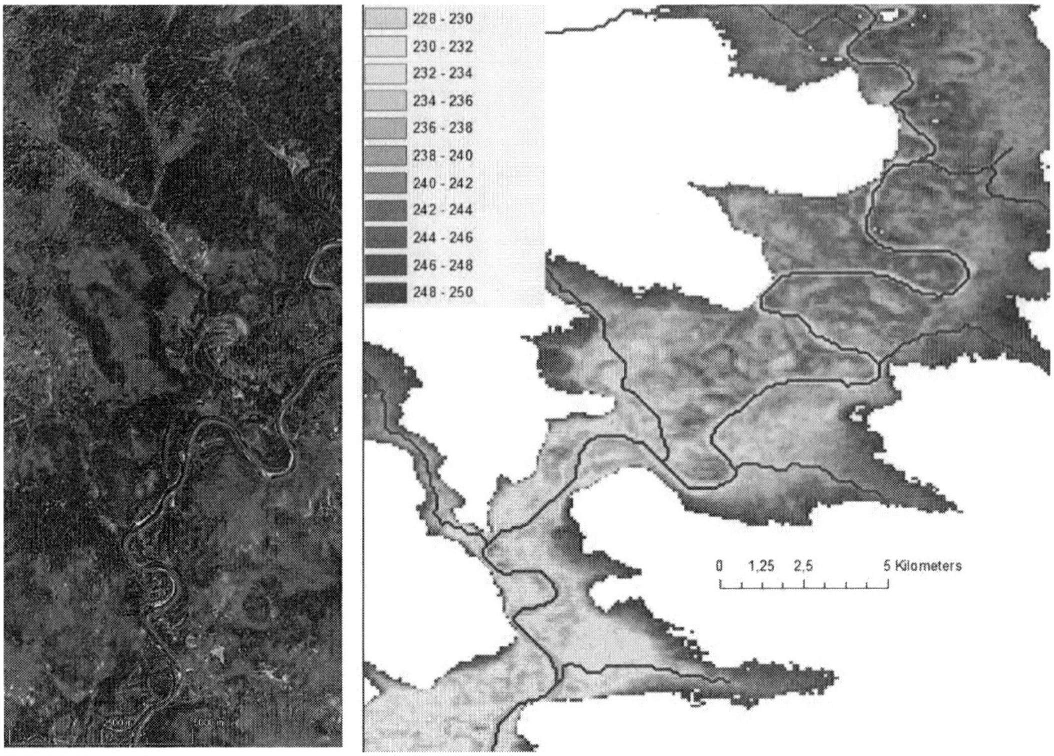

Fig. 5. The basin for research using GIS technologies. On the left: remote sensing data. On the right: digital elevation model (in the legend - elevations in meters).

4 Conclusions

Management of flood discharge of a river basin with distributed DWR with minimal impact on the environment was explained. Mathematical models for operation modes of flood protection DWR on local inflow with unregulated culvert in according with requirements of environmental safety were designed and were implemented.

Operation modes of flood protection DWSs were defined at regulation of peak flows, providing protection of natural flood-inundable processes, as well as biodiversity on temporary flooded areas. Using GIS-technologies the flooding areas were defined in according with water plane by passing of flood of different probability. Detention reservoirs are structural measures for the defense from flash flood events. Proposed approach can solve the problem of flood protection of urban land to ensure sustainable development.

Acknowledgements

This work is supported by the Russian Science Foundation under grant № 16-17-00050

References

1. S. Le Roy, R. Pedreros, C. André, F. Paris, S. Lecacheux, F. Marche, C. Vinchon, Natural Hazards and Earth System Sciences, **15**, 2497-2510 (2015)
2. N.W. Arnell, S.N. Gosling, Climatic Change, **134**, 387-401 (2016)
3. J. Hall, et al, Hydrology and Earth System Sciences, **18**, 2735-2772 (2014)

4. J. Ran, Z. Nedovic-Budic, Environment and Urban Systems, **57**, 68–79 (2016)
5. A.H. Thieken, H. Kreibich, M. Müller, B. Merz, Hydrological Sciences Journal, **52**, 1016-1037 (2007)
6. Z.W. Kundzewicz, N. Lugeri, R. Dankers, Y. Hirabayashi, P. Döll, I. Pińskwar, T. Dysarz, S. Hochrainer, P. Matczak, Mitigation and Adaptation Strategies for Global Change, **15**, 641-656 (2010)
7. Y.L. Vorob'ev, V.A. Akimov, Y.I. Sokolov, *Katastroficheskie navodneniya nachala XXI veka: uroki i vyvody* (DEHKS-PRESS Ltd., Moscow, 2003)
8. S.A. Shlyakov, et al, *Gosudarstvennyj doklad «O sostoyanii zashchity naseleniya i territorij Rossijskoj Federacii ot chrezvychajnyh situacij prirodnogo i tekhnogennogo haraktera v 2013 godu»* (Russian Emergencies Ministry, Moscow, 2014)
9. G.V. Alekseev, et al, *Vtoroj ocenochnyj doklad Rosgidrometa ob izmeneniyah klimata i ih posledstviyah na territorii Rossijskoj Federacii: obshchee rezyume* (Rosgidromet, Moscow, 2014)
10. A.B. Avakyan, M.N. Istomina, Power Technology and Engineering, **35**, 89-93 (2001)
11. C. Butler, N. Pidgeon, Environment and Planning C: Government and Policy, **29**, 533-547 (2011)
12. J. Verschuuren, J. McDonald, Transnational Environmental Law, **1**, 355-379 (2012)
13. V.I. Danilov-Daniliyan, et al, *Reki i ozera mira* (Encyclopedia Ltd., Moscow, 2012)
14. T. Petrow, B. Merz, Journal of Hydrology, **371**, 129–141 (2009)
15. M.P. Fedorov, V.I. Maslikov, Bulletin of the Russian Academy of Sciences, Energetics, **4**, 47-52 (2013)
16. Yu.S. Vasil'ev, S.N. Dobrynin, V.I. Maslikov, T.S. Tikhonova, I.G. Kudryasheva, Hydrotechnical Construction, **34**, 122-129 (2000)
17. R.J. Batalla, C.M. Gomez, G.M. Kondolf, Journal of Hydrology, **290**, 117–136 (2004)
18. M.P. Fedorov, V.V. Elistratov, V.I. Maslikov, G.I. Sidorenko, A.N. Chusov, V.P. Atrashenok, D.V. Molodtsov, A.S. Savvichev, A.V. Zinchenko, Power Technology and Engineering, **49**, 33-39 (2015)
19. V.V. Elistratov, V.I. Maslikov, G.I. Sidorenko, Power Technology and Engineering, **49**, 6-10 (2015)
20. M.P. Fedorov, A.G. Bogolyubov, V.I. Maslikov, Hydrotechnical Construction, **29**, 353-357 (1995)
21. N.V. Aref'ev, V.L. Badenko, G.K. Osipov, Power Technology and Engineering, **32**, 660-663 (1998)
22. N. Arefiev, V. Terleev, V. Badenko, Procedia Engineering, **117**, 39-44 (2015)
23. D. Kurtener, V. Badenko, Journal of the Brazilian Computer Society, **6**, 26-32 (2000)
24. D. Kurtener, V. Badenko, Geomatics Info Magazine, **15**, 76-79 (2001)
25. L. Willuweit, J.J. O'Sullivan, Water Research, **47**, 7206-7220 (2013)
26. M. Deakin, A. Reid, Sustainable Cities and Society, 10, 39-48 (2014)
27. H. Wheater, E. Evans, Land Use Policy, **26**, 251-264 (2009)
28. C. Damodaram, M.H. Giacomoni, C.P. Khedun, H. Holmes, A. Ryan, W. Saour, E.M. Zechman, Journal of the American Water Resources Association, **46**, 907-918 (2010)
29. M. Brilly, M. Polic, Natural Hazards and Earth System Science, **5**, 345-355 (2005)
30. T. Grothmann, F. Reusswig, Natural Hazards, **38**, 101-120 (2006)
31. E.J. Plate, Journal of Hydrology, **267**, 2-11 (2002)

Strengthening and Stabilization of the Weak Water Saturated Soils Using Stone Columns

Leonid Sinyakov[1], *Grigory* Garmanov[1,*] and *Alexey* Melentev[1]

[1]Peter the Great St. Petersburg Polytechnic University, 195251, Polytechnicheskaya 29, St. Petersburg, Russia

Abstract. The article considers innovative modern materials and structures for strengthening of weak soils. In this paper describes a method of strengthening of weak saturated soils using stone columns. The method of calculating the physical-mechanical characteristics of reinforced soil mass is presented. Two approaches to determining the stress-strain state and timeframe of consolidation of strengthened soil foundation using the finite element technique in two-dimensional formulation are proposed. The first one approach it is a modeling of reinforced soil mass, where each pile is represented as a separate 2D stripe. The second approach is to the simulation of the strengthened mass the equivalent composite block with improved physical-mechanical characteristics. The use of the equivalent composite block can significantly reduce the time spent on the preparation of a design scheme. The results of calculations were compared. They show the allowable divergence of results of calculation by two methods were presented, and the efficiency of the strengthening of weak water saturated soils by stone column is proved.

1 Introduction

Weak saturated soil in the foundation of building or structure is one of the most serious problems encountered in the design process. Very important role is played by the correct and possibly more accurate assessment of the impact of weak soil mass for the designed object. Concerning special attention should be also paid to engineering and geological surveys on construction site. Low strength characteristics, strong compressibility and the long duration of building sediment are the key features that must be taken into consideration when buildings on weak saturated soils (especially clay soils) are projected.

Most often weak saturated soil cannot be used as grounds for buildings and structures without prior measures to strengthen the soil mass. From the correctness of the chosen and developed method of strengthening depends on the further process of the construction and operation of the designed object because inaccuracies may lead to significant deformations of the ground and hence malfunction of the building structure and emergencies.

* Corresponding author: gri.garmanov@gmail.com

During the existence of this problem, various methods have been developed in order to solve it: the arrangement of sandy pillows, the replacement of weak soils, the arrangement of sandy piles and different drains with loading mound, fixing of soil by using jet grouting. All the methods have their own characteristics and applicability [1-11].

The paper considers in detail the variant of strengthening of weak saturated soils by modern progressive method of arrangement of soil stone columns what increases strength characteristics of soil mass and also the piles function as drains that greatly accelerates the process of consolidation of the foundation and passing the predicted sediment. This method is not very common today in Russia and a recommendation that regulates use of the stone columns is missing in the normative base. That's why the one of the aim of this paper is the presentation of the method of calculation of physical-mechanical characteristics of the soil. The technique produced using the international experience of employment of the stone columns arrangement technology for strengthening weak grounds [12-15].

2 The technology of arrangement of stone columns

Stone columns are the soil piles which are set through weak layers of soil in order to sealing and transfer loadings on the more strong layers of soil. The piles are formed by immersing in a ground vibration sonde and submission of crushed stone in formed well. Crushed stone substitutes for weak ground, displacing and compacting it, resulting in a significant improvement in strength and deformation characteristics of the soil mass within which the strengthening is carried out.

Also the stone columns work just like normal ground drains resulting in significant reduce of timeframe of consolidation when saturated clay soils lie in the foundation of future building.

This technology is well suited for improving the characteristics of the soil mass, folded by these types of soils like sands, sandy loams, loams and clays.

The technological process of stone columns arrangement consists of the following steps:

the deep vibrator is immersed with the help of vibration and air. If necessary, the deep vibrator is immersed with a minimum flow of water under pressure for passage of the dense layers (Fig. 1, I);

crushed stone is injected through a special drive along the vibrator under the pressure of compressed air (Fig. 1, II);

the deep vibrator is doing reciprocating motion that compacting the crushed stone and gradually forms the body of the pile while moving up (Fig. 1, III).

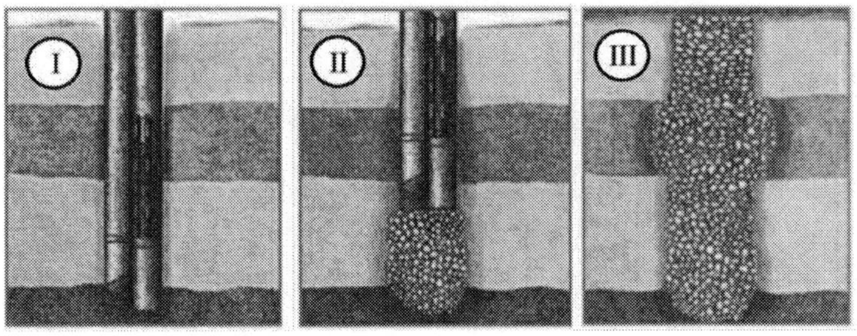

Fig. 1. The sequence of the arrangement stone columns

When the arrangement of stone columns is taken place, the special equipment, so-called vibro-probe – deep vibrator, is used (Fig. 2), and the type of vibro-probe is selected depending on the task. For supplying sealing material under the lower end of the equipment, a special drive is mounted to vibrator.

DOI: 10.1051/matecconf/20167301003

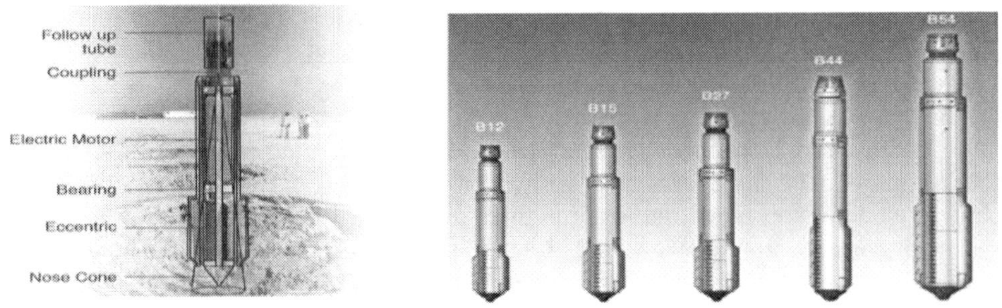

Fig. 2. Vibro-probes

The substructure located on the ground that is strengthened by stone columns should be considered as the foundation on a soil condition taking into consideration physical-mechanical characteristics of a strengthened soil mass.

3 Solution of the problem

Physical and mechanical characteristics of strengthened soil mass are calculated as converted characteristics of natural soil taking into consideration the volume fraction of the stone columns in the soil mass [15].

The determination of the coefficient of internal friction of strengthened soil mass - $\tan\overline{\varphi}$, and its specific weight - $\overline{\gamma}$ of strengthened soil mass is carried out with usage of formulas that are given below:

$$\tan\overline{\varphi} = m' \cdot \tan\varphi_C + (1-m')\tan\varphi_s, \tag{1}$$

$$\overline{\gamma} = m' \cdot \gamma_c + (1-m')\gamma_s, \tag{2}$$

where, φ_c - the angle of internal friction of the stone column material;

φ_s - the angle of internal friction of the soil

γ_c - the specific weight of the stone column material;

γ_s - the specific weight of the soil;

m' - the proportional parameter, that is defined by the formula (3):

$$m' = (n-1)/n, \tag{3}$$

where, n – strengthening coefficient, that is calculated based on the ratio of the squares of stone columns and the strengthened soil mass and defined by the formula (4):

$$n = 1 + \frac{A_C}{A} \cdot \left[\frac{5 - A_C / A}{4K_{aC} \cdot (1 - A_C / A)} - 1 \right], \tag{4}$$

where, A_c - the cross-sectional area of the stone column;

A - the area of influence of the stone column;

K_{aC} - the coefficient of active pressure of the stone columns material, defined by the formula (5):

$$K_{aC} = \tan^2(45^o - \varphi_C / 2), \quad (5)$$

The determination of the adhesion of the of strengthened soil mass $\bar{c}$ is defined by the formula (6), that is given below:

$$\bar{c} = \frac{(A - A_c) \cdot c_s}{A} \quad , \quad (6)$$

where c_s - specified soil adhesion, A and A_c - the same thing like the parameters in the formula (5).

Example.

In the given example, the calculation of the ground of a road embankment performed. The construction of the embankment is considered as stable with the following characteristics: the deformation module (E) – 30 MPa, the angle of internal friction – 30 degrees, the specified adhesion – 1 kPa, the specified weight – 18 kN/m3.

The ground of the road embankment is folded by slowly compacting saturated soil that is clayey silt with the following characteristics: the deformation module (E) – 1 MPa, the angle of internal friction – 2 degrees, the specified adhesion – 12 kPa, the specified weight – 14 kN/m3.

The unpleasant condition of clay soils that are presented in the ground should be noted. The presented sediments have low strength and deformation characteristics and low filtration options and are characterized by a long term seal. The presence of so weak soil in the foundation can leads to long term stabilization of deformations during many years, as well as to uneven deformation during exploitation.

The conversion of building properties of the soil is reached by creation of stone columns with a diameter of 1.2 meters that are arrange to the entire depth of the occurrence of organic soils.

The characteristics of the material of stone columns are the following: the deformation module (E) – 40 MPa, the angle of internal friction – 40 degrees, the specified adhesion – 1 kPa, the specified weight – 20 kN/m3.

The calculated physical-mechanical characteristics of the strengthened soil mass is calculated according to the formulas (1-6) and are provided in Tab. 1.

Table 1. The calculated physical-mechanical characteristics of the strengthened soil mass

Title	Column diameter, [m]	Area of the column, [m^2]	Area of the influence, [m^2]	$\bar{\gamma}$, [kN/m^3]	$\bar{c}$, [kPa]	$\bar{\varphi}$, [deg.]
Strengthened mass	1.2	1.130	4.98	17.6	7.7	27.8

In order to determine the deformation module of the strengthened soil mass when stone column are arranged the series of tests was carried out using the finite element method in three-dimensional statement by means of software package Plaxis 3D (Fig. 3).

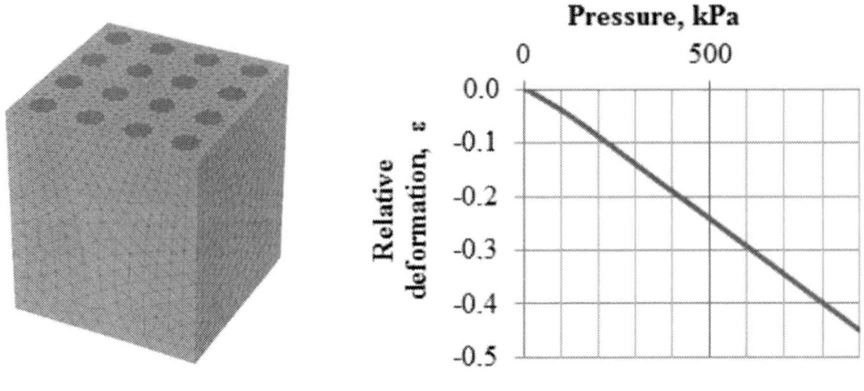

Fig. 3. The general view of the design scheme and the graph of the relative deformation and pressure for the soil mass strengthened by stone columns with a diameter of 1.2 m and a step of 1.9x2.4 m (conditionally the load not shown)

According to the results of calculation the odometer module of deformation of strengthened soil mass is E_{oed} = 1.95 MPa MPa.

During the research process the calculations of subsidence of the mounds and timeframes of consolidation of the soil ground were carried out for the following design schemes:

1) Analysis 1 – 2D FEA without strengthening;
2) Analysis 2 – 2D plane strain FEA with the stone columns modeled as strips (Fig. 4, a);

3)Analysis 3 – 2D FEA with the soil and columns modeled as equivalent composite block (Fig. 4, b).

The calculations carried out by means of software package Plaxis 2D with using Mohr-Coulomb model.

a) b)

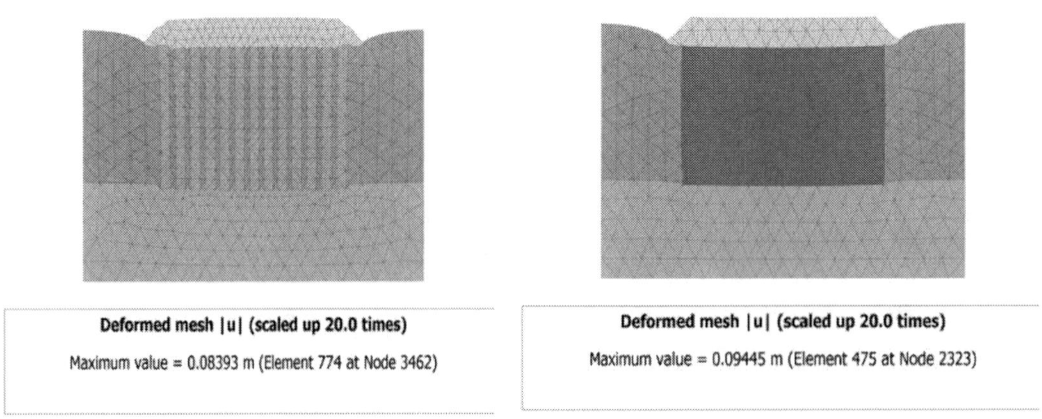

Fig. 4. Embankment deformations schemes
a) FEA with equivalent stone column strips (Analysis 2) b) FEA (Analysis 3) with equivalent composite block

The results of comparative calculations are shown in Tab. 2.

Table 2. Calculation results

Title	Without strengthening (Analysis 1)	With strengthening (Analysis 2)	With strengthening (Analysis 3)
Subsidence of the embankment, [cm]	32.04	8.40	9.44
Time of consolidation, [days]	258	35	38

As can be seen from the calculation results when the ground is strengthened by arrangement of stone columns, the maximal calculated subsidence was 8.40 cm when there were dumping at design height, that is much less than the result without strengthening. The timeframe of the stabilization of deformation of the ground was significantly reduced, from 258 to 35 days. The small differences in the results of the calculation for the full (Analysis 2) and the simplified scheme (Analysis 3) should be noted.

4 Conclusion

1. On the basis of international experience of using of stone columns arrangement technology for strengthening weak grounds was offers the method of calculation physical-mechanical characteristics of a strengthened soil mass.
2. The example of calculation of a weak soil foundation is given, taking in account strengthening by stone columns.
3. The technique of determining odometer module of deformation of strengthened soil mass is displayed.
4. The comparative calculations showed small differences between using the full and simplified scheme. The use of the simplified scheme setting stone columns by soil mass with physical-mechanical parameters calculated using the proposed formulas sufficiently saves time on generating a design scheme and the PC calculation time.
5. The use of the method of weak soils strengthening using stone columns has the following advantages: the decrease of the deformation values of soil mass in several times; a significant reduction in terms of soil consolidation; increased resistance to liquefaction of weak soils under dynamic loading due to an increase in their density and strength; the use in the construction of transport, industrial, civil and hydraulic structures.

References

1. SP 22.13330.2011. *Soil bases of buildings and structures* (Moscow, 2011)
2. SP 24.13330.2011. *Pile foundations* (Moscow, 2011)
3. M. I. Abelev, *Construction of industrial and civil buildings on weak saturated soils* (Stroyizdat, Moscow, 1983)
4. R.A. Mangushev, V.D. Karlov, I.I. Sakharov, *Soil Mechanics: Textbook* (ASV, Moscow, 2009)
5. P.A. Konovalov, *Soil bases and foundation of buildings under reconstruction, fourth ed.* (VNINTPI, Moscow, 2000)
6. V.A. Ilichev, R.A. Mangushev, *Geotechnical manual. Grounds, foundations and underground structures* (ASV, Moscow, 2014)
7. A. G. Malinin, *Strujnaya tsementatsiya gruntov* (Stroyizdat, Moscow, 2010)

8. A. M. Golovanov, V. I. Pashkov, G.A. Revo, D.V. Pashkov, O. V. Nerchinskij, R. I. Turenko, Vestnik MGSU, **8**, 59-67 (2013)
9. B. I. Dalmatov, *Proektirovanie fundamentov zdanij I podzemnykh sooruzhenij: Uchebnoe posobie, second ed.* (ASV, Moscow, 2001)
10. R.A. Mangushev, *Osnovaniya I fundamenty* (ASV, Moscow, 2011)
11. S. G. Bogov, Razvitie gorodov I geotekhnicheskoe stoitel'stvo, **2**, 1-12 (2000)
12. M. Bouassida, S. Ellouze, Z. Ben Salem, M. N. Znaidi, Foundation and Soft Ground Engineering Conference, **1**, 55-66 (2013)
13. J. M. O. Hughes, N. J. Withers, Ground Engineering, **7**, 42-49 (1974)
14. M. R. Shirazi, O. Zarrinkafsh, K. Valipourian, International Journal of Civil, Environmental, Structural, Construction and Architectural Engineering, **9**, 601-605 (2015)
15. Heinz J. Priebe, Ground Engineering, **7-13**, 1-16 (1995)

Protection Parameters against the Cracks by the Method of Volume Compensation Dam

Georgiy Bulatov[1], *Yulia* Ibraeva[1,*], *Philipp* Tarasevskii[1], *Xeniya* Rakova[1] and *Danijela* Zejak[2]

[1]Peter the Great St.Petersburg Polytechnic University, 195251 St. Petersburg, Russian Federation
[2]University of Montenegro, 81000 Podgorica, Montenegro

Abstract. This article provides estimates the parameters of protection from cracking dam due to volume compensation method. This article discusses the method of compensation dam volume. This method allows calculating the settings of security causing cracks the dam. Presents graphs of horizontal deformations of elongation calculated surface along the length of the construction and in time. Showing horizontal stress distribution diagram in the ground around the pile in plan and in section. Given all the necessary formulas for the method of compensation of the dam volume.

1 Introduction

Dam it is an important water works. The damage caused by its failure could be extremely high. Due to the fact that the destruction of direct construction of the dam is often only a small part of the total damage, which included losses from the destruction of associated facilities because the dam is almost always the only part of the waterworks, loss of enterprise. The production of which may be paralyzed as a result of the termination of revenues from hydroelectric power stations, the losses from the devastation produced disastrous spillway downstream of the dam. Global destruction of the dam begins to crack formation. Therefore, of particular relevance is the theme of protecting the dam from the emergence of the weakened areas.

2 Experimental methods

Experimental methods discussed in this article is a method of dam volume compensation. The method consists in the fact that additional volumes in the form of piles or jacks, which periodically introduced into the ground are compensate the deficit of its volume, which arises when dam is lengthening and thus protect it from excessive loosening, and faults – cracks [1-9]. To protect the dam by this method in the extension zone of peak strain produced piling on the crest of the dam for the repayment or reimbursement deformations of the first period. Pile made proactively in time, for example, a month before it reaches peak levels strain the permissible value. Then, with the growth of the deformation of the earth surface at the crest will be set a new high elongation strain, which will also be repaid or compensated stemming piles of the second period of strain. Thus, by plunging into the dam of additional volumes of piles will be compensated strain elongation until their complete decay that will exclude completely the possibility of threatening the dam cracks.

3 Experimental section

It have been known that the relative volume deformation is equal to the amount of linear deformation along the axes

$$\varepsilon_v = \varepsilon_x + \varepsilon_y + \varepsilon_z \tag{1}$$

In the case of displacement soil due to piles in the horizontal plane of the necessary increase (expansion) of the surface at the crest part of the dam determined from expression

$$\varepsilon_s = \frac{\varepsilon_{\text{p}}}{\eta_{\text{н}}} \tag{2}$$

Where ε_{p} – the expected deformation elongation dam from undermining and other effects on settlement dangerous surfaces such as level of filling storage, $\eta_{\text{н}}$ – the degree of orientation of the ground prop (for conventional piles - 0.5 for piles with a special tip for targeted soil prop - 0.7, for piling (tongue) the ranks - 0.9).

$$\varepsilon_p = \varepsilon^* + \frac{\Delta R}{R^*} + \varepsilon_{gp} \tag{3}$$

Where ε^* - the expected deformation elongation of the calculated earth's surface (WHR) from undermining, R^* - radius of its curvature, ΔR – the exceeding of estimated dangerous surface above the WHR, ε_{gp} – deformation elongation from other than the undermining affects to the dam.

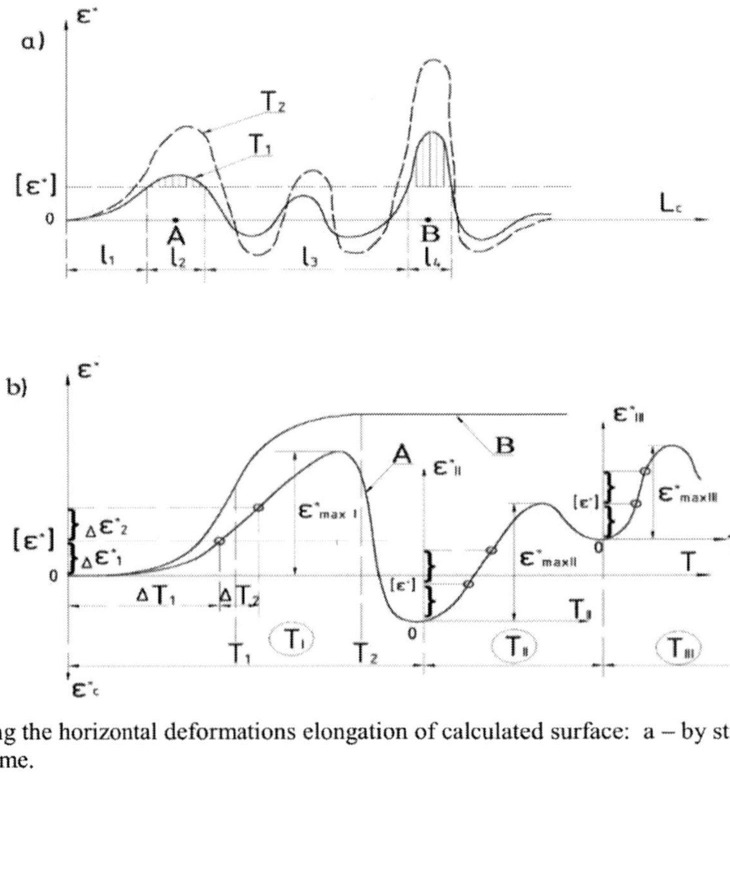

Fig. 1. Changing the horizontal deformations elongation of calculated surface: a – by structures length, b – in time.

Square section piles (or the value of the increment section of jacks or injection) required to secure compensation horizontal extensions the dam, written as

$$S_r = \Omega_r \cdot \varepsilon_s \cdot K_{un} \cdot [K_{s.p}] \quad (4)$$

Where Ω_r – the surface area of the ridge, for one pile (or a jack or an injection well), defined step piles or their rows, K_{un} – coefficient unevenness of sealing between the pile space (approximately equal to 1.5 - for distributed piles and 1.2 - for a pile rows), $[K_{s.p}]$ – admissible generalized coefficient stock for the dam under the deformations of its elongation (approximately equal to 1.5 ... 2.0) [10-14].

Dosed application with the growth of strain elongation the dam crest it is feature of protective measures. For its implementation carried out systematic forecasts and instrumental measurements of deformation.

If now the dashed cut level allowable deformation ε^*, it is possible to select, for example, at the time T_1, portions on which the strain exceeded (shown by vertical hatching) of the permissible level. These sections of dam had to be adequately protected from destruction at the time of T_1. At the time, needed T_2 has other security parameters, and at other sites. From observing practices for the displacement of rocks, it is known that the Earth's surface deformation has pronounced maxima (Fig. 1a) in the form of relative deformations of elongation of the Earth's surface.

On the Figure 1b shows the change in strain in time for the points A and B of considered structures. At the same time, the deformations can change very unevenly, and even change, the sign on the opposite (a curve for point "a"). On the curves can be distinguished several deformation stages $\Delta\varepsilon^*$, each of which will be equal to the allowable strain $[\varepsilon^*]$, and, accordingly, set aside as much time periods ΔT_i development of each of these stages of deformation [15-18].

The onset of each of the periods of strain have been defined due to prognosis and can be monitored during operation of facilities geodesic measurements on two alignments, one of which is laid on the crest of the dam, and the other at the foot of the lower slope. The frames we recommend laying in increments of 5 ... 10 m.

To protect the dam by this method in the peak area of elongation strain (Fig. 1a) produced piling on the crest of the dam for the repayment (reimbursement) deformations of the first period. Pile made proactively in time, for example, a month before it reaches peak levels strain the permissible value.

Then, with the growth of the deformation of the earth surface at the crest will be set a new high elongation strain, which have been also repaid (compensated) stemming piles of the second period of strain [18-21].

Thus, by plunging into the dam of additional volumes of piles will be compensated all strain elongation until the moment of their complete damping that will completely exclude the possibility of appearance the hazardous cracks for the dam [22-26].

In the case of a complex deformation WHR over time as, for example, it is shown in the graph to the point A in Fig. b, the whole period T is divided into cycles of deformation, boundaries are points corresponding to the minimum elongation strain. For points A, there are three such cycles, and total elongation strain, which has to be repaid (offset) of the periodic harvesting, will be

$$\varepsilon^*_{max} = \varepsilon^*_{max\,I} + \varepsilon^*_{max\,II} + \varepsilon^*_{max\,III} \quad (5)$$

Accordingly, the total period of strain will be

$$T = T_I + T_{II} + T_{III} \quad (6)$$

However, more preferred is a separate consideration of each cycle of deformation as a single act (the impact of) of undermining, with a corresponding development of protective measures, which, in principle, in different cycles may be quite different. Wherein it is possible multi directional impact protection measures on the course of deformation WHR in subsequent cycles.

In this case, the strain compensation piles subsequent compression deformation will be difficult. The number of periods piling on the considered deformation area is

$$n_{\mathrm{T}} = \frac{\varepsilon_p}{[\varepsilon_p]} \tag{7}$$

Where $[\varepsilon_p]$ – allowable deformation elongation of ground on the surface settlement.

$$[\varepsilon_p] = \frac{\varepsilon_{cr}}{[K_p]} \tag{8}$$

Where ε_p – the limiting extensibility of the soil, which can be accepted as from the condition of preventing so with taking into account the limited size of cracks assumptions on the estimated dangerous surface.

Sectional area of one pile for immersion in each period of develop and deformation compensation will be

$$S_{pl}^{I} = \frac{S_{pl}}{n_T} \tag{9}$$

The depth of immersion of piles

$$H_{pl} = h_{cr.lim} \cdot [K_{s.p}] \tag{10}$$

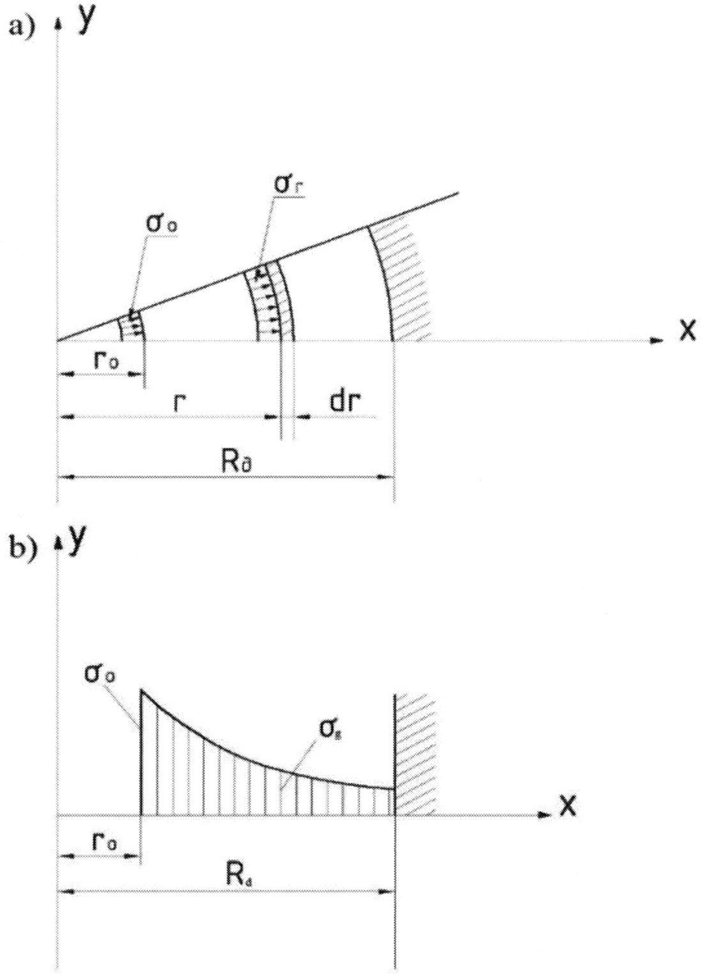

Fig. 2. The distribution of horizontal stress in the soil around the pile: a – in plane, b – a sectional.

Where $h_{cr.lim}$ – maximum depth of crack propagation.

When submerged pile adjacent soil is pushed in the radial direction and compacted. To estimate the horizontal radial stresses in axisymmetric problem use the first part of the formula, shown in the book, which write Yaropolskii I.V.

From this formula is obtained the approximate dependence of the compressive horizontal stress in the soil around the pile (Figure 2)

$$\sigma_r = \frac{r_o}{r} \cdot \frac{E_{com}}{\pi(1-\mu^2)\ln\frac{R_d}{r_o}} \tag{11}$$

Where r_o- sectional radius of the pile, E_{com} and μ – modulus of deformation compression and coefficient Poisson for soil around the pile, r – the distance from the point considered in the soil to the pile axis, R_d – the radius of the deformed zone of soil around the pile at her pile, taken equal to half the maximum distance between adjacent piles [25-27].

Soil compression is considered adequate, provided

$$\sigma_r = \sigma_{Rd} \geq [\sigma_{req}] \quad (12)$$

Where $[\sigma_{\text{req}}]$ – required compressive stress in the ground, that preventing disclosure of through cracks (in the first approximation is taken equal to 0.01 MPa).

4 Results and Discussion

The result of the article is the parameters by which, in the earth dam are determined safe work. All required dependencies have been obtained. Received the formula have been calculated the adequate soil compression.

5 Conclusion

Security is the one of the important criteria for facilities operation. In many ways, it depends on the mechanical processes occurring in the structure. The most common cause is the destruction of the dam it is a cracks. Cracking entails global destruction in dam, thus the prevent appearance of cracks and rendering options cracks that already been formed it is highly important topic. This subject have not been elaborated until the end and need to continue research in this direction.

References

1. R. Usmanov, M. Rakočević, V. Murgul, N. Vatin, Applied Mechanics and Materials, **633–634**, (2014)
2. R. Usmanov, I. Mrdak, N. Vatin, V. Murgul, Applied Mechanics and Materials, **633–634**, (2014)
3. M. Jocovic, B. Melovic, N. Vatin, V. Murgul, Applied Mechanics and Materials, **678**, (2014)
4. Sonoda, Y.,Applied Mechanics and Materials, **566**, (2014)
5. E. Panulinová, S. Harabinová, Advanced Materials Research, **969**, (2014)
6. S. Harabinová, E. Panulinová, Advanced Materials Research, **969**, (2014)
7. M. Sharp, Y. Seda–Sanabria, E. Matheu, Applied Mechanics and Materials, **82**, (2011)
8. Mrdak, M. Rakočević, L. Žugić, R. Usmanov, V. Murgul, N. Vatin, Applied Mechanics and Materials, **633 – 634**, (2014)
9. D. Kubečková, Advanced Materials Research, **1020**, (2014)
10. G.Y. Bulatov, V.I. Teleshev, V.A. Leonov, Fil'tratsionnyye issledovaniya i raschety, **7**, (1983)
11. D. Hilyard, The dam safety minefield, International Water Power & Dam Construction. **5** (2010)
12. S.V. Bortkevich, Power Technology and Engineering, **5**, (2009)
13. A.B. Feringer, N.V. Kabanov, Power Technology and Engineering, **46(5)**, (2013)
14. N. Vatin, N. Lavrov, N. Korzhavin, Applied Mechanics and Materials, **641–642**, (2014)
15. V.B. Vladimirov, Y.K. Zaretskii, V.B. Orekhov, Power Technology and Engineering, **3**, (2003)
16. S.Y. Shkolnikov, I.A. Sekisova, Power Technology and Engineering, **5**, (2008)
17. V.N. Bukhartsev, M.R. Petrichenko, Power Technology and Engineering, **46(3)** , (2012)
18. G.Y. Bulatov, Trudy LPI, **375**, (1981)
19. V.S. Istomina, V.V. Burenkova, G.V. Mishurova, Stroyizdat (1975)

20. A.K. Bougrov, V.F. Markevich, Gidrotekhnicheskoye stroitel'stvo, **4**, (1980)
21. G. Bulatov, N. Vatin, D. Nemova, Y. Ibraeva, P. Tarasevskii, Applied Mechanics and Materials, **725–726**, (2014)
22. V. Deković, A. Andelković, N. Milošević, G. Gajić, M. Janić, Carpathian Journal of Earth and Environmental Sciences, **8 (2)** , (2013)
23. L.N. Rasskazov, M.V. Smirnova, Power Technology and Engineering, **48 (2)** , (2014)
24. L. Svatovskaya, A. Sychova, M. Sychov, V. Okrepilov, MATEC Web of Conferences, **53**, Article Number 01023 (2016)
25. J. Guo, I. Hasan, P.W. Graeber, Springer Series in Geomechanics and Geoengineering, **2015**, (2014)
26. K. Krayushkina, O. Prentkovskis, A. Bieliatynskyi, J. Gigineishvili, A. Skrypchenko, A. Laurinavičius, K. Gopalakrishnan, J. Tretjakovas, Baltic Journal of Road and Bridge Engineering, **11 (1)** , (2016)

Allowance for Sequence of Mass Concrete Dam Erection on Soils During its Stress-strain Analysis

Igor Konstantinov[1], *Sergey* Kuzmin[1], *Alexey* Savchenko[1], *Pavel* Boychenko[1,*] *and Diana* Nagornya[1]

[1]St.-Petersburg State Polytechnical University, Politechnicheskaya St., 29, St. Petersburg, 195251, Russian Federation

Abstract. The following are the research results made both at lengthwise and width directions with and without accounting sequence of a section erection. Calculations were performed in the maths laboratory of «B.E. Vedeneev VNIIG» with software assistance. The main aim of the study is to search for the optimal stress-strain behavior of the dam in terms of forming minimal tension in concrete.

1 Introduction

While designing such critical structures as concrete dams, it is very important to analyze their stress-strain behavior [1-6]. Therefore a selection of a relevant structural design of the dam is also of great importance.

In operative design and construction specifications [7] when choosing structural design of a concrete dam including ones on soils, it is recommended to take into account the sequence of its erection and loads. Rational erection sequence enables better formation of stress-strain behavior in the dam. While designing mass concrete dams on soils (using specific conditions), a choice of an appropriate erection sequence can be done only when a full picture of dam stress-strain behavior at various schemes of installation is well identified.

Nowadays a stress-strain behavior of mass concrete dams on soils at various stages of its erection and loads is quite understudied. The majority of works, concerning this issue were published at the time when calculation methods as well as abilities of PC remarkably limited possibilities of problem solving even in the context of plane elasticity. Today we have more powerful PCs. The finite element method is now widely used for practical design calculations. There is a number of software products (based on finite element method) for solving plane and three-dimensional elasticity used by various scientific institutes and design organizations. These factors remarkably enhances studies in stress-strain behavior of mass concrete dams on soils with consideration of the dams erection sequence and their interaction with foundations. This question was partly considered in the work [8], where in the context of plane elasticity several loading schemes of mass concrete

* Corresponding author: boychenko_pavel@icloud.com

dam (when operated at lengthwise direction) were analyzed. However, the dam erection sequence was not considered in that case.

A section of a dam (transversal and longitudinal sections depicted on Figure1) under inquiry is a monolithic spatial structure with (Figures 1b-c) piers, base slab and crown of inlet on which valve is located. Young's modulus, unit weight, Poisson's ratio of the dam's material are as follows: $E_\delta = 2 \cdot 104$ MPa, γ_δ =24 kN/m3, ν_δ =0.17.

When the valve is closed, basic static loads are following: the dead load of the section, water pressure on both downstream and upstream sides and the counterpressure through the foot of the dam. It is expected that the section bears on foundation with deformation modulus E0 = 20 MPa (fine sand) [11]. Poisson's ratio of v0 =0.25 taken in this study was a quite different that the recommended one [11] v0 =0.3. However, this mismatching slightly influenced stress state and distorted the performance. Practical use of available software for such complicated three-dimensional structure as section of concrete dam on soils is a very difficult task due to the unavailability of automatic boundaries division of the area. That is the reason why the problem of stressed three-dimensional section was approximately solved by analyzing some plane sections. The section included inlet and base slab (Figures 1a-b) was considered for examining stress state in longitudinal section. Furthermore, the task of plane strain of the body interacting with foundation was solved. That method didn't take account of piers influence on stresses in the body.

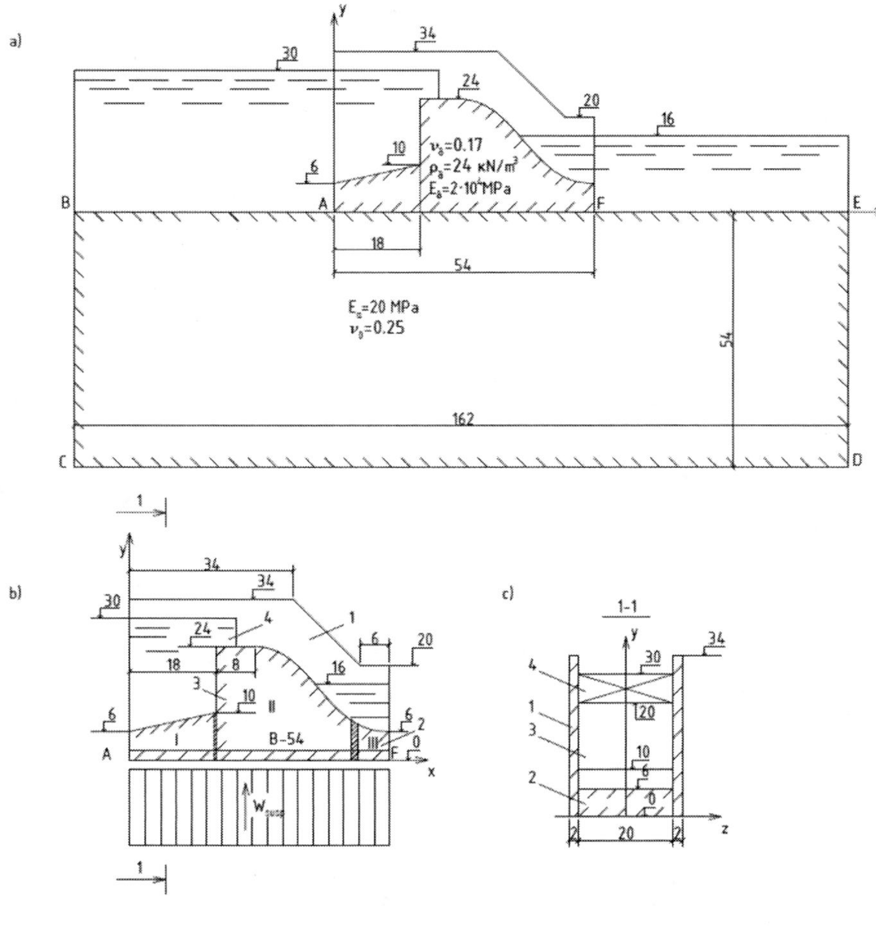

Fig. 1. Structural design: a) design section of the dam and its base zone; b) longitudinal design section of the dam; c) transversal design section of the dam

Stress condition in sections of the body near piers was different from the one found along the centerline. That is why reduced flat stress problem was used in order to roughly take account of piers influence. For this purpose, piers were calculated with density reduced proportionally and following modulus of elasticity: $\rho_{pr} = \rho_{\delta} \cdot t/L$; $E_{pr} = E_{\delta} \cdot t/L$

where t – width of two half-piers in a section, L – overall section width.

After some calculations, tension in pier were increased to $\sigma_{\delta} = \sigma \cdot \sigma_{pr} \cdot L/t$. When calculating, as for account of piers, plane strain problem of body interacting with foundation was solved. Allowance for foundation was made according to [12]. Dimensions of foundation were chosen as follows: AB=BC=DE=EF=B,

where B – longitudinal length of a foot of a dam.

Stresses on BCDE line were calculated according to σ_y and τ_{xy} linearly distributed along the foot of a dam. They were taken based on eccentric compression formulae. Consequently, self-equilibrated forces on CD were not considered. However, that is quite legitimate according to Saint-Venant's Principle.

Calculations were performed with an allowance for two schemes of erection:

1. The dam was believed to be erected by layers. Layer thickness was taken according to recommendations [13]. Hydrostatical pressure and backpressure along the foot of a dam were exerted to the section erected by that time.

2. Dam body was erected layer-by-layer by means of three different parts (Figure 1b, pt. I-III). Only when the parts were erected to the projected height, joints of base slab were concreted. Hydrostatical pressure and backpressure were exerted to the monolithic body along its foot.

Assuming that the dead weight of the structure and hydrostatical pressure exerted to 'momentary' erected structure (Figure 1b), calculations were made to associate longitudinally stress condition of the section with allowance for sequence of structure erection.

A rapid estimate of section's stress condition transversally (crosswise of water current) was considered in conditions of plane strain in section I-I (Figures 1a,c). The following are two options of erection:

1. Base slab and piers were erected separately, then joints were concreted and finally hydrostatical pressure was exerted.

2. Base slab and piers were erected together. Loading was exerted along edges of the slab. Joints between sections were conditionally extended to foundation area. Influence between adjoining sections (same width as a layer) equals to a dam section width (Figure 1c).

Results of calculation are given below on Figures 2-4 in the form of normal stress distribution diagrams σ_x in vertical direction. Analyzing them, we can estimate occurrence of tensile zones of concrete.

1. Figure 2a depicts diagram σ_x (stress distribution in vertical sections of solid part of a dam section). The diagrams marked with '1' were obtained assuming that the only dead weight was applied (without contribution of piers) and that the structure was momentary erected. The diagrams marked with '2'stand for simultaneous actions of dead weight, hydrostatical pressure and backpressure along the foot of a dam. Diagrams 1 and 2 comparison demonstrates that one load from dead weight leads to greater tension σ_x.
The diagrams marked with '3' stand for solution of reduced flat stress problem using basic loads. Piers' effect on stresses in the body were approximately included. A section, as in earlier cases, was considered to be exerted when all loads could be applied. As we can see, allowing for piers' effect massif became stiffer and there were no tensile zones at all. Figure 2b depicts diagrams σ_x (sections 0.5 meters apart from section's base). Let us make notice of diagram 2 and 3. Both of them are different from those introduced in the work [9]. There we did not take into account a backpressure along the foot of the dam.

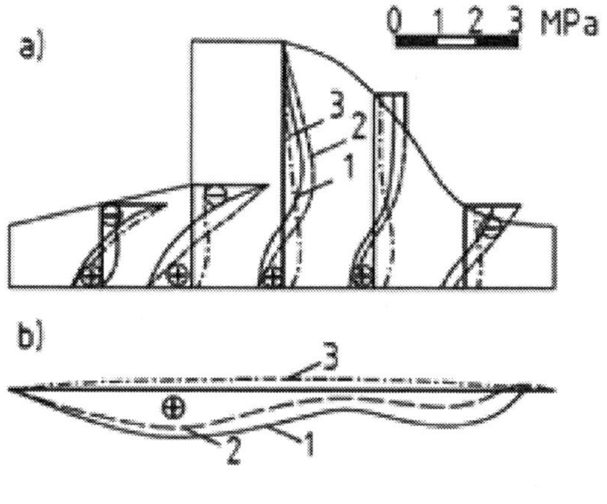

Fig. 2. Diagrams of direct stresses σ_x: a) σ_x across longitudinal sections of the section; b) σ_x 0.5 meters apart from the foot. 1- σ_x when momentary erected and exerted by dead weight without piers' effect; 2- σ_x when momentary erected and exerted by basic loads without piers' effect; 3- σ_x when momentary erected and exerted by basic loads with piers' effect

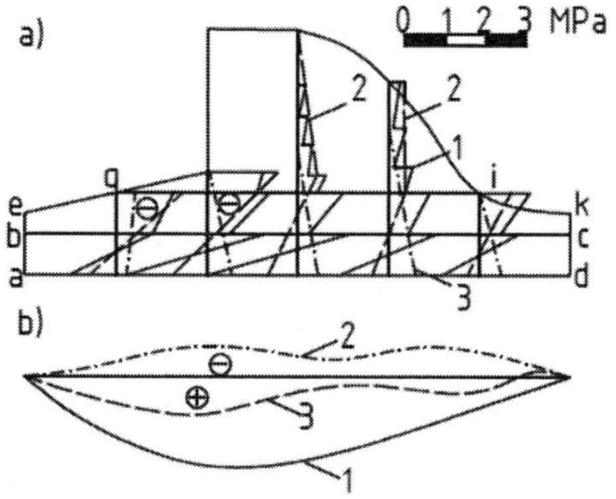

Fig. 3. Diagrams of direct stresses σ_x with consideration of sequence of erection: a) σ_x across longitudinal sections of the section; b) σ_x 0.5 meters apart from the foot. 1- σ_x when erected layer-by-layer; 2- σ_x when erected layer-by-layer with casting joints in the base slab; 3- σ_x when momentary erected

Fig. 3a-b depicts a match of diagrams σ_x (longitudinal sections 0.5 meters apart from the foot; the option without piers' effect and with layer-by-layer erection). It's assumed, that layers were erected 'momentary'. First layer maximum thickness h1 =4.5 m, [13]. The next layer h2 =3 m was placed after the solidification of concrete of the first one. Furthermore, tension σ_x appeared to be a result of bending in abcd layer. Relevant diagrams are approximately linear due to ration of h1 to B. When the second layer with adjustable height gained its strength, the third layer with h3= 3m was casted. When the massif had been erected, a hydrostatical pressure and a backpressure were exerted. Final and conclusive lattice diagrams σ_x resulted from the same loads, but without consideration of sequence of erection (Figure 2a-b). On the contrary, the option with layer-by-layer erection and with casting joints led to diagrams σ_x (dot-dash curve marked with '2', Figure 3a-b).

As it was discussed above, concreting in lifts (layer-by-layer) with casting joints helped to avoid lengthwise tension σ_x in the whole section even without piers' effect.

However, concreting in lifts without vertical joints led to more complicated stress distribution. Furthermore, the smaller the first layer was, the greater stresses σ_x were registered while concreting the next layer. Consequently, a thickness of the first layer should be as great as possible. The dead weight of the first one didn't set up any stresses σ_x across the section. That happened because of strain (due to base slenderness), which took place when the layer had not solidified yet.

First superjacent layers set up the major part of stresses σ_x in the first layer because accumulative section wasn't very thick. Calculation of following layers illustrated interesting constraint. Right after the first layer was loaded by the second one, 36% of maximum tensile stresses σ_x developed across the lower part of the first one. When the first layer was loaded by the third one – 64% were registered etc. The higher the loading layer was, the less it affected the developing of tensile stresses because the thickness of accumulative section was always increasing.

In the process of increasing loads and stresses σ_x the neutral axis position was within the first layer during the whole process of erection. The tensile stresses obtained, were 2-3 times greater than stresses had come from the 'momentary' scheme. Moreover, maximum tensile stresses σ_x developed across the middle vertical crosses of the foot. On the contrary, when we had 'momentary' erection maximum stresses developed in the cross aligned with dam crown of upstream face.

3. Figure 4a-b depicts comparison of diagrams σ_z across vertical sections through upper and lower fibers of the base slab's section in a lateral direction.

The diagrams σ_z marked with '1' were obtained from the base slab and piers separate erection (the 1st scheme of erection) from all basic type of loads. The diagrams σ_z marked with '2' were obtained from layer-by-layer erection of the section as a single whole (the 2nd scheme of erection) without consideration of hydrostatical pressure.

The first scheme of erection diagrams displays that maximum tensile stresses σ_z in the middle of the foot took place at the outset of erection. Tensile stresses on the periphery of the section took place only after hydrostatical pressure was exerted.

When analyzing diagrams obtained from the second scheme of erection (with piers' effect), it turned out that during erection tensile stresses σ_z developed across upper layers of the base slab. Furthermore, compression stresses σ_z across lower layers disappeared. By these means, the most dangerous stress state occurs during construction period.

2 Summary

The carried-out calculations allow to approve the following: significant tensile stresses took place generally by means of surcharge weights of piers. That is the reason of essential base slab reinforcement of both upper and lower parts of it. Calculations and further analysis demonstrates that during the construction period a dam body should be erected detached from piers with casting joints. Those joints should be concreted at the end of the section erection.

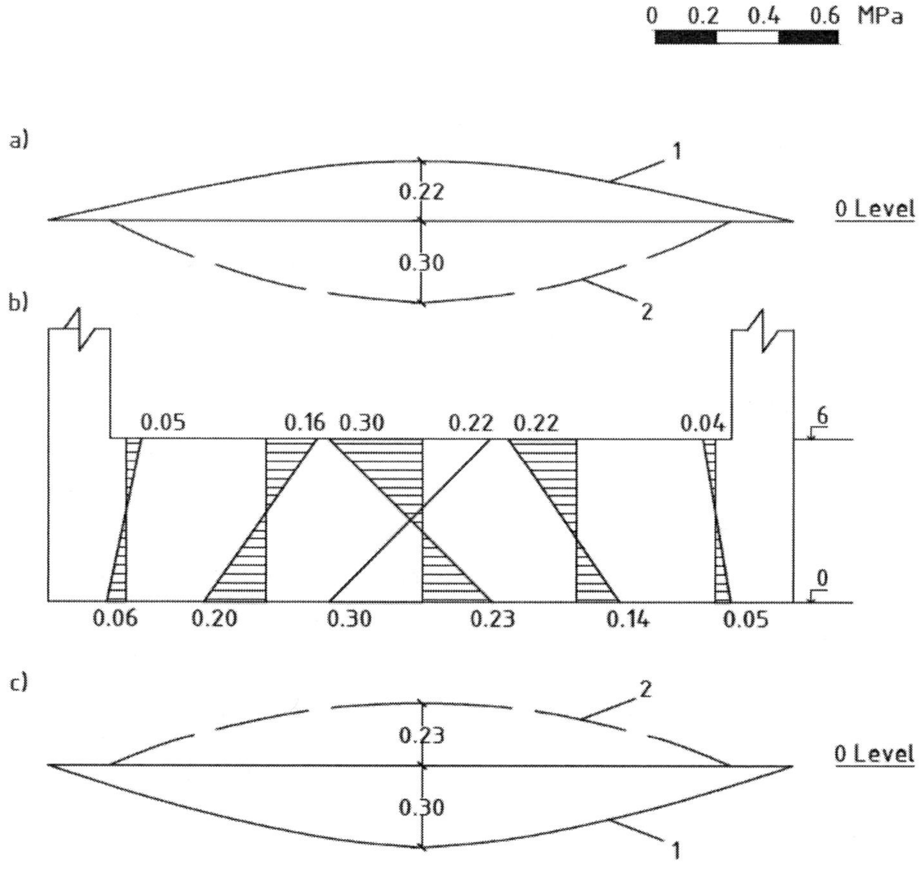

Fig. 4. Diagrams of direct stresses σ_z (section I-I): a) σ_z across vertical areas through upper edge of the base slab; b) σ_z across vertical sections; c) σ_z across vertical areas nearby dam's foot. 1- the base slab and piers separate erection; 2- layer-by-layer erection of the section as a single whole

References

1. Bukhartsev V.N., Petrichenko M.R., Power Techology and Engineering, **46(3)**, 185-189 (2012)
2. Bukhartsev V.N., Petrichenko M.R., Power Techology and Engineering, **46(3)**, 1-3 (2012)
3. Kolosova G.S., Lalin V.V., Kolosova A.V., Magazine of civil engineering, **5**, 76-85 (2013)
4. Benin A.V., Semenov A.S., Semenov S.G., Advanced Materials Research, **831**, 364–369 (2014)
5. Kitaeva D.A., Rudaev Y.I., Ordobaev B.S., Abdykeeva S.S. Applied Mechanics and Materials, **725-726**, 623-628 (2014)
6. Lalin V.V., Savchenko A.V., Zdanchuk E.V., Yavarov A.V., Belyaev D.I., Construction of unique building and structures, **7**, 30-37. (2013)
7. Code Design 40.13330.2012 *Concrete and reinforced concrete dams*. (Moscow, Minregion Russia, 2012)
8. Dyatlovitskiy L.I. *Napryazheniya v gravitatsionnikh plotinakh na neskalnikh osnovaniyakh* [Stresses in gravity dams on the not rocky grounds] (AN YSSR, 1959)

9. Konstantinov I.A., Kuzmin S.A. *Issledovanie napryazhennogo sostoyaniya betonnoy vodoslivnoy plotiny na neskalnikh gruntakh metodom konechnykh elementov* [*Investigation of the stress state of the concrete spillway dam on the not rocky grounds with finite element method*]. (col.book Altay Polytechnic University, 1985)
10. Ymanskiy A.A. Spravochnik proektirovshika [Designer's reference book]. (vol.**2** Moscow, 1973)
11. Efimov YU.N., Kolosova G.S., Konstantinov I.A., Sapozhnikov L.B. *Ychet podatlivosti osnovaniya pri staticheskom raschete plotin metodom konecnykh elementov* [*Accounting for compliance of foundation in static analysis of dams with finite element method*]. (B.E. Vedeneev VNIIG, 1973)
12. Grishin M.M. *Gidrotehnicheskie sooruzheniya* [*Hydrotechnical constructions*] (Vyssh.shk, 1979)
13. Teleshev V.I. *Konstruktivno-tekhnologicheskie meropriyatiya po obespecheniyu treshinostoykosti i monolitnosti massivnyh betonnikh gidrotehnicheskikh sooruzheniy* [*Design and technological measures to ensure the fracture toughness and solidity of massive concrete hydraulic structures*] (LPI, 1983)
14. Katzenbach R., Werner A., Kempfert H.-G., Savidis S., Steffens K., Wesemüller H. Darmstadt Geotechnics, **11**, 109-12. (2003)
15. L. Svatovskaya, A. Sychova, M. Sychov, V. Okrepilov, MATEC Web of Conferences, **53**, Article Number 01023 (2016)
16. Argal É., Korolev V., Kudrin K., Ashikhmen V. Power Technology and Engineering (formerly Hydrotechnical Construction), **43**(6), 337-344 (2009)
17. Mirzabozorg H., Ghaemian M., Noorzad A., Abbasi Zoghi M. International Journal of Earthquake Engineering and Engineering Seismology (EEE), **3**, 52-60 (2007)
18. Mirzabozorg H., Khaloo A.R., Ghaemian M., Jalalzadeh B. International Journal of Earthquake Engineering and Engineering Seismology (EEE), **2**, 48-57 (2007)
19. Brandão J.L.B. Water Resources Management, **24(12)**, 3101-3114 (2010)
20. Kurzin V. B., Seleznev V. S., Journal of Applied Mechanics and Technical Physics, **51(4)**, 590-597 (2010)

Processes at Water Intake from Mountain Rivers into Hydropower and Irrigation Systems

Nikolai Vatin [1,*], *Nikolai* Lavrov [1] and *Gennadi* Loginov[1]

[1]St. Petersburg State Polytechnical University, Politekhnicheskaya, 29, Saint-Petersburg, 195251, Russia

Abstract. In paper, researches of riverbed and hydraulic processes at the water intake from mountain rivers are observed. Classification of designs of the mountain water intake structures, based on continuity signs is offered. Perfecting of base designs of water intake structures of a mountain-foothill zone and means of their hydraulic automation is carried out. The technological, theoretical and experimental substantiation of parameters of basic elements of these designs with a glance of hydromorphometric characteristics of the mountain rivers is given. Complex hydraulic researches of kinematic characteristics and carrying ability of a two-phase stream on water intake structures are executed. Bases of a technique of engineering calculation of the offered designs of water intake structures and the recommendation of their designing and maintenance in various hydrological regimes are developed.

1 Introduction

Now designing and maintenance of water intake structures on the mountain rivers is spent without the sufficient account of demands of creation of necessary riverbed and hydraulic processes on water intake sections that is one of the principal causes of idle times, damage and crashes of constructions. Therefore, there is a necessity of an estimation of rationality of existing lay-out drawings of water intake and regulation structures with definition of their deficiencies. It demands conducting theoretical and experimental researches to raise of characteristics of process of the water intake, by the management of two-phase streams and regimes of formation of bed landforms. In it, the urgency of the given work, devoted to studying and working out of possibilities of use of energy of water streams on sections of regulated riverbeds and in pools of low-head dam water intake structures, consists.

2 Experimental program

The purpose of researches is perfection of the theory, methods of researches and rated validation of necessary characteristics of riverbed and hydraulic processes at the water intake from the mountain rivers. It allows to devide out involved material resources at designing, building and to raise reliability of low-head dam water intake structures during

* Corresponding author: vatin@mail.ru

service. The analysis of characteristics of riverbed and hydraulic processes for the purpose of performance of the basic technical requirements and engineering operations at the water intake, presented in works S. Elsden [1, 2], N. Kanthack [2], R. Krohmer [3], S.T.Altunin [2], K.F.Artamonov [4], N.F.Danelia [2], I.S.Rumjantsev [5], I.I.Levi [2], JA.V.Botchkarev [1], B.I.Melnikov [2], A.S.Ofitserov [1], A.I.Rohman [1], I.K.Rudakov [1], G.V.Sobolin [2], A.V.Filonchikov [6] and other researchers, is made.

Authors of paper prove classification of existing designs low-head water intake structures [7].

This classification defines communication of known lay-out drawings by using of previous rational solutions for working out of the subsequent advanced configurations of structures.

For perfection of the water intake process the following processing methods have been gated out:

- creation of the transverse circulation of water on a section of approach riverbed by means of curvilinear regulated section;
- increasing intensity of the transverse circulation of water before a water intake by creation of the local circulating screw with the help of sediment-protective threshold;
- water selection realisation during the warm period of year from surface horizons of a stream and ground removal of sediments;
- water selection during the winter period from near-bottom horizons of a stream with surface removal and dump ice brash formations;
- creation of depths of water stream which necessary for generation of resistant hydraulic forms before elements of a structure;
- maintenance of the smooth approach of water to devices of the water intake structure and stream outlet;
- definition of a position and parameters of devices of a heading range taking into account conditions of creation of active zones of traffic of hard spots of river sediments in approach regulated riverbed;
- maintenance of rated heads in the upper pool of the structure and stabilization of the water discharge, submitted to outlet canals with use of means of hydroautomatics;.
- at a high water simultaneous removal of excesses of water and a driftwood on all front of the water intake structure and carrying sediments - through river bay apertures is provided;

Performance of these conditions allowed execution during researches the matching of principles of configuration of advanced designs of water intake structures with existing techniques of designing of devices of water intake from the mountain rivers.

Observing the arrangement of a curvilinear approaching riverbed before the heading gate of the water intake structure as the first line of protection against sediments, known representations about the reasons of origin of the transverse circulation have been analysed. M.A.Dementyev, M.V.Potapov, N.F.Danelia, Cecen K, Schucky A, Zolezzi G., Seminara G., Solari. L., [2, 8, 9, 10, 11] as the reasons of formation of the radial secondary currents considered the inertia force - centrifugal.

A.J.Milovich [12] defined as the reason of cross-flows in the radial direction a pressure difference in cross-sections of curvilinear parts.

Passing from the qualitative description to quantitative characteristics, the mathematical model presenting occurring hydraulic processes on curvilinear parts of the rivers has been observed:

$$\left.\begin{aligned} Q_{bi} &= Q_{b(i\text{-}1)} + Q_R \Theta_i - Q_{gr.i} \\ \Theta_i &= \frac{\cos\varphi_{i-1} - \cos\ \varphi_i}{1-\cos\varphi} \\ Q_{gr.i} &= Q_{gr.(i\text{-}1)} + \int_0^{L_{kri}} \omega_i C_i \sqrt{R_i \Delta J}\, dl \\ \omega_i &= k_i dl \\ d_i &= \sqrt{\frac{4Q_{bi}}{\pi v}} \end{aligned}\right\} \qquad (1)$$

Where Q_R - water discharge in a river; Q_{bi} - water flow moving in a rotating cone; φ_i - angle between entrance cross-section of a curvilinear part and observed i - cross-section; $Q_{gr.i}$ - charge of ground currents of the water separating from a cone-shaped stream from its beginning to observed cross-section; Θ_i - coefficient considering a share of a projection of the observed section in the general projection of a concave shore, on a plane placed in the beginning of curvature of a riverbed; ω_i - cross-sectional area of a near-bottom stream separating from cone-shaped roll; k_i - vertical size of a ground stream in observed cross-sections; R_i - design quantity of hydraulic radius; d_i - diameter of cone-shaped roll; v - average speed of water in cone-shaped roll; C_i - coefficient of Shezi for movement of near-bottom currents; ΔJ - increment of the transverse slope of a free surface of a water stream as a result of inertial force act at climb of a stream on a concave shore.

Using the theory of dynamic interacting of solids and a liquid, offered by A.J.Milovich [12], the connection of the longitudinal Vl and transverse Vt components of circulating roll with speed of water movement in the upper pool of a structure and a contour of an external surface of sediment-protective threshold has been defined in an aspect:

$$\frac{V_t}{\sin\gamma} = \frac{V_l}{\cos\gamma} \rightarrow \varphi = arctg\frac{V_t}{V_l} + \pi n \qquad (2)$$

Taking this into account, the system of the equations has been developed for calculation of co-ordinates of a planned configuration of circulating sediment-protective threshold [2] and the planned contour of an external surface of sediment-protective threshold, presented on fig. 1 is gained.

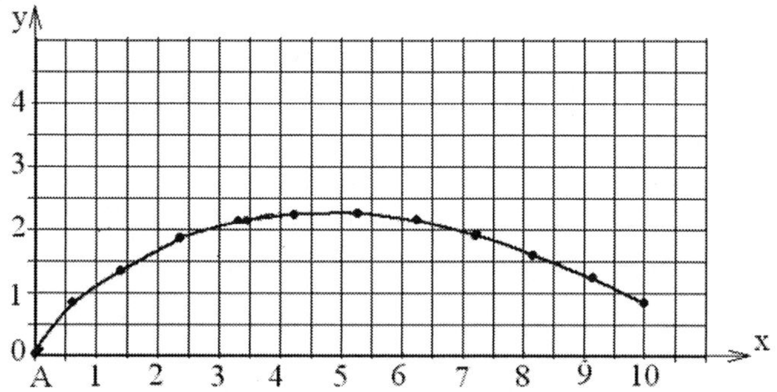

Fig. 1. Contours of an external surface of sediment-protective threshold.

By consideration of alternatives of increase in carrying ability of a sluiceway tract [13, 14, 15], us the arrangement of a horizontal peak on the lower edge of the gate of sluiceway tract was offered. By transformation of the Bernoulli's equation made for cross-sections I-I and II-II (fig. 2), the formula for calculation of speeds of a water stream under a horizontal peak is gained.

$$V_2 = \frac{1}{\sqrt{\alpha_1}}\sqrt{2g(H_R - \varepsilon a - h_\omega) - \alpha_1 V_1^2 + \frac{2(p_1 - p_2)}{\rho}} \tag{3}$$

Where H_R – the water depth in front of the device; p_1 and p_2 - pressure on a stream free surface in cross-sections; ρ - liquid density; g - free fall acceleration; α_1 - coefficient, considering irregularity of stream velocity distribution in initial cross-section; ε - coefficient of jet contraction of a stream; h_ω - stream energy loss on a length of a nozzle.

It allowed to create pressure decline in a near-bottom layer of two-phase stream with increase in speeds of water under a horizontal peak (see fig. 2), and secured raise of removed volumes of sediments.

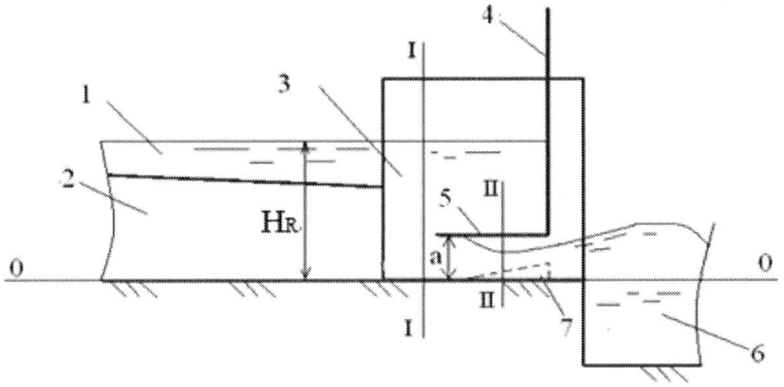

Fig. 2. Scheme of a horizontal peak on the lower edge of the gate of sluiceway tract: 1 - upper pool; 2 – circulating sediment-protective threshold; 3 - sluiceway tract; 4 - flat gate; 5 – horizontal peak; 6 - offtake riverbed; 7 - bottom insert.

Preliminary results of laboratory researches of the gate with a horizontal peak confirmed raise of carrying ability of water stream in a sluiceway tract of water intake structures.

Because of transformation of momentum equation, the formula for definition of the charge of sediments has been gained as [2]:

$$G = \frac{2\Delta K - \rho g b_{st}(2h_2 J \times l_{ac} + J^2 l_{ac}^2)}{2\alpha^{//}(\rho_s - \rho)(v_2 - v_1)} \tag{4}$$

Where ΔK - an increment of a momentum of weight of the two-phase fluid stream removed through a sluiceway tract, accumulating sediments at length l_{ac} in unit of time; b_{st} - width of sluiceway tract, ρ and ρ_s - density of water and carrying sediments; J - hydraulic slope, $\alpha^{//}$- the coefficient of the amount of movement; v_1 and v_2 – movement speeds of carrying sediments in an initial and final range of a tract.

The aspect of the equation (4), allows to define that carrying ability of a water stream increases at growth of value of change of a momentum, and decreases owing to growth of values of following values: width and depths of a stream on length of a sluiceway tract; lengths of a sluiceway tract, density of corpuscles of sediments. Considering these factors, arrangement of a ground insert with an adverse slope at the bottom of a sluiceway tract (fig. 3) was offered.

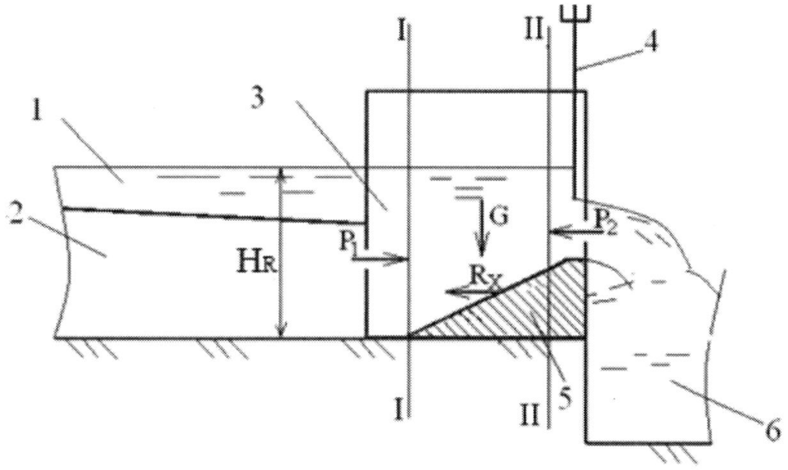

Fig. 3. Scheme of disposing of a ground insert in a sluiceway tract: 1 - upper pool; 2 – sediment-protective threshold; 3 - sluiceway tract; 4 - gate of a sluiceway tract; 5 - ground insert; 6 – offtake riverbed.

For perfection of methods of management of riverbed and hydraulic processes at the water intake with the account of hydro-morphometric features of the mountain rivers on the basis of theoretical and modelling researches three designs of low-head dam water intake structure have been developed (fig. 4): Water intake structure for derivational river plants (patent KR № 607, 2003), Water intake structure for the mountain rivers (patent KR № 920, 2006, Water intake structure from the mountain rivers (the patent of the Russian Federation for useful model №133537, 2013). Elements used in their design and devices allow to create necessary hydraulic forms in a water intake zone.

On three experimental installations in KRSU complex laboratory researches of hydraulic processes at the water intake from the mountain rivers in scale from 1:7 till 1:25 have been executed. Methods of carrying out of physical modelling of two-phase streams (water, sediments) were based on I.I.Levi's recommendations, M.A.Mihalev [16], A.B.Veksler [17], based on conditions of observance of the kinematic similarity. At the first stage of experiments the rational sizes of three-section broken in plan sediment-protective threshold (fig. 4a) have been defined. Criterion was allowable value of hit of sediments in offtake canal. Investigating throughput of four-section broken in plan sediment-protective threshold (fig 4b,), function for definition of discharge coefficient of such surface spillway has been gained:

$$m_{4c.n} = -0{,}0281\left(\frac{H_{spt}}{C_{spt}}\right)^2 + 0{,}171\frac{H_{np}}{C_n} + 0{,}1277 \quad (5)$$

This function is gained for a relative average head of H_{spt}/C_{spt} = 1,3... 3,2, where H_{spt} - an average head value, C_{spt}- thickness of sediment-protective threshold wall.

By results of experimental researches of a sluiceway tract it has been defined that for the least capture of carrying sediments in water intake chamber10 the relative height of intermediate pier 7 (see fig. 4) between a river bay and sluiceway tract should be equal h_{ip}/H_R=0,743…94. Here h_{ip} - height of an intermediate pier, H_R - depth of water in the river before a water intake gate.

DOI: 10.1051/matecconf/20167301006

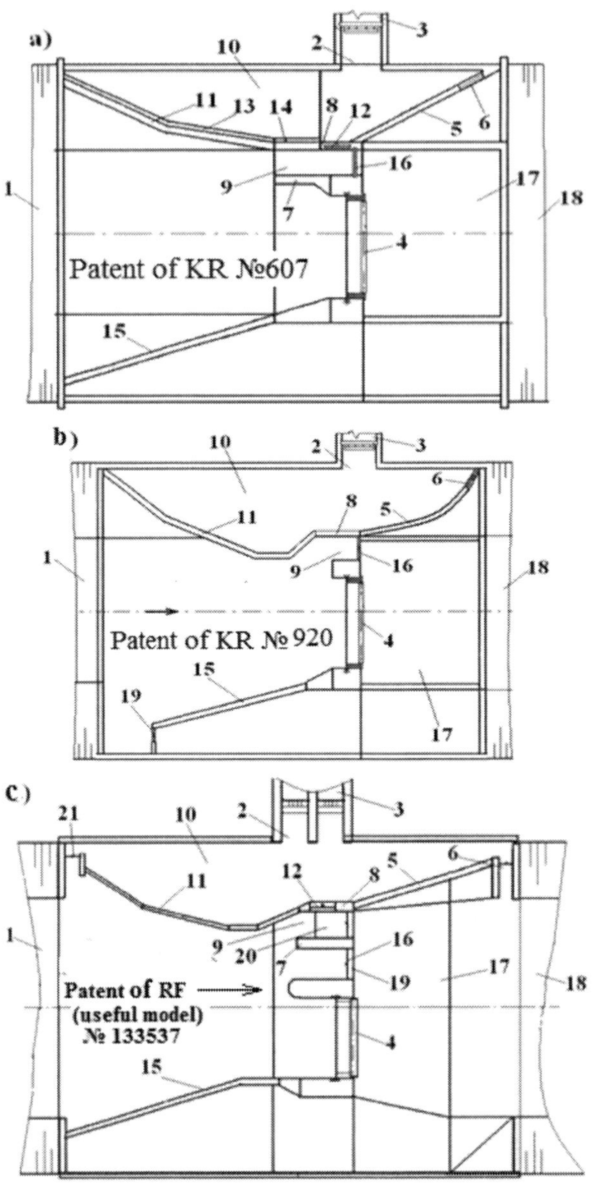

Fig. 4. Layout drawings of offered water intake structures: 1 – approaching riverbed; 2 - water receiver 3 - offtake canal; 4 - automatic gate of level of the upper pool; 5 - autosurface spillway of the water intake chamber; 6 - removal aperture; 7 - intermediate pier; 8 - dividing wall; 9 - sluiceway tract; 10 - water intake chamber; 11 - broken in plan circulating sediment-protective threshold; 12 - aperture of the winter water intake; 13 - inner edge of a threshold; 14 - inset parts for stoplog; 15 - autosurface spillway of upper pool. 16 - twin gate; 17 - hearth; 18 - offtake riverbed; 19 - additional washing aperture; 20 - ground insert; 21 - upper water intake aperture.

Maintenance of a necessary (command) water level in the upper pool of offered types of water intake structures is realized by means of known designs of automatic gates 4 of hydraulic action [1].

Carrying out the analysis of means of hydraulic automation of process of water supply in an offtake canal, the new design of the stabilizer of a water discharge of section parabolic box-shaped gate type (SPBG) has been developed which scheme is shown on fig. 5.

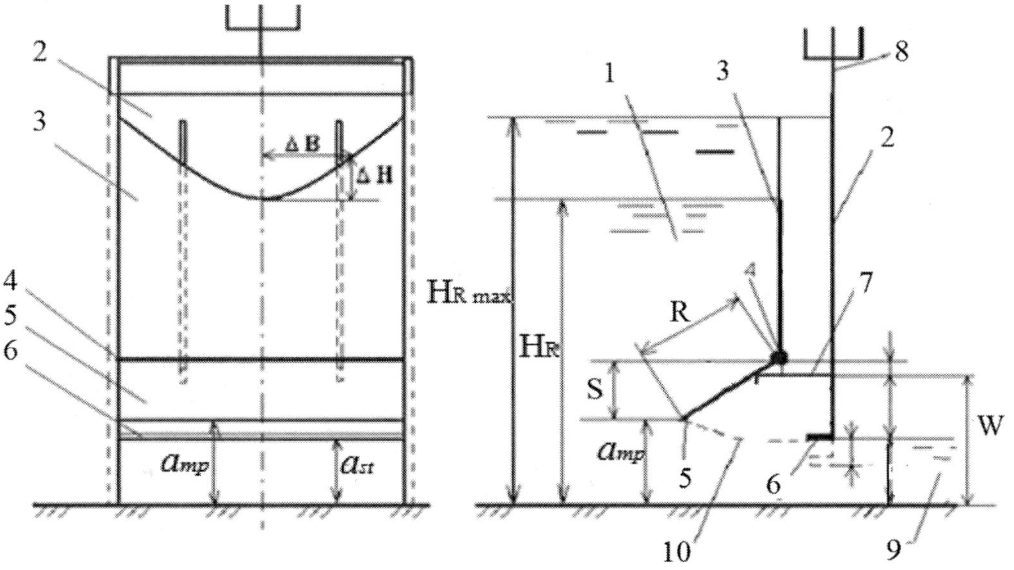

Fig. 5. Designed scheme of SPBG: 1 - upper pool; 2 – flat gate; 3 - head wall; 4 - hinge; 5 - rotary plate; 6 - horizontal peak; 7 - interacting rod; 8 - gate lifting screw; 9 - downstream; 10 - jet contraction path; 11 - path of displacement of the lower edge of the gate at regulating.

Distinctive features of the stabilizer are:

- for exploitation in the conditions of driftwood and trash transport a head wall 3 of discharge stabilizer are arranged separately from the flat gate 2 that allows to make its installation in water receiver before high water passage;

- to raise the accuracy of stabilization of the charge the heckle of the head wall 3 has not the step shape, as at its analogues [1], but the form of parabola which is carried out on the co-ordinates counted by formula [2]:

$$\Delta B = \frac{\mu_{rp} a_{mp} B\sqrt{2g(H_R - a_{st})}}{\mu_{rp} a_{mp}(B/\Delta B - 2)\sqrt{2g(H_R + \Delta H - a_{st})} + 2\mu_{hp} a_{st}\sqrt{2g(H_R + \Delta H - a_{st})}} \qquad (6)$$

Where μ_{rp} and μ_{hp} - water discharge coefficients at the outflow under rotary plate and a horizontal peak; a_{mp} and a_{st} - opening of a moving plate and the stabilizer; B - width of the water receiver; H_R - designed depth in the upper pool. It has been experimentally defined that the lapse in water discharge stabilisation at various opening of gate SPBG does not exceed 4 %. It is confirmed, that the given design of the discharge stabilizer can be used as the water-measuring device.

After definition of rational parameters of sediment-protective elements of water intake structures, we spent a number of modelling researches on perfection of technical characteristics of the devices serving for the winter water intake.

In the course of modelling researches (fig. 6) with a glance of winter hydrological regime of the mountain rivers and deficiencies of the organisation of the winter water intake on the existing water intake structures detected during our researches [1] and researches by A.V. Shipilov [17], the advanced layout drawing of the water intake structure for the mountain rivers (WISMR - 2) is developed.

Fig. 6. Model of water intake structure for the mountain rivers of type WISMR - 2 in winter regime of maintenance.

According to the maintenance circuit design during the winter period for the developed configuration of the water intake structure water levels in the upper pool manually downgraded to a level of a horizontal heckle of sediment-protective threshold [18, 19]. It is made for increasing of superficial speeds of a stream in ice-brash removal tract. Water selection to the water intake chamber is spent from near-bottom layers of a stream through the aperture of the winter water intake 12 (see fig. 4), and also from under the gate of the upper water intaking aperture 21 (fig. 4b). Ice-brash formations which arrive on an approach riverbed, are removed in the downstream through an upper surface of a ground insert 20 at a raising of the gate of the basic sluicing tract 9 and through a heckle of the put down twin gate 16 of additional sluicing tract 19 (fig. 4c). Mathematical modelling of some processes of the winter water intake from the mountain rivers has been executed by A.V.Shipilov [16] and matched with results of our experiments.

Researches of seasonal hydrological regimes influence on change of a landform of a bottom in the upper pool low-head dam water intake structures are executed. Also on model have been studied the conditions at which occurs sedimentation of the water intake structure downstream. Studying of kinematic characteristics and carrying ability of a two-phase stream on investigated water intake structures in various regimes of maintenance [1] is spent.

For the purpose of raise of carrying ability of a stream in structure approaching riverbed, before the winter regime of maintenance, it was recommended to make partial washing of the upper pool on the technique approved on physical models [2].

The designed water intake structure for diversion hydropower plants (WIS for DHPP) was introduced in 2008 at reconstruction of the hydrosystem of Issyk-Ata HPP in Kyrgyzstan and at construction of a diversion HPP-3 on Merke river in Kazakhstan in 2010. On the basis of field research, taking into account the recommendations of [19, 20], at the facilities of the industrial implementation of efficiency and effectiveness for the water intake structure diversion hydropower plants on small mountain rivers [1, 2] confirmed. The deficiency of their exploitation is an adjournment of a prism of river sediments in the upper pool of water intake structure on Merke river and transport of superfluous quantity of ice-brash in the diversion canal of Issyk-Ata HPP.

Detected deficiencies have been considered in 2013 at working out of layout drawing of WIS for DHPP -2 in project of small HPP on the river Usek in Almaty area of Kazakhstan.

3 Conclusion

Conducted researches of hydraulic and riverbed processes of the water intake and calculations of technical and economic efficiency on the specified installations of an industrial implementation confirm expediency of use of the developed structures at the water intake from the mountain rivers in hydropower and irrigational systems. On the basis of carried out researches bases of a technique of engineering calculation of the offered designs of water intake structures and the recommendation about their designing and maintenance in various hydrological regimes [1] are developed.

References

1. N.P.Lavrov, *Hydraulic structures for small hydropower engineering of mountain and foot-mountain area* (KRSU, Bishkek, 2009)
2. G.I. Loginov, *Gidrawlicheskie prozessy pri wodozabore iz gornych reck* (KRSU, Bischkek, 2014)
3. R. Kromer, Gidrosoorugenija, **3**, 11-18 (2009)
4. K.F. Artamonov, A.N. Kroschkin *Voprosy gidrotechnicheckoi klassifikacii ruslovych uchastkov i gidrometrii gornych reck, Morfologia rechnych rusel i ich modelirovanie* (Gidromteoizdat, Leningrad, 1972)
5. I.S. Rumjanzev, R. Kromer, *Prirodopribligennoe vosstanovlenie i expluatacia vodnych objektov* (MGYP, Moscow, 2001)
6. A.V. Filonchikov, *Proektirovanie avtomatizirovannych vodozabornych uzlov na gornych reckach* (Frunze, 1990)
7. G.I. Loginov, N.P. Lavrov, *Klassifikacia vodozabornyx soorugenii. Materialy VIII mezinarodni vedecko-prakticka conference "Dny vedy" po sekcich "Technicke vedu"* (Education and Science, Praha, 2012)
8. G. Zolezzi, J. Fluid Mech., **438**, 183 – 211 (2001)
9. A. Schucky, Thans. Amer. Soc/ Civ. Engr., **115**, 49 (1950)
10. N. Reidar, B. Olsen, *Numerical Modelling and Hudraulics, Department of Hudraulic and Environmental Engineering the Norwegian University of Scienee and Technology*
11. G. Zolezzi, R. Repetto, M. Tubino, M. Toropov, M. Serafini, *Mathematical modeling of silting in Kugart river, Kyrgyzstan, River Costal and EsturineMorphodynamics: RCEM 2007 – Dohmen-Janssen &Hulscher* (Taylor & Francis Group, London, 2008)
12. A.J. Milovich, *Teoriya dinamicheskogo vzaimodeistviya tel i gidkosti. Izdanie – 2, ispravlennoe i dorabotannoe* (Gos. izd-vo literatury po stroitellstvu i architecture, Moscow, 1955)
13. Ch. C. Edson, Civil Engineering, Engineer's Notebook, **1654**, 64-65
14. I.R.D. Francis, Proc. Roy Soc. L, **332**, 443-471 (1973)
15. P. J. Kerssen, A. Urk, J. Hydr. Engrg., ASCE, **112(7)**, 641- 656 (1986)
16. M.A. Michalev *Fizicheskoe modelirovanie gidravlicheskix yavlenii* (isd. Politech. un-ta, SPb, 2012)
17. L. Svatovskaya, A. Sychova, M. Sychov, V. Okrepilov, MATEC Web of Conferences, **53**, Article Number 01023 (2016)
18. J. Giesecke, E. Mosonyi, *WasserkraftanlagenPlanung, Bau und Betrieb4., aktualisierte und erweiterteAuflagemit 368 Abbildungen Springer-Verlag* (Heidelberg, Berlin, 2005)
19. S. Skiena, *The Algorithm Design Manual: 2nd Edition* (Springer, New York, 2008)
20. T. Abbasi, S.A. Abbasi, Renewable and Sustainable Energy Reviews, **15**, 2134-2143 (2011)

Examination of the Basement of Historic Buildings in Investment Activity

Aleksey Ulybin[1,*], *Aleksandr* Lanko[1], *Nikolai* Vatin[1], and *Katerina* Lysnytska[2]

[1]Peter the Great St. Petersburg Polytechnic University, Polytechnicheskaya 29, 195251 St. Petersburg, Russia
[2]National Aviation University, 03058, Kosmonavta Komarova, 1, Kyiv, Ukraine

Abstract. The process and methodology of the survey of basements rarely mentioned in the various construction rules and regulations. Basically describes the procedure of conducting a detailed survey of some of the individual elements. These surveys are fundamental in nature, include a large number of estimates and require significant financial and time costs. Usually the purpose of these surveys is to check the state of the building as a whole, it's safe operation or before starting of reconstruction. In the process of selecting areas of investment activity such large-scale survey is not possible. Needed a quick and inexpensive method intended for decision about investment in a particular object. At the same time, the survey should cover all the elements of the basement significantly affect the cost of reconstruction of the basement associated with his penetration. The article presents the general conception of conducting a rapid survey. The described methods and technologies applicable to the examination for the purpose of making decisions about investments in reconstruction of a basement level rooms. The composition of the works and their sequence. A comparison of the advantages and disadvantages of different methods. The practical examples. Scheme of conducting a rapid survey of the basement. The article analyzes the materials used in the construction of historic buildings in St. Petersburg.

1 Introduction

In today's world investors, specializing in real estate, need to look for new opportunities to find the "extra" square meters. For cities with a long history, particularly cities where the cost of commercial real estate in central, generally historical part of the city is significantly higher than on the outskirts, reconstruction of existing buildings, is relevant. One notable example is St Petersburg, where every year a large number of houses were also renovated in the form of deepening of the basement level rooms. The foundations of such houses are mostly rubble, made from local stone materials, often with the presence of wooden elements - beams, raft foundations, piles. As stone materials were used in the early years of the building of Saint Petersburg rounded rocks (boulders of igneous or metamorphic rocks), in later periods no rounded rock fragments (platy stone from limestone, dolomite,

* Corresponding author: ulybin@mail.ru

sandstone). Sometimes doing the whole foundation with brick of burnt brick. The most widespread rubble foundation as a building material was used mainly platy limestone. The foundations were carried out, as a rule, in the trenches, which were dug out without fixing, with vertical walls. The wooden elements in the foundations of historic buildings in the form of sills from logs with a diameter of 27 - 33 cm, stacked on a bottom of the trench or piles no longer than 6-9 m scored in the soil at the bottom of the trenches, were used in the presence of weak soils and high ground water. Historically, existing basements, as a rule, are not adapted for facilities management in modern conditions, do not meet modern standards and need to increase their height. To investors the problem of determining the volume and composition of works to bring the premises into compliance with the goals and objectives of future use. We need to develop a simple and versatile algorithm to survey the basement for the purpose of decision making about reconstruction. Certainly the results of the survey can only be used for choosing the object of investment, but not to begin work on the development of the project. For the development of the project of reconstruction it is necessary to perform complex research and survey work to identify the actual technical condition of the building and the activities of its recovery [1,2].

Currently you can find a huge number of works devoted to the problem of inspection of buildings [3-15], but mostly they are devoted to a detailed examination. The main objective of this work is to develop the most simple, fast and inexpensive method of examination of different elements of the building for the purpose of making the decision to reconstruct the basement. The novelty in this approach lies in the fact that the survey is not conducted for the purpose of the reconstruction, and for the purpose of making any investment decision, the investor who takes the decision to invest in this space. In contrast to the detailed survey, which devoted the most part of scientific works, the investor has no time and funds to carry out these examination. Reducing the cost of operations in this case are "at the forefront" in the competition for a particular object [16]. Often when buying premises there is no possibility of conducting any survey except visual. A striking example of such scheme of sale of bidding areas are held by the Property Fund of St. Petersburg [17]. During the inspection of the premises at these auctions only allowed a visual inspection of the premises. In this paper we consider the case when the investor can still do the testing, does not violate the integrity and value of the proposed premises. Of course the time factor and the cost of works will continue to solve critical role.

2 Description of the study

The objective of the study is to develop a survey as soon as possible, create a comprehensive survey method based on the combination of various existing methods. The examination should include all necessary elements endemic areas and the building as a whole:

- exterior walls and walls in basement level rooms;
- the foundation of the building;
- the ceiling of the basement level rooms;
- utilities passing through the basement.

Definitely perfect for these parameters, visual method, but in most cases it does not give results sufficient for decision making about investments in a basement. Unfortunately, most of the grounds of the building are not accessible for visual examination method, namely the condition of the grounds of the building is the defining element in assessing the amount of future investments.

The first thing to do would be to visually inspect the walls of the building under which is located the basement. Unfortunately, it is difficult to find a universal algorithm for visual inspection, in the modern normative documents, there is practically no guidance on the procedure for conducting surveys, about the sequence of actions [18]. For the purposes of this survey, it is necessary to inspect both the front and the courtyard section. Should identify the presence and parameters of cracks: the breadth and depth of disclosure, the location, length, angle of inclination, nature of origin. The detection characteristic of cracks, distortions of parts of the building, fissures of walls and other damage and deformation indicates poor base condition. In this case, conducting a detailed survey to identify causes of deformations is required and further conduct a rapid survey is not advisable. To determine the "critical point" after which there is a high degree of probability to argue that the deepening of the basement will require strengthening of bases you can use the tables proposed in VSN 53-86(R) [19] to determine the physical wear of belt stone foundations and brick walls. According to this document depreciation of 40% requires a significant strengthening of structures. This level of damage speaks about the critical condition of basements and foundations of buildings and, with high probability, will require strengthening. Thus, in the first stage of the rapid survey after the visual detection method physical deterioration of the walls and visible portions of the foundations more than 40% discontinue further examination. According to VSN 53-86(R) this wear has the following features: "the buckling and noticeable curvature of the basement, through cracks in the basement with the development of the full height of the building, buckling floors and walls of basement" [19]. Also a quantitative assessment of the degree of wear: "Uneven, precipitate with common deflection walls to 0.02 of its length" [19]. Unfortunately, these parameters determine the physical wear is outdated and requires amendment. For example, decrease the pitch degree of physical deterioration is at least 2, preferably 4 times. At the present time step is 20%. However in this survey they are perfectly applicable.

After visual inspection of the walls and visible portions of the Foundation necessary to proceed to survey the hidden part of the basement. For investor purposes the examination of the grounds of the building, when evaluating investments in a basement are:

- determining the status of the hidden part of the basement;
- definition of the depth of the foundation base;
- determination of groundwater levels and their intensity;
- the presence of a wooden ledge.

The main difficulty of this stage is the need for excavation or drilling operations, which is not always possible when considering the acquisition of a basement. The consent of the current owner to carry out these works. In case of impossibility of the above works, the study will be limited to preliminary visual inspection, which in most cases does not allow to make a reliable prediction about the possible costs of reconstruction of the premises. In this, unfavorable to the investor case, you should stop the rapid survey and to conduct a detailed examination, or to refuse investments. With the consent of the current owner for excavation or drilling operations, you should choose one of the three types of work:

- drilling holes by perforator;
- drilling of cores by diamond drilling machine;
- digging of pit.

Each of these methods has its advantages and disadvantages and choose the type of work should be based on the specific situation.

In the course of examination of a basement in the street Blohina of the Petrograd district of St. Petersburg was a comparison of the advantages and disadvantages of the above methods. 3 dug pit at the base of the Foundation and conducted one trial drilling rig. Methods of examination of grounds of building a drilling rig is described in detail in the work of Kolmogorova S. S. [27]. On the digging of one pit left in an average of about 8 man hours, with most of the work had to be performed below the groundwater level, constantly evacuating the water which was an additional difficulty. Also had to use a heavy jackhammer to overcome the floor plate 30 centimeters wide, the drilling of a single core took about 4 hours, the nature of the work were substantially lighter in nature, less physical labor. Data on the depth of the Foundation and the groundwater level coincided. The difference was less than 5 centimeters.

Examination by excavation of a pit requires large time and cost, but has a number of advantages. Inspection pit allows you to define the cross section of the Foundation from sinking, which is impossible when core drilling. Additionally, the pit is more evident, when it is adopt is exposed about one meter of the Foundation base that allows you to make more accurate measurements. The main disadvantage of the excavation pit is the extent of damage to the cellar. Destroyed not less than 1 square meter of the floor, this disturbed the waterproofing of floor, more expensive to restore.

Drilling of cores is important when large areas of the basement (over 200 m2) or the complex configuration of the basement, when according to GOST 31937-2011 [20] requires a large number (more than 10) cores. In small areas where you can be limited to 2-3 points, especially in the case of ground floors, on the contrary, the digging of pits is preferable.

Drilling holes by perforator is the most quick and inexpensive way to survey the hidden part of the Foundation, but it is the least informative. With the help of drilling of the hole to determine the depth of the Foundation, but the data collected in this way are less accurate than revealed by the excavation of pits or drill core. Accurate determination of current groundwater levels and their degree of intensity is not possible.

The choice of survey method the hidden part of the basement and foundations is influenced by:

- negotiation with the current owner of the particular method;
- the data type that you want to receive with the survey;
- the size and complexity of the configuration of the basement level rooms;
- material of the floors of the basement level rooms.

During practical examination of a basement on Blokhina street in St. Petersburg was drawn up a comparison table (Table 1) of examination methods of the hidden parts of the basement.

Table 1. Comparison methods of excavation the pit and drill core.

	Method of drilling borehole	The method of digging the pit
Speed survey for one object	4 man/hour	8 man/hour
Damage from the survey	Small	Large
Measure the depth of Foundation base	Yes	Yes
Drawing the cross section of the Foundation	No	Yes
Measurements of groundwater level	Yes	Yes
Measurement of the nature of groundwater	No	Yes
Determination of the presence of wooden sills	Yes	Yes

Regardless of the choice of survey method the hidden part of the basement in the first place should determine the depth of the Foundation base. According to calculations by various authors [21,22] when the distance from the sole of the basement floor to the basement at 50 centimeters or more in most cases there is no risk of the lifting soils and additional amplification is not required. This is confirmed by various building codes and regulations for the design of foundations. For example in 50-302-2004 TSN: "the Depth of the base of the foundations from the floor of the underground (basement) shall not be less than 50 cm" [23]. Indirectly this is confirmed by SP 50-101-2004: "allowed to take the depth of the Foundation from the basement floor less than 0.5 m, if the calculation on bearing capacity" [24]. Thus one of the fundamental factors in the rapid method is the distance from the ceiling of the basement (ceiling between the first floor and basement) to the soles of the Foundation. This distance will allow us to determine what the maximum ceiling height of the basement level room is possible to achieve without strengthening the foundations of the building (Eq 1). As a rule the process of strengthening of the bases is so time-consuming and expensive that the reconstruction of the basement coupled with this process is not appropriate.

$$H_{\max} = S - h_{safe} - h_{floor} \quad (1)$$

Where:

H_{max} - the maximum height of a basement level room can be achieved without strengthening the foundations of the building;

S - the distance from the Foundation base to the ceiling between basement and first floors;

h_{safe} - minimal layer of sole Foundation, preventing lifting the soil from under the basement, is taken to be equal 50 centimeters.

h_{floor} - height of floor "pie".

If the estimated height of the ceiling suits the customer based on the intended use and requirements, we should continue the rapid survey. In the case where the estimated height is not enough for operation of the premises, it requires detailed examination for planning of measures on strengthening of the bases. In such unfavorable for the investor case, you should refuse further Express survey.

After determining the calculated height of the basement with the requirements of the customer, it is necessary to determine the groundwater level. This indicator will significantly affect the cost of waterproofing the room. According to the report of Nikolaev and A. S. Savenkova, G. B. ground in St. Petersburg is represented everywhere at depths of from 0.5 to 3.5 meters [25]. The report presents a detailed map of groundwater level in St. Petersburg. Thus, in an ideal for the investor case, there is a possibility that the planned floor of the basement is above the ground water level. The described situation is possible with "high" basements, when a significant portion of the basement level room located above ground level. In such cases, it is possible to confine a light waterproofing that protects against short-term pressures of groundwater. However, you have to take into account seasonal fluctuation of groundwater that can be significant To determine fluctuations in ground waters, in 2008, according to the results of groundwater monitoring of the territorial reach of the state unitary company "SF "Mineral", was built the digital map of maximum predicted groundwater levels in St. Petersburg (the authors Savenkova, G. B., Fedyk, G. N.) [26]. However, the determination of the current level of groundwater directly to the site of the planned reconstruction is important when calculating the cost of works on waterproofing of the room. According to the report of Nikolaev and A. S. Savenkova, G. B. ground water of St. Petersburg are unconfined in nature, which makes possible the waterproofing of basements, but the identification of confined groundwater of the upper pressure aquifer can significantly increase the cost of waterproofing the space and make it a reconstruction is not feasible. Definition of groundwater level can be performed both by the open pit and drilled to the core. However, monitoring of the groundwater level in the pit is preferable to determine the velocity of the water after pumping to detect possible discharge of groundwater. After finding the level of groundwater should determine the presence of a wooden ledge under the base and their status. According to research by Kolmogorsky S. S. wooden sills occur in about 20% of foundations of historical buildings in St. Petersburg [27]. As mentioned previously, the presence of a wooden ledge is a dangerous character by lowering the level of ground water below the ledge. In this undesirable case, the wooden ledge beginning to crumble. S. I. Alekseev in his work [28] says that the process of decay in this case can go so intense that destroys the wood for 3-5 years and inevitably leads to uneven precipitation of additional seals. Thus, the presence of a wooden sills and their location must be considered in the groundwater during the reconstruction of a basement, and if you install the emergency sump pump. Undoubtedly, when it detects the wooden sills should abandon the method of penetration of a basement associated with a long-term lowering of the groundwater level in case if it is below the level of the ledge.

The survey of slabs should be limited to a visual examination. A detailed survey and calculations are expensive and in need of reconciliation. Preliminary examination set the type of overlap (in the form of materials and design features), visible defects and damage, the condition of individual parts of the floors, repaired or enhanced, the existing floor load. When inspecting the slabs record the presence, length and opening width of cracks in the supporting elements or their mates. The deflections of the slabs is determined using methods of geometric and hydrostatic leveling. When carrying out a survey of overlaps detected by the visual method:

For reinforced concrete floors, available:

- deflection in excess of the permitted, possibly with the formation of cracks in the bottom (tensile) zone of reinforced concrete elements caused by exceeding the design load;
- peeling, cracking, blistering or flaking of coatings;
- loss of mortar from joints between precast concrete slabs;
- moisture, efflorescence, rust stains.

To masonry(brick) arches:

- cracks and other damage in the masonry vaults of stone (breaking up of stones of the arches in the castle, the quarters of the span or at the supports, loss of individual stones);
-stratification and a breakdown of the masonry of the vaults caused a significant overload of masonry.

For slabs on girders:

- determine the deflections, possibly with the formation of cracks in small vaults;
- access to beams is determined by the estimated size of the beam, the corrosion of the visible part of the beam (usually the bottom shelf).

The final stage is the examination of utilities that are located indoors. Their status and position can significantly affect to the cost of works on reconstruction of a basement. The examination of heating systems, hot and cold water to produce a survey of the corrosion condition of pipelines and heating devices. The corrosion state is measured by the maximum depth of corrosion of the metal wall and the average value of the narrowing section of pipe corrosion scale deposits in comparison with a new tube. The allowable value of maximum relative depth of corrosion damage of pipes should be taken equal to 50% of the wall thickness of the new pipe. Examination of corrosion condition of pipes allows to estimate the amount of communication needed to be replaced.

The next important element is the examination of engineering communications is to assess the possibility and necessity of migration networks within the basement. This situation often occurs when their location makes it impossible to use this basement for alleged purposes. As a rule of communication pass through the center of the room, occupying a large space, significantly reducing the "useful volume" of the room.In addition, when planning the formation of new openings, exits from the premises, you may also need to transfer communications. At this stage you should assess the amount and type of utilities to be transferred.The General concept of conducting a rapid survey is reflected in the algorithm of Fig. 1.

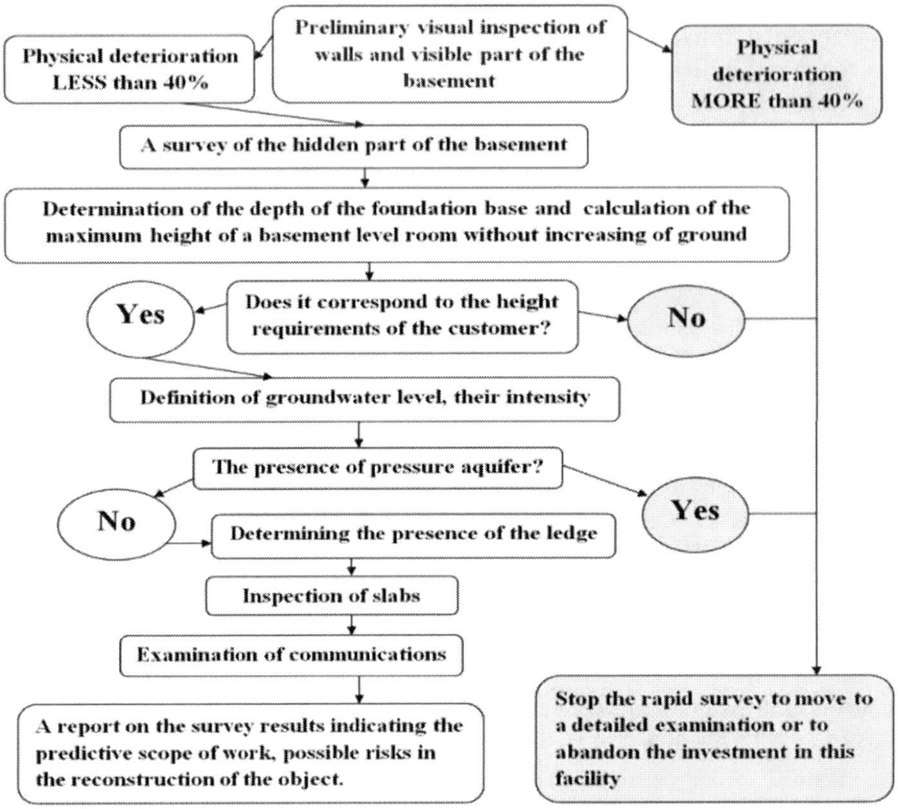

Fig. 1. The algorithm of carrying out a rapid survey of the basement in investment activity.

This algorithm was applied in the examination of a basement on Kazanskaya street in St. Petersburg. The article summarizes the results of this survey. After the initial visual inspection of the walls of the building from the facade and courtyard, walls of a basement are signs of uneven precipitation of the building is not detected. The decision was made to survey the hidden part of the basement. Given the size of the basement (120 square meters), as well as flooring material (clay), the digging pit method of survey was selected. Figure 2 shows a sketch of one of the pits.

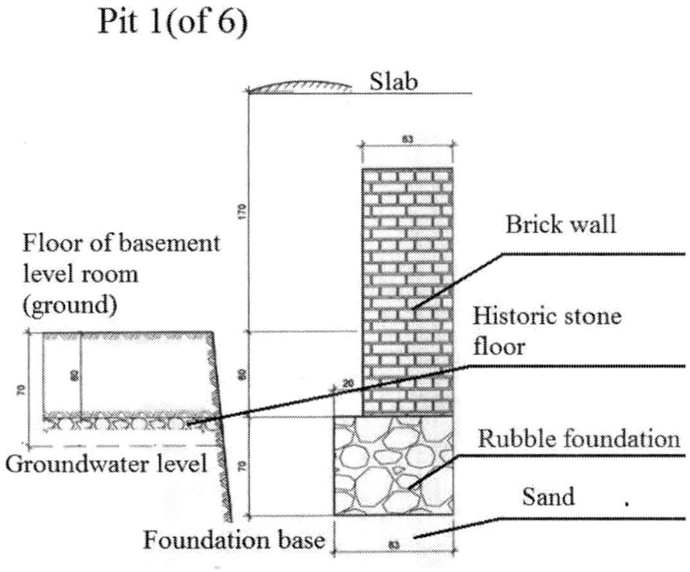

Fig.2. Sketch of test pit

According to the measurement made after excavation the pit, the distance between bottom of Foundation and bottom of the arch is about 300 centimeters. The planned thickness of the floor of the basement is 30 centimeters. Thus, the minimum ceiling height of the room without amplification of the grounds of the building is 220 centimeters to the bottom of the arch ceiling. Groundwater is at 10 centimeters below the planned floor level of the basement. Standard waterproofing with Bentonit plates is recommended. Overlap between the first and basement floors are made as the brick arch. The condition of the arches is satisfactory, significant deformation was not detected.

The results of the survey were established on the materials used in the construction of the Foundation, as well as the material of the base soil of the building. The lower part of the Foundation is made of rubble stone. Masonry made with lime mortar. The upper part of the Foundation is made of brick of high density. On the basis of samples taken from the base of the Foundation established under the Foundation are silty Sands.

3 Conclusion

The findings of the rapid survey of the basement level room are:

1. the Existing base of the building is in satisfactory condition, deepening of a basement to the recommended height possible without strengthening the foundations of the building
2. the groundwater level is small, the waters are unconfined in nature
3. ceiling are in satisfactory condition
4. basement is recommended for purchase and renovation.

As mentioned earlier, this rapid survey is used only for decision making about the purchase and reconstruction of a basement level rooms. For the initial design and main works detailed examination are required. However, the method gives to the investor opportunity to determine the volume of upcoming work, to assess the degree of risk in the reconstruction space. Given the inappropriateness of conducting a detailed survey of each room in the presence of a certain number of alternatives, this method can be used by the investor at the stage of decision making about reconstruction.

References

1. G. Concu, B., De Nicolo, *11th International Conference on Structural Repairs and Maintainance of Heritage Architecture – Tallinn,* 349-358 (2009)
2. G. Leucci, *14th International Conference on Ground Penetrating Radar*, 607-609 (2012)
3. S.V. Semencov, M.M. Orehov, V. I. Volkov, *Metodika provedeniya obsledovaniy i monitoringa tehnicheskogo sostoyaniya zdaniy s ispolzovaniem peredovih tehnologiy* (St-Petersburg, 2013)
4. V.I. Rimshin, V.G. Kazachek, N.V. Nechaev, S.N. Notenko, A.G. Roiman, *Obsledovanie i ispitanie zdaniy i sooryjeniy* (Stroyizdat, Moskow, 2006)
5. V.G. Kazachek, *Obsledovanie zdaniy i sooryjeniy*, (Stroyizdat, Moskow, 2006)
6. A.A. Varlamov, V.B. Gavrilov, U.M. Kruciliak, *Tehnicheskoe obsledovanie konstrykciy zdaniy i sooryjeniy* (RGN, Mignitigorsk, 2007)
7. R.Sh. Adigamov, *Opit stroitelstva, eksplyatacii i obsledovaniya fyndamentov zdaniy* (Inginerno-Economicheskiy Instityt, Cherepovec, 2007)
8. V.S.Abrashitov, *Tehnicheskaya eksplyatacia i obsledovanie stroitelnih konstrykciy* (Stroyizdat, Moskow, 2005)
9. A.I. Bedov, *Obsledovanie i reconstrukcia jelesobetonnih i kamennih konstrukciy ekspluatiruemih zdaniy i sooryjeniy* (Moskow, 1995)
10. A.G. Gindoyan, *Posobie po obsledovaniu stroitelnih konstrukciy zdaniy* (Informreklamizdat, Moskow, 1997)
11. V.T. Grozdov, *Tehnicheskoe obsledovanie stroitelnih konstrukciy zdaniy i soorujeniy* (Stroyizdat, St-Peterburg, 2001)
12. A.N. Dobromislov, *Ocenka nadejnosti zdaniy i soorujeniy po vneshnim priznakam* (Stroyizdat, Moskow, 2004)
13. O.V. Lygin, *Obsledovanie i ispitanie sooryjeniy* (Stroyizdat, Moskow, 1987)
14. V.V. Meshechek, *Posobie po ocenke fizicheskogo iznosa jilih i obshestvennih zdaniy* (Stroyizdat, Moskow, 1999)
15. L. Svatovskaya, A. Sychova, M. Sychov, V. Okrepilov, MATEC Web of Conferences, **53**, Article Number 01023 (2016)
16. A.V. Ulybin, S.V. Zubkov, Magazine of civil ingineering, **(17)**, 53-56 (2010)
17. Information on http:\\www.property-fund.ru
18. A.V. Ulybin, N.I. Vatin, Construction of Unique Buildings and Structures, **10(25)**, 134-146 (2014)
19. VSN 53-86(p) Pravila ocenki fizicheskogo iznosa jilih zdaniy (Stroyizdat, St-Peterdburg, 1986)
20. GOST 31937-2011 Zdaniya i sooryjenia. Pravila obsledovaniya i monitoringa tehnicheskogo sostoyaniya (2012)
21. V.K. Fedulov, L.U. Artemova, *Proektirovanie osnovaniy i fundamentov zdani i soorujeniy* (MADI, Moskow, 2015)
22. A.M. Kornilov, L.I. Cherkasova, *Proektirovanie osnovaniy i fundamentov melkogo zalojenia grajdanskih zdaniy* (MGSY, Moskow, 2005)
23. TSN 50-302-2004 SP 50-101-2004
24. A.S. Nikolaev, G.B. Savenkova, Problemi podtopleniya terrioriy Sankt-Peterberga gruntovimi vodami, information on http://www.infoeco.ru/
25. Environmental portal of St. Petersburg, information on http://www.infoeco.ru/
26. S.S. Kolmogorova, *Ocenka tehnicheskogo sostoyania bytovih fundamentov pri reconstrukcii zdaniy* (Stroyizdat, St-Petersburg, 2007)
27. S.I. Alekseev, *Osadki fundamentov pri reconstrukcii zdaniy* (Stroyizdat, St-Peterdburg, 2009)

Photogrammetry and Laser Scanning for Reconstruction and Restoration of Historical Buildings

Svetlana Golovina[1,*] *and* Svetlana Kanyukova[2]

[1]Saint Petersburg State University of Architecture and Civil Engineering, Vtoraya Krasnoarmeiskaya ul. 4. St. Petersburg, 190005, Russia
[2]Peter the Great Saint-Petersburg Polytechnic University, Politekhnicheskaya, 29, Saint-Petersburg, 195251, Russia

Abstract. Scientific and fundamental metrology concerns the development of new measurement methods. Technology development evaluates in the emergence of new devices, methods and techniques, allowing to obtain spatial and geometric information about constructions with greater accuracy and less labour- and time-consumption. Applied metrology concerns the application of measurement science to their use in society. Such information is of great importance, since the use of these data can provide, for example, a complete view of the construction project documentation accuracy or will allow for further structure and facility monitoring, deviations of structures from the design project, sediment and overall construction state. Another important application is the imprinting objects, in particular listed buildings, in order to preserve their appearance, meaning the use of this data in the restoration or rehabilitation facilities. During the subsequent processing information can be transformed into other forms of fixation like drawings and three-dimensional model. Metrology has thrived at the interface between science and manufacturing. Particularly, Laser scanning and Photogrammetry methods found their wide application in solving practical tasks in architecture and building construction. These are the most common methods of obtaining this information. The article presents the description of both methods and discusses their advantages and disadvantages, as well as aspects, which determine the use of a particular method in different situations.

1 Introduction

The problem of listed buildings and structures fixing methods improvement, in our opinion, is always actual. The traditional form of fixing as an album with photos saves just an object appearance information. It can't give exact information about geometrical properties of object as shooting is conducted on not metric and not calibrated cameras. There are also other known forms of fixing - drawings. They provide geometry, but don't represent the exact appearance of object. Besides, creation of drawings is labour-consuming and expensive. Emergence of new geodetic devices (system of laser scanning) and digital photogrammetry methods development can provide brand new forms of fixing [1].

Technologies of creating three-dimensional models of historical buildings and various construction objects have been developing for more than 20 years in Russia and abroad. Commonly model is created based on results of the land laser scanning (LLS). Many scientists take this approach as a most reasonable one because of the received result accuracy.

Use of laser scanning technologies in 3D design during product creation or modernization allows to create uniform design-technology model, provides effective enterprise designers, technologists and other specialists interaction, and in general considerably reduces terms and increases quality of technological product preparation. 3D design technologies allow to increase work productivity due to a possibility of implementing expeditious changes at any design stage, simultaneous work on model of several people, simplification of design decisions development by means of standardization [2]

A large number of scientific articles confirms efficiency of modern laser scanning methods use, this technology in the few years will become very convenient and technological standard of fixing and documenting of historical objects shapes [3]. Most often LLS is used in restoration projects such as Alexander Nevsky's church in Chelyabinsk [4] or Alexander Nevsky's temple in Novosibirsk [3].

Photogrammetry is generally used for obtaining drawings of buildings facades. This method of inspection has proved itself well in construction branch.

As two various technologies used for the solution of similar tasksPhotogrammetry and LLS were compared many times.

However, techniques allowing to receive 3D models directly from photos using only mathematical algorithms of processing which are carried out on personal computers appeared recently.

Combination of laser scanning data and Photogrammetry is interesting for foreign researchers also, such as Alshawabkeh Y., Haala N., Nuttens T., De Maeyer P., De Wulf A., Goossens R., Stal C., Baltsavias E.P., Ravanel L., Curtaz M., [5-23], who apply these methods in archaeological works and projects of historically important architectural objects archiving. Besides, researchers in other fields of science, such as a mine surveying [24] and mechanical engineering are engaged in studying of this problem [25].

Comparison of the automated models creation methods on the basis of laser scanning and a Photogrammetry data will allow to define strong and weak points of both methods which is necessary for their effective sharing [26-34].

2 Theoretical or experimental methods

The work purpose is comparison of model creation results in program complexes derived from laser scanning and from method of a photogrammetry and consideration of these models possible application for various purposes.

Along with advantages three-dimensional laser systems have also shortcomings: difficulties in irregular shaped objects interpretation; complexity of objects vectorization by "a cloud of points"; slow work with big sized files; it is necessary to work with fragments of an object to increase speed of materials processing. Complex usage of photogrammetric and three-dimensional laser scanning methods opens new opportunities for technological development of architectural constructions measurements and measurement drawings creation [26].

Beside the improvement of a scanning data in particular (a cloud of points method), it is possible to improve a method efficiency by a combination of it with the existing methods. Many modern scanners have the built-in camera which usually doesn't allow to receive the facsimile suitable for photogrammetric processing. But at the same time the modern digital cameras which are available on market for low price are allow to receive high-quality photographic materials. Joint digital facsimiles and a cloud of points processing can also increase data process efficiency considerably as facsimiles are much closer to person's natural object perception than a cloud of points, and also allow to obtain object information in places, which are hard for scanning, - with the presence of obstacles and sharp geometrical changes [27].

The calibrated digital camera, the application for processing of materials of tacheometric survey, a program complex for calculation of elements of orientation and transformation of pictures are necessary for a technology of three-dimensional building model creation with a photogrammetric method.

Listed stages are necessary to create a model by a photogrammetry method. The first stage is field works, including photogrammetric shooting of an object. Next step is a field materials processing which includes affixment results processing and picture correction of result, and also analyse and transformation of pictures in a program complex. The last stage is creation of a three-dimensional model and imposing of textures.

The land laser scanning system consists of LLS (land laser scanner) and the field personal computer with the specialized software. There are laser scanning systems with a built-in control system or support of the wi-fi technology for contact with the personal computer. NLS consists of the laser range finder adapted for work with high frequency and a laser beam development block. The servodrive - a type of the mechanical drive with sensors of provisions, speeds, efforts, etc. acts as the development block in LLS, and the control unit of the drive which is automatically supporting necessary parameters on the sensor and in the device itself, accordingly, depending on the set external value, and also a polygonal mirror or a prism. The servodrive rejects a beam at the set size in the horizontal plane, at the same time the whole upper part of the scanner which is called a head turns. Vertical motion is carried out due to rotation or swing of a mirror. During the scanning the direction of distribution of a laser beam and distance to object points is fixed.

The cloud of points which is a number of object's points with three-dimensional coordinates, from which there was a reflection of a laser signal, is a result of land laser scanning. The cloud of the points covering an object surface forms his spatial dot model.

Processing of the results happens in the program complexes allowing to transform the received cloud of points to other types of fixing: drawings, dot three-dimensional models, orthophotomap, raster development.

3 Experimental section

For model creation methods application the building of Pavlovsky leib-gvardii polka barracks has been chosen – the listed building of the 19th century. It represents a sample of typical Russian classicism and now it is protected by the state. The building is located at the Marsovo pole sq., 1, the central part of a facade facing Millionnaya st. has been chosen.

Fig. 1. Building of Pavlovsky leib-gvardii polka barracks

4 Results and discussion

Laser scanning has been conducted by means of the Leica Scanstation 2 scanner with an 6 mm accuracy of position in space definition , 4 mm distances to object accuracy and the 2 mm accuracy of a surface modeling. While shooting in the automatic mode of polygonal model creation was used, therefore total integrity of model was a little reduced. It has been made for research process acceleration. Usually when scanning architectural objects with manual modeling, which allows to receive more high-precision model. The figure 1 allows to allocate following shortcomings of the automatic mode of a polygonal grid creation:

1) the noise in the lower part which has appeared because of the parked cars and the passing pedestrians;

2) noise in windows – a problem of transparent surfaces scanning;

3) lack of data on shadowed parts of a facade (not visible from points of standing of the scanner) – places behind columns, the lower part of a pediment.

The model of a facade (fig. 2, 3, 4) is constructed of the cloud of points received when scanning by means of the Trimble RealWorks program complex.

Fig. 2. Building facade model in the Trimble RealWorks program

Fig. 3. The left-side view on a facade

Fig. 4. The fragment of façade from the right

The figures show that integrity of object and elements of a facade is preserved, even fine architectural details on model, which sizes can be received with high precision are distinguishable. The labour consisted of 3 man hours for field works and 3 man hours for data processing.

Photogrammetric shooting of a facade by means of the NIKON D90 camera has been executed also. 17 pictures with the resolution of 4288 ×2848 have been taken so that the pictures, next to each other according to an arrangement, overlapped each other not less than for 40%. The model of object has been executed in the Agisoft Photoscan program (fig. 5, 6, 7).

Fig. 5. The building facade Model in the Agisoft Photoscan program

Fig. 6. Left-side view on a facade

Fig. 7. The fragment of façade from the right

The figures show that the received model has a number of deficiencies, only the general form of object and the main architectural elements are preserved, and smaller sites of a facade and elements of the second place (a window behind columns and interiors of apertures) in general remain indiscernible.

Later the overlapping of models on each other has been carried out (fig. 8, 9, 10). It is visible on drawings that the model according to photos is much worse.

Fig. 8. Result of models overlapping

Fig. 9. Left-side view on a facade **Fig. 10.** The fragment of façade from the right

Total labour consisted 0,5 man hours for field works and 2 man hours for data processing.

5 Conclusions

The model received using a photogrammetric method can be used for purposes when high precision is not necessary, for example for the fact-finding purposes or for saving information about the appearance of object. The main advantage of this method is the possibility of receiving the textured model which is most precisely transferring visual characteristics of object (color, invoice). The model received by scanning has much higher precision and degree of specification - this method can be used for control of geometrical parameters compliance of again constructed objects and project documentation on these objects or their adjustments, and also for monitoring of these parameters. Besides, this way can be used for objects restoration, the lost documentation recovery and creation of drawings. At the same time texturing of such model is quite labour-intensive process as it is necessary to carry out overlapping of photos for this purpose which leads to the merge of photogrammetry and laser scanning methods.

References

1. S.V. Tyurin, S.G. Tikhonov, Magazine of Civil Engineering, **7**, 25-30 (2010)
2. I.K. Kiyamov, R.Kh. Mingazov, A.F. Muzafarov, R.A. Ibragimov, A.A. Sibgatullin, Exposition Oil & Gas magazine, **7(32),** 41-43 (2013)
3. A.V. Radzyukevich, M.A. Chernova, V.A. Seredovich, A.V. Ivanov, Interexpo Geo Siberia, **3,** 115-126 (2012)
4. I.G. Shchegolev, A.A. Demin, A.D. Olenkov D.S. Popov, PNRPU Construction and Architecture Bulletin, **1**, 95-98 (2011)
5. M.G. Sinkova, *Razrabotka i issledovaniye fotogrammetricheskikh tekhnologiy obmerov arkhitekturnykh sooruzheniy s ispolzovaniyem lazernykh trekhmernykh skanerov* (2004)
6. V.I. Galakhov, *Razrabotka metodiki sozdaniya vektornykh modeley obyektov po rezultatam nazemnogo lazernogo skanirovaniya i tsifrovoy fotosyemki* (2012)
7. Information on http://www.racurs.ru/www_download/articles/Camaras_digitales_rus.pdf
8. E.P. Baltsavias, ISPRS Journal of Photogrammetry and Remote Sensing, **54**, 83-94 (1999)
9. Y. Alshawabkeh, N. Haala, CIPA 2005 XX International Symposium (2005)
10. Y. Alshawabkeh N. Haala, D. Fritsch, *IAPRS*, **XXXVI,** (2006)
11. Information on http://citeseerx.ist.psu.edu/viewdoc/download?doi=10.1.1.42.6581&rep=rep1&type=pdf
12. Information on http://www.fig.net/pub/fig2011/papers/ts06e/ts06e_nuttens_demaeyer_et_al_5267.pdf
13. Information on: https://www.fig.net/pub/athens/papers/ts26/ts26_4_guarnieri_et_al.pdf
14. Information on: http://www.permanet-alpinespace.eu/archive/pdf/WP6_1_tls_tp.pdf
15. Information on: http://www.diva-portal.org/smash/get/diva2%3A120447/FULLTEXT01.pdf
16. Information on: http://kth.diva-portal.org/smash/get/diva2:139761/FULLTEXT01.pdf
17. Information on: URL: http://etd.fcla.edu/UF/UFE0000526/drake_d.pdf
18. Information on: http://cipa.icomos.org/fileadmin/template/doc/PRAGUE/079.pdf
19. Information on: http://www.isprs.org/proceedings/XXXVI/3-W52/final_papers/Gonzalez_Aguilera_2007.pdf
20. S. Al-Kheder, Y. Al-shawabkeh, N. Haala, Journal of Archaeological Science, **36(2)**, 537–546 (2009)
21. L. Svatovskaya, A. Sychova, M. Sychov, V. Okrepilov, MATEC Web of Conferences, **53**, Article Number 01023 (2016)
22. K. Lambers, H. Eisenbeiss, M. Sauerbier, D. Kupferschmidt, T. Gaisecker, S. Sotoodeh, T. Hanusch, Journal of Archaeological Science, **34(10)**, 1702–1712 (2007)
23. A. Pesci, M. Fabris, D. Conforti, F. Loddo, P. Baldi, M. Anzidel, Journal of Volcanology and Geothermal Research, **162(3–4)**, 123–138 (2007)
24. M. Sturzenegger, D. Stead, Engineering Geology, **106(3–4),** 163–182 (2009)
25. H. Gonzalez-Jorge, B. Riveiro, P. Arias, J. Armesto, Measurement, **45(3)**, 354–363 (2012)
26. A.S. Korchukov, A.V. Leonov, Technology and organization of building production, **1 (2)**,49-51 (2013) (rus)

27. A.V. Leonov, Yu.M. Baturin, I.A. Petropavlovskaya, Voprosy istorii estestvoznaniia i tekhniki (VIET), **3**, 156-170 (2013)
28. V.L. Kvint, V.V Okrepilov, Herald of the Russian Academy of Sciences, **84 (3)**, 188-200 (2014)
29. V.V Okrepilov, V.N. Krutikov, G.I. El'kin, Measurement Techniques, **57 (2)**,109-116 (2014)
30. V.V Okrepilov, A.Yu. Smirnov, Measurement Techniques, **56(1)**, 54-60 (2013)
31. V.V Okrepilov, Studies on Russian Economic Development, **24(1)**, 35-42 (2013)
32. V.V Okrepilov, Standarty i Kachestvo, **10**, 52-55 (2005)
33. V.V Okrepilov, Ivanova, G. Standarty i Kachestvo, **12**, 62-68. (2004)
34. V.V Okrepilov, Standarty i Kachestvo, **3**, 94-96. (2003)

Design Calculation of Drill-Injection Piles with Controlled Broadening and Silty-Clayed Soil Foundation Basic Interaction Parameters

Mikhail Samokhvalov[1], *Juriy* Zazulya [1], *Roman* Melnikov[1,*] and *Victor* Mironov [1]

[1]Industrial University of Tyumen, 625001 Volodarskogo str. 38, Tyumen, Russia

Abstract. The paper describes a new method of drill-injection pile installation formed by the sleeve technology with solution injection using the "fracturing" mode and the controlled broadening at the pile end. This method is aimed at high-quality and effective compaction of the soil under the strip foundation of the reconstructed building. Analyzed are the results of the in-situ tests intended to interaction of drill-injection piles manufactured by the sleeve technology and single and repeated solution injections using the "fracturing" mode and the controlled broadening at the pile end. The radius of the neighboring compacted soil massif zone was determined. The geometric parameters of the pile shaft broadening, and fractures were identified. The algorithm for calculating the basic parameters of the proposed piles was developed in order to forecast their interaction with a silty-clayed soil foundation under static loading. A comparison with classical analytical solutions after soil mechanics, theory of elasticity and plasticity was made.

1 Introduction

There are a large number of buildings and structures worldwide (including cultural heritage) which are in need of renovation, restoration and modernization in accordance with modern requirements which regulate the development of underground space of the buildings in order the objects of social and engineering infrastructure to be housed there [1 - 4]. As a rule, they are located in the constrained urban areas of city centers, in difficult engineering-geological and hydrogeological conditions [5-8]. In this regard, it is necessary to develop an efficient and reliable means for strengthening strip foundations by soil foundation compaction [1, 9-11].

The analysis of different ways of strengthening revealed that the embedded reinforcing elements of different shapes were of high-priority: piles with the controlled broadening at their ends, drill-injection (injection) piles with broadening in the diameter of the pile shaft and the sleeve technology with solution injection using the "fracturing" mode [2, 10, 12, 13]. At the same time, these ways incorporate the main drawback in a varying degree, i.e. unpredictable and uncontrolled expansion of the solution in silty-clayed soil massif, and as a consequence, possible irregular compressibility of the base in plan and other unfavorable effects [10, 14]. As a result, to further improve the workability and effectiveness of the

* Corresponding author: tgasu.melnikov.roman@mail.ru

proposed strengthening methods, the authors suggest the new combined universal method of drill-injection pile installation manufactured by the sleeve technology with solution injection using the "fracturing" mode and the controlled broadening at the pile end. (Fig. 1).

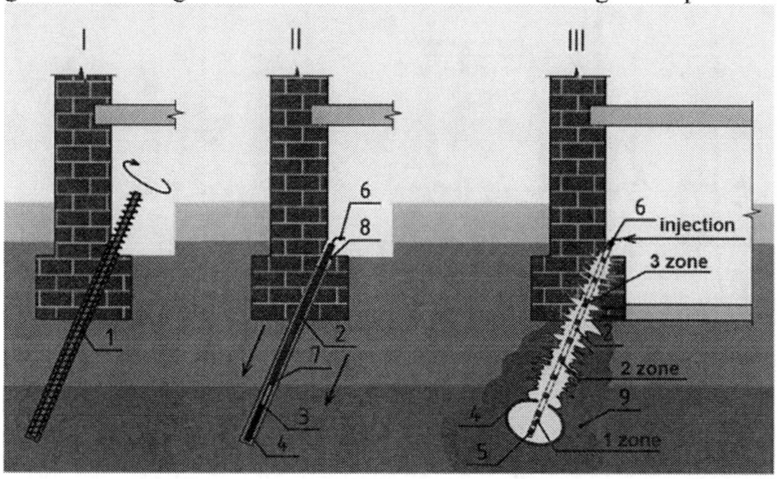

Fig. 1. Schematic illustration of drill-injection pile installation with the controlled broadening: 1-borehole, 2-injector tube, 3-clamp, 4- membrane cup, 5-rubber sleeves, 6 - hose, 7-packer, 8-concrete plug, 9-compressed zone of soil massif. [15]

2 Subject and methods of research

The injector tube with three zones of perforation is used for pile installation. The first zone is located at the end of the membrane cup, the second – beyond the end zone, and the third - in the zone of foundation strengthening. In doing the injection of the first zone the membrane cup is stretched and forms a broadening in the soil massif; the sizes can be controlled with the grout injection pump gauge by the volume flow and solution injection pressure. Then the packer is moved to the second and third zones to form pile shafts; solution injections are done for the foundation members being strengthened using "fracturing" and "healing of cracks" modes. The rubber shell of the membrane fits close the lower end of the injector as a "cup" [15].

3 Design calculation of the proposed pile basic parameters.

3.1 Calculation parameters in controlled broadening development.

Evaluated is the radial movement of the membrane cup wall u1 under injection pressure depending on the required depth of the broadening in the soil massif (pile length), injection pressure and the volume of solution, as well as the stress-strain state of soil under the foundation based on the analytical solutions after O. Hoffman, et al. [16-17]:

$$V = \frac{4}{3} \cdot \pi \cdot (r_b + u_1)^3 \Rightarrow u_1 = \left(\frac{3 \cdot V}{4 \cdot \pi}\right)^{\frac{1}{3}} - r_b \quad (1)$$

in elastic stage with $p_{in1} \leq p_{cr1}$:

$$u_1 = (p_{in1} - \sigma_{01}) \cdot \left(A_{21} + A_{22} \cdot \left(\frac{1+w}{2}\right)\right) \cdot \frac{1}{E} \quad (2)$$

in elastic-plastic stage with $p_{in1} > p_{cr1}$:

$$u_1 = \frac{1}{E} \cdot A \cdot p_{cr} \cdot {k_r}^{\beta} + \frac{1}{E} \cdot M \cdot \frac{2 \cdot (1 - {k_r}^{1-\alpha+\beta})}{1+\beta-\alpha} - \frac{1}{E} \cdot N \cdot \frac{1 - {k_r}^{1+\beta}}{1+\beta} \quad (3)$$

where r_b– initial radius (depends on the screw diameter), r_c=0.04 m;
E – deformation modulus in the broadening zone, kPa;
p_{in1} – solution injection pressure on the membrane cup walls, kPa;

$$\alpha = \frac{12\cdot sin\varphi_0}{3+3\cdot cos\varphi_0}, s = 1;\ \beta = \frac{2\cdot(1-sin\varphi_0)}{1+sin\varphi_0}; \tag{4}$$

$$\beta_1 = A_{11} + \beta \cdot A_{21}, \beta_2 = A_{12} + \beta \cdot A_{22}; A = (A_{21} + A_{22} \cdot \left(\frac{1+w}{2}\right) \tag{5}$$

$$M = \left(\beta_1 + \beta_2 \cdot (1-\alpha)\right) \cdot (p_{cr} + s);\ N = (\beta_1 + \beta_2) \cdot s; \tag{6}$$

σ_{01} – horizontal side deadweight contact soil pressure, kPa;

$$\sigma_{01} = \frac{(2\cdot\sigma_h + \sigma_z)}{3} = \frac{2\cdot\left(\gamma_{gr}\cdot z\cdot\frac{\nu}{1-\nu}\right)+\gamma_{gr}\cdot z}{3} \tag{7}$$

γ_{gr} – gravity weight of the soil massif, kN/m^2; z –depth, m;
ν – Poisson ratio; φ_0 – angle of internal friction;
p_{cr1} – critical pressure kPa of the membrane on the soil massif during solution injection when its elastic strains become plastic ones:

$$p_{cr1} = \frac{4\cdot(\sigma_0\cdot sin\varphi_0 + c\cdot cos\varphi_0)}{(1-w-sin\varphi_0(3+w))} \tag{8}$$

$$w = \frac{A_{12}-2\cdot A_{21}-2\cdot A_{22}}{2\cdot A_{22}} - \frac{\sqrt{4\cdot A_{21}^2-4\cdot A_{21}\cdot A_{12}+A_{12}^2+8\cdot A_{22}\cdot A_{11}+A_{22}^2+2\cdot A_{22}\cdot A_{12}+20\cdot A_{22}\cdot A_{21}}}{2\cdot A_{22}} \tag{9}$$

$$A_{11} = 1, A_{12} = -2\cdot\lambda\cdot\nu, A_{21} = -\nu, A_{22} = \lambda\cdot(1-\nu) \tag{10}$$

λ – multimodulus factor (ratio of compressive deformation modulus E to tensile deformation modulus E_t) allows the elastic-plastic characteristics of silty-clayed soil to be considered; it ranges from 1.0 to 5.0 depending on the type of clay and soil moisture. Fig. 2

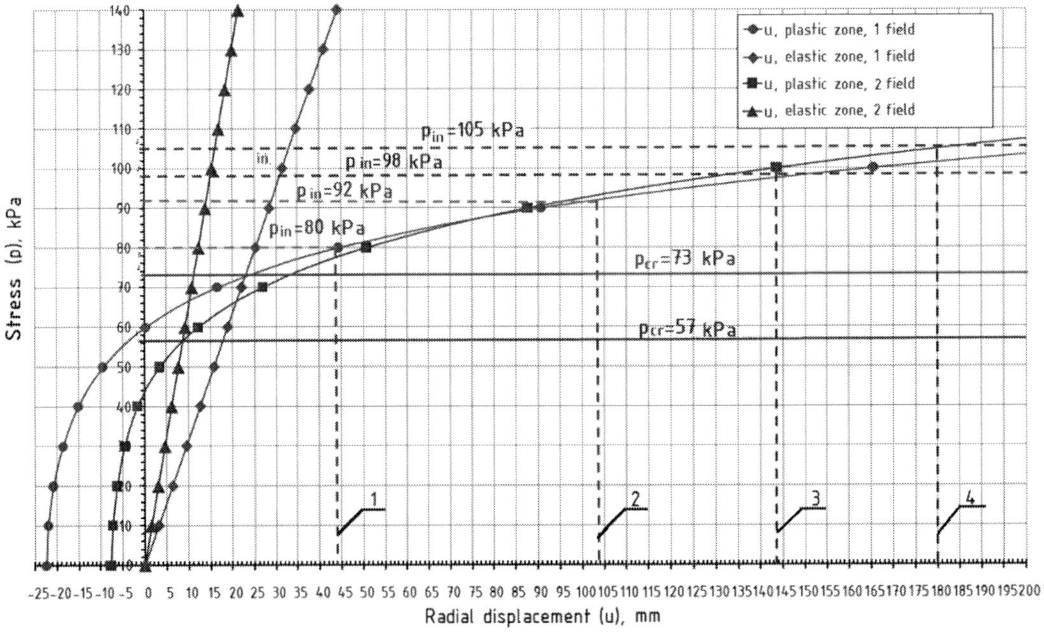

Fig.2. Dependency diagram u_1 from acting solution injection pressure in elastic and plastic zones of the soil massif: 1, 2 – on the 1st site V = 10 and 20 l correspondingly; 3, 4 – on the 2d site V= 30 and 40 l correspondingly.

Evaluated is the radius of the compacted zone depending on u_1:

$$R_{com1} = (r_c + u_1) \cdot k_{r1},\ k_r = \left(\frac{p_{in}^1+s}{p_{cr}^1+s}\right)^{\frac{1}{\alpha}} \quad (11)$$

Evaluated are the residual stresses σ_r^{res} depending on u_1 and R_{com1}:

$$\sigma_r^{res} = \sigma_r - \sigma_r^* \quad (12)$$

$$\sigma_r = -\frac{p_{in} \cdot (r_c + u_1)^n}{R_{com1}^n - (r_c + u_1)^n} \cdot \left(\frac{R_{com1}^n}{r^n} + 1\right), \sigma_r^* = -\frac{p_{in} \cdot (r_c+u_1)^2}{R_{com1}^2 - (r_c+u_1)^2} \cdot \left(\frac{R_{com1}^2}{r^n} + 1\right) \quad (13)$$

where n – factor to be determined from the power law of deformation under shear, $\tau = \tau_0 \cdot \gamma^{\frac{n}{2}}$, n – varies in the range of $0<n<2$ depending on the type of soil and its water saturation; in calculations it is determined on the basis of experimental data –parameters of soil pressure cells when broadening, $n=1.55$.

3.2 Calculation Parameters in pile shaft development.

The radial movement of the borehole wall $u_2$under the action of injection pressure p_{in2}, depending on the pile length of solution injection pressure under the condition $p_{req} > p_{in2} > p_{cr2}$, based on the analytical solutions after Z.G. Ter-Martirosyan, V.G. Fedorovsky, A.I. Polishchuk [18-19] et al. is evaluated by the formulas.:
in elastic stage of broadening with $p_{in2} \leq p_{cr}$

$$u_2 = (p_{in2} - \sigma_{02}) \cdot (A_{21} + \cdot A_{22} \cdot (1 + w)) \cdot \frac{1}{E} \quad (14)$$

in elastic-plastic stage of broadening with $p_{req} > p_{in2} > p_{cr}$

$$u_2 = \frac{1}{E} \cdot A \cdot p_{cr2} \cdot k_{r2}{}^{\beta} + \frac{1}{E} \cdot M \cdot \frac{1 - k_{r2}{}^{1-\alpha+\beta}}{1+\beta-\alpha} - \frac{1}{E} \cdot N \cdot \frac{1 - k_{r2}{}^{1+\beta}}{1+\beta} \quad (15)$$

where p_{in2} –solution injection pressure acting on the borehole walls, kPa;
σ_{02} –side contact soil pressure, kPa:

$$\sigma_{02} = \gamma_{gr} \cdot z \cdot v/(1 - v) \quad (16)$$

$$\alpha = \frac{2 \cdot sin\varphi_0}{1+sin\varphi_0}, s = \frac{C \cdot cos\varphi_0}{sin\varphi_0}, A = A_{21} + (1 + w) \cdot A_{22}, \quad (17)$$

$$\beta = \frac{1-sin\varphi_0}{1+sin\varphi_0}, \beta_1 = A_{11} + \beta \cdot A_{21}, \beta_2 = A_{12} + \beta \cdot A_{22} \quad (18)$$

$$M = \left(\beta_1 + \beta_2 \cdot (1 - \alpha)\right) \cdot (p_{cr2} + s), N = (\beta_1 + \beta_2) \cdot s \quad (19)$$

p_{cr2} – critical pressure, kPa, corresponding to the initial plastic strains is evaluated by the formula:

$$p_{cr} = \frac{-2(\sigma_0 \cdot sin\varphi_0 + c \cdot cos\varphi_0)}{(w+2)sin\varphi + w} \quad (20)$$

$$w = \frac{-A_{21} + A_{12} - 2 \cdot A_{22} - \sqrt{A_{21}^2 - 2 \cdot A_{21} \cdot A_{12} + A_{12}^2 + 4 \cdot A_{22} \cdot A_{11}}}{2 \cdot A_{22}} \quad (21)$$

$$A_{11} = 1 - v^2, A_{12} = -\lambda \cdot v \cdot (1 + v), A_{21} = -v \cdot (1 + v), A_{22} = \lambda \cdot (1 - v^2) \quad (22)$$

p_{req} – hydraulic fracture pressure, kPa, is evaluated by the formula (24);

Evaluated is the radius of the neighboring compacted soil massif zone due to the pile shaft diameter increase under solution injection pressure R_{com2}, depending on u_2:

$$R_{com2} = (r_b + u_2) \cdot k_{r2}, \; k_{r2} = \left(\frac{p_{in2}+s}{p_{cr2}+s}\right)^{\frac{1}{\alpha}} \tag{23}$$

3.3 Calculation parameters in hydraulic fractures development

Evaluated is p_{req}, depending on the depth of the injection horizon z, gravity weight of the soil massif and its mechanical strength characteristics – C и φ, as well as the strength of soil-concrete in compression (tension) σ_c, σ_t, - necessary to be taken into account in repeated solution injection in the "fracturing" mode [10,12]:

$$p_{req} = \gamma \cdot z \cdot \left(1 + \sqrt{3} \cdot tg\varphi\right) + \frac{2}{\sqrt{3}} \cdot C \, , \tag{24}$$

$$\text{где } C = \frac{\frac{2}{\sqrt{3}}\sigma_c \cdot \sigma_t}{\sigma_c+\sigma_t}, f = \frac{\sqrt{3}\cdot(\sigma_c-\sigma_t)}{(\sigma_c+\sigma_t)} \tag{25}$$

Evaluated is the radius of hydraulic fracture expansion $R_{fractur}$ depending on the interval of perforations and their quantity in the injector tube section, as well as the type of injection (single or repeated); due to their random development and uncontrolled expansion, $R_{fractur}$ is taken after the in-situ tests minus R_{com2};

The radius of the neighboring compacted soil massif zone is evaluated with regard to:

hydraulic fractures: $R_{com3} = R_{ycom2} + R_{fractur}$ (26)

change of mechanical characteristics E_{com2},e_{com2}, C_{com3}, proceeding from the radius of their expansion and the solution volume V_s, subjected to injection upon 1rm of the pile length [20-21]:

$$E_{com2} = \frac{E_c+E_a}{2}, \tag{27}$$

where e_a, E_a– air void ratio and modulus of deformation of soil massif reinforced by hydraulic fractures, MPa. It is evaluated by the formula (28);

e_c, E_c – air void ratio and modulus of deformation of soil massif compacted between fractures, MPa.;

e_d, E_d – initial modulus of deformation and air void ratio of soil;

E_s – modulus of deformation of solution injected into soil massif is taken equal to E_s=8000 MPa;

k_a – factor reflecting the proportion of hardened solution in the cracks of hydraulic fracture (minus air void filling) in a given volume of soil massif;

V_p – air void volume in the soil massif under study V_d, m^3;

The coefficient of compaction $C_{com.}$ is determined by SP 50-101-2004 with substitution of value e_{com}.

3.4 Evaluated is the regularity of settlement development in static loading

The settlement-load comparative diagrams (Fig.3) were plotted on the basis of the elastic-plastic model after S.P. Timoshenko and the analytical solutions after Z.G. Ter-Martirosyan [17-18] taking into account the changes in the mechanical properties of the neighboring compacted soil massif zone and preservation of the residual stresses up to the load the pressure of which does not exceed the value σ_r^{res} in the formulas (12, 27).

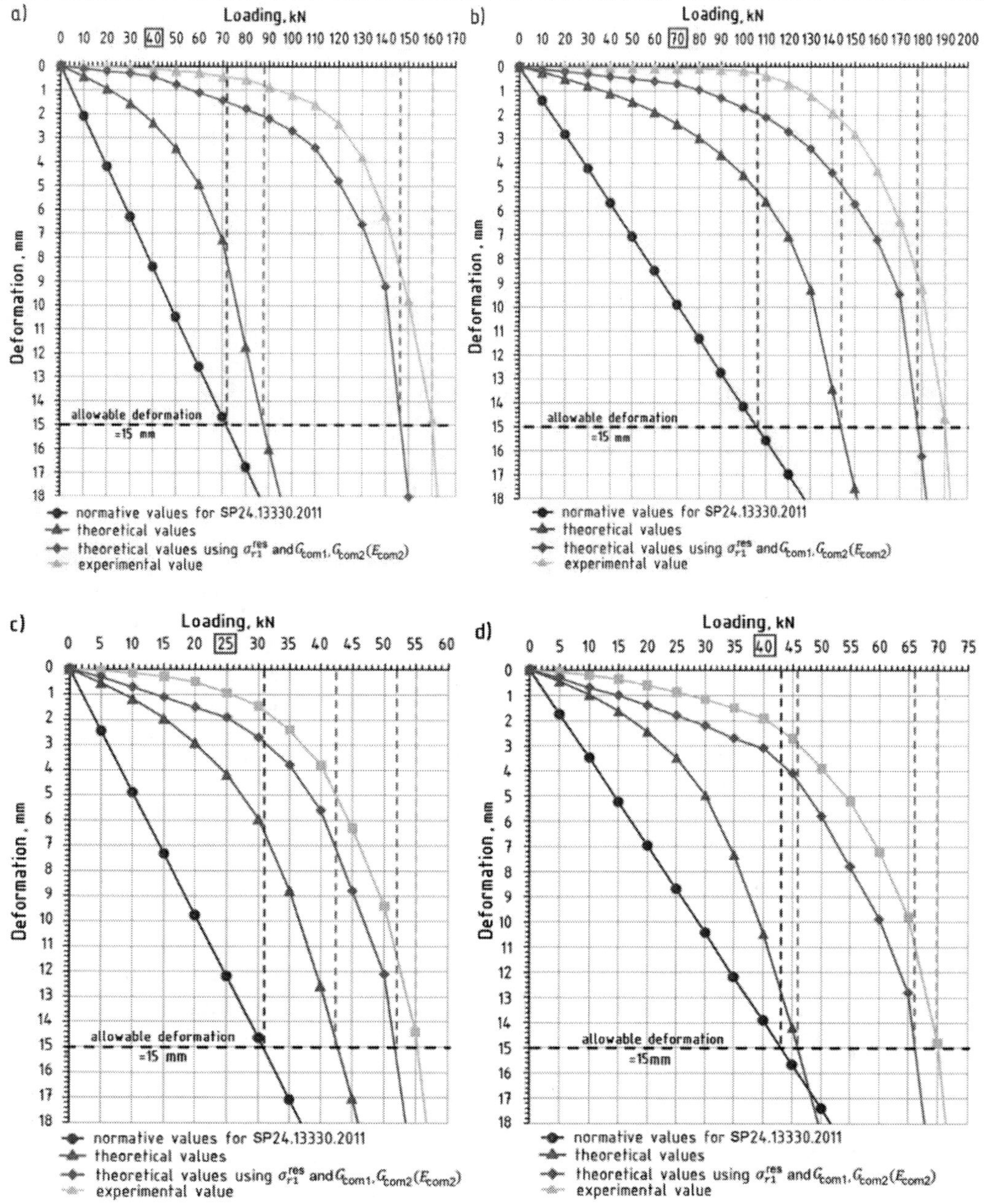

Fig.3. Settlement-load comparative diagrams s=f(p) for drill-injection piles with the controlled broadening and the repeated solution injection in the “fracturing” mode: a), b) on the 1st site V=10-20 l; c), d) on the 2d site V=30-40 l.

$$E_{com1} = \frac{(1+\nu)\cdot\left(1-\frac{\nu}{(1-\nu)}\right)}{\left(\frac{e_{com}\cdot(1+e_0)}{p_{in}-\sigma_0}\right)} \cdot \frac{\sigma_r^{res}\cdot tg\varphi}{\sigma_r^{res}\cdot tg\varphi-\sigma_0\cdot\frac{\left(1-\frac{\nu}{(1-\nu)}\right)}{\sqrt{3}}} \tag{28}$$

$$s = \frac{N}{\pi\cdot(r_b+u_1)\cdot G_{com1}+2\cdot\pi\cdot l\cdot G_{com2}} \cdot \ln\frac{\left(\tau^*\cdot\left(\frac{R_{com1}}{r_b+u_1}+\frac{R_{com2(3)}}{r_b+u_2}\right)-\tau_u\right)}{(\tau^*-\tau_u)} \tag{29}$$

4 The results of in-situ tests

To determine the geometric parameters of the proposed piles and the changes in physical and mechanical characteristics of the soil massif, excavation of piles was carried out layer by layer after in-situ tests and installation of piles. (Fig. 4)

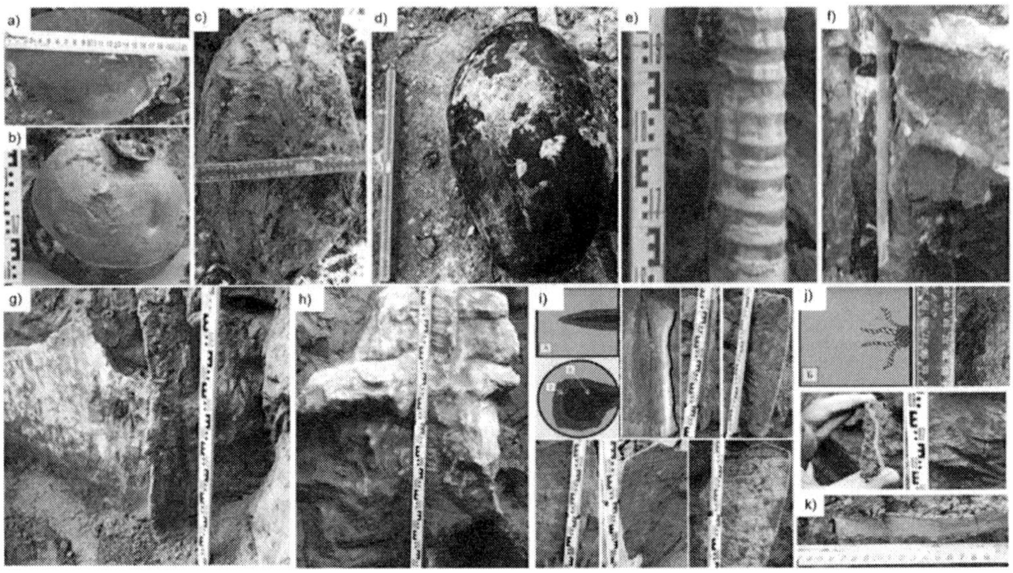

Fig. 4. Geometric parameters of the proposed piles: a, b – controlled broadenings V=10-20 l, c, d – 30-40 l, e – pile shaft diameter increase, f – hydraulic fractures development, g, h – expansion of hydraulic fractures in single and repeated solution injections, i, j – texture of hydraulic fractures of «A» and «B» types, k – horizontal hydraulic fractures.

After the excavation it was found out that a horizontally elongated elliptical controlled broadening of 250-350 mm in diameter with the ratio of d / h = 1.3-1.4 was formed at the pile end in the solution injection of V = 10-20 l into the membrane cup. When the solution injection was V = 30-40 l, the vertically elongated broadening of 360-390 mm in diameter with the ratio of d / h = 0.7-08 was formed. The difference in values was due to the different pile lengths - 2 and 6 m correspondingly. The pile shaft diameter nearly doubled when the drill-injection pile shaft was formed by the sleeve technology with solution injection using the "fracturing" mode due to permanent pressure on the borehole walls with a mean value of 0.2 MPa. Hydraulic fractures up to 40mm in thickness expanded to a considerable distance from the pile, up to 3 m in a single solution injection. Hydraulic fractures localized in the neighboring soil massif up to 130mm in thickness at a distance of 0.5 m from the injector in repeated solution injection. In accordance with the classification of I.I. Sakharov two types of hydraulic fracture textures were found out: type "A" in the form of the greater crack growth with development of solid lens of 10-130 mm in thickness and type "B" – numerous thin cracks up to 5mm in thickness, loosely filled with the solution [10].

Analysis of changes in the physical and mechanical characteristics of the neighboring soil massif showed that the greatest change occurred in the compacted soil massif zone in controlled broadening development: the density increased on the average by 25%, the moisture content decreased on the average by 38%, the modulus of deformation increased by 60%; the density along the pile shaft increased on the average up to 17%, the moisture content decreased up to 28%, the modulus of deformation increased up to 35%. – these must be considered in future in design calculation of the basic parameters of the drill-injection pile and forecast its interaction with soil foundation [19-23].

5 Conclusions

1. Developed is a new method for strip foundation strengthening in buildings and structures erected on silty-clayed soils; it makes it possible to combine drill-injection pile installation with the controlled broadening, sleeve technology with solution injection using the "fracturing" mode (including repeated injections) along any layer of the soil massif and to restore foundation members as an integrated technological unit, thus greatly improving its effectiveness.

2. It was established that in controlled broadening development of 10-40 l in volume, minimum diameter of the broadening was from 250 to 390 mm, the compacted soil massif zone developed with an average radius of expansion of 0.15-0.6 m from the injector tube and a significant change in physical and mechanical properties: the density increased on the average by 25%, the moisture content decreased up to 37%, the modulus of deformation increased by 64%, - these must be considered in future in design calculation of the basic parameters of the controlled broadening.

3. It was established that in drill injection pile shaft development one should do only the repeated solution injection using the "fracturing" mode when the number of hydraulic fractures and their localization increased in the neighboring soil massif at a distance of 0.5 m from the injector tube (at a maximum crack growth up to130 mm) resulting in the pile bearing capacity increase on the average by 20%. Also it was found out that due to the permanent pressure on the borehole wall with a mean value of 0.2 MPa, the diameter of the pile shaft nearly doubled and together with hydraulic fractures resulted in compacted zone development with the mean radius of 0.2-0.4m from the injector tube and the changes in physical and mechanical properties within this zone: the density increased on the average by 17%, the moisture content decreased up to 28%, the modulus of deformation increased by 35%, - these must be considered in future in design calculation of the pile shaft parameters.

4.The results obtained after the in-situ tests made it possible to develop the algorithm for design calculation of the basic parameters allowing the following accuracy: to evaluate the radius of the compacted zone in the controlled broadening and the neighboring soil massif - up to 7%; to evaluate radial movements of the membrane cup walls and borehole – up to 10 and 14% correspondingly; to forecast settlement-load dependency in the proposed piles in silty-clayed soil loaded statically – up to 25%.

References

1. V.F. Baj, T.V. Mal'ceva, A.N. Kraev, Nauchno-tekhnicheskij vestnik Povolzh'ya, **5**, 108-111 (2014)
2. M.N Ibragimov, V.V. Semkin, *Zakreplenie gruntov in"ekciej cementnyh rastvorov. Monografiya* (ASV, Moscow, 2012)
3. V.V, Semkin, M.N. Ibragimov, Ya.Ya. Motuzov, Sbornik nauchnyh trudov NIIOSP, (2006)
4. V.A. Il'ichev, P.A. Konovalov, N.S. Nikiforova, Osnovaniya, fundamenty i mekhanika gruntov, **4**, 19-24 (2001)
5. R.A. Mangushev, A.I. Osokin, *Sovremennye svajnye tekhnologii* (ASV, Moskva, 2007)
6. V.P. Petruhin, O.A. Shulyat'ev, O.A. Mozgacheva, *Novye sposoby geotekhnicheskogo stroitel'stv*a (ASV, Moscow, 2015)
7. H. Brandl, Proceedings of the Twelfth International Conference on Soil Mechanics and Foundation Engineering, Rotterdam, (1992)
8. H.G. Poulos, E.N. Devis, *Pile foundation analysis and design* (Wiley, 2000)
9. Ya.A. Pronozin, Yu.V. Zazulya, R.V. Mel'nikov, M.A. Stepanov, Sovremennye problemy nauki i obrazovaniya, **3**, (2013)

10. I.I. Saharov, M.A, Abbud, Geotekhnika. Nauka i praktika: sb. nauch. tr., 72-76 (2000)
11. Z.G. Ter-Martirosyan, Ya.A. Pronozin, M.A. Stepanov, Soil Mechanics and Foundation Engineering, 1-5 (2012)
12. E.S. Voznesenskaya, V.A. Ermolaev, A.I. Osokin, S.V. Tatarinov, OFMG, **4**, 19-23. (2014)
13. L. Gabrielaitis, V. Papinigis, G. Zarzojus, Procedia Engineering, **57**, 287-293 (2013)
14. Ya.A. Pronozin, M.A. Samohvalov, D.V. Rachkov, Promyshlennoe i grazhdanskoe stroitel'stvo, **3**, 56-60 (2014)
15. Ya.A. Pronozin, Yu.V. Zazulya, M.A. Samohvalov, Sposob izgotovleniya buroin"ekcionnoj svai s kontroliruemym ushireniem, Patent 2522358 Rossijskaya Federaciya, MPK-2014.07 E 02 D 5/46. (2014)
16. O Hoffman, G. Sachs, *Introduction to the theory of plasticity for engineers* (McGRAW-HILL BOOK COMPANY, New York, 1953)
17. Z.G. Ter-Martirosyan, A.S.M. Abdul Malek, OFMG, **6**, 8-11 (2007)
18. Z.G. Ter-Martirosyan, V.S. Avanesov, Vestnik MGSU, 47-54 (2015)
19. A.I. Polishchuk, *Osnovy proektirovaniya i ustrojstva fundamentov rekonstruiruemyh zdanij* (STT, Moscow, 2004)
20. R.H. Karol, *Chemical grouting and soil stabilization, Basel* (Marcel Dekker, Inc., New York, 2003)
21. L. Gabrielaitis, V. Papinigis, J. Sirvydaite, Engineering Structures and Technologies, **4(1)**, 7-15 (2012)
22. L. Epifantceva, O. Poroshin, Y. Pronozin, 10th International Conference on Geosynthetics, ICG (2014)
23. Z.G. Ter-Martirosyan, Y.A. Pronozin, N.Y. Kiselev, Soil Mechanics and Foundation Engineering (2014)

Development of Criteria for Transport Flow Efficiency Assessments of the City Street Road Network

Oleg Danilov[1,*], *Victor* Kolesov[1] and *Denis* Sorokin[1]

[1]Industrial University of Tyumen, 625001 Volodarskogo str. 38, Tyumen, Russia

Abstract. A mathematical model to quantify the efficiency of traffic flow on the road network was developed. The authors have developed a comprehensive criterion considering key performance indicators: the traffic flow performance, traffic safety, environmental safety.Study of the efficiency criterion is based on the analysis of the traffic flow dynamics. The problem of structural and parametric identification of the mathematical model is complex. The solution to this problem is based on the theory of traffic flow, regression analysis, simulation and dynamic programming.

1 Introduction

Modern research shows that the intensive development of motor transport (MT) is supposed to be both positive and negative nowadays. According to the World Health Organization report, road traffic injuries are the eighth leading cause of death in the world. Besides MT is the basic source of adverse environmental impact.

These problems are especially prevalent in large cities, where traffic jams, road traffic accidents (RTF) and smog have become commonplace. Thus, correct road traffic organization, transport safety and environmental risks management are a priority to improve the life quality of population at any modern city. The evaluation criterion of the traffic control efficiency, including the TF performance, the traffic safety level (traffic safety), environmental risk assessment should be the basic principles of traffic control (TC) of the city road network (MAC).

The evaluation criterion can be formed on the basis of dynamic characteristics of the TP study. Study of the TF dynamics, evaluation criteria development, structure and parameters of the model TS identification is of high priority. The final results will allow to efficiently solve the problems of the functioning processes of the TF optimization.

2 Assessment methodology the TF

* Corresponding author: tgasu.danilov.oleg@mail.ru

Contemporary approaches, broadly based on the fundamental study of traffic flows [1], indicate that the effective operation of TF is caused by three main indicators, acting as subtest: TF capacity, traffic safety and environmental risk [2].

Let us consider first subtest. TF performance optimization may provide two options. The first one maximizes the TF intensity, and the second one maximizes the TF kinetic energy. In the city it is advisable to use the first approach, since the main problem is overloaded roads. Thus, the first subtest can be written as

$$g_1 = \frac{N}{N_{\max}} \tag{1}$$

where N is the TF intensity, road vehicle units/h; Nmax is the maximum TF intensity, road vehicle units/h .

As for the second subtest, the traffic safety problems solution is the triad "Intensity of TF – the amount of RTA – the amount of victims" analysis. In Scandinavian researchers works [3] it is noted that relative change of road accident amount is linearly connected with relative change of movement intensity

$$\frac{N_{dtp}}{N_{dtp\max}} = k \cdot \frac{N}{N_{\max}} \tag{2}$$

N_{dtp} is the RTF intensity; N_{dtpmax} – maximum RTF intensity; k – constant (k≈0.8)

The transformation (2) allows to obtain the expression

$$N_{dtp} = \frac{N_{dtp\max} \cdot k}{N_{\max}} \cdot N \tag{3}$$

With this assumption the intensity of the RTF can be interpreted as

$$N_{dtp}(q) = N(q) \cdot RP \tag{4}$$

N(q) is the TF intensity as its density function; RP is the risk of accidents

Inhere it is assumed that the analyzed time interval (for example, year) RP is a fixed value.

The amount of people injured in RTF depends on the risk (probability) of injury (P). Considering the injury rate intensity, it should be noted that the risk is the same both for the driver and for passengers of the vehicle. The data on the probability of the driver's injury due to the driver's vehicle speed at the time of the RTF are given in [4].

The data processing showed that P (V) corresponds to the Weibull law (R2 = 0.99). At the same time, it should be taken into account that among the victims there is a significant proportion of pedestrians. The likelihood of pedestrians injury has been discussed in the report [5]. Identification of model Pped (V) showed that it complies with Weibull law (R2 = 0.93).

The resulting probability should be considered as a weighted sum of P (V) and Pped (V)

$$P(V) = (1-s) \cdot P(V) + s \cdot P_{ped}(V) \tag{5}$$

s - the proportion of injured pedestrians in the overall balance.

DOI: 10.1051/matecconf/20167301010

The results are shown in Fig. 1. Thus, knowing the speed of the TF connection with its density, it is possible to determine the function P (q).

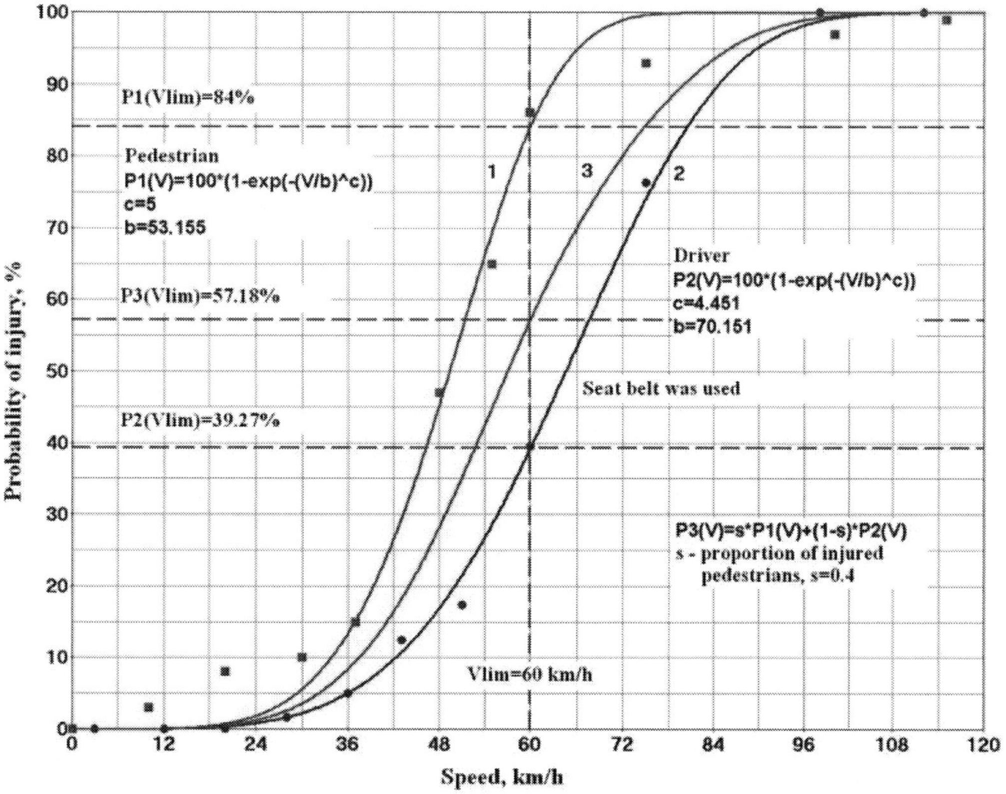

Fig. 1. The probability of injury of RTF participants.

The conditions under which the RTF intensity extremum of involving injury rate (Ntr) is achieved are of most interest. Such RTF include the larger or equal to one quantity of injured people. Intuitively it is clear that the function Ntr(q) is caused by two main indicators: the of the accident (Ndtp) intensity and the risk of injury in case of RTF P(q). This function then can be written in the form of

$$N_{tr}(q) = N_{dtp}(q) \cdot P(q) = RP \cdot N(q) \cdot P(q) \tag{6}$$

It is important to note that the maximum injury rate (Ntrmax) should be achieved in free movement of TF corresponding to high speeds.

Thus, the second subtest is determined by the injury rate intensity, the growth of which reduces the TF efficiency, therefore, it can be represented in the form of

$$g_2 = 1 - \frac{N_{tr}}{N_{tr\,\max}} \tag{7}$$

Let us consider the third subtest. When assessing the environmental risk level generated by urban TF, the approach established in the works [6, 7] is used. These works proved the linear relationship between the concentration of pollutants in the surface layer and TF capacity, ie the product of the TF intensity on its density

$$C_{bb} = k \cdot e \tag{8}$$

where Cbb is concentration of harmful substances in a ground layer of the atmosphere; k is the coefficient depending on average running emission of the vehicle and average speed of a wind stream; e is TF capacity($e = N \cdot q$).

We assume that the environmental risk is proportional to the parameter Cbb, and hence the TF capacity, then the third subtest can be logically written in the form of

$$g_3 = 1 - \frac{e}{e_{\max}} \tag{9}$$

In this case, it is a necessary to identify extrema emax and analyze its dependence on the TF operating conditions.

When assessing the TF efficiency the form of its representation (additive or multiplicative) should be pre-engaged.

Multiplicative assessment, by definition, is equal to zero if at least one of the subtests is zero, but it is hardly consistent with actual practice, therefore, the additive generalized assessment representing the weighed sum of subtests has been chosen. In this case, there is an opportunity either to minimize the negative effect generated by risks or to maximize an advantage due to their decrease. Thus, the criterion function (CF) should be presented in the form of

$$Eff = \sum_{i=1}^{n} \alpha_i \cdot g_i \Rightarrow \max \tag{10}$$

where ai is weight coefficients corresponding to the condition $\sum_{i=1}^{n} \alpha_i = 1$; g_i is a subtest

Formation of subtests is impossible without use of the TF adequate mathematical models allowing to define and to predict TF dynamics of characteristics. Such a model has to consider specifics of functioning of TF in various conditions. Authors have developed a mathematical model of TF dynamics for speed limit [8, 10, 11]. The model is based on longitudinal dynamic dimension of safe driving. The coefficients of the above have been considered by the authors in detail in work (Danilov, etc., 2014).

In work [8] the alternative model of continuous TF allowing to trace dynamics of TF characteristics in the vehicle speed limit for a rather low density has been offered. TF speed - density relationship has a form of [9].

$$V_l(q) = V_{\lim} - K_l \cdot q^2 \tag{11}$$

where Vlim is maximum speed with the limit, km/h; Kl is constant.

When studying TF dynamics in the city it is necessary to consider discrete movement in time. The flow is split internally into groups of cars. Each separate group is formed at the enabling signal of the traffic light which operating mode defines the amount of cars in group. Thus, on roads with the adjustable movement it is advisable to investigate TF dynamics by means of average dynamic characteristics (for a certain period).

Thus, it is advisable to study TF dynamics on roads with traffic control by means of average dynamic characteristics (for a certain period). In this case TF parameters values will be much lower, than in model of continuous TF.

Adaptation of a ratio (11) to city traffic makes it possible to form

$$V(q) = V_{\lim E} - K_E \cdot q^2 \tag{12}$$

where VlimE is the average maximum speed of the TF in a flow; KE is the coefficient considering traffic conditions.

The VlimE parameter is caused by drivers psychology. Theoretically VlimE value has to be equal to speed limit on the studied stage (Vlim), however part of drivers deliberately underreport the speed (VlimE<Vlim), and others exaggerate it (VlimE>Vlim). In the first case the driver *obviously* opts for a more comfortable speed and in the second one he/she violates traffic regulations which extremely negatively affects traffic safety.

The parameter KE should take into account the road surface, its condition, the flow pattern, the conditions of the traffic light regulation and other hindrances.

The first three factors account for the ratio DG m2 [11], and the rest should be described by the empirical laws obtained at TF monitoring in different parts of the Street-road network. Then, the coefficient KE is represented in the form of

$$K_E = F(m_2, K_{sv}, K_{pr}) \tag{13}$$

where Ksv is the coefficient considering an operating mode of the nearest traffic light and distance to it; Kpr is the coefficient considering other factors (the crosswalk, the likelihood of the pedestrian and others).

3 Results

To assess the developed models pilot studies of the TA dynamics in the city of Tyumen have been conducted. a section of Respubliki Street (the area of crossing with Permyakov Street) and a section of Melnikayte Street (the area of crossing with Kotovskogo Street) were investigated as the Street-road network elements. On the chosen sections, 2 one direction lanes have been explored. The database for pilot studies has been received by means of the Traffic-Monitor complex developed by scientific and technological center "Module".

The integral condition of an objective assessment of the TF parameters under studies is collecting sufficient volume of information. Therefore experimental study sample representativeness is significant. At the first stage, histograms of empirical distribution of speed values of the TF have been constructed at various TF density. The analysis of data for two sections (Melnikayte St., Respubliki St.) has shown that the speeds distributions are bimodal. It is obvious that the "peak" of low speeds is caused by higher density of a flow, and the "peak" of high speeds is caused by lower density respectively, which gives a driver a big freedom of choice.

The analysis has shown that distributions of speeds should be considered in two intervals: 0<V<25 km/h and 25<V<65 km/h. Herein in both intervals distributions obey the normal law. Thus it is advisable to study great values of speeds as the error of measurements will be higher: the detector has to record the TF parameters and software process for a smaller period.

Further studies have shown that speeds distributions in the range of 25<V<65 km/h are approximated by the normal law of distribution with an average quadratic deviation of 7.2 km/h. The empirical histogram is illustrated by Fig. 2.

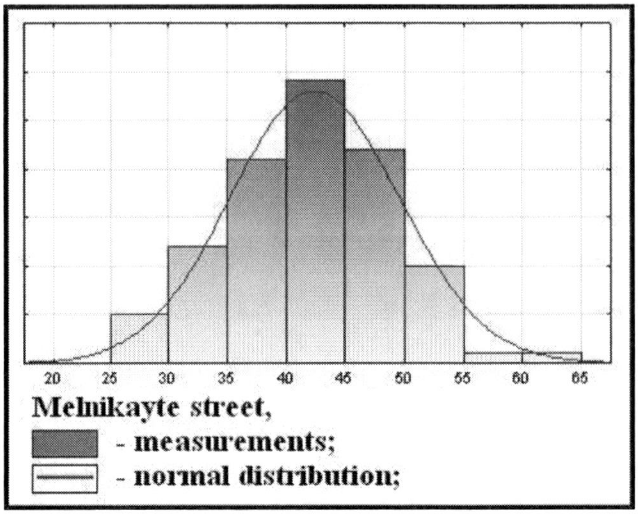

Fig. 2. Histogram of distribution of speeds of the vehicle.

To justify the necessary amount of measurements Student distribution is used at which, having provided the level of confidential probability, it is possible to determine the necessary volume of measurements of n. In Fig. 3 dependence of n from σ for the level of confidential probability 0.9 (at various values of accuracy of measurements Δ) is given. In experimental studies, the value of an average quadratic speed deviation is 7.2 km/h. Accepting the accuracy of measurements Δ=1 km/h, it is possible to claim that the volume of speed selection doesn't exceed 143 values.

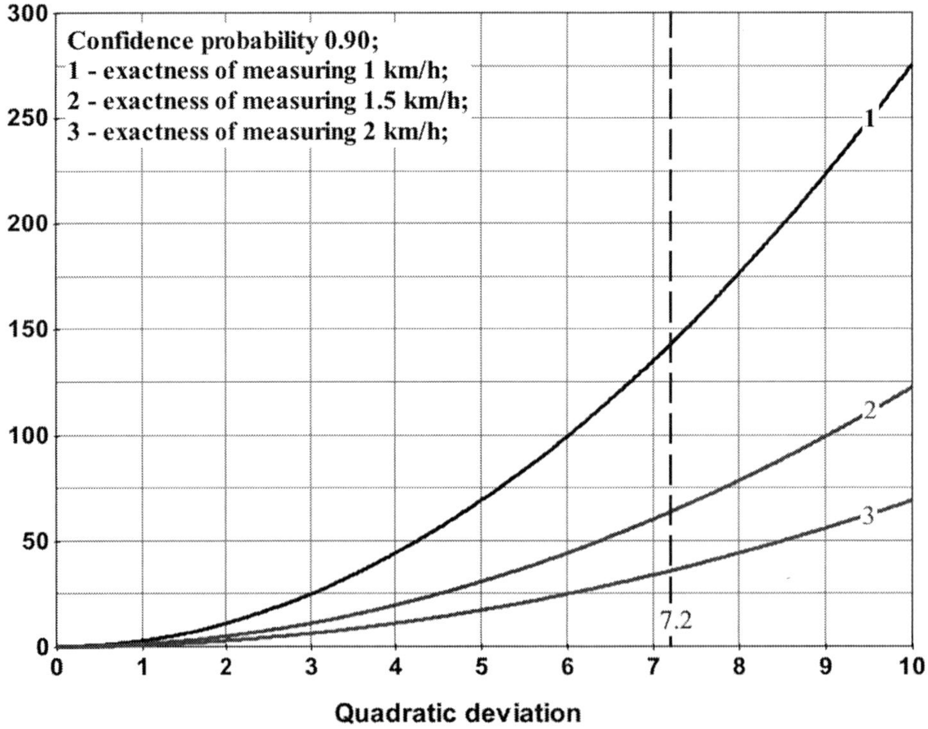

Fig. 3. Definition of necessary amount of speed measurements.

Fig. 4-5 show dependences of intensity of TF on daytime. Experimental data are well described by polynomial regression (with order of n=3, R2>0.75).

a)

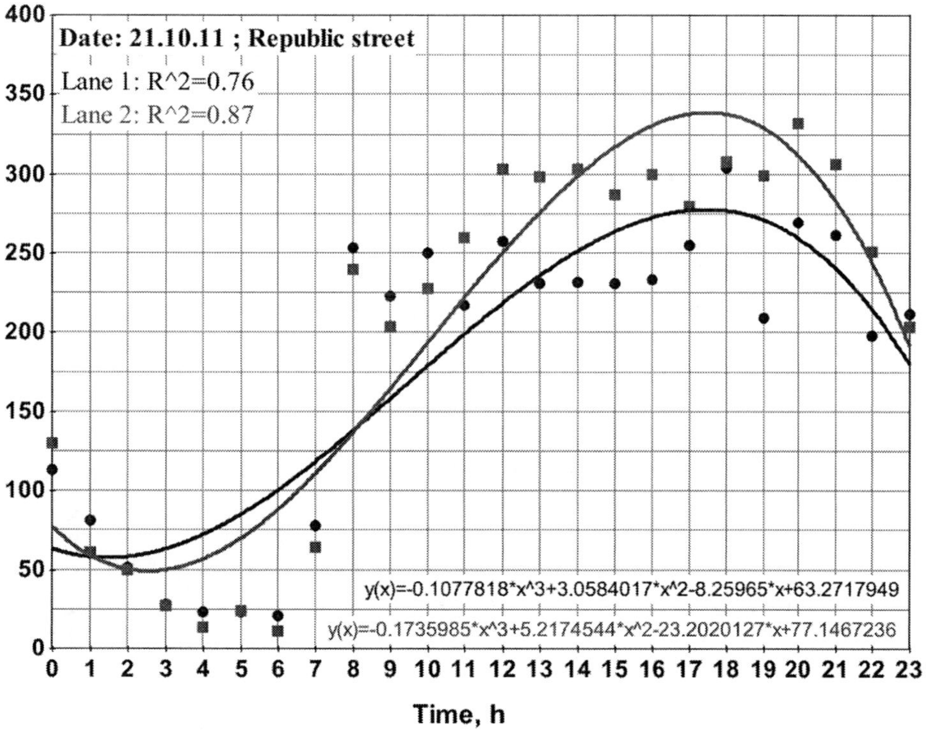

b)

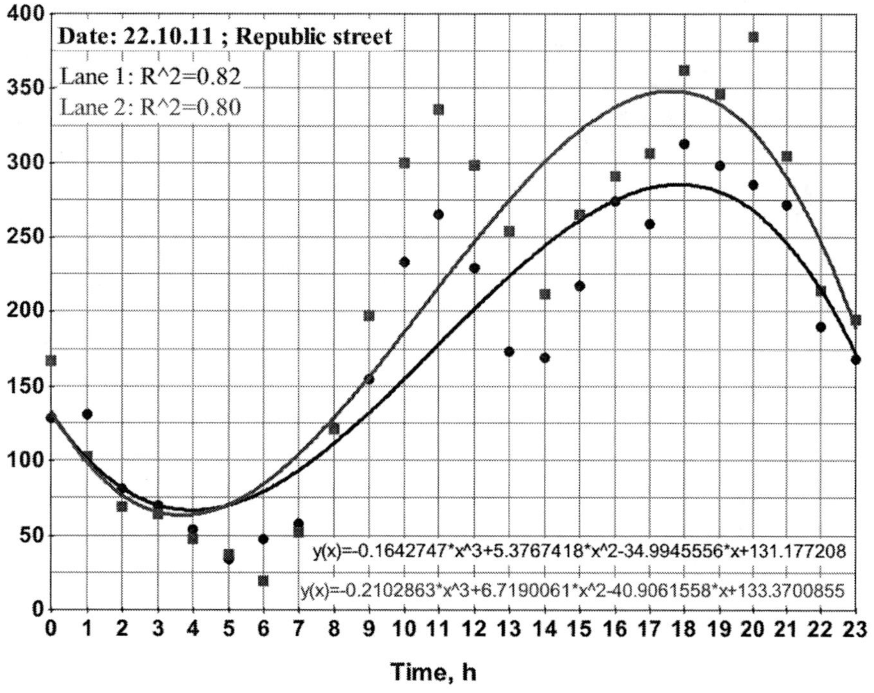

c)

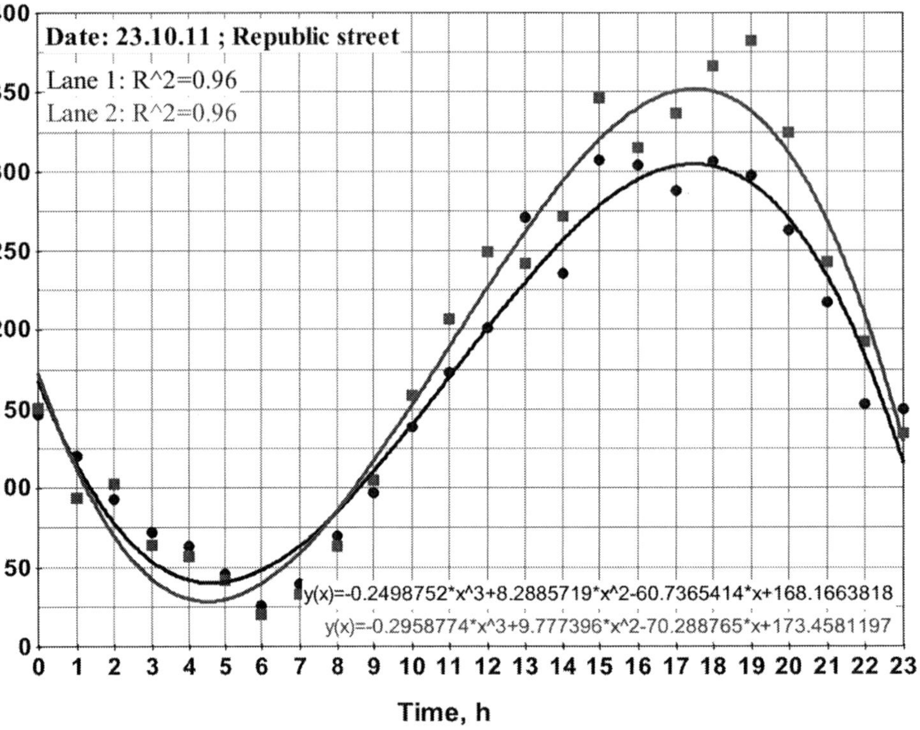

Fig. 4. a), b), c) Dependence of N(t) (Respubliki St.).

a)

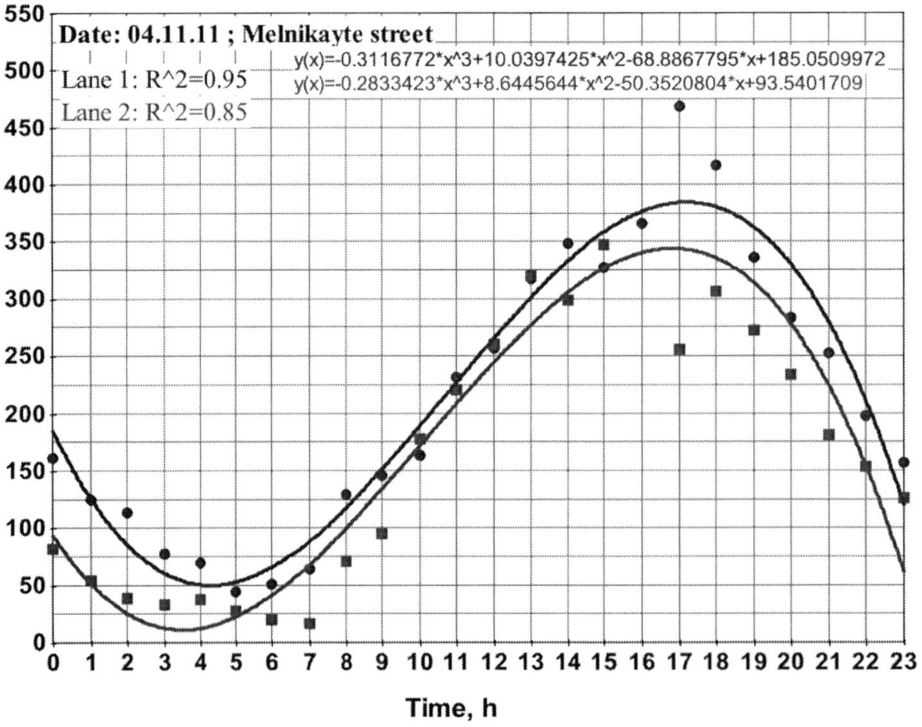

b)

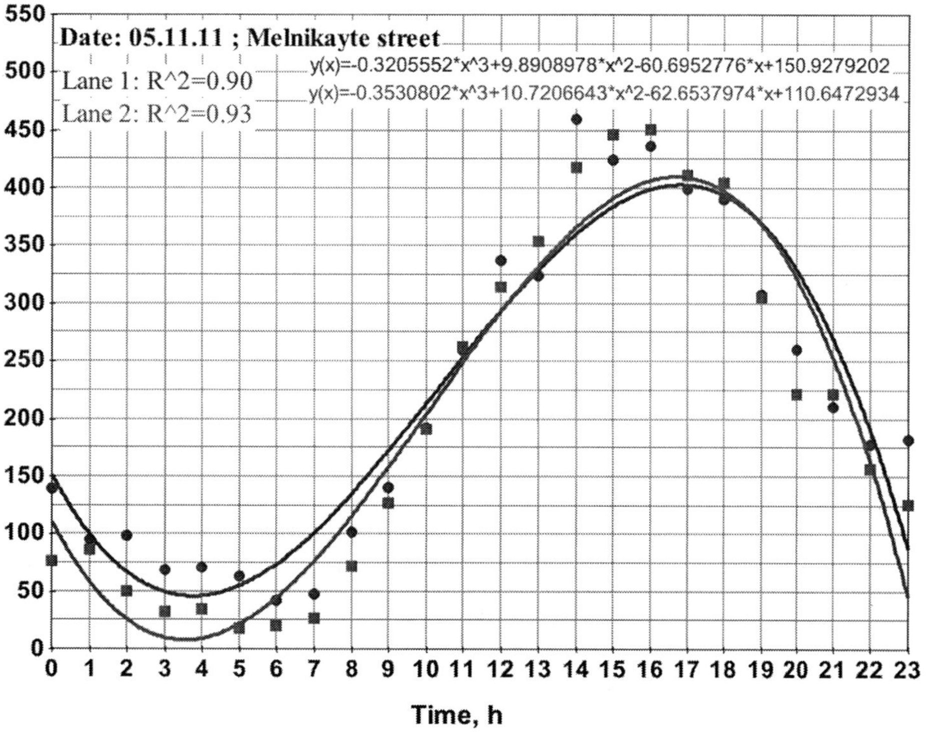

c)

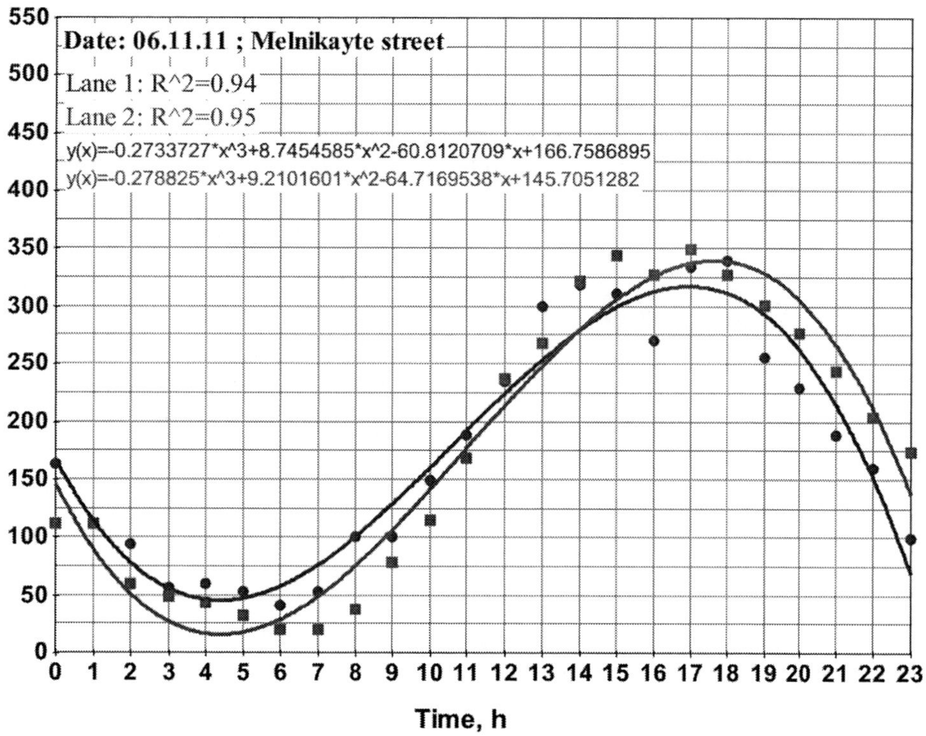

Fig. 5. a), b), c) Dependence of N(t) (Melnikayte St.).

The analysis of Fig. 4-5 suggests that the form of the line N(t) is stable and has both a minimum and a maximum. Thus, the average dynamic TF characteristics vary cyclically with time, obeying similar laws, both on different lanes and at different stages of Street-road network. Approximation of dynamic characteristics interrelations of a real flow provides search of two parameters: VlimE and KE. At the same time, the regression equations have to correspond to the developed models. For automation of the regression analysis, the method of nonlinear assessment of a software package "STATISTICA" has been applied. This method is the universal approximating procedure assessing any kind of dependence between "a response variable" and a set of independent variables. The regression equation corresponding to the developed theoretical models TF dynamics has been put in the program. Theoretical model adequacy is assessed by multiple coefficient of correlation which characterizes narrowness of correlation communication between variables. Fig. 6 illustrates dependences of an average movement interval (ID) on vehicle speed for Respubliki street and Melnikayte street. For Respubliki St. R12=0.85, VlimE1=48.7, KE1=0.154 and R22=0.88, VlimE2=59.5, KE2=0.194 for the first and second lanes respectively, for Melnikayte St. R12=0.71, VlimE1=46.0, KE1=0.068, R22=0.66, VlimE2=52.2, KE2=0.084, with the level of confidential probability 0.95. Actually, ID values are exaggerated because of cyclic stops of cars flow (influence of traffic light regulation). Time of functioning of the main TF part depends on operating mode of the next traffic light.

On the other hand, some drivers keep the greater distance from the vehicle ahead than the theoretical value of confidence limit instructs. There's a simple explanation to it: the driver consciously chooses a greater range of motion. Thus, the regression curve passes through the "cloud" of empirical data and asymptotically tends to the average value of the maximum speed (VlimE) developed in the flow. It is important to note that the values of the coefficient KE for the studied stretch is different for each of the lanes, which indicates the heterogeneous structure of the TF within the considered area of Street-road network. While the KE can vary greatly for the different spans that are caused by the difference in traffic.

a)

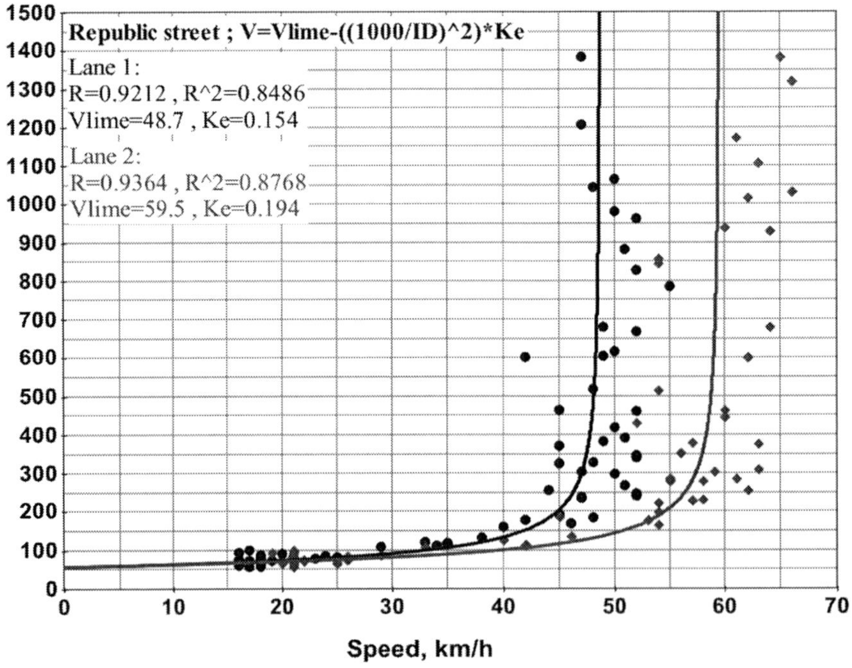

b)

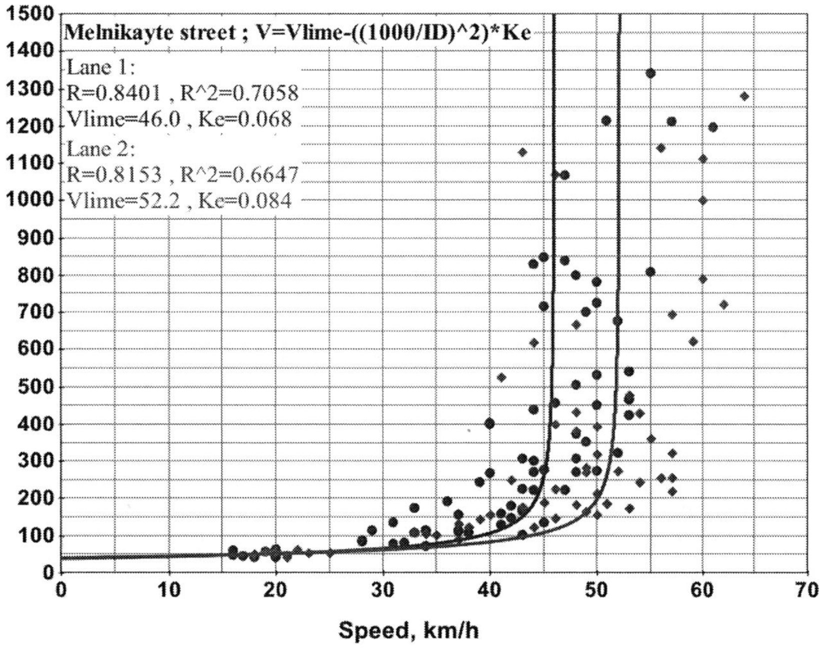

Fig. 6. a), b) Dependence of ID(V).

Fig. 7 illustrates experimental data of traffic intensity from vehicle speed for Respubiliki and Melnikayte streets. Inhere data are approximated by alternative model

$$N(V) = \sqrt{\frac{V_{\lim E} \cdot V^2 - V^3}{K_E}} \tag{14}$$

It is obvious that the extremum of intensity can be reached both at rather high speed of the movement (35-45 km/h), and at low (15-25 km/h).

a)

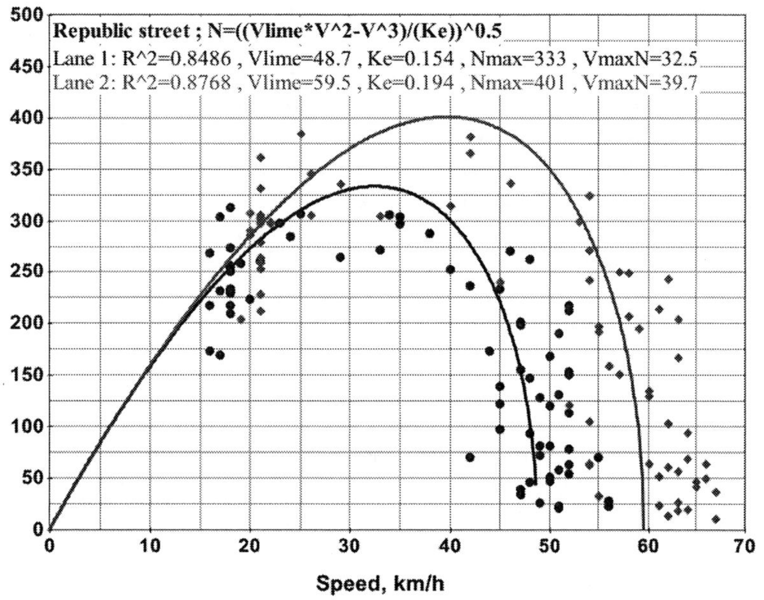

b)

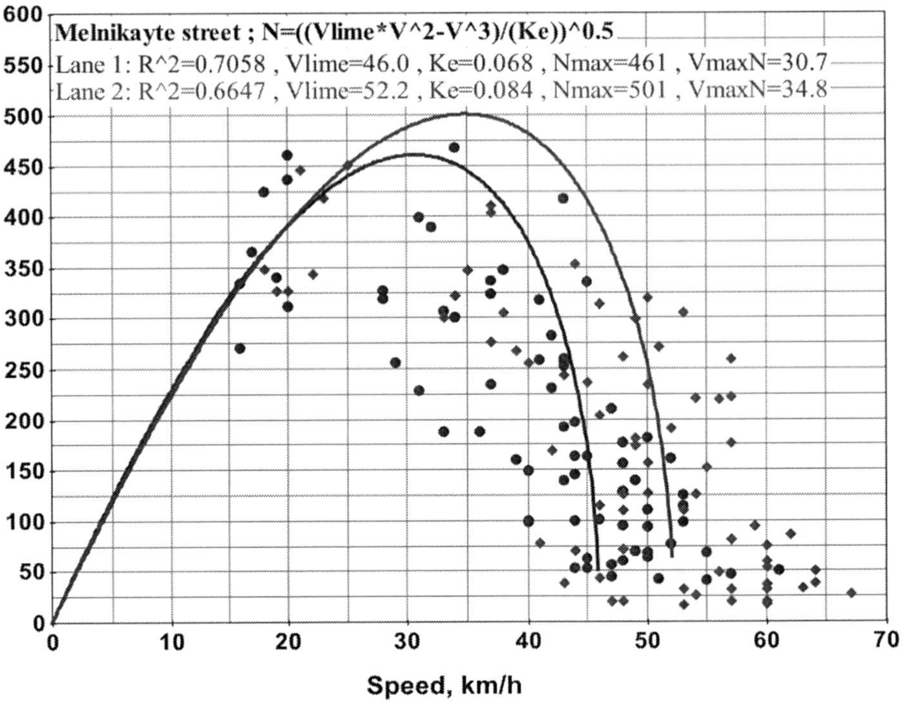

Fig. 7. a), b) Dependence N (V).

To explain this phenomenon, it is necessary to analyze the psychology of the driver at various vehicle speeds. At low speeds the driver obviously keeps lower distance from the vehicle ahead than at high values of V (Fig. 6) as this situation is perceived as safer. Then the average flow density becomes higher than the regression curve directs, whereby the intensity of the movement increases. In the second case (high speed), the situation is more adverse from the traffic safety view point and the driver prefers to keep a longer distance to vehicle ahead, which leads to some decrease of average density. For Respubliki street Nmax1=333 vehicle units/h, VmaxN1=32.5 km/h (first lane) and Nmax2=401 vehicle units/h, VmaxN2=39.7 km/h (second lane). For the Melnikayte street Nmax1=461 vehicle units/h, VmaxN =30.7 km/h, Nmax2=501vehicle units/h, VmaxN2=34.8 km/h. The obtained results allow to assess the first and second subtests of TF efficiency.

Capacity dependences on TF speed have been studied to analyze the third subtest. The greatest capacity values are reached at speeds of 15-30 km/h. For Respubliki street Vmaxe1=24.4 km/h, Vmaxe2=29.8 of km/h. For Melnikayte street Vmaxe1=23.0 km/h, Vmaxe2=26.1 of km/h. Thus, research of real dynamic characteristics of TA creates prerequisites for the analysis of criterion of their efficiency. For an assessment of weight coefficients the following priorities have been placed: a1=0.65, a2=0.28, a3=0.07.

Fig. 8 shows the dependence schedule of Eff (V) created on the basis of average dynamic characteristics models of TF. Maximum efficiency for the front lane (Respubliki St.) makes Effmax1=0.870 (this extremum is expected at a speed of a flow of VmaxEff1=28 of km/h). For the second lane these parameters are Effmax2=0.861, VmaxEff2=33 of km/h. respectively. Parameters for Melnikayte st. are Effmax1=0.871, VmaxEff1=26 km/h, Effmax2=0.868, VmaxEff2=30 km/h. The analysis of fig. 8 allows to claim that flow speed increase in the range VmaxEff<V<Vlim leads to efficiency of his work decrease.

a)

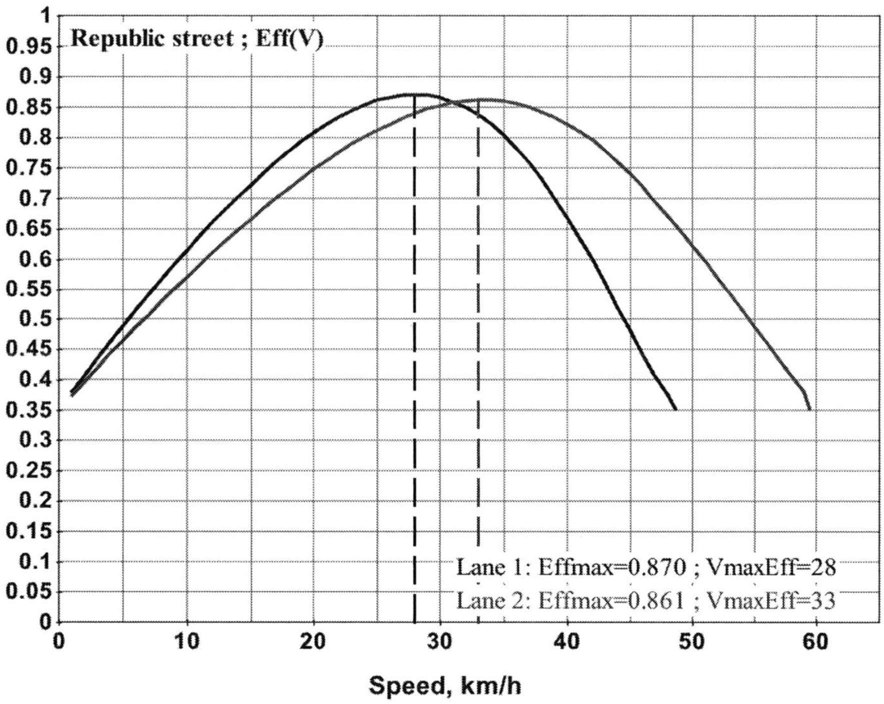

b)

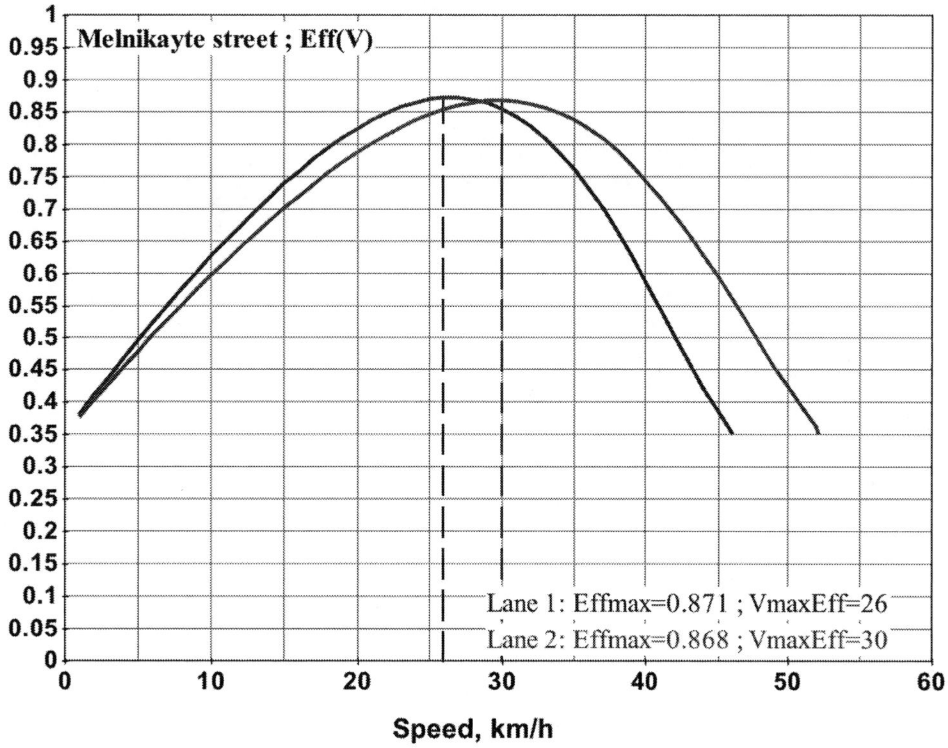

Fig. 8. a), b) Dependence Eff (V).

Fig. 9 illustrates dependence of efficiency on the TF density. Settings for the Respubliki street: qmaxEffl=12 vehicle units/ km, qmaxEff2=12 vehicle units/km. The parameters for the Melnikaite street.: qmaxEffl=17 = 17 vehicle units/ km, qmaxEff2=16 vehicle units/ km.

a)

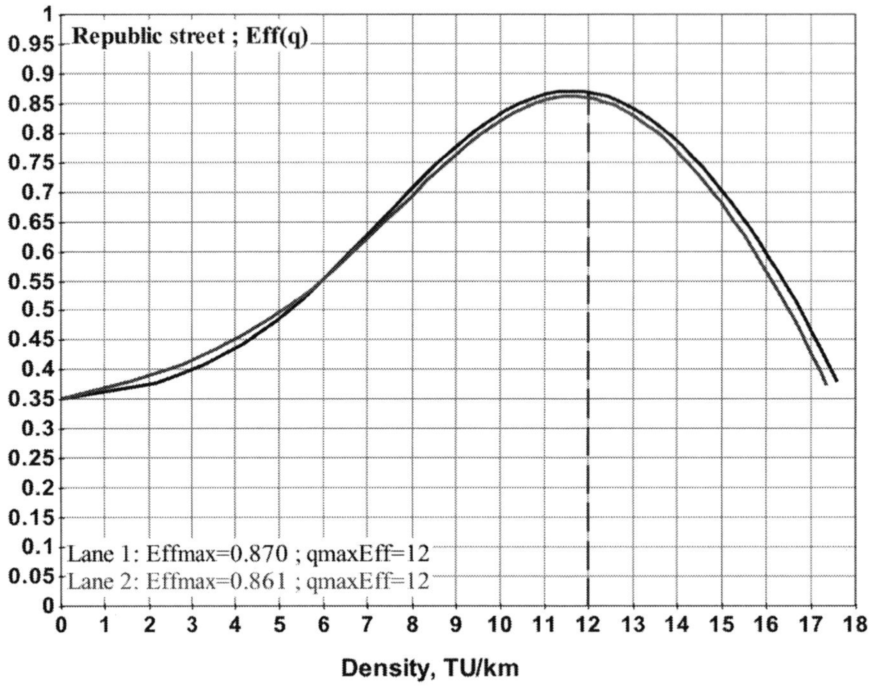

b)

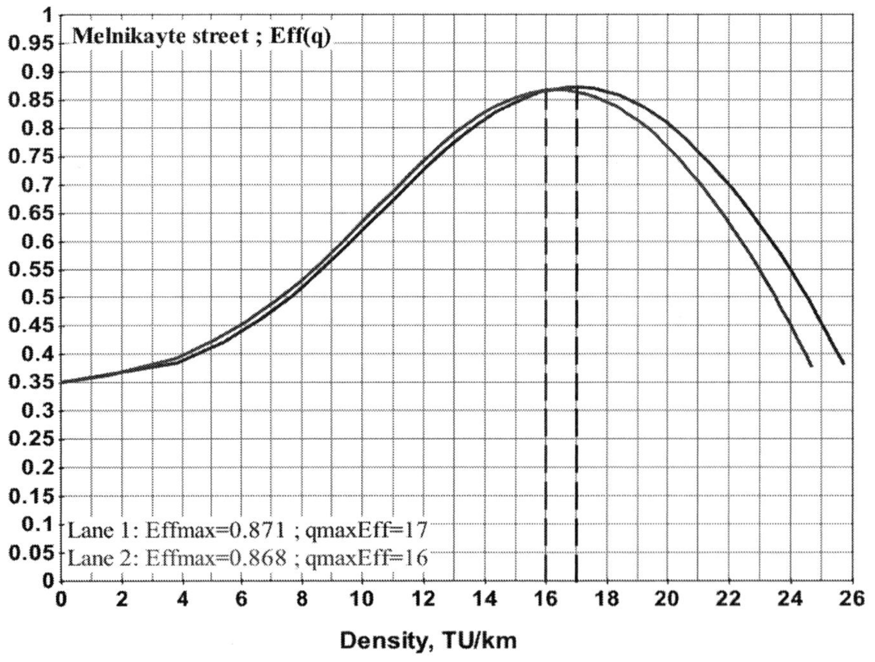

Fig. 9. a), b) Dependence Eff(q).

The analysis of fig. 9 allows to determine the optimum TF density at which maximum efficiency is achieved. However, it is difficult to operate the flow density, as it demands traffic reorganization. It is important to note that the real flow can have three operating modes depending on the actual average density. The first mode provides a flow with its optimum density ($q=q_{max}Eff$).

The second mode is caused by efficiency decrease (Eff<Effmax) due to reduction of actual density (q<qmaxEff). Inhere it is possible to say that the resource of a lane is used incompletely.

The third mode is caused by efficiency decrease due to the density increase ($q>q_{max}Eff$). This situation is most negative as the lane becomes too "overloaded" with cars. Dependences of performance criteria on time are of the greatest value.

Fig. 10 illustrates values of performance criteria within 72 hours for Respubliki and Melnikayte streets respectively (average value in one hour of a flow functioning).

a)

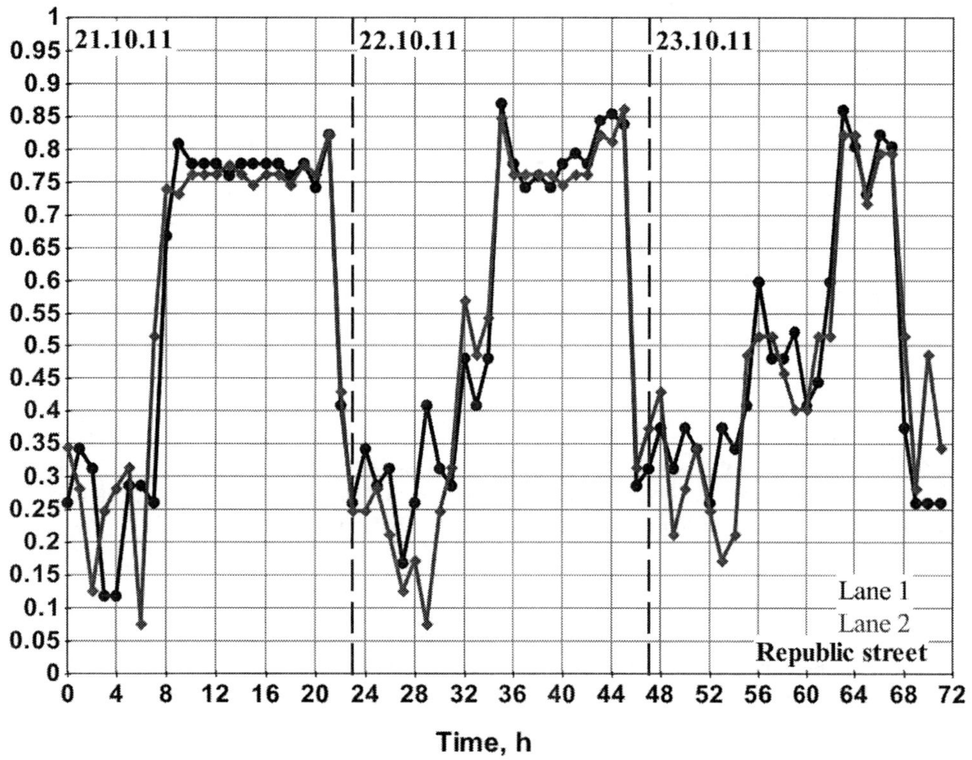

b)

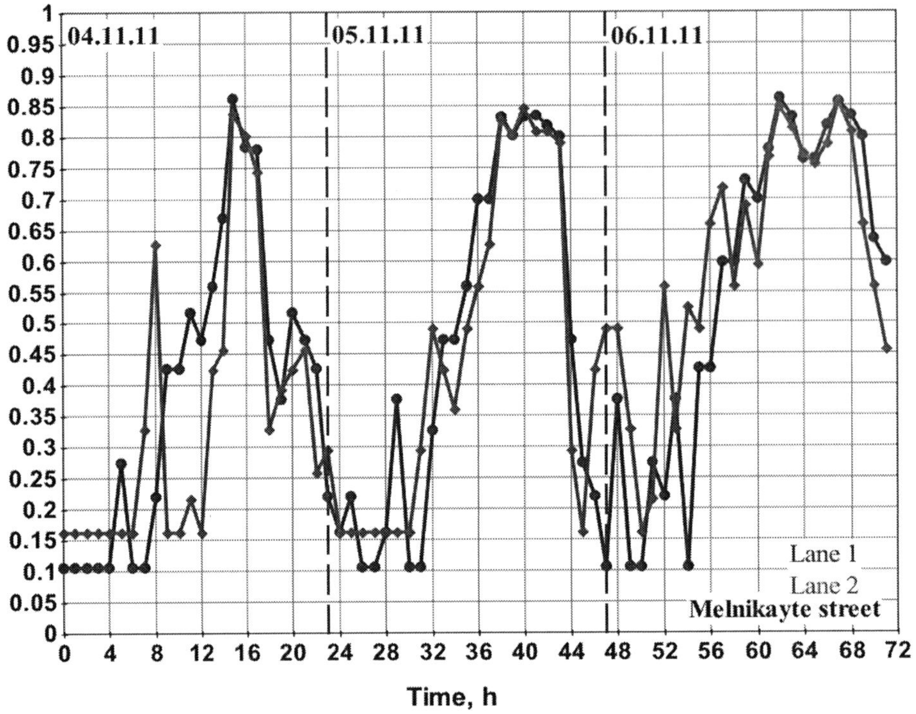

Fig. 10. a), b) Dependence Eff(t).

It is important to note that the chronology of overall any TF performance is caused by two major factors. First, the criterion of Eff significantly depends on dynamics of the flow under study therefore at an initial analysis stage dependences of dynamic characteristics on time have to be known. Secondly, the form of curves of Eff (V) (fig. 8) and Eff(q) (fig. 9) is very sensitive to scales of efficiency subtests (a1, a2, a3).

4 Conclusion

Worldwide researchers have conducted numerous studies of functioning processes of motor transportation flows. Optimization of TF productivity, traffic safety and environmental risks are key tasks against growth of a motor transport role in region the and country economic capacity formation. The results of studies received by authors allow to estimate in a complex the overall TF performance taking into account key efficiency indicators, to mode optimum operating conditions of TF, to develop a program maximizing the subtest taking into account specifics of Street-road network under analyses [11, 12, 13]. All mathematical models have been tested [14]. The program "Effective Transport Stream" allowing to apply various production scenarios on modeling TF dynamics and an assessment of efficiency of its work has been developed.

References

1. D. Drju, *Teorija transportnyh potokov i upravlenie imi* (Transport,Moscow,1972)

2. V. Kolesov, *Transportnye i transportno-tehnologicheskie sistemy. Materialy Mezhdunarodnoj NTK*, 94-103 (2014)
3. R. Jel'vik, B. Anne, V. Trule, Spravochnik po bezopasnosti dorozhnogo dvizhenija: per. s norv. IJeT Norvegii, MADI (GTU), Sovet ministrov severnyh stran. Oslo, Moscow, Hel'sink, 746 (2001)
4. V. Krutov, *Osnovy nauchnyh issledovanij: uchebnik dlja tehnicheskih vuzov* (Vysshaja shkola, Moscow, 1989)
5. Vsemirnyj doklad o preduprezhdenii dorozhno-transportnogo travmatizma: perevod s anglijskogo, Ves' Mir», Moscow (2004)
6. V. Zhdanov, Metod ocenki tehnogennoj opasnosti transportnyh potokov na ulichno-dorozhnoj seti goroda. Avtoreferat dissertacii na soiskanie uchenoj stepeni kandidata tehnicheskih nauk. MADI (Moskovskij Avtomobil'no-Dorozhnyj Institut), Moscow, (2008)
7. V. Zhdanov, *Sovershenstvovanie metodov rascheta jekologicheskih harakteristik gorodskih transportnyh makroistochnikov* (Mashinostroenie, Moscow, 2010)
8. V. Kolesov, D. Sorokin, M. Guljaev, **5**, 89-95 (2012)
9. V. Kolesov, *Neft' i gaz Zapadnoj Sibiri: materialy mezhdunarodnoj NPK – TjumGNGU*, **43**, 26-30 (2013)
10. V. Kolesov, *Materialy 5 Vserossijskoj NTK – TjumGNGU*, 25-31 (2012)
11. O. Danilov, V. Kolesov, *Materialy mezhdunarodnoj nauchno-prakticheskoj konferencii – TjumGNGU*, **44**, 402-405 (2010)
12. A. Moor, O. Danilov, V. Kolesov, *Materialy mezhdunarodnoj NTK.– TjumGNGU*, **45**, 182-193 (2011)
13. A. Moor, O. Danilov, V. Kolesov, Transportnye i transportno-tehnologicheskie sistemy, **4**, 167-171 (2012)
14. D. Sorokin, Vektor Buk, **12**, 126-129 (2010)

Comparison of Electric and Gas Turbine Drive Types at Pump Stations

Vitaly Shpilevoy[1], *Anton* Zakirzakov[1,*] and *Alexander* Shabarov[2]

[1]Industrial University of Tyumen, 625001 Volodarskogo str. 38, Tyumen, Russia
[2]Tyumen State University, 625001 Volodarskogo st. 6, Tyumen, Russia

Abstract. The article analyzes the state of an extensive network of main oil pipelines in Tyumen region on the basis of statistical data, as well as the ways for the efficiency improving of energy-saving policy for the main oil transportation are suggested.

1 Introduction

Issues related to the energy consumption saving are more and more likely to be at the center of the worldwide society attention. The urgency of energy saving [6] focus caused by the limited and slow recovery pace of natural energy sources. It should be noted that the main oil transportation is one of the largest consumers of fuel and energy resources in the Russian Federation [3, 4, 18, 24].

The existing system of oil and gas pipelines in Russian Federation has developed in 60-80 years of the last century, and the greatest development of their structures was caused by the necessity of large volumes of oil and gas transportation from Western Siberia fields to the central regions of the country as well as for export [20, 22].

The state development policy of Russian export main oil pipelines determined the necessity of the creation of two new directions: [13]

- the Northern one: oil export by sea to the countries of North-West Europe and Northern America from Western Siberia fields (the beginning in the Surgut area) as well as Timan-Pechersk oil and gas province to the Barents Sea coast and further;
- the Eastern one: with the prospects of the resources development of Eastern Siberia and Far East, as well as the oil export organization to China, Japan and other countries of the Asian-Pacific region.

Since a significant routes portion of both projects is planned for the northern latitudes, as well as for Siberia area, there is the design and construction experience of such oil pipeline systems under the appropriate climatic conditions in Russia and it can be complemented by the foreign one in some aspects [9, 10].

Main pump units are powerful energy consumers, so their efficient and economical operation is a very important task of operating organizations.

* Corresponding author: zakir1404@mail.ru

Energy consumption for oil transportation can be reduced by choosing the optimal energy drive type of the pump stations as at construction, and at reconstruction of the existing oil pipeline routes, because the main electrical pump drive has no alternative already at the drawing up stage of regulatory and technical documentation [11,12].

Electricity is the main share of energy in the oil pipeline transportation (Figure 1) [21]. GOST for oil pump units, developed in the USSR, only provided the electrical drive for the main pumps, which was caused by the progressive policies of the world powers in the field of the development of a unified power supply system. The standard technical documentation of Russian oil transportation companies took the place of the Soviet GOSTs. It also leaves no choice for a design engineer relative to the main drive units.[15]

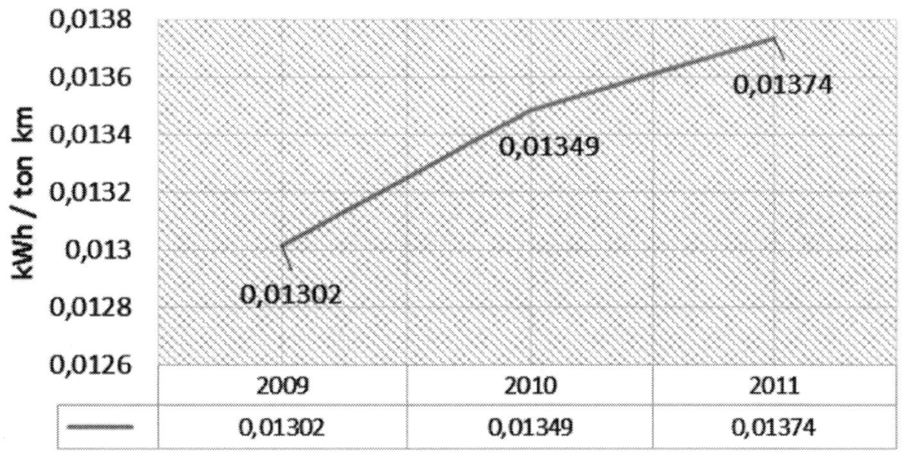

Fig. 1. Specific electricity consumption for oil pipeline transportation in Tyumen region.

2 Comparison of Electric and Gas Turbine Drive Types at Pump Stations

One of the conditions for the dynamic development of the main oil transportation is the pumping cost reducing, which important component the power resources cost is [26]. There are various ways to reduce the energy consumption [25]:

- optimization of oil pumping modes applying the modern methods of pipeline productivity control;
- reduction of energy losses for operating equipment, power transmission lines, distribution networks;
- introduction of modern methods for reducing the pipeline hydraulic resistance;
- the technique optimization for a drive pump units choice because the main pump electrical drive has no alternative already at the drawing up stage of regulatory and technical documentation.

With the development the system of gas supply the oil pump units with gas turbine drive can compete with the existing ones [1]. It is especially important to use them in areas where there is no unified system of power supply. The drive type choice of oil pump system is affected by a number of fundamental factors. First of all there is the availability of powerful energy sources, as well as technical, energy and economic aspects.[14]

In the initial period of the national development of oil pipeline transport a choice of the pump units drive type for the oil pump stations was not difficult. In many cases, there was only one energy source – the combustion energy of the product being pumped. This provided the autonomous reliable and uninterrupted energy supply [5]. This solution is widely used at present for a gas turbine drive in the compressor stations of main gas pipelines, where the energy source is the combustion energy of the pumped gas, and automatic control and regulation system ensures perfect all the requirements for the main gas pipelines operation.[17, 19]

Many years of research and practice of energy drives use at the compressor stations of domestic main gas pipelines determined the preferred appliance of gas turbine units of various capacities and types. It is rational to use them at the pumping stations, located in remote areas from the centralized electricity supply under difficult climatic conditions. It is efficient to use the electrically driven pump stations in areas with large electric power systems with relatively cheap electricity of nuclear and hydro power plants.

For many years, the argument for energy autonomy outweighed the arguments in favor of appliance of other drive types. However, the electrical drive has been widely used in the following and in remote areas as well as areas with complicated environmental climatic and geographical conditions (Western Siberia), for oil pump stations of main oil pipelines. The domestic ideological motivation of priority and the statement that "the electrification is the key element of economic development of the socialist society" has greatly contributed to this. [13]

In addition to this, the planning of the use of each type pumps drive at oil pump stations should be determined by the presence of an energy source and economic aspects of its expediency and necessity connected with the unification of the equipment, different for each energy source (electricity, oil, gas). The influence of factors determining the decision making about a drive type may have a different weight in each case. It is necessary to carry out technical and economic comparisons.

From the energy point of view the products useful factor (KPI) of the drive energy is defined as the product of the efficiency factors of the electric power circuit elements:

$$\eta_E = E^i_I = \eta_P \cdot \eta_{IS} \cdot \eta_{DS} \cdot \eta_L \cdot \eta_n \cdot \eta_{ED}, \quad (1)$$

where ηP — efficiency factor of supplying power plant (for thermal power plants 0.35-0.4, for nuclear and hydro power plants it is significantly higher);

ηIS and ηDS — efficiency factor of raising and lowering substation (0.98);

ηL — efficiency factor of electric power line (0.95);

ηn — efficiency factor of the distribution (power) network (0.96);

ηED — efficiency factor of the drive electric motor (0.85–0.9).

At heat power plants ηE = 0.27–0.31.

The modern gas turbine motor with a gas turbine drive has an efficiency factor of 0.25-0.29 (in the future it will rise up to 0.35-0.4). In addition it has the efficiency use factor of a gas turbine drive ηGTD.

In the energy aspect the electric drive of the heat power plants and the gas turbine drive of the oil pump stations are almost equally effective.

From the economic point of view the efficiency calculation of the compared oil pumping systems is fairly routine [15,16], requiring the synthesis and evaluation of a large number of input data in the current conditions, the previous design decisions, the future operation conditions of the oil pipeline and foreign experience.

Lower costs for buildings and equipment are greatly increased due to the costs for construction of the power supply system when uncontrolled pumps electrical drive. Composition and costs of related equipment (installations for selection, preparation and use of fuel) increase significantly when the pumps gas turbine drive [7, 8].

The estimated prime cost of electric power used at the industrial enterprises of the gas industry, can be made according to the specific fuel consumption spent for 1 kWh energy generation.

Indeed, if the average efficiency factor of heat power plants operating on gas would be taken at the level of 32%, so the specific fuel equivalent consumption for generation of 1 kWh of energy (combustion heat of 1 kg fuel equivalent is taken equal to 29.3 MW/kg) would be equal to kg / (kWh):

$$B_{s\cdot f} = 0,123/0,32 = 0,384 \quad (2)$$

Expressed in terms of natural gas where the fuel combustion heat is taken as 49 MJ/kg, and the gas density under standard conditions – ρST = 0.6-1.205 = 0.72 kg/m, the average specific fuel consumption (of natural gas), released for a 1 kW * h is equal to:

$$b_{ng} = 0,384\text{-}29,3/49 \text{ - } 0,23 \ kg/(kWt\text{-}h) \ or \ 0,319 \ m^3/(kWt\text{-}h) \quad (3)$$

Knowing the numerical value of the gas selling price (P, Rub/kg), it is easy to determine the cost of 1 kWh of electric energy at the power station running on gaseous fuel.

The component of the fuel cost at heat power plants in the total electric power cost is usually at the level of 60 - 65%.

The rest is accounted for by labor and material costs (~ 12 -15%), investments (~ 5 - 7%), taxes (~ 5%), profit (~ 15%), etc.

Thus, the ratio of presented efficiency factors of the gas turbine and electric motor on the pump coupling on the one hand, and the prices relation of the energy carriers in the rapid calculation on the other hand can determine the rational use areas for each drive type (Figure 2).

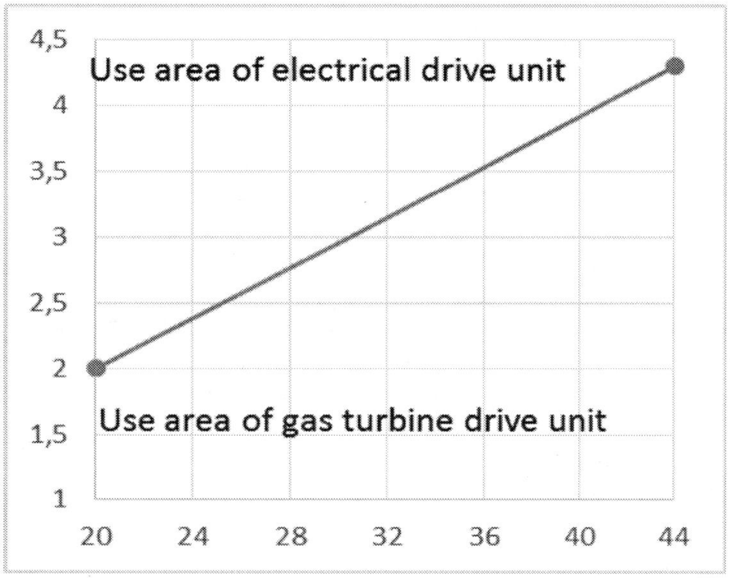

Fig. 2. Rational use areas of gas turbine and electrical drive units.

It is shown that the cost of one kWh of energy from the gas turbine units is about 2.5 - 3 times lower than from the electrical driven ones at current prices for electricity and fuel gas.

The gas turbine drive in the drive of main oil pipelines pumps can be used in areas with different climate conditions. It is preferable to use them in areas with relatively low mean average temperature of the ambient air, that is, in the Far North's and Siberia's areas.

In the world practice there are enough examples of the widespread use of gas turbine drives at the main oil pipelines, including the Trans-Alaska one extending for about 1,300 km under northern conditions close to the Western Siberia's and the Russian Far North's ones.

The question of pumped fuel use takes a special place in the operation of gas turbine drive on the oil pipelines. Such fuel types, as kerosene, gasoline, diesel, natural and associated gas, meet directly the requirements to the gas turbine fuel. Natural gas is effective used when the gas pipeline is laid parallel to the oil one [27]. The associated gas can be used, if it is pumped together with oil. Pumped crude oil can be used after the preparation at special block rectification units. After rectification of the oil 25-30% of natural gas liquids (NGLs) are obtained, and the residue (heavy fuel oil) is pumped into the oil pipeline. Application of such systems is economically advantageous in areas remote from the powerful electric energy supplies. Combustion products disposal solves the problem of heat consumption at oil pumping stations.[17]

The gas turbine drive is the most economical and controlled electrical drive system – the least economical for controlling of the pump unit supply.

There are also isolated examples of evaluation and use of the regulated electrical and gas turbine drive in domestic application area of drive pump units on main oil pipelines.

Developments of OAO "Aviadvigatel" for the "Sakhalin-2" project are of particularly attention [2].

However, further conducting of technical and economic evaluation taking into account the reliability for the specific operating conditions is necessary to fully evaluate the effectiveness of a particular type of oil pump station drive. In areas with a low cost of electricity from nuclear or hydro power plants the electrical drive at oil pump station may be more economical than the gas turbine drive. At the same time the gas turbine drive at the oil pump station may be preferred under conditions of frequent changes in the oil pipeline operation mode.

3 Conclusion

The examples given above show, that the electrical drive is not uncontested drive in oil pipeline transportation systems. There is a need to establish technique of a choice of the pump unit drive, which allows comparing such factors as distance from the power supply sources, the pumped oil composition, the presence of a developed gas supply system, etc.[18]

The implementation of the mentioned measures may be the key factor for significant increase in the efficiency of energy-saving policy on the main oil transportation as well as providing the electric power saving and increase the reliability of the main oil pipelines system.

References

1. V. Durymanov, A. Ivanov, K. Ishbulatova, A. Stepanov, M. Surnikov, Turbines and diesel engines, **3**, (2013)
2. A. Inozemtsev, F. Khairullin, Energy and transport of gas, **17**, (2010)
3. The use of fuel and energy resources in Tyumen region in 2009-2012, Statistical Bulletins
4. The comprehensive program "Energy saving and the energy efficiency enhance in Tyumen region" for 2010-2020
5. A. Smirnov, A. Vengerov, S. Dudin, Yu. Zemenkov, Oil and gas, **3**, 74-77 (2010)
6. I. Tyrylgin, V. Shpilevoy, Yu. Zemenkov, Oil and gas, **6**, 73-81 (2012)
7. V. Shpilevoy, I. Tyrylgin, Yu. Zemenkov, Oil and gas, **5**, 75-78 (2012)
8. A. Kulikov, V. Petryakov, I. Seroshtanov, Business Magazine Neftegaz.RU, **11-12**, 36-40 (2015)

9. A. Kulikov, *New technologies - oil and gas region: Proceedings of the scientific-practical conference of students, graduate students and young scientists,* 102-105 (2010)
10. A. Kulikov, E. Komarova, *Problems of functioning of transport: Proceedings of the International scientific-practical conference,* 188-189 (2010)
11. V. Nekrasov, R. Levitin, Fundamental and applied research: Problems and results, **13**, 223-228 (2014)
12. V. Nekrasov, R. Levitin, I. Tirilgin, Yu. Zemenkov, A device for improving the performance properties of vertical steel tanks. Pat. № 2012125478/05 (RU), declared 10.12.13 (2013)
13. V. Kurushina, Yu. Zemenkov, WIT Transactions on Ecology and the Environment, 190, **2**, 881-888 (2014)
14. V. Antip'ev, A. Nevolin, Yu. Zemenkov, Neftyanoe Khozyaistvo - Oil Industry, **10**, 46-48 (1981)
15. Yu. Zemenkov, V. Shalay, M. Zemenkova, Procedia Engineering, **113**, 254-258 (2015)
16. Yu. Zemenkov, V. Shalay, M. Zemenkova, Procedia Engineering, **113**, 312-315 (2015)
17. A. Pimnev, Thesis of Candidate of Technical Sciences, 169 (2006)
18. E. Kurushina, V. Kurushina, Life Science Journal, **11(11)**, 517-521 (2014)
19. A. Antropova, A. Zakirzakov, *Oil and gas in Western Siberia: Proceedings of the International scientific conference,* 64-66 (2013)
20. A. Zakirzakov A. Egorov, Modern problems of science and education, **1-1**, 309 (2015)
21. Yu. Zemenkov, V. Shpilevoy, S. Podorozhnikov, A. Zakirzakov, *Energotechnological complexes at the design and operation of transport and storage facilities of hydrocarbons* (TSOGU, Tyumen, 2014)
22. Yu. Zemenkov, V. Kurushina, Mining informational and analytical bulletin (scientific and technical journal), **3**, 85-98 (2013)
23. M. Zemenkova, I. Seroshtanov, V. Kurushina, S. Toropov, Yu. Zemenkov, Territory Neftegaz, **10**, 80-86 (2013).
24. Yu. Zemenkov, V. Kurushina, A. Vylegzhanina, V. Barmenkova, N. Khairullina, *Modern management and innovation, science and technology policy* (Publishing House Science and Innovation Center, Ltd., St. Louis, 2013)
25. A. Gumerov, R. Gumerov, A. Akberdin, *Operation of equipment pumping stations* (Nedra-Business Center, Ltd, Moscow, 2001)
26. L. Paleeva, A. Zakirzakov, *Oil and Gas Terminal: Proceedings of the International scientific conference,* 182-184 (2015)
27. I. Kiryanov, S. Toropov, V. Toropov, Neftyanoe Khozyaystvo - Oil Industry, **8**, 48 (2015)

MATEC Web of Conferences 73, 01012 (2016) DOI: 10.1051/matecconf/20167301012
TPACEE-2016

Usage of spatial lattice metal structures as roofing for mechanical equipment rooms of hydroelectric power stations

Mikhail Balzannikov[1], *Vadim* Alpatov[1,*], *Igor* Kholopov[1], *Andrey* Saharov[1] and *Aleksey* Lukin[1]

[1]Samara State University of Architecture and Civil Engineering, Molodogvardeiskaia str. 194, 443001, Samara, Russia

Abstract. The paper focuses on technical analysis of spatial attice structures used in roofing of mechanical equipment rooms of hydroelectric power stations in Russia. The analysis reveals that spatial structures of "Markhi" type are used most frequently there and shows their advantages and disadvantages. The authors put forward their own solution for combining advantages of spatial spatial lattice roofing structures with those of truss frames with gusset plate units. The solution is aimed at enhancing the safety and reliability of hydraulic engineering constructions with no changes in accuracy requirements to their manufacturing. The authors also introduce their model of a spatial lattice roofing unit and give an account of its reliability and strength based on software system calculations.

1 Introduction

Hydroelectric power stations are considered to be hydro-engineering constructions of high level of hazard (1st Degree Amenability according to the classification accepted in Russia). Usually these constructions are unique. The safety and reliability demands for designing, building and operating hydroelectric power stations are very high. One of the most important parts of any hydroelectric power station is a mechanical equipment room. A typical mechanical equipment room is usually of a considerable length with more than 24-meters bays. Spatial lattice roofing structures (as Fig. 1 shows) are used very frequently in mechanical equipment rooms. Metal spatial lattice roofing structures of "Structure" type have been successfully used in mechanical equipment rooms of hydroelectric power stations. They have also been proved effective.

* Corresponding author: avu75@mail.ru

Fig. 1. Spatial spatial lattice roofing structure of a mechanical equipment room, a hydroelectric power station (ANSYS Advantage №15'2011, cover image)

"Structure" has a lot of competitive advantages over other alternative roofing structures. The most important advantages of "Structure" are comparatively little material consumption, the possibility of check assembly in advance and of packaged supply, a block building system, route sequence of mechanical processing, availability of many different parts of one and the same type. "Structure" has no balks, beams and connections typical for other metallic constructions. "Structure" is also characterized by advanced rigidness. The proportion of height to bay in these constructions is about 1/16–1/25. High reliability of "Structure" is based on its multi-indeterminateness. This high reliability (sometimes referred to as "survivability") is mostly due to spatial work of its mechanical system, and also to numerous redundant restraints which give the construction the ability to redistribute forces. "Structure" is a Spatial geometrically invariable structure. Spatial work of "Structure" is evident under non-uniform loads. In such a case overburden pressure on most auxiliary bars (excluding those bars without which the system becomes a kind of a mechanism) does not generally interfere with the normal work of the construction because of its ability to redistribute forces. As far as structural slabs are characterized with advanced rigidness and are of low weight it is reasonable to use them as long span structures. Structural roofing installation can be done as follows: by balanced erection, by erection on design level of trestlework, by erection of the whole construction or its large parts on the ground, by erection of its enlarged blocks. Almost all these types of erection have been tested on different construction sites [1]. Important stage of economic design of large-span metal rod systems is application of methods of optimization [2–6].

2 The model of a spatial lattice roofing unit

Comparative analysis of metal structures used in roofing of mechanical equipment rooms of hydroelectric power stations in Russia showed that only two types of such constructions are usually used. It's either traditional constructions (they are two dimensional trussed rafters of various forms and with different lattice design) or spatial structures of "Markhi" type. "Markhi" was introduced in 1969 by the Department of Engineering Construction of Moscow Architectural Institute. "Markhi" present the next stage in the development of famous German MERO structures (Fig. 2).

The effective service of cross bar constructions depends mostly on the type of joint connection as they contain from 8 to 18 bars connected at different angles. In the most simple structures (which consist of lattice girders intercrossing in one and the same vertical subspace) this connection is usually used to reduce the height of the construction in one level. The girders are interconnected in such unit with the help of cross joint plate, the components of the boom are split in the unit and the joint is covered with a similar gusset sheet joint plate.

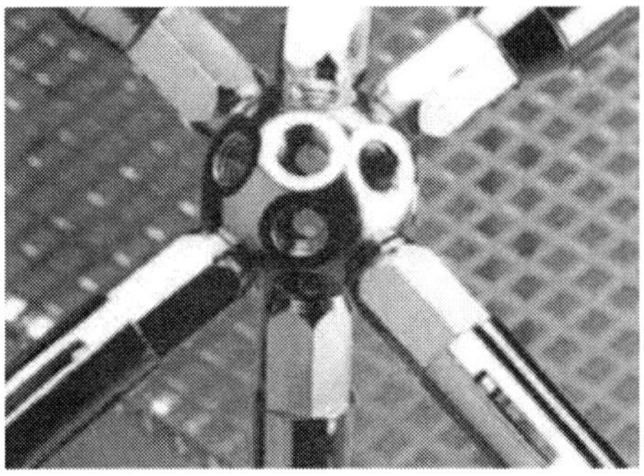

Fig. 2. A unit of MERO system

Both "Markhi" and MERO structures require special unit connectors to be made in the form either of a sphere or of a polyhedral with specially cut tap holes. They also impose strict accuracy requirements to their manufacturing and assembling. Traditional trussed metal rafters have units which are usually made on sheet joint plates and they are able to compensate deviations. The authors here suggest combining positive characteristics of spatial structures and those of trussed rafters on sheet joint plates to design a spatial lattice structure with units on sheet joint plates [7]. Such a structure would combine advantages of spatial structures with the ability to compensate deviations. The authors believe it can be successfully used while roofing building of different types including mechanical equipment rooms of hydroelectric power stations (in designing, building and reconstructing those rooms).

It is believed that separate parts of structural constructions are exposed only to tensile force and direct stress. The situation with unit connectors of these constructions is different. A unit connector in such a construction combines various bars and consequently is under a constant state of tension. Unit connectors of structural constructions should definitely be characterized by enhanced reliability as in case of only one connector failure the whole group of bar fails. That why special attention should be given to these unit connectors and their elements while designing and manufacturing them. To assure it the authors carried out a numerical experiment through simulating the suggested unit connector's stress state. The unit connector is shown below (Figure 3). It is made of sheet joint plates. According to existing recommendations the bulk of the gusset plate should depend on the pressure of the element attached to it (Y.M. Likhtarnikov, Raschet stalnyh constustsiy, Kiev, 1984). However, these recommendations are aimed at flat girder constructions with twin angle joints. There are no recommendations for spatial constructions yet. The authors believe that the stress level of the gusset plate will be reduced and, consequently, the bulk of the gusset plate can be narrowed. The authors of the paper made it their aim to trace the stress pattern in a gusset plate of a spatial construction, to quantify the stress levels in it and to estimate a possibility to make the bulk of the gusset plate narrower than the recommended bulk of a flat girder. The authors have succeeded in doing it by numerical simulation with the help of the modern automated design system (CAD/CAM/CAE).

To determine the rated force applied to a unit connector the authors analyzed design and engineering developments of different spatial constructions for various patterns of loading.

Analyzing the most unfavorable combinations of strain and stress factors which affect the parts of spatial constructions connected with different elements the authors found the most stress-suffering units and calculated their stress-strain state (SSS). Modern software systems of mechanical systems structural analysis – the systems based on finite-element method – were used to analyze SSS of the units in question. All four software systems were used in this analysis to eliminate mistakes in making the simulation model, to eliminate all possible mistakes in the work of those systems, and to verify accuracy of the research.

This paper presents the results of the numerical study of one the units belonging to a lower level. This unit connects eight bars. The loading stress was calculated for these bars. The strain stress in them was simulated with the help of numerical study of bar models of spatial constructions exposed to outer loadings. The scheme of this unit loading is shown on Fig. 4.

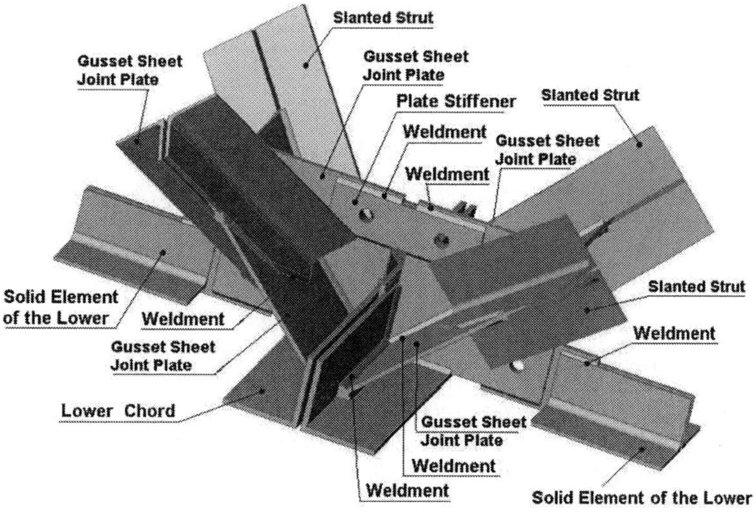

Fig. 3. A unit of spatial bar construction on gusset sheet joint plate

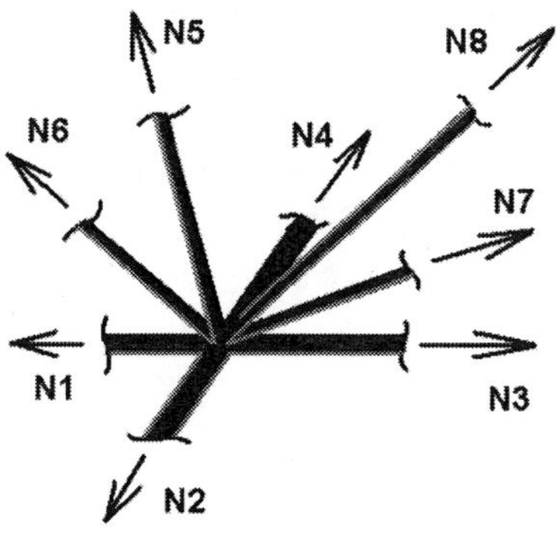

Fig. 4. The scheme of this unit outer loadings

Computational engineering and software of today are capable of calculating almost any of the most advanced systems.

The modern automated design system (CAD/CAE) helps researchers and designers choose appropriate methods and tools [8]. The authors of the paper can mention the following CAD/CAE systems: CosmosM, CosmosWorks, DesignSpace, Ansys, Abaqus, Nastran, Staad, Sap, "Lira", "MIRAZH", SCAD, Ansys.

The units of the structural construction in question were investigated with the help of CosmosWorks, DesignSpace, "Lira", SCAD. Figure 5 show the results of investigation carried out with "Lira".

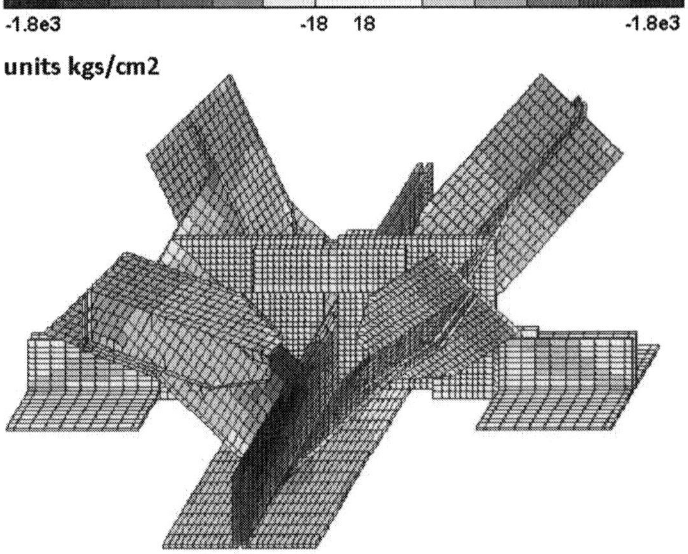

Fig. 5. Stress strain of the unit calculated with "Lira"

Figure 5 shows a calculation model of the unit in question. Finite elements of the calculation model are shown as six- and eight-units spatial finite elements of some form or other. The calculation model contains 14 435 units and 7 733 elements. The characteristics of the material these elements are made of correspond to Steel grade C245 (Ry=240 МПа, E= 2·1011 Н/м2, c=0.3).

Figure 6 demonstrates the results of the investigation. It illustrates stress distribution in the calculation model. The figure clearly shows that gusset plates are used in the so-called region of material elastic work and are 26.6% under loaded. According to the research the other units of the construction have equal excess strength. It leads us to the conclusion that if we follow the recommendations for the bulk of the gusset plates in designing spatial structures it might lead to the excess of the materials used. Stress strain calculations of the same unit with other systems vary from 12% (Cosmos) to 26.6% ("Lira").

3 Conclusions

1. First-hand experience shows that spatial structural constructions, being structures of high "survivability", enhance reliability of potentially hazardous buildings and constructions when they are used as roofing.
2. The authors of the paper introduce the unit connector of a spatial lattice metal construction which makes possible to reduce strict accuracy requirements to its manufacturing with no loss of its reliability or safety.
3. Analysis of the calculation model of the unit showed that the bulk of the gusset plate can be narrowed in comparison with the required bulk of a gusset plate in a flat girder.

References

1. Building and Propety, Internet Journal, №35, (2005), http://www.nestor.minsk.by/sn/
2. Chang-Yu Cui, Bao-Shi Jiang, Engineering Structures, **77**, 109-118 (2014)
3. Lazaros Elias Mavromatidis, Energy and Buildings, **106**, 192-202 (2015)
4. M. I. Balzannikov, I. S. Kholopov, V. Yu. Alpatov, A. A. Lukin, Procedia Engineering, **111**, 72-81 (2015)
5. V. Yu. Alpatov, A. A. Lukin, Procedia Engineering, **111**, 20-29 (2015)
6. V. Yu. Alpatov, Procedia Engineering, **91**, 177-182 (2014),
7. V.Yu. Alpatov, I.S. Kholopov, Sterzhnevaya strukturnaya konstruktsia iz dlinnorazmernykh elementov (Frame Structure made of long dimensional elements), RU Patent 33381. (2003)
8. Y. Gorokhov, V. Mushchanov, I. Romenskii, A. Muschanov, Metal constructions, **21(4)**, 191-206 (2015)

MATEC Web of Conferences 73, 01013 (2016) DOI: 10.1051/matecconf/20167301013
TPACEE-2016

Seismic Isolation of Offshore Pipeline Block Valve Stations

Liudmila Muravieva[1,*]

[1]Peter the Great Saint-Petersburg Polytechnic University, 195251 Polytechnicheskaya str. 29, St. Petersburg, Russia

Abstract. As shown by the actual material analysis of offshore pipeline damage caused by strong earthquakes, the greatest damage is caused to complex assemblies such as branch connections, bends, and various joints .In the seismic resistance analysis of the offshore pipeline a fixed junction of an offshore pipeline section with a block valve station is modeled. To assess the connection assembly for seismic resistance, a number of accelerograms with descriptions of magnitude 8 and 9 earthquakes in the Black Sea region were considered. The calculations have revealed significant pipe wall stresses in the offshore pipeline/block valve station junction. The resulting conclusion is that offshore pipeline block valve.

1 Introduction

Researches M.J O'Rourke and Sh.G. Napetvaridze mention that failure rate of the trunk pipelines increases at the points of connection to the equipment. Seismic resistance of the connections/block valve stations of the buried onshore pipeline is reviewed in the studies of T.R. Rashidov and G.Kh. Khozhmetov.

Calculations of the seismic loads on the linear sections of the pipeline are described in the regulatory documents of STO Gazprom [12], and on the offshore pipeline in the Maritime Register Guidance NDN2-020301-003 [8].

The characteristic feature of the offshore pipelines is that they have only two block valve stations. To control impact of deformations caused by process loads when designing offshore pipelines, pipe bends are used.

Vibration protection issues are addressed in the Manual (RD). It is required that «maximum allowable vibration amplitude of the process pipelines shall be 0.2 mm at a maximum vibration frequency of 40 Hz» [9].

The analysis shows that the detailed seismic resistance study of the offshore pipeline fittings (block valve stations/shutoff valves) shall be further conducted.

2 Landfall section of pipeline

The paper deals with vibration protection of block valve stations /shutoff valves of the offshore pipeline, and includes the trunk pipeline design scheme made with due regard for

* Corresponding author: rfludmia@yandex.ru

damper/elastic supports. Vibration isolation of the offshore pipeline block valve station is also analyzed.

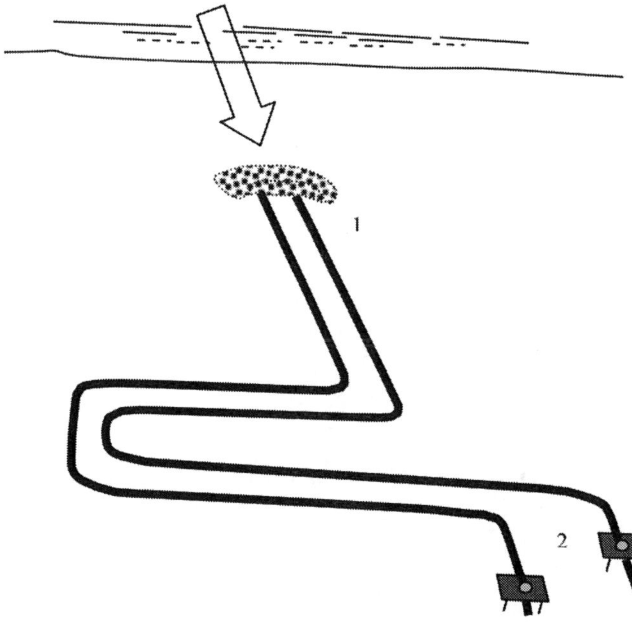

Fig.1. Landfall section of pipeline onshore landfall diagram, 1 – Gas pipelines laid on the seabed have landfall where the shutoff valves 2 are installed, 2 – Ball valves to isolate gas flow.

2.1 Seismic resistance analysis of the offshore pipeline

2.1.1 Calculation between the offshore pipeline section and block valve station

In the seismic resistance analysis of the offshore pipeline (7) a fixed connection between an offshore pipeline section and block valve station is modeled. A number of accelerograms with descriptions of magnitude 7 and 9 earthquakes in the Black Sea and Caspian Sea are considered.
The calculations have revealed significant pipe wall stresses at the connection point between the offshore pipeline section and block valve station. The resulting conclusion is that offshore pipeline block valve stations need to be equipped with dampers.

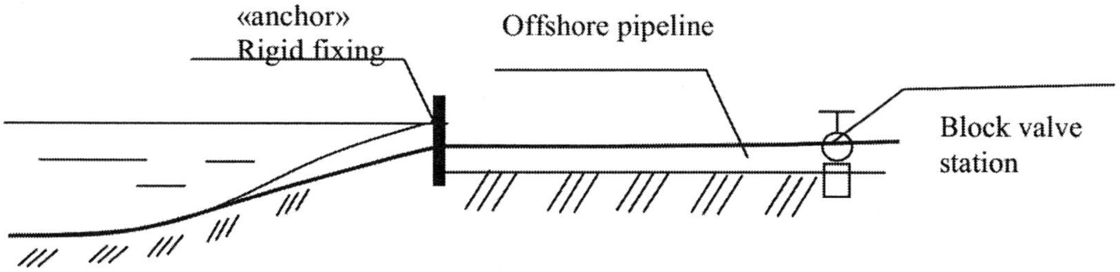

Fig.2. Landfall section of pipeline block valve station.

2.1.2 Seismic resistance analysis of the offshore pipelines

In the course of seismic resistance analysis of the offshore pipelines level of stresses and allowable deformations of the pipe walls are evaluated. Pipe bends are installed to account for the deformations caused by workloads in the offshore pipelines.
Each seismic wave gives earth shocks to buildings. Destructive earthquakes with magnitude of 8 or 9 are characterized by pulse propagation. For example in the northern part of the Izu peninsula (Japan) seismic pulse of 0.015 to 0.12 m/s repeated 18 times within 55 seconds.
In this paper vibration isolation of the offshore pipeline block valve station has been analyzed.

Change in object vibrations from external dynamic loads can be done by different methods: vibration scattering and redistribution of vibration energy. The first case refers to inertial dynamic dampers that are mainly used to suppress monoharmonic or narrow-band random vibrations. In the event of wide-band vibrations it is preferred to connect additional damping elements to an object such as absorbers.

2.2 Equation of the protected object motion

Differential equation of the protected object motion is represented as follows:

$$m\ddot{u} + c_0\dot{u} + f(\dot{u},u) = -m\ddot{u}_g \tag{1}$$

More accurate solutions are observed if behavior of dampers is taken into account. This leads to solving of the nonlinear task. Differential equation of the object motion with consideration of dampers is represented as follows:

$$\ddot{u} + 2\omega_0\left(\xi + \xi_{eq}\right)\dot{u} + \omega_0^2 = -\ddot{u}_g \tag{2}$$

A constant section beam with various end fixings (with one end fixed and the other end semi-fixed relative to angular movement and fixed end relative to transverse movement) is taken as a design model for the offshore pipeline with block valve station. Impacts of shear loads and inertia of cross section are not taken into consideration.

Differential equation of natural bending vibrations of this beam is represented as follows:

$$EI\frac{\partial^4 y(x,t)}{\partial x^4} + m\frac{\partial^2 (x,t)}{\partial t^2} = 0 \tag{3}$$

where y(x,t) is beam deflection.

By introducing $\bar{x} = x/l$ as an independent variable, where l is a beam length, we form the differential equation (3) as follows:

$$\frac{d^4 f(\bar{x})}{\partial \bar{x}^4} - \alpha^4 f(x) = 0 \tag{4}$$

here $\alpha^4 = p^2\frac{ml^4}{EI}$; p is a cyclic frequency of natural vibration, f(x) is a shape function of bending vibrations of the beam.

General solution of the equation (3) is written in the following form:

$$f(x) = AS(\alpha x) + BT(\alpha x) + CU(\alpha x) + DV(\alpha x) \tag{5}$$

where A,B,C and D are arbitrary constants determined from corresponding boundary conditions, $S(\alpha x)$, $T(\alpha x)$, $U(\alpha x)$ and $V(\alpha x)$ are linear combinations of circular and hyperbolic functions introduced by A.N. Krylov [6].

We assume the design model of the linear section of the pipeline with block valve station as a beam which has one end fixed and the other end elastically restrained against angular movement and fixed against transverse movement (Fig. 3)

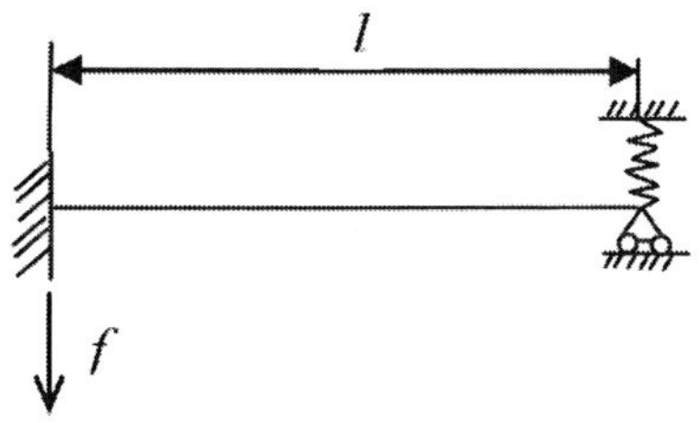

Fig. 3. A beam with one end fixed and the other end elastically restrained against angular movement and fixed against transverse movement.

Boundary conditions are written as follows:

$$\text{- at } x = 0 : f(0) = 0 ; \; f'(0) = 0 ; \tag{6a}$$

$$\text{at } \bar{x} = l \;\; f(l) = 0 ; \; f'(l) = -\frac{hl}{FI} f'(l) \tag{6b}$$

By applying boundary conditions to the functions $f(\bar{x})$и $f'(\bar{x})$ at x =0 we define A and B values.

$$f(x) = AS(\alpha x) + BT(\alpha x) + CU(\alpha x) + DV(\alpha x) \tag{7}$$

$$f'(x) = \frac{\alpha}{l}\left[AV(\alpha x) + BS(\alpha x) + CT(\alpha x) + DU(\alpha x)\right] \tag{8}$$

The result is the following: $A = 0$ and $B = 0$.

With an account of the boundary conditions to the function f($\bar{X}$) and its derivatives '($\bar{X}$) and f''($\bar{X}$) at $\bar{X}$ = 1, we obtain the following system of homogeneous equations:

$$CU(\alpha) + DV(\alpha) = 0;$$

$$C[\alpha S(\alpha) + \bar{h}T(\alpha)] + D[\alpha T(\alpha) + \bar{h}U(\alpha)] = 0 \tag{9}$$

where $\bar{h} = \dfrac{hl}{EI}$ is a relative rigidity of anchorage.

3 Frequency equation of the bending vibrations of the beam

By equating a system determinant (6) to zero we obtain the following frequency equation of the bending vibrations of the beam:

$$U(\alpha)[\alpha T(\alpha)+\bar{h}U(\alpha)]-V(\alpha)[\alpha S(\alpha)+\bar{h}T(\alpha)]=0 \tag{10}$$

We determine the relation between root α of transcendent frequency equation and relative rigidity of anchorage $\bar{h}=\frac{l}{EI}$ based on recommendations [1,6].

Frequency of beam vibration with due account for α is determined from the following formula:

$$\omega=\frac{1}{2\pi}\cdot\frac{\alpha^2}{l^2}\sqrt{\frac{EI}{m}} \tag{11}$$

We suggest that vibrations of the linear section of the pipeline equipped with a block valve station can be damped with the help of an element (vibration isolation) containing elastic members.

We use vibration isolators made from rubber Grade 3311 and dynamic modulus of elasticity of vibration isolator made from rubber E_D=250 N/cm^2 for vibration isolation of the block valve station. We assume height of the deformed part of the vibration isolators as H_p= 3 cm and cross section of the deformed part of the vibration isolator as S_i=14.4 (Fig.1).

Rigidity of the rubber vibration isolators can be calculated from the formula [8], where isolator rigidity is K=11213N/cm^2, Relative rigidity of the elastic anchorage is $\bar{h}$ =52.8 MPa.

$\bar{h}=h\frac{l}{E_D I}$, where E_D is a dynamic rubber modulus and I is a moment of inertia of the isolator.

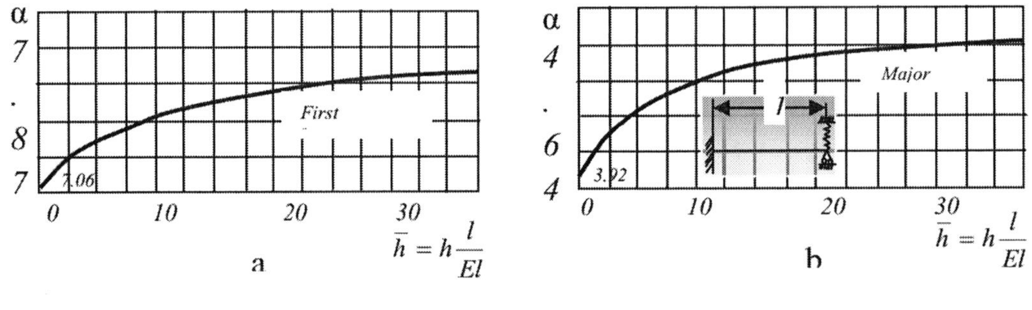

Fig. 4. Relation between root α of transcendent frequency equation and relative rigidity of anchorage $\bar{h}=h\frac{l}{EI}$ for the first harmonics (a) and major pitch (b).

From the curves on the Fig.4 [3] and *h* we define root α of the transcendent frequency equation for the first harmonics, root α value for the first harmonics is α=7.0685 (Fig.4a) and for the main pitch α=3.9266 (Fig.4b).

In practice the effectiveness of the dampers is precisely characterized by equivalent viscous friction.

Natural frequency of the offshore pipeline is determined as per formulas [6,7]. Damping properties of the offshore pipeline are characterized by damping coefficient ξ. Force acting on foundation and block valve station is determined by the deformation of the vibration isolator located between them.

According to [21], the following damping coefficients are adopted; damping coefficient of the pipeline design without concrete coating is $\xi st = 0.005$; damping coefficient of the soil foundation in the vertical direction is $\xi_V = 0.026$; rigidity coefficient of the soil foundation is c_1=230.925 t/m^3 [7]. Non-elastic resistance coefficient of the rubber vibration isolator is ν=0.038.

3.1 Vibration isolator

Vibration isolator is intended to reduce force R(t) generated by the vibration isolator and transmitted to the fixed foundation.

In accordance with [4] design seismic force FE,u,is calculated, taking into account seismic hazard of the pipeline construction area, earth movement rate during earthquakes and design foundation response spectral ordinates.

3.2 Deformation rate of the vibration isolator

In case of force action force R(t) shall be determined by deformation and deformation rate of the vibration isolator, coordinates y and $\dot{y}$:

$$R(t) = k\dot{y} + cy \tag{12}$$

Equation of mass movement *m* may be written as

$$m\ddot{y} + k\dot{y} + cy = F(t) \tag{13}$$

Considering that all variables of the equation (10) vary per the harmonic law, $F(t) = F_0e^{jwt}$, $R(t) = R_0e^{jwt}$, $y(t) = Y_0e^{jwt}$ we obtain a relation to determine amplitude of forced mass oscillations *m*;

$$Y_0 = \frac{F_0}{c}\sqrt{\frac{1+4\xi^2\overline{\omega}^2}{\left[\left(1-\overline{\omega}^2\right)\left(C-\mu\overline{\omega}^2\right)-\overline{\omega}^2\right]^2+4\xi^2\overline{\omega}^2\left[C-\overline{\omega}^2(1+\mu)\right]^2}} \tag{14}$$

where $\overline{\omega} = \omega/\omega_0$ a is relative frequency of "pipeline +block valve station" system (of linear section of the pipeline with block valve station), ω_0 is a natural frequency of the pipeline, c_1 is rigidity coefficient of the soil foundation of the block valve station $c_{rp}=c_1$ N/m^3 and damping coefficients of "pipeline+block valve station" system: $\xi = \xi_{str} + \nu + \xi_{rp}$ [7].

3.3 Force $R_{0,}$ acting on the foundation

We assume seismic vibration model described by V.V. Bolotin. Pursuant V.V. Bolotin we approximate non-stationary random function of seismic acceleration as a product of non-random envelope $A_0e^{-\gamma t}$ and stationary random function $\tilde{X}_0(t)$. Within this system oscillations are induced by seismic vibrations of the foundation that vary per the harmonic law $x(t)=X_0e^{jwt}$.

Coordinates of object vibration *y* vary per the law $y(t)=Y_0e^{jwt}$.

Irrespective of the excitation mode and damping value to ensure vibration protection for amplitude of force R_0, acting on the foundation, the following expression is used,

$$R_0 = \frac{F_0\sqrt{1+4\xi^2\varpi^2}}{\sqrt{\left(1-\varpi^2\right)^2+4\xi^2\varpi^2}} = \frac{F_0\sqrt{1+\eta^2}}{\sqrt{\left(1-\varpi^2\right)^2+\eta^2}} \quad (15)$$

We calculate movement amplitude μ based on the allowable amplitude of foundation displacement Y_D:

$$\mu = \frac{\frac{F_0}{(cY_D)}+1+(1+C)\left(\varpi^2-1\right)}{\varpi^2\left(\varpi^2-1\right)} \quad (16)$$

Amplitude of movement is μ=2.85.

Analysis of various damper types shows that:

- disadvantage of dry friction damper is wear of its mating surfaces which results in potential misalignment and jamming and causes damper failures.
- inconsistent viscous properties of oil due to temperature change causing damper detuning can be considered a problem of many viscous friction dampers.

Calculations of rubber vibration isolator impact on vibration amplitude of the block valve station have been made for the pipeline of the following size: diameter 350 mm, weight 0.148 t/m, coefficient of subgrade reaction is c_{gr}=230.925 t/m^3. Design span length of the pipeline section is 45 m [7].

4 Amplitude of forced vibration

Natural vibration frequency of the linear section of the pipeline present in table 1.

Frequency of the linear section of the pipeline. **Table 1.**

Sr. No	ω(*l*) Frequency, c^{-1}
1	with rigid fixing =5.018

Seismic force equals to F_0=0.464 H [4,7].

Amplitude of forced vibration (of block valve assembly) of mass will be equal to Y_0=0.686 m (c_{gr}=2 t/m^3), amplitude of force R_0, acting on the foundation is 1.008 t.

Study of vibration amplitude variation depending on foundation rigidity.

Force F_0 and allowable amplitude of foundation displacement $\mathbf{Y_D}$ present in table 2.

Force F_0 and amplitude of foundation displacement $\mathbf{Y_D}$ **Table 2.**

c_{onp}, t/m^3	R_0, t	Y_D, m
230.925	1.008	14.334
2	1.008	0.686

5 Conclusions

Calculated R and Y values show insufficient vibration isolation when using only rubber elements. Vibration amplitude of "pipeline+block valve station" system exceeds the required value specified in [10]. Vibration isolation is ensured as a result of combination of mechanical and hydraulic elements. This solution requires installation of hydraulic mount (Fig.5).

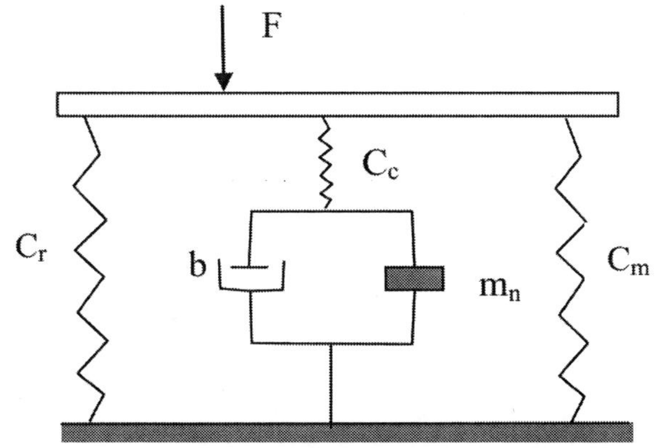

Fig.5. Schematic diagram of the hydraulic mount elastic element.

Vibration isolation of the block valve station of the offshore pipeline can be ensured by hydraulic mount installation. Vibration isolation is made based on interaction of mechanical and hydraulic sub-systems. Mechanical part interrelates with hydraulic part through equivalent resistance C_r, and hydraulic part interrelates with mechanical part through equivalent resistance C_M.

References

1. B.L. Bidermann. *Vibrations and Impact Loads, In: Strength Calculation in Machine Engineering* (Mashgiz, Moscow, 1959)
2. I.A.Birger, *Bar Tension and Bending, Strength and Stability of Vibrations* (M. «Mashinostroenie», Moscow, 1968)
3. *ISO 3010 Basis for Design of Structures – Seismic Actions on Structures. International Standard, Second Revision* (2001)
4. R. Klaff, G. Pensien, *Structure and Dynamics* (Stroyizdat, Moscow, 1979)
5. V.A. Kisilev, *Structural Theory, Special Course* (Stroyizdat, Moscow, 1980)
6. K. Magnus, Vibrations (Mir, Moscow, 1982)
7. *Report on Contract No. PC-13/2015/203-03503. "Development of Seismic Resistance Requirements to Damaged Steel Offshore Pipelines"* (2015)
8. *Register's Rules for Classification and Construction of Subsea Offshore Pipelines. HDN2-020301-003: Russian Maritime Register of Shipping, Saint Petersburg: Russian Maritime Register of Shipping* (2012)
9. *RD 51-3-96 Regulation for Maintenance of Submerged Gas Trunkline Crossings through Water Barriers* (JSC "Giprorechtrans, 1996)
10. *Design Guideline for Vibration Isolation of Machines and Equipment* (Stroyizdat, Moscow, 1972)
11. B.G. Korenev, I.M. Rabinovich, *Guidelines for Structural Dynamics* (Stroyizdat, Moscow, 1972)

12. *Equipment Produced by GERB,* Information on http://www.gerb-vibra.spb.ru/vibfo.html.
13. *STO Gazporm 2-2.1-249-2008 Gas Trunklines*
14. D.E. Chegodaev, Yu.K. Ponomarev. *Damping* (SSAU Publ., Samara, 1997)
15. S. H. Crandall, W. D. Mark. *Random Vibration in Mechanical Systems* (Academic Press, Moscow, 1963)
16. J. P. Den Hartog, *Mechanical Vibrations, 4th Edition* (Dover, 1985)
17. N. M. Newmark, E. Rosenblueth, *Fundamental of Earthquake Engineering* (Prentice Hall, London, 1971)
18. J. Ormondroyd, J. P. Den Hartog, Trans. ASME, J. of Applied Mechanics (1928)
19. A. Preumon, *Vibration Control of Active Structures, Introduction, 3rd Edition* (Springer, London, 2011)
20. A. Preumont, K. Seto, *Active Control of Structures* (Wiley, London, 2008)
21. B. F. Jr. Spencer, M. k. Sain, Special Issue of IEEE Control Systems Magazine on Emerging Technology, **17(6)**, 19-35, (1997)
22. *Recommended Practice DNV-RP-F105 "Free Spanning Pipelines*"(2002)
23. G. B. Warburton, Earthquake Engineering and Structural Dynamics, **10**, 381-401 (1982)

Vertical Displacements Measuring Methods During Bridge Load Tests

Boštjan Kovačič [1,*], *Ante* Marendić [2], *Rok* Kamnik[1], *Mikhail* Volkov[3] and *Vera* Roy[3]

[1]University of Maribor, SI 2000 Smetanova 17, Maribor, Slovenia
[2]University of Zagreb, hr 10000 Kačićeva 26, Zagreb, Croatia
[4] Peter the Great St. Petersburg Polytechnical University, 195251 Politekhnicheskaya str., 29, Saint-Petersburg, Russia

Abstract. This article introduces the use of the GNSS method for the load testings of bridge structures. We know that there area lot of methods by which you can determine vertical displacement. However, as new bridge constructions are constructed over almost impossible and inaccessible gorges and valleys, the classic measurements method do not allow us to determine vertical displacement so reliably any more or they are very time-consuming. The GNSS method is well known but there are difficulties when determining the altitude component of this method, which is rather questionable. For this purpose, the methods were tested as practical examples in which some difficulties were encountered, so subsequently another test recording was performed of altitude changes using this method.

1 Introduction

Load tests are needed in order to assess the behavior of any new building or determine the state of existing facilities before/after any rehabilitation. They enable us to verify the theoretical assumptions, as well as determining use factors which so affect the capacity of the construction they cannot be predicted theoretically. When observing a bridge structure, vertical displacements are very important because they have the biggest impact on the structure and its eventual deformation, while the horizontal displacements are tested especially in the case of rail lines and in more demanding engineering measurements of deformation [1-16].

There are several different methods for performing load tests on bridges but, in general, it involves the placement of sensors on all critical load carrying members and electronic measurements of their individual responses as known vehicles traverse the bridge (Chowdhury and Ray 2003). The method used depends on the type of construction, accessibility, required accuracies, sensitivities and reliabilities of the used instruments.

During scientific research, the newest methods are often used for a link between the results, analysis methods and displaying the results, such as photogrammetry (Guarnieri et. al. 2004, Albert et.al. 2002, Jauregui et.al. 2003, Maas and Hampel, 2006) and laser

* Corresponding author: bostjan.kovacic@um.si

scanning method (Gordon et.al 2004, Schäfer et.al. 2004, Fuchs et.al. 2004). A dynamic testing of bridges using geodetic equipment was considered in Marendić et. al. 2013.

During a load test all characteristic points, namely supports and midpoints of the spans of the bridge and those places where the greatest displacements are expected should be monitored. The dynamic and static loads are investigated during load testing.

Dynamic testing is performed using dynamic loads like shocks, vibrations or dynamic vibrations. For example, in bridge structures, the dynamic load is carried out by driving a fully loaded truck at the facility. According to the former JUS standard, which was valid in Slovenia prior to the introduction of EUROCOD during dynamic loading test, were mandatory for:

- measurements of vertical displacements in the middle of the selected bridge span (mid-of the span - between two supports);
- measuring of the speed of the dynamic loads (speed of the truck);
- measuring deformations in the places of expected maximum tension;
- measurements of other dynamic properties of the structure.

The dynamic part of the load test is usually performed separately, where the left and right parts of the facility is physically separated. In order to better control the speed, the direction of driving the truck at both facilities is the same. In Fig. 1 the dynamic part of the load test can be seen.

Fig. 1. Dynamic load test.

The static part of the load test is carried with an exact number of vehicles and loads, depending on the length and number of fields and supports. Usually both carriage ways burdened at the same time.

According to the former JUS standard during static load testing the following were mandatory:

- measurement of the vertical displacement in the middle of each bridge span;
- measuring the displacement of the support;
- observation of cracks;
- measuring of deformations in foreseen areas, where maximum strains are expected;
- measurement of permanent deformations and deformations after disburden in the bridge.

Static tests are performed for investigating structural behavior under static loads. When investigating this, local and general deformations are measured in order to determine the tensile forces in the structure of a particular element a construction or one of its parts. The geometry of the structure or its components changes under load. These changes are reflected as shrinking or stretching of the structure and as the relative displacement of points. The static part of the load test can be seen in Fig 2.

Fig. 2. Static load test.

2 Materials and methods

2.1 GNSS method

When using GNSS methods for the monitoring of vertical displacements in practice a combination of base-rover is more commonly used (base serves as a reference point). The reference station is placed where no vibrations or other disturbances in the signal are expected and in a sufficiently open place for better reception of the signal and more satellite signals. Fig 3 shows the base or reference station for the measurement.

Fig. 3. Reference station – base.

A mobile receiver or rover is placed at the facility under investigation (Fig 4). It should be placed in the middles of the spans where the greatest displacements can be expected.

Fig. 4. Mobile satellite receiver – rover.

The reference receiver has a permanent coordinate during measurements. The mobile receiver (rover) records the coordinates according to bridge movements. A rover also orients the reference receiver and not only to satellites (with radio link) to obtain better accuracies of the coordinates (Fig 5).

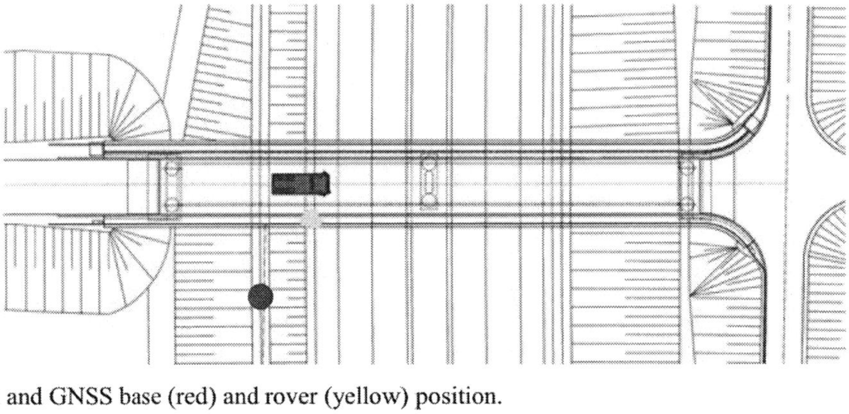

Fig. 5. Terrain and GNSS base (red) and rover (yellow) position.

GNSS data processing follows after completing the measurements as data on the field does not yield the values of displacements. After there are unexpected problems such as insufficient number of readings, signal interference, interrupting of measurements, excessive differences, multipath...

2.2 Levelling method

The movements of the right and left-sides of the bridge were monitored using the levelling method. Three points on each side were monitored. One of them on the embankment, which was the reference point, and two points at the centers of both spans where the load (truck) was standing at a static part of the investigation (Fig 7). DiNi 21 and Leica Sprinter levels were used (Fig 8, 9).

 DOI: 10.1051/matecconf/20167301014

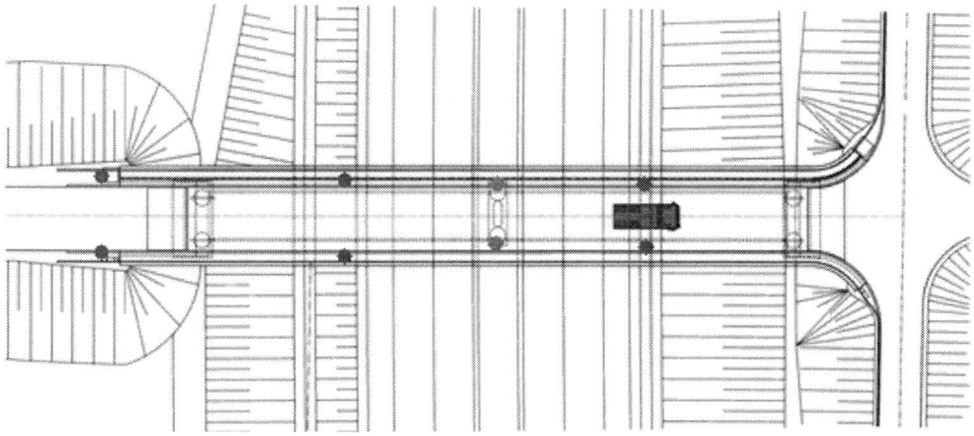

Fig. 6. Levelling points and locations of levels.

- Bench marks
- levels

Fig. 7. DiNi 21 level.

Fig. 8. Leica Sprinter level.

3 Results and discussion

All the results obtained from the analysis should be adequately treated. In such a manner any human error can be eliminated: double observations of the same point, wrong order of the point observation... Each phase is compared with the "zero" state (empty bridge with no burden), which was performed at the beginning of the measurement or with an independent point that is located outside of the deformation zone. For each object (left and right) separately, a comparison is made between the calculated and measured maximal vertical displacements. After the measurements accurate data processing, analysis and correct presentation is made.

In Tables 1 and 2 and Fig. 6 there are examples of the recorded measurements. The receiver records the spatial coordinates x, y and z or h (height). In the case of determining movements in the load test of most interest is the height coordinate at which we know that the method is less accurate.

Table 1. Some of coordinates obtained in the field.

Point	y	x	h
base	552564.529	161175.309	266.322
1	552572.333	161118.975	272.190
2	552572.338	161118. 974	272.191
3	552572.336	161118. 975	272.199
4	552572.339	161118. 977	272.193
5	552572.338	161118. 978	272.188

During the processing procedure the receiver writes the initial 123 points and then the record jumps to the 4724th point. The records of all intermediate points between these two points are been lost.

Table 2. Missing records between 23 and 4724.

Point	y	x	h
22	552572.339	161118.970	272.163
23	552572.337	161118. 967	272.171
4724	552572.329	161118. 978	272.204
4725	552572.332	161118. 980	272.205
4726	552572.330	161118. 978	272.206

The results showed what occurred during the measurements the received losses of the signal or incorrectly written data to the disc. Information about the movements was lost and only 572 points were written. All measurements were 5273 but 4701 of them were not recorded. This means that only about 10 minutes at the 88 minutes of actual measurement were recorded, as the measurement was set to record points every second.

We made a partial analysis of the results using the data. The height of the point varies for one to two centimetres. At the specific load test height differences within the range of 2mm were expected. It can be concluded that this method is insufficiently accurate and reliable for the determination of displacement within such small dimensions.

When measuring vertical displacement suing the GNSS method the equation below tells us how precise all the results we need for quality movements determinations. If centimeter movements are expected, we have to choose such equipment, that has a standard deviation (σf) of measurement equal to a quarter of the expected displacements.

In our case, where a few millimeters of movement were expected we should choose a receiver that detects movements with a standard deviation of one quarter of a millimeter.

$$\frac{\sigma f}{f} < \frac{1}{4}$$

σf – standard deviation
f – expected displacement

The GNSS method at this load test didn't did not provide reliable results. The measurements were too small to conclude anything about the movements during the test.

The results of the measurements can be seen in Table 3.

Table 3. The results of the measurements on left and right side.

side	Left side (Leica Sprinter 50)					Right side (Trimble DiNi 21)				
h (m)	hfl (m)	hfi-hfl (m)				hfl (m)	hfi-hfl (m)			
	reading	2. phase	3. phase	4. phase	5. phase	reading	2. phase	3. phase	4. phase	5. phase
T1 - embank.	3.3384	+0.0003	+0.0003	-0.0003	-0.0006	3.4814	+0.0015	+0.0010	+0.0015	+0.0004
T2 - Span 1	1.6036	+0.0011	-0.0001	0.0000	-0.0006	1.7356	+0.0017	-0.0012	-0.0011	+0.0011
T3 - Span 2	0.7446	0.0000	-0.0001	+0.0010	-0.0003	0.8747	-0.0017	+0.0007	+0.0003	+0.0001

Phase 1	Empty bridge
Phase 2	Burden on span 1
Phase 3	Empty bridge
Phase 4	Burden on span 2
Phase 5	Empty bridge

As can be seen the results for vertical displacements were within the range from 0.3mm to 1.7 mm. These are extremely small displacements which we needed to regard as micro-displacements and should be processed very carefully.

These small amounts can be obtained using reliability of the equipment and the precisions of the measurements, so that measurements with a large redundancy number. Also equipment which allows us to monitor displacement of 0.1mm should be used. GNSS is evidently not yet such equipment.

At load test one should use two mutually independent methods because it could happen that by using only one method one cannot obtain the relevant value.

4 Conclusions

The measurement results showed us how small the shifts on this bridge really were. In such a case it is easy to do inaccurate measurements or errors. While we strive for the best possible performance and accurate measurements, we can see that there are small discrepancies when comparing the left and right sides of the bridge. There are several reasons for that. Deviations usually occur due to the differences in an object's geometry. It could be the case that the object does not have precisely the same characteristics on both sides (different width of the sidewalk, bicycle path just on one side).

Results from both sides were within expectations. The more significant movements were during phase 2 and phase 4 at the moments when the object was loaded in the first and second spans.

The maximal recorded vertical displacement was 2 mm in the second phase of the load test. In other phases these vertical displacements were smaller or not measurable at all. The levelling method for the load tests still appears as a very reliable method for detecting small shifts, where as we have seen that the GNSS method is not an option.

References

1. J. Albert, H.G. Mass, A. Schade, W. Schwarz, *Pilot studies on photogrammetric Bridge deformation measurement* (IAG, Berlin, 2002)
2. M. R. Chowdhury, C. J. Ray, NDT&E, **36**, 237-244 (2003)
3. P. A. Fuchs, G. Washer, S. Chase, M. More, Journal of performance of constructed facilities, (2004)
4. S. J. Gordon, D. Lichti, M. Stewart, J. Franke, *Modelling point clouds for precise structural deformation measurement* (SPRS, Istanbul, 2004).
5. A. Guarnieri, A. Vettore, F. Remondino, PFIG Working Week, may 22-27, Athens, Greece (2004)
6. D. V. Jauregui, R. K. White, C. B. Woodward, K.-R. Leitch, Journal of Bridge Engineering, **8**, 212-222 (2003)
7. H.-G. Maas, U. Hampel, Photogrammetric Egineering & Remote Sensing, **72**, 1 (2006)
8. A. Marendić, Z. Kapović, R. Paar, Geodetski list, **3**, 175-190 (2013)
9. T. Schäfer, T. Weber, P. Kyrinovič, INGEO 2004 and FIG Regional Central and Eastern European Conference on Engineering Surveying, Bratislava, Slovakia, nov 11-13, (2004)
10. N. Harmati, Ž. Jakšić, M. Trivunić, V. Milovanovi, Periodica Polytechnica: Civil Engineering, **58(4)**, 431-444 (2014)
11. K. Krayushkina, O. Prentkovskis, A. Bieliatynskyi, R. Junevičius, Transport, **27(2)**, 129-137 (2012)
12. O. Prentkovskis, J. Tretjakovas, A. Vedas, A. Bieliatynskyi, A. Daninas, K. Krayushkina, Journal of Civil Engineering and Management, **18(5)**, 761-771 (2012)
13. A. Beljatynskij, O. Prentkovskis, J. Krivenko, Transport, **25(4)**, 394-402 (2010)
14. M. Starikov, A. Beljatynskij, O. Prentkovskis, I. Klimenko, Transport, **26(3)**, 255-262 (2011)
15. N. Kuzhel, A. Bieliatynskyi, O. Prentkovskis, I. Klymenko, S. Mikaliunas, O. Kolganova, S. Kornienko, V. Shutko, Transport, **28(4)**, 413-419 (2013)
16. K. Krayushkina, O. Prentkovskis, A. Bieliatynskyi, J. Gigineishvili, A. Skrypchenko, A. Laurinavičius, K. Gopalakrishnan, J. Tretjakovas, Baltic Journal of Road and Bridge Engineering, **11(1)**, 77-83 (2016)
17. I. Mrdak, M. Rakočević, L. Žugić, R. Usmanov, V. Murgul, N. Vatin, Applied Mechanics and Materials, **633-634**, 1069-1076 (2014)
18. J. Ćetković, M. Knežević, M., Žarković, V. Murgul, N. Vatin, Applied Mechanics and Materials, **638-640**, 2465-2470 (2014)

MATEC Web of Conferences 73, 01015 (2016)
TPACEE-2016
DOI: 10.1051/matecconf/20167301015

The Method of Calculating the Settlement of Weak Ground Strengthened with the Reinforced Sandy Piles

Tatyana Maltseva[1,*], *Aleksandr* Nabokov[1], *Yuri* Novikov[1], and *Vladimir* Sokolov[1]

[1]Industrial University of Tyumen, 625001, Volodarskogo str., 38, Tyumen, Russia

Abstract. The paper presents an engineering method for calculating the weak clay base, strengthened with sandy piles reinforced along the contour. The method is based on the principle of layer-by-layer summation, which is used when designing the bases and foundations. The novelty of the suggested method lies in the taking account of the soil reaction along the pile lateral surface and the impact of external vertical loads on the vertical displacement of the base.

1 Introduction

With the advent of geosynthetic materials used as reinforcing elements, it is getting popular to use one of the ways of strengthening the weak clay base with sandy piles reinforced along the contour [1-13]. Problems with weak clay bases are sufficiently presented in [14, 15].

The lack of a unified algorithm for calculating stress and strain states of the base strengthened with sandy piles reinforced along the contour, makes it difficult to extensively use these piles in a low-rise construction [1, 13].

Along with the capacity to predict base settlement by using software package, the practical value can also be seen in engineering-analytical method of calculating the strain state of the reinforced base [11, 12].

The article proposes method of calculation based on the Boussinesq solution in terms of the model of deformable solids. Therefore, the parameter of interstitial water viscosity is not considered explicitly. Deformation modulus is determined by experiment with water-saturated sample, which does not reflect the influence of pore water clearly. The time factor does not appear in the proposed method because his influence on the settlement of the soft clay soils is described in detail earlier by T. Maltseva and A. Minaeva [1, 13, 17].

2 Review

* Corresponding author: tgasu.maltseva.tatyana@mail.ru

A technique for strengthening the bearing capacity and stability of the foundations built on weak water-saturated soils was patented [4]. According to this technique, a sandy pile reinforced vertically along the contour with a geosynthetic permeable element (geogrid) is arranged under the shallow foundation in the ground bed composed of weak soil [5].

The researches [6, 7 and 8] are describing the laboratory experiments, which shows us efficiency of application of reinforced sandy piles. The detailed information [1] is presenting visually in Fig. 1.

The figure shows that the reinforced sandy pile has 55% less settlement compared to natural ground bed and 30% less settlement compared to non-reinforced sandy pile [5, 7].

Full-scale experiments [8, 9] have revealed, that load, which is obtained from the low-rise building is P = 0.111 MPa. On this basis, appropriate settlement of low-rise building is S = 0.0252 m.

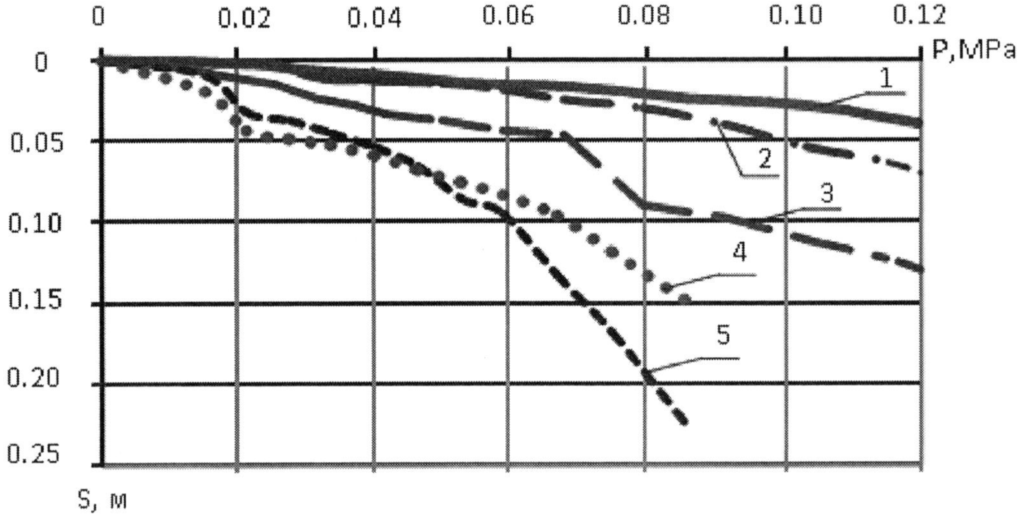

Fig. 1. "Settlement-pressure" diagrams:
1 – foundation soil reinforced with a sandy pile; 2 – foundation soil with a driven sandy pile; 3 – foundation soil reinforced vertically with a geogrid along the contour; 4 – foundation soil reinforced vertically with a permeable element (a film) along the contour; 5 – unstable foundation soil.

3 Problem

Let us analyze a homogeneous cylinder of sand placed in a closed shell (reinforcing element - geogrid), beyond which lies an elastic isotropic material - clay soil (Fig. 2). We replace the effect of clay soil on the cylinder walls with the force of friction Ff. The cylinder is being affected with the static distributed load q. The calculation of sandy cylinder settlement (induced by the external forces) is basing on the principle of layer-by-layer summation, according to which the settlement is defined as a sum of settlements of elementary soil layers. Let us split the reinforced sandy cylinder with height Δz [m] and the cross-sectional area A [m2] into elementary parts (Figure 3).

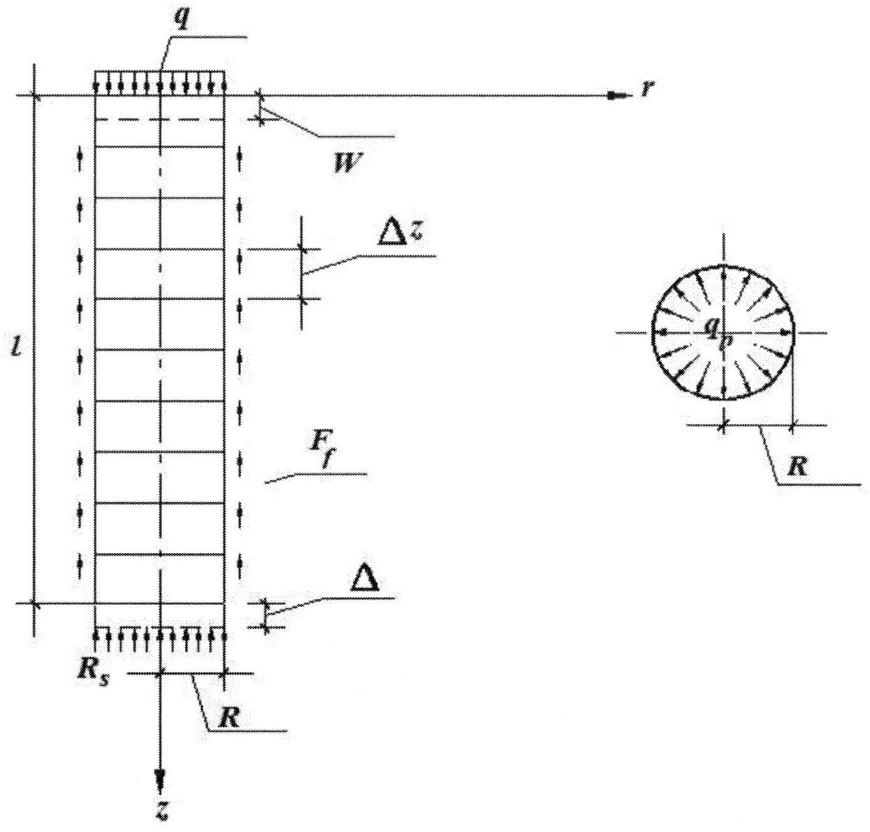

Fig. 2. Structural design

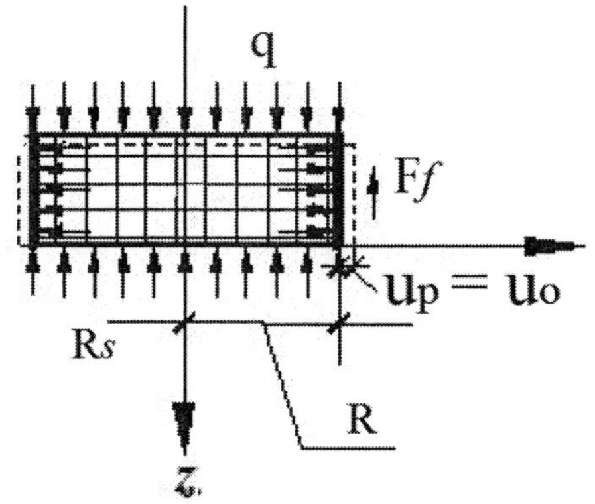

Fig 3. Elementary layer of the sandy pile reinforced along the contour

4 Description of studies

External loads cause the settlement of the sandy pile reinforced along the contour. This settlement is a combination of two components: the settlement of a pile shaft Δw and shift of pile foot Δ.

$$W = \sum \Delta w_i + \Delta, \tag{1}$$

where Δ is also set or determined by layer-by-layer summation method in accordance with construction regulations.

Let us present the settlement of a reinforced sandy pile shaft as a sum of settlements of elementary soil layers $\sum \Delta w_i$, which pile had previously been divided into.

We determine the settlement of the elementary layer of the sandy pile reinforced along the contour as the sum of

vertical displacement of pile shaft (sand) under the impact of the external load - Δw_q,

vertical displacement under the impact (reaction) of the reinforcing element – Δw_{qp},

vertical displacements under the impact (reaction) of soil bearing pressure along the lateral area – Δw_{qs}:

$$\Delta w = \Delta w_q + \Delta w_{qp} + \Delta w_{qs}, \tag{2}$$

The paper [2] presents the calculation of settlement excluding forces of friction generated along the outer lateral surface of the cylinder.

To determine the vertical displacements of the filler Δwq caused by the external load q, we formulate the equilibrium equation of the object:

$$N - Ff = 0, \tag{3}$$

where N is the stress that occurs in shaft of a sandy pile reinforced along the contour. This stress is caused by the external load q and the reaction soil bearing pressure Rs [kPa]; Ff is the force of friction along the lateral surface [kN].

Stress N is

$$N = q \,.\, A - Rs \,.\, A \tag{4}$$

According to Hooke's law, the reaction Rs will be presented in the form of equation:

$$R_s = E_r \cdot \frac{\Delta w_q}{\Delta z}, \tag{5}$$

where Er is a reduced modulus of the filler (sand) deformation [kPa]; Δwq is an absolute deformation of the elementary layer of finite length [m].

The frictional force of the elementary layer along the lateral surface is proportional to the displacement Δz:

$$F_f = k \cdot \Delta z = u \cdot f_i \cdot \Delta z, \tag{6}$$

where k is the coefficient of friction [kN/m]; u is the external perimeter of a pile cross-section [m]; f_i - the calculation resistance of the i-th layer of soil along the lateral surface of the base [kPa] is determined from the equation [16]:

$$f_i = 2 \cdot \pi \cdot r_i \cdot c_i \cdot tg\varphi_i \tag{7}$$

where ri - radius [m]; ci - the adhesion [kPa]; φi - angle of internal friction.

Plugging equations (4), (5) and (6) in the equation (3), and expressing Δwq, we obtain the absolute deformation of the elementary *i* layer:

$$\Delta w_{qi} = \frac{q \cdot A \cdot \Delta z_i}{E_r \cdot A + u \cdot \Delta z_i \cdot f_i} \tag{8}$$

The vertical displacements of the pile shaft with account for impact (reaction) of the reinforcing element Δwqp are defined by the equation:

$$\Delta w_{qpi} = -\frac{\nu \cdot q_{p_i} \cdot \Delta z_i}{E_r} \tag{9}$$

where v is the Poisson's ratio; qpi is a reaction of reinforcing element in the elementary layer of finite length [kN/m2].

Reaction of the reinforcing element qp with account for soil bearing pressure along the lateral area of the pile is determined from the condition of equality of the radial movements of reinforcing element and sand u0=up [13]:

$$q_p = \frac{\nu \cdot q_{zi}}{\dfrac{E_r \cdot R}{C_p \cdot \delta} + 1 - \nu}, \tag{10}$$

where R is a radius of a sandy pile reinforced along the contour [m]; δ - thickness of the reinforcing element [m]; Cp is the equivalent of elasticity modulus of the reinforcing element [kN/m2]; qzi - magnitude of the vertical stresses having impact on the elementary layer [kN/m2].

Reinforcing element modulus of elasticity is determined in [5]:

$$C_p = \frac{E_r \cdot R \cdot (K_a^{ex} - 1)}{\delta \cdot [\nu^2 \cdot K_a^{ex} - (K_a^{ex} - 1) \cdot (1 - \nu)]} \tag{11}$$

where K_a^{ex} is a magnification coefficient of deformation module of the reinforced elementary layer.

Vertical intensity height along the sandy pile reinforced along the outline is approximated to a tolerance of 10% in accordance with the experimental information and is determined in a quadratic dependence [17]:

$$q_z = 0{,}20445714 - 0{,}042892854 \cdot z + 0{,}003794642 \cdot z^2. \tag{12}$$

Vertical displacements Δwqs of the pile shaft from the effects of for soil bearing pressure qsi along the lateral area are determined by:

$$\Delta w_{qs} = -\frac{\nu \cdot q_{s_i} \cdot \Delta z_i}{E_r} \tag{13}$$

Soil surrounding the pile is to be considered as a linearly elastic environment with a coefficient of soil resistance Cz [kN/m3]:

$$q_s = C_z u_s, \quad C_z = \frac{K \cdot z}{\gamma_c}, \tag{14}$$

where K is a constant of proportionality [kN/m4], which is dependent on the type of soil surrounding the pile; z is the depth of the pile section in the ground [m]; γc – coefficient of service factor. Basing on the condition that the radial displacements of the reinforcing element are equal to the radial displacements of the surrounding soil $u_o = u_s$, the reaction of soil bearing pressure on the lateral surface of the elementary layer qs is determined from the equation:

$$q_s = \frac{q_p \cdot R^2 \cdot C_z}{C_p \cdot \delta} \tag{15}$$

5 Conclusion

Results of the analytical and numerical simulation (which were made by FEM model and PLAXIS software) of reinforced base were compared.

Soils with physical and mechanical characteristics similar to soils of on-site field research were taken for calculation by elastic-plastic model with the Mohr-Coulomb strength criterion.

Figure 4 presents "settlement-pressure" diagrams, plotted basing on experimental data calculated by the proposed method and data of numerical simulation.

The diagrams show that the difference between the final settlements of the base (strengthened with the sandy pile reinforced along the outline) calculated by the proposed method and the data of in-situ tests is 22%, the calculation data in "PLAXIS 3D" software package - 37%, the calculation data in "FEM models" program complex - 19 %.

The proposed engineering method of calculation allows to significantly increase the effectiveness in projecting the reinforced soil foundations and thus allows to extend the sphere of application of this method for strengthening the weak clay ground.

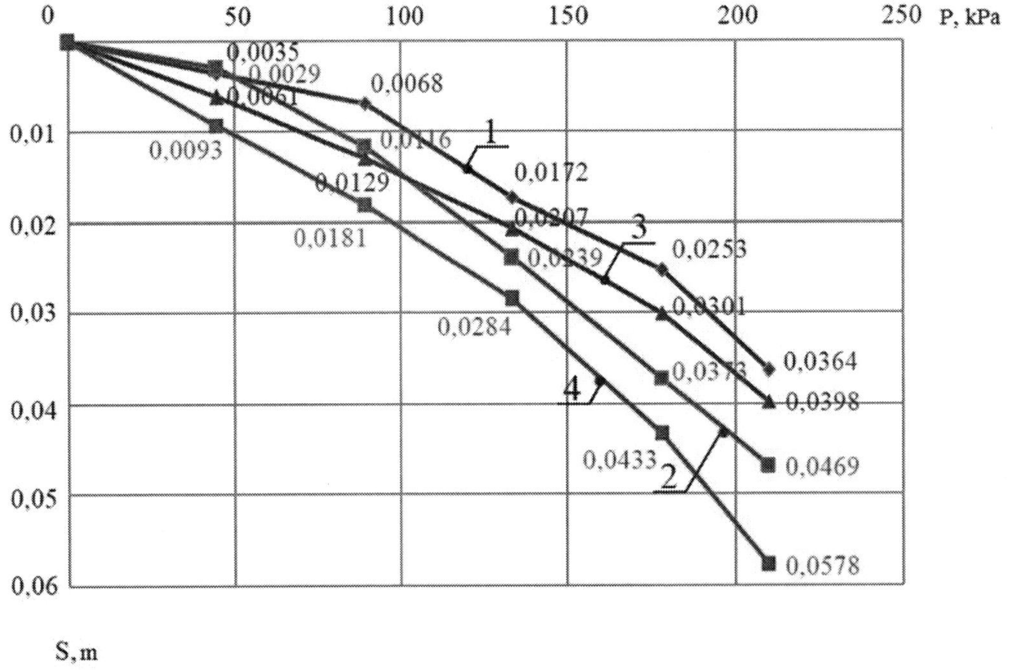

Fig. 4. Comparison of the settlements calculation results:
1 - experimental data; 2 - theoretical calculation according to the method; 3 - the results of numerical simulation in "FEM models" software package; 4 - results of numerical simulation in "PLAXIS 3D" software package.

References

1. T. Maltseva, A. Nabokov, A. Chernykh. Procedia Engineering, **117**, 239-245 (2015)
2. S. Sannikov, Avtoref. dis. na soisk. uch. step. kand. tekh. Nauk, Tyumenskaya gosudarstvennaya arkhitekturno-stroitelnaya akademiya, (2004)
3. A. Kraev, Vestnik Tomskogo arkhitekturno-stroitelnogo universiteta, **4**, 146-151 (2008)
4. Pat. 2361979, MPK51 E 02 D 27/08. Sposob povysheniya nesushhej sposobnosti i ustojchivosti fundamentov na slabyh vodonasyshhennyh gruntah (2009) – Bjul. №20.
5. A. Kraev, Dissertacija na soiskanie uchenoj stepeni kandidata tehnicheskih nauk, Tjumenskaja gosudarstvennaja arhitekturno-stroitel'naja akademija, (2009)
6. V. Baj, A. Nabokov, V. Voroncov, A. Kraev, Vestnik grazhdanskih inzhenerov, **2**, 74–76 (2009)
7. A. Kraev, Nauchno-tehnicheski vestnik Povolzhya, **5**, 221-224 (2013)
8. A. Nabokov, Dissertacija na soiskanie uchenoj stepeni kandidata tehnicheskih nauk (2003)
9. Ju. Novikov, Nauchno-tehnicheskij vestnik Povolzh'ja, **6**, 225-229 (2011)
10. V. Baj, A. Kraev, A. Nabokov, Ju. Novikov, Nauchno-tehnicheski vestnik Povolzh'ja, **2**, 30-33 (2011)
11. V. Baj, T. Maltseva, A. Kraev, Nauchno-tehnicheskij vestnik Povolzh'ja, **5**, 108–111 (2014)
12. V. Baj, A, Kraev, Vestnik grazhdanskih inzhenerov, **3**, 107-110 (2014)
13. L. Maltsev, T. Maltseva, A. Minaeva, A. Nabokov, Nauchno-tehnicheskij vestnik Povolzh'ja, **2**, 234-238 (2012)
14. T. Maltseva, *Matematicheskaja teorija vodonasyshhennogo grunta* (Izdatel'stvo «Vektor Buk», Tjumen', 2012)
15. V. Baj, L. Maltsev, T. Maltseva, A. Nabokov, V. Demin, Jeksperimental'naja ustanovka dlja ispytanija grunta metodom odnoosnogo szhatija (2002): Patent na izobretenie RUS 2213952
16. K. Shadunts, Izd-vo FGOU VPO KubGAU politematicheski setevoi elektronny nauchni zhurnal Kubanskogo gosudarstvennogo agrarnogo unversiteta (2006)
17. V. Baj, T. Maltseva, A. Nabokov, V. Voroncov, A. Minaeva, Izvestija vysshih uchebnyh zavedenij. Neft' i gaz, **1**, 102-106 (2011)
18. Ju. Novikov, *Issledovanie raboty pesceanih armirovannih po konturu svai v slabih glinistih osnovanieah pod lentochnimi fundamentami*, PhD Thesis (2015)

MATEC Web of Conferences **73**, 01016 (2016) DOI: 10.1051/matecconf/20167301016
TPACEE-2016

The Research of Tank Sides Geometrical Forms after their Long-Term Usage in the Oil Transportation System

Alexey Pimnev[1], *Svetlana* Ivantsova[2], *Ruslan* Aginey[3], and *Tatyana* Ponomareva[4,*]

[1,4]Industrial University of Tyumen, 625001, Volodarskogo str., 38, Tyumen, Russia
[2]Gubkin Russian State University of Oil and Gas, Leninsky prospekt 65, Moscow, Russia
[3]JSC "Giprogazcentr", st. Alexeevskaya 26, Nizhniy Novgorod, Tyumen, Russia

Abstract. During the technical diagnosis of vertical steel tanks it is necessary to evaluate the influence of geometry imperfection on deflected mode and resource. One of the most significant characteristics of geometry, definable during the full-scale tank diagnosis, is deflections of tank generatrices from the vertical. In the report, the tank geometry research results are represented on the basis of measuring data of 98 oil transporting system tanks

1 Introduction

The existing tank battery in a system of oil transportation is considerably depreciating. In recent years, the scope of repair, reconstruction, and construction of new storage tanks. It predetermined the increase of scope of technical diagnosis of vertical steel tanks (VST) with attraction of new technical means and methods. VST technical condition's estimation problem can be solved by using two-level technical diagnostics system, whereby frequency and type of inspections depends on the working service. However, the existing system of technical diagnostics reflects modern requirements not to the full [1-8].

In these conditions, it is necessary to take measures increasing the effectiveness of tanks usage by means of extension their safe working service in the process of production [9-13]. As vertical steel tanks are elements of complicated technical systems of extraction, transportation, and storage of oil and oil products, reliability assurance and estimation of safe working service of functional tanks is specifically connected with lifetime improvement of tank battery and all the system of pipeline transport. Besides that, the consequence of reliability enhancement is safety of main funds.

The problem of the reliability of tanks is connected with issues of reduction losses of oil and oil products as a result of leaks and accidents, as well as poor technical condition of a tank shall and bottom. Financial losses in tanks accidents are large and tend to increase due to the increase in reservoir capacity. At the same time, due to numerous indirect expenses, the real damage is higher than direct losses at recovery and the cost of production. Even the

* Corresponding author: feldmarshal@inbox.ru

costs of the fight against environmental pollution caused by leakage of products are higher than direct losses in more than several times.

The problem of estimation of technical condition of vertical steel tanks after long operation service can be solved by using technical diagnostics system. When diagnosing the technical condition of the VST after long-time operation service, we should take into account manufacturing experience in complex tanks surveys during which special attention should be paid to changes in the geometric shape that can occur in the state and operating conditions of the tank batteries of oil pipelines trunks.

2 The object of research

The imperfections of the tank geometric shape leads to significant changes in the stressed-deformed state under the influence of operating loads. Domestic and foreign manufacturing experience enables us to conclude that most of the accident occurs due to rupture of metal structures of the tank. If there are deformations in the areas of welded seams of the wall, the destruction probability increases. Experimental studies carried out on cylindrical shells shown that even a small deviation from the ideal shape leads to a decrease of the critical load. Therefore, we can conclude that the geometrical shape imperfections result in reduced operational reliability of the tank.

Analysis of the geometric shape of the wall of steel vertical cylindrical tanks was made based on the results of field technical diagnosis [8]. As the basis, the deviations of measuring data of 98 oil transporting system tanks. The distribution of the measurements of deviations of tank generatrices from the vertical for different typical sizes of tanks is shown in Figure1.

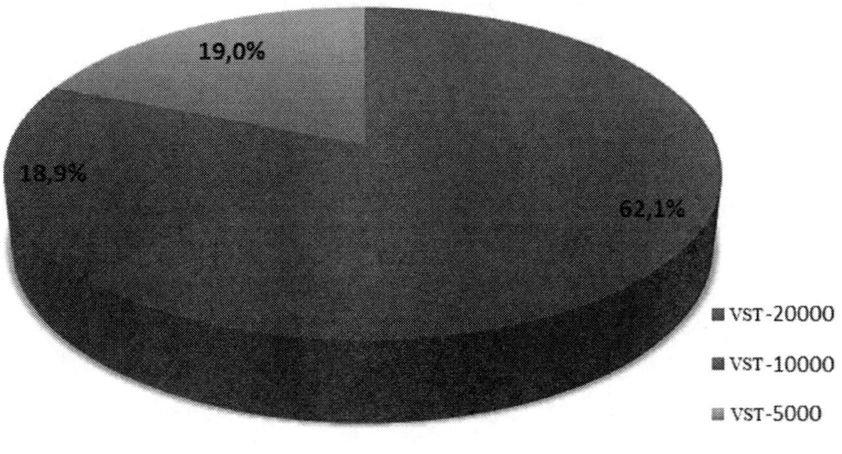

Fig. 1. The distribution of studied tanks by standard sizes

3 Methods

Vertical steel tanks have a cylindrical shape, the geometric dimensions of which are determined by deviations of the tank generatrices from the vertical. During the operation, change of the geometric shape of the tank often occurs due to uneven subsidence of the bottom caused by poor preparation of the base, under the influence of vacuum, hydrostatic load, vibration or overflow.

Measurement of geometric forms of tanks walls was carried out by means of direct measurement of deviations of the tank generatrices from the vertical using the highly accurate theodolite and devices developed at the "Construction of oil and gas pipelines," Department of the Tyumen Industrial Institute (AS.: authors Tarasenko A.A., Galeev V.B., Leshchev N.N.) by the method described in [1].

Goniometer is placed at a tangent to the generatrix at a distance of 20 to 40 meters. From one location of a device, shooting of deviation of maximum two generatrices was made. Depending on the typical size and structural features of tanks, measuring was carried at 24 or 25 generatrices for VST-20000, 18 - for VST -10000 and 12 - for VST-5000. Measurements for each zone, from the 1st to the 8th, were made at points 50 mm below the top horizontal seam of the corresponding zone, as required by current regulatory documentation for diagnostics and maintenance of tanks.

In total, with the participation of the author, 1846 deviations of the tank generatrices from the vertical were measured, which required 14768 measurements. For the analysis, all the measurements were divided into groups depending on the typical sizes and terms of operation of tanks (table 1).

For different typical sizes of tanks, the amount of field joints is equal to the number of rolls, of which the tank side is produced. Thus, for the construction of the VST-5000 one roll is used, that is closed by a vertical field joint, for VST-10000, 2 rolls and 2 field joints respectively, for VST-20000 - 3 vertical field joints. Therefore, the percentage of field joints of tank sides is constant and depends on the typical size of the tank:

- for VST-20000 is 12.5%;
- for VST-10000 is 11.1%;
- for VST-5000 is 8.3%.

Table 1. The number of completed measurements of generatrix deviations from the vertical and distribution by their size and the terms of operation

Terms of operation	Typical size of the tank			Total
	VST-20000	VST-10000	VST-5000	
Up to 20 years	461 (25.0 %)	0	201 (10.9 %)	662 (35.9 %)
Over 20 years	686 (37.1 %)	349 (18.9 %)	149 (8.1 %)	1184 (64.1%)
Total	1147 (62.1 %)	349 (18.9 %)	350 (19.0 %)	1846 (100%)

Table 2 shows the number and percentage of the generatrices, accounted for by the factory and field joints of a tank side. Minor deviations from the theoretical values of field joints of tank sides, defined above, are explained by the inability, in many cases, to make all generatrices measurements because of the presence of the structural elements on the tank side (servicing areas, pipes, stairways, etc.) at the junctions of the tank sides.

Table 2. Percentage of factory and field joints in measurements of generatrices deviations from the vertical

Typical size of the tank	Terms of operation	Generatrices amount		
		Total	incl. joints seams, p. [%]	
			plant	field
VST-20000	Up to 20 years	461	404 (87.6 %)	57 (12.4 %)
	Over 20 years	686	600 (87.5 %)	86 (12.5 %)
	Total	1147	1004 (87.5 %)	143 (12.5 %)
VST-10000	Up to 20 years	0	0	0
	Over 20 years	349	309 (88.5 %)	40 (11.5 %)
	Total	349	309 (88.5 %)	40 (11.5 %)
VST-5000	Up to 20 years	201	184 (91.5 %)	17 (8.5 %)
	Over 20 years	149	136 (91.3 %)	13 (8.7 %)
	Total	350	320 (91.4 %)	30 (8.6 %)

In order to avoid errors and biases in the analysis of the results of measurements deviations of the generatrices, the problem of the representation aggregate sample the obtained values was solved, and interrelation of between the measurement point and the value of the deviations at that point which could be obtained in the study of aggregation was established [16,20].

Since the considered sample is big enough, a hypothesis was advanced that the measured set of random values is subject to a normal distribution law. However, for the final decision on the form of the distribution law, it seems appropriate to check whether this assumption is consistent with reality. At the same time, due to the limited number of observations, an experienced distribution law, even if the assumption about the distribution law is done correctly will still be slightly different from the expected. In this regard, it was necessary to determine: whether the discrepancy between the experimental and anticipated distribution law is a consequence of the limited number of observations or it is significant and connected to the fact that the actual distribution of the random variable is different from the intended.

To verify compliance with the random variable distribution deviations form the wall of the normal distribution law, verification was carried out by one of the most common and important in mathematical statistics fitting criteria - $\chi 2$ goodness of fit test. As an example, a random variable of deviations of geneatrix for the 8th zone of side of VST-20000, which varies from minus 214 to 268 is considered in [8].

As a result of calculations, it was revealed that the discrepancy between the theoretical distribution law and the law, built on the basis of statistical data is irrelevant, accidental, suggesting that the observed value of the test for the 8th zone of tank side of VST-20000 is subject to the normal distribution law. This hypothesis is confirmed for other zones, i.e. in all cases it is subject to the normal distribution law.

For below mentioned groups of tanks, distribution curves of deviation values of generatrices were built at belts. The obtained data were processed by the method described in researches [1, 2]. Axis of abscissas shows the values of deviations form the wall: the outside (positive values) - to the right, the inside (negative values) – to the left, the vertical axis - the accumulated relative frequency of deviations $\Sigma\omega$. Separate frequencies are proportions of the corresponding ranges of values of the measured deviations of generatrices of their total number in this zone.

Connecting each zone accumulated relative frequencies ($\Sigma\omega = 0.5$) to each other, we get the midline, around which there is a change of values of generatrices. This line is a correlation curve between the height of the body and the point value deviations of generatrices (Figure 2).

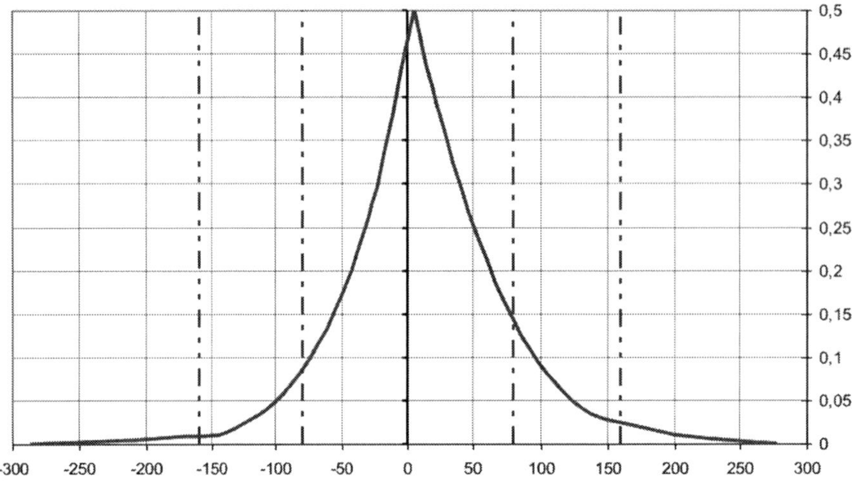

Fig. 2. The distribution curve of generatrices deviations

Analysis of the distribution of values of generatrices deviations of the VST-20000 zones showed that the probability of unacceptable deviations is different at different levels (Figure 3). Analysis of the dependence of the probability of occurrence of unacceptable deviations from the measurement place indicates that maximum probability corresponds to the level of the first zone and is 43%, then it gradually decreases, reaching a minimum at the level of 6-7th zones (18%), and increases again at the level of the 8th zone to 22 %. Thus, special attention in the imperfections of geometric shapes of wall evaluation must be paid to the measurements on the first three and the last zone of the tank.

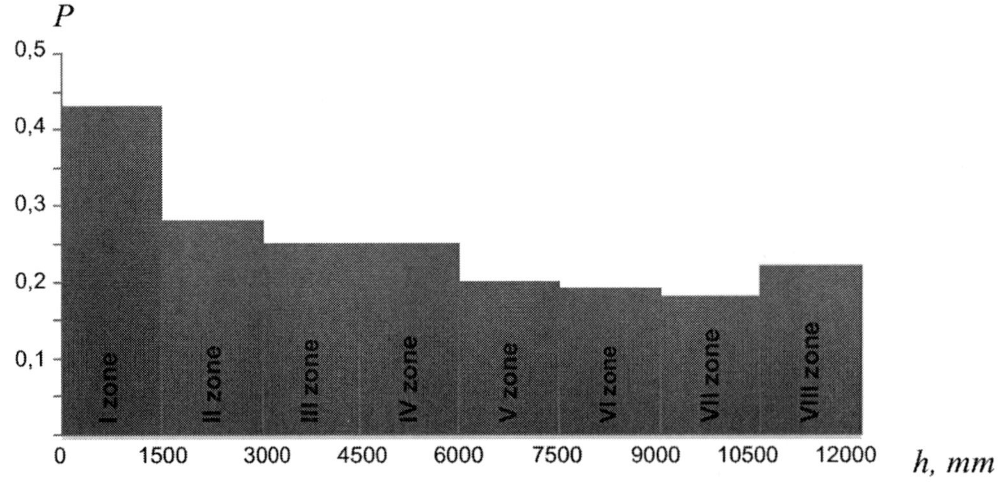

Fig. 3. The dependence of the probability of occurrence of unacceptable deviations on shell height

According to the results of the statistical analysis of deviations measurement data of the generatrices from the vertical, the dependence of the probability of occurrence of unacceptable deviations in height of the shell is obtained. The dispersion coefficient of the approximating curve is 0.91:

$$p(h)=\left(4\cdot 10^{-9}\right)h^2-\left(7\cdot 10^{-5}\right)h+0{,}5 \quad (1)$$

Having interconnected points of the curves for all zones with the accumulated relative frequencies ($\Sigma\omega$ = 0) and ($\Sigma\omega$ = 1), we obtained left and right variation limits of generatrices deviations. The probability of getting the value generatrices deviations in a region bounded by these curves is equal to 1.

If we correspondingly interconnect points of with the accumulated relative frequencies $\Sigma\omega$ = 0,1 and 0,9; $\Sigma\omega$ = 0,2 and 0,8; $\Sigma\omega$ = 0,3 and 0.7; $\Sigma\omega$ = 0,4 and 0.6, we obtain the areas with a probability of getting the generatrices deviations into them, which are equal to: F = 0.8; P = 0.6; P = 0.4; P = 0.2, respectively. Thereby, we obtain the tank wall generatrices deviations, depending on the capacity of the tank.

Table 3. Comparing the characteristics of the studied tanks to tolerance deviations of generatrices from the vertical.

№ of zone	Acceptable deviations with term of operation over 20 years [mm]		Generatrix deviations with probability P=1 [mm]	
	for 75% generatrices	for 25% generatrices	Inside the tank	Outside the tank
I	40	52	74	95
II	60	78	125	130
III	80	104	157	165
IV	100	130	220	205
V	120	156	225	205
VI	140	182	205	210
VII	150	195	248	250
VIII	160	208	270	268

The tolerance limit of deviations of the steel generatrices of the tanks of type VST-20000 from the project verticals should not exceed the values in accordance with the requirements of the current regulatory documentation. In table 3 shows a comparison of characteristics of the studied tanks to the requirements of regulatory documents.

Comparing the data from the survey of tanks to regulatory requirements, it can be concluded that the area of actual values of generatrix deviations goes beyond the area of acceptable values. Thus, the tanks were operated at project conditions with term of operation longer than the standard one in 65% of surveyed of tanks. Therefore, the requirements of the current regulatory documentation can be regarded as overstated, and it is necessary to reconsider the approach to the evaluation of the residual operation life, on the criterion of geometrical shapes, in particular.

In studies [3-7], a method of monitoring the forecasting of indicators of reliability and supplying the transition from "afterfailure" system of service and repair to a "preventive" one on predictive indices of reliability of pipeline transport. Analyzing the abovementioned, we can conclude the necessity of solving problems of estimation the technical state of each tank in accordance with its technical condition [21,22].

4 Results and discussion

Analyzing the distribution curves of the values of generatrix deviations, we can conclude the following:

1. The value of generatrix deviations grows with an increase in height and decrease in the plate thickness of the shell, and this quantity is mostly distributed over a sinusoidal curve.

2. The middle line, around which the value variation of generatrix deviations takes place, is not located in a vertical plane passing through the outer surface of the cylindrical shell, but stands outside of the tank.
3. The largest in absolute magnitude generatrix deviations are observed on the outside face of the tank side.

5 Conclusion

It was found that the maximum probability of unacceptable generatrix deviations corresponds to the level of the first zone and makes up 43%, then it gradually decreases, reaching a minimum at 6-7 zones (18%), and increases again at the level of 8 to 22% zone. Thus, to improve the estimation accuracy of tank side geometric shapes imperfections, it is necessary for the first three and the last zones to increase the density of measurement grid.

The middle line, around which there is the variation of values of generatrix deviations, is not situated in a vertical plane passing through the outer surface of the cylindrical shell, but stands out of a tank. Thus the greatest in absolute magnitude generatrix deviation can be observed at the outer side of the tank.

References

1. A. Tarasenko, *Deflected mode of tanks during repair works* (Nedra, Moscow, 1999)
2. G. Hopersky, *Thesis of Candidate of Technical Sciences* (TSOGU, Tyumen, 1998)
3. M. Zemenkova, *Thesis of Candidate of Technical Sciences* (TSOGU, Tyumen, 2007)
4. V. Kurushina, Yu. Zemenkov, WIT Transactions on Ecology and the Environment, **190(2)**, 881-888 (2014)
5. V. Antip'ev, A. Nevolin, Yu. Zemenkov, Neftyanoe Khozyaistvo - Oil Industry, **10**, 46-48 (1981)
6. Yu. Zemenkov, V. Shalay, M. Zemenkova, Procedia Engineering, **113**, 254-258 (2015)
7. Yu. Zemenkov, V. Shalay, M. Zemenkova, Procedia Engineering, **113**, 312-315 (2015)
8. A. Pimnev, *Thesis of Candidate of Technical Sciences* (TSOGU, Tyumen, 2006)
9. E. Kurushina, V. Kurushina, Life Science Journal, **11(11)**, 517-521 (2014)
10. V. Kurushina, Yu. Zemenkov, E. Kurushina, Fundamental researches, **2-8**, 1632-1636 (2015)
11. E. Kurushina, Canadian Journal of Science, Education and Culture, **2(6)**, 378-384 (2014)
12. E. Kurushina, *Problems of formation of a common economic space and social development in the CIS countries: Proceedings of the International scientific and practical conference*, 27-33 (2011)
13. S. Toropov, Yu. Zemenkov, S. Podorozhnikov, Gas industry, **720(720),** 95-98 (2015)
14. M. Zemenkova, L. Fayzulina, I. Seroshtanov, Gas industry, **720(720)**, 59-64 (2015)
15. R. Mamadaliev, V. Kuskov, Yu. Zemenkov, A. Applied Mechanics and Materials, **770**, 19-22 (2015)
16. A. Shabarov, A. Ibragimov, M. Zemenkova, S. Podorozhnikov, Gas industry, **12(715)**, 50-54 (2014)
17. A. Pimnev, M. Zemenkova, Pipeline transport: theory and practice, **5(51)**, 43-47 (2015)
18. M. Zemenkova, Yu. Zemenkov, A. Pimnev, V. Petryakov, Business magazine Neftegaz.RU, **11-12**, 64-70 (2015)
19. A. Pimnev, M. Zemenkova, Fundamental researches, **11-2**, 292-296 (2015)

20. S. Stepanov, N., Cheremisin, S., Sokolov, A., Altunin, A., Shabarov, Computational method for relative permeability curves based on the Bernoulli generalized porous media equations, ECMOR 2012 - 13th European Conference on the Mathematics of Oil Recovery. (2012)DOI: 10.399/2214-4609.20143270.
21. R. Levitin, Yu. Zemenkov, Neftyanoe khozyaystvo - Oil Industry, **1**, 110-114 (2016)
22. Ya. Gorelik, A. Shabarov, Yu. Sysoyev, Earth's Cryosphere, **12(1)**, 59 (2008)

MATEC Web of Conferences 73, 01017 (2016) DOI: 10.1051/matecconf/20167301017
TPACEE-2016

Theoretical and Experimental Substantiation for Applicability of a Damping Layer in a Foundation Slab Placed on Soil Bed

Nikita Kiselev[1], *Yakov* Pronozin[1], *Maxim* Stepanov[1], *Leonid* Bartolomey[1,*], *Dietmar* Keck[1]

[1]Industrial University of Tyumen, 625001, Volodarskogo str., 38, Tyumen, Russia

Abstract. Authors present the results of studies of innovative foundation structure. The idea of how to increase the operational quality of foundations and reduce the costs due to rational loading of the soil bed is numerically simulated. It is shown that the bending moment in the foundation slab depends on uneven settlements of the soil bed. It is proposed to stabilize the deformable soil bed by the damping layer placed under the slab footing in the zones with minor settlements. Considered is the concept of the damping layer in the foundation slab placed on the soil bed (DLS). The in-situ test for DLS-clayey bed interaction is described. Given are the results obtained after the experiments for DLS performance. The result of DLS implementation in designing the foundation of the 22-storeyed block of flats is considered. The expediency of DLS in comparison to standard foundations is presented.

1 Introduction

The last decade has witnessed a steady growth of residential construction in Russia. In order to create conditions for residential construction and provide population with affordable and comfortable flats, the state program "Residential Construction Development" optimized for 2014-2016 has been launched in Tyumen Region; the program aims at erecting economy-class blocks of flats that meet the requirements of energy efficiency and environmental friendliness. Foundation works greatly influence labor and material costs, i.e. the cost of the foundation is 10-20% from the total cost of the building depending on its structural features. In this regard, applicability of the rational foundations becomes challenging from the point of view of material consumption and technological effectiveness of construction, taking into account sustainability and safety of constructional projects.

Tyumen Region is characterized by complex engineering-geological conditions. As a rule, the soil bed is composed of weak water-saturated floury-clay soils occurring at a considerable stratification depth; they possess low strength and are highly deformable in external loading [1, 2]. The traditional strip foundations are not applicable for 12 … 15-

* Corresponding author: tgasu.bartolomey.leonid@mail.ru

storeyed buildings because of low design strength of the soil bed, weak structural bond between separate strips and their response to uneven settlements. Foundation slabs are characterized by the considerable compressed depth and hence, the excessive design settlements and high materials consumption. Therefore, the idea to expand possible application of shallow foundations due to specified rational loading of the soil bed was formulated [3, 4].

2 Concept for a damping layer in a foundation slab placed on soil bed (DLS)

As is known from "classical" soil mechanics, when a flexible slab is uniformly loaded, the settlement in the center 1.24-1.57 times exceeds the settlements in the edges [5]. As for sustainability of buildings and structures, the value of settlement unevenness is more important than its average absolute value, since it is unevenness that leads to excessive stresses and strains in the slab and above ground structures, and their operational qualities considerably decrease [6, 7, 8]. Let us consider possible loading of the slab, so that minimum unevenness of soil bed settlements was provided.

Numerical computation done in PLAXIS 2D software revealed the design features of soil bed deformation under the load taking into account different distribution laws (Fig. 1a). Flexible slab (l = 6.0 m, t = 0.3 m, E = 30000MPa)-sandy clay soil bed (γ = 18 kN / m3, e = 0.75, φ = 18 °, c = 20 kPa , E = 18 MPa) interaction was considered using elastic-plastic model of Mohr-Coulomb. The slab flexibility value was equal to 6 after the solution of M.I. Gorbunov-Posadov [5], i.e. the slab was considered to possess finite rigidity. In the first computation the slab was uniformly vertically loaded with the intensity of 100 kPa, in the subsequent computations the load was redistributed from the center to the edges 1.0 m in width, saving its total value. The results obtained showed that in the first loading the slab saddled considerably, and in the third loading unevenness of settlements was close to zero (Fig. 1b), in here, the peak value of the bending moment was up to 8 times lower than in the case of the uniform load (Fig. 1c). In further loading of the edge the slab bent in the center, unevenness of settlements and the bending moment increased. The area of the bending moment diagrams (Fig. 1d) affecting material consumption of the foundation was directly dependent upon the settlement differences; the material consumption was 6 times lower for the 3d variant than in the case of the uniform load. Thus, it has been proved that when adapting load distribution over the foundation slab, it is possible to ensure more favorable conditions of its performance.

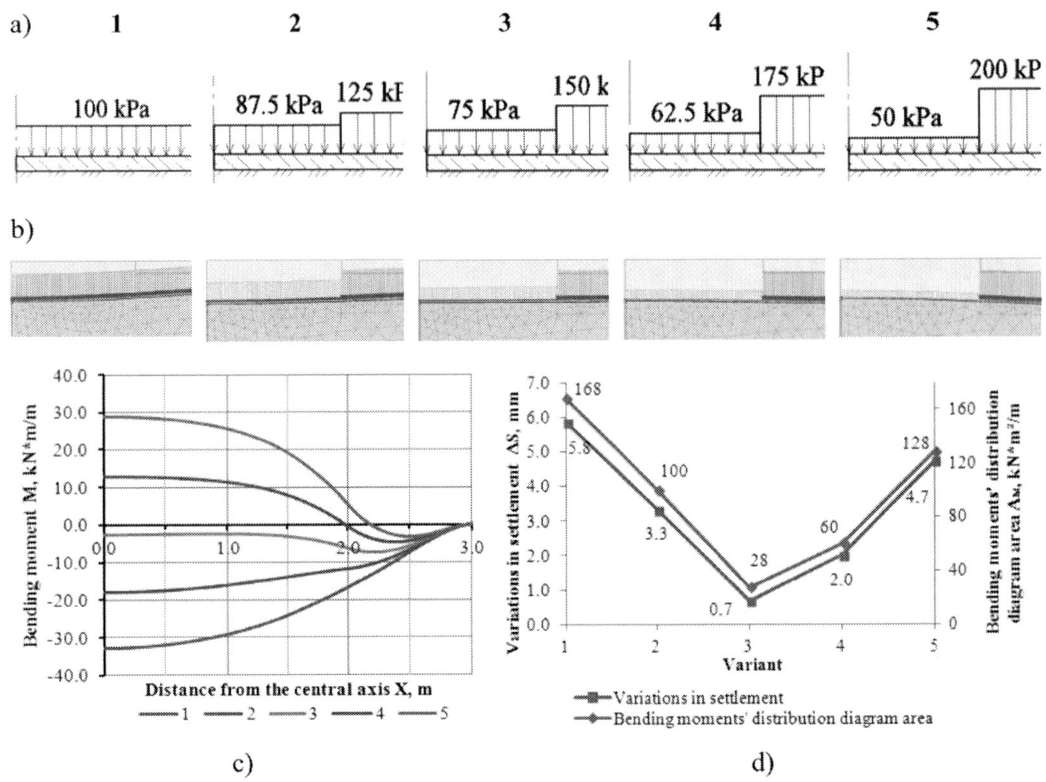

Fig. 1. Plane stress problem with various load distribution on the foundation (variants 1 … 5): a) design diagram; b) strain diagram; c) bending moments in the slab; d) dependency diagram for unevenness of settlements and bending moments' distribution diagram area for different cases of loading

However, loading of the foundation is conditioned by the architectural form of the structure and its technological purpose, i.e. it is "static" and does not offer "plasticity" making it possible to realize the specified loading. The authors [9, 10, 11, etc.] offer to eliminate unevenness of settlements of the foundation slab by correcting soil bed deformability, e.g. by strengthening the soil bed under the center of the slab or weakening it under the slab edges. Taking into account the following hypothesis, let us solve the problem by changing deformability of the soil bed rather than redistribution of the load.

The initial data are given above, but let us introduce the damping layer of b=1.0 m in width and t=0.2m in thickness with the variable modulus of elasticity - E=1.0 … 2.5 MPa under the slab edge; in here, the load remains uniform - 100 kPa (Fig. 2). In general, the revealed dependences are analogous to those mentioned above; thus, the damping layer has made it possible to level the settlements of the soil bed and reduce the bending moments in the slab up to 7 times. Thus, it seems possible to achieve the specified technical result by simpler and more constructable method than in the previous example.

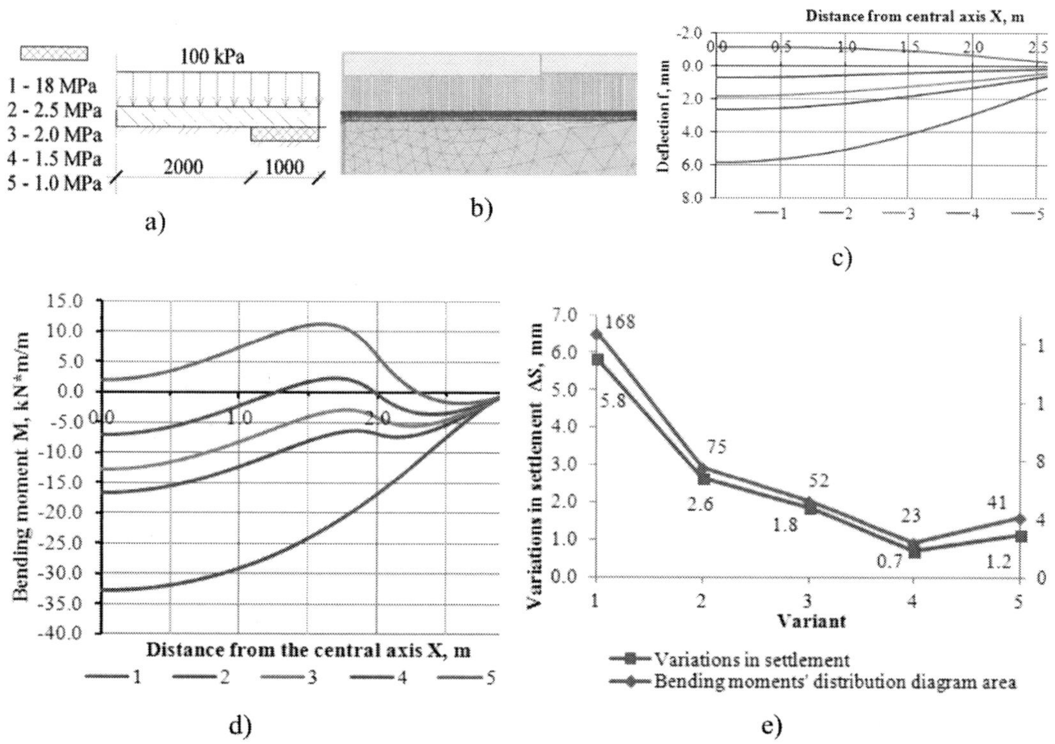

Fig. 2. Plane stress problem with various damping layers under the foundation edge (variants 1 … 5): a) design diagram; b) strain diagram (variant 4); c) deflections in the slab; d)bending moments in the slab; e) dependency diagram of unevenness of settlements and bending moments' distribution diagram area for different cases of loading

A possible way to put this hypothesis into practice is to apply damping layers in slab foundations placed on the soil bed (DLS); this allows the reaction pressures over the slab footing to be rationally distributed proceeding from minimization of stresses in the foundation body or adaptation of the constructional project to specific soil conditions.

The rational use of such foundations is conditioned by the following:

- In case of the flexible slab, e.g. a uniformly loaded foundation of the reservoir, the damping layer in the edges compensates non-uniform compressibility of the soil bed over the foundation footing, decreases deflections and bending moments in the structure. (Fig. 3a).
- In case of the foundation slab of a multistory building, the damping layers in the span zones allow concentrating pressures in the load-bearing axes, unload the span zone and decrease bending moments in the slab (Fig. 3b).
- In case of quite stiff low-compressible covering layers and the underlying thick layer of weak water-saturated soils occurring in the geological structure of the construction site, the decrease of pressures allows reducing the compressible thickness (Fig. 3c) to some extent, the effect on the weak underlying layers, and hence, the settlement of the foundation.

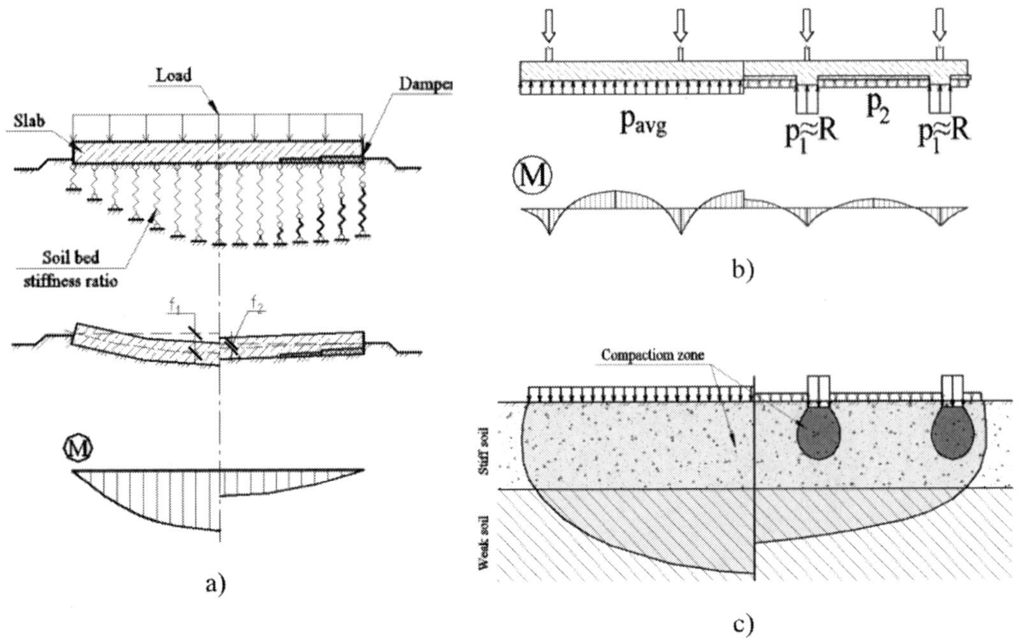

Fig. 3. Rational use of DLS foundations:
a) foundations of reservoirs; b) foundations of buildings; c) weak underlying layer

3 Experimental studies

Complex in-situ tests were conducted to study DLS-soil bed interaction; they were aimed at revealing the features of the deformation process in clayey soil bed loaded with a stamp incorporating the damping layer. Round reinforced concrete stamp models of 1.2m in diameter and 0.2m in thickness were tested on the clayey soil bed composed of lacustrine-alluvial soils ranging from hard to high-plastic ones with the modulus of deformation - E=8 ... 19 MPa (Fig. 4a) : the first stamp incorporated a flat footing, the second one - a foam polystyrene layer (PSB-15), 30 mm in thickness placed between the footing and the soil bed (Fig. 4b). The stamp flexibility value was equal to 0.044 after the solution of Gorbunov-Posadov [5], i.e. the stamps were considered to possess finite rigidity.

The soil bed of the tested stamps was equipped with the screw marks to measure vertical movements of the soil massif and membrane soil pressure cells to determine contact pressures over stamp footings (Fig. 4a) [12, 13, 14]. Movements of the bed and settlements of the stamps were measured by Aistov deflectometers mounted in the stationary benchmark station. Contact stresses were recorded in real-time mode applying Field Point device which transforms a signal received from resistive-strain sensors glued on the membrane of soil pressure cells [15, 16].

The stamps were loaded by a hydraulic cell through the distributive system transferring the load over four points, symmetric to the central axes. The reactive stress from the hydraulic cell was taken by the supporting frame assembled from steel beams and concrete blocks used as the static load (Fig. 4c). The load was applied stepwise after conventional stabilization - 25 kPa for each step, when the rate of the test stamps settlements did not exceed 0.1 mm within the last two hours of observations [17]. Fig. 4d and Fig. 4e show photographic images of stamp making and testing.

Dependency diagrams for stamp settlement-load relation (Fig.5), diagrams of contact stresses over stamp footing (Fig.6) and isolines of vertical movements in stamp beds were plotted after the results of data processing obtained during static tests of the given stamps (Fig.7).

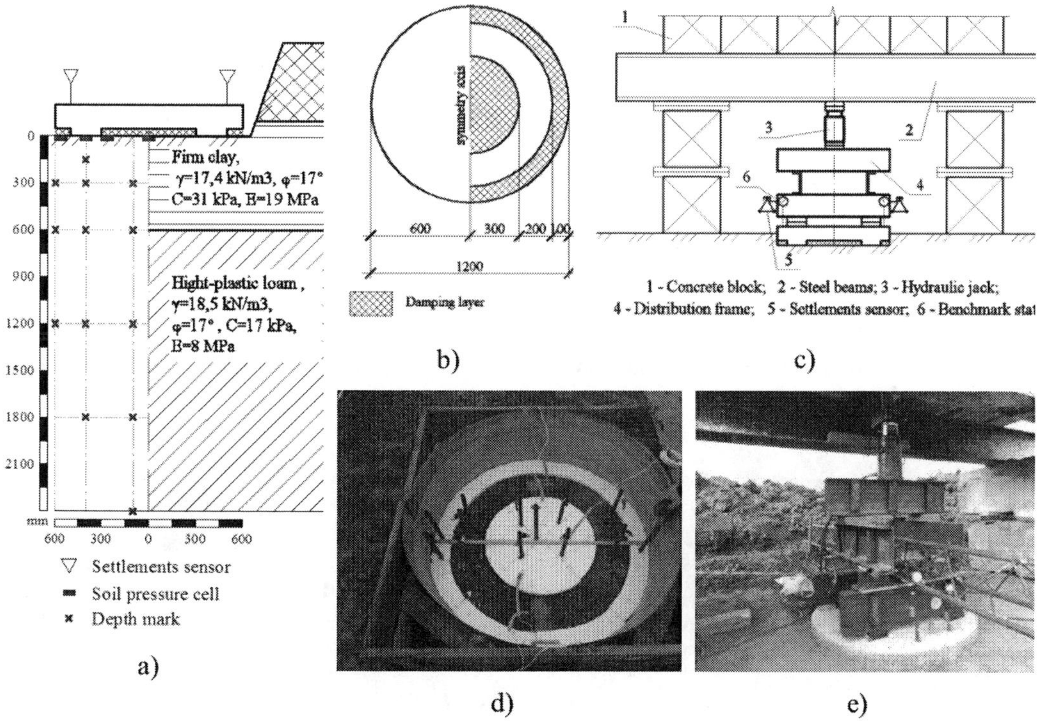

Fig. 4. In-situ tests for DLS:
a) schematic illustration of sensors and geological section of the bed; b) schematic illustration of the test stamps; c) schematic illustration of the test station; d) stamp making; e) stamp testing

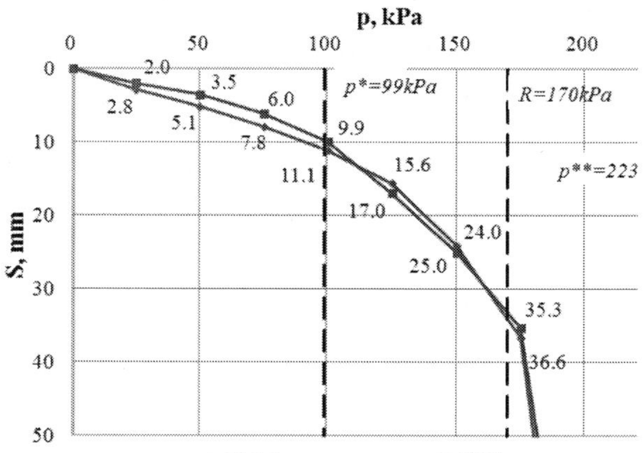

Fig. 5. Dependency diagrams for stamp settlement-load relation

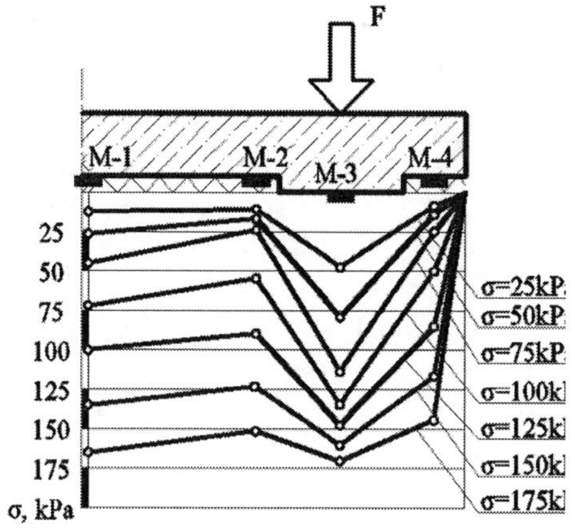

Fig. 6 Contact stresses over DLS footing

One can conditionally point out two characteristic stages of stamp performance on the curves “settlement-load” (Fig. 5): a stage of "minor" quasi-linear settlements, a stage of "major" shear deformations. If the load upon the bed exceeds 175 kPa, soil bed failure occurs resulting in the collapsed unsettled stamp settlements.

The analysis of the test results shows that boundaries of the sites agree with the soil bed bearing capacity values:

- initial critical pressure upon the soil is evaluated by the formula after K.E. Egorov and T.I. Finayeva [18] as for the flat round foundation:

$$p^* = \left(1 + \frac{\sin\varphi}{M_0}\right)\cdot\gamma\cdot h + \frac{\cos\varphi}{M_0}\cdot c = B_0\cdot\gamma\cdot h + D_0\cdot c = 99kPa$$

- soil design resistance evaluated by the formula after SP [19]:

$$R = \frac{\gamma_{c1}\gamma_{c1}}{k}\left[M_\gamma\cdot k_z\cdot b\cdot\gamma_{II} + M_q\cdot d_l\cdot\gamma'_{II} + (M_q - 1)\cdot d_b\cdot\gamma'_{II} + M_c\cdot c_{II}\right] = 170kPa$$

ultimate critical pressure upon the soil evaluated by the formula [20]:

$$p^{**} = (\gamma\cdot h + c\cdot ctg\varphi)\frac{1+\sin\varphi}{1-\sin\varphi}\cdot e^{\pi\cdot tg\varphi} - c\cdot ctg\varphi = 223kPa$$

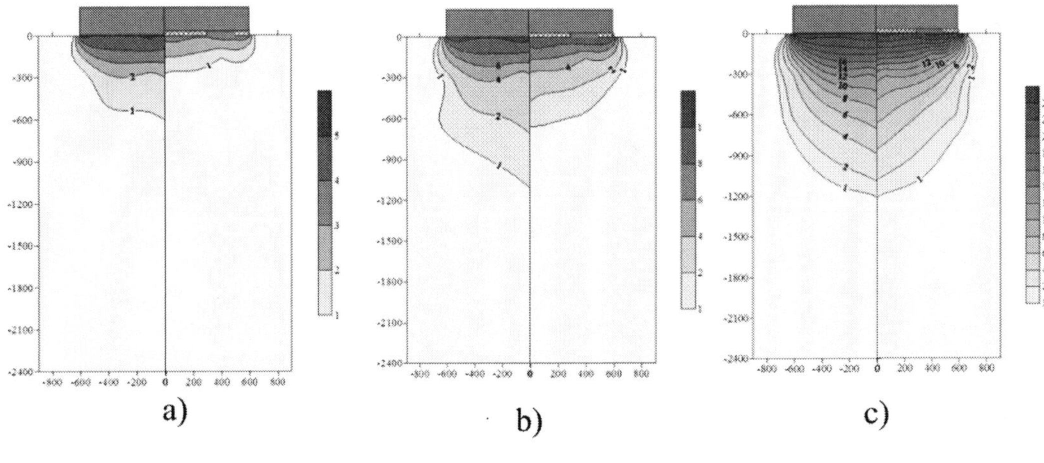

Fig. 6. Isolines of vertical movements in the bed: a) p=50кPa; b) p =100кPa; c) p=150кPa;

At the first stage settlement-load dependency is close to linear if the average pressure is from 0 to 100 kPa; in here, settlements of DLS are up to 25% lower as compared to the flat stamp. At the first stage the contact pressures concentrate under the stamp edge (fig. 6); this occurs due to squeeze reduction of the foam polystyrene damping layer in the span and cantilever zones.

The analysis of layer-by-layer movements of the soil points indicates significant difference in compressibility of separate layers over the soil bed depth of the test stamps. At the first stage one can observe almost absolute loss of movements in the stamp beds at the depths of up to 0.5 of the diameter, although it is to occur at the depths of 1.0 – 1.5m of the diameter after the theoretical solution. In here, soil movements over the central axis of the flat stamp develop to a significantly greater depth in relation to DLS. Vertical deformations of the soil bed under DLS are localized in the upper layer of firm clays, while movements in the flat stamp bed develop in the high - plastic loam subsoil. For instance, in loading with 50 kPa deformation of the near-surface firm clay layer of 0.5 in thickness of the diameter (600 mm) equals to 81% from the total deformation for the flat stamp and 95% - for DLS stamp. The fact makes it possible to expand applicability of DLS in the soil conditions – "stiff upper layer – weak subsoil" or "sandy cushion – weak soil" in case the loads do not exceed p*.

At the second stage, in loads exceeding 100 kPa, DLS performance is characterized by developing ultimate limit state zones under the edges, redistribution of the load over the footing, intensive growth of stamp settlements and deformations in the weak subsoil layer. The growth of settlements is accompanied with transformation of the contact stress distribution diagram caused by extra loading of the foundation in the central and cantilever zones. On the one hand, this is due to the lack of "free movement" of the damping layer, on the other hand – development of plastic deformable zones in the soil bed under the edge. Gradual smoothing of layer-by-layer deformations in DLS bed in relation to the flat stamp is observed. Thus, the efficiency of DLS decreases at the second stage.

In excessive loading over 150 kP, intensive relative deformations in the high - plastic loam subsoil has been recorded resulting in the considerable unsettled settlements. The contact stress values over DLS footing are nearly adjusted, but with their unspecified concentration under the stamp edge. Layer-by-layer vertical movements are almost identical. Pressure increase of up to 200 kPa (in excess of R) is accompanied with continuous unsettled rise of settlements for both stamps indicating linkage of the ultimate limit state zones and loss of soil bed stability. Thus, the range of bearing loads for stamps is actually limited to the design resistance R.

Figure 7 illustrates the dependency diagram for compressible thickness depth-tested stamp load relation. The settlement of the layer – 1mm is taken as a criterion of compressible thickness depth. The given diagram shows that when loading up to 100 kPa, the depth of DLS compressible thickness is nearly twice as low as that of the flat stamp and is actually composed of the layer of firm clays.

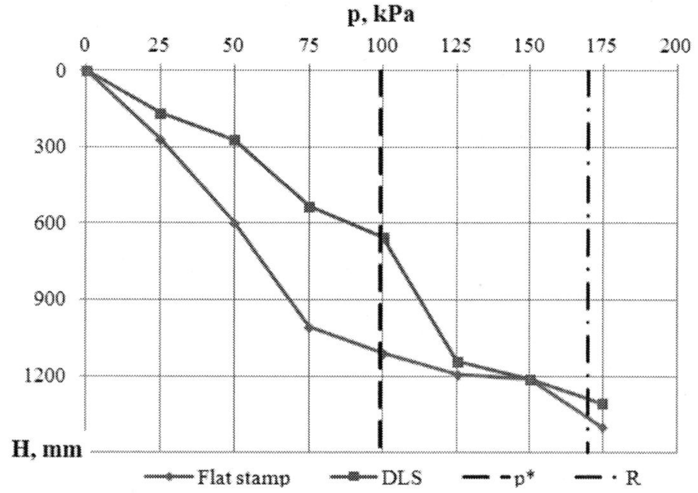

Fig. 7. Dependency diagram for compressible thickness depth-load relation

In here, the DLS diagram before p* is close to linear, and then one can observe the "failure" accompanied with sharp increase of compressible thickness depth; it is likely caused by limit equilibrium zones appearing under the stamp edges and redistribution of load over the footing.

4 Implementation

The results of theoretical and experimental studies for DLS performance were used in foundation works of a block of flats – "Aquarelle" located in 9, Tavricheskaya St., Tyumen (Fig. 8).

Initially the project involved 360 driven piles under a monolithic reinforced concrete slab. After specified calculations the bearing capacity of the pile foundation proved to be excessive in the engineering-geological conditions of the given construction site. It was decided to replace the shallow foundation with more technological one that proved to be suitable in the given conditions.

Fig. 8. General view [21]

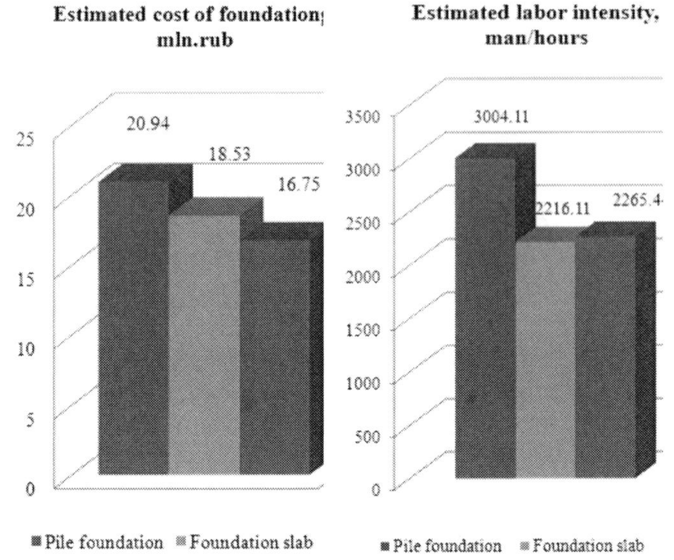

Fig. 9. Technical-and-economic indexes

The following types of foundations were considered during design works:

- Type I – DLS composed of the rigid orthogonal strips 1.2 m in height focused on bearing axes of the building; the slabs – 0.8 m in thickness in span zones; PSB-S-25foam polystyrene damping layers placed between the slab and soil bed (Fig. 4,5).
- Type II – stationary reinforced concrete slab 1.0m in thickness.

The total estimated cost for DLS reduced from 20.94 to 16.75mln. rub (20%); labor intensity – from 3004.11 to 2265.44 man/hours (25%) as compared to the slab-pile foundation. The cost of DLS appeared to be lower than that of the foundation slab, i.e. 1.8 mln. rub. of cost savings in equal labor intensity.

5 Conclusion

1. It has been numerically simulated that due to specified loading of the foundation it seems possible to reduce deflections and uneven settlements of the foundation slab many times and hence, the total distribution diagram of dependable bending moments.
2. The same problem shows that it is possible to achieve the analogous technical result by the damping laying placed under the foundation footing resulting in balance of uneven compressibility of the soil bed and settlements of the foundation slab. The damping layer in the foundation slab placed on the soil bed (DLS) has been analyzed and its rational applications have been highlighted.
3. Scientific research on DLS-soil bed interaction has been done on the model during the in-situ tests. The interaction stages which agree with theoretically proved strength characteristics of soil p *, R and p ** have been determined. At the stage when the loads do not exceed p *, DLS settlements are 25% lower than those of

the foundation slab, and the compressible thickness depth is twice as low as that of the foundation slab, the active DLS bed zone being limited to a layer of firm clays. The fact makes it possible to expand applicability of DLS in the soil conditions – "stiff upper layer – weak subsoil" or "sandy cushion – weak soil".

4. DLS has made it possible to do without 360 driven reinforced concrete piles in construction of 22-storeyed block of flats and reduce the estimated cost of construction by 20% and labor intensity - by 25%.

References

1. T. Maltseva, A. Nabokov, A. Chernykh, Procedia Engineering, **117**, 239-245 (2015)
2. A. Tetior, *Progressivnye konstrukcii fundamentov dlja uslovij Urala i Tjumenskoj oblasti* (Sredne Ural'skoe izdatel'stvo, Sverdlovsk, 1971)
3. Z. Ter-Martirosyan, Y. Pronozin, N. Kiselev,v Soil Mechanics and Foundation Engineering, 51, 157-164 (2014)
4. V. Chikishev, Y. Pronozin, V. Mironov, Internet-vestnik VolGASU. Ser.: Politematicheskaja, **20** (2012)
5. M. Gorbunov-Posadov, T. Malikova, V. Solomin, *Raschet konstrukcij na uprugom osnovanii* (Strojizdat, Moscow, 1984)
6. SP 22.13330.2011 Osnovanija zdanij i sooruzhenij. Aktualizirovannaja redakcija SNiP 2.02.01-83*. Moscow, (2011)
7. M. Abelev, Avarii fundamentov promyshlennyh i grazhdanskih sooruzhenij. FAOU DPO GASIS, (2011)
8. L. Dimov, Geotehnika, **1**, 4-8 (2015)
9. G. Boldyrev, S. Boldyrev, Materialy Mezhdunarodnoy nauchno-tekhnicheskoy konferentsii «Sovremennye problemy fundamentostroeniya (2001)
10. M. Dubina, V. Tselitsom, M. Chukhlatyy, Materialy Mezhdunarodnoy konferentsii «Geotekhnicheskie problemy stroitel'stva krupnomasshtabnykh i unikal'nykh ob"ektov» (2004)
11. V. Krutov, E. Sorochan, V. Kovalev, *Fundamenty melkogo zalozhenija : Racional'nye konstrukcii i tehnologii ustrojstva* (ASV publ., 2009)
12. A. Golli, *Metodika izmerenija naprjazhenij i deformacij v gruntah: uchebnoe posobie* (LISI, Leningrad, 1984)
13. Ju. Murzenko, Jeksperimental'nye issledovanija inzhenernyh sooruzhenij: materialy ko II simpoziumu, Novocherkassk (1999)
14. D. Baranov, *Rukovodstvo po primeneniju prjamogo metoda izmerenij davlenij v sypuchih sredah i gruntah* (CNIISK im. Kucherenko, Moscow, 1995)
15. Z. Ter-Martirosyan, Y. Pronozin, M. Stepanov, Soil Mechanics and Foundation Engineering, **49-4**, 119-123 (2012)
16. A. Kraev, Nauchno-tehnicheskij vestnik Povolzh'ja, **5**, 221–224 (2013)
17. GOST 20276-99. Grunty. Metody polevogo opredelenija harakteristik prochnosti i deformiruemosti, Moscow (2000)
18. K. Egorov, T. Finaeva, Osnovanija, fundamenty i mehanika gruntov, **6**, 26-27 (1984)
19. SP 22.13330.2011 Osnovanija zdanij i sooruzhenij. Aktualizirovannaja redakcija SNiP 2.02.01-83*. Moscow (2011)
20. Z. Ter-Martirosjan, *Mehanika, gruntov* (ASV Publ., Moscow, 2005)
21. Company Group "TIS". Available at: http://ttis.ru/objects/akvarel (accessed 12 May 2016)

MATEC Web of Conferences 73, 01018 (2016) DOI: 10.1051/matecconf/20167301018
TPACEE-2016

Using Linear Spectral Method when Calculating Seismic Resistance of Large-Capacity Vertical Steel Tanks

Alexandr Tarasenko [1], *Petr* Chepur [1,*], and *Alesya* Gruchenkova [1]

[1]Industrial University of Tyumen, 625001 Volodarskogo str. 38, Tyumen, Russia

Abstract. The paper is aimed at determining the possibility of applying the simplified method proposed by the authors to calculate the tank seismic resistance in compliance with current regulations and scientific provisions. The authors propose a highly detailed numerical model for a common oil storage tank RVSPK-50000 that enables static operational loads and dynamic action of earthquakes to be calculated. Within the modal analysis the natural oscillation frequencies in the range of 0-10 Hz were calculated; the results are given for the first ten modes. The model takes into account the effect of impulsive and convective components of hydrodynamic pressure during earthquakes. Within the spectral analysis by generalized response spectra was calculated a general stress-strain state of a structure during earthquakes of 7, 8, 9 intensity degrees on the MSK-64 scale for a completely filled up, a half-filled up to the mark of 8.5 m and an empty RVSPK-50000 tank. The developed finite element model can be used to perform calculations of seismic resistance by the direct dynamic method, which will give further consideration to the impact of individual structures (floating roof, support posts, adjoined elements of added stiffness) on the general stress-strain state of a tank.

1 Introduction

Development of the main oil transport facilities infrastructure and emergence of new supply routes leads to the need for construction of commercial oil and oil products depots in areas with high seismic activity. According to the OSR-97-S maps of seismic zoning of Russia practically all regions of the Far East, Altai, Transbaikal have a 5% likelihood of an increase in earthquakes of 7 degrees intensity on the MSK-64 scale. In the North Caucasus in such conditions large-capacity oil depots are exploited including vertical steel tanks of up to 100,000 m3. To ensure safe operation of large tanks, failure of which may result in environmental and material damage [1-5], it is necessary to develop and improve approaches to calculation of seismic resistance of such facilities.

Currently, there are 3 main methods of calculating structures for seismic effects: static, linear spectral and direct dynamic. The most simple method – static – is based on the assumption that the structure moves with the subgrade in a horizontal plane by a certain law y0(t) without undergoing rotational movements. This method is of little use if you want to

* Corresponding author: chepur@me.com

calculate the seismic resistance of tanks, because it does not take into account features of the hydrodynamic fluid impact on the wall, the base and structures ductility, etc.

Direct dynamic method allows for a highly precise estimation of the vertical steel tank (VST) design stiffness based on the stored liquid, which makes it more attractive for solving specific engineering problems. If there are calculated accelerograms, representing a three-component function of acceleration to time, it is possible to determine deformation, strain and displacement in all points of the structure and fluids for a particular construction site at any moment of time t. Often, however, the results of microseismic zoning of the site are not available, forcing to "pick out" the most suitable existing design accelerograms. For example, in [3] using the direct dynamic method were obtained design calculations of the earthquake impact on the stress-strain state (SSS) of an insulated LNG storage tank, and was proposed a method of reducing the frequency of natural oscillations of the tank by applying damping inserts. However, the results can be extended only on tanks built according to the Chinese project for a specific site with a given impact accelerogram. Experimental studies on seismic action on tanks using reduced models - "vibrating plate-tank-liquid" systems are presented in [1, 3]. These experiments made it possible to obtain high-quality results aimed at further optimizing the design numerical models. Direct dynamic method involves considerable machine time costs (supercomputers of special configuration are required) and complexity of results processing. We propose to use the main provisions of the linear spectral theory of calculating the seismic action for a common in Russia large-sized tank RVSPK-50000. According to this approach, seismic action is considered as a spectrum - dependence of acceleration on frequency. The calculation includes calculating the participation factor of each oscillation mode with the natural frequency of the tank structure, determining the maximum acceleration response, inertial forces and displacements for each mode of the predetermined response spectrum in the initial system. In practice, the main advantage of this method is that the response spectrum can be obtained by synthesizing the accelerograms of real earthquakes. At the same time, the more of them have been taken into account in constructing the action spectrum, the less likely it is that the loading on the tank will exceed the anticipated when the new earthquake occurs.

2 Design scheme

To analyze seismic resistance of the RVSPK-50000 tank, designed by "CNIIPROEKTSTALKONSTRUKCIYA" (D = 60.7 m, H = 17.95 m) the authors carried out a modal analysis to determine the natural frequencies and modes of the structure oscillations. For static problems, the geometric tank model has been developed and verified by the authors in [6-15] using the ANSYS software package, then its optimization was conducted and it was exported to modules «Modal» and «Response Spectrum» to solve dynamic problems.

Let us look at the basic steps of a modal analysis of RVSPK-50000.

- The model has set material properties: 09G2S tank steel with a yield point $\sigma_т$ = 325 MPa, 16G2AF steel with a yield point $\sigma_т$ = 410 MPa and oil with a density of 865 kg/m^3; metal structures are modeled by finite elements SHELL181, BEAM188, SOLID186, oil stored in the tank is modeled by tetrahedral elements FLUID80;
- The calculation is performed for the three cases of VST fill-up: emptied, filled up halfway ($H_{взл}$ = 8.5 m) and filled up to the top level ($H_{взл}$ = 17 m);
- The boundary conditions allow for rigid mounting of the foundation ring, unbroken contact of the bottom and wall with the foundation and the possibility of separation of the central part of the bottom from the subgrade, the properties of which are specified by the coefficient of subgrade reaction k = 200 MN/m^3 (for compacted sand);

- The contact of the walls and bottom with the liquid is determined by the function «no separation», which allows one to simulate the fluid sliding along the wall and bottom during oscillations.

According to [16–19], the results obtained by the modal analysis of RVSPK-50000 and further test calculations, most devastating effects are observed at the oscillation frequency from 0 to 10 Hz. At such frequency the greatest contribution is made by impulsive (inertial) and convective (kinematic) components of hydrodynamic pressure, so the most unfavorable case is the design scheme with a completely filled up oil tank. Figure 1 provides the results of calculating the natural oscillation frequencies of RVSPK-50000 completely filled up to the design level. 10 modes were determined in the calculation; the figure shows general deformations diagrams of the first 6 modes and with the frequency not exceeding 10 Hz.

1st oscillation mode,

natural frequency f= 0.33 Hz

3rd oscillation mode,

natural frequency f= 0.54 Hz

4th oscillation mode,

natural frequency f = 0.94 Hz

5th oscillation mode,

natural frequency f= 0.97 Hz

6th oscillation mode,

natural frequency f= 1.19 Hz

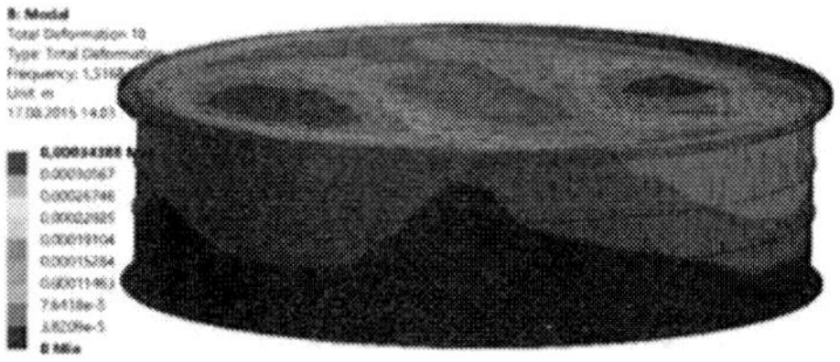

10th oscillation mode,

natural frequency f= 1.52 Hz

Fig. 1. Determining the natural oscillation frequencies of RVSPK-50000, completely filled up to the mark of Нвзл = 17 m.

Table 1 provides the results of calculating the natural oscillation frequencies for the three cases of RVSPK-50000 fill-up at 10 oscillation modes.

Table 1. Results of calculating the natural oscillation frequencies of RVSPK-50000.

RVSPK-50000 fill-up Natural oscillation frequency, Hz	№ of oscillation mode									
	I	II	III	IV	V	VI	VII	VIII	IX	X
Emptied tank	10.04	10.05	10.1	10.16	10.24	10.25	10.26	10.27	10.27	10.28
Filled halfway Нвзл=7.5m	0.27	0.28	0.25	0.84	0.92	1.13	1.22	1.28	1.46	1.54
Completely filled Нвзл=17m	0.33	0.34	0.54	0.95	0.97	1.19	1.26	1.31	1.42	1.52

To perform linear spectral analysis of seismic action the addition of responses calculated from the obtained values of natural modes and oscillation frequencies of the structure was carried out. The SRSS (Square Root of Sum of Squares) method of modes combination was used enabling the input of each oscillation mode in the overall system to be summarized. According to [19], this method yields reliable estimates, if the natural frequencies differ greatly from each other (in our case, the frequencies range from 0.2 to 30 Hz).

To set the external seismic action we used generalized broadband seismic response spectra digitized in [20]. For obtaining acceleration components a formula was used that takes into account the impulse duration, start time and end time, as well as the amplitude:

$$a(t+\tau) = a_m \sin\left(\frac{\pi\tau}{L}\right), \tag{1}$$

where L – the impulse duration; t – the impulse start time; τ – the time within the impulse, $0 \leq \tau \leq L$; a_m – the impulse amplitude.

In Russian design practice it is conventionally recommended to use only one synthesized accelerogram SA-482, established in [18] and underlying the seismic response spectra in this paper. When calculating the structures for seismic resistance within the linear spectral method as a mathematical model of the seismic action a response spectrum is used. It is a graph of the maximum acceleration $A(\omega) = \max(\ddot{x}(t))$ dependence of the oscillators on their natural frequency ω_m at a given value of the oscillation attenuation δ_d due to the action of the accelerograms a(t) (real, synthesized or probabilistic and statistical).

3 Results

Fig. 2 shows the horizontal and vertical ground accelerations dependences on the wave frequency in the range from 0 to 30 Hz used when calculating the seismic action on RVSPK-50000 by the spectral method for earthquakes of 7, 8, 9 intensity degrees on the MSK-64 scale. For areas with the seismic activity of 6 degrees and below the seismic loads are not required to be considered [22]. The logarithmic decrement is assumed to be δ_z = 0.31 for a steel welded structure. The vertical acceleration component value is reduced by a factor of 0.67 from the values of the horizontal components X and Y, in accordance with [17].

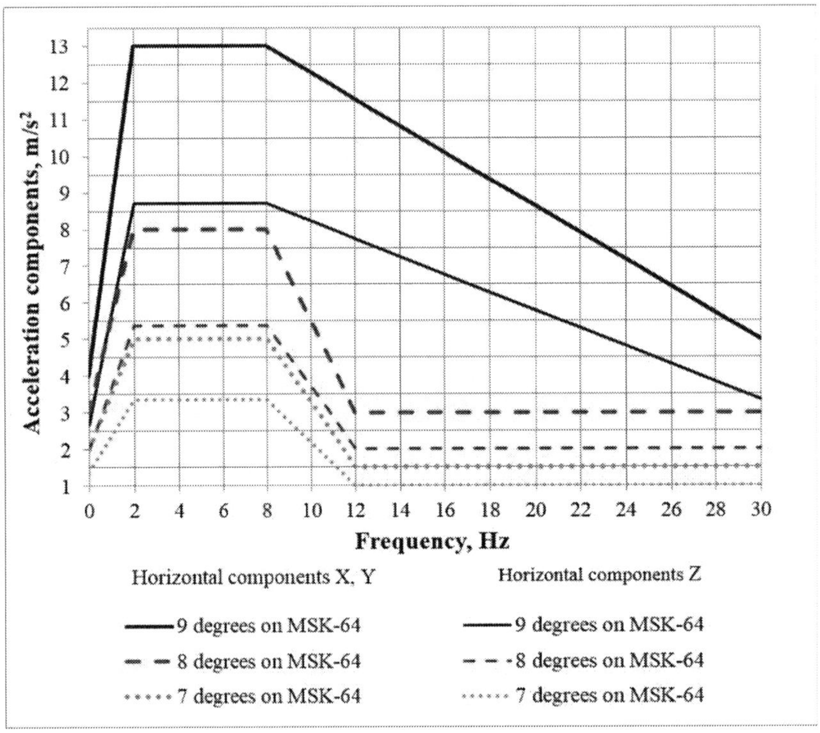

Fig. 2. Seismic response spectra for earthquakes of 7, 8, 9 intensity degrees on the MSK-64. Scale.

A design scheme of the RVSPK-50000 tank filled to the top mark and located on the elastic Winkler base is given in Fig. 3.

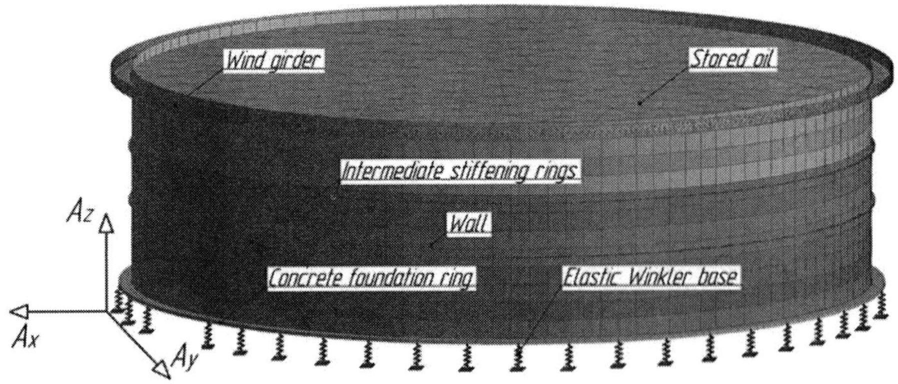

Fig. 3. Design scheme for spectral analysis of RVSPK-50000.Ax, Ay, Az – components of the horizontal and vertical accelerations under the seismic action.

The results of the numerical calculation carried out in the ANSYS «Response Spectrum» software module show that the case when the tank is completely filled up is the most dangerous. A 9-degree seismic action on the MSK-64 scale (substantial destruction of moderate severity, according to the classification) results in considerable hydrodynamic loads, impulsive and convective components of the stored oil cause the appearance of unallowable stresses in the wall that exceed the 09G2S tank steel yield point $\sigma_т = 325$ MPa for the top wall belts in a neutral layer (Fig. 4), with stresses greater than 400 MPa arising in the outer shell layer of the first belt.

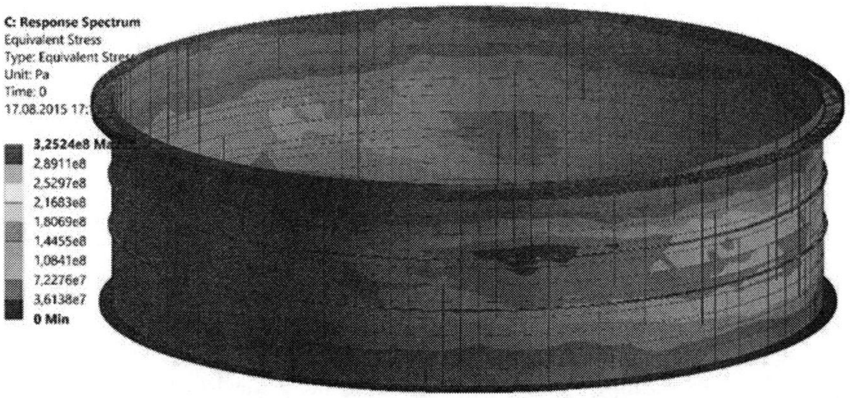

Fig. 4. Distribution of equivalent stresses in RVSPK-50000 structures (neutral layer) at a 9-degree earthquake intensity on the MSK-64 scale.

Reduction in the seismic action intensity and the volume of stored oil decreases the SSS level of the shell. So, in an empty tank the acting equivalent stresses in various metal structures do not exceed 10 MPa at a 9-degree earthquake intensity. The authors have produced a graph (Fig. 5) built on the results of spectral analysis which reflects the dependences of the structure SSS level on the value of the product filling during earthquakes of 7, 8, 9 intensity degrees.

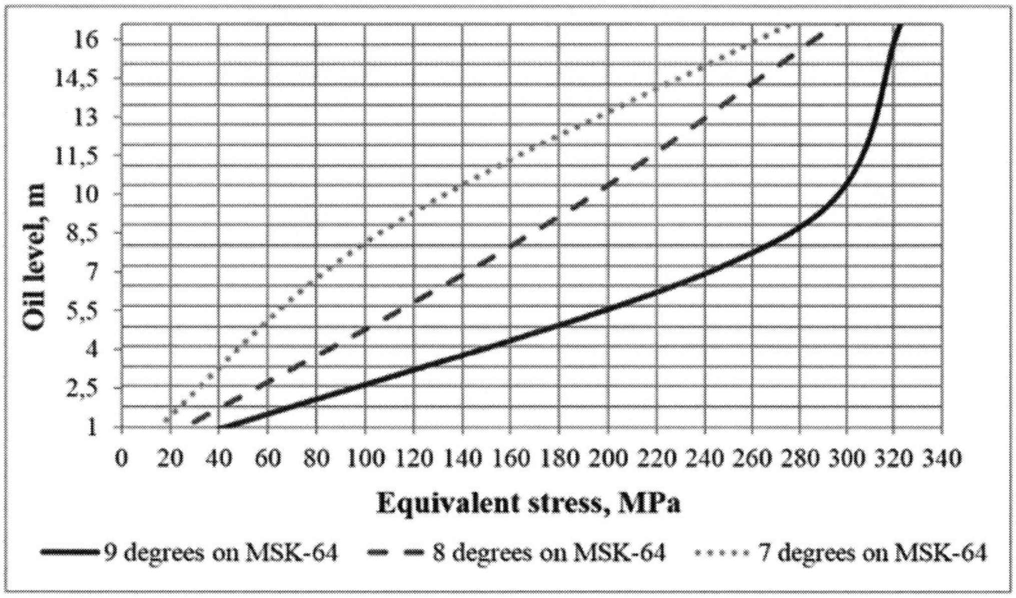

Fig. 5. Dependencies of acting stresses in the neutral wall layer of RVSPK-50000 under seismic actions of 7, 8, 9 intensity degrees.

A high-detail model (the edge length of individual elements is 5 mm), consideration of design geometry and material properties of the tank built on a real project allowed us to obtain the results of the RVSPK-50000 metal structures SSS under seismic actions with an accuracy unattainable by conventional methods.

Linear spectral method with the implementation in the finite element software package actually is a static calculation taking into account seismic forces in the same manner as other static loads. At the same time, it takes into account dynamic characteristics of the calculated structure - natural frequencies, modes and attenuation parameters. When calculating the seismic impact on the oil tank, an advantage of the method is the ability to take into account only a limited number of lower system natural modes caused by the fluid vibration. However, linear spectral method has certain limitations that do not allow taking into account the behavior of the structure during separation (both in case of fixing by anchors, and free support).

4 Conclusions

Analyzing the calculation results, the authors concluded that earthquakes with a large magnitude put at most risk tanks that are completely filled up with product. Thus, it is necessary to develop technological schemes according to which an early warning of a possible earthquake would be possible to redistribute the volume of the stored liquid throughout the tank farm so that each tank would have the lowest level possible [20-22]. This will significantly reduce the probability of failure and structural damage during a possible earthquake. In 2013, for example, after a powerful earthquake in the Chinese province of Sichuan, it was decided to invest more than $300 million in the forecast of seismic shocks: in the most dangerous areas of the country a network of 5,000 observation stations will be set up, the purpose of which is an early warning of severe earthquakes.

A separate issue, which until now has not been fully investigated, is analysis of the influence of the floating roof and pontoon on the tank SSS under seismic actions. Consideration of the contact interaction of the "liquid-pontoon-wall" system is complicated by the fact that a precise modeling of the hydraulic seal, support posts and other structures can solve only a narrow engineering problem for a specific project, without obtaining a generalized solution suitable for amending the existing regulatory and technical documentation. The approach proposed by the authors allows the level of seismic load effect on a tank to be reduced by selecting the damping devices installed between the base and metalwork of the structure.

References

1. G.G. Vasilyev, A.A. Tarasenko, P.V. Chepur, Y. Neftyanoe khozyaystvo, Oil Industry, **10**, 120–123 (2015)
2. A.S. Gorelov, *Inhomogeneous ground base and their influence on the operation of vertical steel tanks: monograph* (Nedra, SPb, 2009)
3. M. De Angelis, R. Giannini, F. Paolacci, *Experimental investigation on the seismic response of a steel liquid storage tank equipped with floating roof by shaking table tests, EarthquakeEngngStruct* Dyn. 2009. DOI: 10.1002.
4. P.A. Konovalov, R.A. Mangushev. S.N. Sotnikov, A.A. Zemlyanskiy, A.A. Tarasenko, *Foundations of steel tanks and deformations of their bases* (Publishing house Associacii stroitel'nyh vuzov, Moscow, 2009)
5. A.A. Tarasenko, *Development of scientific principles of vertical steel tanks repairs methods. Thesis for the degree of Doctor of Technical Sciences* (Tyumen State Oil and Gas University, Tyumen, 1999)

6. A.A. Tarasenko, A.A. Gruchenkova, P.V. Chepur, Pipeline transport: Theory and Practice, **53**, 32-37 (2016)
7. A.A. Tarasenko, P.V. Chepur, A.A. Gruchenkova, Fundamental research, **12-7**, 1418-1422 (2014)
8. A.A. Tarasenko, P.V. Chepur, D.A. Tarasenko, Neftyanoe khozyaystvo, Oil Industry, **4**, 88–91 (2015)
9. A.A. Tarasenko, P.V. Chepur, S.V. Chirkov, Neftyanoe khozyaystvo, Oil Industry, **10**, 121–123 (2014)
10. A.A. Tarasenko, P.V. Chepur, S.V. Chirkov, Neftyanoe khozyaystvo, Oil Industry, **3**, 123–125 (2016)
11. A.A. Tarasenko, P.V. Chepur, Y. Guan, Neftyanoe khozyaystvo, Oil Industry, **4**, 134–136 (2016)
12. P.V. Chepur, A.A. Tarasenko, Fundamental research, **7-1**, 95-100 (2015)
13. P.V. Chepur, A.A. Tarasenko, Fundamental research, **12-1**, 97-102 (2015)
14. P.V. Chepur, A.A. Tarasenko, A.A. Gruchenkova, Fundamental research, **12-3**, 514-519 (2014)
15. S.V. Chirkov, A.A. Tarasenko, P.V. Chepur, Oil and Gas, **5**, 72-78 (2014)
16. *GOST R 53166. Earthquakes. Influence of environmental conditions appearing in nature on the technical products* (2009)
17. *Design regulations NP 031-01. Design regulations for seismic resistant nuclear power plants / Approved by GGTN RF of 19. 10.2001 № 9.*
18. *Regulatory guide RTM 108.020.37-81. Equipment and piping of nuclear power plants. Seismic analysis* (1981)
19. *Design code SP 14.13330.2014, Seismic Building Design Code*
20. A.A. Tarasenko, M.V. Sayapin, Oil and Gas, **1**, 52-56 (1999)
21. A.A. Tarasenko, P.V. Chepur, Fundamental research, **12-1**, 67-84 (2014)
22. P.V. Chepur, A.A. Tarasenko, Fundamental research, **2-8**, 1671-1675 (2015)

Value Justification of Safe Distances from Trunk-Lines by the Methods of Quantitative Risk Assessment

Anton Vengerov [1,*], *Vitaliy* Stroyenko [2], *Aleksey* Gladenko[3], and *Vladimir* Karagusov[3]

[1]Industrial University of Tyumen, 625001 Volodarskogo str. 38, Tyumen, Russia
[2]SIBUR, 016695 Krzhizhanovskogo st. 16/1, Moscow, Russia
[3]Omsk State Technical University, 639548 Mira st. 11, Omsk, Russia

Abstract. The paper discusses the methods of substantiation of safe distances from the pipelines to transport WFLH (Wide Fraction of Light Hydrocarbons) by the methods of quantitative risk assessment. Recommendations are given to maintain the level of probability of occurrence of an accident or failure, as well as their consequences within the given limits of the regulations. Viewed technique was tested on a specific example of the pipeline.

1 Introduction

Trunk-line system (TS), which is unique in extent, performance and security, is being operated and developed in Russia. The data on accidents and injuries indicate that the accident with loss of people`s life in the TS is quite a rare event. However, in the modern urban environment the risk of major industrial accidents involving loss of life increases due to approaching of TS to human locality, production facilities and transport infrastructure. The greatest risk is associated with the operation of the TS, pumping volatile liquid hydrocarbons (liquefied petroleum gases, unstable condensate, broad fraction of light hydrocarbons, with the accidental release of which clouds of the fuel-air mixtures can form that can drift under adverse conditions in the distance of several hundred meters while maintaining a fire capacity [1, 2].

To date, substantial experience in modeling of accidental releases, the calculation of the consequences of accidents at industrial sites and the use of risk analysis methodologies for the assessment of hazards of possible accidents is gained in the works of V.A. Akimova, M.V. Beschastnova, O.M. Ivantsova, A.M. Kozlitina, M.V. Lisanova, S.V. Ovcharov, G.E. Odisharia, A.S. Pecherkina, V.S. Safonov, V.I. Sidorova, S.I. Sumy, S.A. Timasheva, A.A. Shvyryaeva, Y.N. Shebeko, T. Kletz, F. Lees, K. Muhlbauer and other Russian and foreign researchers. Among the ways to ensure the safety of people from accidents the most famous and reliable one is removal of objects from the source of danger at a sufficient distance ("distance protection"). Under the threat of major industrial accidents with the group death of people the issue of establishing safe distances arises particularly acute. After a well-known example of the Ufa catastrophe of 1989 some changes were made in the SBRR (a

* Corresponding author: vengerov_tsogu@mail.ru

Set of Basic Requirements and Regulations) 2.05.06-85 * "Pipelines", which established the extremely pessimistic minimum distance from the LPG product lines. Actually the changes slowed down the design and construction of new objects for decades [3-5].

To establish minimum safety distances to adjacent structures and infrastructure is one of the most important problems of industrial and fire safety in the design and operation of the TS of LPG in connection with the national development plan and gas petrochemical Russia until 2030. The solution to this problem should be based on an analysis of the current state of accidents [6].

2 Research object (Model, Process, Device, Synthesis, Experimental part, etc.)

The wide fraction of light hydrocarbons (WFLH) is a product of processing of associated petroleum gas and gas condensate. It is a mixture of liquefied petroleum gas and heavier hydrocarbons (C2-C6 and above). In general, a typical WFLH composition is the following one: ethane from 2 to 5%; LPG C4-C5 fractions 40-85%; C6 hexane fraction from 15 to 30% and pentane fraction accounts for the remainder. WFLH is not a commercial product, so it is in any case transported to the processing and gas chemical plants for further processing. Russia has huge production facilities for processing of wide fraction of hydrocarbons, but they are located remotely from the fields (Table 1).

Table 1. The main production facilities of WFLH

Company	Date of foundation	Owned by	Raw material	Productivity on raw materials	Productivity on WFLH
Surgut PCS [2].	1984	GAZPROM	De-ethanized gas condensate	12 mln. tons per year	up to 3.3 mln. tons per year
Purovsky PCS [3]	2004	NOVATEK	De-ethanized gas condensate	11 mln. tons per year	3.3 mln. tons per year
Nizhnevartovsk GPP [4]	1974	SIBUR	Associated petroleum gas	16.2 billion m3 per year	up to 1,8 mln. tons per year
Belozerny GPP [4]	1978	SIBUR	Associated petroleum gas	4.2bln. m3 per year	up to 1.26 mln. tons per year
South Balyksky GPP [4]	1974	SIBUR	Associated petroleum gas	2.8 bln. m3 per year	up to 0.9 mln. tons per year
Vyngapurovsky GPP [4]	2011	SIBUR	Associated petroleum gas	2.4 bln. m3 per year	up to 0.64 mln. tons per year
Gubkinsky GPP [4]	1988	SIBUR	Associated petroleum gas	2.1 bln. m3 per year	up to 0.63 mln. tons per year

3 Causes and consequences of accidents on pipelines for WFLH.

According to regulatory documents [6], an accident on the pipeline is defined as the destruction of the pipeline, an uncontrolled explosion and (or) release of liquid or gaseous substances transported by pipeline. It is accompanied by one or more of the following events:

- inflammation of the transported substances, vapor or gas mixture;
- pollution of the water body in excess of the maximum permissible concentrations of harmful substances;
- leakage of transported liquid substance in an amount of 25 tons or more tons;

Classification of accidents and failures in pipelines for WFLH differs little from the traditional one for the majority of the objects of the main transport of hydrocarbons [7]. Among these accidents can be identified:

- depressurization of the pipeline with the spill of pumped medium;
- depressurization of the pipeline with the ignition of the pumped medium;
- depressurization of the pipeline with the explosion of the pumped medium.

The potential danger of accidents on pipelines for WFLH is compounded by the fact that they are prone to instant boiling up, and their clouds can be ignited over several kilometers. There were a lot of accidents on pipelines for petroleum products and volatile hydrocarbons in the history, which led to the loss of billions of dollars and hundreds of deaths. However, most of them took place in countries with warm climate (Kenya, Nairobi, Mexico). It contributed to the rapid ignition of the pumped product when mixed with air. However, in the summer period Western Siberia is hot enough and large seasonal temperature variations contribute to the accumulation of defects in the metal pipes (Table 2).

Table 2. The causes of accidents and failures in TS

Reasons of accidents	% of total
External corrosion	28.9
including SCC (Stress Corrosion Cracking)	22.5
Mechanical damage	19.0
Defect of construction and installation work	21.9
including defect in welding	13.0
Defects of pipes	11.4
Natural disasters	9.5

4 Methods

As with any object oil and gas pipelines for natural gas liquids are especially dangerous. For this reason, safe distances from the MT to the infrastructure must be particularly justified. As any object OF oil and gas industry pipelines for WFLH are especially dangerous. For this reason, safe distances from the ST to the infrastructure must be particularly justified.

However, at the same time there is no clear guidance on the definition of a safe distance from the pipeline for WFLH to industrial and civil objects. Recommendations of SBRR 2.05.06-85 are limited to specific values of minimum distances depending on the group WFLH, pipe diameter and type of object. The specified standards were accepted more than 30 years ago, the last update was made in 2010, but the reason to consider their legacy lies not only in this fact [8,9, etc].

According to the latest revision of standards, WFLH include mixtures containing 5.3% of ethane, whereas in most cases mixtures, which contain10-15% of ethane fraction, are transported through pipelines. The next discrepancy is the result of the first. A large proportion of ethane in the mixture causes a phase of high pressure of its saturated vapor, which in turn creates a need to reduce the inter-station distances in order to prevent the formation of two-phase flow. It is clear that it is possible to achieve it by increasing the number of pumping stations. Thus, there is another factor that affects the safety and the cost of pumping gas liquids.

In addition, with vigorous vaporization arising due to depressurization of the pipeline, there is a significant drop in temperature of the pumped liquid, which entails cooling of the pipe wall. Numerical modeling of lowering temperature of WFLH in the place of pipeline rupture site showed that its value can reach - 45 ° C, resulting in brittle fracture of the pipeline. This problem can be solved by hardening metal pipes alloying elements, but operating specifications does not provide it.

In addition to the actions listed above, SBRR 2.05.06-85 has a number of other inconsistencies in the section relating to the design and operation of pipelines for mixtures of light hydrocarbons. Judging by the aove mentioned information we can state that the reporting guidance document is obsolete and requires new approaches to the formation of rules and recommendations for pipelines that transport unstable mixture of hydrocarbons.

An accident on the main pipelines near Ufa in June 4, 1989 was the most significant one in Russia. There were two passenger trains in the area of the explosion, nearly 600 people were killed and the same number of people was injured. After this incident SBRR 2.05.06-85 * recommended safe distances between populated areas and pipelines were revised in the direction of a significant increase [10].

Some suggestions were made repeatedly and insistently that when designing modern pipelines for WFLH one should turn to the experience of leading foreign companies engaged in the construction and operation of the main product lines. However, analysis of foreign documents indicates that in some countries there are no statutory rules of safe location of pipeline transport facilities on the items with large concentrations of people.

It is possible to rate a safe distance from the water supply to residential or industrial facility by a number of ways.

The analysis of the experience occurred accidents and failures is the most common method. This method is partially reflected in the SBRR 2.05.06-85 *. Sufficiently indicative statistics shows that the size of the affected areas in case of accidents on pipelines for natural gas liquids ranges from 100 to 350 meters away from the axis of the pipeline (thermal radiation, scattering fragments). These boundaries in the first approximation can be determined by diameter and the pressure in the pipeline. This approach does not require the use of any safety factors. However, WFLH ability to fire at a sufficiently large distance from the accident site, requires the repeated increase of these distances.

The second method is based on analyzing the consequences of the maximum hypothetical accident involves the total destruction of the pipeline with the pipeline considering the specific characteristics (geometric parameters, installation conditions, physical properties of the pumped medium, etc.). In this case, the safety factors are implicitly laid in various conventions .

The third method is based on a quantitative assessment of the risk of an accident. The advantage of this method is that one of its principles is the very definition of the probability of an accident or failure, including the maximum. The method allows to evaluate all the accident scenarios in a specific section of the pipeline, the leakage through a defect of any size, as well as all the options on the basis of the outcome of an event tree that allows the fate of the maximum number of factors of impact on the pipeline. The main purpose of risk assessment is to present on the basis of objective evidence the information needed to make informed decisions about risk treatment methods.

Among many methods recommended for the use of quantitative risk assessment the analysis of the causes and consequences [8] is the most appropriate.

The method of analysis of the causes and consequences was originally developed as testing tool system reliability, critical for security, which was used for a more complete understanding of the failures of the system. As well as fault tree analysis method, this method is used to display the failure logic leading to critical events, however, in addition to the functionality of the fault tree, this method allows to analyze the sequence of failures. The method also allows you to take into account the time lag in the analysis of the consequences it is not possible when using the event tree method [9].

The method is used to analyze the different options of the system after the occurrence of a critical event, depending on the behavior of its subsystems (for example, alarm systems). If such options may be described quantitatively, they can be estimated probabilities of possible consequences of critical events.

5 Results and Discussion

As an example of this study we present the results of the risk assessment of accidents on the main reconstructed pipelines of WFLH with diameter of DN = 500 mm [2]. On the route of the product pipeline, there are numerous (over 50 seats) approach to human settlements, production facilities, railways and highways, including deviations from previously existing p. 3.16 of SBRR 2.05.06-85*.

Calculating the weight of leaks in emergency depressurization product pipeline [10-15] (the accident in question with four possible variants of depressurization of the pipeline, "fistula", small, medium cracks and guillotine break) and assessing areas the size of the damaging factors, was built potential risk field, its fragment is shown in Figure 1.

The calculations have shown that the potential level of risk fatalities 10-6 year-1 is achieved at a distance of 50-100 m from the pipeline and only on certain sections of the route; Level 10-7 year-1 - 250-350 m; 10-8 year-1 - 500-700 m.

The effectiveness of technical solutions to ensure the safety of the analyzed pipeline (frequency partition sections with stop valves, the use of modern systems of leak detection, device of protective shafts, relining and stress testing of pipes in the places od dangerous approach , and others.) can also be confirmed by quantitative risk assessment.

For example, the distribution of the homogenized mass leak of WFLH product pipeline around the track before and after the installation of additional valves is shown in Figure1.

The figure shows that the installation of additional valves leads to a significant reduction in leakage volume in some parts of 2-2.5 times.

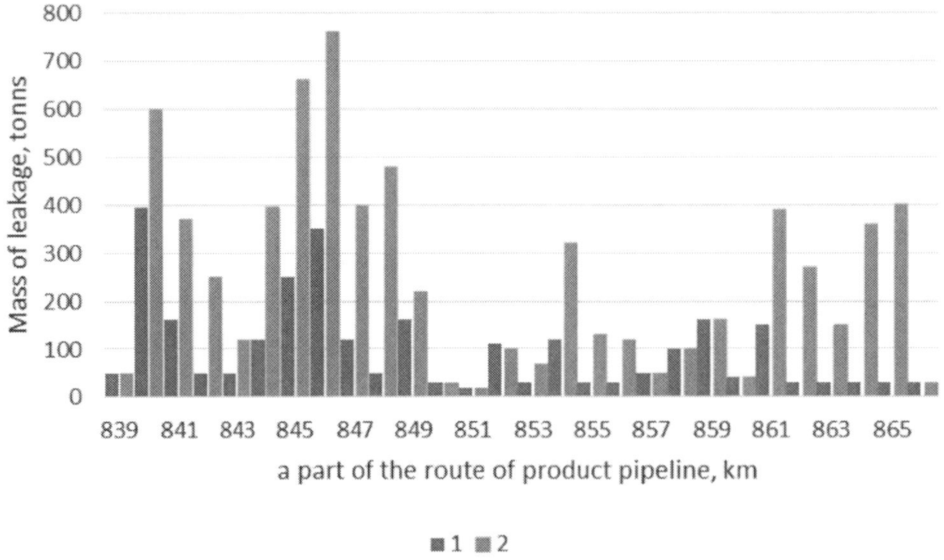

Fig. 1. Distribution of averaged scenarios of mass WFLH product pipeline leak on the highway in the area 839-867 km, 1 and 2 - respectively, before and after the installation of additional valves.

6 Conclusion

The calculations showed that the expected environmental damage caused by oil entering the water bodies, calculated as described in [8], corresponds to the "average" level of risk that may be accepted as admissible on the hard security control condition according to the standard approaches to setting criteria for acceptable risk (for example, the area called according to [16-18] "strict control zone", for which an evaluation of the appropriateness of risk mitigation measures).

In order to reduce risk to the lowest necessary to provide compensatory measures that do not allow the oil to reach the water bodies, i.e. construction of engineering structures and the use of protective covers [19, 20].

As a result of the quantitative risk analysis of the considered the main oil pipeline there were calculated people impact area under various scenarios of oil spills, dispersion and ignition of oil vapor. We analyzed the impact of different design decisions mentioned in [21-23] on risk indicators, including an increase in wall thickness, spacer "pipe in pipe", the construction of dams and other events.

Thus, as a result of the quantitative risk analysis methodology developed and tested to justify a safe distance from the main pipelines to settlements, industrial and agricultural facilities, taking into account the possible design solutions.

References

1. Federal Law Technical Regulations on the safety of buildings and structures from 30.12.2009 N 384-FZ, (2009)
2. S. Sumy, A. Pchelnikov, E. Savin, Occupational safety in the industry, **2**, 48-52 (2007)
3. I. Mazur, O. Ivantsov, *Security of piping systems* (Eslim Publishing, Moscow, 2004)
4. M. Lisanov, S. Sumy, A. Savina, E. Savin, Occupational safety in the industry, **3**, 51-59 (2010)
5. *Methodological guide for risk assessment of accidents on the main oil pipelines,* Ser. 27, V. 1 (Industrial safety", Moscow, 2005)
6. *Declaration of fire safety and fire risk assessment: Coll. Doc. Ser. 19. V. 2. Parts 1-4* (ZAO STC, Moscow, 2009)
7. A. Agapov, I. Lazukin, A. Maruhlenko, Occupational safety in the industry, **1**, 46-52 (2010)
8. The methodology of calculation of the damages caused to water bodies due to violations of water legislation (2009)
9. Federal Law of July 22, 2008. № 123-FZ "Technical Regulations on fire safety requirements" Meeting of the legislation of the Russian Federation (2008)
10. RD 03-418-01. "Guidelines for the risk analysis of hazardous production facilities" approved (Gosgortechnadzor of Russia, 2001).
11. SP 11-113-2002
12. V. Kurushina, Yu. Zemenkov, WIT Transactions on Ecology and the Environment, **190(2)**, 881-888 (2014)
13. R. Levitin, Yu. Zemenkov, Neftyanoe khozyaystvo - Oil Industry, **1**, 110-114 (2016)
14. E. Kurushina, V. Kurushina, Life Science Journal, **11(11)**, 517-521 (2014)
15. Yu. Zemenkov, V. Shalay, M. Zemenkova, Procedia Engineering, **113**, 254-258 (2015)
16. Yu. Zemenkov, V. Shalay, M. Zemenkova, Procedia Engineering, **113**, 312-315 (2015)
17. V. Antipyev, Yu. Zemenkov, N. Kudryavtseva, *A method of calculation of losses of oil at accidents on the main oil pipelines* (UMN Z and SZS, Tyumen, 1985)
18. E. Kurushina, V. Kurushina, Life Science Journal, **11(11)**, 517-521 (2014)
19. R. Mamadaliev, V. Kuskov, Yu. Zemenkov, A. Popova, Applied Mechanics and Materials, **770**, 19-22 (2015)

20. A. Brand, S. Vershinina, A. Vengerov, N. Mostovaya, IOP Conf. Ser.: Mater. Sci. Eng., **93**, 120-125 (2015)
21. E. Kabes, K. Voronin, A. Brand, Sibresurs, (2014) 37
22. I. Kiryanov, S. Toropov, V. Toropov, Neftyanoe Khozyaystvo - Oil Industry, **8**, 48 (2015)
23. A. Vengerov, A. Brand, Proceedings of the International Scientific and Technical Conference, TSOGU, Tyumen, 36-40 (2015)

Upgraded Algorithm for Calculating the Turbo-Expander of Gas Distribution Stations

Mikhail Chekardovskiy [1], *Sergey* Chekardovskiy [1,*], *Konstantin* Ilyukhin [1], and *Alexey* Gladenko[1]

[1]Industrial University of Tyumen, 625001 Volodarskogo str. 38, Tyumen, Russia

Abstract. The article deals with the urgency of adapting computational turbo-expander unit parameters techniques to the conditions of their application at gas distribution stations. Existing computational methods based on the use of air as the working medium yield incorrect data to determine the actual design parameters for operating conditions where the working medium is natural gas. A modernized algorithm of thermogasdynamic calculation of turbo-expanders in order to form the correct initial data for design calculations has been proposed. The objective of calculating turbo-expanders is to identify thermogasdynamic parameters and dimensions of the flow channel, rotational speed, and shaft power. Procedure of thermogasdynamic calculations is shown on the example of a turbo-expander running on natural gas. The result will simplify the process of selecting or designing turbo-expander units for gas distribution stations.

1 Introduction

Currently, obtaining electrical energy using resource-saving, environmentally friendly technologies becomes relevant. One of these areas is using the potential energy of high-pressure natural gas from main gas pipelines with the use of turbo-expander units (TEU). It is known that at gas distribution stations (GDS) and gas control units (GCU) natural gas is reduced. In this case, the potential energy of the compressed gas is lost permanently which can be used to generate electricity and cooling.

For several decades, many European countries (Italy, Germany, etc.) have successfully been using this technology, installing TEU at GDS and GCU pipelines to reduce the gas pressure to the desired pressure for the consumer, both serving as GDS and GCU and simultaneously generating electricity. Wherein the gas is not burned but is used only as a working medium, moving further to the consumer. Thereafter, fuel combustion products do not pollute the environment.

2 Research

The process of gas expansion in the TEU is accompanied by an increase in its temperature. Therefore, in order to prevent hydrate formation, as well as to meet the SNIP temperature standards, at the TEU input gas heating is provided.

* Corresponding author: ldgtd@mail.ru

When designing the TEU at the GDS located in the energy-deficient regions, it is advisable to build a gas turbine power plant (GTP) next to the GDS. Combining a turbo-expander with a heat exchanger of the power plant gas turbine engine which is intended for heating the natural gas at the inlet of the TEU will allow the production of electricity by this complex with a performance coefficient from 65 to 75 % [1, 7, 8]. Therefore, studies aimed at improving the TEU design techniques at GDS are connected with Energy Saving Strategy of Russia and are relevant.

The objective of calculating turbo-expanders is to identify thermogasdynamic parameters and dimensions of the flow channel, rotational speed, shaft power. Procedure of thermogasdynamic calculations is shown on the example of a turbo-expander running on natural gas.

Turbo-expanders of gas distribution stations (GDS) are used to generate electricity by natural gas differential pressure on the turbine stage. Approximate composition and characteristics of the components of natural gas transported through the main gas pipeline and entering the GDS are presented in Table 1.

Table 1. Composition and characteristics of the natural gas components.

Characteristics	**Components**						
	CH4	C2H6	C3H8	C4H10	CO2	H2S	N2
Molar concentration %, ri	98.0	0.4	0.2	-	0.1	-	1.3
Molecular weight of the components, μi, kg/kmol	16.04	30.07	44.09	-	44.0 2	-	28.00
Molecular weight of the mixture	$\mu_m = \frac{1}{100} \cdot \sum r_i \cdot \mu_i = 16{,}34$ kg/kmol						
Gas constant	$R = \frac{\bar{R}}{\mu_m} = 0.509,$ where $\bar{R} = 8.314$ kJ/kg·K						
Density of gas under normal conditions (st)	$\rho_s = \frac{P_s}{R \cdot T_s} = \frac{101{,}33}{0{,}509 \cdot 273} = 0.7292$ kg/m3						

Table 2 shows sample GDS initial data.

Table 2. Main GDS initial data.

Title	**Value**		**Note**
Gas pressure at the inlet	PI = 45 kg/cm	4410 kPa	Density of gas under normal conditions ρst = 0,73kg/m3
Gas pressure at the outlet	PO = 11.9 kg/cm2	1166.2 kPa	
Gas flow rate through the GDS	V=340 thousand n·m3/h	M=69.01 kg/s	
Gas temperature at the inlet	tI=35°C	TI=308 K	
Index of adiabatic gas expansion	Kg =1.2		

Using the initial data in Tables 1 and 2, we made an upgraded algorithm for thermogasdynamic calculation of a turbo-expander in order to form correct input data for design calculations:

1. Gas temperature at the GDS outlet [2, 9, 10], K:

$$T_{OUT} = T_{IN} / \left(\frac{P_{IN}}{P_{OUT}} \right)^{\frac{K_r - 1}{Kr}} \quad (1)$$

2. Average heat capacity of natural gas [3], kJ/kg·K:

$$C_{Pg_1} = \left(0{,}66 + 0{,}34 \cdot \frac{r_{CH_4}}{100} \right) \cdot \left[(0{,}00344 - 0{,}00009 \cdot P_{IN}) \cdot (\frac{t_{IN} + t_{OUT}}{2}) + 0{,}011 \cdot P_{IN} + 2{,}06 \right] \quad (2)$$

3. Available capacity of the GDS, kW:

$$N_{GDS} = M \cdot C_{Pг1} \cdot (T_{IN} - T_{OUT}) \quad (3)$$

As an example, table provides data on the TEU.

Table 3. Initial data on the turbo-expander.

Title	Value		Note
Gas pressure at the inlet	P1 =40 kg/cm2	3920 kPa	Density of gas under normal conditions ρst=0,73 kg/m3
Gas pressure at the outlet	P2=9 kg/cm2	882 kPa	
Gas flow rate	130 thousand n·m3/h	26,386 kg/s	
Gas temperature at the inlet	t1=90 °C	T1=363 K	
Gas temperature at the outlet	t2 = 10 °C	T2 = 283 K	

4. Average heat capacity of natural gas [3, 11, 12], kJ/kg·K:

$$C_{Pg_2} = \left(0{,}66 + 0{,}34 \cdot \frac{r_{CH_4}}{100} \right) \cdot \left[(0{,}00344 - 0{,}00009 \cdot P_1) \cdot (\frac{t_1 + t_2}{2}) + 0{,}011 \cdot P_1 + 2{,}06 \right] \quad (4)$$

5. Available capacity of the GDS, kW:

$$N_{TD} = M \cdot C_{Pг} \cdot (T_1 - T_2) \quad (5)$$

6. Joule-Thomson coefficient, which is a function of pressure, temperature and gas composition, kJ/kg•MPa [4, 13, 14]:

$$(C_P D_i)_m = 10 \cdot \left(1{,}98 - \frac{r_{CH_4}}{100} \right) \cdot \left[(0{,}0000012 \cdot t_1^2 - 0{,}000135 \cdot t_1 + 0{,}00298) \cdot P_A - 0{,}00463 \cdot t_1 + 1{,}119 \right] (6)$$

7. Enthalpy change considering the Joule-Thomson coefficient, kJ / kg:

$$\Delta h_s = C_{Pг} \cdot (T_1 - T_2) - (C_p D_i)_m \cdot (P_1 - P_2) \quad (7)$$

8. Actual capacity of the TEU, kW:

$$N_{TD\,A} = M \cdot \Delta h_s \quad (8)$$

To realize the available capacity of the GDS it is expedient to install two turbo-expanders with a capacity of 4000 kW each.

Let us specify the natural gas parameters by the technique formulas developed by "Gazprom VNIIGAZ", Moscow:

9. Density of gas under standard conditions, kg/m3:

$$\rho_{ST} = \frac{P_{ST}}{R \cdot T_{ST}} \quad (9)$$

where: standard pressure P_{ST} = 101.33 kPa and temperature T_{ST} = 293 K

10. Critical gas pressure, MPa:

$$P_{CR} = 0{,}1773(26{,}831-\rho_{ST}) \quad (10)$$

11. Critical gas temperature, K:

$$T_{CR} = 156{,}24(0{,}564+\rho_{ST}) \quad (11)$$

12. Reduced value of gas pressure:

$$P_R = (P_1 + P_2)/2P_{CR} \quad (12)$$

13. Reduced value of gas temperature:

$$T_R = (T_1 + T_2)/2T_{CR} \quad (13)$$

14. Let us find τ [5, 15, 16]:

$$\tau = 1-1{,}68T_R+0{,}78T^2_R+0{,}0107T^3_R \quad (14)$$

15. Let us define the gas compressibility factor, in a case if a TEU worked as centrifugal gas blower (that is, compressed gas) [6, 17, 18 etc]:

$$Z = 1 - 0{,}0241\frac{P_R}{\tau} \quad (15)$$

16. Available potential work of a turbo-expander, kJ/kg:

$$\omega = Z \cdot R \cdot \left(\frac{T_1 + T_2}{2}\right) \cdot \ln\frac{P_1}{P_2} \quad (16)$$

17. Available potential capacity of a TEU, kW:

$$N_{GDS} = M \cdot \omega \quad (17)$$

18. Polytropic performance efficiency of a turbo-expander [19, 20]:

$$\eta_P = \frac{\Delta h}{\omega} \quad (18)$$

Sample calculation results are summarized in Table 4.

3 Conclusion

Application of a TEU at the GDS produces electricity due to the pressure difference at the input and output of the unit.

Table 4. Calculation results.

№	Parameter symbol	Unit of measurement	Value
1	TOUT	K	246.76
2	Cpg1	kJ/kg·K	2.12
3	NGDS	kW	8877.7
4	Cpg2	kJ /kg·K	2.235
5	NTD	kW	4717.6
6	CpDi	kJ/ kg•MPa	7.036
7	Δhs	kJ /kg	154.12
8	NTDA	kW	4067
9	ρST	kg/m3	0.679
10	PCR	MPa	4.636
11	TCR	K	194.5
12	PR	-	0.518
13	TR	-	1.661
14	τ	-	0.412
15	z	-	0.97
16	ω	kJ /kg	237.78
17	NGDS	kW	6274
18	ηp	-	0.66

This method is adapted for natural gas, since existing techniques are based on calculations for air. The adaptation is made taking into account the heat drop in the TEU; gas density; the degree of expansion as inverse of the compression rate.

Modernization of the existing methods allows generating the correct input data for design calculations of TEU. Calculated by the developed technique thermogasdynamic and design installation parameters for a specific GDS will allow to find the optimal number of existing imported TEU or produce them in Russia, providing jobs and increasing import substitution of the units.

References

1. A. Voronin, O. Malhanov, Energy saving and water treatment, **6**, (2005).
2. M. Chekardovsky, *Methodology for monitoring and diagnostics of the heating system energy equipment* (Nedra, SPb, 2001)
3. G. Krylov, A. Matveev, O. Stepanov, E. Yakovlev, *Operation of gas pipelines in Western Siberia* (Nedra, Leningrad, 1985)
4. O. Stepanov, M. Chekardovsky, S. Chekardovsky, Development of a comprehensive system of control, diagnosis, prediction of technical state of equipment, Scientific report inventory № 02.20. 00391, registration № 01.20. 0001269. TyumGASA, Tyumen, (2000)
5. A. Shekhtman, *Gas-dynamic functions of real gases: Reference book* (Energoatomizdat, Moscow, 1988)
6. V. Gurov, L. Lebedeva, G. Podvidz, E. Shcherbakova, Works of CIAM, **1296**, 259-274 (1996)
7. K. Ilyukhin, B. Moiseev, M. Chekardovskiy, N. Nalobin, S. Chekardovskiy, *Energy technology complexes in designing and operating equipment in heat and gas supply systems* (TyumGASU, Tyumen, 2015)

8. M. Chekardovskiy, K. Ilyukhin, A. Shapoval, S. Chekardovskiy, Upgraded algorithm for calculating the turbo - expanders gas distribution stations, Nauchnoe obozrenie, **12**, 128-134 (2015)
9. M. Malkov, *Handbook of Physical-technical fundamentals of cryogenics* (Energoatomizdat, Moscow, 1985)
10. A. Davyidov, G. Perestoronin, V. Stulov, A. Sherstyuk, *The centripetal turbine expanders* (Kolos-Press, Moscow, 2002)
11. V. I. Epifanova, *Low temperature radial turbine expanders* (Mashinostroenie, Moscow, 1994)
12. S. Mikulin, A. Arharov, *Technique of low temperatures* (Energy, Moscow, 1995)
13. V. Gurov, New generation Turboexpanders, Conversion in engineering, **4**, 33-35 (2006)
14. V. Kirilin, V. Syichev, A. Sheydlin, *Engineering thermodynamics* (Energoizdat, Moscow, 1983)
15. A. Shehtman, *Gas-dynamic function of real gases* (Energoatomizdat, Moscow, 1988)
16. E. Akoev, Conversion in engineering, **5**, 32-39 (1997)
17. Yu. Zemenkov, V. Shalay, M. Zemenkova, Procedia Engineering, **113**, 254-258 (2015)
18. Yu. Zemenkov, V. Shalay, M. Zemenkova, Procedia Engineering, **113**, 312-315 (2015)
19. V. Kurushina, Y. Zemenkov, WIT Transactions on Ecology and the Environment, 190, **2**, 881-888 (2014)
20. V. Antip'ev, A. Nevolin, Yu. Zemenkov, Neftyanoe Khozyaistvo - Oil Industry, **10**, 46-48 (1981)
21. A. Palagin, *Computer-aided design thermal schemes turbines* (Naukova Dumka, Kiev, 1983)
22. A. Stepanets, *Energy Turbine expanders* (Nedra-Businesscenter, Moscow, 1999)
23. A. Boyko, Yu. Govoruschenko, S. Ershov, A. Rusanov, *Aerodynamic calculation and optimal design of a flowing part of turbomachinery* (KhPI, Kharkov, 2002)
24. D. Anderson, Dzh. Tannehil, R. Pletcher, *Computational Fluid Mechanics and Heat Transfer* (Mir, Moscow, 1990)

Increase of a Roadway Covering Durability by Using the Cement-Concrete Base Fragmented with the Geogrid

Sergey Sannikov[1,*], *Sergey* Kuyukov[1], and *Denis* Kubasov[1]

[1]Industrial University of Tyumen, 625001 Volodarskogo str. 38, Tyumen, Russia

Abstract. Presents the results of studies of innovative materials in the field of in road construction. The paper presents an alternative method of increasing the cracking resistance of the roads asphalt-concrete pavement, constructed on the cement-concrete base, due to its fragmentation with the volumetric plastic geogrid while constructing. Theoretical, laboratory and field experimental studies of this design were conducted, as well as the effectiveness of the proposed solution was proved. The use of this design can improve the durability of the roadway coverings and reduce the costs for the roads repair and maintenance.

1 Introduction

In recent years, Russia has significantly increased and altered the vehicle stock, which is associated with the vehicles affordability due to the extended car loans and a variety of purpose-oriented programs. Traffic flow now includes trucks of not only domestic, but of foreign production, which are characterized by a large carrying capacity and significant axial loads. The growth and the composition change of the traffic flow means that there is a necessity to work out the new approaches to the construction and design of the roadway coverings. It also means there is a need to use the new high-strength and durable materials, including cement-concrete pavements and bases.

Abundance of the advanced materials, including synthetic ones, allows to design fundamentally new roadway coverings and road base structures, which helps to reduce the costs for their construction, repair and maintenance, to improve the properties of the traditional materials, and to lengthen the periods between the repair works [1- 5].

Asphalt-concrete is the most common material in the world, and particularly in Russiaб for constructing pavements. According to the data of March 13, 2013, the share of federal roads with asphalt-concrete pavement in Russia is 90.8% [6]. Asphalt-concrete is used due to its positive properties and a relatively simple construction technology.

* Corresponding author: tgasu.sannikov.sergey@mail.ru

DOI: 10.1051/matecconf/20167301021

An extensive range of materials is used to construct a roadway covering base. The choice of the material depends on a number of factors, in particular, on the region of construction and the availability of supply.

In the course of the roadway covering construction, one of the best construction solutions is the structure with the concrete base and asphalt-concrete pavement, which is of high strength and durability. This structure is referred to the roadway coverings of a rigid type [7].

Roadway coverings of a rigid type with the asphalt-concrete pavement and cement-concrete base have approved themselves in rough climatic conditions and on roads with the high traffic density.

Usage of cement-concrete base allows to:

- reduce the need for imported rock material by reducing the thickness of the layer;
- reduce the rutting process;
- reduce the pressure of the vehicle stock on the underlying layers due to the load distribution over a larger area.

Despite the advantages of these roadway coverings, there is also a significant disadvantage - the formation of cracks in the pavement. The cause of this problem lies in the different thermal and physical properties of the base and pavement materials, and also in the absence of the design and construction practice of the structural and technological solutions to reducing the cracking [8, 9].

There are various ways to increase the cracking resistance of the asphalt-concrete pavement in order to lengthen the periods between the repair works [1 - 5]. These include: reinforcement of the asphalt-concrete pavement or asphalt-concrete mix; the use of modified bitumen-based asphalt concrete; constructing the pavements of increased thickness at a single lift; construction of the expansion joints in the pavement, etc. [10 - 17]. These methods have their own significant drawbacks [10, 18].

2 Subjects and methods

In order to increase the cracking resistance of the pavement, the alternative method is suggested – the use of the small-section concrete base. Such base can be obtained in the process of construction by fragmentation of the cement-concrete layer with the volumetric plastic grid (Figure 1).

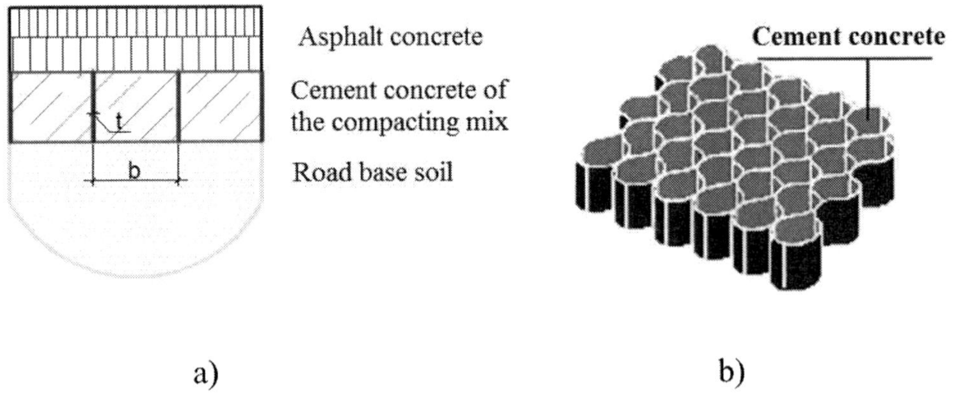

Fig.1. Roadway covering structure with the cement concrete, fragmented with the plastic geogrid:
a) structural scheme of the roadway covering:
b – geocell grid spacing; t – geocell wall thickness;
b) geogrid in expanded form, filled with cement-concrete mix.

The use of the geogrid can reduce thermal stresses in the concrete slab, thus the grid walls, involved in the joint slab work, take up the linear deformations caused by the temperature fluctuations.

Theoretical, laboratory and field experimental studies of the suggested method were conducted at the Highways and Airfields Department of TSUACE. The obtained results demonstrated the effectiveness of the structure.

3 Results

During the research, the theoretical dependences were obtained, which allow to assess the changes of the deformation properties of the concrete slab with the volumetric geogrid [19].

The thickness of the asphalt-cement pavement and concrete base is determined by the condition of tensile strength in bending under the influence of the external loads and temperature fluctuations. [7]

After solving the plane elastic problem, it was found that the final thermal stresses in the concrete slab (fragmented with the grid) depend on the geogrid parameters, in particular, the wall thickness and cells size [19] (Fig. 2).

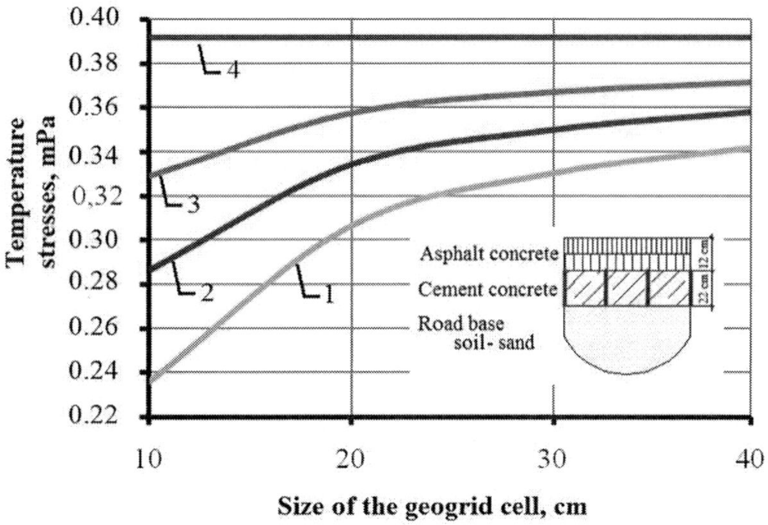

Fig. 2. Concrete slab thermal stresses dependence on the geogrid parameters with the geocell wall thickness (t):
1 – t=2.0 mm; 2 – t=1.35 mm; 3 – t=0.8 mm; 4 – without grid

Thermal stresses reduce by increasing the grid wall thickness and reducing the cell size. Thus, durability coefficient increases, and accordingly, the probability of cracking of the asphalt-concrete pavement on the cement-concrete base reduces.

When the roadway covering is full-strength, the base fragmentation allows to reduce the thickness of the structural layers.

To prove the theoretical results, a series of experiments were set up in the Department laboratories [20].

Initially, cement-concrete slabs of B12.5 grade were made. One slab was fragmented with the plastic grid, and the second one was made with no grid. Then, the slabs were placed into a freezing chamber for the experimental studies.

The purpose of the first stage of the experiment was to determine the stress in the concrete slab arising from changes in ambient temperature. Over thirty observations had been made, each of which was accompanied by lowering the temperature in the freezer down to -170C, followed by disconnecting the chamber and raising the temperature up to + 180C.

To determine the stresses arising from the temperature changes, PL-60 resistance strain gages were used. The resistance strain gages were thin film based. These thin films were attached to the side face of the cement concrete. To determine the temperature, the electronic temperature sensors were used. The "ITC-03p" device was used to determine the stresses while lowering and raising the temperature in the freezing chamber.

The experiment results showed that stress in the concrete slab fragmented with the plastic grid is lower than the one in the slab without the grid. On average, the stress became 1.6 times lower (Fig. 3). This result matches the theoretical values obtained earlier.

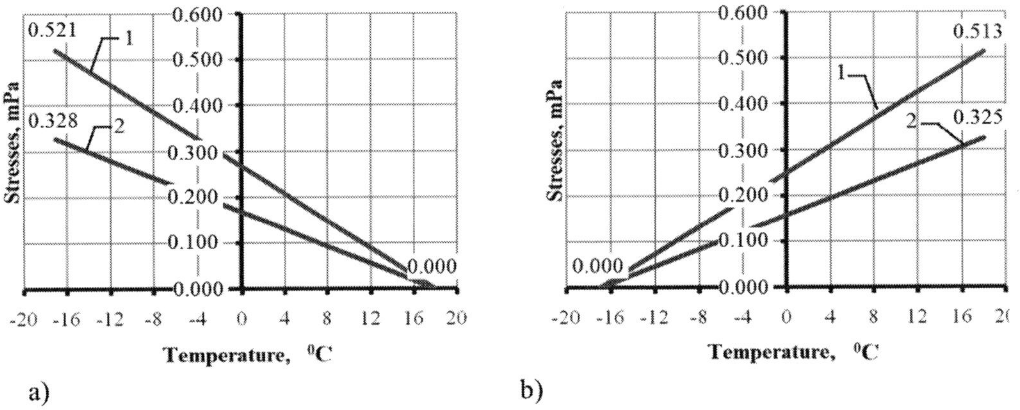

Fig. 3. The measurement results of the thermal stresses in the concrete slab: a) when freezing (transition from +180C to -170C); b) when thawing (transition from -170C to 180C). 1 – without fragmentation; 2 – with fragmentation.

Later on, an experiment was conducted to determine the stresses in the asphalt-concrete layer on the cement-concrete base. To determine the stresses arising from the temperature change, PL-30 resistance strain gages were used.

The measurement results showed that the stress in asphalt-concrete layer on the fragmented cement-concrete slab became 1.3 times lower. (Fig. 4).

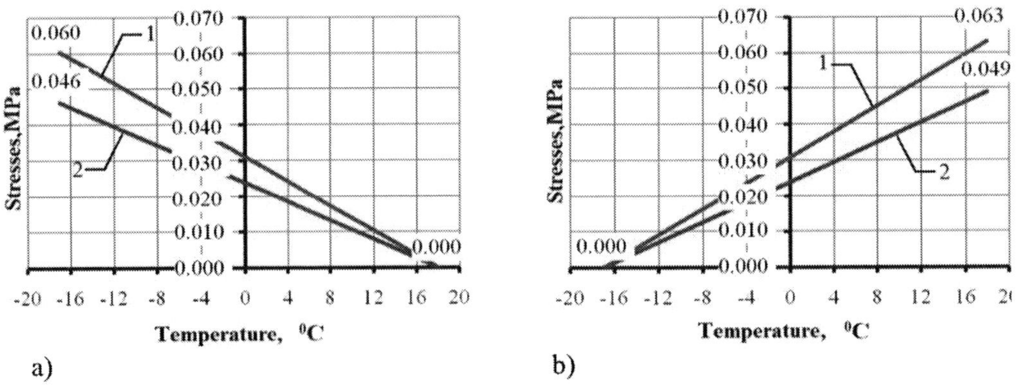

Fig. 4. The measurement results of the thermal stresses in the asphalt-concrete pavement: a) when freezing (transition from +180C to -170C); b) when thawing (transition from -170C to 180C). 1 – without fragmentation; 2 – with fragmentation.

 DOI: 10.1051/matecconf/20167301021

To prove the theoretical calculations and laboratory tests, a construction solution for the concrete base fragmentation was reflected in the working project "Overhaul of the roads with constructing the asphalt-concrete covering in the Leninsky administrative district of Tyumen".

The project passed the State examination and received positive findings.

In summer of 2014, in accordance with the project, the trial lot was constructed in Pyatnitskaya and Venskaya streets in Tyumen.

The roadway covering base was made of cement concrete of B15 grade and with the thickness of 18 cm. The base layer was fragmented with the volumetric perforated geogrid of 15 cm high, and with the protective layer of 3 cm thick. The pavement was constructed of the dense asphalt concrete (mark II) of 6 cm thick.

At the same time, a similar lot with the same structure of a roadway covering was built, but without fragmentation of the concrete base.

Construction included several phases: preparatory works, installation of the volumetric plastic geogrid, construction of the cement-concrete base followed by the compression and construction of the asphalt-concrete pavement (Fig. 5) [21].

Fig. 5. Construction of the trial lot:a) installation of the volumetric plastic geogrid; b) filling the cells with the cement-concrete mix; c) distribution of the mix; d) shaping the base surface;e) compression the base; f) construction of the asphalt-concrete pavement.

Seven days after the constructing completion, for the monitoring of the trial lot, the PL-30 resistance strain gages were attached to the side faces of the asphalt-concrete layer.

Measurements of the stresses in the structure of the roadway covering had been taken during 4 months from July to October, 2014.

Analysis of the results showed that at the lot where the concrete base was fragmented with geogrid, the stresses occurring on the surface of the side faces of asphalt-concrete pavement were 1.33 times lower.

In early spring of 2015, the occurred cracks in the asphalt-concrete pavement were assessed on the trial lot.

The results showed that the number and opening width of cracks is lower in the lot with the fragmented cement-concrete base than in a similar lot without fragmentation (Fig. 6).

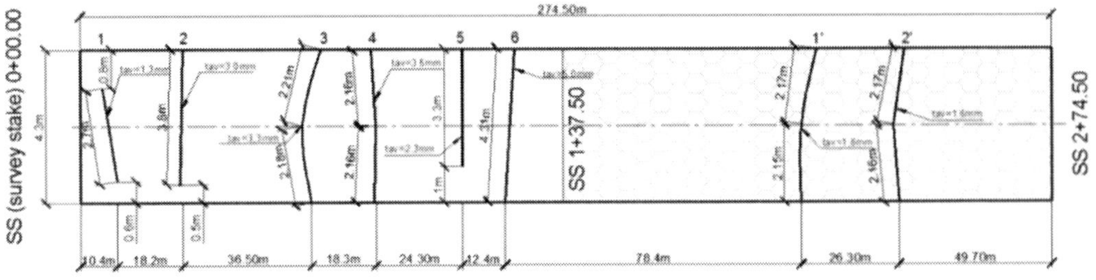

Fig. 6. Scheme of the opening of cracks in the trial lot (lot plan): tav - the average opening width of cracks in mm.

The total extent of cracks in the lot with fragmentation was 8.65 meters with an average opening width of 1.6 mm. In the same lot without fragmentation, the total length of cracks was almost 2.5 times bigger, that is 22.82 m, with an average opening width of 3.0 mm.

Economic efficiency of the suggested solutions was measured in accordance with the reduced total (discounted) costs for: construction, maintenance, repair and overhaul of the roads sections with the fragmented and non-fragmented cement-concrete base. Total costs for the roadway covering with fragmentation is by 13% lower than without fragmentation (Fig. 7).

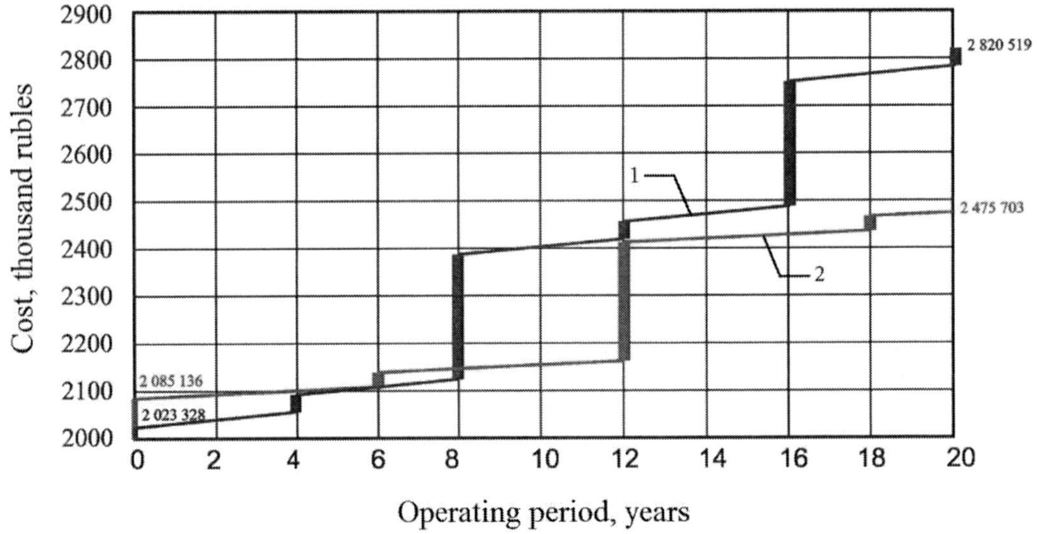

Fig. 7. Diagram of the reduced total costs for the structure of the roadway covering:
1 – without fragmentation; 2 – with fragmentation.

4 Conclusion

After analyzing the results of the theoretical and laboratory research, monitoring the trial lot, one can draw the following conclusions:

- ensuring crack resistance of roadway structures made of cement concrete base and asphalt-concrete pavement is an urgent task at designing and construction;

- method of increasing of crack resistance of road pavements by using small-size cement-concrete blocks for roadway base was proposed;
- theoretical dependencies which allows to calculate the proposed pavement structure were received;
- laboratory experiments which confirms the theoretical studies have been done;
- the proposed method of increasing of crack resistance has been tested at construction of experimental site.

It's found that:

- fragmentation of the cement-concrete base reduces the thermal stresses occurring in the structural layers of the roadway covering by 1.3-1.6 times;
- the use of the small-section concrete base increases the cracking resistance of the asphalt-concrete pavement by 2-3 times;
- in general, the use of the fragmented cement-concrete base, will allow to increase the periods between the repair of the roadway covering by 1.5 times, as well as it will allow to reduce the operating costs for repair and maintenance of roads by 10-20%.

References

1. V.V. Sirotyuk, E.Yu. Krasheninin, Vestnik Sibirskoy gosudarstvennoy avtomobil'no-dorozhnoy akademii, **7,** 31-38 (2008)
2. O.Yu. Moskalev, N.E. Kokodeeva, L.V. Yankovskiy, Transport. Transportnye sooruzheniya. Ekologiya, **2(40)**, 69–78 (2011)
3. V.V. Sirotyuk, G.M. Levashov, Vestnik Sibirskoy gosudarstvennoy avtomobil'no-dorozhnoy akademii, **20**, 21-28 (2011)
4. V.V. Vorontsov, Al.N. Kraev, M.E. Igoshin, Nauchno-tekhnicheskiy vestnik Povolzh'ya, **2**, 119-123 (2014)
5. V.V. Vorontsov, Al.N. Kraev, M.E. Igoshin, Vestnik Sibirskoy gosudarstvennoy avtomobil'no-dorozhnoy akademii, **6(40)**, 67-72 (2014)
6. Information on: http://rosavtodor.ru/information/dorogi_rossii/dorojnoe_hozyaystvo_rossii.htm
7. Metodicheskie rekomendatsii po proektirovaniyu zhestkikh dorozhnykh odezhd (vzamen VSN 197-91), INFORMAVTODOR, Mintrans Rossii, (2004)
8. V.A. Kretov, Nauka i tekhnika v dorozhnom khozyaystve, **1,** 1-9 (1998)
9. V.A. Kretov, *Teoreticheskie osnovy kolichestvennoy otsenki treshchinostoykosti zhestkoy konstruktsii dorozhnoy odezhdy s asfal'tobetonnym pokrytiem* (Informavtodor 5, 1999)
10. V.A. Kretov, *Tekhnologicheskie dorogi neftegazovogo kompleksa Zapadnoy Sibiri iz mestnykh materialov* (Informavtodor, 1999)
11. A.S. Gladkikh, Nauchnyy vestnik Voronezhskogo gosudarstvennogo arkhitekturno-stroitel'nogo universiteta, **4**, 166-173 (2009)
12. Yu. A. Aliver, Trudy Soyuzdornii, **201**, 96-100 (2001)
13. V.A. Kretov, *K voprosu o sposobakh povysheniya treshchinostoykosti pokrytiy* (Informavtodor, 1999)
14. V.A. Kretov, *Vserossiyskoy konferentsii rukovoditeley dorozhnykh nauchnykh i proektnykh organizatsiy*, 45-51 (1998)
15. L.V. Pozdnyaeva, Trudy Rosdornii 10 (2000)
16. P.P. Petrovich, V.V. Savitskiy, A.V. Dmitrichev, V.V. Antoshkin, Konstruktsiya dorozhnoy odezhdy s ispol'zovaniem armirovannogo monolitnogo tsementobetona: Pat. RF № 2003107981/03

17. P. Klekovkina, B.P. Karpov, Sovershenstvovanie konstruktsiy i tekhnologii stroitel'stva dorozhnykh odezhd s betonnym nesushchim osnovaniem, Dissertatsionnaya rabota, Saint Petersburg , (2010)
18. S.P. Sannikov, D.V. Kubasov, *Sb. materialov Vseros. nauch.-prakt. konf. «Aktual'nye problemy stroitel'stva, ekologii i energosberezheniya v usloviyakh Zapadnoy Sibiri».* 89-94 (2011)
19. A.N. Shuvaev, S.P. Sannikov, D.V. Kubasov, Nauchno-tekhnicheskiy vestnik Povolzh'ya, **6**, 440-443 (2012)
20. A.N. Shuvaev, S.P. Sannikov, D.V. Kubasov, Aktual'nye voprosy proektirovaniya avtomobil'nykh dorog, **4(63)**, 138-144 (2013)
21. S.P. Sannikov, D.V. Kubasov, Nauchno-tekhnicheskiy vestnik Povolzh'ya, **1**, 143-145 (2014)

Modelling of Water-Saturated Grounds under a Curved Section of an Oil and Gas Pipeline

Tatyana Maltseva[1,*] and *Tatyana* Saltanova[2]

[1]Industrial University of Tyumen, 625001 Volodarskogo str. 38, Tyumen, Russia
[2]Tyumen State University, 625000 Semakova str. 10, Tyumen, Russia

Abstract. The paper studies the influence of the oil and gas pipeline on the base of water-saturated soil in terms of external forces effect (from the pipeline). The developed mathematical model of stress-strain state of "the oil and gas pipeline and the grounds" system allows us to identify the maximum load resulting from the impact of a curved section of the pipeline on the base as well as the deformations occurring in the grounds.

1 Introduction

Currently, transportation of oil is carried out by one of the most profitable modes of transport – by pipeline.

Its main advantage is a low cost of the transfer. But it has its drawbacks, the main of which is a one-time large capital investment in the construction, as before starting to use the oil pipeline you need to build it from the starting point to its destination.

During the operation, a number of factors influences the pipeline: the pressure caused by curvilinear sections of the pipeline, internal pressure, temperature fields of soil and the pumped medium, the pressure of the overlying soil layer, the deformation of the base associated with the differential settlement of soil under its freezing and thawing. The paper considers a factor of pressure of the pipeline section and the deformation of the base. As shown by calculations, the stresses associated with a curved pipeline axis may exceed the stresses created by the weight of the tube and the pumped product [1].

The biggest part of the territory of oil and gas transportation is made up of weak clay soils. A big number of ways to increase the bearing capacity of foundation soil have been developed and implemented [2-6].

The calculation of water-saturated base is carried out in the framework of the mathematical theory of water-saturated soil, with an account of the effect of residual excess pore pressure in a stable condition when the time factor is absent [7].

The three-dimensional problem of the "oil and gas pipeline + base" system is divided into two sub-problems in the two-dimensional formulations: one problem is related to the horizontal plane and is called the auxiliary one, the other problem deals with the vertical plane x_1Ox_2, is called the principal one.

When addressing the auxiliary two-dimensional problem in the horizontal plane parallel to the ground surface one should define the longitudinal force q3 (with respect to the pipe

* Corresponding author: tgasu.maltseva.tatyana@mail.ru

axis) and transverse force q2. Then one solves the principal two-dimensional problem in the vertical plane, in which the horizontal force q2 is known as a result of the solution of the auxiliary problem. It is found with an account of the action of a longitudinal force q3. Thus, the solution of the three-dimensional problem (task) is replaced with the solution of two two-dimensional problems. Approximate taking account of the action of longitudinal force q3 is carried out through force q2.

2 Problem formulation

A mathematical model of vertical ultimate height (h) of a layer of water-saturated soil with an account of excessive residual pore pressure in a stable condition presents a system of differential equations with constant coefficients G, λ, b, c with respect to the displacement vector u=(u1, u2) of mineral soil particles [8]:

$$-\left(\left((G+\lambda)\frac{\partial\theta}{\partial x_i}+G\Delta u_i+b\frac{\partial^2 u_i}{\partial x_i^2}+c\frac{\partial u_i}{\partial x_i}\right)\right)=F_i,\ i=1,2, \tag{1}$$

$$G=\frac{E_S}{2(1+\nu)},\quad \lambda=\frac{E_S\nu}{(1+\nu)(1-2\nu)},\quad b=\frac{E_l}{\aleph^2},\quad c=\frac{E_l}{\aleph h},\quad \theta=div\,\boldsymbol{u}.$$

E_S - modulus of deformation and ν - Poisson's ratio of mineral soil particles, E_l - mechanical constant of pore water, $\aleph$ – dimensionless quantity $(0<\aleph<1)$ determined from an experiment [9; 10], h – the height of the compressible strata.

The boundary conditions for part of the boundary S_1 – kinematic, for part S_2 - static:

$$\boldsymbol{u}\,|_{S_1}=0,\quad \sigma|_{S_2}=\boldsymbol{q}(x_1,x_2), \tag{2}$$

$\boldsymbol{\sigma}$ - stress tensor in the mineral particle of the soil with the effect of pore water being taken account of only in diagonal elements.

Excessive residual pore pressures vanish on the day (x_1=0) surface of the body S_2 due to the draining coating and only the pressures in the mineral soil particles are left. Load $\boldsymbol{q}(x_1, x_2)$ - resultant of forces: q_1 - the weight per unit length of the pipeline with the product; horizontal force q_2 - the pressure of the curved section of the pipe on the ground. As already noted, it is defined with respect to the action of longitudinal force q_3.

Thus, the solutions of the two two-dimensional problems (the calculation of the curved section of the pipe and the calculation of the water-saturated soil base) have a common link, which is a transverse horizontal component q_2, arising as a result of the movement of oil products along the curved section of the pipeline. The dynamic smoothing method has been applied in order to assess the load rate q_2 [3, 4].

3 Research methods

Let us consider the auxiliary problem. According to the geodetic data the pipeline axis position is known. On the basis of elastic rod equilibrium equations [13; 14] and the use of the d'Alembert principle we obtained homogeneous nonlinear equations of motion of the rod in a Cartesian coordinate system:

$$m\ddot{x}_2 + \mu_{11}\dot{x}_2 + \mu_{12}\dot{x}_3 - \cos\varphi\frac{N' + \varphi' Q}{a} - \sin\varphi\frac{Q' - \varphi' N}{a} = 0,$$

$$m\ddot{x}_3 + \mu_{12}\dot{x}_2 + \mu_{22}\dot{x}_3 - \sin\varphi\frac{N' + \varphi' Q}{a} + \cos\varphi\frac{Q' - \varphi' N}{a} = 0. \quad (3)$$

In these equations the sign « ′ » denotes the derivative with respect to the Lagrangian coordinate of the point of the rod axis; the signs « ˙ and ¨ » stand for the first and the second time derivatives, m – mass per unit length of the pipeline with the product, x_2 and x_3 – Cartesian coordinates; φ – the angle between the pipe axis and the x_3 axis; N – the axial force, $N = D\varepsilon$, $\varepsilon = a - 1$ – axial deformation of the pipe; $D = E{\cdot}A$ – stiffness in tension, E – modulus of the pipe elasticity; A – cross sectional area of the pipe; $Q = H{\cdot}\varphi''$ – shear force, $H = E{\cdot}J$ – pipe bending stiffness, J – second area moment. Values a and φ are determined by known geodetic data – coordinates x_2', x_3' from the equations $x_2' = a\cos\varphi, \quad x_3' = a\sin\varphi.$

Coefficients of resistance μ_{ij} are:
$\mu_{11} = \mu_\tau cos^2\varphi + \mu_n sin^2\varphi,$
$\mu_{22} = \mu_\tau sin^2\varphi + \mu_n cos^2\varphi,$
$\mu_{12} = (\mu_n - \mu_\tau)\ sin\varphi\ cos\varphi,$

where μ_τ and μ_n – known resistivity coefficients for movement along the tangent and the normal to the pipe axis respectively.

Let's introduce the boundary conditions: $x_2(0,t) = x_0$, $x_3(0,t) = y_0$, $x_2(l,t) = x_l$, $x_3(l,t) = y_l$, $\varphi(0,t) = \varphi_0$, $\varphi(l,t) = \varphi_l$.

The initial conditions: at time t = 0 $x_2(0) = x_0(0)$, $x_3(0) = y_0(0)$ - the initial form of a rod, $\dot{x}_2(0) = 0$, $\dot{x}_3(0) = 0$ - the initial velocities.

Transverse and longitudinal forces q_2 и q_3 are determined at any time by the formulas:

$$q_2 = \cos\varphi\frac{N' + \varphi' Q}{a} + \sin\varphi\frac{Q' - \varphi' N}{a}, \qquad q_3 = \sin\varphi\frac{N' + \varphi' Q}{a} - \cos\varphi\frac{Q' - \varphi' N}{a}.$$

The system of equations (3) is solved by the finite difference method with the use of the "cross" explicit difference scheme [15]. The derivatives of x_3 coordinate are approximated by finite-difference formulas of the 2nd order accuracy.

The system of equations (1) is solved with a modified finite element method. Triangular finite elements have been taken as finite elements [8; 16]. The movements inside the triangle elements are described by linear dependencies through the nodal displacements $\boldsymbol{u} = (u_1, u_2)$:

$$u_k = \frac{1}{2\Delta}\left[(p_i + d_i x_1 + n_i x_2)u_k^i + (p_j + d_j x_1 + n_j x_2)u_k^j + (p_m + d_m x_1 + n_m x_2)u_k^m\right],$$

$$p_i = x_1^j x_2^m - x_1^m x_2^j, \quad n_i = x_1^m - x_2^j, \quad d_i = x_2^j - x_2^m, \quad k = 1,2.$$

Nodal displacements u_k^i, u_k^j, u_k^m are the desired values.

When taking the integral over the area of the triangular element there appear definite integrals $\int_S x_1 dS$ and $\int_S x_2 dS$.

Consider taking one of the integrals on the basis of the well-known theorem on the static moment of area according to which the static moment equals the area of a triangle Δ times x_c distance to the center of gravity of the triangle, which coincides with the intersection of its medians:

$$\int_S x_1 dS = x_{1c} \cdot \Delta, \qquad x_{1c} = \frac{x_1^i + x_1^j + x_1^m}{3}, \qquad x_{2c} = \frac{x_2^i + x_2^j + x_2^m}{3}.$$

Here is a system of linear algebraic equations of the finite element method for finding the unknown nodal displacements $\{\delta\} = \left(u_1^i, u_2^i, u_1^j, u_2^j, u_1^m, u_2^m\right)$ [8]:

$$([k_1] + [k_2])\{\delta\} = \{F\},$$

where:

$$[k_1] = \frac{1}{4\Delta}\begin{pmatrix} d_i & 0 & n_i \\ 0 & n_i & d_i \\ d_j & 0 & n_j \\ 0 & n_j & d_j \\ d_m & 0 & n_m \\ 0 & n_i & d_m \end{pmatrix} \cdot \begin{pmatrix} \lambda\frac{1-\nu}{\nu} + b_1 & 2G\frac{\nu}{1-2\nu} & 0 \\ 2G\frac{\nu}{1-2\nu} & \lambda\frac{1-\nu}{\nu} + b_2 & 0 \\ 0 & 0 & G \end{pmatrix} \cdot \begin{pmatrix} d_i & 0 & d_j & 0 & d_m & 0 \\ 0 & n_i & 0 & n_j & 0 & n_m \\ n_i & d_i & n_j & d_j & n_m & d_m \end{pmatrix}$$

$$[k_2] = \frac{1}{4\Delta}\begin{pmatrix} f_i & 0 & 0 \\ 0 & f_i & 0 \\ f_j & 0 & 0 \\ 0 & f_j & 0 \\ f_m & 0 & 0 \\ 0 & f_m & 0 \end{pmatrix} \cdot \begin{pmatrix} c_1 & 0 & 0 \\ 0 & c_2 & 0 \\ 0 & 0 & 0 \end{pmatrix} \cdot \begin{pmatrix} d_i & 0 & d_j & 0 & d_m & 0 \\ 0 & n_i & 0 & n_j & 0 & n_m \\ n_i & d_i & n_j & d_j & n_m & d_m \end{pmatrix},$$

$$f_i = p_i + d_i x_{1c} + n_i x_{2c}, \quad f_j = p_j + d_j dx_{1c} + n_j x_{2c}, \quad f_m = p_m + d_m x_{1c} + n_m x_{2c},$$

$$\int_S dS = \Delta = \frac{1}{2}\begin{vmatrix} 1 & x_1^i & x_2^i \\ 1 & x_1^j & x_2^j \\ 1 & x_1^m & x_2^m \end{vmatrix}.$$

4 Numerical results

The system of equations (1) with the boundary conditions (2) was solved by a modified method of finite elements. The triangular finite elements were chosen as finite element.

The value of the intensity of the horizontal force q_2=0.005 MN/m

Acting at a distance of 10 m along x_3 axis; own weight of the pipe q_1=0.0015 MN/m (Fig. 1)

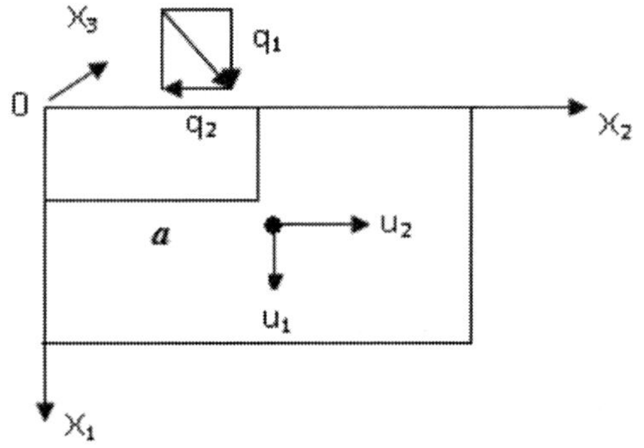

Fig. 1. Forces acting on the water-saturated base.

The mechanical characteristics of the soil

$E_s = 5.2\,MPa$, $E_l = 3.27\,MPa$,
$\nu = 0.3$, $\aleph = 0.52$,

the radius of the pipeline – 0.3 m.
The Fig. 2 shows the horizontal displacement of soil skeleton particles at x_1=0 м, x_2=1 m.

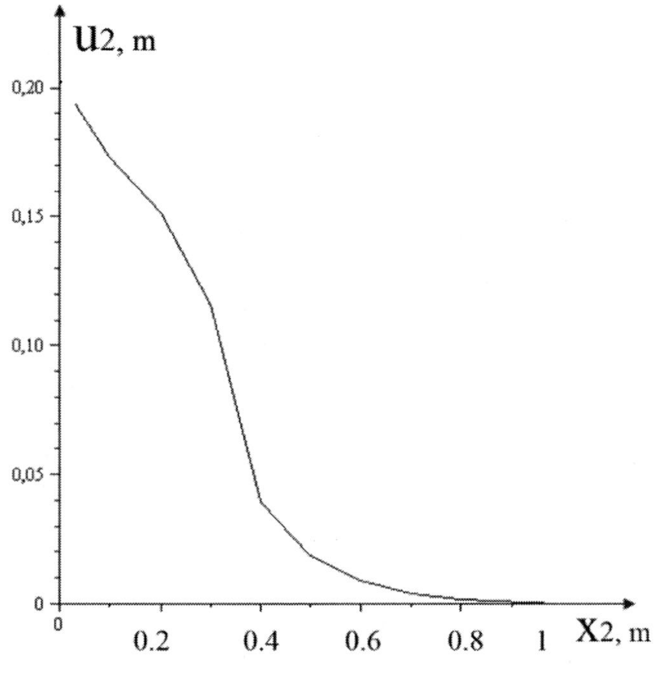

Fig.2. Horizontal displacement of soil skeleton particles

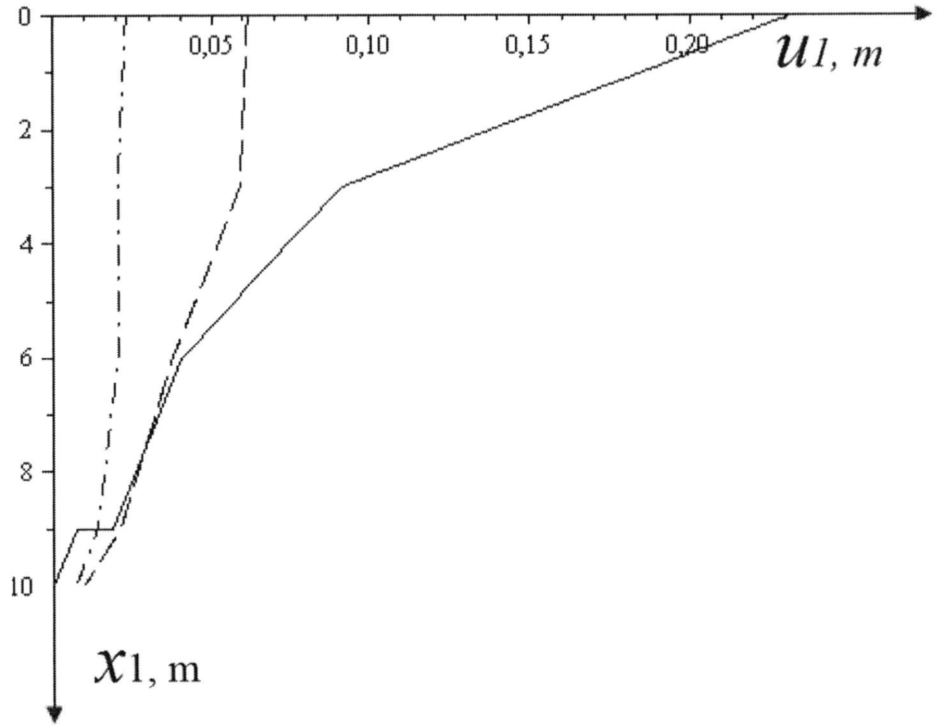

Fig. 3. Vertical displacement for different cross-sections.

Fig. 3 shows the graphs for vertical movement for $x_2 = 0$, at the distance

for $x_2=0$, at the distance $x_2=0.3\ m$, $x_2= 0.9\ m$ from x_1 axis

The maximum horizontal displacements were 0.2 m, the minimum ones – 0.03 m.

The solution method solves the problem of determining the longitudinal and transverse forces of the interaction of the pipeline with the ground for sufficiently long pipeline sections. Transverse bends lead to the loss of stability of plane shape of the bending. The proposed model and solution algorithm allow us to take account of the possibility of the effect of thermal stresses in the pipeline on the strength of the interaction with the ground, but the article does not deal with this question. When solving the problem the maximum load of the pipeline on the ground was determined and it accounted for $q = 0.0166$MN/m

5 Conclusion

The obtained results can be used in the design and the calculation of sediment of engineering pipeline transportation facilities being built on weak grounds in determining the residual life of operated pipelines or in the processing of seismic results. The modified finite element method allows us to determine the stress and strain state of the base of the water-saturated soil under the action of external forces caused by the influence of the pipeline.

The solution to the problem of stress-strain state of a weak base will allow us to predict the maximum permissible values of the base strain and to consider and adopt measures to improve the safety of its construction and operation at the design stage.

References

1. T. Maltseva, S. Dorofeev, T. Selivanov, Izvestija visshikh uchebnikh zavedenij. Neft i gaz, **1**, 59-63 (2012)
2. T. Maltseva, A. Nabokov, A. Chernykh, Procedia Engineering, **117**, 239-245 (2015)
3. A. Kraev, Vestnik Tomskogo arkhitekturno-stroitelnogo universiteta, **4**, 146-151 (2008)
4. V. Baj, T. Maltseva, A. Kraev, Nauchno-tehnicheskij vestnik Povolzh'ja, **5**, 108–111 (2014)
5. V. Baj, A. Kraev, Vestnik grazhdanskih inzhenerov, **3**, 107-110 (2014)
6. L. Maltsev, T. Maltseva, A. Minaeva, A. Nabokov, Nauchno-tehnicheskij vestnik Povolzh'ja, **2**, 234-238 (2012)
7. T. Maltseva, *Matematicheskaja teorija vodonasyshhennogo grunta* (Vektor Buk, Tjumen', 2012)
8. L. Maltsev, T. Maltseva, T. Saltanova, Problemi prochnosti i plastichnosti, **68**, 181 (2006)
9. V. Baj, L. Maltsev, T. Maltseva, A. Nabokov, A. Demin, R.F. Patent 2213952. (2002)
10. L. Maltsev, T. Maltseva, A. Demin, Izvestija vysshih uchebnyh zavedenij. Neft' i gaz, **2**, 96-102 (2001)
11. S. Dorofeev, N. Nesterovich, Sbornik nauchnikh trudov TumGU, **11**, 65-70 (2009)
12. S. Dorofeev, N. Nesterovich, Sbornik nauchnikh trudov TumGU, **11**, 71-75 (2009)
13. S. Toropov, S. Dorofeev, E. Sapozhnikov, Izvestiya vysshikh uchebnykh zavedenij. Neft' i gaz, **3**, 49-53 (2003)
14. E. Grigolyuk, V. Shalashilin, *Problemy nelinejnogo deformirovaniya*, (Nauka, Moscow, 1988)
15. N. Kalitkin, *Chislennie metodi* (Nauka, Moscow, 1978)
16. T. Maltseva, T. Saltanova, Vestnik TumGU, 49-56 (2007)

Moving of Inline Cleaning Units Along Submerged Crossing in Main Pipelines

Sergey Toropov[1], *Sergey* Podorozhnikov[1,*], *Evgeniy* Toropov[1], and *Larisa* Markova[1]

[1]Industrial University of Tyumen, 625001 Volodarskogo str. 38, Tyumen, Russia

Abstract. The article analyzes the forces acting on the inline cleaning units of new design, operating on a bailer principle. In this case all cleaning products are collected in the internal cavity of the cleaning unit, which eliminates the formation of the ground plug and can significantly reduce the friction of the cleaning unit on the inner surface of the pipeline. The moving of the cleaning units of the given design within the space of a curved steel pipeline is investigated, the solution of the obtained equations and their analysis related to pipeline submerged crossing are shown.

1 Introduction

Cleaning the inner cavity of main and field pipelines is one of the main technological processes during the construction period of the oil and gas pipelines as well as their operation [1, 2, 3]. Periodic cleaning can reduce the water resistance of the pipeline and thereby significantly reduce the cost for the product pumping. This paper presents the results of the dynamics test of new design inline cleaning units taking into account their service particularities. In contrast to the previous papers, considering the moving of inline cleaning units like a motion of a material point, the cleaning units in this paper are considered as the contiguous flexible objects with distributed loads [4, 5]. Such formulation corresponds to a constructive solution of the cleaning unit data and differs significantly from the models of the previous authors [7]. The obtained results allow within the accepted assumptions to evaluate the impact of the design parameters of the materials of the cleaning units and their geometric characteristics on the moving process of the cleaning units along a spatially curved pipeline [9].

2 Research object (Model, Process, Unit, Synthesis, Experimental procedure, etc.)

Being developed method of cleaning the internal cavity of the pipelines is based on the use of cleaning units like a "bailer". The design is based on plastic pipes, which ensures the friction lowering on the internal surface of the pipeline and the possibility for a large structure length to pass sections with a small curvature [21]. In the adopted model the

* Corresponding author: sergey_urevih@mail.ru

assumption for the constancy of the mass of the cleaning unit and the uniform distribution of the cleaning products on the cleaning unit was made [10].

3 Method

Cleaning unit is a plastic pipe of the length about 5 – 10 diameters, closed at one end, Figure 1.

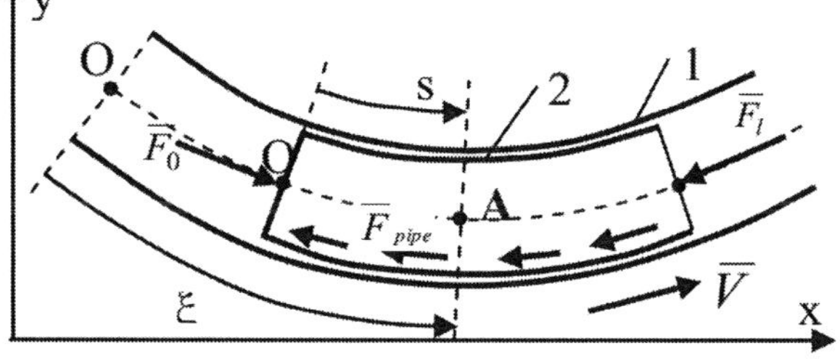

Fig. 1. Scheme of the clearing unit.

In the figure are indicated: 1 – pipeline, 2 – cleaning unit. The cleaning unit moves at a rate $\overline{V}$ under the acting force $\overline{F}_0$, generated by extreme pressure of the air injected into the pipeline. A force $\overline{F}_l$ acts on the right end of the cleaning unit (Figure 1), caused by the resistance of the being removed from the internal surface of the pipeline deposit products $\overline{F}_c$, and extreme pressure of air displaced from the pipeline [11]. The distributed over the length of the unit friction force $\overline{F}_{pipe}$, weight force and reaction forces act on the cleaning unit while moving.

The position of any arbitrary point A of the cleaning unit is defined by its coordinate s about a point O of the cleaning unit – "beginning" or coordinate ξ – about a point O′ – the beginning of the pipeline, Figure 1. The coordinate ξ depends on the time t, as well the coordinate s i.e. $\xi = \xi(s, t)$.

To derive the motion equation of the cleaning unit an element of the length ds is considered, Figure 2.

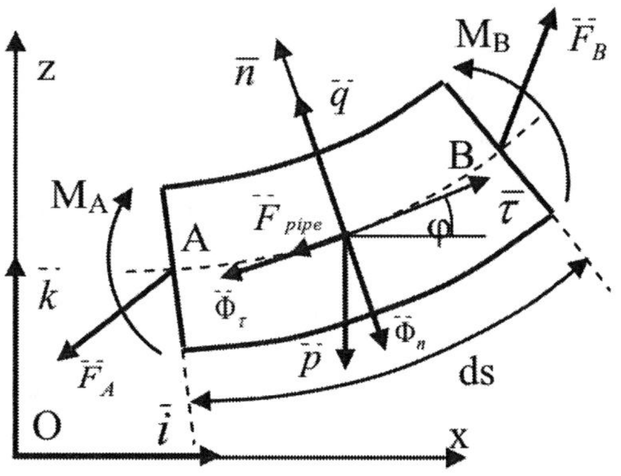

Fig. 2. An element of the clearing unit.

In Figure 2 is indicated:

$\overline{i}$, $\overline{k}$ – the unit vectors of the Cartesian coordinate system XOZ;

$\bar{\tau}$, $\bar{n}$ – ordinary vectors directed tangentially and normally to the pipe axis in the concavity direction

$\bar{F}_A$, $\bar{F}_B$ – forces acting in sections A and B, where $\bar{F} = T\bar{\tau} + N\bar{n}$;

T and N – tangential and shearing forces, respectively;

MA, MB – bending moments in the cross section A and B;

$\bar{p}$, $\bar{q}$ –weight and reaction forces acting on the element in question;

Φ_τ, Φ_n – tangential and normal inertia forces;

$\bar{F}_{pipe}$ – friction force.

Derivation of the motion equations is carried out as usual. We assume

$$\overline{F_B} = -\overline{F_A} + d\bar{F}, \quad d\bar{F} = \frac{d\bar{F}}{ds}ds = \frac{d}{ds}\left(T\bar{\tau} + N\bar{n}\right)ds. \tag{1}$$

Derivatives of vectors $\bar{\tau}$ and $\bar{n}$ are equal respectively [6]:

$$\frac{d\bar{\tau}}{ds} = \chi\bar{n}, \quad \frac{d\bar{n}}{ds} = -\chi\bar{\tau}, \tag{2}$$

where $\chi = \frac{1}{R}$ – the curvature of the pipe axis.

Taking into account these designations we obtain:

$$d\bar{F} = \left(\frac{dT}{ds} + \chi N\right)\bar{\tau}ds + \left(\frac{dN}{ds} - \chi T\right)\bar{n}ds \tag{3}$$

The bending moment M is a scalar value, so

$$M_B = -M_A + \frac{dM}{ds}ds, \tag{4}$$

$$\bar{Q} = qds\bar{n}, \tag{5}$$

where q – the distributed reaction force. The weight of element of the length ds is equal

$$\bar{P} = -gdm\bar{k} = -g\gamma\, ds\bar{k}, \tag{6}$$

γ - mass of a length unit.

The vector $\bar{k}$ is decomposed into components, directed along the axes $\bar{\tau}$ and $\bar{n}$:

$$\bar{k} = \sin\varphi \cdot \bar{\tau} \pm \cos\varphi \cdot \bar{n}, \tag{7}$$

The "+" – if the bulge is upward, "-" – if it is down, then the vector

$$\overline{P} = -g\gamma \sin\varphi ds\overline{\tau} \pm g\gamma \cos\varphi ds\overline{n}, \tag{8}$$

friction force

$$\overline{F}_{pipe} = -f|q|ds\overline{\tau}, \tag{9}$$

f – the friction coefficient.

The considered element of the repairing insert moves with acceleration. Therefore, the inertia force acts on it in accordance with the principle of d'Alembert [12]. Taking this into account we obtain

$$\begin{cases} -\gamma a_\tau - \gamma g \sin\varphi + \dfrac{dT}{ds} - \chi N - f|q| = 0 \\ -\gamma a_n \pm \gamma g \cos\varphi - \chi T + \dfrac{dN}{ds} + q = 0 \\ \dfrac{dM}{ds} + N = 0. \end{cases}, \tag{10}$$

The unknowns in these equations are accelerations $a\tau$ and an, forces T and N, and the reaction force q. The bending moment M can be considered as known, because the shape of the curved pipe axis is considered as acquainted [14]. The total number of unknowns – more than three, therefore, the problem in this formulation is statically indeterminate.

We denote u(s, t) – moving the point of the repairing insert along the axis of the pipeline with coordinate s at the time moment t. Axial deformation at the point s:

$$\varepsilon = \frac{\partial u}{\partial s}, \tag{11}$$

then the tangential force T

$$T = EA \cdot \varepsilon = EA\frac{\partial u}{\partial s}, \tag{12}$$

where A – the sectional area of the pipe of the cleaning unit; E –elasticity modulus of the pipe material.

Bending moment is

$$M = EJ\chi, \tag{13}$$

where J – the inertia moment of the cleaning unit section, the curvature χ is assumed to be acquainted at each point of the pipeline.

The rate of the point with the coordinate s

$$v(s,t) = \frac{\partial u}{\partial t}, \tag{14}$$

Tangential and normal accelerations are

$$a_\tau = \frac{\partial^2 u}{\partial t^2}, \quad a_n = \chi v^2 = \chi\left(\frac{\partial u}{\partial t}\right)^2. \tag{15}$$

The moving rate of the cleaning unit is about 3-5 m / s. At such rates, the normal inertial force has no significant impact on the moving of the repairing insert, so it will not be taken into account [13].

Expressing the shear force N taken into account the expressions (11) – (15) from the last equation of the system (10) we obtain:

$$\begin{cases} \gamma\dfrac{\partial^2 u}{\partial t^2} + \gamma g \sin\varphi - EA\dfrac{\partial^2 u}{\partial s^2} - EJ\chi\dfrac{d\chi}{ds} + f|q| = 0, \\ \pm \gamma g \cos\varphi - EA\chi\dfrac{\partial u}{\partial s} + EJ\dfrac{d^2\chi}{ds^2} - q = 0. \end{cases}$$

Finally, expressing q from the second equation and substituting it in the first one, we obtain:

$$\gamma\frac{\partial^2 u}{\partial t^2} - EA\frac{\partial^2 u}{\partial s^2} - EJ\chi\frac{d\chi}{ds} + \gamma g \sin\varphi + f\left|EJ\frac{d^2\chi}{ds^2} - EA\chi\frac{\partial u}{\partial s} \pm \gamma g \cos\varphi\right| = 0 \tag{16}$$

The values φ and χ in the equation are different at different points of the trajectory, and they should be considered as a function of ξ (but not s). Therefore, the derivatives dx/ds and d2x/ds2 in the equation (16) should be replaced by dx/dξ and d2x/dξ2. The coordinate ξ is expressed in terms of s and u as follows:

$$\xi = s + u(s,t) \tag{17}$$

Therefore

$$\varphi = \varphi(s + u(s,t)), \quad \chi = \chi(s + u(s,t)). \tag{18}$$

The obtained equation (16) is a nonlinear differential equation in partial derivatives [15]. The initial and boundary conditions are for the considered problem as follows:

$$\begin{cases} u(s,0) = 0, \quad v(s,0) = 0, \\ T(0,t) = EA\dfrac{\partial u}{\partial s}\Big|_{S=0} = T_0, \; T(l,t) = EA\dfrac{\partial u}{\partial s}\Big|_{S=l} = -T_l, \end{cases} \tag{19}$$

where l – length of the cleaning unit, T0 = PL·S, T1 = PR·S, PL – the pressure on the left of the cleaning unit (Fig. 1), PR – the pressure on the right, S – the cross-sectional area of the pipeline.

The pressure on the left PL and PR depends on the moving rate of the cleaning unit v, on the length LL = u(0,t) of the pipeline on the left from the cleaning unit and on the length LR = L – u(0,t) on the right from the cleaning unit. Furthermore, the pressure drop depends on the thermodynamic air parameters [16]. Loss of pressure on the left and on the right from the cleaning unit is determined by expressions [8]:

$$P_L = \frac{P_{IN}}{\sqrt{1+\lambda \frac{L_L}{d}\cdot\frac{v^2}{RT}}}, \qquad P_R = \frac{P_0}{\sqrt{1-\lambda \frac{L_R}{d}\cdot\frac{v^2}{RT}}} \tag{20}$$

where PIN – the inlet pressure in the pipeline; P0 – the outlet pressure in the pipeline (atmospheric pressure); λ – coefficient of hydraulic friction; d – the inner pipeline diameter, R – gas constant for air; T – absolute temperature. The coefficient of hydraulic friction is determined by the formula of Altshul [18].

Solution of the motion equations. The analytical solution of the problem (16), (19) is not possible. The finite difference method for the solution is applicable.

We write the motion equation (16) as a system of two differential first-order equations with the variable t:

$$\begin{cases} \dfrac{\partial v}{\partial t} = \dfrac{EA}{\gamma}\dfrac{\partial^2 u}{\partial s^2} + \dfrac{EJ}{\gamma}\chi\dfrac{d\chi}{ds} - g\sin\varphi - f\left|\dfrac{EJ}{\gamma}\dfrac{d^2\chi}{ds^2} - \dfrac{EA}{\gamma}\chi\dfrac{\partial u}{\partial s} \pm g\cos\varphi\right|, \\ v = \dfrac{\partial u}{\partial t}. \end{cases} \tag{21}$$

We apply the simplest difference scheme. Approximation of the derivatives is taken as follows

$$\frac{\partial v_j^{i+1}}{\partial t} \approx \frac{v_j^{i+1} - v_j^i}{\Delta t}, \quad \frac{\partial u_j^i}{\partial s} \approx \frac{u_{j+1}^i - u_{j-1}^i}{2\Delta s}, \quad \frac{\partial^2 u_j^i}{\partial s^2} \approx \frac{u_{j+1}^i - 2u_j^i + u_{j-1}^i}{(\Delta s)^2}.$$

Difference equations will be as follows:

$$\begin{cases} v_0^{i+1} = v_0^i + \dfrac{\Delta t \cdot EA}{\gamma}\dfrac{2(u_1^i - u_0^i)}{(\Delta s)^2} + \dfrac{\Delta t \cdot EJ}{\gamma}\chi(u_0^i)\dfrac{d\chi(u_0^i)}{ds} - \Delta t \cdot g\sin\varphi(u_0^i) - f\Delta t \cdot \left|\dfrac{EJ}{\gamma}\dfrac{d^2\chi(u_0^i)}{ds^2} \pm g\cos\varphi(u_0^i)\right| \\[2ex] v_j^{i+1} = v_j^i + \dfrac{\Delta t \cdot EA}{\gamma}\cdot\dfrac{u_{j+1}^i - 2u_j^i + u_{j-1}^i}{(\Delta s)^2} + \dfrac{\Delta t \cdot EJ}{\gamma}\chi(j\Delta s + u_j^i)\dfrac{d\chi(j\Delta s + u_j^i)}{ds} - \Delta t \cdot g \cdot \sin\varphi(j\Delta s + u_j^i) - \\ \qquad - f\Delta t \cdot \left|\dfrac{EJ}{\gamma}\cdot\dfrac{d^2\chi(j\Delta s + u_j^i)}{ds^2} - \dfrac{EA}{\gamma}\chi(j\Delta s + u_j^i)\dfrac{u_{j+1}^i - u_{j-1}^i}{2\Delta s} \pm g\cos\varphi(j\Delta s + u_j^i)\right|, \quad j = 1, 2, \ldots n-1; \\[2ex] v_n^{i+1} = v_n^i + \dfrac{\Delta t 2EA}{\gamma}\cdot\dfrac{u_{n-1}^i - 2u_n^i}{(\Delta s)^2} + \dfrac{2\Delta t T}{\gamma\Delta s} + \dfrac{\Delta t EJ}{\gamma}\chi(n\Delta s + u_n^i)\dfrac{d\chi(n\Delta s + u_n^i)}{ds} - \Delta t \cdot g \cdot \sin\varphi(n\Delta s + u_n^i) - \\ \qquad - f\Delta t \cdot \left|\dfrac{EJ}{\gamma}\cdot\dfrac{d^2\chi(n\Delta s + u_n^i)}{ds^2} - \dfrac{T}{\gamma}\chi(n\Delta s + u_n^i) \pm g\cos\varphi(n\Delta s + u_n^i)\right| \quad . \end{cases}$$

From the stability conditions $\Delta t \le \Delta s\sqrt{\dfrac{\gamma}{EA}}$ should be chosen.

From these equations for i = 0 we find v_j^1, then we find u_j^{i+1} from the equation $u_j^{i+1} = u_j^i + \Delta t \cdot v_j^{i+1}$ $(j = 0, 1, \ldots, n)$ and return again to the equations (22).

4 Results and Discussion

The obtained equations allow us to calculate the moving of the cleaning unit for pipeline submerged crossing of any profile. To do this, it is necessary to determine the spatial position of the pipeline axis and to approximate the axis by any functions.

As an example, the results of calculation of the submerged crossing profile are given, shown in Figure 3. The submerged crossing length is conditionally divided into three sections [17].

Sections L_1, L_3 are submerged crossing areas with distinctive variable curvature, characterized by the presence of differential of the elevation marks. The section L_2 is a straight portion of the river bed of the submerged crossing. It is characterized by the absence of the curvature of the pipeline axis.

The cleaning element starts moving from the start-receiver pig chamber, (initial point of section L_1) passes the area L_1, where the rolling force $F_{ск}$ accelerates the moving. The rolling force is absent on a straight section. The rolling force has a retarding impact on the moving of the repairing insert on the section L_3.

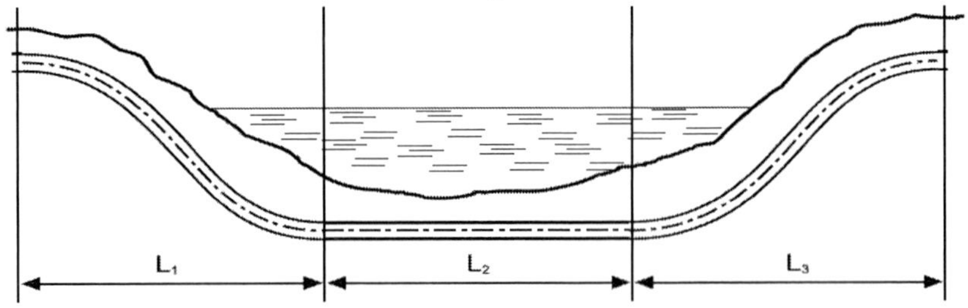

Fig. 3. Profile of submerged crossing.

The moving of the cleaning unit is carried out by the force T_B created by air pressure acting on the section of the cleaning unit. The air supply Q is carried out by the compressor, while it remains constant in the throughout time interval of moving [20].

Typical graphs on the calculation results of the cleaning unit moving dynamics ($v(t)$ – change in moving speed; $P(t)$ – change in the air pressure) for the submerged crossing corresponding to the profile Figure 2.

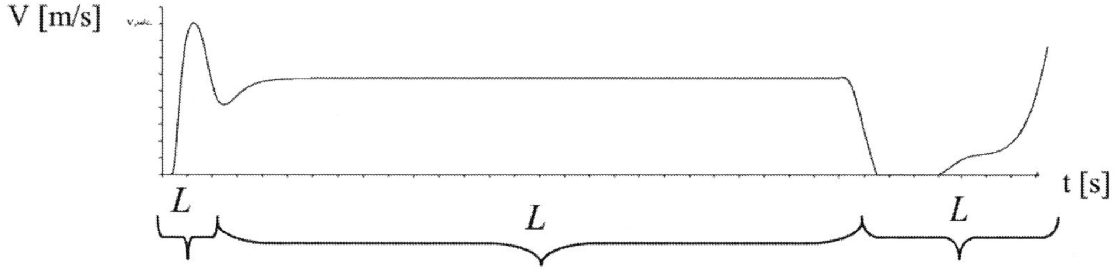

Fig. 4. Change in moving rate of the cleaning element in the submerged crossing.

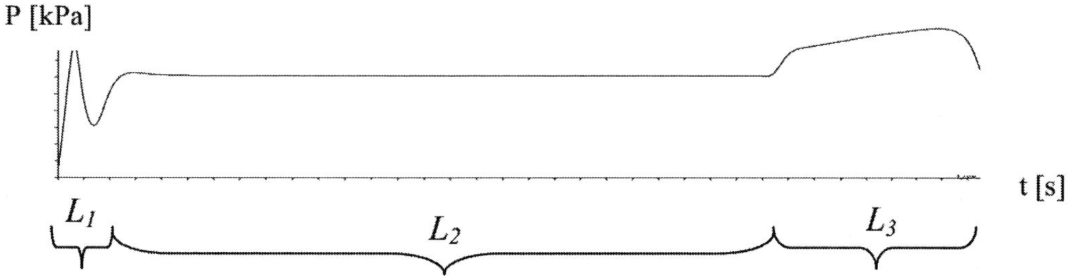

Fig. 5. Time pressure changes character.

From the graphs the moving parameters character of the cleaning unit can be seen.

The initial time t = 0 corresponds to the air feed release. Until the time moment t1 the overpressure increases from 0 up to values at which the force acting on the cleaning unit reaches a friction force value. After that, it begins to move. When this pressure continues to increase because the speed of the leaning unit increases slower than the pressure increases [19]. But the acceleration of the cleaning unit also increases (because the pressure increases). At the time moment t2 rate of the cleaning unit reaches such value that the volume on the left of the cleaning unit starts to increase faster than the air is injected. At this time, the pressure Pmax is maximal. The speed continues to increase because the rolling force acts on the cleaning unit together with the pressure. At the point t3 the rate reaches a maximum value and the pressure – the minimum one, i.e. the speed and pressure change are shifted in phase. After some fluctuations in the section L1 the cleaning unit enters the straight portion L2. The rate and pressure in the straight section are almost constant. While reaching the curved section L3 the rate decreases sharply, because as the resistance force increases sharply and the rolling force occurs, retarding moving. At the time t4 a stop occurs, continuing until the pressure reaches the value necessary to start the moving.

Thus, in general, the moving rate of the cleaning unit has an uneven character and is determined by the profile of the submerged crossing.

3 Conclusion

It should be noted that basic elements of HSC DSS have been implemented in the software product the System of Remote Monitoring of Parameters (SRMP), there is a Certificate of state registration of the software (Appendix No. 1), co-authors are Razboynikov A.A. and Vashchilin V. V. Within a calculation method approbation of a residual resource pump stations' equipment, on the basis of stationary monitoring systems of vibration parameters, on the CPS-4 object of the Ust-Balyksky field, the implementation measures of the residual resource calculation method on three pump units SRP-240 on the basis of the permanently installed monitoring systems of vibration characteristics (PIMSVC) "Vibrobit-300" of LLC OPP "Vibrobit" production were held. The protocol of successfully carried out results after pilot tests has been signed (Appendix No. 2). According to the conducted researches, a comprehensive application of the developed methodology and software operation system of the main equipment of pumping stations by the technical state described in this work will allow to:

- to provide accident-free operation of an oil and gas complex equipment (in particular-pumping stations);
- to withdraw the defective aggregate from operation promptly, thereby preventing an emergency situation and serious damage to the equipment units;
- to define an aggregate residual resource effectively;
- to optimize an aggregate operating mode;
- to reduce the repair cost, through targeted funding aggregates according to the actual state;
- to increase the efficiency of repair work planning by identifying defects at an early development stage;
- to estimate the preliminary cost of repair work.

The expected results from implementation of the above-stated developments and improvements fully corresponds to the Strategy of innovative development of the Russian Federation theses till 2030.

References

1. R. Sprague, A *Framework for Research on Decision Support Systems. Decision Support Systems: Issues and Challenges* (Pergamon Press, Oxford, 1980)
2. GOST 27.002-89 *Reliability in equipment. Basic concepts. Terms and definitions*
3. *MOOR 98 The method of residual resource determination of the process equipment of oil refineries, petrochemical and chemical industries* (Volgograd, 1998)
4. V. Voyevodin, *Numerical methods of algebra. Theory and algorithms* (Science, Moscow, 1966)
5. *RD 153-39TH-008-96 Guide for operations and technology maintenance and equipment repair and structures of oil pumping stations* (Ufa, 1997)
6. B. Antonyuk, *The development of expert systems of artificial intelligence in the USA* (RRIS, Moscow, 1986)
7. A. Borisov, E. Vilums, L. Sukur, *Dialog systems for decision making based on minicomputers: Information, mathematical and software* (Zinatne, Riga, 1986)
8. B. Denning, G. Essig, S. Maas, D*ialog systems "Man-computer". Adaptation to user requirements* (Mir, Moscow, 1984)
9. O. Larichev, A. Petrovskiy, *Decision support system. Modern state and development prospects. Science and technology results* (AUISTI, Moscow, 1987)
10. E. Trakhtengerts, *Computer decision support* (SINTEG, Moscow, 1998)

11. V. Kostrov, M. Zemenkova, S. Sokolov, A. Ryabkov, S. Chekardovskiy, Fundamental researches, **10**, 274-278 (2015)
12. M. Chekardovskiy, K. Ilyukhin, S. Chekardovskiy, *Actual problems of architecture, construction, ecology and conservation in Western Siberia: Proceedings of the international scientific-practical conference*, 217-221 (2015)
13. V. Vaschilin , A. Starovoytov, S. Chekardovskiy, Oil and Gas, 2, 93-99 (2014)
14. M. Chekardovskiy, K. Ilyukhin, Yu. Stavitskiy, S. Chekardovskiy, Collection of scientific papers «Megapaskal», **3**, 35-36 (2009)
15. M. Chekardovskiy, K. Ilyukhin, A. Razboynikov, S. Chekardovskiy, *Proceedings of the international scientific conference dedicated to the 80th anniversary of the Ural power system, UPI-UGTU, Ekaterinburg*, 106-111 (2003)
16. A. Razboynikov, M. Chekardovskiy, Yu. Zemenkov, S. Chekardovskiy, *Diagnostics and elimination of vibration equipment oil and gas facilities* (TSOGU, Tyumen, 2014)
17. Yu. Zemenkov, [and others] *Electromechanical equipment of pumping stations oil pipelines* (TSOGU, Tyumen, 2014)
18. Yu. Zemenkov, [and others] *Operation of pumping equipment at the facilities of pipeline transport* (TSOGU, Tyumen, 2010)
19. Yu. Zemenkov, V. Shalay, M. Zemenkova, Procedia Engineering, **113**, 254-258 (2015)
20. Yu. Zemenkov, V. Shalay, M. Zemenkova, Procedia Engineering, **113**, 312-315 (2015)
21. V. Kurushina, Yu. Zemenkov, WIT Transactions on Ecology and the Environment, **2**, 881-888 (2014)
22. V. Antip'ev, A. Nevolin, Yu. Zemenkov, Neftyanoe Khozyaistvo - Oil Industry, **10**, 46-48 (1981)
23. A. Bacherikov, M. Pashkov, Yu. Bogatenkov, Yu. Zemenkov, V. Shalay, G. Vasiliev, A. Prokhorov, S. Sidelnikov, S. Podorozhnikov, V. Kehl, *Diagnosis in the maintenance of pipeline transportation* (Nedra, St.Peterburg, 2007)
24. R. Levitin, Yu. Zemenkov, Neftyanoe khozyaystvo - Oil Industry, **1**, 110-114 (2016)
25. R. Mamadaliev, V. Kuskov, Yu. Zemenkov, A. Popova, Applied Mechanics and Materials, **770**, 19-22 (2015)
26. J. Broch, *Mechanical Vibration and Shock Measurments* (Bruel & Kjear, 1984)

Perspective Application Directions of Fiber Optic Lasers in the Repair and Renewal Operations on the Linear Part of Main Gas Pipelines

Vadim Ivanov[1], *Maxim* Zykov[2], *Vasiliy* Rybin[3,*], and *Semen* Plotnikov[4]

[1]Industrial University of Tyumen, 625001 Volodarskogo str. 38, Tyumen, Russia

Abstract. Large number of main pipelines requires today an immediate repair. Equipment producing entertainments supply the market with more and more new and advanced devices. However, their use does not always allow reducing the repair work period. That way it is appropriate to develop the alternative repair methods that allow reducing the labor intensity as well as the cost of repair and renewal operations. One of the promising areas is the laser technologies application in the repair process.

1 Introduction

The gas industry is one of the most important economy sectors of the Russian Federation. Therefore the increase of productivity and quality of manufacturing processes of this industry will boost the qualitative and quantitative development of not only modern industry, but also economy of our country.

The main pipelines are the key objects in the processing chain of the gas pipeline transportation. Consequently, the performance of the entire Russian gas transportation system as well as the profitability of the energy industry in general depends on the quality of their construction and technical condition [1-10].

During the construction of modern hydrocarbon transportation facilities a great importance is attached to compliance with rules and regulations of work performance, as well as to the modern equipment appliance, ensuring the reliable operating throughout its life time [11, 14-16, 18, 20-22, 32].

However, in the epoch of Russian oil and gas industry, in the middle of the twentieth century, it was necessary to provide uninterrupted transportation of hydrocarbon raw materials from the extraction spots to the final consumer [12] as short as possible. In turn, such factors as the construction advancing deadlines, inaccessibility of oil and gas production facilities, lack of quality equipment and materials affected the work performance quality in oil and gas networking construction, which operate through present. It should be noted that the construction of the linear part of the main oil and gas pipelines is

* Corresponding author: vtec11@mail.ru

an independent subsystem in the pipeline constructions system, which is accounted for approximately 50% of the total volume of construction and installation jobs, and therefore its quality factors largely determine the industry efficiency level as a whole [13, 17, 19, 23, 24, 25].

Thus, the actual task is to improve existing repair methods of the linear part of main gas pipelines, as well as the substitution of the obsolete repair equipment for modern and potential analogues.

One of such areas is the development of automated facilities, using laser units as an operating device. The implementation of the fiber optic lasers will allow quality and performance improving of such operations as the pipes surface cleaning, the insulation and removing of coating, the pipeline route preparation in the mountainous areas, the welding and repair of the main gas pipelines linear part [16, 17, 18, 19].

As it is known, welding is one of the main technological operations, both in the construction and the repair of pipelines of any diameter [20, 21].

Currently, arc welding is the main way for pipe sections connecting. This process has a number of significant disadvantages:

low productivity;

burning and spatter metal losses;

formation of heat affected zone, which reduces the steel mechanical properties and the pipe joint operational characteristics.

The use of fiber optic lasers for pipeline joints welding (Figure 1) allows to avoid the above mentioned problems and obtain weld metal by 90-95% identical to the base metal (welding arc ranges in this rate between 70-80%). Consequently, there is a real opportunity to get the pipe joint, which is almost as good as the base metal by its characteristics, which guarantees their longer and trouble-free operation.

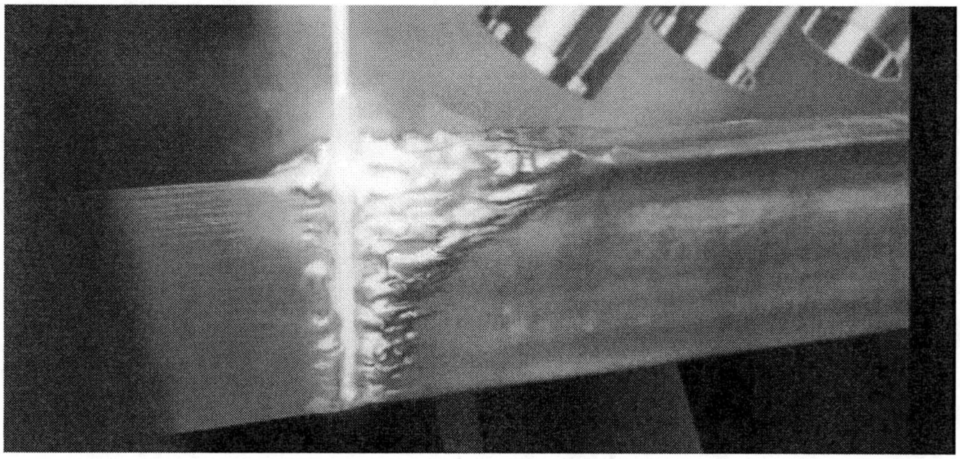

Fig.1. Laser welding process.

It is also possible to use a laser beam for the orbital welding (Figure 2), the welding speed varies at that from 3 to 12 m/min (at the arc welding this rate hardly reaches 20-25 m/h).

Fig. 2. Orbital laser welding of the pipe joint.

Application of laser units is not limited to only welding. It is also possible to perform the weld facing, for example, for the defects repair of the type such as metal loss, etc. It is known that the mentioned defects are one of the most important problems of modern pipeline transportation. The emergence and development of this type of defects leads to a thinning of the pipe wall and as a consequence the loss of the bearing capacity of the pipeline linear part.

The main repairing methods of such defects are:

- general repair related to the pipeline network stoppage;
- coupling structure installation
- weld facing of the defective part with a pumping stop or non-stop.

The repair of the defective part by weld facing without stoppage the hydrocarbons transportation is the most effective from the technological and economic points of view. However, in this case, the repair area is subjected to temperature softening, which requires a reduction of the operating pressure to the repairing one. That is, the greater the pipe wall heating is, the less the load bearing capacity is due to the metal transition in plastic condition. Therefore, reducing the heating effect made it possible to fold increase the steel strength properties on the repairing area, which means to increase the pressure in the repaired pipeline.

The appliance of laser technology will allow to reduce the heat impact on the metal surface, as well as achieve a high quality of metal alloying. At that the weld facing rate may exceed 7-10 m/min and the dimensions of the heat affected zone as well as structural changes degree in the welded area is ten times less than those for arc surfacing (Figure 3).

DOI: 10.1051/matecconf/20167301024

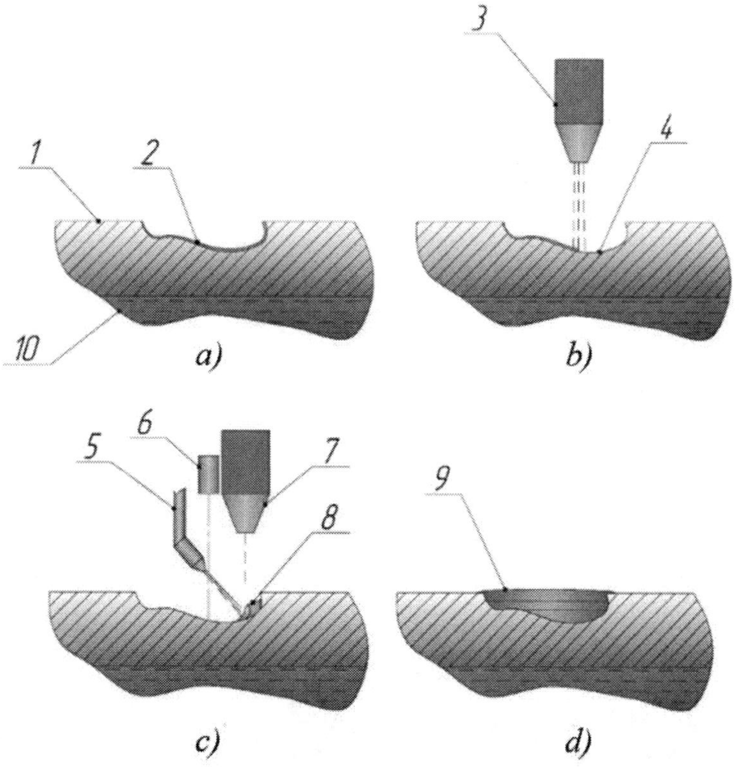

Fig. 3. Repair technology of the defect type "Metal loss": 1 – pipeline wall; 2 – corrosion damage; 3 – the cleaning laser; 4 – the cleaned surface; 5 – nozzle for supplying by the metallic powder; 6 – the laser level; 7 – the welding laser; 8 – the weld metal; 9 – the repaired area.

Protection of the main pipeline against corrosion is achieved by applying a high-quality waterproof coating and functioning of electrochemical protection system throughout its service life.

There are the following necessary conditions for metal long-term protection against corrosion using protective coatings: the insulating material must have sufficient strength, long-term moisture resistance and good adhesion to metal; the metal surface must be cleaned out of the insulation residues, corrosion products, dirt, salt and other deposits. The roughness should be created on the surface to ensure the required bondability.

Modern pipeline systems are constructed of pipes with factory waterproof coating based on extruded polyethylene, which provides the reliable protection for the entire service period, but their share in the total length of the gas transportation system is undoubted small. However, the most of gas pipelines are those of the old fund, with a service life longer than 20-25 years.

For the re-isolation of such pipelines areas the tape sealing compound coating ("PAM") and the boundary compound coating "Transkor-Gas" are currently used. In route conditions, the pipe surface preparation for the coating application is performed by means of automated mechanical cleaning facilities. The working devices of the facilities are scrapers and wire brushes. When cleaning in this way the surface preparation degree "3" on the Russian Industry Building Code 008-88 is achieved. This preparation is sufficient for the application of the above mentioned coatings. However, a disadvantage of coatings of this class is their short effective service life which is 5-15 years.

For application of more reliable coatings, such as the polyurethane ones, the surface preparation is required up to degree Sa 2½ according to ISO 8501-1 as well as the

 DOI: 10.1051/matecconf/20167301024

roughening of at least Rz = 60 microns. Such cleaning quality may be achieved only by using of sandblasting or shot-blasting units. The use of sandblasting units in route conditions is difficult and inefficient because of the low productivity and high sand consumption without its re-use possibility. Air shot blasting and airless shot blasting cleaning is mainly used in the factory conditions.

One of the potential alternative methods is a surface cleaning using the pulse laser systems of the company IPG-Photonics (Figure 4).

Fig. 4. Metal laser cleaning.

At present, the experimental research on the metal cleaning out of old insulation coating residues as well as rust and scale based on the adhesive tapes were made (Figure 5).

Fig. 5. The cleaned pipe sample 1 – uncleaned portion with melted insulating coating and corrosion.

As a result of the surface cleaning by the pulse laser with the capacity of 50 W the surface has been brought to result shown in the picture after 5 passes. Each pass took an average 1 minute. The average productivity has made 1.5 m2/hour which is certainly less than even the weakest sandblasting unit (its capacity ranges from 10 up to 30 m2/hour). However, taking into account the possibility of increasing the power of the laser unit up to 5-10 kW its capacity can be increased tenfold. Furthermore another advantage of laser cleaning appliance is controlled required surface roughening. Then the surface does not heat when laser cleaning due to the short interaction time – of 120 ns. However, if necessary the pipe surface may be heated under control up to the required temperature for the application of new insulation coating.

In addition to the insulating coating removing, lasers can also be used for its application (see Figure 5, 7).

DOI: 10.1051/matecconf/20167301024

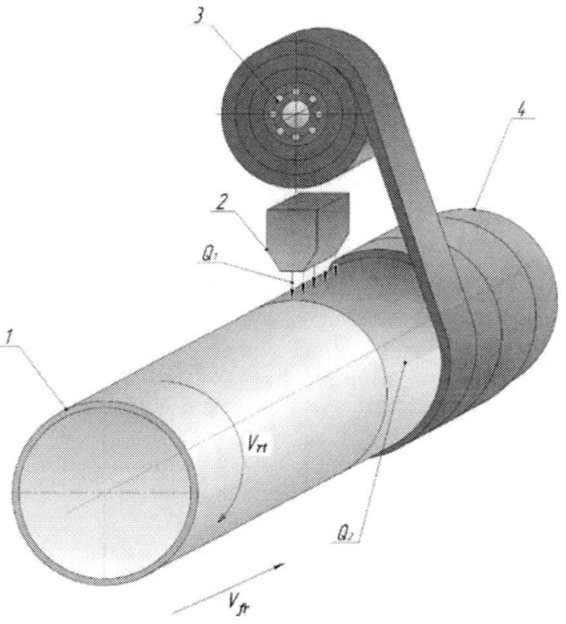

Fig. 6. Factory coating process:1 – the pipe section; 2 – laser heat cleaning emitter; 3 – spool with polymer; 4 – polymer on the pipeline; Vrt – the pipeline rotation speed; Vfr – forward speed of the pipeline section; Q1 – the heat flow from the laser; Q2 – the residual heat flow in the pipe section wall.

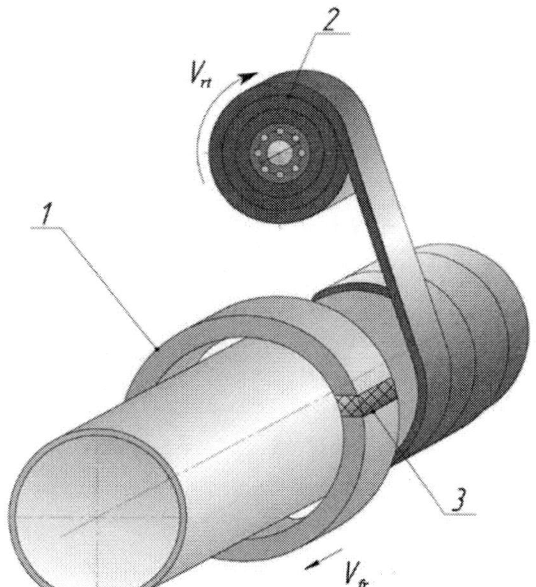

Fig. 7. Route coating process.

1. Laser ring emitter; spool on winding machine (the winding machine bonding is not shown, it is ordinary); Vrt – the spool rotation speed; Vfr – the laser head speed.

Another focus of laser technology application for the repair of the main pipelines is the polymer coatings spraying.

The principle of this method is as follows (Figure 8): a finely dispersed polymer powder is supplied to the zenith of the laser emitter (active spot) through the nozzle under the pressure. Entering the active spot of the laser emitter, the powder is heated up to the melting point and its adhesion to the weld surface emergences.
In case of the weld facing onto the polymer surface the upper edge of the product surface is also subjected to melting (depending on the laser wavelength).

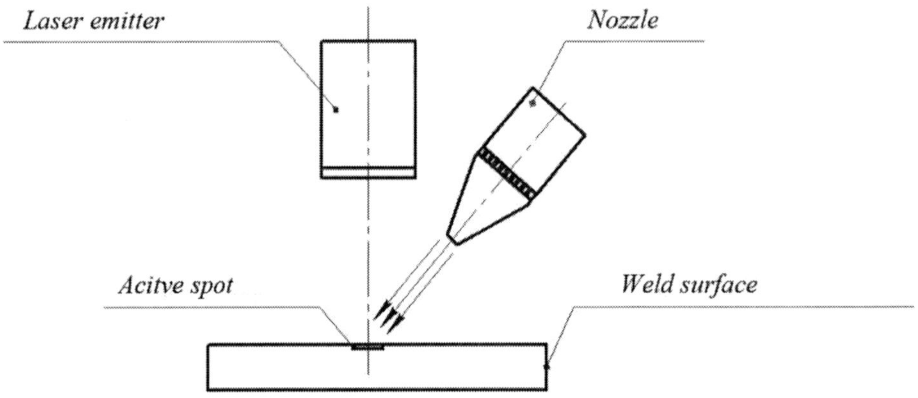

Fig. 8. Weld facing scheme.

Thus it is possible to carry out the surface recovery of plastic and composite pipes, machine components and devices.
To enhance the fusion layer strength (Figure 9) it is possible to apply the reinforced elements such as nets, etc.

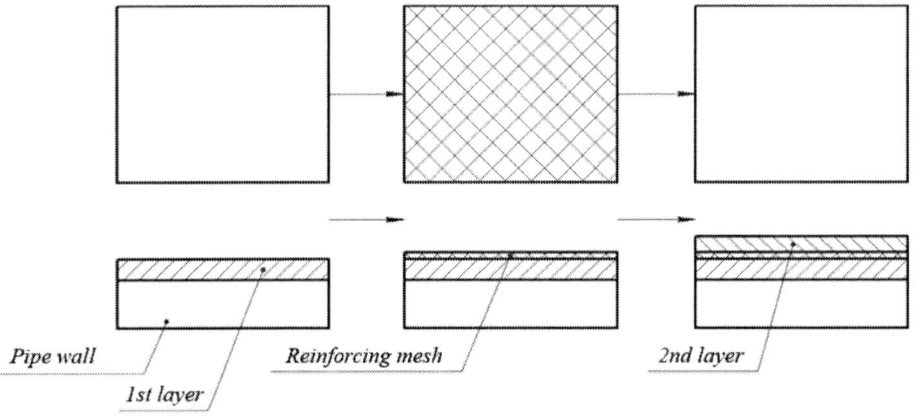

Fig. 9. Adaptation of the reinforced coating in surfacing.

Thus, further work on testing and implementation of the optic fiber laser technologies in pipeline maintenance and construction process will allow introducing of a new approach to the performing technology of construction and assembly jobs. Supply of construction and repair jobs by laser devices will allow achieving high quality and performance results as well as reducing the risks of human factor influence. However, before the implementation of these technologies in real conditions it is necessary to carry out more detailed experimental studies.

References

1. C.S. Sirimanna, A.C. Manalo, W. Karunasena, S. Banerjee, L. McGarva, Rehabilitation of Pipelines Using Fiber-reinforced Polymer (FRP) Composites, 267-285 (2015)
2. T, S. Khan, M, S. Al-Shehhi, Journal of Natural Gas Science and Engineering, **25**, 66-(2015)
3. S. Hertelé, A. Cosham, P. Roovers, Engineering Structures, **124(1)** , 429-441 (2016)
4. J. Fink, Guide to the Practical Use of Chemicals in Refineries and Pipelines, 25-29 (2016)
5. L.P. Djukic, W.S. Sum, K.H. Leong, A.G. Gibson, Rehabilitation of Pipelines Using Fiber-reinforced Polymer (FRP) Composites, 237-265 (2015)
6. R. Shahandeh, H. Showkati, Journal of Constructional Steel Research, **121**, 237-252 (2016)
7. O. Shabarchin, S. Tesfamariam, Journal of Loss Prevention in the Process Industries, **40**, 479-495 (2016)
8. P. Hopkins, Underground Pipeline Corrosion, 62-84 (2014)
9. N. Saeed, H. Ronagh, A. Virk, Composites Part B: Engineering, **58**, 605-610 (2014)
10. A. Shahriar, R. Sadiq, S. Tesfamariam, Journal of Loss Prevention in the Process Industries, **25(3)**, 505-523 (2012)
11. M. Zykov, V. Ivanov, Oil and Gas,**4**, 29-35 (2014)
12. V. Ivanov, Yu. Savinykh, M. Zykov, Fundamental researches, **2**, 3709-3712 (2015).
13. V. Ivanov, M. Zykov, V. Rybin, Gas industry, **7**, 49-51 (2015)
14. V. Ivanov, B., Gas industry, **3**, 12-14 (2011)
15. V. Ivanov, I., Volynets, B., Gas industry, **692**, 64-66 (2013)
16. V. Ivanov, M., Gas industry,**7**, 49-51 (2015)
17. V. Ivanov, S. Burakovsky, Oil and Gas, **1**, 61-66 (2011)
18. V. Ivanov, N. Savchenko,. Oil and Gas, **2**, 57-61 (2011)
19. V. Ivanov, E. Mikhalenko, S. Sokolov, Oil and Gas, **1**, 76-80 (2012)
20. V. Ivanov, M. Zykov, Oil and Gas, **4**, 29-35 (2014)
21. V. Ivanov, S. Sokolov, Exposition Oil Gas, **2(34)**, 91-92 (2014)
22. V. Ivanov, A. Kirsch, Fundamental research, **9**, 256-261 (2014).
23. V. Ivanov, A. Davydov, V. Berg, D., Mining information and analytical Bulletin (scientific and technical journal), **4**, 169-173 (2014)
24. V. Ivanov, A. Davydov, D. Serebrennikov, Exposition Oil Gas, **6(38)**, 100-101 (2014)
25. V. Ivanov, A. Davydov, D. Serebrennikov, Exposition Oil Gas, **6(31)**, 25-26 (2013)
26. V. Ivanov, V. Rybin, Exposition Oil Gas, **7(32)**, 55-59 (2013)
27. V. Ivanov, V. Rybin, Exposition Oil Gas, **7(32)**, 63-66 (2013)
28. V. Ivanov, V. Rybin, Exposition Oil Gas,**7(32)**, 60-62 (2013)
29. V. Ivanov, V. Rybin, Exposition Oil Gas, **7(32)**, 52-54 (2013).
30. V. Ivanov, A. Akulov, V. Rybin, Improving safety in the repair of main pipelines by welding method, Technosphere safety, **3(4)**, 5-7 (2014)
31. V. Rybin, Exposition Oil Gas,**1**, 52-53 (2015)
32. V. Rybin, Exposition Oil Gas, **6**, 102-103 (2014)

MATEC Web of Conferences 73, 01025 (2016) DOI: 10.1051/matecconf/20167301025
TPACEE-2016

Study of Deformations in a Large-Capacity Oil Storage Tank in the Presence of Subgrade Inhomogeneity Zones

Alexandr Tarasenko [1], *Petr* Chepur [1,*], and *Alesya* Gruchenkova [1]

[1]Industrial University of Tyumen, 625001 Volodarskogo str. 38, Tyumen, Russia

Abstract. Characteristics of a joint action of a subgrade and a large 50000 m3 capacity storage tank have been overviewed. The maximum allowable values of the RVSPK-50000 base immersion in the presence of the inhomogeneity zone have been determined given the stiffness of metal structures. To simulate the inhomogeneity zone we applied the Drucker–Prager model – a linear elastoplastic material implemented in the finite element software package ANSYS. The dependences of the maximum design value of the outer tank bottom contour immersion on the inhomogeneity zone sector length have been obtained (in the range of 10 to 95 meters). It has been found that 95% of all cases of uneven immersion occurring in practice fall within this range according to data on diagnostics of 40 vertical steel tanks.

1 Introduction

Current trends in main oil transportation can be characterized by a constant increase in the unit of storage capacity in commercial storage tanks with subsequently remote regions with harsh engineering and geological construction conditions. Despite the fact that vertical steel tanks of more than 50,000 m3 capacity refer to I-class hazard structures and requirements for design decisions and quality of construction and installation works are at a very high level, there are cases of inhomogeneous immersion of RVSPK-50000 [1-6]. Research on the stress-strain state (SSS) of tanks during immersion is, as a rule, limited to strength analysis of the wall, bottom and annular plate structure. At the same time, simplified subgrade models are introduced that do not reflect the real nature of its deformation under the influence of operational and off-design loads. This approach ultimately leads to high errors in determining the SSS of the structure as a whole, and the lower circumferential joint in particular.

The authors propose to consider the joint work of the subgrade and the RVSPK-50000 tank as a thin-walled shell structure having finite bending stiffness. Based on the analysis results of diagnosing more than 40 RVSPK-50000 tanks [7-8], it has been shown that any outer bottom contour immersion can be represented as a combination of the wall surface arc value L along which the immersion develops and the immersion value u. Let us assume that the base properties correspond to the design. Then, for the most unfavorable case of loading

* Corresponding author: chepur@me.com

the tank metal structures in the immersion zone it is necessary to cut out a base sector and determine to what values the immersion can develop if the soil at the site of its development is completely absent. It is obvious that the value of immersion for the sector with a combination of operational loads is finite and will be determined by the elastic properties of the tank. Knowing the full range of immersion values encountered in practice and changing the value of the arc along which they occur, we can obtain the range of possible values of the outer tank bottom contour deformations. Immersion cannot exceed these values. The authors propose to use a research tool available in the ANSYS package [9-12] and replace the sector without the foundation base with the one with minimal strength properties, and perform the analysis of the interaction between the base and the structure using a model based on the Drucker-Prager strength criterion.

Introduction of powerful finite element analysis packages provided the opportunity to obtain reliable numerical solutions based on the actual geometry of the system and high detail of all interacting elements of the system.

In international practice there are cases of trouble-free operation of vertical steel tanks with uneven immersion of more than 1 meter; however, according to current design standards in the Russian Federation [21-22] the permissible uneven outer bottom contour immersion is limited to 40 mm for new vertical steel tanks and 80 mm for those being in service for more than 20 years. Permissible values of immersion in tank construction have significant differences from country to country, due to the type of construction materials used and design approaches in the regulatory documents (GOST, RD, SNiP, STO – Russia; API – the USA, EN – the European Union, British Standard – Great Britain).

In most cases, experts encounter the problem of immersion development when the reservoir has already been built and in operation for a long time. Considering that main oil transportation infrastructure is located in regions with harsh engineering and geological construction conditions designers are required to develop technological solutions for a particular project that would ensure normal lifetime operation conditions of the tank. An important condition for quality design is sufficient survey operations, the volume of which must be determined individually for each construction site depending on the characteristics of the structure and the construction properties of local soils. However, the amount of survey operations and the number of wells to be drilled are often designated incorrectly. Therefore, the customer's desire to maximize the reduction of costs (on survey operations) leads to an incorrect assessment of the geological conditions of the site. For example, according to regulatory documents, for a tank with the diameter of more than 60 meters and the design filling level of 17 m (base pressure under the central part of the bottom qbot = 144,2 kPa, under the foundation ring qbase = 203 kPa) only 5 wells are allowed to be drilled: one in the center and four along the contour at diametrically opposite points. This approach leads to the fact that design organizations apply design decisions based on incorrect information about the real properties of the soil forming the natural base. There are cases when there is no information about the presence of weak layers of organic soil in the base. Thus, errors at survey and design stages, violations of the construction and assembly operations technology, deviations in the hydraulic testing mode, changes of the oil transmission process scheme after the commissioning of the object (as a consequence, the emergence of off-design loads) may be the reasons for the emergence of inhomogeneity zones and uneven tank immersion.

The authors propose to apply the ANSYS finite element software package and analyze the RVSPK-50000 structure behavior features considering the joint operation of a reinforced concrete ring foundation, tank metal structures and a base with inhomogeneity zones composed of weak compressible soils. From [13-20] we can deduce that correct solution of such a problem requires a most accurate tank model to be developed that takes into account not only the stiffness of the structure, but also the contact interaction at the interface of the elements in the "soil-foundation-bottom-wall" system. When solving the contact problem it is important to consider the possibility of separation of the contact surfaces in immersion modeling, such as disconnection of the RVSPK metal structures from the foundation ring or foundation sagging over the inhomogeneity zone.

2 Tank model development. Description of the design scheme

A finite element model of RVSPK-50000 tank has been developed according to the design project of «CNIIPROEKTSTALKONSTRUKCIYA», with its verification reviewed in [17]. The tank diameter is 60.7 m; height – 17.95 m; wall thickness for belts I–XII ranges from 17 to 8 mm aligned with the inner surface; on wall belts V and VIII are located stiffening rings from roll-formed sections – 100x300 angles, 8 mm in thickness; the wind girder is an L-shaped structure of conjugated sheet and beam elements reinforced by braces located at 2.5 m intervals, welded to the XII wall belt via intermediate assembly plates; the foundation is circular, reinforced concrete, from rectangular section dimensioned 1.5x0.4 m; the bottom 9 wall belts and annular plates are made from 16G2AF steel with a yield point σ_T = 440 MPa, other structures are from 09G2S steel with a guaranteed yield point σ_T = 325 MPa. To simulate the wall, bottom, annular plate and stiffening rings 4-node finite elements SHELL181 are used with six degrees of freedom per node, taking into account the membrane tension, compression and bending. For the foundation ring a 20-node element SOLID186 is applied with three degrees of freedom per node. The subgrade is modeled using 10-node finite elements SOLID187 with three degrees of freedom per node and support of large deformations. When simulating immersion of the RVSPK-50000 foundation in a deformable soil volume the Drucker-Prager law was applied. In accordance with the Drucker-Prager strength criterion it is assumed that the yield surface does not change with the increase of plastic deformation; hence, there is no hardening effect, and the material is ideally elastoplastic. According to Drucker-Prager, the equivalent stress above which the material experiences plastic deformation is determined by taking into account the hydrostatic material squeeze reduction: the higher the mean stress, the greater the strength of the material. This condition is a modification of the Mises strength condition and better reflects the work of the base soils. The design scheme of RVSPK-50000 considering the acting load is given in Fig. 1.

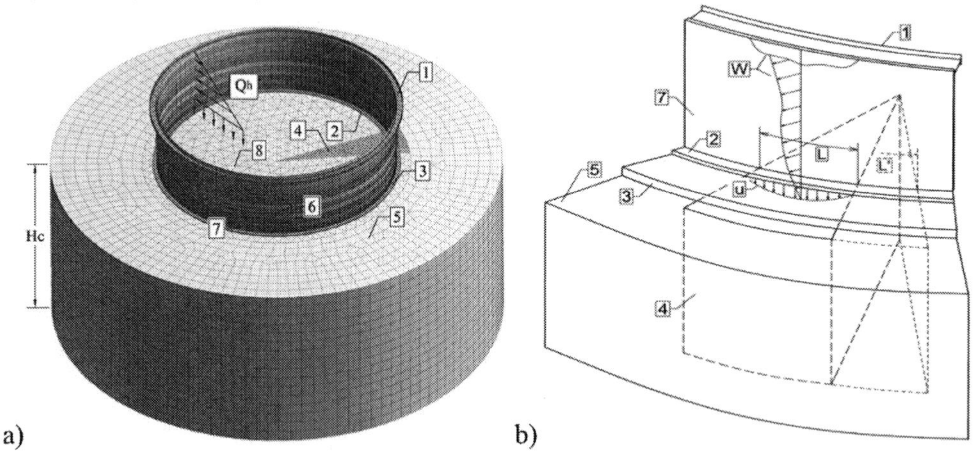

a) b)

Fig. 1. Design scheme (a, b). 1 – wind girder; 2 – annular plate; 3 – reinforced concrete ring foundation; 4 – highly compressible soil of inhomogeneity zone; 5 – subgrade with design characteristics; 6 – stiffening ring on wall belts V and VIII; 7 –RVSPK-50000 wall; 8 – central bottom section; Qh – hydrostatic load; R – tank radius; L – inhomogeneity zone sector length (along the outer contour of the bottom); Hc – compressed thickness; W – wall radial movements; u – bottom outer contour immersion.

Previous diagnostic tests [8] confirm that the inhomogeneity zone under the VST bottom and annular plate tends to form a triangular sector. The inhomogeneity zone in this research is given as a triangular sector with the arc size L along the VST outer contour. This design scheme was proposed for the first time by I.V. Slepnev [13], where the design loading was produced by cutting a segment of the VST foundation ring. However, in [11] a section of the wall, annular plate and bottom "sagged" above the inhomogeneity zone, i.e. soil and foundation within the specified sector limits were non-existent. This was justified by the fact that such a loading scheme reflects the most unfavorable operating conditions of VST metal structures in the presence of the inhomogeneity zone.

When performing test calculations, as well as the analysis of [12] and diagnostic survey data [8], intervals that determine the minimum and maximum values of the inhomogeneity zone sector of the RVSPK-50000 base were set (along the outer contour of the bottom). The minimum value was L_{min} = 10 meter, maximum - L_{max} = 95.3 meters (which corresponds to the RVSPK-50000 roll). The authors set the properties of a highly compressible soil in the inhomogeneity zone using the Drucker–Prager model, taking into account the most unfavorable case ever encountered according to diagnostic tests of real tanks. Table 1 shows the physical and mechanical characteristics of the soil for the modeled base sections.

Table 1. Basic physical and mechanical characteristics of soils for the modeled base sections.

Characteristics	Unit	Weak high-plasticity clay soil of the inhomogeneity zone (4 – fig. 1)	Artificially compacted sand and clay soil (5 – fig. 1)
Elastic modulus, E	MPa	5	30
Poisson's ratio, μ	–	0.43	0.3
Density, ρ	kg/m3	1800	1650
Cohesion, c	kPa	0	4
Angle of internal friction, φ	degree	14	36

The thickness of the model's active base zone is taken to be H_c = 40 m. This value was determined in accordance with the recommendations [21] based on the calculation results of the foundation plate on the elastic base for a highly compressible soil in the inhomogeneity zone (4). The area of the active base zone is taken to be 11304 m^2 and its diameter is 120 m.

The boundary conditions of the model are defined by a rigidly fixed lower edge of the subgrade at the level of "-40 m", as well as by the exclusion of the lateral displacement of soil along the perimeter of the computational domain. A nontrivial problem of model development is consideration of contact interaction between loaded structures. So, contacts of the bottom metal structures, wall and annular plate (surface–surface, edge–surface) connected by welds are modeled as deformable bonded, but without the possibility of separation – command «bonded». However, contact of the bottom, annular plate and foundation with the subgrade, the annular plate with the foundation ring cannot be set to «bonded», because for large deformations in the base inhomogeneity zone the contact area may change: there can be disconnection of metal structures from foundations, raising and lowering of the foundation ring and the bottom, whereby both interpenetration of contact surfaces (the reinforced concrete foundation into the soil) and their separation at any point is possible. So, the authors used an algorithm, the augmented Lagrange method, which takes into account the sliding friction force proportional to the normal reaction in the contact model. In this case, the contact area can be changed and includes both areas of adhesion, sliding and complete separation. This technique allows controlling the amount of introduction of the contact surfaces, also, the contact stiffness values in the normal direction based on the Young's modulus - E and the size of adjacent elements are calculated. The calculation results confirm the validity of applying the approach used. Thus, at great values of the inhomogeneity zone the circumferential joint of RVSPK-50000 is actually "hanging" over the highly compressible soil base section. The model consists of 301647 finite elements and takes into account the physical and geometric nonlinearities.

3 Results

Fig. 2–3 show the subgrade deformations diagrams and RVSPK-50000 metal structures with a minimum considered inhomogeneity zone L = 10 m and maximum – L = 95.3 m (along the outer contour of the bottom), in which case there is tank roll with a break at the interface of the weak soil and soil with design characteristics. The RVSPK-50000 design model with the inhomogeneity zone considers hydrostatic load Q_h = 144.2 kPa, oil innage height $H_{взл}$ = 17 m (oil density ρ = 865 kg/m^3). For visualization purposes, on the diagram soil and metal structures deformations are presented using a scale factor x3.

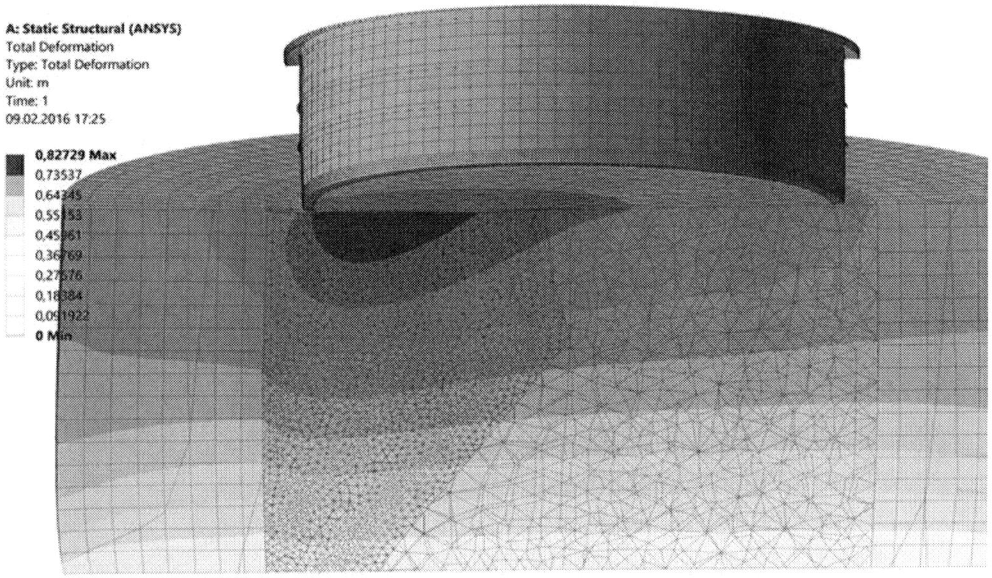

Fig. 2. Diagram of RVSPK-50000 movements with the inhomogeneity zone L = 10 m (section view).

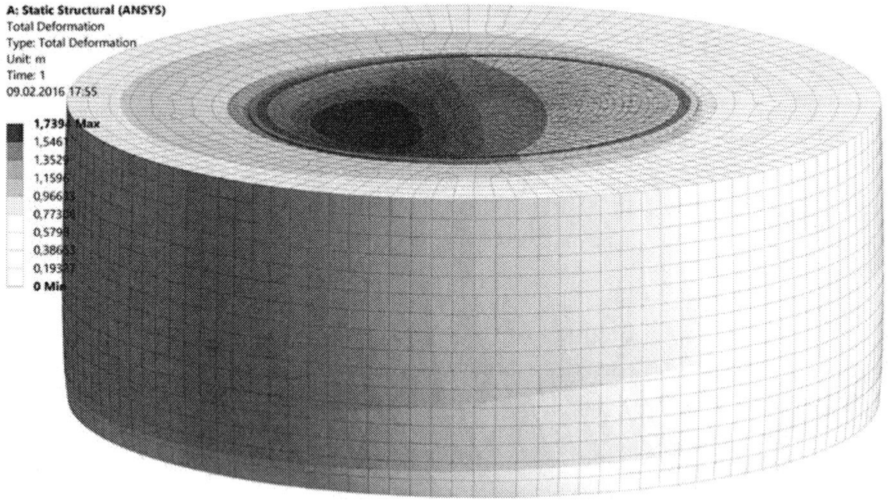

Fig. 3. Diagram of RVSPK-50000 base and foundation movements with the inhomogeneity zone L = 95.3 m (tank removed for clarity).

The authors conducted interpretation of the data obtained from finite element modeling and found that when the tank is filled to the maximum the bottom outer contour immersion above the inhomogeneity zone with the length L = 95 m (1/2 of the RVSPK-50000 perimeter) can reach 325 mm. For the inhomogeneous immersion zone with the sector length L = 10 m the immersion of the bottom outer contour is 29 mm. The greatest values of the bottom outer contour immersion are observed in the middle of the inhomogeneity zone regardless of the sector length. A graph has been built (Fig. 4) that shows the dependence of the maximum design value of the RVSPK-50000 bottom outer contour immersion on the inhomogeneity zone sector length.

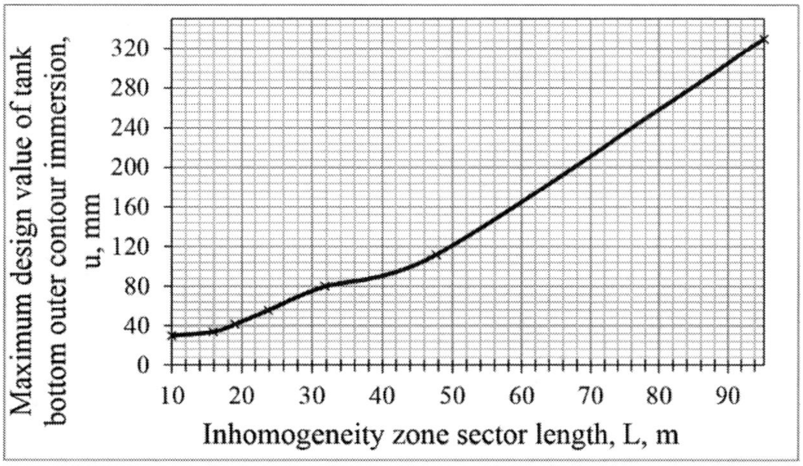

Fig. 4. Dependence of the maximum design value of the RVSPK-50000 bottom outer contour immersion on the inhomogeneity zone sector length.

Note that structures of this type have a large cylindrical stiffness that prevents the development of uneven immersion. However, it should be considered that if the bottom outer contour immersion is more than 40 mm when the inhomogeneity zone sector length is 20 m there occurs an increased stress-strain state in metal tank structures, which confirms the requirements of the regulatory documents.

The developed model considering the contact interaction of the base with the structure has allowed establishing the acting stresses in the tank metal structures and identifying worst case deformations in the presence of inhomogeneous immersion [22-23]. At the inhomogeneity zone sector length of 30-50 meters the maximum stresses in all load-bearing elements of RVSPK-50000 exceed the yield point of steel. However, at maximum values of the inhomogeneity zone sector (L = 70–95.3 m) the level of SSS decreases. This is due to the stiffness characteristics of the structure and forms of possible deformations under nonaxisymmetric loads. Note that the greatest stresses occur in the three areas: the wind girder above the inhomogeneity area, in places of wall break at the interface of two subgrade types, as well as in the additional stiffening ring on the V wall belt.

Dependences of the acting equivalent stresses in the RVSPK-50000 metal structures on the inhomogeneity zone sector length for the metal structures of the I–XII wall belts, stiffening rings on the V and VIII belts, as well as the wind girder are given in Fig. 5.

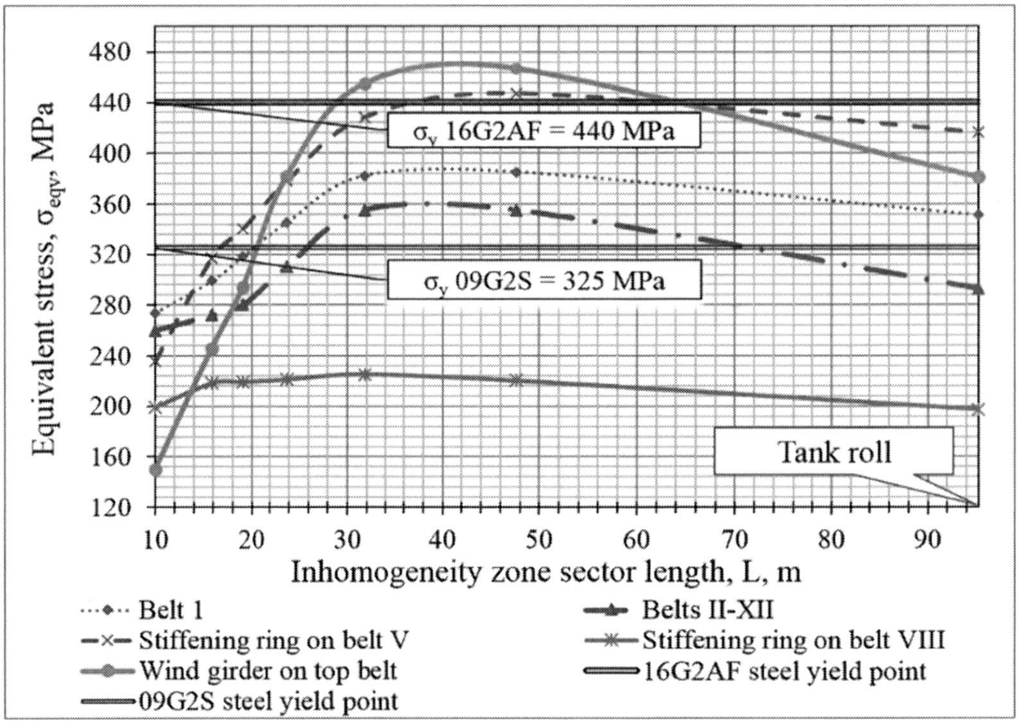

Fig. 5. Dependences of the acting equivalent stresses in the RVSPK-50000 metal structures on the inhomogeneity zone sector length.

4 Conclusions

1. A finite element model of RVSPK-50000 has been developed that allows us to determine the stress-strain state of the structure and soil mass in presence of the inhomogeneity zone taking into account the physical Drucker–Prager model, contact interaction of the "base-foundation-bottom-tank" system.

2. It has been established that if there is inhomogeneous immersion, separation of the reinforced steel ring from the annular plate and bottom results in a sharp increase in the stress-strain state of the entire structure. This highlights the need to develop and improve solutions to enhance the design of the foundation ring.

3. The dependences of the maximum design value of the outer tank bottom contour immersion on the inhomogeneity zone sector length have been obtained (in the range of 10 to 95 meters). It has been found that 95% of all cases of inhomogeneous immersion occurring in practice fall within this range according to data on diagnostics of 40 vertical steel tanks.

References

1. G.G. Vasilyev, A.A. Salnikov, Neftyanoe khozyaystvo, Oil Industry, **2**, 106-108 (2015)
2. V.B. Galeev, *Stress-strain state of tanks built on weak waterlogged soils, DSc. Thesis* (Tyumen, 1987)
3. A.A. Tarasenko, A.A. Gruchenkova, P.V. Chepur, Pipeline transport: Theory and Practice, **53**, 32-37 (2016)
4. V.A. Ilyichev, R.A. Mangushev, *Geotechnics reference. Bases, foundations and underground structures* (Publishing house ASV, Moscow, 2014)

5. A.A. Tarasenko, P.V. Chepur, Fundamental research, **12-1**, 67-84 (2014)
6. A.S. Gorelov, *Inhomogeneous ground base and their influence on the operation of vertical steel tanks: monograph* (Nedra, SPb, 2009)
7. P.A. Konovalov, R.A. Mangushev. S.N. Sotnikov, A.A. Zemlyanskiy, A.A. Tarasenko, *Foundations of steel tanks and deformations of their bases* (Publishing house Associacii stroitel'nyh vuzov, Moscow, 2009)
8. A.A. Tarasenko, M.V. Sayapin, Oil and Gas, **1**, 52-56 (1999)
9. G.G. Vasilyev, A.A. Tarasenko, P.V. Chepur, Y. Guan, Neftyanoe khozyaystvo, Oil Industry, **10**, 120–123 (2015)
10. A.A. Tarasenko, P.V. Chepur, D.A. Tarasenko, Neftyanoe khozyaystvo, Oil Industry, **4**, 88–91 (2015)
11. A.A. Tarasenko, P.V. Chepur, S.V. Chirkov, Neftyanoe khozyaystvo, Oil Industry, **10**, 121–123 (2014)
12. A.A. Tarasenko, P.V. Chepur, S.V. Chirkov, Neftyanoe khozyaystvo, Oil Industry, **3**, 123–125 (2016)
13. I.V. Slepnev, *Stress-strain elastic-plastic state of steel vertical cylindrical tanks with irregular immersion of bases* (MISI, Moscow, 2014)
14. A.A. Tarasenko, P.V. Chepur, A.A. Gruchenkova, Fundamental research, **12-7**, 1418-1422 (2014)
15. P.V. Chepur, *Stress-strain state of the tank in the development of irregular immersion of its foundation: thesis for the Candidate of technical sciences degree* (Gubkin Russian State Oil and Gas University, Moscow, 2015)
16. P.V. Chepur, A.A. Tarasenko, Fundamental research, **2-8**, 1671-1675 (2015)
17. P.V. Chepur, A.A. Tarasenko, Fundamental research, **7-1**, 95-100 (2015)
18. P.V. Chepur, A.A. Tarasenko, Fundamental research, **12-1**, 97-102 (2015)
19. P.V. Chepur, A.A. Tarasenko, A.A. Gruchenkova, Fundamental research, **12-3**, 514-519 (2014)
20. S.V. Chirkov, A.A. Tarasenko, P.V. Chepur, Oil and Gas, **5**, 72-78 (2014)
21. A.A. Tarasenko, P.V. Chepur, Y. Guan, Neftyanoe khozyaystvo, Oil Industry, **4**, 134–136 (2016)
22. *GOST R 52910-2008. Vertical cylindrical steel storage tanks for oil and petroleum products* (Standartinform, Moscow, 2007)
23. *RD-23.020.00-KTN-283-09. Rules of repair and reconstruction of oil storage tanks 1000-50000 cubic meters in capacity* (2009)

Effective Structures of Tamped Foundations of Frame Buildings

Leonid Bartolomey [1,*], Alexey Glushkov[2], and *Vyacheslav* Glushkov[2]

[1]Industrial University of Tyumen, 625001 Volodarskogo str. 38, Tyumen, Russia
[2]Volga State University of Technology, 424000 Lenin Sq. 3, Yoshkar-Ola, Russia

Abstract. The application results of foundations in a tamped ditch in complex geotechnical conditions are strongly considered. The article informs about the experience of the effective foundation application instead of pile foundations. Also, a special attention is given to stress-strain analysis in the active zone of the soil basement of the foundation in a tamped ditch. The article exposes innovative technology of transformation the soil materials structure. Presents the results of studies of innovative soil materials state in the field of geotechnics and foundation engineering. This article seems to be interesting for those who work in the field of building construction and geotechnics.

1 Introduction

Recently in construction practice the quick-mounting houses with a metal framework and reinforced concrete overlapping and covering find application. As walls heat-effective sandwich panels are used. The traditional solution of the frame building foundations in complex geotechnical conditions are pile foundations made from big length piles. The available domestic designing of the foundations in similar soil conditions indicates on an opportunity and advisability of the fullest use of the bearing capacity of tamped foundations [3, 4, 5]. The most reasonable structures of the frame building foundations in complex geotechnical conditions are tamped foundations [1, 2, 3].

Experimental results demonstrate that tamped foundations have the bearing capacity by 2.0-2.5 times bigger in comparison with the prismatic pile of equal volume [6]. The significant increase in the bearing capacity is explained by a large soil compression and the other nature of spatial interaction with the foundation basis in comparison with prismatic piles [7, 8].

2 Technological features, experimental and theoretical results

2.1 Technological features of the foundation making

Tamped foundations are made by the pile engine installation equipped with the diesel hammer and the conical operating element (fig. 1a). At a tamping of ditches a metal tamper

* Corresponding author: tgasu.bartolomey.leonid@mail.ru

by the size of 900 mm on top, 200 mm low, 2.5 m long, working by the principle of a wedge with spring-return was used (fig. 1b). The optimum size of the ditch lateral face inclinations to a vertical makes 0.14-0.20 [9].

The spring mechanism is used to avoid the exception of jamming, suction of a tamper in the basis and its extraction from the soil. The principle of the device operation while tamping foundations is shock-mechanical.

The conical tamper is fastened to the guideway with the bracket, and the impact action from the diesel hammer is transferred to the operating element through the cap. During the work of the diesel hammer as a result of the repeated impact action the conical tamper plunges into the soil basis. When the reaction pad achieves the surface of the soil, pad springs begin to contract, and at the moment when the cumulated compressive force of the pad springs become bigger than friction forces and the surface coherence of the operating element upon the ditch walls, there is a separation of a conical tamper from the soil.

Due to the work of the baseplate which is located along the perimeter of a tamped ditch, the ground softening, soil loosening and its uplift is excluded in the area next to the surface of the ditch. (Fig. 2, 3)

a b

Fig. 1. (a) pile engine; (b) metal tamper.

a b

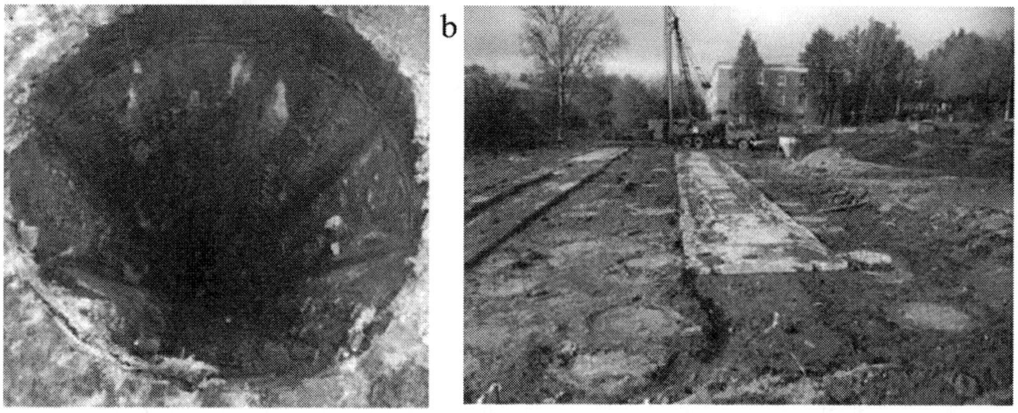

Fig. 2. Ditch (a) after tamping; (b) after concreting.

Fig. 3. (a) metal frame; (b) with reinforced concrete overlapping and covering.

The process of tamping is followed by soil compression with the formation of the compaction zone around the sidewalls and is lower than the plane of a metal tamper connected with the creation of a new structure and the reduction of the void ratio e [10].

At a tamping of a ditch there is an elliptic compaction zone which increases the soil density and improves strength and deformation properties. The control of the soil density in the basis of shallow foundations is determined by the soil resistance to penetration of a micropenetrometer tip in it.

The thickness of a packed layer under the certain foundation in a tamped ditch makes (1.6-2.2)bm below the plane of the cone edge, the width of the compaction zone makes (2.2-2.6)bm respectively, where bm is a diameter of a tamped ditch on an average depth section. The density of the packed soil at a depth of 20-30 cm from the bottom of the tamped ditch makes ρd = 1.70-1.75 g/cm3 at a moisture degree of the packed soil Sr = 0.6-0.7, and ρd > 1.75 g/cm3 at Sr < 0.6. In a packed soil volume after the tamping of ditches in the basis the specific cohesion of soil C increases by 5-7 times, the module of deformation E0 increases by 2-5 times. The depth where the density of the dry soil reaches the value is usually ρd = 1.60 g/cm3, it is accepted as the lower bound of the compaction zone. The greatest efficiency of packing is reached in case of tamping of ditches at the optimum moisture content ω0, and is determined by the formula:

$$\omega_0 = \omega_p - (0.01 \div 0.03) \tag{1}$$

where ωp is a plastic limit.

2.2 Experimental and theoretical investigation of the bearing capacity

Full-scale tests of the deformable foundation basis in the tamped ditches reflect the aspects of spatial interaction of frame buildings and the basis more precisely. In-situ experiments in full-scale conditions were made on three grounds for the purpose of a reliable assessment of the bearing capacity of tamped foundations.

The experimental ground No. 1 of an apartment frame house is located in Ufa. According to the geomorphological investigations the site is related to the III left-bank terrace valley of the river White. The geological section of a site of researches up to the explored depth of 15.0 m is presented by quaternary, neogen sediments. On top there is loam (adQ) from semisolid to stiff consistence with an opened power of 3.5-7.8 m. Loam is spread by semisolid clay (N23-Q1) of a thick layer of 2.1-10.2 m and sand of average fineness with the power up to 2.0 m. From the depth of 8.0-12.0 m clay is semisolid, neogen (N2ak).

The experimental ground No. 2 of a hostel where 277 people live is located in Nizhny Novgorod. In the geomorphological relation the ground (platform) is in the centre of the East European Plain, on the contact point (interface) of the Western part of the Volga Highlands, the Volga-Vetluzhsky Plain and the Mari Lowland. Sediments of the Perm (P2) and quaternary (Q) systems take part in a geological structure of the working area up to the depth of 60.0 m. In the soil profile of the explored ground (platform) up to the explored depth the following engineering- geological elements are met: IGE 1 – technogenic formations (tIV) – are presented by the mixture of resedimented sand and sandy loam of grey colour; IGE 2 – blanket layer of loams (pr, d, II-III); IGE 3 – alluvial fluvioglacial sediments (and, f, IIdn) – fine-grained sand, clay with fine gravel of metamorphic rocks. Ground waters are assumed at a depth of 2-5 m from the earth's surface.

The experimental ground No. 3 of a dwelling house is located in the town Reshetikha of the Nizhny Novgorod Region. In the top part of the alluvial layer of loams under filled-up soil lie quartz (adQIII), fine-grained sands and sands of average density with the power up to 7.7 m. Quarternary layer loams are spread by the formations of Urzhumsky tier of Central Perm and put by stiff clays and siltstones. Below at the depth of 56 m lie karstic carbonate rocks and sulphatic breeds of the Kazan and Sakmarsky tiers of Central and Low Perm.

Experimental investigation of the bearing capacity of tamped foundations were carried out according to State Standard 5686-2012 Soils. "Field test methods by piles" [11]. For the ultimate strength of the tamped foundation there is a loading under the pressure of which the tested foundation has received the settlement equal to s and is determined by the formula (2):

$$s = \zeta \cdot S_{u,mt} \tag{2}$$

where $S_{u,mt}$ – the limiting value of the foundation settlement of the designed frame building is established according to the BC 22.13330, $S_{u,mt}$ = 15 cm (civil buildings with metal frame) [12];

ζ – transfer factor from the limiting value of the foundation settlement of the building Su, mt to the settlement of the tamped ditch, $\zeta = 0.3$.

A structural design load on the pile foundation is determined by the formula:

$$N = \frac{F_d}{\gamma_k} \tag{3}$$

where γ_k is the safety factor on soil for the bearing capacity of the tamped foundation according to the results of field test methods by the static load, $\gamma k = 1.0$.

Thus, taking the results of static tests, the calculated load on the tamped foundation of frame buildings is defined at settlement S within 45 mm.

Theoretical studies of the bearing capacity of tamped foundations are made with the use of the geotechnical complex Plaxis. The spatial elastoplastic analysis was carried out by the method of final elements with the simultaneous accounting of strength and deformation properties of the basis [13, 14]. As yield condition while carrying out the analysis the limit equilibrium equation of Mohr-Coulomb was accepted [15, 16]. The design model of the basis considered the formation of the compaction zone while tamping a ditch.

The dead load of soil was carried out in the form of initial stresses, the deformed condition of the basis was defined only from the external load on the foundation [17].

Matching conditions of the soil basis are 8.0×8.0×12.0 m (fig. 4). Under the loads on the frame column more than 750 kN the foundation is made from two combined tamped ditches with the monolithic reinforced concrete capping.

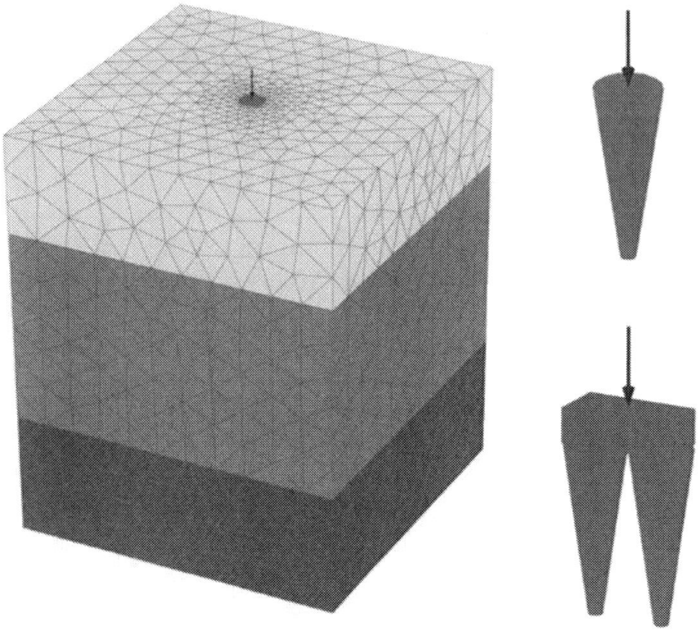

Fig. 4. A design model of the method of final elements.

Comparison of the results of experimental and theoretical investigations of the bearing capacity of tamped foundations on experimental grounds No. 1, 2, 3 is given in fig. 5-10.

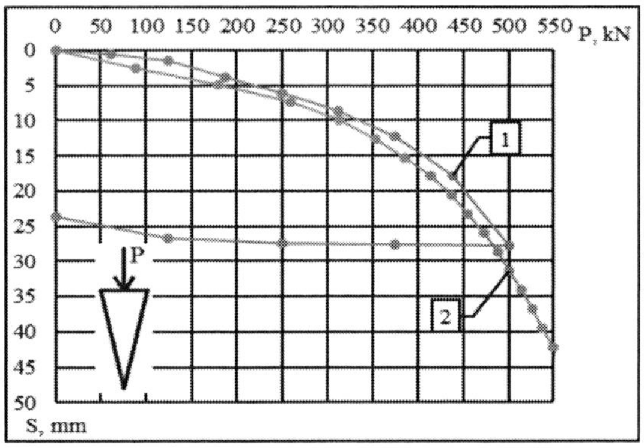

Fig. 5. Dependence of S=f (P) for platform No. 1:
1 – experiment; 2 – calculation.

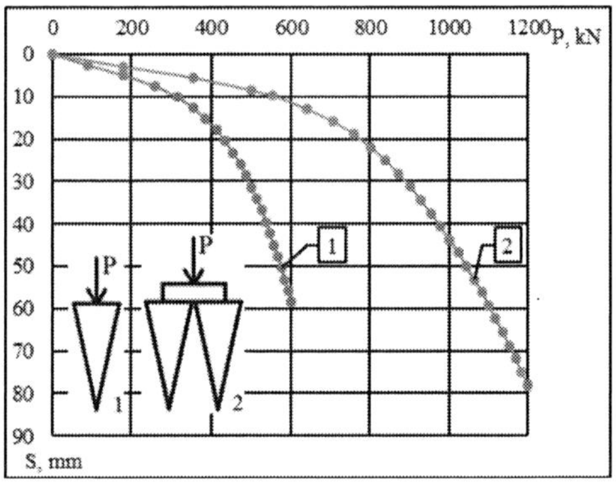

Fig. 6. Results of the calculation of tamped foundations:
1 – single; 2 – double (platform No. 1).

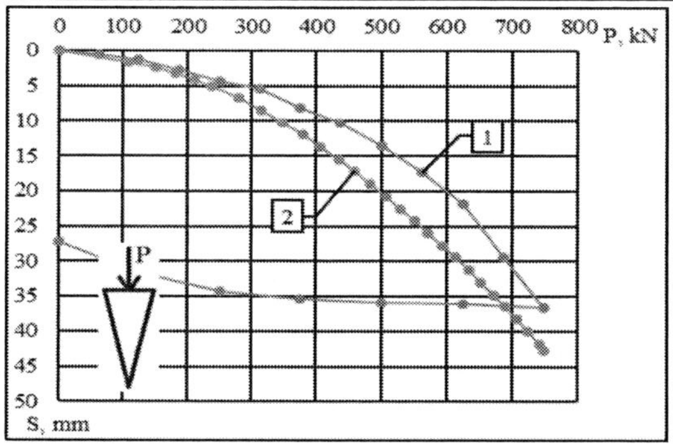

Fig. 7. Dependence of S=f (P) for platform No. 2:
1 – experiment; 2 – calculation.

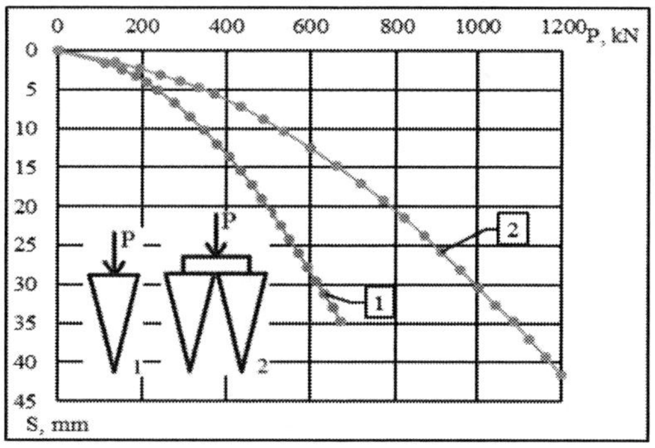

Fig. 8. Results of the calculation of tamped foundations:
1 – single; 2 – double (platform No. 2).

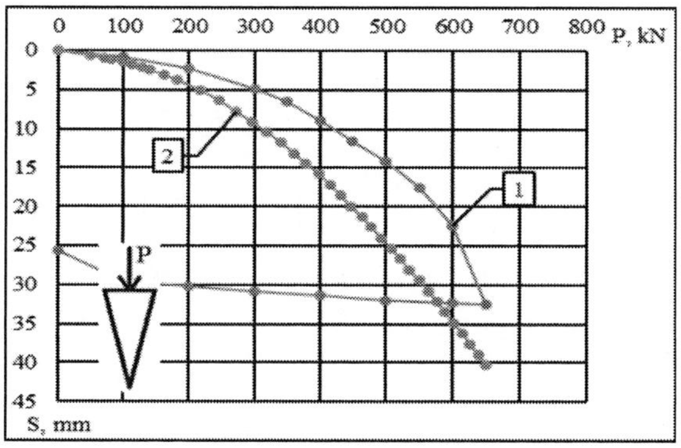

Fig. 9. Dependence of S=f (P) for platform No. 3:
1 – experiment; 2 – calculation.

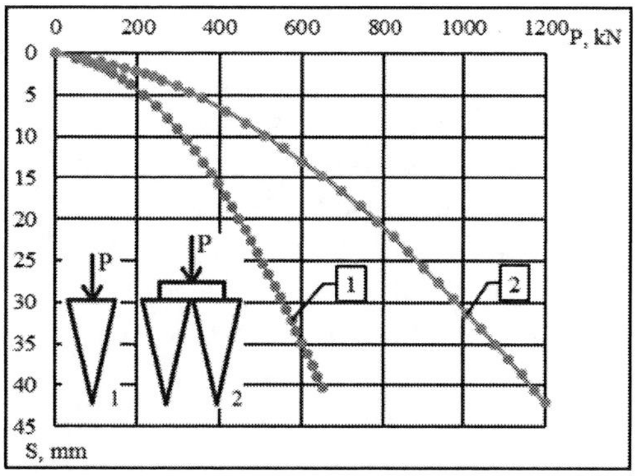

Fig. 10. Results of the calculation of tamped foundations:
1 – single; 2 – double (platform No. 1).

For the increase of the bearing capacity of the tamped foundation a solid material is tamped into the foundation ditch. It allows to increase the area of the spread foundation below the plane of the cone edge and leads to the increase of the bearing capacity by 1.2-1.5 times (fig. 11) [18, 19].

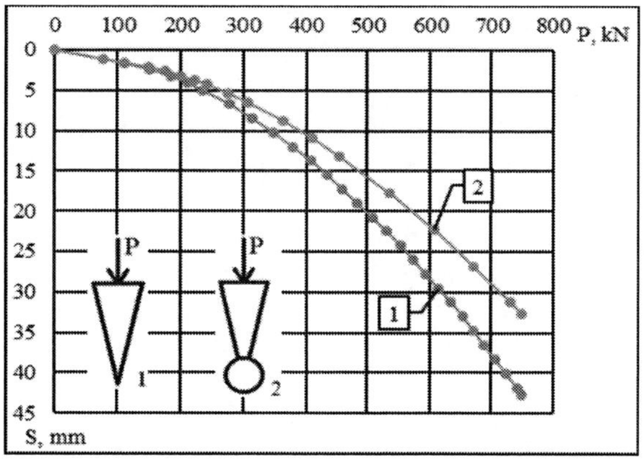

Fig. 11. Dependence of S=f(P) of the ditch on platform No. 2 (1) without tamping of crushed stone and (2) with tamped crushed stone in the bottomhole of the cone edge

3 Conclusion

The use of tamped foundations for frame buildings allows to reduce material capacity of a zero cycle by 1.5-2.0 times, to reduce the volume of excavation and shuttering works. The load carrying capacity of the foundations in tamped ditches on experimental grounds No. 1, 2, 3 makes 820-1120 kN/m3 that is significantly bigger than for traditional foundation designs. The use of nonlinear mechanics of soil allows to estimate the stress-strain state of tamped foundation bases in the high dynamic range of changing the loads up to their limits [20]. For the practical application the engineering method is allowed to determine the tamped foundation settlement taking into account the strength and deformation properties of soil. As the mathematical model, connecting the value of settlement with the initial parameters the multifactor nonlinear dependence is accepted. It allows to design tamped foundations by using maximum permissible deformations.

References

1. A.A. Bartolomey, V.E. Glushkov, I.V. Glushkov, Combined foundation, in: 52 World exhibition of innovation, research and new technology, Brussels.
2. V.E. Glushkov, I.V. Glushkov, Tamped foundation with the central pile, Patent 12148 RF, in: Rospatent, Bul. 12. (1999)
3. D.D. Ahmetov, A.P. Konovalov, L.O. Epaneshnikov, Soil Mechanics and Foundation Engineering, **3**, 21–25 (2005)
4. V.I. Krutov, Yu.A. Bagdasarov, I.G. Rabinovich, *Fundamenty v vitrambovannykh kotlovanakh* (Stoyizdat, Moscow, 1985)
5. Iu.E. Ponomarenko, M.V. Maksimov, V.I. Krutov, L.V. Erofeev, Mechanics and Foundation Engineering, **6**, (1991) 19–21.
6. *Manual on design and construction of foundations in tamped pits* (Stroiizdat, Moscow)
7. I.V. Glushkov, trudy Mejdunarodnogo nauchno-praknicheskogo seminara, Perm, **2**, (2005) 13–18
8. A.A. Bartolomey, I.V. Glushkov, Sbornik nauchnykh trudov Kubanskogo gosudarstvennogo agrarnogo universiteta. Krasnodar, (2003) 174–181
9. V.E. Glushkov, Equipment for tamped foundation making, Patent 4307 RF, in: Rospatent, Bul. 6, (1997)
10. A.L. Gotman, Iu.M. Shemenkov, Vestnik PNIPU. Perm, **3**, (2015) 23–40
11. *State Standard 5686-2012 Soils. Field test methods by piles* (Moscow, 2013)
12. *Building Code 22.13330.2011. Soil bases of buildings and structures* (Moscow, 2011)
13. A. Lyav, Matematicheskaya teoriya uprugosti (Math elasticity theory), Moscow, ONTI, 1935
14. V.G. Korotkin, Sb. Gidroenergoproekta, **4**, 52-85 (1938)
15. V.A. Florin, *Osnovy mekhaniki gruntov* (Gosstroyizdat, Leningrad, 1959)
16. H.W.Wei, Y.Z. Wu, Z.H. Yu.Design, Journal of Central South University, **19(6)**, 1753–1763 (2012)
17. H.S. Yu, Springer Science + Business Media, 2000
18. V.Ia. Shishkin, A.A. Anikeev, Zhilizhchnoe stroitel'stvo, **9**, 33–37 (2012)
19. S. Van Baars, Soils and Foundations, **54(1)**, 985–992 (2014)
20. M.E. KHarr, *Osnovy teoreticheskoy mekhaniki gruntov* (Stoyizdat, Moscow, 1971)
21. L. Epifantceva, O. Poroshin, Y. Pronozin, Building technology for strip foundations combined by membranes in heavy geotechnical conditions of Siberia, ICG (2014)
22. Z.G. Ter-Martirosyan, Y.A. Pronozin, N.Y. Kiselev, Soil Mechanics and Foundation Engineering (2014)
23. T. Maltseva, A. Nabokov, A. Chernykh, Reinforced Sandy Piles for Low-Rise Buildings. Procedia Engineering, **117**, 239-245 (2015)

Research of Spined Heat-Exchanging Pipes

Kirill Akulov [1,*], *Vasiliy* Golik [1] , *Mikhail* Chekardovsky [1], and *Anatoly* Serebrennikov [1]

[1]Industrial University of Tyumen, 625001 Volodarskogo str. 38, Tyumen, Russia

Abstract. Work is devoted to a research of spined heat-exchanging pipes that are assumed to use in air-cooler exchangers (ACE). The proposed new geometry of finning allows intensifying heat exchange and improving the efficiency of air coolers. It is caused by the increased area of finned surface with a value of finning ratio (the ratio of the area of the smooth pipe to a finned one) to 42.7, while in the commercially available ACE, the figure is 22. Besides, the geometrical arrangement of the pin fins turbulizes the airflow. It should be mentioned that an easier method of manufacturing of heat exchanging pipes is proposed to use, which will reduce their costs. The proposed heat exchange pipes are made by winding cut aluminum strip to the supporting pipe or stretching stamped blanks on it. To increase the efficiency of the heat exchange surface pin fins should be as thin and long as possible; however, their strength should be sufficient for deformation-free operation. Fins should be staggered to maximize the distance between them. Spined heat-exchange pipes are designed to operate in a commercially produced ACE and their service is carried out similarly to commercially produced transversely finned pipes.

1 Introduction

Gas industry at the present stage of development, is characterized by the fact that its resource base is located in remote northern regions of the country, and key gas consumers - in the central, industrial areas and abroad. This causes the necessity to transport natural gas at very long distances [1-23]. Under such circumstances, the final price for natural gas includes a significant percentage of transportation costs [4, 7, 8, 16]. It should be noted that at compressor stations with gas turbine-driven superchargers (which have the vast majority) most of the electricity spent on gas transportation is consumed on its cooling. Therefore, improvement of the cooling system is one of the primary challenges facing the gas industry to reduce energy consumption for gas transport and its cost.

2 The object of research

In this study we focus on one of the directions of intensification of the cooling of natural gas associated with convection on the surface of heat-exchanging pipes of air coolers (ACE) (Figure 1.).

* Corresponding author: akukir@gmail.com

Fig. 1. Air coolers on compressor station.

In ACE, commercially available transversely finned heat-exchange pipes are used (Figure 2.). The main way to increase the efficiency of a heat exchange surface is the increase of finning area [2, 10, 14, 15, 18, 19]. This metal-intensive method is characterized by the fact that with an increase in finning ratio, increases the relative depth of intercostal gap, root zone gradually excluded from the heat and the heat emission growth slows down. This problem was solved using a pin fining [1, 12, 13, 17] (Figure 3a., Figure 6.). This type of finning is almost not been studied and it was researched only with small finning coefficients not applicable in industrial air coolers. An advantage of heat exchanging pipes is that fins in the transverse plane are disposed at an angle to each other and by increasing of the length of the fin an intercostal spacing increase. Therefore, such a fin may have considerable height of fins without the formation of dead zones. The location and shape of fins contributes to the turbulence flow while turbulization in the intercostal cavities of transversely finned pipes is weak. It is noteworthy that the finned convective surface is not inclined to clogging and has an improved clean ability [20].

Fig. 2. Heat exchange pipes with wound spiral finning.

A technological way to the supporting pipe finning with pin fins was developed. For this purpose, the supporting tube of winding tape should make transverse cuts or incisions at intervals of several millimeters, depending on the thickness of the pins. Received petals will form pins of the radiator when pulling on the blank (Figure 3b.). The process is similar to the production of standard spiral-wound fins. However, there becomes no need in rolling rollers, flattening the ribbon to give it bending (Figure 3a.). Thus, the ribbon winding process can be produced without special equipment on a conventional lathe. Simplification of the production process will inevitably lead to products cheapening.

3 Methods

Experimental setup study of pipes designed and manufactured for the study of thermal characteristics of single finned tubes in air flow speed range up to 6 m / s. Installation diagram of the experimental research of finned tubes is shown in Figure 4. It represents a low-pressure open-loop wind tunnel with a width of the flow-cross section of 0.23 m and a height of 0.34 m. The air temperature is defined by two benchmark platinum resistance thermometers Pt100 class 1, one of which is a dry thermometer and another is wet. To adjust the air flow a section of centrifugal fans with variable speed is used. For alignment and smoothing the flow of air pressure pulsations after the fan section, the narrowing element was installed. Volumetric flow rate was measured by the Venturi nozzles, built into the flowmeter.

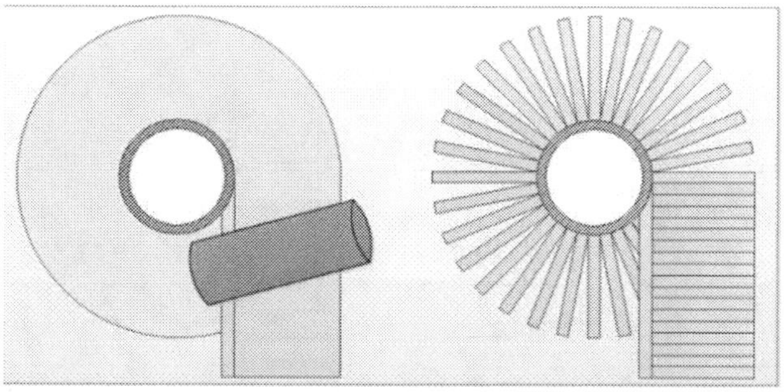

Fig. 3. Formation schemes: a) scheme of bending formation of the spiral fin by rolling rollers. b) scheme of formation of pin fins from ribbon with notches.

Studied finned tubes had pin finning of a round profile with geometrical parameters shown in Table 1. Finned tubes impaled on tubular electrical heaters (TEHs) with interference at nominal fit diameter of 10 mm. The length of finned tubes and the length of the active part of electrical heating element = 230 mm. The voltage at the electric heating elements U = 220 V, and their active resistance RU = 120 - 130 ohm. The ends of the ribbed pipes are heat-insulated with fluoroplastic caps.

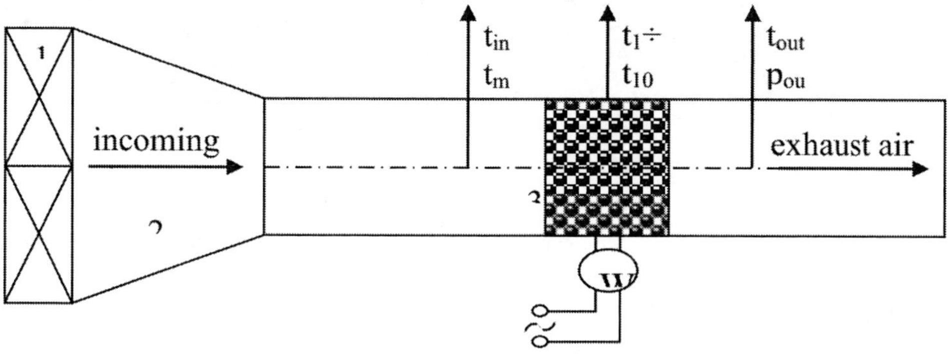

Fig. 4. Schematic scheme of the installation of experimental studies of finned tubes. 1. Section of fans with variable frequency of rotation; 2. narrowing element; 3. The studied finned pipe.

Experimental studies of heat emission of finned tubes consisted of 6 sets (the number of finned tubes). In each set, one finned pipe was studied. The finned pipes were heated by means of electric heaters built into them. To reduce the TCR between the finned pipes and heating elements, the inner surface of the pipe is pushed over the outer surface of the electrical heating element with an interference fit.

The heat transfer coefficient on the surface of the finned tube was determined by averaging over the entire outer surface of the finning, which is the sum of surfaces pits, tops and sides of fins. To calculate the heat transfer coefficient, the value of the forced convective component of the heat flow as the difference between the electric power heating finned tube of electric heaters and radiant heat flow component, as well as natural convective component of the heat flow was used [21]. Here radiant heat flow component was determined by the surface temperature of the finned tube and the walls of the channel of the wind tunnel. The ends of the finned tubes were heat-insulated, so heat loss from the ends was neglected.

Table 1. The geometrical finning parameters of the samples of finned pipes.

№ of sample	finning coefficient ψ	intercostal distance δ [mm]	fin thickness δf [mm]	The fin height hf [mm]	The diameter of the support tube d [mm]	The outer diameter of the finning D [mm]
1	41.7	2	3	95	35	22.5
2	37.4	2	3	85	35	20.5
3	33.1	2	3	75	35	18.5
4	28.9	2	3	65	35	16.5
5	24.6	2	3	55	35	14.5
6	20.3	2	3	45	35	12.5

Fig. 5. The appearance of machine of the experimental studies of finned tubes.

The temperature of finned surface elements was measured by copper-nickel (C-N) thermocouples. The temperature of peaks was measured by fins glued on a hot melt adhesive Alsil-2 to the ends of fins thermocouples. The temperatures of fins foundation was measured by fins, glued on a hot melt adhesive Alsil-2 to the base thermocouples. The rate of air flow sequentially took values from 0.5 to 6 m / s. The time of exit of a machine on a nominal mode at steady-state value of air flow rate and temperature of the surface of the finned tube is about 1 hour. After the installation of the nominal mode at steady-state value of air flow rate, such parameters as air speed, current and voltage of the heater, the average temperature of the sample, the temperature of the incoming and outgoing flow were measured. For each value of air flow rate, all values were measured by alarm-control system for 70 minutes at intervals of 30 seconds. For the user, the averaged values were displayed and used for further processing.

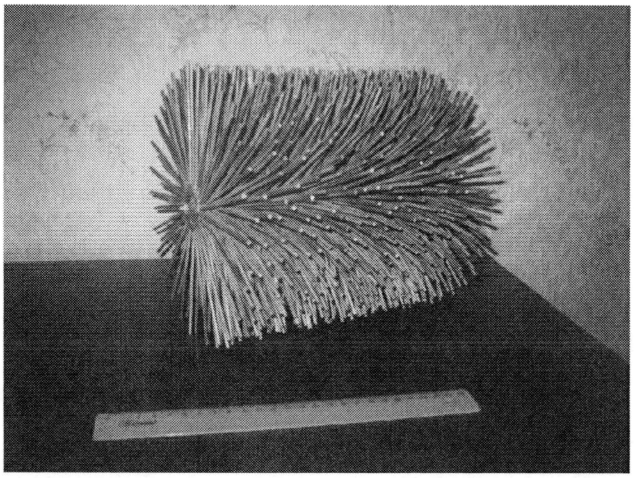

Fig. 6. Appearance of the sample №1.

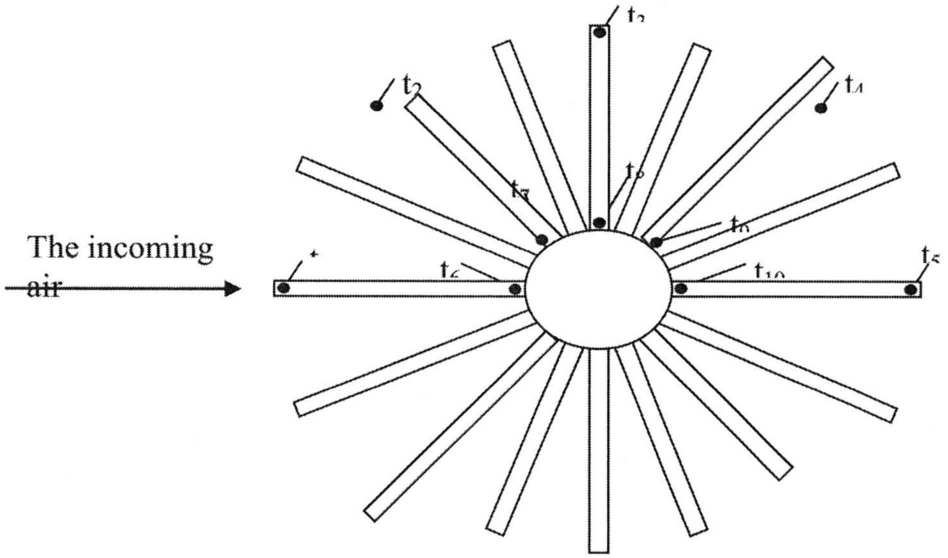

Fig. 7. The layout of thermocouples on the test sample.

To simplify the calculations, program was designed, the algorithm of which is presented in Figure 8. The program was written in tabular processor Microsoft Excel of Microsoft Office. The resulting calculation values of Nusselt and Reynolds numbers were approximated by the equation [5,11,20]:

$$Nu = c \cdot \mathrm{Re}^{m} \tag{1}$$

where Nu - Nusselt number from the air; Re - Reynolds number from the air; c and m - constant coefficients.

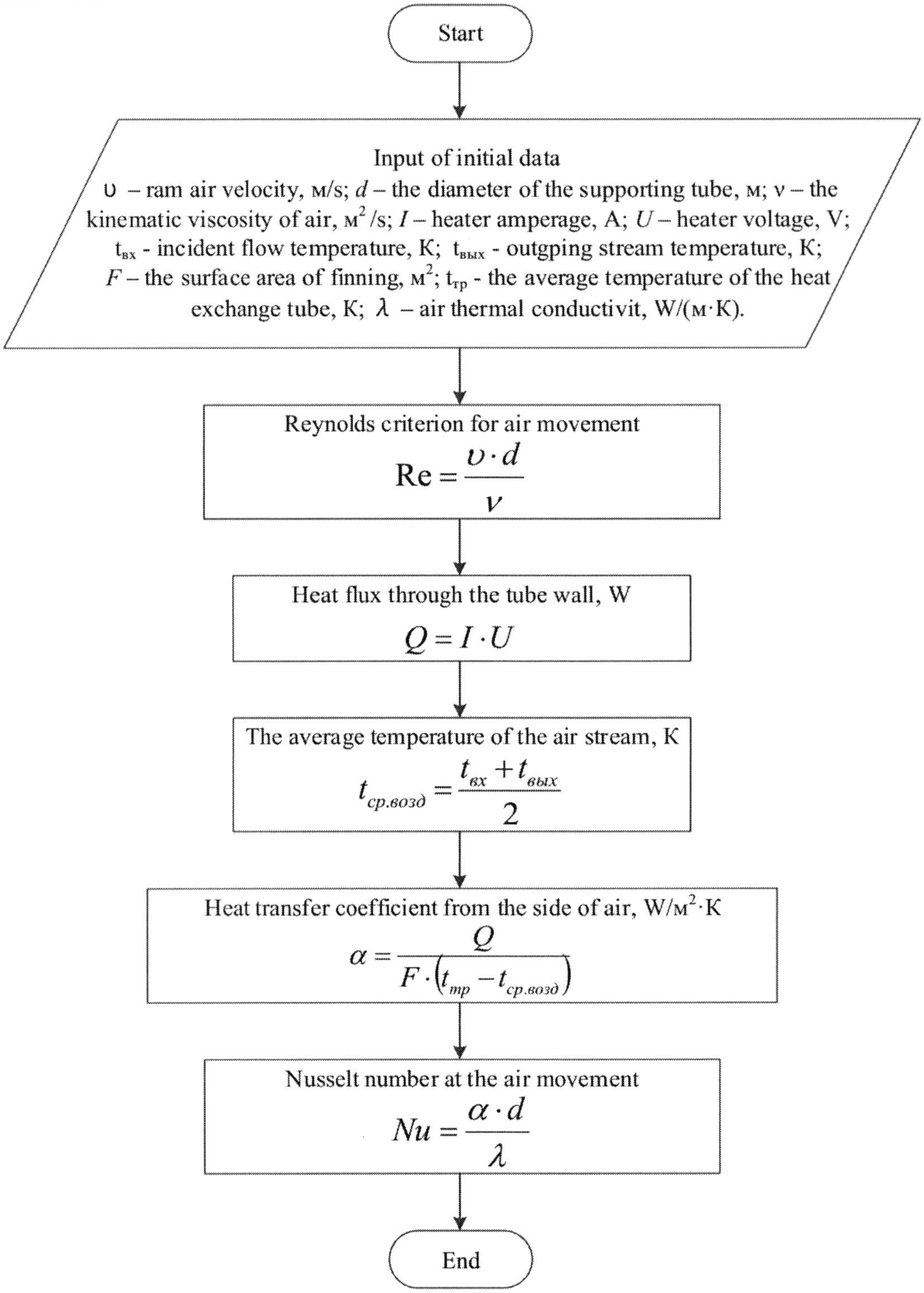

Fig. 8. Algorithm of processing of experimental data.

The generalized equation for the sample №1 is:

 DOI: 10.1051/matecconf/20167301027

$$Nu = 0,1454 \cdot Re^{0,541} \quad (2)$$

The generalized equation for the sample №2 is:

$$Nu = 0,1528 \cdot Re^{0,541} \quad (3)$$

The generalized equation for the sample №3 is:

$$Nu = 0,1602 \cdot Re^{0,541} \quad (4)$$

The generalized equation for the sample №4 is:

$$Nu = 0,1676 \cdot Re^{0,541} \quad (5)$$

The generalized equation for the sample №5 is:

$$Nu = 0,175 \cdot Re^{0,541} \quad (6)$$

The generalized equation for the sample №6 is:

$$Nu = 0,1824 \cdot Re^{0,541} \quad (7)$$

4 Results and Discussion

The dependence of the Nusselt number for total heat exchange surface of the Reynolds number is shown in Figure 9. For comparison, the figure shows the characteristics used in the ACE heat exchanging pipes coefficients with finning 22, 14.6 and 9 [3,6,9] (corresponding pipe 1, 2 and 3 in Figure 9.). For the reasons of clarity, characteristics of a smooth pipe are also shown.

Characteristics of spined pipes are more gentle than the transversely finned pipes. Heat transfer of spined pipes with finning factor 22 is more than transversely finned at 77% in the area Re = 5000 and 41% at Re = 40000. Spined pipe with a coefficient of finning 41.7 has heat transfer more by 190% and 131%, respectively.

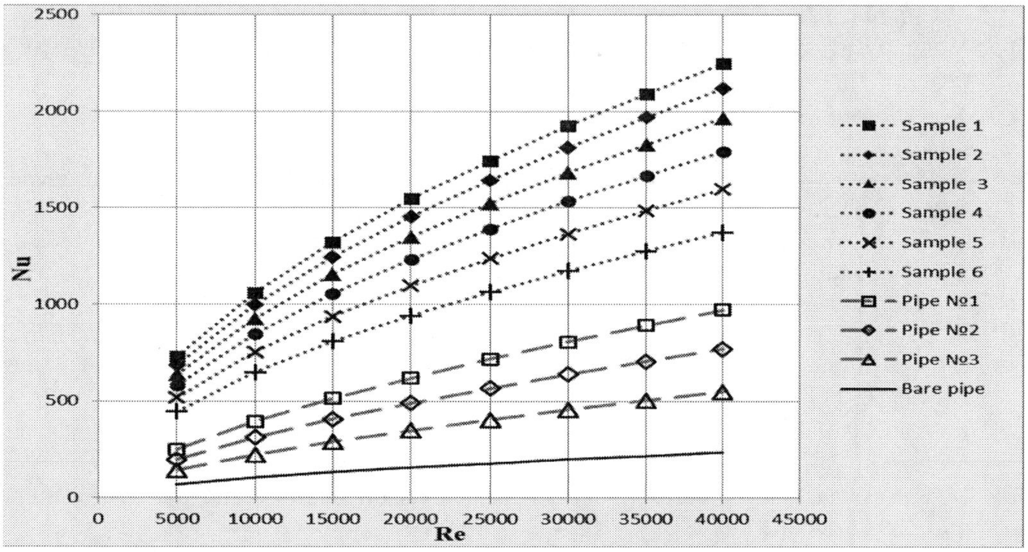

Fig. 9. Heat transfer of the researched and existing heat-exchange pipes.

5 Conclusion

Compared to standard, heat exchanger pipes studied in research possess enhanced heat transfer. Their manufacturing is simpler and allows doing without special equipment; the process can be established on a conventional lathe. Spined convective surface is not prone to clogging, and has an improved clean ability. These advantages make it possible to reduce the cost and increase the efficiency of heat exchange of air coolers.

References

1. K. Akulov, Problems of Functioning Transport Systems: Proceedings of the scientific-practical conference of students, graduate students and young scientists, TSOGU, Tyumen, 42-44 (2012)
2. K. Akulov, V. Golik, Problems of Functioning Transport Systems: Proceedings of the scientific-practical conference of students, graduate students and young scientists, TSOGU, Tyumen, 27-32 (2015)
3. K. Akulov, A. Istomina, New technologies - oil and gas regions: Proceedings of all-russian scientific-practical conference, TSOGU, Tyumen, 44-45 (2011)
4. K. Akulov, P. Matantsev, Oil and Gas Terminal: Proceedings of the International scientific and technical conference, TSOGU, Tyumen, 15-18 (2015)
5. K. Akulov, P. Matantsev, New technologies - oil and gas regions: Proceedings of all-russian scientific-practical conference, TSOGU, Tyumen, 45-49 (2011)
6. K. Akulov, P. Matantsev, A. Istomina, Oil and gas in Western Siberia: Proceedings of the international scientific conference, TSOGU, Tyumen, 5-8 (2011)
7. K. Akulov, V. Golik, T. Ponomareva, Fundamental researches, **12-3**, 453-456 (2015)
8. G. Bakhmat, N. Eremin, O. Stepanov, *Air cooling devices at gas compressor station* (Nedra, SPb, 1994)
9. A. Bessonny, G. Dreytser, V. Kuntysh, ed. V. Kuntysh, *Basement of calculation and designing of heat exchangers of air cooling: Dictionary* (Nedra, SPb, 1996)
10. K. Degtyarev, A. Soloviev, Young scientist, 1(8), 107-112 (2011)
11. A. Gabdrahmanov, N. Garris, Oil and Gas Business, **1**, 44 (2003)
12. *Handbook of Heat Exchangers, V. 1* (Energoatomisdat, Moscow, 1987)
13. V. Kuntysh, A. Suhotskiy, S. Filatov, A. Zhdanovich, Chemical equipment, **11**, 40-43 (2013)
14. V. Nekrasov, S. Dudin, Yu. Zemenkov, R. Levitin, Territory Neftegaz, **11**, 79-82 (2014)
15. V. Nekrasov, R. Levitin, A. Mayer, Higher Educational Institutions News. Oil and Gas, **5**, 63-65 (2014)
16. V. Nekrasov, R. Levitin, Yu. Zemenkov, Journal of the Tajik National University. Series of Natural Sciences, **1-2**, 140-144 (2015)
17. V. Nekrasov, Yu. Zemenkov, Pipeline transport: Theory and Practice, **6**, 24-26 (2012)
18. R. Levitin, Yu. Zemenkov,. Neftyanoe khozyaystvo - Oil Industry, **1**, 110-114 (2016)
19. V. Antip'ev, A. Nevolin, Yu. Zemenkov, Neftyanoe Khozyaistvo - Oil Industry, **10**, 46-48 (1981)
20. R. Mamadaliev, V. Kuskov, Yu. Zemenkov, A. Popova, Applied Mechanics and Materials, **770**, 19-22 (2015)
21. V. Kurushina, Yu. Zemenkov, WIT Transactions on Ecology and the Environment, 190, **2**, 881-888 (2014)
22. M. Omelyanyuk, Neftepromyslovoe delo, **6**, 62-65 (2008)
23. M. Chekardovsky, V. Ivanov, A. Khamidov, S. Chekardovsky, K. Ilyukhin, Higher Educational Institutions News. Oil and Gas, **1**, 44-49 (2008)

Development of Mathematical Model for Traffic Flow within Speed Limitation

Oleg Danilov[1,*], *Victor* Kolesov[1], and *Denis* Sorokin[1]

[1]Industrial University of Tyumen, 625001 Volodarskogo str. 38, Tyumen, Russia

Abstract. The paper discusses a symbiotic model of the traffic flow, based on the classical model of the longitudinal dynamic gauge of the automobile and an alternative model developed by the authors. The presented mathematical model makes it possible to quantify the dynamic characteristics of the traffic flow for various levels of speed limitation and for a wide range of road conditions. The problem of structural and parametric identification of the mathematical model is challenging. A solution to this problem is based on the traffic flow theory, regression analysis and dynamic programming.

1 Introduction

Traffic management, road-traffic safety and environmental risks are priorities for improving the standards of living. It is possible to meet the challenges provided the traffic flow management principles (TF) are realized on the road network (RNW). In turn, traffic flow management is intimately associated with the innovative approaches in the field of mathematical modeling. Thus, the TF dynamics analysis, identification of the structure and parameters of the TF models are the basis for sustainable development of the motor industry.

"Car-Following" or "Follow-the-Leader" models are widely used for road-traffic safety; they are based on the longitudinal dynamic gauge (DG) of the automobile. In 1963 Tanaka proposed some new models based on DG [1]. Updating of the proposed models, development of new alternative solutions will make it possible to complete the tasks related to the study of the TF dynamics [1-19].

2 Technique

The classical model of the longitudinal dynamic gauge is as follows:

$$L(V) = m_2 \cdot V^2 + m_1 \cdot V + m_0, \qquad (1)$$

L – longitudinal dynamic gauge, m; V – speed of the traffic flow, m/s; m2 – coefficient, c2/m; m1 – coefficient, c; m0 – coefficient, m.

* Corresponding author: tgasu.danilov.oleg@mail.ru

The TF density (q) is evaluated by the number of vehicles on the lane with a fixed length and is inversely proportional to the DG

$$q(V) = 1/L(V) = 1/(m_2 \cdot V^2 + m_1 \cdot V + m_0). \quad (2)$$

Intensity (N) is evaluated by the basic equation of the TF connecting the intensity with its density and speed

$$N(V) = V \cdot q(V) = V/(m_2 \cdot V^2 + m_1 \cdot V + m_0). \quad (3)$$

DG coefficients (m2, m1, m0) were carefully analyzed by the authors.

The classical model of Tanaka (1) does not take into account the factor of speed limitation on different sections of the road network. The solution to the problem of the given model adaptation to any speed rates must meet two requirements:

Firstly, the vehicle speed cannot exceed the given limit Vlim.=60 km/h (chosen is speed limitation for a city);

Secondly, the longitudinal DG appropriate for safe driving must not be violated, i.e. drivers follow the DG with regard to TF density growth.

Here, two issues are of practical interest:

How to describe speed-density ratio analytically (Vl(q)) in low values of q (0<q<qa);

What are the coordinates of the function transition point Vl(q) in transition to function VL(q) corresponding to the DG model (A(qa,Va)).

For rational choice of the function Vl(q) it is necessary to take into account a number of conditions [3]. Firstly, when q=0 Vl(q)=Vl(0)=Vlim. secondly, when q> 0 the function must steadily decrease and be differentiable, thirdly, when q=qa, the condition Vl(qa)=VL(qa) must be performed and fourthly, the derivatives of both functions must be equal in the point A. The following function is chosen as Vl(q):

$$V_l(q) = V\lim. - K_l \cdot q^2, \quad (4)$$

Vlim. – maximal speed with regard to limitation, km/h; Kl – coefficient, km3/h.

The value of Kl is evaluated from the condition of equality of speeds Vl(qa) и VL(qa) in the point A. Taking into account that Vl(qa)=VL(qa)=Va, let us find Kl from (4), expressing the speed in m/s, and density in Vehicles/m

$$K_l = \frac{3.6 \cdot [V\lim - V_a]}{1000^2 \cdot q_a^2}. \quad (5)$$

The function VL(q), characterizing TF performance in keeping to the longitudinal DG can be presented in the following way

$$V_L(q) = \frac{-m_1 + \sqrt{m_1^2 - 4 \cdot m_2 \cdot (m_0 - 1/q)}}{2 \cdot m_2}. \quad (6)$$

Let us find the derivatives of the functions Vl(qa) and VL(qa) to meet the conditions of their conjugation in the point A. In the first case, when applying the equation (4) and transforming V into m/s, and q – into Vehicles/m, we obtain

$$\frac{dV_l}{dq}=-\frac{2\cdot K_l\cdot 1000^2\cdot q_a}{3.6}=-\frac{2\cdot K_l\cdot 1000^2}{3.6\cdot L_a}, \tag{7}$$

La – the value of the dynamic gauge in the point A.
Taking into account (1) let us find the derivative of the second function

$$\frac{dV_L}{dq}=-\frac{(m_2\cdot V_a^2+m_1\cdot V_a+m_0)^2}{m_1+2\cdot m_2\cdot V_a}=-\frac{L_a^2}{m_1+2\cdot m_2\cdot V_a}. \tag{8}$$

Making (7) and (8) equal, we obtain

$$K_l=\frac{3.6\cdot L_a^3}{2\cdot 1000^2\cdot(m_1+2\cdot m_2\cdot V_a)}. \tag{9}$$

In here, the condition imposed by the ratio (5), where qa=1/La is needed to be met. Simultaneous solutions (5) and (9) make it possible to note

$$\frac{3.6\cdot(V\lim-V_a)\cdot L_a^2}{1000^2}=\frac{3.6\cdot L_a^3}{2\cdot 1000^2\cdot(m_1+2\cdot m_2\cdot V_a)},$$

$$5\cdot m_2\cdot V_a^2+(3\cdot m_1-4\cdot m_2\cdot V_{огр})\cdot V_a+(m_0-2\cdot m_1\cdot V_{огр})=0,$$

$$V_a=\frac{-(3\cdot m_1-4\cdot m_2\cdot V_{огр})+\sqrt{(3\cdot m_1-4\cdot m_2\cdot V_{огр})^2-4\cdot 5\cdot m_2\cdot(m_0-2\cdot m_1\cdot V_{огр})}}{2\cdot 5\cdot m_2} \tag{10}$$

The equation (10) allows calculating the speed in the conjugation point A, when drivers follow the longitudinal DG.

When studying the association of the TF with its speed, one should consider two intervals. For 0≤V≤Va the function N(V) is determined by the ratio (3), and for Va≤V≤Vorp it is conditioned by the formula (4), which implies

$$N(V)=\sqrt{\frac{V\lim\cdot V^2-V^3}{K_l}}. \tag{11}$$

The DG coefficients should be chosen to develop the basic models of the TF. To do this, the completed algorithms have been used (Danilov et al., 2014). The values of m0=8 m, m1=1,6 s have been taken. The coefficient m2 is calculated for the rough asphalt concrete pavement and models for different operation conditions of the TF are made: dry pavement, wet clean, wet dirty, compacted snow, soft snow, clear ice (status of the pavement condition is r = 1..6 respectively). Table 1 shows the parameters of the model.

Table 1. Model parameters for the rough asphalt concrete pavement.

Status of the pavement condition r	m_2	V_a, km/h	q_a, Vehicles/km	K_l
1	0.016159	36.02	39.02	0.015748
2	0.020056	36.44	38.09	0.016235
3	0.026390	37.09	36.65	0.017057
4	0.038427	38.17	34.15	0.018721
5	0.069727	40.28	28.88	0.023651
6	0.312132	45.25	12.92	0.088421

3 Results

Fig. 1 illustrates the speed-density ratio of the traffic flow at various states of the road pavement. The conjugation points are highlighted in red. The left parts of the graphs evaluated by (4) are highlighted in blue, and the right ones corresponding to the DG model – in black. It is obvious that for the relatively favorable conditions of road traffic (r = 1.3) the curves are hardly ever distinguishable, and conjugation points are concentrated in a small area. In this case it seems reasonable to combine such curves into a single group. In fact, taking into account the applied modeling of the TF, the road conditions are to be differentiated into three types: normal, snowy and icy.

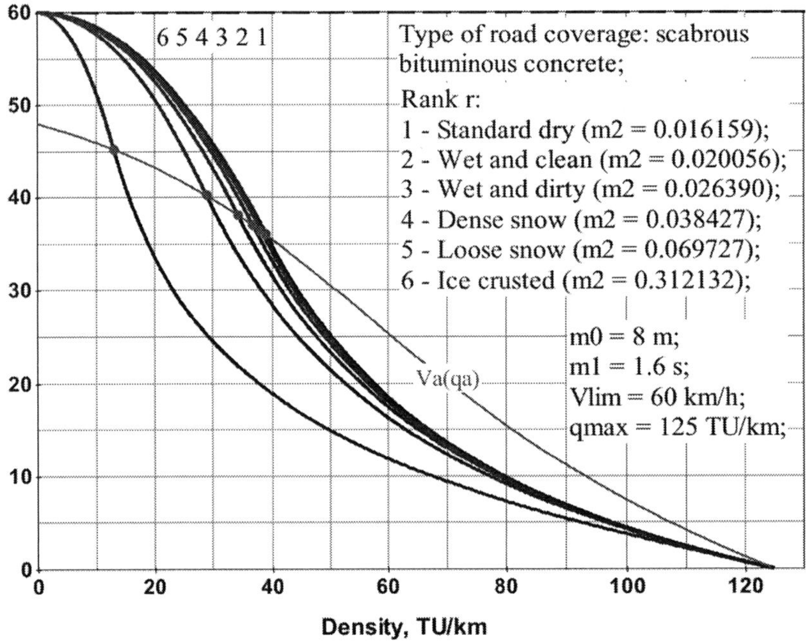

Fig. 1. Speed-density ratio of the traffic flow.

At the same time, the coefficient m2 determining the shape of the line V (q) depends on the structure of the TF [4]. Therefore, the calculated values of m2 (Fig. 1) can be considered conventional complying with the particular case. Thus, needed is the identification of the drift trajectory of the point A, the position of which is caused by the parameter m2.

Let us express the coefficient m2 using the simultaneous solution of the equations (5) and (9)

$$m_2 = \frac{2 \cdot m_1 \cdot V\,\text{lim} - 3 \cdot m_1 \cdot V_a - m_0}{5 \cdot V_a^2 - 4 \cdot V\,\text{lim} \cdot V_a}. \tag{12}$$

On the other side, as it follows from (3)

$$m_2 = \frac{1}{N \cdot V} - \frac{m_1}{V} - \frac{m_0}{V^2}. \tag{13}$$

Making (12) equal to (13) with regard to V=Va, we obtain the solution

$$N_a = \frac{5 \cdot V_a^2 - 4 \cdot V\,\text{lim} \cdot V_a}{2 \cdot m_1 \cdot V_a^2 - 2 \cdot m_1 \cdot V\,\text{lim} \cdot V_a + 4 \cdot m_0 \cdot V_a - 4 \cdot m_0 \cdot V\,\text{lim}}. \tag{14}$$

The trajectory of the point A(Va,Na) is shown in Fig. 2 (highlighted in red).

As for the position of the point of extremum intensity, here it is necessary to consider two situations. In the interval of 0≤V≤Va the rate of Nmax advances is determined by finding the derivative of the function (3) making it possible to obtain the formulas

$$V_{\max N} = \sqrt{m_0 / m_2}, \tag{15}$$

$$m_2 = \frac{m_0}{V_{\max N}^2}, \tag{16}$$

VmaxN – the rate of maximum intensity advances.

Using the simultaneous solutions of (16) and (13) we obtain the following formula taking into account that the actual intensity corresponds to the maximum one:

$$N_{\max} = \frac{V_{\max N}}{m_1 \cdot V_{\max N} + 2 \cdot m_0}. \tag{17}$$

In the interval of Va≤V≤Vlim the speed VmaxN is in the cross point of the functions Na(Va) and Nmax(VmaxN) (marked as B(Vb,Nb)). Point B is highlighted in dark blue in Fig. 2. The coordinates of B can be determined analytically making (14) equal to (17) taking into account that Va=VmaxN=Vb. Then, after mathematical transformations we obtain

$$V_b = \frac{-(6 \cdot m_0 - 2 \cdot m_1 \cdot V\,\text{lim.}) + \sqrt{(6 \cdot m_0 - 2 \cdot m_1 \cdot V\,\text{lim.})^2 - 4 \cdot 3 \cdot m_1 \cdot (-4 \cdot m_0 \cdot V\,\text{lim.})}}{2 \cdot 3 \cdot m_1}. \tag{18}$$

The analysis (18) suggests that within the interval of Va≤V≤Vlim. the point of extremum intensity is attained at the fixed rate of VmaxN=Vb (Fig. 2). In here, the value of Nmax is evaluated by the formula (11), which implies

$$N_{\max} = \sqrt{\frac{V\,\text{lim.} \cdot V_{\max N}^2 - V_{\max N}^3}{K_l}}. \tag{19}$$

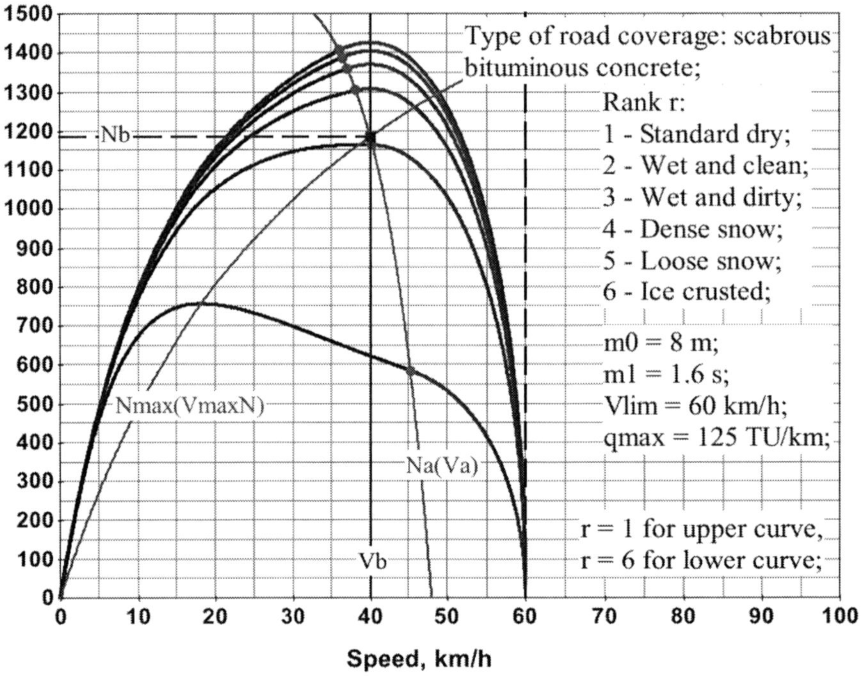

Fig. 2. Intensity-speed ratio of the traffic flow.

In this case, Vb = 40 km / h, and Nb = 1184 Vehicles / h. Substituting the data obtained in (13) we find the value of m2, where the curve N (V) passes through point B (the condition Nmax = Na is met). We obtain m2 = 0.064849. Thus, the developed models with the parameter of $m2 \leq 0.064849$ will have the maximum intensity of the TF at the speed of VmaxN=Vb=40 km/h, and the models with $m2 \geq 0.064849$ - at the speed evaluated by the formula (15).

It is obvious that the distance between the vehicles in the density range of $0<q<qa$ exceeds the value of DG (Fig. 1). Therefore, further research is needed to introduce the following concept - the average interval of vehicle movement. The average interval of movement (ID) is the distance between the vehicles in traffic environment (TE). For the density interval $qa \leq q \leq qmaxID$ corresponds to the value of the longitudinal DG L(V), but the analysis of Fig. 1 suggests that in the range of $0<q<qa$ the condition ID(V)>L(V) is met. The evolution of the interval of vehicle movement is illustrated in Fig. 3. The black colour shows the function L(V) to the left of the conjugation point A (a dotted line marks its theoretical extension). The blue colour shows the function ID(V) which describes the interval of vehicle movement. Moreover, Fig. 3 illustrates the curve La(Va) which describes the drift of point A within the changing conditions of the TE (for different values of m2).

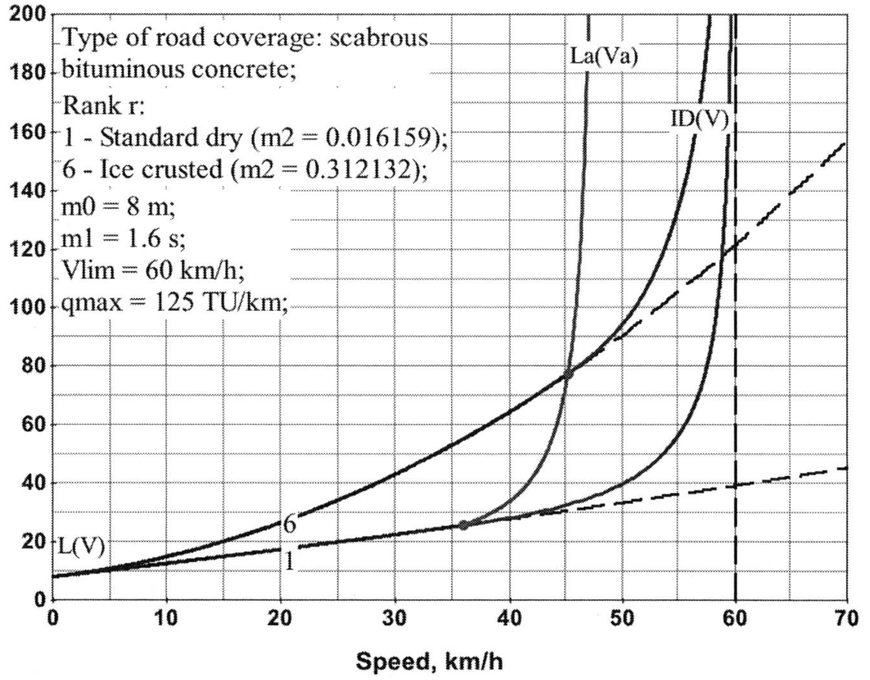

Fig. 3. Dynamics of average interval of movement ID.

It is obvious that the interval of movement in the TE gradually increases if the speed increases within the constraints of (V≤Vlim.). This fact radically distinguishes the developed alternate function (4) from the classical model of the DG. It has a logical explanation. Having passed from the associated TF mode to the free one, the driver does not tend to increase the speed of the vehicle to the maximum. In this case, the behaviour of the driver is still conditioned by the distance to the "leader" vehicle. The gist is simple: the greater the distance between the vehicles, the more freely a person feels, and the more subjected he is to speeding. In here, the driver accelerates to the maximum in the absence of vehicles in his visibility distance, or if they are at a great distance.

It should be noted that the developed model makes it possible to track the dynamics and other parameters of the TF, e.g. its capacity. Space-time capacity of the TF (e) is evaluated as the product of TF intensity and density. Proved is the linear interrelation between the levels of the environmental risks arising from the TF on the RNW and the parameter e [5]. The capacity value (taking into account the basic equation of the TF) is evaluated by

$$e = qN = \frac{N^2}{V} = q^2 V. \tag{20}$$

For speed intervals 0≤V≤Va (the DG model) the equation (20) should be transformed with regard to (2); then we obtain the formula

$$e(V) = \frac{V}{L^2(V)} = \frac{V}{(m_2 \cdot V^2 + m_1 \cdot V + m_0)^2}. \tag{21}$$

For the range of Va≤V≤Vlim. (the alternative model) the capacity will be determined by the ratio between (4) and (11) which implies

$$e(V) = \frac{V \cdot (V\lim - V)}{K_l}. \tag{22}$$

Fig. 4 illustrates the capacity-speed ratio at two different states of the road pavement (dry road and ice). Black colour illustrates the curve obtained after the DG model; in here, a dotted line marks the theoretical extension to the right of point A. The curve determined by (22) is highlighted in blue (reflecting the dynamics of the TF capacity under speed limitation).

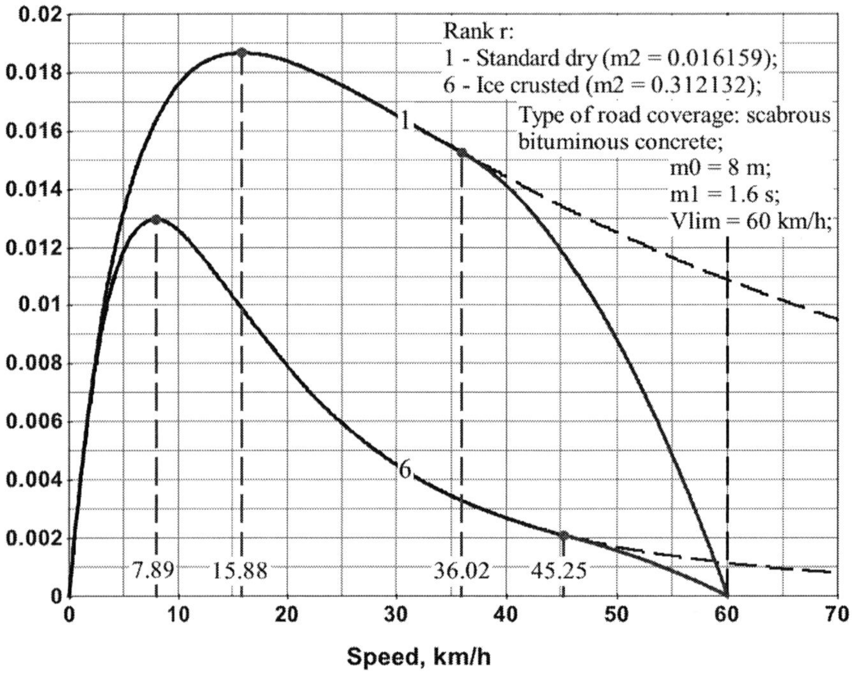

Fig. 4. Capacity-speed ratio of the traffic flow.

The position of the TF capacity extremum is marked in green dot in Fig.4. Obviously, the maximum value of e is achieved when de(V)/dV=0. Then, applying the function (21) let us find its derivative and equate it to zero. After mathematical transformations we obtain the formula for the TF speed achieved in maximum capacity.

$$V_{\max e} = \frac{-m_1 + \sqrt{m_1^2 + 12 \cdot m_2 \cdot m_0}}{6 \cdot m_2} \tag{23}$$

Vmaxe – speed in attaining TF maximum capacity.

The ultimate value of emax can be evaluated when applying the formula (21) and if the condition V=Vmaxe is met.

$$e_{\max} = \frac{V_{\max e}}{(m_2 \cdot V_{\max e}^2 + m_1 \cdot V_{\max e} + m_0)^2}. \tag{24}$$

4 Conclusion

Different types of models that form the basis of modern algorithmization of the subject-matter have been applied when studying the TF dynamics. Numerous regression curves of the TF speed-density ratio are available to be found in scientific books, but one can fail to find analytical description of the function V(q) in various ranges of the allowable speed. The developed mathematical model of the TF combines both classical and innovative approaches for the comprehensive analysis of the TF dynamic parameters.

The proposed model is a symbiosis of the DG model and the alternative model making it possible to track the dynamics of the TF parameters in terms of speed limitation. The coordinates of the fixed point, where the functions of models conjugate, are determined by the DG coefficients and speed limitation. The point allows determining the boundary of TF phase transition from the "free" state into the "bound" one.

The whole of the mathematical models have been analyzed by soft testing. The results obtained are widely used in experiments aimed at studying transport flows in the city of Tyumen. Based on the approaches given, the authors have developed the mathematical model allowing the effectiveness of the TF performance to be evaluated taking into account the key parameters: TF productivity, road traffic safety and environmental risks.

References

1. H. Inosje, T. Hamada, *Upravlenie dorozhnym dvizheniem* (Transport, Moscow, 1983)
2. V. Kolesov, *Materialy 5 Vserossijskoj NTK – TjumGNGU*, 25-31 (2012)
3. D. Sorokin, V. Kolesov, *Materialy Vserossijskoj NPK (Nauchno-Prakticheskoj konferencii). TjumGNGU*, 388-395 (2011)
4. V. Zhdanov, Metod ocenki tehnogennoj opasnosti transportnyh potokov na ulichno-dorozhnoj seti goroda. Avtoreferat dissertacii na soiskanie uchenoj stepeni kandidata tehnicheskih nauk. MADI (Moskovskij Avtomobil'no-Dorozhnyj Institut), Moscow, (2008)
5. V. Zhdanov, *Sovershenstvovanie metodov rascheta jekologicheskih harakteristik gorodskih transportnyh makroistochnikov* (Mashinostroenie, Moscow, 2010)
6. V. Kolesov, D. Sorokin, M. Guljaev, *Materialy mezhdunarodnoj NTK – TjumGNGU*, 89-95 (2012)
7. V. Kolesov, *Materialy mezhdunarodnoj NPK – TjumGNGU*, 26-30 (2013)
8. V. Kolesov, *Materialy 5 Vserossijskoj NTK – TjumGNG*, 25-31 (2012)
9. O. Danilov, V. Kolesov, *Materialy mezhdunarodnoj nauchno-prakticheskoj konferencii – TjumGNGU*, 402-405 (2010)
10. A. Moor, O. Danilov, V. Kolesov, *Materialy mezhdunarodnoj NTK.– TjumGNGU,* 182-193 (2011)
11. A. Moor, O. Danilov, V. Kolesov, *TjumGNGU*, 167-171 (2012)
12. D. Sorokin, *Bezopasnost' na dorogah radi bezopasnosti zhizni: materialy vtorogo oblastnogo foruma, «Vektor Buk»*, 126-129 (2010)
13. V. Kolesov, *Materialy Mezhdunarodnoj NTK. TjumGNGU*, 94-103 (2014)
14. V. Krutov, *Osnovy nauchnyh issledovanij: uchebnik dlja tehnicheskih vuzov* (Vysshaja shkola, Moscow 1989)
15. S. Ognjenovic, Z. Krakutovski, N. Vatin. Procedia Engineering, **117(1)**, 564-572 (2015)
16. S. Ognjenovic, R. Donceva, N. Vatin. Procedia Engineering, **117(1),** 556-563 (2015)
17. S. Ognjenovic, R. Ristov, N. Vatin. Procedia Engineering, **117(1)**, 573-578 (2015)
18. S. Ognjenovic, R. Donceva, N. Vatin. Procedia Engineering, **117(1)**, 549-555 (2015)
19. S. Ognjenovic, Z. Zafirovski, N. Vatin. Procedia Engineering, **117(1)**, 579-584 (2015)

MATEC Web of Conferences 73, 01029 (2016) DOI: 10.1051/matecconf/20167301029
TPACEE-2016

Modeling Parameters of Reliability of Technological Processes of Hydrocarbon Pipeline Transportation

Viktor Shalay[1], *Maria* Zemenkova[1,*], *Yuri* Zemenkov[1], and *Sergey* Toropov[2]

[1]Omsk State Technical University, st. Mira 11, Omsk, Russia
[2]Industrial University of Tyumen, 625001 Volodarskogo str. 38, Tyumen, Russia

Abstract. On the basis of methods of system analysis and parametric reliability theory, the mathematical modeling of processes of oil and gas equipment operation in reliability monitoring was conducted according to dispatching data. To check the quality of empiric distribution coordination , an algorithm and mathematical methods of analysis are worked out in the on-line mode in a changing operating conditions. An analysis of physical cause-and-effect relations mechanism between the key factors and changing parameters of technical systems of oil and gas facilities is made, the basic types of technical distribution parameters are defined. Evaluation of the adequacy the analyzed parameters of the type of distribution is provided by using a criterion A.Kolmogorov, as the most universal, accurate and adequate to verify the distribution of continuous processes of complex multiple-technical systems. Methods of calculation are provided for supervising by independent bodies for risk assessment and safety facilities

1 Introduction

In modern conditions of operation of technical systems of oil and gas industry an important task is to ensure the safety of objects by selecting the optimal operating conditions, timely and high-quality maintenance and repair [1,2,4,7,8]. The current system of assessing the reliability, based on the statistical evaluation of reliability index does not fully meet modern requirements [9], since, by definition, involves the fact of the occurrence of failure [11,13-16,19,20]. Operation of oil and gas companies provide preventive maintenance system with conditions to minimize any incidents, so the statistical information is unsuitable for classical reliability analysis. On-line monitoring models of reliability of the equipment and technologies of predicting failures are taking particular relevance and therefore require modern methods of analysis of the parametric reliability [10,17,18]. For high-quality mathematical reliability evaluation and selection of models and techniques, it is necessary to observe the main stages of preliminary data analysis based on the application of

* Corresponding author: muzemenkova@mail.ru

engineering data; graphical methods of estimation; nonparametric methods and evaluation criteria to determine the physical and mathematical models of reliability and features techniques. To obtain an adequate model of reliability it is very important to choose the type of distribution of failures or events corresponding to the actual, which depends largely on the accuracy of the results.

2 Methodological foundation

We'll define the aspects of mathematical modeling of processes for monitoring the reliability of oil and gas equipment according to dispatching data. To check the quality coordination of the empirical distribution agreement with the hypothetical use of fitting criteria. The most common and rather simple criteria are χ - square, Kolmogorov-Smirnov criteria. Kolmogorov-Smirnov test is applied in cases where the estimated distribution is continuous (particularly at tests for reliability).

Analysis of publications suggests that the following types of distributions are among the many most applicable distribution for parameters and characteristics of reliability of technical systems: exponential, normal, Weibull, Rayleigh, and logarithmic.

In most cases, gradual failures of complex equipment are subjected to the normal law; the laws of exponential form are accepted in the simulation of sudden failures.

Despite the fact that in some cases the use of the exponential law allows to obtain a good approximation, to construct correct models in any particular case justification is required. It should be noted that the use in the calculation of the reliability of the exponential distribution law is possible only under the condition that the failure rate is constant, i.e. the system is operated in the normal operation period. In other cases, it is necessary to use other laws, although this fact complicates the task of determining reliability. Analysis of the physical laws of cause and effect relationships between the key factors and changes in parameters of engineering systems of oil and gas facilities [3,4,5,6] has allowed to identify the main types of distributions - the most suitable are the normal law and Weibull. The normal law can be seen as the limit to which other laws approach with common typical conditions. The law allows describing parameters that are formed in real conditions under the influence of numerous mutually independent and weakly dependent factors or are the sum of some random variables. Distribution gives the opportunity to describe wearing processes and change of properties of the system taking into account working conditions (loading, speed, and temperature), personnel qualification, quality of material, hardness, purity of surface, precision manufacturing etc. When sudden changes of operating conditions occur (overload, violation and incomplete execution of modes, changing the equipment and parts for repairs etc.), application of the normal aw is not correct. The advantage of Weilbull theoretical distribution law is shown in a model of a "weak chain"when the system consists of a group of independent elements, where each failure leads to the failure of the entire system, and allows you to define the properties of the system, characterizing the rate of reaching the system limit state. Weilbull distribution gives the possibility to simulate processes of sudden failures (when distribution form parameter b is close to a unit, i.e. $b \approx 1$) and failures due to wearing ($b \approx 2.5$), and also when the reasons cause both these failure. Distribution also finely describes gradual failures caused by aging of the material, e.g. by corrosion. The advantage in application of the developed Weilbull model of distribution is that the law is rather universal and for certain values of the parameter form and dynamics of change in the failure it takes the form: exponential (for $b = 1$, i.e, the failure rate is constant); normal (when $b = 3.3$); Rayleigh (for $b = 2$) the failure rate - a linear function); for $b > 1$ and $b < 1$ the failure rate increases and decreases, respectively. Weibull distribution allows to vary the distribution of parameters and apply it to a large number of cases of probability changes. As a logarithmic distribution, it describes satisfactorily the operating time to failure of various units of machines, bearing temperature sensors. In IUT it was defined, that the reliability of oil and gas equipment is also subject to the Weibull law. Thus, we assume that the predominant use of the law of the Weibull model in the development of safety model and in the description of the intensity of parameter changes of technical condition of the equipment and facilities of oil transport systems consisting of multiple elements with a variety of cause-and-effect relationships. Table 1.

Table 1. Distribution function of index of persistence.

Distribution function	Type of distribution function	Basic parameters	Function chart
Normal	$P(y)=F\left(\frac{y-m}{\sigma}\right)+0{,}5$	m, σ	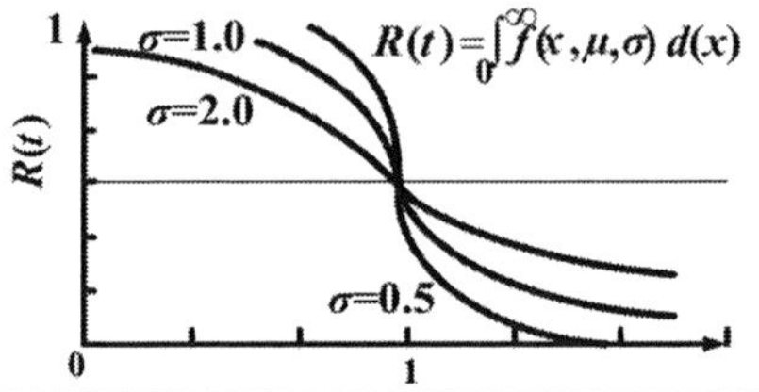
Weibull	$P(y)=e^{-\left(\frac{y}{a}\right)^{b}}$	A, B	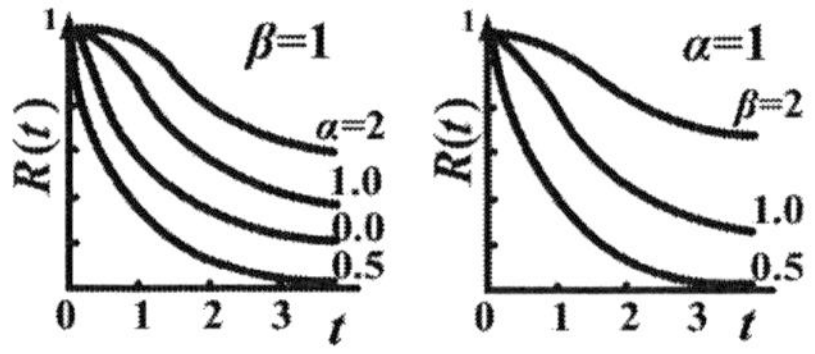
Exponential	$P(y)=e^{-(\lambda y)}$	λ	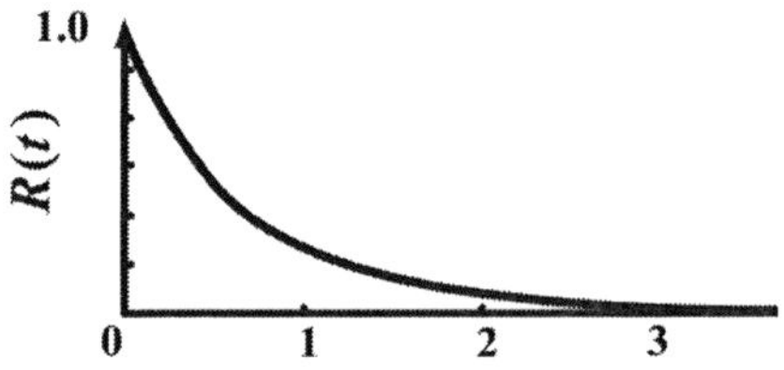
Gamma	$P(y)=\int_{0}^{y}\frac{\lambda^{c}}{G(c)}y^{c-1}e^{-\lambda y}$	λ, c	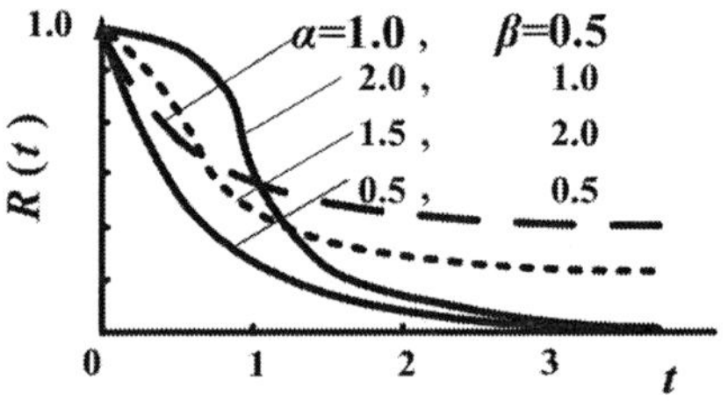
Releigh	$P(y)=e^{-\frac{1}{2}\left(\frac{y}{\sigma}\right)^{2}}$	σ	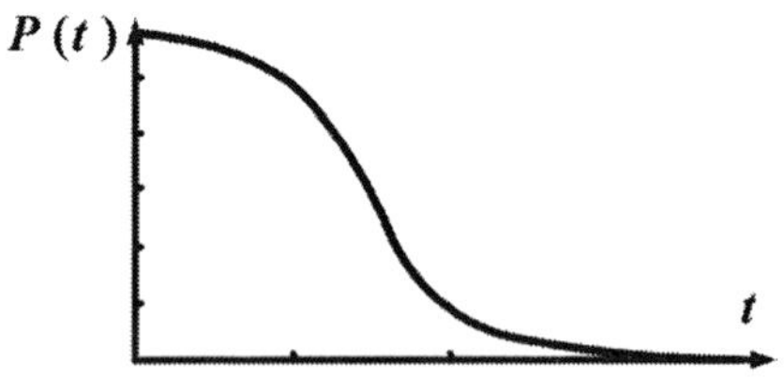

The type of distribution function is proposed to determine, applying the criteria of acceptance of preliminary engineering assessment of technological features of systems.

Analysis of experience in application of methods of probability theory in assessing the reliability of oil transportation equipment and facilities shows that possible criteria of consent to assess the types of distributions are criteria of K. Pearson, V. Romanovsky, Mises, etc.

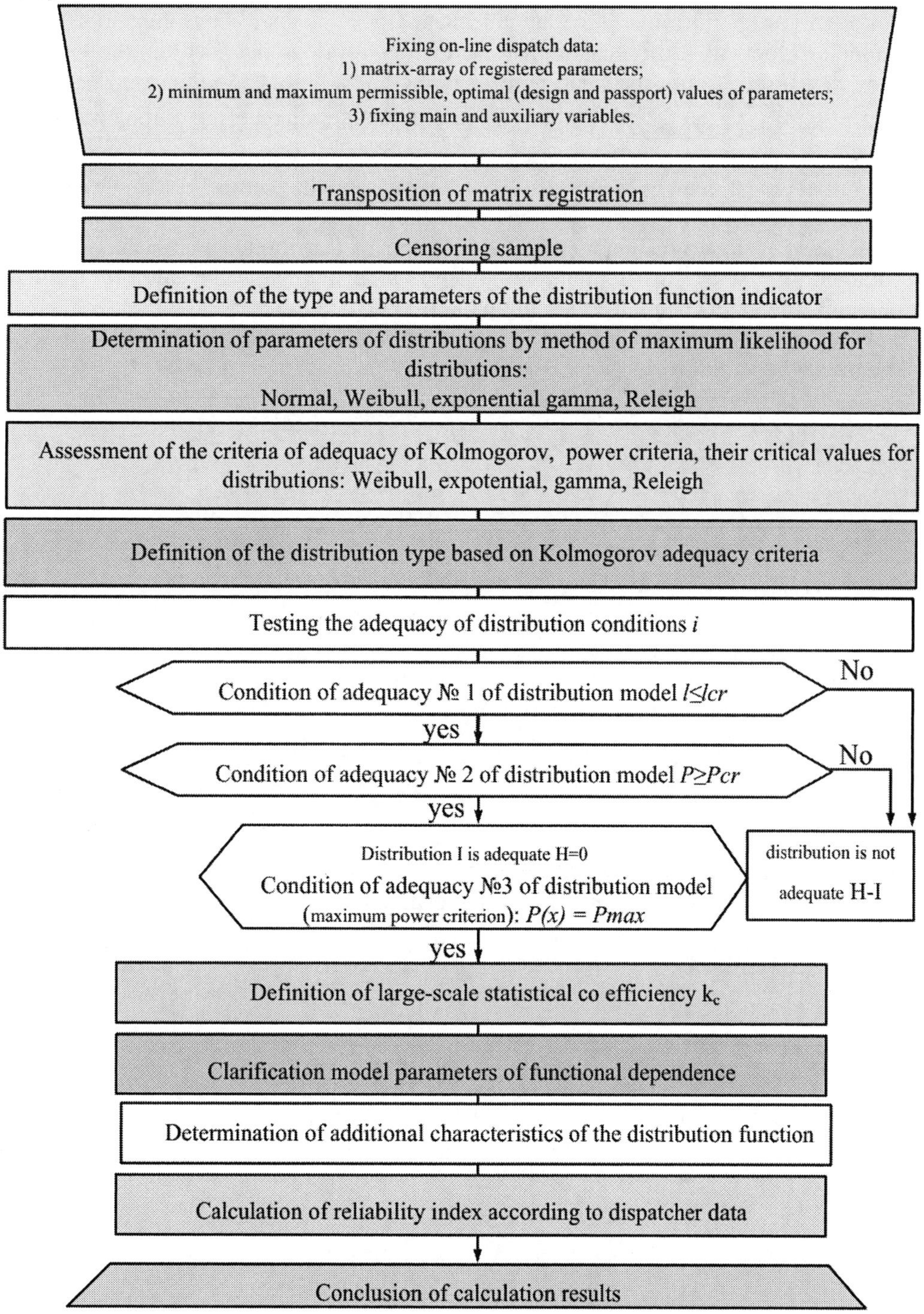

Fig. 1. Algorithm of program realization of the model of analysis of technological parameters.

It should be noted that an important stage in the developed technique is the necessity to check the quality of coordination of the empirical distribution with a hypothetical.

However, the evaluation of adequacy of type of distribution of the analyzed parameters we can carry out using the most appropriate criterion of A. Kolmogorov, as the most versatile, accurate and adequate to check the distributions of continuous processes of complex multi-element technical systems.

For comfort of application of mathematical apparatus the distribution function should be determined for deviation of technological parameter from optimal according to the model. If the deviation from the optimum parameter is not in the middle of the interval $y_i \in \left(y_{\min Cr}; y_{\max cCa}\right)$, the deviation will be determined using scaling methods.

Thus, in accordance with the method by A. Kolmogorov as a measure of discrepancy between theoretical and experimental distributions, we'll use the maximum value of difference between empirical and theoretical functions.

Since the developed technique of an estimation of distribution parameters is specific depending on the type of distribution, we'll take as an example the algorithm for estimating the Weibull parameters using the maximum likelihood method: .

1. Probability density is determined (derivative of the probability distribution function), assuming that Weibull distribution is the three-parameter:

$$f(y,\alpha,\beta,\gamma) = \frac{\beta}{\alpha}(y-\gamma)^{\beta-1} e^{-\frac{(y-\gamma)\beta}{\alpha}},$$

$$y \geq \gamma, \gamma \geq 0, \beta > 0, \alpha > 0, \tag{1}$$

where α, β, γ - scale parameter, forms, position respectively.

2. Position parameter y is different from zero if it is known that for real parameter values of y value argument does not specify the function value.

Position parameter in the analysis of the reliability of technical systems is determined to be zero, if Weibull distribution is two-parameter:

$$f(y,\alpha,\beta) = \frac{\beta}{\alpha}(y)^{\beta-1} e^{-\frac{y^\beta}{\alpha}} \tag{2}$$

3. Maximum likelihood estimations of the parameters α, β are obtained using the equations:

$$n\alpha - \sum_{i=1}^{n} y_i^\beta = 0 \tag{3}$$

$$\frac{n}{\beta} + \sum_{i=1}^{n} \ln yi - \frac{1}{\alpha}\sum_{i=1}^{n} y_i^\beta \ln y_i = 0 \tag{4}$$

$$\bar{y} = \alpha^{\frac{1}{\beta}} G\left(1 + \frac{1}{\beta}\right) \tag{5}$$

4. There is the first approximation of the parameter according to the formulas (7) and (9) using the method of successive approximations. 5. After the evaluation β α is determined:

$$\alpha = \frac{\sum_{i=1}^{r} y_i^{\beta} + (n-r)y_r^{\beta}}{r} \tag{6}$$

$$r = \frac{N_{df}}{2}, \tag{7}$$

where N_{df} – the number of degrees of freedom .

6. Distribution parameters are elaborated in real time as data become available and increase of volume of statistics.

7. Probability is determined (reliability index of the parameter) in Weibull distribution:

$$P(y) = e^{-\frac{y^{\beta}}{\alpha}} \tag{8}$$

or $$P(y) = e^{-\left(\frac{y}{a}\right)^{b}} \text{ when } a = \frac{\alpha}{\beta}; b = \beta \tag{9}$$

8. An evaluation of the correctness of the analysis of reliability is carried out for the following quantities (it should be noted that similar functios (12) and (13) and accepted designations are used in modern programming environments such as Matlab):

- intensity of parameter changes or failures:

$$\lambda(y) = \frac{\beta}{\alpha} y^{\beta-1}; \tag{10}$$

-mathematical expectation:

$$\mu_y = \alpha^{1/\beta} G\left(1 + \frac{1}{\beta}\right); \tag{11}$$

-dispersion:

$$\sigma^2{}_y = \alpha^{2/\beta}\left[G\left(1+\frac{2}{\beta}\right) - \left\{G\left(1+\frac{1}{\beta}\right)\right\}^2\right]; \tag{12}$$

-coefficient of variation:

$$k_v = \sqrt{\frac{G\left(1+\frac{2}{\beta}\right)}{G\left(1+\frac{1}{\beta}\right)^2} - 1} = \frac{\sigma_y}{\mu_y} \quad (13)$$

9. The values of the coefficients of variation kv, are checked , which should be within the following limits:

normal law $k_v \leq 0.4$;
the law of the Weibull distribution, $k_v = 0.35 \ldots 0.8$;
the exponential law, $k_v > 0.8$.

10. The conclusion is made (eg, based on the results of experimental data): analysis of changes in the y values for the object revealed that the change in absolute values of the operational parameters is subject to the normal law and deviations of these parameters - to Weibull distribution).

Since the definition of the form of the distribution function and assessment of the adequacy of its parameters are based on probabilistic-statistical method of Kolmogorov, in Matlab environment it is possible to implement the methods to use functions KSTEST of Kolmogorov-Smirnov (non-contradiction of general combination to law, the law with fixed parameters). The following conditions are checked as a result of calculation:

$$P \geq P_{cr} \quad \text{and} \quad l \leq l_{cr} \quad (14)$$

When H=0 probabilistic-statistical model is adequate. Among the alternative hypotheses about the form of distribution law the hypothesis with the greatest test of significance P is accepted (power of test).

3 Conclusion

Thus, the developed mathematical model and algorithms enable us to obtain a set of indicators that adequately characterize system reliability for time-varying parameters of the technical condition of oil and gas equipment.

The obtained mathematical models enable operating organizations, Federal services of supervision, independent expert organizations to assess and predict the development of reliability on a register block control data, in real time, taking into account the dynamics of the operating conditions of the object.

References

1. M. Zemenkova, Thesis of Candidate of Technical Sciences (TSOGU, 2007)
2. V. Kurushina, Yu. Zemenkov, WIT Transactions on Ecology and the Environment, **190(2)**, 881-888 (2014)
3. V. Antip'ev, A. Nevolin, Yu. Zemenkov, Neftyanoe Khozyaistvo - Oil Industry, **10**, 46-48 (1981)
4. A. Bacherikov, M. Pashkov, Yu. Bogatenkov, Yu. Zemenkov, V. Shalay, G. Vasiliev, A. Prokhorov, S. Sidelnikov, S. Podorozhnikov, V. Kehl, *Diagnosis in the maintenance of pipeline transportation* (Nedra, St.Peterburg, 2007)
5. S. Dudin, Yu. Zemenkov, A. Shabarov, N. Saranchyn, Oil and Gas, **6**, 63-68 (2010)

6. A. Shipovalov, Yu. Zemenkov, S. Toropov, M. Zemenkova, S. Podorozhnikov, I. Tyrylgin, V. Pavlov, Aspects of technological reliability and economic efficiency of operation of underground storages of natural gas of Western Siberia (2012)
7. Yu. Zemenkov, V. Shalay, M. Zemenkova, Procedia Engineering, **113**, 254-258 (2015)
8. Yu. Zemenkov, V. Shalay, M. Zemenkova, Procedia Engineering, **113**, 312-315 (2015)
9. S. Toropov, Yu. Zemenkov, S. Podorozhnikov, Gas industry, **720(720)**, 95-98 (2015)
10. Yu. Zemenkov, B. Moiseyev, N. Nalobin, S. Oil and Gas, **5**, 106-110 (2015)
11. Yu. Zemenkova, I. Seroshtanov, V. Shantarin, Yu. Zemenkov, S. Toropov, Fundamental researches, **10-2**, 259-263 (2015)
12. Yu. Zemenkov, B. Moiseyev, S. Dudin, N. Nalobin, Neftegaz territory, **9**, 112-117 (2015)
13. V. Pavlov, K. Voronin, Yu. Zemenkov, *Oil and Gas Terminal: Proceedings of the International scientific conference*, 169-173 (2015)
14. A. Pimnev, M. Zemenkova, Pipeline transport: theory and practice, **5**(51), 43-47 (2015)
15. A. Pimnev, *Oil and Gas Terminal: Proceedings of the International scientific conference*, 178-179 (2015)
16. M. Zemenkova, Yu. Zemenkov, A. Pimnev, V. Petryakov, Business magazine Neftegaz.RU, **11-12**, 64-70 (2015)
17. S. Stepanov, N., Cheremisin, S., Sokolov, A., Altunin, A., Shabarov, *ECMOR 2012 - 13th European Conference on the Mathematics of Oil Recovery* (2012) DOI: 10.399/2214-4609.20143270
18. S. Dudin, A. Shipovalov, S. Podorozhnikov, Yu. Zemenkov, Oil and Gas, **2**, 65-69 (2011)
19. V. Ostreykovsky, *Theory of reliability* (High school, Moscow, 2003)
20. E. Yasin, V. Berezin, K. Rashchepkin, *Reliability of the main pipelines* (Subsoil, Moscow, 1972)

Assessment of the Energy - Efficient Modernization of Residential Historical Buildings in Kiev

Vera Murgul[1,a]

[1]St. Petersburg State Polytechnical University, Politekhnicheskaya, 29, Saint-Petersburg, 195251, Russia

Abstract. Most of residential historical buildings in Kyiv, Ukraine are in need of energy efficiency modernization. The present research exposes the assessment of warming effectiveness for the certain enclosing structures in historical residential buildings, considering the overall thermal balance of building. Heat losses through certain enclosing structures considering actual thermal resistance were estimated, also in compliance with standard requirements for thermal resistance of enclosing structures according to the standard DBN B.2.6-31:2006. It turned out that the warming of the whole building envelope and the replacement of transparent enclosing structures allows to reduce the heat flux of the heating system for compensation of thermal losses through the building envelope in 3 times.

1 Introduction

Improved energy efficiency of historic buildings is one of the most urgent problems of construction science of Ukraine. The following researches [1-9] expose the problem of energy efficiency in residential historical buildings.

In this article the possibility of energy-efficient modernization of residential historical buildings in Kiev was considered in accordance with modern standards [10].

Residential building built in 1909 on the Malaya Zhitomirskaya street, 12-A, Kiev was taken as an example for the assessment (historical and architectural monument of local importance [2].

It should be noticed that the design of walling (external walls, attic slabs and basement ceiling) in residential historical buildings of St. Petersburg and Kiev have the common features [2-7, 11-13].

For example, exterior walls have a variable thickness and become thinner with height: the wall thickness varies in the range from 2 to 5 bricks. Therefore, the methodology of energy efficiency modernization can be quite similar for such buildings.

[a] Corresponding author: october6@list.ru

2 Thermo-technical calculation of building envelope

Design conditions for Kiev and numerical values used for further calculations presented in Tables 1, 2

Table 1. Design conditions for Kiev

Object: Malaya Zhitomirskaya street, 12-A, Kiev	Requirements DBN B.2.6-31:2006	Was in 80-s
Design ambient air temperature t_{ext} °C	-22	-22
The average outdoor temperature during the heating season, $t_{h..s.}$, °C	-0.1	-0.1
Duration of the heating season, z_{ht} , day	176	176
Design interior temperature, t_{int}, °C	20	18
Degree-day heating season	3538	3186

Table 2. Numerical values used for further calculations

Enclosing structure	External wall		Attic slabs		Basement ceiling		Windows and doors	
	DBN B.2.6-31:2006	Was in 80-s	DBN B.2.6-31:2006	Was in 80-s	DBN B.2.6-31:2006	Was in 80-s	DBN B.2.6-31:2006	Was in 80-s
Coefficient taking into account the dependence of the enclosing structure position in relation to ambient air, n	1	1	0.9	0.9	0.4	0.4	-	-
Normalized temperature difference between the interior temperature and the temperature of enclosing structure inner surface, $\Delta t_{н}$, °C	4	6	3	4	2	2	-	-
Heat transfer coefficient of the enclosing structure inner surface, α_{int}, W/m²*°C	8.7	8.7	8.7	8.7	8.7	8.7	-	-
Required thermal resistance of enclosing structure according to hygiene requirements $R = n\ (t_{int} - t_{ext}) / \alpha_{int}\ \Delta t_n$ (m² •°C)/W	**1.2069**	**0.7663**	**1.4483**	**1.0345**	**0.9655**	**0.9195**	**0.7500**	**0.42**
Normalized heat transfer resistance of enclosing structure, (m² •°C)/W	3.3	-	**4.9500**	-	**1.5**	-	-	-
Heat transfer coefficient of the enclosing structure outer surface, α_{ext}, W /(m ²*°C)	23	23						

The facade and section of the building on Malaya Zhitomirskaya st., 12A is illustrated in Figure 1.

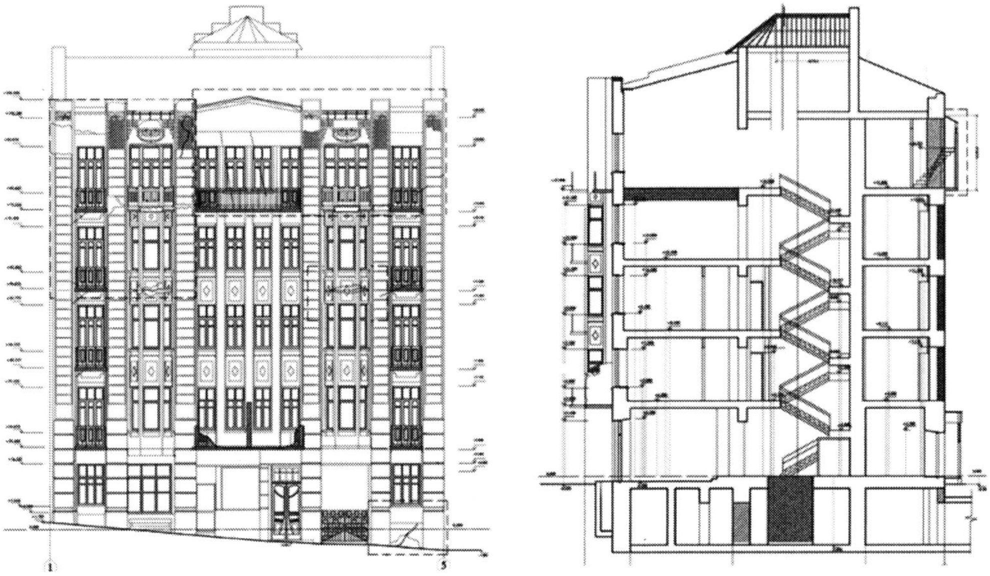

Fig. 1. The façade and section of the building on Malaya Zhitomirskaya st., 12A, Kyiv

As already mentioned above, the design of walling (external walls, attic slabs and basement ceiling) in residential historical buildings of St. Petersburg and Kiev have the common features. Exterior walls have a variable thickness and become thinner with height: the wall thickness varies in the range from 2 to 5 bricks.

The results of calculation for the bricks wall thickness presented Table. 3.

Table 3. The results of calculation for the bricks wall thickness.

Material	Layer thickness δ_i, m	Coefficient of heat conductivity, λ_i, [W / (m·°C)]	R_i, [(m^2•°C)/W]
External wall of the 3rd floor			
Brickwork of ordinary loam bricks. (Sand-cement mortar), 1800 kg/m^3	0.62	0.81	**0.76543**
$1/\alpha_в$			0.1149
$1/\alpha_н$			0.0435
r (heat transfer performance uniformity factor)			0.698676
Thermal resistance of enclosing structure R (m^2*°C)/ W			**0.6455**
External wall of the 2nd floor			
Brickwork of ordinary loam bricks. (Sand-cement mortar), 1800 kg/m^3	0.86	0.81	**1.06173**
$1/\alpha_в$			0.1149

$1/\alpha_н$			0.0435
r (heat transfer performance uniformity factor)			0.698676
Thermal resistance of enclosing structure R (m^2*°C)/ W			**0.7441**
External wall of the 1st floor			
Brickwork of ordinary loam bricks. (Sand-cement mortar), 1800 kg/m^3	1.06	0.81	**0.76543**
$1/\alpha_в$			0.1149
$1/\alpha_н$			0.0435
r (heat transfer performance uniformity factor)			0.698676
Thermal resistance of enclosing structure R (m^2*°C)/ W			**0.7860**

Dependence diagram of heat transfer performance uniformity factor to the thickness of brick wall presented Figure 2.

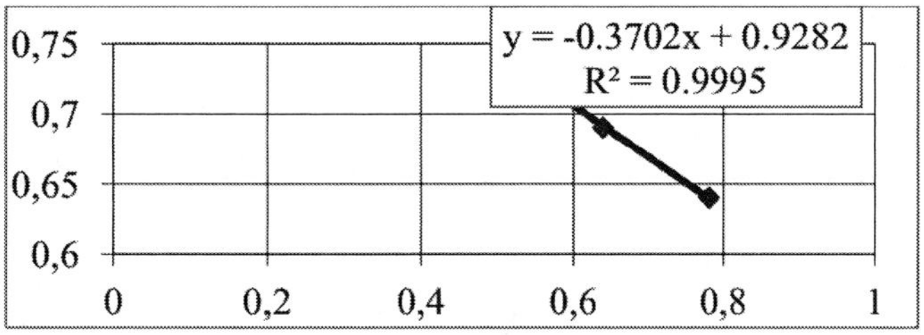

Fig. 2. Dependence diagram of r (heat transfer performance uniformity factor) to the thickness of brick wall.

The calculation results of design and standard values of structural thermal resistance for a residential building are shown in Table 4.

Table 4. Showings of the thermal resistance of enclosing structures in residential buildings

Type of enclosing structures	Thermal resistance, [(m^2·°C)/W]		
	Design	The minimum required for the sanitary and hygiene standards	Standard
External walls: - 1st floor; - 2nd floor; - 3rd, 4th and 5th floors	1.467 / 0.786 1.22 / 0.744 0.924 / 0.646	1.207	3.3
Attic slab	1.035	1.448	4.95
Overlap above the unheated basement	2.3/0.92	2.415/0.966	3.75/1.5
Transparent façade systems	0.42		0.75

Notes: 1. The actual thermal resistance for the exterior walls which is specified in the denominator was determined taking into account the coefficient of thermotechnical homogeneity (r - heat transfer performance uniformity factor);
2. The values of thermal resistance for the overlap above the unheated basement were specified in the denominator with coefficient n=0.4, which was taken for the overlap above the unheated basement below the level.
3. The minimum required thermal resistance for the sanitary and hygiene standards was defined by the standard temperature differential between the inside air temperature and temperature of the inner surface of the building envelope [1];
4. Specified indicators were adopted for the 1st temperature zone of Ukraine [1].

2.1 Heat losses calculation. Variant 1

Considering the first type according to actual terms of thermal resistance of enclosing structure (Table 5).

Table 5. Thermal resistance of enclosing structure:

External walls 1st floor	0.7860	$(m^2 \cdot {}^0C)/W$	Basement ceiling	0.9195	$(m^2 \cdot {}^0C)/W$
External walls 2nd floor	0.7441	$(m^2 \cdot {}^0C)/W$	Attic slabs	1.0344828	$(m^2 \cdot {}^0C)/W$
External walls 3rd floor	0.6455	$(m^2 \cdot {}^0C)/W$	Windows (double glazing in wooden separate bindings)	0.420	$(m^2 \cdot {}^0C)/W$
Doors	0.290	$(m^2 \cdot {}^0C)/W$			
t_{int}	20	°C			

The results of calculation of thermal losses are presented in Table 6.
Conventions are as follows:
Ff – Front façade facing the Zhitomirskaya st.
Fy – Yard facade.
Fg – gable facade (adjoins with the Murashko)
EW – External walls
BC – Basement ceiling
AS – Attic slabs
WD – windows and doors

Design interior temperature: $t_{int} = 20$ [°C]
Design ambient air temperature: $t_{ext} = -22$ [°C]

Table 6. The results of calculation of thermal losses

Room number and its purpose	Characteristic of walling					n	Walling heat transfer coefficient, k, [W / (m·°C)]	General heat losses, Qo=k·F· (t_{int} - t_{ext}) · n , [W]	Quantity of additional heat losses due to orientation, β	Heat losses including extra losses, Q=Qo·(1+β), [W]	Total losses, [W]
	Title	Orientation	Geometry, [m]		Area size, F, [m²]						
			a	b							
1st floor Ff	EW1	SW	21.58	4.60	99.27	1	1.2722	3921	-	3921	3921
	WD	SW			25.88	1	2.3810	2588	-	2588	2588
1st floor Fy	EW1	NE	21.58	4.60	99.27	1	1.2722	4144	0.1	4558	4558
	WD	NE			21.72	1	2.3810	2172	0.1	2389	2389
	BC		21.58	21.41	462.03	0.4	1.0875	8441	-	8441	8441
											21898
2nd floor Ff	EW1	SW	21.58	3.90	84.16	1	1.3439	3138		3138	3138
	WD	SW			28.56	1	2.3810	2856		2856	2856
2nd floor Fy	EW	NE	21.58	3.90	84.16	1	1.3439	3932	0.1	4325	4325
	WD	NE			14.50	1	2.3810	1450	0.1	1595	1595
											11915
3rd floor Ff	HC1	SW	21.58	4.00	86.32	1	1.5492	5617		5617	5617
	WD	SW			28.56	1	0.8317	998		998	998
3rd floor Fy	EW	NE	21,58	4.00	86.32	1	1.5492	5617	0.1	6178	6178
	WD	NE			14.50	1	0.8317	507	0.1	557	557
											13350
4th floor Ff	EW	SW	21.58	3.90	84.16	1	1.5492	3618		3618	3618
	WD	SW			28.56	1	2.3810	2856		2856	2856
4th floor Fy	EW	NE	21.58	3.90	84.16	1	1.5492	4533	0.1	4986	4986
	WD	NE			14.50	1	2.3810	1450	0.1	1595	1595
4th floor Fg	EW	NE	21.41	3.90	83.50	1	1.5492	5433	0.1	5976	5976
											19031
5th floor Ff	EW	SW	21.58	4.26	91.93	1	1.5492	4298		4298	4298
	WD	SW			25.88	1	2.3810	2588		2588	2588
5th floor Fy	EW	NE	21.58	4.26	91.93	1	1.5492	4569	0.1	5025	5025
	WD	NE			21.72	1	2.3810	2172	0.1	2389	2389
5th floor Fg	EW	NE	21.41	4.26	91.21	1	1.5492	5935	0.1	6528	6528
	AS		21.58	21.41	462.03	0.9	0.9667	16882		16882	16882
											37711

Total heat losses – 103 905 W

As a result,
Overall heat losses through the whole walling – 103905 W
Heat losses through all basement ceilings – 8441 W
Heat losses through all attic slabs – 16882 W
Heat losses through all glazed surfaces and doors – 20411 W
Heat losses through all external walls – 58170 W

Total building heat losses before warming of building envelope is illustrated in Figure 3

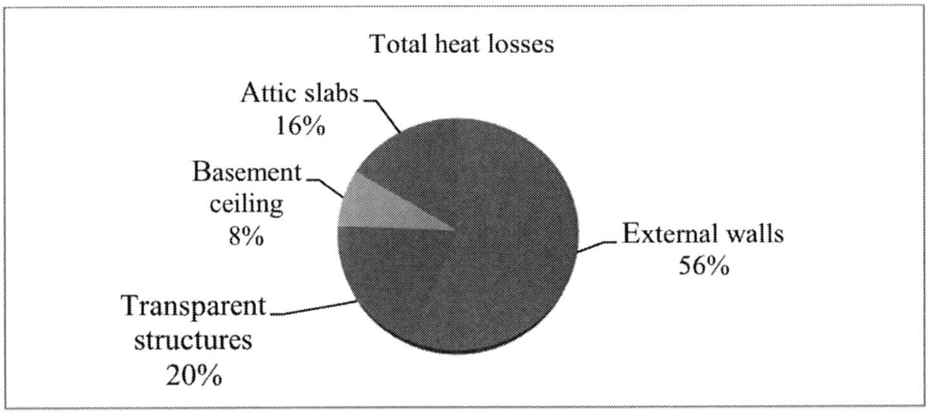

Fig. 3. Total building heat losses before warming of building envelope

2.2 Heat losses calculation. Variant 2

We consider the option of insulating the whole walling to normalized values, according to DBN B.2.6-31:2006 (Table 7)

Table 7. Thermal resistance of enclosing structure:

External walls 1st floor	3.3	$(m^2 \cdot {}^0C)/W$	Basement ceiling	1.50	$(m^2 \cdot {}^0C)/W$
External walls 2nd floor	3.3	$(m^2 \cdot {}^0C)/W$	Attic slabs	4.95	$(m^2 \cdot {}^0C)/W$
External walls 3rd floor	3.3	$(m^2 \cdot {}^0C)/W$	Double window glazing	0.750	$(m^2 \cdot {}^0C)/W$
Doors	0.290	$(m^2 \cdot {}^0C)/W$			
t_{int}	20	°C			

Calculation results:
Overall heat losses through the whole walling – **33151 W**
Heat losses through all basement ceilings – 5175 W
Heat losses through all attic slabs – 3528 W
Heat losses through all glazed surfaces and doors – 12486 W
Heat losses through all external walls – 11962 W

Total building heat losses after warming of building envelope according to DBN B.2.6-31:2006 requirements is illustrated in Figure 4.

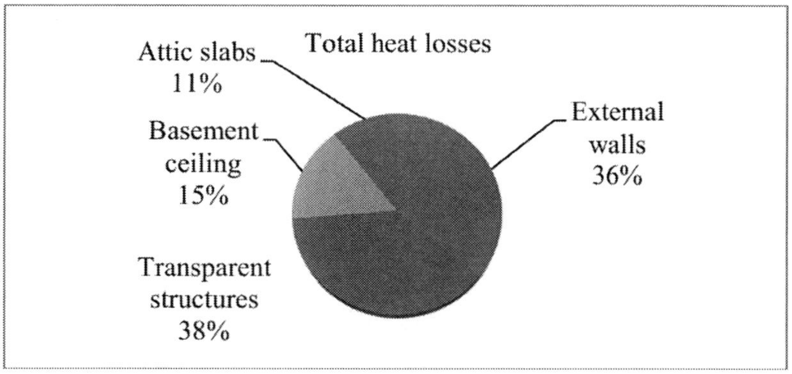

Fig. 4. Total building heat losses after warming of building envelope according to DBN B.2.6-31:2006 requirements

2.3 Heat losses calculation. Variant 3

We consider the option of insulating external walls only (Table 8)

Table 8. Thermal resistance of enclosing structure.

External walls 1st floor	3.3	(m^2 •^{0}C)/W	Basement ceilings	0.9195	(m^2 •^{0}C)/W
External walls 2nd floor	3.3	(m^2 •^{0}C)/W	Attic slabs	1.0344828	(m^2 •^{0}C)/W
External walls 3rd floor	3.3	(m^2 •^{0}C)/W	Double window glazing	0.420	(m^2 •^{0}C)/W
Doors	0.290	(m^2 •^{0}C)/W			
t_{int}	20	°C			

Calculation results:
Overall heat losses through the whole walling – **60027 W**
Heat losses through all basement ceilings – 8441 W
Heat losses through all attic slabs – 16882 W
Heat losses through all glazed surfaces and doors – 22741 W
Heat losses through all external walls – 11962 W

2.4 Heat losses calculation. Variant 4

We consider the option of replacing old windows with the glass units (Table 9)

Table 9. Thermal resistance of enclosing structure.

External walls 1st floor	0.7860	(m^2 •^{0}C)/W	Basement ceilings	0.9195	(m^2 •^{0}C)/W
External walls 2nd floor	0.7441	(m^2 •^{0}C)/W	Attic slabs	1.0344828	(m^2 •^{0}C)/W
External walls 3rd floor	0.6455	(m^2 •^{0}C)/W	Windows	0.750	(m^2 •^{0}C)/W
Doors	0.290	(m^2 •^{0}C)/W			
t_{int}	20	°C			

Calculation results:
Overall heat losses through the whole walling – **93650 W**
Heat losses through all basement ceilings – 8441 W
Heat losses through all attic slabs – 16882 W
Heat losses through all glazed surfaces and doors – 10156 W
Heat losses through all external walls – 58170 W

2.5 Heat losses calculation. Variant 5

We consider the option of warming the attic slabs only (Table 10)

Table 10. Thermal resistance of enclosing structure.

External walls 1st floor	0.7860	$(m^2 \cdot {}^0C)/W$	Basement ceilings	0.9195	$(m^2 \cdot {}^0C)/W$
External walls 2nd floor	0.7441	$(m^2 \cdot {}^0C)/W$	Attic slabs	4.95	$(m^2 \cdot {}^0C)/W$
External walls 3rd floor	0.6455	$(m^2 \cdot {}^0C)/W$	Double window glazing	0.420	$(m^2 \cdot {}^0C)/W$
Doors	0.290	$(m^2 \cdot {}^0C)/W$			
t_{int}	20	°C			

Calculation results:
Overall heat losses through the whole walling – **90551 W**
Heat losses through all basement ceilings – 8441 W
Heat losses through all attic slabs – 3528 W
Heat losses through all glazed surfaces and doors – 20411 W
Heat losses through all external walls – 58170 W

2.6 Heat losses calculation. Variant 6

We consider the option of warming the basement ceilings only (Table 11)

Table 11. Thermal resistance of enclosing structure.

External walls 1st floor	0.7860	$(m^2 \cdot {}^0C)/W$	Basement ceilings	1.50	$(m^2 \cdot {}^0C)/W$
External walls 2nd floor	0.7441	$(m^2 \cdot {}^0C)/W$	Attic slabs	1.0344828	$(m^2 \cdot {}^0C)/W$
External walls 3rd floor	0.6455	$(m^2 \cdot {}^0C)/W$	Double window glazing	0.420	$(m^2 \cdot {}^0C)/W$
Doors	0.290	$(m^2 \cdot {}^0C)/W$			
t_{int}	20	°C			

Calculation results:
Overall heat losses through the whole walling – **100638 W**
Heat losses through all basement ceilings – 5175 W
Heat losses through all attic slabs – 16882 W
Heat losses through all glazed surfaces and doors – 20411 W
Heat losses through all external walls – 58170 W

General and additional heat losses through the building envelope at the design outdoor temperature minus 22°C are as follows:
- considering actual terms of thermal resistance - 103910 W;
- considering the observance of regulatory requirements [10] for whole walling - 33950 W;
- considering the observance of regulatory requirements only for external walls - 60030 W;
- considering the observance of regulatory requirements only for transparent structures – 93650 Вт;
- considering the observance of regulatory requirements only for the attic slabs – 90550 Вт;
- considering the observance of regulatory requirements only for the basement ceiling – 100640 Вт.

Conclusion

Thus, implementation of energy saving measures such as insulating for whole walling and replacing transparent walling allows to reduce heat flux of the heating system which compensates the heat loss through the building envelope in 3.1 times. At the same time the external walls facing Malaya Zhitomirskaya st. have to be insulated from the inside, and in other cases the external warming is possible.

During the carrying-out of the regulatory requirements [1] for the thermal resistance of the individual walling structures, the reduction of heat losses will be as follows:

- for external walls - ΔQ=103910 - 60030 = 43880 W (42.2%);
- for transparent structures - ΔQ=103910 - 93650 =10260 W (9.9%);
- for attic floor slab - ΔQ=103910 - 90550 =13360 W (12.9%);
- for the basement ceiling - ΔQ=103910 - 100640 =3270 W (3.1%).

References

1. K. Fabbri, B. Brunetti, Energy Procedia, **78**, 1281-1286 (2015)
2. F. Robertia, U. F. Obereggera, A. Gasparellaba, Energy and Buildings, **108**, 236–243 (2015)
3. E. Arumägia, M. Mändelb, T. Kalamees, Energy Procedia, **78**, 1027-1032 (2015)
4. C. D. Şahin, Z. D. Arsan, S. S. Tunçoku, T. Broström, G. G. Akkurt, Energy and Buildings, **96(1)**, 128-139 (2015)

A. L. Pisello, A. Petrozzi, V. L. Castaldo, F. Cotana, Applied Energy, **162(15)**, 1313-1322 (2016)

5. Ü. Alev, L. Eskola, E. Arumägi, J. Jokisalo, A. Donarelli, K. Siren, T. Broström, T. Kalamees, Energy and Buildings, **77**, 58-66 (2014)
6. F. Ceroni, F. Ascione, R. F. De Masi, F. de' Rossi, M. R. Pecce, Energy Procedia, **75**, 1325-1334 (2014)
7. D. Milone, G. Peri, S. Pitruzzella, G. Rizzo, Energy and Buildings, **95**, 39-46 (2015)
8. A. Gagliano, F. Nocera, F. Patania, M. Detomaso, V. Sapienza, Energy Procedia, **62**, 62-71 (2014)
9. Gavrilyuk L.O., Denisenko G.G., Katargina T.I., et.al. *Pam'yatki istorii̊ ta kul'turi Ukraïni: Katalog-dovidnik. Zoshit 2: Katalog-dovidnik pam'yatok istoriï ta kul'turi Ukraïni*, (Kiïv, Kiev, 2007).
10. DBN B.2.6-31:2006
11. O.P. Oliynik, N. Gizhko, Tekhnichna estetika i dizayn, **9**, 203-210 (2011)
12. G.N. Ageeva, Conference Proceedings,Tret'yey Vserossiyskoy konferentsii, 5-9 (2007)
13. G. N. Ageeva, *Rekonstruktsiya zhitla: nauk.-virobn.vid,* (Logos, Kiev, 2012)

Experimental Assessment of Moisture Transfer in the Vertical Ventilated Channel

Tatyana Musorina[1], *Vyacheslav* Olshevskyi[1], *Anastasia* Ostrovaia[1] and *Elena* Statsenko[1,*]

[1]Peter the Great St. Petersburg Polytechnic University, 195251 Polytechnicheskaya str. 29, St.Petersburg, Russia

Abstract. This article considers the moisture transfer phenomenon in the vertical ventilated channel. Also the article considers innovative structures for the vertical ventilated channel.The negative factors connected with moisture accumulation and excessive moistening of a heat insulation are given. Optimum design parameters for definition of the most intensive moisture transfer are discovered: systems with and without grooved lines. Influence that existence of technological gaps (grooved lines) has on the speed of air flows in a gap is investigated. Dependence of speed on width of the ventilated gap for a design with the opened and closed grooved lines is established empirically. The dehumidification properties of the vertical ventilated channel are described, as well as dependence of mass evaporation rate on time is established. Processes of drying-out in various conditions are compared.

1 Introduction

Energy and resource conservation is the central focus area of modern technical policy of the Russian Federation in the construction industry. The increase of thermal insulation of the enclosure counts for a great deal in a set of measures for energy conservation. In this regard multilayered enclosure structures with the use of porous thermal insulators gained widespread in modern construction. These materials offer the increased cold-resisting properties and allow to reduce thickness and weight of the enclosure structures. At the same time the durability and service property of such structures substantially are defined by their moisture conditions.

The control of moisture level is a necessary requirement for maintenance of the internal microclimate of buildings. Well-timed definition of excessive moistening sources also allows to support an optimum work level of the structures with the ventilated gap. That allows to improve service properties of the enclosing structures and air condition systems.

When considering this point the moisture transfer phenomenon is extremely essential. It is especially actual in climate conditions of the Northwestern District and similar regions since the increased precipitation level leads to the excessive moistening of systems.

* Corresponding author: staclena@mail.ru

Moisture transfer is a process of the movement of moisture which represents transfer of vacant and physically-bounded water under the influence of gravitational and sorption (molecular and capillary) forces.

Thus, quantitative calculation of moisture transfer is one of the most important in the multilayered enclosure structure designing.

This article considers influence of parameters of the ventilated airgap on air velocity speed which moisture transfer directly depends on.

2 Literature References

A significant contribution to the promotion of the subject "Moisture Transfer Assessment in the Vertical Ventilated Channel" was made by [1-35].

The article [1] considers the conditions under which there is a cold air filtration in the ventilated gap. The publication [2] considers features of work of building faces with and without grooved lines. In the work [6] the optimum sizes of the airgap are defined. The author of the work [10] reveals the key problems of structures with the ventilated airgap. In the text of the work [11] the moisture transfer phenomenon in disperse and porous materials is estimated. In the article [14] authors reveal adequate existence conditions of the free convection stream in the vertical conduit. The authors of the publication [23] consider issues of a natural ventilation of the vertical channel. The work [28] considers the application of the HVF Hinged Ventilated Facades) constructions in northern climate areas. The author of the work [29] investigates the wall building with the ventilated building face thermal protection taking into account a longitudinal air filtration.

3 Setting goals and objectives

- Influence of grooved lines on air exchange and moisture transfer in the HVF gap
- Research of the moisture transfer phenomenon in the vertical channel
- Comparison of the dehumidification properties under various conditions

4 Materials and Methods

The majority of the humidity testing methods are focused on reducing moisture intake into the airgap. However the imperfection of building structures leads to their excessive moistening. In this regard more and more attention is given to the moisture accumulation ability of materials and the possibility of their drying out.

Moisture evaporates while drying from a surface of the dried-up materials. At the same time there is a moisture movement from within of a surface. The heat, necessary for moisture evaporation, is transferred from environment by means of convective surface heat transfer. If the environment has the increased temperature, the material heats up when drying and there is a movement of heat in consequence of heat conductivity. Thus, the main physical phenomena of convective drying are the following:

- heat transfer from a gaseous fluid to a surface of the dried-up samples, i.e. heat transfer between a surface and the environment by means of convection;
- movement of heat from external into internal layers in consequence of heat conductivity;
- moisture evaporation from a surface of the dried-up materials - moisture exchange with the environment
- moisture movement from the central layers to a surface; it is accepted to be called *moisture transfer*.

Moisture transfer is the most difficult and important of all the listed phenomena. It has significant effect on drying behavior.

Moisture transfer, i.e. moisture movement, happens under the influence of three main forces: drop of humidity, temperature drop and pressure drop of vapor-air mixture in a material.

Moisture transfer rarely happens under the influence of any of the listed mechanisms. All types of moisture transfer which can lead to emergence of the problems caused by moisture can also help when construction materials and surfaces are drying.

The saved-up experience of research of heat and moisture transfer in the enclosure structures gives the grounds to claim that a zone of the greatest moistening is the layer of a mineral wool heat insulation adjoining the gap. Therefore, the degree of water vapor saturation in the ventilated gap is defining rather than average humidity on heat insulation thickness.

One of the most important characteristics of air exchange for moisture transfer calculation is air speed in the air layer. For its definition we will consider the stylized scheme of the ventilated gap between the "hot" plane y = 0 (with T_h=67°*C*) and the cold plane y = h (with Tc=23°C). Heating is provided with three heating elements located throughout the height of the installation. For equal distribution of heat elements attach to a tin sheet, therefore T_h=const. The height of a gap is equal to L=204 cm, and L ≫h.

While producing a facade sample we consider the temperature difference of external and internal layers proved by performance of heating systems. The model considers wall heating and heat input from it in the air layer. The external air is supposed to come into the air layer through the lower air hole. It rises through the ventilated channel and leaves through the top air holes. Air temperature increases and its relative moisture changes during the rise. At the same time moisture increases throughout the height of the layer.

While researching moisture transfer in the air layer it is necessary to know air velocity in the gap that depends on the chosen structure of the ventilated building face – with open or pressurized grooved lines.

Let us carry out an experiment. We take the speed and air temperature in the air gap depending on its width in the structure with and without grooved lines. We use thermal anemometer for measurement.

The main calculatory dependence for the average speed of the free convection stream (FCS) in the ventilated gap:

$$\vartheta = \varphi\sqrt{1-\frac{T_c}{T_h}}\sqrt{2gL}, \qquad (1)$$

Where T_c is the temperature of a cold wall, T_h is the temperature of a hot wall, L is the building face height, φ is the velocity coefficient which is determined by a formula:

$$\varphi = \frac{\vartheta}{\sqrt{2gL(1-\frac{T_c}{T_h})}}, \qquad (2)$$

By results of experiment we construct the schedule of dependence on the relation of L to h where the result is presented by red color at the closed grooved lines and blue at the open grooved lines:

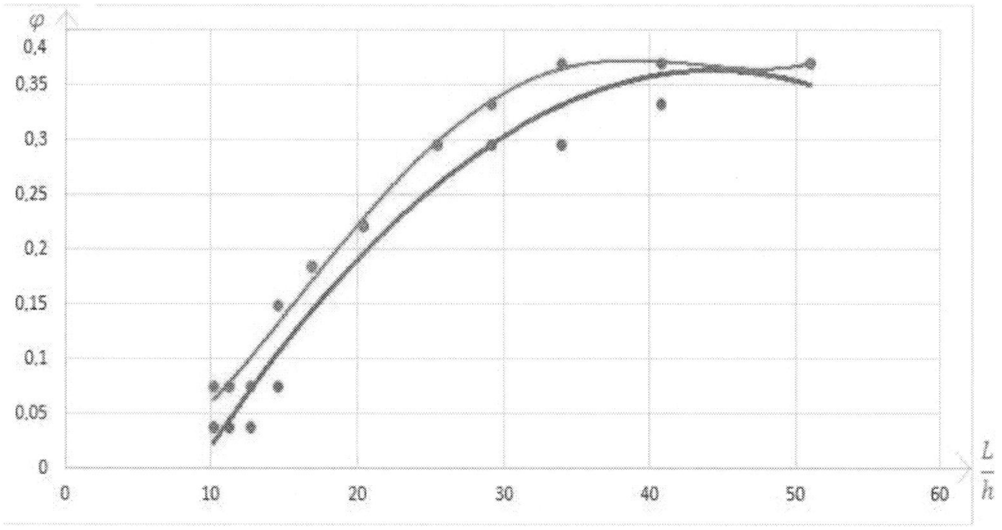

Fig. 1. Dependence φ on L/h.

On the basis of the data available from experiments it is possible to draw the following conclusions:

- at great L/h values (≥25) the vertical ventilated gap behaves as the channel with coefficient $\varphi \leq 0{,}4$
- at small L/h values (<25) the vertical ventilated gap behaves as the heated plate in free-air space.
- at equal L/h values the coefficient of velocity when the grooved lines is closed is more, than when ones are open. It testifies to higher velocity values in a structure with closed grooved lines

For simulation of the channel it is necessary to have L/h ≥ 25, i.e. $h \leq \frac{L}{25} \approx 8$ cm. Thus for determination of mass evaporation rate we take a gap equal to 8 cm and a structure without grooved lines.

For definition of the influence of air velocity on moisture transfer it is reasonable to conduct the following experiment: to place a humidified heat insulation in the ventilated gap of a structure and to measure weight under various conditions of drying-out: in the ventilated channel and in natural environment.

For definition of air velocity influence on moisture transfer we use a mineral-cotton sample 50 mm thick.

The size of the sample is 95x210x50 mm. The mass is 140 g. Let us humidify the sample. The mass of the saturated sample is 154 g.

One prototype is located in the airgap of HVF, and the identical one we will leave in natural environment. Then the time of drying-out of samples is measured and compared.

Using the obtained data we find the mass evaporation rate:

$$m^{\cdot} = \frac{\Delta m}{\Delta t}, \tag{3}$$

where Δm - the mass difference of two consecutive measurements, Δt - the interval between measurements, Δt=600 s.

Mass rate of vaporization in the structure:

$$m^{\cdot} = \frac{\Delta m}{\Delta t} = \frac{14}{5100} = 0.0027;$$

In natural environment: $m^{\cdot} = \frac{\Delta m}{\Delta t} = \frac{6}{5100} = 0.0012$.

Let us give the dependence of mass speed on each interval from the time where the result is presented by red color under natural conditions, blue - in the air gap:

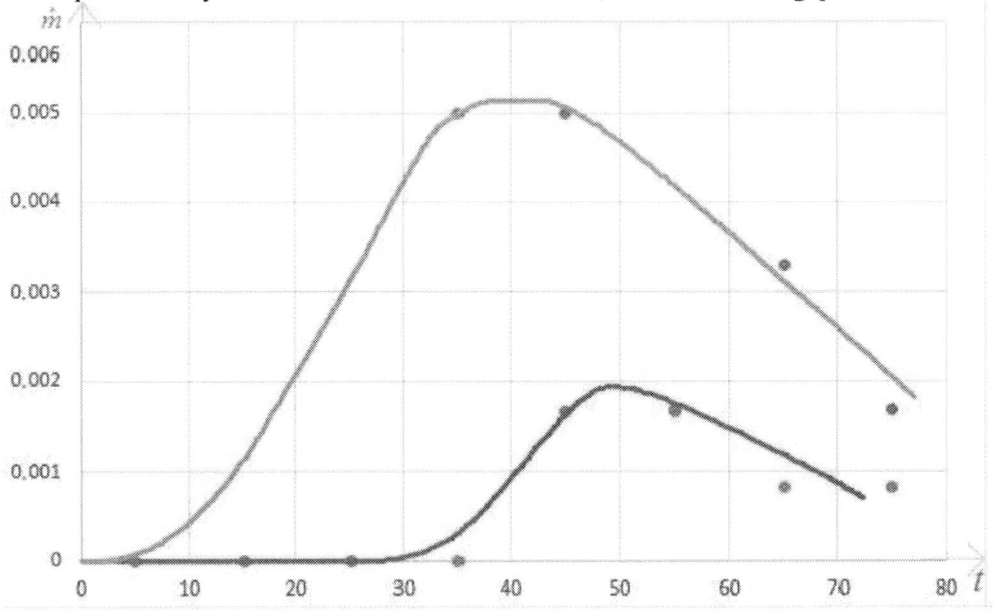

Fig. 2. Dependence of $m^{\cdot}$ on *t*.

The figure shows that the hidden heat of evaporation takes place at the initial stage of time, there is no evaporation. Evaporation begins when the sample warms up to the saturation point at this pressure.

5 Summary

Building humidity control is a key condition of increase of their durability, effective use and a healthy microclimate. Knowing of moisture sources and moisture transfer mechanisms in buildings and in enclosure structures allows to improve design choices of buildings and ventilation systems. The strategy of humidity control must be based on this knowing.

On the ground of the experiments we draw the following conclusions:

• a structure without grooved lines represents the "ideal" gap since velocity values are hare higher in it.

• the average moisture evaporation rate from a physical model is more, than in fixed air since the saturation temperature is higher in the ventilated air gap.

References

1. V.L. Kvint, V.V Okrepilov, Herald of the Russian Academy of Sciences, **84 (3)**, 188-200 (2014)
2. A.F. Ostrovaja, E.A. Stacenko, V.Ja. Olshevskyi, T.A.Musorina, MATEC Web of Conferences, **53**, (2016)
3. V.V Okrepilov, V.N. Krutikov, G.I. El'kin, Measurement Techniques, **57 (2)**,109-116 (2014)

4. M.R. Petrichenko, Soviet research, **23**, 703-715 (1991)
5. A.F. Ostrovaja, M.R. Petrichenko, E.A. Stacenko, Applied Mechanics and Materials, 725-726 (2015)
6. M.R. Petrichenko, Second Edition St. Petersburg State Polytechnical University Journal. Physics and Mathematics, 143–149 (2012)
7. M.R. Petrichenko, M. Petrichenko, E.B. Yevtushenko, Magazine of Civil Engineering, **37,** 35-40 (2013)
8. V.V Okrepilov, A.Yu. Smirnov, Measurement Techniques, **56(1)**, 54-60 (2013)
9. V.V Okrepilov, Standarty i Kachestvo, **10**, 52-55 (2005)
10. N.I. Gamajunov, S.N. Gamajunov, Vestnik TvGTU, **22,** 46-54 (2012)
11. S.V. Kornienko, Inzhenerno-stroitel'nyj zhurnal, **2,** 18-23 (2012)
12. V.V Okrepilov, Ivanova, G. Standarty i Kachestvo, **12**, 62-68. (2004)
13. E.B. Yevtushenko, M.V. Petrochenko, Magazine of Civil Engineering, **8,** 38-45 (2013)
14. V. Murgul, Procedia Engineering, **117**, 808-818 (2015)
15. V.V Okrepilov, Standarty i Kachestvo, **3**, 94-96. (2003)
16. V.V Okrepilov, Studies on Russian Economic Development, **24(1)**, 35-42 (2013)V. G. Gagarin, V. V. Kozlov, I. A. Mehnecov, AVOK, **8**, 71-80 (2005)
17. J.Li, Y.Chow, Centre for Sustainable Energy Technologies (CSET), 198-203 (2014)
18. Jun Hana, Lin Lua,Jinqing Penga, Energy and Buildings, **56,** 204–209 (2013)
19. V.N. Bukhartsev, M.R. Petrichenko, Power Technology and Engineering, **5**, 374-377 (2011)
20. Alin V. Rosca, Ioan Pop, International Journal of Heat and Mass Transfer, **60**, 355-364 (2013)
21. G. Barrios, G. Huelsz, R. Rechtman, J. Rojas, Energy and Buildings, **43**, 219-223 (2011)
22. H. Yang, X. Feng, G. Xia, Q. Wan, ISHVAC, 384-396 (2013)
23. L. Gaillard, S. Giroux-Julien, C. Menezo, H. Pabiou, Chair Habitats and Energy Innovations, 54-67 (2012)
24. N. Harmati, R. Folić, Z. Magyar, J. Dražić, N. Kurtović-Folić, Thermal Science, Nuclear Institute Vinča, Belgrade (2015)
25. N. Harmati, Z. Magyar, Energy Procedia, **78**, 2458 – 2463 (2015)
26. V. Pukhkal, A. Bieliatynskyi, V. Murgul, Journal of Applied Engineering Science, **14**, 93 – 101 (2016)
27. N. Harmati, Ž. Jakšić, N. Vatin, Procedia Engineering, **117**, 791 – 799 (2015)
28. N. Harmati, R. Folić, Z. Magyar, *Thermal Science*, **19(3)**, 865-880 (2015)
29. V.V. Okrepilov, V.N. Krutikov, G.I. El'kin, Measurement Techniques, **57 (2)**, 109-116 (2014)
30. V.L. Kvint, V.V. Okrepilov, Herald of the Russian Academy of Sciences, **84 (3)**, 188-200 (2014)
31. V. Murgul, MATEC Web of Conferences, **53**, 01046 (2016)
32. A. Gorshkov, V. Murgul, O. Oliynyk, MATEC Web of Conferences, **53**, 01045 (2016)

Efficient Thermal Insulation of Passive House with Curved Façades in Cold Climate

Askar Aznabaev[1,*], *Ivan* Pshuk[1] , *Daria* Likhaia[1], *Semen* Bondarenko[1], *Kirill* Gureev[1] and *Sergey* Amelchenko[1]

[1]Peter the Great St. Petersburg Polytechnic University, 195251 Politekhnicheskaya, 29, Saint-Petersburg, Russia

Abstract. In this paper there is exploration of the possibility of creating effective sealed building envelope in zero energy building (ZEB) with unique façades. This article describes passive and active measures of decrease of thermal loss through enclosure structures, usage of modern insulation materials; positive effect of rationalization of the façade orientation and form of a building in a plan; on the base of existing architecture concept structural solutions of building envelope are suggested. Using thermal FEA the enclosure structure was optimized. The results of modeling of units and elements of building envelope are: heat transfer coefficient U-value for heterogeneous multilayer structure, picture of temperature distribution and magnitude of the thermal conductivity and temperature of internal wall surface.

1 Introduction

According to Kyoto Protocol [1], more than 140 industrialized countries have committed themselves to substantially reduce CO_2 emissions. So countries have decided that the use of energy-saving technologies is a priority in order to preserve our natural resources.

The building sector is responsible for 40% of total energy consumption and CO_2 emissions in the world. That is why there is a necessity of a new approach to the design of each new project and each of the reconstructed building. E. Enache-Pommer, R. Mayer and G. Parsons write [2], that all over the world emissions of hydrocarbons can be reduced by 715 million TFOE by improving energy efficiency in buildings (27% of the projected increase in greenhouse gas emissions).

The most rapidly developing direction of the construction was the design of buildings with low energy consumption in the 1990's, according to paper [3]. In 1988, W. Feist and B. Adamson formulated in their book [4] theory of passive houses, which describes the principles of energy-efficient buildings. The articles [5, 6] mention that main requirement is the creation of a highly efficient building envelope. This is achieved through detailed design of exterior walls with a large layer of insulation. The tightness of the building envelope is provided by low air permeability and the lack of thermal bridges. This is

* Corresponding author: askar.spbstu@gmail.com

noticed in the articles [7-9] written by T.S. Egorova, S.G. Sheina, A.A. Fedotova and others authors.

One of the best complex approaches for the design of a passive house is the use of passive measures, which rationalizes the use of insulating materials. According to V. Murgul,
N. Vatin, E.H. Varzaneha [10-12], it includes: increased a compact shape of the building, the preferred orientation of windows and building façades to the south to provide solar heat gains. All these activities and recommendations for the design joint of building envelopes by the Passive House Institute restrict architects and designers. From an architectural point of view, the vast majority of buildings with zero energy consumption has a simple shape like a square or cylinder. The theory of the passive house limits future building projects in their forms of expression and a variety of architectural façades.

Recently, there is a complex task for the team of architects and designers. The complexity of its implementation consists in two aspects. On the one hand there is a need to create an unusual religious architecture of the building. On the other hand, the project shall conform to the requirements of energy efficiency. It is mentioned in article [13]. Thus the use of non-standard design solutions for passive house is not possible. Project teams need to find a balance between architectural concepts and solutions related to energy efficiency.

In this paper we consider the possibility of creating energy efficient sealed envelope for a unique building. This article describes the features of solutions used in a building with architectural expressive and complex façades of the nZEB.

The object of study is a concept architectural project, created by students of the SPbPU for the international competition "Multi-comfort House ISOVER-2015". It is noted in article [14] that the project involves a high degree of architectural study and detailed design of the building envelope.

The paper has the following goals and objectives:

- analysis of the shape and envelope of an existing architectural project from the viewpoint of energy efficiency and the definition of geometrical parameters,
- offer passive measures to reduce heat loss through the building envelope with rationalization of its shape and orientation of the façades,
- selection of the optimal type of building envelope and wall structure and the construction of individual nodes and elements of building enclosure with a continuous insulation layer,
- thermal modeling, analysis and determination of thermal properties of previously designed nodes of the heterogeneous building envelope.

Nomenclature

nZEB	near-zero energy building
FEA	finite element analysis
TFOE	tonnes of fuel oil equivalent
f-factor	form factor (perimeter to area ratio)
LSTWC	light steel thin-walled construction

2 Methods

2.1 Passive action to reduce heat loss through the building envelope

The initial data for the study are presented in the form of digital conceptual 3D-building model created in Revit. In addition, the projected building is located in Astana, Kazakhstan.

This area has a sharply continental climate and as a result strong winds. Number of windy days is 280-300 per year. These factors can have a significant impact on the overall thermal resistance. High wind pressure on the façade is the cold air infiltration into the room. These factors must be considered when the architecture of the building façades is designed.

Exploration of the initial architectural model has revealed negative factors which increase the heat loss through the building envelope:
- non-compact form caused by multisection structure of the building (Fig. 1a),
- vertical irregularity of enclosure construction (Fig. 1b),
- increased area of translucent structures as part of the façades,
- large number of terraces with accessible roof (Fig. 1a).

Fig. 1. General view of the building: (a) plan, (b) south façade, (c) perspective from North.

To minimize the negative impact of these factors the measures that do not change the original architecture of the project have been suggested. Among them the following are:
- horizontal rotation of major sections of the building to reduce the total façades' area oriented to the north (Fig. 1c),
- orientations of terraces and the extended side of façades to the southern direction for maximizing solar heat gains and insolation (Fig. 1a).

Also, during the analysis the positive moments in a curvilinear configuration of façades were revealed. In the plan the triangle deprived of corners due to bionic modification (Fig. 1a). Firstly, such form has f-factor = 0,93 that is 12-14% less, than for traditional rectangular in plan of a building. Secondly, rounded building shapes chosen due to the aerodynamic properties reduce the wind pressure upon a wall (Fig. 2a). Therefore, the air infiltration through a building cover is passively reduced. Thirdly, trilateral configuration of façades provides their natural day lighting. Thus the relative area of the unlit façades is 25% less (Fig. 2b) than for traditional. In summary, the cover of a façade has no corners, thereby excludes undesirable thermal bridges.

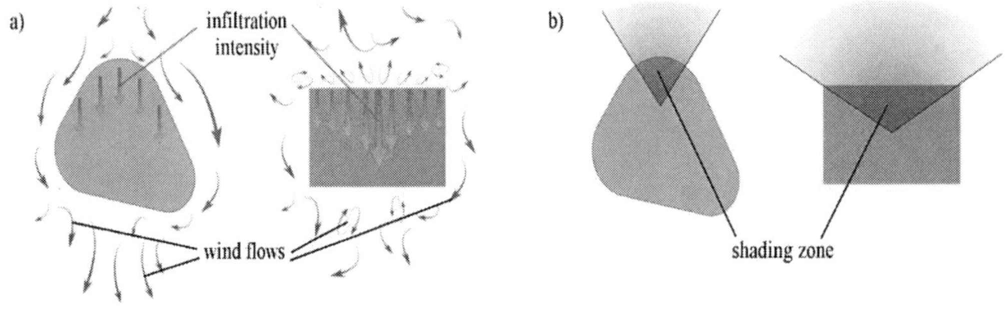

Fig. 2. Comparison of building forms in the plan of: (a) the perception of wind pressure, (b) the level of daylight illumination.

2.2 Special aspects of designing the effective building envelope

To choose the type of enclosure walls and to appoint its level of thermal performance it is needed to consider two main factors:

- climatic conditions of the area,
- features of the architecture and shape of the façades.

Long heating period (216-229 days/year) is one of reasons for constructing walls with heat transfer coefficient $U \leq 0.1$ W/(m^2·K). On this basis framed type of walls were chosen for designing building. Space frame of walls is formed with flat trusses [15] that repeat geometry of concept architecture project (Fig. 3a). Applying LSTWC allows to minimize weight of enclosure constructions (Fig. 3c).

Besides that, fulfilling the frame with effective insulation gives a wall with high thermal resistance [16]. The important point in designing of insulated frame is to form ongoing layer of insulation through the whole cross-section of building. It can be achieved by complete isolation of supporting reinforced concrete elements of the environment and by setting up mutual arrangement of 4-ply insulation with the displacement. The frame of enclosure walls is designed as whole curtain and does not have thermal bridges at the junction of adjacent floors and in interface assembly of roof and wall (Fig. 3b).

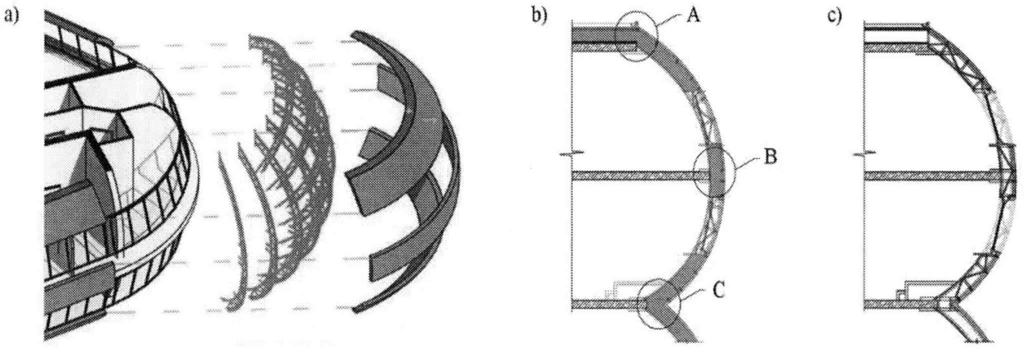

Fig. 3. (a) spatial model of the building envelope, (b) arrangement of insulation layer in cross section, (c) profile of bearing framework (truss).

Obtaining the most adequate value of coefficient of heat transfer is reached by detail analysis of processes of thermal conductivity in heterogeneous and complicated assembles of construction [17].

This research focuses on the development of FEA and of thermal 2D modeling . Construction of assembles was created in AutoCAD. After that thermal valuations were calculated in THERM software.

For analyzing boundary conditions were set (Tab. 1) which are regulated by National standards of buildings thermal protection and microclimate in internal building spaces [19].

DOI: 10.1051/matecconf/20167302003

Table 1. Boundary conditions for thermal analysis of enclosing structures.

Performance	Notation	Units of measurements	Value
Heat transfer coefficient of wall surfaces: 1. inside 2. outside	 α_{in} α_{out}	 [W/(m^2·K)] [W/(m^2·K)]	 8.7 23.0
Optimum relative humidity of the indoor air in the cold season	W_{in}	[%]	45.0
Outdoor air humidity	W_{ext}	[%]	65.0
Optimum indoor temperature	t_{in}	[°C]	+21.0
Outdoor temperature (average temperature of the coldest five-day period per year)	t_{out}	[°C]	-33.0
Absolute minimum temperature	t_{min}	[°C]	-52.0
Absolute maximum temperature	t_{max}	[°C]	+45.0
Average outdoor temperature of the heating season	T_{ave}	[°C]	-8.4
Dew point temperature is taken on the basis of the abovementioned preset conditions	t_d	[°C]	+8.6

3 Results

3.1 Effect of passive measures

At the base of project documentation (Fig. 4a) and of simulation of solar activity using Autodesk Ecotect Analysis software for Astana city, Kazakhstan, energy optimum location of a building was determined. Analysis is based on average daily incident radiation on a vertical surface. Major solar heat gains are reached by orientation of façades on the angle of 160° (direction SSE) (Fig. 4b). Due to obtained data and rotation of building sections (Fig. 4c) optimal orientation of the building as regards energy rationality was achieved.

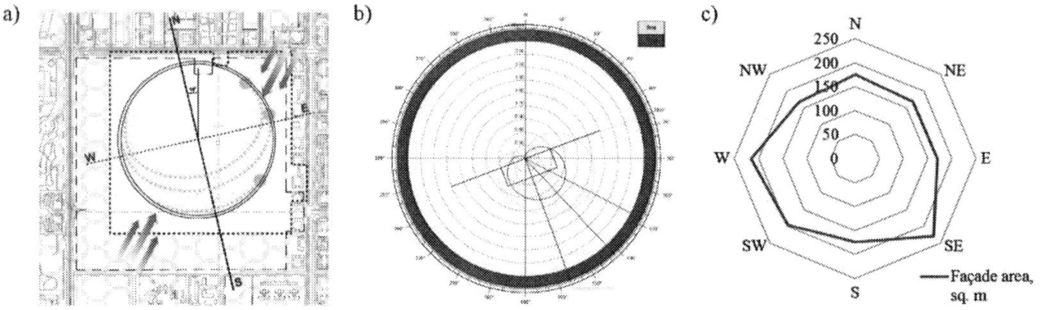

Fig. 4. (a) climatic conditions of building site, (b) diagram of optimum building orientation, (c) ratio of the façade areas to the cardinal directions.

3.2 Thermal analysis of interface nodes of enclosure structures

Structural solutions of building envelope that were demanded were optimised to exclude through-the-thickness thermal bridges. Three assembles of roof, wall and slab – A, B and C (Fig. 3b) were exposed to comprehensive thermal analysis. For some of them there are following pictures:

- structure scheme and composition of optimized (Fig. 5a,c),
- picture of temperature distribution in envelope (Fig. 5b),

• minimal temperature in interior surface of enclosure structure (Fig. 5d).

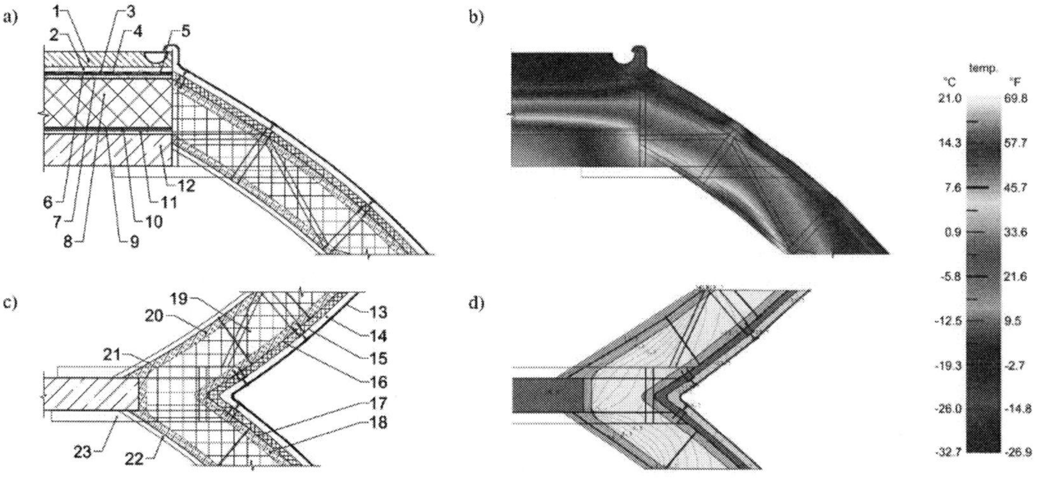

Fig. 5. (a,c) constructive drawings of nodes A and C, (b) color infrared picture of the temperature distribution in node A, (c) temperature distribution in node C.

It has to be admitted, that relative thickness of effective insulation composes ~ 90% or 390 mm. It indicates low material consumption of building and rationality of design solution. Materials that are used in enclosure structures and its thickness are restricted in appendix A.

Modeling shown that in unfavorable climate conditions temperature of interior wall surface t_{int}^{surf} = +16.3°C (indoor air temperature t_{int} = +20.0°C). It satisfies requirements ISO 7730 [20, 21].

Main characteristic of energy efficiency and structure comfort is airtightness of building envelope. Its numerical rating is air permeability, occurrence of thermal bridges and intensity of thermal loss through it. According to multistage thermal analysis and optimization of envelope construction, the fact that it is better to locate all supporting parts inside a building was determined. There are only metal profiles of LSTWC as thermally conductive inclusions. Thus due to high thickness and thermal lag of walls and thinness of metal profiles influence of heat transfer through metal frame is insignificant (Fig. 6).

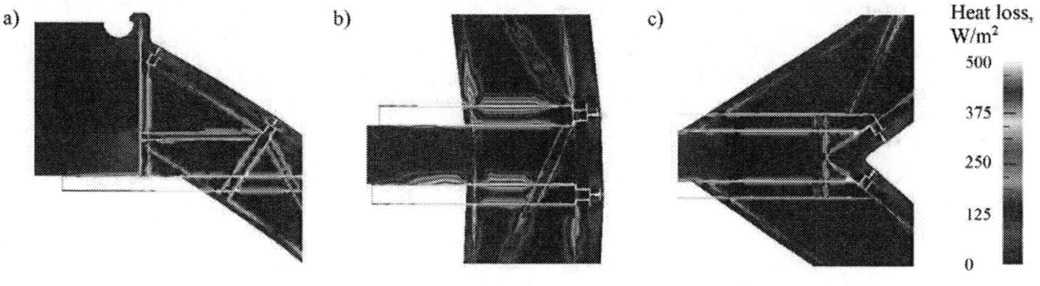

Fig. 6. Magnitude of heat transfer through envelope structures in assembles: (a) A, (b) B and (c) C.

Values of U-factor that was calculated for assembles of heterogeneous multilayer structure and translucent constructions in THERM software are described in table 2.

On the base of taken wall construction and concept building model energy balance of the object was calculated in Multi-Comfort Designer software that was certified by Passive House Institute. The results are shown in table 3.

Table 2. Results of calculating U-value of enclosure structures.

Thermal transmittance	U, [W/(m²·K)]
Walls	0.089
Roof	0.098
Slab above the basement	0.109
Windows	0.769
Doors	0.714

Table 3. Numerical energy balance characteristics of building.

Performance	Units of measurements	Value
Total heated space area	[m²]	3 478.0
Total heated space volume	[m³]	11 477.0
Total enclosure structures area	[m²]	1 496.0
Transmission heat losses	[kW·h/year]	120 094.6
Ventilation heat losses		25 634.1
Total heat losses		145 728.7
Internal heat losses		39 089.9
Available solar heat gains		76 627.2
Total heat gains		105 677.1
Annual heat demand		40 051.6
Specific annual heat demand	[kW·h/(m²·year)]	11.52
Energy Efficiency Class	—	A+

4 Discussion

The fundamental principle to design energy efficient unique building is a comprehensiveness of approach [22]. Detailed analysis of building form is necessary for creating of a passive house. Modeling of building geometry lets to considerate climatic factors that influence indirectly on energy efficiency of future structure. Form compactness, rational façades orientation, simple building geometry in plan are passive factors that reduce energy costs of building operation. Active measure is a designing of effective building envelope. Individual approach is necessary for designing of enclosure structures of buildings with unique architectural façades. It is necessary to consider each interface assemble of elements in details. However, it makes building process longer and more expensive. The solution of this problem can be using CAD and FEA technologies and creation of BIM-model also according to articles [23, 24]. Existence of BIM allows to control energy consumption and efficiency during whole building life cycle [25]. Besides, it will make certification of building with green standards (LEED, BREEAM) easier. Total obtainment of high energy efficiency class is possible with using of complex of engineering solutions: usage of renewable energy sources, heat pumps, reuse of water, smart house systems [26-33].

5 Conclusion

The base of unique buildings is an architectural component that creates just a shaping envelope, because of that the main aim of engineers and project designers becomes competent integration of energy efficiency standards and regulations with initial concept of future structure.

This paper includes actual researches in the area of thermal insulation of houses with unique façades. As a result, the suggestion of insulation of such constructions was offered.

Detailed designing and thermal analysis allowed to create an effective thermal insulation of a building that is located in cold climate. Specific annual heat demand reached the value near 11.5 kW·h/(m^2·year). It permits to classify this building to A+ class (less than 15 kW·h/(m^2·year) for passive houses), despite on unfavorable climatic conditions.

Eventually, complex of engineering solutions permitted to change from just architectural concept to complex system that creates comfort and safety of interior space.

6 Appendix. Composition of enclosure structures

Table 1. Materials used in the nodes A, B and C.

№	Name of enclosure structure layer	Thickness, mm	№	Name of enclosure structure layer	Thickness, mm
1	reinforced coating	95.0	13	Eco-polymer composite façade panel	7.0
2	Sand cement screed	30.0	14	Air gap	40.0
3	Double-layer polyethylene film	1.5	15	House wrap (Isover HB)	1.0
4	Waterproof bitumen coating	2.0	16	Light gauge steel framing	2.5
5	Primer	1.0	17	Fiber-glass thermal insulation (Isover VentFacade-Top)	40.0
6	Sand cement screed	20.0	18	Fiber-glass thermal insulation (Isover VentFacade-Optima)	50.0
7	Roofing felt	4.0	19	Fiber-glass thermal insulation (Isover VentFacade-Bottom)	250.0
8	Mineral wool thermal insulation (Isover OL-TOP+OL-Pe)	30+300	20	Fiber-glass thermal insulation (Isover VentFacade-Optima)	50.0
9	Sealant compound	2.0	21	Vapour seal covering (Isover Vario/VS 80)	1.0
10	Vapour seal covering	1.0	22	Drywall	24.0
11	Primer	1.0	23	Wall covering	5.0
12	Reinforced concrete slab	200.0			

References

1. *United Nations Framework Convention on Climate Change* (United Nations, Kyoto, 1997)
2. E. Enache-Pommer, R. Mayer, G. Parsons, ASHRAE Pap, **28**, 1-9 (2013)
3. A. Gorshkov, X. Rakova, Proc. Eng, **100**, 1505-1514 (2015)
4. W. Feist, *Main provisions of the designing of passive houses* (Ass. of Const. Univ, Moscow, 2008)
5. N. Vatin, D. Nemova, P. Rymkevich, A. Gorshkov, Mag. of Civ. Eng, **8**, 4-14 (2012)
6. M. Tanic, D. Stankovic, V. Nikolic, M. Nikolic, D. Kostic, A. Milojkovic, S. Spasic, N. Vatin, Proc. Eng, **117**, 924-937 (2015)
7. T. Egorova, V. Tscherkas, Bull. MGSU, **3-1**, 421-428 (2011)
8. S. Sheina, A. Minenko, Analysis and calculation of thermal bridges for improvement of residential building's energy efficiency, Engineering Journal of Don. 4-1 (2012) 131.
9. T. Egorova, A. Fedotova, V. Tscherkas, P. Belogurov, Roof. and insul. Mat, **4**, 26-32 (2013)
10. E. Varzaneha, M. Aminib, M. Bemanian, Proc. Eng, **94**, 25-32 (2014)

11. N. Kobelev, S. Emelyanov, V. Kretova, A. Morzhavin, V. Amelin, V. Kobelev, Proc. Eng, **117**, 186-190 (2015)
12. V. Murgul, N. Vatin, I. Zayats, Proc. Eng, **117**, 824-829 (2015)
13. A. Aksamija, Proc. Eng, **118**, 72-80 (2015)
14. A. Nefedova, D. Chernyshev, D. Tseytin, Cons. of Un. Build. and Str, **10**, 73-87 (2014)
15. S. Ovsyannikov, T. Vyazova, Roof. and insul. Mat, **1**, 28-31 (2014)
16. L. Long, H. Ye, Ener. Proc,**75**, 1850-1855 (2015)
17. A. Volkov, A. Sedov, Proc. Eng, **91**, 362-367 (2014)
18. A.A. Aznabaev, A.V. Ovsyannikova, A.O. Povzun, Z.A. Gaevskaya, Cons. of Un. Build. and Str., **4(43)**, 103-116 (2016)
19. Building regulations *SP 50.13330-2012 Thermal performance of the building* (Scientific-Research institute of building physics of Russian architectural and building science, Moscow, 2013)
20. K. Lomas, R. Giridharan, Build. and Envir, **55**, 57-72 (2012)
21. National code *GOST R ISO 7730-2009 Ergonomics of the thermal environment. Analytical determination and interpretation of thermal comfort...* (ANO "NIC KD", Moscow, Russia, 2010)
22. K. Gureev, A. Kazimirova, V. Avvakumov, G. Kafidov, P. Shaybacovich, A. Aznabaev, Sc.-meth. Elec. Jour. "Concept", **5**, 76-80 (2014)
23. V. Bazjanac, *IFC BIM-based methodology for semi-automated building energy performance simulation* (Lawr. Berk. Nat. Lab., CA, USA, 2008)
24. M. Royapoor, T. Roskilly, Ener. and Build, **94**, 109-120 (2015)
25. T. Lu, X. Lu, M. Viljanen, Ener. Proc, **75**, 1825-1831 (2015)
26. N. Klueva, S. Emelyanov, V. Kolchunov, A. Bukhtiyarova, App. Mech. and Mat, **725-726**, 1423-1429 (2015)
27. V.L. Kvint, V.V Okrepilov, Herald of the Russian Academy of Sciences, **84 (3)**, 188-200 (2014)
28. V.V Okrepilov, V.N. Krutikov, G.I. El'kin, Measurement Techniques, **57 (2)**,109-116 (2014)
29. V.V Okrepilov, A.Yu. Smirnov, Measurement Techniques, **56(1)**, 54-60 (2013)
30. V.V Okrepilov, Studies on Russian Economic Development, **24(1)**, 35-42 (2013)
31. V.V Okrepilov, Standarty i Kachestvo, **10**, 52-55 (2005)
32. V.V Okrepilov, Ivanova, G. Standarty i Kachestvo, **12**, 62-68. (2004)
33. V.V Okrepilov, Standarty i Kachestvo, **3**, 94-96. (2003)

MATEC Web of Conferences **73**, 02004 (2016) DOI: 10.1051/matecconf/20167302004
TPACEE-2016

LED Lightbulbs as a Source of Electricity Saving in Buildings

Akhtar Zeb[1], *Marcelo* de Andrade Romero[2], *Dinar* Baiguskarov[3,*], *Sanzhar* Aitbayev[3], *Ksenia* Strelets[3]

[1]University of Engineering & Technology Peshawar, Khyber Pakhtunkhwa, Pakistan.
[2]University University of Sao Paulo, Rua da Reitoria, 109 - Butanta, Brazil
[3]Peter the Great St. Petersburg Polytechnic University, Politechnicheskaya, 29, St. Petersburg, 195251, Russia

Abstract. Today a number of renewable energy technologies are available for power generation, but fossil fuels are still providing a dominant share nevertheless. In order to decrease the electric bill and save our environment, energy conservation is always crucial. In this paper a very interesting idea is presented which is economically viable to reduce electricity usage in our buildings. An effort has been made to estimate the amount of energy that could be saved in the dormitory section of the IMOP building in Russia. Although there are many ways to reduce the consumption of electricity in this building but here the emphasis is on changing light bulbs inside the rooms, kitchen, toilet and bathroom of each apartment. The scope of the study is to figure out monthly electricity saving by replacing traditional light bulbs by LED light bulbs in the building under consideration. The total investment required and the payback period is also presented.

1 Introduction

Lighting represents almost 20% of global electricity consumption. This consumption is similar to the amount of electricity generated by nuclear power. The latest IEA estimates show the total savings potential in residential and services lighting at more than 2.4 EJ per year by 2030 [1]. The energy savings potential in the buildings sector is massive. Globally, the wide deployment of best available technologies and efficiency policies could yield annual savings in building final energy use in the range of 53 EJ by 2050 (a 29% reduction in projected building energy consumption relative to a business-as-usual scenario in 2050) – equivalent to the combined energy use of buildings in China, France, Germany, Russia, the United Kingdom, and the United States in 2012 (IEA, 2015a). Capturing this energy savings potential would deliver a range of benefits: lower electricity and fuel costs for businesses and households; greater reliability in meeting energy demand without costly infrastructure and disruptions; and reductions in emissions of heat-trapping GHGs and other pollutants that pose a threat to human health [2].

* Corresponding author: dinarbay@gmail.com

With the emerging concept of retrofitting, it is desirable to spare as many kilowatts of power as possible. This not only results in better utilization of energy inside buildings but also helps us to cut our electricity bill. It is quite feasible and economically viable by just making few changes inside our buildings or homes, for instance, changing paint color of the walls, choosing windows that prevent unnecessary heat transfer and even the simplest trick is to use LED light bulbs of suitable illuminance and wattage instead of traditional light bulbs.

In almost every working area appropriate lighting is the key to higher efficiency. It is also associated with educational field especially the student's dorms. The student's dormitory is an important part of the education area and should be provided with suitable lighting conditions. It should be equipped according to the students visual comfort, however, should be made available in a way to reduce the energy consumption that would result in lower rent charges. In order to achieve this goal it is recommended to replace the traditional light bulbs by LEDs.

In this report a case study is made to represent the importance of changing LED lamps in the dormitory of international students known as IMOP located in Saint Petersburg, Russia.

The aim of this paper is to present:

- The current electricity consumption;
- Total investment required for changing LED light bulbs in all apartments;
- The payback period;
- The amount of electricity saving on monthly basis.

The current scenario of the building focusing on the calculations of daily electricity utilization and monthly electricity bill, which served as the basis for annual projections. The total investment required for LEDs light bulbs is estimated, the payback period is calculated and the amount of electricity that could be saved is also reported.

2 Light Emitting Diodes (LEDs) light bulbs

Traditional incandescent bulbs use a lot of energy to produce light and are no longer manufactured. 90% of the energy is given off as heat. That lost energy is money we are throwing away. Three of the most common energy-efficient lighting types include halogen incandescents, CFLs, and LEDs. LEDs use only 20%–25% of the energy and last up to 25 times longer than the traditional incandescent bulbs they replace. LEDs use 25%–30% of the energy and last 8 to 25 times longer than halogen incandescent [3]. LED bulbs are currently available in many products such as replacements for 3W, 4W, 5W, 6W, 10W, 12W, 40W, 60W, and 75W traditional incandescents, reflector bulbs often used in recessed fixtures, and small track lights [4]. The LED lamps have very high efficiency and can be used easily in place of traditional light source [5]. Also, it acts as an environmental light source from manufacturing to disposal [6]. There are many different applications, for instance; modern household equipment [6], head light and brake lights in automobiles [8], street lights [9], interior household illumination [10], hospital diagnostic and treatment room [11]. LED lighting is very different from other lighting sources such as incandescent bulbs and CFLs.

Key differences include the following [12]:

- Light Source: LEDs are the size of a fleck of pepper, and a mix of red, green, and blue LEDs is typically used to make white light.
- Direction: LEDs emit light in a specific direction, reducing the need for reflectors and diffusers that can trap light. This feature makes LEDs more efficient for many uses such as recessed downlights and task lighting. With other types of lighting, the

light must be reflected to the desired direction and more than half of the light may never leave the fixture.
- Heat: LEDs emit very little heat. In comparison, incandescent bulbs release 90% of their energy as heat and CFLs release about 80% of their energy as heat.

Some of the distinctive features of LEDs are [13]:

1. A light-emitting diode, or LED, is a type of solid-state lighting that uses a semiconductor to convert electricity into light. Today's LED bulbs can be six-seven times more energy efficient than conventional incandescent lights and cut energy use by more than 80 percent.
2. Good-quality LED bulbs can have a useful life of 25,000 hours or more -- meaning they can last more than 25 times longer than traditional light bulbs. That is a life of more than three years if run 24 hours a day, seven days a week.
3. Unlike incandescent bulbs – which release 90 percent of their energy as heat -- LEDs use energy far more efficiently with little wasted heat.
4. From traffic lights and vehicle brake lights to TVs and display cases, LEDs are used in a wide range of applications because of their unique characteristics, which include compact size, ease of maintenance, resistance to breakage, and the ability to focus the light in a single direction instead of having it go every which way.
5. LEDs contain no mercury, and a recent Energy Department study determined that LEDs have a much smaller environmental impact than incandescent bulbs. They also have an edge over compact fluorescent lights (CFLs) that expected to grow over the next few years as LED technology continues its steady improvement.
6. Since the Energy Department started funding solid-state lighting R&D in 2000, these projects have received 58 patents. Some of the most successful projects include developing new ways to use materials, extract more light, and solve the underlying technical challenges. Most recently, the Energy Department announced five new projects that will focus on cutting costs by improving manufacturing equipment and processes.
7. The first visible-spectrum LED was invented by Nick Holonyak, Jr., while working for GE in 1962. Since then, the technology has rapidly advanced and costs have dropped tremendously, making LEDs a viable lighting solution. Between 2011 and 2012, global sales of LED replacement bulbs increased by 22 percent while the cost of a 60-watt equivalent LED bulb fell by nearly 40 percent. By 2030, it's estimated that LEDs will account for 75 percent of all lighting sales.
8. In 2012, about 49 million LEDs were installed in the U.S. – saving about $675 million in annual energy costs. Switching entirely to LED lights over the next two decades could save the U.S. $250 billion in energy costs, reduce electricity consumption for lighting by nearly 50 percent and avoid 1,800 million metric tons of carbon emissions.

3 What's a lumen?

Lumens measure how much light you are getting from a bulb. More lumens means it's a brighter light; fewer lumens means it's a dimmer light.
Lumens are to light what:

- Pounds are to bananas;
- Gallons are to milk.

Lumens let you buy the amount of light you want. So, when buying your new bulbs, think lumens not watts.
We typically buy things based on how much of it we get, right? When buying milk, we buy it by volume (gallons). So, why should light be any different ? For decades, we have been buying light bulbs based on how much energy they consume (watts) – no matter how much

light they give us (lumens). When you're shopping for light bulbs, compare lumens to be sure you're getting the amount of light, or level of brightness, you want.

4 Current scenario

In this work the dormitory portion of the building is considered. There are total nine floors. The first floor consists of many different offices and other facilities for the students like medicine room, laundry, etc. Therefore, only 8 floors are examined in this study. There are total of 15 apartments on a floor. Each apartment consists of two bedrooms, kitchen, toilet and a bathroom. Figure 1 depicts the number of lamps inside an apartment.

Fig. 1. Lamps in apartment.

The current total power requirement of an apartment is presented in Table 1 (the study lamps are excluded).

Table 1. Current consumption of an apartment.

Apartment	Number of lamps	Watt	Total Wattage
Bedroom 1	4	60	240
Bedroom 2	4	60	240
Kitchen	2	50	100
Toilet	1	50	50
Shower	2	50	100
Total current power requirement of an apartment			730 W

The daily electricity usage and monthly bill of an apartment is calculated with the assumptions that the light bulbs are turn on for the following periods of time:

Bedrooms: 8h/day;
Kitchen: 16h/day;
Toilet & bathroom: 4 h/day.
Current electricity consumption and monthly & annual bill of an apartment:
Electricity consumption per day: 6.04 KWh;
Total consumption per month: 181.2 KWh;
Monthly electricity bill: 543.6 rubs (1 KWh = 3 rub);
Annual electricity bill: 6523.2 rub.

5 After retrofitting

In order to reduce the electricity consumption, appropriate LED light bulbs were chosen having the same amount of lumens but low wattage. In our case the LED light bulbs of 6 and 10 W are the suitable ones. With the usage of these lamps, the total power requirement of an apartment would be as under as shown in Table 2.

Table 2. Total power requirement of an apartment after retrofitting.

Apartment	Number of lamps	Watt	Total Wattage
Bedroom 1	4	10	40
Bedroom 2	4	10	40
Kitchen	2	6	12
Toilet	1	6	6
Shower	2	6	12
After Retrofitting the total power requirement of an apartment			110 W

After retrofitting, electricity consumption and monthly bill of an apartment:
Electricity consumption per day: 0.904 KWh;
Total consumption per month: 27.12 KWh;
Monthly electricity bill: 81.36 rub (1 KWh = 3 rub);
Annual electricity bill: 976.32 rub.

6 Results

With a vacancy rate of 25%, there are 100 occupied and 30 vacant flats in the building (the vacancy rate may change). Therefore, for the 100 occupied flats the total investment required for changing lamps, the electricity saving and payback period would be as under:
Total investment for changing lamps: 273.000 rub;
Amount of electricity saving per month: 15.408 KWh.
Annual Economy by reducing electricity bill: 554.688 rub. Payback period: 6 months.

7 Conclusion

In this study an idea of electricity saving inside buildings has been presented. The current electricity consumption with traditional lights is calculated. The LED lights are used replacing the traditional lights and the electric consumption is recalculated. The amount of electricity saving is figure out. The total investment required is estimated and thus the payback period is presented. It is concluded that replacing the present traditional lights by

LED lights in our buildings would not only result in low electric bills but also in energy conservation. Furthermore, the process is quite feasible and economically viable.

References

1. International Energy Agency
2. Information on http://iea.org/topics/energyefficiency/subtopics/lighting
3. International Partnership for Energy Efficiency Cooperation (IPEEC) & International Energy Agency (IEA), 'Building Energy Performance Metrics', Supporting Energy Efficiency Progress in Major Economies
4. Information on www.mrs.680org/bulletin
5. Information on http://energy.gov/energysaver/choices-save-you-money
6. Safaa Alzubaidi and Prashant kumar Soori, J.Light & Vis. Env.,**36(1)**, 23-31 (2012)
7. R. Devonshire, J. Light & VisEnv., **32**, 275-287 (2008)
8. N. Zheludev, Nature Photonics, **1**, 189-192 (2007)
9. S. Liu, A. Minato, S. Ozawa and M. Nakagawa, J. Light & Vis.Env, **31**, 65-69 (2007)
10. M.A.Dalla Costa, L.Schuch, L.Michels, C.Rech, J.R. Pinheiro and G.H.Costa, "Autonomous Street Lighting System based on Solar Energy and LEDs", 1143-1148 IEEE, (2010)
11. Wen-Shing sun, Chih-Hsuan Tsuei, and Yi Han huang, "Simulating the Illuminance and Efficiency of the LEDs used in General Household Lighting", Science Direct, Physics Procedia, **19**, 244-248 (2011)
12. P. Sathya & Dr. R. Natarajan, "Design of Energy Efficient lighting system for Educational Laboratory".
13. L. Svatovskaya, A. Sychova, M. Sychov, V. Okrepilov, MATEC Web of Conferences, **53**, Article Number 01023 (2016)
14. Information on http://energy.gov/articles/top-8-things-you-didn-t-know-about-leds
15. Information on http://energy.gov/energysaver/lumens-and-lighting-facts-label

MATEC Web of Conferences 73, 02005 (2016) DOI: 10.1051/matecconf/20167302005
TPACEE-2016

Combined Thermal Insulating Module of Mounted Vented Facades

Svetlana Ryabukhina[1,*], *Tatjana* Simankina[1], *Marina* Koshkarova [2], *Nikita* Sokolovskii[1] and *Oleg* Ryzhkov[3]

[1]St.-Petersburg State Polytechnical University, 195251 Politechnicheskaya St. 29, St. Petersburg, Russian Federation
[2]Ltd «Teploizolyatsiya ograzhdayushchikh konstruktsiy», 197342 Torshkovskaja St. 1, St.Petersburg, Russian Federation
[3]Ltd «URSA EURASIA», 196191 Leninskij Ave., 168, St.Petersburg, Russian Federation

Abstract. In order to define an optimum type of mounted vented facades among the existing ones, comparative analysis of two façade modules has been conducted. The first module type is a widespread standard module of hinged vented facade and the second type is less applicable combined thermal insulating module. Those two technologies were compared thermal engineering and energy efficiency parameters. It was defined that the application of a thermal insulating module with combined insulation system improves thermal engineering parameters of the building as well as leads to a substantial savings. This article exposes innovative materials and structure of vented facades which can be applied in modern construction.

1 Introduction

System of mounted vented facades (MVF) have found it's wide application in the construction market [1], although, despite the wholesale application, there are some aspects which require modernization [2]. At the current developmental stage of construction the increasing of energy efficiency and energy saving indexes [3] of buildings and structures with different purposes remains the major challenge.

As discussed elsewhere [4, 5, 6] it is possible to figure out the solution of those tasks only by the means of complex approach for the building construction itself, not by examining it's separated components. The other problem is defining the economic effectiveness. This task also demands for a complex approach, taking into account the whole cycle of construction and mounting works, and also the logistic costs, expenses for the storage and cargo works, not to take the distinct costs of construction materials for the unit of measure.

Authors Haase, Pukhkal, Penic [7, 8, 9] evaluates different design and configuration ventilated facade in respect to the amount of energy saved.

The optimal solution has to be found among the existing options of mounted ventilated facades (MVF), knowing the problems of a particular structure.

* Corresponding author: snowflake2203@yandex.ru

Let us compare a wide-applicable constructional solution of a MVF – the standard module and a less common one, though selectively applicable combined thermal insulating module for the thermal engineering and economic efficiency parameters.

2 Comparison of existing and optimize thermal insulation modules

Today the popular solution of MVF is a double-layered heat-insulating module made of mineral wool plates obtained from basalt rocks, where both layers are mechanically fastened to base – the wall. The whole structure itself fastens to the suspension system made of corrosion-resistant galvanized steel with polymer coating. Facing is made of porcelain stoneware slabs [10, 11, 12]. Along with the obvious advantages, such as high-class fire safety construction design (R0), which is allowing you to apply this structure with no restrictions for the building height, there is also a number of indicators that needs to be refined.

One of the constructional optimization options can be the application of the combined heat- insulating module, for example, the MVF PARURS structure. The structure consists of thermal insulating module with internal layer made of fiberglass staple plates (with a proper thickness according to thermal engineering calculations for the application region), which are fixed on the cantilevers of mounted system without and additional bracing. It also consists of a blanket with a thickness of more than 30 mm made of mineral wool plates obtained from basalt rocks, which is fixed on cantilevers of the mounted system with an additional bracing. As a windproof layer the materials with a high-class fire safety construction design (R0) are used [13]. Otherwise the construction has the standard MVF construction system.

Benefits of that kind of structure are the following parameters: minimization of thermal bridges, the improvement of layer uniformity index, improved heat-resistance index, which leads to increased energy efficiency as well as economic benefits during construction and installation works, storage and logistics.

Let us compare those types of heat-insulating modules.

One of the demands during the mount of heat-insulating plates in two layers is a fixing the first layer plates with a poppet head dowels with 110 mm head diameter or with poppet head dowels with additional 140 mm shims irrespectively to the attachment of the second layer [14]. On the basis of the technical solutions album ATS-01-2014, inner insulating layer of the PARURS structure is hung directly on brackets of the suspension system without additional mechanical fixing, which clearly reduces the amount of thermal bridges in the heat-insulating unit [15].

It should be noted that the outer walls of the constructions, unfortunately, don't always have a flat, smooth structure (Figure 1).

Fig. 1. The roughness on the outer wall construction

Comparing modules for the fit force of the inner heat insulating layer in enclosing structure revealed that mineral wool plates obtained from basalt rocks cannot always provide the required density and elasticity of the fit [16], while the use of a combined heat-insulating module shows a snug fit of insulation to the wall with a roughness and fills the irregularities of the boundary wall (Figure 2).

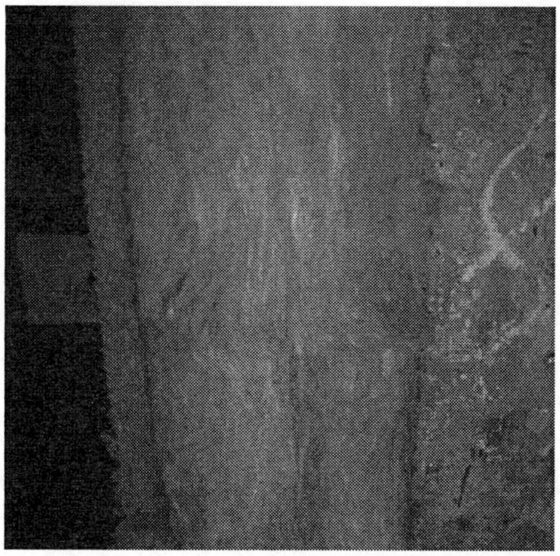

Fig. 2. The fit of a combined heat-insulating module to the wall with roughness

Heat transfer resistance of the outer walls in accordance with the thermal calculations, particularly for Saint-Petersburg, is 3.83 ($m^2 \cdot {}^\circ C / W$). Heat-insulating module made of two layers of mineral wool plates obtained from basalt rocks equals this parameter with the use of 150 mm heat-insulating layer. Combined heat-insulating module with the same thickness of the heat- insulating layer shows a 4.03 ($m^2 \cdot {}^\circ C / W$) index, which indicates more efficient heat-insulation in the structure [17].

In order to compare economic efficiency estimates were made for each type of construction with equal areas of the facade (Table 1, Table 2). Calculations during the mount of 1000 m^2 of the facade showed that combined system of insulation is 4400 euros cheaper than the standard one.

Table 1. Cost estimates of mounting 1000 m^2 of a combined module structure

№	Name	Measure units	Laboriousness, man-hour/ machine hour	Amount	Unit value, rub.	Total cons, rub.
	Mounting of vented facades with a porcelain facing and thermal insulation layer Overhead cost: 105% of Wages fund Estimated profit: 55% Wages fund	m^2 of facing		1000	263.29	263288.12
1	Manning of workers	man- hour	2.400	2399.865	60.13	144291.88
2	Manning of machinists	man- hour	0.258	258.16	72.50	18716.60
	Machinery					
3	Lifts with a capacity up to 500 kg, single-masted, elevating height 35 m	machine hour	0.258	258.16	111.63	28817.11
4	Screwdriver	machine hour	0.131	130.575	18.75	2 448.28
5	Drills: electric	machine hour	0.384	383.55	12.19	4 674.52
6	Electric grinding machines	machine hour	0.070	69.6	32.06	2 231.55
7	Electric 1.5 kW hammer with impact energy up to 18 J	machine hour	0.378	378.42	164.13	62 108.18
					Total:	263 288.12
	Overhead cost	105%				171 158.91
	Estimated profit	55%				89 654.67
					Total:	524 101.70
					VAT 18%:	94 338.31
4	Screwdriver	machine hour	0.131	130.575	18.75	2 448.28
5	Drills: electric	machine hour	0.384	383.55	12.19	4 674.52
				Overall with VAT 18%:		618 440.00

Table 2. Cost estimates of mounting 1000 m^2 of a standard module structure

№	Name	Measure units	Laboriousness, man-hour/ machine hour	Amount	Unit value, rub.	Total cons, rub.
	Mounting of vented facades with a porcelain facing and thermal insulation layer Overhead cost: 105% of Wages fund Estimated profit: 55% Wages fund	m^2 of facing		1000	390.35	390 347.40
1	Manning of workers	man-hour	3.6921	3692.1	60.13	221 987.51
2	Manning of machinists	man-hour	0.3688	368.8	72.50	26 738.00
	Machinery					
3	Lifts with a capacity up to 500 kg, single-masted, elevating height 35 m	machine hour	0.3688	368.8	111.63	41 167.30
4	Screwdriver	machine hour	0.1741	174.1	18.75	3 264.38
5	Drills: electric	machine hour	0.5114	511.4	12.19	6 232.69
6	Electric grinding machines	machine hour	0.0696	69.6	32.06	2 231.55
7	Electric 1.5 kW hammer with impact energy up to 18 J	machine hour	0.5406	540.6	164.13	88 725.98
					Total:	390 347.40
	Overhead cost	105%				261 161.79
	Estimated profit	55%				136 799.03
					Total:	788 308.22
					VAT 18%:	141 895.48
				Overall with VAT 18%:		930 203.70

4 Results and Discussion

The comparison of a widely applicable standard module of MVF to the optimized less applicable combined heat-insulating module structure solution based on the standard structure on the basis of thermal engineering and economic efficiency parameters, as well as sound absorbing characteristics of the heat-insulating layer, it was concluded that the application of the heat- insulating unit with combined insulation system not only improves the thermal performance of buildings, but also leads to a significant economic effect. The future is about neoteric technologies. Even if it seems utmost was invented earlier, we possess a wide range of activity lines for implementation of something new, original and more effective. For today combined module PAPURS is only coming to market and hasn't been tried on practice yet. This construction was patented in Russia this year. Although, many specialists have already estimated it's properties and are ready to implement it.

5 Conclusion

Based on the conducted comparison methodology authors obtained following data:
- heat-insulating module of a combined type is 4.03 ($m^2 \cdot {}^\circ C / W$) combined to 3.83 ($m^2 \cdot {}^\circ C / W$) of a standard system [17];
- from the point of economical effectiveness the usage of combined module leads to the significant economic effect (mounting 1000 m^2 of combined module is 4400 euros cheaper comparing to standard type).

References

1. Rynok ventiliruemyh fasadov, Ventiliruemye fasady «SK Divart» - Jelektronnyj resurs: http://sk-divart.ru/rynok- ventiliruemyh-fasadov
2. A.S. Gorshkov, D.V. Nemova, N.I. Vatin, Stroitel'stvo unikal'nyh zdanij i sooruzhenij, **7**, 49-63 (2013)
3. D.V. Nemova, Jenergojeffektivnye tehnologii v ograzhdajushhih konstrukcijah, Stroitel'stvo unikal'nyh zdanij i sooruzhenij, **3**, 78-82 (2012)
4. A. Milajić, D. Beljaković, N. Davidović, N.I. Vatin, V. Murgul, Procedia Engineering, **117**, 916-923 (2015).
5. N. Harmati, Ž Jakšić, N.I. Vatin, Procedia Engineering, **117**, 791-799 (2015)
6. M. Tanic, D. Stankovic, V. Nikolic, M. Nikolic, D. Kostic, A. Milojkovic, S. Spasic, N.I. Vatin, Procedia Engineering, **117**, 924-937 (2015)
7. M. Haase, F. Marques da Silva, A. Amato, Energy and Buildings, **41**, 361-373 (2009)
8. V. Pukhkal, M. Tanić, N.I Vatin, V. Murgul, Procedia Engineering, **117**, 864-869 (2015)
9. M. Penić, N.I. Vatin, V. Murgul, Applied Mechanics and Materials, **680**, 534-538 (2014).
10. L. Svatovskaya, A. Sychova, M. Sychov, V. Okrepilov, MATEC Web of Conferences, **53**, Article Number 01023 (2016)
11. D. Terentyev et al, Fasadnye teploizoljacionnye sistemy s vozdushnym zazorom. *Rekomendacii po sostavu i soderzhaniju dokumentov i materialov, predstavljaemyh dlja tehnicheskoj ocenki prigodnosti produkcii* (first ed., Gosstroj Rossii, Moscow, 2004)
12. Information on - URL: http://www.minplita.biz/solutions/50/.
13. № 123-FL (Federal Law) *«Tehnicheskij reglament o trebovanijah pozharnoj bezopasnosti»*
14. Information on http://www.minplita.biz/solutions/50/
15. Al'bom tehnicheskih reshenij ATR – 01 – 2014 *«Sistema ventiliruemyh fasadov PARURS» (2014)*
16. Information on : http://stoydiz.ru/pokraska-i-otdelka/montazh-ventiliruemyh-fasadov.html.
17. CP (Code of Practice) 50.13330.2012 *«Teplovaja zashhita zadnij»*

The Efficiency of the Ventilated Gap of the Double-Skin Facade Systems Using Fire Crosscuts

Aleksandra Barabash[1,*], *Elizaveta* Naumova[1], *Oksana* Zhuvak[1], *Darya* Nemova[1], and *Vyacheslav* Olshevskiy[1]

[1]Peter the Great St. Petersburg Polytechnic University, 195251 Politechnicheskaya Str. 29, St. Petersburg, Russia

Abstract. Double-skin facade is an excellent material for decoration of buildings. Despite the large number of advantages over other types of finishes, ventilated facades also have their drawbacks. In particular, a violation of installation conditions and fire safety regulations can lead to serious consequences. Innovative materials and structures in the field of fire safety are analyzed in this paper. The article deals with fire protection methods, special attention is given to the unit of fire crosscuts in the air gap of double-skin facade. Analysis of experimental data has revealed some drawbacks of this method and the necessity of design improvements.

1 Introduction

Recently, double-skin facades (DSF) are widely distributed due to their numerous advantages over traditional methods of facade finishing. The advantages are: energy efficiency, the flexibility of design decisions, noise protection and simplicity in installation at building's construction and reconstruction [1]. The structure of the curtain wall system is shown on (Fig. 1). The design is a steel frame connected with the insulant plate and facing panels. It is set on the outer part of the wall.

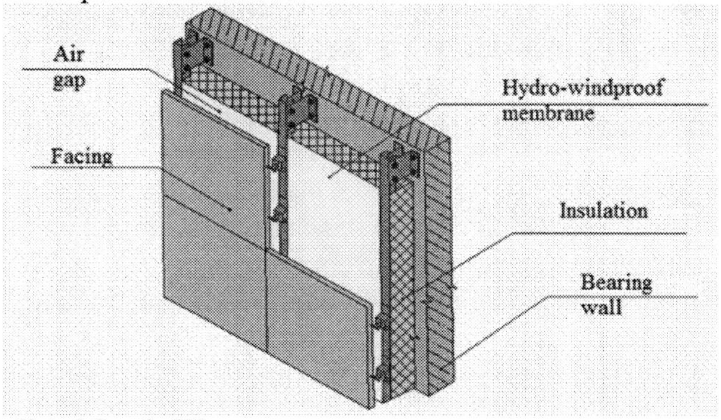

Fig. 1. The device of ventilated facade [2]

* Corresponding author: aleksandra17au@yandex.ru

Ventilated facades are mostly used in decoration or renovation of urban, office and residential buildings [3]. According to HVAC system analysis involving the use of the simulation software Computational Fluid Dynamics (CFD), Autodesk Simulation CFD and Energy Plus®, it is found that ventilated facade is an element of energy-efficient building. The construction keeps the fixed temperature inside the building regardless of the outside temperature changes that can significantly reduce the costs for air conditioning and heat [4-6]. The main difference of such facades is the presence of the air gap between the insulation and the facing protection material. The main purpose of the gap is to provide a thermal protection of the building, playing the role of the air buffer, reducing heat loss. One of the important properties is the removal of moisture: the condensation zone is shifted into the insulation layer which is bordered by a ventilated air interlayer [7]. In gap installation it is necessary to take into account technical characteristics and performance requirements of the building. If the preset thickness of the gap is excessive, the facade will buzz and whistle because of the impact of wind flow. Otherwise, in case of deficient air gap, the convective flow will be reduced, the moisture removal will be impossible; the insulation will damp and collapse eventually. In this situation the function of the air gap is reduced to zero.

For obtaining the qualitative and durable system it is necessary to comply with all installation and exploitation requirements and use high-quality materials only. Thus, the main problems of hinged ventilated facade systems are [8]:

- Calculating the necessary ventilated gap
- Choice of high-quality materials
- Qualitative installation execution
- Provision of the required heat transfer resistance
- Fire safety

The demand for these systems are explained by their functionality: DSF systems give not only an impressive architectural appearance to building, but also carry out the thermal protection functions and contribute to the normalization of the building's heat and humidity mode which achieved due to the design of these systems [8].

Fire safety is the most important parameter which is worth to take into account when installing the DSF. Basic requirements for the facade systems are established by The Federal Law "Technical Regulations on Fire Safety Requirements" [9]. The various aspects of fire safety of materials in the construction of facade systems are described in [10-12], which contain the classification of flammable materials on the smoke-forming ability and the toxicity of combustion products.

The most dangerous area of ventilated facade is a metal frame. It is responsible for ensuring the integrity and stability of the whole structure. Therefore, the use of fusible metals in the construction of the frame are not recommended. Otherwise, at the fire, parts of the facade will break down and fall, injuring passers-by and damaging nearby facilities. Thus, the most reliable are steel frames [13].

The composition of the facade also includes waterproof film which can easily catch fire in case of violation of assemble and exploitation conditions. No less common cause of fires is the use of low-quality or flammable cladding materials. In particular, the use of aluminum composite panels (ACP) was the cause of the rapid spread of fire on the facade of the building Address Downtown, Dubai; in high-rise building "Transport Tower", Astana; in high-rise building complex "Grozny City", Grozny.

In order to protect the structure during a fire, the following measures are used:

- The use of high-quality, fire-resistant materials, the relevant The Federal Law N 123
- The use of fire box

- The use of fire crosscuts

One of the measures of fire safety, in addition to the use of non-flammable materials, is installing the fire-protective boxes. The fire-protective box is a metal structure on the perimeter of window and door openings, attached to the wall by anchors. It can be done as a single structure or as a composite, interconnected by the metal ware. The space between the metal elements is filled with heat-resistant mineral wool. It does not allow the flame to penetrate into the facade. However, this system is not effective in conditions where the fire source is located outside the premise.

Therefore, fire crosscuts are set with definite step, spanning the whole thickness of the air gap. In case of fire, they prevent the spread of combustion of windproof membrane and falling of burning droplets film out of the gap [14], as well as the spread of flammable gases by perimeter wall.

Fire crosscut is a metal plate (solid or perforated) made of sheet steel with a minimum thickness of 0.6 mm and at a distance of not less than 6 m from each other. Since crosscuts should fit snugly to the facade when installed, function of the gap is reduced to zero. Therefore, perforation in a crosscut is performed in order not to disturb the movement of the air flows. The degree of punching is defined by calculations. According to [9] fire crosscuts can be attributed to a fire barrier, but the specific design requirements are only for structures that prevent the spread of fire from one room to another, and therefore are not covered.

(Fig. 2-3) presents assemblies of fire crosscuts by Kraspan and Sial systems.

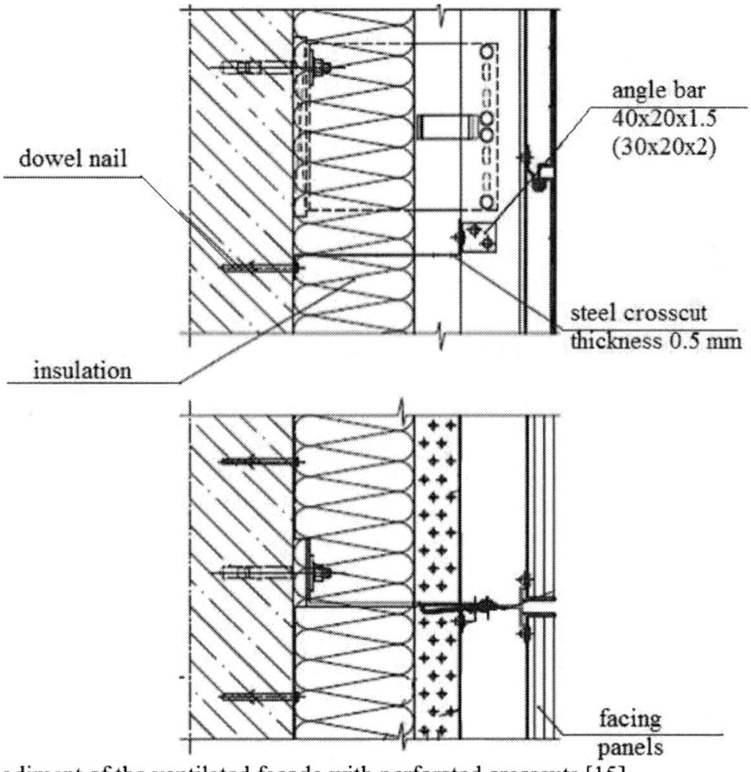

Fig. 2. An embodiment of the ventilated facade with perforated crosscuts [15]

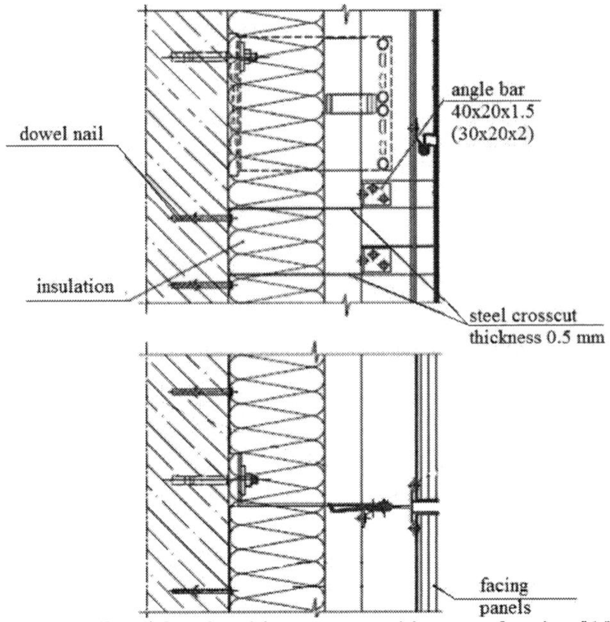

Fig. 3. An embodiment of the ventilated facade with crosscuts without perforation [15]

In the Kraspan system the perforated metal crosscut completely covers the gap. In the Sial system it spans gap partially, until the edge of the vertical profile. Also, there is embodiment of crosscut performance consisting of two continuous sheets, raised in height with overlap.

2 Publications review

Research of a number of famous scientists is devoted to this theme [16-20, 34-35]. In article [21] the importance of the use of fire crosscuts to ensure fire safety facilities is considered, as well as the need for measures to prevent fire and falling of combustion products.

Great attention should be paid to the material of crosscuts, as it must meet all the requirements of fire safety, the most important of which is do not let the fire spread across the facade.

A.D. Zhukov, A.V. Chugunkov set task to produce a material for fire crosscuts homogeneous structure [22]. The experience of products' bench manufacturing of flexible membranes is taken into account. These membranes allow to adjust the excess tension, and the material retains its structure with minimal variation in density. The consequence of this method is application of aerated concrete as expanding fire crosscuts. In this case, aerated concrete poured into the mould is sandwiched between sheets of insulation and formwork. The mixture distends, forming stress condition and cellular mixture fills all voids and leakages, thus relaxing own excess tensions [22].

If the building is not equipped with fire crosscuts, it belongs to a class K3 fire danger according to [23].

In [24] authors make demands to the material, targeted for a product with desired properties (primarily low thermal conductivity) and maintaining these properties over time in specific operating conditions. In this case two kinds of constraints are taken into account: safety of material in terms of fire impacts and minimizing negative impacts on the environment.

Geir Jenssen leads in his article the European and American testing standards of ventilated facades on the impact of the fire. In particular, comparative evaluation of the use of perforated and solid cut-offs according to European standard tests E2912-13 is performed [25]. David Crowder has reviewed the methods of experiments based on Fourier transform (FTIR) to analyze the effect of toxic gases on the rate of spread of a fire. The tests have been produced in special flue boxes BS EN ISO 5659 [26].

The conclusion based on reviewed papers is that fire-fighting crosscuts are the best way of protection of the facade against the spread of flame, however, it is necessary to identify how this system will affect the functionality of the DSF.

Purpose: to assess the impact of fire crosscuts on the efficiency of a ventilated gap.

Objectives:

- To conduct a theoretical assessment of the impact of fire crosscuts on the air flow in the ventilated fence of DSF;
- To make the experiment on a physical model of the ventilated gap in the laboratory

3 Advantages of double-skin facades

Ventilated facades are increasingly used by builders. The removal of atmospheric moisture and condensate from the building envelope happens due to the air gap between the insulation and facade cladding in ventilated facade of the building [27].

Specific points of ventilated facade:

1. This facade can significantly reduce the thickness of the bearing wall, which leads not only to direct savings of wall materials, but also allows to soften the requirements to the bearing capacity of the Foundation [28];
2. With the right setup of the facade the danger of moisture condensation inside walls of elements is eliminated [28, 29];
3. In the event of a catastrophic destruction of the insulating layer the functionality of the facade can be quite easily recovered by pouring the resulting cavity with air-polymerizing foam composition or filling lightweight backfilling — expanded clay aggregate or foamed glass [30, 31];
4. For facade masonry is not necessary to use "warm" masonry mortars [32];
5. As the material for the top layer of the facade could be used a cost-effective ceramic and clinker bricks of "American type" [28].

Advantages:

1. High workability: preparation of a bearing wall is almost not required; all components have a high degree of prefabrication;
2. High speed of installation;
3. The absence of "wet" processes;
4. Installation of systems of ventilated facades is simple, but requires qualification and training of workers;
5. The device does not require scaffolding; installation can be carried out with cradles;
6. Facing the elements if necessary can be customized by size directly on the site [32];
7. The possibility of carrying out installation work at any time of the year;
8. Design features allowing to increase the service life of building elements, the estimated lifetime can reach 50 - 100 years;
9. Low operating costs;
10. A wide range of coating materials allows to realize almost any design decision [27].

4 An experiment on a physical model

In the experiment, the perforated fire crosscuts shown on (Fig.4) block most of the ventilated gap.

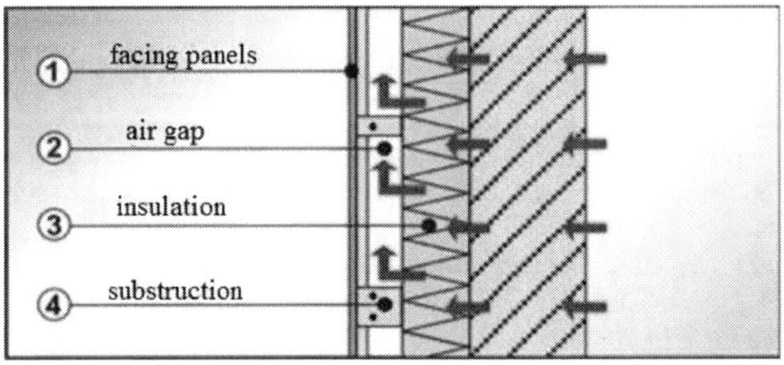

Fig. 4. Scheme of a double-skin facade [33]

There are installations in which the air flow velocity changings in the gap can be studied, as well as temperature, depending on the width of the air gap under the influence of the crosscut and without it.

The physical model of a vertical ventilated layer (Fig.5). is an installation 2040 mm in height, 600 mm in width, there are two faces bounding the air gap and simulating the facing layer (non-heated) and the wall of the building. One of the faces of the installation is heated and fixed, the second face, simulating the facing layer, is unheated and can move free, creating a different air gap width h in the range 20-300 mm. The edge which imitates the facing layer (unheated), consists of sheets 300 mm high, separated by 5 mm gaps, imitating rustication. There is also an opportunity providing to control access of air from the bottom of the channel. Over the entire height of the unit a heating element is located which allows to supply the heat for heating the motionless face.

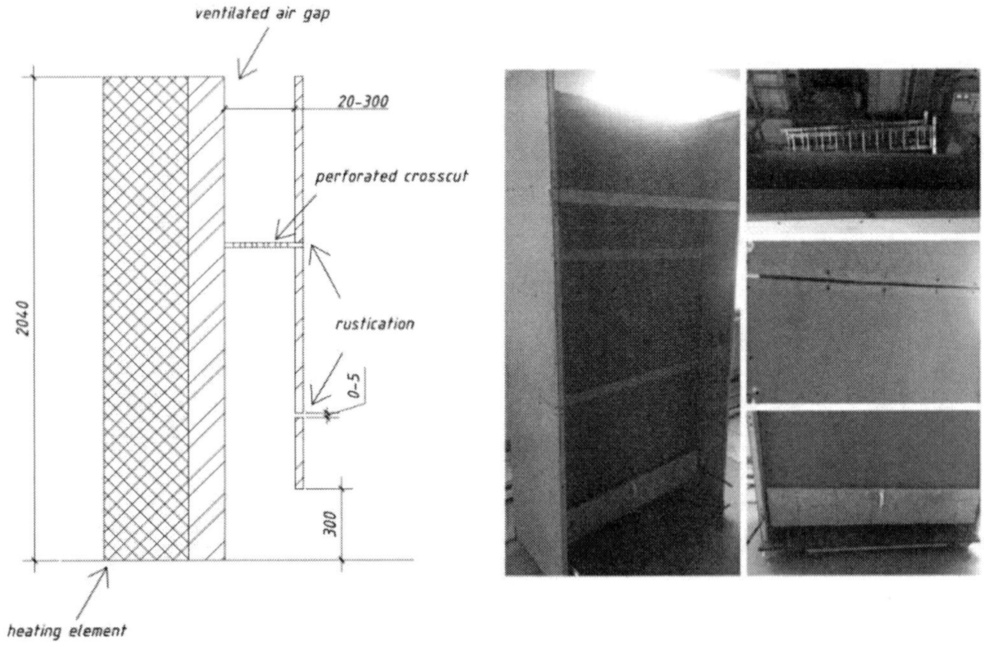

Fig. 5. Installation for physical experiments [8]

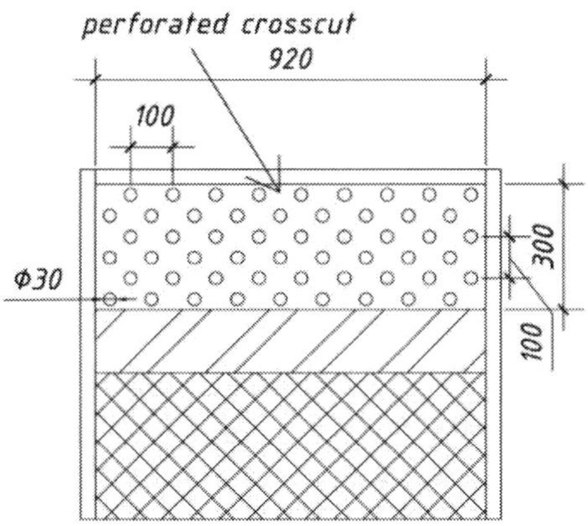

Fig. 6. Top view of the installation.

Thus, the installation (Fig.6). allows measurements by adjusting the geometrical parameters of the air gap.

A series of experiments have been carried out under the following sizes and installation configurations on a physical model to confirm the theoretical calculations and the numerical experiment and also studying the effect of crosscuts on the role of the air gap and air access:

1st case: the crosscut is not used, free flow of air in the gap is organized

2nd case: the perforated crosscut is across the width of the gap

In both cases, the width of the air gap was variable and ranged from 20 to 300 mm. Height of the air gap was constant and amounted to 2040 mm.

Before a series of experiments, the temperature was measured in the laboratory, which amounted to 293 K. During the experiment, in both cases, the average speed and the temperature of the air flow in the channel at each width of the air gap was measured using a hot-wire anemometer.

The air velocity varied from 0 m/s to 0.7 m/s, the temperature varied from 296.8 K to 309.7 K.

In the case of use of fire crosscut the imitation (Fig.7). was taken, which represented a piece of chipboard the size of 920x300x5 mm, with 13% degree of perforation (the hole diameter is 30 mm, the interval between them is 100 mm).

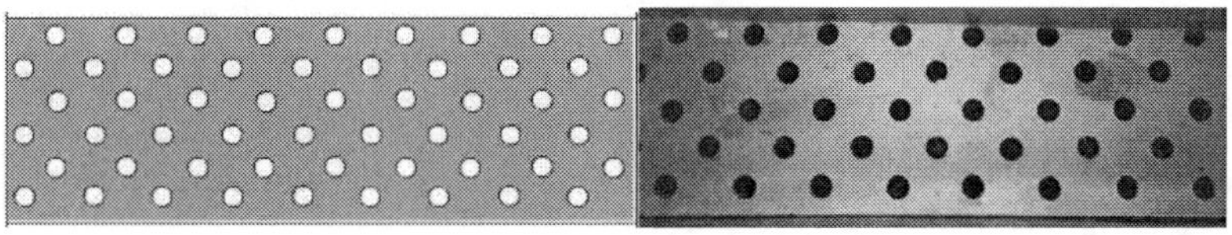

Fig. 7. The view of the crosscut from the top.

Processing of experimental data was performed for two series of experiments. L/h and φ were used as the variables.

5 The calculated dependencies

1. The General formula for average speed:

$$v = \varphi\sqrt{1-\frac{T_c}{T_h}}\sqrt{2gL} \text{ [8]} \tag{1}$$

where:

$$\varphi = \frac{1}{\sqrt{1+\lambda\frac{L}{h}+\zeta}} \text{ - speed coefficient} \quad (2)$$

λ - the coefficient of hydraulic friction,
ζ - coefficient of local pressure loss,
L – the height of a ventilated gap,
h – the width of the gap,
T_c – the room temperature in Kelvin,
T_h - the temperature of the heating element in Kelvin;

2. With crosscuts ($v \approx 0.1$):

$$v = \varphi\sqrt{1-\frac{T_c}{T_h}}\sqrt{2gL} \rightarrow \varphi \rightarrow +0 \quad (3)$$

The installation of crosscuts increases the channel resistance, reducing φ to zero.

$$\varphi = \frac{1}{\sqrt{1+\frac{\lambda L}{h}}} \quad (4)$$

$$1 + \lambda\frac{l}{h} = \frac{1}{\varphi^2} \quad (5)$$

$$\varphi_p = \frac{1}{\sqrt{1+\frac{\lambda L}{h}+\zeta_p}} \quad (6)$$

$$\rightarrow \zeta_p = \frac{1}{\varphi_p^2} - \frac{1}{\varphi^2} \quad (7)$$

$$\varphi = \frac{v}{\sqrt{1-\frac{T_c}{T_h}}\sqrt{2gL}} \quad (8)$$

The maximum coefficient of local pressure loss of crosscuts was calculated when flow velocity was 0.1 m/s: $\zeta_p = 566.67$

The graph of speed dependence φ from the ratio of the height of the ventilated gap L to its width h is shown on (Fig. 8).

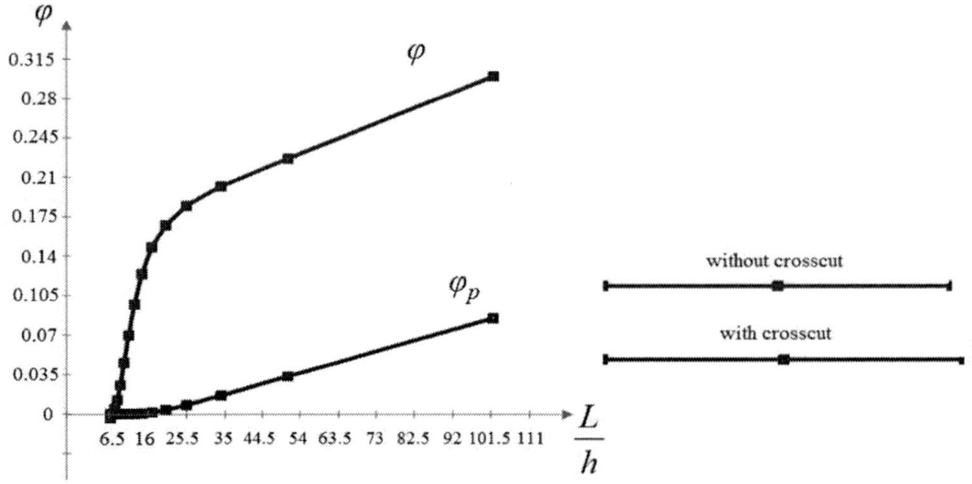

Fig. 8. A graph of dependence φ from L/h

6 Conclusions

During the carried-out tests and calculated results, it was found that while using the crosscuts (even perforated), the air gap's role is reduced to zero, because the facade stops working as ventilated. With a relatively small gap size of 60 mm the air flow velocity drops to zero, air circulation with further increase of channel doesn't occur. On the one hand, this means that the crosscut actually works as a means of preventing the spread of hot and flammable vapors along the facade. But on the other hand, it interferes with the normal air convection inside the facade. It can be concluded that the rational solution is the use of non-combustible materials or installation the system of automatic "ejection" of crosscuts in case of fire. It could be implemented as follows: in the design of the facades to provide thermal anemometers, which will register a temperature increase above a certain limit. In case of fire the signal from the anemometer is transmitted to a remote control system of crosscuts and they fall across the entire air gap. The scheme of the device is shown in (Figs. 9-10).

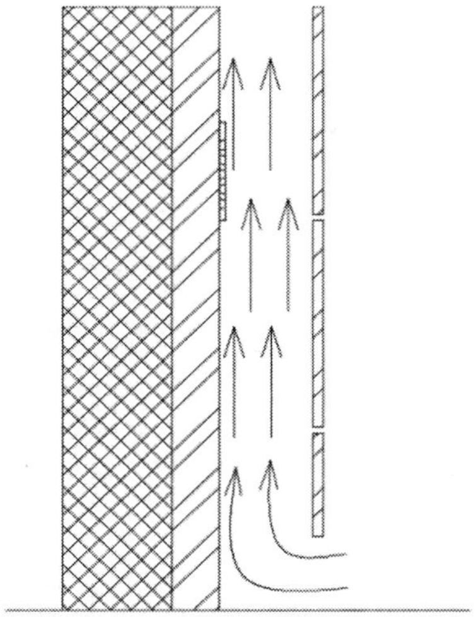

Fig. 9. Normal position of the crosscut - along the facade

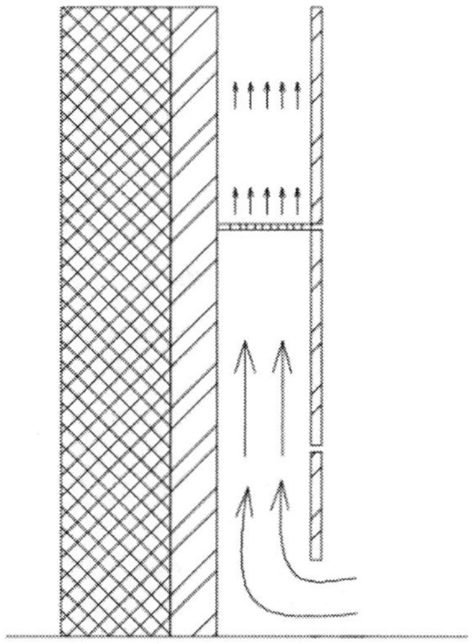

Fig. 10. Case of fire – crosscut closes air gap

References

1. C. Suarez, P. Joubert, J. L. Molina, F. J. Sanchez, Ener. and Build., **43**, 3696–3703 (2011)
2. Information on http://vfasade.blogspot.ru/2013/06/no-fire.html
3. D.V. Nemova, Mag. of Civil Eng., **5**, 7-11 (2010) (rus)
4. B. Gaujena, A. Borodinecs, J. Zemitis, A. Prozuments, *IOP Conference Series: Materials Science and Engineering,* **96 (1)** (2015)
5. N. Harmati, Z. Jakšić, N. Vatin, Proc. Eng., **117 (1)** , 791-799 (2015)

A. Borodinecs, J. Zemitis, A. Prozuments, *Proceedings of World Renewable Energy Forum (WREF 2012)* 4181-4186 (2012)

6. A.D. Zhukov, SCE, **1**, 1-15 (2012) (rus)
7. D.V. Nemova, *PhD Tesis: 05.23.16, Peter the Great St.-Petersburg Polytechnic University, Saint-Petersburg,* 18 (2015) (rus)
8. *The Federal Law N 123* (Moscow, 2008) (rus)

9. *National Standard of Russian Federation 30244-94* (Moscow, 1995)
10. *National Standard of Russian Federation 30402-9* (Moscow, 1996)
11. *National Standard of Russian Federation 12.1.044-89* (Moscow ,1996)
12. I.R. Khasanov, I.S. Molchadsky, K.N. Goltsov, AV. Pestritsky, FS, **5** (2006) (rus)
13. V. N. Vorobev, (Vladivostok, 2012) (rus)
14. Information on http://vfasade.blogspot.ru/2013/07/uzel-ventfasad.html
15. S.A. Isaev, N.I. Vatin, S.V. Guvernyuk, V.G. Gagarin, B.I. Basok, Y.V. Zhukova, High Temp., **53 (6)**, 873-876 (2015)
16. G.I. Grinfeldi, A.S. Gorshkov, N.I. Vatin, Mat. Res., **941-944**, 786-799 (2014)
17. N.I. Vatin, J. Havula, L. Martikainen, A.S. Sinelnikov, A.V. Orlova, S.V. Salamakhin, Adv. Mat. Res., **945-949**, 1211-1215 (2014)
18. M.R. Petrichenko, D.W. Serow, J. NPCS. **17 (3)**, 336-339 (2014).
19. Nefedova, J. Bykova, S. Kosov, M. Petrichenko, Proc. Eng., **117 (1)**, 1102-1111 (2015)
20. I.A. Lobaev, A.J. Bazilevich, A.O. Andreev, *Materialy dvenadcatoj nauchno-tehnicheskoj konferencii «Sistemy bezopasnosti»*, 47 – 249 (Moscow, 2003.) (rus)
21. A.D. Zhukov, A.V. Chugunkov, Proc. MSUCE, **5**, 128—132 (2012) (rus)
22. *National Standard of Russian Federation 31252-2003* (Moscow, 2003) (rus)
23. B.M. Rumyantsev, A.D. Zhukov, T.V. Smirnova, **4 (35)**, 3 (Volgograd, 2014) (rus)
24. G. Jensen, Proc. of matec web of conf., **9**, 1-11 (2013)
25. D. Crowder, *BRE Trust Review 2012*, 46-47 (2013)
26. A.E. Kiryudcheva, V.V. Shishkina, Const. of Un. Build. and Struct., **4 (31)**, 248-262 (2015) (rus)
27. E. Cykanovski, V. Gagarin, A. Granovski, M. Pavlova, Build. Tech., **6**, 28-33 (2002) (rus)
28. *Centralny nauchno-issledovatelski i proektny institut jilyh i obscestvennyh zdanii (CNIIEP zhilishha),* (Pravitelstvo Moskvy, Moskomarhitektura, 2002), Dmitriev, En-sber., **6** (2001) (rus)
29. A.S. Gorshkov, Mag. of Civil Eng., **1**, 9-13 (2010) (rus)
30. M.A. Platonova, N.I. Vatin, D.V. Nemova, S.A. Matoshkina, D. Iotti, I. Togo, Const. of Un. Build. and Struct., **4 (19)** ,83-95 (2014) (rus)
31. Information on http://pfkmsk.ru/
32. M. Penić, N. Vatin, V. Murgul, App. Mech. and Mat., **680**, 534-538 (2014)
33. A. Gorshkov, V. Murgul, O. Oliynyk, MATEC Web of Conf., **53** (art. no. 01045, 2016)

MATEC Web of Conferences 73, 02007 (2016) DOI: 10.1051/matecconf/20167302007
TPACEE-2016

Impact of Rustication Joints on Lightweight Insulation in Ventilated Facade Systems

Mikhail Petrichenko[1], *Darya* Nemova[1], *Elisaveta* Reich[1*], *Svetlana* Subbotina[1], *Faina* Khayrutdinova[1], *Rudolf* Schilling[2], *Vyacheslav* Olshevskyi[1]

[1]Peter the Great St. Petersburg Polytechnic University, 195251 Polytechnicheskaya Str. 29, St. Petersburg, Russian Federation
[2] Technical University of Munich, D-80333 Arcisstraße 21, Munich, Germany

Abstract. One of the problems of ventilated facades systems is the emission of heat-insulating layer. Emission is a process of separation or removal of material's particles from insulation by virtue of different factors. In this article research of emission of contemporary light weight insulation in system of hinged ventilated façade (VF) with various design is carried out. Impact of design and hydraulic factors on insulation destruction is estimated. Quantitative characteristic of blast air through rustication joints in a vertical channel is given and, consequently, it is possible to determine control point of recirculation zone formation – the beginning of turbulent air flow. Particle's removal in insulation and decline of its properties are most probable in case of tangential stress, that conform to turbulent air flow.

1 Introduction

In modern construction using of hinged ventilated facades as an enclosure structure has increased. This type of systems is characterized by high heat-shielding properties in comparison with other popular types of walling. One of the aspects of study is the air gap of the system, which characteristics affect heat transfer and durability. [5, 6]

Insulation with a high density, which fiber's emission is relatively low is often used to correct this effect. [7,10] However, this type of insulation is more expensive. Another way to reduce emission of insulation layer is using of wind-shelter membranes. [8] But these membranes also have some deficiencies, such as complication of erection procedure, high ignitability.

Consequently, the question of the possibility of applying lightweight insulation without a wind-shelter membrane in VF systems becomes relevant.

The aim of this study is determination of impact of rustication joints on lightweight insulation in VF systems. [1,3]

Accomplish this aim such challenges were figured out:

1. determination of functional connection between tangential stresses on lightweight insulation's surface and height for VF with various designs.

* Corresponding author: lisa_reich@mail.ru

2. theoretical estimation of filtration velocity of an air flow in a lightweight insulation.
3. determination of amount of the blast air through rustication joints in a vertical channel and control point of recirculation zone formation.

2 Literature References

A large amount of papers are deals with studying of VF systems [1-31]. Characteristics of an air flow in interlayer of VF were analyzed by Diego Angeli, Alessanro Dama [1]. They have examined precision of performance of simplified analytical model of air motion in an air layer in case of free ventilation. Due to this model it is possible to estimate heat transfer in façade systems.

Alexander Meneylyuk, Igor Babij, Ivan Meneylyuk, Artem Gladischuk in their paper [2,7] selected the most appropriate insulation for free-convective air motion in the interlayer of VF. Basing on experimental and static modelings, necessity of applying of compact insulation for its longevity was determined.

3 Building Characteristics

Heat conservation in VF system is possible due to its critical part – insulation. In case of its destruction heat loss increase and energy efficiency of the building degrades. Therefore, it is vital to research sources and methods of avoidance of insulation's destruction.

Insulation's emission can be caused by plenty of reasons. Hydraulic factors, such as external air flow and filtration flow, are able to impact on destruction of insulation. In this paper possibility of applying of lightweight spun glassy insulation membrane in VF systems is analyzed. Insulation has a following thermotechnical characteristics:

Dry density $\rho_0 = 2.2$ kg/m^3;

Vapor permeability rate $\mu = 0.389$ mg/(m·h·Pa) ;

Vapor permeability coefficient $s = 10^{-10}$ m^2.

For the estimation apartment building with a maximum allowable height according to Snip 21-01-97 [32] was used without additional agreement (75 meters). It has a following characteristic:

Temperature of the cold surface $T_c = 293$ K;

Temperature of the heat surface $T_h = 298$ K;

Head loss coefficient of the rustication joint $\cdot\xi_r = 4$;

Head loss entrance coefficient $\xi_{in} = 0.45$;

Width of the air interlayer h=0.1 m;

Width of the rustication joint $\delta_r = 0.008$ m;

Width of the cladding panel $b_p = 0.6$ m.

4 Estimate of Filtration Velocity

For checking purposes of impact of filtration flow, filtration velocity in insulation is required to determine:

$$V_{filtr} = k \cdot J \tag{1}$$

where:

k - filtration coefficient,

$$k = s \cdot g / \nu \tag{2}$$

where:

v -air viscosity;
J - hydraulic gradient,

$$J = \Delta P/(\rho \cdot g \cdot L) \tag{3}$$

where:
L - height of the channel,
ΔP - differential head heightwise,

$$\Delta P = \rho \cdot g \cdot L. \tag{4}$$

Therefore,

$$V_{filtr} = k = s \cdot g/v \tag{5}$$

Estimated filtration velocity for insulation is negligible quantity, that is equal to 10^{-4} m/s. (Figure 1).

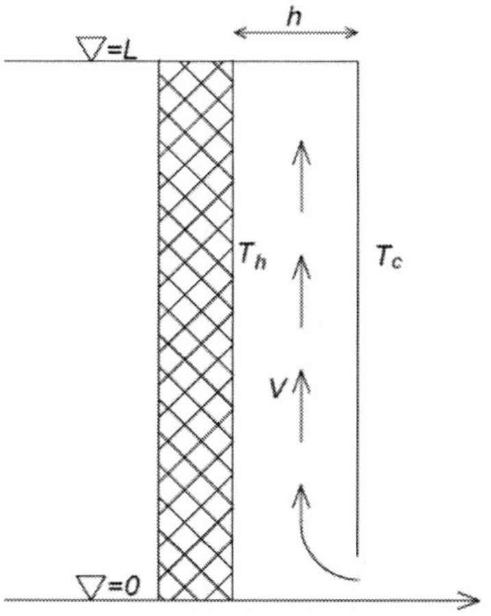

Fig. 1. Diagram of the ideal channel.

4.1. Ideal Channel

Ideal channel is a vertical channel that has no rustication joints (Fig.1.). Air flow in this type of construction operates at right angles to temperature head. Air velocity remains constant along the height.

For estimation of the impact of the height on insulation's emission functional connection between tangential stresses and rustication joint's height is determined:

$$\tau_0 = \rho \cdot g \cdot R_{hydr} \cdot J \tag{6}$$

where:
ρ - ambient density,
g - gravitational acceleration,
R_{hydr} - hydraulic mean depth, incidentally $R_{hvdr} = h$,

$$J = \Delta P_T/(\rho \cdot g \cdot L) \tag{7}$$

where:
ΔP_T - thrust pressure, that is calculated according to the formula:

$$\Delta P_T = \xi \cdot \rho \upsilon^2/2 \tag{8}$$

where:
υ - average velocity of the air flow in the air gap, it is calculated as follows:

$$v = \varphi\sqrt{1 - {}^{T_C}\!/_{T_h}}\sqrt{2gL} \tag{9}$$

where:

$\varphi = 1/\sqrt{1 + {}^{\lambda L}\!/_{h} + \zeta}$ - velocity coefficient,

λ - pipe friction number,
ξ - coefficient of local head losses, in the present case $\xi = \xi_{in}$,
Therefore:

$$\Delta P_T = \frac{\xi_{in}}{1+\xi_{in}+(\lambda \cdot L/h)} \cdot \gamma \cdot L \cdot (1 - T_c/T_h), \quad (10)$$

where γ - relative weight,
end formula for determination of tangential stresses in the air interlayer of an ideal channel is as follows:

$$\tau_o = h \cdot \frac{\xi_{in}}{1+\xi_{in}+(\lambda \cdot L/h)} \cdot \gamma \cdot (1 - T_c/T_h), \quad (11)$$

Consequently, basing on the connection, it can be said that in case of ideal channel gap's height influences on tangential stresses merely by the velocity coefficient. Therefore, tangential stress near the insulation's surface is constant along the air layer.

For the selected building in case of applying VF system without rustication joints tangential stress near the insulation's surface is equal to $9.9 \cdot 10^{-4}$ Pa

Velocity fluctuation and increasing of tangential stresses in 10 times are specific to turbulent flow.

Motion mode is determined by Grasgoph number:

$$Gr_h = \left(g \cdot h^3 / v^2\right) \cdot \left(\Delta T / T_h\right), \quad (12)$$

$Gr_h \approx \mathrm{Re}_n^2$.

For laminar flow there is criteria as follows:

$$\mathrm{Re}_n^2 = \sqrt{\left(g \cdot h^3 / v^2\right) \cdot \left(\Delta T / T_h\right)}, \quad (13)$$

where v - kinematic viscosity coefficient.

When $\mathrm{Re}_n \leq 2300$ – flow mode is laminar.

For the selected channel $\mathrm{Re}_n = 1290$, that conforms to laminar flow of the air.

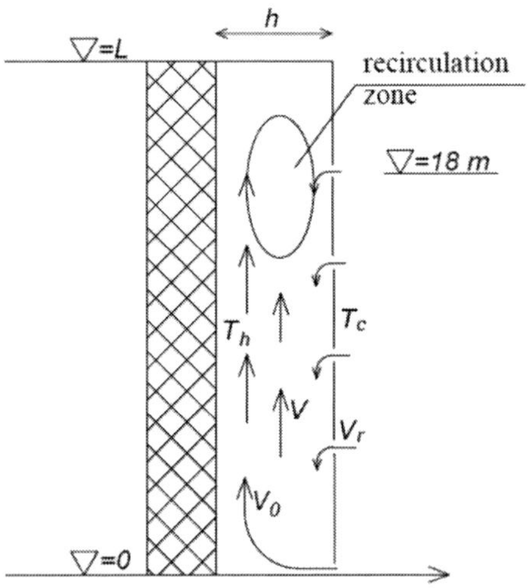

Fig. 2. Diagram of the channel with rustication joints

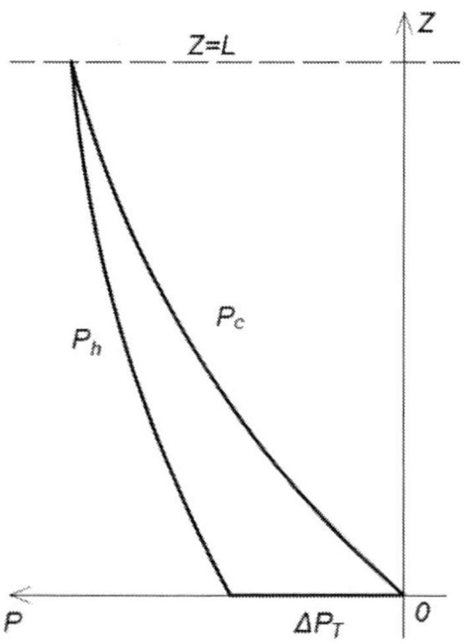

Fig. 3. Pressure difference.

4.2. Channel with Rustication Joints

Practically, in most cases channel with rustication joints (gaps between cladding panels) are applied (Fig.2.). They are necessary for temperature deformation compensation. Presence of rustication joints in VF system decrease Froude number almost twice, therefore, velocity coefficient φ decreases proportionally. Due to air indraft in joints pressure is not constant along the height, and, with the height extension, the average velocity diminishes until setting to zero. In this moment recirculation zones occur and motion mode becomes turbulent.

Pressure difference (Fig.3.), that is equal to thrust pressure, between hot and cold surfaces at any height is calculated as follows:

$$\Delta P_T = (1/n - 1/k)\cdot\left(\frac{g\cdot(L-Z)}{R\cdot T_o}\right)^2 / 2, \tag{14}$$

where:
n,k – polytropic index,
Z – calculated point,
T_0 - outdoor air temperature, $T_0 = T_c$,
R – absolute gas constant.

$$\Delta P_T = P_o\cdot((k-1)/k)\cdot\left(\frac{g\cdot(L-Z)}{R\cdot T_o}\right)^2 / 2, \tag{15}$$

where:
P_0 – static pressure at Z=0,

$$\Delta P_T / P_o = ((k-1)/k)\cdot\left(\frac{g\cdot(L-Z)}{R\cdot T_o}\right)^2 / 2, \tag{16}$$

Graph (Fig.4.) depicts quadratic dependency between thrust pressure and height of the channel.

It stands to reason, that maximum pressure difference occurs at *Z=0,* when *Z=L* pressures are equal.

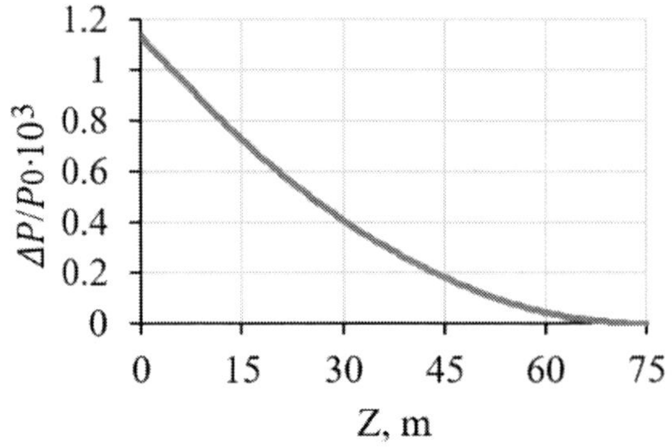

Fig. 4. Dependence ΔP/P0 from the vertical point Z.

Functional connection between thrust pressure and velocity is as follows:

$$\Delta P_T = \zeta \cdot \rho \cdot \upsilon^2 / 2 \tag{17}$$

Therefore:

$$\upsilon_r = \sqrt{2 \cdot \Delta P_T / \zeta_r \cdot \rho} \tag{18}$$

where:

υ_r - velocity in rustication joint,

Consequently, at *Z=0* velocity in rustication joint is extreme, dependence between average velocity and vertical coordinate is linear.

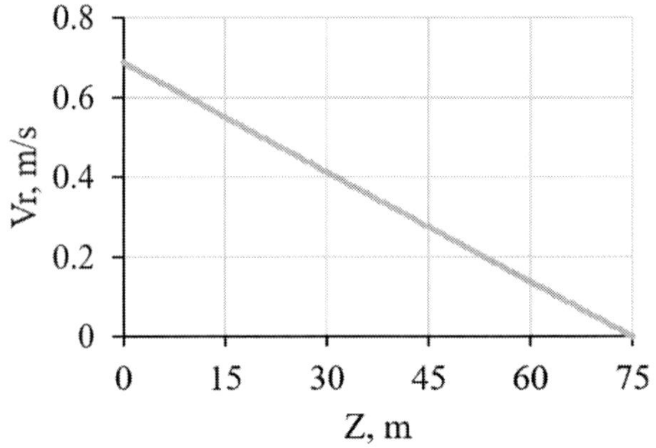

Fig.5. Graph of variance of the average velocity in the joint in relative to the height Z.

In the picture (Fig.5.) graph of variance of the average velocity in the joint in relative to the height Z is represented.

Average velocity without rustication joints can be calculated by the formula:

$$\upsilon_o = \varphi \cdot \sqrt{(1 - T_c / T_h)} \cdot \sqrt{2gL}\,. \tag{19}$$

For selected façade average velocity without rustication joints is equal to 1.5 m/s.

With the height increasing, velocity will be decreasing, due to penetration of the cold air through rustication joints, which motions contradirectionally to the major flow.

Air flow rate is determined as follows:

$$Q = h \cdot \upsilon, \tag{20}$$

where:
υ – air velocity.

Consequently, blast air flow rate through rustication joints can be calculated by the formula:

$$Q_c = \delta_r \cdot \sum_{i=1}^{x} \upsilon_{ri}, \tag{21}$$

where:
X – number of joints in the channel.

Blast cold air flow rate through rustication joints in the channel for selected building is equal to 0,35 m^2/s.

Hot air flow rate in the channel can be described as follows:

$$Q_h = h \cdot \upsilon_0 \tag{22}$$

Therefore, hot air flow rate in the channel of VF systems is constant and equal to 0.15 m^2/s.

Height, at which blast cold air flow rate through rustication joints is equal to hot air flow rate, conforms to the beginning of recirculation zone's formation:

$$\delta_r \cdot \sum_{i=1}^{x} \upsilon_{ri} = h \cdot \upsilon_o \tag{23}$$

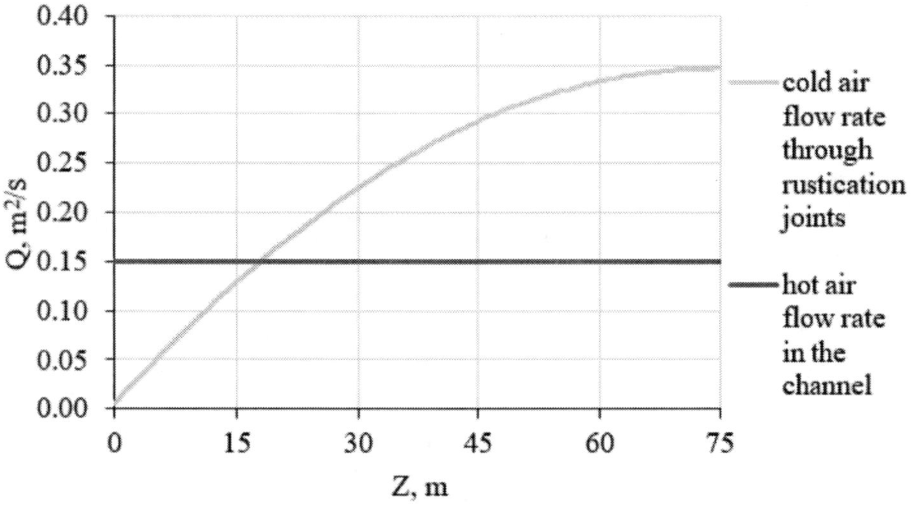

Fig. 6. Dependence between air flow rate and height point of the channel Z.

It can be seen, that at the height $Z \approx 18$ m cold air income through rustication joints derails updraft air movement. By virtue of this penetration, recirculation zones occurs, motion mode is turbulent and tangential stresses near the surface increase abruptly.

Value of decreasing of an average velocity in the air gap of every joint can be estimated from flow equality:

$$\upsilon_r \cdot \delta_r = \Delta\upsilon \cdot h, \tag{24}$$

$$\Delta\upsilon = \upsilon_r \cdot \delta_r / h, \tag{25}$$

where:

δ_r - width of rustication joint.

Therefore, total velocity is as follows:

$$\upsilon = \upsilon_o - \sum_{i=1}^{x} \Delta\upsilon_i \,. \tag{26}$$

After point of 18 meters velocity reverse the direction, air motions downward. By virtue of it, façade seizes perform correctly – moment of channel closure (Fig.7).

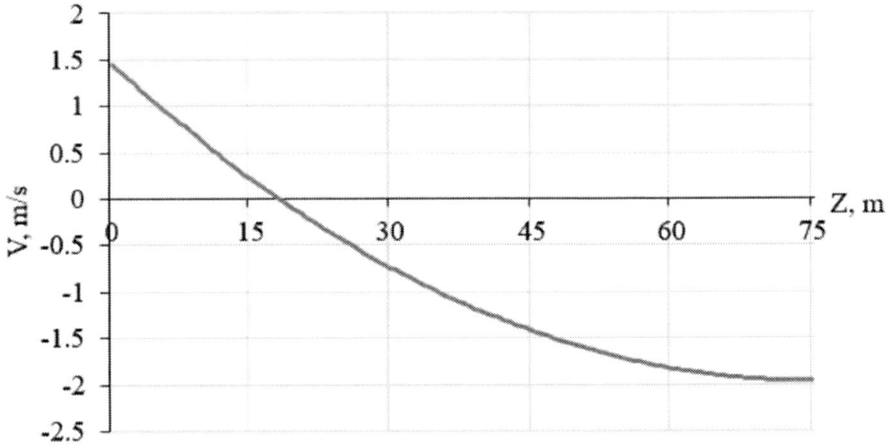

Fig. 7. Dependence between air velocity and height point of the channel Z.

With notice of velocity, tangential stresses for laminar flow can be estimated:

$$\tau_L = \rho \cdot \upsilon^2 / 2 \cdot \pi \,. \tag{27}$$

For turbulent flow tangential stresses are approximately 10 times as much as in case of laminar flow (Fig. 8):

$$\tau_t \approx 10 \tau_L \,. \tag{28}$$

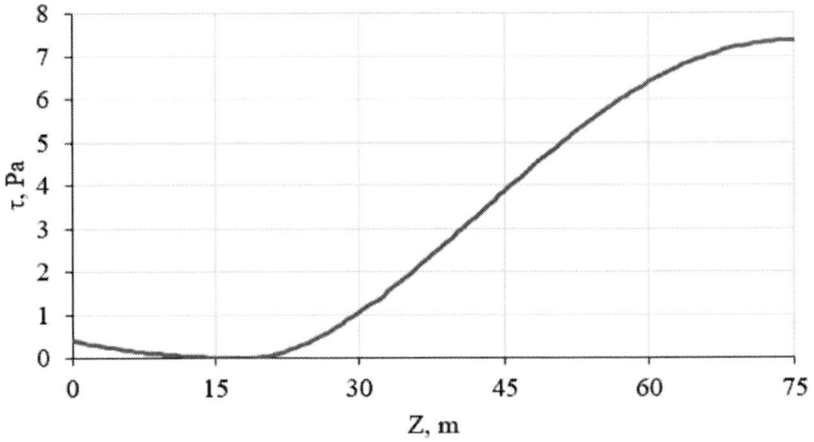

Fig. 8. Dependence between tangential stresses in the channel with joints and height point Z.

Tangential stresses in selected channel decrease until point of 18 meters, after that they become increase attaining its maximum of 7,4 Pa at the height of 75 meters (Fig.8.).

5 Conclusions

Based on these studies, the following connections were identified:

1. Filtration velocity for insulation of VF system was estimated and its value is negligible.
2. For ideal channel tangential stresses near the insulation's surface are constant along the height of the façade, due to laminar flow. For the channel with rustication joints tangential stresses decreases until definite height, and after that, extremely increase by virtue of substitution of the motion mode for turbulent flow.
3. Quantity of the blast cold air through rustication joints of the vertical channel was determined. It afforded to estimate height of the formation of recirculation zones, that was equal to 18 m.

Recirculation zones substitute laminar flow for turbulent motion mode of the flow, thereby increasing tangential stresses in times, that negatively affects insulation in the system.

References

1. D. Angeli, A. Dama, Energy Procedia, **78**, 1537–1542 (2015)
2. A. Meneylyuk, I. Babij, I. Meneylyuk, A. Gladischuk, Suchasni Tekhnolohiyi, Materialy i Konstruktsiyi v Budivnytstvi, **19**, 28-33 (2015)
3. C. Popa, D. Ospira, S. Fohanno, C. Chereches, Energ. Buildings, **50**, 229–233 (2012)
4. N. Vatin, M. Petrichenko, D. Nemova, Appl. Mech. Mater, **633-634**, 1007-1012 (2014)
5. M. Petrochenko, K. Strelets, M. Petrichenko, E. Yavtushenko, Appl. Mech. Mater, **672-674**, 567-570 (2014)
6. I. Babij, I. Meneylyuk, Suchasni Tekhnolohiyi, Mat. i Konstr. v Bud., **17**, 26-31 (2014)
7. I. Smirnov, E. Shesterova, D. Nemova, M. Petrichenko, Appl. Mech. Mater, **725-726**, 130-137 (2015)
8. N. Mingotti, T. Chenvidyakarn, A. Woods, Build. Environ, **46**, 807-823 (2011)
9. A. Korjenic, J. Zach, J. Hroudová, Energ. Buildings, **116**, 45–58(2016)
10. H. Elarga, M. Carli, A. Zarrella, Energ. Buildings, **104**, 97–107 (2015)
11. T. Zadvinskaya, A. Gorshkov, Adv. Mat. Res, **953-954**, 1570-1577 (2014)
12. A. Girgidov, Power Tech. Eng., **5** (2015)
13. A. de Gracia, L. Navarro, A. Castell, L.F. Cabeza, Energ. Buildings, **91**, 37–42 (2015)
14. M. Nizovtsev, V. Belyi, A. Sterlygov, Energ. Buildings, **75**, 60–69 (2014)
15. C. Marinosci, G. Semprini, G.L. Morini, Energ. Buildings, **72**, 280–287 (2014)
16. N. Vatin, I. Pestryakov, D. Avdeev, Appl. Mech. Mate.r, **725-726**, 454-460 (2015)
17. M. Penić, N. Vatin, V. Murgul, Appl. Mech. Mater., **680**, 534-538 (2014)
18. F.M. da Silva, M.G. Gomes, Energ. Buildings, **40**, 1553–1559 (2008)
19. A. Hazem, M. Ameghchouche, C. Bougriou, Energ. Buildings, **102**, 225-236 (2015)
20. V. Pukhkal, V. Murgul, N. Vatin, Appl. Mech. Mater., **633-634**, 1077-1081 (2014)
21. N. Parfentyev, O. Samarin, K. Lushin, S. Paulauskaite, Pap. of conf. of VGTU, **854—858** (2008)
22. G. Grinfeld, A. Gorshkov, Vatin N., Adv. Mat. Res, **786-799** (2014)
23. A. Milajić, D. Beljaković, N. Čulić, N. Vatin, V. Murgul, Appl. Mech. Mater., **725-726**, 1544-1551 (2015)
24. M. Petrichenko, N. Vatin, D. Nemova, V. Olshevskiy, Appl. Mech. Mater., **725-726**, 93-99 (2015)
25. T. Zadvinskaya, A. Gorshkov, Adv. Mat. Res., **953-954**, 1570-1577 (2014)
26. N. Vatin, O. Gamayunova, Appl. Mech. Mater., **672-674**, 550-553 (2014)
27. N. Harmati, Ž. Jakšić, N. Vatin, Appl. Mech. Mater., **725-726**, 1572-1579 (2015)
28. N. Vatin, O. Gamayunova, Adv. Mat. Res., **1065-1069**, 2159-2162 (2015)
29. Snip 21-01-97 *Fire Safety of buildings and structures*

MATEC Web of Conferences 73, 02008 (2016) DOI: 10.1051/matecconf/20167302008
TPACEE-2016

Airways in Apartment Buildings as a Method of Thermal Control

Anna Suslova[1,*], *Michail* Petrichenko[1] *and Aleksandr* Sivokhin[1]

[1]St. Petersburg State Polytechnical University, 19525 Polytecknycheskaya str. 29, St. Petersburg, Russia.

Abstract. In general, the majority of modern apartment buildings are rather high. Altitude of such structures attains 50 meters. It is clear that for such high structures every extra meter of elevation costs a lot. For this reason, architects are trying to avoid adding attics above the last floor of the buildings. However, attic is not only an architectural element. It is an important part of the thermal control process of the entire building, especially of the apartments located on the last floor. In this article, construction of airways under the roof is suggested and discussed in detail. Airway acts as an attic, but has a significantly lower construction cost due to the lower height. Application of this technology allows providing comfortable microclimate on the living quarters in an economical way.

1 Introduction

Construction of traditional attic means presence of a big amount of free space for communications under the roof. Airway acts as a garret in temperature control field due to the working-section equality. Whole area of the attic functions like a heat insulator because of the air inside. But by imposing motion to this air and reducing its temperature, similar thermal effect can be achieved with garret of lower height.

The idea of not using attic is not new. Since the beginning of 20th century a lot of new buildings avoid additional construction elements like garrets. It gives an economical effect but causes some problems, such as icicle formation on the edges of housetop, high moisture of the last floor apartments, strong dependence of the temperature inside the flats from weather conditions. To avoid these problems, airway can be applicated. Air moving under the housetop makes the lowest layer of the roof construction cooler. It prevents snow from melting on the roof during the winter. At the same time, temperature of this air is higher than outside, therefore providing more comfortable temperature in the dwelling. During the summer, airway prevents excessive heating of the building.

There are some articles about attic ventilation from all around the world [1-8, 14-17, 26-29], particularly from Russia [9-13, 18-21, 30]. They are made not only by professionals, but also by ordinary private house owners. Comprehensive research work was performed by Mark Graham [14, 15]. HVAC-systems and the branches of knowledge connected with it are widely explored. Topics are various: from preventing ice damming by different

* Corresponding author: suslovanya@mail.ru

measures [12, 18, 20, 31-32], to economic effects of modern materials and technologies usage [21]. The main problem of all those researches is that they do not take into account the altitude of the building. For high apartment buildings speed of the wind on the top of them is big enough to use it without any fans. Wind speed influence on temperature inside the airway is considered in this paper.

2 Theory or experimental methods

The main goal of this article is to estimate the optimal cross section of an airway and evaluate its efficiency. The criteria the conclusion is based on is the temperature under the roof and over the covering slab of the last floor. Temperature should be high enough to provide the sufficient amount of heat to the last floor and low enough to exclude the icicle formation on the roof. Optimal temperature is calculated according to the first law of thermodynamics. Based on this temperature, the necessary cross section can be estimated by using boundary problem solutions.

3 Research description

First of all, optimal height of an attic should be estimated. It can be made, according to the governmental standard in Saint-Petersburg [22]. The medium temperature outside during the winter is -4.6°C. Temperature in bedrooms and living rooms is 20°C, according to sanitary regulations [23]. Temperature inside the roof space should be at least -1.6°C to prevent icicle formation [24]. Using the first law of thermodynamics, the heat balance of the garret can be written:

$$0 = Q_{slab} - Q_{walls} - Q_{roof}; \tag{1}$$

According to the Newton-Rixman law:

$$q = \alpha \cdot \Delta T; \tag{2}$$

Where q - heat-flow rate, α- heat-transfer coefficient, ΔT- temperature difference. Heat flow for all the filler structures is:

$$\begin{cases} q = \alpha_h \cdot (T_h - T_0) \\ q = \dfrac{\lambda \cdot (T_0 - T_1)}{\delta} \\ q = \alpha_c \cdot (T_1 - T_c) \end{cases}; \tag{3}$$

Where α_h and α_c are heat-transfer coefficients of the material inside and outside the building, T_h-inside temperature T_h=20°C, T_c–outside temperature T_c=-4.6°C, T_0 and T_1 are the walls surface temperatures, λ is a thermal conductivity coefficient, δ is a thickness of the walling. Solving given system, the following results have been obtained:

$$q = \frac{T_h - T_c}{\dfrac{1}{\alpha_h} + \dfrac{\delta}{\lambda} + \dfrac{1}{\alpha_c}}; \tag{4}$$

Estimate the amount of heat per time by heat-flow rate:

$$Q = q \cdot A; \tag{5}$$

Where A is an area of the surface.

$$Q = \frac{T_h - T_c}{\frac{1}{\alpha_h} + \frac{\delta}{\lambda} + \frac{1}{\alpha_c}} \cdot A; \tag{6}$$

Similar equations for all the filling structures are written below:

$$Q_{wall} = 2\frac{T_{attic} - T_c}{\frac{1}{\alpha_h} + \frac{\delta_{wall}}{\lambda_{wall}} + \frac{1}{\alpha_c}} \cdot l \cdot h; \tag{7}$$

$$Q_{slab} = \frac{T_h - T_{attic}}{\frac{1}{\alpha_h} + \frac{\delta_{1slab}}{\lambda_{1slab}} + \frac{\delta_{2slab}}{\lambda_{2slab}} + \frac{1}{\alpha_c}} \cdot l \cdot a; \tag{8}$$

$$Q_{roof} = \frac{T_{attic} - T_c}{\frac{1}{\alpha_h} + \frac{\delta_{roof}}{\lambda_{roof}} + \frac{1}{\alpha_c}} \cdot l \cdot a. \tag{9}$$

Put equations Eq.7 – Eq.9 into Eq.1:

$$\frac{T_h - T_{attic}}{\frac{1}{\alpha_h} + \frac{\delta_{1slab}}{\lambda_{1slab}} + \frac{\delta_{2slab}}{\lambda_{2slab}} + \frac{1}{\alpha_c}} \cdot l \cdot a = \frac{T_{attic} - T_c}{\frac{1}{\alpha_h} + \frac{\delta_{roof}}{\lambda_{roof}} + \frac{1}{\alpha_c}} \cdot l \cdot a + 2\frac{T_{attic} - T_c}{\frac{1}{\alpha_h} + \frac{\delta_{wall}}{\lambda_{wall}} + \frac{1}{\alpha_c}} \cdot l \cdot h. \tag{10}$$

All the parameters of the walls, the slab and the roof are supposed to be typical for residential buildings: $\alpha_h = 23\frac{W}{m^2 \cdot K}$, $\alpha_c = 8.7\frac{W}{m^2 \cdot K}$, $\delta_{1slab} = \delta_{walls} = 0.2m$; the walls and the slab are made from reinforced concrete: $\lambda_{walls} = \lambda_{1slab} = 2.04\ \frac{W}{m^2 \cdot K}$, the slab is also winterized by a mineral wool insulation: $\delta_{2slab} = 0.15m$, $\lambda_{2slab} = 0.048\frac{W}{m^2 \cdot K}$, the roof itself is made from the mineral wool insulation: $\delta_{roof} = 0.15m$, $\lambda_{roof} = 0.048\frac{W}{m^2 \cdot K}$. Only meaning layers are taken into account. The attic section dimensions are: l=20m, a=10m. According to those data, h can be determined.

After the calculations, it is achieved that h = 2.1m is enough to prevent the icicle formation on the housetop of a typical building. It is an ordinary height of the attic, but it seems that to achieve this result it is necessary to use a large amount of heater both on the roof and the slab. Another option is to provide a natural ventilation for the under-roof space. It can be achieved by using the airways.

Mean wind speed on the top of the building is governed by governmental regulations. Regulatory wind speed, according to geographic position, is 27m/sec. Coefficient of correction is equal to 1 because of the type of the region and the height of the building.

Pressure losses take place on entrance, exit and along the airway.

$$\Delta P = \Delta P_{in} + \Delta P_{out} + \Delta P_{fr}; \tag{11}$$

$$\Delta P = \xi_{in} \cdot \rho \cdot \frac{v_\infty^2}{2} + \xi_{out} \cdot \rho \cdot \frac{v_{in}^2}{2} + \lambda \frac{L}{d} \cdot \rho \cdot \frac{v_{in}^2}{2} = \rho \cdot \frac{(v_\infty - v_{in})^2}{2}; \tag{12}$$

$$\xi_{in} = \frac{1-\frac{S_{in}}{S_{out}}}{2}, \xi_{out} = \left(1-\frac{S_{in}}{S_{out}}\right)^2, \qquad (13)$$

Since $\frac{S_{in}}{S_{out}} \rightarrow 0$, then $\xi_{in} = 0.5$ and $\xi_{out} = 1$.

$$\lambda = 0.11 \cdot (\Delta_{eq})^{0.25} = 0.11 \cdot \left(\frac{k \cdot P}{4 \cdot A}\right)^{0.25} = 0.11 \cdot \left(\frac{0.002 \cdot 2}{4 \cdot 0.3 \cdot 0.3}\right)^{0.25} = 0.036; \quad (14)$$

$$P_1 = \Delta P. \qquad (15)$$

$$\frac{27^2}{2} = \frac{1}{2} \cdot \frac{27^2}{2} + \frac{v_{in}^2}{2} + 0.036 \cdot \frac{20}{0.5} \cdot \frac{v_{in}^2}{2}; \qquad (16)$$

$$\mathrm{v_{in}} = 12.2\ \mathrm{m/sec.}$$

As soon as wind speed inside the airway is known, it is possible to determinate the cooling effect of an air circulation under the roof.

Energy conservation law, according to the heat power looks like:

$$\rho \cdot v_{in} \cdot A \cdot c_p \cdot dT = \alpha \cdot (T_h - T) \cdot dA, \qquad (17)$$

On the Fig.1 cross section of the attic is shown. Calculations are made on the plate along the wind direction under the housetop. It is supposed that parameters of the roof space are constant in width, so the plane problem can be estimated.

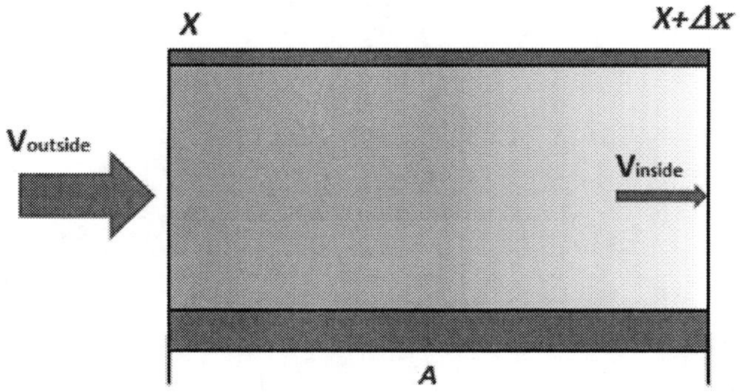

Fig. 1. Attic cross section.

Where A is the cross section of an attic, c_p is the transfer characteristic.

$$dA_w = \chi \cdot dx, \qquad (18)$$

Where χ is the coordinate increment in horizontal surface.

$$\rho \cdot v_{in} \cdot A \cdot c_p \cdot dT = \alpha \cdot (T_h - T) \cdot \chi \cdot dx; \qquad (19)$$

$$\frac{dT}{dx} = \frac{\alpha \cdot \chi}{\rho \cdot v_{in} \cdot c_p} \cdot (T_h - T). \qquad (20)$$

Add new quantities: Stanton number and hydraulic radius.

$$St = \frac{\alpha}{\rho \cdot v_{in} \cdot c_p} \text{ and } Rg = \frac{A}{\chi}. \tag{21}$$

Put them into the original equation Eq.1:

$$\frac{dT}{dx} = St \cdot \frac{(T_h - T)}{R_g}; \tag{22}$$

Numerical value of them can be estimated:

$$R_g = \frac{h^2}{4 \cdot h} = \frac{h}{4} = \frac{0.3}{4} = 7.5 \cdot 10^{-2}, \tag{23}$$

$$St = \frac{\alpha}{\rho \cdot c_p \cdot v_{in}} = \frac{15.85}{1{,}293 \cdot 1000 \cdot 12.2} = 1 \cdot 10^{-3}. \tag{24}$$

Make the original equation suitable for integration:

$$\frac{dT}{(T_h - T)} = \frac{St}{R_g} \cdot dx. \tag{25}$$

Integrate Eq.20 all along the airway:

$$-ln(T_h - T) = \frac{St}{R_g} \cdot x + c; \tag{26}$$

If x=0, then:

$$-ln(T_h - T_c) = c; \tag{27}$$

$$ln(T_h - T) = ln(T_h - T_c) - \frac{St}{R_g} \cdot x. \tag{28}$$

Exponentiate the equation to segregate the decision variable.

$$T_h - T = e^{\frac{St}{R_g} \cdot x - ln(T_h - T_c)}; \tag{29}$$

$$T = 20 + 2.73^{\frac{1 \cdot 10^{-3}}{7.5 \cdot 10^{-2}} \cdot 10 - ln(20+4.6)} = -1.53°C. \tag{30}$$

The estimated temperature under the roof is -1.53°C. It means that the icicle formation on the housetop will be prevented.

4 Results

Calculations show that the effect from the airway with the cross section 0.3x0.3 meter is equal to effect of the attic construction with height of 2.1 meter. It is clear that the cost of airway is less than the cost of attic. Saving time and finances during the construction phase is the foreground task for every building project. The airways are easy-making and their effect is proved. It makes the application of such kind of construction optimal. Only rectangular attic has been considered. According to other researches, it is possible to produce calculations for all types of roof. In [3] triangular roof is considered and air behavior within the roof is described. Combination with other methods of energy-saving can improve the effect from suggested measures. Some of them are offered in [24]. Generally, the result is highly corresponded with other researches data. Close result was achieved by W. Tobiasson, J. Buska, A. Greatorex [9]. They came to conclusion that icings can be avoided by maintaining an attic temperature of −1 °C when the outside temperature is −5.5 °C.

5 Conclusions

The present article has shown the possibility of airway application as a method to prevent the icicle formation on the roof of the apartment buildings during the winter. Evidently, that only the most general case has been considered. For every particular building the calculations should be performed independently.

There are a few factors which significantly increase the overall value of the described method. First of all, implementation of airways can potentially bring tangible monetary benefits due to the fact that this technology is much cheaper than attic construction. Additionally, airways are much easier to plan, design and implement than adding fans and other ventilation systems to a particular building. Finally, airways can be designed, measured and integrated into most of the buildings. Notably, considering the relatively low cost of airways and the overall simplicity of their design and integration, the method described in this article can potentially be found beneficial by construction industry workers.

References

1. S.C. Saha, J.C. Patterson, C. Lei, International Journal Of Thermal Sciences, **49**, 1899-1910 (2010)
2. I. Suwan,S.C. Saha, J.C. Pattersonb, C. Leib, Energy and Buildings, **42**, 1192-1204 (2010)
3. S.C. Saha, Energy and Buildings, **43**, 695–703 (2011)
4. E.F. Kent, Mechanics Research Communications, **36**, 497–508 (2009)
5. T. Basaka, S. Royb, S.K. Babub, A.R. Balakrishnana, International Journal of Heat and Mass Transfer, **51**, 4496–4505 (2008)
6. El H. Ridouanea, A. Campob, M. Hasnaouia, International Journal of Heat and Fluid Flow, **27**, 476-489 (2006).
7. V.A. Akinsete T.A. Coleman, International Journal of Heat and Mass Transfer, **25**, 991-998 (1982)
8. A.D. Suslova, A.D. Sivokhin, Construction of Unique Buildings and Structures, **5**, 54-64 (2014)
9. W. Tobiasson, J. Buska, A. Greatorex, AVOK: ventilation, heating , air conditioning, heating and Building Thermal Physics, **3**, 20-25 (2011)
10. N.I. Shamrikov, V.A. Volnukhin, T.N. Makarova, *Gable Roof With Unheated Attic*, (R.F. Patent 2263749, 2005)
11. O.A. Sizenko, A.P. Prokhorenko, Stroitel'stvo I Arkhitektura, **4**, 46-51 (2009)
12. I.I. Staroverova, *Thermal Conditions Of Warm Attics And Adjacent Spaces In Modern Residential High-Rise Buildings In The Cold Season (On The Example Of Construction In Moscow)* (Dissertation 1984)

13. M. Graham, National Roofing Contractors Assn., **7**, 1-12 (2015)
14. M.Graham, National Roofing Contractors Assn., **11**, 1-12 (2012)
15. J. Straube, Building Science Press. Building Science Digest, **135**, 23-28 (2006)
16. P.V. Nielsen, *Displacement Ventilation: Theory and Design* (Denmark: Aalborg University, 1993)
17. M.R. Petrichenko, V.N. Bukhartsev, *Extreme Problems Of Filtration Stream* (Palmarium Academic Publishing, 2012)
18. M.R. Petrichenko, Nauchno-Tekhnicheskiye Vedomosti SPBGPU ,**1**, 108-112 (2012)
19. V.I. Lipko, Magazine of Polotsk State University, B Series : Industry, Applied Science, **9**, 40-45 (2006)
20. V.S. Belyaev, V.F. Tikhonova, Residential Buildings Construction, **3**, 34-45 (2014)
21. SNiP (Construction Rules and Regulations) № 41-01-2003. Otopleniye, ventilyatsiya i konditsionirovaniye.
22. GOST (Russian National Standard) № 30494-96 Zdaniya zhilyye i obshchestvennyye. Parametry mikroklimata v pomeshchenii
23. D.R. Miftakhova, D.V. Nemova, Construction of Unique Buildings and Structures, **43**, 34-45 (2016)
24. A.S. Gorshkov, D.V. Nemova, N.I. Vatin, Construction of Unique Buildings and Structures, **12**, 2458 – 2463 (2013)
25. N. Harmati, R. Folić, Z. Magyar, J. Dražić, N. Kurtović-Folić, *Thermal Science* (Nuclear Institute Vinča, Belgrade, 2015)
26. N. Harmati, Z. Magyar, Energy Procedia, **78**, 2458 – 2463 (2015)
27. N. Harmati, Ž. Jakšić, N. Vatin, Procedia Engineering, **117**, 791 – 799 (2015)
28. N. Harmati, R. Folić, Z. Magyar, Thermal Science, **19**, 865-880 (2015)
29. V. Pukhkal, A. Bieliatynskyi, V. Murgul, Journal of Applied Engineering Science, **14**, 93 – 101 (2016)
30. V. Okrepilov, Ivanova, G., Standarty i Kachestvo, **12**, 62-68 (2004)
31. V. Okrepilov, Standarty i Kachestvo, **3**, 94-96 (2003)
32. V. Okrepilov, Standarty i Kachestvo, **10**, 52-55 (2015)

MATEC Web of Conferences **73**, 02009 (2016) DOI: 10.1051/matecconf/20167302009
TPACEE-2016

Energy efficiency in multi-story buildings

Anastasiia Staritcyna[1,*], *Eric* Pollock[2], *Ekaterina* Sokolova[3] and *Elena* Martynenko[4]

[2]Helsinki Metropolia University of Applied Sciences, Vanha maantie 6, 02650 Espoo, Finland

[1,3,4]St. Petersburg State Polytechnical University, Politekhnicheskaya, 29, Saint-Petersburg, 195251, Russia

Abstract. In this project a research on energy efficiency of Malta house was provided, it is a residential multi-story building in Helsinki, Jätkäsaari area. This project describes introduction with a new heating system for residential dwellings, which uses only heated air. To maintain air temperature in comfort level heat recovery and district heating is used in the same system. The task was to research efficacy of the enclosure structures. For research the 3D model has been created in the program the Revit 2015 and Lumion 13. Thermotechnical calculation for three types of a design has been executed in the program U-value.net.

1 Introduction

Energy efficiency problem gets more actual due to prices rise for energy resources, as well as the reduction of non-renewable energy sources. In addition, to that we should take into account the amount of carbon dioxide are increasing in the atmosphere. Human being has to reduce the power consumption to a minimum value [1, 2]. One of the areas to reduce energy consumption is housing, which consumes approximately 30 to 40% of all energy resources. While the existing old buildings consume from 200 to 400 [kWh / (m^2 • Year)] thermal energy for heating, buildings of the future generation would be expected to consume from [20 to 50 kWh /(m^2 • Year)] [3]. Some European countries already have similar standards of energy consumption. Solution to this problem is the construction of buildings with zero energy consumption, or close to zero [4]. Such buildings do not have heating expenses because the heat losses are fulfilled by renewable energy sources, as well as household and solar energy [5].

The heat camera was used to monitor thermal radiation from building's surfaces, such as exterior walls, windows and floors, to estimate thermal resistance of the house. For testing the efficiency of the heating system, the Wirepas (Wireless building monitoring) system was used in a single apartment.

Minimal requirements set for multi-storey buildings in Finland are described in the National Building Code of Finland. Part D2 of the code addresses the requirements aimed for the indoor environment, ventilation and air conditioning of residential spaces [6].

* Corresponding author: a.staritsyna@mail.ru

Table 1. Guideline values for room temperatures for different room types during the heating season for spaces where the room temperature design value is not 21 °C

Room type	**Room temperature [°C]**
Stair well	17
Bathroom, washroom	22
Drying room	24
Room type	Room temperature [°C]
Shop	18
• fixed work station in a shop	21
Gymnasium	18

When using the guideline values the level of comfort of adjoining rooms must be ensured.

Table 2. Air flow, air movement and sound.

Space/use	Outdoor air flow [(dm³/s)/per-son]	Outdoor air flow [(dm3/s)/m²]	Extract air flow [dm³/s]	Sound level [$L_{A.eq.T}$/$L_{A.max}$] dB	Air velocity winter [m/s]	NB
Dwe-lling spaces:	6					
Dwe-lling rooms		0.5		28 / 33*	0.20	°C1 regula-tion
Kit-chen		#S	8 #A	33 / 38*	0.20	°C1 regula-tion
impro-vement of occu-pancy period		#S	25	33 / 38	0.20	
Cloak room, store-room		#S	3	33 / 38		
Bathroom		#S	10 #B	38 / 43	0.20	
Impro-vement of occu-pancy period		#S	15	38 / 43	0.20	
Toilet		#S	7 #B	33 / 38		
impro-vement of occu-pancy period		#S	10	33 / 38		
Utility room		#S	8	33 / 38	0.30	
Impro-vement of occu-pancy period		#S	15	33 / 38	0.30	
Sauna in the flat		2 #C	2/m² #C	33 / 38		
Com-mon spaces:						

Stair-well		0.5 1/h	0.5 1/h	38 / 43		
Store-rooms		0.35	0.35/m^2	43 / 48		
Cold cellar (also cold stores in apart-ments if area is > 4m^2)		0.2	0.2/m^2	43 / 48		
Dres-sing room		2	2/m^2	33 / 38	0.20	
Wash-room		3	3/m^2	43 / 48	0.20	
Hot room in sauna		2	2/m^2	33 / 38		
Laun-dry room in the buil-ding		1	1/m^2	43 / 48		
Drying room		2 #D	2/m^2 #D	43 / 48		
Hobby room, club room		1#E	2/m^2 #E	33 / 38	0.20	

Addresses the guidelines for air flow, air movement and sound levels in residential spaces. If there are insufficient grounds for the dimensioning of the air flows, according to number of occupants, the surface area is used for the dimensioning. Extract air flows in large dwellings are generally designed to be higher than the guideline values so that the outdoor air flow for each space is in accordance with the guideline value. The air change coefficient of the space must be at least 0.5 of the total volume per hour. Part D3 of the building code discusses the requirements and guidelines of energy efficiencies. The code sets out the specific values to be used while calculating the efficiencies. The code considers total energy consumption, air-tightness of the building shell, heat losses, ventilation system efficiency and needed heating power. New apartment buildings are required to have an E-ratio (kWh/M^2 per year) of no more than 130 [7].

2 Literature review and formulation of research problems

Great contribution to the possible solutions of energy saving and energy efficiency was made by the following researchers: D.I. Markov, A.O. Pogonin, A.S. Gorshkov, D.V. Nemova, N.I. Vatin and many others.

Markov's article describes the experience of the design and construction of energy efficient buildings, talks about the principles of formation for such buildings (architectural, urban planning and engineering solutions energy saving). The present work focuses on the experience of designing «passive» houses in Darmstadt. It describes the experience of designing, a model of a residential complex of energy efficient buildings for the conditions of the Gatchina District, Leningrad Region, architectural planning, design, engineering solutions of the complex and technologies of renewable energy in this model [8-12].

In this article Pogonin considers the peculiarities of residential buildings stand-alone type of renewable energy, analyzes historical and contemporary experience of energy-efficient housing and cites, the example of a pilot project of autonomous energy-efficient residential unit [13].

The article A.S. Gorshkov, D.V. Nemova, N.I. Vatin in conceptual form is considered the major architectural and planning, three-dimensional and constructive solutions to save energy and ensuring energy efficiency of residential buildings. In this work was a comparative analysis of thermal costs energy for heating during the heating season for a private single-family house designed, according to the minimally acceptable level of regulatory requirements for thermal insulation standards in Russia and Finland. It has been proven that the determining factors for low-level consumption in buildings thermal energy for heating are: a high level of thermal insulation of external building envelope (walls, windows, roofs, etc.), tightness of the outer shell of the building and its compactness (low coefficient of compactness) [14-20].

Despite the large number of publications, that discusses advantages of consumption with zero and low power consumption, it has not yet been developed in such construction. The main objective of this work is to provide avoidance which shows modern Finland energy efficiency issues.

3 The object of work

Brand new building with low energy consumption was built on the territory of Helsinki, Jätkäsaari area. The building is equipped with modern energy-saving technologies and equipment that meets the highest international trends of construction Passive House type. Malta building uses decentralized forced air heating and ventilation system [21-28].

Throughout the operational phase was providing energy monitoring, evaluation of thermal characteristics, the determination of actual performance energy consumption. To provide this, the building was equipped with Wirepas wireless building monitoring system and heat camera.

4 Indoor environment quality

For the Malta building we used the Wirepas wireless building monitoring system. This system uses a closed wireless network to collect data from nodes placed around the building. This system can be used to monitor many aspects of the indoor environmental quality, for example temperature, relative humidity, carbon dioxide, and luminance. This information is uploaded to a wireless server and can be monitored in real time from a computer or mobile device. For our short time of monitoring we placed four sensors in apt. A21. This allowed us to observe both the building HVAC system and the Wirepas system in action. The time and scale allowed for our time and monitoring was not enough to get a good overview of the indoor environment, but the apartment which we did monitor seemed quite good. The Wirepas system is quite sensitive, showing quite small variations in the carbon dioxide and relative humidity.

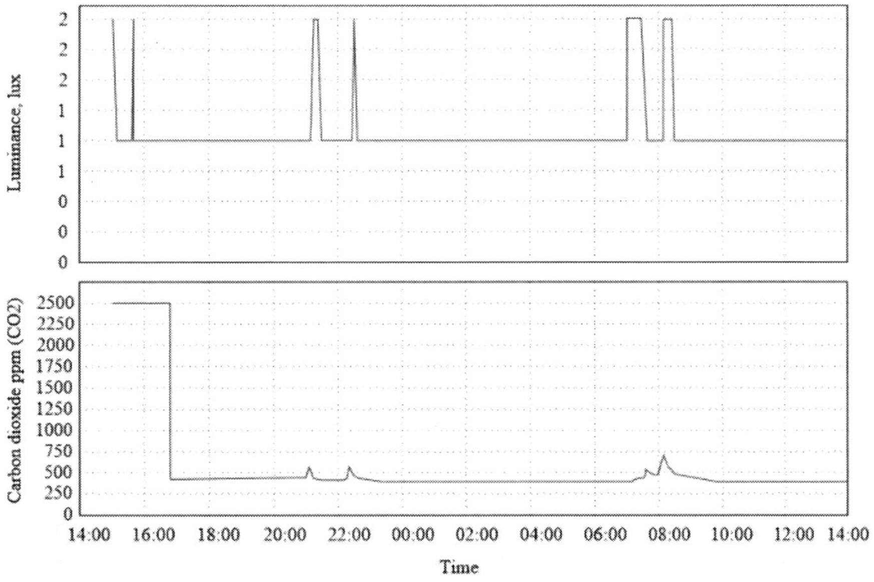

Fig. 1. Carbon dioxide and luminance values.

For example in the above data collected in the WC of apt. A21 we can see that the occupant used the WC five times during the monitoring period. At each occurrence of luminance, or use, we can see a small spike in the carbon dioxide corresponding to it. This however drops again as soon as the occupant left. This tells us that the ventilation system is working very well to circulate the air and keep it fresh. Another interesting data point we observed was from the sensor we left in place directly in front of the supply air vent in the main living space.

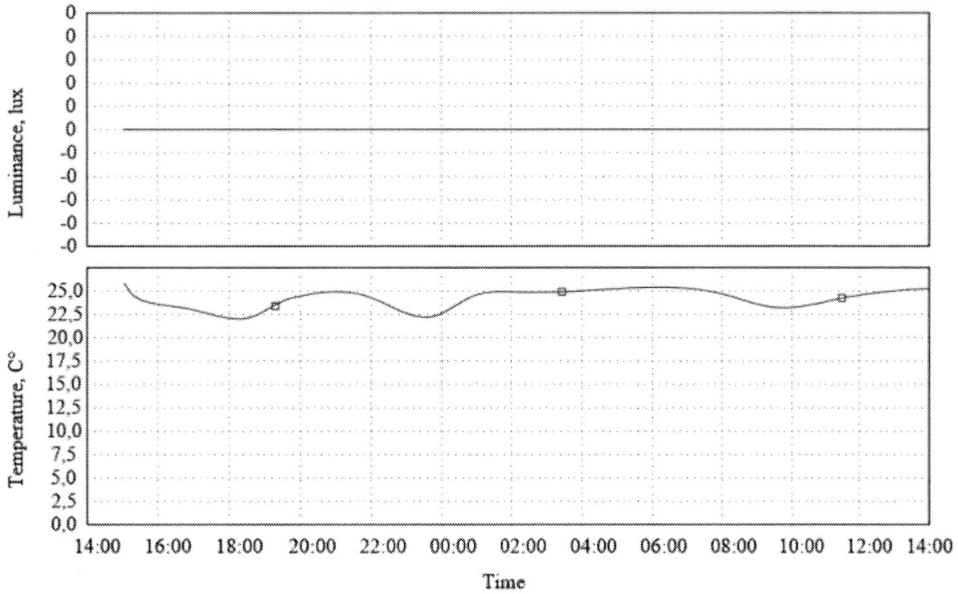

Fig. 2. Temperature and luminance values.

Here in the temperature data taken from 30 cm in front of the living room supply air vent we can see quite large temperature changes over the course of the evening and night. However, when we examine the data from sensors which were farther out into the space we find something quite different.

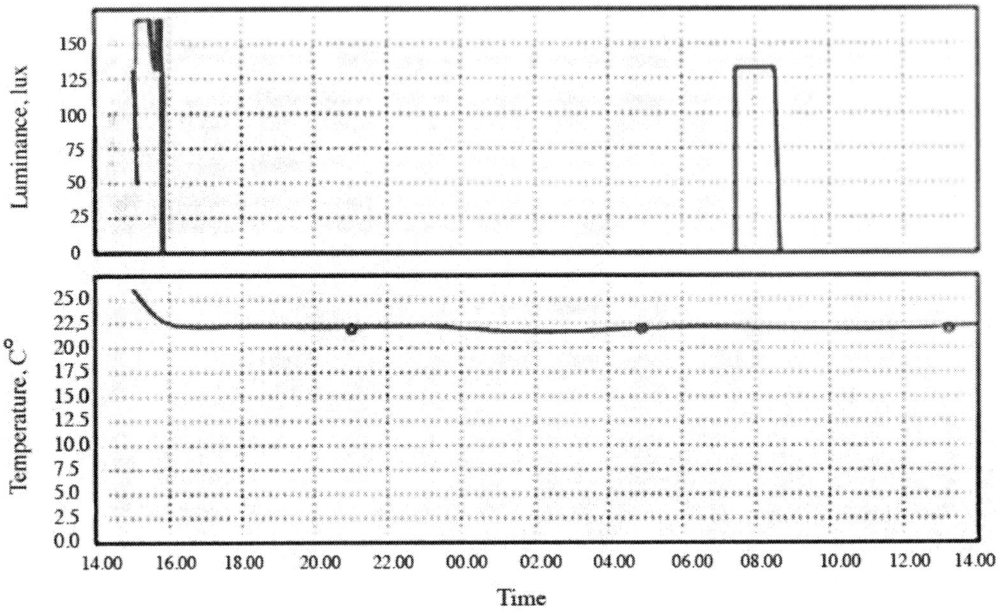

Fig. 3. Temperature and luminance values.

Here farther out in the room we notice that the temperature remains quite steady without major fluctuations in temperature. From this we determine that the system works quite well at reading the temperature of the room and responding in an acceptable time to avoid fluctuations.

In this exercise we experienced two main challenges. The first was a lack of available data, due to the limited time and scope we had available to us we were not able to make any broad range conclusions about the building as a whole. This however would be quite simple to fix. A more in depth study over a longer time would give much more relevant data. The second challenge we experienced was in setting up the Wirepas system. The system itself is quite intelligent, it actively searches for connections as each node is turned on, and if a connection is lost it will seek a new connection back to the main hub, scanning the nodes near it for an open connection. However since this process is automated, there is nothing that we can do if one node fails to work. We experienced some difficulty in starting to set up the system, struggling to achieve the first connections. However it must be said that once the network is up and running, there were no difficulties and the data was easy to collect and compare using Java script software.

5 Examples

The most weak and vulnerable sections of the building, which often formed "thermal bridges" are the following:

- joints of foundation walls to the floor of the first floor;
- joints of foundation walls with bearing exterior walls of the building;
- joints external load-bearing walls with a roof;

- angles of the exterior walls;
- window openings;
- doorways;
- balcony slabs;
- architectural elements of the building;
- any site of attachment of various structural parts to the building.

Let's have a look at the different examples of thermal bridges in case of Malta.

1. Problem with doorways.

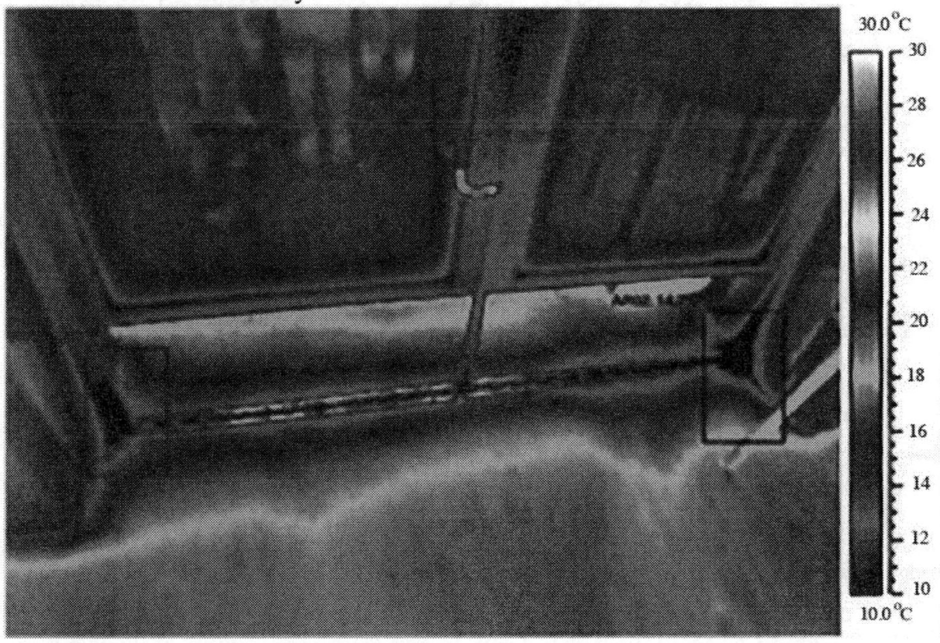

Fig. 4. Cold bridges thru door.

Normal wear possibly has caused damage to the weather stripping seals on the doors. These should be inspected and if necessary replaced.

Some spaces require maintenance.

2. Problem with the construction (Fig. 5)

Fig. 5. Cold bridges thru wall and door.

Possible empty space or error during the construction process. Unknown cause which requires further inspection.

6 Issues

There are some problems which should be pointed out. First of all, the frame in figure has faulty installation and there is high heat loss around the frame. Second of all, on the bottom of the door is a crack between door surface and frame. In addition, in most of door that was inspected team found issues with locks. Their temperature is too low, what makes them cold bridges. (Fig. 6, 7)

Fig. 6. Cold bridges thru door frame.

Next figure represents issue with seal which might be worn off or faulty installation during construction process.

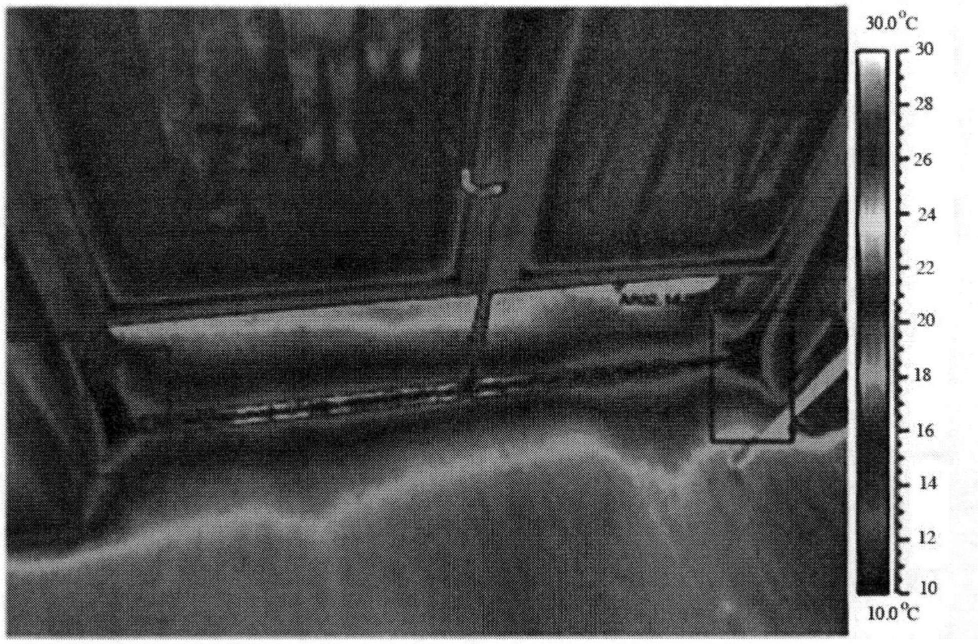

Fig. 7. Cold bridges thru door frame.

Another issue within corners of front door is shown in figure. In these spaces temperature is too low which indicates mechanical wear or faulty installation. However, mechanical damage is likely in this situation.

7 Thermomechanical calculation of the multilayered enclosure structure u-value.net

Design conditions and heat energy parameters of the buildings for the climatic conditions of Helsinki, Finland presented at the table 3.

Table 3. The design conditions.

The characteristics	Unit of measure	Value
Temperature of inside air	°C	20
Temperature of outside air	°C	-7
Humidity of inside air	%	50
Humidity of outside air	%	80

According to initial data, let's execute thermal calculation of the three types of multilayered enclosure structure.

For research the following parameters of the enclosure structures have been taken for achievement of thermal transmission coefficient value U=0,23.

Table 4. Structures of walls and its characteristic

Type of structure	Structures of walls	λ, W/mK	ρ, kg/m3	δ, mm
	Concrete	2.0	2400	500
Type I	Foam polystyrene URSA XPS	0.033	35	120
	Air (ventilated layer)	0.024	0,001	30
	Granite tile	1.5	2600	12
	Brick	0.960	2000	250
Type of structure	**Structures of walls**	**λ, W/mK**	**ρ, kg/m3**	**δ, mm**
Type II	Rock wool facade insulation	0.035	100	150
	Air (ventilated layer)	0.024	0,001	30
	Granite tile	1.5	2600	12
	Light weight concrete	1.3	1800	300
Type III	Polyurethane foam MUSTOVO	0.022	50	200
	Air (ventilated layer)	0.024	0,001	30
	Granite tile	1.5	2600	12

The following is a report on thermal calculations of the first type of the wall (fig.8-11).

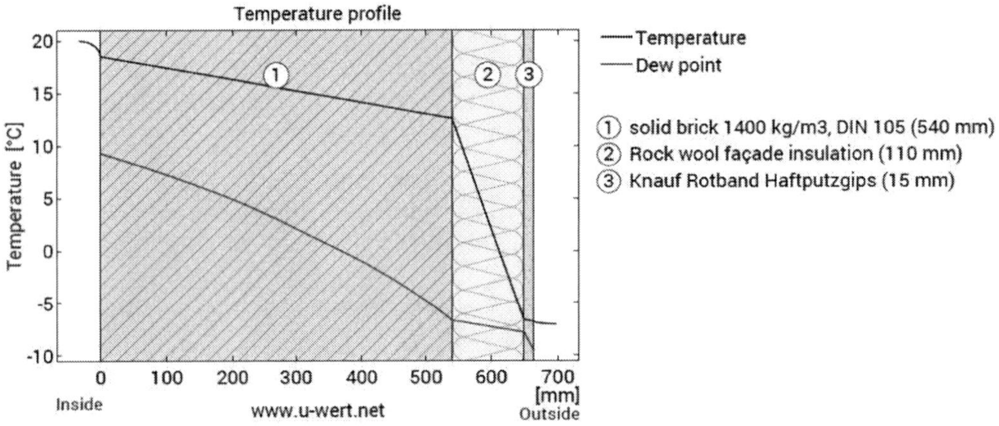

Fig. 8. Temperature gradient. Condensing zone.

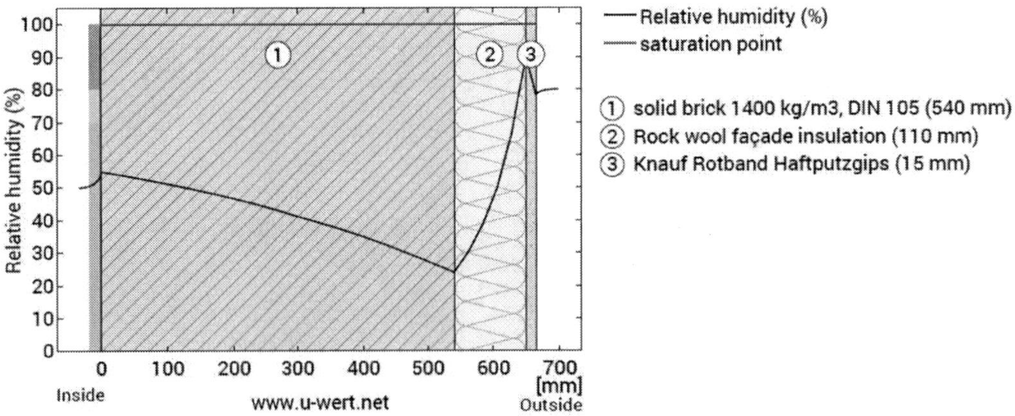

Fig. 9. Humidity schedule.

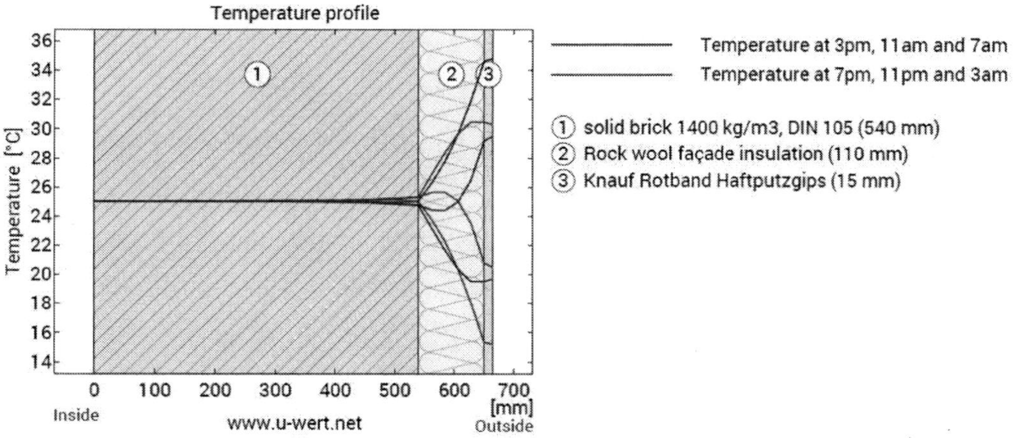

Fig. 10. Temperature schedule.

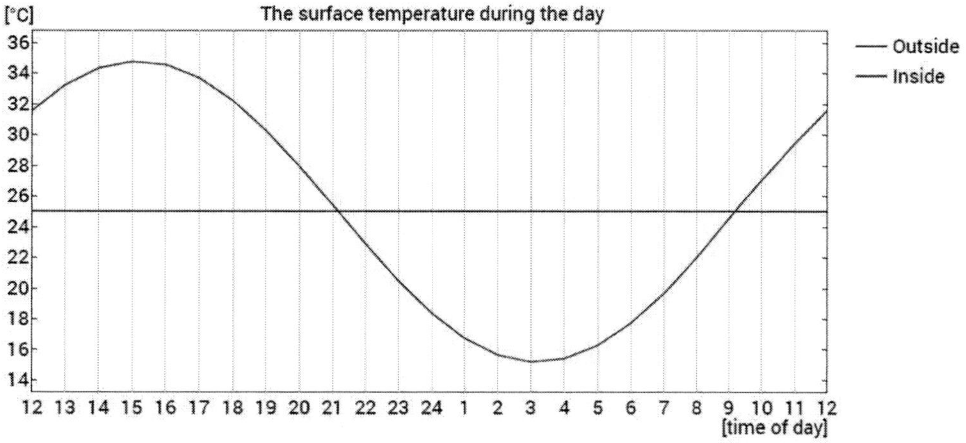

Fig. 11. The surface temperature during the day.

The report on thermal calculations of the second type of the wall (fig.12-15).

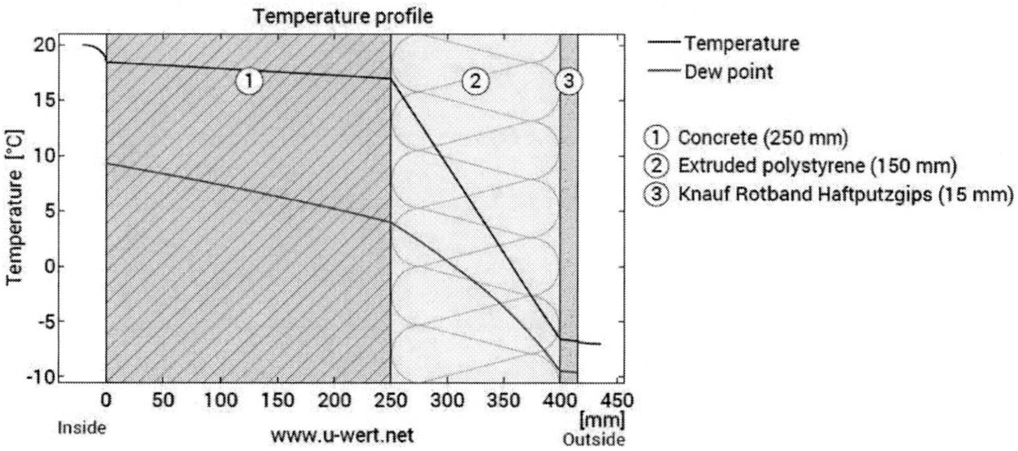

Fig. 12. Temperature gradient. Condensing zone.

 DOI: 10.1051/matecconf/20167302009

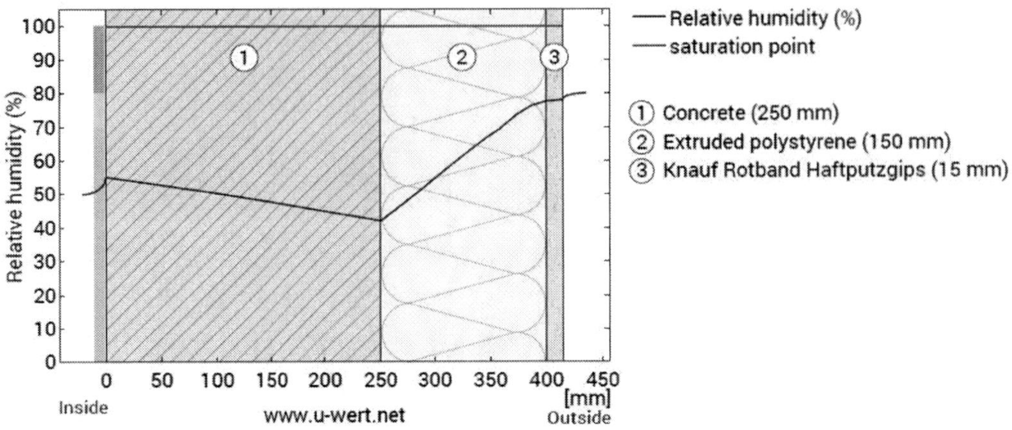

Fig. 13. Humidity schedule.

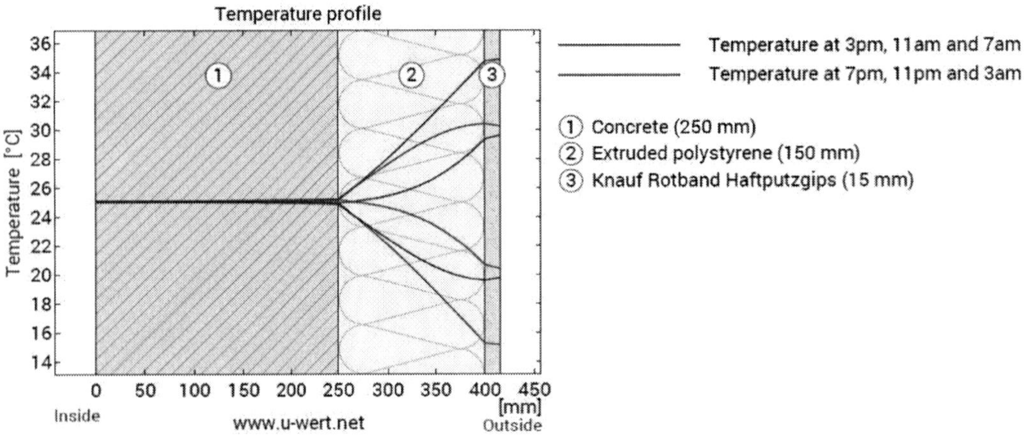

Fig. 14. Temperature schedule.

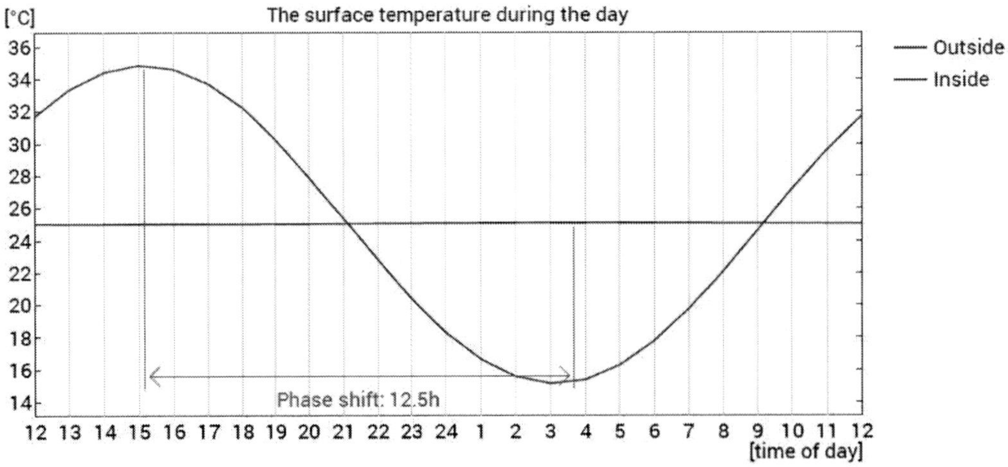

Fig. 15. The surface temperature during the day.

The report on thermal calculations of the third type of the wall (fig.16-19).

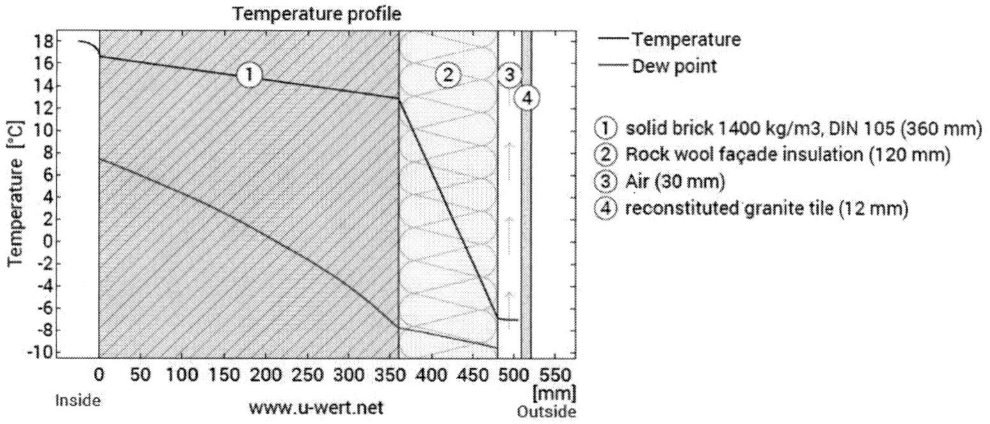

Fig. 16. Temperature gradient. Condensing zone.

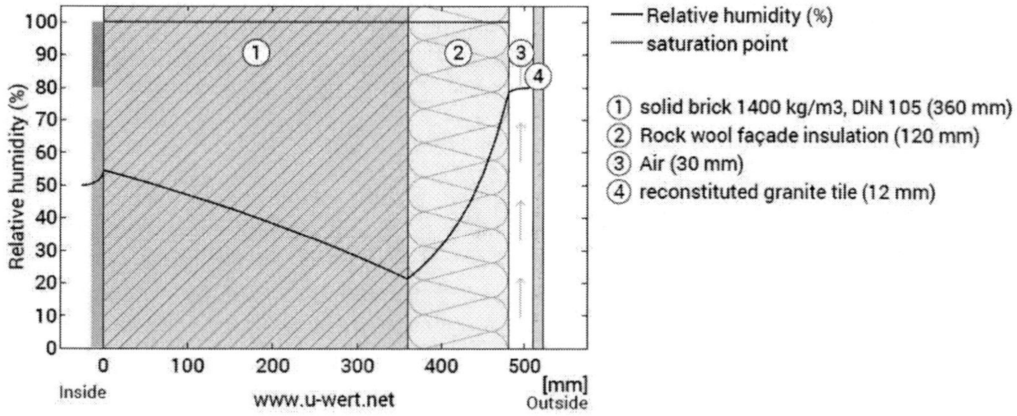

Fig. 17. Humidity schedule.

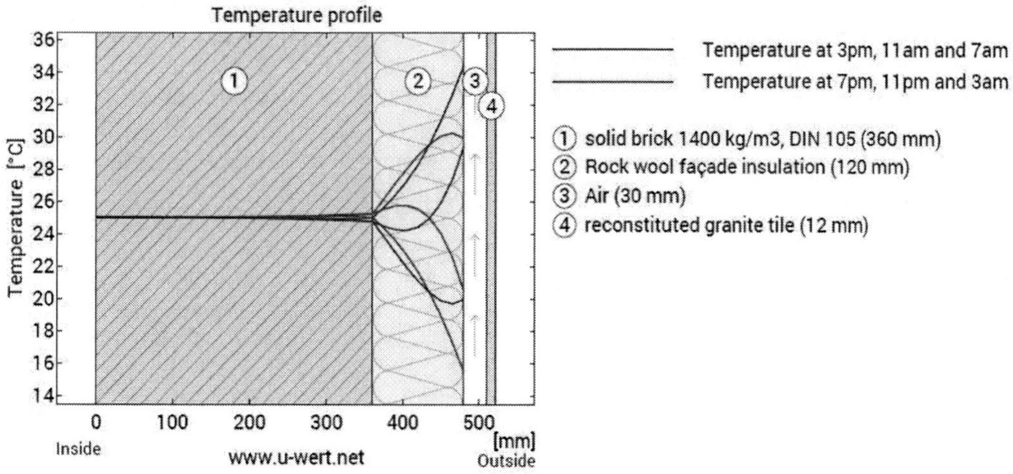

Fig. 18. Temperature schedule.

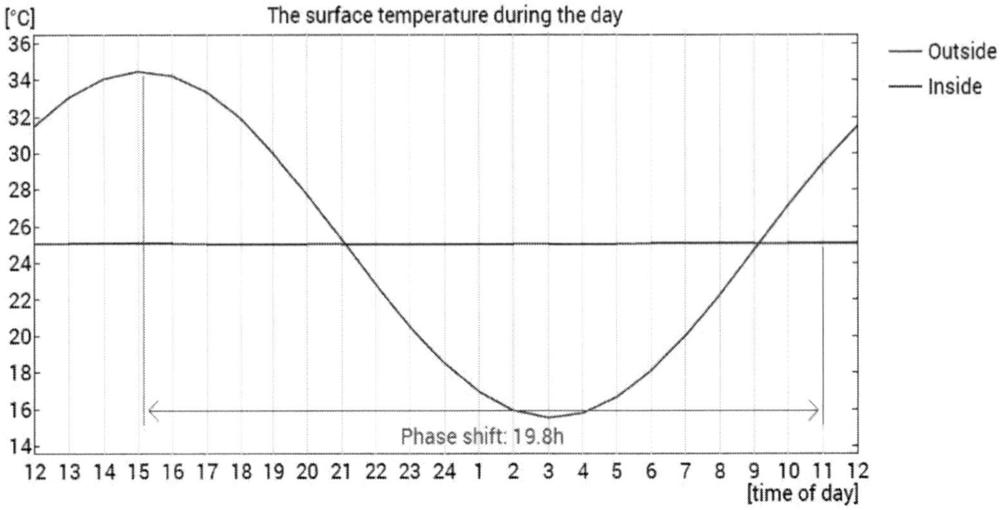

Fig. 19. The surface temperature during the day.

8 Conclusion

Buildings account for 40% of energy consumption in most countries, yet their potential for energy efficiency is huge. By using adapted insulation and energy savings techniques, up to 80% of a building's energy consumption for heating or cooling can be saved. In a building that is insufficiently or not insulated, heat flows through the roof or the façades thus leading to increased energy needs for heating (in winter) or for cooling (in summer).

According to the building inspection, we can make the following conclusions: The main heat leaks occur at the location of windows, doors and joints of window frames with walls, also joints of doors with floor and field installation of keyholes.

At purpose of identical the thermal transmission coefficient the most massive design is Type 1 (665 mm) after comparison of three types of the enclosure structures in the program U-value.net. In addition, the thinnest enclosure structure is Type 2 (415 mm). Therefore it brings influence to bear on architectural appearance and weight of the building in general that allows to reduce sections of the bearing designs and to reduce construction cost. By results of research the construction consisting of concrete (250 mm), Extruded polysterene (150 mm) and Gypsum Plaster (15 mm) is the most power effective type of an enclosure structure at this value of the thermal transmission coefficient U=0.23.

References

1. N. Vatin, D. Nemova, A. Kazimirova, K. Gureev, Advanced Materials Research, 1537-1544 (2014)
2. N. Vatin, A. Gorshkov, P. Rymkevich, D. Nemova, D. Tarasova, Applied Mechanics and Materials, 365-369 (2014)
3. N. Vatin, A. Gorshkov, D. Nemova, O. Gamayunova, D. Tarasova, Applied Mechanics and Materials, 349-354 (2014)
4. I. Marcov, Architecture and Modern Information Technologies, **1(18)**, 13-18 (2014)
5. Xinhong Zhaoa, Congyu Mab, Pingdao Gub, Energy Procedi, 1523–1527 (2012)
6. National Building Code of Finland
7. N. Vatin, O. Gamayunova, Applied Mechanics and Materials, 972-976 (2014)

8. N. Vatin, A. Gorshkov, D. Nemova, A. Staritcyna, D. Tarasova, Advanced Materials Research, 905-920 (2014)
9. A. Pogonin, Architecture and Modern Information Technologies, **4 (13)**, 13 (2010)
10. A. Gorshkov, D. Nemova, N. Vatin, Construction of Unique Buildings and Structures, **7 (12)**, 49-63 (2013)
11. O. Averyanova, Magazine of Civil Engineering, **5**, 53-59 (2011)
12. D. Nemova, V. Murgul, A. Golik, E. Chizhov, V. Pukhkal, N. Vatin, Journal of Applied Engineering Science, **12 (1)**, 37-44 (2014)
13. S. Ginestet, D. Marchio, O. Morisot, Energy Conversion and Management, **76**, 368-376 (2013)
14. R. Alihodzic, V. Murgul, N. Vatin, E. Aronova, V. Nikolić, M. Tanić, D. Stanković, Applied Mechanics and Materials, **624**, 604-612 (2014)
15. N. Vatin, O. Gamayunova, Advanced Materials Research, 2159-2162 (2015)
16. N. Vatin, A. Gorshkov, D. Nemova, A. Staritcyna, D. Tarasova, Advanced Materials Research, 905-920 (2014)
17. D. Nemova, V. Murgul, A. Golik, E. Chizhov, V. Pukhkal, N. Vatin, Journal of Applied Engineering Science, **12 (1)**, 37-44(2014)
18. N. Vatin, D. Nemova, L. Khazieva, D. Chernik, Applied Mechanics and Materials, 2057-2062 (2014)
19. N. Vatin, A. Gorshkov, D. Nemova, D. Tarasova, Mechanics and Materials, 991-996 (2014)
20. Y. Iwafune, K. Nishio, Journal of Environmental Engineering, **650**, 371-379 (2010)
21. M. Bassanino, T. Fernando, J. Masior, M. Kadolsky, Fouchal R. Scherer, Fouchal, R., Firth, T. Hassan, K. Klobut, 10th European Conference on Product and Process Modellin, 863-870 (2015)
22. A. Gorshkov, Energosberezhenie, **4**, 12-19 (2014)
23. M. Petrichenko, D. Nemova, A. Staritcyna, D. Tarasova, Applied Mechanics and Materials, 2308-2315 (2014)
24. N. Vatin, D. Nemova, D. Tarasova, A. Staritcyna, 3rd International Conference on Energy and Environmental Protection, 854-870 (2014)
25. N. Vatin, D. Nemova, L. Khazieva, D. Chernik, Applied Mechanics and Materials, 2057-2062 (2014)
26. K. Petrov, E. Zolotareva, V. Volodin, N. Vatin, E. Zhmarin, Magazine of Civil Engineering, **2**, 59-64 (2010)
27. V. Murgul, Procedia Engineering, **117**, 808-818 (2015)
28. V. Murgul, D. Vuksanovic, V. Pukhkal, N. Vatin, Applied Mechanics and Materials, 977-981 (2014)

Possibility of Thermal Storage System Use with Different Accumulating Material in SPbSTU

Anna Nefedova[1,*], *Julia* Bykova[1], *Maxim* Kukolev[1], *Sergei* Kosov[2], *Aleksandrs* Zajacs[3], and *Anatolijs* Borodinecs[3]

[1]Peter the Great St-Petersburg State Polytechnical University, Polytechnicheskaya, 29. St. Petersburg, 195251, Russian Federation
[2]Ltd "Ecomatic SPb", Kurskaya, 27 B. St.Petersburg, 192007, Russian Federation
[3]Riga Technical University, Kalku Street, Riga LV-1658, Latvia

Abstract. In this article the possibility of heat accumulator use in one of the buildings of SPBSTU was analyzed. The use of this accumulator could lower the hot water demand. To complete the goal it was assumed that there is an additional café with defined working hours.
Different kinds of heat collecting materials were studied: water, red brick, concrete and granite. Depending on charging time of the heat accumulator, the results of the heat gaining abilities of different materials were made.
As a result it was calculated that single-phased accumulating materials should not be used as heat collectors in public buildings.

1 Introduction

Nowadays, energy efficiency is a leading topic in civil engineering. It saves money, resources so in the long run it helps economics and ecology.

To achieve high energy efficiency the heating system distribution should be modernized. One of the variants is to install heat accumulators for hot water and heating systems. In Russia it is popular in small houses outside cities.

In this article one of SPBSTU campuses is studied when it is assumed that there is an additional heating source along with higher hot water demand.

2 Literature review

Energy efficiency has become world tendency and now is going to be the first thing to consider during the designing process of all constructional elements [2-4]. In many countries there is legislation for energy efficiency, in Russian Federation it is Federal Law No. 261-FZ of November 23, 2009 "On Energy Saving and Increase of Energy Efficiency and Introduction of Changes into Separate Legislative Acts of the Russian Federation" [5].

During the design development designers in the first place try to apply energy saving space-planning concepts - construction of wide-bodied houses [6, 7] allows to reduce specific area of enclosing structures. They also use energy saving construction systems like

* Corresponding author: anyanefedova94@mail.ru

heat insulation of exterior walls and hinged ventilated facades [8, 9]. Projects of "Green houses", partly or fully isolated from energy sources, appear [10-12]. As practice shows, energy saving can be highly increased by implementing new building services systems or by renovating them, for example, by installing automated heating unit [13-15]. Lately the appliance of heat accumulators became a very popular approach. They can store some amount of heat energy to use it during day periods when charges are higher or during the hot water consumption peaks to reduce the load on the heating unit [16-17].

3 Purposes and tasks

The purpose of this article and our research was to choose the most suitable type of heat accumulator for help to smooth the peak of hot water consumption or to define that under existing conditions it is inappropriately.

Following problems were solved:

- Calculation of hot water demand after the reconstruction of café;
- Calculation of possible amount of accumulated heat for different types of single-phased accumulating material (SPAM) in the accumulator;
- Analysis of the results and conclusion about possibility of application of different types of heat accumulators for smoothing the hot water consumption peak in considered problem.

4 Research description

Nowadays, there is a serious problem with cafes in SPBSTU because there is not enough place for all the people. Let us assume that to solve the task an additional café was working during peak loads on the existing café. Changes of the number of water-supply points can be seen in Table 1.

Table 1. Comparison of number of water-supply points in existing café and in designing café.

		Existing café	Additional café
1	Number of seats	18	50
2	Number of was-basins for clients	0	2
3	Number of sinks	1	3
4	Number of was-basins for stuff	0	1

There are two ways to lower the maximal hot water use:

- Recalculation of the substation and an expensive modernization afterwards;
- Installation of the heat collector as an additional equipment.

The heat collector will be used during when occupation of the café is maximal. From 12.00 to 16.00 there is the highest use of hot water. Outside this period an additional café is not needed.

These changes will highly increase the load on hot water supply system and the existing heat unit will not be able to bear it. Nominal capacity of the heat unit is 200 W and maximum hot water flow is 0.96 l/s. In order to smooth the consumption peak it was decided to install heat accumulator in series, when the accumulator is connected into power circuit after the energy source.

Main element of every heat accumulator is accumulating material. Hot water in hot water supply system transfers the heat to the SPAM, which, due to its high heat capacity, can accumulate a huge amount of heat during the charging time. Due to good quality of thermal insulation the heat energy may be stored for a long time and be used when needed – in our case during the consumption peak since 12 a.m. to 4 p.m.

In order to find the most profitable accumulator we considered several types of relatively cheap and non-toxical SPAMs (Fig.1):

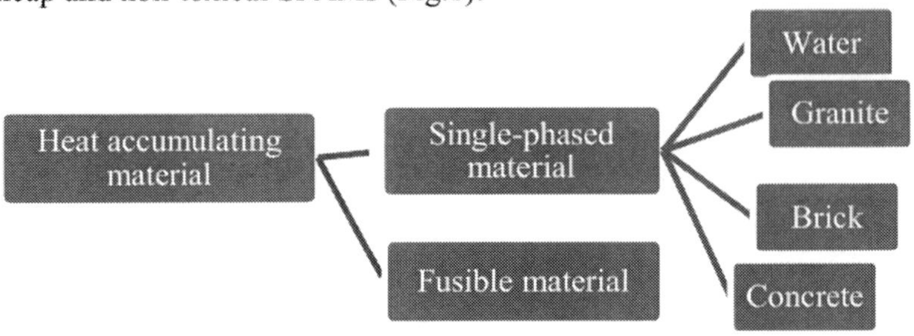

Fig. 1. Types of heat accumulating materials.

In this article we considered heat accumulators with single-phased SPAM.

Let's have a closer look to the working process of heat accumulator.

Heat accumulator charge

Single-phased SPAM (Fig. 2a) is located around the pipe in which heat-transfer medium flows. Tci — input medium temperature, Tco — output medium temperature. Passing through the channel heat-tranfer medium is getting colder (Tci < Tco), transferring heat to the SPAM, thus the temperature of accumulating material Tn is increasing.

Heat accumulator discharge

The most common system is used: cold heat-transfer medium with input temperature Tdi flows through the channel and warms up to output temperature Tdo (Fig.2b).

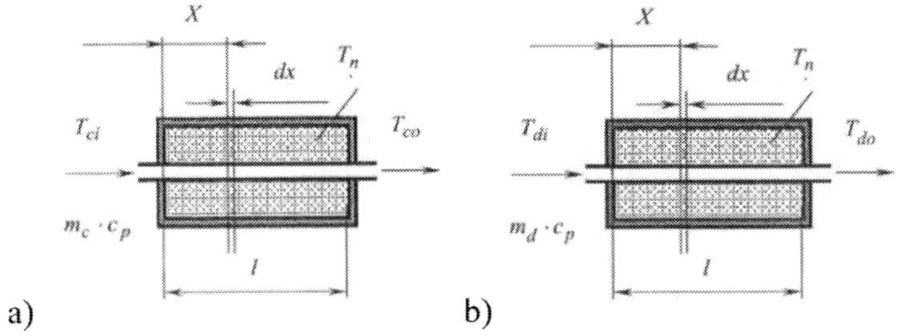

Fig. 2. Heat accumulator with single-phased SPAM: a. Charge b. Discharge.

By changing charging time we can choose the most profitable variant of using the heat accumulator for every type of SPAM.

5 Calculations

Firstly, we calculate the number of conditional dishes consumed per hour and per one working shift. Café is designed for 50 seats.

Number of dishes consumed per hour can be calculated with the formula (1):

$$u = n \cdot m \cdot u_o = 50 \cdot 3 \cdot 2.2 = 330 \tag{1}$$

Where

n — number of seats, n=50;

m — number of different customers seating at one seat per hour, m=3 (for students' café);

u_o — number of conditional dishes consumed by one customer (u_o = 2.2).

Café is open for 4 hours in a day. Daily quantity of dishes consumed can be calculated with the formula (2):

$$U = \frac{u \cdot T}{K} = \frac{330 \cdot 4}{1.5} = 880 \tag{2}$$

Where
u — calculated number of conditional dishes consumed per hour;
T — working time of café;
K — hourly non-uniformity coefficient (K=1.5).
Therefore, for the calculated number of conditional dishes hot water flows were calculated (using SNiP 2.04.01-85 «Domestic water supply and plumbing system"):
q_h = 0.65 l/s — maximal design flow;
q_{hhr} = 1.28 m3/h — maximal hourly flow.
Heat flow during the hour of maximal water consumption Q_{hr} = 76800(ccal/h) = 321 546.240 (kJ/h) = 89.3 kW.
So, we can calculate how much heat is needed to be stored in the heat accumulator, if we multiply heat flow per hour by working time.
Then, amount of required heat is Q=1 286 184.96 kJ.
Physical characteristics of materials, calculated on the base of parameters found, are presented on the Table 2.

Table 2. Physical characteristics of different types of SPAM

SPAM	Spec.heat capacity [J/kg·K];	Density [kg/ m3] liquid/solid	Mass flow [m3/h].	Time of discharge [h]	Mass [t]
Water	4200	998,2	1277,7	4	9.34
Granite	750	2800	3200		14,3
Brick	840	1100	1408		5,6
Concrete	1000	2500	3584		12,8

As a result of modeling of heat accumulators with single-phased SPAM in series we have following functional connections:
For the temperature of SPAM (3),(4):
Charge:

$$T_{nc}(t) = T_o + (T_{ci} - T_o) \cdot (1 - e^{-y_c \cdot \theta_c}) \tag{3}$$

T_o — initial temperature of SPAM (To = 283 K);
T_{ci} — input temperature of heat-transfer media entering the accumulator during charge (T_{ci} = 338 K).
Discharge:

$$T_{nd}(t) = T_{nc} - (T_{nc} - T_{di}) \cdot (1 - e^{-y_d \cdot \theta_d}) \tag{4}$$

T_{di} — input temperature of heat-transfer media entering the accumulator during discharge (T_{di} = 283 K – in summertime, T_{di} = 278 K – in wintertime)
For the temperature of heat-transfer media in channel as it leaves heat accumulator (5), (6):
Charge:

$$T_{co}(t) = T_o + (T_{ci} - T_o) \cdot (1 - y_c \cdot e^{-y_c \cdot \theta_c}) \tag{5}$$

Discharge:

$$T_{do}(t) = T_{co} - (T_{co} - T_{di}) \cdot (1 - y_d \cdot e^{-y_d \cdot \theta_d}) \tag{6}$$

To find the temperature difference for the beginning and the end of charge and discharge it's needed to calculate coefficients with formulas (7)-(9):

a) Non-dimensional transfer numbers

$$N_c = \frac{K_c \cdot F(x)}{m_c \cdot c_p} \qquad N_d = \frac{K_d \cdot F(x)}{m_d \cdot c_d} \tag{7}$$

b) Non-dimensional process times:

$$\theta_c = \frac{m_c \cdot c_p \cdot \eta_c}{M_\Sigma \cdot c_n} \cdot t_c \qquad \theta_d = \frac{m_d \cdot c_p}{M_\Sigma \cdot c_n \cdot \eta_d} \cdot t_d \tag{8}$$

c) Ratios

$$y_c = 1 - e^{-\frac{N_c}{\eta_c}} \qquad y_d = 1 - e^{-N_d \cdot \eta_d} \tag{9}$$

As soon as we've calculated these coefficients as a function of charging time tc, we can calculate temperatures for charge and discharge with formulas (3)–(6). Let's assume that charging time vary between 1 and 20 hours. Calculation results for single-phased SPAM can be presented in table form (Table 3).

Table 3. Dependence of the temperature of heat-transfer media on the charging time.

	T [K]	**t_c [h]**								
		1	**2**	**3**	**4**	**5**	**6**	**7**	**8**	**20**
Water	$T_{co}(t)$	289.7	295.5	300.7	305.3	309.2	312.7	315.8	318.5	*333.9*
	$T_{do}(t)$	286.7	290.1	292.9	295.4	297.7	299.6	301.3	302.9	*311.5*
Granite	$T_{co}(t)$	303.5	316.4	324.4	329.5	332.7	334.6	335.9	336.6	*337.9*
	$T_{do}(t)$	285.5	287.0	288.0	288.6	289.0	289.3	289.4	289.5	*289.7*
Brick	$T_{co}(t)$	319.3	331.6	335.8	337.2	337.7	337.9	337.9	337.9	*338*
	$T_{do}(t)$	287.4	288.9	289.4	289.6	289.7	289.7	289.7	289.7	*289.7*
Concrete	$T_{co}(t)$	300.9	313.1	321.2	326.7	330.4	332.8	334.5	335.6	*337.9*
	$T_{do}(t)$	285.2	286.6	287.6	288.3	288.8	289.1	289.3	289.4	*289.7*

With the formula (10) we can calculate possible amount of heat we can accumulate in dependence of charging time of heat accumulator and can compare the results (Table 4) with the required amount of heat for café.

$$Q = M \cdot \Delta T \cdot C_n \tag{10}$$

Table 4. Dependence of heat amount on the charging time

		t_c [h]								
		1	**2**	**3**	**4**	**5**	**6**	**7**	**8**	**20**
Water	Q [MJ]	115.7	217.4	306.6	385.0	453.8	514.2	567.3	613.9	*879.1*
Granite		193.4	314.6	390.5	438.2	467.7	486.4	498.0	505.3	*517.5*
Brick		149.9	200.8	218.1	224.0	226.0	226.6	226.9	227.0	*227.0*
Concrete		202.0	337.9	429.4	491.0	532.4	560.3	579.1	591.7	*617.6*

Dependence of heat amount on the charging time is shown on the fig.3.

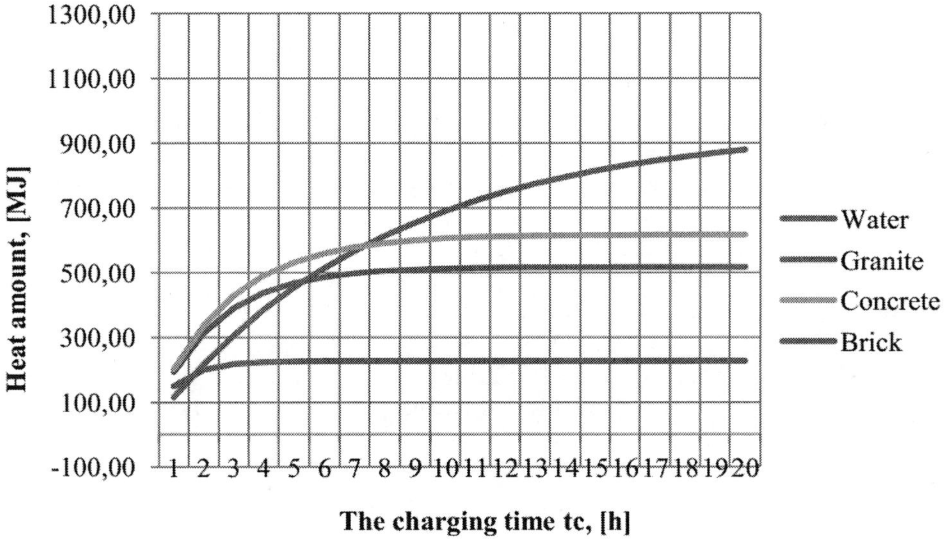

Fig.3. Dependence of heat amount on the charging time.

As it can be seen from the graph, even during the maximal charging time – 20 hours – SPAM can't accumulate the required amount of heat (1286 MJ).

6 Conclusions

1. Hot water flow in "Hydrotechnical building-2" of campus of SPbSTU during the working hours of designing café has increased almost in two times and became 1.61 l/s instead of 0.96 l/s
2. Applying of heat accumulator is unpractical for full and autonomous supply of café for 50 seats.
3. Heat accumulator with SPAM-water can be used as additional power source.

4. Heat accumulators with SPAM-granite/brick/concrete don't accumulate a lot of heat when charging time is increased maximally
5. Recalculation of heat unit with nominal capacity of 289.2 kW and hot water flow of 1.61 l/s is needed.

References

1. Government program "Energy efficiency and energetic development", (approved by Government Resolution of Russian Federation N 321, 2014)
2. D. Verneke, Stroitel'nye materialy, oborudovanie, tehnologii XXI veka, **10**, 26-27 (2008)
3. T. Zadvinskaya, A. Gorshkov, Adv Mat Res, **953-954**, 1570-1577 (2014)
4. N. Zaguskin, Problemy sovremennoy ekonomiki, **4 (48),** 314-319 (2013)
5. *About energy saving and energy efficiency improvements and introduction of amendments to certain regulatory act of the Russian Federation,(Federal Law N 261-FZ, 2009)*
6. E. Semenova, D. Kosheleva, Nauchnyj vestnik voronezhskogo gosudarstvennogo arhitekturno-stroitel"nogo universiteta, **1**, 150-153 (2011)
7. E. Semenova, L. Amirkhanyan, Nauchnyj vestnik voronezhskogo gosudarstvennogo arhitekturno-stroitel"nogo universiteta, **1**, 142-144 (2011)
8. S. Maloedov, V. Vyguzov, Stroitel'nye materialy, **5**, 24 (2001)
9. R. Zorin, I. S'yanov, **1**, 139-142, (2010)
10. Nefedova, D. Chernyshev, D. Tseytin, Constr of U Build and Struct, **10(25),** 73–87 (2014)
11. E. Bazhenova, J. Bykova, D. Bryus, D. Tseytin, Ap Mech and Mat, **725-726**, 1445-1456 (2015)
12. V. Murgul, J of Ap. En. Sc., **12 (2),** 121-128 (2014)
13. N. Koval'nogov, A. Rtishheva, E. Cynaeva, Izvestija vysshih uchebnyh zavedenij. Problemy jenergetiki, **3-4**, 100-107 (2007)
14. V. Afanas'ev, V. Kovalev, V. Tarasov, V. Tarasova, D. Fedorov, Vestnik chuvashskogo universiteta, **2**, 10-18 (2014)
15. Nefedova, J. Bykova, S. Kosov, M. Petrichenko, Proc Eng, **117**, 1102 – 1111 (2015)
16. A.Nefedova, J. Bykova, S. Kosov, Constr of U Build and Struct, **5 (32)**, 50-62 (2015)
17. M. Kukolev, Ju. Kukelev, L. Lucenko, CSME FORUM'98, 584-588 (1998)
18. V.Borovkov, M. Kukolev, V. Chahovskij, Ju. Kukelev, Nadezhnost' i bezopasnost' jenergetiki, **2**, 56-58, (2008)
19. Mohamad, Sol En, **3(61),** 211-218 (1997)
20. Z. Yuwen, A. Faghri, Int. J of H and M Tr, **15(39),** 3165-3173 (1996).

MATEC Web of Conferences **73**, 02011 (2016) DOI: 10.1051/matecconf/20167302011
TPACEE-2016

Solutions to Improve the Thermal Protection of the Administrative Building

Valeriya Kostenko[1,*], *Nailya* Gafiyatullina[1] , *Gafur* Zulkarneev[1], *Aleksandr* Gorshkov[1] , *Marina* Petrichenko[1] and *Sara* Movafagh[2]

[1] Peter the Great St. Petersburg Polytechnic University, 195251 Politechnicheskaya str. 29, St. Petersburg, Russia
[2] University of Guilan, 4199613776 Khalij Fars highway, Ghazvin road, Rasht, Iran

Abstract. At the end of the 90s, with the introduction of changes in the regulatory documents of the Russian Federation №3 to SNiP II-3-79*, regulatory requirements for thermal protection of buildings were revised towards increase. For this reason, the buildings built till 2000 don't conform to modern requirements. The actual solution of this problem is to carry out works on renovation of facades of the existing buildings with the use of innovative materials. As object of research one of educational cases of Peter the Great St. Petersburg Polytechnic University has been chosen, where by practical consideration the size of the actual thermal resistance of external walls has been determined by heat flux meter, the numerical value of which was significantly lower than the standard values. Based on the obtained data two modern ways of thermal insulation of facades (Ventilated Façade System (VFS) and External Thermal Insulation Composite System (ETICS)) have been analyzed, the assessment of energy saving potential and the discounted payback period of the investments directed to warming of facades has been made.

1 Introduction

About 95% of the buildings of the total amount existing housing stock in Russia do not correspond to modern requirements for thermal insulation, i.e. they are obsolete. The physical state of the building facades is also often faulty. In this regard, measures for heat insulation of external building envelope should be recognized as reasonable.

In addition to improving the appearance of buildings, raise thermal insulation of the building envelope leads to decrease of transmission heat losses [1-6]. The lower the heat loss in the building, the lower the amount of thermal energy is required to bring to the building from a heat supply source to compensate for transmission losses. Thus, the thermal insulation leads to decrease energy consumption in the building and consequently to reduction of payments for heating [3-6].

One of the important characteristics of thermal insulation performance of the building envelope is the resistance to heat transfer. The methods of determining it based on

* Corresponding author: kostenkolera@gmail.com

laboratory tests of material samples or fragments of building structures in climatic chambers, or researches are conducted directly in a natural external environment.

The latter method is more suitable because it does not require destruction of the structure and enables to determine the characteristics of the actual design and not the model. The test allows to quantify the thermotechnical characteristics of the building envelope (heat transfer resistance, thermal conductivity, thermal flux density) and to check their compliance with the regulatory requirements established in accordance with the rules [10], to determine the real heat loss through exterior walls.

The aim of the article is a quantitative analysis of the thermotechnical characteristics of the building envelope, which are carried out by comparing the actual heat transfer resistance values with normative data, offer methods of the building envelope insulation with their technical and economic comparison, a choice of the most effective technical solutions.

2 Initial design data

The building was designed in 1970 that is the reduced value of the heat transfer resistance of the building envelope R_0^r was determined according to the standard [7] by climatic parameters of Leningrad and was equal 0.98 m^2·°C/W.

3 The results of field observations

Experimental research of exterior walls R-value of Peter the Great St. Petersburg Polytechnic University Gidrokorpus-2 (Fig. 1) was carried out using ITP-MG4.03/5(I) "POTOK" heat flux density meter. The test was conducted from 15 to 20 January 2016. During this period, an outdoor air temperature varied in the range from minus 23 °C to minus 2 °C, an inner wall surface temperature – from 14 °C to 16 °C. The building walls are made of solid brick 510 mm thick with ceramic plating.

Fig. 1. Peter the Great St. Petersburg Polytechnic University Gidrokorpus-2.

Fig. 2 is a graph showing changes in heat fluxes throughout the entire study period. The actual heat transfer resistance has been calculated in the area of heat fluxes stationary thermal conditions ranging from 30.6 W/m^2 to 33.95 W/m^2. Further, the test was discontinued, since a sharp warming began from 19 January 2016, the outside air temperature became higher than minus 10 °C.

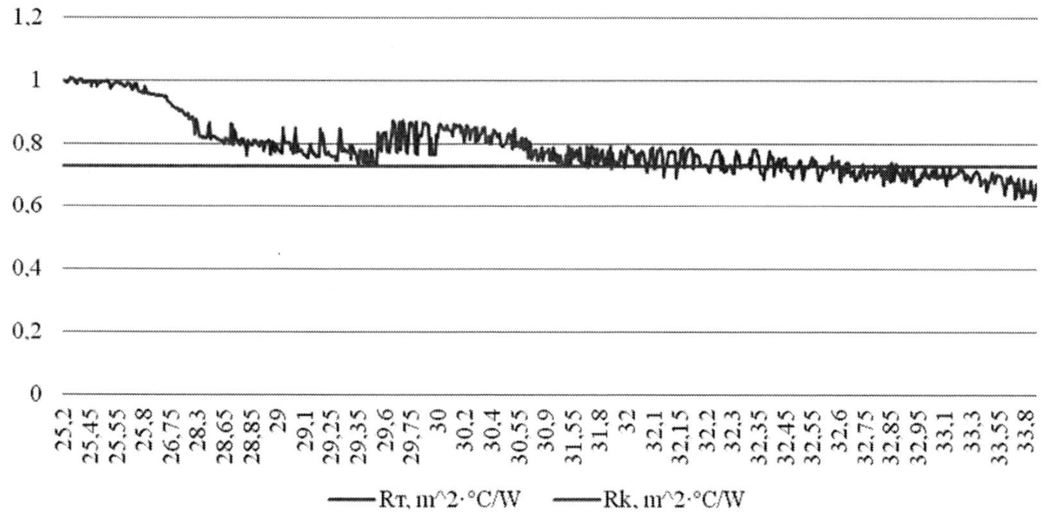

Fig. 2. Graph of heat flux changes.

Fig. 3 is a graph of temperature changes on the wall surfaces indoors (τ_1) and outdoors (τ_2).

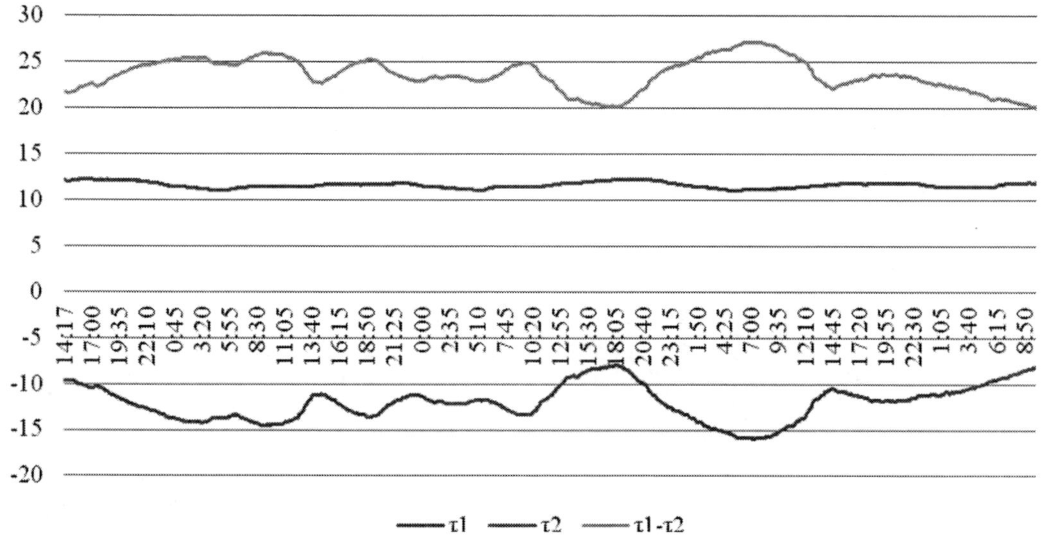

Fig. 3. Graph of temperature changes: τ_1 – the inner wall surface temperature, τ_2 – the outer wall surface temperature, τ_1-τ_2 – temperature difference.

In according with the standard [9] the building envelope R-value R_k is defined as the difference between the average temperatures of the building envelope inner and outer surfaces during the test period to the average heat flux density passing through the enclosure. According to what it is equal to 0.73 m^2·°C/W. Thermal resistance calculated by taking into account additional factors, equal to 0.89 m^2·°C/W.

For the purpose of comparison of the thermal conductivity actual values of the materials used in construction, with project values, masonry U-value was determined: λ_k = 0.7 W/m·°C.

4 Current regulatory requirements

According to the requirements [10, 12] in St. Petersburg the basic value of the reduced total thermal resistance of the building envelope is $R_0^r = 2{,}5\text{m}^2\cdot°\text{C/W}$, the normalized value of the reduced total thermal resistance of the building envelope is $R_0^{norm} = 1{,}57\text{m}^2\cdot°\text{C/W}$.

5 Comparative analysis of the façade solutions

Compare the walls R-values when warming by ETICS (External Thermal Insulation Composite System) – R_{ETICS}, and VFS (Ventilated Façade System) – R_{VFS}. (Fig. 4)

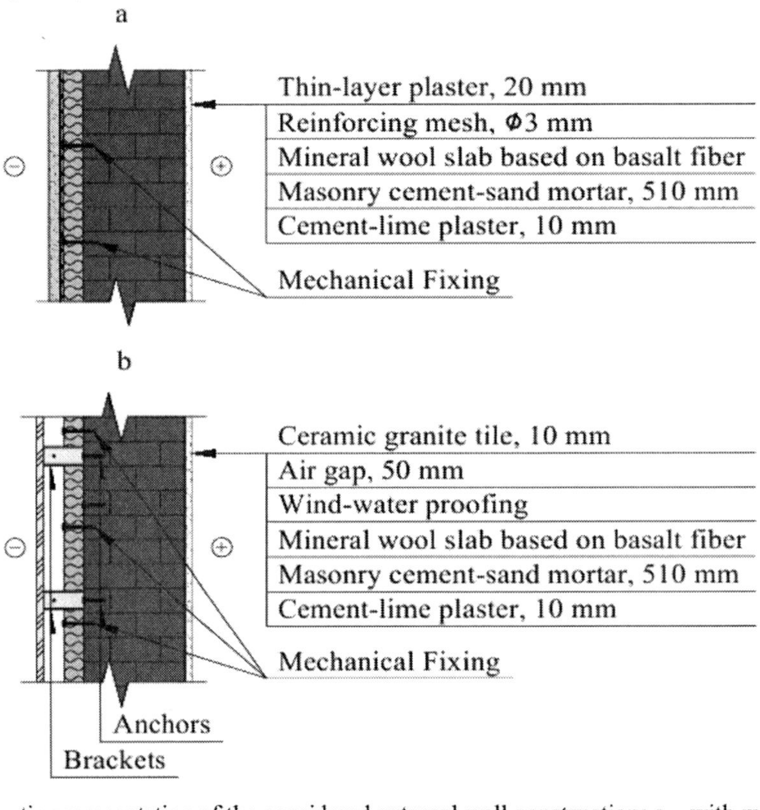

Fig. 4. Schematic representation of the considered external wall construction: a – with mineral wool thermal insulation with plaster finish; b – with a ventilated suspended facade system.

Temperature analyze of the wall with mineral wool thermal insulation based on basalt fiber ($\rho = 100$ kg/m^3 semirigid) with thin layer plaster finish (Fig. 4a) [13]:

1 – cement-lime plaster: $\lambda_1 = 0.87$ W/m·°C, $\delta_1 = 0.01$ m;

2 – brick masonry: $\lambda_2 = 0.7$ W/m·°C, $\delta_2 = 0.51$ m;

3 – mineral wool slab: $\lambda_3 = 0.045$ W/m·°C, $\delta_3 = 0.1$ m;

4 – protective layer from thin layer plaster: $\lambda_4 = 0.87$ W/m·°C, $\delta_4 = 0.045$ m.

Rated thermal resistance was determined according to the standard [13], taking into account the structural features of the facade systems.

The ETICS R-value is equal to $R_{ETICS} = 2.7$ m^2·°C/W > 1.57 m^2·°C/W.

Temperature analyze of the wall with suspended facade system with ventilated air-space and with mineral wool thermal insulation based on basalt (ρ = 100 kg/m^3 semirigid) (Fig. 4b):

1 – cement-lime plaster: $\lambda_1 = 0.87$ W/m·°C, $\delta_1 = 0.01$ m;

2 – brick masonry: $\lambda_2 = 0.7$ W/m·°C, $\delta_2 = 0.51$ m;

3 – mineral wool slab: $\lambda_3 = 0.045$ W/m·°C, $\delta_3 = 0.1$ m;

4 – air gap: $R_{a.g.} = 0$, because air-space is not closed;

5 – ceramic granite tile (heat-insulating quality of decorative screen is not taken into account).

The VFS R-value is equal to $R_{VFS} = 2.06$ m^2·°C/W > 1.57 m^2·°C/W.

6 Determination of transmission heat losses

The calculation of transmission heat losses during the heating period Q_{be} is performed by the overall heat transfer equation. The value of heat loss during the heating period:

– actual: $Q_{be} = 548.5$ Gcal/year;

– ETICS: $Q_{be\ ETICS} = 180.8$ Gcal/year;

– VFS: $Q_{be\ VFS} = 237$ Gcal/year.

Thus, the heat energy savings will be:

– $\Delta Q_{ETICS} = 367.7$ Gcal/year after ETICS application,

– $\Delta Q_{VFS} = 311.5$ Gcal/year after VFS application.

Methods of calculating capital costs for additional insulation, operating costs before and after thermal insulation, as well as the payback period of energy-saving measures are described in detail in the works [15-19].

Calculate the economic effect in relation to the two selected variants of facades heat insulation:

– $\Delta E_{ETICS} = 520663.2$ rub/year;

– $\Delta E_{VFS} = 441084$ rub/year.

Capital expenditure on insulation of external walls of the existing building:

– $\Delta C_{ETICS} = 14\ 742\ 323.8$ rub.

– $\Delta C_{VFS} = 40\ 983\ 113.0$ rub.

Simple payback period of insulation the outer wall of the existing building insulation with a thickness of 100 mm is:

– $T_{ETICS} = 28.3$ years;

– $T_{VFS} = 92.9$ years.

Note that the calculated payback period was obtained without taking into account:

– tariffs raising for thermal energy;

– interest on the loan (in the case of the application of loan funds to carry out events for the insulation of building external walls);

– discounting of achieved future cash flows as a result of implementation of energy-saving measures under consideration and of reduce the heat energy loss for heating.

For this reason, the calculated values of the projected payback period of the investment can be regarded only as estimates.

Heat tariffs increase annually. This means that the annual cost savings will increase with each subsequent year (heating period).

Taking into account the above-mentioned factors, the projected payback period of investment in additional facades insulation is determined by a logarithmic equation [15-20]. According to the equation the payback period of energy-saving measures was calculated taking into account the total capital expenditure for its implementation, the rising cost of tariffs for thermal energy (r), discounting future cash flows (i), achieved by cost-cutting as a result of the implementation of energy-saving measures.

Values r and i are fickle and change every year. Possible options for payback, depending on the changes of the variables are presented in Table 1 for ETICS and in Table 2 for VFS.

Table 1. Correlation analysis results of the payback period of ETICS taking into account the changes in the average annual rate of the rising cost of tariffs for thermal energy (r) and in the future cash flows discount factor on the key rate of the Central Bank of the Russian Federation (i).

i \ r	0.05	0.06	0.07	0.08	0.09	0.1	0.11	0.12	0.13	0.14	0.15	0.16
0.05	-	24.70	22.46	20.70	19.27	18.08	17.08	16.21	15.45	14.79	14.20	13.67
0.06	-	-	24.72	22.49	20.74	19.32	18.13	17.13	16.26	15.51	14.84	14.25
0.07	-	-	-	24.75	22.53	20.78	19.37	18.18	17.18	16.32	15.56	14.90
0.08	-	-	-	-	24.77	22.57	20.83	19.41	18.23	17.23	16.37	15.61
0.09	-	-	-	-	-	24.80	22.60	20.87	19.46	18.28	17.28	16.42
0.1	-	-	-	-	-	-	24.82	22.64	20.91	19.51	18.33	17.33
0.11	-	-	-	-	-	-	-	24.84	22.67	20.96	19.55	*18.65*
0.12	-	-	-	-	-	-	-	-	24.86	22.71	21.00	19.60
0.13	-	-	-	-	-	-	-	-	-	24.88	22.74	21.04
0.14	-	-	-	-	-	-	-	-	-	-	24.91	22.77
0.15	-	-	-	-	-	-	-	-	-	-	-	24.93
0.16	-	-	-	-	-	-	-	-	-	-	-	-

Table 2. Correlation analysis results of the payback period of VFS taking into account the changes in the average annual rate of the rising cost of tariffs for thermal energy (r) and in the future cash flows discount factor on the key rate of the Central Bank of the Russian Federation (i).

i \ r	0.05	0.06	0.07	0.08	0.09	0.1	0.11	0.12	0.13	0.14	0.15	0.16
0.05	-	63.84	51.91	44.41	39.17	35.26	32.21	29.75	27.71	26.00	24.53	23.26
0.06	-	-	63.99	52.08	44.59	39.34	35.42	32.36	29.90	27.85	26.13	24.66
0.07	-	-	-	64.13	52.25	44.76	39.51	35.59	32.52	30.04	27.99	26.27
0.08	-	-	-	-	64.28	52.42	44.93	39.68	35.75	32.67	30.19	28.13
0.09	-	-	-	-	-	64.42	52.59	45.10	39.84	35.90	32.82	30.33
0.1	-	-	-	-	-	-	64.56	52.76	45.27	40.01	36.06	32.97
0.11	-	-	-	-	-	-	-	64.70	52.92	45.44	40.17	*37.35*
0.12	-	-	-	-	-	-	-	-	64.83	53.09	45.61	40.33
0.13	-	-	-	-	-	-	-	-	-	64.97	53.25	45.77
0.14	-	-	-	-	-	-	-	-	-	-	65.10	53.41
0.15	-	-	-	-	-	-	-	-	-	-	-	65.24
0.16	-	-	-	-	-	-	-	-	-	-	-	-

To solve the problem of estimating the projected payback period of investment in energy efficiency magnitude r was adopted of the average over the last 5 years (16%), and the value i was for 2016 (11%).

In this case the payback period of investment will be:

– T_{ETICS} = 18.7 years;

– T_{VFS} = 37.4 years.

On results of calculations in Tables 1 and 2, it can be assessed how the investment payback period varies depending on the differentiation of the coefficient the average annual tariff raising for thermal energy (r) and the discount rate of future cash flows on the key rate of the Central Bank of the Russian Federation (i). It can also be concluded that the facade renovation of obsolete buildings is possible, if $r \geq i$, and is not justified, when $r < i$.

7 Results

According to the experience results:

– thermotechnical characteristics and envelopes of the building were evaluated with regulatory requirements [10]. Required design heat transfer resistance is $R_0^r = 2{,}5\text{m}^2\cdot°\text{C/W}$, actual heat transfer resistance is $R_0 = 0.89$ m$^2\cdot$°C/W;

– options for renovation of facades are considered in the most widespread ways. Heat transfer resistance values of facade systems were founded: $R_{0\ ETICS}$ = 2.7 m$^2\cdot$°C/W at ETICS, $R_{0\ VFS}$ = 2.06 m$^2\cdot$°C/W at VFS;

– transmission heat losses for heating were defined: actual Q_{be} = 548.5 Gcal/year; with ETICS $Q_{be\ ETICS}$ = 180.8 Gcal/year; with VFS $Q_{be\ VFS}$ = 237 Gcal/year;

– capital costs for performance of work upon insulation of administrative building facades and investment payback period for both insulation methods, taking into account the average annual tariff raising for thermal energy and discounting future cash flows on the key rate of the Central Bank of the Russian Federation were calculated: these are ΔC = 14 742 323,80 rub, T_{ETICS} = 18.7 years at ETICS; these are ΔC = 40 983 113,0 rub, T_{VFS} = 37.4 years at VFS.

8 Conclusion

By results of experience it was revealed that the actual condition of external walls of the academic building of Peter the Great St. Petersburg Polytechnic University does not meet current regulatory requirements for thermal protection. In order to eliminate the revealed discrepancy it is necessary to carry out additional work on the building envelopes renovation by one of the proposed methods of warming: ETICS or VFS. The least expensive method is SFTK, which is characterized by lower estimated payback period of investments. Its application can reduce the heat losses through the walls by about 57%

Implementation of the aforementioned recommendation will allow to bring closer the building into compliance with modern requirements for thermal protection, which will increase the building energy performance class.

References

1. V.I. Livchak, AVOK, **3**, 62-67 (2010)
2. V.I. Livchak, AVOK, **6**, 14-21 (2012)
3. N.I. Vatin, D.V. Nemova, P.P. Rymkevich, A.S. Gorshkov, Mag. of Civ. Eng, **8**, 4-14, (2012)
4. A.S. Gorshkov, D.V. Nemova, P.P. Rymkevich, Roof. and Ins. Mat., **2**, 34-39 (2013)
5. A.S. Gorshkov, D.V. Nemova, P.P. Rymkevich, Energ. Sav., **2**, 26-32 (2014)
6. N.I. Vatin, D.V. Nemova, A.S. Gorshkov, Build. Mat., Eq., Tech. of XXI cen., **1**, 36-39 (2013)
7. *SNiP II-A.7-62* Therm. Eng. Des. Reg.*

8. *SNiP II-A.6-62 Build. Clim. and Geoph.*
9. *GOST R 54853-2011 Build. and. Cons.*
10. *SP 50.13330.2012 Therm. Prot. of Build.*
11. *SP 131.13330.2012 Build. Clim.*
12. *GOST 30494-2011 Res. and Pub. Build.*
13. *GOST R 53786-2010 Therm. Ins. Com. Fac. Sys.*
14. *GOST R 54851-2011 Cons. Build. Env. Het.*
15. D.V. Nemova, N.I. Vatin, A.S. Gorshkov, A.V. Kashabin, P.P. Rymkevich, D.N. Tceitin, Cons. of Un. Build. and Str., **8**, 93-115 (2014)
16. L. Svatovskaya, A. Sychova, M. Sychov, V. Okrepilov, MATEC Web of Conferences, **53**, Article Number 01023 (2016)
17. A.S. Gorshkov, N.I. Vatin, P.P. Rymkevich, AVOK, **3**, 32-36 (2014)
18. A.A. Romanova, P.P. Rymkevich, A.S. Gorshkov, Tech. and Tech. Ser. Pr., **4**, 68-74 (2014)
19. A.S. Gorshkov, MSUCE, **12**, 136-146 (2015)
20. A.S. Gorshkov, V.I. Livchak, *Building envelope: review and analysis of the regulatory requirements* (Lap Lambert Academing publishing, Saarbruecken, 2015)

Energy-Efficient Features to Upgrade Attic Slabs of Historical Buildings (Case Study of Saint-Petersburg)

Vera Murgul[1,*]

[1]St. Petersburg State Polytechnical University, Politekhnicheskaya, 29, Saint-Petersburg, 195251, Russia

Abstract. Thermal insulation of attic slabs is a conventional measure during the energy-efficiency improvement of residential buildings. A specific problem is warmth-insulation of attic slabs in historical buildings, which are the objects of cultural heritage, and for them the main modernization rule is the preservation of historical interiors (ceiling is a protection subject). The article exposes design of reconstruction of attic slabs while preserving the existing historical ceiling and warmth-insulation design is proposed.

1 Introduction

Traditionally, energy efficiency improvement of old buildings includes warming of attic slabs (in case when the mansard is absent). The historical building in St. Petersburg of pre-industrial development period was chosen as the object of this study.

The peculiarity of the historical buildings in St. Petersburg is the existence of protected interiors, particularly ceilings with rich stucco decoration. Reconstruction of the attic slab in such buildings is hampered by the need of preserving historical ceiling. At the same time load bearing capacity of wooden beams in historical attic slabs is already exhausted.

Studying the Energy Performance of Historic Buildings is presented in the number of studies [1-5]. Review of structures and materials of attic floors in apartment houses of St. Petersburg, which were used for the construction in different periods, is given in [6, 7]. As a general rule, for now, a significant part of historical buildings of pre-industrial period have already been reconstructed with partial or complete replacement of the slabs due to material wear.

Numerous technical surveys of pre-industrial period buildings, which were performed by the author, revealed the diversity and uniqueness of the designs made in 18-19th centuries in St. Petersburg. Nevertheless, it is possible to allocate the established basic designs of floors in historical buildings of St. Petersburg in 18-19th centuries.

This article will expose the main types of attic slabs in historical buildings before and after renovation due to material wear.

The project of reconstruction including warmth insulating of attic slabs in historical building while preserving the invaluable ceiling was proposed.

*Corresponding author: october6@list.ru

2 Features of attic slabs in historical pre-industrial period buildings of Saint-Petersburg

A survey of structures in old residential housing was conducted in Saint-Petersburg during 1968-1972 by the means of special Lenzhilproekt institute methodology [8].

The historical St. Petersburg residential buildings constructed before 1915 can be divided on five types of floors (Table 1) [8].

Table 1. Types of slabs in historical residential buildings

Floor type	Area, [million m^2]	The average standard terms of service, [years]
1. Plastered wood	13	60
2. Plastered wood on steel beams	3.5	80
3. With brick and concrete filling on steel beams	0.5	125
4. Circular brick vaults	0.5	-
5. Precast, made of reinforced concrete, set after complex overhaul	4.5	125

Note: Survey data for 1931. Concrete floors are made during complex overhaul.

2.1 Wooden slabs

Wooden slabs can have diverse cross-section of beam: beams with a selected slot (Fig. 1a), with the selected "skull" (Fig. 1b), hewn on two edge for the attic slabs (Fig. 1, c). [7]

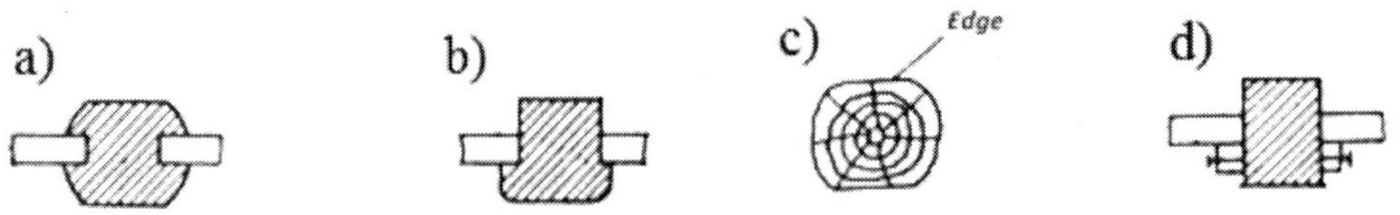

Fig. 1. The cross sections of wooden slabs beams. a) beam with a selected slot; b) a beam with hewn "skulls"; c) beam with two cut corners; d) from a of timber bar nailed with cranial bars

"Grease" is used for heat and steam insulation of ceilings. The lubricant must be lightweight in order to not burden the beam, refractory to serve as at least some obstacle to the rapid spread of fire from the upper floors to the bottom, and finally, should not contribute to the decay or deterioration of beams and subfloors [9]. Lubricant is based on clay and brick. To maintain the lubricant there is subfloor. Wooden subfloors can be divided to sectional and laid ones (Fig. 2). They have different beam joining.

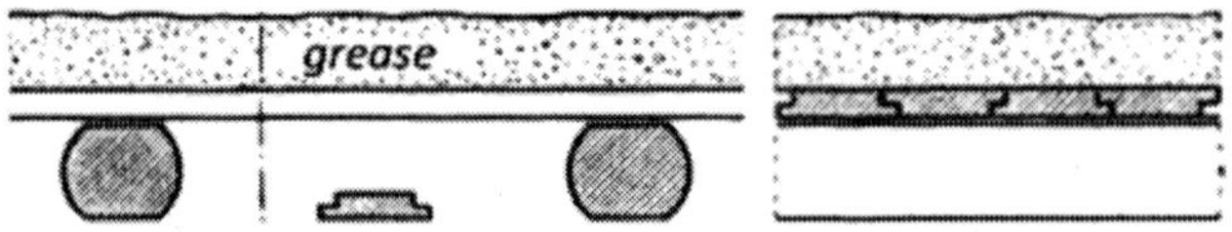

Fig. 2. Attic slab design with laid subfloor

Laid subfloors are arranged of 6-inch wooden boards, set entirely on the beams and combined in the quarter. To install sectional subfloors (or conjunctions) skulls are cut in the beams or wooden bars are nailed to them.

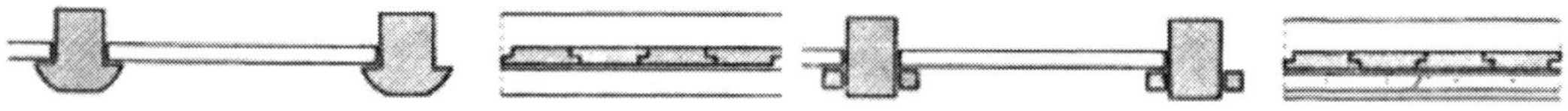

Fig. 3. Floor design with sectional subfloor.

Debris were put in a layer of 8-10 cm directly over the subfloor conjunction. When wooden bars began to be used as beams, the connection of boards in conjunctions was made back to back and covered with one layer of roofing felt before filling with bebris.

Rarely met hardwood slabs of separate structure where the load from the the ceiling structure and the floor is perceived by the individual supporting beam systems. The height of these slabs is more than 0.5 meters (Fig. 4).

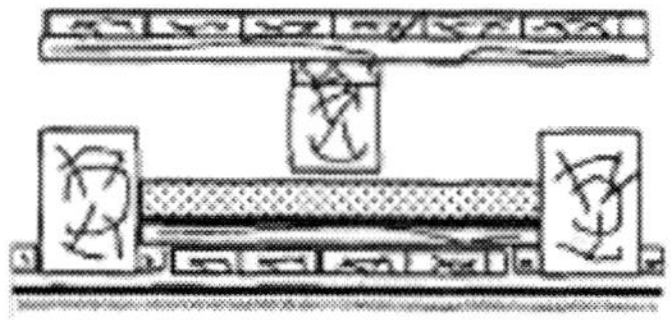

Fig. 4. Wooden slabs of separate structure

Brick grease was used for thermal insulation of the attic space. The kneaded clay layer of 2-4 cm was pressed by a half-brick, so that individual bricks abut each other. The clay squeezed during that filled joints.

After drying seams for filling the voids were spilled with lime liquid mortar. Area of wooden floors in the old residential houses in St. Petersburg at the beginning of the 80s. of XX century was about 13 million m^2, about 2% of this area have valuable architectural decoration.

2.2 Floors on steel beams

Table 2 lists the specifications which are affecting the heat-insulation parameters of building envelope and which are required for builders during installation of steel beams. There were also numerous recommendations, which in practice were often neglected. For example, recommendations said to plaster the ceiling over the hem. But in houses with cheap apartments boards in conjunctions were cut, then sutured with shingles and plastered ceilings without the hem. It all depended on what goals the owner of the house pursued and of a durability presumed during construction. Many old houses can not be repaired because of the condition of main structures.

Table 2. Technical Requirements for overlaps on the metal beams

Requirements	Causes
Minimum sealing of the ends of beams with spans up to 4 m - 15 cm; 6 m - 18 cm; up to 9 m - 20 cm	These dimensions provide a reliable connection to the support beam
The minimum distance from the beam end to the front surface of the outer wall should be not less than 40 cm	Such distance prevents end of the beam from freezing
When making attic overlaps setting of laid floors is required	Setting laid floors in attics with steel beams is more important than with wooden ones, because board row and a grease layer prevent steel from freezing

Thus, considering historical buildings of pre-industrial period as an object of study, 3 main types of attic floors can be considered: wooden floors on the wooden beams, wooden floors on steel beams, stone vaulted ceiling on steel beams .

3 The results of field surveys of historical buildings

The building at the following address was surveyed: St. Petersburg, Nevsky avenue, 41 (palace of Beloselsky-Belozersky) [10]. Figures 5, 6.

Year of construction - 1800; year of reconstruction – 1848.

Attic floor of mixed type.

- On the metal beams of the I-sections № 18, 22 (b = 18.98 mm), packed in increments of 90-100 cm. Insulation is a clay-brick grease on the wooden filling and construction debris on the concrete filling. Wooden boards filling and concrete vaulted filling between the beams;

- On the wooden beams made of hewn logs with a diameter of 20-40 cm and beams with cross-section 7x15 cm, packed in increments of 70-150 cm: insulation - clay-brick grease on laid floor. Part of the beams are one end embedded in the masonry of interior walls, the other end is rested on the beams of hewn logs with a diameter of 32 cm, arranged along the exterior walls.

Fig. 5. Attic overlap on steel beams.

Fig. 6. Attic overlap on wooden beams; insulation - slag, construction debris

Administrative building in Vasilyevsky District, St. Petersburg was surveyed. [7] Time of construction is before 1917. In 1989 the building was overhauled with rearrangement of overlaps. The site is located within the boundaries of regulating built-up areas and economic activity of objects of cultural heritage.

The attic floor is flat on the metal beams, arranged with 800-900 mm increment in the axes 2-3 / A-I and 1100-1200 mm in axes 6-7 / A-I, with a filling of small-sized concrete ribbed slabs (Fig.7).

Floor beams in different spans of the building are I-section beams №22, №30; №24 [8] and the compound welded beams which consist of channel №24 and 63x6 corner; compound welded beams made of channel №22 and 63x6 corner.

The space between the beams is filled with expanded clay gravel and slag. In some places above the concrete ribbed slabs vapor barrier was designed made of bitumen-based materials and mineral wool.

The layers of the attic slab:

- Filling with expanded clay with density $\rho = 900$ kg / m3; thickness $\delta = 0.17$ m;
- Concrete ribbed slabs - $\rho = 900$ kg / m3; $\delta = 0.06$ m;
- Plaster - $\rho = 1600$ kg / m3; $\delta = 0.03$ m.

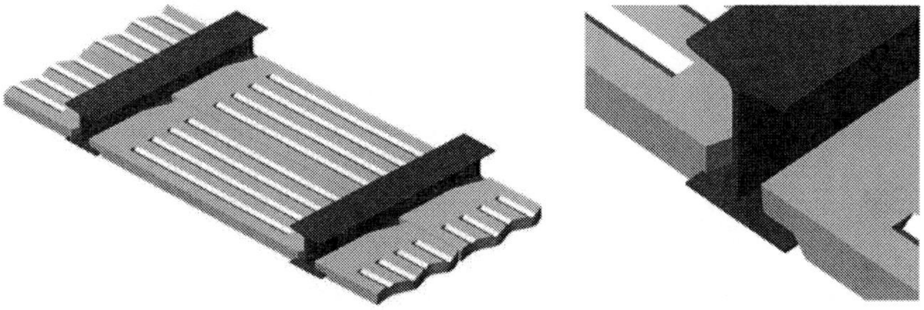

Fig. 7. Reinforced concrete ribbed slabs on steel load-bearing structures

The attic slab above the staircase is presented as a lightweight vault which is 1 m height, made of hollow ceramic tubes shaped as corrugated cylinder (Fig. 8). The gaps were filled with the debris (slag, broken bricks, expanded clay gravel, etc.), with density ρ = 1200 kg / m3.

Fig. 8. Vault overlap over the staircase

As shown by field researches, wooden beams are often in good condition and can be preserved. In the opposite case, the reconstruction of overlapping is made with full or partial replacement of the load-bearing structure.

4 Reconstruction of the floors of historical buildings with preservation of the ceiling

We choose the worst case for the study, when the bearing capacity of the wooden ceiling is exhausted, but its replacement is not possible because of the need to preserve the historical ceiling (as part of the interior in the protected building - object of cultural heritage).

The design choice may be to preserve combustible ceiling and create reinforced concrete diaphragm above it which is 120 mm thick. Firstly, the old filling and insulating layers of strainer are removed, then wooden boards are laid on old beams (as permanent formwork) and monolith concrete slab-diaphragm is made. During monolithic works it is necessary to ensure an insertion of steel anchors (for the suspension of the old wooden beams of the ceiling) and overall stability of the entire structure (additional supports are made while mounting if necessary). By the means of anchors the load from combustible overlap is transmitted to the reinforced concrete slab. The design is shown in Fig. 9 [11].

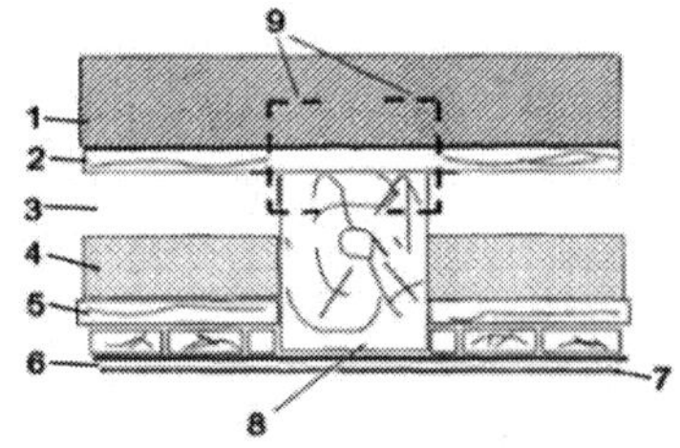

1 – concrete slab - 120 mm;
2 – formwork of boards - 34 mm;
3 – air gap - 120 mm;
4 - filling (sand, crushed bricks) + insulating layer (clay coating) - 80 mm;
5 – double-layered boards - 80 mm;
6 - supporting boards - 25 mm;
7 – plaster on lath - 25 mm;
8 – wooden floor beams 200x280 mm;
9 – steel anchors.

Fig. 9. The design of overlap reconstruction with the preservation of valuable ceiling

5 Thermal calculation of warming layer (for the climatic conditions of St. Petersburg)

For calculation following conditions were determined: internal air temperature +20 °C, outside air temperature -24 °C, the number of degree-days in heating season is 4537 °C • d, required thermal resistance of the attic slab is determined as 4.4685 m2 • °C / W.

The values are taken as for roof structure. The calculation considers outdoor temperature, i.e. the "n" factor was not considered.

The calculation is performed according to CP 50.13330.2012 "Thermal performance of the buildings". According to this newly integrated document, the outdoor air temperature during the heating period and the duration of the heating period is changed, so the number of degree-days of heating period changed too (CP 131.13330.2012 "Building climatology").

DDHS (degree-days in heating season) = (20+1.3) •213 = 4537;

R_{req}= 0.0005•4537+2.2 = 4.6585 – for roof structure;

R_{req}= 0.00045•4537+1.9 = 3.94 – for attic overlap.

The temperature field of the original design is shown in Fig. 10. Thermal resistance of the attic overlap with the concrete diaphragm which is 120 mm thick and formwork thickness 34 mm – 1.283 m^2·°C/W.

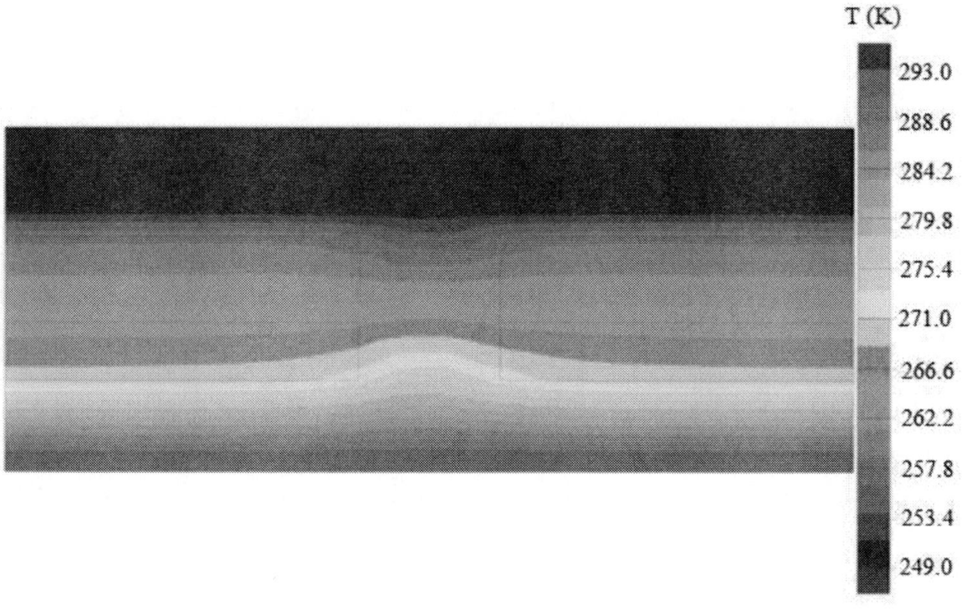

Fig. 10. Thermal field. Initial design.

As the insulation rock wool plates were selected, with density of 180 kg / m3 and 160 mm thick. The temperature field is shown in Fig. 11. The thermal resistance of the attic slab with insulation – 4.617 m2 • ° C / W.

Fig. 11. Thermal field of attic slab with insulation.

6 Conclusions

The maximum preservation principle of historical structures of pre-industrial period buildings should be fundamental during the energy modernization of the old housing. Resolved design of reinforcing attic slab with its subsequent insulating allows the use of modern, efficient insulation materials without the risk of condensation in the layers of preserved wooden floor, which often occurs during the immediate installation of effective heat insulation on existing wooden slabs.

References

1. E. Arumägia, M. Mändelb, T. Kalamees, Energy Procedia. **78,** 1027 – 1032 (2015)
2. C. D. Şahin, Z. D. Arsan, et al. Energy and Buildings, **96**, 128-139 (2015)
3. L. L. Pisello, A. Petrozzi, V. L. Castaldo, F. Cotana, Applied Energy, **162**, 1313-1322 (2016)
4. F. Ceroni, F. Ascione, R. F. De Masi, et al. Energy Procedia. **75**, 1325-1334 (2015)
5. P. Leardini, M. Manfredini, M. Callau, Energy and Buildings. **95**, 211-218 (2015)
6. K. Fabbri, B. Brunetti, Energy Procedia, **78**, 1281-1286 (2015)
7. S.G. Golovina, S.V. Sementsov, *Istoriya razvitiya konstruktsiy zdaniy zhiloy istoricheskoy zastroyki na primere Sakt-Peterburga*, (SPbGASU, Saint-Petersburg, 2012)
8. G.T. Popov, L.Ya Burak, *Tekhnicheskaya ekspertiza zhilykh zdaniy staroy zastroyki*, (Stroyizdat, Leningrad, 1986)
9. V. Statsenko. *Chasti zdaniy. Grazhdanskaya arkhitektura*, (Gosudarstvennoe izdatel'stvo, Leningrad, 1930)
10. Information on: http://dl.unilib.neva.ru/dl/694.pdf
11. D. A. Slepukhin, PhD thesis. Moscow, 2001, 154 p.

Optimization of Energy Performance of Windows by Applying Self-Adjustable Shadings

Vadim Starkov[1], *Petr* Ovchinnikov[1,*] *and Temur* Dzampaev[1]

[1] Peter the Great St.Petersburg Polytechnical University, Polytecknycheskaya st., h. 29, St. Petersburg, 195251, Russia

Abstract. The paper studied the thermo-physical properties of the windows with self-adjustable jalousie and consequent cooling load reduction for the buildings in 60-degree North latitude throughout the year. We focused on using different types of innovative materials and window structures and specified possible ways of improvement for modern and conservative windows. The paper discussed main factors that induce choosing specific window type. First, we investigated quantity of solar radiation energy that is taken into account while calculating the cooling load. Then this dependence was analyzed with different scenarios for specific window structures facing to corresponding cardinal direction. The results provide useful information for optimizing the thermal response of an enclosure to periodic solar heat gains.

1 Introduction

Fenestration has always been regarded as the weak element in the building envelope due to its role in determination of the energy balance of a building. At the same time, cooling performance is becoming a significant parameter for evaluating the energy performance of buildings [1]. Windows bring air and light into a room and connect the room with the world outside [2, 3].

This happens since thermal transmittance of the ordinary toughened glass is just 0.76, $\left[\frac{W}{m^2K}\right]$, when the thickness of the glass is 6 mm. Heat transfer coefficient of 6 mm thick ordinary toughened glass is 5.80, $\left[\frac{W}{m^2K}\right]$. Even though heat transfer coefficient of the 24 mm thick low-emissivity insulating glass is 1.80, $\left[\frac{W}{m^2K}\right]$, it is still much higher than the same coefficient for a simple wall with insulation layer [4,5].

* Corresponding author: pshenichca@hotmail.com

Table 1. Thermophysical properties of glasses.

Glass Type	U Value, $\left[\frac{W}{m^2 K}\right]$	Emissivity rate
Single Clear Glass	5.80	0.8400
Low E Single Glass	3.70	0.0600
Low E Double Glazing	1.80	0.0036

Numerous experiments and building inspections clearly shows that the cause of the biggest thermal losses are window structures. Even if total area of windows in building is significantly lower than the area of walls, the share in total thermal losses usually lies in the rate of 25-40 % [6]. The distribution of total thermal losses of a building during the heating season can be shown as following:

1. Losses through external walls – 23-25 %;
2. Losses through windows – 25-40 %;
3. Losses through infiltration – up to 40 %.

On the opposite side, during the cooling season window structures should operate without creating unreasonable heat gain in order to reduce possible negative influence on the air-conditioning system [4]. Thus, improving window structures can optimize overall energy performance of a building in demand for cooling and heating. A distinct improvement in comfort can be achieved by selecting a beyond-compliance solution to glazing for a new or retrofitted residential, commercial or institutional development. The main factors are dimensions, orientation of the windows, amount of shading and type of glass that is used.

2 Determining of Solar Radiation Heat Influence

The rigorous calculation of the cooling load due to solar heat gains through the glazing is quite complex, as it involves several phenomena [4]. Incident global solar radiation includes direct beam, sky, and ground-reflected radiation as it shown on the Figure 1.

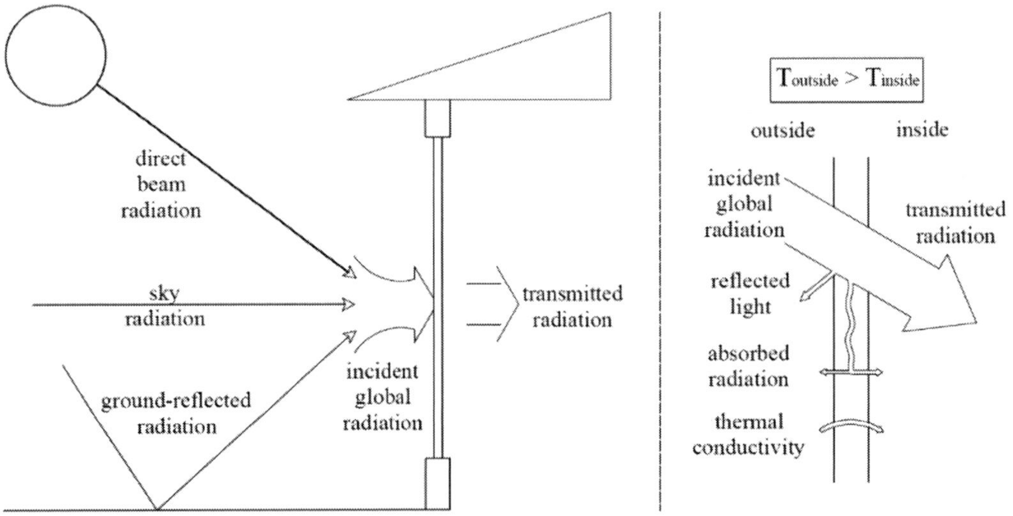

Fig. 1. Solar radiation and transmitting heating process.

The solar radiation transmitted into a living space is less than the radiation that strikes the outside of a window, because of the losses associated with radiation reflected off and absorbed by the glass [7]. Transmitted radiation depends on the emissivity level that crucially differs for conventional single-glass window and modern low-e glass (that is shown in Table 1).

Ultimately, there are number of ways to improve energy performance of window structures, some of them are:

1. Low emissivity glass (low-E glass), has a high transmittance in the visible region, a high reflectance in the near infrared region and a low emittance in the far infrared region. It has been conventionally used in energy efficient windows because it not only prevents near infrared radiation (known as a source of heat) from penetrating the glass, but also blocks the emission of far infrared radiation of building interiors [8,9].

2. Electrochromics (EC) are glazings whose optical transmittance can be controlled via a small applied electric potential. This is potentially useful in managing the admission of daylight into buildings, since they can be darkened when too much of it causes glare and be made lighter when more is needed to illuminate the interior space. With hourly blind position adjustments, electrochromics showes significantly higher (62% and 53%, respectively without and with overhang) lighting energy consumption than clear glass [10-12].

3. Applying self-adjustable external or internal shadings or jalousie may also improve heat resistance of windows. Using these blinds in conventional way inside the building is less effective since they block solar radiation after it transmits through glazing. These blinds are more effective outside a building but cost of exploiting such a system is higher than of the same system that is placed in-between two layers of glass [13].

Furthermore, self-adjustable shadings in-between external and internal glasses may significantly improve conventional windows in achieving better results in energy performance during both heating and cooling season.

3 Structure of a Window with Self-Adjustable Shading

Construction of a window is more complex than of a wall. Window consists of light, sealing profile with chambers, insulation and gas. The typical width of window frame is 125 mm. Free space between glasses is already being used for manually adjustable jalousie in a few European countries [6,14] but putting this scheme together with automatization system, self-powered by solar radiation, we can get the structure, that is shown on the Figure 2.

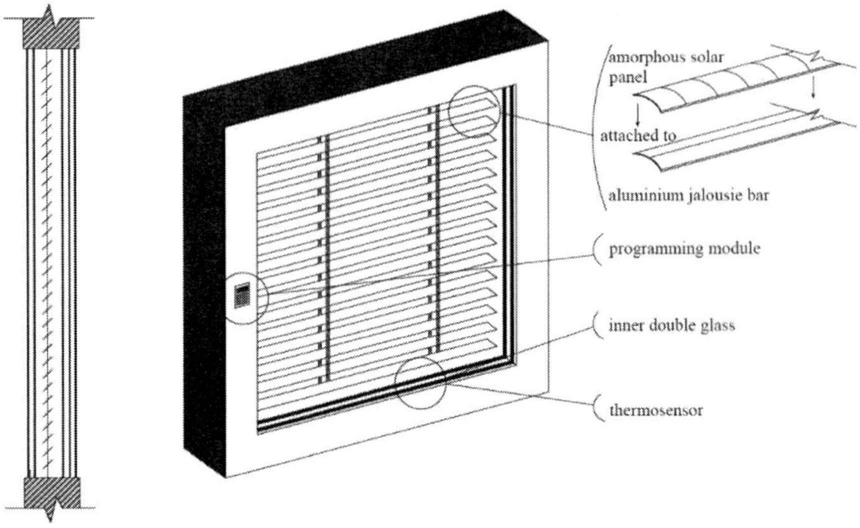

Fig. 2. Schematic image of the structure.

To make them self-adjustable we can attach amorphous solar panels to the jalousie bars, that will produce energy to meet the needs of the whole structure [15]. Jalousie can be adjusted with respect to temperature inside and outside which will be determined by two thermal sensors inside and outside of a frame. These panels will change their inclination to block extra heat gain or pass it free depending on temperature and irradiance. At the same time, the internal part of jalousie has to be covered with heat reflecting membrane to keep heat inside of a building during night times.

4 Method of Determining the Energy Output and Heat Blocking Rate

To estimate energy output of our system the following procedure was applied:

1. In order to assess the feasibility of solar installations the analyses of solar radiation has been conducted on the differently oriented surfaces with geographic 60^0 North latitude [16]. Getting data of total radiant exposure H, $\left[\frac{W \cdot h}{m^2}\right]$[17], we can calculate irradiance rate E, $\left[\frac{W}{m^2}\right]$, for every month and cardinal direction using Formula 1:

$$E = \frac{H}{t}, \quad (1)$$

where t – hours of daytime during the month, [h].

Table 2. Irradiance rate.

Month	Hours of daytime, [h]	Irradiance on a vertical surface, $\left[\frac{W\cdot h}{m^2}\right]$				
		N	NE/NW	E/W	SE/SW	S
Jan	6	-	-	102	371	505
Feb	8	-	-	192	446	598
Mar	8	-	133	347	617	734
Apr	12	86	183	383	481	528
May	18	91	163	272	323	294
Jun	18	122	213	287	298	263
Jul	18	109	179	276	285	272
Aug	12	97	196	360	446	441
Sep	12	-	139	275	428	469
Oct	10	-	68	187	416	523
Nov	8	-	-	125	342	475
Dec	6	-	-	97	269	441

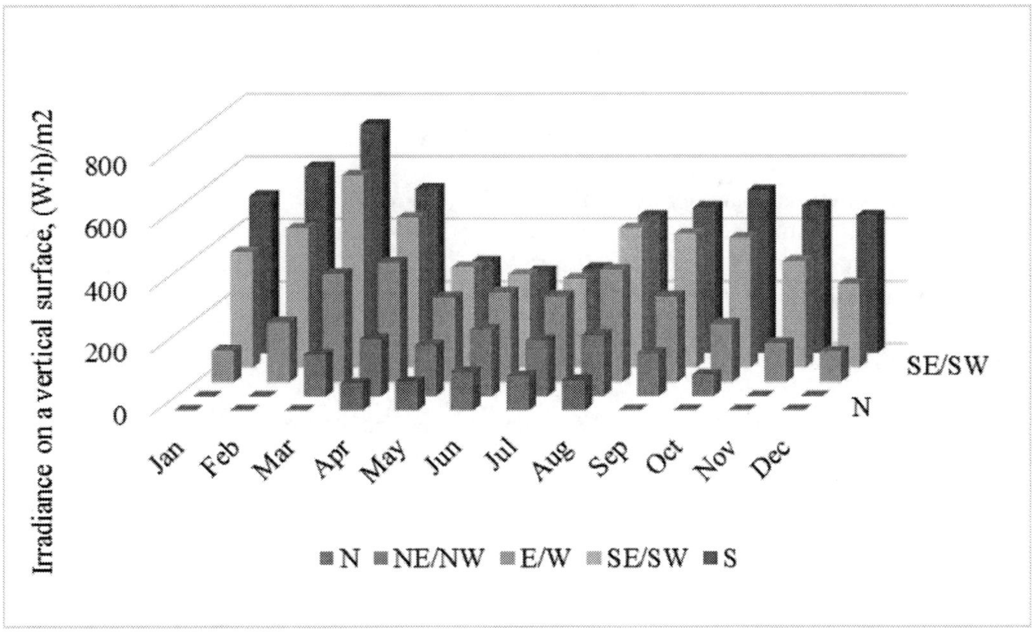

Diagram 1. Irradiance on a vertical surface.

2. The particular amorphous flexible solar panels made of steel foil and active ceramic membrane with the following characteristics were chosen:
   a. dimensions of a panel are 180 x 39, [mm];
   b. voltage 1.5, [V];

c. maximum output power for 1 element is 0.45, [W].

Maximum output of a single solar panel, [W], depends on solar irradiance rate E, $\left[\frac{W}{m^2}\right]$, and can be shown on the Diagram 2 [18]:

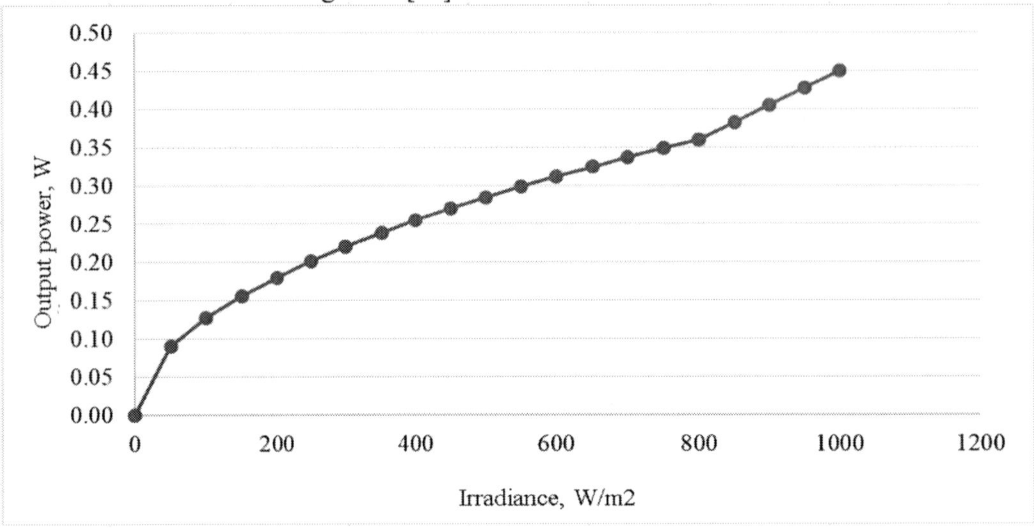

Diagram 2. Output power of solar panel.

According to the Diagram 2, if the irradiance rate is less than 800 $\left[\frac{W}{m^2}\right]$, the Formula 2 can be used for energy output of 1 vertical panel:

$$P_0 = \sqrt{0.000162 \cdot E} \cdot t, \tag{2}$$

where P_0 – output power for 1 solar panel per day, $[kW \cdot h]$; E – average irradiance rate per day, $\left[\frac{W}{m^2}\right]$,
t – number of daytime hours per day, [h].

Giving to emission rate of single clear glass at 0.84 data from the Diagram 1, Table 2 and Table 3, we can calculate mean output power for 1 m^2 of solar panels. Since 1 m^2 of panels consists of 142.45 panels, and they will be inclined by 45^0. The Formula 3 is used:

$$P = 142.45\sqrt{0.000162 \cdot E \cdot K \cdot 0.84} \cdot t, \tag{3}$$

where P – output power for 1 m^2 of solar panels per daytime, $[kW \cdot h]$; K – panel inclination coefficient, 0.84 – emissivity rate of a single glass.

To estimate the heat-blocking rate of the glazing we applied the following procedure:

1. the reflected solar heat for different types of structures was estimated by the Formula 4:

$$= r \cdot E \cdot t, \tag{4}$$

where R - reflected solar heat by 1 m^2 of glazing, $\left[\frac{kW \cdot h}{m^2}\right]$, r – reflection coefficient.

2. To estimate the reflected solar heat by 1 square meter of simple glazing with jalousie Formula 4 is to be rewritten to the Formula 5:

$$R = 0.16 \cdot E \cdot t + 0.84 \cdot E \cdot t \cdot \sin(\alpha), \quad (5)$$

where α – inclination of panels, 0.16 – reflection coefficient for simple glazing.

5 Results and Discussion

Accordingly, to 1 m^2 of attached solar panels we can get the following numbers for the angle equal to 45^0 [19].

Table 3. Mean output power for 1 m^2 of solar panels per day (45^0 inclination).

Month	Panel inclination coefficient	Mean output power for 1 m^2 of solar panels per day (45^0 inclination), $[W \cdot h]$				
		N	NE/NW	E/W	SE/SW	S
Jan	0.835	-	-	92	175	205
Feb	0.886	-	-	173	264	306
Mar	0.996	-	153	247	330	359
Apr	1.286	210	306	443	496	520
May	1.537	355	474	612	666	636
Jun	1.637	423	558	648	661	621
Jul	1.600	396	506	629	639	624
Aug	1.444	236	336	455	506	503
Sep	1.215	-	259	364	455	476
Oct	1.018	-	138	229	342	383
Nov	0.896	-	-	141	233	274
Dec	0.826	-	-	89	149	190

Looking at the data from this table, we can notice the following:

1. System meets their demands in energy;
2. Applying this system on the northern facades is unreasonable;
3. Excess power can be used for the lighting system inside;
4. The most proper orientation in terms of gaining energy is south.

Applying Formula 5 we calculated the heat-blocking rate. The results are represented in the Table 4.

Table 4. Reflected solar heat by 1 square meter of glazing per day.

Month	Clear glass, $\left[\frac{kW \cdot h}{m^2}\right]$					Low-e glass, $\left[\frac{kW \cdot h}{m^2}\right]$				
	N	NE/NW	E/W	SE/SW	S	N	NE/NW	E/W	SE/SW	S
Jan	-	-	0.10	0.36	0.49	-	-	0.58	2.09	2.85
Feb	-	-	0.25	0.57	0.77	-	-	1.44	3.36	4.50
Mar	-	0.17	0.44	0.79	0.94	-	1.00	2.61	4.64	5.52

Month	Clear glass, $\left[\frac{kW\cdot h}{m^2}\right]$					Low-e glass, $\left[\frac{kW\cdot h}{m^2}\right]$				
	N	NE/NW	E/W	SE/SW	S	N	NE/NW	E/W	SE/SW	S
Apr	0.17	0.35	0.74	0.92	1.01	0.97	2.07	4.32	5.42	5.95
May	0.26	0.47	0.79	0.93	0.85	1.55	2.76	4.61	5.46	4.97
Jun	0.35	0.61	0.83	0.86	0.76	2.07	3.60	4.86	5.05	4.45
Jul	0.32	0.52	0.80	0.82	0.79	1.85	3.03	4.67	4.82	4.61
Aug	0.19	0.38	0.69	0.86	0.85	1.09	2.21	4.06	5.03	4.97
Sep	-	0.27	0.53	0.82	0.90	-	1.57	3.10	4.83	5.30
Oct	-	0.11	0.30	0.67	0.84	-	0.64	1.76	3.91	4.91
Nov	-	-	0.16	0.44	0.61	-	-	0.94	2.57	3.57
Dec	-	-	0.09	0.26	0.42	-	-	0.55	1.52	2.49

Low-e glazing works efficiently during the cooling season, when no extra heating is required. During the heat season, the heat it reflects can be useful. By using adjustable jalousie we can improve the performance of the structure.

Table 5 shows the results of simple use of this automatic system – 45-degree closure during the cooling season and completely open during heating season. Even this simple method may improve the overall performance of glazing.

Table 5. Reflected solar heat by 1 square meter of glazing per day.

Month	Clear glass and jalousie , $\left[\frac{kW\cdot h}{m^2}\right]$				
	N	NE/NW	E/W	SE/SW	S
Jan	-	-	0.098	0.356	0.485
Feb	-	-	0.246	0.571	0.766
Mar	-	0.170	0.444	0.790	0.939
Apr	0.165	0.352	0.736	0.923	1.013
May	1.240	2.213	3.697	4.378	3.989
Jun	1.659	2.890	3.896	4.046	3.569
Jul	1.484	2.432	3.746	3.867	3.697
Aug	0.876	1.775	3.259	4.037	3.989
Sep	-	1.257	2.488	3.870	4.247
Oct	-	0.108	0.299	0.666	0.836
Nov	-	-	0.160	0.437	0.608
Dec	-	-	0.093	0.258	0.423

Higher heat blocking during summer could be achieved. Table 6 shows the difference between heat reflected by clear glass and the system with jalousie within a day. The calculation Formula 6 is used:

$$\Delta = R_1 - R_2, \tag{6}$$

where R_1 - reflected solar heat by 1 square meter of clear glass and jalousie per day, $\left[\frac{kW \cdot h}{m^2}\right]$; R_2 - reflected solar heat by 1 square meter of clear glass per day, $\left[\frac{kW \cdot h}{m^2}\right]$;

Table 6. Blocking rate of transmitted solar radiation.

Month	Difference between reflected solar heat in accordance with surface orientation, $\left[\frac{kW \cdot h}{m^2}\right]$				
	N	NE/NW	E/W	SE/SW	S
May	0.977	1.744	2.912	3.449	3.142
Jun	1.307	2.277	3.069	3.188	2.811
Jul	1.169	1.916	2.951	3.046	2.912
Aug	0.690	1.399	2.567	3.181	3.142
Sep	-	0.990	1.960	3.049	3.346

Due to reduction of transmitting solar heat by 73.5 %, cooling load decreases by 24 %. The total improvement depends on specific dimensions and parameters of room.

Using advanced approaches, simulation of a computer model or construction of a prototype combined with thermography [20], may entail even higher result. Analyses provided by P. Pfrommer [21] can also be applied for more elaborated results.

6 Conclusion

Series of studies have been made, and according to data presented in the article, the following statements were proposed:

1. self-adjustable heat-blocking jalousie on the base of solar panels are feasible;
2. extra solar energy gained by solar panels may be used for powering inner light;
3. the most proper orientation for this system, in terms of gaining energy, is south;
4. using low-e glass is more effective way to block solar heat but it cause unreasonable heat blocking during the heating season;
5. applying self-adjustable heat-blocking jalousie with low-e glass improve heat blocking up to 0.38 %, that can not be considered as an effective combination;
6. the most effective way of exploiting this system is to combine it with clear glass – that would decrease cooling load by 24 %;
7. total transmitted solar heat blocking rate for the system is 73.5 % for combining with clear glass and approximately 98% for combining with low-e glass.

References

1. K. Tsikaloudaki, K. Laskos, T. Theodosiou, D. Bikas, Energy and Buildings, **49**, 192-199 (2012)
2. V. Murgul, N. Vatin, I. Zayats, Procedia Engineering, **117**, 824-829 (2015)
3. M. Tanic, D. Stankovic, V. Nikolic, M. Nikolic, D. Kostic, A. Milojkovic, S. Spasic, N. Vatin, Procedia Engineering, **117**, 924-937 (2015)
4. G. Evola, L. Marletta , Applied Energy, **160**, 431-441 (2015)

5. Z. Huifen, F. Yingchao, Y. Fuhua, T. Hao, Z. Ying, Mathematical Problems in Engineering, 1-9 (2013)
6. V.N. Starkov, *Povyshenie energoeffektivnosti zdaniy i sooruzheniy* (SPBSTU, St.Petersburg, 2014)
7. W. Marion, S. Wilcox, *Solar radiation data manual for buildings* (NREL, 1995)
8. A. Jonsson, A. Roos, Solar Energy, **84**, 1370-1375 (2010)
9. J. Qu, J. Song, J. Qin, Z. Song, W. Zhang, Energy and Buildings, **77**, 1-10 (2014)
10. R. Baetens, B.P. Jelle, A. Gustavsen, Solar Energy Materials and Solar Cells, **94**, 87-105 (2010)
11. L.L. Fernandes, E.S. Lee, G. Ward, Energy and Buildings, **61**, 8-20 (2013)
12. B.P. Jelle, Solar Energy Materials and Solar Cells, **116**, 291-323 (2013)
13. B. Anderson, *Solar energy: fundamentals in building design* (Total Environmental Action, 1977)
14. G. Nair, K. Mahapatra, L. Gustavsson, Applied Energy, **89**, 329-338 (2012)
15. S. Guha, J. Yang, High-Efficiency Amorphous Silicon Alloy Based Solar Cells and Modules, **130**, (2005)
16. G. Radović, V. Murgul, M. Cvetkovska, E. Aronova, N. Vatin, Journal of Applied Engineering Science, **12**, 277-284 (2014)
17. *SP 131.13330.2012 Stroitel'naya klimatologiya (aktualizatsiya SNiP 23-01-99) (SR 131.13330.2012 Building Climatology)*
18. M. Pagliaro, G. Palmisano, R. Ciriminna, *Flexible Solar Cells* (WILEY-VCH Verlag GmbH & Co. KGaA, Weinheim, 2008)
19. W. Beckman, J. Duffie, S. Klein, *Simulation of solar heating systems* (ASHRAE GRP N.Y., 1977)
20. G. Baldinelli, F. Bianchi, Applied Energy, **136**, 250-258 (2014)
21. P. Pfrommer, K.J. Lomas, C. Seale, Chr. Kupke, Solar Energy, **54**, 287-299 (1995)

Thermal Stability Influence of the Enclosure Structure on the Building's Energy Efficiency

Daria Zaborova[1,*], *Marina* Petrochenko [1] and *Ludmila* Chernenkaya[1]

[1]Peter the Great Saint-Petersburg Polytechnic University, 195251 Polytechnicheskaya str. 29, St. Petersburg, Russia

Abstract. Thermal stability of the enclosure structures is one of the most important objective in the building design. Thermal processes in the wall depend not only on the internal and external air temperature, but also on many other factors. Therefore, complexity of this process make heat conservation in the room to be an actual problem. This paper presents dependence between thermal stability of the enclosure structure and its design. It was shown that thermophysical characteristics of materials directly affect the thermal processes in the wall. For the research, three frequently used types of enclosure structure in Russia were taken. For each wall was found the average temperature and cooling time. As a result, it was found that the higher values of thermal conductivity, specific heat and material density are, the higher average temperature of the wall is.

1 Introduction

Nowadays the problem of energy efficiency is one of the most discussed issues [1-5]. Especially for Russia, effective use of the fuel and natural resources is the main task. The main object of energy consumption is residential building. Approximately, energy consumption for heating buildings is about 40% of all extracted fuel [6]. The main energy task in the construction industry is design of buildings and constructions with effective use of energy [7, 8]. It is known that the biggest heat loss is through the walls. Therefore, much attention is paid to design the enclosure structures. Works of many people are devoted to the solution of this problem [9-13].

Operating principle of the enclosure structure is to keep heat in the house as long as possible [14-16]. For this reason, it is so important that material can be able to resist the stresses caused by change of temperature.

Thermophysical properties of structures and materials were investigated in works of G.M. Kondratyev, V. N. Bogoslovsky, G.N. Dulnev [17], Y. A. Tabunshchikov [18], A. S. Gorshkov [19, 20]. These materials show the importance of heat transfer problem in structures.

1.1 Purposes and objectives

The purpose of the study is to establish the connection between the energy efficiency of the enclosure structure and its thermal stability. Thus, the following tasks have been set:

1. To estimate dynamic of changes the average stationary temperature in multi-layer walls of different design;
2. To evaluate influence of thermal resistance on the enclosure structure average temperature.

1.2 Research description

Thanks to introduction of new rules for heat resistance of enclosure structures and increases cost of energy, in Russia is becoming more and more popular use of lightweight concrete in construction [21, 22]. In addition, such materials as reinforced concrete and brick do not maintain the requirements for thermal resistance any more. To compare what wall accumulates heat better we chose three different materials: lightweight concrete, brick and reinforced concrete.

For this study were selected three types of enclosure structure which are frequently used in construction (Fig. 1, 2, 3).

Fig. 1. Wall №1.

Fig. 2. Wall №2.

Fig. 3. Wall №3.

First type of the enclosure structure consists of three layers and has a total thickness 595 mm (Fig.4). The structure of the wall is shown below.

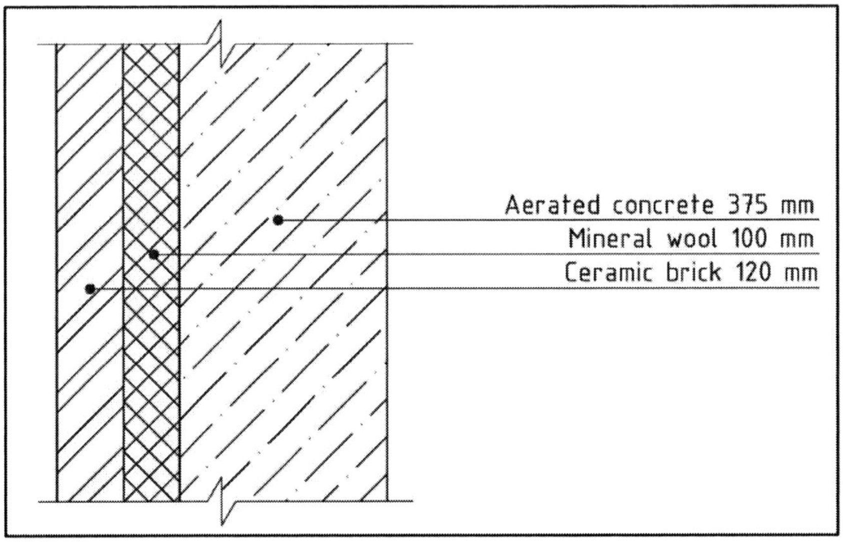

Fig. 4. The section view of the wall №1 by layers. Outdoor is on the left.

The thermal characteristics of materials are given in table 1.

Table 1. Thermal characteristics of materials.

№ of layer	Layer of material	Layer thicknes s, δ	Coefficient of thermal conductivity, λ	Specific heat, c_p	Density, ρ	Coefficient of thermal absorption, β
		[m]	[W/m·°C]	[J/kg·°C]	[kg/m^3]	[J/$\sqrt{s}$·m^2·°C]
1	Aerated concrete	0.375	0.14	1000	500	264.57
2	Mineral wool	0.1	0.04	920	150	71.46
3	Ceramic brick	0.12	0.64	880	1600	949.27

Second type of the enclosure structure consists of three layers and has a total thickness 470 mm (Fig.5). The structure of the wall is shown below.

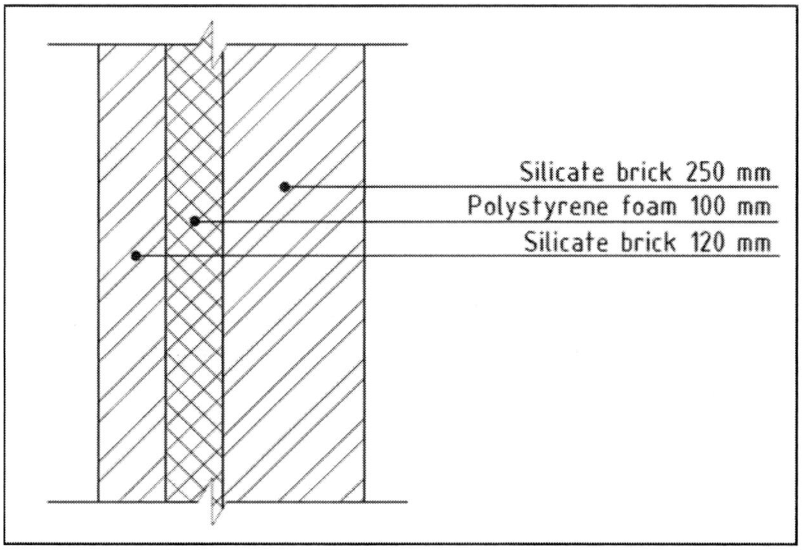

Fig. 5. The section view of the wall №2 by layers. Outdoor is on the left.

The thermal characteristics of materials are given in table 2.

Table 2. Thermal characteristics of materials.

№ of layer	Layer of material	Layer thickness, δ	Coefficient of thermal conductivity, λ	Specific heat, c_p	Density, ρ	Coefficient of thermal absorption, β
		[m]	[W/m·°C]	[J/kg·°C]	[kg/m^3]	[J/$\sqrt{s}$·m^2·° C]
1	Silicate brick	0.25	0.81	800	1500	985.90
2	Polystyrene foam	0.10	0.04	1340	150	89.67
3	Silicate brick	0.12	0.81	800	1500	985.90

Third type of the enclosure structure consists of three layers and has total thickness 470 mm (Fig.6). The structure of the wall is shown below.

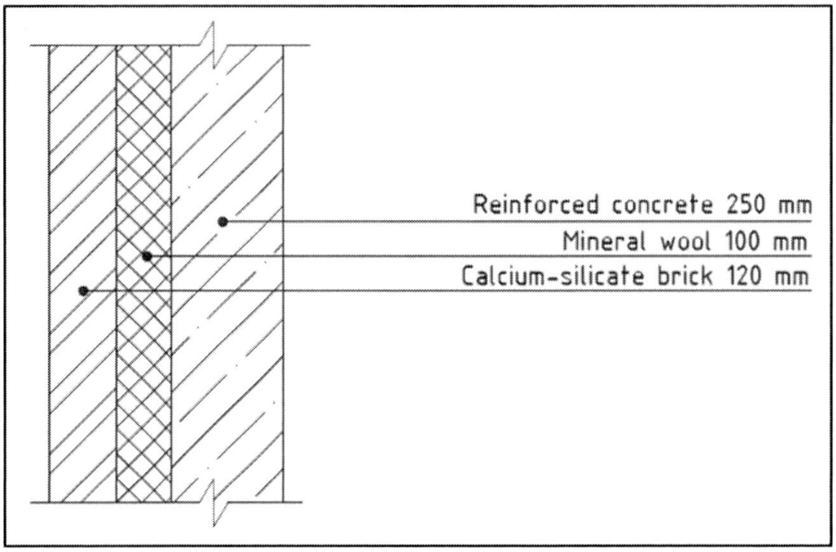

Fig. 6. The section view of the wall №3 by layers. Outdoor is on the left.

The thermal characteristics of materials are given in table 3.

Table 3. Thermal characteristics of materials.

№ of layer	Layer of material	Layer thickness, δ	Coefficient of thermal conductivity, λ	Specific heat, c_p	Density, ρ	Coefficient of thermal absorption, β
		[m]	[W/m·°C]	[J/kg·°C]	[kg/m^3]	[J/$\sqrt{s}$·m^2·°C]
1	Reinforced concrete	0.25	1.70	840	2500	1889.44
2	Mineral wool	0.1	0.04	920	150	74.30
3	Silicate brick	0.12	0.81	880	1600	1067.93

Internal coefficient of heat transfer α_{int}, W/ (m^2·°C) demonstrates the ability of material to absorb heat when temperature fluctuates on its surface and can be found by the formula:

$$\alpha_{int} = \sqrt{\frac{\lambda \cdot \rho \cdot c_p}{t_0}}, \quad (1)$$

where t_0 - fluctuation period of a thermal steam, s;

$\beta = \sqrt{\lambda \cdot \rho \cdot c_p}$ - coefficient of thermal absorption, J/($\sqrt{s}$·m^2·°C).

Thus, the heat absorption of construction depends on physical properties of material.

2 Calculations of the wall average temperature

Because the multilayer walls consist of different layers, the coefficients λ, ρ, c_p are changed discretely. With great number of layers, the discrete distribution can be replaced by a continuous distribution, and from summation turn to integration of continuous distributions.

$$R = \sum \frac{\delta_i}{\lambda_i} = \int_0^{\delta} \frac{dx}{\lambda(x)}, \qquad (2)$$

where R wall heat transfer resistance, $(m^2 \cdot °C)/W$.

We determined the average stationary temperature for each type of wall by the formula:

$$T_\infty = \frac{\alpha_c \cdot T_c + \alpha_h \cdot T_h \cdot \left(1 + \alpha_c \cdot \int_0^\delta \frac{dx}{\lambda(x)}\right)}{\alpha_c + \alpha_h \cdot \left(1 + \alpha_c \cdot \int_0^\delta \frac{dx}{\lambda(x)}\right)} + \frac{\alpha_c \cdot \alpha_h \cdot (T_h - T_c)}{\alpha_c + \alpha_h \cdot \left(1 + \alpha_c \cdot \int_0^\delta \frac{dx}{\lambda(x)}\right)} \cdot \frac{1}{\delta} \cdot \int_0^\delta \frac{(\delta - \xi) d\xi}{\lambda(\xi)}, \qquad (3)$$

where $\alpha_c = 23\ W/(m^2 \cdot °C)$ - outer surface heat transfer coefficient of the building envelope;

$\alpha_h = 8.7\ W/(m^2 \cdot °C)$ - inner surface heat transfer coefficient of the building envelope;

T_h - the internal air temperature in the premises of a residential building in St. Petersburg, taken in accordance with GOST 30494 at 20 ° C;

T_c - the average monthly outdoor air temperature in January, taken in accordance with SNIP 23-01-99* at -8.7 ° C;

δ- thickness of the wall, m;

λ- thermal conductivity, $W/(m \cdot °C)$. The calculation results are summarized in Table 4.

Table 4. Average temperature of enclosure structure.

Average stationary temperature	Sample №1	Sample №2	Sample №3
T_∞, °C	5.22	8.34	8.89

Calculations have shown that the wall with the highest thermal conductivity has the highest average temperature. Influence of wall heat conductivity on its average temperature is shown in Figure 7.

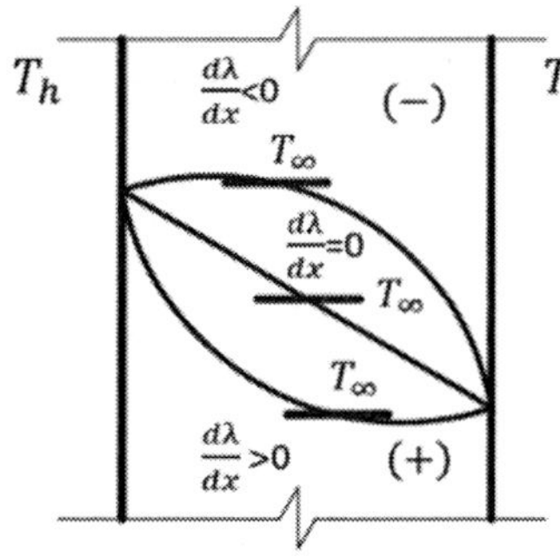

Fig. 7. Dependence of average temperature on heat conductivity coefficient.

Now we determine the instantaneous average temperature of the wall:

$$\overline{T} = (T_h - T_\infty) \cdot \theta + T_\infty. \qquad (4)$$

Temporary measure of thermal stability - θ can be found by the formula (4).

$$\theta = \exp\left(-\frac{t}{t_h} - \frac{t}{t_c}\right), \qquad (5)$$

where t - cooling time of the wall, s.

The value of t_h and t_c were determined from the following equations:

$$t_h = \frac{\rho \cdot c_p \cdot \delta}{\alpha_h}, \qquad (6)$$

$$t_c = \frac{\rho \cdot c_p \cdot \delta}{\alpha_c}. \qquad (7)$$

The results of calculation for each construction type are summarized in table 5.

Table 5. Instantaneous average temperature.

Time, hours	Wall № 1	Wall № 2	Wall № 3	Time, hours	Wall № 1	Wall № 2	Wall № 3
0	20	20	20	13	6.26	9.18	9.98
1	17.27	17.86	18.18	14	6.07	9.03	9.80
2	15.04	16.12	16.66	15	5.91	8.90	9.65
3	13.23	14.69	15.39	16	5.78	8.80	9.53
4	11.75	13.53	14.33	17	5.68	8.71	9.42
5	10.54	12.58	13.44	18	5.59	8.64	9.33
6	9.56	11.80	12.70	19	5.53	8.59	9.26
7	8.76	11.17	12.07	20	5.47	8.54	9.20
8	8.10	10.65	11.55	21	5.42	8.51	9.15
9	7.57	10.22	11.12	22	5.39	8.48	9.11
10	7.10	9.88	10.75	23	5.35	8.45	9.07
11	6.78	9.59	10.45	24	5.33	8.43	9.04
12	6.5	9.37	10.19				

Time-average temperature transformation in all free types of enclosure structure are presented in Figure 8.

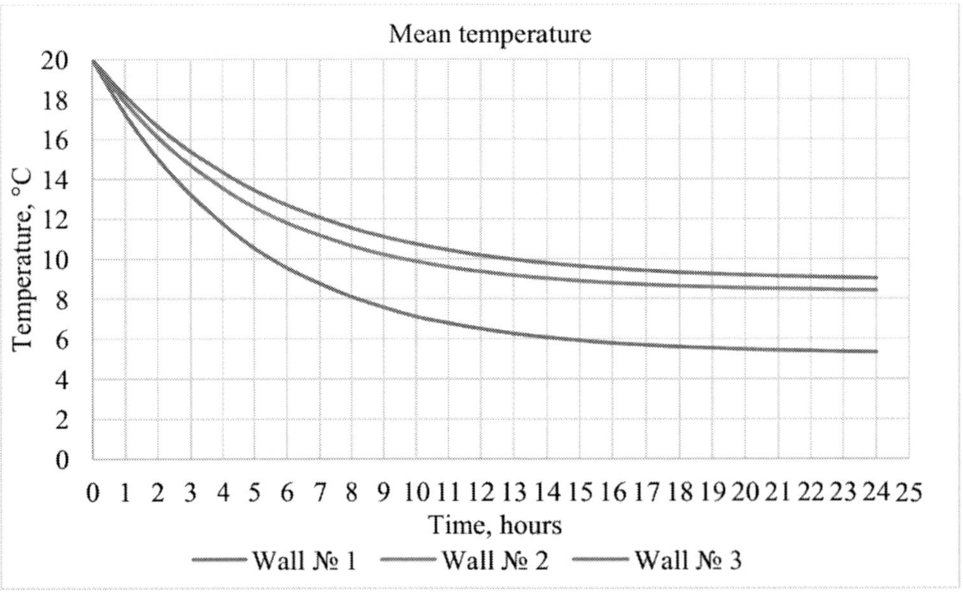

Fig. 8. Cooling time of the walls.

3 Conclusions

1. The results of calculations showed that reinforced concrete wall has the highest average temperature 8.89 °C. The aerated concrete wall has the lowest average temperature 5.22 °C. These values confirm that wall with big coefficient of thermal absorption β has a higher average temperature than wall with small coefficient of thermal absorption.
2. Aerated concrete wall cools much faster than wall made from brick or reinforced concrete.
3. Heat accumulation is very important for energy saving and creation the optimal climate in the house. However, despite the small coefficient of thermal conductivity λ, heat storage capacity of aerated concrete is less than in reinforced concrete wall.

From the standpoint of energy efficiency, reinforced concrete wall is unreasonable. However, from the standpoint of thermostability this is reasonable wall.

References

1. V. Murgul, JAES, **12 (1)**, 1-10, (2014)
2. M.V. Knat'ko, M.N. Efimenko, A.S. Gorshkov, MCE, **2**, 50-53, (2008)
3. D. Nemova, V. Murgul, A. Golik, E. Chizhov, V. Pukhkal, N. Vatin, JAES, **12 (1)**, 37-44, (2014)
4. A. Gorshkov, N. Vatin, D. Nemova, A. Shabaldin, L. Melnikova, K. Paramonov, Procedia Eng., **117**, 1080-1089, (2015)
5. L. Svatovskaya, A. Sychova, M. Sychov, V. Okrepilov, MATEC Web of Conferences, **53**, Article Number 01023 (2016)
6. E. Perlova , S. Karpova, X. Rakova, E. Bondarenko, M. Platonova, A. Gorshkov, AMM, **725-726**, 1580-1588, (2015)
7. A. Lupíšek, M. Bureša, M. Volfa, J. Hodkováa, J. Nováčeka, P. Hejtmáneka, J. Tywoniak, Energy Procedia, **78**, 285-290, (2015)
8. J. Tywoniak, K. Sojková, Energy Procedia, **78**, 1069-1074, (2015)
9. D.V. Petrosova, N.M. Kuz'menko, D.V. Petrosov, MCE, **8**, 31-37, (2013)
10. E. Kossecka, J. Kosny, J. Therm. Envelope Build. Sci., **7**, 721-731, (1998)
11. S.V. Korniyenko, MCE, **8**, 25-37, (2014)
12. J. Carmeliet, H. Hens, C. Vermeir, *Research in building physics*, (2003)
13. Medved, A. Trnik, *Thermophysics and Mass Transfer in Materials Science and Construction*, (2015)
14. M. R. Hall, *Materials for energy efficiency and thermal comfort in buildings*, (2010)
15. S. Kubba, *Green building design and construction*, (2012)
16. P. Glover, *Building surveys 7 edition*, (2008)
17. G.N. Dul'nev, D.P. Volkov, V.I. Malarev, V.I. Malarev, J. Eng. Phys. Thermophys., **56 (2)**, 198-206, (1989)
18. G.P. Vasilyev, V.A. Lichman, M.V. Kolesova, N.V. Peskov, M.M. Brodach, Y. A. Tabunshchikov, Energy Build., **86**, 803-807, (2015)
19. N.I. Vatin, A.S. Gorshkov, D.V. Nemova, A. A. Staritcyna, D. S. Tarasova, Adv. Mater. Res., **941-944**, 905-920, (2014)
20. N.I. Vatin, D.V. Nemova, P.P. Rymkevich, A.S. Gorshkov, MCE, **8**, 4-14, (2013)
21. A.S. Gorshkov, P.P. Rymkevich, N.I. Vatin, MCE, **8**, 38-48, (2014)
22. N.I. Vatin, A.S. Gorshkov, D.V. Nemova, Constr. of Unique Build. and Str. **3 (8)**, 1-11, (2013)

Assessing Energy Efficiency of Compression Heat Pumps in Drying Processes when Zeotropic Hydrocarbon Mixtures are Used as Working Agents

Alexander Shurayts[1], *Alexander* Rulev[2,*], and *Elena* Usacheva[3]

[1]Yuri Gagarin State Technical University of Saratov, 410054, st. Politechnicheskaya, Saratov, Russian Federation
[2]Public Limited Company "Giproniigaz", 410012, st.Kirova, Saratov, Russian Federation
[3]Scientific Technologies Ltd, 410062, st. Moscow highway, Saratov, Russian Federation

Abstract. Presents the results of studies of innovative materials in the field of renewable energy.The paper proposes a design and a formula for assessing energy efficiency of the heat pump air dryer, which uses zeotropic hydrocarbon mixtures of saturated hydrocarbons as a working agent and applies the principle of a counter-current heat exchanger with a variable temperature of both the working and the drying agents. Energy efficiency of the heat pump is achieved by means of obtaining a greater part of heat from renewable energy sources, in this case by cooling the air and condensing the water vapors in the heat pump. A conducted analysis identified correlations in establishing the marginal real coefficient of performance of the compression heat pump dryer running on zeotropic hydrocarbon mixtures and operating a cycle with variable temperatures of both the working and the drying agent in the evaporator and the condenser of the heat pump. According to the established correlations, the marginal real coefficient of performance of the compression heat pump dryers running on zeotropic hydrocarbon mixtures of 40 mol% of R600a and 60 mol% of R601 is 1.92 times higher than that of the same dryers running on only R600 (n-butane).

1 Introduction

Drying agricultural and other types of produce, including grassy forage, vegetables, fruits, grains, timber, pottery and other materials, accounts for a significant energy consumption.

A literature review [1-5] demonstrates that grassy forage, grains, vegetables, fruits, different types of timber, and other materials are dried most properly and effectively with heat pumps at low temperatures using the air as the drying agent with its complete recirculation and dehydration. Energy efficiency of the heat pump is achieved by means of obtaining heat from a renewable energy source, in this case by cooling the air and condensation of water vapors in the heat pump. According to [1,5,6], the average rate of

* Corresponding author: ritamitrofanova@yandex.ru

return of such installations is 3-4 years. An optimal type of engine for the compressor of the industrial heat pump dryer is a gas engine [1], which allows for a considerable increase in the combustion efficiency of the primary fuel by means of useful heat extraction from flue gases, condensation of water vapors in them, and the use of heat from cooling systems and engine lubrication. Existing types of heat pumps use substances with a constant boiling point as a working agent, which is not energy efficient when low-temperature heat sources, such as ground and outdoor bodies of water, have a high volumetric heat capacity [1,2,4], and therefore their temperature remains constant or varies slightly.

At the same time, when a source or sink for heat has limited volumetric heat capacity, for example, the drying air [7], whose temperature changes significantly while it is being dehydrated in the evaporator and being heated in the condenser, using a substance with a constant boiling point as a working agent [8,9] is characterized by reduced efficiency of non-renewable energy (electricity from thermal power plants, gas, etc.) used to drive the compressor.

2 Heat pump dryer design concept and description

The proposed design of the heat pump dryer that applies a complete recirculation of air and runs a cycle with variable temperatures of both the working and the drying agents that flow in opposite directions, is an energy efficient drying solution (fig. 1).

The design concept is as follows. The working agent chosen is a two-component zeotropic mixture 4, which, when boiled in evaporator 1, increases its temperature from the initial value at the inlet of evaporator tubes unit 3 to the final temperature value at the outlet. Contrary to that, humid air 5, while flowing throw shell side 2 in the opposite direction, reduces its temperature from the initial value at the outlet to the final temperature value at the inlet of evaporator tubes unit 3. During its transition from vapor to liquid in condenser 9, the zeotropic mixture reduces its temperature from the initial value at the inlet to condenser tubes unit 11 to the final value at the outlet. Contrary to that, dehumidified air 5, flowing throw shell side 10 in the opposite direction, increases its temperature from its initial value, at the outlet, to the final value at the inlet to condenser tubes unit 11. The change in the temperature of boiling and condensation of the zeotropic mixture, as well as the change in the value of the temperature difference of the beginning and end of boiling or condensation, can be achieved by changing the percent composition of one of its components. Figure 1 shows the design and the operating principles of the gas engine heat pump.

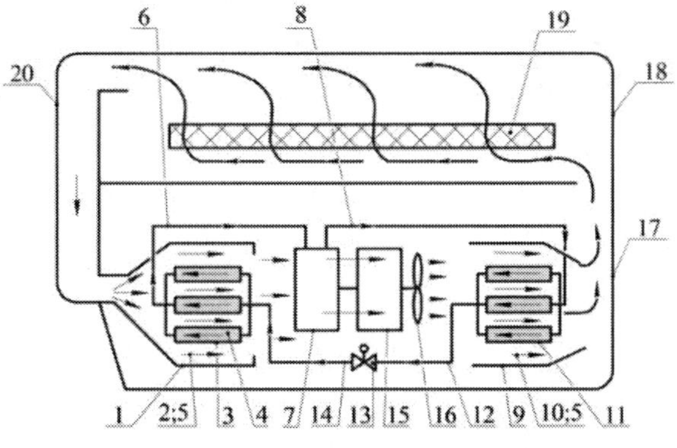

Fig. 1. Design of the heat pump dryer running on a zeotropic hydrocarbon mixture with a gas engine drive: 1 — evaporator; 2 — evaporator shell side; 3 — evaporator tubes unit; 4 — working agent made from a two-component zeotropic mixture; 5 — air; 6— tubes of the vapor phase of working agent 4 used to connect to the outlet of evaporator tubes unit 3; 7 — compressor; 8 — tubes of the vapor phase of working agent 4 used to connect to the outlet of compressor 7; 9 — condenser; 10 —

condenser 9 shell side; 11 — condenser tubes unit; 12 — tubes of the liquid phase of working agent 4 used to connect to the outlet of condenser tubes unit 11; 13 — temperature regulator reducing the temperature of the saturated liquid phase of working agent 4; 14 — tubes of the liquid phase of working agent 4 connecting regulator 13 with the inlet of evaporator tubes unit 3; 15 — gas engine to drive the compressor; 16 — fan; 17 — heat pump station; 18 — drying station; 19 — drying products; 20 — recirculation duct.

3 Selection of the working agent

Working substances used as components of a zeotropic mixture in heat pump dryers should not contribute to the depletion of the ozone layer of the atmosphere [10] and climate change [11], should be fireproof and cost-efficient, available for use and have no negative impact on the human body, they should have variable values of boiling points and condensation, which are the most suitable for the drying process of grass, grain and similar products. Currently, there are no working agents that can perfectly satisfy all these requirements. For example, freons R11, R12, R22 widely used in the past, which satisfy the low-temperature drying conditions of agricultural and other similar products and could be used as components of a zeotropic mixture in heat pump dryers, have negative effects on the ozone layer of the atmosphere and contribute to climate change. In this respect, the choice of components for the zeotropic mixture is carried out individually, relevant to the low-temperature heat pump air dryer considering all positive and negative factors of their effect.

Based on the results of the conducted analysis, zeotropic mixtures of saturated hydrocarbons, consisting of R600a (isobutane) and R601 (n-pentane) meet these requirements more fully. R600a and R601 are considered to be completely safe in terms of their effect on the ozone layer of the Earth. These gases do not cause the greenhouse effect, do not have affect climate change or the human body, are extracted directly from natural gas, and are much cheaper than other working agents.

However, using zeotropic mixtures can result in a reduced heat emission coefficient in "the inner surface of the heat exchange tubes unit — the working agent" system of the evaporator and the condenser of the heat pump (fig. 1) due to reduced vaporization centers and decreased bubble departure diameter, compared to each of the components of the mixture [12]. The calculations, carried out in accordance with [12-14], demonstrate that the reduction of the overall heat emission coefficient does not exceed 5.5%, since the main specific weight in the total balance of its value takes the heat emission coefficient in "the outer surface of the finned heat exchanger tubes unit — the drying air" system, which is characterized by relatively low heat emission in case of forced air flow. Such heat emission reduction rate and a subsequent increase in capital investment in heat exchange surfaces of the evaporator and the condenser in this case are as not as significant as the reduced operating costs due to an increased coefficient of compression heat pump dryer performance, which is explained below.

The drawback of saturated hydrocarbons, when used as components of zeotropic mixtures, is that they pose a fire hazard. In this respect, industrial heat pump dryers should be equipped with smoke detectors. At the same time, when used as a drive for the compressor of the gas engine, which is most preferred in the drying process, a smoke detector is mandatory, based on the security requirements for gas supply items [15,16]. Apart from that, in the case of modern heat pump dryers with a small filling capacity (35-50 g), even if the entire working agent leaks out, its concentration in a space of 15 m3 will be ten or more times lower than 1.7-1.35%, the lowest inflammability limit of R600a and R601 mixtures [17,18]. Given the overall balance of advantages and disadvantages, currently more than 35% of household refrigerators in Europe run on isobutane R600a. The low-boiling component of the mixture, R600a has the boiling point of $t = -11.73°C$ in the atmospheric pressure and the high-boiling component R601 has the temperature of t = +36.07 °C.

Now we estimate the change of temperature conditions for the heat pump, designed for air drying, operating on a zeotropic mixture of R600a (isobutane) and R601 (n-pentane), containing 40 mol% of R600a.

4 Estimating temperature conditions of the working and the drying agents

In order to estimate temperature conditions of the heat pump, we need to know the following:

1) change in the current temperature of the saturated zeotropic mixture of R600a and R601 in the evaporator and the condenser of the heat pump, including its initial and final values, in relation to the degree of dryness X in the variation range from 0 to 1;

2) change in the temperature of the drying air in the evaporator and the condenser of the heat pump, including its initial and final values, in relation to the temperature and flow volume of the saturated zeotropic mixture.

The nature of temperature t changes in the evaporator and the condenser of the heat pump, including its initial and final values for the zeotropic mixture of R600a and R601, for which Dalton's law, Raoult's law [17,18], Konovalov's first law [19], the Antoine equation [20,21] for saturated vapor pressure for pure components are applicable, can be estimated depending on the degree of dryness X, according to [20]:

$$X = P_m\left(\frac{\psi}{\beta} + \frac{1-\psi}{\xi}\right) \quad (1)$$

$$\beta = \left(P_m - 10^{A_{R601} - \frac{B_{R601}}{C_{R601}+t}}\right) \quad (2)$$

$$\xi = \left(P_m - 10^{A_{R600a} - \frac{B_{R600a}}{C_{R600a}+t}}\right) \quad (3)$$

where: Pm is the absolute total pressure of the mixture, Pa 105; ψ is the R600a content in the initial liquid and vapor phases of the mixture, with which it enters the evaporator and the condenser of the heat pump, mol.%; AR600a, BR600a, CR600a , AR601, BR601, CR601 are coefficients typical for saturated hydrocarbons R600a and R601 within a certain temperature range; t is the boiling point of an individual hydrocarbon (R600a and R601),°C.

Based on the most accurate experimental data on the saturated hydrocarbons R600a and R601, [21, 22] gives the coefficients AR600a, BR600a, CR600a, AR601, BR601, CR601 in formulas (2) and (3). Differences, resulting from applying calculation formulae (1)–(3) based on Dalton's laws, Raoult's law and the Antoine equation for zeotropic mixtures of saturated hydrocarbons at the absolute pressure up to 1.0 MPa, to the actual measurement data, are 4.5% [23].

Changes in the temperature of evaporated and condensed mixture in the range of the dryness degree X from 0 to 1.0 result in a change in the temperature of the heated or cooled drying air. The initial temperature value of a limited heat capacity environment, in this case the air, depending on the temperature changes, dryness factor and the zeotropic mixture flow rate, for the case of the counter-current heat exchanger, is determined from the balance equation between the working and drying agents in the evaporator and the condenser of the heat pump.

For the heat pump evaporator:

$$T^{v}_{ai.i}=\frac{G_{ai}c^{v}_{ai}T^{v}_{ai.f}+G_{m}\left[r^{v}_{m}X^{v}_{m}+c^{v}_{m}(T^{v}_{m.f}-T^{v}_{m.i})\right]}{c^{v}_{ai}G_{ai}}. \quad (4)$$

For the heat pump condenser:

$$T^{c}_{ai.i}=\frac{G_{ai}c^{c}_{ai}T^{c}_{ai.f}-G_{m}\left[r^{c}_{m}X^{c}_{m}+c^{c}_{m}(T^{c}_{m.i}-T^{c}_{m.f})\right]}{c^{c}_{ai}G_{ai}}. \quad (5)$$

where: $T^{v}_{ai.i}, T^{v}_{ai.f}$ are the initial and final temperatures of the drying air in the evaporator, K; $T^{v}_{m.f}, T^{v}_{m.i}$ are the initial and final temperatures of the zeotropic mixture in the evaporator, K; $T^{c}_{ai.f}, T^{c}_{ai.i}$ are final and initial temperatures of the drying air in the condenser, K; $T^{c}_{m.i}, T^{c}_{m.f}$ are initial and final temperatures of the zeotropic mixture in the condenser, K; G_m, G_{ai} are the mole discharge of the zeotropic mixture and the drying air, mol/h; r^{v}_{m}, r^{c}_{m} are the average values of latent heat of evaporation (vaporization) of the zeotropic liquid-vapor mixture in the condenser and the evaporator, kJ/mol; c^{v}_{ai}, c^{c}_{ai} are mean specific heat of the drying air in the evaporator and the condenser, kJ/(mol·K); c^{v}_{m}, c^{c}_{m} are mean specific heat of zeotropic liquid-vapor mixture in its boiling and condensation temperature intervals in the evaporator and the condenser, kJ/(mol·K); X^{v}_{m}, X^{c}_{m} are the dryness factor of the zeotropic mixture in the evaporator and the condenser, unit fractions.

Estimating the temperatures of the zeotropic mixture and the drying air at given X is carried out using the formulae (1)–(3) and applying the method of successive approximations in the following manner.

1. We shall assume that the concentration value of R600a and R601 in the zeotropic mixture equals 40 mol.%.

2. The temperature of the drying air in the condenser outlet considering low temperature drying conditions equals $t^{c}_{ai.f}$ = 65 °C. The initial and final temperature values are shown in the figure "temperature (T) — entropy (S)", (fig. 2). Since the temperatures in the T-S diagram are indicated in Kelvin degrees, the following notation and value will be applied: $T^{c}_{ai.f}$ = 338 K.

Assume the initial temperature drop between heat exchange streams at the inlet of the counter-current condenser is at a rate of Δt = 7 °C. Then the initial temperature for zeotropic mixture of R600a and R601 in the condenser will be $t^{c}_{m.i}=t^{c}_{ai.f}+\Delta t=65+7=72$ ^{0}C ($T^{c}_{m.i}$=345 K). Then, according to the formulae (1)–(3), taking the values of X = 1 (beginning of condensation of the saturated vapor phase), $t^{c}_{m.i}$=72 ^{0}C and ψ=40 mol% of

R600a, by the trial method we estimate the absolute pressure, whose value, based on the results of calculations, equals $P^{c}_{m} = 0.4$ MPa.

3. The final temperature value of the mixture in the condenser is estimated using the formulae (1)–(3) and applying the trial method with the value of X = 0 (end of condensation of the saturated vapor phase) and $P^{c}_{m} = 0.4$ MPa, whose value equals $t^{c}_{m.f} =$ 55 0C ($T^{c}_{m.f} = 328K$) (fig. 2).

4. Further, according to the formula (5), for the case of the counter-current heat exchanger (fig. 2), we estimate the drying air temperature in the condenser inlet, in relation to the temperature change and the zeotropic mixture flow, which is $T^{c}_{ai.i} = 317K$, based on the results of calculations.

5. Assume the temperature of the drying air at the evaporator outlet equals its temperature at the condenser inlet, i.e., $T^{v}_{ai.f} = T^{c}_{ai.i} = 317K$.

6. Assume the absolute pressure in the evaporator is P^{v}_{m} =0.25 MPa. Then, according to the results of calculations based on the application of the formulae (1)–(3) using the trial method given the value of X=0 and absolute pressure P^{v}_{m} =0.25 MPa, the initial temperature value for the zeotropic mixture of R600a and R601 in the evaporator of the heat pump that contains ψ=40 mol.% of R600a, will be $t^{v}_{m.i}$ =36 0C ($T^{v}_{m.i}$ =309 K). The value X=0 corresponds to a state of an equilibrium system, when the mixture contains only the saturated liquid phase and, vice versa, if x=0 then the mixture contains only the saturated vapour phase. Similarly, determine the final temperature value of the mixture in the evaporator $t^{v}_{m.f}$ =54 0C ($T^{v}_{m.f}$ =327 K) given the value of X=1 and P^{v}_{m} =0.25 MPa. The temperature curve from point 4 to point 1 for the mixture containing ψ=40 mol.% of R600a, given the absolute pressure P^{v}_{m} =0.25 MPa, is shown on the temperature-entropy diagram (fig. 2.)

7. After that, according to the formula (4), determine the temperature of the drying air at the outlet of the evaporator, which is $T^{v}_{ai.i}$ =332 K based on the results of calculations for the case of the counter-current heat exchanger. Apply the obtained values of the drying air temperatures $T^{v}_{ai.i}$ =332 K; $T^{v}_{ai.f}$ =317 K and the direction of the air flow to the T-S diagram (fig. 2.)

8. For comparison, the dashed line in the diagram shows temperatures in the condenser and the evaporator of the heat pump when it runs on the working agent that consists of one substance $T^{c}_{s.mt}$; they equal, respectively, $T^{c}_{s.mt} = T^{c}_{m.i}$ =345 K and $T^{v}_{s.mt} = T^{v}_{m.i}$ =309 K.

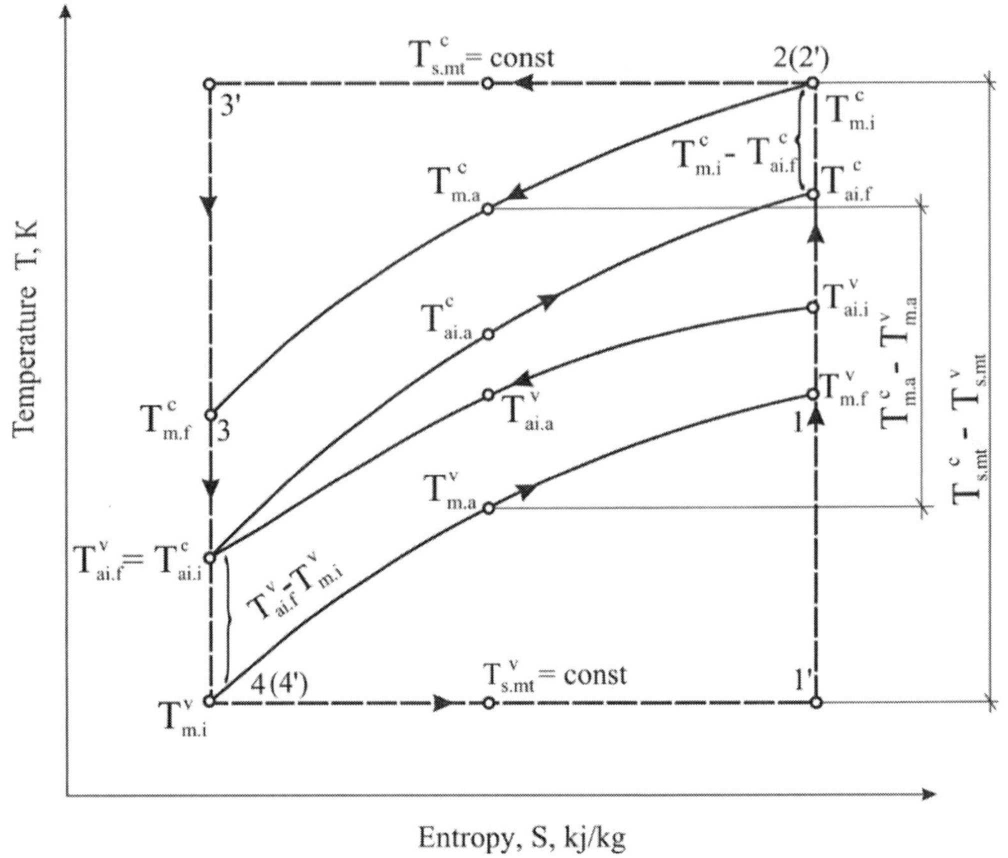

Fig.2. Comparison of a cycle 1-2-3-4-1 of the heat pump running on a mixture of R600a and R601 with a cycle 1′-2′-3′-4′-1′ of the heat pump running on pure R600
1-2 (1′-2′) is the vapor phase compression in the compressor; 2-3 (2′-3′) is the vapor phase condensation in the condenser; 3-4 (3′-4′) is throttling of the liquid phase; 4-1 (4′-1′) – is boiling of the liquid phase in the evaporator.

5 Assessing energy efficiency of the heat pump dryer

Evaluate the effectiveness of the heat pump dryer cycle that runs on the working agent with a fluctuating temperature of 1-2-3-4-1 compared to the cycle 1′-2′-3′-4′-1′ for the dryer that operates on a pure working agent of R600. The real coefficient of performance of the heat pump, recommended by [24] for a preliminary assessment of these cycles, is estimated using the formula:

$$\varphi_r = [0.74 \cdot T^{v}_{m.a} /(T^{c}_{m.a} - T^{v}_{m.a}) - (0.0032 \cdot T^{v}_{m.a} + 0.765 \cdot T^{v}_{m.a} / T^{c}_{m.a})] + 0.9. \quad (6)$$

where $T^{v}_{m.a}$, $T^{c}_{m.a}$ are the average temperatures of the working agent that consists of a zeotropic mixture in the evaporator and the condenser, K.

According to [24], in case of variable temperatures, the performance evaluation is conducted with the help of the equivalent Carnot cycle. In this case, the average temperature in the condenser $T^{c}_{m.a}$ (line 2-3) and the evaporator $T^{v}_{m.a}$ (line 4-1) are estimated, respectively:

$$T^{c}_{m.a} = (T^{c}_{m.i} - T^{c}_{m.f}) / \ln \cdot (T^{c}_{m.i} / T^{c}_{m.f}). \quad (7)$$

$$T^{v}_{m.a} = (T^{v}_{m.f} - T^{v}_{m.i}) / \ln \cdot (T^{v}_{m.f} / T^{v}_{m.i}). \quad (8)$$

Putting the corresponding temperature values $T^{v}_{m.a}$=321.4 K and $T^{c}_{m.i}$=340.1 K, calculated using the formulae (7) and (8) for the cycle 1-2-3-4-1, in the formula (6), we deduce the marginal real coefficient of performance φ_r=13.3. For the cycle 1′-2′-3′-4′-1′- given the constant temperature in the evaporator $T^{v}_{s.mt}$ equal to the initial temperature of the mixture $T^{v}_{m.i}$ for the cycle 1-2-3-4-1, i.e. $T^{v}_{s.mt}$ = $T^{v}_{m.i}$=309 K, and given the constant temperature in the condenser $T^{c}_{s.mt}$ equal to the initial temperature of the mixture $T^{c}_{m.i}$ for the cycle 1-2-3-4-1, i.e. $T^{c}_{s.mt}$=$T^{v}_{m.i}$ = 345 K, the marginal real coefficient of performance according to the formula (6) is φ_r'=6.94.

Thus, the marginal real coefficient of performance of compression heat pump dryers running on zeotropic mixtures that contain 40 mol.% of R600a, is 1.92 times higher (φ_r / φ_r'=13.3/6.94) than that of the same installation operating on a pure working agent of R600.

6 Conclusions

1. The paper present a design of a heat pump air dryer that uses a zeotropic mixture of saturated hydrocarbons consisting of 40 mol.% of R600a and 60 mol.% of R601 as a working agent, and applies the principle of a counter-current heat exchanger with a variable temperature of both the working and the drying agents. The proposed design guarantees a reduced mean temperature difference between the air and the zeotropic hydrocarbon mixture in the evaporator and the condenser and, consequently, an increases coefficient of performance of the heat pump.

2. The correlations (1)–(3) were suggested for calculating the current temperature values of the saturated mixture of R600a and R601 of the given composition □ in the evaporator and condenser of the heat pump, including the initial and final temperature values, depending on the relative amount of evaporated or condensed mixture, i.e. the vapor dryness degree X.

3. The equations (4)–(5) were devised to estimate the initial and final values of the drying air temperature in the evaporator and the condenser of the heat pump, depending on the flow volume of the zeotropic mixture and the drying air, and temperature conditions of boiling or condensation of the mixture.

4. The correlations (6)–(8) were worked out to roughly estimate the marginal real coefficient of performance of compression heat pump dryers running on a zeotropic mixture and operating a cycle with variable temperatures in the evaporator and the condenser of the heat pump.

According to the correlations (6)–(8), the marginal real coefficient of performance of the compression heat pump dryer running on the zeotropic mixture of 40 mol.% of R600a, is φ_r=13.3, which is 1.92 times higher than that of the same installation running on a pure working agent of R600 with the value φ_r'=6.94, i.e. φ_r / φ_r'=13.3/6.94 =1.92.

Acknowledgements

This work has been prepared with the support of the "Yuri Gagarin State Technical University of Saratov" and Public Limited Company "Giproniigaz".

References

1. D. Reay, D. Macmichael, *Heat Pumps, Design and Applications, translation from English* (Energoizdat, Moscow, 1982)
2. E.I. Yantovskiy. L.A. Levin, *Industrial Heat Pumps* (Energoatomizdat, Moscow, 1989)
3. International Simposium on the Industrial Application of the Heat Pump (1982)
4. V.G. Gorshkov, Heat Pumps, Analytical review, An industrial equipment guide, **2**, 47-80 (2004)
5. O.Sh. Vezirishvili, G.I. Chogovadze et al., Teploenergetika, **12**, 45-47 (1981)
6. B.N. Kuritsin, *Fundamentals of energy saving in heating and ventilation technology* (Publishing house Nadezhda, Saratov, 1996)
7. P.D. Lebedev, *Calculation and design of the drying equipment* (Gosenergoizdat, Moscow-Leningrad, 1973)
8. E.V.Romanova, A.Yu. Orlov, Vestnik of Tambov State Technical University, **14(3)**, 591-596 (2008)
9. V.I. Konovalov, E.V. Romano)va, N.Ts. Gatapova, Vestnik of Tambov State Technical University, **17(1)**, 153-178 (2001)
10. The Montreal Protocol on Substances That Deplete the Ozone Layer, adopted by the government of the USSR in November (1988), enacted on January 1 (1989)
11. The Kyoto Protocol to the United Nations Framework Convention on Climate Change. Ratified by the Federal Law of the Russian Federation N 128-FZ of November 4 (2004) It came into force on February 16 (2005)
12. A.M. Kutepov, L.S. Sterman, N.G. Styushin, *Hydrodynamics and heat transfer in vaporization: a textbook for universities* (Vysshaya shkola, Moscow, 1977)
13. M.A. Mikheev, I.M. Mikheeva, *Foundations of heat transfer* (Energiya, Moscow, 1973)
14. A.P. Usachev, A.V Rulev, Teplovye processy v tehnike, **5(8)**, 343-353 (2013)
15. SP 42-101-2003). General provisions for the design and construction of gas distribution systems made of metal and plastic pipes, Stroyizdat, Moscow (2003)
16. PB 12–529–03 Safety rules of gas distribution and gas consumption systems, JSC STC Industrial Safety, Moscow (2003)
17. N.I. Preobrazhenskiy. *Liquefied gases* (Nedra, Leningrad, 1975)
18. N.L. Staskevich, D.Ya. Vigdorchik, *Handbook of liquefied petroleum gases* (Nedra, Leningrad, 1986)
19. M.Kh. Karapetyants, *Chemical thermodynamics* (Khimiya, Moscow, 1975)
20. A.P. Usachev, A.L. Shurayts, A.V. Rulev, T.A. Usacheva, *System research on increase of heat exchange intensity of liquefied petroleum gas regasificators* (Saratov State Technical University, Saratov, 2010)
21. B.N. Kuritsin, A.P. Usachev, B.P. Bogdanov, *The use of gas in the national economy: collection of articles of the Griproniigaz institute* (Saratov University Publishing House, Saratov, 1974)
22. M.D. Tilicheev, *Physicochemical properties of individual hydrocarbons* (Gostoptehizdat, Moscow-Leningrad, 1947)
23. C.H. Nysewander, B.H. Sage, W.N. Lesey, Industrial and Engineering Chemistry, **32(1)**, 118-123 (1940)
24. V.S. Martynovskiy, *Cycles, schemes and characteristics of thermotransformers* (Energiya, Moscow, 1979)

Centralized Heating System with Heat Pumps

Oleg Stepanov [1,*] and *Polina* Tretyakova [1]

[1]Industrial University of Tyumen, 625001 Volodarskogo str. 38 Tyumen, Russia

Abstract. The authors introduce the centralized heat supply system of remote consumers which is based on the heat pump systems using secondary energy resources of a thermal power plant as a source of low-grade heat. To assess the effectiveness of the proposed system, the indexes of thermal efficiency are given. The authors describe a mathematical model of determining the dependence of the discounted net profit of consumers' remoteness, the pipe diameter, and the thermal input of the consumer groups located in the city of Tyumen as a result of the introduction of heat pumps in the centralized heating based on the combined heat and power plant.

1 Introduction

In fact, the most important task is to select main energy sources while designing and constructing energy-efficient heating systems in residential and public buildings, because the main purpose of the Federal Law of 23.11.2009 № 261-FL [1] is to reduce the use of resources consumed in the heating needs while maintaining the consequent benefits from their use.

When considered thermodynamically, the combined production of electricity and heat at the combined heat and power plant (CHPP) is much more efficient than separate production of electricity at condensing thermal power plants and heat boilers [2]. However, the use of centralized heat supply systems has disadvantages:

1) a high level of heat energy losses in heat systems (across northern latitudes);

2) an increase in the cost of fuel and energy resources;

3) depreciation of heat systems and equipment, and thus a high level of thermal networks operating costs;

4) violation of hydraulic control of heat systems which leads to underheating and overheating [3].

These unresolved technical and economic problems have a negative impact on the quality and energy efficiency of centralized heat supply in Russia.

The Decree of the Government of the Russian Federation of November 13, 2009 № 1715-p approved the Energy Strategy of Russia up to 2030. According to it, the aim of the energy policy of Russia is to establish the most efficient use of natural energy resources and ensure ecological safety of Russia's energy sector. A significant role is assigned to the development of renewable energy sources [4]. Their usage will save fossil fuels; will balance energy demand due to the needs of consumers located either away from centralized

* Corresponding author: kpt@tgasu.ru

heat systems or close to them, and will reduce the environmental pressure on the part of energy enterprises.

The proposed technical solution is aimed to increase the efficiency of the centralized heat supply of remote heat consumers using heat pump systems utilizing low-grade thermal energy of thermal power plants (TPP).

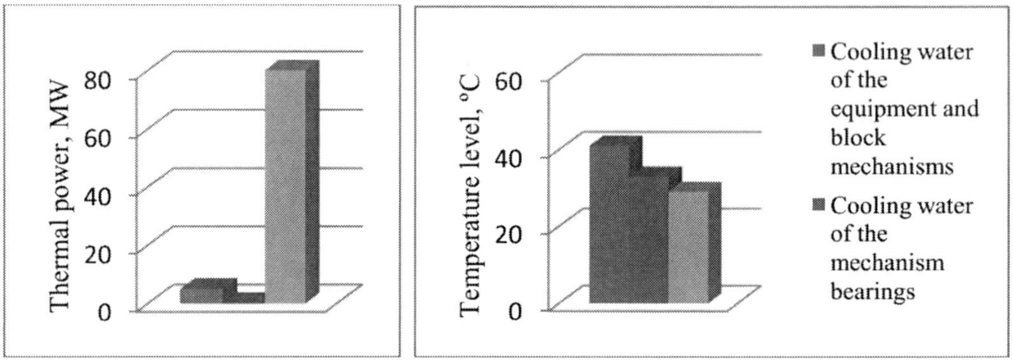

Fig. 1. Thermal power and temperature of renewable energy resources at the combined cycle cogeneration plant (CCCP).

Figure 1 shows thermal power and temperature level of thermal secondary energy resources of a combined-cycle plant (CCP), as exemplified by CCP-110 [5]. The cooling water of the equipment and block mechanisms is of the highest temperature. Its usage as a source of low-grade heat increases the transformation rate of a heat pump system. But water consumption is low and in this case it is more efficient to use it for hot water supply or heating water for cold water supply (CWS).

Taking into account the consumption of heating areas of the city, it is possible to use cooling water thermal power of a vacuum-condensation system [6].

The heat pumps improve the existing technologies of electric and thermal energy and reduce the pressure on the environment by reducing emissions of the pollutants produced due to combustion of natural gas [7]. In addition, the share of alternative energy sources in the supply-demand balance will increase.

2 Subjects and methods

Currently, there are a large number of works devoted to the modernization of the existing system using heat pump systems aimed to improve energy efficiency and the development of combined systems, which combine structural elements of centralized and decentralized heat supply systems. Further, we will consider three main categories of heat pumps application:

- as a peak source [8, 9];
- to use reversed heating system water as low-grade heat [10, 11];
- to reduce a temperature chart of a heat supply system [12, 13].

The integration of heat pump technology in the combined production of heat and electricity increases the production of electricity by reducing losses in heating mains while reducing the temperature of the heat-transfer fluid in the lines, and increases thermal transfer capacity at the same flow rate. More efficient use of fuel energy reduces CO2 emissions [14]. The transition to a low-temperature schedule of heating mains increases the service life, reduces the breakdown rate and material costs by using cheaper heat insulating materials. In the long run, it is quite realistic to introduce different heat tariffs from CHPP—depending on the capacity [15].

To provide remote consumers, we suggest using the centralized heat supply system. In this system, CHPP transfers a low-grade heat-transfer fluid (cooling water thermal power of a vacuum-condensation system) to a central heating station (CHS) which is equipped with heat pumps for heating and hot water supply (HWS).

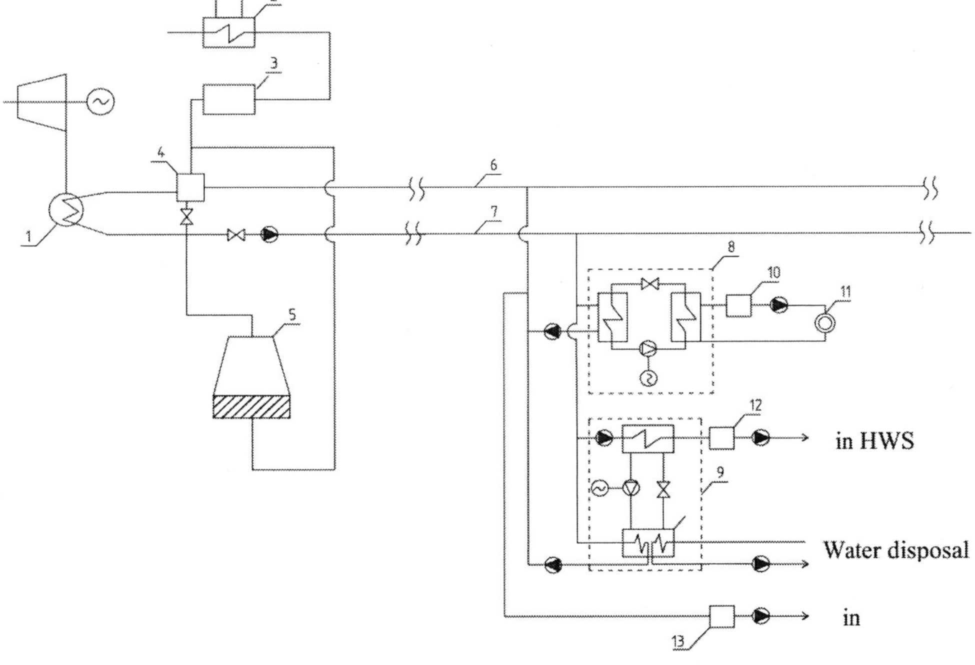

Fig. 2. The heating system using a heat pump: the steam-turbine condenser of CHPP 1, the heat exchanger of cooling system of units and mechanisms 2, the water purification plant 3, the water storage tank 4, the cooling tower 5, straight 6 and 7 reverse system water pipes and remote heat points and heat consumers 11. Each remote heat point includes a heat pump plant of a heating system, a heat pump of hot water system, a hot water storage tank 12, a cold water storage tank 13, a hot water storage tank of the heating system 10 [17].

One part of heat-transfer fluid giving up heat to a heat pump system, returns to CHPP through the return pipeline of heating-system water, and another part is taken by hot water system of consumers. Domestic sewage water is directed to the evaporator inlet of the heat pump of the hot water supply system, and heating water is directed to the evaporator only in case of absence or shortage of domestic sewage heat [16].

Figure 2 shows one of the possible centralized heating systems implementing the proposed method.

Source water is supplied through a pipeline by the heat exchanger 2 in the water purification plant 3 and added to the water storage tank 4, which also receives water from the reverse system water pipes 6. The heating water from the water storage tank 4 is heated in the steam-turbine condenser 1 and is transported by a pipeline network 3 to remote heat points equipped with compression heat pumps and the unclaimed water is cooled in the cooling tower 5.

In the heat pump of the heating system, the water low-grade heat of the heating water pipe is transported to the working fluid of the evaporator heat pump. The cooled water is fed into the reverse pipe. The condensation of the working fluid of the heat pump in the heating system is carried out by return water from the heat consumer 11.

In the heat pump of the hot water system, the domestic sewage heat and / or straight reverse system water pipe are transferred to the working fluid of the heat pump in the evaporator. As a result, the cooling water is fed into the reverse system water pipe and waste water goes down the drain. The condensation of the working fluid of the heat pump of the hot water system is carried out by the water supplied from the straight system water pipe 3. The water in the condenser is heated up to the temperature of hot water system and then it is directed to the hot water storage tank 12.

If necessary, the cold water from the storage tank 13 is collected from the reverse system water pipe, and then it is directed by a pump to the consumers.

The evaluation of the proposed solution of heat pump usage is based on the electric / heat output ratio E [MWt / Gcal] [MWtEE / MWtHE] and the coefficient of fuel heat utilization (CFHU) [18, 19].

3 Results

Heat pumps produce the load required by the consumers. In addition, pipe losses are reducing and as a consequence, steam flow is decreasing which increases electricity generation. But a part of the load is spent on the compressor drives of heat pump systems.

The electric / heat output ratio can be represented as a ratio of underproduced electrical energy to generated thermal energy [MWt / Gcal] [MWtEE / MWtHE].

$$E = \frac{\Delta N}{Q_{determ}} \tag{1}$$

Where: ΔN - electricity additionally produced by the work of condensing cycle CHPP, MWt

Q_{determ} – thermal energy of network heaters, MWt

The heat generation / electric output ratio can be considered as a reverse value of electric / heat output ratio (equivalent of the conversion ratio of heat pumps):

$$\varphi_{hppchs} = \frac{1}{E} = \frac{h''_{exir} - h'_{exir}}{\eta_{oi}\,\eta_m\,\eta_e\left(h''_{exir} - h''_{condenser}\right)} \tag{2}$$

Where: h''_{exir}, h'_{exir} – enthalpy of steam input and enthalpy of water leaving network heater, kj/kg °C;

η_{oi} – relative internal turbine's efficiency;

$\eta_M\, \eta_E$ – electromechanical turbine's efficiency;

$h_{condenser}$ - enthalpy of steam in the condenser.

The electric power is extra expensed due to the use of heat pumps (HP) in the CHS to the compressor drive:

$$N_{condenser} = \frac{\sum Q_{hpp\ chs}}{\varphi_{hpp\ chs}} \tag{3}$$

Where: $\varphi_{hpp\ chs}$ – coefficient of heat pump performance of CHS;

$\sum \varphi HPP\ CHS$ – amount of thermal energy generated by the heat pump, MWt.

The total thermal capacity of HP consists of the amount of heat coming from the source of low-grade heat and the compressor's electric power. Since the heat pump uses the rejected heat, energy saving will be:

$$\Delta N_{transportosser} = \frac{\sum Q_{hpp\ chs}}{\varphi_{chpp}} \tag{4}$$

Electric power generated optionally due to reducing transport losses.

$$\Delta N_{pipe} = \frac{Q_{pipe} - Q_{pipe}^{'}}{\varphi_{chpp}} \tag{5}$$

Where: $Q_{pipe} - Q_{pipe}{}'$ – loss of thermal energy while transporting the heat transfer fluid from the CHPP to the CHS without heat pumps and with them in the CHS.

Changes in electricity generation by means of the heat pump plant (HPP) in CHS:

$$\Delta N_{hppchs} = \Delta N_{transportosser} - N_{condenser} = \frac{\sum Q_{hpp\ chs}}{\varphi_{chpp}} - \frac{\sum Q_{hpp\ chs}}{\varphi_{hpp\ chs}} = \sum Q_{hpp\ chs}\left(\frac{1}{\varphi_{chpp}} - \frac{1}{\varphi_{hppchs}}\right) \tag{6}$$

Changing of HPP electrical power:

$$\Delta N_{hpp+hpp} = \Delta N_{hppchs} + \Delta N_{pipe} = \sum Q_{hpp\ chs}\left(\frac{1}{\varphi_{chpp}} - \frac{1}{\varphi_{hpp\ chs}}\right) + \frac{Q_{pipe} - Q_{pipe}^{'}}{\varphi_{chpp}}$$

$$= \frac{\Delta Q_{pipe} + \sum Q_{hpp\ chs}}{\varphi_{chpp}} - \frac{\sum Q_{hpp\ chs}}{\varphi_{hpp\ chs}} \tag{7}$$

Then the change in power generation at thermal consumption for HP – CHPP with HP will be:

$$\Delta R = E_1 - E = \frac{\Delta N + \Delta N_{chpp+hpp}}{Q_{determ}} - \frac{\Delta N}{Q_{determ}} = \frac{\Delta N_{chpp+hpp}}{Q_{determ}} =$$

$$= \frac{1}{Q_{determ}}\left(\frac{\Delta Q_{pipe} + \sum Q_{hpp\ chs}}{\varphi_{chpp}} - \frac{\sum Q_{hpp\ chs}}{\varphi_{hpp\ chs}}\right) \tag{8}$$

Changing of the coefficient of fuel heat utilization (CFHU) – the proportion of the fuel heat, useful electricity and heat gain. In other words, the ratio

$$CFHU = \frac{W^{\Sigma}_{sustem} + Q^{\Sigma}_{sustem}}{B_{district}} = \eta^{e}_{district} + \eta^{t}_{district} \tag{9}$$

where $B_{district} = B^{\Sigma}_{e} + B^{\Sigma}_{t}$ – fuel consumption

B^{Σ}_{e} – fuel consumption for the production and supply of electricity;

B^{Σ}_{t} – the same for heat;

$W^{\Sigma}_{sustem}, Q^{\Sigma}_{sustem}$ – electricity and heat consumption of the district

Changing of the coefficient of fuel heat utilization – additionally generated electricity to heat of the fuel burned in combustion chambers of gas turbines (Q_t) Q_{TOT}):

$$\Delta CFHU = \frac{\Delta N}{Q_t} \tag{10}$$

If HPP uses rejected heat in CHS, then

$$\Delta CFHU = \frac{1}{Q_t}\left(\sum Q_{hpp\ chs}\left(\frac{1}{\varphi_{chpp}} - \frac{1}{\varphi_{hpp\ chs}}\right) + \frac{Q_{pipe} - Q_{pipe}'}{\varphi_{chpp}}\right) \tag{11}$$

When the use of heating systems with heat pumps is effective from the thermodynamic point of view, the minimum value of heat transformation ratio can be calculated according to the conditions: electricity underproduced as a result of heat energy in a standard schema must be greater than the share of electricity underproduced in the schema with HPP.

$$\frac{1}{\varphi_{chpp}} \rangle \frac{1}{\varphi_{hpp\ chs}} - \frac{\Delta \eta_{pipe}}{\varphi_{chpp}} \tag{12}$$

$$\varphi_{hpp\ chs} \rangle \frac{\varphi_{chpp}}{1 + \Delta \eta_{pipe}} \tag{13}$$

where: $\frac{1}{\varphi_{hpp\ chs}}$ – the share of electricity spent on the compressor drive;

$\frac{\Delta \eta_{pipe}}{\varphi_{chpp}}$ – the share of electricity saved by reducing pipe losses.

The analysis of the presented above equations shows that the system is effective only at high conversion effectiveness of HP (use of low-temperature heating systems) and high pipe losses in the existing system.

We want to evaluate the effectiveness of the proposed heating system for a conventional settlement with the building area of 700 km2 and consumed heat input of 5.6 mln Gcal. When CHPP is located as close as possible to the center of the settlement, the most remote consumer is located in 15 km from the source of thermal energy. The total heat loss is 1.68 Gcal/year. When $\varphi_{hpp\ chs} = 3.9$ and $\varphi_{chpp} = 4.73$ $\varphi_{HPP\ CHS} = 3.9\ and\ \varphi_{CHPP} = 4.73$we get an efficiency limit of HPP across 7 km (Fig. 3).

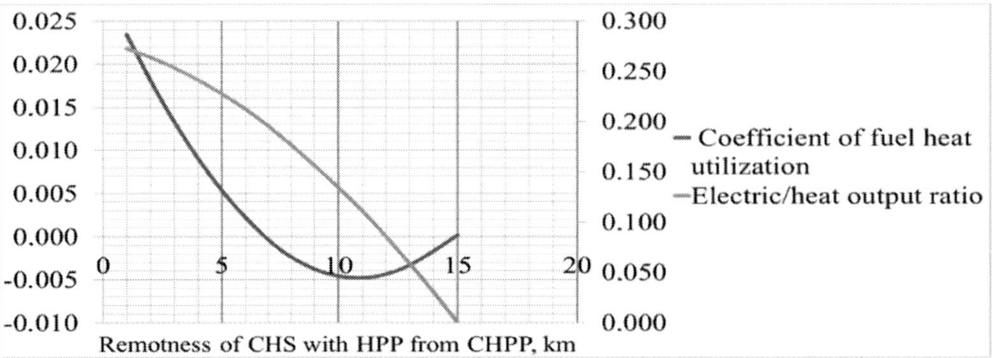

Fig. 3. Dependence of the coefficient of fuel heat utilization and electric/heat output ratio on the consumer remoteness from CHPP.

Technical and economic indicators of the heat supply systems with heat pumps depend on the consumption and electricity cost, capital costs for the systems. In its turn, the electricity cost is reducing while the speed reduction of heat-transfer fluid in the main pipeline. But the capital expenditures on the heat supply system increase, since the pipeline diameter increases:

$$NPV = f\left(l_{pipe}, d_{pipe}, q_{consumer}\right) \tag{14}$$

To determine the dependence of net discounted profits on the pipeline length (remoteness of customers) $\left(l_{pipe}\right)$, pipeline diameter (d_{pipe}) and consumers' heat input ($q_{consumer}$), we built a mathematical model based on the method of orthogonal central compositional planning [20, 21] applicable for the climatic parameters of Tyumen, and calculated net discounted profits on the basis of the model. As a result, we have the dependence $NPV = f\left(l_{pipe}, d_{pipe}, q_{consumer}\right)$ in the form of an equation:

$$\begin{aligned} y &= -57953586 + 477314 \cdot d - 1080373 \cdot l + 1482485 \cdot \mathrm{q} + 9829 \cdot \mathrm{dl} + \\ &+ 31028 \cdot \mathrm{dq} - 410045 \cdot \mathrm{lq} - 1303 \cdot \mathrm{d}^2 - 118871 \cdot \mathrm{l}^2 - 196220 \cdot \mathrm{q}^2 \end{aligned} \tag{15}$$

These results allow us to determine the optimal size of the diameter, the remoteness of consumers and their input. For example, the optimum diameter of the pipeline from the CHPP is 380 mm if the input of consumers is 5 Gcal/h and the remoteness from the CHPP is 7 km, while the NPV is 38.091 thousand rubles when replacing an existing source of heat boiler (Fig. 4).

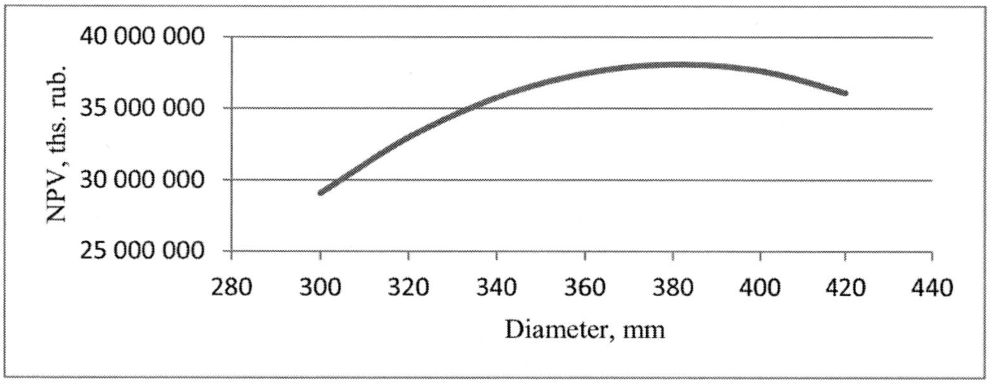

Fig. 4. $NPV = f\left(l_{pipe}, d_{pipe}, q_{consumer}\right)$ dependencies.

We can determine the optimum temperature for the calculation of the heating system of a developing area with 380 mm pipeline diameter (Fig. 5, Fig. 6).

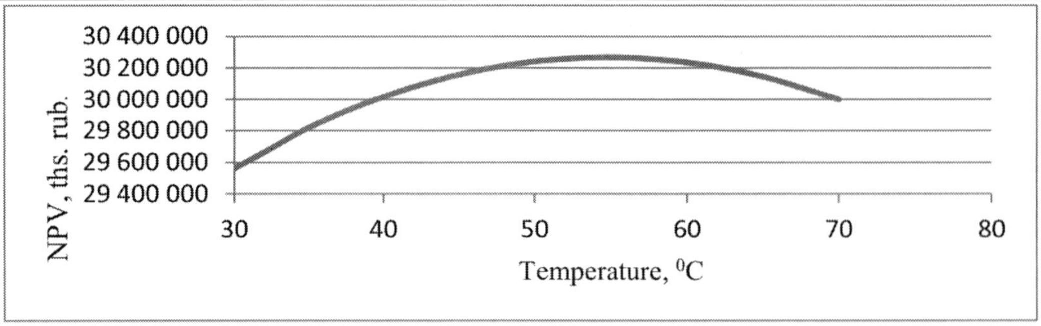

Fig. 5. The optimum temperature in the supply pipeline.

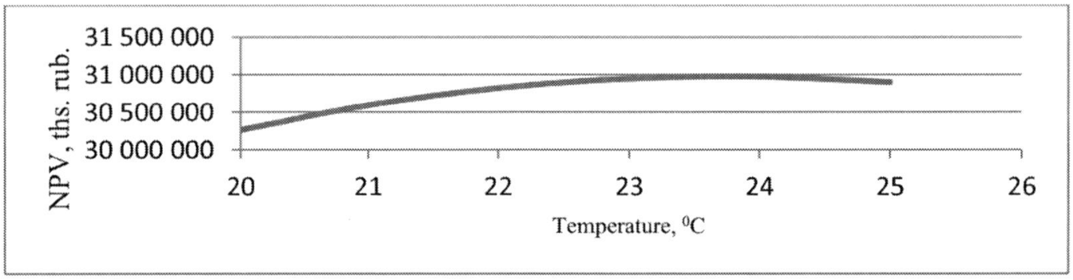

Fig. 6. The optimum temperature in the return pipeline.

4 Summary

We have developed a heat supply system for remote heat stations of a low-grade heat-transfer fluid from CHPP (it is obtained after heat recovery from steam-turbine condenser and cooling water of the equipment and mechanisms of the electrical power unit). Each heat station is equipped with vapor compression heat pumps of the heat supply system and hot water supply. A part of heat-transfer fluid returns to CHPP through the return pipeline of heating-system water, and another one is taken by the hot water system of consumers.

We have proposed heat efficiency indicators of CCCP—heating main—CHS to assess the effectiveness of the proposed heat supply system: changing of the coefficient of fuel heat utilization and changing of the electric/heat output ratio.

We have built a mathematical model to determine the dependence of net discounted profits as a result of the introduction of heat pumps in the centralized heat supply system based at Tyumen CHPP. The profits depend on the consumers' remoteness, heat input and pipeline diameter.

We have given the examples of application of these indicators for the consumers with a given heat input at a given distance. The optimum pipeline diameter, the optimum water temperature in the straight and reverse system water pipes of the consumers' heat supply system are determined.

References

1. *Federalnyy zakon ob energosberezhenii i o povyshenii energeticheskoy effektivnosti i o vnesenii izmeneniy v otdelnye zakonodatelnye akty Rossiyskoy Federacii: feder. zakon ot 23.11.2009 № 261-FZ* (Rossiyskaya gazeta, 2009)
2. A. Sokolov, *Sistemnaya effektivnost otopitelnyh PGU TEC v sistemah teploenergosnabzheniya : Dissertaciya kand. tehn. Nauk* (2004)

3. N.M. Chikisheva, Tyumeni Topical issues of construction, ecology and conservation in Western Siberia, **1**, 188–192 (2014)
4. Information on http://minenergo.gov.ru/aboutminen/energostrategy
5. P. Tretyakova, Materialy mezhdunarodnoy nauchno-tehnicheskoy konferencii studentov, aspirantov, uchenyh, Chelybinsk (2013)
6. P. Tretyakova, Energosberezhenie na kurganskoy tec-2 za schet utilizacii nizkopotencialnoy teploty, Omskiy nauchnyy vestnik (2014)
7. A. Andryushchenko, *Kombinirovannye sistemy energosnabzheniya* (Tepoenergetika, Moscow, 1997)
8. V. Danilov, V. Slavin, Patent 2170885 Rossiyskaya Federaciya, MPK7F24D15/04, F01K17/02. Sistema teploenergosnabzheniya (2001)
9. M. Orlov, V. Sharapov, P.v Chaukin, V. Mordovin, Patent 2467257 Rossiyskaya Federaciya, MPK7 F 24 D 3/08. Sposob teplosnabzheniya (2012)
10. V. Stenin, Patent 2239129 Rossiyskaya Federaciya, MPK7 F 24 D 3/08. Sposob teplosnabzheniya (2003)
11. V. Stenin, Patent 2266479 Rossiyskaya Federaciya, MPK7 F 24 D 3/08. Sposob teplosnabzheniya (2004)
12. A. Bogdanov. Vseros. nauch.-prakt. konf, Krasnoyarsk, **1**, 20–23 (2009)
13. V. Tomilov, Y. Pugach, G. Nozdrenko, L. Pugach, Yu. Ovchinnikov, P. Shchinnikov, Patent 2163327 Rossiyskaya Federaciya, MPK7 F 24 D 3/08. Sposob centralizovannogo teplosnabzheniya (2001)
14. E. Sheremet, A. Seminenko. *Primenenie teplovyh nasosov vsistemah centralizovannogo teplosnabzheniya v celyah povysheniya ekonomichnosti i energoefektivnosti teplovyh setey* (Tepoenergetika, Moskva, 2013)
15. V. Berzan, S. Robu, M. Shyt, Problemy regionalnoy energetiki, **1**, 91–94 (2011)
16. P. Tretyakova, Omskiy nauchnyy vestnik, 178–182 (2014)
17. P. Tretyakova, Pat. 2571361 Rossiyskaya Federaciya, MPK7 F 24 D 11/02. Sistema centralizovannogo teplosnabzheniya, goryachego i holodnogo vodosnabzheniya (2015)
18. A. Batukhtin, A. Kalugin, *Modelirovanie sovremennyh sistem centralizovannogo teplo snabzheniya* (Moscow, 2011)
19. P. Shchinnikov, Vestnik SGTU, **1**, 139–149 (2004)
20. A. Batukhtin, V. Makkaveev. Primenenie optimizacionnyh modeley funkcionirovaniya sistem teplosnabzheniya dlya snizheniya sebestoimosti teplovoy energii i uvelicheniya raspolagaemoy moschnosti stancii, **1**, 7–8 (2010)
21. V. Suzdaltsev, E. Takhavova, M. Shleymovich, *Predstavlenie znaniy v informacionnyh sistemah, Matematicheskoe modelirovanie: uchebnoe posobie* (KGTY, Kazan, 2005)

MATEC Web of Conferences 73, 02017 (2016) DOI: 10.1051/matecconf/20167302017
TPACEE-2016

About the Territorial Potential of the Construction of Battery-Charging Stations for Autonomous Electric Motor Vehicles in the Regions

Lyubov Shilova[1,*], *Dmitry* Soloviev[2], *Vassily* Timatkov[2], and *Aleksey* Adamtsevich[1]

[1]Moscow State University of Civil Engineering, 129337 Yaroslavskoe sh. 26, Moscow, Russia
[2]Joint Institute for High Temperatures of the Russian Academy of Sciences, 125412 Izhorskaya str. 13/2, Moscow, Russia

Abstract. The article describes the main current trends in the development of electric motor vehicles with "zero emission" as well as the battery-charging stations concerned. The study is based on a preliminary comparative analysis of the RF regions with respect to five indices (average per capita income, number of private cars in the region, air pollution level, provision of the region with power supply and the potential use of local renewable energy resources), and it gives some recommendations on the prospects of possible construction of battery-charging stations in the regions.

1 Introduction

A steady progress in the electrification of the economy is one of the main trends in the development of economies and power engineering. Despite the fact that the electric energy is less than the half of the current energy consumption, the aforesaid trend (with consideration of the urban population growth, new electronic technologies, the automated industries and others) steadily leads the global economy to the so-called "electric world" situation when the electricity is the dominant type of energy consumed [5,13,15].

At present, one of the signs of the economy electrification is the development of electric motor vehicles with hybrid power units or completely electric ones for both the individual and commercial use and the public transport lines [2,16]. The key attention here is paid to the electric motor cars.

Many experts consider a technological break-through in the development of electric motor cars by 2020-2030 quite possible. It will be provided for by new light-weight and durables batteries (able to produce more than 5000 recharge cycles) as well as new electric motors (preferably without rare earth metals) and ultra-light construction materials [6,7].

* Corresponding author: shilovaLA@mgsu.ru

According to the plans of developed world countries supporting the production of electric motor vehicles, there will be 5 to 7 mln of electric motor cars in the megalopolises to say nothing of the millions of hybrid cars. At first, the demand basis may be formed by the taxi companies, the logistic companies and others changing their fleets due to the support of their states [4].

The leading European countries and the USA declared their intention to abandon the use of internal combustion engines by 2050. In Russia, at present, we also have taken some programmes on introduction of electric motor transport which are put into practice in some regions. Here, the analysis of possible engineering solutions for such projects as well as of their competitiveness with consideration of natural climatic conditions and other factors in very important.

An active use of electric motor transport is impossible without a developed battery-charging system, which should be efficient from point of view of both economy and power engineering, and which should be able to be installed into both the existing power engineering systems and the prospects ones. In this situation, the mass use of electric motor cars will cause the growth of total electric energy consumption, and it will require the introduction of new power plants. In the process of integration of electric motor cars into transport system in Russia, the question of the influence of battery-charging system on the existing power grid as well as on the power engineering services for the other consumers will be very important. It is also important to consider a possible use of renewable energy sources as additional capacities for battery-charging systems. Taking into consideration both the solar energy potential in Russia and the promising regions concerned, we may conclude that it is quite logical to use the solar energy for the battery-charging systems in southern and central RF regions.

The average electric motor power is 3 to 5 kW and more, the battery capacity is not less than 2 to 5 kWh for hybrid cars and 20 kWh and more for electric motor cars. It provides for a considerable increase in the demand for electricity and stimulates the development of the power system which acquires a new active consumer. On the other hand, the power system can use the batteries for some time periods and so to smooth the energy consumption peaks etc. [1]

The electric motor vehicles and renewable energy sources (RES) are the most promising technologies for the improvement of urban environment in megalopolises [3,5,10]. The motor transport is the main pollutant in cities. In our country, one of the most important unsolved problems is the development of the battery-charging system including local renewable energy sources efficiently integrated into the power grid [2, 8, 9, 14]. The promising regions for the development of the battery-charging stations should be chosen; it will allow us to plan an efficient development of both the battery-charging system and the electric motor transport.

2 How to choose the appropriate regions for the development of electric motor transport and battery-charging systems

To promote such an unusual product for the consumers as the electric motor vehicle, it is necessary to choose the appropriate regions in the RF. Considering the special features of this motor transport type, we suggest the following criteria for the estimate of regions under consideration:

– average per capita income;
– number of private cars;
– air pollution due to the fossil fuel combustion;
– provision of the region with electric energy;

– available local renewable energy sources.

The electric motor cars will be more expensive in the near future than the conventional motor cars, so the regions with a higher per capita income are preferable.

The RF regions with the highest wages are the Central region, the North-West one, the Urals and the Far-East one [19]; the RFsubjects with higher wages are Moscow, St.-Petersburg, Khanty-Mansi autonomous region-Yugra, Moscow Region, Tatarstan autonomous republic. The aforesaid regions have rather developed economies. The top ten rigions also include Yamal-Nenets autonomous region, Tyumen region, Sakhalin region, Samara region and Bashkortostan autonomous republic.

The introduction of a new product into the market requires a certain critical demand volume determined by the economic expediency of the process. Ceteres paribus, the estimated demand volume for the motor transport with "zero emission" will be proportional to the number of motor cars in the region. Therefore, the regions with the largest motor car fleets are preferable.

The motor car fleet in Russia includes 39 349 300 cars ("Avtostat Info" Co., 01.01.2015) [18,20].

The number of private motor cars in Russia is calculated as a specific parameter "number of motor cars for 1000 people". At the beginning of 2015 this parameter was equal to 274 (without Crimea) in comparison with 258 an year earlier (6.2% year's increase). Thus, the country motorization goes on despite the economic crisis caused by international punitive sanctions and the reduction of oil price. The Central federal region has the largest car fleet (30% with the 6.2% year's increase), the Volga region has the second largest one (20.1% with the 7.6% year's increase). The top ten regions are presented in Fig.1.

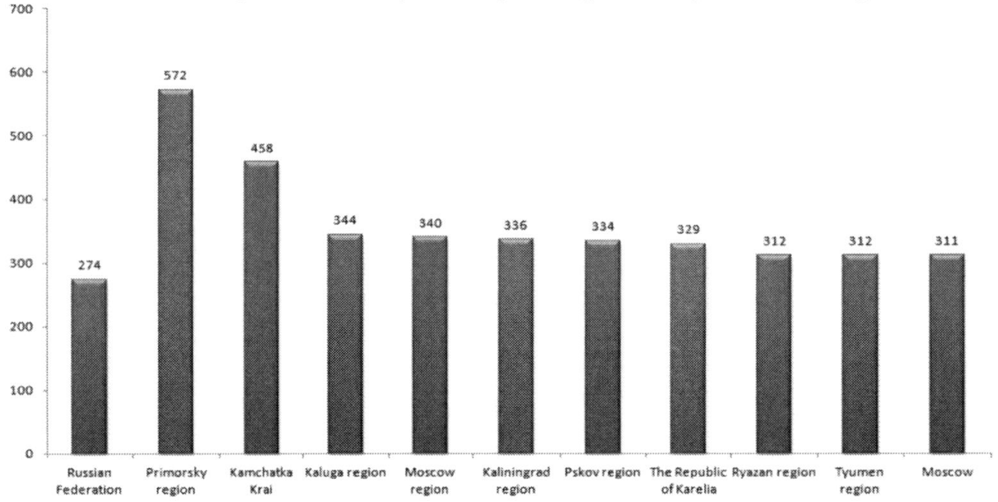

Fig.1. The RF regions with the highest values of the "number of motor cars for 1000 people' in 2014 [18].

The motor transport contributes much to the air pollution, and its improvement can play a considerable role in reduction of air pollution. In regions with high air pollution levels, the vehicles with "zero emission" will be welcome by both the population and the local authorities.

The estimated mass of the pollutants released into the air is not quite informative as it is not proportional to the pollutant concentration in megalopolises and large industrial centres. The data on air pollution in the regions are necessary, but this statistics in beyond the scope of this study.

It is expedient to estimate the fuel expenditures as they considerably determine the total emission of greenhouse gases. On account of the quotas on greenhouse gas emissions, the electric motor cars may be considered as a means of reduction of these emissions due to the state subsidies for the purchase of vehicles with "zero emission".

According to the "Autostat Info" Co. [17], 15 largest regions in Russia consume 47% of the fuel. The other 68 regions consume 53% of that. The European part of the country consumes more than 70% of the fuel. The regions with the highest consumption of fossil fuels: Moscow (8.36%), Moscow region (5.58%), Krasnodar region (3.81%), St.-Petersburg (3.42%), Primorsky region (3.12%), Rostov region (2.77%), Tatarstan (2.68%), Sverdlovsk region (2.67%), Chelyabinsk region (2.21%), Samara region (2.19%).

The electric energy is relatively cheap, and the maintenance of electric motor cars is cheaper as compared with conventional motor cars. However, the mass use of electric motor cars may cause a considerable increase in the maximum electric load in the region, and it would require either the introduction of additional electric generators or the improvement of carrying capacity of power systems. Both variants are rather expensive, and they may be rejected by the local authorities.

Several regions already have considerable negative energy balance due to the ever-increasing consumption of electric energy by both the private households and the industry. So the preference should be given to the regions with positive or slight negative electric energy balance. An exception may be the regions with irregular daily schedule of electric energy consumption, as the reserve energy can be used at battery-charging stations.

The data on production and consumption of electric energy in the RF regions are available in the open Rosstat statistics issues [21].

Top ten RF regions with positive electric energy balance: Saratov region, Tver region, Leningrad region, Kursk region, Smolensk region, Krasnoyarsk region, Kostroma region, Stavropol region, Irkutsk region, Amur region.

Among the regions with decentralized power supply, it is necessary to determine the regions with considerable renewable energy resources and to evaluate their potential use at the battery-charging stations.

Fig.2 presents the engineering capacity of the RF regions with the highest local wind and solar energy resources. The engineering capacity of the use of renewable energy sources is the wind and solar energy available for the conversion into the electric energy in compliance with the environment protection regulations [12]. This criterion is very important for the development of battery-charging systems for electric motor cars in decentralized regions.

The use of renewable energy sources at battery-charging stations in the RF regions with developed power systems may be expensive and inefficient. In regions with dominant nuclear power plants and hydroelectric power plants, the environmental benefits may be questionable as these technologies themselves do not produce considerable heat energy emissions. Moreover, the NPP have the surplus energy at night, when the energy consumption in less. At the same time, the renewable energy sources may produce considerable environmental effect in the regions with dominant thermal power plants and even economic effect in the regions with undeveloped power systems.

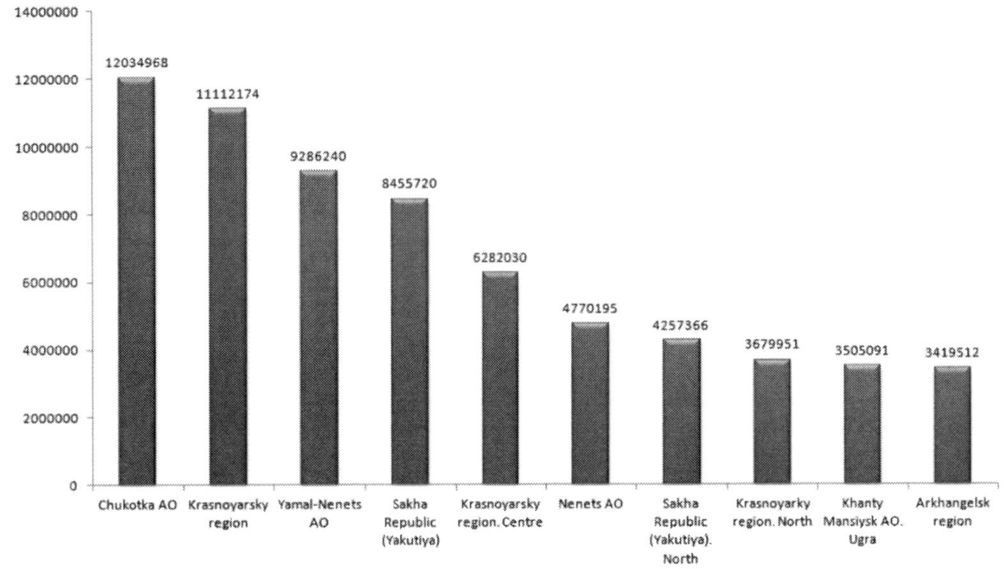

Fig.2. The RF regions with the highest local wind and solar energy resources, billion kWh [11].

On the basis of the aforesaid study, we can conclude that the most promising regions for the development of electric motor vehicles and the battery-charging systems are Moscow, Moscow region, St.-Petersburg and Leningrad region which head the top ten regions with respect to the aforesaid parameters. The following regions are good for these projects as well: Kaliningrad region, Krasnoyarsk region, Primorsky region, Sakhalin region, Tyumen region, Tatarstan autonomous republic. The most promising for the use of renewable energy sources at the battery-charging stations are the following regions: Yamal-Nenets region, Sakha region (Yakutiya), Krasnoyarsk region and Arkhangelsk region.

3 Conclusion

Despite the success in the field of development of autonomous electric motor transport, one of the main problems in the field will be the undeveloped battery-charging system in Russia. This problem is rather an economic one than a technological one. There are still not many battery-charging service users, and so the investors are reluctant to invest into the projects concerned, and the sale of electric motor vehicles is growing slowly due to the lack of battery-charging services.

The use of electric motor vehicles promises both benefits and some expenses. Currently, for the end users, the lower operating costs do not compensate a higher cost of electric motor vehicles. In this situation, the encouragement of introduction of electric motor vehicles may be carried out not only due to the state subsidies to the purchase of electric motor cars (as is the case in the EU countries) but also due to a professional planning and the investments into the development of an efficient system of battery-charging stations.

The electric motor vehicles help to diversify the fuel consumption balance reducing the dependence on oil products. The electric energy in our country is not expensive, and the production of electric motor vehicles may help to develop new high-tech industries, to create new working places and to increase the economic growth. In the long term, the use of electric motor vehicles will allow us to reduce the level of harmful emissions in the cities. The study based on a preliminary comparative analysis of the RF regions with respect to five indices (average per capita income, number of private cars in the region, air pollution level, provision of the region with power supply and the potential use of local renewable energy sources) allowed us to choose the most promising regions from the point of view of possible construction of battery-charging stations necessary for the use of electric motor vehicles with "zero emission". The most important of the aforesaid regions are Moscow and Moscow region. Ten more regions in Russia are promising as well. The use of renewable energy sources at the battery-charging stations will be expedient only in Northern and Eastern regions, where the most part of electric energy is produced at thermal power plants and the power systems are not properly developed yet.

The environmental effect of construction of battery-charging stations is obvious only for megalopolises. The total environmental effect will depend on the engineering characteristics of battery-charging stations, the charging speed, the climatic conditions of their operation, the type of energy generation (thermal, nuclear, hydraulic, etc.) and the fuel type. The various combinations of the aforesaid parameters in the process of the use of electric motor cars may result in both the reduction and the increase of the greenhouse gas emissions in the country. At the same time, the reduction of greenhouse gas emissions and the sale of quotas at the foreign market is nowadays the most obvious and simple way of monetary evaluation of the environmental effect of the development of electric motor transport.

The further studies will include a thorough analysis of daily and seasonal irregularity of the consumption of electric energy, applied to the possible use of the power reserves at the battery-charging stations. It is also necessary to determine some additional factors influencing the development of battery-charging systems with the use of renewable energy sources.

References

1. P.H. Andersen, J.A. Mathews, M. Rask, Energy Policy, **7(37),** 2481–2486 (2009)
2. E. Azadfar, V. Sreeram, D. Harries, Renewable and Sustainable Energy Reviews, **42**, 1065–1076 (2015)
3. A.R Bhatti, Renewable and Sustainable Energy Reviews, **54**, 34–47 (2016)
4. M.A. Hannan, F.A. Azidin, A. Mohamed, Renewable and Sustainable Energy Reviews, **29**, 135–150 (2014)
5. L. Liu, Renewable and Sustainable Energy Reviews, **51**, 648–661 (2015)
6. S. Manzetti, Mariasiu, L. S. Florin, M.F.E. Manzetti, Renewable and Sustainable Energy Reviews, **51**, 1004–1012 (2015)
7. I. Rahman, Renewable and Sustainable Energy Reviews, **58**, 1039–1047 (2016)
8. S. Speidel, T. Bräunl, Renewable and Sustainable Energy Reviews, **40**, 97–110 (2014)
9. K.M. Tan, V.K. Ramachandaramurthy, J.Y. Yong, Renewable and Sustainable Energy Reviews, **53**, 720–732 (2016)
10. A. Verma, Energy, **86**, 548–559 (2015)
11. T.I. Andreenko, T.S. Gabderahmanova, O.V. Danilova, *Atlas of Russian renewable energy resources: scientific edition* (MUCTR. Mendeleev, Moscow, 2015)
12. P.P. Bezrukikh, *Russian renewable energy resources directory* (Energia, Moscow, 2013)
13. V.V. Bushuyev, V.V. Timatkov, ELECTRO: electrical engineering, power engineering, **3**, 2-5 (2012)
14. S.V. Kiseleva, O.S. Popiel, A.A. Tarasenko, International Scientific Journal for Alternative Energy and Ecology, **11(133)**, 16-24 (2013)
15. V.V. Timatkov, Energy policy, **6**, 58-67 (2014)
16. V.V. Timatkov, V.V. Bushuyev, *Electric transport as part of the electrical world. Facts and Forecasts* (Energy, Moscow, 2015)
17. Information on https://www.autostat.ru/infographics/10178

18. M*arket of passenger cars in Russia. The results of the 2014, trends and outlooks* (, Autostat, Togliatti, 2015)
19. Information on http://www.gks.ru/free_doc/new_site/population/trud/sr-zarplata/t2.xlsx.
20. Information on http://avtostat-info.com/Article/130.
21. Information on http://www.gks.ru/wps/wcm/connect/rosstat_main/rosstat/ru/statistics/enterprise/industrial

MATEC Web of Conferences 73, 02018 (2016) DOI: 10.1051/matecconf/20167302018
TPACEE-2016

Sustainable Building Life Cycle Design

Alexander Ginzburg[1,*]

[1]Moscow State University of Civil Engineering, 129337 Yaroslavskoe sh. 26, Moscow, Russia

Abstract. The current building life cycle management system in the Russian Federation is a family of discrete subsystems that exist independently for different building life cycle stages. In this situation building reliability and sustainable functioning are out of the question. The implementation of a united information model (BIM-model) intended to describe building entire life cycle will allow to raise the sustainability, but this will happen only if goals and concerns of all participants of the project process are properly coordinated. An important figure of process sustainability is the organizational and technological reliability (OTR) that describes the possibility of a system to reach a goal. In case of building life cycle design, the economical efficiency of a building can be considered as the goal. The required technical, ecological, organizational, and other parameters form a complex of constraints that determine the area of allowable values for building functioning. In its broad meaning, OTR may be understood as the probability of receiving an economical effect based on the value of organizational and economical reliability (OER).

1 Introduction

Any physical object goes through different stages of existence. The "life cycle" term is intended to cover all these stages: from the concept design of a building to its removal and waste disposal. During its life cycle, a building can go as well through rebuilding, refurbishment, and even purpose changing.

At the concept design stage, the parameters of a future building are specified. The customer (Construction Investor) defines building expediency and specifies financing terms. If necessary, a detailed analysis and a product/service demand forecast are done. Their purpose is to ensure funding for the planned building.

The main concern for the investor is the conditions, the time of investment return and the profit margin from project realization. Even at this stage, i.e. before the beginning of any works serious contradictions of interests appear. The investors planning to sell the building immediately after construction will apply design requirements totally different from those of the investors planning to use the building by themselves. If a customer is not interested in a cost-saving and efficient use of a building, he will not order, for example, any energy and resource saving materials and technologies because they are usually more

* Corresponding author: ginav@mgsu.ru

expensive and therefore disadvantageous at the design and construction stage. Such customer is not interested in the stage where energy saving costs begin to pay back.

This is only one of the most obvious examples. In fact, there are much more contradictions that affect the entire future life cycle of a building, particularly sustainability and duration.

Although as a rule people address all their complaints regarding buildings to builders, in fact, the builders build according to the project documentation developed by designers. What are designers' interests then? Here we should mention that the common practice in the Russian Federation is that design cost is calculated as a percent from construction and installation works. This means that the more expensive the construction object is, the better it is for the designer. Design firms are not interested at all in using innovative cost-saving materials, accelerated construction technologies etc. in their projects.

This can be well demonstrated by a comparative analysis of project designs made by design and construction entities, that is, by the companies doing both design and construction (in simple words, those who design for themselves) and usual design companies, that design for others. Even for similar buildings, the discrepancies are very significant.

2 United informational base. BIM technologies

The modern building life cycle management technologies are based on the creation and support of a building information model (BIM), the goal of which is to ensure the existence of a united informational base that would comprehensively describe a building.

Since the implementation of BIM technologies requires investment outlays, it can be considered as a realization of an investment project. This, in its turn, implies the necessity to justify the investment attractiveness of their implementation through a complex of assessment figures reflecting their economic efficiency and non-economic effects.

Here we have a question again: Economic efficiency for whom? The answer to this question determines not only the informational body of a BIM, but even the necessity of its application.

BIM technology application experience shows that their maximum pay off begins at the operational stage. As well, BIM is very effective at the stage of building removal and disposal, since the builders need information about the structure of the building, materials it was made of etc. As a rule, such analysis of a building takes a lot of time and money. But a building may serve for 100 years before it is removed and disposed of, while BIM requires investment at the design stage. Is the investor interested in it? He is, if he invests in a building intended for himself. If the building will be used by others, the modern business-models do not encourage the investor to spend extra money. Under current conditions in the Russian Federation, the investor as a rule cannot warrant a BIM by proving it to the buyer that though a building costs more, it will pay off faster.

Is BIM advantageous to design companies? That is, are design companies ready to apply BIM without additional funding (at their own cost)? As a matter of principle, BIM can give them additional advantage when they bid for tenders, they will be able to select and match a lot of building forming components more effectively, and to issue the required graphic and calculation documentation faster. Most Russian designers point out the effectiveness of BIM technologies when there is a need to reveal design internal mismatches and discrepancies (this especially applies to the design of complex utility systems), and note a significant time saving in case any modifications are needed. The software systems used for designing in BIM interface automatically replace one or another unit, assembly or material throughout the entire package of design documentation, which allows not only save time, but also avoid any engineering faults that often occur when high scale changes take place.

But we should keep in mind that to apply the BIM technology, a design company will have to purchase corresponding (and quite expensive) software and equipment, train their specialists to work with BIMs, and involve additional specialist managers into their staff to provide technical support to the system.

If assessed solely at the design stage, the effectiveness of the BIM technology does not look definitely justified. The increase of investments pay off is much more obvious when the generated model turns out to be of demand at the following stages of building life cycle.

If a BIM is intended to ensure building sustainability during its entire life cycle, then it should be filled with the figures that characterize the building at its different life cycle stages. Here we face the need to assess effectiveness again. For designers, the extension of mass data in a BIM in fact means the increase of overhead costs.

The modern BIMs do not imply that information on a building is entered into the united storage once and forever. In fact, the talk is about not one, but many databases interconnected by means of different interfaces and software systems. It is clear that during building lifetime, the data and the composition of the data change many times depending on the changing conditions, goals, etc. But practice shows that an originally correct data composition (composition of the data entered when a BIM was created at the design stage) significantly increases BIM effectiveness at the following stages.

In theory, BIM technologies can be applied, for example, from the construction stage if no BIM was created by designers. Having received design documentation in the electronic form the designers, a construction company in certain cases can export the received data into a software system intended for BIM based designing and use it for effective construction management, including coordination of works and materials, structures, and building products supplies in time.

Still, as well as in the case of designers, the construction company should take into account that application of BIM technologies in construction will imply significant material, time and organizational costs. The analysis of BIM application at the construction stage in the Russian Federation clearly shows that the implementation of these technologies is successful when the original BIM has been created at the design stage, and when designers entered the data taking into account it's future use at the construction stage. As a rule, this is possible in design and construction firms only, where designers and builders are united in one holding. In other cases, the investor should be seriously interested in such BIM and ready not only to pay for such close interaction of different participants of the process, but also organize it.

The next and the most important stage of a building life cycle is its use. As of today, a significant experience in the creation and use of smart buildings and even their conglomerates (micro-districts if not cities) has been accumulated. Technologies for monitoring the parameter that characterize building state and behavior have been developed, the mechanisms of human activity parameters adjustment and adaptation to different conditions have been proposed. However, practice shows that at least in the Russian Federation the building use stage is absolutely isolated from those information modelling achievements that are applied at the design and construction stage. In fact, there are no examples of BIM technology application at the stage of building use in the Russian Federation. Though, it is that very stage where companies might make use of all advantages of BIMs, because BIMs describe the history of building life cycle in details.

This can be partially explained by the fact that the wide application of BIM technologies is relatively young in the Russian Federation. So, the buildings for which BIMs were developed are either still being built, or have been built recently and their models have not been put into practice by the companies that use the buildings. It may be so, but the problem of economic interest in the creation and informational support of BIM technologies steps forward again. What can motivate a construction company to spend time and efforts on filling a BIM with data needed to someone else, not them?

3 Content of BIM

It should be mentioned that at present there is no complete and formalized mathematical formulation for building life cycle. This means that at different stages different isolated models are used. Different executors use different data sets, solve different problems and, by the way, not always correctly or successfully. Of course, all of these impacts building life cycle sustainability.

When speaking of BIMs, many people imply different informational content. The most spread opinion is that the main feature of a BIM design is its 3D representation. In fact, this achievement of BIM technologies is the most questionable one. The future experts are taught to read 2D drawings from the first course of the institute. A 3D picture is, of course, good when you do a presentation of your project to a non-professional customer, but it gives little in content respect.

The real advantage of BIM technologies emerges when 3D models are complemented with the technical parameters of structures, construction materials data, prices, load curves, delivery time, outsources' e-mail addresses, etc. At the use stage, all this information is as well complemented with the information on building components' technical state, repairs, upgrades, replacements, service life, expenses, and responsible persons and organizations. This is why today experts speak of N-D design, that is, of buildings described in general by N-parameters needed for the analysis and mathematical formulation of its entire life cycle.

The more N-parameters are used, the more complex the mathematical model is. And this does not always imply that it is more effective. For example, we can collect a lot of data describing the current state of a building. The technological infrastructure of modern buildings and structures well allows to do it. But it is not always clear what should one do with these data. The modern construction science by no means always allows to interpret them correctly. A signal of temperature raising does not always indicate that a fire is spreading; and the information on cracks or micro-damages in separate structures is not always a sign of alarm. This is why the issue of reasonable sufficiency of BIM information should be discussed.

The main criteria for sufficiency or non-sufficiency of BIM size is its economical advisability, that is, the result of the comparison between the costs required for BIM creation, support at all stages of building life cycle and the advantages provided by the received informational and mathematical model. The fact that a lot of participants with different interests are involved in BIM creation process increases the complexity of the problem. In addition to this, it often happens so that ones are to bear the costs, and the others are to get the advantage.

Speaking formally, we have to solve a multi-attribute task with many contradicting criteria. In general cases, such tasks do not have an exact one-valued solution. In theory, a number of approaches to mathematical analysis of such situations is used to solve the task. As a rule, the approaches come down to trying to choose one most important criteria and optimize it. The other criteria are considered as a set of constraints.

In construction system engineering, there is a possibility to reject the analysis of contradicting criteria and go over to generalized, so called system engineering integrated criteria that describe the behavior of a system as a whole. The main idea is, that proper functioning of the system in fact results in reaching a balance of contradicting characteristics.

4 Organizational and technological reliability

Organizational and technological reliability (OTR) is one of the important system engineering integrated criteria. This figure describes the probability of complex system object's reaching a goal. As opposed to classical reliability, here, in addition to system failures (capable to cause system breakdown) system disturbances (that impair life cycle figures but do not cause breakdown) are taken into account.

Construction companies operate in the stochastic environment. Being influenced by different random factors, course of the construction production objectively deviates from an established one. As a dynamic system the construction production demonstrates a strong probabilistic character. Not only the external factors, which influence the whole system, are probabilistic but also the behavior of each particular element in this system, beginning with human communities.

The main feature of construction systems, unlike technological systems, is their organizational character. In production process technical and social systems are united. The interaction between these two stochastic systems has not been taken into account neither in the produced organizational and technological documentation, nor in normative and reference basis (the construction norms and rules etc.). Failure to consider the stochastic character of the process of construction/reconstruction of manufacturing enterprises leads to insufficient adequacy and poor reliability of many organizational and technical, economical, management decisions, based on classical models.

The complexity of constructional technological processes, the complexity and the increasing numbers of constituent parts of organizational structures, the complexity of planning, economical and especially management decisions has led to the necessity of using entirely new, systematic approaches to the analysis of construction production as a system.

The solution to the problem of improving the organizational and technical reliability (OTR) holds big possibilities of improving the effectiveness of construction. The organizational and technical reliability can be viewed as the ability of organizational, technological and economical solutions to achieve the given objective of construction production in conditions of random shocks, usual in complex stochastic systems such as construction.

Man-machine (organizational, technological, management) systems, which include, not only technological but economical and social aspects, are characterized by a certain level of reliability, which decreases considerably with the complexity of the system. To define OTR of the system, the methods of the reliability theory, which is based on the analysis of distribution of random measures' sets - reliabilities of separate elements of a complex, are used. The likelihood of the index to achieve the projected size can be estimated with the help of these models and methods.

The increase of OTR can be achieved in different ways:

- by decreasing the size of factors, responsible for reliability of construction organization performance;
- by designing systems, functioning reliably in the presence of the above factors.

The realization of the first method is not always possible due to difficulties in determining the origins and causes of these factors. The second approach as well as the combination of the two are the most realistic ones.

In its broad meaning, OTR may be understood as a probability of receiving an economical effect. In this case, organizational and economical reliability (OER) will be the main figure describing building life cycle sustainability.

5 Conclusion

Thus, the conditions of ecologically acceptable development of regional investment - construction complex are: minimization of construction products are nature-based sustainable use of natural resources within the ecological life cycle; maximization of the generalized indicator of the ecological-economic efficiency of investment projects based on application of the best available technologies and environmental initiatives; modeling the equilibrium state, in which the increase of volume of performed works, manufactured products and the cost of ecological support of investment - construction activity does not violate the financial stability of economic entities.

The operation of construction companies as ecological - economic system necessitates evaluation of the effectiveness of activities based on ecological-economic analysis for structured components: an analysis of organizational aspects of environmental support productive activities; analysis of impact on components of the regional ecological-economic system; analysis of the ecological and nature intensity of work; analysis of the use of resources, providing ecological support of investment-construction activities; analysis of the impact on economic and financial performance of enterprises; investment analysis [2,3,5].

The practical significance of the study lies in the feasibility of using proposed scientific approaches, techniques in the practice of management, organization and planning of activity of the enterprises of investment-building complex, taking into account environmental aspects of development, in the activities of state administration bodies in the development of strategic plans and programs of development of investment and construction sector, relevant social and environmental priorities of the region development, the implementation of ecological-economic monitoring of construction industry.

The author believes that mainstreaming environmental knowledge in the regulation of investment and construction activities is possible through the expansion of the range of integration of ecoinstitut in the institutional environment of investment and construction sector and the institutional environment of ecological and economic regulation of investment and construction activities. The completeness of the integration processes is determined by the degree of greening of the totality of tools and methods of regulation of investment and construction activities.

6 Summary

The current building life cycle management system in the Russian Federation is a family of discrete subsystems that exist independently for different building life cycle stages. In this situation building reliability and sustainable functioning are out of the question. The implementation of a united information model (BIM-model) intended to describe building entire life cycle will allow to raise the sustainability, but this will happen only if goals and concerns of all participants of the project process are properly coordinated.

References

1. V.V. Talapov, *BIM Technology. Essence and features of building information modeling introduction* (DMK Press, Moscow, 2015)
2. A.V. Ginzburg, E.I. Nesterova, Vestnik MGSU, **5**, 317 (2011)
3. A.V. Ginzburg, *The automation of the organizational technological construction reliability designing* (SEP RIA, Moscow, 1999)
4. A.V. Ginzburg, P.B. Kagan, CAD and graphics, **9**, 32-34 (1999)

5. A. Ginzburg, *Processing Models In Organizational and Technological Design. In Abstract Volume, 14th International Conference on Computing in Civil and Building Engineering (14th ICCCBE)*, 518-519 (2012)
6. A. Ginzburg, Applied Mechanics and Materials, **405-408**, 3352-3355 (2013)
7. A. Ginzburg, A. Khripushin, Applied Mechanics and Materials, **411-414**, 2313-2316 (2013)
8. A. Ginzburg, A. Ryzhkova, Applied Mechanics and Materials, **672-674,** 2221 (2014)
9. A. Ginzburg, A. Ryzhkova, International Journal of Applied Engineering Research, **10(21)**, 41828-41830 (2015)
10. A. Ginzburg, A. Ryzhkova, International Journal of Applied Engineering Research, **10(21)**, 42410-42411 (2015)
11. A. Ginzburg, A.Skiba, Applied Mechanics and Materials, **584-586**, 507-511 (2014)
12. A. Ginzburg, S. Kachanov, International Journal of Applied Engineering Research, **11(3)**, 1660 (2016)
13. A. Ginzburg, S. Kachanov, International Journal of Applied Engineering Research, **10(20)**, 40869-40872 (2015)
14. A.A. Gusakov, *System engineering in construction* (Stroyizdat, Moscow, 1983)
15. S.A. Sinenko, *Automation of organizational technological design in construction: The manual* (Publishing company ASV, Moscow, 2002)
16. A.A. Volkov, S.N. Petrova, A.V. Ginzburg, *Information management systems and data processing technologies in civil engineering* (MGSU, Moscow, 2015)
17. A. Volkov, Applied Mechanics and Materials, **838-841**, 2973 (2014)
18. A. Volkov, V. Chulkov, R. Kazaryan, R. Gazaryan, Applied Mechanics and Materials **584-586**, 2685 (2014)
19. A. Volkov, Civil Engineering, **3**, 54 (2012)
20. A. Volkov, A. Sedov, P. Chelyshkov, Applied Mechanics and Materials, **409-410**, (2013) 630
21. V.V. Garyaeva, N.A. Garyaev, *Computing in Civil and Building Engineering Proceedings International Conference*, 1336 (2014)

Energy-Efficient Devices for Transporting and Feeding Bulk Materials in the Construction Industry

Alexander Ishkov[1,*]

[1]Moscow State University of Civil Engineering, 129337 Yaroslavskoye sh.26, Moscow, Russia

Abstract. Only in the construction industry millions of tons of bulk materials that need to be transported to the place of processing, storing and evenly or dosed feeding are recycled annually. Decreasing the costs of these processes will significantly reduce the cost of the finished product. The article presents a review of studies conducted in the field of storage, transport and feed bulk materials, and it describes the innovative design of energy-efficient disc vibrating feeder bulk materials.

1 Introduction

In the construction industry millions of tons of bulk materials that need to be transported to the place of processing, storing and evenly or dosed feeding are recycled annually. Thus, reducing the cost of transportation, storage and flow of granular materials significantly affects the cost of the finished product. Therefore, many researchers around the world are looking for ways to improve energy-efficiency of the above processes.

2 Storage of Bulk Materials and Transportation in Special Conditions

From review papers, you can highlight the article by A. W. Roberts [1], which presents a historical overview of the development of the processes of storage, transport and processing of bulk materials, as well as directions of future strategic studies.

In the field of storage of bulk materials the study of R. Selva Ganapathy and B. Pitchumani [2] can be identified. They studied the outflow of coal out of the bunker, depending form the bunker, angle of inclination of the walls and size of the outlet opening.

Bulk materials transportation is performed by various devices and in different conditions. Therefore, much research is devoted to studying these issues.

In the work of A. De Visscher [3] the article discusses environmental pneumatic unloading of wood pellets from barges to power plants in the Netherlands, and P. Mukhopadhyay and S. Chandrasekar [4] propose solutions aimed at improving the efficiency of loading coal on the trains in India. The system of loading and unloading bulk

* Corresponding author: aishkov@gmail.com

materials from ships operating in the most severe weather conditions is presented in publication [5].

The Australian company Sea Transport Corporation and the Australian Maritime College at the University of Tasmania has developed new technology transshipment of bulk materials (iron ore and coal) from small feeder vessels to large bulk carriers [6], which gives a high environmental and economic effect, and in some cases completely eliminates to costly dredging operations. In this paper the issues of hydrodynamic interactions of multiple floating close to each other ships during the seaway is study.

J. Doller [7] describes the technology of pneumatic transport of bulk materials for the construction of a tunnel on the Swiss project NEAT.

Design, installation and operation the longest (8.2 km) in the world conveyor system in Peru is described in the article by J. Wiedenroth [8]. And in the work of S. Hoheisel [9] the commissioning of the longest in Europe belt conveyor for French cement factory in Montalieu is described. Its length is about 13 km, and the total power of three motors is 750 kW.

3 Conveyor Systems Transport Bulk Materials

The widespread use of conveyor systems pushes researchers to search for ways to improve their energy efficiency.

C. Wheeler, P. Munzenberger and D. Ausling [10] conducted a large studies to reduce the energy consumption of belt conveyors. Studies have helped to develop theoretical models that can be used in the design process of energy efficient systems of belt conveyors.

In publications [11] and [12] an energy-efficient samples of conveyor belts with lower rolling resistance to reduce energy consumption are proposed. And in the article by G. Lodewijks [13] discusses options of conveyor belts, made of new rubber compounds that can significantly reduce the consumed energy of the conveyor.

A wide range of produced rubber conveyor is presented in by R. Alles and K. Steinbrueck report about the Hannover Fair [14]. The authors mark trend toward standardization and automation of conveyors, as well as the wide use of plastics as load-bearing elements and protective coatings.

Studies conducted in recent years in the field of conveyor systems are mostly aimed at increasing the efficiency of transporting bulk materials. S. Ramjee and P. Staples [15] achieve this by improving the design of the conveyor. G. Lodewijks и Y. Pang [16] studied all possible energy-saving options for continuous transport systems. C. A. Wheeler [17] has developed a theoretical method of calculating the resistance to rotational motion of idler rolls in a conveyor belt system and a device to measure the strength of this resistance. G. Hu and colleagues [18] propose to use a discrete element method (DEM) to study the behavior of the equipment during transportation of powder and bulk materials.

4 Systems Pneumatic Conveying of Bulk Materials

Pneumatic conveyor systems of bulk material received are widespread in the industry. Therefore justified the attention, which researchers direct at increasing the energy-efficiency of pneumatic transport installations.

Publication [19] describes the energy-efficient transporting system of cleaning products with compressed air, which helps in protecting the environment by reducing emissions. N. Barlmeyer [20] proposes to transport the metal particles using negative pressure. C. Saccani, C. Panciroli and C. Ligabue [21] substantiate the advantages of pneumatic conveying installations for transporting granular particles and dust.

T. Schuster, M. S. A. Bradley and A. R. Reed [22] investigated the impact properties of the bulk material, the diameter of pipelines and flow regime on the power consumption of pneumatic conveying systems.

In the work of S. Magister, J. Wildfoerster and P. Hilgraf [23] the results of studies of 10 pneumatic conveyor systems for transporting bulk materials at a lime plant are described. It helped to minimize the cost of electricity and compressed air.

L. Frye [24] developed a qualitative three-level model of the attrition process involving stress mode, aimed at determining the dominant mechanism of the abrasion in the process pneumatic transport of bulk materials.

5 Transporting Bulk Materials by Means of Screw Device

Screw conveyors have not lost their popularity.

M. Rackl and W. A. Günthner [25] conducted a pilot study of the effect of different varieties of wood chips on the performance of a screw conveyor. The results of the study allowed exclude jamming during transporting wood chips by a screw conveyor.

V. A. Evstratov, V. A. Rud and K. Y. Belousov [26] developed a mathematical model of the process of vertical screw transport of bulk material flow, this allows to reduce consumption energy of the conveyor, by means of the choice of optimal values of geometric, kinematic and dynamic parameters of working bodies.

6 Systems Vibration Conveying of Bulk Materials

Transportation of bulk and powder materials with the help of vibration is applied in many production processes in various industries, because it has a simple design, high reliability and efficiency.

Different designs of vibratory conveyors with electromagnetic drive are presented in the paper by R. Alles and K. Steinbrueck [27], devoted to the Hannover Fair. In the paper [28] K. Steinbrueck describes the benefits of vibration conveyors, in particular, the ability to solve many technological problems due to the change of the angle of inclination of the vibrating conveyor, the amplitude and frequency of its oscillation.

The work [29] is aimed at studying the impact of the transported material on heavy-loaded vibrating conveyors and finding ways to prevent the occurrence of resonant frequency. The results of the study can be applied in the design of vibrating conveyors with electromagnetic drive.

The method of selection of vibratory conveyors depending on the kind of the bulk material and the operating conditions is presented in article [30]. S. Bhuvaneswari and S. Ravikumar [31] have developed a system of calculating the optimum power vibrating feeder, which allows to reduce energy consumption and increase equipment life.

7 Vibratory Feed System Bulk Materials

The quality of the finished product and the efficiency of many technological processes are often determined by the feeders – devices that provide a uniform flow of bulk materials to machines for their processing or transporting.

Vibration feeders [32] are widespread among the various types of feeders, since they have a simple design, small dimensions, have the reliability, accuracy and smoothness of control flow. Vibratory feeders bulk materials are used in the manufacture of building materials and other industries, for example, in the production of cement, building mixtures and etc.

Disk vibratory feeder [33, 34] is widely used in the construction industry. It is placed coaxially underneath the exhaust pipe 1 of conical hopper 2 (Fig. 1). In the traditional embodiment the disk vibratory feeder is a horizontal spreading disk 3, which is installed with a gap under the exhaust pipe 1 and connected with the electromagnetic drive 4.

Feeder operates as follows.

Bulk material from the hopper 2 under the action of gravity poured through the exhaust pipe 1 on the spreading disc 3, where self-closes, forming on the disk 3 the annular mound with the natural angle of repose.

After switching the electromagnetic drive 4 the spreading disk 3 starts to vibrate. Bulk material, located on the spreading disk 3, goes into a state of vibrating liquefaction. In this case the angle of repose of the granular material reduces and it begins to expire with spreading disk 3. After turning off the electromagnetic drive 4, the vibration spreading disk 3 is stopped and the bulk material self-closes, forming on the spreading disk 3 the sustainable annular mound with the natural angle of repose.

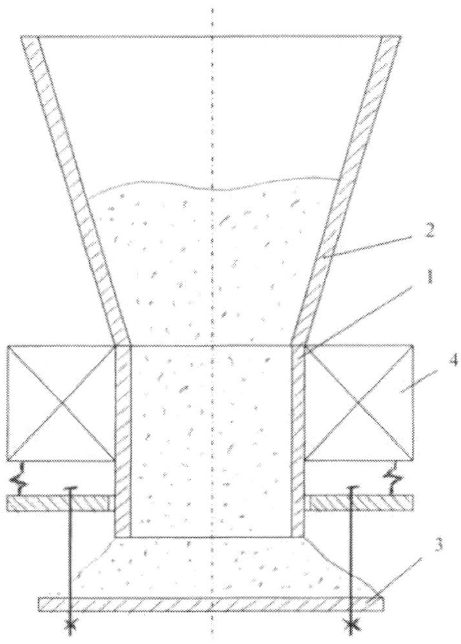

Fig. 1. Disc vibratory feeder of bulk materials.

8 Results

In the work [35] the energy-efficient design of the disc vibratory feeder is proposed. The spreading disc is formed with a central hole, overlapped with elastic membrane. The researchers were able to reduce specific energy consumption per unit of supplied granular material due to a sharp (5-6 times) increase in performance of the feeder. According to the authors, this effect is due to increased driving force of the expiry of the material (pressure of the column of material in the hopper) and the increase of the oscillation amplitude of the spreading disk due to the elastic properties of the membrane.

In the future the design of energy-efficient disc vibratory feeder of bulk materials was improved (Fig. 2). The elastic membrane, overlapping the central hole in the spreading disk 3 is removed, and under the spreading disk 3 is mounted with a gap coaxially fixed disk 5, rigidly secured to the hopper 2.

The advanced design of the energy-efficient disc vibratory feeder allows you to eliminate wasteful expenditure of electricity on useless vibration of the central part of the pillar of bulk material, located in the hopper 2.

9 Conclusion

The advanced design of the energy-efficient disc vibratory feeder reduces the cost of electricity for 16-27%. Research in this direction continues, but the disc vibratory feeder is already in use at the enterprises of the construction industry.

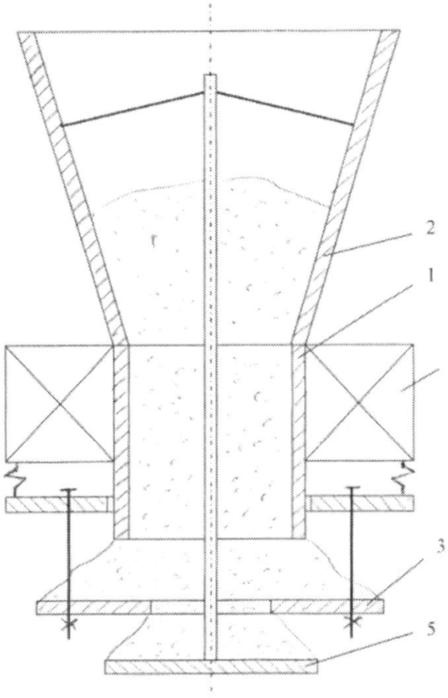

Fig. 2. Advanced disc vibratory feeder of bulk materials.

References

1. A.W. Roberts, Bulk Solids Handling, **26-6**, 392-414 (2006)
2. R. Selva Ganapathy, B. Pitchumani, Bulk Solids Handling, **29-4**, 218-222 (2009)
3. A. De Visscher, Bulk Solids Handling, **33-4**, 32-33 (2013)
4. P. Mukhopadhyay, S. Chandrasekar, Bulk Solids Handling, **35-5**, 30-32 (2015)
5. *Bulk handling in extreme conditions* (2007)
6. G.J. Macfarlane, N.T.M. Johnson, L.J. Clarke, R.J. Ballantyne, K.A. McTaggart, , Proceedings of the International Conference on Offshore Mechanics and Arctic Engineering – OMAE, **7**, (2015).
7. J. Doller, ZKG International, **58-6**, 67-71 (2005)
8. J. Wiedenroth, Cement International, **7-4**, 54-63 (2009)
9. S. Hoheisel, Bulk Solids Handling, **35-1**, 17-19 (2015)
10. C. Wheeler, P. Munzenberger, D. Ausling, Bulk Solids Handling, **35-5**, 40-50 (2015)
11. M. Hager, A. Hintz, Bulk Solids Handling, **13**, 749-758 (1993)
12. A. Greune, M. Hager, Bulk Solids Handling, **10**, 245-248 (1990)
13. G. Lodewijks, Bulk Solids Handling, **32**, 52-56 (2012)

14. R. Alles, K. Steinbrueck, Foerdern Heben, **21-10**, 571-581 (1971)
15. S. Ramjee, P. Staples, Bulk Solids Handling, **35- 4**, 20-25 (2015)
16. G. Lodewijks, Y. Pang, *Energy saving options for continuous transport systems, an exploration, ICBMH 2013 – 11th International Conference on Bulk Materials Storage, Handling and Transportation* (Newcastle, NSW; Australia, 2013)
17. C.A. Wheeler, Journal of Manufacturing Science and Engineering, Transactions of the ASME, **138-4**, (2016)
18. G. Hu, M. Liu, Y. Liu, H. Wan, L. Liu, 2010 International Conference on Mechanic Automation and Control Engineering, MACE2010, 923-926 (2010)
19. Powder Handling and Processing, **18-1**, 44-46 (2006)
20. N. Barlmeyer, Bulk Solids Handling, **33-1**, 28-30 (2013)
21. C. Saccani, C. Panciroli, C. Ligabue, Powder Handling and Processing, **13-3**, 295-299 (2001)
22. T. Schuster, M.S.A. Bradley, A.R. Reed, Proceedings of the Institution of Mechanical Engineers, Part E: Journal of Process Mechanical Engineering, **209-E1**, 69-75 (1995)
23. S. Magister, J. Wildfoerster, P. Hilgraf, ZKG international, **41-12**, 601-606 (1988)
24. L. Frye, Handbook of Powder Technology, **12**, 1149-1218 (2007)
25. M. Rackl, W.A. Günthner, Biomass and Bioenerg, **88,** 106-115, (2016)
26. V.A. Evstratov, A.V. Rud, K.Y. Belousov, Procedia Engineering, **129**, 397-402 (2015)
27. R. Alles, K. Steinbrueck, Foerdern Heben, **21-10**, 582-589 (1971)
28. K. Steinbrueck, AEG-Telefunken Progress, **4**, 102-107 (1981)
29. K. Steinbrueck, Aufbereitungs-Technik, **22-8**, 410-417 (1981)
30. K. Steinbrueck, Technische Mitteilungen AEG-Telefunken, **71-3**, 87-93 (1981)
31. S. Bhuvaneswari, S. Ravikumar, International Journal of Applied Engineering Research, **10-7**, 16441-16449 (2015)
32. A.D. Ishkov, D.A. Semernin, S.V. Miloradov, I.V. Voronina, Applied Mechanics and Materials, **741**, 500-503 (2015)
33. G.G. Arsoev, R.U. Patent 2,340,530. (2008).
34. A.D. Ishkov, A.V. Kosjakov, A.N. Cetovich, F.M. Filipkov and O.Ju.Grigor'ev, S.U. Patent 1,590,414 (1990)
35. A.D. Ishkov, A.V. Stepanov, S.V. Miloradov, I.V. Voronina, Applied Mechanics and Materials, **670-671**, 458-461 (2014)

MATEC Web of Conferences **73**, 02020 (2016) DOI: 10.1051/matecconf/20167302020
TPACEE-2016

Specifics of Building Envelope Air Leakage Problems and Airtightness Measurements

Anatolijs Borodinecs[1,*], *Jekaterina* Nazarova[1], *Aleksandrs* Zajacs[1], *Alexandr* Malyshev[2] *and Vladimir* Pronin[2]

[1]Riga Technical University, LV – 1658 Kalku str. 1, Riga, Latvia
[2]ITMO University, 197101 Kronverksky Ave. 49, St.Petersburg, Russia

Abstract. In addition to transmission heat loses the infiltration of outdoor air can cause significant heat losses. The external building envelope should be airtight in order to prevent uncontrolled cold air infiltration. The article analysis modern building materials and structures influence on airtightness. The practical measurements of renovated buildings' airtightness are presented and compared to non-renovated buildings. In addition paper presents data on airtightness measurements of whole multi apartment building and single apartment in analyzed building taking inco accout properties of building materials. The airtightness of single apartment was evaluated with support pressure in neighbor apartments. The results show that the airtightness measurements of multi apartment building can be evaluated by measuring single apartment on last floor with support pressure in neighbor apartments. The practical measurement of renovated buildings had shown the air leakage rate q50 of typical Latvian construction after renovation is between 2.5 and 2.9 $m^3/(m^2 \cdot h)$. Since the building envelope has to minimize the heat loses (transmission and infiltration) and ventilation system either mechanical or natural has to provide necessary air exchange, the building envelope airtightness shouldn't be dependent on type of ventilation systems.

1 Introduction

General, building energy consumption depends on transmission heat losses, infiltration heat losses and ventilation heat losses. While the U-value of building envelope has reached reasonable minimums the requirements for air permeability of building envelop are still under constant review across the EU. The strong requirements for air permeability reduce amount of infiltrated outdoor air, thus minimizing the heat losses. However in some cases the air infiltration is considered as part of natural ventilation. Nowadays Belgium, Germany and some other countries has different air-tightness requirements for building with natural and mechanical ventilation [1]. For example in Germany air permeability of building mechanical ventilation is 1.5 (h^{-1}) and 3(h^{-1}) for building with natural ventilation. It should be noted that natural ventilation as well as mechanical ventilation should ensure necessary air exchange rate independently to outdoor air conditions such as wind and temperature

* Corresponding author: anatolijs.borodinecs@rtu.lv

difference. Modern energy efficient ventilation systems also have to provide possibilities airflow regulation according to room occupancy [2-6]. The air flow through the buildings envelope couldn't be controlled and additional heat consumption for infiltrated air will be needed even in case of human absence in room. This issue is especially actual in office building, schools and kindergartens, where during the night and weekends the ventilation should ensure only minimal air exchange.

Air tightness of building envelope can be expressed as air permeability q50 $m^3/(m^2 \cdot h)$ as well as air change rate coefficient n50 (h^{-1}). The air permeability value q50 directly characterizes properties of building envelope. The air change rate coefficient n50 (h^{-1}) should be evaluated with the respect to building surface area to volume ratio.

2 Evaluation of building envelope air leakage

It is clear that building envelope of modern buildings should be well insulated as well as airtight. The airtight building has both – positive and negative properties. The positive impact of air tightness is possibility to create controlled air supply scheme and to reduce energy consumption for supply air heating [7-11]. This positive effect could be reinforced by use of the ventilation systems with heat recovery. The negative property is lack of natural air supply and additional costs for equipment and exploitation of mechanical ventilation. The negative property of non-airtight buildings is uncontrolled cold air infiltration which requires also additional heat in time than room is not occupied. Existing researches [12, 13, 14] have proved positive effect of renovation works done in kindergartens and schools. However measurements done in Latvia [15, 16, 17] have shown significant increase of indoor air CO_2 concentration after renovation, which has negative impact on human health.

Almost in all cases the level of building envelope airtightness could not be calculated during design stage. The level of airtightness much more depends on quality of buildings construction works. Especially quality of window/wall joints, quality of brick or block walls and roof/wall joins. The junction solution of prefabricates elements should be carefully evaluated during the design stage. There is no common developed evaluation methodology for building envelope airtightness building at design stage. At the moment there is a strict gradation for windows air leakage [18]. But for external elements such as walls and roofs structures, theoretical evaluation of air leakage is problematic due to the lack of information on building materials air permeability. The study presents data only for limited number of materials [19, 20].

Typically dwelling buildings up to 16 floors are built from concrete panel with mineral wool insulation and with concrete external finishing. The resistance to infiltration through wall's materials itself can be calculated on the data [20] basis:

$$R_{\inf} = R_{\inf}^{concrete} + R_{\inf}^{insulation} + R_{\inf}^{ext.fin}, [m^2 hPa/kg] \quad (1)$$

where: $R_{\inf}^{concrete}; R_{\inf}^{insulation}; R_{\inf}^{ext.fin}$ – resistance to infiltration of concrete panel, insulation and external finishing, $[m^2hPa/kg]$.

The resistance to infiltration of a wall: $R_{\inf} = 39242.[m^2 hPa/kg]$.

Such high resistance to infiltration does not have any impact on total building's energy performance and quality of indoor air.

The air tightness of windows still should be taken into account. The windows with resistance to air leakage $R_{\inf} = 0{,}65$ $kg/(m^2 \cdot h)$ or 0,3 m^3/hm^2 are typically used for dwelling building construction.

The level of infiltrated air through the windows can be calculated [21]:

$$G_0 = \frac{1}{R_{\inf}}\left(\frac{\Delta p}{\Delta p_o}\right)^{\frac{2}{3}}, [kg/(m^2 h)] \quad (2)$$

For the first floor the level of infiltrated air will be: $G_0 = \frac{1}{0.65}\left(\frac{23}{10}\right)^{\frac{2}{3}} = 2.68, [kg/(m^2 h)]$

For the 9th floor: $G_0 = \frac{1}{0.65}\left(\frac{1}{10}\right)^{\frac{2}{3}} = 0.33, [kg/(m^2 h)]$

The Comparison of air tightness requirements in different European countries are shown in Table 1 [22].

Table 1. Comparison of air tightness requirements in different EU countries.

EU Member State	Air tightness requirements at 50 Pa pressure	
	Natural ventilation	Mechanical ventilation
Czech Republic	4.5 [1/h]	w/o heat recovery: 1.5 [1/h] with heat recovery: 1.0 [1/h]
Germany	3.0 [1/h] or 7.8 m³/h per m² floor area	1.5 1/h or Germany 3.9 m³/h per m² floor area
Denmark	1.5 [l/s per m² floor area]	
Norway	3.0 [1/h]	
The Netherlands	Dwellings: 200 dm³/s (at 10 Pa) Non-residential buildings: 200 dm³/s per 500 m³ (at 10 Pa)	
United Kingdom of Great Britain	New dwellings and new commercial and public buildings over 500 m²: 10 m³/m²h (stated as reasonable limit for the design air permeability in building regulations 2000 L1A and L2A)	

3 Evaluation of the buildings air tightness

After the renovation works the thermal performance of external building envelope significantly increases. The U-values of walls after thermal insulations is 0.25 – 0.30W/(m2·K), while for roofs 0.15 – 0.20W/(m2·K). Such thermal bridges as external wall/floor junctions, internal corners and external corners are fully prevented after heat insulation works.

In order to get more objective information of renovation quality, including also changes in ventilation rate, the airtightness tests for kindergarten were done. Figure 1 shows preparation work done for airtightness test. In order to test air leakage rate of building envelope, all ventilation shafts were sealed and all windows and external doors were closed while the internal doors were opened to prevent closed spaces. Preparation works where done in compliance with the requirements of standard EN 13829 [23]. Thermal Performance of Buildings – Determination of Air Permeability of Buildings – Fan Pressurization Method. Brussels: European Committee for Standardization method B.

Figure 1. Preparation work for Blower Door test.

Air leakage test was done using two Blower Door installations. Blower Door installations were placed on opposite facades. The total capacity of Blower Door installations was 28000m3/h. The 3D model of kindergarten and dimensions are shown in Figure 2. Verification was performed in two modes:

- by creating underpressure in range between 40Pa and 70Pa;
- by creating overpressure in range between 25Pa and 60Pa

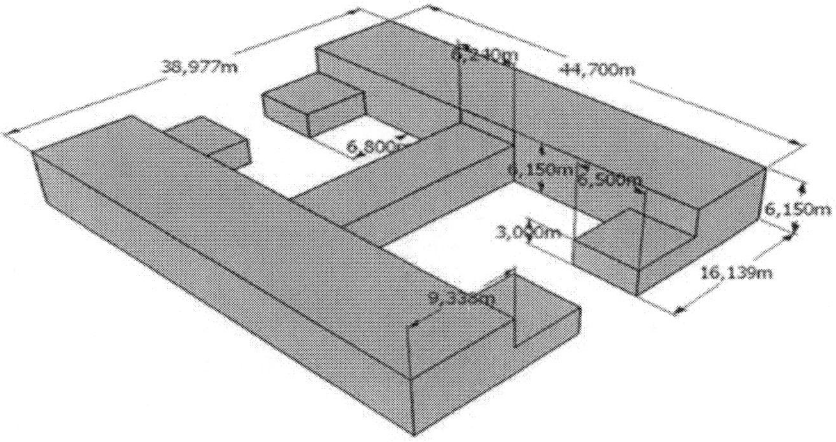

Figure 2. 3D scheme of kindergarten.

Air permeability verification was performed in compliance with standard EN 13829 (methods A and B). Assessment of verification results was performed in compliance with standard LBN 002-01. According to the requirements of LBN 002 – 01 air permeability q50 of the buildings envelopes being tested should not exceed $3m^3/(m^2{\cdot}h)$, with pressure difference 50Pa, where:

$$q50=V50/A_E,\ [m^3/(m^2{\cdot}h)] \quad (3)$$

V50 - the measured air flow that flows through the construction with pressure difference 50Pa $[m^3/h]$;

A_E - area of building's envelope $[m^2]$.

Additionally, air change rate coefficient n50 $[h^{-1}]$ with pressure difference 50Pa was calculated,

$$n50= V50/V,[h^{-1}] \quad (4)$$

where:

V50 - the measured air flow that flows through the construction with pressure difference 50Pa $[m^3/h]$;

V - building volume [m^3].

The results of Blower Door test after renovation are presented in Table 2. In all cases the air leakage rate is in conformity with requirement of LBN 002-01. Air leakage rate according to this code should not exceed 3.0 $m^3/(m^2 \cdot h)$ for residential buildings, hospitals and kindergartens.

Table 2. Airtightness of buildings envelope in kindergartens after renovation [24].

Case study	Volume,[m3]	Area of building envelope,[m2]	Air leakage rate [m3/(m2·h)] q50	Air change rate, [h-1]n50
1	4045	5358	2.89	2.19
2	6375	4044	2.57	1.63
3	3895	6228	2.68	1.68

Unfortunately there are no precise data available on air leakage rates of kindergarten buildings before renovation. However measurement of one, partly renovated, kindergarten has shown that the air leakage was 6.5 $m^3/(m^2 \cdot h)$. In addition to building envelope airtightness evaluation the ventilation systems were also checked in kindergarten "Kamolitis". The ventilation rate under pressure difference 50Pa created 4.62 h^{-1} but under 4Pa – 1.29 h^{-1}. In case of complex architecture of some buildings for example as private houses, calculation of area of the envelope and q50 [$m^3/(m^2 \cdot h)$] could be challenging, therefore another normative as DIN 4108-7:2011-01 defines the maximum air exchange rate (Tab. 3).

Table 3. Recommended building preparation and recommended maximum air change rates for the airtightness measurement at 50 Pa pressure difference according to DIN 4108-7:2011-01 [25].

<table>
<tr><th colspan="2">Ventilation system</th><th>Type of outdoor air aperture</th><th>Preparation of outdoor air aperture and exhaust air grilles</th><th>Maximum air change rate at 50Pa, n50, max [h-1]</th></tr>
<tr><td rowspan="7">Natural ventilation</td><td>Via windows only</td><td>-</td><td>Not applicable</td><td>3.0</td></tr>
<tr><td rowspan="3">Cross ventilation via outdoor air apertures</td><td>Not closable</td><td>No measures</td><td>3.0</td></tr>
<tr><td>Closable, without self-regulation</td><td>Closure of outdoor air aperture</td><td>3.0</td></tr>
<tr><td>With self-regulation</td><td>Sealing of outdoor air aperture</td><td>1.5</td></tr>
<tr><td rowspan="3">Shaft ventilation</td><td>Not closable or no outdoor air aperture</td><td>No measures at outdoor air aperture, sealing of exhaust air grilles</td><td>1.5</td></tr>
<tr><td>Closable, without self-regulation</td><td>Closure of outdoor air aperture, sealing of exhaust air grilles</td><td>1.5</td></tr>
<tr><td>With self-regulation</td><td>Sealing of outdoor air aperture and exhaust air grilles</td><td>1.5</td></tr>
<tr><td rowspan="3">Mechanical ventilation</td><td rowspan="2">Exhaust system</td><td>Closable, without self-regulation</td><td>Sealing of outdoor air aperture</td><td>1.0</td></tr>
<tr><td>With self-regulation</td><td>Sealing of outdoor air aperture</td><td>1.0</td></tr>
<tr><td>Supply and exhaust system</td><td>-</td><td>Sealing of exhaust, exit, supply and outdoor air ducts</td><td>1.0</td></tr>
</table>

Current study shows that the measured values of the air change rate in the number of private houses differs a lot and depends mostly on the quality of the construction work (Tab. 4).

Table 4. Air change rate in one family houses.

Case study	**Volume, [m3]**	**Air change rate, [h-1] n50**	**Description of ventilation system**
1	310	4.78	MEH
2	345	0.82	NAT
3	420	8.9	NAT
4	467	3.6	NAT
5	571	3.16	NAT
6	572	4.98	NAT
7	610	4.2	NAT
8	635	6.1	NAT
9	640	6.7	NAT
10	725	10.75	NAT
11	810	8.4	MEH
12	854	2.95	MEH
13	1026	5.3	NAT
14	1055	3.08	NAT
15	1060	2.77	MEH
16	1070	1.94	MEH
17	1319	8.2	MEH

4 Conclusions

1. The heat consumption after applying thermal insulation during refurbishment process of buildings can be realistically reduced down to 70kWh/m^2 in typical Latvian climate conditions.
2. The optimal U-values of walls after thermal insulations is 0,25 – 0,30W/(m^2·K), while for roofs 0,15 – 0,20W/(m^2·K).
3. Due to ventilation systems usually are not renovated, the IAQ measurements in renovated buildings have shown unsatisfactory situation with increased CO_2 concentration and high relative humidity.
4. Measurement airtightness of partly renovated kindergarten has shown that the air leakage was 6.5 m^3/(m^2·h).
5. The practical measurement of renovated buildings had shown the air leakage rate q50 of typical Latvian construction after renovation is between 2.5 and 2.9 m^3/(m^2·h).
6. On average the share of energy consumption for ventilation needs in renovated building without heat recovery is 50% of total heat consumption for space heating and ventilation. The IAQ measurements in renovated buildings have shown slightly unsatisfactory situation with increased CO_2 concentration higher relative humidity while the energy consumption can almost be reduced to level of energy efficient buildings.

This work partially financially supported by Goverment of Russian Federation (Grant 074-u01).

References

1. Information on on: http://www.stb.rwth-aachen.de/projekte/2008/ETHICS/docs/ETX-136_Deliverable_WP1_a.pdf
2. V. Pukhkal, M. Tanić, N. Vatin, V. Murgul, Procedia Engineering, **117**, 864-869 (2015)
3. V. Murgul, D. Vuksanovic, N. Vatin, V. Pukhkal, Applied Mechanics and Materials, **635-637**, 370-376 (2014)
4. V. Pukhkal, N. Vatin, V. Murgul, Applied Mechanics and Materials, **680**, 529-533 (2014)
5. V. Murgul, D. Vuksanovic, N. Vatin, V. Pukhkal, Applied Mechanics and Materials, **680**, 524-528 (2014)
6. V. Murgul, D. Vuksanovic, V. Pukhkal, N. Vatin, Applied Mechanics and Materials, **633-634**, 977-981 (2014)
7. V. Murgul, MATEC Web of Conferences, **53**, 01046 (2016)
8. V. Pukhkal, A. Bieliatynskyi, V. Murgul, Journal of Applied Engineering Science, **14(1)**, 93-101 (2016)
9. A. Milajić, D. Beljaković, N. Davidović, N. Vatin, V. Murgul, Procedia Engineering, **117**, 916-923 (2015)
10. A. Gorshkov, V. Murgul, O. Oliynyk, MATEC Web of Conferences, **53**, 01045 (2016)
11. V. Pukhkal, N. Vatin, V. Murgul, Applied Mechanics and Materials, **633-634**, 1077-1081 (2014)
12. N.I. Vatin. D. V. Nemova, A. S. Kazimirova, and K. N. Gureev, Advanced Materials Research, **953-954**, 1537-1544 (2014)
13. N. I. Vatin, D. V. Nemova, D. S. Tarasova, and A. A. Staritcyna, Advanced Materials Research, **953-954**, 854-870 (2014)
14. N. Vatin, M. Petrichenko, D. Nemova, A. Staritcyna, and D. Tarasova, Applied Mechanics and Materials, **1**, 633-634 (2014)
15. A. Borodinecs, A. Kreslins, E. Dzelzitis, A. Krumins, Ventilation and Energy Conservation in Buildings: Sustainable Built Environment, 361-368 (2007)
16. A. Borodinecs, B. Gaujena, J. Zemitis, A.Kreslins, 12th International Conference on Indoor Air Quality and Climate, 973-978 (2011)
17. A. Borodinecs, Z. Budjko, *Indoor air quality in nursery schools in Latvia 9th International Conference and Exhibition - Healthy Buildings* (2009)
18. *LVS EN 12207:2001 Windows and doors - Air permeability – Classification* (2001)
19. Information on: http://www.cmhc-schl.gc.ca/publications/en/rh-pr/tech/98109.htm
20. *SP 23-101-2004 Thermal performance design of buildings* (Moscow, 2008)
21. E.G. Malyavina, *Building Heat Losses* (Avok-Press, Moscow, 2007)
22. H. Erhorn, H. Erhorn-Kluttig, F.R. Carrié. *Airtightness requirements for highperformance buildings, Conference paper, AIVC, Kyoto, Japan*, (2008) Information on http://www.asiepi.eu/wp-5-airtightness/related-papers.html
23. EN 13829 *Thermal Performance of Buildings–Determination of Air Permeability of Buildings–Fan Pressurization Method* (European Committee for Standardization, Brussels, 2000)
24. A. Borodinecs, A. Rodriguez-Gabriel., O. Tatarchenko, etc. *Handbook on Buildings Renovation in Central Baltic Region* (Riga Technical University, Riga, 2013)
25. DIN 4108-7:2011-01 *Thermal insulation and energy economy in buildings - Part 7: Air tightness of buildings - Requirements, recommendations and examples for planning and performance* (2011)

MATEC Web of Conferences **73**, 02021 (2016)
TPACEE-2016
DOI: 10.1051/matecconf/20167302021

Modeling Hydrodynamic State of Oil and Gas Condensate Mixture in a Pipeline

Sergey Dudin[1], *Konstantin* Voronin[1,*], *Svetlana* Yakubovskaya[1] and *Sharifjom* Mutavaliev[1]

[1]Industrial University of Tyumen, 625001 Volodarskogo str. 38, Tyumen, Russia

Abstract. It is known that energy costs to pump a unit mass of hydrocarbon mixture in the form of gas 2÷3 times exceed energy costs to pump hydrocarbon mixture in the form of liquid. As far as energy conservation during energy carrier transportation is concerned, an important task is development and application of a method to calculate gas-liquid hydrocarbons flow and heat and mass transfer in process and trunk pipelines during their design and operation.
The authors have developed a physical-mathematical model of a hydrocarbon flow in the gas condensate line based on balance equations of conservation of mass, impulse and energy of the transported medium within the framework of a quasi-1D approach.
Based on the developed model a calculation method was obtained which is used to analyze hydrodynamic state and composition of hydrocarbon mixture in each ith section of the pipeline when temperature-pressure and hydraulic conditions change.

1 Introduction

Transportation of hydrocarbons is one of the most energy-intensive industries. From the literature it is known that energy costs to pump a unit mass of hydrocarbon raw in the form of gas 2÷3 times exceed energy costs to pump it in the form of liquid. [10] As far as energy conservation during energy carrier transportation is concerned, an important task is development and application of a method to calculate gas-liquid hydrocarbons flow and heat and mass transfer in process and trunk pipelines during their design and operation [21].

Gas condensate fields have a number of thermodynamic characteristics [26] that must be considered both in their field development and during transportation and processing of gas condensate. A complicated phase behaviour of the gas condensate system, as well as the dependence of the extracted raw materials on the phase state of the deposit other conditions being equal, is a key aspect. Therefore, when designing gas condensate lines the crucial task is to select the most appropriate methods of calculating thermophysical properties and phase equilibrium of the transported gas condensate [22-25].

* Corresponding author: voronin_tsogu@mail.ru

2 Research object

Methods of calculating thermophysical properties (TPP) and phase equilibria are reviewed in quite a lot of scientific publications and reference books [2, 3, 9, 11, 13, 14, 18, 19 etc.]. However, when it comes to such complex systems as oil, gas condensate, natural gas and their by-products, using even the known techniques is a difficult task. It should also be taken into account that large part of calculation methods is designed for individual hydrocarbons and mixtures of known composition, that is why using them to calculate TPP of complex mixtures of unknown composition is not always possible or can even results in additional, hard-to-estimate errors [2].

At the department of Transportation of Hydrocarbon Resources in Tyumen State Oil and Gas University (TSOGU) the authors have developed a physical-mathematical model of a hydrocarbon mixture (HCM) flow in the gas condensate line considering phase transitions and changes in the phase composition.

3 Methods

A detailed description of the model is presented in [4, 6, 7, 8] according to which the main model closing relations are the following.

1) The dependence of the gas condensate mixture density on pressure and temperature.

Given the gas condensate line upstream initial conditions (mass flow rate of component – Gi, true density of gas condensate – ρsm0, temperature – T0, pressure – p0, mass fractions of component — Nmi) let us find the density of a group of hydrocarbons hexane+ at the inlet of the rated pipeline section by the formula:

$$\rho_{C6+} = \frac{Nm_n}{\frac{1}{\rho sm_0} - \sum_{i=1}^{n-1} \frac{Nm_i}{M_i} \upsilon_i}, \qquad (1)$$

wher υi – molar volume of other components calculated for both gas and liquid by a single Peng-Robinson equation of state.

Having determined the density of a group of hydrocarbons hexane+ by (1) we find boiling temperature TbC6+, molar weight MC6+, critical pressure and temperature TcC6+, pcC6+ and eccentricity factor ωC6+ using techniques from [2, 17]. Using these data, mass fractions zip of HCM components are calculated. Therefore, a group of hydrocarbons hexane+ is included in the gas condensate mixture as a single pseudo-component.

2) The viscosity of the condensate mixture is calculated at temperature of 20°C and pressure of 1 atm by the formula:

$$\mu sm_0 = \frac{\sum_{i=1}^{N} z_i \sqrt{M_i}\, \mu ind_i}{\sum_{i=1}^{N} z_i \sqrt{M_i}} \qquad (2)$$

Recalculation of viscosity on thermobaric conditions in different sections of the pipeline is carried out according to the formula:

$$\mu sm = 10^{\lg\left(\mu sm_0\right) + 1.4503 \cdot 10^{-3} \cdot \left(\frac{p}{10} - 0.101325\right) \cdot \left(5.2054 \cdot \mu sm_0^{0.278} + 0.0239\right) - 3} \qquad (3)$$

3) The isobaric heat capacity of the individual components in the mixture is calculated using a database from [11]. Isobaric heat capacity of a group of hydrocarbons hexane+ is determined using the Watson-Nelson formula. The heat capacity of the entire mixture was calculated according to the additivity rule:

$$Cpsm = \sum_{i=1}^{N} z_i \frac{Cp_i}{M_i} \quad (4)$$

4) The soil heat transfer is found by:

$$\alpha_s = \begin{cases} \dfrac{2\cdot\lambda_s}{D\cdot\ln\left(\dfrac{2\cdot H}{D}+\sqrt{\left(\dfrac{2\cdot H}{D}\right)^2-1}\right)}, & if\ \dfrac{H}{D} > 3(\text{Forchheimer - Vlasov}) \\ \dfrac{2\cdot\lambda_s}{D\cdot\ln\left(4\dfrac{H_{pr}}{D}+\dfrac{1}{Nu_s}\right)}, & if\ \dfrac{H}{D} < 3(\text{Arson - Kutateladze}) \end{cases}$$

where H – the depth of the pipeline occurrence in the soil, m; H_{pr} is found by:

$$H_{pr} = H + H_e; \qquad H_e = H_{SN}\frac{\lambda_{SO}}{\lambda_{SN}},$$

and also λ_{sn} and λ_{so} – the thermal conductivity coefficients for snow and soil:

$$\lambda_{SN} = 0{,}465\frac{W}{m\cdot K}, \qquad \lambda_{SO} = 1{,}4\frac{W}{m\cdot K},$$

Nu_{so} – the Nusselt number for soil:

$$Nu_{SO} = \frac{\alpha_{air}\cdot D}{\lambda_{SO}},$$

where α_{air} – the heat exchange coefficient for air:

$$\alpha_{air} = 11{,}6\frac{W}{m^2\cdot K}.$$

5) The distribution coefficient or equilibrium constant of the ith component of the hydrocarbon mixture Ki is the relation of its molar fraction in a vapor phase yi to its molar fraction in a liquid phase xi.

Composition and quantitative ratio of the equilibrium vapor and liquid phases is found using equations of phase concentrations of the mixture components [1, 20].

$$x_i = \frac{z_i}{V(K_i - 1) + 1} \tag{5}$$

$$y_i = \frac{z_i K_i}{V(K_i - 1) + 1} \tag{6}$$

Equations (5) and (6) are called equations of phase concentrations of the mixture components. They allow us to determine the molar fractions of components in vapor and liquid phases of the mixture composition zi at given values of distribution coefficients Ki and molar fraction of the vapor phase V.

6) By definition, the actual volumetric gas content is equal to:

$$\alpha_g = \frac{V_g}{V_{sm}} = \frac{V_g}{V_g + V_L} = \frac{1}{1 + \frac{V_L}{V_g}}$$

If gas and liquid volumes are found as a product of the molar volume of gas and liquid (υ_g, υ_L – calculated by the equation of state given the known phase compositions) and the number of moles of the mixture in the vapor and liquid state, respectively (N_V, N_L),

$$\alpha_g = \frac{1}{1 + \frac{\upsilon_L}{\upsilon_g}\frac{N_L}{N_V}} = \frac{1}{1 + \frac{\upsilon_L}{\upsilon_g}\left(\frac{1}{V} - 1\right)} \tag{7}$$

7) The flow regime of the gas-liquid mixture is determined by one of the known techniques, for example, by the methods shown in [12, 15, 16].

4 Results and discussion

Based on the developed model a calculation method was obtained which is used to analyze hydrodynamic state and composition of hydrocarbon mixture in each ith section of the pipeline when temperature-pressure and hydraulic conditions change. The developed technique was tested on the hydrocarbon mixture of de-ethanized condensate and oil transported from northern oil and gas condensate fields via the gas condensate line to the refinery. Results of a computational [9] and parametric study of thermophysical parameters of oil and gas condensate mixture in various product pipeline sections are discussed in [5, 8] and are shown in Figure 1, 2.

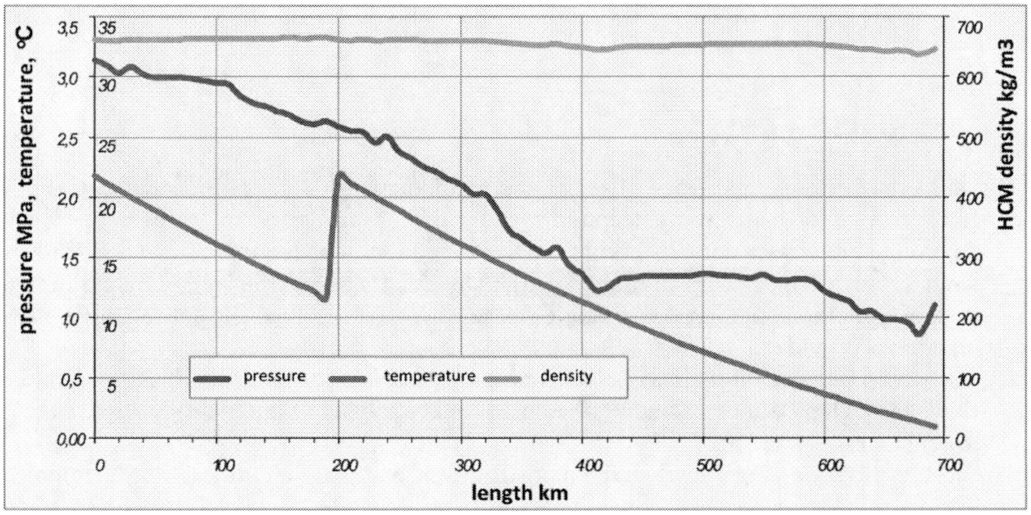

Fig. 1. HCM parameters along the gas condensate line.

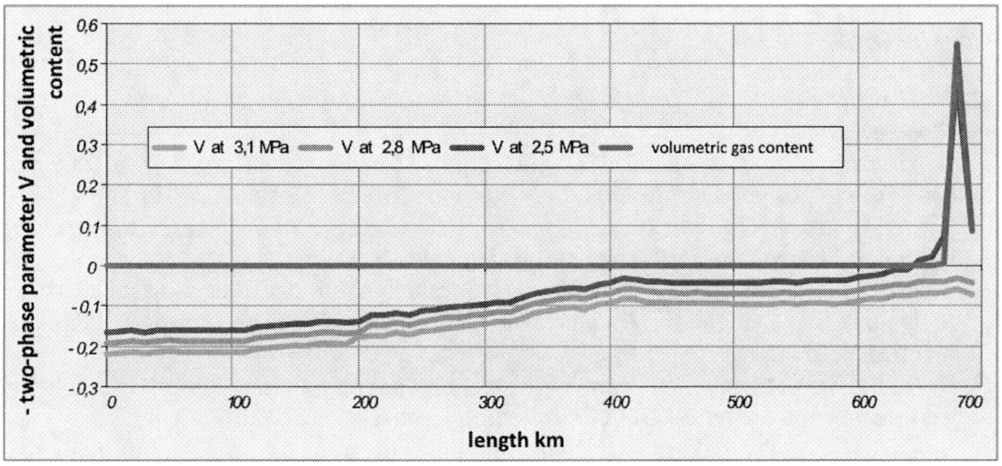

Fig. 2. Changes in HCM phase state parameters along the gas condensate line at differential product pressure in the inlet section.

Figure 1 shows the results of calculating HCM parameters along the gas condensate line. By comparing the calculated and actually observed differential pressure and temperature, we can conclude on the adequacy of the selected computational physical and mathematical model of the condensate line (relative deviation of the calculated differential pressure from actually observed is 4.6% and deviation of the estimated temperature change from the experimental was 1.5%).

Figure 2 shows a computational and parametric study of HCM phase state along the gas condensate line, with the curves showing changes in the two-phase parameter V and the true volumetric gas content of hydrocarbons when lowering the operating pumping pressure by0.6 MPa at the beginning of the pipeline. It follows from these graphs that with a decrease in operating pumping pressure on the condensate line may arise areas where hydrocarbons enter the gas-liquid two-phase state (molar fraction of the vapor phase is $0 < V < 1$ and acquires a definite physical meaning). In this case, the subsequent calculation of hydrodynamic parameters of the condensate line was carried out taking into account the two-phase flow of hydrocarbons.

5 Conclusion

1) On the basis of the developed physical and mathematical model of the gas-liquid media flow in a condensate line a technique and calculation algorithm were composed to determine the phases composition in an arbitrary section of the pipeline and to find the valid values of mass concentration of light hydrocarbons in the inlet condensate line section from the condition that a single-phase flow regime along the entire length of the pipeline is provided.
2) The developed technique can be used when managing unstable regimes of pumping fluids through pipelines, including the operational analysis of the hydrodynamic state of the condensate pipeline.
3) Results of a computational and parametric study show that in case of incomplete loading of the condensate line capacity there are areas with a two-phase flow. Therefore, in order to reduce energy consumption for transportation of condensate it is necessary to provide a single-phase flow regime along the entire length of the pipeline.

References

1. A. Brusilovsky, *Phase transformations in the development of oil and gas* (Graal, Moscow, 2002)
2. B. Grigoryev, A. Gerasimov, G. Lanchakov, *Thermophysical properties and phase equilibria of gas condensates and their fractions* (MPEI, Moscow, 2007)
3. Yu. Zemenkov, *Diagnosis of damage and leakage during pipeline transport multiphase hydrocarbons* (Vector Book, Tyumen, 2002)
4. S. Dudin, Yu. Zemenkov, A. Shabarov, N. Saranchyn, Oil and Gas, **6**, 63-68 (2010)
5. S. Dudin, Yu. Zemenkov, A. Shabarov, N. Saranin, Pipeline transport: theory and practice, **5**, 42-45 (2010)
6. S. Dudin, V. Nekrasov, Yu. Zemenkov, Oil and gas: a Separate edition of the Mining information and analytical Bulletin (research journal), 53-62 (2013)
7. S. Dudin, Yu. Zemenkov, Electronic journal "Oil and gas" (2007) Information on: http://www.ogbus.ru/authors/Dudin/Dudin_1.pdf.
8. S. Dudin, M. Zemenkova, *Oil and gas terminal: Proceedings of international scientific-technical conference "Transportation and storage of hydrocarbons"*, **7**, 86-91 (2015)
9. S. Dudin, A. Shapovalov, *Oil and gas terminal: Proceedings of international scientific-technical conference "Transportation and storage of hydrocarbons"*, **8** (2015)
10. A. Mirkin, V. Usins, *Piping systems: reference edition* (Chemistry, Moscow, 1971)
11. R. Reid, J. Prausnitz, T. Sherwood, *Properties of gases and liquids: a reference guide* (Chemistry, Leningrad, 1982)
12. A. Guzhov, V. Titov, V. Medvedev and others. *Collection, transport and storage of natural hydrocarbon gases: a Training manual* (Nedra, Moscow, 1978)
13. Yu. Zemenkov, *Technical and parametric diagnostics in pipeline systems*, (Vector Book, Tyumen, 2002)
14. Yu. Zemenkov, *Monitoring of hydrodynamic and specifications piping systems: a tutorial* (Vector Book, Tyumen, 2008)

15. A. Shabarov, N. Saranin, V. Kutrunov, *Proceedings of a Roundtable "Education through science and innovation", The modernization of education in conditions of globalization* (2005)
16. V. Kurushina, Yu. Zemenkov, WIT Transactions on Ecology and the Environment, **190(2)**, 881-888 (2014)
17. D. Peng, D. Robinson, Ind. Eng. Chem. Fundam, **15**, 59-64 (1976)
18. S. Stepanov, N., Cheremisin, S., Sokolov, A., Altunin, A., Shabarov, *ECMOR 2012 - 13th European Conference on the Mathematics of Oil Recovery* (2012) DOI: 10.399/2214-4609.20143270
19. R. Levitin, Yu. Zemenkov, Neftyanoe khozyaystvo - Oil Industry, **1**, 110-114 (2016)
20. E. Kurushina, V. Kurushina, Life Science Journal, **11(11)**, 517-521 (2014)
21. R. Mamadaliev, V. Kuskov, Yu. Zemenkov, A. Popova, Applied Mechanics and Materials, **770**, 19-22 (2015)
22. Yu. Zemenkov, V. Shalay, M. Zemenkova, Procedia Engineering, **113**, 254-258 (2015)
23. Yu. Zemenkov, V. Shalay, M. Zemenkova, Procedia Engineering, **113**, 312-315 (2015)
24. V. Antipyev, Yu. Zemenkov, N. Kudryavtseva, *A method of calculation of losses of oil at accidents on the main oil pipelines* (UMN Z and SZS, Tyumen, 1985)
25. G. Wilson, *Modified Redlich-Kwong EOS Application to General Physical Data Calculations* (Paper 15C presented at the 1968 AIChE Annual Meeting, Cleveland, Ohio, 1969)

The Testing Equipment to Study Heat Transfer through a Frame-Panel Enclosure Structure Fragment

Vladimir Borzykh[1,*], *Pavel* Mikhailovt[2], and *Igor* Shalagini[3]

[1,3]Industrial University of Tyumen, 625001, Volodarskogo str., Tyumen, Russia
[2]Tyumen State University, 625003, Lenin str., Tyumen, Russia

Abstract. Thermal properties of the panel - frame enclosure structures used for construction purposes in the city of Tyumen have been studied. The effect of filtration on the change in the thermal flow rate has been revealed.

1 Introduction

Panel-frame enclosure structures are widely used in present - day construction. Thermal losses in such products mainly depend on the air and/or steam filtration speed rate through insulation layer [1, 2, 3].

However, existing methods of thermotechnical calculations do not take into account this fact [4, 5, 6]. Nevertheless, for energy saving it is important to estimate the heat amount transported through the enclosure structure depending on filtrating degree (mass rate of gas flow). In turn, it is evident that the filtering degree will depend on the pressure difference in the layered construction, which is the frame-panel enclosure structure. Filtrating degree has become especially important after the new insulating layers with selectivity in gas-transmission were invented. Experimental study approach here seems preferable, as it allows to evaluate the thermophysical properties of heat and mass transfer and mathematical model adequacy of the process under study.

It became possible to obtain the temperature distribution on the surface, as well as the actual heat flux density passing through the test structure at various differential pressures that are generated by various modes of built-in fans by using the presented automated research tool set.

2 The objects and methods of study

The aim of this study was to determine some thermophysical properties on the enclosure structure surfaces, such as temperature fields, surface density of heat flow, heat transfer dynamics and its dependence on the mass transfer degree through the layered structure.

[*] Corresponding author: tgasu.borzykh.vladimir@mail.ru

The testing equipment was climate chamber heat-cold-moisture REOCAM TCM – 1000, allowing to use objects in the temperature range -60 ÷ 100°C and relative humidity 20 ÷ 98% in the programmable mode. The main technical specifications are illustrated in Tab. 1.

Table 1. Technical characteristics of REOCAM TCM-100.

Parameter	units	value
Volume capacity Dimension capacity: Width Depth Height	l mm mm mm	1000 1000 1000 1050
Outline dimensions: Width Depth Height	 mm mm mm	 2330 2015 1490
Mass	kg	1100
Thermal insulation thickness of the chamber	mm	100
Insulation material	Foamed polyurethane	
Temperature range	°C	- 60 ... + 100
Accuracy of temperature by time	°C	from ± 0.01 to ± 0.5
Accuracy of temperature in volume (temperature homogeneity over the working volume, uneven temperature distribution.	°C	from ± 0.5 to ± 1.8
Relative humidity range (at temperatures from 20 to 60 ° C)	%	20…98
Humidity error on time (the uneven moisture distribution), not more	%	± 3
The heating rate	°C /min	1…3
The average rate of cooling to - 40°C, not less	°C / min	2
The average cooling rate -40°C to -70°C, not less	°C / min	1
The speed of air circulation	m/c	2...3
The maximum power consumption	kW	14
Power supply network access	~ 380 V/ 50 Hz, 3f	
Used refrigerants/oils: - the refrigeration unit of the upper stage - the refrigeration unit of the lower stage - refrigerating unit of a dehumidifier/cooler	 R404a/Suniso SL22 R23/Suniso SL22 R134a/Suniso SL22	
Interface and protocol of data transmission	RS232, ModBUS-RTU	

The fragment of the enclosure structure under analysis has a layered construction with the frame (see Figure 1, 2, PL. 2).

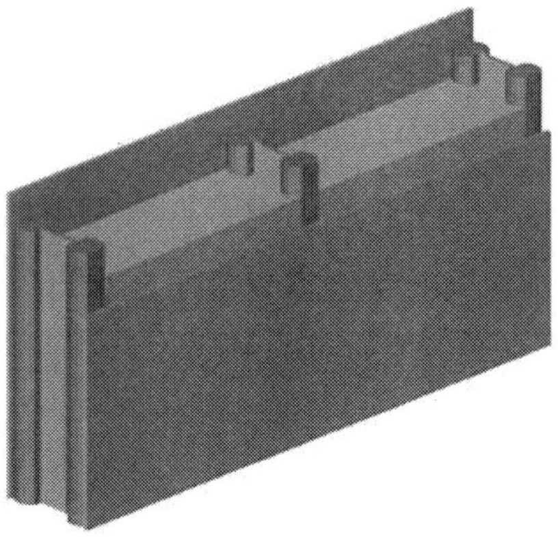

Fig. 1. The enclosure structure

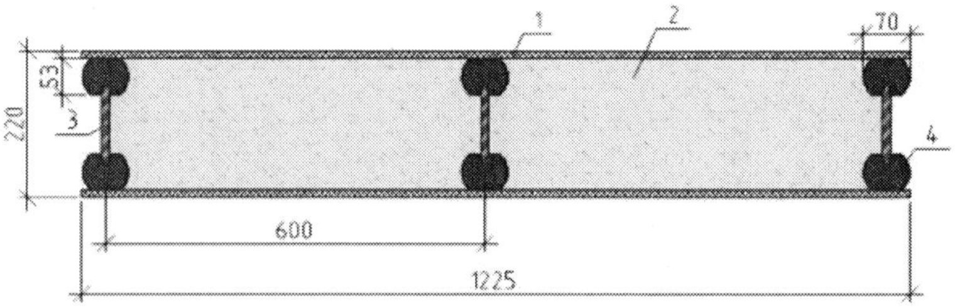

Fig. 1. Horizontal section of the enclosure structure

Table 2. Enclosure structure materials specifications.

Sr.No.	Name title	Thickness, mm	Heat capacity W / (m °C)	Density, kg/m3
1	gypsum-chip board (moisture-resistant)	8	0.41	800
2	slag wool	-	0.046	50
3	oriented standard board	9.5	0.16	600
4	wood (pine)	-	0.229	600

Testing procedure:

1. Thermoelectric sensors (TXK 0006) were attached to internal and external surfaces of the enclosure structure under testing by KPT heat-conducting paste – 8. The sensors were connected to the IT-2 calculator. Temperature sensors placement is shown in fig. 3, 4, 7, 8.

The thermoelectric sensor No. 15 was attached between the external layer of gypsum-chip board and slag wool, and the sensor No. 16 was attached between an inside layer of gypsum-chip board and slag wool to measure the insulation surface temperature both outside, and inside the enclosure structure.

Such geometrical position of sensors and measuring instruments is caused by simultaneous indicating on as large area as possible. Stiffening ribs placement has heat conductivity coefficient other than insulation conductivity coefficient. Specifications of the TXK 0006 thermoelectric sensors and the IT-2 calculator are given in table 3, 4.

Measuring instruments.

Temperature and heat conductivity sensors (multichannel IT-2) consist of an IT-2 block and a set of connection box (compensation devices) UK-4 (1 piece per each of 16 canals). IT-2 are connected to the computer via the RS-232 interface. Each UK-4 contains the integrated temperature sensor of the thermocouple cold ends.

Temperature multichannel IT-2 calculators take temperature sensors readings of the thermocouple cold ends on each of channels and transfer received data to the computer. The computer defines values of surface density of a thermal flow (q) or temperature (t) on the obtained data if necessary.

The received values are displayed on the computer in the form of voltage value (U), q or t. Temperature-cycle test is conducted. During each temperature-cycle test the IT-2 block tests all channels and transfers data to the computer. Calculators can test continuously ie temperature-cycle tests are conducted one by one, or they can conduct the required amount of temperature-cycle test. Calculators also conduct temperature-cycle tests with a programmable delay between cycles from 1 second till 60 min.

Table 3. Specifications of the thermoelectric chromel-copel sensors

Technical specifrications	chromel-copel thermocouple 0006
The measured temperatures range, °C	-40…+600
Nominal static properties	chromel-copel (L) (XK (L))
Tolerance class	2
Thermal inertia, c	0.5 ... 12
The protection against dust and water degree	IP51
Protective reinforcement material	steel 12X18H10T
Junction filling	isolated, non-isolated
Conditional pressure range, MPa	0.2...2.0

Table 4. Specifications of the IT-2 calculator

Technical specifrications	IT-2
Amount of measuring channels (Depending on model)	16, 32, 48, 64, 80, 96
Voltage measuring range of the sensor output, mV	-99.999…+99.999
Type of display	on the computer screen
The display capacity resolution:	
Voltage U, mV	0.001
Heat flow density q, W / m2	0.01
- Temperature T, ° C	0.01
Main absolute error, mV, not more, where U is the changed modulus value of the measured voltages, mV	± (5 + 1.5 ·10-4 ·Uchanges)
Temperature error of thermocouples cold ends, °C ()	±0.5
Access to PC	RS-232
Degree of protection against solid objects and water ingress according to GOST 14254-96	IP20
Air Temperature, ° C	20±5
Power supply	~ 220 V; 50 Hz
Current consumption, mA, not more	50
Overall dimensions, mm, not more:	
- -IT-2 blocks	250x110x355
- UK-4 complete set device	190x40x80
Weight, kg, no more:	
- IT-2 block	3.5
-entire set	8

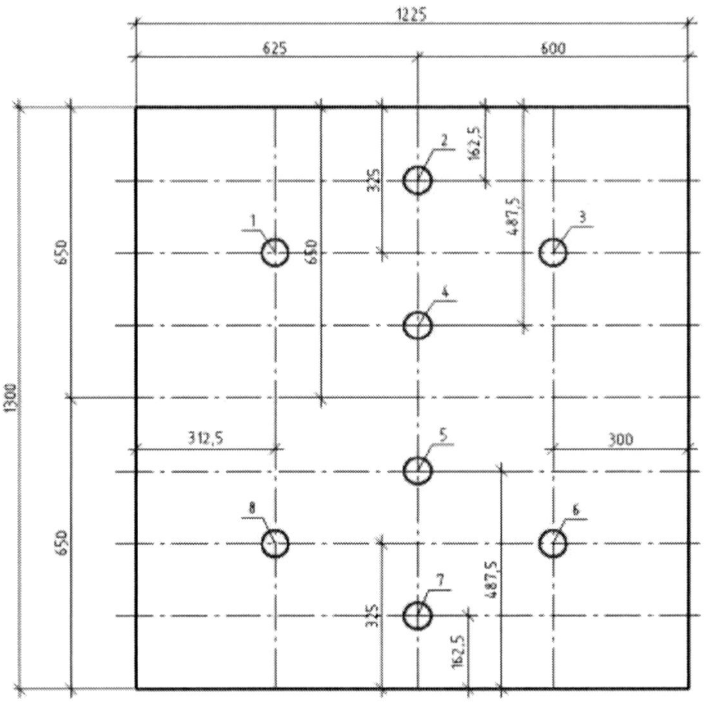

Fig. 3. Placement of the CCT (chromel-copel thermocouple) 0006 thermoelectric sensors on the enclosure structure external surface.

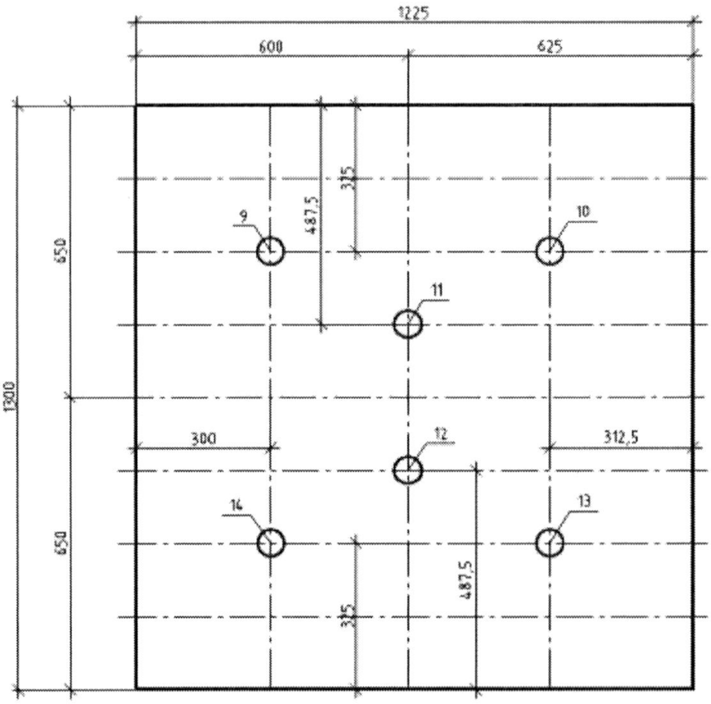

Fig. 4. Placement of the CCT (chromel-copel thermocouple) 0006 thermoelectric sensors on the enclosure structure external surface.

2. Sensors of temperature and thermal flow of ITP of MG 4.03/X(Y) "STREAM" covered with the heat-conducting (organic-silicon Heat-Conducting Paste) SHCP-8 paste were placed on an internal and external surfaces. Placement of temperature and a thermal flow sensors is shown in figures 5, 6, 7, 8. Structurally the calculator is made in the form of the electronic block and the module connected by cables. 4 thermal flow sensors and 6 temperature sensors are connected to the module by cables. The calculator is intended for testing and registration of a thermal flow density and temperatures on surfaces and / or surrounding gaseous and loose media.

Maintenance conditions:

For thermal flow and temperature sensors:

- air temperature is-30 ÷ 70 °C;
- atmospheric pressure is 84,0 ÷ 106,7 kPa (630 ÷.800 mm hg);
- relative air humidity 95% at 35 °C.

For the electronic block:

- air temperature - 20 ÷ 50 °C;
- atmospheric pressure is 84,0 ÷ 106,7 kPa (630 ÷ 800 mm hg);
- relative air humidity 95% at 35 °C.

Specifications of small thermal conductivity meter (STCM) 4.03/X(Y) "STREAM" are given in Tab. 5.

Table 5. Specifications of STCM 4.03/X(Y) "STREAM"

Name of the specification	The value of the specification
1. Testing range: heat flow density channels, W/m2 temperature channel, °C	 from 10 to 999 from - 30 to 100
2 The permissible main relative error limits when testing the heat flow density, % The permissible basic absolute error limits when temperature testing , °C	± 6 ± 0.2
3. Complementary relative error limits at thermal flow density measurement caused by a temperature deviation of a thermal flow sensors from 20 °C, %	±0.5
4 Allowed complementary error limits of channels of temperature measurements caused by a temperature deviation of the electronic block and modules from 20 °C, °C	± 0.05
5 The maximum total amount of thermal flow and temperature sensors connected to the sensor, not less, than ___pieces	100
6. The voltage supply of the calculator and modules	from 1.7 to 3.5
7. The conversion factor of the heat flow sensors, W/(m2•mV), not more	50
8. Sensors thermal resistance, m2•K/W, not more than the temperature heat flow density	0.005 0.001
9. The current consumed by the calculator, not more than, mA	28
10. The current consumed by the module, not more, mA	7
11. Overall dimensions, mm, not more: calculator calculator with larger display screen module	175x90x30 250x350x100 120x75x35 ø12x3

temperature sensors heat flow sensors (rectangular)* heat flow sensors (round)*	from 10x10x1 to 100x100x3 from 0 18x1.5 to 0 100x3
12. Weight, kg, not more electronic unit electronic unit with larger display screen module with ten sensors (cable length 5 m) a single temperature sensor (cable length 5 m) a single heat flux sensor (cable length 5 m)	0.25 1.70 1.20 0.3 0.3
13. The maximum cable length connecting each heat flow and temperature sensor to calculator , m, min.*	50
14. Mean time between failures, hour, not less	20000
15. Average service lifetime, years	10

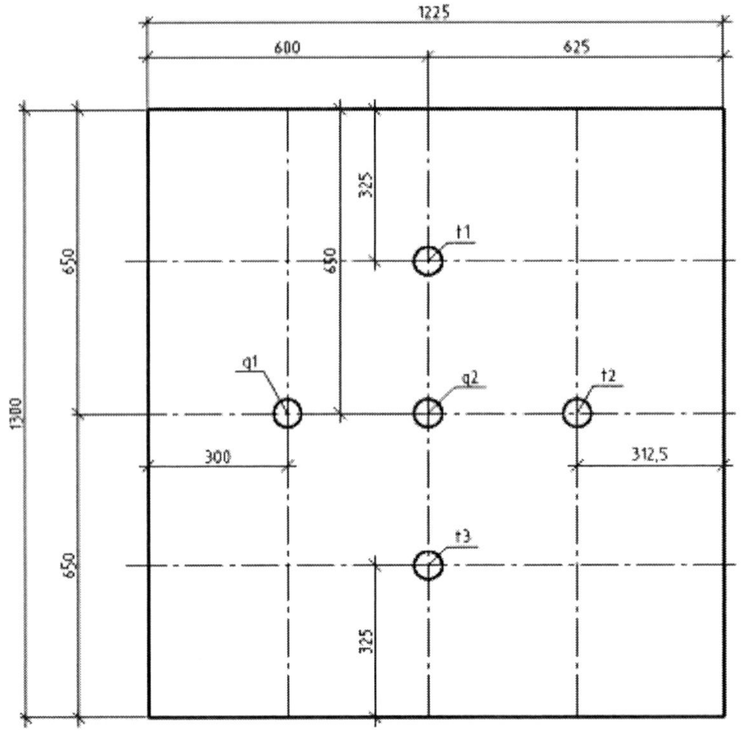

Fig. 5. Placement of temperature and heat flow sensors on the inner surface of the enclosure structure.

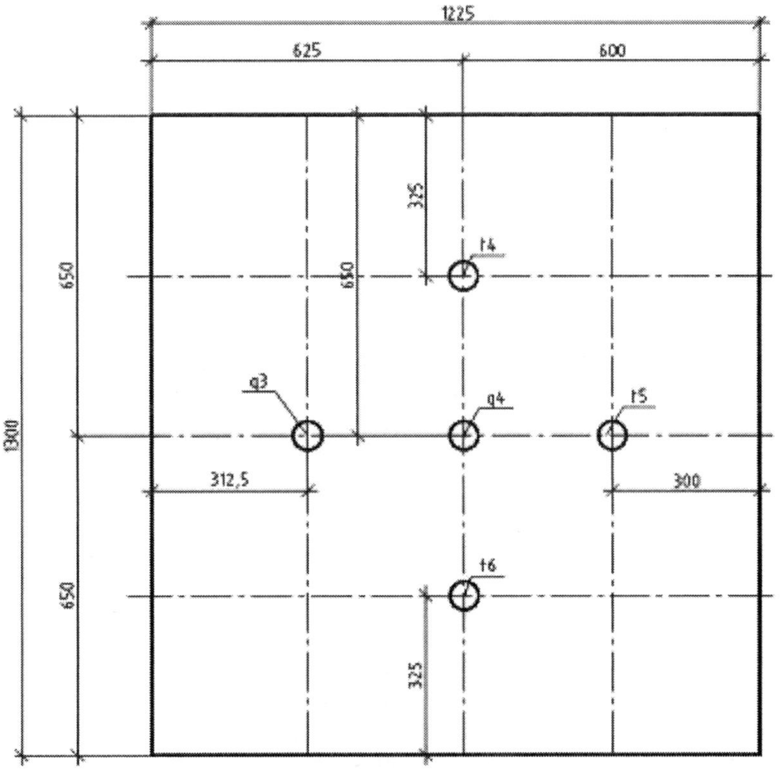

Fig. 6. Placement of temperature and heat flow sensors on the outer surface of the enclosure structure.

Fig. 7. Placement of temperature and heat flow sensors, thermoelectric sensors 0006 CCT (chromel-copel thermocouple) on the outer surface of the enclosure structure.

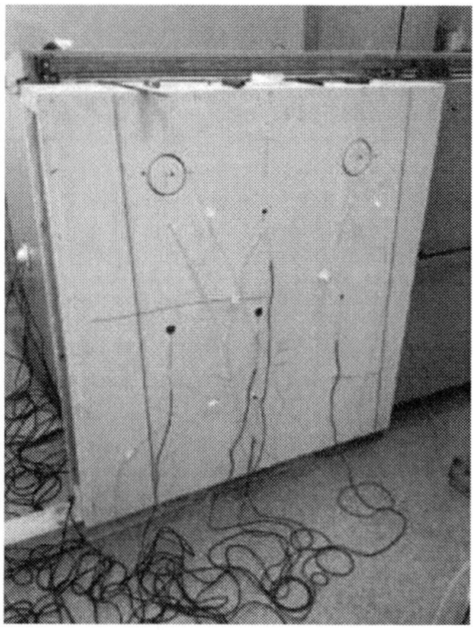

Fig. 8. Placement of temperature and heat flow sensors, thermoelectric sensors 0006 CCT (chromel-copel thermocouple) 0006 on the inner surface of the enclosure structure.

1. The sample of a enclosure structure under testing was attached to a working chamber by ropes.
2. The mode (cycle) of the camera operation was set. The schedule of the camera operation is given in fig. 9.
3. Gas filtration level through a layered construction was set by method of excess pressure or discharge from outer side.
4. The fragment under testing was exposed continuously to three cycles. Typical thermograms of a cycle are shown in fig. 10; surface densities of a thermal flow and temperature are given in fig. 11.
5. Test results were mathematically processed in the normal extension using Student's test.

3 Results

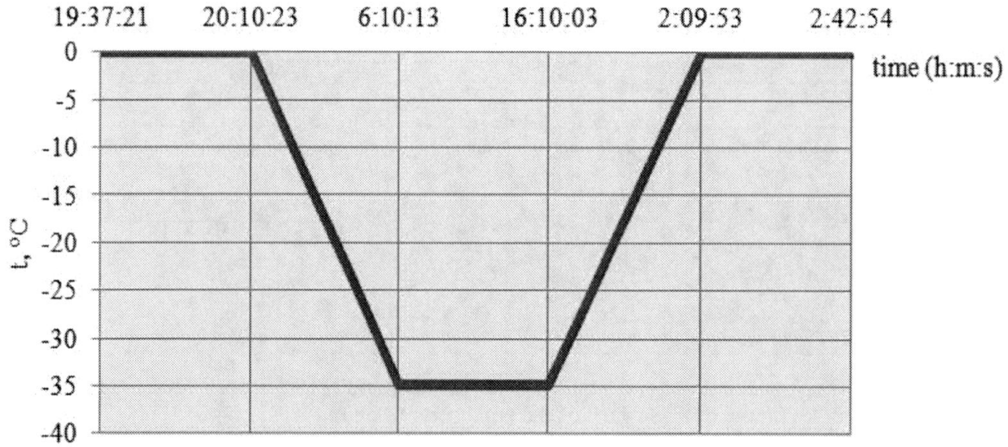

Fig. 9. Mode (cycle) of the camera operation.

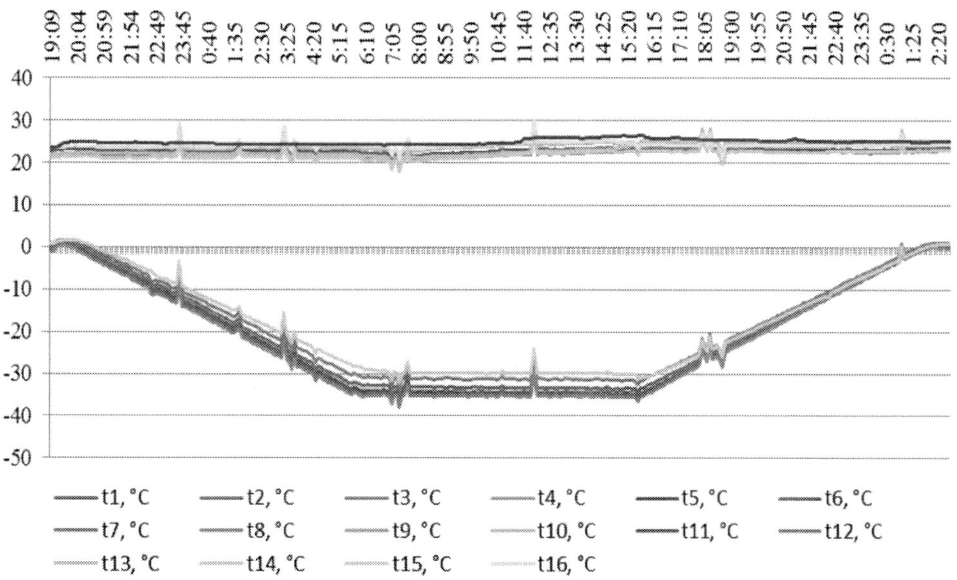

Fig. 10. Mode (cycle) of the camera operation.

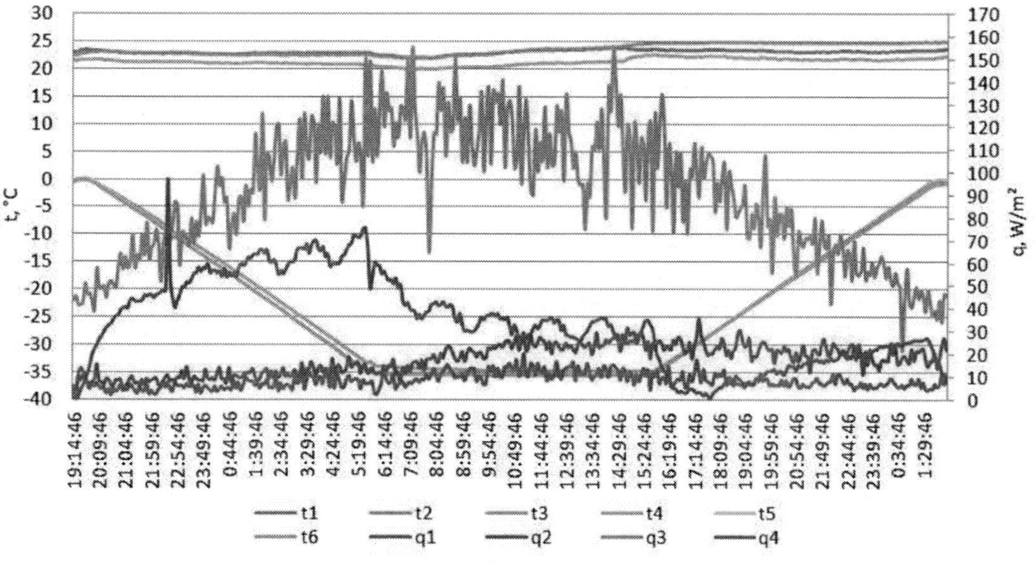

Fig. 11. Surface densities of thermal flows (q) and surface temperatures (t).

4 Conclusions

The presented experimental complex allows to carry out a wide range of heat-physical properties studies necessary to assess the enclosure structure heat saving efficiency.

The testing equipment is completely computer-aided which allows to conduct testing continuously saving results. Statistical processing is also computer-aided which increases the studies efficiency and excludes mistakes.

It is now possible to carry out a comparison of experimental study and simulation (in the measurement points) results in real time. The testing equipment is universal, as it allows to study the heat and mass transfer processes in multilayer enclosure structures without readjustment (or with small readjustment).

The automated research's stand, which was described in the article, allows to make real-time study of the interrelated processes of heat and mass transfer with air filtration, the level of measuring's equipment automation allows us to simulate the impact of external factors such as temperature, barometric pressure difference and to conduct a statistical analysis of the measured data.

New experimental information about the interrelation of distribution of temperature, pressure differential, heat flux density of studied construction of the exterior wall panel in a non-stationary thermal and mass transfer processes occurring in the building materials were received. The results of our research are applicable in the validation of computational models by quantitative comparison of the calculated temperatures distribution, pressure differences and heat flux density.

References

1. D. Krajnov, R. Sadykov, *Materialy X mezhdunarodnogo simpoziuma «Jenergojeffektivnost' i jenergosberezhenie* (2009)
2. D. Krajnov, R. Sadykov, Zhilishhnoe stroitel'stvo, **6**, 10-12 (2012)
3. V. Savin, Jenergosberezhenie, **6**, 63-65 (2002)
4. F.V. Ushkov, *Stroitel'naja teplofizika*, (Gosstroj, Moscow, 1979)
5. F. Ushkov, Stroitel'naja promyshlennost', **8**, 16-19 (1951)
6. F. Ushkov, Issledovanie temperaturnyh polej stykov trehslojnyh panelej. Gos. kom. po grazhd. stroitel'stvu i arhitekture pri Gosstroe SSSR. Nauch.-issled. in-t teorii, istorii i perspektivnyh problem sovetskoj arhitektury (1966)
7. N. Kurylenko, L. Mikhaylov, P. Artamonov, MATEC Web of Conferences (2015)
8. G. Kuznetsov, N. Kurilenko, G. Mamontov, L. Mihajlova, *EPJ* Web of Conferences (2015)
9. N. Kurilenko, G. Mamontov, L. Mihajlova, *EPJ Web of Conferences* (2015)
10. I. Shalagin, N. Kurilenko, R. Davlyatchin, *MATEC Web of Conferences* (2015)
11. G. Kuznecov, N. Kurilenko, V. Maksimov, G. Mamontov, T. Nagornova, Journal of Engineering Physics and Thermophysics (2013)
12. N. Kurylenko, D. Zverev, A. Idrisov, Gas industry (2001)

MATEC Web of Conferences **73**, 02023 (2016) DOI: 10.1051/matecconf/20167302023
TPACEE-2016

Mathematical Modeling of Temperature Field of Multilayer Enclosure Structures

Boris Aksenov[1,*], *Svetlana* Karyakina[1], *Oleg* Stepanov[1], *Anatoly* Shapoval[1], and *Mikhail* Bodrov[2]

[1]Industrial University of Tyumen, 625001 Volodarskogo str. 38, Tyumen, Russia
[2]Nizhny Novgorod State University of Architecture and Civil Engineering, 603950 Ilyinskaya str. 65, Nizhny Novgorod, Russia

Abstract. The mathematical model of heat exchange process through multilayer enclosure structure is constructed. Herein infiltration and heat-conducting inclusions are taken into account. Heat exchange process modeling in the stationary and non- stationary conditions is considered.The method of temperature field constructing based on the suggested model is offered. The three-dimensional boundary problem of non- stationary heat conductivity is solved numerically using the implicit method of alternating directions. The temperature field constructing of the enclosure structures in the non- stationary condition allows to track dynamics of thermal flows through enclosing structures and to receive the thermal resistance considering influence of heat-conducting inclusions.

1 Introduction

There are problems with indoor microclimate when using the thermally insulated cladding panel with aluminum alloy skin.

Modeling of heat exchange processes in the panel is required to work out the premises design with the required positive temperature indoors at a negative external air temperature. Moreover the inside temperature differentials should not exceed the set norms.

Work [1] offers the heat exchange model in aluminum panels at the following assumptions: stationary condition of the process, constancy of thermophysical properties of materials, independence of heat exchange process on mass transfer. This model represents a system of equations describing heat exchange on cladding panel internal and external skins and through panel joints.

It is possible to calculate temperature distribution on the panel skin by a two-dimensional boundary problem of stationary heat conductivity.

The stationary model is insufficient for accurate thermal structural calculations when operating the factory-assembled buildings in environments with large diurnal temperature variation and strong winds. During the time required to establish stationary state both the outside air temperature and the indoor heat balance may change.

* Corresponding author: tgasu.aksenov.boris@mail.ru

2 Modeling of heat exchange processes in the enclosure structures in non-stationary conditions

The physical and mathematical model is based on the following assumptions. Heat is transferred from indoor air to the inner skin plate, then through the insulation layer to the outer siding and then into the atmosphere. Besides there is a heat flow through the "thermal bridge" having a high thermal conductivity. It should be noted that the standard panel thickness is an order less than its height and length.

Thermophysical properties of materials are represented by functions depending on time and coordinates. Mass transfer is not taken into account.

The temperature distribution on internal and external panel skin surface is characterized by heat exchange at a panel skin surface. The equations describing the distribution of temperature on an internal and external panel skin surface are derived from the equation of thermal balance, and contain heat and moisture exchange coefficients which are defined by the intensity of a convective heat transfer, radiation heat exchange of internal and external enclosure structure surfaces.

Temperature on a side plate is not calculated. Heat exchange through panel joints is taken into account when formulating boundary conditions. The thermal flow arriving to a joint along the plate consists of the heat loss flow through heat-conducting inclusion from the premises into environment, and the cold air infiltration flow heating conditioned by more preheated panel.

The thermal flow in a thermal insulation layer is considered directed strictly perpendicular to the panel plane as near the panel edge the flow through insulation is small enough in comparison with the flow through a metal panel skin.

Let us introduce a three-dimensional Cartesian coordinate system with the origin at the center of the internal panel skin size $2a \times 2b \times d$. The direction of the axis Z from the inner panel skin to the outer one is then $-a \le x \le a, -b \le y \le b, 0 \le z \le d$.

To derive the equation describing the temperature distribution t_1 on the inner panel skin surface, let us consider the surface element $dx \times dy \times \delta$, where δ is the panel skin thickness.

Let us express the heat flowing into surface element along the axes X, Y, Z respectively at time τ by $Q_x(\tau, x, y)$ $Q_y(\tau, x, y)$ $Q_z(\tau, x, y, z)$. Based on Fourier's law mathematical expression of these flows is as follows:

$$Q_x(\tau,x,y) = -\lambda_a \cdot \delta \cdot dy \frac{\partial t_1(\tau,x,y)}{\partial x} \cdot d\tau,$$
$$Q_y(\tau,x,y) = -\lambda_a \cdot \delta \cdot dx \frac{\partial t_1(\tau,x,y)}{\partial y} \cdot d\tau, \qquad (1)$$
$$Q_z(\tau,x,y,z)\big|_{z=0} = \alpha_1 (t_w(\tau) - t_1(\tau,x,y,z)) dx \cdot dy \cdot d\tau.$$

The expressions for the resulting heat flow can be written as follows:

$$Q_x(\tau,x+\Delta x,y) = -\lambda_a \cdot \delta \cdot dy \frac{\partial t_1(\tau,x+\Delta x,y)}{\partial x} \cdot d\tau;$$
$$Q_y(\tau,x,y+\Delta y) = -\lambda_a \cdot \delta \cdot dx \frac{\partial t_1(\tau,x,y+\Delta y)}{\partial y} \cdot d\tau; \qquad (2)$$
$$Q_z(\tau,x,y,z)\big|_{z=\delta} = -\lambda_s \cdot dx \cdot dy \frac{\partial t_s(\tau,x,y,z)}{\partial z}\Big|_{z=\delta} \cdot d\tau.$$

where λ_a, λ_s are panel skin and insulation thermal conductivity, α_1 is heat exchange coefficient on an internal panel skin, t_s is isolation temperature; t_w is average air temperature indoors.

In the non-stationary condition the heat balance equation is:

$$\begin{aligned}&\left(Q_x(\tau,x,y)+Q_y(\tau,x,y)+Q_z(\tau,x,y,z)\big|_{z=0}\right)-\\&-\left(Q_x(\tau,x+\Delta x,y)+Q_y(\tau,x,y+\Delta y)+Q_z(\tau,x,y,z)\big|_{z=\delta}\right)=dH,\end{aligned} \qquad (3)$$

where dH is the enthalpy change of the panel skin element

$$dH = c_a \frac{\partial t_1(\tau,x,y,z)}{\partial \tau} \cdot d\tau \cdot dV, \qquad (4)$$

where c_a is thermal capacity of panel skin unit volume, dV - panel skin element volume

After rearrangement the thermal balance equation (3) with regard to (1), (2), (4) will change to:

$$\begin{aligned}&\lambda_a\left(\frac{\partial^2 t_1(\tau,x,y)}{\partial x^2}+\frac{\partial^2 t_1(\tau,x,y)}{\partial y^2}\right)+\frac{\lambda_s}{\delta}\frac{\partial t_s(\tau,x,y,z)}{\partial z}\Big|_{z=\delta}+\\&+\frac{\alpha_1}{\delta}\left(t_w(\tau)-t_1(\tau,x,y)\right)=c_a\frac{\partial t_1(\tau,x,y)}{\partial \tau},\\&-a<x<a,-b<y<b,\end{aligned} \qquad (5)$$

The equation describing thermal balance for the panel external surface is derived similarly and has a form:

$$\begin{aligned}&\lambda_a\left(\frac{\partial^2 t_2(\tau,x,y)}{\partial x^2}+\frac{\partial^2 t_2(\tau,x,y)}{\partial y^2}\right)-\frac{\lambda_s}{\delta}\frac{\partial t_s(\tau,x,y,z)}{\partial z}\Big|_{z=d-\delta}+\\&+\frac{\alpha_2}{\delta}\left(t_u(\tau)-t_2(\tau,x,y)\right)=c_a\frac{\partial t_2(\tau,x,y)}{\partial \tau},\\&-a<x<a,-b<y<b,\end{aligned} \qquad (6)$$

where t_2 is external panel skin temperature, t_u is the average air temperature outdoors, α_2 is heat exchange coefficient on an external panel skin.

For the insulation layer the heat flows only in the direction Z according to the accepted assumptions. $Q_x = Q_y = 0$ is considered. Let us consider the balance of heat flows for an insulation element $dx \times dy \times dz$.

The inward flow

$$Q_z(\tau, x, y, z) = -\lambda_s \cdot dx \cdot dy \cdot \frac{\partial t_s(\tau, x, y, z)}{\partial z} \cdot d\tau \qquad (7)$$

where t_s is temperature insulation, λ_s is thermal conductivity of the insulation material.

The outward flow

$$Q_z(\tau, x, y, z + \Delta z) = -\lambda_s \cdot dx \cdot dy \cdot \frac{\partial t_s(\tau, x, y, z + \Delta z)}{\partial z} \cdot d\tau . \qquad (8)$$

The equation of thermal balance is:

$$\lambda_s \cdot dx \cdot dy \cdot d\tau \left(\frac{\partial t_s(\tau, x, y, z + \Delta z)}{\partial z} - \frac{\partial t_s(\tau, x, y, z)}{\partial z} \right) = c_s \cdot \frac{\partial t_s(\tau, x, y, z)}{\partial \tau} \cdot d\tau \cdot dV, \qquad (9)$$

where c_s is heat capacity of insulation material unit volume, dV is the insulation unit volume.

After the rearrangement (9) will take a form:

$$\frac{\lambda_s}{c_s} \frac{\partial^2 t_s(\tau, x, y, z)}{\partial z^2} = \frac{\partial t_s(\tau, x, y, z)}{\partial \tau},$$
$$-a < x < a, -b < y < b, \delta \le z \le d - \delta. \qquad (10)$$

Boundary conditions are set similarly [1]:

$x = -a$:

$$\frac{\partial t_1(\tau, x, y)}{\partial x} = -\frac{\partial t_2(\tau, x, y)}{\partial x} =$$
$$\frac{(t_1(\tau, x, y) - t_2(\tau, x, y))}{l_2} + \frac{MC_p(t.(\tau) - t_{cw}(\tau, x, y))}{\delta \lambda_a}. \qquad (11)$$

$x = +a$:

$$-\frac{\partial t_1(\tau, x, y)}{\partial x} = \frac{\partial t_2(\tau, x, y)}{\partial x} =$$
$$\frac{(t_1(\tau, x, y) - t_2(\tau, x, y))}{l_1} + \frac{MC_p(t.(\tau) - t_{cw}(\tau, x, y))}{\delta \lambda_a}. \qquad (12)$$

$$y=-b:$$
$$\frac{\partial t_1(\tau,x,y)}{\partial y}=-\frac{\partial t_2(\tau,x,y)}{\partial y}=$$
$$\frac{(t_1(\tau,x,y)-t_2(\tau,x,y))}{l_4}+\frac{MC_p(t_i(\tau)-t_{cw}(\tau,x,y))}{\delta\lambda_a}. \tag{13}$$

$$y=+b:$$
$$-\frac{\partial t_1(\tau,x,y)}{\partial y}=\frac{\partial t_2(\tau,x,y)}{\partial y}=$$
$$\frac{(t_1(\tau,x,y)-t_2(\tau,x,y))}{l_3}+\frac{MC_p(t_i(\tau)-t_{cw}(\tau,x,y))}{\delta\lambda_a}. \tag{14}$$

where l_1, l_2, l_3, l_4 are length dimension parameters characterizing heat-conducting inclusions, M is the discharge of infiltrating air per unit length per second, t_{cw} is temperature of the cold air coming inside through a joint; C_p is specific air heat capacity.

The left part of boundary conditions characterizes a heat flow due to heat conductivity of a panel skin, the right one is proportional to a temperatures difference of external and internal part of a joint.

It is advisable to accept

$$t_s(\tau,x,y,\delta)=t_1(\tau,x,y),\quad t_s(\tau,x,y,d-\delta)=t_2(\tau,x,y). \tag{15}$$

Initial conditions:

$$\begin{aligned} &t_1(0,x,y)=t_w(0),\\ &t_2(0,x,y)=t_u(0),\\ &t_s(0,x,y,z)=t_w(0)-z\cdot\frac{t_w(0)-t_u(0)}{d},\\ &-a\le x\le a,\quad -b\le y\le b,\quad \delta\le z\le d-\delta. \end{aligned} \tag{16}$$

The simultaneous solution of (5), (6), (10), (11) - (16) are used to construct the temperature field of the panel.

Numerical solution of the boundary value problem of non-steady heat conduction

For the numerical solution of the problem (5), (6), (10), (11) - (16) we use the implicit method of alternating directions. We introduce the grid

$$\omega_{h\tau}=\omega_h\times\omega_\tau=\left\{\begin{matrix}(i_1h_1,\, i_2h_2,\, i_3h_3,\, j\Delta\tau),\\ i_1=0,1,\dots,N_1, i_2=0,1,\dots,N_2,\\ i_3=0,1,\dots,N_3, j=0,1,\dots,N_\tau\end{matrix}\right\}$$

With grid steps $h_1 = \dfrac{2a}{N_1}, \quad h_2 = \dfrac{2b}{N_2}, \quad h_3 = \dfrac{d}{N_3}, \quad \Delta\tau = \dfrac{T}{N_\tau}.$

For the equation of an internal panel skin we receive two boundary problems. The index i_3 is omitted, since we get $i_3 = 0$ for points of the inner panel skin.

The first problem is

$$0,5\beta_1 t_{1\,i_1-1,i_2}^{\,j+1/2} - k_1 t_{1\,i_1,i_2}^{\,j+1/2} + 0,5\beta_2 t_{1\,i_1+1,i_2}^{\,j+1/2} = -F_{i_1,i_2}^{j+1/2},\ i_1 = 1,2,...,N_1-1\ (0 < i_2 < N_2),$$

$$\beta_1 = \frac{\lambda_a\Delta\tau}{c_a(h_1)^2},\ \beta_2 = \frac{\lambda_a\Delta\tau}{c_a(h_2)^2},\ k_1 = 1+\beta_1 - \frac{\Delta\tau}{2\delta c_a}\left(\alpha_1 + \frac{\lambda_s}{h_3}\right),$$

$$F_{i_1,i_2}^{j+1/2} = 0,2\beta_2\left(t_{1\,i_1,i_2-1}^{\,j} + t_{1\,i_1,i_2+1}^{\,j}\right) + (1-\beta_2)t_{1\,i_1,i_2}^{\,j} - \frac{\lambda_a\Delta\tau}{2c_a} f^{j+1/2}, \tag{17}$$

$$f^{j+1/2} = \left(\frac{\lambda_s}{\lambda_a\delta h_3} t_{s\,i_1,i_2}^{\,j} + \frac{\alpha_1}{\lambda_a\delta} t_w(\Delta\tau(j+0,5)\right).$$

The boundary conditions are:

$$t_{10,i_2}^{j+1/2} = t_{11,i_2}^{j+1/2} + h_1\left(\frac{(t_{10,i_2}^{j} - t_{20,i_2}^{j})}{l_2} + L\right),$$

$$t_{1N_1,i_2}^{j+1/2} = t_{1N_1-1,i_2}^{j+1/2} + h_1\left(\frac{(t_{2N_1,i_2}^{j} - t_{1N_1,i_2}^{j})}{l_1} + L\right), \tag{18}$$

where $L = \dfrac{MC_p\left(t_u^{\,j} - t_{cw}^{\,j}\right)}{\delta\lambda_a}$.

The second problem is:

$$0,5\beta_2 t_{1\,i_1,i_2-1}^{\,j+1} - k_2 t_{1\,i_1,i_2}^{\,j+1} + 0,5\beta_2 t_{1\,i_1,i_2+1}^{\,j+1} = -F_{i_1,i_2}^{j+1},\ i_2 = 1,2,...,N_2-1\ (0 < i_1 < N_1),$$

$$\beta_1 = \frac{\lambda_a\Delta\tau}{C_a(h_1)^2},\ \beta_2 = \frac{\lambda_a\Delta\tau}{C_a(h_2)^2},\ k_2 = 1+\beta_2 - \frac{\Delta\tau}{2\delta c_a}\left(\alpha_1 + \frac{\lambda_{из}}{h_3}\right),$$

$$F_{i_1,i_2}^{j+1} = 0,5\beta_1\left(t_{1\,i_1-1,i_2}^{\,j+1/2} + t_{1\,i_1+1,i_2}^{\,j+1/2}\right) + (1-\beta_1)t_{1\,i_1,i_2}^{\,j+1/2} - \frac{\lambda_a\Delta\tau}{2c_a} f^{j+1/2}, \tag{19}$$

$$f^{j+1/2} = \left(\frac{\lambda_{из}}{\lambda_a\delta h_3} t_{из\,i_1,i_2}^{\,j} + \frac{\alpha_1}{\lambda_a\delta} t_w(\Delta\tau(j+0,5)\right).$$

The boundary conditions are:

$$t_{1i_1,0}^{j+1} = t_{1i_1,0}^{j+1} + \mathrm{h}_2 \left(\frac{(t_{1i_{1,0}}^{j} - t_{2i_{1,0}}^{j})}{l_4} + L \right),$$
$$t_{1i_1,N_2}^{j+1} = t_{1i_1,N_2-1}^{j+1} + h_2 \left(\frac{(t_{2i_1,N_2}^{j} - t_{1i_1,N_2}^{j})}{l_3} + L \right); \tag{20}$$

The scheme of alternating directions for the equation of an internal panel skin is derived similarly.

For the insulation layer the differential analog of the equation (10) will have a form:

$$\frac{\lambda_s \Delta\tau}{c_s (h_3)^2} t_{s\ i_1,i_2,i_{3-1}}^{\ j+1} - \frac{2\lambda_s \Delta\tau}{c_s (h_3)^2} t_{us\ i_1,i_2,i_3}^{\ j+1} + \frac{\lambda_s \Delta\tau}{c_s (h_3)^2} t_{s\ i_1,i_2,i_{3+1}}^{\ j+1} = -t_{s\ i_1,i_2,i_3}^{\ j+1},\ i_3 = 1,2,\dots,N_3 - 1. \tag{21}$$

The boundary conditions are

$$t_{s\ i_1,i_2,0}^{\ j+1} = t_{1\ i_1,i_2}^{\ j},\ t_{s\ i_1,i_2,N_3}^{\ j+1} = t_{2\ i_1,i_2}^{\ j}. \tag{22}$$

The received systems allows to use the tridiagonal matrix algorithm to solute them.

The constructed scheme of alternating directions is stable and converges at a speed of

$$O\left(\sum_{i=1}^{3} h_i^2 + (\Delta\tau)^2 \right).$$

To calculate the temperature of infiltrated air, the methodology described in [1] has been used. When computer modelling the temperature field the symmetry was taken into account. All information was stored for a quarter of the fragment and restored due to the symmetry for other sections of the construction. Besides, variable step toward the panel edges was used, for the temperature field at the fragment edges has very strong heterogeneity. On each time step heat rejection ratio of a vertical and horizontal joints surface, the cold air temperature arriving through joints, and calculation of temperature at an internal and external panel skin surface as well as in the insulation layer were defined. To check the calculations accuracy by the proposed method additional calculations were made, the results of which were compared with the calculations results by stationary model, for which there is a comparison with in - situ measurements [1]. The comparison showed almost total agreement between the calculated temperatures. Figure 1 shows the results of the calculated steady temperature fields of the inner and outer surfaces according to [1].

The temperature field calculation by the above method allows to calculate the heat flow through the panels and through the joints and assess their contribution to the overall heat loss at any time.

Heat flow through panels

$$Q(\tau) = \int_{-(a-\delta)}^{a-\delta} dx \int_{-(b-\delta)}^{b-\delta} \alpha_1 (t_w(\tau) - t_1(\tau, x, y)) dy.$$

Heat flow through a panel joint $x = a$:

$$Q(\tau)\big|_{x=a} = -\lambda_a \cdot \delta \cdot \int_{-b}^{b} \left[\frac{\partial t_1(\tau, x, y)}{\partial x} \Big|_{x=a} \right] dy,$$

similarly $x = -a,\ y = \pm b$.

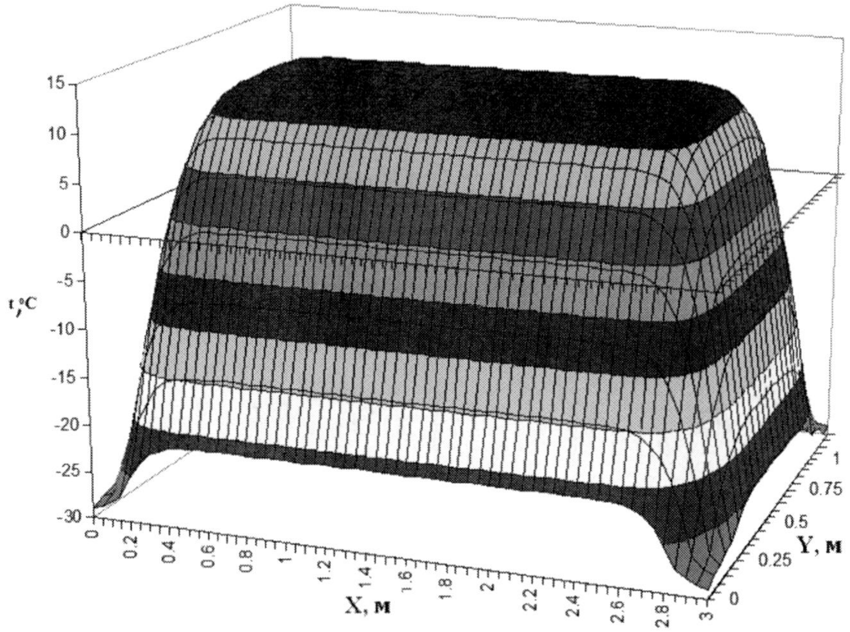

a)

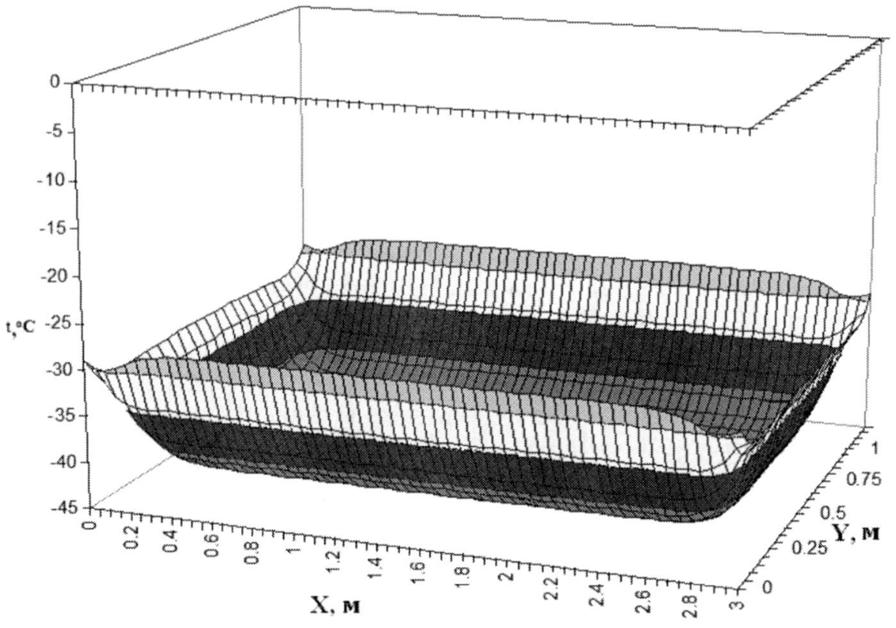

b)

Fig. 1. Temperature field of the panel skin; a) is an internal surface; b) is an external surface.

3 Results

The physical and mathematical model of heat exchange through the enclosure structure is developed. The model considers non-stationarity of the process and heat exchange connected with heat-conducting inclusions and infiltration of air in panel joints.

This model can be rather effective to design a multilayered heat insulation panels as the above calculations have shown that at non-stationary heat exchange process heat loss depends on sequence of layers along cross section.

4 Conclusion

When using the enclosure structure it takes from 15 to 20 hours to establish the stationary condition. Thus, the stationary condition practically does not occur. Accurate thermo - physical calculation can be made only with the help of the non-stationary model.

In lightweight enclosure structure up to 75% of heat loss falls on the joints between panels. And about 30% is lost through infiltration and 45% through the joints formed in "thermal bridge". To reduce heat loss through the enclosure structure it is more efficient to seal joints and destroy "thermal bridge" but not increase the panel thickness or use a better thermal insulator.

References

1. V.V. Remizov, A.F. Shapoval, B.V. Moiseev, B.G. Aksenov, *Osobennosti stroitel'stva ob'ektov v neftegazodobyvajushhih rajonah Zapadnoj Sibiri* (Nedra, Moscow, 1996)
2. A.N. Tihonov, A.A. Samarskij, *Uravnenija matematicheskoj fiziki* (Nauka, Moscow, 1977)
3. S.Ya. Kushnir, A.I. Gorkovenko, A.S. Gorelov, Soil Mechanics and Foundation Engineering, **42(4)**, 147-151 (2005)
4. G.V. Kuznetsov, M.A. Sheremet, Izv. RAN. Mehanika zhidkosti i gaza, **6**, 29-39 (2006)
5. V.N. Bogoslovskij, *Teplovoj rezhim zdanija* (Strojizdat, Moscow, 1979)
6. E.G. Maljavina, *Teplopoteri zdanija* (AVOK – PRESS, Moscow, 2007)
7. L. Banhidi, *Teplovoj mikroklimat pomeshhenij* (Strojizdat, Moscow, 1981)
8. V.N. Bogoslovskij, *Stroitel'naja teplofizika* (Vysshaja shkola, Moscow, 1982)
9. L.D. Boguslavskij, *Snizhenie rashoda jenergii pri rabote sistem otoplenija, ventiljacii i kondicionirovanija vozduha* (Strojizdat, Moskva, 1982)
10. E.A. Nasonov, *Sb. nauchnyh trudov NIIST*, 161-169 (1966)
11. SP 60.13330.2012 SanPiN 2.2.4.548 – 96
12. SNiP 23-02-2003.
13. Tri «Je», ESCO Jelektronnyj zhurnal jenergoservisnoj kompanii «Jekologicheskie sistemy», **1**, (2007)
14. Ju.A. Surinov, Izvestija vuzov. Chernaja metallurgija, 58-66 (1997)
15. A.N. Skanavi, *Konstruirovanie i raschet vodjanogo i vozdushnogo otoplenija zdanij* (Strojizdat, Moscow, 1983)
16. S.N. Shorin, Teploperedacha. V sbornike: Sovremennye voprosy otoplenija i ventiljacii, (1949)
17. Ju.A. Tabunshhikov, *Raschety temperaturnogo rezhima pomeshhenija i trebuemoj moshhnosti dlja ego otoplenija ili ohlazhdenija* (Strojizdat, Moscow, 1981)
18. S.A. Krupennikov, Izv. vuzov, Chernaja metallurgija, **5**, 46-49 (1995)
19. T.A. Nagornova, *Vserossijskaja nauchno-prakticheskaja konferencija s mezhdunarodnym uchastiem «Fedorovskie chtenija»*, 165-166 (2013)

20. G.Ja. Mamontov, T.A. Nagornova, N.I. Kurilenko, V.I. Maksimov, Izvestija Tomskogo politihnicheskogo universiteta, **323(4),** 66-70 (2013)

Influence of a Cylindrical Thermally-Conductive Inclusion on the Heat Flow through the Light Building Envelopes

Konstantin Afonin [1], *Tatyana* Zhilina [1,*], *Anatoly* Shapoval[1], and *Boris* Moiseev [1]

[1]Industrial University of Tyumen, 625001 Volodarskogo str. 38, Tyumen, Russia

Abstract. Paper introduces a mathematical model describing the distribution of temperature field around the cylindrical thermally-conductive inclusion. It also presents an experimental setup based on REOCAMTCM-1000 climate chamber, as well as the obtained results of the temperature distribution on the surface of a light building envelope under the influence of a cylindrical thermally-conductive inclusion.

1 Introduction

Western Siberia has all kinds of fuel and energy resources and is a major oil and gas producing region. Fuel and energy complex of Western Siberia is one of the main factors of the country economy development. When considering the development of the North, one should consider its specific features:

- distance from developed regions of the country,
- absence of a stable transport links within the area and of the developed construction materials industry,
- random placement of its plants,
- high labor costs.

Harsh climatic conditions of the North, marked by low outside temperatures, large diurnal changes in temperature, strong winds, snow drifts and intense rains, make it necessary to consider all these factors in the design and construction of buildings [1].

Construction in the North is imposed with strict requirements defining the appropriate indoor climate, provision of the essential conveniences for living, as well as economic indicators - reducing the labor intensity, increasing the level of industrialization of construction, lasting quality, reduction of capital and operational costs. Climatic region of construction largely determines the requirement for a reliable thermal protection of buildings and provision of a comfortable microclimate in residential facilities in the North [1,2].

The main direction of the industrialization of the oil and gas extraction facilities arrangement is a factory-assembled method, which is indispensable in the construction of energy complex surface facilities in the northern regions of Western Siberia [3].

* Corresponding author: tgasu.zhilina.tatyana@mail.ru

Oil and gas facilities refer to the small-volume and distributed objects, which are currently built in factory-assembled way. Outer light envelopes are constructed of metal panels with an effective insulation [4,5].

The main feature of light envelopes is the necessity to study their properties over time, due to presence of thickening, which eventually leads to a loosening of the hermetic sealing and increase of the air permeability through the outer envelopes and as a result - to increase of the energy consumption for climate control.

Due to specific design features, zones with different thermally-conductive inclusions occur in building envelopes. They can be through, perforated, etc., and have different shapes. In northern conditions there is a presence of a large temperature gradient between the outer and inner walls of the building envelope. The combination of the temperature gradient and specific design features cause the heavy increase of heat loss and breach of the facility comfort conditions.

2 Subjects and methods

The authors analyze the distribution of temperature field around the cylindrical thermally-conductive inclusion [6,7,8].

Figure 1 presents the scheme that has been selected for calculation the stationary distribution of thermal field on the building envelope surface, containing thermally-conductive inclusion of cylindrical shape.

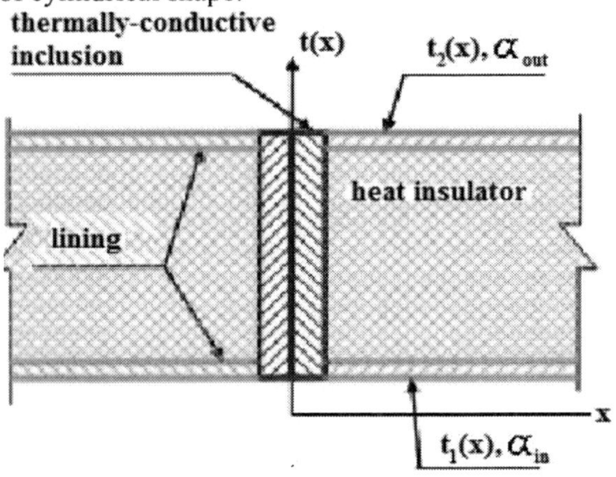

Fig. 1. Scheme of the cylindrical thermally-conductive inclusion.

Building envelope consists of two metal layers between which there is a layer of material with high thermal resistance. The envelope parameters are going to be designated as following:

- δ_1,δ_2 - thickness of the inner and outer lining,
- R_0 - thermal resistance of the lining material,
- R - thermal resistance of the thermally-conductive inclusion,
- t_{in}, t_{out} - indoor and outdoor air temperature,
- $\alpha_{in},\alpha_{out},\lambda_1,\lambda_2$ – coefficients of surface heat exchange and thermal conductivity for inner and outer surfaces accordingly,
- r_0 – radius of the thermally-conductive inclusion.

If we consider the problem as the spatial one, there arises a problem of solution of the Laplace's equation with various spatial physical properties which makes it necessary to reconcile the obtained results in different areas with one another. This leads to unnecessary complication of the solving the problem process.

The authors put forward a different approach. Its feature is that the temperature distribution inside the envelope layer is not so significant when calculating of heat loss through this structure. Temperature fields on the mating surfaces, which are involved in major heat exchange, are more important. Spatial temperature distribution along the *Z* axis and the angular dependence are also ignored, as the problem has cylindrical symmetry by condition [6,9,10]. For example, to balance the heat flow in a cylindrical reference frame, we get the following equation in partial derivatives [11,12]

$$\frac{\partial^2 t}{\partial r^2}+\frac{\partial t}{r\partial r}+\frac{\partial^2 t}{r^2\partial\theta^2}+\frac{\partial^2 t}{\partial z^2}=0 \tag{1}$$

where $t(r, \theta, z)=t(M)$ – temperature at any point of the envelope.

If we get down to the description of the heat flow based on the assumptions for inner and outer surfaces of the envelope set out above, the following components can be defined. Let us assume that $t_1(r)$- temperature distribution on the inner surface of the envelope, and $t_2(r)$ - temperature distribution on the outer surface of the envelope.

Then:

$\frac{d^2 t_1}{dr^2}+\frac{dt_1}{rdr}$ - reduced density of a heat flow distributing along the uniform inner surface;

$\frac{\alpha_{in}}{\lambda_1\delta_1}(t_{in}-t_1)$ - reduced density of a heat flow ingoing due to a heat exchange between surface and environment;

$\frac{t_1-t_2}{\lambda_1\delta_1 R_o}$ - reduced heat flow, outgoing from the surface due to flows through the material inside the envelope.

Taking into account these flows, differential equations for each of the surfaces can be written. For further convenient manipulations, we introduce the following designations:

$$m_1=\frac{\alpha_{in}}{\lambda_1\delta_1};\ n_1=\frac{\alpha_{out}}{\lambda_2\delta_2};\ m_o=\frac{1}{R_o\lambda_1\delta_1};\ n_o=\frac{1}{R_o\lambda_2\delta_2};\ m_{10}=m_1+m_0;\ n_{10}=n_1+n_0;$$

$$p_{11}=\frac{r_0}{2\lambda_1\delta_1}(\alpha_{in}+\frac{1}{R});\ p_{22}=\frac{r_0}{2\lambda_2\delta_2}(\alpha_{out}+\frac{1}{R}); \tag{2}$$

$$p_{12}=\frac{-r_0}{2R\lambda_1\delta_1};\ p_{21}=\frac{-r_0}{2R\lambda_2\delta_2};\ q_1=-\frac{\alpha_{in}r_0t_{in}}{2\lambda_1\delta_1};\ q_2=-\frac{\alpha_{out}r_0t_{out}}{2\lambda_2\delta_2}$$

Using new designations the system of equations for determining the temperature fields will be presented as follows:

$$\begin{aligned}&\frac{d^2t_1}{dr^2}+\frac{dt_1}{rdr}-m_{10}t_1+m_0t_2+m_1t_{in}=0\\&\frac{d^2t_2}{dr_2}+\frac{dt_2}{rdr}-n_{10}t_2+n_0t_1+n_1t_{out}=0\end{aligned} \tag{3}$$

where $t_1(r)$ and $t_2(r)$ enter each equation symmetrically. Determination of their functional dependence is interrelated.

Boundary conditions along the boundary of a thermally-conductive inclusion with radius r_0 and thermal resistance R can be written as follows:

$$\lambda_1\delta_1 2\pi r_0 \frac{dt_1}{dr} + \alpha_{in}\pi r_0(t_{in} - t_1) = \frac{t_1 - t_2}{R}\pi r^2{}_0$$
$$-\lambda_2\delta_2 2\pi r_0 \frac{dt_2}{dr} + \alpha_{out}\pi r_0^2(t_2 - t_{out}) = \frac{t_1 - t_2}{R}\pi r^2{}_0 \qquad (4)$$

In this equation, the first term is associated with the description of amount of heat outgoing to the surface due to heat exchange along the boundary of a thermally-conductive inclusion, the second term describes the loss of heat due to heat exchange between the surface of a thermally conductive inclusion with the outer air. The right side of the equation presents the heat flow going through the thermally-conductive inclusion from one side of the surface to another with a thermal resistance R, area πr_0^2 with the temperature difference t_1-t_2.

For the equations describing the boundary conditions for the temperature fields in new designations we get the following system:

$$\left.\frac{dt_1}{dr}\right|_{r=r_0} = p_{11}t_1 + p_{12}t_2 + q_1$$
$$\left.\frac{dt_2}{dr}\right|_{r=r_0} = p_{21}t_1 + p_{22}t_2 + q_2 \qquad (5)$$

The solution of the initial system of equations (and each of them is the Bessel equation) can be found in the form of a linear combination of Macdonald functions, bounded at infinity, modifications of Bessel functions [13,14]. To eliminate the constant component in each of the equations (4), we will seek the solution in the following form:

$$t_1 = t_{10} + t_1^*(r)$$
$$t_2 = t_{20} + t_2^*(r) \qquad (6)$$

After substituting the equations solutions of this form into the equations of the system (4) and setting the constant component of each equation to zero, we obtain the following system of equations for determining the constant components t_{10} and t_{20}:

$$-m_{10}t_{10} + m_0 t_{20} + m_1 t_{in} = 0$$
$$-n_{10}t_{10} + n_0 t_{20} + n_1 t_{out} = 0 \qquad (7)$$

After applying the corresponding quantities instead of the previously introduced designations (2) and after solving the received system (7), we obtain the corresponding values for these constants:

$$t_{10} = t_{in} - \frac{t_{in} - t_{out}}{1 + R_0\alpha_{in} + \dfrac{\alpha_{in}}{\alpha_{out}}}$$
$$t_{20} = t_{out} + \frac{t_{in} - t_{out}}{1 + R_0\alpha_{out} + \dfrac{\alpha_{out}}{\alpha_{in}}} \qquad (8)$$

The variable component of the temperature field $t_1^*(r)$ и $t_2^*(r)$ will be sought in the following form:

$$t_1^* = A_{11}K_0(\nu_1 r) + A_{12}K_0(\nu_2 r)$$
$$t_2^* = A_{21}K_0(\nu_1 r) + A_{22}K_0(\nu_2 r) \tag{9}$$

In these equations A_{ij} and ν_i are undetermined constants that will be chosen to satisfy both the equations (3) and the boundary conditions (4). $K_0(\nu_i r)$ function is a Macdonald function of zero order. After applying the solution (9) into the first equations of the system (4), we obtain:

$$A_{11}\left(\frac{d^2K_0(\nu_1 r)}{dr^2} + \frac{1}{r}\cdot\frac{dK_0(\nu_1 r)}{dr} - \nu_1^2K_0(\nu_1 r)\right) + A_{11}\nu_1^2K_0(\nu_1 r) +$$
$$+ A_{12}\left(\frac{d^2K_0(\nu_2 r)}{dr^2} + \frac{1}{r}\cdot\frac{dK_0(\nu_2 r)}{dr} - \nu_2^2K_0(\nu_2 r)\right) + A_{12}\nu_2^2K_0(\nu_2 r) - m_{10}A_{11}K_0(\nu_1 r) - m_{10}A_{12}K_0(\nu_2 r) + \tag{10}$$
$$+ m_0A_{21}K_0(\nu_1 r) + m_0A_{22}K_0(\nu_2 r) = 0$$

We obtained equations in brackets with coefficients A_{11} and A_{12}. Macdonald functions satisfy these equations, so these terms of equations become zero. By combining the remaining terms, we obtain:

$$\left(A_{11}\left(\nu_1^2 - m_{10}\right) + A_{21}m_0\right)K_0(\nu_1 r) + \left(A_{12}\left(\nu_2^2 - m_{10}\right) + A_{22}m_0\right)K_0(\nu_2 r) = 0 \tag{11}$$

By requiring coefficients to become zero in ahead of the Macdonald functions in this equation, we get:

$$A_{11}(\nu_1^2 - m_{10}) + A_{21}m_0 = 0;$$
$$A_{12}(\nu_2^2 - m_{10}) + A_{22}m_0 = 0. \tag{12}$$

The received requirement means that there is a link between the coefficients included in the solution. The link is determined by the equation, that is:

$$A_{21} = -A_{11}\frac{\nu_1^2 - m_{10}}{m_0};$$
$$A_{22} = -A_{12}\frac{\nu_2^2 - m_{10}}{m_0}. \tag{13}$$

This means that we can seek the solution to this equation for the same constant values ν, but with different coefficients, namely:

$$t_1^* = CK_0(\nu\, r);$$
$$t_2^* = DK_0(\nu\, r). \tag{14}$$

We use this type of solution (14) in both equations (4) and obtain:

$$C\nu^2K_0(\nu r) - m_{10}CK_0(\nu r) + m_0DK_0(\nu r) = 0;$$
$$D\nu^2K_0(\nu r) - n_{10}DK_0(\nu r) + n_0CK_0(\nu r) = 0; \tag{15}$$

Or

$$(C(v^2 - m_{10}) + Dm_0)K_0(vr) = 0;$$
$$(D(v^2 - n_{10}) + Cn_0)K_0(vr) = 0. \quad (16)$$

From the condition of equation satisfaction it follows that the coefficient ahead of the function K_0 *(vr)* must become zero. Therefore:

$$C(v^2 - m_{10}) + Dm_0 = 0;$$
$$D(v^2 - n_{10}) + Cn_0 = 0. \quad (17)$$

The received system of equations is homogeneous and has non-zero solutions only if determinant of the system becomes zero. Then:

$$\begin{vmatrix} v^2 - m_{10} & m_0 \\ n_0 & v^2 - n_{10} \end{vmatrix} = 0. \quad (18)$$

By expanding the determinant, we obtain a biquadratic equation as regard to v, or

$$v^4 - v^2(m_{10} + n_{10}) + (m_{10} \cdot n_{10} - m_0 \cdot n_0) = 0 \quad (19)$$

Positive roots of this equation can be presented in the following form:

$$v_{1,2} = \sqrt{\frac{(m_{10} + n_{10}) \pm \sqrt{(m_{10} - n_{10})^2 + 4m_0 n_0}}{2}}. \quad (20)$$

From the analysis of the values included in this determinant, we can get some useful relations:

$$(v^2 - m_{10})(v^2 - n_{10}) - m_0 n_0 = 0 \Rightarrow \frac{v_{1,2}^2 - m_{10}}{m_0} = \frac{n_0}{v_{1,2}^2 - n_{10}}. \quad (21)$$

After introducing the auxiliary constants M and N, the relations between them will be presented in the following form:

$$M = \frac{m_{10} - v_1^2}{m_0} = \frac{n_0}{n_{10} - v_1^2}; N^{-1} = \frac{m_{10} - v_2^2}{m_0} = \frac{n_0}{n_{10} - v_2^2};$$
$$\frac{M}{N} = \frac{m_{10} - v_1^2}{m_0} \cdot \frac{n_0}{n_{10} - v_2^2} = \frac{m_{10} - v_1^2}{m_0} \cdot \frac{n_0}{n_{10} - m_{10} - n_{10} + v_1^2} = -\frac{n_0}{m_0}, \quad (22)$$

where the relation between the roots of a quadratic equation $v_1^2 + v_2^2 = m_{10}^2 + n_{10}^2$ is used.

Using the relation between the coefficients for the desired solution, we can now present the desired solution in the following form:

$$t_1 = t_{10} + A_{11}K_0(v_1 r) + NA_{22}K_0(v_2 r);$$
$$t_2 = t_{20} + MA_{11}K_0(v_1 r) + A_{22}K_0(v_2 r). \quad (23)$$

The remaining unknown coefficients of solutions A_{11} and A_{22} will be determined from the fact that this solution must satisfy the boundary conditions (5). It will also be regarded that the derivative of the Macdonald zero order function is equal to the Macdonald first order function, that is:

$$\frac{dK_0(vr)}{dr} = -vK_1(vr). \tag{24}$$

After using the solution (23) in the equations describing the boundary conditions, we obtain:

$$\begin{aligned}
&A_{11}\frac{dK_0(v_1r)}{dr} + NA_{22}\frac{dK_0(v_2r)}{dr} = p_{11}t_{10} + p_{12}t_{20} + q_1 + p_{11}A_{11}K_0(v_1r) + p_{11}NA_{22}K_0(v_2r) + \\
&+ p_{12}MA_{11}K_0(v_1r) + p_{12}A_{22}K_0(v_2r); \\
&MA_{11}\frac{dK_0(v_1r)}{dr} + A_{22}\frac{dK_0(v_2r)}{dr} = p_{21}t_{10} + p_{22}t_{20} + q_2 + p_{21}A_{11}K_0(v_1r) + p_{21}NA_{22}K_0(v_2r) + \\
&+ p_{22}MA_{11}K_0(v_1r) + p_{22}A_{22}K_0(v_2r);
\end{aligned} \tag{25}$$

By grouping terms beside unknown values A_{11} and A_{22} we obtain:

$$\begin{aligned}
&A_{11}(-v_1K_1(v_1r_0) - p_{11}K_0(v_1r_0) - p_{12}MK_0(v_1r_0)) + A_{22}(-Nv_2K_1(v_2r_0) - p_{11}NK_0(v_2r_0) - \\
&- p_{12}K_0(v_2r_0)) = p_{11}t_{10} + p_{12}t_{20} + q_1; \\
&A_{11}(-Mv_1K_1(v_1r_0) - p_{21}K_0(v_1r_0) - p_{22}MK_0(v_1r_0)) + A_{22}(-v_2K_1(v_2r_0) - p_{21}NK_0(v_2r_0) - \\
&- p_{22}K_0(v_2r_0)) = p_{21}t_{10} + p_{22}t_{20} + q_2.
\end{aligned} \tag{26}$$

If we introduce new designations, then (26) can be rewritten in a shorter form:

$$\begin{aligned}
&A_{11}P_{11} + A_{22}P_{12} = Q_1; \\
&A_{11}P_{21} + A_{22}P_{22} = Q_2,
\end{aligned} \tag{27}$$

where new designations were obtained by substituting the values $r=r_0$в (26).

$$\begin{aligned}
P_{11} &= -v_1K_1(v_1r_0) - p_{11}K_0(v_1r_0) - p_{12}MK_0(v_1r_0) \\
P_{12} &= -Nv_2K_1(v_2r_0) - p_{11}NK_0(v_2r_0) - p_{12}K_0(v_2r_0) \\
P_{21} &= -Mv_1K_1(v_1r_0) - p_{21}K_0(v_1r_0) - p_{22}MK_0(v_1r_0) \\
P_{22} &= -v_2K_1(v_2r_0) - p_{21}NK_0(v_2r_0) - p_{22}K_0(v_2r_0) \\
Q_1 &= p_{11}t_{10} + p_{12}t_{20} + q_1; \\
Q_2 &= p_{21}t_{10} + p_{22}t_{20} + q_2.
\end{aligned} \tag{28}$$

Thus, the values of the unknown coefficients that determine the solution of this problem can be represented as follows:

$$\begin{aligned}
A_{11} &= \frac{Q_1P_{22} - Q_2P_{12}}{P_{11}P_{22} - P_{12}P_{21}}; \\
A_{22} &= \frac{Q_2P_{11} - Q_1P_{21}}{P_{11}P_{22} - P_{12}P_{21}}.
\end{aligned} \tag{29}$$

If we use the identical equation for Bessel functions:

$$\frac{d(r^nK_n(r))}{dr} = r^nK_{n-1}(r), \tag{30}$$

then we can present the final form of the equation for the heat flow outgoing, for example, from the inner surface of building envelope:

$$Q = \int_{0}^{2\pi} d\varphi \int_{r_0}^{r_1} \alpha_{in}(t_{in} - t_1) r dr \text{, or}$$

$$Q = 2\pi \int_{r_0}^{r_1} \alpha_{in}(t_{in} - t_{10} - A_{11}K_0(\nu_1 r) - NA_{22}K_0(\nu_2 r)) r dr$$

Using the relation of the following type:

$$\frac{d(\nu_1 r K_1(\nu_1 \nu r))}{d(\nu_1 r)} = -\nu_1 r K_0(\nu_1 r) \tag{31}$$

equation for Q can be presented as follows, after the integral is taken over the length:

$$Q = 2\pi\alpha_{in}((t - t)(\frac{r_1^2 - r_0^2}{2}) - (\frac{A_{11}}{\nu_1}(r_1 K_1(\nu_1 r_1) - r_0 K_1(\nu_1 r_0)) + \\ + \frac{A_{22}N}{\nu_2}(r_1 K_1(\nu_2 r_1) - r_0 K_1(\nu_2 r_0))). \tag{32}$$

To confirm that developed technique works, a series of experiments was conducted and obtained data were processed.

3 Results

For the experiment a 1000x1000 mm panel fragment was taken, in the center of which a metal rod (rod diameter varied) was installed. The panel was attached to a bore of the climate chamber with the retaining fasteners. REOCAMTCM-1000 climate chamber was used. General view of the chamber is shown in Figure 2. [6,10,15]

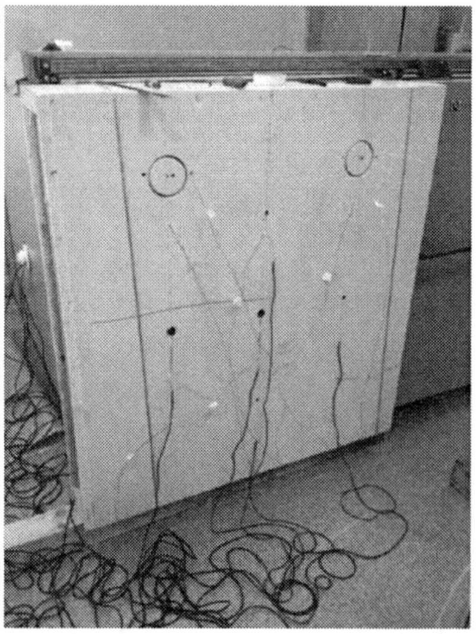

Fig.2. General view of the climate chamber with a fragment of the panel.

To create a hermetically sealed space inside the climate chamber, the perimeter of the chamber bore was treated with a sealant. After that the panel fragment with thermally conductive inclusion was attached to the chamber bore with the help of retaining fasteners. Imperviousness in the chamber was controlled by a pressure gauge 516.11 (accuracy class 0.15). Measurement design is shown in Figure 3.

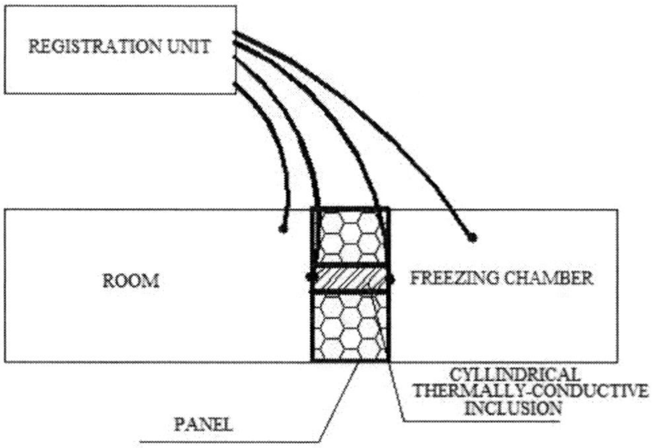

Fig. 3. Circuit diagram of measurement.

For measuring the temperature at the distinguished points the copper-constantan thermocouples were used as defined in GOST 50431-92 standard. A microammeter M903/1 was used as a secondary device. Total amount of the thermocouples installed for the experiments was 5. Thermocouple measurement accuracy -0.1 °C.

Results of the experiment and calculations are shown in Table 1.

Initial data for the calculations:

1. Indoor air temperature t_{in}=16; 17; 18; 20 °C;
2. Outdoor air temperature t_{out}=-20; -30; -35; -46 °C;
3. Coefficient of heat transfer of the building envelope inner surface α_{in}=8.7 W/(m^2 ^{0}C);
4. Coefficient of heat transfer of the building envelope outer surface α_{out}=23.2 W/(m^{20}C);
5. Thermal resistance of the insulator material R=0,00057 (m^2 ^{0}C) /W;
6. Thermal resistance of the thermally-conductive inclusion R_0= 1.923 (m^2 ^{0}C) /W;
7. Thickness of lining δ_1=δ_2=0.001 m;
8. Thermal conductivity coefficient of lining λ=175 W/(m°C);
9. Radius of thermally-conductive inclusion r_0=8; 10; 12; 16; 20 mm.

Table 1. Results of the experiment and calculations

N	Rod diameter, mm	Indoor air temperature tin, °C	Outdoor air temperature tout, °C	In practice		In theory	
				Temperature at the outer end of the rod t2, °C	Temperature at the inner end of the rod t1, °C	Temperature at the outer end of the rod t2, °C	Temperature at the inner end of the rod t1, °C
1	8	18	-30	-19.2	11.2	-22.0	8.9
2	10	18	-30	-17.9	8.9	-20.1	6.5
3	12	18	-30	-16.3	6.8	-18.4	4.5
4	16	18	-30	-13.9	3.5	-16	1.4
5	20	18	-30	-12.3	1.5	-14.4	-0.8
6	8	17	-46	-32.9	7.2	-35.1	4.9
7	10	17	-46	-30.2	4.2	-32.6	1.9
8	12	17	-46	-28	1.6	-30.5	-0.8
9	16	17	-46	-24.8	-2.4	-27.3	-4.8

10	20	17	-46	-23.0	-5.4	-25.2	-7.6
11	8	16	-20	-11.7	11.4	-14.01	9.1
12	10	16	-20	-10.3	9.7	-12.54	7.4
13	12	16	-20	-9.0	8.2	-11.31	5.9
14	16	16	-20	-7.2	5.8	-9.5	3.5
15	20	16	-20	-6.0	4.2	-8.3	1.9
16	8	20	-35	-23.5	11.8	-25.9	9.5
17	10	20	-35	-21.4	9.2	-23.6	6.8
18	12	20	-35	-19.5	6.9	-21.7	4.5
19	16	20	-35	-16.6	3.2	-18.9	1
20	20	20	-35	-14.8	0.8	-17.1	-1.5

4 Conclusion

In course of the theoretical studies by means of mathematical physics methods we obtained analytical equations of temperature field distribution around the through cylindrical thermally-conductive inclusion in the light building envelopes of block boxes.

The analysis of the experimental data and the data obtained by the developed and refined physical-mathematical model of a stationary heat transfer through the building envelope, shows that the difference between the data varies between 10% and 25%, which proves the accuracy of the developed model.

References

1. A.I. Kruglova, *Klimat i ograzhdayushchie konstruktsii* (Stroyizdat, Moscow, 1990)
2. Yu.P. Batalin, V.G. Chirskov, G.I. Shmal, *Komplektno-blochnoe stroitel'stvo ob"ektov neftyanoy i gazovoy promyshlennosti* (Nedra, Moscow, 1996)
3. Yu.P. Batalin, S.N. Vasserman, Ya.N. Kagan, V.L. Trofimov, *Inzhenerno-stroitel'nye osobennosti zastraivaemykh territoriy neftegazodobyvayushchikh rayonov Zapadnoy Sibiri* (Moscow, 2001)
4. A.Sh. Dekhtyar, *Oblegchennye konstruktsii metallicheskikh sten promyshlennykh zdaniy* (Stroyizdat, Moscow, 2009)
5. F.F. Tamplon, *Metallicheskie ograzhdayushchie konstruktsii dlya zdaniy, vozvodimykh v surovykh klimaticheskikh usloviyakh* (Stroyizdat, Moscow, 2008)
6. K.V. Afonin, Sbornik materialov nauchno-prakticheskoy konferentsii prepodavateley, molodykh uchenykh, aspirantov TyumGASA, 3-6 (2002)
7. A.I. Gorkovenko, T.S. Zhilina, V.S. Starikov, Stroitel'ny vestnik Tyumenskoy oblasti, **4**, 48-51 (1999)
8. A.I. Gorkovenko, T.S. Zhilina, Oil and gas, **1**, 108-112 (2000)
9. K.V. Afonin, A.F. Shapoval, K.N. Ilyukhin, M.N. Chekardovskiy, Sbornik dokladov nauchno-prakticheskoy konferentsii, 487-491 (2000)
10. T. Zhilina, Applied Mechanics and Materials, **729**, 224-228 (2015)
11. V.N. Bogoslovsky, *Stroitel'naya teplofizika* (Stroyizdat, Moscow, 2013)
12. V.N. Bogoslovsky, *Teplovoy rezhim zdaniya* (Stroyizdat, Moscow, 1999)
13. B.G. Aksenov, *Stefurak, Matematika* (RIO TyumGASU, Tyumen, 2013)
14. L.Z. Rumshinskiy, *Matematicheskaya obrabotka rezul'tatov eksperimenta* (Nauka, Moscow, 1971)
15. I.Yu. Shalagin, Inzhenernyy vestnik dona magazine, **2-2**, (2015)
16. N.I. Kurilenko, G.Ya. Mamontov, L.Yu. Mikhaylova, EPJ Web of Conferences, **82**, (2015)
17. G.V. Kuznetsov, N.I. Kurilenko, V.I. Maksimov, G.Ya. Mamontov, Journal of Engineering Physics and Thermophysics, **86**, 519-524 (2013)

Dielectric Method for Diagnosing Formation of Gas Hydrates in Gas Pipelines

Sergey Podorozhnikov[1,*], *Alexander* Shabarov[2], *Albina* Shirshova[2], and *Artyom* Berdyshev[2]

[1]Industrial University of Tyumen, 625001, Volodarskogo str. 38, Tyumen, Russia
[2]Tyumen State University, 625001 Volodarskogo st. 6, Tyumen, Russia

Abstract. Experimental study of low-frequency dependence of the dielectric properties of gas hydrate in the process of growth and dissociation in the range 50- 1000 kHz is conducted. It is found that the dielectric loss tangent maximum value is observed at a frequency of 260 kHz and depends on the amount of gas hydrate in the measuring cell. Comparison of degrees of gas-hydrate formations obtained by two independent methods (PVT-method and dielectric method) showed good agreement. It allows us to offer a simple method of controlling the process gas-hydrate formations in the pipes based the measurement of the dielectric loss tangent at a certain frequency.

1 Introduction

It is known that natural gas in certain thermodynamic conditions enter into combination with water to form hydrates, which are accumulating in the field and main pipelines, significantly increase its hydraulic resistance and hence reduce capacity. The problem of combating the hydrates formation becomes particularly important while developing deposits in Western Siberia and the Far North. Low temperatures and severe climate conditions in these areas provide favorable conditions for hydrate formation in wells and pipelines (see. Figure 1).

For the destruction of hydrates accumulations, the method of applying electromagnetic waves of microwave range is very efficient [2]. For example, in [3] the results of field tests of this method at a special test section (length 16 m) flowline connected to one of the wells Efremovsky deposits of Kharkov GPU are shown.

2 Object of research (model, process, configuration, synthesis, experimental part, etc.)

Diversity of wells and pipes equipped for production and transportation of oil and gas requires the development of a number of fundamentally different methods and devices for

* Corresponding author: sergey_urevih@mail.ru

HF EMF actuation in pipes and wells, in which the different obstructions are formed. Depending on the design and parameters of wells and pipelines, it is necessary to choose a specific technological scheme of electromagnetic exposure and the necessary range of the electromagnetic equipment. However, their physical basis is made by transformation of electromagnetic energy part either in intertube space, or inside the tubes into heat due to the finite conductivity of pipes materials and dielectric losses of the area of filling intertube space and pipes.

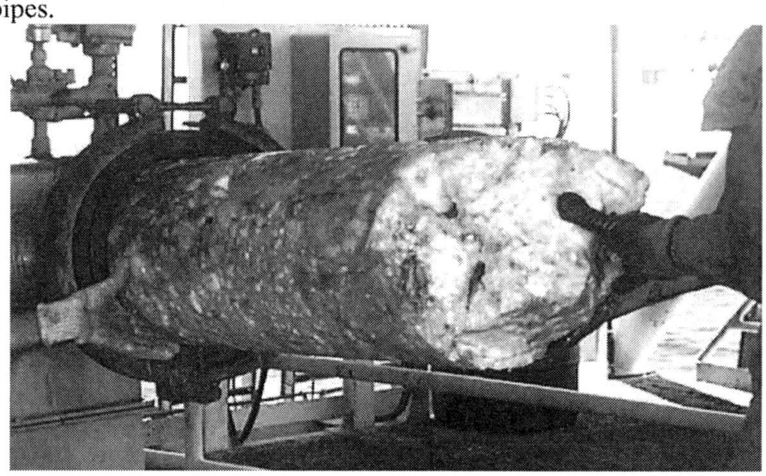

Fig. 1. Hydrate plug in the pipeline [1].

Researches [4], [5], [6], [7], [8], [9] are devoted to theoretical and experimental investigation of the dielectric properties of gas hydrates. However, these works mainly represent the results of researches in high-frequency electromagnetic field, where energy is converted to heat, sufficient for decomposition of gas hydrate.

On the other hand, the electronic response to the impact of low frequency electromagnetic field of low power and gas hydrate can provide useful information about the beginning of the process of gas-hydrate formation in pipe, thus avoiding accidents and pipeline downtime. Furthermore, this information can quantify the degree of gas-hydrate formation, not destroying the formed gas hydrate.

3 Methods

The objective of this paper is to provide a method of control the gas hydrate formation process using dielectric method.

Tasks:

1. Experimental study of hydrate dielectric parameters in the process of its formation and dissociation in the low-frequency area from 50 to 1000 kHz and at low power of electromagnetic field.
2. Studying the possibility of determining gas-hydrate formation beginning and measuring this process by means of a dielectric the parameters.

Description of the experimental mechanism. Experimental studies were carried out at the mechanism, which block diagram is shown in Figure 2. The main part of the mechanism is gas hydrate thermostatic chamber (reactor-crystallizer) with controlled gas supply system, as well as temperature and pressure control measuring instruments. Mechanism is designed for a pressure of about 1.6 MP and temperatures down to -300C. The main element of mechanism is a reactor (6) of high pressure where gas hydrates are either formed or dissociate under controlled conditions. Cylindrical condenser has a capacity about 88 sm3. Reagent is placed and gas is pumped into it.

For the formation of hydrates in the reactor cell, it is necessary to maintain a certain range of temperatures and pressures. To modify and maintain the temperature thermostat KRIO-VT-01 is used (1). The pressure is maintained by means of the cylinder and valves (2, 12a, 12b). Pressure control is carried out with a sensor, which is attached to the reactor in a separate pipe. The temperature is monitored using a thermistor (5). After the experiment, the gas is vented from the system by opening the tap 12a, which also allows smoothen pressure release in the reactor for dissociation experiments.

To study the dielectric parameters of gas hydrates, electric circuit is hermetically mounted in the mechanism, consisting of a capacitor of cylindrical shape (7), mounting wires and kumetra (11) - measuring the quality factor - all this in general is an oscillation circuit.

The methodology of the experiment. In the experiments we used process water, as the hydrate-forming gas - propane. The stability of the hydrate formation process flow was supplied by maintaining temperature and pressure of certain conditions in reactor. In particular, in our experiments, the temperature was kept constant (1.60 C) and a pressure was pumped cyclically to a constant maximum value exceeding the equilibrium pressure of the gas - gas hydrate - water system, at a given temperature. One cycle lasted 24 hours. Detailed methodology of increase kinetics and dissociation of gas hydrate is described in research [10]. In this work, a methodology measuring the dielectric parameters of gas hydrate in the process of growth and dissociation is presented for the first time.

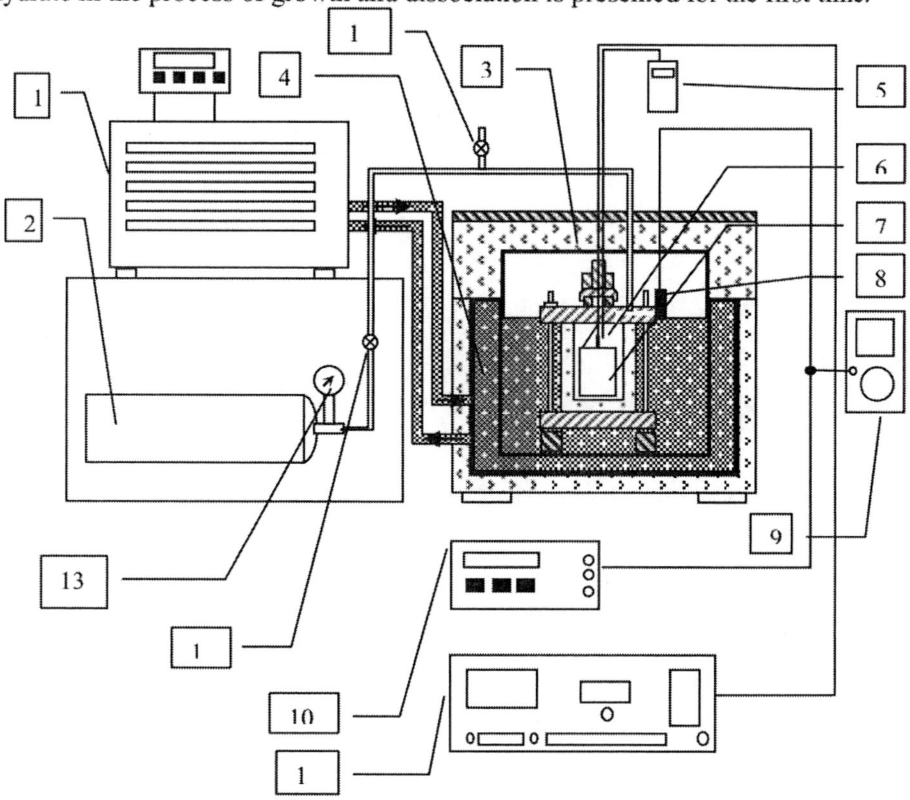

Fig. 2. The block diagram of mechanism
1 - thermostat KRIO-VT-01, 2 - a gas cylinder, 3 - cover of the heat exchanger, 4 - the first circuit of the heat exchanger, 5 - thermistor 6 - reactor 7 - cylindrical capacitor 8 - pressure sensor, 9 - multimeter 10 - power source for the pressure sensor, 11 - Qmetr, 12a, 12b - acicular valves for pressure control, 13 -reduktor.

The measurements of dielectric parameters were performed after each cycle using hydrate kumetra VM 560 in the range from 50 kHz to 1 MHz. Measuring the dielectric loss tangent $tg(\delta)$ and dielectric permittivity ε' were performed according to standard procedures [11]. In the jump of capacitance and quality factor of the oscillatory circuit, it can be judged on start of the formation of gas hydrate in the measuring cell. This determines the induction time of the beginning of the gas-hydrate formation and time of beginning measuring the dielectric parameters of gas hydrates. After the 1st cycle of hydrate formation, no signal was observed at kumetre, which is connected with the presence of significant amounts of liquid water and a small amount of gas hydrate in the measuring cell. After the second cycle of, the signal was reliably recorded. Since the temperature in the reactor is positive, and there is no ice in the measuring cell, the electromagnetic wave is mainly damping in gas hydrates. In subsequent cycles, with the accumulation of gas hydrate (dielectric) between the measuring capacitor plates, losses of electromagnetic energy are increasing. The greater the power dissipates in the dielectric, the larger the dielectric losses angle δ and the value of the function $tg(\delta)$ are. Hence, can be judged not only on the presence of gas hydrate, but also its quantity at the value.

In the process of measuring the oscillation circuit parameters, the gas hydrate formed in the measuring cell heats. As experts' estimation shows, the rise in temperature must not exceed $0{,}8\ ^{\circ}C$ [11] - in this case, the condition of the existence of gas hydrate complies. Therefore, the question arises about the temperature change in the measuring cell during the dielectric measurements. Our estimates of temperature change at kumetra power VM 560 and the length of the dielectric parameters measurements for 1.5 hours showed that the heat generated in the measuring cell (in the adiabatic approximation), is not enough to heat the gas hydrate even to decomposition temperature, not to mention the decomposition itself. Hence, gas hydrate decomposition does not occur in the dielectrical measurements.

Calculation of the mass of the formed gas hydrate in a cycle was determined experimentally from the ideal gas equation (PTV-method) according to the formula:

$$\Delta m_i = \frac{\Delta P_i V \mu_H}{R T_\gamma} \quad (1)$$

where ΔPi – pressure drop in a reactor during one cycle of hydrate formation, Pa, V – the volume of the reactor, R – the gas constant; T - reactor temperature, K; i – cycle number.

The mole fraction of hydrating agent in gas hydrates:

$$\gamma = \frac{v_{G-H}}{v_H}$$

where υG-H – the number of moles of gas, which passed into gas hydrate; υH – the number of moles of gas hydrate. For propane, γ = 0.0555 [14].

4 Experiment Results and discussion

Figure 3 shows photographs of gas hydrate grown in a cylindrical capacitor for 8 days (cycles).

Fig. 3. Photos gas hydrate.

Figure 4 shows the dependence of the dielectric loss tangent of the frequency and number of a cycle at growth and dissociation of propane gas hydrate. The graph shows that with the growth of gas hydrate from cycle to cycle, the dielectric loss tangent increases, while dissociation - decreases. The frequency at which the kumetre signal is maximum, depends on the scheme, and the oscillating circuit parameters used in the experiment. In our case, the frequency is f = 260 kHz.

Experimentally obtained values of dielectric loss tangent of gas hydrate allow to judge the degree of hydrate formation α, which was determined experimentally, as the ratio of the dielectric loss tangent i-th cycle tg(δi) to the maximum tangent of the loss tg(δmax), where the water is completely converted to a gas hydrate:

$$\alpha = \frac{tg(\delta_i)}{tg(\delta_{max})} \tag{2}$$

In order to ensure consistency of the dielectric methods for quantifying the degree of gas-hydrate formation, it is necessary to compare it to another, independent from it, method. In this case, the simplest and most convenient method of comparison seems PVT method. Using the formula (1) we can calculate the degree of hydrate formation as:

$$\alpha^* = \frac{\sum_{i=1} m_i}{m_{max}}, \tag{3}$$

where mi - the gas hydrate mass, formed after i-th cycle, mmax - the total mass of the formed gas hydrate.

Figure 3 shows a comparison of dependencies of the degree on the time of hydrate formation obtained by two independent methods. It is clear from the graph, that the dielectric data are in good agreement with the results of the calculation by PVT method (difference does not exceed 10%).

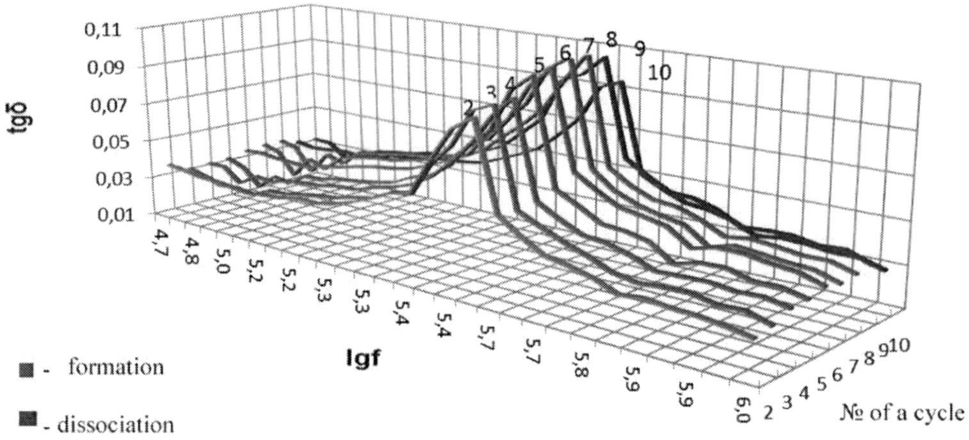

Fig. 4. The dependence of the dielectric loss tangent on the frequency and numbers of cycle in the formation and dissociation of gas hydrate.

In practical application [16-22] of surface-dielectric sensor, "dangerous" in terms of the start of hydrate formation, pipe section can be determined by the method developed in [15].

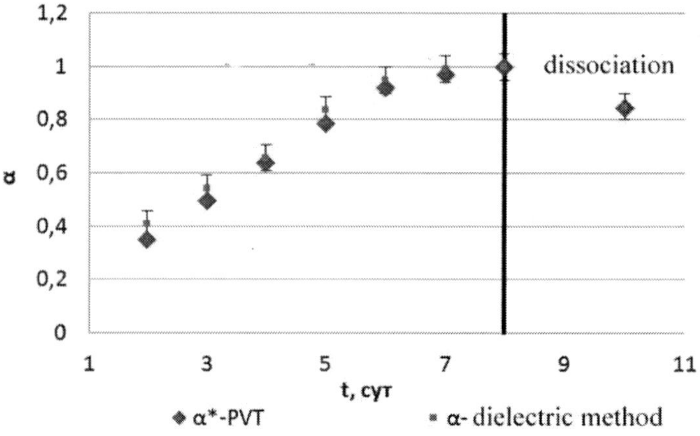

Fig. 5. The dependence of the degree hydrate formation on the time, obtained by two independent methods.

5 Conclusion

The results obtained allow offering a simple method of monitoring the gas hydrate formation process. This method is based on measuring the loss tangent at a certain frequency in the low frequency range of the electromagnetic field. The suggested method makes it possible to predict the time of safe operation of process equipment and avoid accidents.

References

1. Yu. Makogon, Geology and mineral resources of the oceans, **2**, 9 (2010)
2. B. Degtyarev, B. Buchgalter, *Fighting hydrates in the operation of gas wells in the northern regions* (Nedra, Moscow, 1976)

3. B. Beshevli, V. Savchenko, A. Kasyan, G. Kovalenko, Gas Industry, **2**, 21-22 (1975)
4. T. Jakobsen, K. Folgerøzk. Meas. Sci. Technol, **8**, 1006–1015 (1997)
5. M. Fatykhov, N. Bagautdinov, Oil and Gas Business, 15 (2006)
6. N. Bagautdinov, Dissertation of Doctor of Technical Sciences (2007)
7. S. Gough, Journ Phys Chem, **77(25)**, 2969 - 2976 (1973)
8. D. Davidson, *Clatherate hydrates* (NY Plenum Press, 1973)
9. D. Davidson, Can Journ Chem, **81**, 248-253(1977)
10. A. Shabarov, A. Shirshova, S. Gasheva, Oil and gas, **2**, 88-93 (2014)
11. Manual. Q-Meter VM 560. USSR, Brno TESLA Brno, Nat. entrepr. (1978)
12. D. Davidson, *Clatherate hydrates* (NY Plenum Press, 1973)
13. A. Groisman, *Thermal properties of gas hydrates* (Nauka, Novosibirsk, 1985)
14. D. Carroll, *Natural gas hydrates* (JSC "Premium Engineering", Moscow, 2007)
15. S. Buchinskiy, A. Shabarov, A. Burbasov, Vestnik TSU, **6**, 28-33 (2008)
16. S. Stepanov, N., Cheremisin, S., Sokolov, A., Altunin, A., Shabarov, *ECMOR 2012 - 13th European Conference on the Mathematics of Oil Recovery* (2012) DOI: 10.399/2214-4609.20143270
17. V. Antip'ev, A. Nevolin, Yu. Zemenkov, Neftyanoe Khozyaistvo - Oil Industry, **10**, 46-48 (1981)
18. V. Pavlov, K. Voronin, Yu. Zemenkov, *Oil and Gas Terminal: Proceedings of the International scientific conference*, 169-173 (2015)
19. S. Dudin, A. Shipovalov, S. Podorozhnikov, Yu. Zemenkov, Oil and gas, **2**, 65-69 (2011)
20. Yu. Zemenkov, V. Shalay, M. Zemenkova, Procedia Engineering, **113**, 254-258 (2015)
21. Yu. Zemenkov, V. Shalay, M. Zemenkova, Procedia Engineering, **113**, 312-315 (2015)
22. R. Levitin, Yu. Zemenkov, Neftyanoe khozyaystvo - Oil Industry, **1**, 110-114 (2016)

Determination of Glazing Material's Influence on the Energy Performance of Office Buildings Using Dynamic Simulation Techniques

Norbert Harmathy[1,*]

[1]Budapest University of Technology and Economics, Műegyetem rkp. 3. K. Build. 231, H-1111 Budapest, Hungary

Abstract. This investigation presents a detailed analysis in an effort of building energy performance improvement from the aspect of building envelope and glazing influence on the annual heating and cooling energy demand. This work is the continuance of our previous envelope optimization investigation and elaborates the influence of five different glazing types on the annual heating and cooling energy demand of the reference office building in compliance with energy efficiency regulations. Total annual energy demands were determined according to the building envelope's thermal performance and glazing characteristics. Findings from the dynamic simulations indicated the influence of glazing parameters (U-value, Solar heat gain coefficient) on the annual heating and cooling demand of the multi-zone building model. From the five simulated envelopes it was concluded that with adequate window application the annual heating demand could be reduced by 27% and the annual cooling demand by approximately 40%.

1 Introduction

Our previous investigations cover building envelope optimization in the function of indoor daylight quality in offices in order to determine preferable window to wall ratios and window geometries. This work is the continuance of the envelope optimization investigation and elaborates the influence of five different glazing types on the annual heating and cooling energy demand of the reference office building in compliance with energy efficiency regulations [1-5].

Thermal and lighting simulations from the energy perspective were investigated in previous researches applying energy modeling. Envelope glazing's transmittance dependence on the solar radiation in order to reduce building energy demand was investigated respectively from wavelength and economic aspects respectively. Daylight analysis has been a widespread topic investigated in numerous papers using daylight coefficient concept, daylighting schemes, window properties, building design and climate conditions [6 - 10]. The research was conducted on a typical not rehabilitated 10 level reference office building which is part of the Faculty of Technical Sciences complex in

* Corresponding author: harmathy@egt.bme.hu

Novi Sad. The adopted best case indoor illumination performance model from the previous research [11-16] was simulated in EnergyPlus with the aim to explore and determine the annual heating and cooling energy demand for five different glazing types. Double and triple pane, low-E coated windows with Argon filling, were simulated on an annual basis with hourly time steps. The improved reference building's heating and cooling energy demand was determined and analyzed for five glazing types, for the previously inherited window to wall ratio and window geometry. The research goal was to conduct the influence of window construction and parameters, such as solar heat gain coefficient (SHGC) and overall heat transfer coefficient (U-value) on the annual heating and cooling demand. The possibility of application could be enabled in the decision making process in early design stages and during building envelope rehabilitation stages of existing office buildings. Throughout the research, the simulated exterior wall's thermal properties and window parameters complied with the Serbian and EU regulations [17-20].

Materials and methodology

The research was conducted on a reference multi-level office building which is part of the Faculty of Technical Sciences complex in Novi Sad, Serbia. The total building area is 3430 m2 consisting of 10 levels. An aerial view with the building is presented in Fig. 1.

Fig. 1 Multi-level reference office building of the Faculty of Technical Sciences

Energy expenses for district heating and electricity were gathered for the year 2013 in order to compare them with the results from the simulation, as shown in Table 1 and Table 2. The building used in total 338 MWh/a (100 $kWh/m^2/a$) for district heating on an annual basis, and 204 MWh/a (59 $kWh/m^2/a$) for electricity; combined cooling, lighting and equipment.

Table 1. Heating energy consumption

Month	Heating [kWh]	Month	Heating [kWh]
Jan	115993	Oct	9551
Feb	63473	Nov	45003
Mar	42323	Dec	61030
Apr	1415	Annual Sum	338788

Table 2. Electricity consumption

Month	Cooling, lighting and equipment electricity [kWh]	Month	Cooling, lighting and equipment electricity [kWh]
Jan	19214	Aug	16652
Feb	17478	Sep	14113
Mar	18519	Oct	17245
Apr	16918	Nov	15282
May	14375	Dec	20230
Jun	16706	Sum [kWha^{-1}]	203810
Jul	17078	[kWhm^{-2}a^{-1}]	59

The location and climate data for Novi Sad city were imported from the global climatological database Meteonorm (~150 000 parameters, 15 min intervals) [23] as shown in Table 3 and Fig. 2.

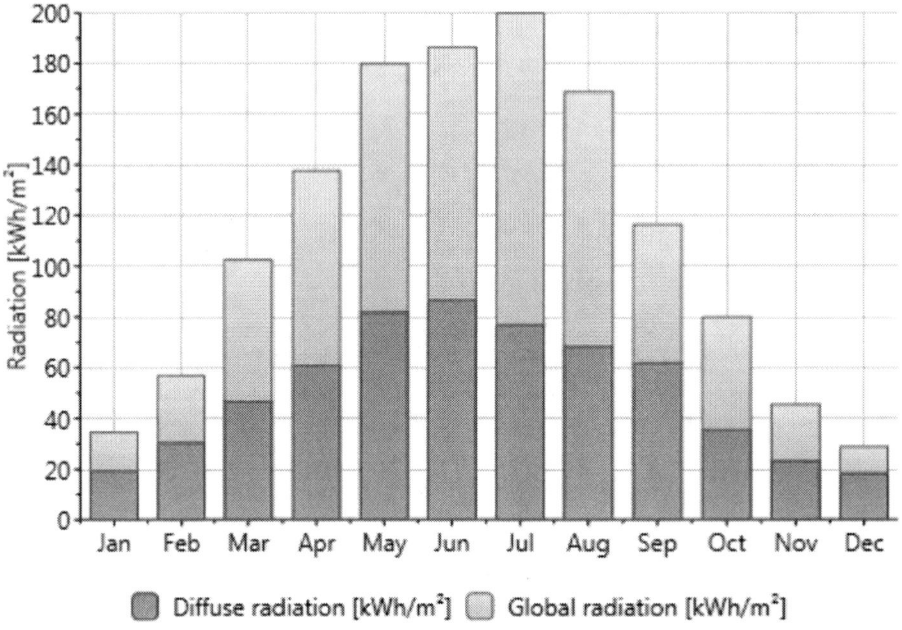

Fig. 2 Solar radiation and air temperature data for the city of Novi Sad

Abbreviations: Ta: Air temperature, G_Gh: Mean irradiance of global radiation horizontal, RH: Relative humidity, G_Dh: Mean irradiance of diffuse radiation horizontal, FF: Wind speed, SD: Sunshine duration, PAR: Photosynthetically active radiation

Table 3. Climate data – monthly average values for Novi Sad.

Month	Ta	G_Gh	RH	G_Dh	FF	SD	PAR
Jan	0.4	46.3	81.9	26	2.6	70	20.1
Feb	2.3	84.5	76.8	41.1	2.8	89	36.2
Mar	7.3	137.7	65	60.5	3.1	145	58.7
Apr	12.7	191	62.7	93.5	2.9	180	82.3
May	18	241.8	63.3	105.5	2.4	230	104.7
Jun	20.8	258.8	65.9	118.4	2.1	251	112.9
Jul	22.4	268.5	64.2	100.2	2.1	289	117
Aug	22.2	226.8	63.3	98.3	1.9	269	99.6
Sep	16.9	161.5	68.6	77.1	2	207	71.5
Oct	12.6	107.5	73.6	58.8	2.3	170	47.9
Nov	7.1	63	78.7	35	2.6	87	28.2
Dec	1.7	38.7	83.8	23.2	2.6	60	17.3
Year	12	152.2	70.7	69.9	2.5	2047	66.4

In order to determine annual heating and cooling loads a multi-zone thermal model was constructed using OpenStudio plug-in for Sketchup. Each thermal zone was assigned with internal load properties typical for a large office building. The thermal zones were formed and named according to their function and position in the building. According to the investigation phases and complexity of the model and simulation processes, three programs were applied in this research, which are the following:

- Google Sketchup (Multi-zone thermal model construction) [34];
- OpenStudio (multi-zone model properties; construction, materials, occupancy, internal loads and schedules) [35] and,
- EnergyPlus (dynamic energy simulation) [36].

The overall energy performance of five multi-level office building models (tot. area per model 3430m^2) was simulated in EnergyPlus engine. Factors influencing the energy demand were modeled together with the various glazing types. Energy demands were determined according to the energy efficiency regulations of the building construction's thermal performance.

Inherited envelope properties from previous Radiance illumination research

The daylight quality was evaluated according to three criteria; spatial illumination dispersion, average daylight factor during occupied hours, and photo-electric lighting simulation for electricity reduction. The daylight intensity analysis and daylight dispersion required numerous simulations which depended on the analyzed period, time, sky conditions and zone orientation. 720 simulations were performed in Radiance (3 WWR x 3 WG x 4 orientations x 6 months x 3 intervals = 648 and 72 simulations for the reference model consisting of 4 orientations x 6 months x 3 intervals).

All output images were evaluated according to the illumination intensity between 350 and 500 lx. The calculation of average daylight factor (DF) was performed in zone centre points as BRE DF calculation for WWR of 20%, 25%, 30%, and base case 50%. Results closest to 2.0 DF were adopted since it satisfies the minimal illumination quality in an office environment. DF simulations were applied for previously selected vertical rectangular window geometry and the total number of conducted calculations was 16 (4 orientations x 4 WWR's) [31].

From the comparative analysis the vertical rectangular window geometry presented the most preferable results due to window height which contributed to deeper daylight dispersion in the indoor environment resulting in qualitative natural illumination in offices [31]. Illumination intensity detectors were setup and adjusted that if lighting level in the center zone point falls below 350 lx the sensors automatically turn on electric lights or switch to dimming mode. Findings indicated the best performance if dimming mode is applied. In conclusion, considering the average daylight factor (DF) and annual percentage of unnecessary usage of electric lighting the adopted models of WWR for vertical WG are presented in Table 4 [31].

Table 4. Daylight factor [-], adopted WWR [%] and photoelectric dimming

East Offices	Percentage working year lighting OFF (%)	
DF 1.97	69	WWR **30%** / 30° rotated floor plan
South Offices	Percentage working year lighting OFF (%)	
DF 2.05	70	WWR **25%** / 30° rotated floor plan
West Offices	Percentage working year lighting OFF (%)	
DF 1.78	66	WWR **30%** / 30° rotated floor plan
North Corridor	Percentage working year lighting OFF (%)	
DF 1.89	67	WWR **25%** / 30° rotated floor plan

Energy simulation - construction, occupancy and operation schedules

The building envelope applied in the simulation was selected according to the thermal insulation requirements of the Serbian Directive - Official Gazette RS no. 61/2011 and EU Standard [37-39]. The overall heat transfer coefficient of existing exterior walls have 2.32 W/(m2K) and existing exterior glazing has 2.78 W/(m2K). The U-value of the modified exterior wall is significantly reduced to 0.22 W/(m2K) by adding 14cm of expanded polystyrene. Further, internal gains from occupants were assigned in OpenStudio in the "people definition" dialog. The number of occupants and internal gains were implemented in the energy simulation setup by the following steps:

- Expectable number of occupants was calculated;
- Occupied office areas were calculated;
- Unoccupied areas were calculated.

The expectable number of occupants on building levels is shown in Table. 5.

Table 5. Occupant number and approximated office areas

No. of occupants	Building level	Office area approx. [m^2]	
(18 x 6) 108 pers. 8 pers. 12 pers. 16 pers. 10 pers. Rarely occupied	4th – 9th level 3rd level 2nd level 1st level Ground level Basement	(196 x 6) 1176 196 196 196 133 0	Office area: 1897 m^2 Other: 1533 m^2 (Entrance, hall, corridor, staircase, elevators, WC, sub-station spaces, installation spaces, archive)
Total 154 (adopted 160 pers.)	Total no. 11 levels	Total area: 3430 m^2	

Occupancy is defined according to the occupancy intensity in the function of occupied period, and people activity in the function of the occupied period.

The "Run Period Profiles" were formulated as a Priority 1 profile for weekdays (8 hours) followed by Priority 2 for Saturday (4 hours) and Priority 3 (0 hours) for Sunday. Electric lighting, electric equipment and thermostat schedules for heating and cooling were also assigned according to the occupancy intensity and building operation hours.

Applied glazing types and parameters

Glazing types were applied according to window properties (parameters: U-value, Solar Heat Gain Coefficient, Visible Transmittance) as shown in Table 6.

Various double and triple pane window constructions were applied in the simulations. U-values were in the range from 1.5 W/(m^{2}K) to 0.7 W/(m^{2}K) high performance tri-pane Pilkington windows with low-E [40].

The Guardian Clima Guard window construction was simulated respectively, which is mainly applied in cold climate conditions [41].

Table 6. Window properties

Model	Windows	Parameters
W1	Dual pane; Pilkington, Optifloat clear	U-value 1.30 W/(m^{2}K); SHGC 0.50; Visible transmittance 0.73
W2	Tri-pane; Pilkington, One pane with Sun-Stop coating & Ag	U-value 1.05 W/(m^{2}K); SHGC 0.34; Visible transmittance 0.63
W3	Tri-pane; Pilkington, Planar + Optifloat + Optitherm glass	U-value 0.70 W/(m^{2}K); SHGC 0.26; Visible transmittance 0.52
W4	Tri-pane; Pilkington, Planar + Optifloat + K Glass	U-value 0.90 W/(m^{2}K); SHGC 0.34; Visible transmittance 0.57
W5	Dual pane; Guardian Clima-Guard 80/70	U-value 1.53 W/(m^{2}K); SHGC 0.69; Visible transmittance 0.81

From the dynamic energy simulation the heating and cooling demands were assessed and the influence of window parameters was determined.

Results and evaluation – heating and cooling demands

Prior to heating and cooling demand determination the building envelope glazing was improved according to the aforementioned multi-criteria analysis in the previous research. The annual heating and cooling energy demands of the five models are shown in Table 7, Fig. 3 and Fig. 4.

Model W2 presented the highest heating energy demand, in total 39 MWh/a. The lowest heating energy demand was determined in case of W3 model.

The heating energy demand reduction of model W3 was 27% lower compared to model W1 and W2, 24% lower compared to W4, and 19% lower compared to W5.

Table 7. Simulated monthly heating and cooling energy demands

Mon.	Heating Energy W1 [kWh]	Cooling Energy W1 [kWh]	Heating Energy W2 [kWh]	Cooling Energy W2 [kWh]	Heating Energy W3 [kWh]	Cooling Energy W3 [kWh]	Heating Energy W4 [kWh]	Cooling Energy W4 [kWh]	Heating Energy W5 [kWh]	Cooling Energy W5 [kWh]
Jan	14349	0	14387	0	10445	0	13737	0	13477	0
Feb	9052	0	9264	0	6924	0	8816	0	8143	1
Mar	984	146	1434	37	834	55	1218	59	384	922
Apr	0	6073	0	4553	0	4881	0	4914	0	9970
May	0	18661	0	16083	0	14197	0	16440	0	24580
Jun	0	26459	0	23719	0	20685	0	24002	0	32703
Jul	0	32210	0	29105	0	24960	0	29355	0	39079
Aug	0	29602	0	26644	0	22917	0	26891	0	35761
Sept	0	15045	0	13371	0	12369	0	13662	0	18927
Oct	2	4782	5	3918	1	4127	3	4158	0	6938
Nov	1190	0	1396	0	631	0	1170	0	715	30
Dec	12468	0	12478	0	8939	0	11873	0	11618	0
Sum	38045	132978	38963	117430	27773	104191	36819	119481	34337	168912
Max	14349	32210	14387	29105	10445	24960	13737	29355	13477	39079

The cooling demand in all models presented significantly higher values compared to heating due to specifically high internal gains and highly insulated building envelope. Model W5 presented the highest cooling energy demand, in total 169 MWh/a. The lowest cooling energy demand, 104 MWh/a, was determined in case of W3 model. The cooling energy demand reduction of model W3 was 21% lower compared to model W1, 11% lower compared to model W2, 12% lower compared to W4, and 38% lower compared to W5. The highest SHGC value 0.69 in W5 model contributed to high cooling demands of the building. The cooling demand was highly influenced by the SHGC value since the cooling demand was approximately 40% higher compared to W3 model with 0.26 SHGC.

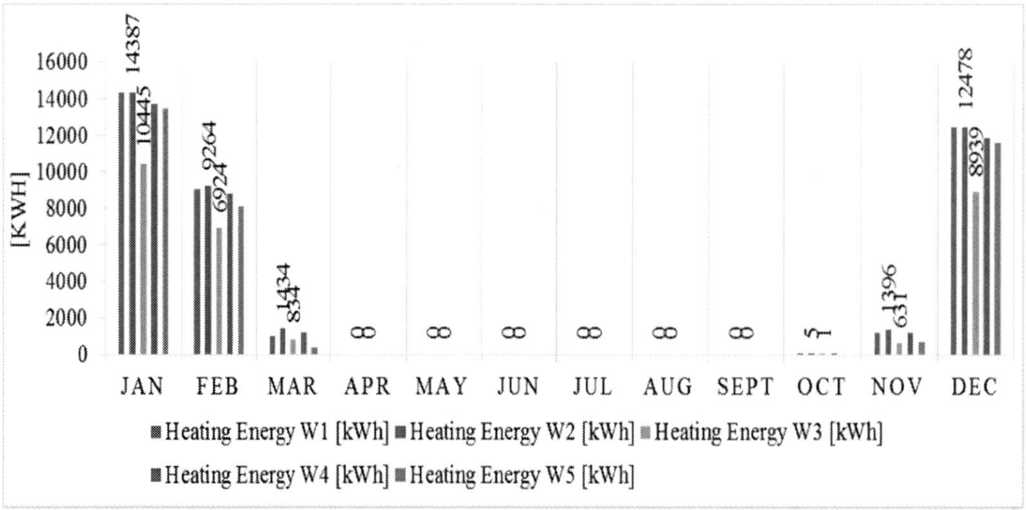

Fig. 3 Annual heating energy demand (W1-W3)

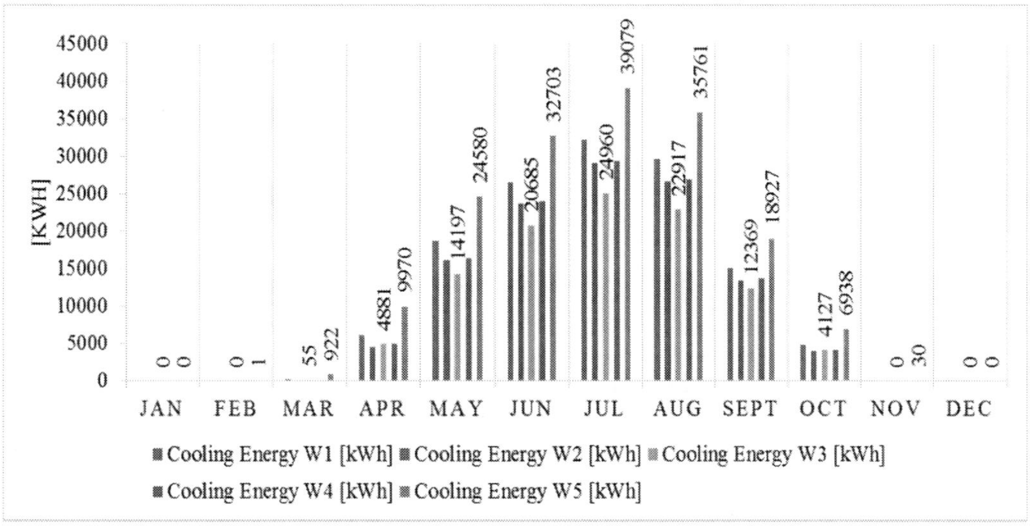

Fig. 4 Annual cooling energy demand (W1-W3)

The W3 model was adopted as the most preferable among the simulated and was compared with the reference office building's energy performance as presented in Table 8, below. The total energy demand of the Best Case Model (W3) could be reduced roughly by 80% in case of annual heating. Simulated cooling energy demand was 40% higher compared to annual heating due to internal heat gains of occupants and electric equipment, which are specific for office environments.

The calculation of annual heating and cooling demand for the Best Case Model was performed according to the EN 15251 Annex B; Basis for the criteria for indoor air quality and ventilation rates; B.1 Recommended design ventilation rates in non-residential buildings [39], as seen in Table 8.

Table 8. Energy performance comparison

Reference FTS office tower energy expenses (2013)			Best Case Model energy demand – W3		
	Heating energy [kWh]	Cooling, lighting and equipment electricity [kWh]	Heating energy [kWh]	Cooling energy [kWh]	Energy demand for lighting and equipment [kWh]
Sum	338788	203810	27773	104191	
			EN 15251; air ventilation amount + 37325 (for heating) +7330 (for cooling)		106330
[kWh/m^2/a]	100	59	19	32	31

According to the climatic conditions of Novi Sad and EN 15251, 37 MWh/a, was added to the simulated heating energy and 7 MWh/a, for the cooling energy, because an ideal air load system was simulated in EnergyPlus, without taking into account the air preparation procedure.

Conclusion

The investigation presented the significance on the reduction of annual energy performance of building envelope's thermal properties and the application of adequate windows in the function of climate conditions and building type and compliance to energy efficiency regulations.

The improved reference building's heating and cooling energy demand was analyzed in case of five glazing types with various parameters, using the previously inherited window to wall ratio and window geometry. From the dynamic energy simulations the annual heating and cooling demands were determined and the influence of window parameters was assessed. During the simulation the main criteria was the maintenance of indoor occupant comfort according to EN 15251 in compliance with the building envelope's thermal properties according to the Serbian Regulations RS no. 61/2011. From the five simulated envelopes it was concluded that with adequate window application the annual heating demand could be reduced by 27% and the annual cooling demand by approximately 40%. If compared to the reference building's energy expenses the annual heating demand could be reduced drastically by 80%.

Further directions of investigation will cover the thermal comfort parameters in the function of minimizing annual heating and cooling loads. In order to find a reasonable solution for cooling energy demand reduction, further research will include the simulation of night time ventilation to determine the cool air accumulation capacity of the building. The simulation and analysis of solar shading systems will cover further directions of investigation respectively.

References

1. G. Radovic, V. Murgul, N. Vatin, E. Aronova, Applied Mechanics and Materials, **627**, 357-364 (2014)
2. M. Penić, N. Vatin, V. Murgul, Applied Mechanics and Materials, **680**, 534-538 (2014)
3. V. Murgul, D. Vuksanovic, N. Vatin, V. Pukhkal, Applied Mechanics and Materials, **635-637**, 370-376 (2014)
4. V. Pukhkal, N. Vatin, V. Murgul, Applied Mechanics and Materials, **633-634**, 1077-1081 (2014)
5. V. Murgul, N. Vatin, E. Aronova, Applied Mechanics and Materials, **635-637**, 2029-2035 (2014)
6. V. Pukhkal, N. Vatin, V. Murgul, Applied Mechanics and Materials, **680**, 529-533 (2014)
7. P. Polina, N. Vatin, V. Murgul, Applied Mechanics and Materials, **680**, 510-516 (2014)
8. R. Alihodzic, V. Murgul, N. Vatin, Applied Mechanics and Materials, **680**, 494-498 (2014)
9. V. Murgul, D. Vuksanovic, N. Vatin, V. Pukhkal, Applied Mechanics and Materials, **680**, 524-528 (2014)
10. G. Radović, V. Murgul, M. Cvetkovska, E. Aronova, N. Vatin, Journal of Applied Engineering Science, **12 (4)**, 277 – 284 (2014)
11. D. Vuksanovic, Y. Nikitin, V. Murgul, N. Vatin, V. Pukhkal, Applied Mechanics and Materials, **680**, 499-503 (2014)
12. G. Radovic, V. Murgul, N. Vatin, Applied Mechanics and Materials, **641-642**, 634-638 (2014)

13. V. Murgul, D. Vuksanovic, V. Pukhkal, N. Vatin, Applied Mechanics and Materials, **633-634**, 977-981 (2014)
14. M. Penića, S. Golovina, V. Murgul, Procedia Engineering, **117**, 883-890 (2015)
15. V. Murgul, V. Pukhkal, Procedia Engineering, **117**, 891-899 (2015)
16. V. Pukhkal, M. Tanić, N. Vatin, V. Murgul, Procedia Engineering, **117**, 864-869 (2015)
17. V. Murgul, N. Vatin, I. Zayats, Procedia Engineering, **117**, 824-829 (2015)
18. V. Murgul, Procedia Engineering, **117**, 808-818 (2015)
19. V. Murgul, MATEC Web of Conferences, **53**, Article number 01046 (2016)
20. A. Gorshkov, V. Murgul, O. Oliynyk, MATEC Web of Conferences, **53**, Article number 01045 (2016)
21. F. Goia, M. Haase and M. Perino. Applied Energy, **108,** 515–527 (2013)
22. J.T. Kim and M.S. Todorovic. Energy and Buildings, **63**, 108–118 (2013)
23. M.S. Mayhoub and D.J. Carter. Building and Environment, **46**, 698-710 (2011)
24. S. Paunović-Žarić, V. Radulović, E. Alihodžić-Jašarović, V. Murgul. Journal of Applied Engineering Science, **14(1)**, 140 – 147 (2016)
25. V. Pukhkal, A. Bieliatynskyi, V. Murgul. Journal of Applied Engineering Science, **14 (1),** 93-101 (2016)
26. V. Pukhkal, V. Murgul, M. Garifullin, Procedia Engineering, **117**, 624-627 (2015)
27. A. Milajić, D. Beljaković, D. Davidović, N. Vatin, V. Murgul, Procedia Engineering, **117**, 916-923 (2015)
28. Nabil and J. Mardaljevic. Energy and Buildings, **38**, 905–913 (2006)
29. N. Harmati and Z. Magyar. Proc. of the Fifth German-Austrian IBPSA Conference BauSim. pp. 115-122 (2014)
30. N. Harmati, Z. Magyar and R. Folić. Proc. of the Int. Congress E-nova. pp. 179-189 (2014)
31. N. Harmati and Z. Magyar Proc. of the Int. Congress E-nova. pp. 313-319. FH Burgenland. Pinkafeld (2015)
32. N. Harmati and Z. Magyar. Energy Procedia, **78,** 2458-2463 (2015)
33. Information on: http://meteonorm.com/en/downloads
34. Information on: http://www.sketchup.com/buy/education-licenses
35. Open Studio 2013: http://openstudio.nrel.gov
36. Energy Plus 2013: http://apps1.eere.energy.gov/buildings/energyplus
37. Official gazette RS no. 61/2011. Rules on conditions for the contents and manner of certificate issuance of energy performance for buildings. 2011.
38. Directive 2012/27/EU
39. European Standards: EN 15251 2007
40. Information on: http://www.pilkington.com/europe/germany/german/products/bp/downloads/byproduct/glasssystems/default.htm
41. Information on: https://www.guardian.com/residential/WindowSolutions/EnergyEfficiency/index.htm

MATEC Web of Conferences **73**, 03001 (2016) DOI: 10.1051/matecconf/20167303001
TPACEE-2016

Mathematical Modeling the Hydrological Properties of Soil for Practical Use in the Land Ecological Management

Vitaly Terleev[1] , *Ekaterina* Petrovskaia[1,*] , *Aleksandr* Nikonorov[1] , *Vladimir* Badenko[1] , *Yulia* Volkova[1] , Sergey Pavlov[1] , *Natalia* Semenova[2] , *Kirill* Moiseev[3] , *Alex* Topaj[3] ,and *Wilfried* Mirschel[4]

[1]Peter the Great St. Petersburg Polytechnic University, Polytechnicheskaya, 29, St. Petersburg, 195251, Russia

[2]St. Petersburg State University, Universitetskaya naberezhnaya, 7/9, St. Petersburg, 199034, Russia

[3]Agrophysical Research Institute, Grazhdanskii pr. 14, St. Petersburg, 195220, Russia

[4]Leibniz-Centre for Agricultural Landscape Research, Eberswalder Straße 84, Müncheberg, 15374, Germany

Abstract. An original and convenient (from a practical point of view) method to estimate the supply of productive moisture in the soil is offered. The method is based on a physically adequate mathematical model of the soil hydrological properties considering the hysteresis of the water-retention capacity. The computation of the irrigation rates, which is based on such estimates, minimizes the water wastage if the excess of the gravitational water is formed and this water percolates out of the moisturized soil profile under watering conditions. The practical applying of the method is able to optimize the crop irrigation techniques, eliminates any inefficient losses of irrigation water and nutrients (and other agricultural chemicals), promotes the rational usage of the water resources as well as provides developing effective solutions of urgent problems of the land ecological management.

1 Introduction

The ecological management of lands is rather relevant under the climate change and the intensification of anthropogenic impact [1-4]. The assessments of weak soils and drained lands are quite important [5-7]. The researches on the use of water intakes and protecting designs of hydraulic structures, as well as for the protection of areas from floods, are very urgent [8-10]. Limitation of natural resources, in particular water resources, raises the question of their protection and rational use of the designers of ameliorative systems, as well as the specialists operating. The problem of preventing admission of nutrients and other agriculture chemicals to the natural groundwater and surface water applies to the topical issues of the environment management.

* Corresponding author: kate-petro@yandex.ru

The economic and ecological significance of this problem is obvious: the removal of nutrients and other agricultural chemicals out of the root zone of the soil by means of a downward water flux creates unproductive losses of mineral fertilizers, pesticides and land reclamation chemicals. This poses a threat of eutrophication and pollution of natural waters. For this reason, in the construction of water management objects as well at the irrigation agriculture there is a challenge to develop energy-saving technologies of crops irrigation, ensuring the protection of natural waters and the rational water resources usage. The development of such technologies involves the use of a significant amount of data on the reclaimed lands in particular – data on hydrological properties of the soils. The direct measurements of the soil hydrological indicators are the rather labor-consuming procedures. Therefore, for the indirect quantitative estimation of these indicators and for prediction of the state of ecosystems it is widely applied mathematical modeling techniques [11-18].

Consider one of such techniques on the example for water-retention capacity (WRC) of the soil. This property is usually described in the form of the functional dependence of the volumetric water content θ [$cm^3 \cdot cm^{-3}$] on the capillary pressure (capillary-sorption potential) of the soil moisture ψ [cm H_2O] [19-21]. The $\theta(\psi)$ dependence is widely used in the calculations of the soil moisture dynamics [12, 22]. To the characteristic features of soil water-retention capacity a phenomenon of hysteresis refers. In the world literature quite a lot of research in relation to this phenomenon is represented. They revealed multiple significant differences between the wetting and drying branches of the WRC hysteretic loop [23-26]. Due to hysteresis, one ψ value corresponds to an infinite number of θ values, which, depending on the «background» of the wetting-draining cycles of the soil, may differ essentially. For complete information on WRC, it is required the direct measurements of $\theta(\psi)$ on a set of all possible wetting-draining «scenarios» of the soil. This requirement is impossible practically.

2 Theory and method

An identification of θ for a given Ψ is a rather urgent task, for example when irrigation rates are calculating. In most practical situations this calculation is reduced to determining the potential supply of the productive moisture in the soil. As is known, such supply is determined by the difference between the two so-called soil hydrological indices (SHIs), namely: «field capacity» (FC is equal to θ value at pF $\approx$ 2.5) and «permanent wilting point» (WP is equal to θ value at pF $\approx$ 4.2). Recall that pF is the decimal logarithm of the absolute ψ value measured in cm H_2O.

The numerical values of the SHIs are usually obtained by direct measurements of the main drying branch of $\theta(\psi)$ dependence. In practical irrigation agriculture agronomists are trying to prevent drying of the soil till WP, leading to plant death. Therefore, irrigation is implemented at slightly moist soil, and irrigation rates are calculated as the difference between FC and value of the present soil water content, that is controlled. As the allowable lower border of θ value, it is accepted SHI, which is called «the moisture of capillary break» (CB is equal to θ value at pF $\approx$ 3.48). Usage of CB as such is explained due to the fact that this index characterizes the transition between two categories of the soil moisture: «readily-available for plants water» and «difficult-available for plants water». The numerical value of CB is usually obtained by direct measurements of WRC main drying branch.

Note that the availability of water for plants is not determined by its specific content per unit volume of the soil, but is characterized by the energy state of water, which depends on the action of capillary sorption forces that retain moisture in the soil. As the quantitative characteristics of the energy state of water in the soil, ψ and pF are used usually. However irrigation rate should be expressed in units of volume (mass) of water. Therefore, it is now widely practiced, the following approach to the calculation of irrigation rates. If the present soil water content is θ_{in}, hence the amount of irrigation water per cm^3 of the soil V is calculated by the formula: $V = FC - \theta_{in}$. But this seemingly obvious approach leads to an overestimation of irrigation rates and, as a result, unproductive losses of irrigation water. We prove this statement.

It is known that for a given pF any θ value on the main drying branch of WRC exceeds all relevant values of volumetric water content on other branches of the hysteresis loop. In particular, for pF ≈ 2.5 FC index, which characterizes θ on the main drying branch of WRC, is the maximum of all the possible θ values on the remaining branches of the hysteretic $\theta(\psi)$. Furthermore, if we compare the branches of WRC hysteresis loop using pF value, corresponding to θ = FC, the maximum of the pF value refers to the main drying branch of $\theta(\psi)$ dependence. Consequently, the application of the irrigation rate $V = FC - \theta_{in}$ causes that soil «goes» along the WRC wetting branch to condition, which is characterized by pF value that is less than pF, corresponding to the measured FC on the main drying branch of the hysteresis loop. This means that an excess of free moisture, which the soil is not able to retain, will flow out by gravity from the analyzed soil layer. It should be noted that, in addition to the unnecessary loss of irrigation water, some undesirable phenomena, such as leaching of nutrients and other agricultural chemicals from the root zone of the soil as well as them percolation into ground water with concomitant negative consequences of eutrophication of natural waters on the areas of the watershed, occur.

Of course, the hysteresis phenomenon of soil WRC is well-known to specialists of hydromelioration. But accounting of this phenomenon in calculating the irrigation rates with usage of data on WRC direct measurement is not possible for the following reason. Suppose, for the soil, on which the irrigated crops are cultivated, the main drying branch of WRC has been measured, as well as FC and pF value, which corresponds to FC, have been identified. For example, taking into account the peculiarities of the physiology of cultivated plants, the pF maximum threshold value (for instance, corresponding CB) has been determined as well the primary wetting branch of WRC has been measured from this pF value to pF that corresponds to FC. With these data, it would be possible to use fixed cycles of wetting and drying of the soil using the well-defined irrigation rates.

However, in real conditions of agricultural field such an «ideal» scenario will likely be disturbed by atmospheric precipitations. Under such conditions, the soil can «go» along the wetting branch of WRC, about which there are no data; further it will change on the draying branch of WRC, which may not coincide with the main drying branch of the hysteretic $\theta(\psi)$ dependence. The intensity of the precipitation and the duration of their falling may differ essentially at different vegetation periods, so it is not possible to measure the diversity of «configurations» for WRC hysteresis loop. For this reason, accounting of the hysteresis phenomenon in the calculating the irrigation rates is possible only with use of physically adequate mathematical models. Of course, the parameters of such models have to be identifiable using of a limited data set of $\theta(\psi)$ direct measurements, for example – data about the main drying branch as well as about one of the primary wetting branches of the hysteresis loop. This condition is satisfied if the model parameters have a physical meaning.

3 Results and discussion

A model of water-retention capacity of soil as a capillary-porous media is described in [27]. For the parameters of the model, it is proposed the physical interpretation in some concepts about interphase abiotic interactions in soil with involvement of water as a liquid, as well as about the configuration space of soil pores. A computer program «Hysteresis», which allows carrying out computational experiments using the mathematical model of hysteretic $\theta(\psi)$, is developed at the laboratory of mathematical modeling of agroecosystems (Agrophysical Research Institute, Russia).

The experiments with the oscillations of the capillary pressure of moisture within the various ranges of values showed no negative «pump effect». Fig. 1 shows the result of calculation of WRC hysteresis loop using the model parameters typical for loamy soil: $n_d = 2.8$; $n_w = 2.8$; $\theta_s = 0.6$ [cm^3·cm^{-3}]; $\theta_r = 0.1$ [cm^3·cm^{-3}]; $\psi_{ae} = -20$ [cm H_2O]; $\psi_{we} = 150$ [cm H_2O]; $\alpha_d = 0.0020$ [cm H_2O^{-1}]; $\alpha_w = 0.0025$ [cm H_2O^{-1}]. The vertical dashed line intersects several drying and wetting branches of hysteretic WRC at $\psi = -330$ [cm H_2O], corresponding to FC.

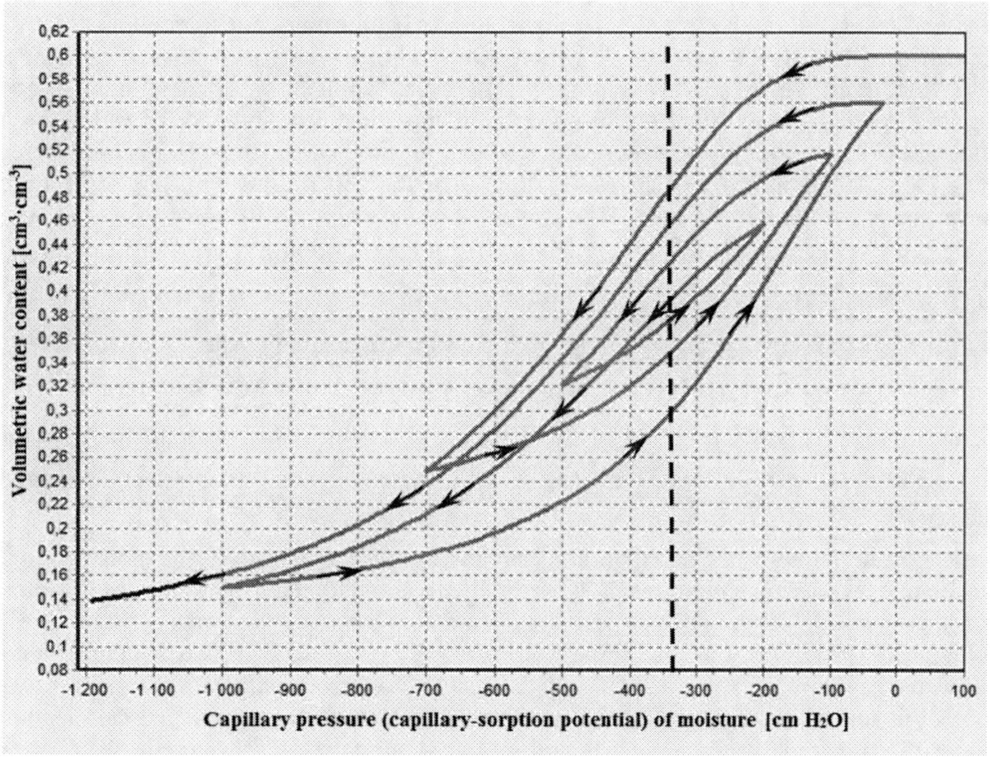

Fig. 1. A possible set of the drying and wetting branches of the hysteretic WRC.

Fig. 2 shows the results of calculating the irrigation rates in two variants. In the calculations it is used the set of typical for loamy soil SHI, such as: FC = 0.495 [cm^3·cm^{-3}] at $\psi = -330$ [cm H_2O]; CB = 0.165 [cm^3·cm^{-3}] at $\psi = -1000$ [cm H_2O]. In the *a*) variant after the first watering free gravitational water flows out of the analyzed soil layer, and the capillary pressure of soil moisture takes the value $\psi = -330$ [cm H_2O], corresponding to the «new FC», which takes the value: $\theta = 0.415$ [$cm^3 \cdot cm^{-3}$]. The loss of water in this case is $0.495 - 0.415 = 0.08$ [$cm^3 \cdot cm^{-3}$]. In the *b*) variant there is not any loss of irrigation water. The second watering in the *a*) variant begins at $\theta = 0.165$ [$cm^3 \cdot cm^{-3}$], while the «new CB» at $\psi = -1000$ [cm H_2O] takes the value: $\theta = 0.135$ [$cm^3 \cdot cm^{-3}$]. The second watering in the *b*) variant begins at $\theta = 0.135$ [$cm^3 \cdot cm^{-3}$]. As a result of watering the volumetric water content of soil takes the value 0.270 [$cm^3 \cdot cm^{-3}$] $\psi = -330$ [cm H_2O]. In this case, it is additionally used the readily available for plants moisture, which is characterized by the difference: $0.165 - 0.135 = 0.03$ [$cm^3 \cdot cm^{-3}$].

Besides, whole irrigation water is used without any loss. In the *a*) variant the irrigation rates for the second watering and for subsequent ones are 2.44 times higher than in the *b*) variant $\{(0.495 - 0.165)/(0.270 - 0.135) = 2.44\}$. Moreover, in the *a*) variant the watering is consumed only at 75.8% $\{(0.415 - 0.165)/(0.495 - 0.165) \cdot 100 = 75.8\}$, whereas in the *b*) variant the irrigation rate is fully used. Unproductive losses of irrigation water can be rather impressive: to increase water content in the soil layer by 30 cm thickness (with the above parameters) from the value $\theta = 0.165$ [$cm^3 \cdot cm^{-3}$] till a value $\theta = 0.495$ [$cm^3 \cdot cm^{-3}$] such loss for every watering can reach about 240 [$m^3 \cdot ha^{-1}$].

Along with the above example of the advantages of the proposed model, which is the basis for the computer program «Hysteresis», earlier there was described an example of the practical using the model in [27]. To justify the irrigation rates for agricultural crops the specific hydrophysical indices, which are typical of the loamy chernozem (FC = 0.35 [$cm^3 \cdot cm^{-3}$], CB = 0.26 [$cm^3 \cdot cm^{-3}$]), were applied. The maximum effective supply of productive moisture in this soil (that is initially fully saturated) after draining the gravitational water out of the water-saturated soil profile is characterized by the difference FC – CB = 0.09 [$cm^3 \cdot cm^{-3}$]; such water supply for the soil profile with the depth of 1 m corresponds to 90 mm of the H_2O layer. The hysteretic WRC of the soil under consideration has the following parameters: $\theta_s = 0.54$ [$cm^3 \cdot cm^{-3}$]; $\theta_r = 0.13$ [$cm^3 \cdot cm^{-3}$]; $\psi_{ae} = -5.0$ cm H_2O; $\psi_{we} = 0.0$ cm H_2O; $\alpha_d = 0.0029$ [cm H_2O^{-1}]; $\alpha_w = 0.0032$ [cm H_2O^{-1}]; $n_d = 2.7$; $n_w = 2.7$. For this example, the following θ values correspond to ψ value at FC (– 330 [cm H_2O]): $\theta_d(-330) = FC = 0.35$ [$cm^3 \cdot cm^{-3}$] and $\theta_w(-330) = 0.32$ [$cm^3 \cdot cm^{-3}$]. The difference between the mentioned θ values: $\theta_d(-330) - \theta_w(-330) = 0.03$ [$cm^3 \cdot cm^{-3}$] fully agrees with the measured data for chernozem [28].

Hence, to minimize the percolation of the gravitational water out of the moisturized soil profile under watering conditions, it is preferable to compute the irrigation rates as follows. For this purpose, it should be used the difference between the prospective values of volumetric water content in soil on the wetting curve of hysteretic WRC (that correspond to the water capillary pressure at FC) and the real pre-watering values of volumetric water content in soil on the drying curve of hysteretic WRC. However, in practice the mentioned values of volumetric water content in soil on the wetting curve of hysteretic WRC are not always available; therefore, the irrigation rates are computed from the difference FC – CB or from the difference between FC and the pre-watering value of volumetric water content in soil. It is obvious that such practice results in the formation of the excess of gravitational water.

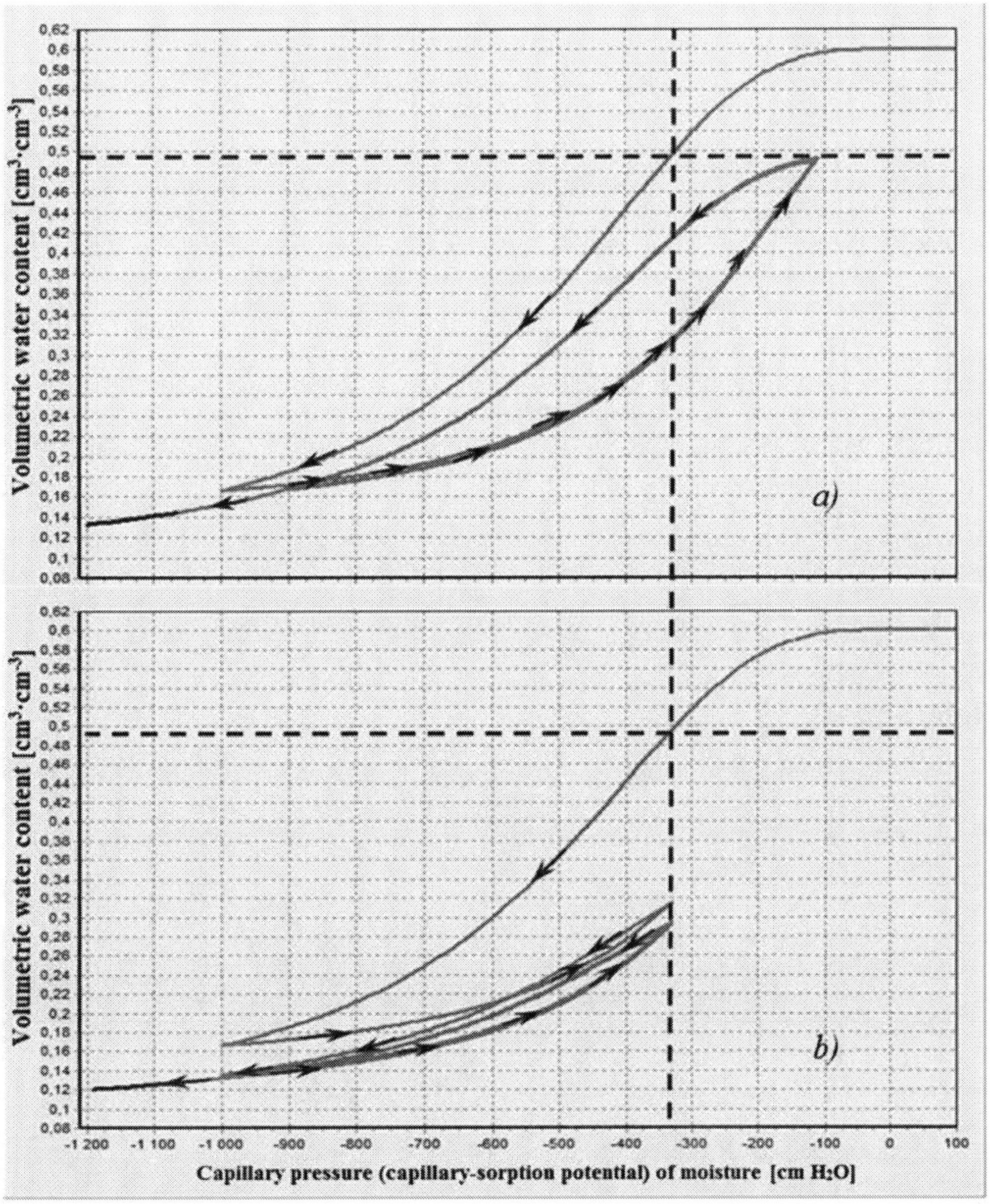

Fig. 2. *a*) Variant: two cycles of «drying-wetting» of the soil with watering from CB to FC, which are defined by the «classical» WRC curve (without hysteresis); *b*) Variant: two cycles of «drying-wetting» of the soil with watering from ψ, corresponding to CB, till ψ, corresponding to FC (taking into account the hysteresis).

The prevention of this negative process and the minimization of water wastage seem to be quite realizable at the presence of reliable data concerning the hysteretic WRC. However, to obtain such data, some rather labor-intensive soil-hydrophysical investigations (that consider the great set of drying-wetting variants) should be carried out. In this paper it was considered the adequate model of hysteretic WRC with physically interpreted parameters which can be identified using the limited set of experimental data (for example, with usage of SHIs and the data on the wetting curve of hysteretic WRC) as an alternative to labor-intensive measurements.

4 Summary

The original mathematical model of the soil hydrological properties has been considered. This model has been developed in the framework of physical concepts on the phenomenon of the hysteresis of soil water-retention capacity. The «Hysteresis» software has been employed for carrying out the computer experiments with this model. Usage of this software allows minimizing the time consuming calculations and improving of their accuracy. This is especially important for reclamation issues solving and for rational using of water resources. The approach to calculate the supply of productive moisture and the irrigation rates has been proposed.

The calculation is based on the values of capillary pressure at FC and CB, taking into account the hysteretic properties of the soil. The rates of watering, which are computed by facilities of this approach, are able to minimizing the losses of irrigation water, mineral fertilizers, pesticides and land reclamation chemicals caused by the percolation of the excess of free water by gravity out of the moisturized soil profile. Practical usage of the approach proposed here allows solving the urgent challenges of the land ecological management. This approach favors the optimization of crop irrigation techniques, as well as provides the rational usage of water resources.

The research was supported by DAAD (A/10/01103), DFG (MI 526/3-1) and Russian Foundation for Basic Research (#16-04-01473-a).

References

1. M.G. Ryzhakova, V.I. Maslikov, A.N. Chusov, V.V. Korablev, AMM, **675-677**, 761-769 (2014)
2. J. Lednova, A. Chusov, G. Gogoberidze, *IEEE/OES Baltic International Symposium 2014*, Article number 6887841, (2014)
3. J. Lednova, A. Chusov, M. Shilin, *Proceedings of the 10th Global Congress on ICM: Lessons Learned to Address New Challenges, EMECS 2013 – MEDCOAST 2013 Joint Conference*, **2**, 1024-1034 (2013)
4. N. Arefiev, V. Garmanov, V. Bogdanov, Y. Ryabov, V. Terleev, V. Badenko, Procedia Eng. **117**, 26-31 (2015)
5. R.A. Usmanov, N.I. Vatin, V.A. Murgul, AMM, **680**, 474-480 (2014)
6. N. Arefiev, V. Badenko, A. Nikonorov, V. Terleev, Y. Volkova, Procedia Eng. **117**, 20-25 (2015)
7. V. Chechevichkin, N. Vatin, AMM, **641-642**, 409-415 (2014)
8. N. Arefiev. M. Mikhalev, D. Zotov, K. Zotov, N. Vatin, O. Nikonova, O. Skvortsova, S. Pavlov, T. Chashina, T. Kuchurina, V. Terleev, V. Badenko, Y. Volkova, V. Salikov, K. Strelets, M. Petrochenko, A. Rechinsky, Procedia Eng., **117**, 32-38 (2015)
9. A. Nikonorov, S. Pavlov, V. Terleev, N. Arefiev, V. Badenko, Y. Volkova, Procedia Eng. ,**117**, 258-263 (2015)
10. A. Makarov, A. Mihailova, N. Arefiev, S. Pavlov, T. Chashchina, V. Terleev, V. Badenko, Procedia Eng., **117**, 225-231 (2015)
11. R.A. Poluektov, S.M. Fintushal, I.V. Oparina, D.V. Shatskikh, V.V. Terleev, E.T. Zakharova, Archives of Agronomy and Soil Science, **48**, 609-635 (2002)
12. R.A. Poluektov, V.V. Terleev, K.Ch. Kersebaum, J.-M. Hecker, W. Mirschel, M. Wegehenkel (Eds), *Leibniz Centre of Agricultural Landscape Research (ZALF) in Müncheberg/Germany*, 75-89, (2007)
13. V. Badenko, V. Terleev, A. Topaj, AMM, **635-637**, 1688-1691 (2014)
14. S. Medvedev, A. Topaj, V. Badenko, V. Terleev, IFIP AICT, **448**, 252-261 (2015)

15. V. Badenko, V. Terleev, N. Arefiev, J. Volkova, O. Nikonova, Proceedings of the AASRI International Conference on Industrial Electronics and Applications (IEA 2015). *Book Series: AER-Advances in Engineering Research* ,**2**, 452-455 (2015)
16. N Arefiev, V. Terleev, V. Badenko, Procedia Eng., **117**, 39-44 (2015)
17. V. Terleev, E. Petrovskaia, N. Sokolova, A. Dashkina, I. Guseva, V. Badenko, Y. Volkova, O. Skvortsova, O. Nikonova, S. Pavlov, A. Nikonorov, V. Garmanov, W. Mirschel, MATEC Web of Conferences, **53**, 1-6 (2016)
18. V. Terleev, A. Nikonorov, V. Badenko, I. Guseva, Y. Volkova, O. Skvortsova, S. Pavlov, W. Mirschel, Advances in Civil Engineering 2016, **8176728,** 1-7 (2016)
19. R. Poluektov, V.Terleev, Russ meteorol hydro+, **11**, 70-75 (2002)
20. R. Poluektov, V.Terleev, Russ meteorol hydro+, **12**, 73-77 (2005)
21. V.Terleev, W. Mirschel, U. Schindler, K.-O. Wenkel, INT AGROPHYS, **24**, 381-387 (2010)
21. R. Poluektov, I. Oparina, V. Terleev, Russ meteorol hydro+, **11**, 61-67 (2003)
23. Y. Mualem, Water Resources Research, **10**, 514-520 (1974)
24. J.-Y. Parlange, Water Resources Research, **12**, 224-228 (1976)
25. P.S. Scott, G.J. Farquhar, N. Kouwen, American Society of Agricultural Engineers, St. Joseph, MI, Publ., **11**, 163-170 (1983)
26. J.B. Kool, J.C. Parker, Water Resources Research, **23**, 105-114 (1987)
27. V.V. Terleev, A.G. Topaj, W. Mirschel, Russ meteorol hydro+, **40**, 278-285 (2015)
28. V.I. Sukharev, *the Doctor's Thesis in Agriculture, Kursk State Agricultural Academy, Kursk [in Russian]* (2006)

Reclamation of the illegal dump for sustainable development the environment in Sverdlovo of Leningrad Oblast', Russia

Maria Bukova[1,*], *Andrey* Bondal[1], *Olga* Skvortsova[1], *Olga* Nikonova[1], *Artem* Kholodiakov[1], *Inna* Guseva[1], *Tatiana* Makarova[1], and *Wilfried* Mirschel[2]

[1]Peter the Great St. Petersburg Polytechnic University, 195251 Polytechnicheskaya str. 29, St. Petersburg, Russia
[2]Leibniz-Centre for Agricultural Landscape Research, 15374 Eberswalder Straße 84,Müncheberg, Germany

Abstract. Illegal dumping is dumping of any waste such as oil, furniture, appliances, trash, litter or landscaping cuttings, upon any land of state, city, village or private ownership without consent of the owner. Illegal dumping has a great negative and fatal impact on our environment and all living organisms both fauna and flora. It also exposes people to various risks of chemicals (fluids or dust) and is a big threat to all under-ground and surface water resources. Illegal dumps also attract all kinds of bugs such as rodents and insects. For example, illegal dumps with waste tires provide a practically perfect place for mosquitoes to breed. Mosquitoes can multiply 100 times faster than normal in the warm, stagnant water in waste tires. Exemplary for the illegal dump in Sverdlovo of Leningrad Oblast' the main purpose of this article is to offer a possible option for the remediation of contaminated area.

1 Introduction

Nowadays ecological situation in Sverdlovo, which is situated in Vyborgsky District of Leningrad Oblast', is deteriorating because of the dump, which has illegally appeared there. Earlier, when this territory was clean, there were some clay pits, where people got clay for brick production. The clay pits appeared there at the beginning of the XIX centaury, and during the 200 years after establishing, they were filling up with water. Landscape, which had transformed due to the clay-extraction, turned to the recreation area for many people. Unfortunately, the territory has recently become operating as an illegal dump: it is filling up with construction and household wastes (Fig. 1).

The relevance of cleaning the territory is obvious, because the dump has significantly deteriorated the environment and the surrounding area. Today we can verify harmful gases (CH_4, CO_2) from the decomposition of organic wastes and an increasing number of rodents and insects. Periodic burning of the landfill pollutes the atmosphere with toxic substances. In addition, it has a harmful impact on the quality of the nearby ponds and streams. The

* Corresponding author: miss-kant@yandex.ru

water research, which was conducted by members of the FBU «Center of laboratory analysis and technical measurements», revealed the presence of a high content of phenol, iron and oil products [1].

The main purpose of this article is to offer a possible option for the remediation of contaminated areas. As the state of the environment at the place of the dump becomes environmentally dangerous, it is necessary to carry out measures to restore the favorable ecological situation in this area.

Fig. 1. Illegal dump in Sverdlovo of Leningrad Oblast′, Russia.

2 Problem of urgency concerning illegal dumpsite

A dumpsite (also tip, landfill, rubbish dump, garbage dump or dumping ground) is a site for the disposal of waste materials by burial and is the oldest form of waste treatment [2]. Historically landfills have been the most common method of waste disposal organization. Moreover, they remain so in many countries around the world. Some dumps are also used for waste management purposes, such as the temporary storage, consolidation and transfer, or processing of waste material (sorting, treatment, or recycling). A dump also may refer to ground that has been filled in with rocks instead of waste materials, so that it can be used for a specific purpose, such as for building houses. Unless they are stabilized, these areas may experience severe shaking or liquefaction of the ground during an intensive earthquake.

Dumpsites have the possible potential to cause a number of issues. Infrastructure disruption, such as damage to access roads by heavy vehicles, may occur. Pollution of local roads and water courses from wheels on vehicles when they leave the dumpsite can be significant. Pollution of the local environment, such as contamination of groundwater or water horizon or soil pollution may occur, as well. Extensive efforts are made to capture and treat leachate from landfills before it reaches groundwater horizon, but engineered liners always have a lifespan, though it may be 100 years or more. Eventually, every dumpsite liner will leak, allowing the leachate to contaminate the groundwater [3].

Methane is naturally generated by decaying organic wastes in a dump or landfill. It is a potential greenhouse gas which can destroy the atmospheric ozone, and can itself be a risk or danger because it is flammable and potentially explosive. In properly managed landfills, gas is collected and utilized. This could range from simple flaring to landfill gas utilization [4]. Poorly run dumpsites may become nuisances because of vectors such as rats and flies which can cause infectious diseases. The occurrence of such vectors can be mitigated through the use of daily cover. Other potential issues include wildlife disruption, dust, odor, noise pollution, and reduced local property values.

3 Method and proposed solutions

The method for land reclamation described here, are implemented in three stages: preparatory, technical and biological.

3.1 Preparatory stage

At the preparatory stage, scientists are exploring the disturbed areas and determine the direction of reclamation. The major activities of mechanical cleaning of territories and removal of surface contamination are carried out before the technical stage. The preparatory phase of the water (drainage and water barrage) facilities construction involves a preliminary assessment of the hydrological conditions of the recultivated object.

Indicators of soil hydrophysical properties: water-retention capacity and hydraulic conductivity of the soil are among the important indicators that characterize the hydrological conditions of recultivated lands. It is necessary to conduct soil-reclamation engineering research in order to receive these indicators. As part of these investigations appropriate laboratory studies are carried out. These studies are very time- and labour-consuming. Using the technology of indirect estimation of indicators of soil hydrophysical properties with respect to the available data is advisable to reduce the costs of carrying out soil-reclamation research. For example, the water-retention capacity of the soil can be estimated using so-called soil hydrological constants, such as Field Capacity and Wilting Point [5], and applying pedotransfer functions [6].

Hydraulic conductivity of the soil can be assessed using an improved method of Mualem-Van-Genuchten [7-10]. The method based on the fact that the water-retention capacity and soil hydraulic conductivity are described by mathematical functions with a single set of parameters. Physical interpretation is proposed for these parameters. This interpretation is based on the idea of the soil as a capillary-porous media. In accordance with these representations distribution of soil pores by its size obeys a lognormal distribution [11-14]. The value of the capillary pressure (capillary-sorption potential) of the soil moisture Ψ [cm H_2O] is used to describe the interaction of water with the surface of the soil particles. The volumetric water content θ [$cm^3 \cdot cm^{-3}$] is used to describe the moisture content of the soil. Function $\theta(\psi)$ describes WRC (water-retention capacity) of the soil as:

$$\theta = \begin{bmatrix} \theta_r + (\theta_s - \theta_r)(1/2)\mathrm{erfc}\{(n\sqrt{\pi}/4)\ln[-\alpha(\psi - \psi_{ae})]\}, \psi < \psi_{ae} \\ \theta_s , \psi \geq \psi_{ae} ; \end{bmatrix} \tag{1}$$

where: θ_s is volumetric saturated soil water content; θ_r is volumetric residual soil water content; $\psi_{ae} = -\beta/r_{\max}$ is bubbling pressure; $\alpha = r_{max}\boldsymbol{r_0}/\beta$ and $n = 4/(\sigma\sqrt{2\pi})$ are position and shape parameters respectively (further explanation: $\boldsymbol{r_0} = r_0/(r_{\max} - r_0)$; σ is standard deviation of random variable $\ln \boldsymbol{r}$; $\boldsymbol{r_0}$ is pore radius that corresponds to the most probable value $\ln r_0$ of random variable $\ln r$; $r = (r - r_{\min})/(r_{\max} - r)$; r is pore radius; $r_{\min}$ is radius of the smallest pore; $r_{\max}$ is radius of the largest pore; $\beta = 2\gamma\cos\varphi/(g\rho_w)$ is coefficient of proportionality; γ is measure the surface tension of soil moisture; φ is angle of the contact of the surface of soil particles; g is gravity acceleration, ρ_w is density of liquid water).

Hydraulic soil conductivity is described as following:

$$k = \begin{cases} k_S \sqrt{\operatorname{erfc}\{(n\sqrt{\pi}/4)\ln[-\alpha(\psi - \psi_{ae})]\}}\left(\operatorname{erfc}\{(n\sqrt{\pi}/4)\ln[-\alpha(\psi - \psi_{ae})] + 2/(n\sqrt{\pi})\}\right)^2 /(4\sqrt{2}), \\ k_S, \psi \geq \psi_{ae}, \end{cases} \quad (2)$$

where k_s - coefficient of filtration of soil moisture.

3.2 Technical stage

Technical stage includes the planning, the installation of hydraulic and drainage structures, the covering potentially fertile soil layer and other works, which establish necessary conditions for further use of the land for the intended purpose. The technical stage has several steps and may include an intermediate step - chemical remediation.

The first step involves the selective mining, storage fertile soil layer and non-toxic species, which will be used after finishing the work. The second stage includes the planning and formation of the surface area. Pits, quarries and ravines are used for mounds, with the creation of a relief in accordance with the sanitary standards of the chemical and mineralogical composition of stored rocks. The third stage involves works on the formation of the potentially fertile soil layer for the subsequent stage of the biological remediation.

Formation of the root zone depends on their rocks; if they contain more than 20% of toxic compounds, chemical reclamation is carried out; then a layer of parent rock and on top of the root zone - potentially fertile layer. At the same stage, there are works to protect areas from storms and flood waters, water and wind erosions. These works include construction of drainage and water protective structures [15].

After the completion of the technical stage of recultivation biological stage occurs, whose task is to improve the properties of the soil. During this stage the physical and chemical properties of formed soils should be taken into account. The soil may be fertilized or sowed tree strengthen mixtures for fixing the surface layer of soil to prevent erosion, if it is necessary.

3.3 Biological stage

The biological reclamation stage includes a set of agronomic and phyto-reclamation activities using different bioremediation techniques [15]. Agrotechnical activities contribute to the creation of soils with favorable for the development of plant properties and include:

- loosening the soil to accelerate the physical, chemical and biochemical processes;
- creation of an artificial micro-relief, which is made of alternating grooves and mounds on soils with excess moisture;
- implementation of necessary fertilizers.

Recultivated soil and the surrounding area must be ecologically balanced sustainable landscape after the completion of all works. According to the instructions of land reclamation [16], connected with hygiene, the following requirements must be complied:

- a choice of means preserving disturbed land, which depends on the condition, composition and properties of rocks, weather conditions, technical and economic indicators;
- activities for technical and biological reclamation disturbed lands with preserving, are agreed in accordance with the sanitary and epidemiological service;
- the use of binding materials for fixing the surface of disturbed lands, that do not have a negative impact on the environment and have sufficient water resistance and resistance to temperature fluctuations;
- covering a shielding layer of soil from the potentially fertile soils on the surface of industrial mounds, made of unsuitable for biological remediation substrate;
- implementation of land reclamation;
- conservation of the tailings, slurry tanks, ash dumps and industrial waste dumps containing toxic substances;
- fixing the industrial waste dumps using technical, biological and chemical methods.

The phytoremediation method is used to cleanse the area from harmful gases. The method is based on the ability of plants to absorb different compounds, split them and use as a food. Using the method of phytoremediation, territories become cleaner from many kinds of pollution, including the salts of heavy metals, pesticides and petroleum hydrocarbons. Moreover, plants prevent the transfer of pollutants from the soil surface, contaminated sites in water ecosystems and other environmental components [17, 18]. As a result, after preparatory and technical works on the reclamation, it is needed to use plants that are typical for the Sverdlovo region: red clover, timothy grass, meadow fescue, white clover.

4 Results and discussions

The mathematical model of soil hydrophysical properties mentioned above is used in the following way. The first step is to measure the dependence $\theta(\psi)$ or to assess this relationship using soil hydrological constants. Further, using the measured data $\theta(\psi)$ the hydrophysical parameters of the functions (1) and (2) can be defined. The second step is to measure the ratio k_s and then calculate the values of the hydraulic conductivity of the soil. Hydrophysical functions (1) and (2) with estimated parameters could be used in calculations of soil moisture dynamics [19-22].

These calculations allow taking reasonable engineering solutions during the development of measures for land reclamation, as well as the design of reclamation systems. If reclamated lands, which are very important to protect [23, 24], involved in agricultural use, so the calculations of the soil moisture dynamics are relevant during the forecasting of crop yields [25-29].

Further, the other stages of reclamation are realized, with strict adherence to the instructions. It is also important to implement individual approach to every case of described issue. It is necessary to calculate correctly the applied engineering solutions such as drainage or hydraulic structures and especially arrangement of new soil layer [30].

5 Conclusion

Illegal dumping is a dangerous affair and puts people health and safety at risk, especially when hazardous chemicals are involved and can threat the environment by polluting land and water resources [31]. Illegal dumping is also dangerous for land owners who could be breaking the law by knowingly or unknowingly allowing waste to be dumped on their property without the necessary approvals [32-35]. Even if a land owner unwittingly allows waste to be dumped on their property it can still pollute land and water, and clean-up cost may be very expensive [36]. In this paper the method for reclamation of the polluted areas as the complex way is proposed. The problem urgency is considered. A step-by-step draft of land reclamation is offered, preliminary phase is considered attentively, the connection with the hydrophysical properties of the soil researches is approved, and the importance of them is founded. The technical and biological stages of case are also investigated, all researches are done for the example of the illegal dumping site in Sverdlovo of Leningrad Oblast', Russia.

Acknowledgements

The research was supported by DAAD (A/10/01103), DFG (MI 526/3-1) and Russian Foundation for Basic Research (#16-04-01473-a).

References

1. Information on http://eco.lenobl.ru/programm/news?id=38660
2. V.I. Smetanin, *Reclamation and improvement of disturbed lands*, 94 (2003).
3. H. Ødegaard, Water Science and Technology, **53**, 17-33 (2006)
4. A. Kholodiakov, A. Bondal, M. Zubkova, A. Chusov, Ability to use the methane component of the modern WWTP biogas as an alternative energy source: submitted to Materials Science Forum (2016)
5. H. Vereecken, M. Weynants, M. Javaux, Y. Pachepsky, M.G. Schaap, M.Th. Van Genuchten, Vadose Zone J., **9**, 795-820 (2010)
6. V.V. Terleev, W. Mirschel, U. Schindler, K.-O. Wenkel, *Estimation of soil water retention curve using some agrophysical characteristics and Voronin's empirical dependence*, 381-387 (International Agrophysics 24, 2010)
7. Y. Mualem, Water Resour. Res., **12**, 513-522 (1976)
8. M.Th. Van Genuchten, Soil Sci. Soc. Amer. J., **44**, 892-989 (1980)
9. V. Terleev, V. Badenko, I. Guseva, W. Mirschel, Applied Mechanics and Materials, 725-726 (2015)
10. V. Terleev, A. Nikonorov, V. Badenko, I. Guseva, Yu. Volkova, O. Skvortsova, S. Pavlov, W. Mirschel, Advances in Civil Engineering, Article ID 8176728, 7 (2016) (doi:10.1155/2016/8176728)
11. L. Svatovskaya, A. Sychova, M. Sychov, V. Okrepilov, MATEC Web of Conferences, **53**, Article Number 01023 (2016)
12. K. Kosugi, Water Resour. Res., **32**, 2697-2703 (1996)
13. K. Kosugi, J.W. Hopmans, Soil Sci. Soc. Am. J., **62**, 1496-1505 (1998)
14. K. Kosugi, Soil Sci. Soc. Am. J., **63**, 270-277 (1999)
15. A.V. Motorina et al., *Industrial and land reclamation*, 240 (1975).
16. D.V. Gorlov, *Land reclamation in the quarries*, 260 (1981).
17. A.D. Gorbovskay, *Ecology of soils*, 170 (2015).
18. B. Baranova, P. Manko, T. Jaszay, Ecological Engineering, **81**, 1-13 (2015)
19. N Arefiev, V. Terleev, V. Badenko, Procedia Engineering, **117**, 39-44 (2015)
20. R.A. Poluektov, V.V. Terleev, Russian Meteorology and Hydrology, **11**, 70-75 (2002)
21. R.A. Poluektov, V.V. Terleev, Russian Meteorology and Hydrology, **12**, 73-77 (2005)
22. R.A. Poluektov, I.V. Oparina, V.V. Terleev, Russian Meteorology and Hydrology, **11**, 61-67 (2003)
23. N. Arefiev, V. Badenko, A. Nikonorov, V. Terleev, Y. Volkova, Procedia Engineering, **117**, 20-25 (2015)
24. A. Nikonorov, S. Pavlov, V. Terleev, N. Arefiev, V. Badenko, Y. Volkova, Use of Procedia Engineering, **117**, 258-263 (2015)

25. R.A. Poluektov, S.M. Fintushal, I.V. Oparina, D.V. Shatskikh, V.V. Terleev, E.T. Zakharova, Archives of Agronomy and Soil Science, **48**, 609-635 (2002)
26. K.Ch. Kersebaum, J.-M. Hecker, W. Mirschel, M. Wegehenkel (Eds), *Proceedings of the Workshop, Leibniz Centre of Agricultural Landscape Research (ZALF) in Müncheberg/Germany,* 75-89 (2007)
27. V. Badenko, V. Terleev, A. Topaj, Applied Mechanics and Materials, 635-637 (2014)
28. S. Medvedev, A. Topaj, V. Badenko, V. Terleev, IFIP Advances in Information and Communication Technology, **448**, 252-261 (2015)
29. V. Badenko, V. Terleev, N. Arefiev, J. Volkova, O. Nikonova, *Proceedings of the AASRI International Conference on Industrial Electronics and Applications (IEA 2015). Book Series: AER-Advances in Engineering Research,* **2**, 452-455 (2015)
30. V. Chechevichkin, N. Vatin, Applied Mechanics and Materials, 641-642 (2014)
31. M.G. Ryzhakova, V.I. Maslikov, A.N. Chusov, V.V. Korablev, Applied Mechanics and Materials, 675-677 (2014)
32. N. Seror, S. Hareli, B. Portnov, Waste Management, **34**, 1436-1445 (2014)
33. K. Ishii, T. Furuichi, Y. Nagao, Waste Management, **33**, 445-455 (2013)
34. D. Ichinose, M. Yamamoto, Resource and Energy Economics, **33**, 79-93 (2011)
35. R. Chaubey, S. Sahu, O.O. James, S. Maity, Renewable and Sustainable Energy Reviews, **23**, 443-462 (2013)
36. N. Arefiev, V. Garmanov, V. Bogdanov, Y. Ryabov, V. Terleev, V. Badenko, Procedia Engineering, **117**, 26-31 (2015)

MATEC Web of Conferences 73, 03003 (2016) DOI: 10.1051/matecconf/20167303003
TPACEE-2016

Ecological Aspect of Dam Design for Flood Regulation and Sustainable Urban Development

Vladimir Badenko[1], *Nikolai* Badenko[1], *Aleksandr* Nikonorov[1,*], *Dmitry* Molodtsov[1], *Vitaly* Terleev[1], *Julia* Lednova[1] and *Vladimir* Maslikov[1]

[1]Peter the Great St. Petersburg Polytechnic University, 195251 Polytechnicheskaya str. 29, St. Petersburg, Russia

Abstract. Many floodplains are excluded from urban development because the floods cause considerable damage to people's lives and properties. This requires the development of new approaches to flood management and mitigation for support sustainable urban development. In present study as the measures for mitigation of flash floods, the regulation of river flow by the system of detention reservoirs for flood diversion with dams, which do not need any operation management, are analyzed concerning of Far East region of Russia. The main objective of this paper is to develop a method for analysis how the dam site selection meets the environmental criterion. The method to justify a selection of self-regulated flood dam parameters, primarily a height of a dam and its location on a water stream, providing minimization of impact on the environment have been developed. The result for Selemdzha river basin in Far East monsoon region of Russian Federation is analyzed. The result shows the robustness of the method.

1 Introduction

Floods are one of the most dangerous natural disasters. Flooding is an issue that affects all countries and each year they cause considerable damage to people's lives and properties. This leads to that many floodplains are excluded from sustainable urban development. Therefore the development of new approaches to flood management and mitigation are required for sustainable urban development. Thus with the Directive 2007/60/EC of the European Parliament and of the Council of 23 October 2007 on the assessment and management of flood risks the strategy changed from Flood Protection to Flood Risk Management [1, 2]. Furthermore, climate change is likely to increase the frequency of flood events, thus increasing the associated hazard [3]. Therefore, the likelihood of the overuse of land in flood-prone areas and the risk of devastating floods grows with population growth, particularly in urban areas.

One of most dangerous type of flood events is sudden (flash) flood. These phenomena are caused by a specific type of intensive rain (rain storm) over a river basin. As a result, a catastrophic rise of water levels and flows in catchment streams are observed. There are many publications, which are devoted to preparing of appropriate flood-protection measures [4-7]. Some of these measures are detention reservoir for flood diversion with dams, which do not need any operation management and could regulate water level automatically [8]. Such types of structural measures for mitigation of flash floods are analyzed in this paper. In addition, it must be pointed out that floods cause the most damage in urban areas, and the main method of researches is modeling [9-11].

The results in [12] based on economic evaluation in trading off between the construction costs and expected value of damage reduction as the benefits have shown that detention dams are the best alternatives of flood mitigation methods. The work [13] shows the possibility of urban flood modelling applications using the high resolution of topographic data that can nowadays be easily acquired. In addition, flood mitigation measures must taking into account that for local authorities achieving sustainable urban development remains one of the most important goals [14, 15]. Therefore, authorities usually choose low impact mitigations as a sustainable solution for urban storm water management [16]. European Union recognizes that member states may face significant challenges in implementing the Flood Directive [17] and solutions are likely to require sustainable flood detention basins [18]. Some comparison of national and local policy tensions in flood risk management have been presented in [19]. The structural measures for flood mitigation are reviewed and compared to each other in [20] according to the type, duration of application, and the return period of the design flood event, how the project measures are integrated into spatial planning that occur. A wetland park in Zhangzhou City, China is taken as an example to show how to apply the concepts and measures of resource saving and environment friendliness [21]

For Russian Federation, the flash floods are a serious problem too [22]. For example, heavy rain falling in Far East areas of Russia, can create floods with runoff about 12-20 m^3/s per km^2 in small river basins.

For comparison, the runoff of melting of snow during the spring flood does not exceed 4 m^3/s per km^2 in the same river basins. An increasing of volume and frequency of extreme precipitation, particularly in areas with a monsoon climate in the Far East of Russia are forecasted by experts [23]. Preliminary evaluation gave the following results: the frequency of floods caused by heavy rains in the Far East will increase by 1.2 - 1.5 times. So, if now they are repeated 1 time to 10 - 15 years in the future, they will be repeated with time intervals to 7 - 12 years. The number of floods will increase by 45 - 60% in rivers of Far East region of Russia that is a big challenge for sustainable urban development in the region [24].

In present study as the measures for mitigation of flash floods, the regulation of river flow by the system of Self-Regulated Flood Dams (SRFD) are analyzed concerning of Far East region of Russia. These dams are distributed on the river network and can save additional water volume from flash floods in temporary detention reservoirs created by SRFD. The selection of the best sites for the SRFD on specific branch of the river network of the river basin is a multicriteria task with economical, ecological, social and technological criteria on the top level of the criteria hierarchical structure [25, 26]. It should take into account that a modern feature of flood risk management is integration of ecological, economic and social factors in urban flood risk prevention and mitigation. During sustainable urban development risk-reducing measures for flood events can be in conflict with ecosystem functions and complicate upstream/downstream relations. Flood risks are also influenced by processes in the catchment, such as changes in land-use and increases of vulnerable urban areas for example in floodplains [27].

The main objective of this paper is to develop methods for analysis how the site selection of SRFD ("cross-section" or "point" on a river) meets the environmental criterion. Ecological research is the preliminary analysis before final decision [28]. Consideration of all criteria in complex from the beginning is both labor-consuming and financial expensive arrangement, which only complicates the process. The methodological basis of analysis is the modeling in GIS environment of hydrological processes in a river basin based on the basin-landscape approach [29] for the selection of modeling objects.

2 Materials and methods

The structural measures are usually used for flood mitigation [8, 12, 17, 20]. SRFD may be considered as an example of the structural measures. However, this is a special kind of dam with a hole, which in a normal situation does not resist to flows but in ideal situation, SRFD transforms the flash flood into a 10% probability flood. For this reason, SRFD has minimal impact on the environment. SRFD is used primarily for solving hydrological problems by changing the shape of the catastrophic flood hydrograph, which become smoother. If SRFD is distributed on branches of mainstream of river basin, then resulting hydrograph in a basin control section will have a smooth shape without peaks, which are characteristics of catastrophic floods. Therefore, flood risks for population and sustainable urban development will decrease. The objective of this work is to develop a method to justify a selection of SRFD parameters, primarily a height of a dam and its location on a water stream, providing minimization of SRFD impact on the environment.

Because ecological criterion is in the focus of the research, following tasks were analyzed in this paper:

- selection of appropriate SRFD parameters in specific location for SRFD;
- calculation of duration of flooding time for each ecosystem/landscape for the particular SRFD parameters during flash floods.

Based on the comparison of duration of flooding time for each ecosystem/landscape at the fixed location of the dams with different heights and width of holes at different points on the river, the best SRFD parameters choosing for specific locations can be performed.

During modeling of hydrological process on river basin rainfall-runoff models [31] are used with following equations. For the velocity of surface runoff under assumption that the Manning equation can be used the following equation is used:

$$v = k_r H^{2/3} \sqrt{I} \tag{1}$$

where k_r is roughness (according land use), I – is a slope, H is a depth of water at a soil surface. For a pulse inflow, outflow peaks after a time given by the time lag, and then decays exponentially as follow:

$$Q_o = (1 - x\,/\,(dt/K))\,Q_i \ ; \ x = 1 - e^{(-dt/K)} \tag{2}$$

where Q_o is outflow, dt – is a time step, Q_i is inflow, K is a delay parameter.

As the selection of appropriate dam height is considered for SRFD at current stage of the research a model with virtually unlimited height is used. This approach makes it possible to get a time for landscapes flooding in river basin for all possible values for a dam height with given flood hydrograph. The calculation of the flood volume $V_s\,(t)$ that is accumulated in a reservoir of SRFD at time t is calculated using balance method:

$$V_s\,(t) = V_s(t-1) + (Q_i(t) - Q_o(t))dt \tag{3}$$

where $Q_i(t)$ – inflow to SRFD, $Q_i(t)$ – outflow through SRFD, dt – time step.

For modeling in GIS environment raster model is used for a studied basin. The basic raster layers of the GIS database are following:

- landscapes (land use types) - $\{LS\}$;
- Digital Elevation Model (DTM) - $\{DT\}$.

In these layers, the pixels of the same size are used. The main resulting layer is a raster layer with the attribute - flood time - $\{Tf\}$.

Let X is the location of SRFD on the river; w is the width of the SRFD hole, Hd - the height of SRFD. In the basin there are N landscapes. For every j-th landscape, a critical TS_j flooding time, beyond which for the j-th landscape will be an environmental hazard, is known. Let $HYD(t)$ is the hydrograph of specific flood. $H(t)$ is the water level in the SRFD upstream, before flood $H(t)=Const$. Flooding begins at time $t = 0$. At each time step t is computed $H(t)$:

$$H(t) = f(X, w, Hd, \{DT\}, HYD(t-1), H(t-1)) \tag{4}$$

On the basis of $H(t)$ can define a set of pixels which are flooded at time t: $\{LS\}_t$.Then for each landscape:

$$\{Tf\}_j = \textstyle\sum_t \{LS\}_t \tag{5}$$

Then the condition to select the best location for SRFD can be written as:

$$\textstyle\sum_j \{Tf\}_j \rightarrow min, \ \{Tf\}_j < TS_j \tag{6}$$

A method scheme is based on experience of development of a method for estimation of gross hydropower potential in Russia [30]. The scheme of the proposed method and required databases based on [30] are shown in Figure 1.

On the base of the method proposed a comparison of the hydrological situation in the local watershed during a flash flood in natural conditions (without SRFD) and with SRFD is performed by modeling in a GIS environment.

Also it must be pointed out, that it is necessary to begin with the preliminary sites selection for SRFD to except the wide areas, which have high ecological value. Building of the mathematical model begins with the creation of land use GIS database. Knowledge of the ecological value of the determined river floodplain areas, land use regime specifics are necessary for the definition of main parameters, such as flood covered area and critical time duration of flooding. As initial conditions, flood hydrograph and size of hole (non-operational) should be considered. It is important to notice, that during the single calculation parameters of the hole would be fixed, and then, selecting these parameters, we could determine the required conditions. In addition, it is important, that the flood hydrograph at this stage of the research is hypothetical. However, the method proposed with real flood event hydrograph can be used. For the realistic calculations, every hydrographs for every scenario will be received, but this kind of hydrograph also can take place.

For modeling, raster GIS representation of river basin landscapes is used. The Python language is used for modelling in GIS environment.

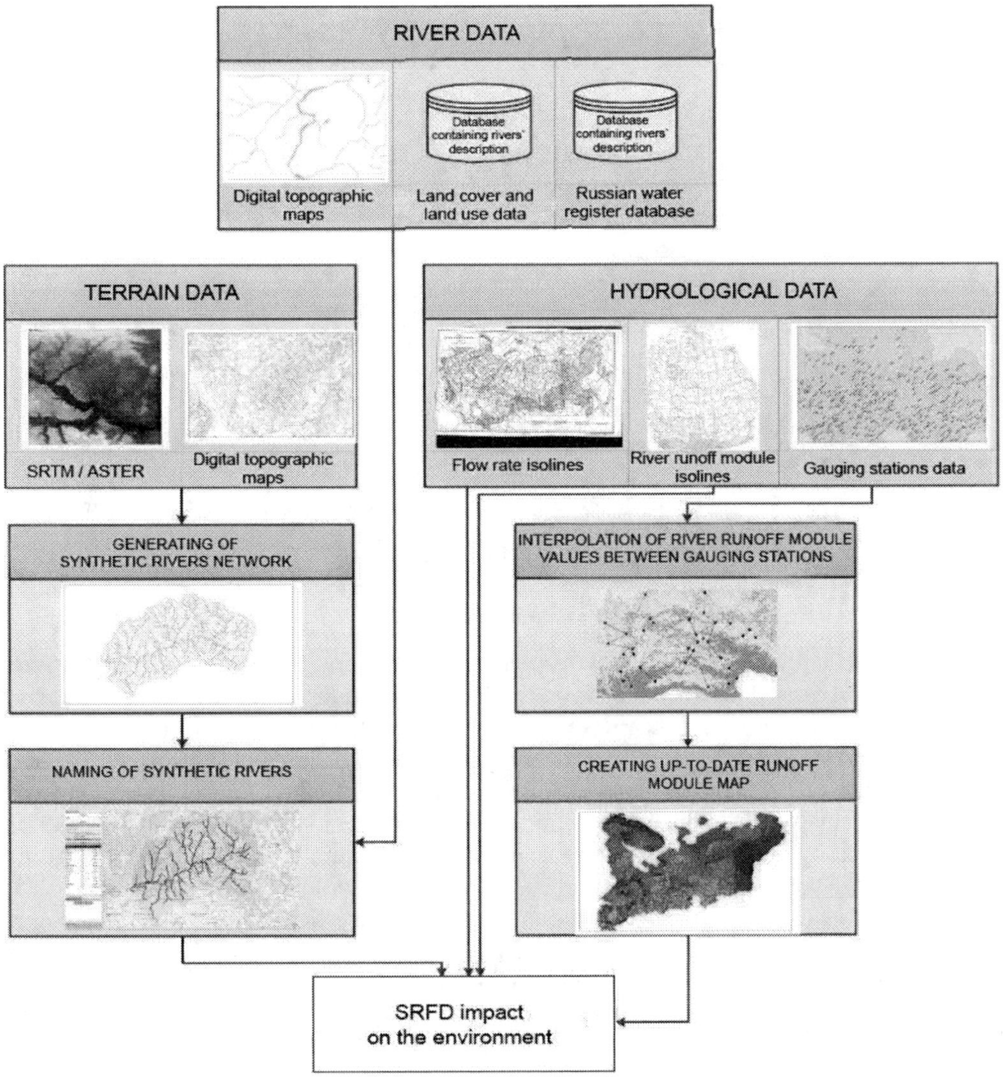

Fig. 1. The scheme of the proposed method.

3 Results and discussion

Selemdzha river basin in Far East monsoon region of Russian Federation is considered. Remote sensing data for part of the region analyzed shows in Fig. 2. An urban territory can be found in the middle of the picture. A width floodplain with unstable riverbed can be recognized. This is why the synthetic rivers network using thalweg of digital elevation model are created for modeling.

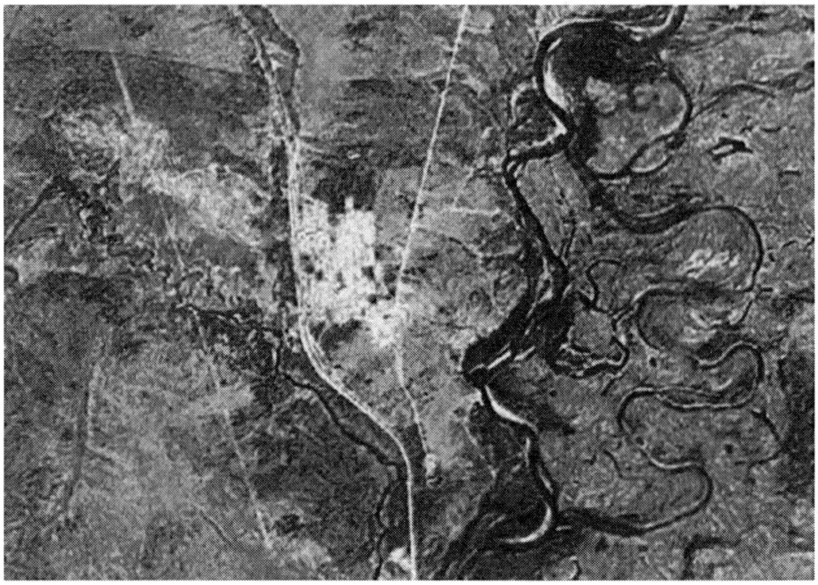

Fig. 2. Remote sensing data for part of the region analyzed.

According to the scheme in Figure 1 and the relations (1) - (6) some scenarios for branches were carried out for the river Selemdzha – Nora basin. The vector and raster layers in GIS database have been created to simulate the hydrological processes on the basin territory. DTM ($\{DT\}$) for the territory was obtained using SRTM dataset.

The results of calculations by the program using the Python language in the form of graphs and thematic GIS maps were obtained. At the current stage of research, in modeling scenarios flood hydrographs of varying triangular shape were used. At the next stage more realistic hygrograph obtained by rainfall/runoff modeling will be use.

Some results for one scenario of flood event in the specific locations shown in Fig.3. A model for the incoming hydrograph with triangular shape is shown in Fig.3A. Also outflow SRFD hydrographs for different SRFD holes (w = 3, 5 and 8 meters) is shown in Fig.3A. Fig. 3B shows a thematic GIS map for flooding for the specific locations of SRFD ($\{LS\}_t$.for using in (5) formula) Flooding time duration for *w=5m* is shown with different color. Pixel size for the raster GIS database is 100 meters.

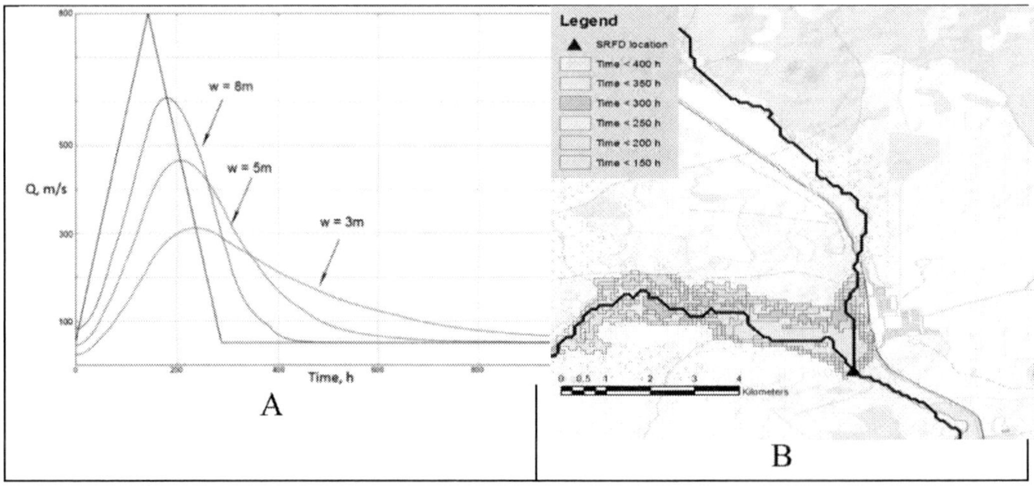

Fig. 3. Results of modelling in GIS environment.

Raster GIS database layer in Fig.3B was combined with the landscape layer and critical times to fit (6) formula. This allowed a comparison of locations with others and chooses a location that has a minimal impact on the environment.

If ecological management concerning risk-based flood management define as a system approach that assesses and compares the structural and non-structural measures to pursue the best ameliorating effects, than the method proposed can solve some sustainable urban development problem in flood-prone areas taking into account environmental impact [31-34]. The method proposed adapts the principles supporting structural measures as diverting water away from urban area to making room for water and save natural landscapes.

4 Conclusions

As the measures for mitigation of flash floods, the regulation of river flow by the system of detention reservoirs for flood diversion with dams, which do not need any operation management, are analyzed concerning of Far East region of Russia. The method for analysis how the dam site selection meets the environmental criterion is developed. This method used to justify a selection of self-regulated flood dam parameters, primarily a height of a dam and its location on a water stream, providing minimization of impact on the environment have been developed. The method proposed adapts the principles supporting structural measures as diverting water away from urban area to making room for water and save natural landscapes. The result for Selemdzha river basin in Far East monsoon region of Russian Federation shows the robustness of the method.

Acknowledgements

The research was supported by Russian Science Foundation (grant #16-17-00050).

References

1. C. Butler, N. Pidgeon, Government and Policy, **29**, 533-547 (2011)
2. J. Ran, Z. Nedovic-Budic, Computers, Environment and Urban Systems, **57**, 68-79 (2016)
3. L. Yao, W. Wei, L. Chen, Ecological Indicators **60**, 893-905 (2016)
4. N. Arefiev, V. Badenko, A. Nikonorov, V. Terleev, Y. Volkova, Procedia Engineering, **117**, 20-25 (2015)
5. A. Nikonorov, S. Pavlov, V. Terleev, N. Arefiev, V. Badenko, Y. Volkova, Procedia Engineering, **117**, 258-263 (2015)
6. K.M. Debusk, T.M. Wynn, Journal of Environmental Engineering, **137**, 800-808 (2011)
7. L. Ahiablame, R. Shakya, Journal of Environmental Management ,**171**, 81-91 (2016)
8. D. Nijssen, A. Schumann, M. Pahlow, B. Klein, Natural Hazards and Earth System Science, **9**, 1349-1363 (2009)
9. V.V. Elistratov, V.I. Maslikov, G.I. Sidorenko, Power Technology and Engineering, **49**, 6-10 (2015)
10. A.H. Thieken, H. Kreibich, M. Müller, B. Merz, Hydrological Sciences Journal, **52**, 1016-1037 (2007)
11. M. Guan, N. Sillanpää, H. Koivusalo, Hydrological Processes, **29**, 2880-2894 (2015)
12. A. Heidari, Natural Hazards and Earth System Science, **9**, 61-75 (2009)
13. J. Leandro, A. Schumann, A. Pfister, Journal of Hydrology, **535**, 356-365 (2016)
14. A.M. Hassan, H. Lee, Land Use Policy, **48**, 199-212 (2015)

15. M.P. Fedorov, V.V. Elistratov, V.I. Maslikov, G.I. Sidorenko, A.N. Chusov, V.P. Atrashenok, D.V. Molodtsov, A.S. Savvichev, A.V. Zinchenko, Power Technology and Engineering, **49**, 33-39 (2015)
16. H. Qin, Z. Li, G. Fu, Journal of Environmental Management, **129**, 577-585 (2013)
17. Z.W. Kundzewicz, N. Lugeri, R. Dankers, Y. Hirabayashi, P. Döll, I. Pińskwar, T. Dysarz, S. Hochrainer, P. Matczak, Mitigation and Adaptation Strategies for Global Change, **15**, 641-656 (2010)
18. M. Scholz, Q. Yang, Landscape and Urban Planning, **98**, 190-199 (2010)
19. S. Beucher, Environmental Hazards, **8**, 101-116 (2009)
20. A. Kryžanowski, M. Brilly, S., Rusjan, S. Schnabl, Natural Hazards and Earth System Sciences, **14**, 135-142 (2014)
21. L. Qiao, M. Li, Y. Zhang, X. Zhao, Nature Environment and Pollution Technology, **15**, 221-226 (2016)
22. A.B. Avakyan, M.N. Istomina, Power Technology and Engineering, **35**, 89-93 (2001)
23. P.C.B. Milly, R.T. Wetherald, K.A. Dunne, T.L. Delworth, Nature, **415**, 514-517 (2002)
24. S.G. Dobrovolskii, M.N. Istomina, Water Resources, **36**, 491-506 (2009)
25. N. Arefiev, V. Terleev, V. Badenko, Procedia Engineering, **117**, 39-44 (2015)
26. D. Kurtener, V. Badenko, Journal of the Brazilian Computer Society, **6**, 26-32 (2000)
27. L. Nyberg, M. Evers, M. Dahlström, A. Pettersson, Aquatic Ecosystem Health and Management, **17**, 331-340 (2014)
28. V. Bauni, F. Schivo, V. Capmourteres, M. Homberg, Remote Sensing Applications: Society and Environment, **1**, 50-60 (2015)
29. N.V. Aref'ev, V.L. Badenko, G.K. Osipov, Power Technology and Engineering, **32**, 660-663 (1998)
30. N. Arefiev, O. Nikonova, N. Badenko, T. Ivanov, V. Oleshko, *"ENVIRONMENT. TECHNOLOGY. RESOURCES" Proceedings of the 10th International Scientific and Practical Conference*, 41-50 (2015)
31. M.P. Fedorov, A.G. Bogolyubov, V.I. Maslikov, Environmental safety of power plants using renewable sources of energy, Hydrotechnical Construction, **29**, 353-357 (1995)
32. V. Terleev, A. Nikonorov, V. Badenko, I. Guseva, Y. Volkova, O. Skvortsova, S. Pavlov, W. Mirschel, Advances in Civil Engineering, **2016**, 7 (2016)
33. O. Skvortsova, A. Dashkina, E. Petrovskaia, V. Terleev, A. Nikonorov, V. Badenko, Y. Volkova, S. Pavlov, MATEC Web of Conferences, **53**, 7 (2016)
34. V. Terleev, E. Petrovskaia, N. Sokolova, A. Dashkina, I. Guseva, V. Badenko, Y. Volkova, O. Skvortsova, O. Nikonova, S. Pavlov, A. Nikonorov, V. Garmanov, W. Mirschel, MATEC Web of Conferences ,**53**, 6 (2016)

MATEC Web of Conferences **73**, 03004 (2016) DOI: 10.1051/matecconf/20167303004
TPACEE-2016

Assessment of Natural Resource Potential of a Territory for Planning of Investment Development and Construction in Suburban Areas

Marina Usoltceva[1,*], *Yulia* Volkova[1], *Artyom* Radchenko[1], *Inna* Guseva[1], *Tatiana* Makarova[1], *Vitaly* Garmanov[2], and *Karl-Otto* Wenkel[3]

[1]Peter the Great St. Petersburg Polytechnic University, 195251 Polytechnicheskaya str. 29, St. Petersburg, Russia
[2]St. Petersburg State Agrarian University, 196601 Peterburgskoye shosse 2, St. Petersburg, Russia
[3]Leibniz-Centre for Agricultural Landscape Research, 15374 EberswalderStraße 84, Müncheberg, Germany

Abstract. The paper describes the types of assessment and assessment process of the natural environment. A review of judgments about the ecological and geo-ecological assessment of the environment by known academics, environmentalists, physical geographers, geobotanists is made here. In this paper the method of the environmental assessment of natural resource potential within the North-West region of Russia is proposed. Authors divide this method into blocks, which are described in the paper. A produced method for evaluating territory used to calculate the cadastral value of land.

1 Introduction

Evaluation of natural bodies and phenomena is a form of reflection of the interaction between society and nature. Due to the fact that in themselves natural bodies and phenomena have neither useful nor harmfulness should clearly indicate what is the object of evaluation, and in the interests of what subject it is produced. Proceeding from the above, the assessment is the process of establishing conformity assessment facility requirements to it by the subject expressed in the form of regulatory or other assessment categories. In view of the fact that each assessment is subjective, the proximity of the two estimates, we can speak only in the event that there is similarity of the positions of subjects of evaluation.

According to the process of estimating the evaluation can be divided into direct and indirect ones. Assessments can be divided into private, comprehensive and multi-criterial by the manifestation of systemic criteria. Private assessments are separate comparison of baseline characteristics with the accepted background levels and standards. This usually exploded assessment determining a positive or negative significance of the object. Private assessments are traditional geographical and environmental studies. They form the basis of

* Corresponding author: slcvmrn@rambler.ru

integrated and multicriterial assessment of natural and economic systems. Integrated assessments are to establish the significance of the object on the set of indicators. Comprehensive assessment as a rule is not systemic, since the concept of "complex" and "system" is not identical to each other. This is the most perspective, but the least developed and standardized assessment. The founder of an integrated approach to the study of the natural environment is V.V. Dokuchaev. His ideas had been enhanced by L.S. Berg, A.I. Voeikov and other scientists.

Multicriterial (synthetic) assessment establish the importance of an object by constructing a vector of values of normalized parameters of the object or by the convolution of information on the state of the object as a function of desirability. These assessments are systemic; they suggest the need for a specific procedure clotting information, which is a goal-oriented process, based on clear principles of selecting or designing the most informative variables of the system - the index of its condition. These indexes can be represented by a single number (one-dimensional code), or a set of numbers (multivariate index) [1, 2].

Multicriterial assessment of a territory includes determining the usefulness of its properties for production non-economic and economic ways. Non-economic method is based on determining the value of the territory, is not expressed in economic terms, and economic way - by determining its social usefulness in satisfying human needs for food security through agricultural production [3, 4]. On this basis, with the multicriterial assessment territory as non-economic way we use aeco-geographical approach, and as economic way we use an inventory approach.

Results of the assessment usually expressed in the form of two basic forms: verbal and numerical. The verbal form is a qualitative characteristic of the assessment results, divided into comparative gradations, such as "suitable", "relatively suitable", "average fitness", "relatively suitable", "unfit" and the digital form is a record of verbal forms scores, grades, or categories of direct calculation indices.

2 Eco-geographical and cadastral approaches to estimating the land resources

The geographical assessments of environment is focused on a comprehensive study of nature, to establish the nature of the interaction between human society and the geographical environment [4].

Objects of the geographical environment assessments are natural or anthropogenic changes in geosystems of different hierarchical levels. In assessing characteristics such as geosystems: their structure, spatial and temporal variability, particularly peculiar geosystems biogeochemical cycles of chemical elements, spatial relationships between the natural and social objects within geosystems, interconnection and interoperability of geographic phenomena occurring in the ecosystems and others are determined.

Along with geosystem in geographical studies are widely used term environmental systems (TES), which refers to a space-time system of geographic components, interrelated in their placement and developing as a whole [5, 6]. However, it should be noted that compared to geosystem of natural-territorial complex (NTC) has a broader interpretation, since the term NTC applicable only to certain geographical parts of the shell, but does not apply it as a whole.

Thus, the assessment of the geographical environment is focused on the analysis of its present state, identifying ways of its development, to determine the degree and direction of human impacts on nature, the study of the deep bonds of abiotic factors and determining the importance of natural ingredients for the society. The main content of this assessment is a complex physical-geographical regionalization of the earth's surface or the properties of geosystems different hierarchical levels, zoning largest integral indicators of the natural environment, the identification of conditions for rational use and protection of natural resources.

The objective of the environmental assessment of the environment according to Odum [7] is the answer to the question: how many organisms inhabit this area, when and where they can be found and why. At the same time, in his opinion, as the subject of the evaluation should serve biological systems superorganism level. N.F. Reimers considers environmental assessment of the environment as a process of determining the state of the environment or the life of the impact on it of any factors (taking into account also the dynamics of the impact) [4].

V.V. Dmitriev under the environmental assessment of the environment understands: 1) determination of the suitability of natural systems and their components for living organisms; 2) parametric definition of the natural environment, ensure the existence of communities living organisms characteristic of these states in terms of a natural or man-made conditions for their development; 3) obtaining based on multicriterial "portrait ecosystems" and relating it to the "standard portrait" of the same ecosystem; 4) subject object criterial assessment of a natural object from a sustainable functioning of ecological communities, save them in the natural course of successional processes or from the standpoint of the degree of suitability (potential usefulness) for his people (society) [1, 2]. A.G. Isachenko states that any environmental assessment is biocentric, it examines only one side of the interaction in the life environment, i.e., the ties that focus on living, and relationships in the abiotic parts of the system and feedbacks directed from the living to the inanimate essentially remain outside research interests [8]. In addition, the author emphasizes the weakness of the environmental assessment of the environment, which, in his opinion, is "one size fits all" major categories of assessments, that is, uncertainty in its spatial boundaries and no territorial "bindings".

Thus, the subject of the environmental assessment of the environment is a living systems superorganism level existing in natural or modified conditions, and is focused on the assessment of management and optimal operation of these systems with the needs of society. Recently intensively developing geo-ecological, geographical and environmental assessments of the environment are used. The relationship of geography and ecology in the study of the natural environment first is stressed by Sochava [9].

Geoecology tasks in the process of its development became more sophisticated, so if at the end of the 70s in German Geoecology considered only as the doctrine of the natural landscape of the budget, which is qualitatively and quantitatively investigated through the cycling of matter, by the 80s it was formed in the scientific field that studies spatial and functional aspects of social and natural systems of different ratings. The transfiguration geoecology, unlike bio-ecology, studying not only the paired connection "living system - environment", but also network-mediated relationships between the components of geographical systems included in the scope of his study mechanisms of adaptation of human society to natural and anthropogenic environmental changes [10].

Thus, geo-ecological assessment of a territory is focused on diagnosis of the natural and anthropogenically modified geosystems different hierarchical levels; on the analysis of environmental impacts, including human beings; on geosystems; to clarify the stability of geosystems to these impacts; to determine the trajectory of geosystems behavior for different types of impacts; the determination of the natural ecological potential of geosystems, etc. Consequently, geo-ecological assessment is a spatial parametric analysis capability of natural and anthropogenic changes in terrestrial systems in different types of human impacts steadily to serve life support (environment and resource reproduction).

Cadastral assessment of land has arisen as a result of the introduction of private ownership of land and payment for land use, land market development, reforming the tax system and collection of land rent for the land. It is applied economic analysis of supply and demand of land for a specific functional use and forecast of the likely offer price areas in the land market. This task is designed to solve the massive state cadastral valuation of lands conducted within the administrative-territorial units.

State cadastral assessment of land is based on a statistical analysis of market prices and other information about the objects of real estate, as well as on methods of mass appraisal of real estate.State cadastral assessment of agricultural land outside of urban and rural settlements and forest land is based on the capitalization of the estimated rental income, as well as other categories of land on the basis of the estimated rental income capitalization, and on the basis of costs to reproduce or preserve and maintain the value of their natural potential [11-14].

The results of state cadastral assessment of land are in the form of evaluation zoningscheme. Evaluation of the land area is recognized, uniform for the intended purpose, since the functional use and close to the value of cadastral value of land state cadastral valuation of land is carried out based on the data of the land, urban, forest, water and other inventories. The results of state cadastral valuation of land approved by the executive authorities of the Russian Federation and entered into the State Land Cadaster. Consequently, the state cadastral assessment of land is a set of legal, administrative and technical measures aimed at establishing the cadastral value of land as a certain date.

3 Procedure of the geo-resource potential assessment in Russian North-West Region

Based on the basic provisions of the system analysis can be formed the following procedure of the geo-resource potential assessment within the North-West region of Russia (Fig. 1).

In view of the fact that in this region consists of several major administrative entities having a certain economic independence, each of them should be regarded as an independent object of ecological and geographical land evaluation. Accordingly, the factors involved in the evaluation, for all administrative units included in the study region should be united, but their qualitative scale is individual. A briefly description of each of the blocks method used is given in the following:

The first block includes the collection and study of scientific research, literature references, statistical and cartographic materials that characterize the region under study, and especially its development.

The list of indicators is determined while realizing the second block on the base of analysis of collected materials. These indicators characterize nature-agricultural potential of the research region. Then indicators are chosen which will be participated in the assessment. After that "tree of properties" is constructed with usage of choosing models.

The third block is advisable to use expert methods of pare comparisons for determination of irregular coefficients of coefficient significance which are included in "the tree of properties".

This choice was caused by simplicity of expertize holding and good results of expert analysis.

Scientific and methodological providing of expert survey includes development of formalized questionnaires and guidance on how to fill them.

One of the most important questions, which are considered on this stage, is question of expert number determination. This is due to the fact that big number of experts can raise cost of expertise but their little number can't provide to get objective expert data of analysis. The optimal number of experts is 8-12 persons in dependence of expertise difficulty and from experience of providing.

While realizing this method pares of alternatives are presented to experts, in which is suggested to choose more preferred consideration [15]. As a result-matrix of pare comparisons is formed. Then components of it eigenvector are determined on lines with usage of dependences. After that irregular vectors of priority are considered with usage of dependences.

$$\sqrt[4]{(W_1/W_1)\times(W_1/W_2)\times(W_1/W_3)\times(W_1/W_4)}=a \qquad (1)$$

$$\sqrt[4]{(W_2/W_1)\times(W_2/W_2)\times(W_2/W_3)\times(W_4/W_4)}=b \qquad (2)$$

$$\sqrt[4]{(W_3/W_1)\times(W_3/W_2)\times(W_3/W_3)\times(W_4/W_4)}=c \qquad (3)$$

$$\sqrt[4]{(W_4/W_1)\times(W_4/W_2)\times(W_4/W_3)\times(W_4/W_4)}=d \qquad (4)$$

$$S=a+b+c+d \qquad (5)$$

$$P_1=a/S; \qquad (6)$$

$$P_2=b/S; \qquad (7)$$

After calculation of irregular priority vectors the consistency of expert's opinions is considered by following dependences:

$$F_1=(W_1/W_1)+(W_2+W_1)+(W_3/W_1)+(W_4/W_1), \qquad (8)$$

$$F_2=(W_1/W_2)+(W_2+W_2)+(W_3/W_2)+(W_4/W_2), \qquad (9)$$

$$F_3=(W_1/W_3)+(W_2+W_3)+(W_3/W_3)+(W_4/W_3), \qquad (10)$$

$$F_4=(W_1/W_4)+(W_2+W_4)+(W_3/W_4)+(W_4/W_4), \qquad (11)$$

$$\lambda_{max}=F_1\times P_1+F_2\times P_2+F_3\times P_3+F_4\times P_4, \qquad (12)$$

$$IC=(\lambda_{max}-n)/(n-1), \qquad (13)$$

$$RC=IC/Z, \qquad (14)$$

Where IC is the index of the consistency of expert's opinions; OC is the relatively consistency of expert's opinions; Z is the random consistency; n is the number of compared indicators. (Figure 1)

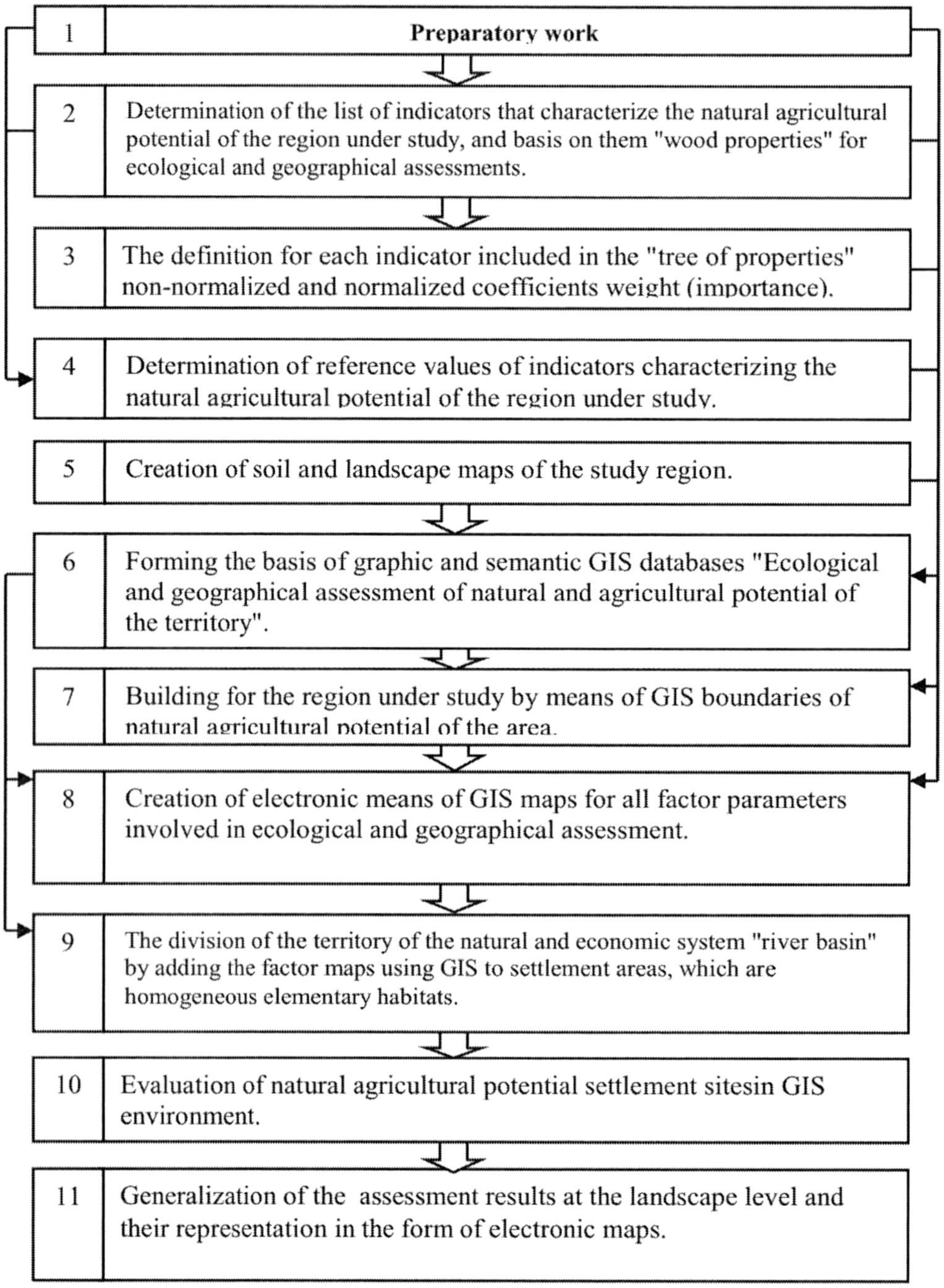

Fig. 1. Block diagram of the methodology of ecological and geographic assessments of natural and agricultural potential of the area.

The relative consistency of expert's opinions is considered acceptable if it is in the range from 0 to 10%. Otherwise, the experts need to revise their judgments.The normalized coefficients of the importance of each indicator which participate in the evaluation are calculated according to the following dependence:

$$P_j = \overline{P}_{j1} \times \ldots \times \overline{P}_{jn} \tag{15}$$

Where P_j is the normalized importance of the j-th elementary (quasielementary) properties; $\overline{P}_{j1}$, $\overline{P}_{jn}$ is average unstandardized coefficients of importance of the i-th and n-th level of the tree properties of hierarchically related within the same branch with the j-th elementary (quasielementary); n is the number of levels in "the tree of properties".

According to the dependences of the normalized weight (importance) elementary (quasielementary) properties, P_j is calculated by multiplying the average unstandardized coefficients of importance of the individual properties $\overline{P}_{j1} \ldots P_{jn}$ hierarchically related to each other in "the tree of properties".

The fourth block. While implementing of this unit there are defines the standardized, i.e. the best for each administrative region, values of indicators of natural and agricultural potential of landscapes. For this purpose stock, regulatory materials, data of field surveys and aerospace information are used.

The fifth block. Landscape and soil maps for research region are developed by typical methods in scale of base material [16, 17-19].While the preparation of landscape maps objects of mapping are natural-territorial complexes in the rank of the individual tracts or typologically similar groups [20-23].Using and development of soil maps are impossible without pervasive soil researches, especially when the case is connected to the soil moisture [24-28].

The sixth block. It includes the creation of digital models of basic cartographic principles, soil and landscape maps in the research region, their entry in the created GIS and design on the adopted conditional signs. In parallel to this there is performed the semantic content of the database information which is necessary for the production of eco-geographical assessment of natural and agricultural potential [29].

The seventh block. The construction of the boundaries of the natural-economic systems called "river basin" within the limits of the studied region is carried out in a GIS environment on a digital landscape model of basic cartographic foundation and automated with the help of a special software.

The eighth block. The graphical display of spatially referenced information about the level of demonstration in limits of natural and commercial system "river basin" of researching indicator is mean as a factor map. Factor maps are prepared in GIS with usage of standard or special developed software [30].

The ninth block. There is the summation of the factor maps in the process of implementation of this unit using the "overlay" method. Ultimately, the study area is divided into design areas which are represented by homogeneous elementary areas within of which each of the analyzed indicator has only one value.

The tenth block. The assessment of natural and resource potential of designed areas is carried out in the following sequence.

1) For all elementary and quasielementary properties within "the tree of properties" the absolute values are determined. In that case, if the indicator is amenable to the methods of physical measurements, its values are expressed in any physical units, and otherwise in points.

2) For all elementary and quasielementary properties within "the tree of properties" with usage of the relations (16) and (17) the relative values of their indicators are determined as follows:

$$R_{pj} = W_{pj} / W_{jэ} \text{ while } W_{pj} < W_{jэ}, \tag{16}$$

$$R_{pj} = W_{js} / W_{pj} \text{ while } W_{pj} > W_{jэ} \tag{17}$$

Where R_{pj} is the relative value of indicator characterizing the j-th elementary (quasielementary) property of the p^{th}design area, R_{pj} is changed in limits from 0 till 1 ($0<R_{pj}<1$); W_{pj} is the absolutevalue of indicator characterizing the jth elementary (quasielementary) property of the p^{th}designarea; W_{je} is the standard absolutevalue of indicator characterizing the jth elementary (quasielementary) property.

3) The value of the composite indicator of natural and agricultural potential design area is determined using the following relationship:

$$, \quad K_p^0 = \sum_{j=1}^{n} R_{pj} \times P_j \tag{18}$$

where K_p^0 is the composite indicator describing the natural-resource potential of the pth design area.

The high natural resource potential has the current design area in which the indicator K_p^0 has the greatest value. The index "0" in the indicator K_p^0 means that the assessment is given in relation to the very low (zero) level in the hierarchy of properties, i.e. in relation to the quality of natural resources. If the objects are compared, in comparison variants the best quality has the object which indicator K_p^0 will have the biggest difference. The ratio of quality assessments will accurately reflect the reality only in the case when the assessment is made.

The eleventh block. Natural and resource potential of the landscape is a weighted value that characterizes the potentials of design areas located within it:

$$L_m^0 = \sum_{d=1}^{n} K_d^0 \times Q_d, \tag{19}$$

where L_m^0 is natural and resource potential of the mth landscape; Q_d is share in the landscape of design areas with d^{th} natural resource potential; K_d^0 is score characterizing the dth natural resource potential of design areas.

The twelfth block of methodology is intended to represent the results of natural and agricultural [31, 32] potential assessment of the territory in the form of a series of electronic cards, comprising a set of factor maps and the final map describing the distribution of values of natural and agricultural potential within the study region. Agricultural potential calculation is tightly connected to the soil properties researches and experiments [20-22, 26-28].

4 Conclusions

On the base of conducted analysisis we can make further conclusions. The assessment of natural bodies and phenomena is a form of reflection of the interaction between society and nature. Direct assessments establish the importance of the object in the set of properties that define it. Indirect assessments establish the importance of the object on functional indicators and correlations between measured properties. The manifestations of assessment systemic can be divided into private, comprehensive, and multi-criteria assessments. Multi-criteria assessment of land includes the determination of the usefulness of its properties for the production of non-economic and economic ways.

There are two approaches to the assessment of land resources: ecological and geographical and cadastral. Geo-ecological assessment is parametric spatial analysis of capabilities of natural and anthropogenicallymodified geosystems at the various types of anthropogenic impacts steadily to perform life functions. State cadastral assessment of land resources is a complex of legal, administrative and technical measuresaimed at establishing the cadastral value of land as at a certain date [33].There is a method of geo-ecological assessment of natural resource potential within the North-Western region of Russia, which consists of 12 blocks. This methodology can be used to assess the natural and resource potential of the territory for the calculation of the cadastral cost of the land when planning investment development of territories [32-36].

The research was supported by DAAD (A/10/01103), DFG (MI 526/3-1) and Russian Foundation for Basic Research (#16-04-01473-a).

References

1. V.V. Dmitriev, *Diagnostics and modeling of aquatic ecosystems*, Saint Petersburg: Publisher SPbSU (1995)
2. V.V. Dmitriev, Vestnik SPbSU, **14**, 60-70 (1994)
3. N. Arefiev. M. Mikhalev, D. Zotov, K. Zotov, N. Vatin, O. Nikonova, O. Skvortsova, S. Pavlov, T. Chashina, T. Kuchurina, V. Terleev, V. Badenko, Y. Volkova, V. Salikov, K. Strelets, M. Petrochenko, A. Rechinsky, Procedia Engineering, **117**, 32-38 (2015)
4. N.F. Reymers, *Natural resources: dictionary* (Misl', Moscow,1990)
5. A.G. Isachenko, Izv. VGO, **6**, 97-101 (1991)
6. A.G. Isachenko, Izv. VGO, **4**, 45-54 (1991)
7. U.P. Odum, *Fundamentals of ecology* (Mir, Moscow, 1975)
8. A.G. Isachenko, *Ecological geography of the North-West of Russia* (Russian geographic society, Saint Petersburg, 1995)
9. V.B. Sochava, *Geography and ecology* (Geographical society of USSR, Novosibirsk, 1970).
10. V.S. Preobrazhensky, New conceptions in geography and forecasting, 8-21 (1993)
11. N. Arefiev, V. Garmanov, V. Bogdanov, Y. Ryabov, V. Terleev, V. Badenko, Procedia Engineering, **117**, 26-31 (2015)
12. S.I. Nosov, *State cadastral valuation of lands of agricultural purpose: practice manual* (Publisher house "Russkaya otsenka", Moscow, 2006)
13. *Resolution of the Government of the Russian Federation from 25.08 1999 No. 945 "On the State cadastral valuation of lands"* (Moscow, 1999)
14. N.V. Arefiev, *Management of natural and manmade systems: introduction to ecoinformatics* (Publisher of SPbSPU, Saint Petersburg, 2000)
15. N. Arefiev, V. Terleev, V. Badenko, Procedia Engineering, **117**, 39-44 (2015)
16. A.G. Isachenko, *Methods of field of landscape research and landscape-ecological mapping* (Publisher of SPbSU, Saint Petersburg, 1999)
17. N.S. Chochia, *Summer field practice of landscape science* (Publisher Leningrad State University, Saint Petersburg, 1969)
18. V.M. Chupahin, *The basics of landscape learning* (Agropromizdat, Moscow, 1987)
19. M.G. Ryzhakova, V.I. Maslikov, A.N. Chusov, V.V. Korablev, Applied Mechanics and Materials, **675-677**, 761-769 (2014)
20. V. Badenko, V. Terleev, A. Topaj, Applied Mechanics and Materials, **635-637**, 1688-1691 (2014)

21. R.A. Poluektov, S.M. Fintushal, I.V. Oparina, D.V. Shatskikh, V.V. Terleev, E.T. Zakharova, Archives of Agronomy and Soil Science, **48**, 609-635 (2002)
22. R.A. Poluektov, V.V. Terleev, *Proceedings of the Workshop, Leibniz Centre of Agricultural Landscape Research (ZALF) in Müncheberg/Germany*, 75-89 (2007)
23. R.A. Usmanov, N.I. Vatin, V.A. Murgul, Applied Mechanics and Materials, **680**, 474-480 (2014)
24. V. Badenko, V. Terleev, N. Arefiev, J. Volkova, O. Nikonova, *Proceedings of the AASRI International Conference on Industrial Electronics and Applications (IEA 2015). Book Series: AER-Advances in Engineering Research*, **2,** 452-455 (2015)
25. R.A. Poluektov, V.V. Terleev, Russian Meteorology and Hydrology, **11**, 70-75 (2002)
26. R.A. Poluektov, V.V. Terleev, Russian Meteorology and Hydrology, **12**, 73-77 (2005)
27. V.V. Terleev, W. Mirschel, U. Schindler, K.-O. Wenkel, International Agrophysics, **24**, 381-387 (2010)
28. R.A. Poluektov, I.V. Oparina, V.V. Terleev, Russian Meteorology and Hydrology, **11**, 61-67 (2003)
29. A. Nikonorov, S. Pavlov, V. Terleev, N. Arefiev, V. Badenko, Y. Volkova, Procedia Engineering, **117**, 258-263 (2015)
30. N. Arefiev, V. Badenko, A. Nikonorov, V. Terleev, Y. Volkova, Procedia Engineering **117**, 20-25 (2015)
31. S. Medvedev, A. Topaj, V. Badenko, V. Terleev, IFIP Advances in Information and Communication Technology, **448**, 252-261 (2015)
32. V.V. Terleev, A.G. Topaj, W. Mirschel, Russian Meteorology and Hydrology, **40**, 278-285 (2015)
33. V. Chechevichkin, N. Vatin, Applied Mechanics and Materials, **641-642**, 409-415 (2014)
34. A. Makarov, A. Mihailova, N. Arefiev, S. Pavlov, T. Chashchina, V. Terleev, V. Badenko, Procedia Engineering, **117**, 225-231 (2015)
35. O. Skvortsova, A. Dashkina, E. Petrovskaia, V. Terleev, A. Nikonorov, V. Badenko, Y. Volkova, S. Pavlov, MATEC Web of Conferences, **53**, 7 (2016)
36. V. Terleev, E. Petrovskaia, N. Sokolova, A. Dashkina, I. Guseva, V. Badenko, Y. Volkova, O. Skvortsova, O. Nikonova, S. Pavlov, A. Nikonorov, V. Garmanov, W. Mirschel, MATEC Web of Conferences, **53**, 6 (2016)

Methods of Improving Water Treatment Systems for Individual Residential Houses

Sergey Pavlov[1], *Alexander* Novikov [1,*], *Alexander* Pavlov[1], *Olga* Skvortsova[1], *Olga* Nikonova[1], *Ekaterina* Semanina[1], *Ruslan* Zafarov [1] and *Karl-Otto* Wenkel[2]

[1]Peter the Great St. Petersburg Polytechnic University, 195251 Polytechnicheskaya str. 29, St. Petersburg, Russia
[2]Leibniz-Centre for Agricultural Landscape Research, EberswalderStraße 84, Müncheberg, 15374, Germany

Abstract. One of hot topics for ecological management is sewage treatment today in places where there is no sewerage. The volume of country construction in territories, which are not connected to the public sewage system increasing nowadays. Therefore, problem of wastewater treatment take place. Currently, there are a lot of different designs of local waste treatment plants is offered to consumers. However, a large number of negative reviews indicate serious shortcomings in most of the local plants offered in the market. The purpose of this paper is the proposal of improvement of the most common local treatment plants in Russia.

1 Introduction

For the reason of treatment plants development and improvement, it seems necessary to solve the following problems:

1. Offer classification of local waste treatment plants because of their various operation and serious constructive differences.

2. If it is possible, identify the advantages and disadvantages of the local waste treatment plants. For this purpose, take one or several units of each group. Creators present the following disadvantages: a) lack of necessary or desirable elements to provide high-quality process of clearing; b) elements in the system which are almost not influence on the quality of clearing, make the price of the clearing system higher; c) lack of official document confirming the compliance of the treated wastewater accepted by sanitary standards in this region. For example, in the Russian Federation such standards are SanPiN 2.1.5.98-00 "Hygienic requirements for surface water protection" ; d) the absence of well-composed guide which describes the device and operation of the cleaning plant.

3. Analysis of advantages and disadvantages offers how to improve design and make the cleaning systems of this group cheaper.

* Corresponding author: coolhabit@yandex.ru

2 Method

Classification of treatment plants is based on standard wastewater treatment procedure. This scheme includes the following steps: clarification (gratings, grit chambers, presettling tanks), biological treatment (aerotanks or trickling filters also possible anaerobic treatment in septic tanks), final settling tank (the systems of moving of recurrent and surplus sludge in case of aerotanks, system of processing waste activated sludge or biofilm, disinfection of treated waste water, removal of treated waste water into the intake channel or in the soil [1, 2].

On the basis of data about the treatment plants available to the authors, we propose the following classification.

2.1 Group 1 - Plants, including the aeration tank and secondary settler

Installation represents aerotank only with a secondary settler; primary settler is missing. There are no trash-rack structure and disinfection of treated wastewater. Division into excess and return sludge is not presented. All activated sludge from the secondary settling tank is fed back into the aeration tank. If the concentration of waste activated sludge above the limit, plant stops and the most part of sludge is taken out by cesspool transport. The purification steps: 1) supply the waste water to the plant; 2) inflow of waste water in the aeration tank and the processing by activated sludge; 3) settling of activated sludge and contaminants in the secondary settling tank which have passed through the aeration without changes; 4) export the purified organic and supernatant water by gravity into the ground.

There is an example of the installation related to this group "Independent Sewerage SANI"(Fig. 1).

This plant represents concentric volumes. The inner volume represents the conical chamber with a hole at the bottom, filled with the elements covered by micro-organisms that consume organic matter dissolved in the waste water. Air moves into this capacity. External volume serves (according to the description) as a settler where a sedimentation of the fractions which are not exposed to biological cleaning take place. This plant represents the simplified version of the biofilter by the principle of action [3, 4].

The efficiency of the plant is questionable. Waste water on loading with a biofilm has to be pumped by a drop irrigation, but not a continuous stream in the classical version of the biofilter which provides a 95-percent cleaning of organic chemistry. Primary settler is required for biofilter. The pores between the loading elements quickly clogged with large fractions of waste which is contained in the waste water without primary settler. The settler does not be able to release the waste water from the slurry. The stream moving the fractions intended to precipitation (they have to move down) directed upwards to an interceptor [5]. The code certificate «736 The main center of state sanitary and epidemiological supervision of the Russian Federation Ministry of Defense» of the waste water treatment plant is questionable.

It is specified that biological sorption of oxygen after cleaning decreases from 300 mg of O_2/l to 2 mg of O_2/l, and concentration of the weighed substances decreases from 300 mg/l to 3 mg/l. How it occurs without of primary settler and with so small volume of a secondary settler is unclear.

SanPiN 2.1.5.98-00 "Hygienic requirements to protection of a surface water" is not used as a hygienic requirement. It is substituted by the unknown document according to which purified waste water shouldn't include any suspended solids at all that is impossible [6]. There is no information about bacterial pollution of the purified waste water. Also there is a question about installation design. In the technical passport of the device there is no description of the process of intake of waste water from aeration chamber in a secondary settler.

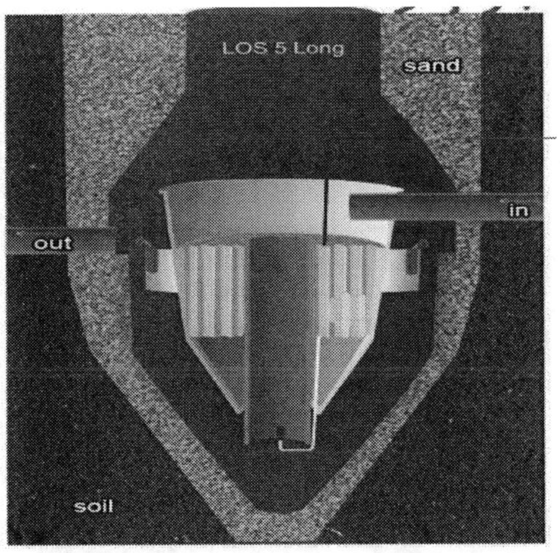

Fig. 1. Independent Sewerage SANI.

2.2 Group 2 - The installation consisting of a uniform chamber

The plant consists of only the aeration tank, which operates periodically, and in breaks between stages of processing waste by activated sludge, it performs the functions of the secondary settler (Fig. 2). There are no grids and disinfection is absent. These systems does not have division into excess and returnable sludge because the plant represent a single chamber and the entire sludge remains in it. If the concentration of sludge above the limit, the plant stops. A large part of the sludge is transported by Cesspool transport.

The purification steps: 1) supply of waste water to the plant; 2) faltering aeration during10 hours; 3) stage of settling during 2 hours; 4) export the purified organic and supernatant water.

The example of 2nd group is the ONE2CLEAN plant (Fig. 2). Setting represents the uniform chamber divided by a small partition. In this plant the aerator works, oxygenating water and activating aerobic process of purification of waste water with active sludge during 10 hours. After that aeration stops and process of settling during 2 hours begins. At this time active sludge accumulates on an installation bottom, and the purified waste water is removed in the soil or to the district. According to the manufacturer, sewage can inflow in the plant during of all process. The efficiency of this plant is highly questionable. According to the manufacturer, wastewater activated by acid for elimination of a smell.

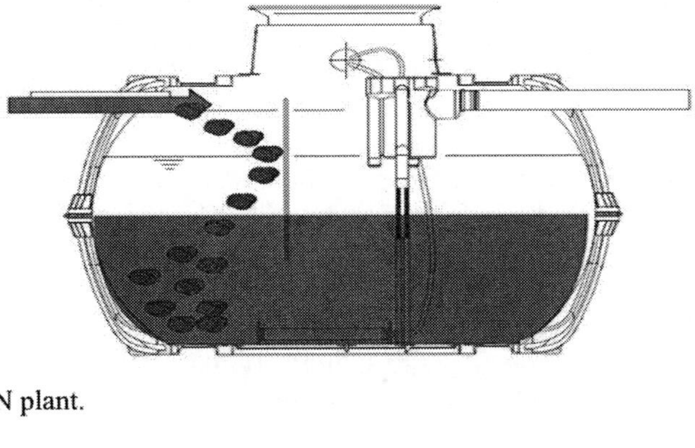

Fig. 2. ONE2CLEAN plant.

It completely contravenes the basic principle of sewage treatment. Acid inhibits microorganisms, and as a result, sewage treatment is not performed [7]. The fact that waste water may enter the plant during settling process is doubtful. Purified water will be mixed with waste water because the chamber presents uniform capacity. The waste water received as a result of mixing can't be removed from operation [8]. The Finnish Biosetti plant also belongs to the same group (Fig. 3). The difference is only that the general amount of a chamber is divided by an impenetrable partition, and waste water is pumped from the first chamber in the second. In this plant process of purification of waste water is identical to process in the ONE2CLEAN plant. Although the movement of water and sediment is carried out at least by 3 pumps.

Fig. 3. Biosetti plant.

2.3 Group 3 - Installation with the most complete set of cleaning elements

It includes: a primary settling tank, aeration tank, secondary sedimentation tank, disinfection chamber. There are no trash screen. The camera acts as a primary clarifier and settling tank, and a septic tank. When the concentration of activated sludge in the aeration tank reaches a limit value 10.000 mg/L [9], the bulk of the activated sludge is removed. In the absence of flow of sewage in the aeration tank, the biomass contained in the sludge, according to the manufacturer, is subjected to aerobic mineralization, substantially reducing in volume. The purification steps: 1) inflow of wastewater to be treated in a primary settling tank; 2) the passage of wastewater through aeration; 3) after aeration stripped of organics but with more activated sludge water passes advocacy; 4) purified from organic and from activated sludge water passes disinfection by chlorination; 5) conclusion decontaminated water from the treatment plant to the soil or ditch.

There is "5 VOCs" is analyzed as an example (Fig. 3). Set is a complex of series-connected blocks, in which there is stepwise purification of waste water. First cleaning unit - primary sedimentation tank, which the role of the septic tank. In it there is a mechanical and microbiological treatment of wastewater. In the settler there is a smooth natural motion of the fluid from one chamber to another sediment through the overflow openings. Suspended solids contained in the waste water settle to the bottom and from a slang. It is subjected to slow the process of anaerobic fermentation, in which part of impurities is dissolved in water and the other at the bottom of the settler accumulates as insoluble minerals. Second cleaning unit - aeration chamber, a secondary settling tank and the disinfection chamber. In the aeration tank waste water using submersible aerator is saturated with oxygen. Simultaneously, there is intense mixing of aerated water. During aeration the activated sludge consumes organic compounds. After the aeration tank the water enters the secondary settling tank where activated sludge is precipitated and purified by allocating organics and water from the active sludge. Activated sludge from the secondary clarifier and primary sludge mixed with anaerobic sludge must be removed periodically. Unfortunately, the frequency of cleaning in the instructions to the VOC is not specified, which may create difficulties in the operation. The last stage of purification is disinfection of treated wastewater. Disinfection takes place at the contact of water with the tablet slow release of chlorine.

After disinfection the treated wastewater comes in ready-dispensing well, where they are displayed in the ground, if possible, or discharged off-site. Installing such plants, it should pay attention to temperature control area. The manufacturer points out that for the effective operation of the primary sump (used as a septic tank) needs to make its insulation. It is also necessary to make pipe insulation; the temperature in the waste water was between 15-30 C^0. Datasheet waste water treatment plant VOC 5 up with great care. In it painted in great detail the waste water treatment process. Each element of the installation is painted in detail, and how it is involved in the cleaning process. Also, manufacturers are given very detailed instructions on the installation and operation of the facility, as well as instructions in case of malfunction. Certificate of compliance for treated waste water sanitary norms and rules beyond doubt.

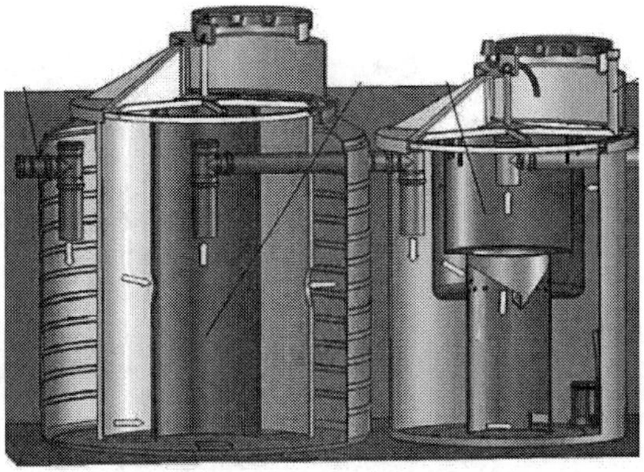

Fig. 4. LOS 5 plant.

"YUNILOS Mega M-30" refers to the same group of installations (Fig. 5). YUNILOS "MEGA-M-30" plant consists of a series-connected units: 1) block the mechanical purification of waste water; 2) the unit of biological treatment in aeration tanks and the bioreactor, with the participation of attached and free-floating microorganisms; 3) block additional cleaning and disinfection; 4) block sludge dewatering. First, the waste water flows into the sewage pumping station, after that they enter the mechanical grates, where the wastewater treatment of the coarse inclusions of size greater than 1-2 mm. Apparently (this is in the "Description" is not mentioned), lattice are equipped with mechanical rakes to remove the dirt from the rods. Then drain enters the biological treatment unit. In this unit, it first enters the homogenization tank where they are mixing and the averaging effluent occurs then it enters the denitrifies wherein biological coagulation occurs.

On this occasion, you will notice that there is a process of bio-coagulation under aeration of wastewater (without oxygen activated sludge is not working [10]), while denitrification is a process of anaerobic [11, 12]. As a single device can be carried out two different nature of the process - is unclear.

Then, the mixture of sewage and activated sludge flows into a biological purification unit, which in turn provides sequential process deeper mineralization of organic substances. In the first section by maintaining anoxic oxygen regime and high loads on the mud to develop a culture of microorganisms, as a result of vital activity which occurs enzymatic hydrolysis of organic matter [13]. In fact, this element is a septic tank [14].

Next brightened waste water enters the aeration tank, in which the biochemical oxidation of organic substances of activated sludge microorganisms using dissolved oxygen. Aeration tank includes an anaerobic denitrification zone (in the diagram in the "Description" in the anaerobic denitrification zone somehow fed air) and the aerobic nitrification zone.

Aerobic nitrification zone allows the oxidation of organics. The mixture of purified water and activated sludge flows into the aeration tank after the secondary clarifier. Il settles in a secondary settling tank; part of it through the air lift system is given, as stated in the "Description" in the septic chamber primary settling tank.

Primary settler is not mentioned in the text above and in Scheme absent. Apparently it meant "block of deep mineralization" (septic tank). Another part of the activated sludge is returned to the aeration tank. Secondary clarifier effluent from the bioreactor fall with floating on the surface of which is formed and fixed biofilm microorganisms contributing to the ultimate oxidation of the organic contaminants. The effectiveness of this element is questionable, as in classical biofilters for quality treatment of wastewater from organic drip irrigation elements required to boot from a biofilm and intensive air blowing [15].

Fig. 5. YUNILOS Mega M-30 plant.

After bioreactor clarified juice enters the tertiary settling tank for the final settling and going on to post-treatment advanced treatment and disinfection unit. To do this, apply a filter and sorption floating load. Purified water is supplied for the installation of ultraviolet disinfection and diverted to the intake channel. In biological purification coagulant is added to the sump for removing phosphate solution.

The excess activated sludge system is going to collect and remove sludge and is removed in the mineralizer provided for the collection and stabilization of primary sludge and secondary settling tanks in the anaerobic mode.

After mineralizer pellet fed to the dehydration. As part of the documents attached to the YUNILOS, there is a certificate confirming the purification of waste water to sanitary norms and rules standards. However, the document states that provided for the analysis of the original waste water is quite clear: it is the water BOD5 = 21.1 mg/l while the standard wastewater BOD5 has close to 200 mg/l.

2.4 Group 4 - Installation with the anaerobic principle of cleaning

Treatment plants of the Group are conducted mainly anaerobic treatment of waste water, which is followed by settling. After the anaerobic treatment of waste water can be supplied to the primitive camera aerobic treatment.
The purification steps: 1) the inflow of wastewater to be treated into the first chamber, where it takes place the anaerobic treatment; 2) bypass water in the well, which, under certain conditions, can work as a filter well; 3) the challenge of waste water from the well by gravity or pump. As an example, analyzes the Finnish plant (septic tank) Sakosetti (Fig. 6).

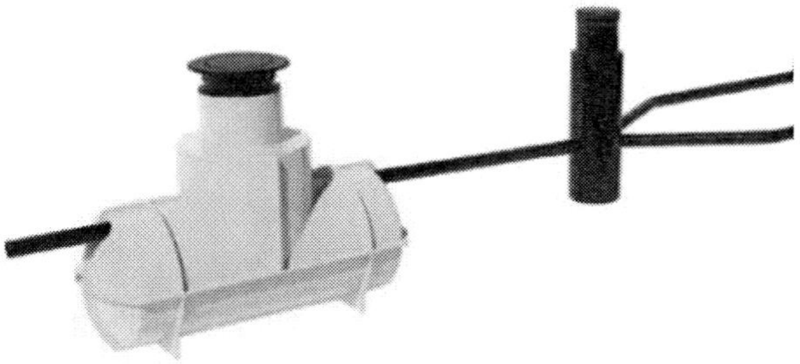

Fig. 6. Sakosetti plant.

Installation is a septic tank, connected in series with a distribution well, from which treated waste water is discharged to the filter box or off-site. The first treatment phase is deposition coarse-particulate matter, which passes inside the settling tank. From the start operation the septic tank is filled with clean water, according to the authors "Description", which allows reduce effluent concentrations which fall into it. Technologically septic tank is made so that it passes through the air.

According to the manufacturer, this is as well as reduced effluent concentrations, made for a more active involvement of the bacteria in the process of cleaning and more qualitative origin of the fermentation process. As anaerobic treatment and fermentation takes place in the septic tank of oxygen access - it is unclear [16]. The second phase of the clearing - conditionally clean water enters the septic tank after the distribution well, after that it enters the spray tube and is absorbed in a layer of crushed stone.

In it, in the oxygen-free environment it is a final post-treatment wastewater. Authors and experts in the field of wastewater treatment believe that in the soil occurs aerobic post-treatment [17], all the more that, according to the scheme "Descriptions", perforated pipe filter fields are equipped with aerators [25-28].

3 Results and discussion

Table 1 below systematized proposals to improve the treatment plants of different groups. Improvements are: a) addition elements that will improve the quality of treatment; b) renouncement of devices available to install, but does not improve the cleaning; c) constructive changes of existing elements in the installation.

Table 1. Recommendations for improvement of sewage treatment plants

	Installation example	Recommendations for improvement
Group 1	Autonomous sewerage Sani	Primary settler apparatus, increase in the secondary clarifier, the disinfection device camera.
Group 2	Installation ONE2CLEAN, Biosetti	Separation of the aeration tank and secondary sedimentation chamber into two successive element, the addition chamber primary settling tank, disinfection chamber device.
Group 3	Installation LOS 5 YUNILOS Mega-M 30	In the apparatus the LOS 5 to use primary settling tank and an anaerobic treatment in the chamber. For this purpose it is necessary to insulate and seal the hatches, the waste water fed through the elements impeding the flow of air from the inlet pipe. Enlarge secondary settling tank and replace the chlorination system disinfection (chlorine adversely affects the chemical composition of the soil) to irradiation with ultraviolet light. Cut from the system YUNILOS Mega-M30 air supply to the denitrification chamber. purification systems in a bioreactor with floating loading replace a traditional biofilter with drip irrigation fixed crushed stone or plastic boot and forced ventilation.
Group 4	Installation Sakosetti	Sealing chamber sump and remove it from the air supply system to provide him with a septic tank functions. Do not dilute the wastewater with clean water, which will provide enough food for anaerobic bacteria and accelerate their development. To improve the quality of the treated effluent (septic removes not more than 70% organic matter [18]) can be successively put two units. If there is no filtering of treated waste water into the soil, the installation must always be complemented by the disinfection system

General recommendations relating to the settings of all the groups in the case of operation in the North-West region is their thorough insulation. As known, for the normal functioning of aerobic (and especially anaerobic) bacteria as required sufficiently high temperature [19, 20].

Treated wastewater must satisfy the requirements in order that the local clearing waste treatment system would be accepted by official organizations. Authors realize that it is difficult to achieve these stringent requirements for inhabitants of country houses. As stated above, authors offer how to improve plant operation. For this purpose, authors made the analysis of the local waste treatment plants sold in the market. Due to the large number of models of clearing systems for the analysis of all models will take a lot of time. In addition, not all submitted a certificate confirming compliance with the declared and valid characteristics of wastewater treatment.

Technical passport does not contain drawings and operation of water treatment system. Also authors, describing the functional efficiency of the plant in technical passport, confirm that the plant removes up to 98% of harmful substances, what is wrong! Therefore, it is offered to divide local clearing plants into groups. Each group of units have one operation and similar design features. According to mentioned above, this work has the following results: 1) For the first time suggested the division into groups according to structural features of local treatment facilities; 2) The scheme of construction of each group was analyzed, the shortcomings identified plants; 3) The measures to improve facilities were offered.

4 Conclusion

All of the sewage treatment plants are characterized by a number of design flaws that reduce the quality of treated wastewater. According to this, the researches of soil properties, which may be affected by this pollution, is also important [21-24, 29-33]. The documents attached to products, often have significant inaccuracies and errors in the descriptions of structures and processes that occur at different stages of water treatment. The proposed measures will allow authors without significant additional costs to improve the quality of waste water treatment, coming from a small village, or from an individual country house. Also received results will help to solve some problems of ecological management on purification of waste water and to improve a state of all ecological environment.

References

1. M. Henze, C.P. Leslie Grady Jr., W. Gujer, G.V.R. Marais, T. Matsuo, Water Research, **21**, 505-515 (1987)
2. C.D. Beal, E.A. Gardner, N.W. Menzies, Soil Researches, **43**, 781-802 (2005)
3. R. Tenno, R. Vilu, H. Oit, Automatica, **25**, 739-742 (1989)
4. J.V. Voronov, S.V. Yakovlev, *Vodootvedenie i ochistka stochnyh vod. Uchebnik dlya vyzov* (Publisher assoxtsiatsii building schools, Moscow, 2006)
5. S.V. Yakovlev, J.A. Karelin, *Kanalizaciya. Uchebnik dlya vyzov. Pyatoe izdanie* (Stroyizdat, Moscow, 1975)
6. M.B. Yates, Groundwater, **23**, 586-591 (1985)
7. M. Wagner, A. Loy, Current Opinion in Biotechnology, **13**, 218-227 (2002)
8. P.J.A. Withers, H.P. Jarvie, C. Stoate, Environmental International, **37**, 644-653 (2011)
9. M. Hummer, *Tehnologiya obrabotki prirodnyh i stochnyh vod* (Stroyizdat, Moscow, 1979)
10. G. Dobele, N. Bogdanovich, G. Telysheva, U. Viesturs, Applied Biochemistry and Biotechnology, **57**, 857-867 (1996)
11. Y.J. Chan, M.F. Chong, C.L. Law, D.G. Hassell, Chemical Engineering Journal, **155**, 1-18 (2009)
12. R.F. Hickey, W.-M. Wu, M.C. Veiga, R. Jones, Water Science & Technology, **24**, 207-255 (1991)
13. O. Legrini, E. Oliveros, A.M. Braun, Chemical Review, **93**, 671-698 (1993)
14. L.J. Arnade, Seasonal Correlation of Well Contamination and Septic Tank Distance, Groundwater, **37**, 920-923 (1999)
15. A.J. Balkema, H.A. Preisig, R. Otterpohl, F.J.D. Lambert, Urban Water, **4**, 153-161 (2002)
16. L.S. Philippi, R.H.R. da Costa, P.H. Sezerino, Water Science and Technology, **40**, 125-131 (1999)
17. G. Kassab, M. Halasheh, A. Klapwijk, M. Fayyad, J.B. van Lier, Bioresource Technology, **101**, 3299-3310 (2010)
18. J.V. Voronov, E.V. Alekseev, *Vodootvedenie* (INFRA-M, Moscow, 2012)
19. S. Luostarinen, W. Sanders, K. Kujawa-Roeleveld, G. Zeeman, Bioresource Technology, **98**, 980-986 (2007)
20. H.E. Muga, J.R. Mihelcic, Journal of Environmental Management, **88**, 437-447 (2008)
21. V. Badenko, V. Terleev, A. Topaj, Applied Mechanics and Materials, **635-637**, 1688-1691 (2014)
22. S. Medvedev, A. Topaj, V. Badenko, V. Terleev, IFIP Advances in Information and Communication Technology, **448**, 252-261 (2015)

23. V.V. Terleev, A.G. Topaj, W. Mirschel, Russian Meteorology and Hydrology, **40**, 278-285 (2015)
24. V.V. Terleev, V. Mirschel, U. Schindler, K.-O. Wenkel, International Agrophysics, **24**, 381-387 (2010)
25. V. Chechevichkin, N. Vatin, Applied Mechanics and Materials, **641- 642**, 409-415 (2014)
26. N. Arefiev, V. Garmanov, V. Bogdanov, Y. Ryabov, V. Terleev, V. Badenko, Procedia Engineering, **117**, 26-31 (2015)
27. A. Beljatynskij, O. Prentkovskis, J. Krivenko, Transport, **25**, 394-402 (2010)
28. V.V. Okrepilov, M.V. Leonidovich, Asian Social Science, **11**, 312-325 (2015)
29. V. Terleev, A. Nikonorov, V. Badenko, I. Guseva, Y. Volkova, O. Skvortsova, S. Pavlov, W. Mirschel, Advances in Civil Engineering, **2016**, 7 (2016)
30. O. Skvortsova, A. Dashkina, E. Petrovskaia, V. Terleev, A. Nikonorov, V. Badenko, Y. Volkova, S. Pavlov, MATEC Web of Conferences, **53**, 7 (2016)
31. V. Terleev, E. Petrovskaia, N. Sokolova, A. Dashkina, I. Guseva, V. Badenko, Y. Volkova, O. Skvortsova, O. Nikonova, S. Pavlov, A. Nikonorov, V. Garmanov, W. Mirschel, MATEC Web of Conferences, **53**, 6 (2016)
32. N. Arefiev, V. Badenko, A. Nikonorov, V. Terleev, Y. Volkova, Procedia Engineering, **117**, 20-25 (2015).,
33. A. Nikonorov, S. Pavlov, V. Terleev, N. Arefiev, V. Badenko, Y. Volkova, Procedia Engineering, **117**, 258-263 (2015)

Performance based evaluation of rainwater harvesting system in public buildings

Ksenia Strelets[1], *Petr* Ovchinnikov[1*] and *Temur* Dzampaev[1]

[1] Peter the Great St.Petersburg Polytechnical University, 195251 Polytecknycheskaya str. 29, St. Petersburg, Russia

Abstract. Harvested rainwater has become an inseparable part of sustainable development a long before term 'sustainability' appeared. Even though domestic systems have been investigated to some extent, there are few recent studies of larger non-domestic systems used in public and commercial buildings, especially in Russia. This paper aims to describe a combined performance evaluation of a configured rainwater harvesting system for a specific study case of public building located in St. Petersburg. The careful analyses of empirical and theoretical data set was applied while calculating water demand, optimal storage sizing and subsequent economical juxtaposition of scenarios. The results provide useful information for optimizing the economic response of every specific scenario in terms of payback period, initial costs and end use.

1 Introduction

The total usable freshwater supply for ecosystems and humans, according to the United Nations water organization, UN-Water [1], is only about 200 000 km^3 of water – less than one percent (<1%) of all freshwater resources. And, water use has been growing at more than twice the rate of the population increase in the last century. Based on the map published by the Consultative Group on International Agricultural Research (CGIAR), the countries and regions suffering most water stress are North Africa, the Middle East, India, Central Asia, China, Chile, Colombia, South Africa and Australia. Water scarcity is also increasing in South Asia. With this, good management of water resources and corresponding water infrastructure can jointly manage human water security [2].

The domestic use of freshwater accounts for approximately 10% of the total global freshwater consumption [3]. The use of rainwater harvesting systems (RHS) and rainwater tanks specifically is an old practice, which corresponds to these environmental and social issues. RHS stores rainwater from impervious catchments, usually a rooftop for non-potable use. Water is typically used for outdoor irrigation or indoor non-potable uses like flushing toilets. Components usually include a roof catchment, a filter to remove the initial fraction of roof runoff, a storage tank (also known as a cistern), and a pump to supply system demand.

[*] Corresponding author: pshenichca@hotmail.com

Currently, RHS are widely used for non-potable applications in countries such as Australia, South Africa, the USA, Italia and Spain, where funding is provided for the construction of rainwater harvesting systems [4], or where significant water scarcity is explicitly manifested. A thorough inspection of RHS studies also demonstrates that, although historically used in areas where the water supply was limited by climate or infrastructure, the practice of rainwater harvesting has recently been used in humid and well developed regions, namely in response to the increased interest in green building practices which support the smart use of water [5,6]. This is concerned of such countries as England, Germany, France and some others.

With this, Russian Federation has never been regarded as a part of a European vanguard in the matter of sustainability and St. Petersburg, being an article's case study, has never been lack of fresh potable water. But, giving to world's increasing demand, the use of RHS in St. Petersburg area seems to us as a very provident move.

Common practices show that performance of RHS model is predominantly influenced by the following:

1. Identification of rainfall trends in location. It can be analyzed by taking data from the last years and calculated as mean [7] or data horizon can be broaden and calculated as a trend [8]. Otherwise, more representational rainfall years can be taken for further analyses;
2. Roof area, tank sizing, piping, co-related dimensions and end uses scenarios affect financial feasibility of a system [9];
3. The selection of time scale for modeling RHS is important. Although time step selection often depends on the objective of the analysis and on the availability of data, a sensitivity analysis conducted by Mitchell [10] has shown that it may influence the estimation of the tank volumetric reliability (up to about 8%) depending on the algorithm used to run the water balance simulation. A variety of time steps have been used for RHS, ranging from 5 min [11] to daily [10] and monthly. Early results by Fewkes and Butler [12] point out that simulations with monthly time steps may provide inaccurate evaluation of the RHS system water saving performance and suggest to use the daily time step resolution for such an evaluation.

2 Materials and Methods

2.1 Case Study

The RHS system was analyzed for Hydrocampus – 2 of Peter the Great St.Petersburg Polytechnic University located in St. Petersburg, Russian Federation. This is a public building which was built in 70's and currently passes the stage of a major overhaul. The 3D model of the existing building is presented on Fig. 1.

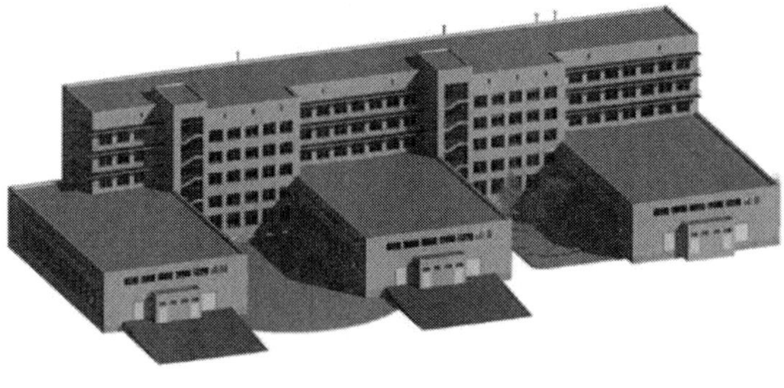

Fig. 1. Hydrocampus – 2.

Hydrocampus – 2 was selected in this study because it is representative of other higher education campuses. It is a five-story building with a total area of 8559 m^2, roof area of 4050 m^2, and lawn area of 3992 m^2. It has 3 embedded lab buildings and 10 lavatory rooms. All the area-volume data set is presented in the Table 1.

Table 1. Area data.

№	Parameter	Value
1	Overall height of the building	23.40 m
2	Total floor area	8 559.00 m^2
3	Total volume	41 609.00 m^3
4	Total lawn area	3992.00 m^2
5	Parking area	3780.00 m^2
6	Total collecting area of the roof	4049.62 m^2

To provide thorough analyses, the RHS was modelled for this building (Figure 2). Inspection of the object revealed that existing rainfall drainage system (marked in green) is in a sufficiently good state so to reduce initial costs we decided to leave it as it is. Storage tank was decided to place in the middle laboratory building. This building formerly included a pool for hydraulic tests and now this space is unoccupied.

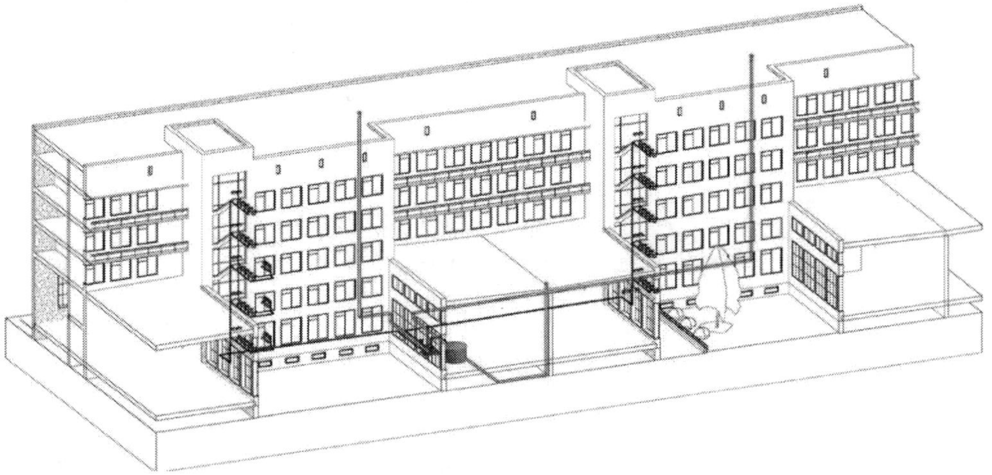

Fig. 2. RHS model.

For the hydraulic head reaches the height of 18 meters it is impossible to omit hydraulic pumps – 2 of them were placed in the middle laboratory building along with storage tank. Water supply pipes (marked in blue) provides lavatory rooms on every floor with non-potable water for flushing. In every lavatory room, a small water storage has to be placed in order to ensure continuous water flow.

2.2 Rainfall Data for St. Petersburg

The rainfall data was obtained from the Hydrometeorological Research Centre of Russian Federation (Hydrometcentre of Russia). Average monthly and annual rainfall data set in the considered period (1970–2005) was collected; however due to some missing data the years of 1999 and 1986 were not included in the analysis. Thus, the studied overall record period was 33 years (Table 2).

Table 2. Average monthly rainfalls.

Month	Average rainfall, [mm]	Minimum rainfall, [mm]	Maximum rainfall, [mm]	Daily maximum, [mm]
Jan	44	2 (1838)	82 (2011)	23 (1955)
Feb	33	3 (1886)	92 (1990)	23 (19900
Mar	37	0.7 (1923)	90 (1763)	26 (1971)
Apr	31	2 (1850)	99 (1764)	29 (1991)
May	46	2 (1978)	127 (2003)	56 (1916)
Jun	71	8 (1889)	199 (1742)	44 (2004)
Jul	79	5 (1919)	166 (1979)	69 (2002)
Aug	83	1 (1955)	197 (1869)	76 (1947)
Sep	64	2 (1851)	190 (1767)	34 (1912)
Oct	68	5 (1987)	150 (1984)	37 (2003)
Nov	55	2 (1993)	118 (2010)	31 (2010)
Dec	51	4 (1852)	112 (1981)	28 (2009)
Year	661	395 (1882)	861 (2012)	76 (1947)

2.3 Occupants

Using the design occupancy of the building in estimating demand lead to an over-sizing of the storage tank, when the actual occupancy of the building, as a rule, ends up being significantly lower [6]. In order to assess sufficiently precise demand in water, the number of students and employees for the building was asked to the institute's management team. This number was carried out through study plan and divided among weekdays and time zones (Table 3).

Table 3. Estimated occupants.

Semester:	**Winter semester**					
Day	Monday	Tuesday	Wednesday	Thursday	Friday	Saturday
Visitors	1265	1099	1205	1204	1175	1085
Semester:	**Summer semester**					
Day	Monday	Tuesday	Wednesday	Thursday	Friday	Saturday
Visitors	1102	936	1042	1041	1012	922

These figures were reduced for the summer semester by the number of graduating master students for it they do not visit institute during that semester on a daily basis.

2.4 Specific Water Demand

Specific end use of non-potable water was divided into 3 parts. Frequency of toilet and urinals flushing was assessed according to data derived from Heaton and Radvan [13]. Day consumption can be calculated as:

$$C_{tf}\left(\frac{m^3}{day}\right)=\frac{\sum n^i F^i C_u^i \cdot pop}{1000}, \quad (1)$$

where
n^i – estimated number of flushes per unitary;
F^i – estimated percentage of people using toilet/urinal flush, [%];
C_u^i – unitary consumption of the toilet/urinal flush, [liters];
Pop – number of visitors.
Floor/pavement washing is calculated as:

$$C_{pw}\left(\frac{m^3}{day}\right)=\frac{A \cdot C_u \cdot NW}{1000}, \quad (2)$$

where
A – floor/parking area, [m^2];
C_u – unit consumption for washing floors, as recommended by the Technical Specification ANQIP (ETA 0701) [14], [liters/m^2];
NW – number of washes provided. For irrigation of territory this formula is to be rewritten as:

$$C_I\left(\frac{m^3}{day}\right)=\frac{A \cdot C_u \cdot d}{1000}, \quad (3)$$

where
A – area of garden [m^2];
C_u – consumption unit for garden irrigation as recommended by the Technical Specification ANQIP (ETA 0701) [14], [liters/m^2];
NI – number of watering provided.
Total monthly consumption per end use is presented in the Table 4.

Table 4. Consumption in non-potable use.

Month	Toilet flushing, [m^3]	Floor washing, [m^3]	Parking washing, [m^3]	Irrigation, [m^3]
Jan	42.20	3.42	0.00	0.00
Feb	63.30	3.42	0.00	0.00
Mar	127.15	3.42	0.00	0.00
Apr	145.32	3.42	75.60	239.52
May	108.99	6.85	75.60	319.36
Jun	108.99	6.85	75.60	239.52
Jul	24.22	6.85	75.60	319.36
Aug	24.22	6.85	75.60	239.52
Sep	168.79	6.85	75.60	319.36
Oct	168.79	3.42	75.60	239.52
Nov	189.89	3.42	0.00	0.00
Dec	189.89	3.42	0.00	0.00

Four scenarios were developed to investigate the best way to utilize rainwater harvesting at the Hydrocampus – 2 (Table 5). Currently, the potable municipal water is the only source of water used. The scenarios varied with respect to water demand, end use and initial costs:

— Basic scenario: City supplied potable water is used for both flushing toilets and irrigation.

— Scenario 1: Drainage system was modeled as being renovated to collect roof runoff for its use in flushing and floor washing.

— Scenario 2: Drainage system was modeled as being renovated to collect roof runoff for its use in toilet flushing, floor and parking washing.

— Scenario 3: Drainage system was modeled as being renovated to collect roof runoff for its use in toilet flushing, floor and parking washing and irrigation.

Total daily non-potable consumption for j-scenario:

$$C_w^j = \sum_{i=1}^{i} C_i, \tag{4}$$

where

C_w^j - total daily consumption for j-scenario, [m^3/day];

C_i – specific daily consumption (Eq. 1, Eq. 2, Eq. 3), [m^3/day];

Table 5. Scenarios.

Scenario	Rainwater re-use				Additional supply	
	Toilet flushing	Floor washing	Parking washing	Irrigation	Municipal water	Electricity for pumps
Basic scenario	-	-	-	-	+	-
1	+	+	-	-	-	+
2	+	+	+	-	-	+
3	+	+	+	+	-	+

2.5 Water Tank Dimensioning

The water storage unit is usually the most expensive part of the rainwater harvesting system, so an accurate analysis and optimization is critical to create an overall efficient water supply. Therefore, the capacity and all other construction parameters should be determined properly, among others, to minimize the payback period.

There are various methods to estimate the design of storage tank. The Ripple method [15] involves calculating the volume of water that needs to be stored throughout the year to compensate the lack of water during the dry period. According to the Ripple method, the reservoir volume corresponds to the maximum of the positive accumulated differences between the consumption of non-potable water and the collectable rainwater volume, observed within an evaluation period. This method tends to oversize the reservoir volume because it is prepared to meet the water demands of critical drought periods [16]. Nevertheless, it has the benefit of checking the upper limit for the volume of a storage reservoir [4], corresponding to the ideal situation of 100% efficiency.

The average rainfall distribution is enlisted in Table 2. Average daily precipitations are shown on Figure 3. Despite the fact, that recent studies [19] have shown the influence of time scale resolution for modeling the consumption, for high-demand applications, like in the current study.

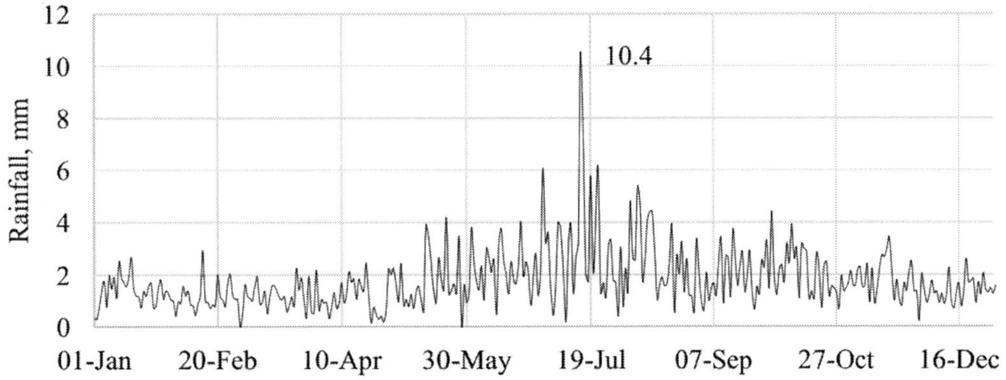

Fig. 3. Average precipitations distribution throughout the year.

The roofs of building are covered with black sheeting rolls. A collecting surface [A_r, m^2] is 4049.62 m^2. According to Roebuck [17], the runoff coefficient [C, dimensionless] of these surfaces is around 0.8 – 0.95. The value used in the present design was set to 0.8.

The volume of water, available for collecting, was calculated using the equation:

$$V_m = C \cdot (10^{-3} \cdot P) \cdot A_r \cdot n_f, \tag{5}$$

where
V_m – available volume of rainwater, [m^3];
P – daily average of precipitations, [m^3];
n_f – hydraulic filtration efficiency of the system, [dimensionless].

Hydraulic filtration efficiency is the ratio of the quantity of filtered rainwater arriving at the reservoir and the amount of rainwater that reaches the filter. In case the maintenance and cleaning of the filters are ensured, the hydraulic filtration efficiency may be considered constant and equal to 0.95 [14].

Further, the sequences of positive accumulated differences [S_t, m^3] were calculated as follows:

$$\begin{cases} S_t = 0, & C_{w,t} - V_{w,t} \le 0 \\ S_t = C_{w,t} - V_{w,t}, & C_{w,t} - V_{w,t} > 0 \end{cases}, \quad t = 1 \tag{6}$$

$$\begin{cases} S_t = 0, & S_{t-1} + C_{w,t} - V_{w,t} \le 0 \\ S_t = C_{w,t} - V_{w,t}, & S_{t-1} + C_{w,t} - V_{w,t} > 0 \end{cases}, \quad t > 1 \tag{7}$$

where
S_t – positive accumulated difference of the time period t, [m^3];
S_{t-1} – positive accumulated difference of the time period t-1, [m^3];
$C_{w,t}$ – total daily non-potable consumption of the time period t, [m^3];
V_{wt} – volume of water, available for collecting of the time period t, [m^3].
Results are enlisted in Table 6.

Table 6. Reservoir volume according to different scenarios and years.

		Maximum reservoir volume, m^3		
Year	Precipitations, [mm]	Scenario 1	Scenario 2	Scenario 3
1992	504	138	254	1692
2000	713	122	245	1085
2001	685	103	157	1143
2002	610	103	272	1416
2003	842	110	185	616
2004	699	141	254	1053
2005	650	152	303	1306
Average year	657	28	91	1150

The average rainfall data was proven inaccurate, as the difference between the calculated reservoir volume between average and realistic models was significant.

Therefore, in order to create a sustainable and relevant system, real precipitation among the most recent data should be used for determining the volume of the tank.

The Ripple approach considers only the environmental benefit, and does not reveal the most economically efficient result. So it should be improved with some modeling techniques [19], and in this study, performance evaluation approach is suggested.

To estimate the efficiency of the system, energy consumption should be calculated first.

2.6 Booster Pumps

Booster pumps are required to transport the rainwater for flushing toilets and urinals. The sizing of the pumps and their annual energy requirements were estimated using the standard pump power equation:

$$N_{real} = \frac{Q \cdot \gamma \cdot (h_e + h_p)(1 + \alpha)}{\eta}, \tag{8}$$

where,

N_{real} – power input to pump [W],

η – combined mechanical and hydraulic efficiency of the pump,

Q – flow rate [m3/s],

γ – specific weight of water [N/m3],

α – percentage of energy lost to friction,

h_e – elevation head provided by pump [m],

h_p – pressure head provided by pump [m].

Obtained results are shown in Table 7.

Table 7. Demand of pumps.

Parameter	1st pump	2nd pump
Q, [m^3/sec]	0.00338	0.00274
H, [m]	18	18
γ, [N/m^3]	9810	9810
η	0.72	0.72
$1+\alpha$	1.5	1.5
N_{useful}, [kWt]	0.60	0.48
N_{real}, [kWt]	1.24	1.00

2.7 Performance Evaluation

The resulting parameter, that was used to determine the efficiency, is payback period [PB, years]. It was calculated as follows:

$$V_t^{wT} = \begin{cases} V_{w,t} + V_{t-1}^{wT} - C_{w,i}, & V_{w,t} + V_{t-1}^{T} - C_{w,i} < V^T \\ V^T, & V_{w,t} + V_{t-1}^{T} - C_{w,i} \geq V^T \end{cases} \quad (9)$$

$$V_t^{ru} = \begin{cases} C_{w,t}, & V_{w,t} + V_t^{wT} - C_{w,t} \geq 0 \\ V_{w,t} + V_t^{wT}, & V_{w,t} + V_t^{wT} - C_{w,t} < 0 \end{cases} \quad (10)$$

$$V^{ru} = \sum_{t=1}^{n} V_t^{ru} \quad (11)$$

$$P_r = P_m \cdot A_r + P_w \cdot A_r \quad (12)$$

$$A = f(V^T) \quad (13)$$

$$P_s = P_r + P_o \quad (14)$$

$$S = (P_{wu} + P_{ww}) \cdot V^{ru} - P_e \cdot N_{real} \cdot t_d \cdot N \quad (15)$$

$$PB = \frac{P_s}{S} \quad (16)$$

where
V_i^{wT} – volume of water in a tank at the time period t, [m^3];
V_t^{ru} – volume of collected rainwater, that was used during the time period t, [m^3];
P_r – price of a reservoir, [rub.];
P_m – price of materials for a reservoir, [rub.];
P_m – price of construction of a reservoir, [rub.];
P_s – price of the rainwater harvesting system, [rub.];

P_o – price of other components of the system, [rub.];
S – amount of savings after applying the system, [rub./year];
P_{wu} – price of water utilization, [rub./m^3];
P_{ww} – price of water supply, [rub.];
N_{real} – power demand for the pump, [W];
t_d – working hours in a day, [h];
N – number of working days in a year, [days].

Reservoir volume changes from 0 to V_{max}, and that affects whole model in an unpredictable way, therefore the simulation was carried through all years and scenarios. Results of simulation for the first scenario are shown on Figure 4, for the second – on Figure 5.

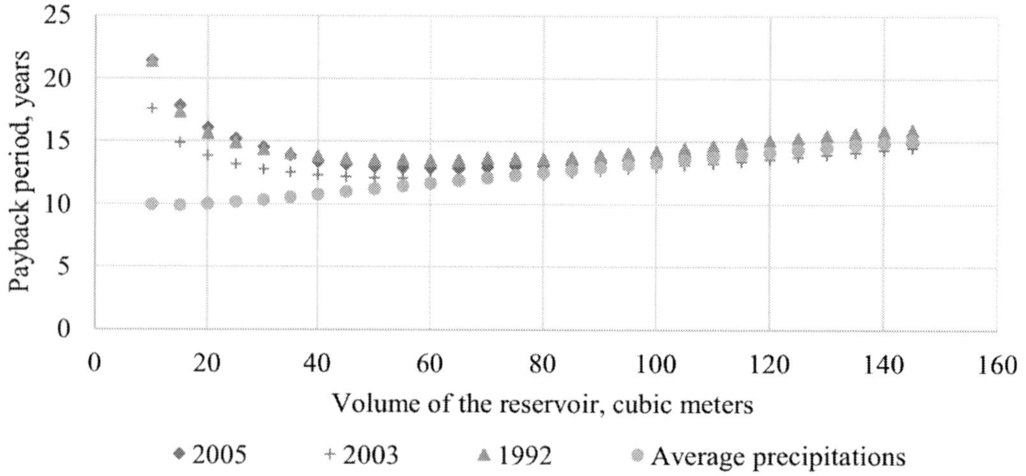

Fig. 4. Payback period depending by the volume of the reservoir for the year 2005, 2003, 1992.

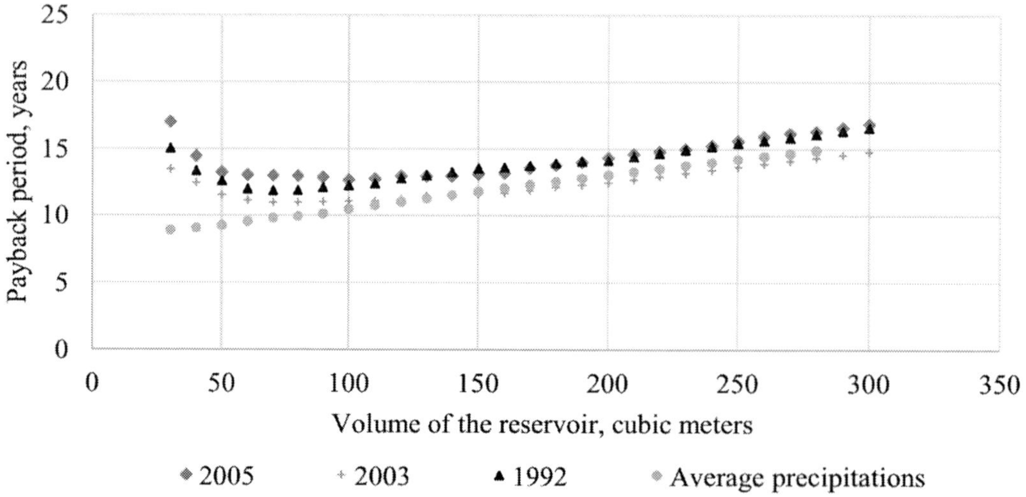

Fig. 5. Payback period depending by the volume of the reservoir for the year 2005, 2003, 1992.

3 Results and Discussion

3.1 Optimum Tank Sizes

Performance evaluation revealed that optimal storage volume for the first scenario is 60..65 cubic meters. With exception for average precipitations model. This confirms the fact, that average data can not be used for designing RHS.

The average payback period for the first scenario in case of applying the 60 m^3 storage is 12.77 years. Spread of performance evaluation results for the second scenario is higher, as it is shown on Figure 5. But the average payback period for 80 m^3 reservoir is 11.94 years. Therefore, the second scenario is more beneficial.

Overall, this technique allows us to choose both main parameters of the system – for what purposes rainwater can be used to maximize efficiency, and how big should the storage be.

Performance evaluation also revealed that volume of a tank depends not on the total amount of rainfall in a year, but on the distribution within a year, which is confirmed by longer payback period for 2005 precipitation model.

Figures 6 and 7 show the comparison of Ripple approach and suggested performance evaluation approach. These graphs show the volume of water in the tank at every day of the year, with the second consumption scenario and 2005 precipitation model.

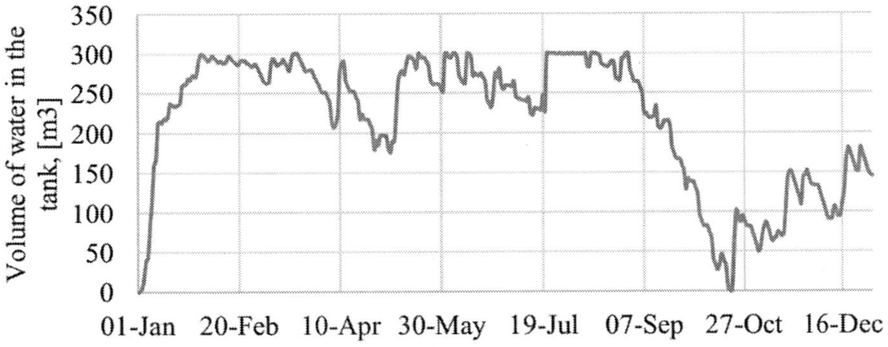

Fig.6. Water balance in the tank for the year 2005, second scenario and Ripple approach.

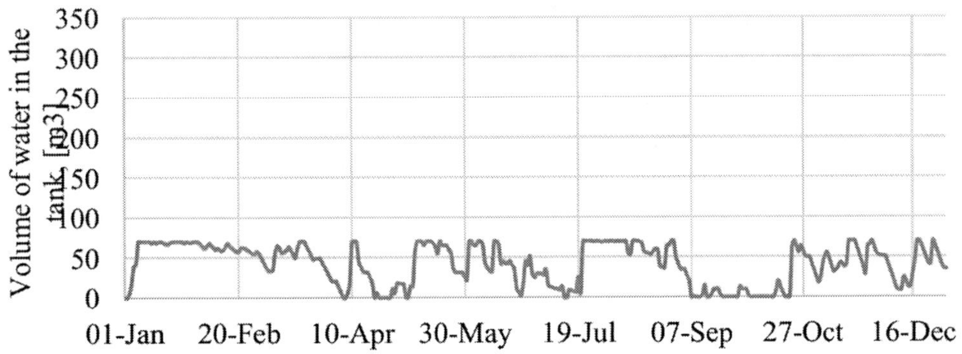

Fig. 7. Water balance in the tank for the year 2005, second scenario and economic approach.

Despite the fact, that Ripple approach allows to save more water, it is economically inefficient, which is proven by performance evaluation. Even taking into account periods, when the system can not supply the building, it is more efficient to reduce the volume of the tank to 80 m^3.

Another valuable detail, that these graphs show, is the most appropriate time for system installation. In this particular application the best timing will be from January to February, or from the middle of July to the beginning of September, where graphs for all the three chosen years behave similarly and are at their peaks, as it is shown on Figure 8.

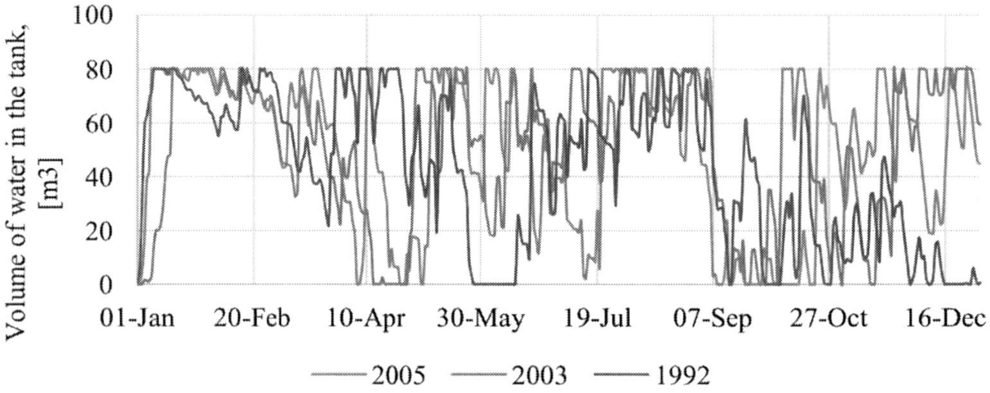

Fig. 8. Water balance in the 80 m^3 tank for the 2005, 2003, 1992, second scenario

3.2 Economic Effect

According to the results of performance evaluation, maximum economic effect was achieved when rainwater is used for toilet flushing, floor cleaning and parking washing. That consumption scenario, combined with the tank of 80 [m^3], reduces the payback period of the system to approximately 12 years.

4 Conclusions

The empirical and theoretical assessment of the modelled RHS system for the existing public building has produced the following conclusions and recommendations:

1. RHS system, that was modelled for the Hydrocampus – 2, ensures significant level of fresh water savings;
2. According to calculations of initial costs and related to RHS electricity consumption, payback period for the 2nd scenario is equal to 12 years. Energy consumption of pumps, however, can be further reduced by supporting innovations in RHS system design, for example around gravity-based systems;
3. Results of RHS modelling, carried by Ripple method, may not be accurate to proper size the system. Ripple method maximizes the ecological impact, and is good for determining modeling limits. It should be improved with additional techniques, like performance evaluation, that leads to determining the optimum volume;
4. The performance evaluation approach provides users with significantly more accurate results, than any other technique, if the algorithm stands on daily data. Use of monthly or weekly averages increase inaccuracy;
5. Application of the developed technique to existing university campus gives the optimal storage volume of 80 m^3.

References

1. Information on http://www.unwater.org/
2. Information on http://www.worldbank.org/
3. A. Bocanegra-Martínez, J.M. Ponce-Ortega, F. Nápoles-Rivera, Resources Conservation Recycling, **84**, 44–56 (2014)
4. P. Tomaz, *Aproveitamento de água da chuva para áreas urbanas e fins não potáveis. Navegar* (São Paulo, Brazil, 2005)
5. S. Ward, S. Barr, D. Butler, F.A. Memon, Technological Forecasting and Social Change, **79**, 1354-1361 (2012)
6. S. Ward, F. Memon, D. Butler, Water Resources, **46**, 5127-5134 (2012)
7. C. Matos, C. Santos, S. Pereira, I. Bentes, International Journal of Sustainable Built Environment, **2**, 109-118 (2013)
8. D. J. Sample, J. Liu, Journal of Cleaner Production, **75**, 174-194 (2014)
9. J. Devkota, H. Schlachter, D. Apul, Journal of Cleaner Production, **95**, 311-321 (2015)
10. V.G. Mitchell, D.T. McCarthy, A. Deletic, Environmental Modelling Software, **23**, 782–793 (2008)
11. P. J. Coombes, G. Kuczera, D. Jetse, R. Kalma, Urban Water, **4**, 307-320 (2002)
12. A. Fewkes, D. Butler, Building Services Engineering Resources and Technology, **21**, 99-106 (2000)
13. K. Heaton, J. Radvan, H. Cripps, R. A. Mountford, F. Braddon, A. Hughes, Gut, **33**, 818-824 (1992)
14. ANQIP, 2009. ETA 0701 — Sistemas de Abastecimento de Águas Pluviais em Edifícios [SAAP], Version 7. Portuguese National Association for Quality in Building Services – ANQIP, Lisbon, Portugal
15. E.J. Ripple, Minutes Procedia Inst. Civil Engineering, **71**, 270–278 (1883)
16. F. Luís, P.S. Daniela, Science of the Total Environment, **529**, 91–100 (2015)
17. R. Roebuck, *A Whole Life Costing Approach for Rainwater Harvesting Systems, in School of Engineering* (Design and Technology, University of Bradford, Bradford, UK, 2007)
18. A. Campisano, C. Modica, Procedia Engineering, **70**, 218-227 (2014)
19. C. Onyeka Okoye, O. Solyalı, B. Akıntuğ, Res., Conservation and Recycling, **104**, 131-140 (2015)

Research of Methods, Technologies and Materials for Drainage Water Treatment at the Municipal Solid Waste Landfill in Salaryevo

Elena Gogina [1,*], and *Alexey* Pelipenko[1]

[1]Moscow State University of Civil Engineering, 129337 Yaroslavskoe sh. 26, Moscow, Russia

Abstract. The article deals with innovative methods, technologies and materials intended to reduce the adverse ecological impact of human waste and various industrial waste situated in municipal solid waste landfills (MSW), on water bodies, soil, and atmosphere. The existence of these factors makes the region less attractive for urban development. A comparison has been made of the methods intended to reduce the damage caused to the environment, in order to provide for sustainable development of cities, using the example of an actual landfill situated in the territory of Moscow. A scheme of reconstruction is recommended for the drainage water treatment plant at this landfill, which will lead to improvement of the environmental situation and contribute to the development of territories in the adjacent districts, and to reduction of pollution load on the river and atmosphere.

1 Introduction

It is known that human waste is removed to municipal solid waste landfills. Officially, the total number of registered landfills is 865, of which 37 are located in the Moscow region [1]. The largest operating landfills include Dmitrovsky, Khmetyevo, Levoberezhny, and Saburovo landfill. Decommissioned landfills and those in the process of reclamation include Nekrasovka and Salaryevo.

The main problem is the treatment of MSW landfill drainage flows with elevated concentrations of organic and mineral contamination as well as heavy metals [2].

There are three major sources of leachate formation at solid domestic waste landfills:

- Precipitation infiltrating through the landfill body in contact with the surface of solid waste (the main source of leachate formation);
- Initial moisture content of certain types of waste;
- The moisture released from the section as a result of biochemical processes involving the formation of water at the anaerobic decomposition of organic components [3].

In practice, there is "young" and "old" leachate. "Young" leachate is formed at the initial stage of landfill operation after 2-7 years of storage and burial of solid domestic waste and lasts for 5-10 years. This leachate has an average value of pH, high values of COD and BOD, and high content of ammonia nitrogen and iron; the composition of organic

* Corresponding author: goginaes@mgsu.ru

compound is represented by volatile organic fatty acids. "Old" leachate is formed mainly on the post-operational stage of landfill life [4-6].

The aim of the research work is the reconstruction of the existing drainage water treatment plant at the Salaryevo landfill with minimal capital investments both for the refurbishment of the existing buildings, and for the construction of new buildings and the purchase of new equipment. The main task is to find the optimal scheme, which will ensure the proper indicators of active substances for their subsequent discharge to the river Setun. At the moment, the discharge is made to the relief. As of recent, the need to reduce the environmental impact of leachate, both on the relief and on ponds, has been the most urgent. The Russian government has been carrying out a set of measures aimed at improving the quality of treating various wastewaters by increasing the fees for discharge of polluted waters both to the municipal sewage system, and to ponds; in addition, measures have been implemented to refuse using relief as the site for wastewater discharge.

Implementation of the authorized discharge of solid domestic waste landfill leachate to the relief has the following negative effects:

- Change in the degree of soil salinity
- Change in the chemical composition
- Possibility of toxic pollutants entering the ground water, thus contributing to their change
- Formation of dangerous gases released into the atmosphere during decomposition of organic contaminants [7-8].

Disposal of drainage water into fishing industry waters contributes to the change of water chemical composition, resulting in:

- Threat to human life upon further use as a source of drinkable water;
- Increased level of water pollution causing water body eutrophication;
- Changes in the microbiological composition of water;
- Reduction of pond's self-cleaning ability;
- Possibility of toxic pollution spreading in the range of 50-60 km;
- Disposal of unfiltered contaminants into the pond [9-10].

The landfill is currently under reclamation; however, the drainage water will continue to flow during the next few decades, which is why it was decided to renovate the existing treatment plant.

The current diagram of drainage water treatment consists of the following facilities: reception chamber combined with bar screen building, primary sedimentation tanks, bag filters, sand filters, cartridge filters, and reverse osmosis plant.

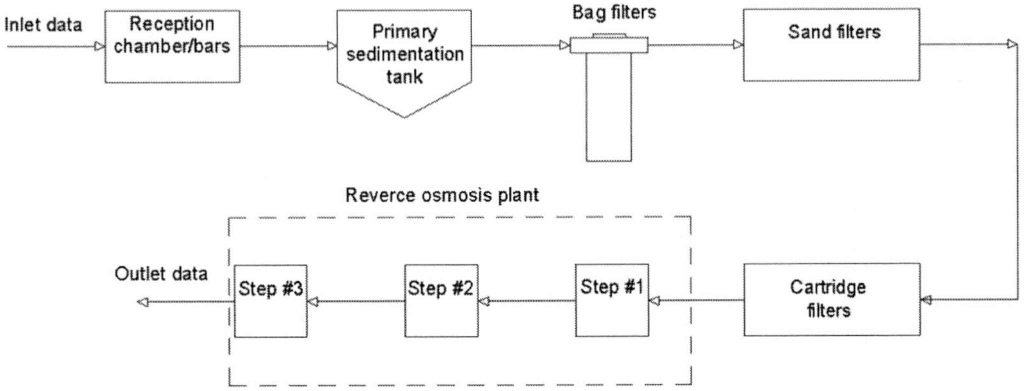

Fig. 1. Diagram of Leachate Treatment Plant before Reconstruction.

The actual contaminant indicators obtained during laboratory tests were used during the research work. The results show that the drain water is characterized by high values of COD, suspended solids, and ammonia nitrogen.

2 Theory

At the moment, there is a large number of SDW landfills leachate treatment methods on the market. Current trends have formed two main areas on which all the developments are based. There is a variety of modular systems, which are placed in the shipping containers or the existing facilities. In addition, there is stratification by the treatment methods. Thus, there are biochemical (aerobic and anaerobic) and physicochemical (oxidation, ion exchange, adsorption) methods [11-12].

For example, Osmotics Company proposes a method based on the reverse osmosis membrane technology in combination with ozonization and electrochemical machining technologies [13].

Moscow State Environmental Engineering University presented the MSW landfill leachate chemical treatment and detoxication technology, under which the active substances are treated in the following order: chemical treatment with lime milk to pH 11-12; ammonia air stripping; use of coagulant along with feeding into a thin-layer elements sedimentation tank; filtration through the contact with silica sand (particle size 1-3 mm); electrofrothcoagulation; 30% hydrogen peroxide treatment for organic dissolved substances decomposition; heavy metals adsorption on the natural sorbent (tripoli powder, particle size 200-300 mkm) with final additional purification at the membrane module. The limited version of the plant with the use of deep desalination is available at Dmitrovsky Solid Domestic Waste Landfill, Moscow.

BMT CJSC (Vladimir) has developed a technology comprising of the following stages:

- pre-treatment – chemical treatment, electrochemical oxidation, sedimentation, ultrafiltration treatment
- deep purification and desalination of waste water using two-stage reverse osmosis and subsequent disposal of reverse osmosis concentrate to the landfill body;
- sorption purification from low-molecular organics;
- UV disinfection of purified water (ultraviolet spectroscopy) [14].

The following method developed by MIS Company presents wastewater filtration using ultrafiltration vacuum plant (without pretreatment) with subsequent ozone treatment. In this case, the concentrated precipitation from the ultrafiltration step is fed back to the landfill. The initial ultrafiltration uses surface treated (to reduce sludge production) membranes based on ether sulfones. Water ozonation occurs in the vortex reactor. At the low volumes of inlet drainage water, this method can be carried out in standard sea containers.

Ekologicheskaya gruppa LLC (Kaliningrad) offers its own vision of the effective drainage water purification technology. Its development is based on electrical chemisorption method, which includes various stages of other processes (flotation, filtration, electrooxidation, and chemisorption). At the core of the reaction is the mechanism of electro-initiated free radical chain oxidation. The initiation occurs through the generation of electrons with electrodes in an aqueous medium. In contrast to other processes, both anode and cathode function in this method.

Due to the high oxidation potential, the reaction occurs at a high speed, thus catalyzing other purification processes of organic matter; as a result, the organics oxidize to carbon dioxide and water [15].

Another development by BMT CJSC is based on the use of sequential purification process comprising of the following steps: mechanical cleaning of mechanical and colloidal particles, fine purification unit on a mechanical barrier filter with retention potential of up to 20 microns, deep purification unit, two-stage desalination with a membrane module, and final purification using ion exchange filters with the concentrate return to the landfill body for participation in the biochemical processes [14].

In addition, there were found various advanced methods that are currently being developed, but have not been applied at the actual landfills, so there is no realistic assessment of their effectiveness . This includes active substance flow evaporation method (Kimo-business, Kiev), laser beam water treatment (Bauman MSTU, Kaluga Branch), contaminant extraction using biological sorption filters (PermGTU, Perm), as well as the biological method of drainage water treatment in a system of cascade biofilters and slowly flowing ponds (Ecoprom CJSC, Samara).

Institute of Colloid Chemistry and Water Chemistry (NIKTI GH, Kiev) has developed a scheme based on the use of reverse osmosis method with a preliminary two-stage anaerobic-aerobic biological treatment, chemical treatment, setting-out, disinfection with sodium hypochlorite, and filtration through sand and charcoal filters. This scheme also includes ion exchange steps with Na-cation exchange filters and electrodialysis.

However, it is worth noting that the current technologies are mainly based on the foreign experience, which is often not adapted to the climatic conditions and a specific site, and is often economically unjustified in view of the more stringent requirements of the Russian legislation on the quality of purified active substances sent for drainage into the waterways [16-18].

3 Experimental part

Based on the review results of the existing methods, we selected three methods that constitute complete solutions for further work on MSW landfill drainage water purification and are the most optimal in terms of the applied technologies.

Method #1 is the development of BMT CJSC. It includes membrane technology with pretreatment of heavily contaminated drainage water.

Method #2 – Alpha mobile complex by Ekologicheskaya gruppa LLC.

Method #3 by BMT CJSC consisting of several purification steps aimed at removing pollutants of various sizes.

For each of the selected schemes we found published results on the leachate treatment of solid domestic waste landfills with contaminant indicators on the inlet and outlet. Based on this, the contaminant removal efficiency for each method was found. Based on the efficiency values, the purification results with respect to the Salaryevo landfill were calculated through analytic determination.

We also evaluated the environmental damage from the discharge of untreated and treated leachate into the pond. When conducting the analysis, we calculated the fee for the discharge of pollutants by four parameters: BOD, suspended solids, petroleum products and ammonia nitrogen. The comparison on the four indicators owed to the fact that there were results for a variety of substances for the selected schemes in the public domain; however, it was on these indicators that two methods coincided.

4 Results and discussion

By the results of the outlet pollutant indicators simulation, and after applying each method, two methods with the best results of solid domestic waste landfill leachate treatment were selected (methods #2 and #3). There were requested commercial offers and conducted economic comparison of each method.

The cost of method #2 amounted to RUB 10 million; that of method #3 – RUB 14 million.

Method #3 includes capital costs for the construction works and purchase of new equipment. Therefore, it is fair to reduce the cost of this method to RUB 10 million.

By the results of comparing the fee for discharge of pollutants into a water body, method # 3 provides the lowest payment. It is 5 times less than that for the second method and 220 times less than the fee for discharge without reconstruction.

Upon selection of method #3, the economic effect on the four contaminants will amount to 49,341.7-224.41=49,117.29 RUB/year.

It is expected to reconstruct the current treatment plants using method #3. It allows the fullest use of the available facilities and includes the following steps during the reconstruction:

- Reception chamber combined with the bar screen building are not reconstructed; however, it is necessary to take into account the bar spacing in the existing grids, as well as the basic requirements for the reconstruction, such as the availability of engineering systems (water supply, ventilation, plumbing, heating, electricity)
- Primary sedimentation tanks are converted to storage tanks for matching the incoming wastewater and leachate flow offset
- Bag filters are decommissioned
- As part of reconstruction, sand filters are upgraded by removal of the old medium and its replacement by a larger medium, i.e. granular bed, for the implementation of rough mechanical treatment stage
- Cartridge filters are decommissioned and replaced with mechanical filters for the fine purification of waste water
- Reverse osmosis plant is subjected to the reconstruction only in terms of membrane replacement and stage #3 decommissioning.
- Ion exchange filters are added to the scheme after the reverse osmosis plant stage #3.

The diagram of leachate treatment plant before and after the reconstruction is shown in Figures 2 and 3.

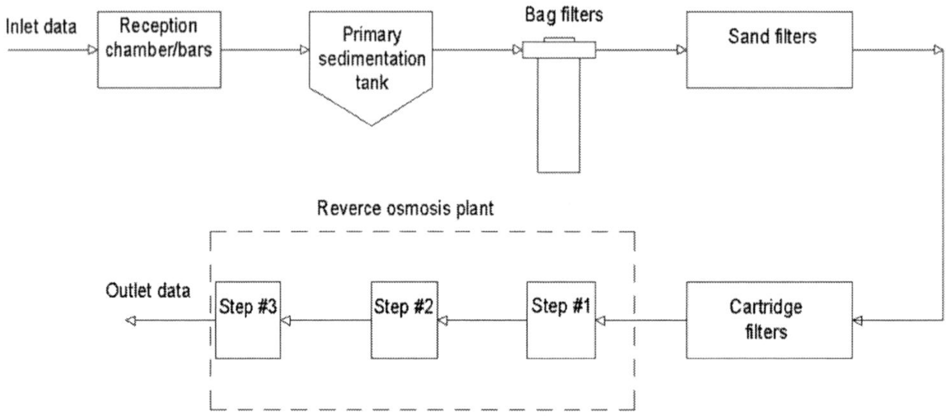

Fig. 2. Leachate Treatment Plant before Reconstruction.

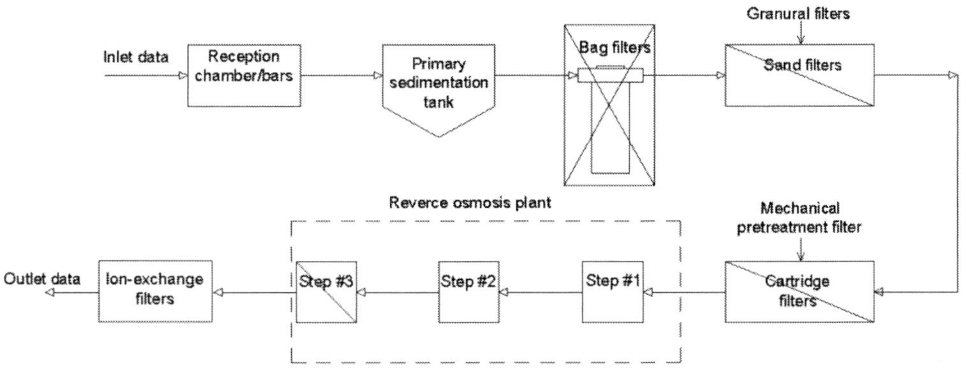

Fig. 3. Leach ate Treatment Plant after Reconstruction.

5 Conclusion

By the results of economic comparison and the analytical results of drainage water purification by different methods, the best solution is method #3, which is a sequence of various leachate treatment stages aimed at obtaining the best treatment results and requiring minimal capital expenditures for the reconstruction due to the possibility of using the existing facilities.

References

1. Information on http://rpn.gov.ru/node/853
2. C. Brown, M. Milke, Conversation and Recycling, **106**, 21-32 (2016)
3. D. Han, X. Tong, M. J. Currell, G. Cao, M. Jin, C. Tong, Journal of Geochemical Exploration, **136**, 24-39 (2014)
4. Y.Voronov, *Vodootvedenie: uchebnik* (INFRA-M, Moscow, 2011)
5. E.Gogina, N. Makisha, Applied Mechanics and Materials, **361-363**, 628-631 (2013)
6. E.Gogina, N. Makisha, Applied Mechanics and Materials, **587-589**, 636-639 (2014)
7. A. Trivedi, A. Singh , A. Chauhan, International Journal of Disaster Risk Reduction, **14**, 527-535 (2015)
8. B. Gworek, A. Hajduk, E. Koda, A. Grochowalski, A. Jeske, Chemosphere, **92**, 753-759 (2013)
9. A. Kulako, Water and ecology: problems and solutions, **1**, 26-40 (2015)
10. E. Gogina, O. Ruzhitskaya, O. Yantsen, Advanced Materials Research, **919-921**, 2145-2148 (2014),
11. N. Makisha, O. Yantsen, Applied Mechanics and Materials, **587-589**, 640-643 (2014)
12. M. Hrad, M. Huber-Humer, B. Wimmer, T. G. Reichenauer, Waste Management, **32**, 2324-2335 (2012)
13. Information on http://osmotics.ru
14. Information on http://zaobmt.ru
15. Information on http://ecoalfa.ru
16. I. Gulshin, A. Kuzina, International Journal of Applied Engineering Research, **10**, 42618-42623 (2015)
17. N. Zaletova, Y. Voronov, N. Makisha, International Journal of Applied Engineering Research, **10**, 42454-42455 (2015)
18. E. Gogina, O. Ruzhitskaya, O. Yantsen, Advanced Materials Research, **919-921**, 2145-2148 (2014)

Analysis of Qualities of Construction Wastes as Secondary Material Resources from the Position of the Theory of Stability of Disperse Systems

Vadim Bespalov[1], *Oksana* Gurova[1], *Oksana* Paramonova[1,*], *Natalia* Samarskaya[1], *Ekaterina* Lysova[1], and *Natalia* Yudina[1]

[1]Architectural and Construction Academy Don State Technical University, 344022, Sotsialisticheskaya str. 162, Rostov-on-Don, Russia

Abstract. Each type of wastes of different aggregate states coming into the environment from various sources is characterized by certain set of parameters of qualities determining in the future the possibility of their secondary use. The article represents results of innovative research in the sphere of broadening of opportunities of use of construction wastes as secondary material resources. Parameters of qualities of construction wastes created in the process of technological processes at the enterprises of construction industry have been researched and classified. Construction wastes and their qualities are represented from the position of the theory of stability of disperse systems. Physical essence of processes and events was chosen as the main classification criteria in distribution of parameters of qualities of construction wastes by groups. Parameters of qualities of construction wastes are represented for disperse phase and dispersion medium.

1 Introduction

Purpose of the research is analysis of qualities of construction wastes on basis of application of theory of disperse systems stability from the point of view of broadening of the sphere of their repeated use in economic activity.

For achievement of the set goal the following tasks have been solved:

– generalized parameters of qualities of construction wastes have been systematized;

– their energetic parameters have been researched;

– construction wastes stability has been researched as a resulting criteria representing the waste's capability to resist external influence.

Presently construction of buildings and constructions is carried out with a rather high speed which leads to increase of rates of production of construction wastes, growth of the number

* Corresponding author: paramonova_oh@mail.ru

of demolished old buildings and, as a consequence, the volume of the produced construction wastes grows [1, 2].

Nomenclature of wastes created in the construction area of industry includes tens of waste items characterized by various volumes of creation, quantitative and qualitative structure, hazardous characteristics in relation to environment, which complicates their classification, accounting for, collecting and processing [1, 2-4].

Existing classifications of construction wastes are rather different, but one-sided, they do not review the peculiarities of behavior of construction wastes in the environment on each stage of their handling, which could lead to changes of their qualities.

Different approaches to classification of wastes are based on the following classification features: sources of wastes creation; stage of technological process; type of waste; volumes of creation; degree of damage to the environment and human health; direction and efficiency of use; degree of examination and development of the utilization technology [1, 2, 4-10].

Many construction wastes should be utilized, it depends on various parameters allowing to identify 3 categories of wastes:

1 category – wastes created, as a rule, in the beginning of construction (heavy bulk waste, created during the building demolition).

2 category – wastes, created during construction (construction materials packing).

3 category – wastes created during finishing works [8-13].

Besides the classification represented above construction wastes could be divided into:

- raw wastes, source materials wastes being remnants of source materials not changing their main qualities (bark, road metal, keramzit wastes, etc.);
- construction production wastes (scrap brick, ceramic tile scrap, congealed solution, PWM tare, glass scrap, etc.) [1, 4-7, 12].

From the point of view of possibility of use of wastes, as a rule, all wastes, including construction ones, are divided into two flows:

- wastes directed for use as secondary material resources;
- wastes directed finally for burial.

In the majority of cases various trends were developed for use of construction wastes which are determined by physical and chemical qualities of the created wastes, technical opportunities and other factors [1-16]. That is why the issue of examination of qualities of construction wastes is a rather pressing one.

2 Theory or experimental methods

Substantial dimensions of creation and accrual of construction wastes lead to environmental pollution of cities and pose pressing issues for the community to lower pollution, as well as develop innovative methods of wastes utilization [1, 2, 4-13, 15-18]. In connection with that description of qualities of wastes should be an important information, preceding the search for solution of these issues.

Existing approaches to examination of qualities of wastes (including construction ones) practically do not reveal the interrelation of characteristics of construction materials, wastes created from them with the characteristics of the environment [14, 17-21].

Processes of formation, division and distribution of construction wastes, as it is known, are explaining the environmental pollution and in the actual conditions could happen either consequently or simultaneously in the space and time [1, 3, 12-14, 17, 18, 20, 22-24]. Thus, they should be reviewed in aggregate, but not separately from each other. Unity of nature of such processes underlines the close interconnection between them. In this case we could speak only of prevalence of the influence of one of the processes. This influence depends on the number of parameters, such as qualities of disperse phases (d.p.) and dispersion medium (d.m.) of the waste, characteristics of the environment, etc. [25-29].

Recently number of researches have been suggesting to review any pollutant as a dispersion system [25-29]. Such an approach realizes the theory of dispersion systems based on classical ideas of colloid and physical chemistry. As that in accordance with this approach construction wastes, in our view, in the majority of cases represent a dispersion system made of solid dispersion phase (or several dispersion phases) and gas dispersion medium.

3 Experimental section

Methods of examination are based on the main statements of the theory of dispersion systems, system analysis, analytical generalization of the known scientific and practical results.

4 Results and Discussion section

Reviewing construction wastes as a dispersion system allows to make a conclusion that the description of their behavior in the environment could be built on basis of complex consequent examination of their state and ordered analysis of parameters, characterizing the qualities of the dispersion phase and dispersion medium.

With this purpose the parameters determining the qualities of dispersion phase (d.p.) and dispersion medium (d.m.) should be groupped taking into consideration physical essence of processes and events in the construction wastes. Analysis of such processes allows to identify the following main groups of parameters (Tab. 1) [2, 3, 5, 6, 10, 14, 17, 18, 25, 27-30].

Table 1. Main parameters, identifying qualities of dispersion phase and dispersion medium

Ref. No.	Parameter determining qualities of d.p. and d.m.		
	Name	Denomination	Unit of measurement
1	**2**	**3**	**4**
Parameters of qualities of dispersion phase of a construction waste			
Geometric parameters			
1	dispersity of waste particles	$d_{(ч)}$	[mkm]
2	specific surface of waste particles	$S_{уд}$	[m^2]
Physical-chemical parameters			
3	pour density of particles	$\rho_{нас\ (ч)}$	[kg/m^3]
4	rupture strength	$P_{разр\ (ч)}$	[Pa]
Hydrodynamic parameters			
5	wetting angle	$\Theta_{(ч)}$	[degree]
6	relative humidity	$\varphi_{(ч)}$	[%]
Aerodynamic parameters			
7	particles surface roughness factor	$k_ш$	-
8	dynamic factor of the particles form	$k_ф$	-
Electromagnetic parameters			
9	specific electrical resistance	$R_{уд(ч)}$	[Ohm·m]
10	Material magnetic susceptibility	χ	-
11	Internal electrical field intensity	E	[V/m]

12	average charge	q	[Kl]
Optic parameters			
13	refraction index	n	-
Parameters of qualities of dispersion medium of construction waste			
Geometric parameters			
14	d.m. volume	V	[m^3]
Physical-chemical parameters			
15	d.m. density	ρ_z	[kg/m^3]
Hydrodynamic parameters			
16	d.m. humidity (absolute and relative)	φ_z	[%]
Aerodynamic parameters			
14	dynamical air viscosity	μ_z	[Pa·s]
Thermophysical parameters			
15	temperature	t_z	[°C]

Result of the carried out analysis of parameters determining qualities of d.p. and d.m. could be expressed by aggregate of parameters determining qualities of construction wastes in general (PQ_{CW}). In general obtained data could be represented as a functional dependence between groups of parameters of phase components of construction wastes:

$$PQ_{CW} = f_1\left(PQ_{d.p.}, PQ_{d.m.}\right) \quad (1)$$

where $PQ_{d.p.}$ – is the aggregate of parameters determining qualities of d.p. of construction wastes; $PQ_{d.m.}$ – is the aggregate of parameters determining qualities of d.m. of construction wastes.

The represented aggregate allows to amend in the process of formation of theoretical bases each group of parameters of d.p. and d.m. with new characteristics and carry out purposeful and consequent evaluation of all sides of the dynamics of creation, accrual, distribution and demolition of construction materials and wastes [25, 27-29].

Purchasing, distribution and consumption of energy of construction materials and wastes is qualitatively characterized by energy parameters which are determined by an interrelated complex of groups of parameters of qualities of d.p. and d.m. of construction wastes.

Energy state of construction wastes will be described by corresponding energy parameters which are characterized by groups of parameters of qualities.

By arrangement of parameters of qualities of d.p. and d.m. of construction wastes from the separated groups the following main energy parameters of construction wastes are formed: kinetic energy $W_{kin.}$; energy of optical interaction W_{opt}.; energy of adhesive interaction $W_{adh.}$; energy of hydrodynamic interaction $W_{hd,}$ heat energy $W_{h.}$; electromagnetic energy W_{elect}.

The reviewed types of energy in aggregate determine summarized free energy (activation energy) of construction wastes $(W_{S\,CW.})$, which functionally depends on certain energy parameters of d.p. and d.m. of construction wastes which assumes the opportunity of re-distribution of certain types of energy reflecting peculiarities of behavior of construction wastes [25, 27-29].

Having a certain stock of summarized free energy (activation energy) construction materials and wastes show in the environment peculiarities of their behavior which, finally, influences its stability:

$W_{S\,CW.} \sim U_{CW.}$

According to modern ideas, stability is a resulting characteristics, determining peculiarities of behavior of dispersion systems, i.e. parameters of their working life.

Stability is a characteristics of intensity of transformation of parameters of qualities and energy parameters of dispersion system for the certain time interval under the conditions of external influences.

Decrease of stability of dispersion systems is demonstrated either in strengthening of particles of d.p. (re-crystallization), or in adhesion and enlargement of these particles (creation of aggregates is happening) [25, 27-29].

Analysis of parameters characterizing the state of construction wastes as dispersion systems allows to conclude that for description of behavior of construction wastes in the environment as elements of its pollution, as well as for choosing or development of possible technologies for their secondary use, it is necessary to review construction materials and wastes from the position of theory of dispersion systems.

5 Conclusion

As the result of the research carried out we have solved the following issues:

– parameters of qualities of construction wastes have been researched and distributed into groups;

– energy parameters of construction wastes have been reviewed;

– stability of construction wastes has been researched.

It will allow in the future to manage the behavior of construction wastes with the purpose of decreasing of their stability and, finally, decreasing of environmental pollution.

References

1. E. V.Kondrashchenko, A. A. Kachura, Komunal'ne gospodarstvo mist, **107**, 150-155 (2013)
2. V.V. Berdus, Stroitel'nye materialy, **2**, 31-32 (2001)
3. B.Bossink, H. Brouwers, J. Constr. Eng. Manage, **122(1)**, 55-60 (1996)
4. C. S.Poon, T. W. Ann, H. Ng L. Resources, conservation and recycling, **32(2)**, 157-172 (2001)
5. T.B. Tkachenko, Promyshlennoe i grazhdanskoe stroitel'stvo, **8**, 35-37 (2011)
6. E.Nikolaeva, Khimicheskiy zhurnal, **4**, 34-38 (2003)
7. Kh. S. Vorob'yev, Stroitel'nye materialy, **10**, 6-7 (1985)
8. A.G. Yudin, Ekologiya i promyshlennost' Rossii, **3**, 229-235 (2010)
9. V. W. Y.Tam, C. M. Tam. A review on the viable technology for construction waste recycling (2006) Resources, Conservation and Recycling, 47 (3), pp. 209-221.
10. P.P.Oleynik, V.I. Brodskiy, Visnik PDABA, **10 (187)**, (2013) URL: http://cyberleninka.ru/article/n/organizatsiya-upravleniya-pererabotkoy-stroitelnyh-othodov (data obrashcheniya: 07.05.2016)
11. N. G. Chumachenko, Vestnik SGASU. Gradostroitel'stvo i arkhitektura, **1**, 112-116 (2011)
12. T. Napier, National Institute of Building Science, **25(05)**, 2013 (2012)
13. A.J Morrissey, J Browne, Waste Management, **24(3)**, 297–308 (2004)
14. N. G.Chumachenko, Ekologiya i zdorov'ye cheloveka, 201-203 (2001)
15. V. D. Kotlyar, A. V. Ustinov, Internet-zhurnal Naukovedenie, **3(16)**, (2013) URL: http://cyberleninka.ru/article/n/effektivnaya-stenovaya-keramika-na-osnove-opok-i-othodov-ugleobogascheniya (data obrashcheniya: 07.05.2016)
16. Yu. V.Selivanov, V. I.Vereshchagin, A. D.Shil'tsina, Izvestiya TPU, **1**, (2004) URL: http://cyberleninka.ru/article/n/poluchenie-i-svoystva-poristoy-stroitelnoy-keramiki (data obrashcheniya: 07.05.2016)

17. T.O. Kondratenko, A.V. Saybel', IVD, **4** **(2)**, (2012). URL: http://cyberleninka.ru/article/n/otsenka-vozdeystviya-stroitelnogo-proizvodstva-na-okruzhayuschuyu-sredu (data obrashcheniya: 07.05.2016)
18. A.S. Grinin, V.N. Novikov, *Promyshlennye i bytovye otkhody: Khranenie, utilizatsiya, pererabotka* (FAIR-PRESS, Moskva, 2002)
19. O.N. Paramonova, S.O. Koryagin, *Stroitel'stvo-2011: Materialy Mezhdunarodnoy nauchno-prakticheskoy konferentsii,* 137-139 (2011)
20. L.N. Bel'deeva, Yu.S.Lazutkina, L.F. Komarova, *Ekologicheski bezopasnoe obrashchenie s otkhodami,* (Azbuka, Barnaul, 2006)
21. I.A.Ryb'yev, *Stroitel'noe materialovedenie* (Vysshaya shkola, Moskva, 2003)
22. T. A.Sharkov, S. Yu. Efremova, Izvestiya PGU im. V.G. Belinskogo, 29, (2012) URL: http://cyberleninka.ru/article/n/upravlenie-ekologicheskoy-bezopasnostyu-v-oblasti-snizheniya-obrazovaniya-othodov-proizvodstva (data obrashcheniya: 07.05.2016)
23. T.O. Kondratenko, A.V. Saybel', IVD, **4** **(2)**, (2012) URL: http://cyberleninka.ru/article/n/ekologicheskaya-otsenka-pri-vybore-stroitelnyh-materialov-dlya-novogo-stroitelstva-rekonstruktsii-i-restavratsii (data obrashcheniya: 07.05.2016)
24. N. A. Babak, Izvestiya Peterburgskogo universiteta putey soobshcheniya, **3**, (2011) URL: http://cyberleninka.ru/article/n/preventivnyy-metod-otsenki-sostoyaniya-zaschity-prirodno-tehnogennyh-sistem-i-upravleniya-imi-pri-osuschestvlenii-stroitelnoy (data obrashcheniya: 07.05.2016).
25. E.P. Lysova., L.A. Lisutina, *Malootkhodnye, resursosberegayushchie khimicheskie tekhnologii i ekologicheskaya bezopasnost',* 98-100 (2013)
26. S. V. Belov, F. A. Barbinov, A. F. Koz'yakov i dr., *Okhrana okruzhayushchey sredy* (Vysshaya shkola, Moskva, 1991)
27. V.I. Bespalov, L.Z. Ganicheva, N.V.Yudina, T.L.Pirozhnikova, I.N.Gevorkyants, Biosciences, Biotechnology Research Asia, **11**, 27-35 (2014)
28. V.I. Bespalov, O.S. Gurova, N.S. Samarskaya, E.P. Lysova, A.N. Mishchenko, Biosciences Biotechnology Research Asia, **11(3)**, 1615-1620 (2014)
29. V.I. Bespalov, Izvestiya SKNC VSH. Estestvennye nauki, 37-47 (1995)
30. V.V.Privalenko, I.N.Kuznetsov, S.G. Demchenko, YuNTs RAN, Rostov-na-Donu, (2009)

Radiation and Geochemical Assessment of the Soil State in the South of Tyumen Region

Leonid Skipin[1,*], *Elena* Zakharova[1], *Elena* Gaevaya[1], *Vasilya* Burlaenko[1], and *Aleksandr* Mitrikovskiy[1]

[1]Industrial University of Tyumen, 625001 Volodarskogo str. 38, Tyumen, Russia

Abstract. The article presents a study of the radiation and geochemical background state of soil in the south of Tyumen region. According to the obtained data it was revealed that within the study areas, the power values of gamma radiation exposure dose ranged from 6 to 12.5 mcR/h, due to the natural gamma background of the area. The research results of the soils radiation status have shown that the main contribution to the radiation activity of soils is made by the natural radionuclide potassium-40 and man-made strontium-90. Elevated concentration of strontium-90 was observed in grey forest soils and amounted to 102 Bq/kg. The study of soil radiation contamination in the "Tavda" explosion epicenter (Nizhnetavdinsky district) revealed the dynamics of migration of radionuclides along the soil profile, which takes place at a considerable depth (80-100cm). The maximum concentration of strontium-90 was observed in the area "South-200" at the soil layer of 10-20 cm depth and was 1680 Bq/kg. The presence of heavy metals mobile forms in the soils of the Tyumen region south is within the MPC. The article presents a mathematical processing of the obtained results.

1 Introduction

Soil cover of the biosphere is one of the main components in nature, in which radionuclides and heavy metals are located due to the fact that man discharges them into the environment as a result of his anthropogenic activities [1-21]. When it comes to the artificial radionuclides and heavy metals, soil has a unique sorption capacity that implies an ambivalent meaning for their migration in the biosphere [3, 16, 17]. Specific features of radionuclides sorption by soil absorbing complex provide their strong retention in the terrestrial compartment with a long half-life [1, 14, 15, 19].

In biological chain, soil is one of the main ways by which radionuclides and heavy metals get into the feeding and food plants, and through them – into animals and humans [4, 18, 20]. Due to the construction of an underground oil storage tank in Nizhnetavdinsky District in 1959, as well as the consequences of the East Ural Radioactive Trace and the accident at "Mayak", there is a necessity to check the possibility of artificial radionuclides

* Corresponding author: tgasu.skipin.leonid@mail.ru,

entering the territory of Tyumen region. It is possible that radionuclides cesium-137 and strontium-90 will enter Tyumen region. [10,12].

Technological impact of long-lived radionuclides and heavy metals located in natural objects caused contamination of large areas. Consequently, to assess the ecological status of the ecosystem units, there occurred a need of monitoring research of the artificial radionuclides and heavy metals presence in the region [2, 8, 9, 10, 13].

2 Subjects and Methods

The subject of the research is the soil selected on the benchmark areas located in the territory of the administrative districts at a depth of 0-20 cm in the upper horizontal layer of soil. In the area of Nizhnetavdinsky district, in the center of underground tank construction, sampling was carried out at a depth of 0-10, 10-20, 40-60, 80-100 cm. The samples are represented by the following soil types: sod-podzolic, deep dark grey podzolized, leached chernozem, sod-podzolic gley, grey forest podzolized, grey forest, grey forest podzolized of medium depth, light grey forest.

Sampling was carried out in the south of Tyumen region for the further analysis in accordance with GOST 17.4.3.01-81, GOST 17.4.402-84, GOST 28168-89. The samples were tested in the "Radiation monitoring", "Environmental monitoring" laboratories at the Safety of Technosphere Department of TSUACE.

Radiometry of soil samples was measured with the "Progress" spectrometric complex. The presence of cesium-137 radionuclides was revealed by the scintillation gamma-ray spectrometry, and strontium-90 presence was revealed by scintillation beta-ray spectrometry in the native material. Detection of heavy metals in soil samples was carried out by stripping voltammetry at the "CTA" voltammetry complex and "Temos-Express" complex of sample preparation. EDR (exposure dose rate) of gamma radiation was measured with DRG 01 T1 dosimeter.

3 Results

The level of natural radiation background depends on the concentrations of natural of radionuclides (radium-226, thorium-232 and potassium-40) in the Earth surface, water, air and other elements of the biosphere. Technologically changed natural radiation background is generated from natural sources of ionizing radiation as a result of radionuclides extraction from the earth interior, along with minerals or building materials [21]. Self-purification of the atmosphere upper layers from natural and man-made radionuclides is a slow process, which continues to be a global source of radioactive contamination of the environment. The source of contamination of surface layer atmosphere is the wind-driven rise of radioactive substances from the soil surface in areas contaminated due to nuclear weapons tests and accidents at enterprises of nuclear fuel cycle [11, 12].

The exposure dose rate (EDR) values of gamma-radiation in the southern areas of Tyumen region ranged from 6 to 12.5 mcR/h, due to the natural gamma background of the territory which does not exceed the established reference levels of 15mcR/h for Tyumen region (Fig.1). The maximum EDR values of gamma radiation have been observed in Vagaysky Disrtict (12.27 mcR/h), Omutinsky District (10.82 mcR/h) and Aromashevsky District (10,10mcR/h). It should be noted that increased EDR values were observed in the regions bordering the Omsk region and Kazakhstan. This can be explained by the possible impact of the Semipalatinsk test site and the high concentration of NRN in the earth surface. Minimum rates were recorded in Yalutorovsk (6.23 mcR/h) and Tobolsk Districts (6.46 mcR/h).

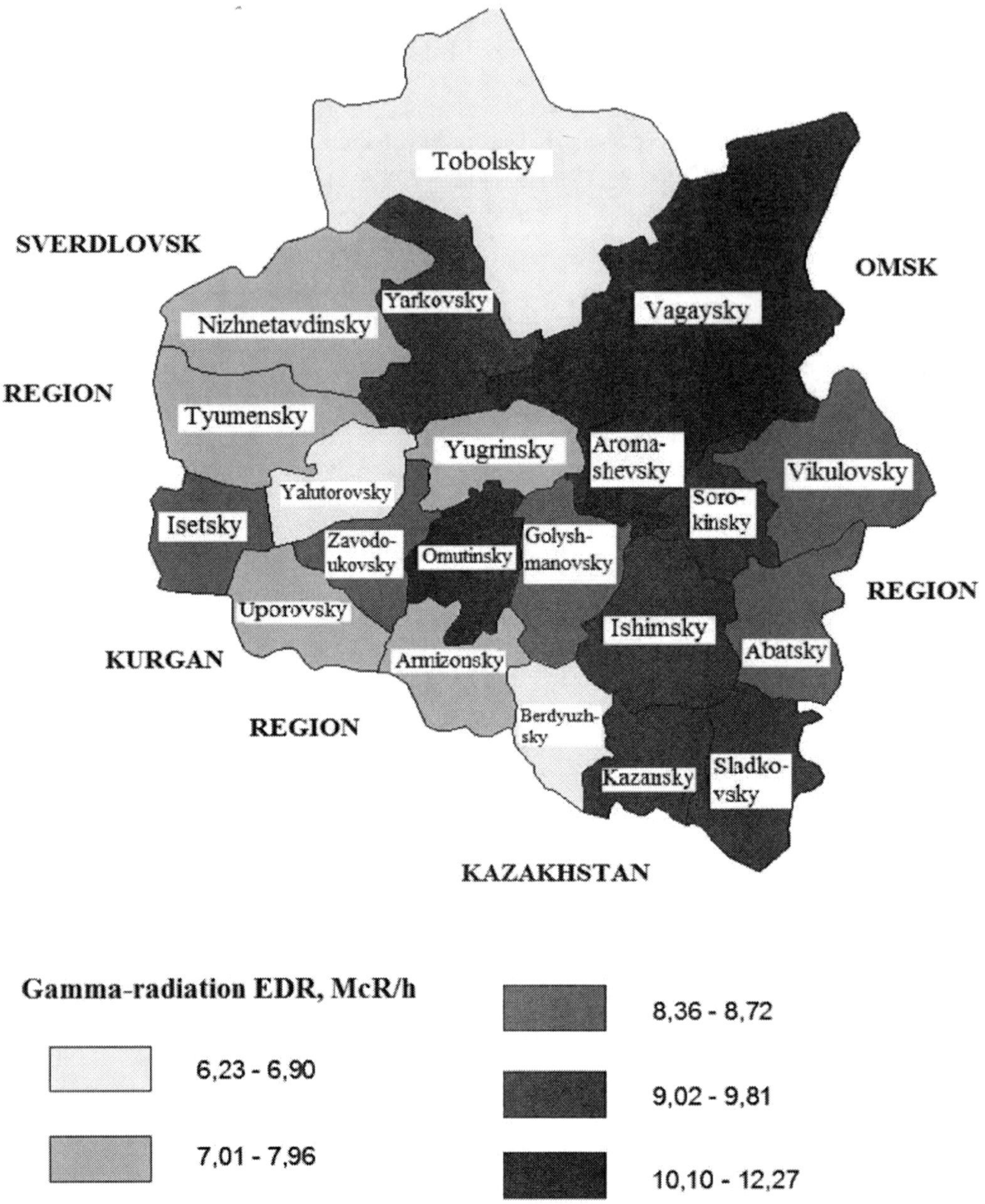

Fig. 1. Exposure dose rate of gamma-radiation on the south of Tyumen region (average values).

Fig. 2 presents the research of soils for the presence of natural and man-made radionuclides in the upper horizontal layer (0-20 cm) in 6 control districts in the south of Tyumen region.

The diagram shows that the natural radionuclide potassium-40 and artificial strontium-90 mostly contribute to the radiation state of soils. The maximum values of the potassium-40 concentration were observed in Vagaysky (926 Bq/kg) and Nizhnetavdinsky (719Bk/kg) Districts. Presence of strontium-90 was 102 and 95 Bq/kg accordingly. Thus, the maximum strontium-90 concentration of 66.4 Bq/kg was recorded in Vagaysky and Isetsky Districts, the minimum - in Ishimsky District (36,3Bk/kg). Thorium-232 and cesium-137 had the minimum values of concentration that amounted to 18.2 and 4.5 Bg/kg in Ishimsky and Tyumensky Districts.

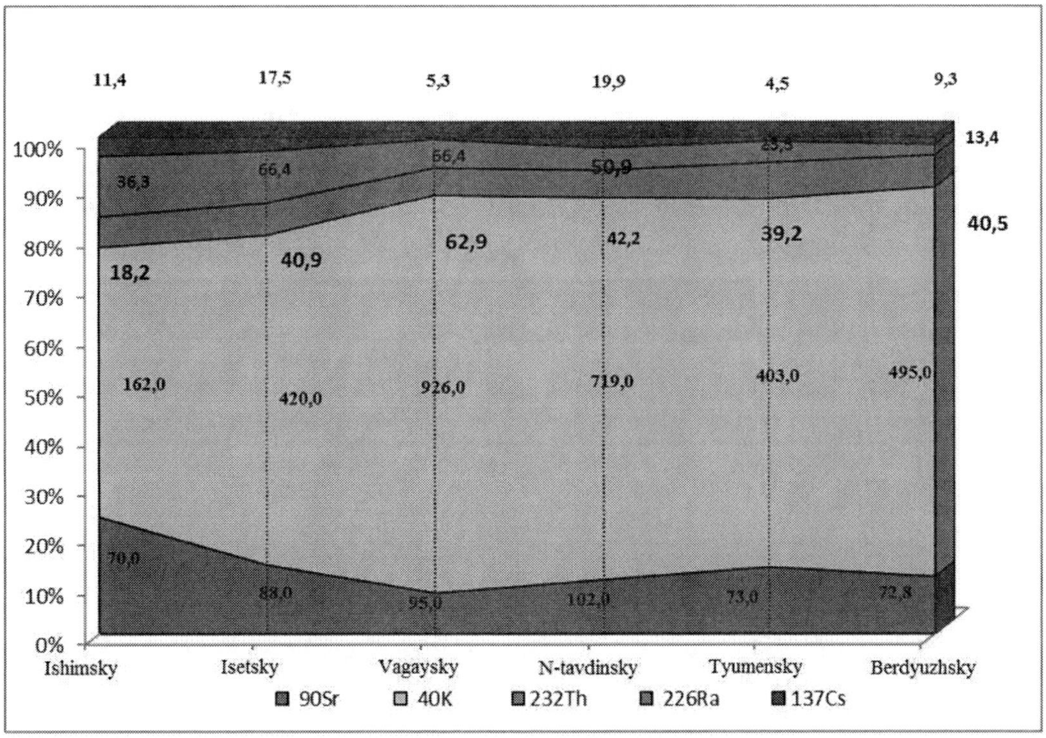

Fig. 2. Presence of natural and man-made radionuclides at a depth of 0-20 cm in control districts.

Figures 3 and 4 graphically represent the presence of cesium-137 and strontium-90 radionuclides in the soils of the control points.

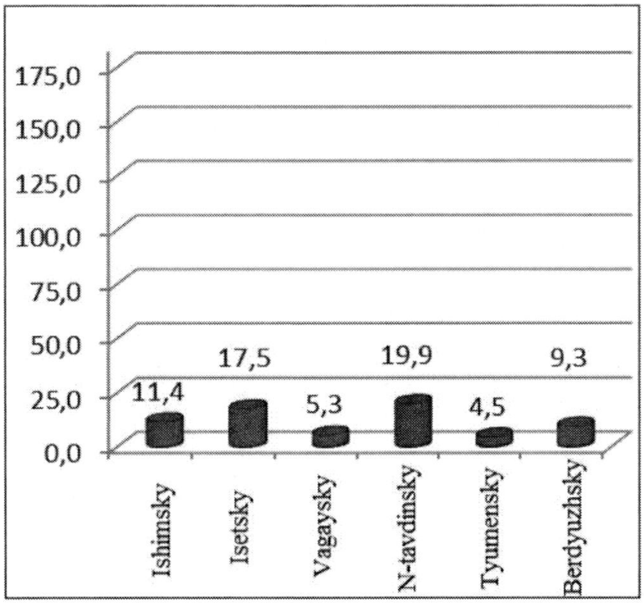

Fig. 3. graphically represent the presence of cesium-137 radionuclides in the soils of the control points.

DOI: 10.1051/matecconf/20167303009

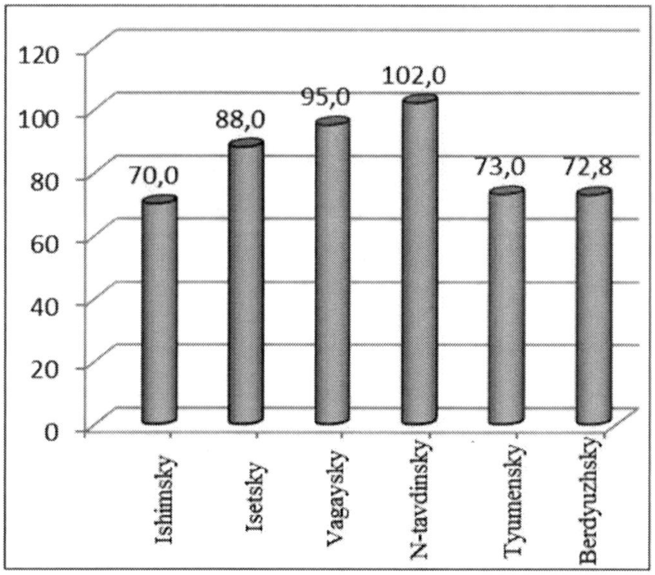

Fig **4.** graphically represent the presence of strontium-90 radionuclides in the soils of the control points

Accumulation of radiocesium in soils is less intense than accumulation of radiostrontium. The highest concentration of both radionuclides was detected in Nizhnetavdinsky District.

The presence of cesium-137 in grey forest soils of Nizhnetavdinsky District was 19.9 Bq/kg, of strontium-90 – 102 Bq/kg. The minimum amount of these elements was recorded in Tyumensky and Ishimsky Districts, which was 4.5 Bq/kg and 70 Bq/kg, accordingly.

Research of the soils radiation state on the territory of the underground peaceful nuclear explosion "Tavda" gave the results, basing on which the concentrations of cesium-137 and strontium-90 were determined in a variety of soil profiles at a depth of 0-10, 10-20, 40-60, 80-100 cm (Fig. 5, 7), as well as the density of radionuclides entering the soil was calculated - distance of 100m and 200m from the epicenter of the underground tank (Fig. 6, 8).

Distribution of cesium-137 in a vertical soil profile is characterized by its maximum accumulation in the topsoil (0-20 cm), which is averagely more than 60% of its total presence in the root zone (0-60 cm).

Maximum lateral migration of radiocesium is observed in an easterly direction at a distance of 200 meters from the explosion.

The study of the vertical distribution of the gross amounts of cesium - 137 along the soil profiles showed that since the moment of soil contamination, radionuclides in these areas migrated to a considerable depth (80-100cm).

The maximum concentration of this element was recorded at the site "East-200" in the soil of 10-20 cm depth and was 296.5 Bq/kg.

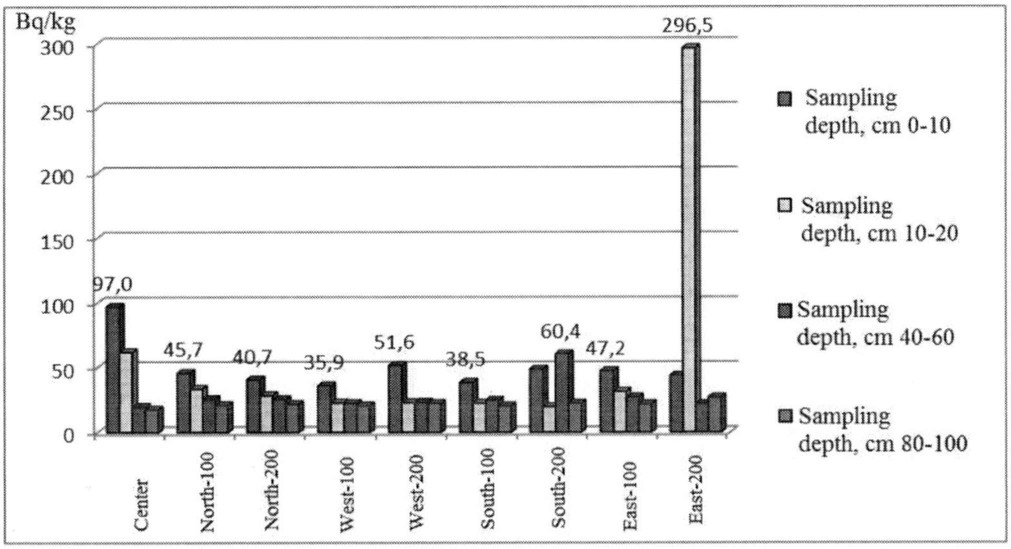

Fig. 5. Concentration of cesium-137 in the soil of Nizhnetavdinsky District at a distance of 100 and 200 meters from the explosion epicenter.

The density of the cesium-137 deposition into the soil (Fig. 6) is of steady rate in both vertical and horizontal directions, and varies from 0.146 to 0.141 Ci/km2.

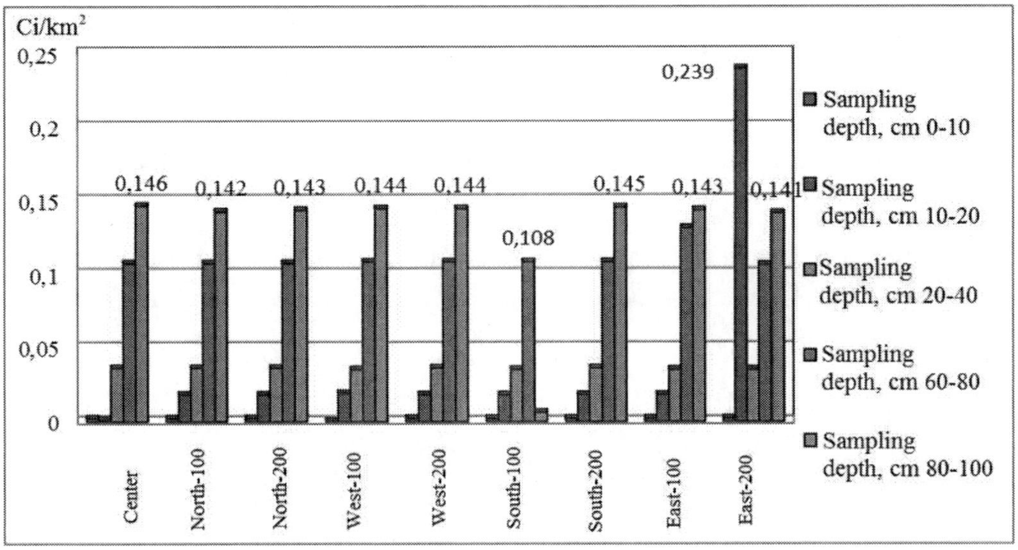

Fig. 6. The density of the cesium-137 deposition into the soil of Nizhnetavdinsky District at a distance of 100 and 200 meters from the explosion epicenter.

The basic mass of radionuclides is concentrated at a depth of 80-100 cm in all directions. The highest density of deposits was 0.239 Ci/km2 at a distance of 200 meters in an easterly direction and at a depth of 10-20 cm, the lowest –0.02 Ci/km2 in the layer of 0-10 cm in all directions.

The accumulation of strontium-90 in grey forest soils is much more intense than accumulation of cesium. All soil profiles are characterized by the consistent pattern of increased concentration of a man-made element (Fig. 7). The maximum radionuclide concentration was observed at a depth of 10-20 cm in a southerly direction at a distance of 200m from the epicenter, and was 1680 Bq/kg.

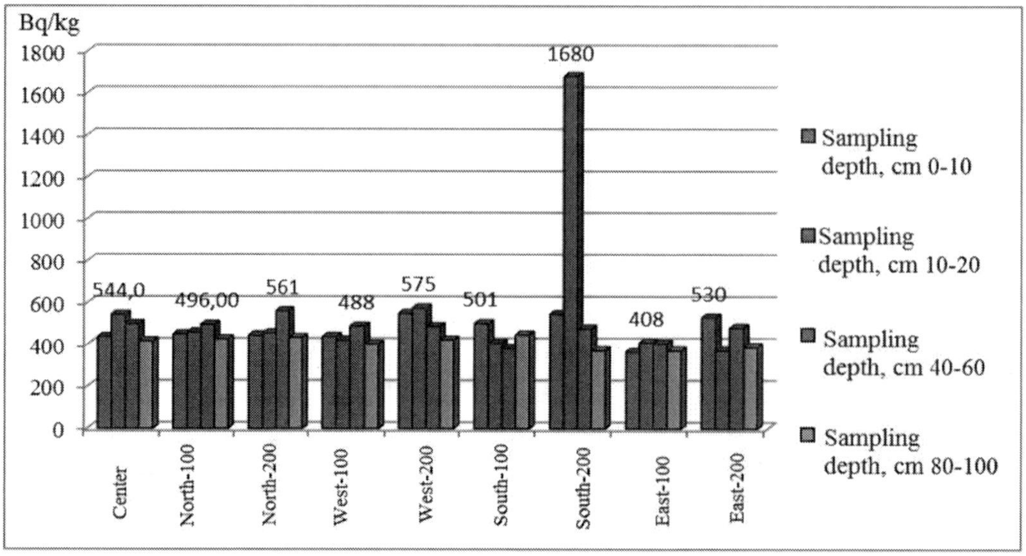

Fig. 7. Concentration of strontium-90 in the soil of Nizhnetavdinsky District at a distance of 100 and 200 meters from the explosion epicenter.

Density of strontium-90 deposition tends to increase with depth of the soil profiles (Fig. 8). In the upper layers of the soil (0-10cm) inactivation of strontium-90 was more intense, and deposition density here does not exceed 0.04 Ci/km2. This fact can be explained by the increased aeration of the upper level, the filtration capacity and biological activity. The maximum density of strontium deposition was observed in the epicenter of the explosion at the depths of 40-100cm. The layer of 80-100cm at the epicenter of the explosion is mostly subjected to contamination, the deposition density here reached 3.464 Ci/km2. In the points North-100, West-100, South-100, East-100, the indicators of soil contamination density amounted to 2.906; 2.827; 0.126; 2.432 Ci/km2 accordingly. This dynamics maintained in samples taken at a distance of 200 meters from the epicenter of the explosion, and concentration of strontium-90 in the layer of 80-100cm decreased by 0.2-0.4 Ci/km2 on average.

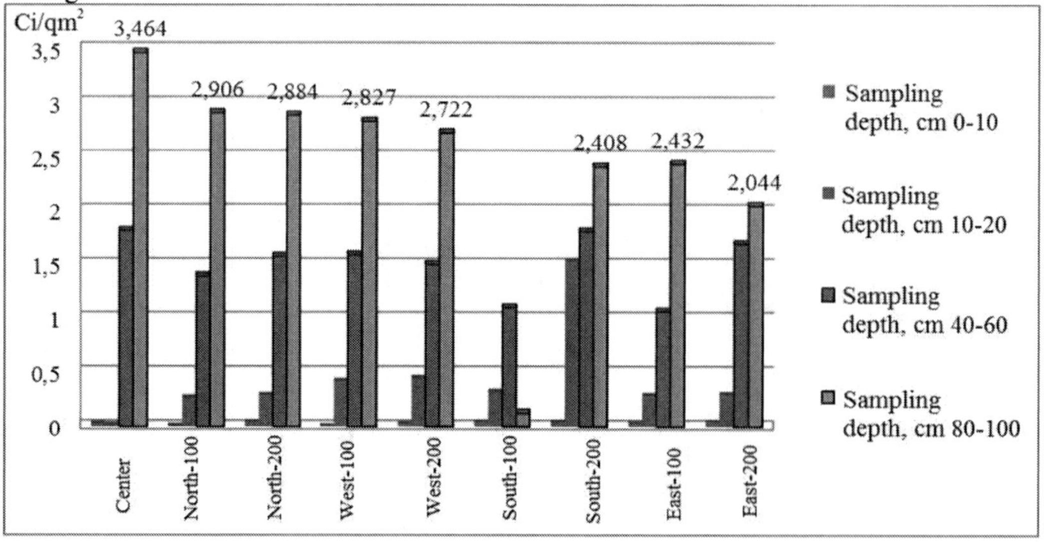

Fig. 8. Density of the strontium-90 deposition into the soil of Nizhnetavdinsky District at a distance of 100 and 200 meters from the explosion epicenter.

Results of the research of the "Tavda" explosion venue raise the possibility of cesium-137 going out of the mouth of the well drilled into the blast zone.

In accordance with the classification of territories ranged in the degree of ecological trouble (Table 1), the soil in the epicenter of the "Tavda" underground explosion can be attributed to the areas of the environmental emergency.

Table 1. Criteria for environmental assessment of the soils state.

Criteria	Ecological disaster, [Ci/km^2]	Environmental emergency, [Ci/km^2]	Satisfactory situation, [Ci/km^2]	Exposure time, [years]	Area of the assessed territory, [thousand hectares]
Radioactive contamination of soils with cesium-137.	Over 40	15-40	Less than 1	1	0.1
Radioactive contamination of soils with strontium-90.	Over 3	1-3	Less than 0.3	1	0.1

Radioactive contamination of soil in layers of 40-60, 60-80, 80-100 cm with strontium-90 was recorded in the range of 1 to 3 Ci/km2. In the central point of sampling the concentration of strontium-90 at a depth of 80-100 cm was 3.464 Ci/km2.

Taking into account the classification of areas ranged in the degree of ecological trouble, this soil can be characterized as a zone of ecological disaster.

In the upper layers of soil (up to 40 cm) the density of contamination with strontium-90 does not exceed 1 Ci/km2, which means that this zone can be attributed to the territory with a satisfactory situation.

Concentration of the mobile forms of heavy metals, acidity and humus in the soil samples was studied in the laboratories of TSUACE. Basing on the results it can be stated that the presence of the mobile forms of heavy metals in the soils of Tyumen region south is within the MPC and can be assessed as "low".

These indicators show that southern territory of Tyumen region is ecologically clean and is favorable for the cultivation of various crops.

Currently, volumes of soils liming are being reduced (in these conditions soil are acid and slightly acid, pH=5.59 averagely), which will lead to the increased mobility of heavy metals.

Soil organic matter (humus) contributes to the binding of the mobile forms of heavy metals and to their transforming into insoluble compounds.

While heavy metals are firmly bound with the corresponding soil organic matters, they are remote for plants, and thus have less detrimental effect on the soil and plants.

Loss of humus in the soil is also able to induce the mobility of heavy metals. This point is proved by some results of the research (Tab. 2).

Table 2. Average data of the mobile forms of heavy metals, humus and acidity in the soils of the southern part of Tyumen Region, 2012-2015.

District	[Zn]	[Cu]	[Cd]	[Pb]	Humus. [%]	[pH]
Taiga-forest zone		South Taiga				
Tobolsky	1.14±0.34	0.43±0.13	0.09±0.03	1.79±0.54	4.20	5.30
Vagaysky	1.49±0.45	0.34±0.10	0.12±0.04	1.51±0.45	3.60	5.40
		Sub-Boreal Forest				
Nizhnetavdinsky	0.82±0.25	0.31±0.09	0.08±0.02	2.46±0.74	3.70	5.20
Yarkovsky	1.20±0.36	0.51±0.15	0.07±0.02	1.77±0.53	4.10	5.60
Yurginsky	0.57±0.17	0.30±0.09	0.06±0.02	3.04±0.91	4.30	5.40
Aromashevsky	<0.01	<0.05	<0.01	<0.10	4.50	5.40
Vikulovsky	<0.01	<0.05	<0.01	<0.10	4.80	5.50
Sorokinski	<0.01	<0.05	<0.01	<0.10	4.00	6.00
Forest-steppe zone		North Forest Steppe				
Tyumensky	1.39±0.41	0.48±0.14	0.08±0.02	1.66±0.50	5.50	5.70
Isetsky	0.98±0.29	0.42±0.13	0.13±0.04	1.86±0.56	6.30	5.40
Yalutorovsky	1.74±0.52	0.64±0.19	0.18±0.05	1.20±0.36	5.40	5.80
Zavodoukovsky	1.07±0.32	0.37±0.11	0.12±0.04	1.20±0.36	6.50	5.30
Uporovsky	0.47±0.14	0.23±0.07	0.07±0.02	1.18±0.35	5.80	5.40
Omutinsky	0.90±0.27	0.38±0.11	0.06±0.02	1.26±0.38	6.00	5.50
Golyshmanovsky	<0.01	<0.05	<0.01	<0.10	4.30	5.70
Ishymsky	<0.01	<0.05	<0.01	<0.10	5.80	5.70
Abatsky	<0.01	<0.05	<0.01	<0.10	5.30	5.80
		South Forest Steppe				
Armizonsky	1.07±0.32	0.57±0.17	0.08±0.02	1.62±0.49	6.10	5.40
Berdyuzhsky	<0.01	<0.05	<0.01	<0.10	4.00	6.00
Kazansky	<0.01	<0.05	<0.01	<0.10	5.40	5.90
Sladkovsky	<0.01	<0.05	<0.01	<0.10	5.80	5.90
MPC/APC	23.00	3.00	0.50	6.00		

When the presence of humus is 4.3% (Yurginsky District), the concentration of mobile forms of lead is 3.04 mg/kg. If humus is 6.1% (Armizonsky District), 6.5% (Zavodoukovsky Disrtict), then lead - 1.62 and 1.20 mg/kg, accordingly. Rather significant correlation dependence was observed only between the humus presence and the mobile lead (r=-0.54). Zinc and copper show insignificant dependence on the presence of organic matter in the soil. This can be explained by the fact that these chemical elements intensively form tightly bound complexes with mineral colloidal fractions. It is important to note that soils of the southern part of Tyumen region are characterized as heavy-textured.

After analyzing the correlation dependence between the heavy metals and agrochemical parameters (humus and soil acidity) it can be stated that mechanisms of the metals retention in soil include: adsorption on the surface of the humus, formation of complex compounds with humus, and formation of insoluble compounds, especially within remediation, when reaction of the environment plays an important role. There is an inverse relation (r = from -0.30 to -0.90) between the humus presence and the mobility of heavy metals which means that with increase of humus, the mobility of heavy metals decreases. The data of correlation analysis of soil acidity and mobile forms of heavy metals have different characteristics. There is a direct dependence between the acidity and the presence of cadmium mobile forms (r = 0.68); an inverse relation - between acidity and lead (r = -0.69). Positive coefficient of correlation presumes that with the increase of the first characteristic, the second one increases as well. Negative coefficient presumes that with the increase of acidity, the mobility of zinc and lead decreases.

Mathematical processing of the results (Table 1) showed a significant variability, the coefficient of variation of all the elements is over 20%. Lead has the greatest degree of dispersion (V = 88%), copper is the least susceptible to variations (V = 60%). According to the degree of variation heavy metals form the following decreasing series: Pb> Cd> Zn> Cu.

Beyond that, the presence of some heavy metals determines the presence or, on the contrary, absence of the other. The increase of the lead mobility is directly proportional to the zinc mobility (r = 0.70) and inversely proportional to the mobility of cadmium (r = -0.92). Change in the presence of copper is directly proportional to the change in the zinc presence (r = 0.44).

4 Conclusion

After analyzing the concentration of cesium-137 and strontium-90 radionuclides in soils of the southern parts of Tyumen region, it can be concluded that soil contains small quantities of radionuclides, and the natural gamma background is within the norm.

The dose rate of gamma radiation at a height of 1m from the ground surface varied from 6 to 12.5 mcR/h, which corresponds to the natural background radiation, and does not exceed the norm of 15 mcR/h for Tyumen region. These results do not raise serious environmental concerns. Besides, by taking the agrotechnical measures, one can significantly reduce the concentration and proximity of these radionuclides in the treated soils.

Analysis of the data on concentrations of the man-made radionuclides in the soil of the "Tavda" underground explosion epicenter (Nizhnetavdinsky District) refers to a local crisis in the deep layers of the soil profile, as well as the need for continuous radiation monitoring of the area.

Soils examination for heavy metals concentration showed that they do not exceed the MPC, although they are unevenly distributed in the surveyed areas. As for mobility in the soil, metals form the following row: Pb> Cd> Zn> Cu. With the increase of soil acidity, about 50% of cadmium evolves into mobile form, and about 50% of lead, on the contrary, is bound by the soil organo-mineral complex in the forms inaccessible for plants. Beyond that, with increasing zinc mobility, the mobility of lead increases as well.

With the increase of the lead proximity, the concentration of cadmium mobile forms reduces.

The dependence between the humus concentration and heavy metals is weak, so it is impossible to say for sure whether this characteristic does or doesn't influence on the mobility of heavy metals in soil. Geochemical situation in the south of Tyumen region can be described as weakly exposed to the technogenic contamination.

References

1. A.G. Berseneva, Sbornik materialov XV nauchno-prakticheskoy konferentsii molodykh uchenykh, aspirantov, soiskateley i magistrantov TyumGASU, 16-21 (2015)
2. A.G. Berseneva, Aktual'nye problemy stroitel'stva, ekologii i energosberezheniya v usloviyakh Zapadnoy Sibiri: sbornik materialov Mezhdunarodnoy nauchno-prakticheskoy konferentsii, 153-156 (2014)
3. L.A. Bol'shov, Rol' radiatsionnykh faktorov v ekologicheskikh riskakh dlya naseleniya Rossii: TSNII atominform, 29-32 (2001)
4. A.A. Vaymer, DSc Thesis (Tyumen, 2006)
5. E.V. Gaevaya, Vestnik Krasnoyarskogo gosudarstvennogo agrarnogo universiteta, **11**, 149-153 (2013)
6. E.V. Gaevaya, Visnik Dnipropetrovs'kogo derzhavnogo agrarnogo universitetu, **2(34)**, 99-102 (2014)
7. G.P. Gamzikov, Plodorodie pochv i resursosberezhenie v zemledelii: materialy vseros. nauch.-prakt. konf., 20-27 (2003)
8. A.N. Guseynov, *Ekologiya goroda Tyumeni: sostoyanie, problem* (Slovo, Tyumen, 2001)
9. G.A. Demidenko, N.V. Fomina, *Monitoring okruzhayushchey sredy* (Krasnoyarsk, 2013)
10. V.A. Logachev, *Mirnye yadernye vzryvy: obespechenie obshchey i radiatsionnoy bezopasnosti pri ikh provedeni* (IzdAt, Tyumen, 2001)
11. E.V. Zakharova, Plodorodie, **6(81)**, 46-48 (2014)
12. E.V. Zakharova, Agroprodovol'stvennaya politika Rossii, **9**, 88-92 (2013)
13. Kh. Zigel', A. Zigel', Nekotorye voprosy toksichnosti ionov metallov (Mir, 1993)
14. M.A. Kiseleva, Sbornik materialov XV nauchno-prakticheskoy konferentsii molodykh uchenykh, aspirantov, soiskateley i magistrantov TyumGASU. 43-49 (2015)
15. M.A. Kiseleva, Aktual'nye problemy arkhitektury, stroitel'stva, ekologii i energosberezheniya v usloviyakh Zapadnoy Sibiri: Sbornik materialov mezhdunarodnoy nauchno-prakticheskoy konferentsi, 227-232 (2015)
16. A.O. Oznobikhina, Sbornik materialov XV nauchno-prakticheskoy konferentsii molodykh uchenykh, aspirantov, soiskateley i magistrantov TyumGASU. 96-100 (2015)
17. A.O. Oznobikhina, Sbornik materialov XV nauchno-prakticheskoy konferentsii molodykh uchenykh, aspirantov, soiskateley i magistrantov TyumGASU, 92-96 (2015)
18. Yu.A. Sapozhnikov, R.A. Aliev, S.N. Kalmykov, *Radioaktivnost' okruzhayushchey sredy. Teoriya i praktika* (BINOM, Tyumen, 2006)
19. L.N. Skipin, Agroprodovol'stvennaya politika Rossii, **4(16)**, 29 (2014)
20. V.D. Starkov, *Radiatsionnaya ekologiya* (IPP "Tyumen"', Tyumen, 2001)
21. A.T. Khusainov, L.N. Skipin, L.I. Sofronova, *Vliyanie otkhodov uranopererabatyvayushchikh predpriyatiy na sostoyanie komponentov ekosistem Severnogo Kazakhstana* (Kokshetau, Tyumen, 2012)

Environmental Safety for Chernozem Soil Fertilized with Phosphogypsum and Ash for Spring Wheat Cultivation in North Kazakhstan

Ainur Muhanbet [1,*], *Abilzhan* Khusainov [2] , and *Sagintai* Elubaev [3]

[1]Kazakh National Agrarian University, 050010 Abay Avenue 8, Almaty, The Republic of Kazakhstan
[2]Academy "Kokshe", 020000 Kaptsevich street 247, Kokshetau, The Republic of Kazakhstan
[3]Kokshetau Abay Myrzakhmetov University, 020000 Auezov street, Kokshetau, The Republic of Kazakhstan

Abstract. Currently, year by year the industrial waste stockpiles increase, but at the same time its recycling is not more than 10-15% of the production. There are several methods of recycling of such waste as phosphogypsum and ash, among which one of the perspective directions is considered using them in agriculture as fertilizer and ameliorator. The limiting factor of their widespread use in agriculture is the presence of heavy metals and radionuclides in them. This article presents data about environmental safety of using phosphogypsum and ash for cultivation spring wheat on the chernozem soils of Northern Kazakhstan. According to the results of research, using phosphogypsum and ash doesn't give negative impact on the environment, the presence of heavy metals and radionuclides in the soil and grain do not exceed the maximum allowable concentrations.

1 Introduction

The accumulation of industrial waste is a global environmental problem. Industrial waste, the production of which is growing every year, includes ash and phosphogypsum. The accumulation of ash adversely affects the environment, it occupies large areas, due to its sparseness it is hard for transporting, light particles are carried by wind and pollute nearby water courses and land and cause damage to the local population's health by penetrating into the body through the respiratory system[1].

Annual production of ash and its utilization varies greatly in different countries. Annual production of ash in China is about100 million tons, 75 million tons in USA, 40in Germany, the UK and Canada and15 and 6 million tons respectively. If the disposal of ash and in China and the USA is 38.45, and 65% of the annual production, in Germany, England and Canada they utilized 85, 50and 75% of the annual production of ash[2].

Phosphogypsum is a by-product of the production of phosphate fertilizers [3]. The chemical composition of phosphogypsum is represented by CaO, sulfates (such as SO3), SiO2, Al2O3, Fe2O3, P2O5, and F, contains heavy metals such as lead, cadmium, chromium, mercury and selenium [4] Only 15% of the total production of phosphogypsum

* Corresponding author: muhanbet.ainur@mail.ru

in the world is used mainly in the construction and agricultural sectors. The remaining 85% are not recycled, and of course it causes great harm to the environment, occupies vast areas, polluting the environment with chemicals and radiation. One of the promising areas of application of industrial wastes is their using as fertilizer or ameliorator. Phosphogypsum in agriculture is used to improve soil structure and increase productivity [5], reducing erosion and increasing soil available presence of S and P forms [6].

In turn, the using of ash for various crops helps not only to increase concentration of such elements required for plants such as K, Na, Ca, and Mg, but also reduce the presence of heavy metals such as Mo, Se, Al [7]. Ash adding helps to reduce the presence of heavy metals in plant tissues such as Cd, Cu, Cr; The researchers attribute this to improving reaction of the soil environment [8].

In Kazakhstan, the annual output of ash and ash mixtures from the combustion of coal is about 19 million tons, and in the ash dumps, nowadays, there has been accumulated more than 300 million tons of waste, the volume of their use does not exceed 10% in our country. [9]. Only in Akmola region as a result of TPP and boiler rooms work in 2011, according to the Department of Natural Resources and Environmental Control ,there was formed 599, 372 thousand tons of ashes representing 29% of the total volume of the produced ones in Northern Kazakhstan (202 457.6 thousand tons) [10]. In general, in Northern Kazakhstan, according to the Department of Natural Resources and Environmental Control of Akmola, Kostanai, Pavlodar and North Kazakhstan regions, there were formed 202.5 millions of tons of ash waste, the amount of formation of phosphogypsum only in Stepnogorsk is 9 million tons [11].

That means, in the country for a long time there is an acute issue of utilization of stockpiled ash and phosphogypsum. However, despite the above mentioned numerous studies demonstrating the possibilities and prospects for the use of phosphogypsum and ash in agriculture [5-8], information about the studies on phosphogypsum and ash in Northern Kazakhstan is very small. It is also necessary to take into account the fact that a number of foreign scientists have warned about the dangers of the use of phosphogypsum because of the presence of toxic metals and radionuclides, which, as a result of steeping can contaminate the soil and plants [12,13]. In addition, Page's studies show that the use of ash increases the presence of boron, which in turn inhibits microbial activity [14].

Thus, for the use of phosphogypsum and ash as fertilizer a thorough research about the environmental safety of industrial waste in soil and climatic conditions of Northern Kazakhstan is necessary. On this basis, field experiments and laboratory studies have been conducted to study the use of ash and phosphogypsum as the fertilizers on chernozem soils. The results of the environmental impact on the safety of industrial waste are presented in this article.

2 Material and Methods

The experiments were conducted on the experimental field of LPO "North-Kazakhstan Research Institute of Agriculture", Chaglinka village, Akmola region (53 ° 10 '9.12 "N, 69 ° 7' 37.57" E 53.1692 °, 69.127103 °), during 2014 -2015. Spring wheat of local selection "Astana" was cultivated in the experiment and zonal cultivation technology was used. Soil from the experimental area - ordinary black loam moderately low humus content. Ashes for the experiments were taken from gold shaft district in Kokshetau and pre-sowing treatment for 10 days before planting was done. Phosphogypsum was taken from dumps LPO "Kazphosphate" Stepnogorsk. Experiments have been carried down according to systematic method in triplicate. The scheme of experiments with ash includes 6 following options: (1) control, (2) superphosphate (3) Ashes from 0.2 t / ha (4) Ashes from 0.3 t / ha (5) Ashes from 0.4 t / ha (6) ash and 0.5 t / ha. The scheme of experiments with phosphogypsum is also presented in six options: (1) control, (2), superphosphate (3) phosphogypsum 1 t / ha (4) phosphogypsum 2 t / ha (5) phosphogypsum 3 t / ha (6) and phosphogypsum 4 t / ha.

The determination of zinc, copper, lead and cadmium in soil and plant samples was conducted on atomic absorption spectrometry (AA 6200, Japan) in atomic adsorption mode. The definition of the presence of mobile forms of heavy metals in soil samples was carried out in 0.1N HNO3 extract; and the determination of heavy metals in plant samples was performed after mineralization by dry ashing to constant weight.

Results and their discussion

Among the pollutants on the scale and effects of pollution on biological objects, heavy metals take special place. In principle, many of them are essential for living organisms, but as a result of intensive atmospheric dispersion in the biosphere and a significant concentration in the soil, it becomes toxic to biota.

Migration and accumulation of contaminants such as heavy metals and radionuclides by the complex is closely related with the processes as leaching, capillaries size, sorption, pH, humidity of the root system and etc. [15,16].

It is also known that heavy metals are found in nature, in the soil organic material, clay, in the Fe or Mn oxides, carbonates, etc., without disturbing the ecological balance [17-19].

It is known that, the term "maximum permissible concentration" (MPC) is used to assess major environmental soil pollution standards, such as contained element in the soil, which for constant contact or interaction for a certain period of time has no effect on the animals, plants, microorganisms and humans.

According to modern conception, the MPC of chemical substances in the soil is a complex indicator of harmless to human of chemicals in the soil, as used in its justification criteria reflect the impact of the possible ways of contacting elements on the environment, soil biological activity and its self-purification processes.

The laboratory tests confirm the sanitary and environmental safety of introducing ash into spring wheat on chernozem soils of Northern Kazakhstan. The use of different doses of ash entering from 0.2 to 0.4 tons / ha does not exceed MPC of heavy metals in the soil and wheat grain (Table 1).

According to the research, during the year of application of ashes the excess of Pb in the soil was observed on all variants of the experiment and ranged from 4.08 (Ashes from 0.4 t / ha) to 4.51 mg / kg (Ashes from 0.2 t / ha) that exceed the control at 3.65 and 4.08 mg / kg respectively. However, it did not affect the content of Pb in the grain of wheat, where there was noted a slight excess - in the range 0.03-0.1 mg / kg. Noteworthy is the fact that in the following year after the introduction of ash, Pb content in soil was reduced by half, and Pb content in the grain and does less control. We suppose that this is due to the improvement of soil reaction and absorption ability stipulated with the introduction of ash, that is confirmed by Petruzelli's data[8].

Significant increase of Cu and Zn content was not observed in the grains. The highest rate of Cu in 2014 noted at the option Ashes 0.3 t / ha (4.8 mg / kg) and the excess Zn in the form of ash 0.5 t / ha (10.7 mg / kg) does not exceed the maximum permissible concentration. In 2015, the content of these elements in wheat grains did not differ from the control variant. However, there was a significant excess of Cu and Zn in the soil of the ash entering year and in subsequent years. Compared with the control, an increase in the content of Cu to various doses of ash entering varied slightly and averaged 5.2 mg / 2014 and 1.47 to 2015. Content of Zn is also increased in 2014 and in 2015 and the total amount was 17.08 mg / kg and 3.75 mg / kg, respectively. This is largely due to the high content of these elements in the chemical composition of the ash. It is necessary to point out that both Zn, and Cu are also trace elements, essential for the normal growth and development of plants [20]. Thus, we can conclude that the introduction of ash improves soil nutrient regime, since the use of ash does not exceed the maximum permissible concentration.

Table 1. The effect of dose application of ash on the content of heavy metals in the soil and in grain of spring wheat, mg / kg.

Experiment variants	Pb		Cd		Cu		Zn	
	2014y.	2015y.	2014y.	2015y	2014y.	2015y.	2014y.	2015y.
Soil*								
Control	0.43	0.47	0.0	0.078	0.5	0.21	0.12	0.93
P20	0.50	2.13	0.0	0.038	0.32	1.24	0.38	1.82
Ash 0,2 [t/ha]	4.51	0.84	1.7	0.054	5.1	1.26	17.0	3.28
Ash 0,3 [t/ha]	4.20	1.84	1.5	0.071	5.4	1.84	17.2	3.95
Ash 0,4 [t/ha]	4.08	2.86	1.2	0.042	5.7	1.68	17.1	4.68
Ash 0,5 [t/ha]	4.12	2.74	1.3	0.028	5.9	1.89	17.8	4.75
Wheat**								
Control	0.25	0.44	0	0	0.09	1.6	7.7	6.3
P20	0.28	0.46	0	0	0.16	1.9	3.5	2.8
Ash 0,2 [t/ha]	0.14	0.094	0	0	4.2	0.35	8.8	1.8
Ash 0,3 [t/ha]	0.28	0.091	0	0	4.8	0.95	9.2	2.0
Ash 0,4 [t/ha]	0.31	0.06	0	0	2.8	0.87	10.1	2.3
Ash 0,5 [t/ha]	0.35	0.13	0	0	0.87	0.76	10.7	2.7
MPC*	Pb – 32		Cd -3,0		Cu – 33		Zn – 23	
MPC**	Pb – 0.50		Cd – 0.10		Cu – 10		Zn – 50	

The excess of Cd to within a 2.7-4.71 mg / kg observed in the soil in the year ash was added, while the content was not found in the wheat grain. The following year as compared to the control, there was not exceeding the level of Cd, in any of the variants of the experiment.

As a result, despite a slight excess on the deposit options ash content of mobile heavy metals did not exceed the maximum allowable concentrations or in the soil or in the grain of wheat. The results of our study are confirmed by the experience of foreign scientists. In the experiments made by Nileshusing of ash did not exceed Cd and Pb in the grain of wheat, beans and mung slightly (within MPC) exceeding Zn [21] content.

As mentioned above, one of the problems hindering its widespread in agriculture is the content of radioactive Radium and radon. However, according to Mays and Mortved, introduction of 112t / ha phosphogypsum containing 25 pCi 226Ra in loamy soils helped to grow such cereals successfully(maize, spring wheat) and oilseeds (safflower). As they note, the application rate of 112 t / ha application rate exceeds the plaster under the peanuts more than 200 times, but it was not observed any negative effect on the excess of the radioactivity of any grain or in the soil [22].

At the same time some scientists warn about the dangers of the use of phosphogypsum because of the content of toxic metals and radionuclides, that as a result of leaching can contaminate the soil and plants [12,13].

Our studies have shown that the use of phosphogypsum in doses of 1-4 t / ha contributed to a slight increase in the content of heavy metals in soil and wheat grain, but did not exceed the maximum allowable concentrations (Table 2). According to Table 2, in the year of application of phosphogypsum as compared to the control there was no excess of heavy metals in the soil in all experimental variants. However, in the following year noted an excess of Cu Zn and Pb. Partly, it can be explained by poor solubility of phosphogypsum in soil, and therefore in the second year the strong aftereffect of phosphogypsum was observed. Comparing different doses of phosphogypsum application, an excess of the highest Pb, Zn and Cu are found to form phosphogypsum 4t / ha; in comparison with the control, excess Pb in this embodiment made - 2.55 mg / kg: Zn 4.58 mg / kg and Cu 3.93 mg / kg. However, these data, is the greatest indicator of increasing data content items, are within the acceptable concentrations.

Regarding the impact of phosphogypsum for heavy metal content in the grain must be noted that the application rates of phosphogypsum had not significant impact on their content, except Zn, the content of which in the wheat grain in 2014 exceeded the control on an average of 4.18-6.88 mg / kg . However, already in the following year the Zn content in the wheat grain has come back to normal, and all indicators do not exceed their maximum allowable concentrations.

O. V. Dubravina notes that introduction of phosphogypsum at the norm of 3 t/hectare doesn't cause accumulation in the soil of heavy metals, but increases in the humus of the Kamennaya steppe the presence of P2O5 on 5.6 mg/kg of the soil [23].

N. A. Protasova, N. S. Gorbunova claim that prolonged using of phosphogypsum and composts on its basis doesn't lead to essential change of HM presence in the plain chernozem soil and agricultural production, thanks to high buffering and ecological stability of humus [24].

Table 2. Influence of application of phosphogypsum on the presence of heavy metals in the soil and in a grain of a spring wheat, mg/kg.

Experience options	Pb		Cd		Cu		Zn	
	2014	2015	2014	2015	2014	2015	2014	2015
Soil*								
Control	0.48	0.41	0.00	0.052	0.7	0.27	0.13	0.8
P20	0.52	2.28	0.00	0.034	0.5	1.8	0.35	1.74
Phosphogypsum1 [t/hectare]	0.44	1.27	0.00	0.061	0.8	2.3	0.02	3.8
Phosphogypsum 2 [t/hectare]	0.45	1.82	0.00	0.037	0.7	2.5	0.03	4.1
Phosphogypsum 3 [t/hectare]	0.41	1.93	0.00	0.083	0.6	3.2	0.02	4.8
Phosphogypsum 4 [t/hectare]	0.42	2.96	0.00	0.054	0.6	4.2	0.02	5.38
Grain**								
Control	0.28	0.34	0	0	0.12	1.8	7.3	6.0
P20	0.23	0.45	0	0	1.9	1.4	3.6	3.0
Phosphogypsum1 [t/hectare]	0.08	0.043	0	0	0.84	2.8	4.2	8.5
Phosphogypsum 2 [t/hectare]	0.09	0.045	0	0	4.3	1.2	5.0	9.3
Phosphogypsum 3 [t/hectare]	0.10	0.17	0	0	6.8	0.9	6.1	9.4
Phosphogypsum 4 [t/hectare]	0.12	0.16	0	0.033	7.0	1.1	7.2	10.1
MPC*	Pb – 32		Cd –3,0		Cu – 33		Zn – 23	
MPC**	Pb – 0,50		Cd – 0,10		Cu – 10		Zn – 50	

3 Conclusion

Researches have shown ecological safety of application of waste of the local industry as fertilizer on black soils of Northern Kazakhstan. Introduction of doses of phosphogypsum till 4/t and ash till 0.4 t/hectare hasn't exceeded maximum-permissible concentrations of heavy metals and radionuclides in the plain chernozem soil and in a grain of a spring wheat.

References

1. J Jankowski, C.R. Ward, D. French, S. Groves, Fuel, **85**, 243-256 (2006)
2. Information on http://www.tifac.org.in
3. A.H. Hunter, *Use of phosphogypsum fortified with other selected essential elements as a soil amendment on low cationexchange soils* (Florida Institute of Phosphate Research, 1989)
4. R. Taha, R.K. Seals, Proceedings of Utilization of Waste Materials in Civil EngineeringConstruction New York, NY. American Society of Civil Engineering (1992)
5. G.L. Mullins, C.C. Mitchell, *Use of phosphogypsumto increase yield and quality of annual forages* (Florida Institute of Phosphate, Bartow, 1990)
6. X. Zhang, W. Miller, M. Nearing, L.D. Norton, Transactions of ASAE, **41(4)**, 989-994 (1998)
7. S.K. Sharma, N. Kalra, J SciInd Res, **65(5)**, 383–90 (2006)
8. G. Petruzzelli, L. Lubrano, S. Cervelli, *Heavy metal uptake by wheatseedling grown on fly ash amended soils. J Environ Qual* (London, 1986)
9. A.U. Akhmedyanov, K.Zh. Kirgizbayeva, G.I. Turekhanova, Eurasian National University named after L.N. Gumilyov, Kazakhstan.
10. Information on http://www.rusnauka.com/19_AND_2012/Tecnic/10_114203.doc.html
11. *The information bulletin about condition of environment of the North Kazakhstan region for 2011* (2011)
12. A.T. Khusainov, A.A. Sarsenova, Achievements of modern Natural Sciences: Materials of VI All-Russian scientific conference with the international participation. Sochi, Russian Federation, **2**, 73-75 (2005)
13. P.M. Rutherford, M.J. Dudas, R.A. Samek, Science of theTotal Environment, **99**, 1-38 (1994)
14. M.S. Al-Masri, A. Ali, M. Keitou, Z. Al-Hares, Environmental radiochemical analysis. London: Royal Society of Chemistry (1999)
15. A.L. Page, A.A. Elseewi, I.R. Straughan, Residue Rev, **71**, 83-120 (1979)
16. J.J. Mortverdt, Journal of Environmental Quality, **23**, 643–650 (1994)
17. C. Papastefanou, S. Stoulos, A. Ioannidou, M. Manolopoulou, Journal ofEnvironmental Radioactivity, **89**, 188-198 (2006)
18. A. Tessier, P.G. Campbel, M. Bisson, Analytical Chemistry, **51(7)**, (1979)
19. A.M. Ure, P. Quevauviller, H. Muntau, B. Griepink, International Journal of Environmental Analytical Chemistry, **51**, 135-151 (1993)
20. B.J. Alloway, *Heavy metals in soils (2nd ed.)* (Blackie, London, 1995)
21. C. Zaccone, C. Cocozza, V. D'Orazio, C. Plaza, A.K. Cheburkin, T.M. Miano, Talanta, **73**, 820-830 (2007)
22. N. K Mahale, S. D Patil, D. B Sarode, S. B. Attarde, Pol.J.Environ.Stud, **21(6)**, 1713-1719 (2006)
23. D. Mays, J. Mortvedt, Journal of Environmental Quality, **15**, 78-81 (1986)
24. O.V. Dubrovina, *Influence of fertilizers and phosphogypsum on productivity, quality of grain of corn and fertility of the plain black soil in the conditions of the Southeast Central Humus Reserve. Candidate of agricultural sciences dissertationauthor's abstract, Kamennaya steppe* (2004)
25. N.A. Protasova, N.S. Gorbunov, Agrochemistry, **7**, 52-61 (2010)
26. T. Maltseva, A. Nabokov, A. Chernykh, Procedia Engineering, **117**, 239-245 (2015)

AUTHOR INDEX

AUTHOR INDEX

AUTHOR INDEX

AUTHOR INDEX

XV International Conference "Topical Problems of Architecture, Civil Engineering, Energy Efficiency and Ecology – 2016"

MATEC Web of Conferences Volume 73 (2016)

Tyumen, Russia
27 - 29 April 2016

Part 2 of 2

Editors:

Tatiana Maltseva

ISBN: 978-1-5108-2871-1

Printed from e-media with permission by:

Curran Associates, Inc.
57 Morehouse Lane
Red Hook, NY 12571

Some format issues inherent in the e-media version may also appear in this print version.

Printed by Curran Associates, Inc. (2016)

For additional information, please contact EDP Sciences – Web of Conferences
at the address below.

EDP Sciences – Web of Conferences
17, Avenue du Hoggar
Parc d'Activité de Courtabœuf
BP 112
F-91944 Les Ulis Cedex A
France

Phone: +33 (0) 1 69 18 75 75
Fax: +33 (0) 1 69 28 84 91

contact-edps@webofconferences.org

Additional copies of this publication are available from:

Curran Associates, Inc.
57 Morehouse Lane
Red Hook, NY 12571 USA
Phone: 845-758-0400
Fax: 845-758-2633
Email: curran@proceedings.com
Web: www.proceedings.com

TABLE OF CONTENTS

PART 1

PART 2

The Determination of Stress State of Structures Considering Sequence of Construction and Load Application

Vladimir Kuroedov[1], *Luka* Akimov[1,*], *Artem* Frolov[1], *Alexey* Savchenko[1], *Anton* Kuznetsov[1], and *Aleksandr* Kostenko[1]

[1]Peter the Great Saint-Petersburg Polytechnic University, 29 Polytechnicheskaya st., St.Petersburg, 195251, Russian Federation

Abstract. For more accurate calculation of structure it is required to apply loads in the process of its construction instead of load application after structure has taken its final form. This action is necessary in order to observe changes in the stress-strain state of the structure under study. This circumstance is important for the massive hydraulic structures, such as hydroelectric dams. It is required to determine tensions, deformations and displacements in solving the building-up tasks. It is necessary to consider the problem for constructions of linear and nonlinear materials and check the principle of superposition on which the method of solving nonlinear problems is based. Also it is necessary to consider the solution of finite element schemes with the help of various iterative methods, such as the method of additional loads and the method of variable.

1 Introduction

It is usually assumed in the static calculation of structures that load application is carried out after the structure has taken its final form [1]. Changes of the stress-strain state of structures are mostly investigated only with the appearance of various defects in the construction that could occur during the construction process [2-6].

However, a number of authors show that it is necessary to take into account the following circumstance: the load is applied during the erection of structure, the stress-strain state changes until the construction gets its final form and loads get their final quantity [7-11].

It is particularly important to pay attention to this circumstance during the calculation of massive structures [12]. These requirements have been included in EC [13, 14].

* Corresponding author: lukas-ak@mail.ru

2 Materials and Methods

The solution of the building-up tasks [8, 9] is leaded to the determination of tension $\bar{\sigma}$, deformation $\bar{\varepsilon}$ and displacement $\bar{u}$ in the finally formed body depending on its configuration changes, characterized by the k-phase of the building-up task by the quantities $\overline{\Omega}^{(k)}$, $\overline{S}^{(k)}$ and the sequence of the load application $P^{(k)}$. Thus, in case of linear task (when $\sigma = D\varepsilon$), for the weighty body:

$$\overline{\Omega}^{(N)} = \sum_{k=1}^{N} \omega^{(k)} \tag{1}$$

which is constructed by layers of finite thickness $\omega^{(k)}$, expressions for the quantities $\overline{\sigma_i}$, $\overline{\varepsilon_i}$, $\overline{u_i}$, appearing in i-th layer, would have the following form:

$$\overline{\sigma_i} = \sum_{k=1}^{N} \sigma_i^{(k)} \tag{2}$$

$$\overline{\varepsilon_i} = \sum_{k=1}^{N} \varepsilon_i^{(k)} \tag{3}$$

$$\overline{u_i} = \sum_{k=1}^{N} u_i^{(k)} \tag{4}$$

Here: N – the number of building-up layers; k – the number of stage of the building-up when the structure gets the $\overline{\Omega}^{(k)}$ form and the forces of tension, deformation and displacement come into effect $P^{(k)}$: $\sigma_i^{(k)}$, $\varepsilon_i^{(k)}$, $u_i^{(k)}$. These forces appear in the i-th layer in the process of solution of the tasks with fixed borders $\overline{S}^{(k)}$ of the linear theory of elasticity for the computational region $\overline{\Omega}^{(k)}$ during the action of dead load of the layer $\omega^{(k)}$ – $G^{(k)}$, as $G^{(k-1)} = G^{(k-2)} = ... = G^{(k)} = 0$. At the same time, the other components of the load $P^{(k)}$ are calculated in the same way and their quantities and points of application must be known for every stage of the building-up.

We shall consider the solution of the same problem for the structure made of nonlinear material. The law of elasticity of such material in the general case is written as:

$$\sigma = D(\varepsilon)\varepsilon \tag{5}$$

where $D(\varepsilon)$ is the matrix of elasticity which components are situated on the basis of the adjusted link between the invariants of the tensor of tensions σ_0, σ_i and the tensor of deformations e_0, e_i:

$$\sigma_0 = f_1(e_0, e_i) \tag{6}$$

$$\sigma_i = f_2(e_0, e_i) \tag{7}$$

The solution of the tasks that were mentioned above is complicated by the following circumstance: the principle of superposition could not be used in this case.

3 Result and Discussion

It is necessary to derive the equilibrium equation in terms of displacements. The solution of the u_k task with fixed boundaries on the k-th stage of the calculation for the area $\Omega^{(k)} = \Omega^{(k-1)} + \omega^{(k)}$ with fixed boundaries $S^{(k)}$ should satisfy this equilibrium equation. In this case we should assume the following statement: the final quantities of tensions that were received on the (k-1)-th and k-th stages of the calculation must satisfy the law of elasticity Eq.5 and the balance equation:

$$A'\bar{\sigma} + P = 0 \tag{8}$$

The expressions for the tension $\bar{\sigma}_k$ could be represented as follows:

$$\bar{\sigma}_k = D(\bar{\varepsilon}^*)\bar{\varepsilon}^* + \bar{D}(\varepsilon, \bar{\varepsilon}^*)\varepsilon_k \tag{9}$$

where $\bar{\varepsilon}^*$ – the deformation which has occurred in the building-up body before the installation of the $\omega^{(k)}$ layer, $\varepsilon_k = Au_k$; $\bar{D}(\varepsilon, \bar{\varepsilon}^*)$ – the reduced matrix of elasticity which components depend not only on the solution of the $u_k$task, but also on the deformations $\bar{\varepsilon}^*$ which have already taken place in the building-up body before the installation of the layer.

Substituting Eq.9 into Eq.8 that corresponds to the k-th stage of the erection of array and considering that a similar equation holds for the (k-1)-th stage of the building-up task, we could receive the equilibrium equation in terms of displacements which must satisfy the solution of the task with fixed boundaries $S^{(k)}$:

$$A'\bar{D}(\varepsilon, \bar{\varepsilon}^*)Au_k + P_k = 0 \tag{10}$$

In this expression P_k corresponds to the load of the re-stacking layer $\omega^{(k)} \cdot \sigma^{(k)} \cdot (\sigma^{(k-1)} = \ldots = \sigma^{(1)} = 0)$ and to the forces applied to the construction at this stage of building-up.

Substituting Eq.5 into Eq.9 and considering that $\bar{\varepsilon}_k = \bar{\varepsilon}^* + \varepsilon_k$, we could obtain the expression for the calculation of components of the reduced matrix of elasticity:

$$\overline{D}(\varepsilon,\overline{\varepsilon}^*) = \left[D(\overline{\varepsilon}^* + \varepsilon) - \left[D(\overline{\varepsilon}^*) - D(\overline{\varepsilon}^* + \varepsilon)\right]\varepsilon^* \cdot \frac{\varepsilon'}{|\varepsilon|^2} \right] \qquad (11)$$

where $|\varepsilon| = \sqrt{\varepsilon'\varepsilon}$.

Thus, the solution of the building-up task for the body $\overline{\Omega}$ constructed from the material, which behavior is described by the nonlinear law Eq.5, reduces to the solution of the tasks with fixed boundaries $S^{(k)}$ (k=1,2,…,N) concerning to the Uk sequence. The behavior of the material in each task is described by the nonlinear law:

$$\sigma = \overline{D}(\varepsilon,\overline{\varepsilon}^*)\varepsilon \qquad (12)$$

with the asymmetrical matrix $\overline{D}(\varepsilon,\overline{\varepsilon}^*)$, defined according to Eq.11.

In the case of simple connection between tension and deformation (for instance, for incompressible media) the law of elasticity Eq.11 could be interpreted as the original Eq.5 and written in the new system of coordinates $\overline{\sigma} - \overline{\varepsilon}$ with the beginning that corresponds to quantities $\overline{\sigma}^*$ and $\overline{\varepsilon}^*$. Similar considerations could be used for more general cases of the construction of buildings.

In solving tasks by finite-element method [15] we also solve the nonlinear task with the fixed boundaries Sk on the each stage of the building-up:

$$\overline{K}(q)^{(k)} q^{(k)} = P^{(k)}, \text{ k=1,2,…,N} \qquad (13)$$

The expression for the calculation of the final values of the stresses in the elements $r \in \omega^{(i)}$ would take the following form:

$$\{\overline{\sigma}^r\}_i = D(\varepsilon) \cdot \{\overline{\varepsilon}^r\}_i \qquad (14)$$

where

$$\{\overline{\varepsilon}^r\}_i = B^r \{\overline{q}^r\}_i \qquad (15)$$

and

$$\{\overline{q}^r\}_i = \sum_{k=i}^{N} \{q^r\}_i^{(k)} \qquad (16)$$

It is necessary to mark that using similar considerations in case of the linear formulation we would come to the known scheme of solution of the building-up tasks.

Problems of type Eq.13 are usually solved using various iterative methods. In this case the method of the additional loads is more suitable, because the methods of the variable parameters are unsuitable due to the peculiarities of the matrix $\overline{D}(\varepsilon,\overline{\varepsilon}^*)$.

Let us consider one option of the iterational process when the system of linear algebraic equations Eq.13 with the symmetric stiffness matrix of the element system is solved $r \in \Omega^{(k)}$ [16]:

$$K^{(k)} q_{j+1}^{(k)} = P^{(k)} - \overline{L}_j^{(k)} \quad , j=1,2,\ldots,n \tag{17}$$

Here $\overline{L}_j^{(k)}$ is the vector of the generalized nodal forces, caused by the nonlinear coupling Eq.12 between the tension and deformation:

$$(\overline{L}_m)_j^{(k)} = \sum_{r \in m} \{(\overline{L}_m)_j^r\}^{(k)} \tag{18}$$

$$\{(\overline{L}_m)_j^r\}^k = \int_{V'} \{B^r\}' \cdot \left\{ \{\overline{D}(\varepsilon, \overline{\varepsilon}^*)_j^r - D^r\} \varepsilon_j^r \right\}^{(k)} dv \tag{19}$$

where D^r is the stiffness matrix of the element r, that corresponds to the initial approximation (j-0).

4 Conclusions

By using this method it becomes possible to solve the task considering the geometrical nonlinearity. This method allows to solve stability problems of elements and constructions [17-20].

The method of solving tasks with the loads application to a building in the process of its construction or the method of up-building has demonstrated sufficient efficacy in the calculation of structures [21]. In solving these tasks it is necessary to take into account both the nonlinear behaviour of the material and the sequence of construction of buildings [22, 23]. It is especially important to consider these two factors in the calculation of massive hydraulic structures in which the stress-strain state changes as long as the structure has not obtained its final form. It would be especially relevant by using this method in a variety of computer-aided calculation complexes of building design in solving planar linear and physical nonlinear problems of up-building. In these tasks the load will be applied during the construction of buildings whereby it will be possible to observe the change of the stress-strain state of the facilities. It is also appropriate to explore this circumstance in a calculation of other industrial and civil buildings in which it is necessary to take into account the sequence of construction and the application of loads.

References

1. Yousuf Dinar , Munshi Md. Rasel , Muhammad Junaid Absar Chowdhury , Md. Abu Ashraf, The Int. J. Of Eng. And Sc, **3(1)**, 52-63 (2013)
2. O. V. Kabantsev, S. V. Gorbatov, K. O. Pesin, Vestnik TSUAB. English ver. app. to No. 6, **1-2**, 11-25 (2015)
3. O.V. Kabantsev, Proc. 3rd All-Rus. Conf. on Conc. and Reinf. Conc., **1**, 282–292 (2014)
4. O.V. Kabantsev, A.V. Karlin, J. Ind. and Civil Eng., **7**, 33–35 (2012)
5. O.V. Kabantsev, A. Perelmuter, Proc. Eng., **57**, 479–488 (2013)

6. N.D. Sergeev, Stroit. mekh. i raschet sooruzhenii., **6**, 11–16 (1975)
7. S. Albanesi, N. Antolini, S. Biondi, *The influence of building phases in the post-elastic behavior of multistoried reinforced concrete frames* (11th World Conference on Earthquake Engineering, Acapulco, 1996)
8. S.G. Shulman, *Raschety gydrotehnicheskih sooruzhenyi s uchetom posledovatelnosty vozvedeniya* (Energia, Moscow, 1975)
9. L.I. Dyatlovickiy, A.I. Raynberg, *Formirovaniye napryazheniy v gravitacuonnih plotinah* (Naukova Dumka, Kiev, 1975)
10. H.L. Yip, F.T.K. Au, S.T. Smith, Proc. Eng., **14**, 1384-1391 (2011)
11. V. V. Kuroedov, Trudy LPI. **363**, 70-73 (1997)
12. Bondarev, A., Yugovb, A., Proc. Eng., **117**, 948-958 (2015)
13. D. Zejak, N. Vatin, V. Murgul, Appl. Mech. and Mat., **725-726**, 111-117 (2015)
14. Z. Zafirovski, N. Vatin, Appl. Mech. and Mat., **725-726**, 214-219 (2015)
15. L.A. Rozin, *Raschet gydrotehnicheskih sooruzhenyi na ECVM. Metod konechnih elementov* (L: Energia, Leningrad, 1971)
16. V.V. Kuroedov, Trudi LPI., **349**, 43-47 (1976)
17. V.V. Lalin, D.A. Kushova, Int. j. for comp. civ. and str., **4(9)**, 178-185 (2013)
18. V.V. Lalin, A.V. Yavarov, **5**, 51-54 (2013)
19. V. Lalin, E. Zdanchuk, Int. J. Of Math. Mod. and Meth. In Appl. Sc., **8**, 208-123 (2014)
20. V. Lalin, D. Kushova, *Variational formulations of the nonlinear equilibrium and stability problems of elastic rods* (International Conference „Innovative Materials, Structures and Technologies", Tallinn, 2014)
21. M. Dabaon, E. Ellobody, K. Ramzy, J. of Constr. Steel Res., **110**, 16-28 (2015)
22. Sayed Behzad Talaeitabaa, Minoo Halabiana, Mohammad Ebrahim Torkib, Thin-Walled Struct., **95**, 389-407 (2015)
23. O.A. Rosenboom, T.F. Paret, G.R. Searer, *Chronological Construction Sequence, Creep, Shrinkage, and Pushover Analysis of an Iconic 1960s Reinforced Concrete Building* (15th World Conference on Earthquake Engineering, Lisboa, 2012)

MATEC Web of Conferences **73**, 04002 (2016) DOI: 10.1051/matecconf/20167304002
TPACEE-2016

Effect of Internal Pressure on Parametric Vibrations and Dynamic Stability of Thin-Walled Ground Pipeline Larger Diameter Connect with Elastic Foundation

Vladimir Sokolov[1], *Igor* Razov[1,*] and *Yulia* Ogorodnova[1]

[1]Industrial University of Tyumen, 625001 Volodarskogo str. 38, Tyumen, Russia

Abstract. The article describes the research methodology of parametric vibrations and dynamic stability of ground thin-walled large-diameter pipeline. The equation of motion of the middle surface of the cylindrical shell element based on elastic resistance of ground. To determine the frequency of free oscillations using the assumptions of the semimomental theory of cylindrical shells and get a complete system of equations for the case of an articulated fixing of the cylindrical shell. Solving the resulting system in analytical form the expression for determining the square of the frequency of free oscillations of a thin-walled ground gas pipeline of large diameter. The methodology of the study of the dynamic stability of a gas pipeline under unsteady effects of internal operating pressure, and the parameter of the longitudinal compressive force, using Mathieu equations. With the help of the Bogolyubov-Mitropolsky techniques built area of dynamic instability in the form of Ince-Strutt diagram for different values of the internal pressure of the pipeline section, the parameter of the longitudinal compressive force, and parameter of thin-walled h / R.

1 Introduction

Nowadays pipelines are the main means of delivery of the product from the production site to the consumer. There is currently a construction of new and expansion of existing trunk pipeline networks. The linear part of main pipelines is the most expensive metal content and engineering construction, durability and reliability that is necessary to provide for the entire period of operation.

It should be noted the works of the authors based on the classical theory of the beam: S.P.Timoshenko, Y.G.Panovko, V.V. Bolotin [1], V.I. Feodosiev, H. Jeshli, N.A. Alfutov, G.V. Hauzner, A.P. Kovrevskij, V.A. Svetlickij и др. From the position of the theory of cylindrical shells: M.A. Ilgamov, M.P. Paidoussis [2, 3], A.S. Volmir [4], B.K. Mihajlov, Je.I. Ivanjuta, S.N. Kukudzhanov, V.P. Ilyin [5, 6], O.B. Haleckaja, A.A. Efimov [7], A.V. Bereznev [8] и др. It should be noted the works deal with pipelines connect with elastic foundation: P. Djondjorov [9], O. Doare [10], T.V. Maltseva [11], S.V. Lilkova–Markova

* Corresponding author: tgasu.razov.igor@mail.ru

[12], I. Lottati [13] и др. There are many works connect with determine of the free oscillation. But the Problem deal with dynamic calculation of ground pipeline wasn't solved, because in free source not enough information of method which to take into account the effect of soil radial pressure on the outer surface of the pipe on its own and parametric oscillation.

2 Review

The study of parametric oscillations of pipelines with a flowing has researched liquid based on the beam theory. One of the first works in this direction was an article I.I. Goldenblat [14] which deals with the parametric oscillations of a vertical twin pipe with flowing fluid directed in opposite directions. The analysis showed that at certain critical fluid flow velocities, the straight pipe becomes unstable. H.A. Kartvelishvili [15] solved this problem has already been considered closer to the needs of the practice.

There was formulated and solved the problem of parametric vibrations of a horizontal straight pipe with a pulsating fluid flow. It was found Mathieu equation and investigated the oscillation frequency corresponding to the parametric resonance, that is dynamic stability of the pipeline. In that paper it is shown that the fluid flowing in the pipeline with the flow rate of the pulsation, the load acts as a parametric with respect to transverse oscillations of the tube.

The most detailed parametric vibrations and dynamic stability of the straight pipelines with unsteady fluid flow studied in S.S. Chen [16] and M.P. Paidoussis [2,3] within beam theory. The authors of these works have received systems are not separable differential equations Mathieu - Hill, a solution which was made by using numerical methods, the method of S.S. Hsu [17].

The parametric vibrations and dynamic stability of thin-walled pipes of large diameter with a stream flowing pulsating fluid was investigated on the theory of thin cylindrical shells. One of the first works in this field were published by E.I Grigolyuk, which set out a solution of the problem of parametric vibrations of a cylindrical shell, loaded pulsating axial force [18]. The most complete foundations of the general theory of dynamic stability of elastic systems are described in the monograph VV Bolotin [1], which as a special case considered parametric vibrations of a cylindrical shell, loaded periodically changing forces.

A.C. Volmir [4] studied parametric vibrations of a cylindrical shell with pulsed fluid, he is based on the potential flow of an ideal fluid theory, and then he received a solution to determine the hydrodynamic fluid flow pressure on the pipe wall. A system of coupled differential equations of Mathieu and are built boundaries of areas of dynamic instability.

The most detailed study of parametric vibrations of cylindrical shells with unsteady fluid flow, followed by the construction of the upper and lower boundaries of the main and secondary areas of dynamic instability, carried out M.P. Paidoussis [2, 3]. In these studies was used a simplified assumptions of the theory of shells, and the dynamic instability of research carried out based on a complex numerical solution of coupled systems of differential equations of Mathieu.

In all the works devoted to the evaluation of dynamic stability of thin-walled pipes of large diameter with a flowing liquid, based on the theory of thin cylindrical shells, mostly obtained systems did not share, that is coupled differential equations Mathieu. Therefore, the solutions of these equations is necessary to use of numerical methods with subsequent constructions areas of dynamic instability for each particular case. This difficulty is overcome in the works of V.P. Ilyin, V.G. Sokolov [5], AA Efimov [7], which solved the problem of parametric vibrations and dynamic stability of large diameter pipe sections with a pulsating flow of liquid on the basis of a geometrically nonlinear variant semimomental theory of cylindrical shells and potential flow of an ideal fluid theory. In these papers received Separating Systems of differential equations Mathieu, are built and analyzed Ince - Strett diagrams modified for practical use.

3 Problem

Let us consider the straight section of the ground pipeline finite length, lying on the elastic foundation, or on a sand bedding (Fig. 1). To account for the influence of the elastic foundation to the free vibration frequency, as the design scheme of the pipeline circuit adopted homogeneous isotropic cylindrical shell, closed cross-section, the final length L, and radius R of the middle surface, and h means wall thickness of the shell. The shell is exposed to the action of a stationary internal operating pressure p0, longitudinal compressive force F, and the influence of the elastic foundation is $\Psi(\theta)$ (Fig. 2).

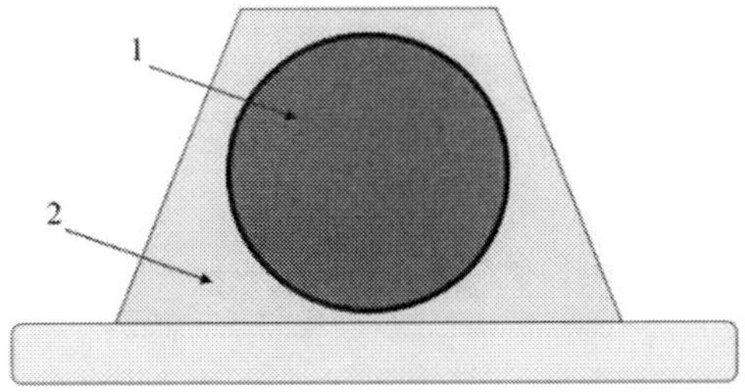

Fig. 1. Ground pipeline: 1 – pipeline, 2 – sand bedding и external covering soil.

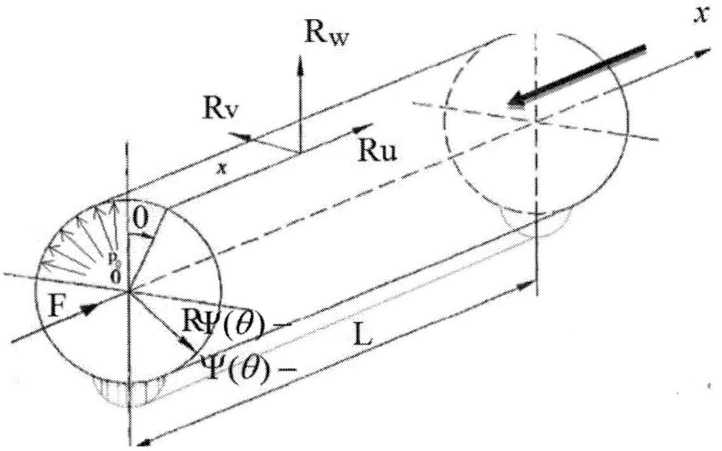

Fig. 2. Ground pipeline on elastic foundation.

In the figure the following notation: Ru, Rv, Rw – components move longitudinally, circumferentially and radially, x – coordinate along the longitudinal axis of the shell, $\Psi(\theta)-$ soil pressure on the radial an external surface of the pipe, θ – polar angle of rotation in the circumferential direction, p0 – internal operating pressure.

In order to solve the problem of parametric vibrations and dynamic stability of the pipeline is necessary carry out:

- Get the equation of motion of ground thin-walled large-diameter pipeline under unsteady effects of internal operating pressure, and the parameter of the longitudinal compressive force;

- Get the equation in analytic form for determining the square of the frequency of free oscillations of an ground pipeline;
- Using solutions N.N. Bololyubova - YA Mitropolsky define and to explore areas of dynamic instability ground pipeline with the influence of the elastic soil foundation;
- To study the influence of the internal operating pressure, and the parameter of the longitudinal compressive force on the size and location of areas of dynamic instability;
- To investigate the effect of the thin-walled parameter h / R ground pipeline in the dynamic zone of instability.

4 Description of studies

Based on the geometrically nonlinear semimomental theory of cylindrical shells of the average bending and using semimomental assumptions of the theory, we obtain the equation of motion in displacements [20]:

$$\frac{\partial^3 u}{\partial \xi^3}+h_v^2\frac{\partial^3}{\partial\theta^3}(\vartheta_2+\frac{\partial^2\vartheta_2}{\partial\theta^2})+2\frac{\partial^2}{\partial\theta^2}(\frac{\partial^2 w}{\partial\xi^2}\varepsilon_0)-\frac{R}{Eh}p_0\frac{\partial^3\vartheta_2}{\partial\theta^3}+$$
$$+\frac{R}{Eh}\frac{\partial^2}{\partial\theta^2}[\Psi(\theta)w(\xi,\theta,t)]-\frac{R^2\rho}{E}(\frac{\partial^3 u}{\partial\xi\partial t^2}-\frac{\partial^3 v}{\partial\theta\partial t^2}-\frac{\partial^4 w}{\partial\theta^2\partial t^2})=0, \tag{1}$$

were u,v,w – moving components middle surface attributed to the radius R; ϑ_2 – rotation angle of the tangent to the median cross-sectional surface; E – the elastic modulus of the shell material;

$\tilde{v}$ – Poisson's ratio; $h_{\tilde{v}}=\dfrac{h}{R\sqrt{12(1-\tilde{v}^2)}}$ – parameter of the relative thickness of the shell; h – thickness of the shell; $\varepsilon_0=\dfrac{F}{EA}$ - original longitudinal deformation caused by the action of the longitudinal compressive force, which is determined on the assumption of non-deformable cross section, ρ – the density of the shell material;

$$\Psi(\theta)=\frac{kR}{\pi}\left[\frac{\varphi_0^3}{3}+\sum_{m=1}^{\infty}\beta_m\cos m\theta\right];$$
$$\beta_m=(-1)^m m^{-3}\left[2\varphi_m\cos\varphi_m-(\varphi_m^2-2)\sin\varphi_m\right];\quad \varphi_m=m\varphi_0, \tag{2}$$

where $\Psi(\theta)$– soil pressure on the radial an external surface of the pipe [21], k - soil ratio in accordance with the model Fuss-Winkler, $\varphi_0=\sqrt[3]{\dfrac{3Q_{общ}}{2kR^2}}$, $Q_{общ}=Q_y$, – linear pipe weight, Q_y – foundation reaction, N/m, R – the radius of the middle surface.

The equations of motion in displacements (1) has four unknowns u,v,w, и ϑ_2. For complete system of equations we add assumptions of semimomental theory:

$$\frac{\partial v}{\partial\theta}+w=0;\ \frac{\partial v}{\partial\xi}+\frac{\partial u}{\partial\theta}=0;\ \vartheta_2=\frac{\partial w}{\partial\theta}-v, \tag{3}$$

As a result, we obtain a complete system of equations (1), (3). Solve the resulting system by separation of variables (Fourier method) represent the function of the radial displacement $w(\xi,\theta,t)$ for the case of an articulated fastening at a [20]:

$$w=\sum_m\sum_n f(t)\sin(\lambda_n\xi)\cos(m\theta) \tag{4}$$

где $\lambda_n=\dfrac{n\pi R}{L}$; m,n=1,2…– wavenumbers in the circumferential and longitudinal directions,

The other components of displacement and rotation angle of the tangent are determined from the semimomental assumptions of the theory, which have the form:

$$u=-\sum_m\sum_n\frac{\lambda_n}{m^2}f(t)\cos(\lambda_n\xi)\cos(m\theta),$$

$$v=-\sum_m\sum_n\frac{1}{m}f(t)\sin(\lambda_n\xi)\sin(m\theta), \tag{5}$$

$$\vartheta_2=-\sum_m\sum_n\frac{m^2-1}{m}f(t)\sin(\lambda_n\xi)\sin(m\theta).$$

The result of the differentiation of the term $\dfrac{\partial^2}{\partial\theta^2}\left[\Psi(\theta)w(t,\xi\theta)\right]$ represented as a product of the partial sums. First we multiply the ranks, and then calculate the derivative. We introduce two partial sums:

$$S_1=\sum_m a_m\cos(m\theta);\quad m=1,2,3,\dots\quad a_0=0,\ S_2=b_0+\sum_m b_m\cos(m\theta);\quad b_0=\frac{\varphi_0^3}{3};\quad b_m=\beta_m.$$

Next we have:

$$Z=S_1S_2=z_0+\sum_{k=1}^{\hat{m}}z_k\cos k\theta,$$

Differentiation sums reduces to differentiation of terms:

$$\frac{\partial}{\partial\theta}(Z)=-\sum_{k=1}^{\hat{m}}z_k k\sin k\theta,\qquad \frac{\partial^2}{\partial\theta^2}(Z)=-\sum_{k=1}^{\hat{m}}z_k k^2\cos k\theta.$$

Further will take the notation k=m:

$$\frac{R}{Eh}\frac{\partial^2}{\partial\theta^2}\left(\Psi(\theta)w(\xi,\theta,t)\right)=\frac{k_sR^2}{\pi Eh}f(t)\sum_{m=1}^{\infty}m^2z_m\cos m\theta, \tag{6}$$

где k – soil ratio [Н/м3], m – волновое число в окружном направлении. Calculating z_m with different wave numbers m=1,2,3… represented in the form:

$$z_1=a_0b_1+a_1b_0+\frac{1}{2}(a_1b_2+a_2b_1+a_2b_3+a_3b_2+a_3b_4+a_4b_3+a_5b_4+a_4b_5);$$

$$z_2=a_0b_2+a_2b_0+\frac{1}{2}(a_1b_3+a_3b_1+a_2b_4+a_4b_2+a_3b_5+a_5b_3+a_1b_1);$$

$$z_3=a_0b_3+a_3b_0+\frac{1}{2}(a_1b_4+a_4b_1+a_2b_5+a_5b_2+a_1b_2+a_2b_1); \tag{7}$$

$$z_4 = a_0b_4 + a_4b_0 + \frac{1}{2}(a_1b_5 + a_5b_1 + a_1b_3 + a_3b_1 + a_2b_2);$$

$$z_5 = a_0b_5 + a_5b_0 + \frac{1}{2}(a_1b_4 + a_4b_1 + a_2b_3 + a_3b_2);$$

$$a_m = 1;$$

$$b_0 = \frac{\varphi_0^3}{3}; \quad b_m = \beta_m,$$

$$\beta_m = (-1)^m m^{-3}\left[2\varphi_m \cos\varphi_m - (\varphi_m^2 - 2)\sin\varphi_m\right]; \quad \varphi_m = m\varphi_0.$$

Substituting (4), (5) в (1) and equating coefficients of the same trigonometric functions, we obtain a system of equations of separable:

$$\left\{\tilde{\lambda}_n^4 + m^4(m^2-1)(m^2-1+p^*) + k^*m^4 - \tilde{\lambda}_n^4 m^4 P/n^2\right\} f(t) + \rho^* Rh[\tilde{\lambda}_n^4 h_v + m^2 + m^4] f''(t) = 0, \quad (8)$$

where $\tilde{\lambda}_n = \frac{n\pi R}{L\sqrt{h_{\tilde{v}}}},\ p^* = p_0\frac{R}{Ehh_v^2},\ \rho^* = \rho\frac{R}{Ehh_v^2},\ k^* = \frac{R^2k}{\pi Ehh_v^2}z_m.$

Considering that the free oscillations are harmonic, represent the function of time $f(t)$ in form:

$$f(t) = \sin\omega_{mn}t,\ f''(t) = -\omega_{mn}^2 \sin\omega_{mn}t, \quad (9)$$

The resulting system is a solution to the problem of free oscillations of a thin-ground large diameter pipelines at steady impact. Substituting (9) in (8) transforming an equation for determining the square of the frequency of free oscillations of a ground pipeline:

$$\omega_{mn}^2 = \frac{\tilde{\lambda}_n^4 + m^4(m^2-1)(m^2-1+p^*) + k^*m^4 - \tilde{\lambda}_n^4 m^4 P/n^2}{\rho^* Rh[\tilde{\lambda}_n^2 h_v + m^2 + m^4]}. \quad (10)$$

The expression (10), allows you to define a wide range of frequencies for enveloped forms at wave numbers m и n=1,2,3…., considering the internal operating pressure parameter of the longitudinal compressive force and elastic soil foundation, at different values of the geometric characteristics.

In actual practice the pipeline is exposed during operation of various kinds dynamic effects and vibrations, including unsteady, caused by the work of compressor stations with a predetermined frequency. In this mode, the internal operating pressure is unsteady changing the law:

$$p(t) = p_0(1 + \mu\cos\gamma t), \quad (11)$$

where γ – frequency of excitation; μ – parameter of excitation, $\mu \leq 0,5$; p0 – internal

operating pressure.

As a result of the work, the unsteady internal operating pressure leads to the excitation of a stationary longitudinal compressive force to the non-stationary:

$$F(t) = F_0(1 + \mu\cos\gamma t), \quad (12)$$

where F_0 - Longitudinal compressive force.

Substituting (11), (12) in system (8) after transformations, we obtain a system of equations with separable unsteady exposure:

$$\left\{\widetilde{\lambda}_n^4 + m^4(m^2-1)(m^2-1+p^*) + k^*m^4 - \widetilde{\lambda}_n^4 m^4 P/n^2 - \mu\cos\gamma t\cdot m^4\widetilde{\lambda}_n^4[P/n^2 - p^*(m^2-1)/\widetilde{\lambda}_n^4]\right\}f(t) +$$
$$+\rho^* Rh[\widetilde{\lambda}_n^2 h_v + m^2 + m^4]f''(t) = 0. \quad (13)$$

Solving and converting (17) we obtain a separable system of equations of Mathieu:

$$f''(t) + \omega_{mn}^2(1-\delta_{mn}\cos\gamma t)f(t) = 0, \quad (14)$$

where ω_{mn}^2 - (10), and δ_{mn} - excitation coefficient:

$$\delta_{mn} = \frac{[P/n^2 - p^*(m^2-1)/\widetilde{\lambda}_n^4]\cdot m^4\widetilde{\lambda}_n^4}{\widetilde{\lambda}_n^4 + m^4(m^2-1+p^*) + k^*m^4 - \widetilde{\lambda}_n^4 m^4 P/n^2}\cdot\mu \quad (15)$$

Evaluation of the dynamic stability of the ground pipeline is to construct dynamic instability areas in the parameter plane γ, P for given values of the internal operating pressure, parameter h/R and coefficient of soil. The area of dynamic instability is determined like:

$$\omega_{mn} = \frac{\gamma}{2}k,\ k=1,2,3... \quad (16)$$

Where is the most wide area is realized with k=1, is called the main area of instability. Secondary region of instability for k>1 are narrower in width and is usually the main area of overlap.

Using the method of Bogolyubov - Mitropolsky defined upper and lower bounds of the dynamic instability, with the help of modified diagrams Ince-Strutt:

$$1-\frac{\delta_{mn}}{2} < \left[\frac{2\omega_{mn}^2}{\gamma}\right] < 1+\frac{\delta_{mn}}{2}. \quad (17)$$

Where the upper and lower limits are determined by equations:

$$\gamma^2 = 4\omega_{mn}^2[1-\frac{\delta_{mn}}{2}]^{-1},\quad \gamma^2 = 4\omega_{mn}^2[1+\frac{\delta_{mn}}{2}]^{-1} \quad (18)$$

Accordance with obtained values of the above procedure, which are summarized in Table 1,2 and illustrated in the form of graphs in Figures 3-11.

Table 1. Definition of the boundaries of dynamic instability of ground pipelines, depending on the parameter P.

Pipe	MPa	*k=1.0*10⁷N/m³; L/R=10; h/R=1/31*					
		P					
		0.101	0.15	0.20	0.25	0.30	0.35
1220x20 (*h/R=1/31*)	*p₀=2*	93.72	80.24 / 71.93	67.49 / 47.46	– / 11.00	-	-
		P					
Pipe	*MPa*	0.376	0.40	0.50	0.55	0.60	0.65
1220x20 (*h/R=1/31*)	*p₀=8*	93.80	88.22 / 85.05	64.72 / 36.90	-	-	-
		*k=1.0*10⁷N/m³; L/R=10; h/R=1/41*					
		P					
Pipe	*MPa*	0.126	0.15	0.20	0.25	0.30	0.35

1220x15 $(h/R=1/41)$	$p_0=2$	81.99	75.36 / 71.89	59.22 / 46.70	-	-	-
		P					
Pipe	*MPa*	0.502	0.55	0.60	0.65	0.70	0.80
1220x15 $(h/R=1/41)$	$p_0=8$	82.65	68.25 / 60.94	49.94 / 31.36	-	-	-
The numerators are given excitation frequency values γ upper boundary of the dynamic instability, and the denominator, the excitation frequency corresponding to the lower boundary of the dynamic instability							

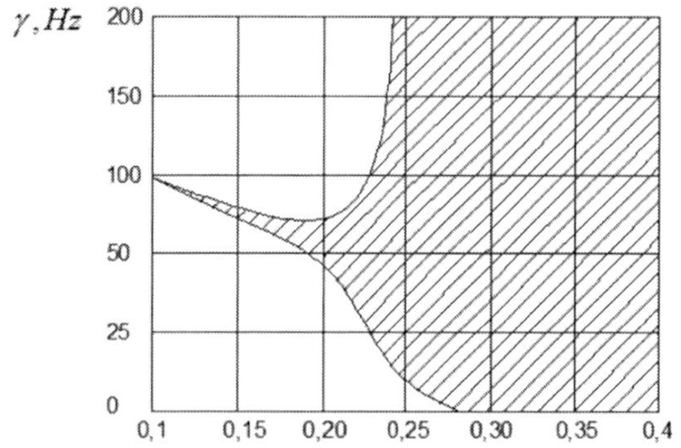

Fig. 3. The area dynamic instability ground pipeline with different values of the parameter *P* with $p_0=2$ *MPa; h/R=1/31* и $k = 1*10^7 N/m^3$, *L/R=10.*

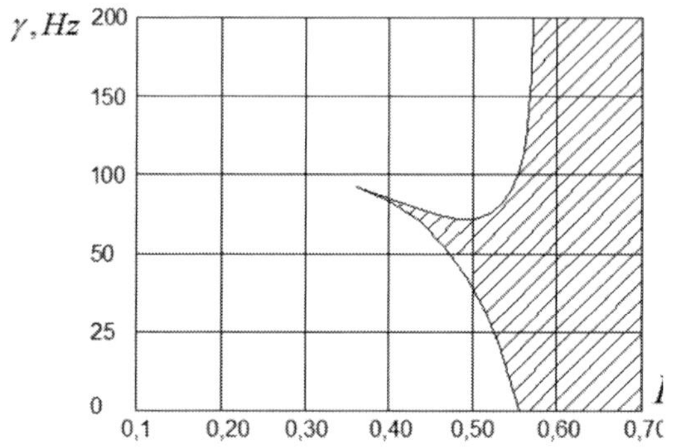

Fig. 4. The area dynamic instability ground pipeline with different values of the parameter P with p0=8 MPa; h/R=1/31 и k = 1*107N/m3, L/R=10.

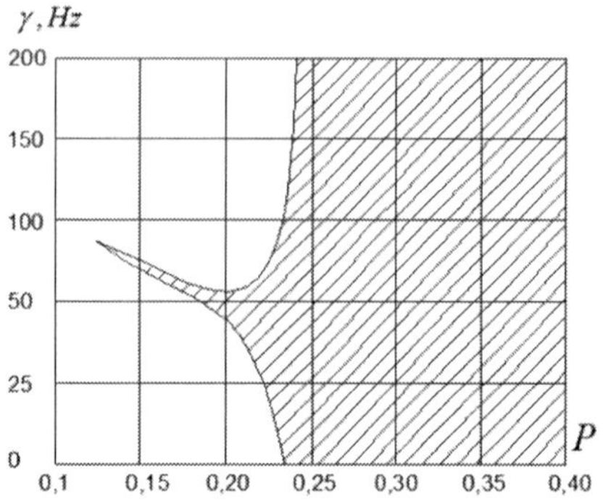

Fig. 5. The area dynamic instability ground pipeline with different values of the parameter P with p0=2 MPa; h/R=1/41 и k = 1*107N/m3, L/R=10.

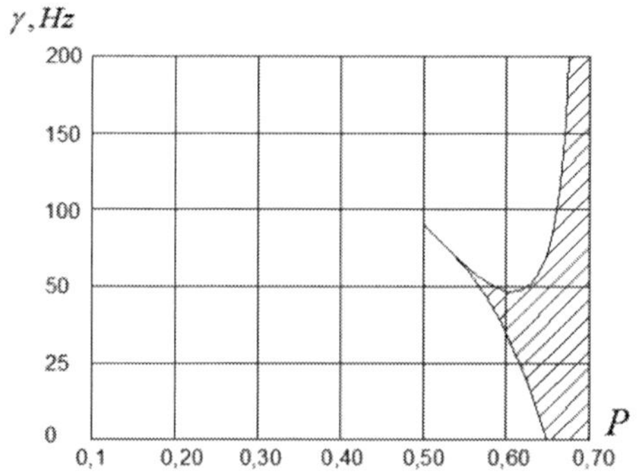

Fig. 6. The area dynamic instability ground pipeline with different values of the parameter P with p0=8 MPa; h/R=1/41 и k = 1*107N/m3, L/R=10.

Table 2: Definition of the boundaries of dynamic instability of ground pipelines, depending on the size of the internal operating pressure p0.

Pipe	*P*							
		$k=1.0*10^7 N/m^3$; $L/R=10$; $h/R=1/31$;						
		p_0, *MPa*						
		1	2	3	4	5	6	6.5
1220x20 (*h/R=1/31*)	*0.3*	-	-	— / 7.19	66.25 / 44.09	77.97 / 68.04	89.61 / 87.27	95.16
		p_0, *MPa*						
Pipe	*P*	1	2	3	4	4.2	-	
1220x20 (*h/R=1/31*)	*0.2*	— / 14.26	67.49 / 47.46	79.48 / 70.64	90.99 / 89,45	94.22	-	
		p_0, *MPa*						
Pipe	*P*	1	2	2.1	-	-	-	
1220x20 (*h/R=1/31*)	*0.1*	80.89 / 73.18	92.91 / 91.59	93.42	-	-	-	
		$k=1.0*10^7 N/m^3$; $L/R=10$; $h/R=1/41$						
		p_0, *MPa*						
Pipe	*P*	1	2	3	4	4.8	-	
1220x15 (*h/R=1/41*)	*0.3*	-	-	43.83 / 20.01	67.93 / 60.44	82.45		
		p_0, *MPa*						
Pipe	*P*	1	2	3	3.2	-	-	
1220x15 (*h/R=1/41*)	*0.2*	-	59.23 / 46.71	78.95 / 77.35	82.34	-	-	
		p_0, *MPa*						
Pipe	*P*	1	1.6	-	-	-	-	
1220x15 (*h/R=1/41*)	*0.1*	71.59 / 66.11	82.25	-	-	-	-	
		$k=1.0*10^7 N/m^3$; $L/R=10$; $h/R=1/51$						
		p_0, *MPa*						
Pipe	*P*	1	2	3	3.8	-	-	
1220x15 (*h/R=1/41*)	*0.3*	-	-	47.02 / 39.08	65.88	-	-	
		p_0, *MPa*						
Pipe	*P*	1	2	2.6	-	-	-	
1220x15 (*h/R=1/41*)	*0.2*	-	54.51 / 49.41	67.51	-	-	-	

		p_0, *MPa*					
Pipe	*P*	1	1,3	-	-	-	-
1220x15 (*h/R=1/41)*	*0.1*	$\frac{60.88}{58.43}$	65.01	-	-	-	-

The numerators are given excitation frequency values γ upper boundary of the dynamic instability, and the denominator, the excitation frequency corresponding to the lower boundary of the dynamic instability

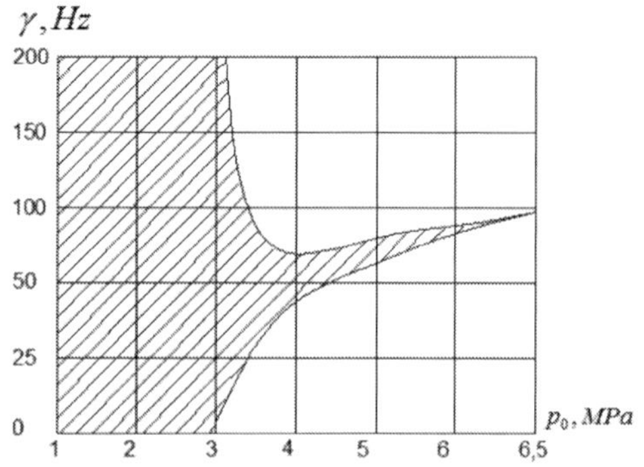

Fig. 8. The area dynamic instability ground pipeline with different internal operating pressure p0 with h/R=1/31 и

k=1.0*107N/m3, L/R=10; P=0.3.

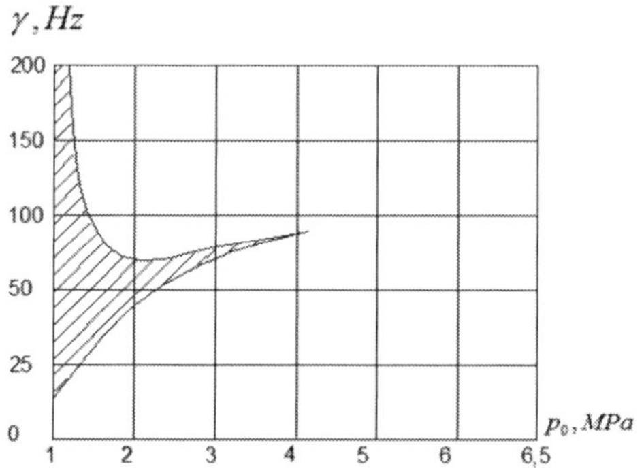

Fig. 9. The area dynamic instability ground pipeline with different internal operating pressure p0 with h/R=1/31 и

k=1.0*107N/m3, L/R=10; P=0.2.

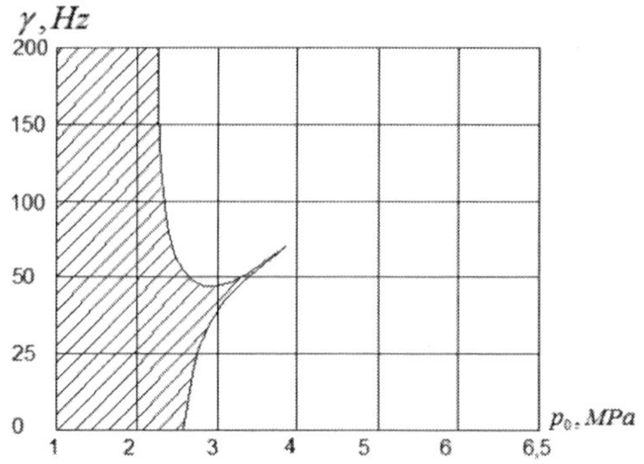

Fig. 10. The area dynamic instability ground pipeline with different internal operating pressure p0 with h/R=1/51 и k=1.0*107N/m3, L/R=10; P=0.3.

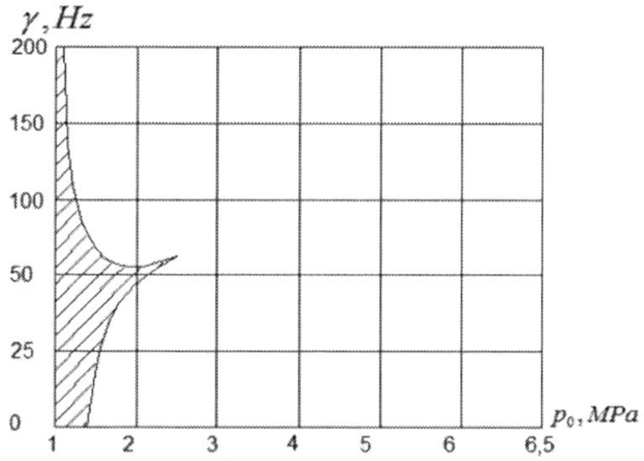

Fig. 11. The area dynamic instability ground pipeline with different internal operating pressure p0 with h/R=1/51 и k=1.0*107N/m3, L/R=10; P=0.2.

5 Conclusion

Based on the results it follows that an increase in the value of the internal operating pressure of the dynamic instability of the region is shifted upward parameter P values testify to the values in Table 1 and the graph in Figure 3-6. Such a change in the dynamic of instability zones means that with an increase in the loss of dynamic stability of internal operating pressures possible for large values of P. Next we note that for thinner pipes h/R=1/41 influence of the internal operating pressure considerably more than for thicker pipe h/R=1/31. This effect is observed when comparing such areas in figures 4 и 6, with p0=8 MPa.

Further study is aimed at studying the effect of the internal operating pressure on the size and boundaries of areas of dynamic instability, the values obtained (Table 2) and a built area of dynamic instability in Figures 9-14. According to the results, it follows that an increase in the value of the internal working pressure the area of dynamic instability is narrowed, and for certain values of p0 refers to the point. It should be noted that an increase in the parameter of the longitudinal compressive force, the dynamic instability area expands as it was proved in [19]. This increase the parameter P values increases the likelihood of dynamic stability losses several times.

It was further found that the internal operating pressure significantly effect for thinner pipe, for example h/R=1/51. This is because the internal operating pressure prevents the deformation (ovalization) of the cross section, increasing the rigidity of the pipeline section.

The results obtained in the paper can be used for optimal selection of the thickness of the pipe wall and can help for the detuning from the resonance of the linear part of pipe. This approach ensures the reliability of the pipeline on the entire period of operation. Building of dynamic instability zones, depending on the tested parameters, allow an assessment of the dynamic stability of the pipeline. In the case when the value of (γ, P) enters the shaded area, this means that the pipeline is located in a dynamically unstable position, which significantly increases the amplitude of the oscillations, which contributes to the appearance of fatigue fracture of the metal. Otherwise, when the value falls into the free the area of the hatching dynamic stability provided by pipeline.

References

1. V.V. Bolotin, *Dinamicheskaja ustojchivost' uprugih system* (Gostehizdat, Moscow, 1956)
2. M.P. Paidoussis, N.T. Issid, Journ. of sound and vibr, **33(3)**, 264-294 (1994).
3. M.P. Paidoussis, N.T. Issid, *Trans of ASME*, 198-202 (1996)
4. Volmir, A.S. *Obolochki v potoke zhidkosti i gaza, Zadachi gidrouprugosti*)Nauka, Moscow, 1999)
5. V.P. Ilyin, V.G. Sokolov, *Voprosy mehaniki stroitel'nyh konstrukcij i materialov. Mezhvuzovskij tematicheskij sbornik,* 6-10 (1987)
6. V.P. Ilyin, Vestnik rossijskoj akademii Arhitektury i stroit, nauk, **1**, 15-21 (1996)
7. A.A. Efimov, V.G. Sokolov, Izv. vuzov, Neft i gaz, **4**, 47-51 (2007)
8. A.V. Bereznev, Vestnik grazhdanskih inzhenerov, **3(50)**, 101-104 (2015)
9. P. Djondjorov, Journal of Theoretical and Applied Mechanics, **31(3)**, 101-112 (2001).
10. O. Doare, E. de Langre, Journal of Fluids and Structures, **16**, 1-14 (2002)
11. T.V. Maltseva, S.M. Dorofeev, T.V. Saltanova, Izvestija vysshih uchebnyh zavedenij. Neft i gaz, **1**, 59-63 (2012)
12. S.V Lilkova–Markova and D.S. Lolov, Sadhana, **29(3)**, 259-262 (2004)
13. I. Lottati, A. Kornecki, Journal of Sound and Vibration, **2(109)**, 327-332 (1996)
14. I.I. Goldenblat, *Issledovatel'skie raboty po inzhenernym konstrukcijam*, 4-12 (1998)
15. N.A. Kartvelishvili, Izv. Vsesojuznogo NII Gidrotehniki, **49**, 31-53 (1998)
16. S.S. Chen, Journ. Of the Eng. Mech. Division, **97**, 1469-1485 (1991)
17. C.S. Hsu, J. Appl. Mech., **3(30)**, 367-372 (1993)
18. Je.I. Grigoljuk, V.V. Kabanov, *Ustojchivost' obolochek* (Nauka, Moscow, 1998)
19. V.G. Sokolov, , I.O. Razov, Vestnik grazhdanskih inzhenerov, **2**, 65-68 (2014)
20. V.G. Sokolov, I.O. Razov, Vestnik grazhdanskih inzhenerov, **1**, 29-32 (2013)
21. I.O. Razov, Vestnik grazhdanskih inzhenerov, **3**, 58-62 (2015)

MATEC Web of Conferences **73**, 04003 (2016) DOI: 10.1051/matecconf/20167304003
TPACEE-2016

A FEM Modeling of the Concrete Pavement Made of the Recycling Material

Miloš Šešlija[1], *Nebojša* Radović[1] and *Issa* Togo[2,*]

[1]Faculty of Technical Sciences. 21000 Trg Dositeja Obradovića 6, Novi Sad, Serbia
[2]St. Petersburg State Polytechnical University, 195251 Politekhnicheskaya str. 29, Saint-Petersburg, Russia

Abstract. Paper is a brief review of the research focused on formulation an numerical model for the concrete pavement which is made by the recycling material. For numerical modeling the finite element model (FEM) and the 3D finite element model were applied. The software EverFE 2.25, was used. The results of FEM analysis is in a chapter shape showing move value change, strees and deflections for all layers a construction road model. In the next phase of the research was provided by FEM software with appropriate general purpose non-linear models, which allows the analysis of the real behavior of solid pavement under load.

1 Introduction

The concrete pavement consists of four layers, including a subgrade. The first layer is concrete pavement, the second layer is made of crushed stone with fraction 0/31.5, and third layer is made of crushed stone 0/63. Each layer takes over the load and transmits it to the subgrade. The goal of this study is to find out maximal/minimal principal stress values per slab. We used a bus, very frequent on that section of the road. The model is carried out with the software EverFE 2.25 which gives us the model results with 3D FE element [1-11].

2 3D fe analaysis

To model a stress notation of the special systems 3D solid FE are used as the most complex and ensure the most precise approximation. Applying these FE we get models with high degree of freedom, therefore their application is limited to analysis of specific construction systems [1].

The software Ever FE version 2.25, has been used for modeling. The software is equipped with the qualitative tools for creating 3D FE models. The first step in creating a model of the concrete pavement is to determine its dimensions and to define the number of layers. Each layer has a specific modulus of elasticity (E), Poisson`s ratio (v) and layer thickness (d) [2]. All the values are presented in Table 1.

* Corresponding author: issatogo@mail.ru

Table 1. Pavement structure with characteristic data

Layers	Layers thickness (cm)	Modulus of elasticity (MPa)	Poisson's ratio
Concrete pavement layer	24	33000	0.35
Crushed stone road-base 0/31.5 (base)	15	480	0.35
Crushed stone road-base 0/63 (sub-base)	25	400	0.35
Clay subgrade	10	50	0.45

Ever FE is equipped to allow the user to specify a wide range of realistic loads that may be applied to a solid pavement through a vehicle's tire.

The first step in selecting a load is to choose the type of tire contact which can be rectangular or circular. After selecting the type of tire contact, the user must define the rest of the load parameters. The wheel type can be selected as single or dual. The axle type can be selected as single or tandem.

The test bus station has tree concrete pavements that are made of recycling material and they have 15 dowels. The dowels durations are 600mm and diameters are 22mm. For testing load was used the bus which has 120kN on front axle and 120kN on rear axle (Fig 1, 2).

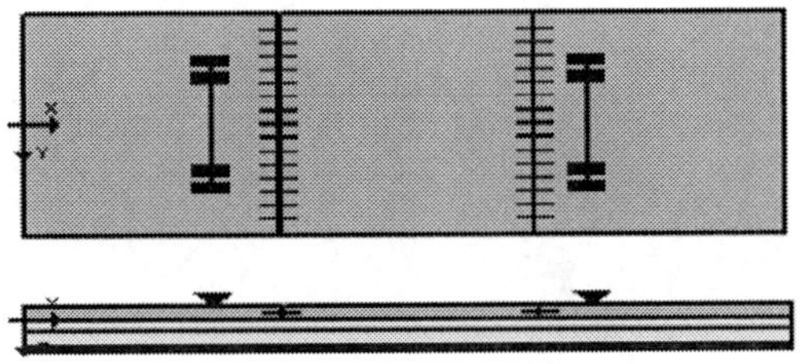

Fig. 1. Bus load scheme.

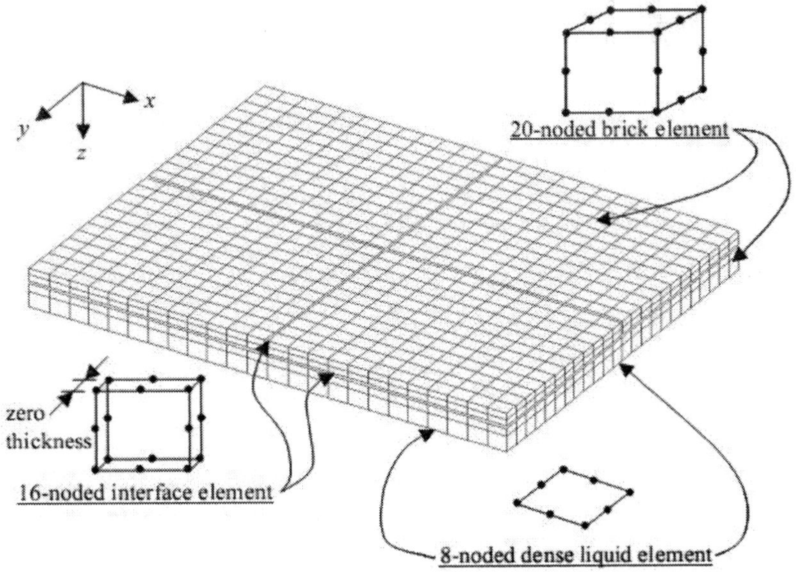

Fig. 2. The Basic Finite Element Discretization [2].

3 Research results and discussion

After defining the parameters the above menitoned, the research results are obtained. EverFE allows the results of the finite-element analysis to be presented using any of four options available from the visualization menu: Stresses (Maximal Principal, Minimal Principal, Stresses in the x-x plane, y-y plane, z-z plane, x-y plane, y-z plane and z-x plane), Displacements, Results for Points, and Results for Dowels [2].

Displaced shapes of slabs and the base layer can be viewed for any project for which a solution exists. The results for this points are showing the maximum and minimum principal stresses and their locations in the corresponding slab; the red text show the highest principal stress on the slab.

Selecting dowels from the Visualize menu show all moments for any selected dowel. The five force components that can be viewed: Nq (axial force), the shears Fr and Fs, and the moments Mr and Ms. The subscripts q, r and s refer to the dowel local coordinate system, and correspond to the global model (x,y,z) axes [2]. The maximum values of shear and moment are displayed for the currently selected dowel at the bottom of the dowel results panel. These maximum values are computed as the magnitude of the “r” and “s” components, which is appropriate for a dowel with a circular cross-section. The following charts present the characteristic cross-sections of the stresses, displacements, results for points and results for dowels. (Fig. 3)

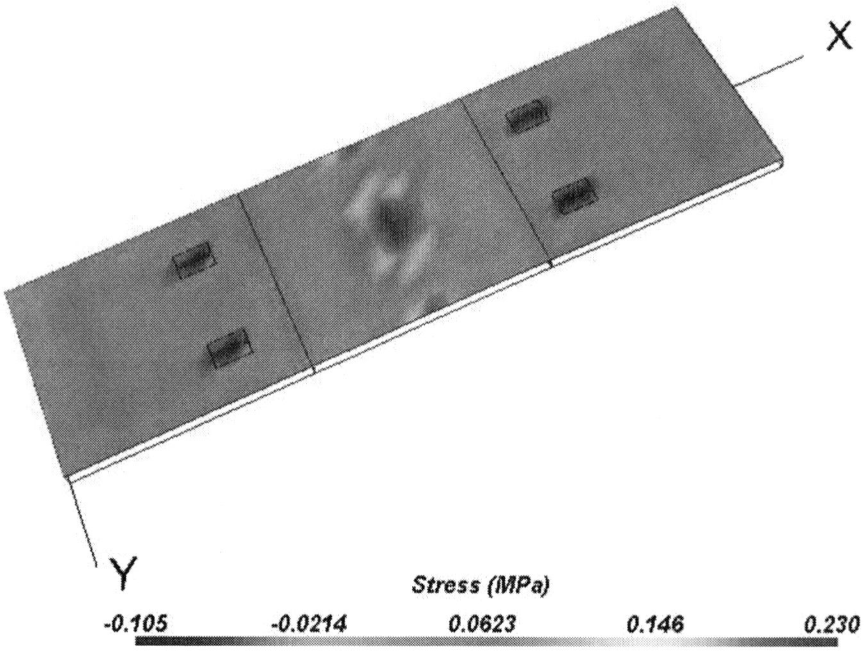

Fig. 3. Maximal Principal.

The maximal principal stress is in the center concrete pavement, so the major impact is noticeable in the upper layers (Fig. 4).

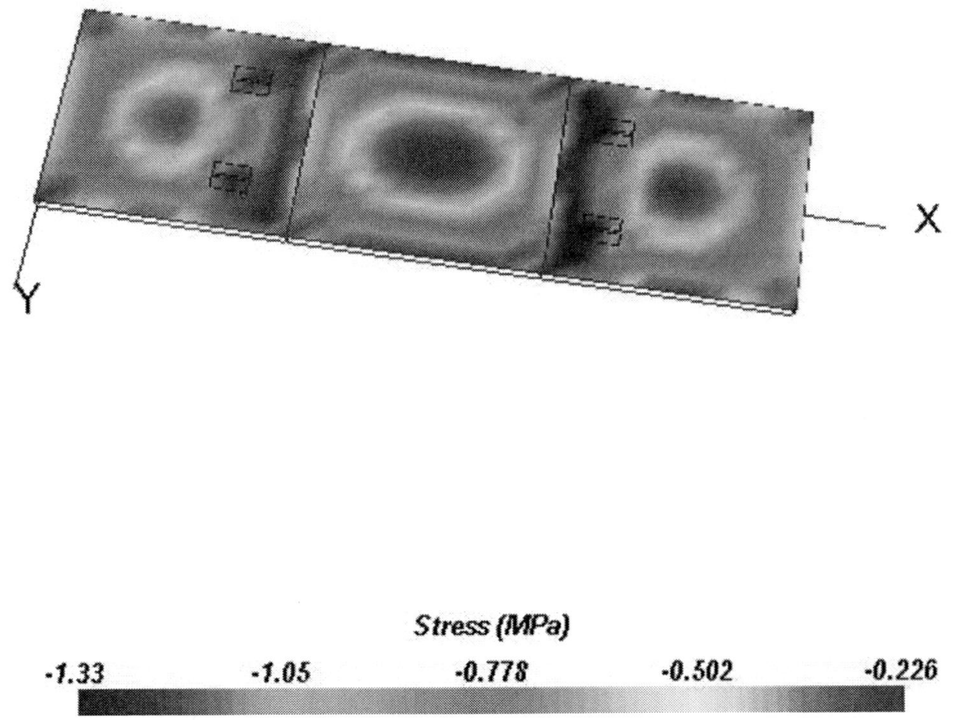

Fig. 4. Minimal Principal.

The minimal principal stress is in the end of the first pavement and the start of the third pavement. (Fig. 5)

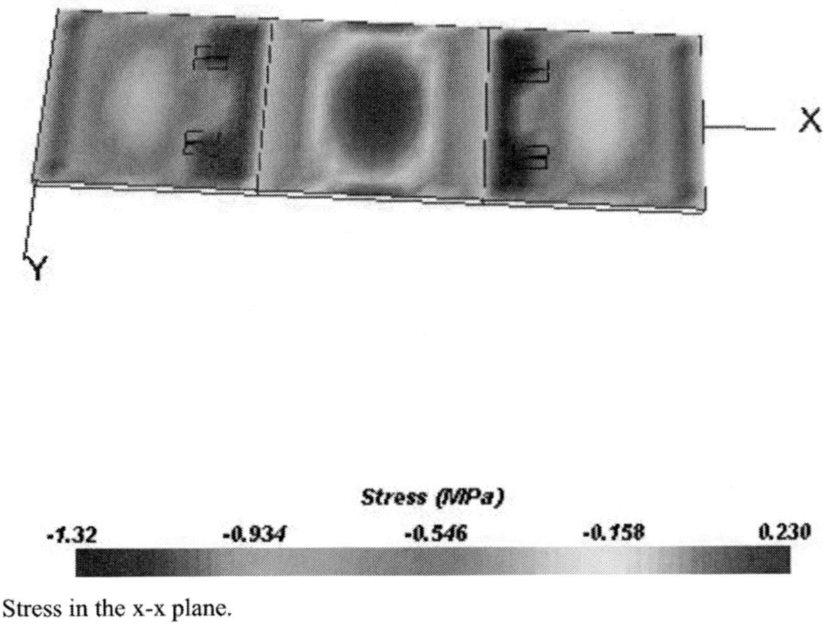

Fig. 5. Stress in the x-x plane.

The maximal stress is in the center of middle concrete pavement, and the minimum is in the end of the first pavement and the start of the third pavement (Fig. 6, 7).

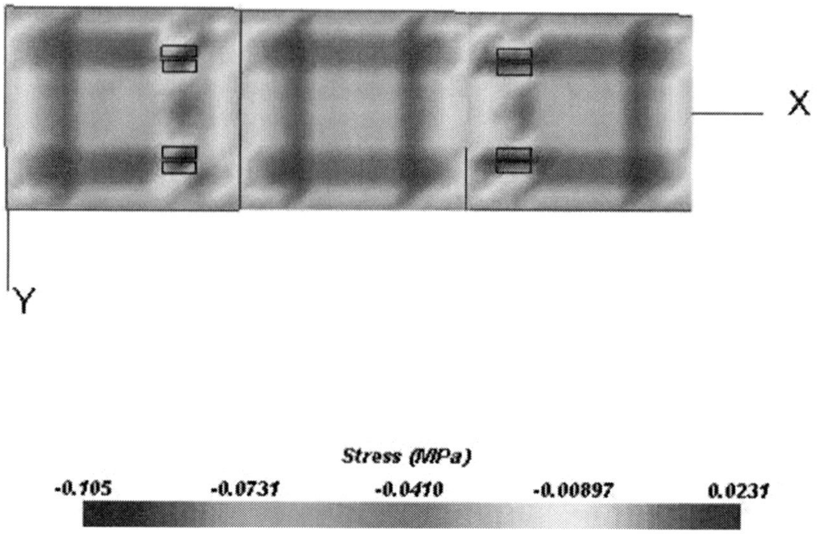

Fig 6. Stress in the z-z plane.

The maximal stress is in the all slabs, but the minimum is under the wheels.

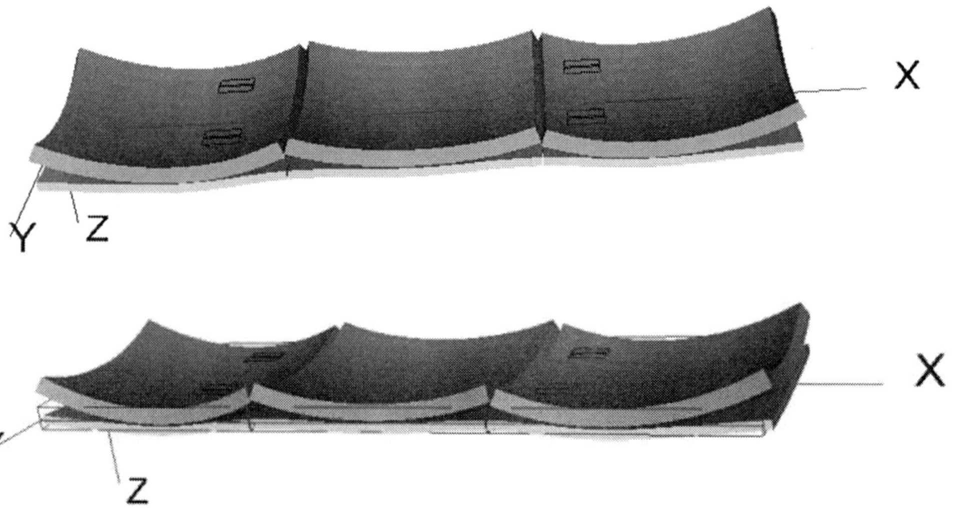

Fig. 7. Deflections of the all slabs and the base.

All slabs are deflected, even thought the loads are placed on first and third slab, because it has dowels that link all slabs and loads.

MAX: 1.22587MPa	MAX: 1.22607MPa	MAX: 1.22691MPa
X: 3666.63mm	X: 7679.32mm	X: 8358.77mm
Y: -1458.29mm	Y: -1458.29mm	Y: -1458.29mm
Z: -119.99mm	Z: -119.99mm	Z: -119.99mm
MIN: -1.72333MPa	MIN: -2.2094MPa	MIN: -1.80472MPa
X: 1666.68mm	X: 6012.7mm	X: 10025.4mm
Y: 2.27374e-013mm	Y: 2.27374e-013mm	Y: 2.27374e-013mm
Z: -0.01mm	Z: -0.01mm	Z: -0.01mm

Fig. 8. Maximal/Minimal Principal Stress Values per Slab.

The numbers in each rectangle shows the maximum and minimum principal stresses and their locations in the corresponding slab; the red text show the highest principal stress on the slab (Table. 2).

Table 2. Maximal moment and maximal shear for dowels

slab	Dowel	Max.Moment (Nmm)	Max.Shear (N)	slab	Dowel	Max.Moment (Nmm)	Max.Shear (N)
First and second slabs, link with dowels	1	48200.800	1165.410	Second and third slabs, link with dowels	**16**	**48895.200**	**1294.790**
	2	45975.100	1051.580		17	46927.500	1193.800
	3	44134.201	957.750		18	45292.000	1109.750
	4	42560.701	878.386		19	43801.500	1034.140
	5	41339.901	816.631		20	42556.501	971.100
	6	40385.001	768.949		21	41486.401	917.526
	7	39776.300	738.882		22	40761.700	881.268
	8	39494.200	725.150		23	40428.600	864.761
	9	39774.000	738.758		24	40763.600	881.366
	10	40385.301	768.965		25	41486.101	917.513
	11	41342.801	816.785		26	42554.101	970.977
	12	42556.501	878.163		27	43805.100	1034.310
	13	44132.301	957.647		28	45293.600	1109.830
	14	45975.600	1051.610		29	46927.000	1193.780
	15	**48203.600**	**1165.560**		30	48892.700	1294.670

The bold text values maximal moments both on first and second slabs and second and third slabs. Fig. 9-12.

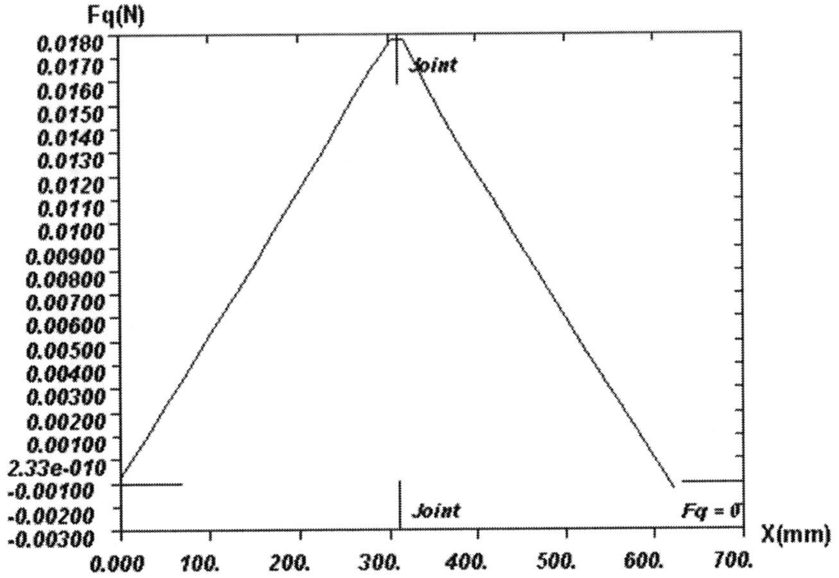

Fig. 9. Axial Forces (Nq).

They have an upward linear dependence on the end of second pavement and another linear decreasing dependence at the beginning of third pavement.

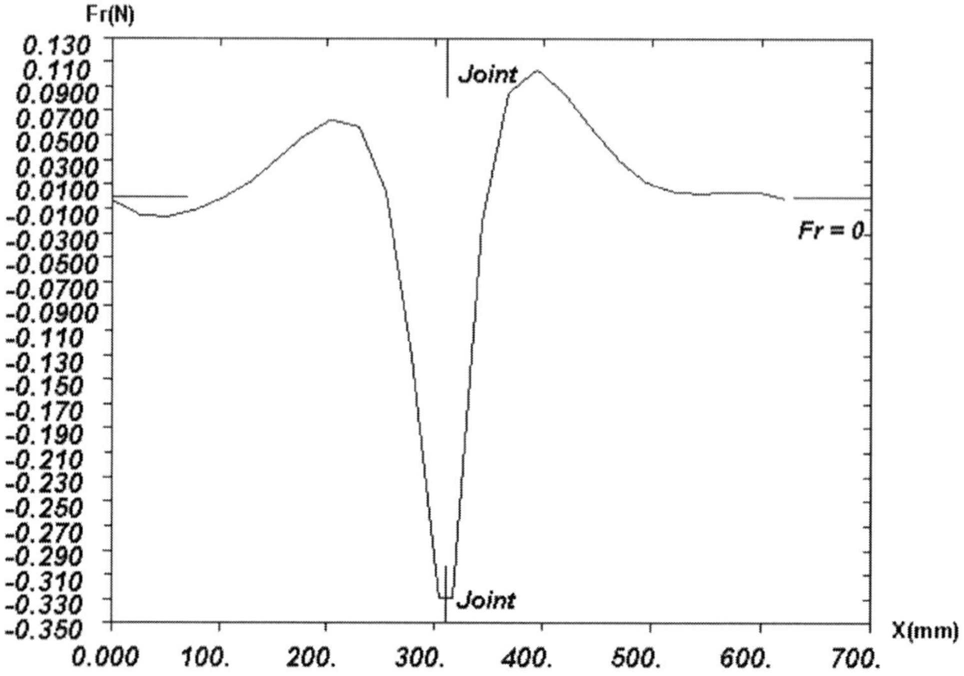

Fig. 10. Shear Forces (Fr).

Between second and third concrete pavement there are big leaps for Shear Forces (Fr).

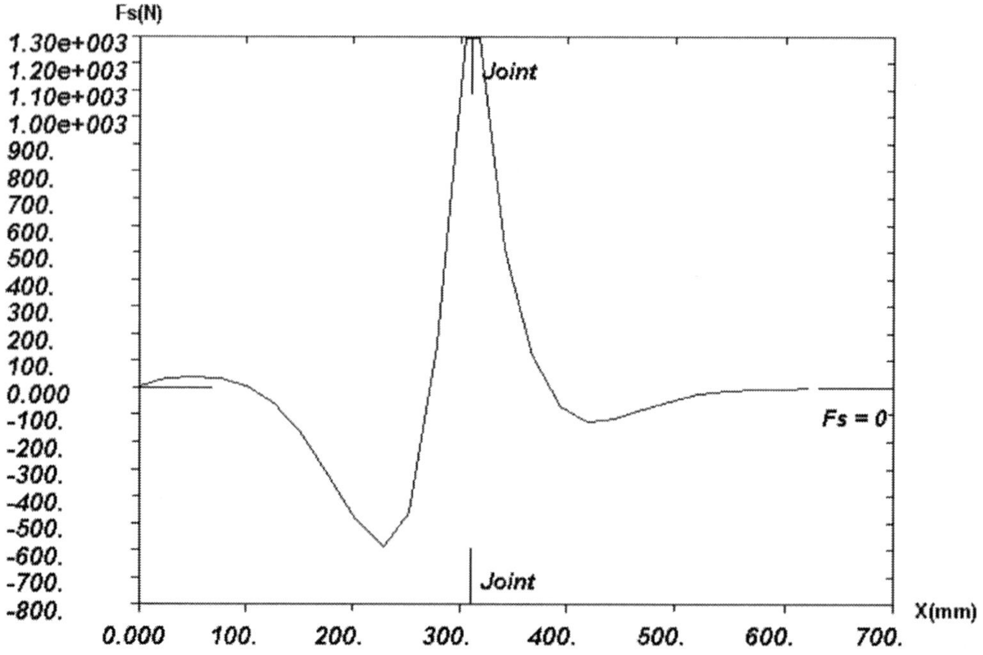

Fig. 11. Shear Forces (Fs).

Between second and third concrete pavement there are big leaps for Shear Forces (Fs).

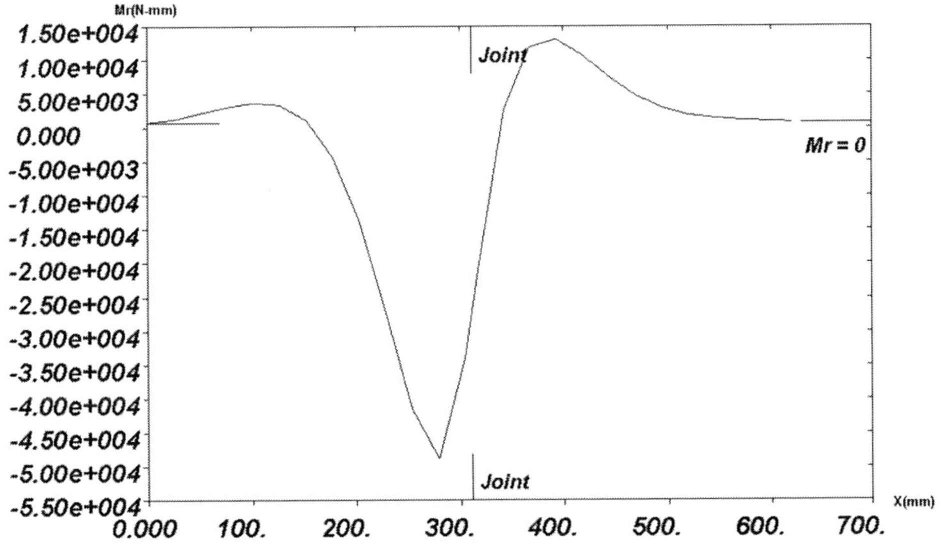

Fig. 12. Moment of dowel (Mr).

Between second and third concrete pavement there are big leaps for Moment (Mr).

4 Conclusion

After the research was carried out, we have got the results in the form of a model and graphics which ensure an insight in all their changes and deviations. The conclusions of the study are following:

- there are substantial deviations in the stresses, displacement and results for dowels. The cause of these changes is the traffic loading,
- the loads are changeable in all three concrete pavements.
- The direction of further research:
- test influence to use reinforce concrete pavement,
- testing joints with the big diameter at the ends of the concrete pavement,
- test influence with angle on the first and the final pavement.

5 Acknowledgements

The work reported in this paper is a part of the investigation within the research project TR 36017 "Utilization of by-products and recycled waste materials in concrete composites in the scope of sustainable construction development in Serbia: investigation and environmental assessment of possible applications", supported by the Ministry for Science and Technology, Republic of Serbia. This support is gratefully acknowledged.

References

1. D. Kovacevic, *"FEM Modeling in structural analysis" (in Serbia)* (Građevinska knjiga, Belgrade, 2006)
2. F.E. Ever, *Software for the 3D Finite Element Analysis of Jointed Plain Concrete Pavements* (2015) http://www.civil.umaine.edu/EverFE/theory_manual.pdf

3. K. Krayushkina, O. Prentkovskis, A. Bieliatynskyi, R. Junevičius, Transport, **27(2)**, 129-137 (2012)
4. O. Prentkovskis, J. Tretjakovas, A. Vedas, A. Bieliatynskyi, A. Daninas, K. Krayushkina, Journal of Civil Engineering and Management, **18(5)**, 761-771 (2012)
5. K. Krayushkina, O. Prentkovskis, A. Bieliatynskyi, J. Gigineishvili, A. Skrypchenko, A. Laurinavičius, K. Gopalakrishnan, J. Tretjakovas, Baltic Journal of Road and Bridge Engineering, **11(1)**, 77-83 (2016)
6. M. Arsenovic, Z. Radojević, Ž. Jakšić, L. Pezo, Ceramics International, **41(3)**, 4899-4905 (2015)
7. M. Arsenovic, Z. Radojević, Ž. Jakšić, L. Pezo, Ceramics International, **41(3)**, 4890-4898 (2015)
8. V.V. Okrepilov, M.V. Leonidovich, Asian Social Science, **11(7)**, 312-325 (2015)
9. V.V. Okrepilov, V.L. Makarov, A.R. Bakhtizin, S.N. Kuzmina, Economy of Region, **2**, 301-313 (2015)
10. V.L. Kvint, V.V. Okrepilov, Herald of the Russian Academy of Sciences, **84(3)**, 188-200 (2014)
11. V.V. Okrepilov, V.N. Krutikov, G.I. El'kin, Measurement Techniques, **57(2)**, 109-116 (2014)

MATEC Web of Conferences **73**, 04004 (2016) DOI: 10.1051/matecconf/20167304004
TPACEE-2016

Technological Decision to Renewable Energy Usage Biogas for Off-grid Systems Consumption

Marina Zubkova[1,*], *Vladimir* Maslikov[1], *Dmitry* Molodtsov[1], *Alexander* Chusov[1], *Viacheslav* Zhazhkov[1], and *Alexander* Stroganov[2]

[1]Peter the Great St.Petersburg Polytechnic University, 195251, Polytechnicheskaya st. 29, St. Petersburg, Russia
[2]Lumex Instruments Canada, Unit 207, 31510 Gill Avenue, Mission, B.C., V4S0A1 Canada

Abstract. This paper presents the results of the energy experiments based on electrochemical researches and the thermodynamic calculations, which are carried out on the hydrogenous fuel with the residual content of methane obtained from biogas selected organic waste. Energy indicators are examined in comparison to electrolysis hydrogen. The use of technical and electro physical indicators together with parameters of the fuel operating allowed assessing energy efficiency the module reformer - fuel cell running on a non-standard hydrogenous fuel. Numerical characteristic the efficiency of workflows fuel system reformer – fuel cell is about 39%. To operate the power installation with a predetermined capacity amount used of hydrogenous fuel is comparable to required electrolysis hydrogen amount. Shown the possibility of creation the systems of power supply based on new hydrogen technologies using renewable energy resources local waste. Confirms the relatively high efficiency the usage of hydrogenous fuel for the tasks of off-grid systems consumption.

1 Introduction

One of the most important directions of energy saving is creation of autonomous and economic systems of electric power supply of the consumers, allowing to receive electric and thermal energy in a place of its consumption for decrease in losses and increase reliability of energy supply [1]. It is promising to use for the purpose of fuel cells, offering a direct conversion of the energy of the fuel into electrical energy [2-4]. They are characterized by high energy- economic indicators, reliable performance, quiet operation, mobility, lack of harmful emissions, etc. Fuel cells can be used as stationary and emergency backup power supply in autonomous systems [3-7]. For loads from a few kilowatts to several hundred kilowatts (consumers in a low housing estate) it is preferred application of low-temperature fuel cells that run on hydrogen.

In the European Union countries, the USA, Japan, etc. works on receiving hydrogen from biogases and its use in energy supply of autonomous consumers. Identified

* Corresponding author: marina19zubkova@list.ru

weaknesses of installations - the low content of hydrogen in the reformate (70-80% vol.) and existence in it harmful impurity. It demands use of expensive and difficult systems in operation of its cleaning that leads to increased power consumption and reduced process effectiveness. Therefore, it is urgent is the development of technologies for producing biogas from relatively cheap enough "clean" hydrogen and methods of its direct use in low-temperature hydrogen fuel cells [7-11].

Efficiency of operation the autonomous system with fuel cells in many respects will be defined by the cost of receiving hydrogen. The cheapest way by far is the production of hydrogen from natural gas. The hydrogen received from biogas of agricultural waste is rather cheap. In the long term, it is necessary to expect increase of cost of natural gas owing to what production of hydrogen from biogas can be cheaper, than from natural gas [11-14].

The task to determine the possibility of using hydrogenous fuel with residual methane content in fuel cells, and carrying out the calculations for assessing the energy efficiency of non-standard fuel, gives the opportunity fundamentally changed the approach to technologies considering secondary sources [10, 12-14].

Biogas can be obtained practically everywhere as organic-containing waste is constantly formed in the course of economic activity of the person, meanwhile resources of natural gas are limited, and its use is limited. There is a unique possibility of creation of systems of energy supply of autonomous consumers based on the new hydrogen technologies using as an energy resource local waste [8, 11-15].

In Peter the Great St. Petersburg Polytechnic University experimental system for studying the processes of obtaining and using biogas consisting of biogas generating modules, its accumulation and purification, methane reforming to hydrogen production is created [11]. Stable hydrogen production of high purity (98.5% vol.) is experimentally confirmed at biomethane supply (about 95%) in the wide range of loadings of a reformer (30-100% from nominal) [12,13]. In a reformer there is an internal utilization of CO_2, and its contents in reformate also does not exceed 10 ppm. The experiments on the use of "weak" biogas containing methane 30-45 % vol. were also carried out [14-16]. The concentration of hydrogen in the reformate remains high (more than 93% vol.) [12]. This extends the use of the considerable potential "weak" biogas with a relatively low concentration of organic components for autonomous power engineering in the direct use of hydrogen-derived fuels in the energy system to produce electricity and heat. Testing of a fuel element during multiple cycles (supplying-reset) allows to define its energy efficiency and optimum operating modes and to estimate opportunity and efficiency of use of hydrogen fuel with the methane component, received from biogases by two-steps conversion, in a low-temperature fuel cell

Principle possibility of the fuel cell operation on gas mixture with methane not oxidizing on the anode [12,15,16] and formation of a stream of purge of methane-hydrogenous mixture is confirmed. Modes of return of the fuel cell to the baseline characteristics after steady work for a non-standard hydrogenous fuel are fulfilled [14-18].

2 Experimental studies

We set the task based on the experimental results, the data of the material balance and thermodynamic evaluations on the effectiveness of the use of non-standard hydrogenous fuel, carry out the model calculation of fuel consumption derived from biogas required for the efficient operation of the unit conversion fuel – "reformer-fuel cell" to meet the optimal resource capacity and the time duration.

The studies were carried out using the unique scientific unit "The laboratory complex for study of processes for obtaining and converting biogas from organic waste" in Peter the Great St.Petersburg Polytechnic University.

Experimental studies were carried out on a special stand [11] in the mode of direct supply of hydrogenous fuel to the fuel cell with parallel testing of the fuel cell on hydrogen and oxygen obtained by electrolysis, which allows to make comparisons of the electrochemical characteristics of the oxidation the fuel and the number of purges of unfinished fuel [11-17].

Results of experiments at step change of the set potential on the fuel module FC-50 (dead end) with connection of purge system for unfinished fuel from the anode of a fuel cell for hydrogen obtaining by electrolysis (mode 1) and non-standard hydrogenous fuel (mode 2) are presented in Fig. 1. As an oxidizer air oxygen with a necessary excess expense was used. Excessive pressure of fuel supply on a cell maintained within the limits 0.9 bar. Analysis of plots (Fig. 1) shows that when running on electrolysis hydrogen, fuel purges does not occur on the interval considered a step change in capacity. Consequently, the fuel feeding rate and the resulting number of protons at the anode sufficient to maintaining a stable electrochemical process in the fuel cell. For a hydrogen fuel gas mixture purges occur at high current (or power) as not discharged molecules of CH_4 interferes with access of necessary amount of hydrogen to a catalytic surface of the anode.

Characteristics of studies for modes 1 and 2 are shown in Table 1.

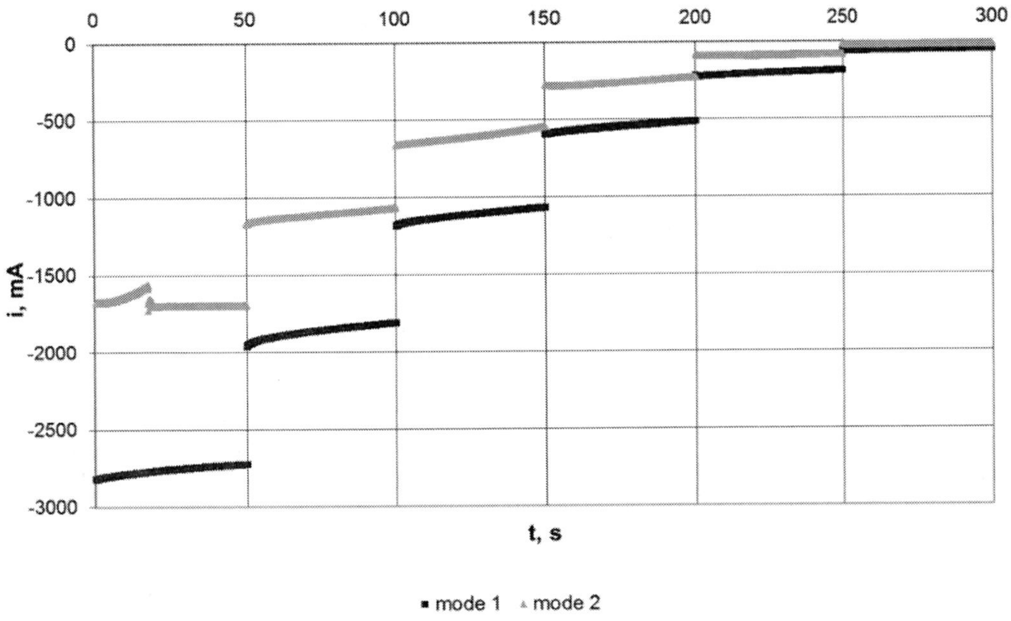

Fig. 1. The change of current over time at step change of potential. Mode 1: H2/air, P= 0.9 bar; mode 2: hydrogenous fuel/air, P=0.9 bar.

It should be noted that the amount of gas consumed per unit of produced electricity and power comparable to study modes. This confirms the ability effectively module work on hydrogenous fuel derived from biogas, and achieving fuel-efficient modes.

Joint analysis of plots (Fig. 2) and in Table 2 suggests that while working on electrolysis hydrogen fuel discharge does not occur on the considering duration the fuel cell operation. Consequently, the fuel feeding rate and the resulting number of protons at the anode sufficient to sustain a stable electrochemical process in the fuel cell. For a gas mixture of CH_4 fuel discharges occur at high current (or power) as not discharged molecules CH_4 prevent access required amount of hydrogen to the catalytic surface of the anode. Increasing the excess fuel pressure reduces the number of purges due to increased fuel supply rate, that is, increase the importance of mass transfer stages providing delivery of H_2 to the catalyst surface, and thus to quantity of the particles which are discharged on the anode [17-19].

Table 1. Basic parameters the step change in the potential for module FC-50

E, V	t, s	mode 1 H_2/air, P= 0.9 bar		mode 2, hydrogenous fuel/air, P=0.9 bar	
		V_1, ml	V_2, ml	V_1, ml	V_2, ml
6.0	50	148.3		99.2	
6.5	100	103.3		65.8	
7.0	150	60.8	0	36.7	
7.5	200	31.7		17.5	8.0
8.0	250	13.8		8.3	
8.5	300	6.2		4.6	
		Σ =364.1	purges have not been	Σ = 232.1	1 purge
		Σ= 364.1		Σ = 240.1	

Where:
E – the potential;
t – the time period of step change of potential;
V_1 – the operative volume of the spent gas;
V_2 – gas volume in purges.

The statistical analysis of data shows that on each site of the cycle "work-purge" fall of current and power characteristics at supply of electrolysis hydrogen makes 7%, in case of use of hydrogenous mixture with residual methane falling of these indicators makes 4÷6 of % depending on an operating mode. Use of non-standard fuel demands frequent updating of a surface, i.e. the number of purges for removal from the anode of not oxidized methane molecules together with which also not discharged hydrogen molecules are removed that reduces fall of current and power characteristics. During the work on electrolysis hydrogen purges happen much less often, however fall of these characteristics says that on the anode there is an accumulation of the molecules of hydrogen which aren't discharged to a proton interfering access of again arriving fuel to a catalytic surface.

It should be noted that amount of the spent gas on unit of the made electricity and power are commensurable for all studied modes. It confirms a possibility of effective operation of the module on hydrogenous fuel with the residual content of methane received from biogases and achievements of the modes of effective use of fuel.

Using the relation (1) was determined by the amount of consumed gas for a coulomb electricity (N_q).

$$N_q = \frac{n \cdot G_s \cdot t_s + \Delta t \cdot G_r}{I \cdot t_r}, \quad (1)$$

where:
n – the number of fuel purges on the selected plot of time;
G_s – gas flow rate at the moment purge, ml;
t_s – the time discharge of fuel, s;
Δt – the time interval stable operation of the fuel cell, s;
G_r – flow rate gas to the fuel cell operation, ml;
I – amperage (output characteristic), A;

t_r – the duration the selected area of the fuel cell operation (400 ÷ 1200), s.

The corresponding quantitative indices on considered modes are presented in Table 2.

Table 2. Data of fuel cell operation for fuels of different composition.

mode	fuel	I, A	n	Gs, ml/s	Gr, ml/s	Nq, ml/C
1	H_2	2.0÷2.2	1	12.0	2.5	1.2
2	98% H_2 + 2% CH_4	1.8÷1.9	12	12.0	2.7	1.6

The amount of consumed gas for retrieving 1 coulomb electricity under feeding the electrolysis hydrogen and hydrogen-containing mixture with the residual methane content at an identical oxidizer and pressure of supply of fuel are commensurable. This confirms the possibility of efficient operation of the module on the hydrogenous fuel with a residual methane derived from biogas, and achieving fuel efficiency modes [17-20].

The FC-50 module may be considered as the scale model of more difficult and power-intensive system. The presented results are the basis for calculations for an assessment of overall performance of the 500 W fuel module which can be considered as basic element for off-grid system consumption.

Presented results form the basis of calculations to evaluate the effectiveness of the fuel module 500 W, which can be considered as a basic element for autonomous consumer. The results of estimating calculations the required amount of energy (electrolysis hydrogen and hydrogenous fuel) for the module operation during 2000 hour/year are presented in Table 3.

Table 3. Estimated parameters of necessary expenses energy for operation of the power installation with a capacity of 500 W for t=2000 hours/year

Title indicators	Designation, dimensionality	hydrogen electrolysis	Hydrogenous fuel
Efficiency	η, %	46.3	44.0
Rated Power Energy System	P, kW	0.5	0.5
Power created by the consumable	Pg, kW	1.08	1.01
Electric power generation	W, kW h/year	1000	1000
Mass of fuel consumed	m, kg/year	64.85	75.50

Comparison of the obtained values of mass of fuel consumed to ensure stable operation of the installation within the specified period shows that the consumption of hydrogenous fuel is comparable with the required flow rate of hydrogen electrolysis.

Table 4 shows the results of calculation of the methane and biogas volumes required for obtaining hydrogenous fuel in an amount providing work of the power plant with capacity of 500 W for the module operation during 2000 hour/year.

Table 4. Estimated parameters fuel consumption in the reformer for power installation capacity of 500 W for t=2000 h/year

Title indicators	mass of methane	volume of methane	volume of biogas
Designation, dimensionality	m_{CH4}, kg/year	V_{CH4}, m^3/year	V_{biogas}, m^3/year
Digital value	142.0	197.9	328.0

The efficiency of the tandem system "reformer-fuel power unit" is defined by Eq. 2 [21]. The initial data used normalized values of a tandem system reformer-fuel cell and workflows.

$$\eta_{\Sigma} = \eta_{ref} \cdot \left[2 - u + \left(1 - \frac{1}{\lambda_{H2}} \right) \cdot \eta_{ref} \right] \cdot \frac{1}{\lambda_{H2}} \cdot \frac{U}{u_{ref}}, \quad (2)$$

where:

η_{ref} =0.7 — efficiency of the reformer;

u=0.95 — coefficient of methanization;

λ_{H2}=1.25 — output ratio of hydrogen to one mole of methane;

U=0.720 V — the voltage on the unit cell;

u_{ref} =1.25 V — the reference voltage.

Estimated the effective efficiency of the tandem system based on the reformer with steam reforming of methane and power plant with capacity of 500 W amounted ~ 39 %. This confirms the relatively high efficiency of non-standard hydrogenous fuel use.

3 Conclusions

1. A promising direction in the creation of autonomous power supply systems in low-rise residential development is the use of fuel cells running on hydrogenous fuel derived from biogas.
2. Expenses of hydrogen fuel to operate the power installation with a predetermined capacity comparable with the required amount of hydrogen obtained by electrolysis.
3. The calculated value of the efficiency of the system reformer – fuel cell, based on data on the required quantity of biogas and methane, as well as numerical characteristics of workflows fuel system, nearly to 39 %, confirms the relatively high efficiency of the use of hydrogenous fuel for the tasks of autonomous power supply.
4. Usage of relatively low-cost hydrogen-containing fuel obtained from biogas from local secondary renewable resources will contribute to the creation of autonomous economic systems of power supply.

References

1. N.I. Vatin, D.V. Nemova, D.S. Tarasova, A.A. Staritcyna, Advanced materials research, **953-954**, 854-870 (2014)
2. D.O. Dunikov, V.I. Borzenko, S.P. Malyshenko, D.V. Blinov, A.N. Kazakov, Teploenergetika [Thermal Engineering], **60(3)**, 202-211 (2013)
3. S.P. Malyshenko, V.I. Borzenko, D.O. Dunikov, O.V. Nazarova, Teploenergetika [Thermal Engineering], **59(6)**, 468-478 (2012)
4. G. Gahleitner, International Journal of Hydrogen Energy, **38(5)**, 2039-2061 (2013).
5. K.C. Surendra, D. Takara, A.G. Hashimoto, S.K. Khanal, Renewable and Sustainable Energy Reviews, **31**, 846-859 (2014)
6. J. Havukainen, V. Uusitalo, A. Niskanen, V. Kapustina, M. Horttanainen, Renewable Energy, **66**, 232-240 (2014)
7. R. Chaubey, S. Sahu, O.O. James, S. Maity, Renewable and Sustainable Energy Reviews, **23**, 443-462 (2013)
8. H.J. Alves, C. Bley Junior, R.R. Niklevicz, E.P. Frigo, M.S. Frigo, C.H. Coimbra-Araújo, International Journal of Hydrogen Energy, **38(13)**, 5215-5225 (2013)
9. I. Rodriguez-Verde, L. Regueiro, M. Carballa, A. Hospido, J.M. Lema, Science of the Total Environment, **497**, 475-483 (2014)
10. F. Bauer, T. Persson, C. Hulteberg, D. Tamm, Biofuels, Bioproducts and Biorefining, **7(5)**, 499-511 (2013)

11. M.P. Fedorov, V.I. Maslikov, A.N. Chusov, D.V. Molodtsov, St. Petersburg State Polytechnical University Journal, **4(135)**, 35-41 (2011)
12. M.P. Fedorov, M.Yu. Zubkova, A.A. Thomasov, D.V. Molodtsov, N.K. Zelenina, *Proc. Reports. 7th Russian Conference "Physical Problems of Hydrogen Energy." St. Petersburg: FTI. Joffe*, 160-162 (2011)
13. M.Yu. Zubkova, V.I. Maslikov, D.V. Molodtsov, A.A. Thomasov, *Proc. Report of the Eighth Russian Conference "Physical Problems of Hydrogen Energy", St. Petersburg. FTI. Joffe*, 151-153 (2012)
14. A.N. Chusov, M.Yu. Zubkova, V.V. Korablev, V.I. Maslikov, D.V. Molodtsov, St. Petersburg State Polytechnical University Journal, **4-1(183)**, 78-85 (2013)
15. M.Yu. Zubkova, V.V. Korablev, V.I. Maslikov, D.V. Molodtsov, A.N. Chusov, *Abstracts. International Conference "Innovative materials, Structures and Technologies", Riga. RTU*, 54 (2013)
16. M.P. Fedorov, A.N. Chusov, V.I. Maslikov, M.Yu. Zubkova, D.V. Molodtsov, *Abstracts, VIII International Conference, Fundamental problems of electrochemical energy, Saratov*, 436-439 (2011)
17. M.Yu. Zubkova, V.I. Maslikov, D.V. Molodtsov, A.N. Chusov, Advanced Materials Research, **941-944**, 2107-2111 (2014)
18. M.Yu. Zubkova, V.I. Maslikov, D.V. Molodtsov, A.N. Chusov, Applied Mechanics and Materials, **587-589**, 330-337 (2014)
19. M.Yu. Zubkova, M.P. Fedorov, A.N. Chusov, Innovations, **196(2)**, 70-76 (2015).
20. M.Yu. Zubkova, V.I. Maslikov, M.P. Fedorov, A.N. Chusov, *Proceedings Sardinia 2015, 15-th International Waste and Landfill Symposium, Forte Village, Santa Margherita de Pula, Italy; 5 9 October 2015*, 229 (2015)
21. R. Schmersahl, E. Janine, S. Volkhard, available at: http://brennstoffzellen.pitcom.net/upload/dokument139.pdf (accessed October 05, 2015)

Applicability Problem in Optimum Reinforced Concrete Structures Design

Assedeq Ashara [1,*], *Aleksandar* Milajić [1] , *Dejan* Beljaković [1] and *Issa* Togo [2]

[1]Union Nikola Tesla University, Faculty of Construction Management, Belgrade, Serbia
[2]Peter the Great St. Petersburg Polytechnic University, 195251 Polytechnicheskaya 29, St. Petersburg, Russia

Abstract. Optimum reinforced concrete structures design is very complex problem, not only considering exactness of calculus but also because of questionable applicability of existing methods in practice. This paper presents the main theoretical mathematical and physical features of the problem formulation as well as the review and analysis of existing methods and solutions considering their exactness and applicability.

1 Introduction

Search for an effective and applicable method for optimum design of the reinforced concrete structures is not a new subject, but the great majority of solutions proposed in literature is aimed at finding minimal weight of a given structure, although the decision making process is usually, if not always, aimed at minimal price. Therefore, material and labour costs should also be important issues in design and construction of the reinforced concrete structures. Besides that, obtained solution should not be only mathematically acceptable, but also applicable in reality, i.e. at the building site.

Common approach in optimum design of the reinforced concrete structures is focused on optimizing the cross-sectional dimensions and total quantity of reinforcement but usually does not include more detailed analysis of the reinforcement pattern and possibility of its proper placing and fixing at the building site. However, in reality, reinforcement design includes detailed specification of many data beyond the determination of area of steel, such as the selection of bar diameters and the number of bars, as well as longitudinal distribution of bars, positioning of bars within critical sections, determination of curtailment points, specification of the size and spacing of stirrups, etc. As a consequence, proposed theoretical methods are usually adequate only for the preliminary analysis but not sufficiently applicable in engineering practice.

The aim of this paper is to present structural, mathematical and practical aspect of optimal reinforced structures structural design, as well as to provide concise and clear overview of the solutions proposed in literature and their basic features in order to enable other researchers to easily find adequate benchmark problems and to develop proper criteria for optimality assessment.

[*] Corresponding author: aleksandar.milajic@gmail.com

2 Literature Overview

Development of optimization methods in the field of reinforced structures design was closely connected with available computation possibilities. Until the information technology was not developed enough to support very complex calculus models and procedures, problem of optimal reinforced structures design was usually solved by considering only basic variables such as cross-sectional dimensions and total amount of the reinforcement, while the problem of reinforcement bars placing within a concrete members remained almost untouched or was avoided by introducing too generalized assumptions [1]. Friel [2] derived an equation for determining optimal ratio of steel to total concrete area in a singly reinforced beam, while Chou [3] used Lagrange multiplier method for minimizing total cost of the T-shaped beam. Kirsch [4] presented iterative procedure in three levels of optimization for minimizing the cost of continuous girders with rectangular cross section, in which the total amount of the reinforcement is minimized at the first level, cross-sectional dimension are minimized at the second level, while the third level of optimization iz minimizing the design moments. Lakshmanan and Parameswaran [5] derived a formula for direct determining of optimal span to cross-sectional depth ratio so the iterative trial and error procedure can be avoided, while Prakash et al. [6] based their cost-minimization method on Lagrangian and simplex methods. Kanagasundaram and Karihaloo [7,8] introduced the crushing strength of concrete as an additional variable along with cross-sectional dimensions and steel ratio to optimize the cost of simply supported and multi-span beams with rectangular and T-sections using sequential linear programming and convex programming. Chakrabarty [9,10] presented cost-optimization method for rectangular beams using the geometric programming and Newton–Rapson method, while Al-salloum and Siddiqi [11] proposed optimal design of singly reinforced rectangular beams by taking the derivatives of the augmented Lagrangian function with respect to the area of steel reinforcement. Coello et al. [12] proposed the cost optimal design of singly reinforced rectangular beam using Genetic Algorithms by considering cross-sectional dimensions and the reinforcement area as variables. More detailed overview of literature on cost-optimization of reinforced concrete structures up to 1998 can be found in [13].

Further development of the information technology enabled researchers to develop innovative methods of design by including the optimality aspect more thoroughly and to include more realistic requirements and optimality criteria, as well as the reinforcement detailing. One of the first papers that deal with reinforcement placing details was presented by Koumousis and Arsenis [14]. This method is based on multi-criterion optimization using Genetic Algorithms for finding a compromise between minimum weight, maximum uniformity and the minimum number of bars for a group of members. After that, researchers have started to introduce reinforcement detailing data as variables in optimization methods, usually by using one of two basic approaches. In the first one, reinforcement spacing demands are included into calculus as constraints [15-18], while the other one uses previously developed data base of possible reinforcement patterns [19-22]. Constraints in the first approach are based on maximum allowable number of reinforcement layers (usually one or two) and maximum allowable number of bars per layer (usually up to four or five). The second approach is in fact simplification of the first one because the data-base of allowable reinforcement patterns is developed by introducing the same limitations and demands proposed by a given code of practice. Only a few papers, such as [23] have no previously adopted arbitrary limitations.

It can be observed that the main problem in comparing efficiency and applicability of different approaches is the fact that they are based on different codes of practice, i.e. on different reinforcement placing rules and restrictions. Because of that, and as opposite of the steel structures, there is no standard benchmark problems for testing a given method so the parametric sensitivity analysis is the only available tool for applicability assessment.

3 Optimization Problem

The first variable studied in optimum structural design is the cross-sectional shape for all members that the structure is composed of that will result in the most suitable stress distribution. In order to obtain mostly uniform forms within one building, the cross-sectional shapes are usually selected from a relatively limited list of possible or available sections. When the shape is determined and fixed, the second and possibly the most challenging task is the placement of reinforcing bars within concrete members, often called detailing. From the optimization point of view, this task generally belongs to the field of topology optimization, where the number of bars, their shape and space position are searched for. The type and form of a chosen parameterization of the shape will determine the computational complexity of a given problem.

If the optimal design of a given reinforced structure is cost-oriented, i.e. if the aim is to minimize the total price of a structure, than the objective function can be defined as:

$$F_{(x)} = V_c P_c + W_s P_s + A_f P \tag{1}$$

where Vc is the volume of concrete, Ws is weight of steel, Af is total area of formwork and Pc, Ps and Pf are unit price of concrete per m^3, of steel per kg and of formwork per m^2, respectively. Prices of the materials include material, fabrication and labor.

Since the structural geometry (spans and supports positions), material properties and prices and loads (except self-weight) are usually predefined in the designing process, variables in this problem are the cross-sectional dimensions, i.e. width b and depth h. Basically, there is no need to include the total steel area in critical cross sections in the variables because it can be calculated according to a given code of practice. However, this aspect of design should not be totally neglected since the complete solution of the problem should also include details about bars diameters and placing scheme, which can be quite a demanding task, especially considering the great variety of possible patterns and possibility of combining bars with different diameters or forming bundles made of same or different bars. It should be noted here that although codes of practice can vary more or less between different countries, they all generally come down to the same set of requirements because what is obligated in one country usually is accepted as a rule of thumb in another and vice versa.

4 Applicability Problem

The main problem in all approaches mentioned in the second section is detailing, i.e. forming the reinforcement pattern with sufficient amount of data and details that would be applicable for placing and fixing in situ. Having in mind that reinforcement bars come in more than ten different available diameters, this task is not as easy as it is usually considered. Consequently, there are numerous possibilities for achieving any given required total area of steel. The level of complexity gets even higher if we consider the possibility of using different bars diameters in the same cross section and the fact that reinforcement can also be grouped in bundles consisting of two, three or four same or different bars.

This variety of feasible solutions faces a designer with several further problems. The first and the most obvious one is how to take all possible reinforcement patterns into account. Complexity of this problem can be illustrated by a relatively simple example shown on Figure 1. For a given rectangular cross-section b/h = 25/35 cm calculated minimal area of steel is 9.30 cm^2. All presented patterns are theoretically acceptable, but the computer would always opt for the third one (9.36 cm^2), although an experienced designer would dismiss it as too complicated. Note that solutions with bundles and combinations of different bundles or of bundles and single bars are not presented, because in that case there would be more than 20 feasible solutions.

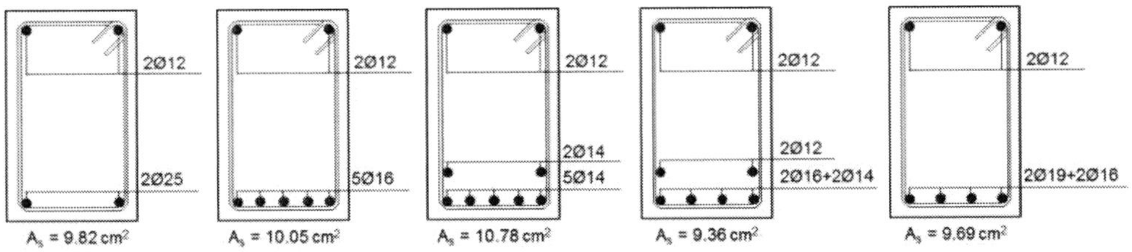

Fig. 1. Feasible reinforcement patterns for Ar=9.30 cm^2.

5 Possible Solutions

One possible solution that can be found in the literature is to use previously developed database consisting of all possible patterns for all available bar diameters placed in cross-sections with different widths. However, proposed databases are usually extremely generalized, simplified and used only to show the basic principle and possibilities of a given methods. The only one that might be considered sufficiently detailed is presented in papers by Govindaraj and Ramasamy [19,20], but it should be noted that their database covers only very limited cross-sectional width span of 10 cm and provides almost 10.000 possible reinforcement patterns. This means that for greater span of allowable widths, for example 20–50 cm, database would consist of hundreds of thousands possible patterns and development of such database would be meticulous and demanding task. Besides that, even if such database would exist, the computer would always opt for the solution that would demand the minimal amount of steel, regardless of it applicability at the building site. But in reality, when faced with a variety of possible and theoretically acceptable solutions, a designer is supposed to find a compromise between the exactness and the applicability at the building site. Because of that, it is recommendable to use some additional criteria for estimating the applicability of a given reinforcement pattern, such as the one proposed in [24].

The second possible solution, and the more applicable one, is to divide the reinforcement problem into two separate and generally independent steps. The first one would be to find only the optimum amount of steel and its centrum, and it would be a part of a chosen optimization method, usually some metaheuristics. After the optimal solution is obtained, the second step would be to find feasible reinforcement pattern that would provide calculated amount of steel and that also could be placed within optimum cross-sectional width. This approach has several benefits. First of all, it does not affect the exactness of calculus because, due to a great number of possible reinforcement patterns, any given steel area can be sufficiently obtained, without compromising exactness of the calculus. Besides that, it would decrease number of variables and restrictions in the optimization process and thus would make it faster and more efficient.

6 Conclusion

Theoretical methods for optimum reinforced concrete structures design that can be found in the literature usually consider this problem only as the mathematical one, regardless of applicability of obtained solutions in practise. Purpose of this paper is an attempt to abridge a gap between theory and practise in this field by emphasizing importance of assessing obtained solutions from the practical point of view. The most critical points and problems are emphasized, as well as the possible solutions that would lead to more applicable solutions.

References

1. G. Steven, FENet Newsletter, January, (2003)
2. L. Friel, ACI Journal, **71(11)**, 556–558 (1974)
3. T. Chou, Journal of the Structural Division ASCE, **103(8)**, 1605–1617 (1977)
4. U. Kirsch, Engineering Optimization, **6(4)**, 207–212 (1983)
5. N. Lakshmanan, V.S. Parameswaran, , IE (I) J, **66**, 92–98 (1985)
6. A. Prakash, S.K. Agarwala, K.K. Singh, Computers and Structures, **30(4)**, 1009–1011 (1988)
7. S. Kanagasundram, B.L. Karihaloo, Computers and Structures, **41(6)**, 1357–1364 (1991)
8. S. Kanagasundram, B.L. Karihaloo, Computers and Structures, **41(3)**, 509–518 (1991)
9. B.K. Chakrabarty, Computers and Structures, **42(3)**, 447–510 (1992)
10. B.K. Chakrabarty, Journal of Structural Engineering ASCE,**118(11)**, 3238–3242 (1992)
11. Y.A. Al-Salloum, G.H. Siddiqi, ACI Structural Journal, **91(6)**, 647–655 (1994)
12. C.A. Coello, A.D. Christiansen, F. S. Hernandez, Engineering and Computers, **13**, (1997) 185–196.
13. K.C. Sarma, H. Adeli, ASCE Journal of Structural Engineering, **124(5)**, 570-578 (1998)
14. V.K. Koumousis, S.J. Arsenis, Comput Aided Civil Infra-structural Engineering, **13**, 43–52 (1998)
15. M. A. Alqedra, I. Mohammed, Journal of Artificial Intelligence, **4**, 76-88 (2011)
16. A. Kaveh, O. Sabzi, International Journal of Civil Engineering, **9(3)**, 193-206 (2011)
17. A.F.M. Barros, M.H.F.M. Barros, C.C. Ferreira, Structural and Multidisciplinary Optimization, **45**, (2012) 845–860.
18. G. Bekdas, S.M. Nigdeli, Proceedings of the 3rd European Conference of Civil Engineering (ECCIE '12), 122-126 (2012)
19. V. Govindaraj, J.V. Ramasamy, Computers and Structures, **84**, 34-48 (2005)
20. V. Govindaraj, J.V. Ramasmy, Engineering Optimization, **39(4)**, 471-494 (2007)
21. M.M. Jahjouh, M.H. Arafa, M.A. Alqedra, Structural and Multidisciplinary Optimization, **47(6)**, 963-979 (2013)
22. S.T. Yousif and R.M. Najem, International Journal of Applied Sciences and Engineering Research, **2(1)**, 79-92 (2012)
23. A. Milajić, D. Beljaković, G. Pejičić, Technics Technologies Education Management, **8(2)**, 533–540 (2013)
24. A. Milajić, Technical Gazette, **22(1)**, 119-124 (2015)

MATEC Web of Conferences 73, 04006 (2016) DOI: 10.1051/matecconf/20167304006
TPACEE-2016

Forecast Jointed Rock Mass Compressive Strength Using a Numerical Model

Anatoliy Protosenya[1], and *Pavel* Verbilo[1,*]

[1]National Mineral Resources University, 199106 St Petersburg, Russia

Abstract. The method of forecasting the strength of the jointed rock mass by numerical modeling of finite element method in ABAQUS was described. The paper presents advantages of this method to solve the problem of determining the mechanical characteristics of jointed rock mass and the basic steps of creating a numerical geomechanical model of jointed rock mass and numerical experiment. Numerical simulation was carried out with jointed rock mass in order to obtain the ratio of strain and stress while loading the numerical model, determining parameters of quantitative assessment of the impact of the discontinuities orientation on the value of the compressive strength, compressive strength anisotropy. The results of the numerical experiment are compared with the data of experimental studies investigations. Innovative materials and structures are analyzed in this paper. The results that were obtained by calculation show qualitative agreement with the results of laboratory experiments of jointed rock mass.

1 Introduction

The rock mass is a system of discrete blocks of intact rock that are separated by fractures, structural weakening zones, different defects that have a significant influence on mechanical characteristics of the jointed rock mass [1]. Methods of determining mechanical characteristics of the rock mass are divided into direct and indirect. Direct methods are used in the laboratory or field conditions, where the rock mass is loaded with the help of press installations and then the value of strength in uniaxial compression is determined. Laboratory or field experiments for testing of the rock mass is the most reliable way to determine the mechanical characteristics, but in a laboratory conducting tests on samples of large dimensions are associated with difficulties and high cost [2-3]. Other approaches to determine the strength and deformation characteristics of the rock mass are indirect methods, which include: empirical, analytical and numerical methods. The empirical approach [4-6] for determining the mechanical characteristics of the rock mass is based on case history, which is widely used in the design of tunnels by using the classification systems of the rock mass. The main disadvantage of this method is the lack of a mathematical framework for the establishment of geomechanical model. At present in the domestic industry in order to determine the strength and deformation properties of jointed rock mass empirical coefficient of structural weakening is used that takes into account only

* Corresponding author: pashaverbilo@mail.ru

one characteristic of jointed rock mass, the distance between discontinuities. In most cases this method is not scientifically justified and therefore can leads to inaccurate results. Analytical methods 11 are very convenient to use in geomechanics, but their application to determine the mechanical properties of the rock mass does not allow to get solution for system, which including a complex system of discontinuities and intact rock.

A big impulse for the development of numerical approaches to determine the strength and deformation properties of jointed rock mass was the development of computer technology. Development of numerical methods and their integration with existing analytical methods for solving problems, which are related with the determination of influence on the value of the rock mass mechanical properties, the scale effect, is the subject of serious research, because there are uncertainties, which are associated with the transition from the test results of intact rock to the physical and mechanical properties of the rock mass. The main advantage of numerical methods is the ability to create a geomechanical model of the rock mass on the basis of accounting of discontinuities system and interaction conditions of intact rock along of discontinuity 12.

The finite element method is one of the most widely used numerical methods for determining the mechanical characteristics of jointed rock mass, as it allows explicitly take into account fractures of rock mass and conditions of interaction of rock blocks 13. Recently a lot of researches are conducted using the finite element method under jointed rock mass to describe its mechanical behavior 14.

2 Materials and Methods

Construction of numerical model of scale-inhomogeneous rock mass requires singling out and analysis of typical structures of discontinuities by geotechnical data, because the structure of discontinuities reflect the structural features of the rock mass and determine form of blocks of intact rock. The size of the rock block, inside which the conditions quasicontinuous and quasihomogeneity are satisfied, is representative elementary volume. Results of laboratory experiments on the compression press equipment well describe the mechanical behavior of the elementary intact rock block. The behavior of the rock within the representative elementary volume in the numerical model is described by results of laboratory experiments. Results of laboratory experiments is achieved by compilation of a model of intact rock and consistent quality correction to match the results obtained in laboratory tests with the possibility of quantitative differences within the margin of error. The mechanical behavior of intact rock in the numerical model will correspond to reality, if we get results that are similar to the experimental results.

We must always remember that in the rock mass there are several levels of typical structures of discontinuities and defects. Each sample of every structure consists of a finite number of previous level structures, which leads to the requirement of compliance of conditions of quasihomogeneity and quasicontinuous which are calculated on each level. This requirement is set by the need to eliminate the influence of the scale effect on the final result.

The subsequent development of geomechanical numerical model requires a similar sequential execution of numerical experiments by modeling of interaction of blocks of intact rock along the contact between them in order to achieve result that is not quantitatively different from the laboratory experiment on direct shear. This step is the most important, because carried out research with rock mass indicates that discontinuities, as the main factor, mostly affect the quantitative value of mechanical characteristics of the rock mass. Based on the above steps we construct a common geomechanical model of the rock mass, taking into account the features of the behavior of blocks of intact rock and their interaction on the contact during loading.

Changing the size of the numerical models of rock mass allows reasonable calculation if scale effect. The model can be applied for numerical experiments with different systems of discontinuities.

3 Formulation of the problem and description of the study

Aspects of modeling of rock mass sample impaired by discontinuities system during uniaxial and biaxial compression testing using the finite element method in ABAQUS to study the anisotropy of uniaxial compression strength of jointed rock mass are considered in this paper, the procedure of simulation is described consistently. Graphs of the stress-strain dependence and graphs of the dependence of the compressive strength on the angle of discontinuities in the rock mass built on the results of the calculations. We use numerical geomechanical model of the rock mass to study the behavior of the rock mass under the influence of loads and for determination of mechanical characteristics of jointed rock mass. This model reflects actual mechanical characteristics of the rock mass and laws of its changes in the space by interconnected mathematical and formal logic expressions. The task is to build a numerical model of jointed rock mass and model's test under uniaxial and biaxial compression in order to obtain the limit values of strength and to obtain an influence estimation of discontinuities parameters in conjunction with an applied lateral load on the value of compression strength.

It is assumed that the destruction in the numerical experiment of a sample of the rock mass will occur on the surfaces of the weakening. Destruction of blocks of intact rock in the model is not provided. In the numerical model interaction on the contact between the blocks is defined by friction, as it is assumed that the cohesion between blocks of intact rock is no taken into consideration.

Actually, cohesive bonds in jointed rock mass along the contact interaction of blocks of intact rock have small values, so we can ignore cohesion in modeling of jointed rock mass 17. Representative elementary volume of intact rock in the study under consideration is form of parallelepiped. Therefore, system of the rock mass discontinuities is a system of vertical and horizontal parallel lines with a constant distance between them.

Basic information about mechanical characteristics of the intact rock and the discontinuities that was used in the modeling is given in 0.

Table 1. The mechanical properties of the intact rock and discontinuity.

Intact rock	Density, (kg/m^3)	Young's modulus, E (GPa)	Poisson's ratio
	2700	2	0,3
Discontinuity	Joint friction angle, (^{o})	Joint cohesion, [MPa]	
	21	0	

The most time-consuming step in the construction of numerical geomechanical model of the rock mass is the geometric construction of the rock mass, as it is formed by a plurality of independent units that must be simulate.

To solve this problem two options scheme application loads on the numerical model of the rock mass have been chosen. In the numerical model vertical load in combination with various lateral loads (0, 10, 20, 30 MPa) were set. Lateral load were applied to the side surface uniformly over the whole area. Typical design schemes of load application to the numerical model during the numerical experiment are shown in 0, wherein 0a shows a scheme of uniaxial compression. 0b shows biaxial compression test.

Vertical load to the numerical model of the rock mass was provided through a specified move to absolutely rigid plate located on the top surface of the rock mass. Lower plate in the numerical experiment was also absolutely rigid. Any possible move prohibits to support plate. Thus, in the blocks of intact rock and contract between them is occurring the emergence and subsequent redistribution of stresses during the motion of the plate on the ordinate. The loss of the bearing capacity of the model occurs after reaching the limit state of numerical model of jointed rock mass.

Models of jointed rock mass with angles to the horizontal plane of the discontinuities of 0, 30, 45, 60, 90 degrees were built for research strength anisotropy of uniaxial and biaxial compression of jointed rock mass. For each models carried out of simulation tests in uniaxial and biaxial compression. Several simulation sets had been described above. Black lines within the test sample of the jointed rock mass in 0 indicate the discontinuity, and the intersection of the lines with each other forms blocks of simulated intact rock.

The results are analyzed after carried out virtual experiments. Results of virtual experiment are summarized in the graphs of the relative value of the stress to the relative magnitude of strains of jointed rock mass where the angle of discontinuities is 45 degrees (0). Also below is a graph of relative magnitude of the strength on uniaxial and biaxial compression on the angle of inclination of discontinuity to the horizontal plane (0).

As you can see, the orientation of discontinuities in the rock mass has a significant impact on its strength and deformation characteristics and the process of deformation and failure of the rock mass in volume state of stress occurs in the form of shear and breakout, which is confirmed by experimental studies [18-19].

The greatest value of the uniaxial compression strength belongs to the model of the rock mass with an angle of discontinuities 0 degrees to a horizontal plane (0). Further, there is gradual decline in the value of the compression strength of the jointed rock mass by increasing the angle of discontinuities to a horizontal plane, this dependence can be described by a linear relationship. The biggest difference from the compression strength of intact rock and compression strength of a rock mass belongs to rock mass with the angle of discontinuities to a horizontal plane 55 degrees. This ratio is typical for this system of discontinuities. The subsequent increasing of the angle of inclination of discontinuities to the horizontal plane is accompanied by a gradual increasing in the value of the strength of the rock mass under uniaxial and biaxial compression.

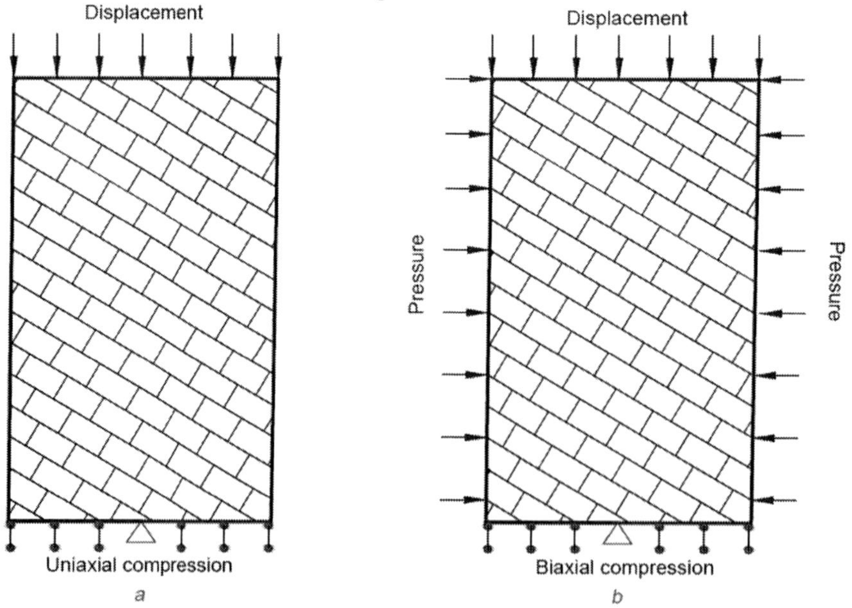

Fig. 1. Schemes of loads to models in the numerical experiment: (a) scheme of uniaxial compression; (b) scheme of biaxial compression.

The simulation results (0) show that the increase in load leads to compaction of jointed rock mass and rigidity increased. With further load increasing blocks are displaced relative to each other and form voids within the array. When the shear occurs it leads to separation of intact rock along the discontinuity 17. But shear occurs when the shear stresses on the micro areas reach a critical value. Similar shears take place all over the jointed rock mass, which leads to increase of the size of model, that is to say dilatancy effect is observed. It should be noted that rock mass, without applied lateral load, has the lower value of compressive strength. Rock mass has brittle deformation, this fact is confirmed by realized numerical experiments.

In 0 the mechanism of rock mass deformation is considered under biaxial compression. Diagrams of stresses distribution in the rock mass during deformation are shown in 0. Micro shears occur during the initial deformation that leads to the formation of micro discontinuities of separation. Hardening occurs in the next step of deformation, which leads to an increase of stresses in the rock mass. The contact area of interaction between the intact rock blocks (the area that takes the load) is reduced with the hardening of the rock mass. Further, the shear deformation is accompany by the formation of voids and increase of volume of the model. These two processes determine the value of the jointed rock mass compressive strength [18-19].

Thus, at the beginning of the loading deformation of the rock mass has is linear character. After reaching the elastic limit of irreversible deformation occurs that extends to the maximum of compression strength. At this stage in the rock mass shears are taken place along of contacts between blocks of intact rock. It results in formation of main shear surface in the model. This section of the diagram which continues to the maximum magnitude of compressive strength is characterized by homogeneous deformations over the whole volume of the jointed rock mass. On the maximum of compressive strength deformation of this character stops. After passing through the maximum deformation occurs on area of residual deformations, which continues until the limit of residual strength. The result is the destruction of the rock mass along the main sliding surface, where maximum shear stresses have been formed. The above-described behavior of jointed rock mass during deformation experimentally was determined in laboratory conditions on press equipment [18-19].

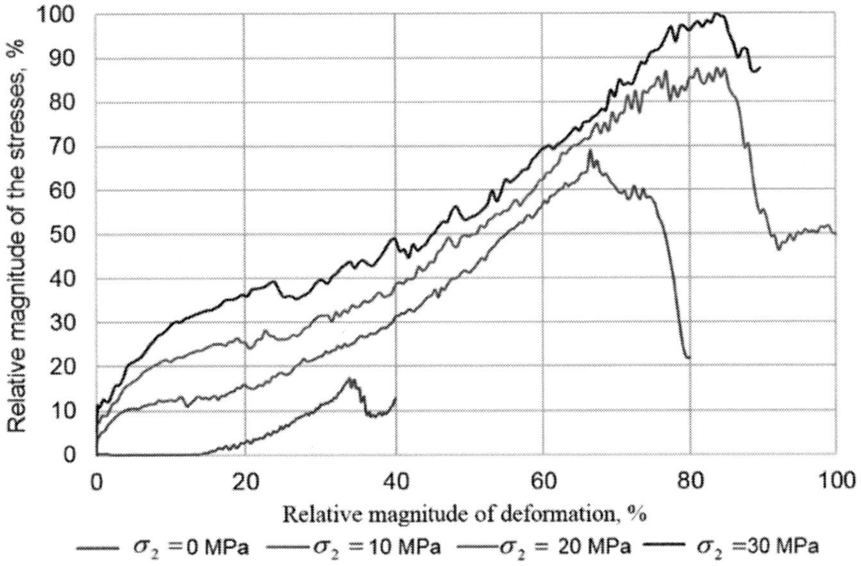

Fig. 2. Graphs of dependences of the relative magnitude of the stresses on the relative magnitude of deformation of the rock mass with an angle of discontinuities to the horizontal plane 45 degrees.

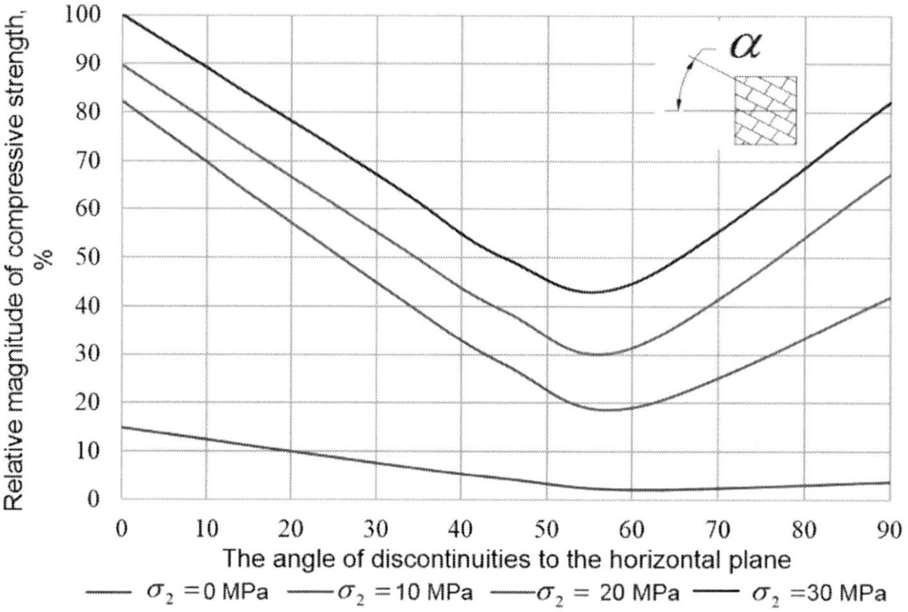

Fig. 3. Graphs of dependences of the relative magnitude of compressive strength on the angle of discontinuities to the horizontal plane at different lateral loads.

Increased strength with increasing lateral pressure is explained based on the assumption that the jointed rock mass has an internal friction which increases with increasing pressure. By analyzing the results of the numerical experiment we can make the following conclusions: jointed rock mass had destroyed by slide. The magnitude of the failure stress, the magnitude of the residual strain and the angle of orientation of the slide plane depend on state of stress and angle of discontinuities in the rock mass to the horizontal plane; the increase of lateral stress leads to increased shear strength at the contact between the intact rock blocks. Discontinuities are the cause of discontinuities of separation during deformation of the jointed rock mass. Concentration of stresses on the surfaces of defects leads to emergence of discontinuities of separations.

4 Comparison of analysis results

The article describes a numerical method for determining the strength and deformation characteristics of fractured rock mass using the finite element method. Numerical experiments were done with jointed rock mass of constant size by varying lateral loads and angle of discontinuities to horizontal plane. As a results dependence of strength on biaxial and uniaxial compression on the direction of force application to the sample of jointed rock mass was derived. The results were compared with similar studies with jointed rock mass, which were done by other researches 20-22, where numerical simulation was carried out on basis of experimental data obtained by testing samples of the jointed rock mass in the laboratory. The results obtained in this study have a good match convergence with experimental results 17. The experimental results show that the structural defects have such as discontinuities in the rock mass has a significant impact on the value of the strength and deformation characteristics of the jointed rock mass and strength, and deformation characteristics depend on the direction of the load in relation to major system of discontinuities [20-26].

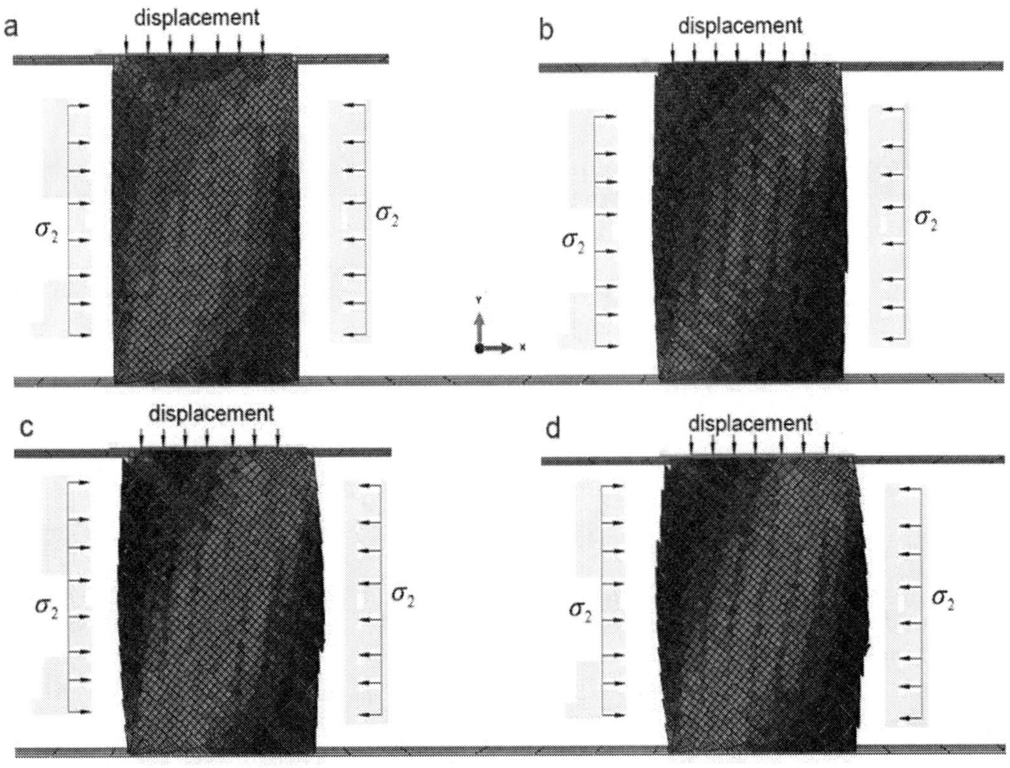

Fig. 4. The diagram of the stresses distribution in the numerical model during deformation: (a) first stage; (b) second stage; (c) third stage; (d) forth stage.

5 Conclusions

The most difficult tasks for geomechanics are related with the definition of mechanical characteristics of jointed rock mass. Application of numerical methods allows to solve such problems. The main goal when we using numerical modeling application is to define of accuracy of results. Comparison with results of experiment allows to determining the accuracy of the numerical model and its adequacy to real conditions. Though the possibilities of numerical modeling methods are very high and they allow to process and solve large number of complex systems of equations.

Quantitative accuracy of the results is still an important issue and requires further research. Thus, application of numerical methods allows to solve complex problems by linking scientific achievements in geomechanics with actively developing computer technologies. If in the laboratory conditions large size sample of jointed rock mass can not be studied, the use of numerical methods, in this case, the finite element method, allows to take into account the discontinuities and obtain different mechanical parameters which depend on the direction of the load in relation to the basic system of discontinuities of the jointed rock mass.

Also numerical experiments with models of jointed rock mass of different sizes allow to determine the reasonable limits of homogeneous rock mass size and scale effect. Anisotropy of strength is characteristic for jointed rock mass, this fact is explained by the presence of defects in the form of discontinuities in the rock mass. Quantifying such effect of anisotropy is possible to obtain with the help of the finite element method. It is required to perform because it is a very important stage in the design of underground structures.

Further researches on the development of numerical geomechanical model of jointed rock mass requires detailed study of the modeling of the mechanical behavior of intact rock and the mechanical behavior of the intact rock along contact of interaction between them. These studies require both high qualitative and convergence of results. Therefore its necessary to use experimental results in the numerical simulation.

According to the results of performed work, we have arrived the following conclusions which are presented in next sentences. The combination of analytical and empirical methods in the numerical approach (in the finite element method) of modeling allow to take into account discontinuities and different defects of the jointed rock mass. With the created geomechanical model of jointed rock mass it is possible to obtain of stress-strain graph and the anisotropy value of the of the strength and deformation properties of jointed rock mass.

References

1. E. Hoek, M.S. Diederichs, International Journal of Rock Mechanics and Mining Sciences, **43(2)**, 203-215 (2006)
2. N. Barton, Journal of Rock Mechanics and Geotechnical Engineering, **5(4)**, 249-261 (2013)
3. L. Jing, K.B. Min, A. Baghbanan, Rock mechanics new research, 109–165 (2009)
4. E. Hoek, E.T. Brown, International Journal of Rock Mechanics and Mining Sciences, **34(8)**, 1165-1186 (1997)
5. Özkan, B. Erdem, A. Ceylanoğlu, International Journal of Rock Mechanics and Mining Sciences, **73**, 28-41 (2015)
6. A. Palmstrom, R. Singh, Tunnelling Underground Space Technol, **16**, 115–31 (2001)
7. A. Trivedi, Journal of Rock Mechanics and Geotechnical Engineering, **7**, 540-565 (2015)
8. M. Sanei, L. Faramarzi, A. Fahimifar, S. Goli, A. Mehinrad, A. Rahmati, International Journal of Rock Mechanics and Mining Sciences, **75**, 119-131 (2015)
9. H.W. Jing, S.Q. Yang, M.L. Zhang, G.A. Xu, K.F. Chen, Tunnelling and Underground Space Technology, **43**, 184-197 (2014)
10. J.C. Jaeger, N.G.W. Cook, R.W. Zimmerman, *Fundamentals of rock mechanics: 4th edition* (London, 2007)
11. N. Halakatevakis, A. Sofianos, Int J Rock Mech Min Sci, **47(4)**, 568–582 (2010)
12. A. Khani, A. Baghbanan, S. Norouzi, H. Hashemolhosseini, International Journal of Rock Mechanics & Mining Sciences, **60**, 345-352 (2013)
13. Y. Ji. Ping, C. W. Zhong, Y. D. Sen, Y. J. Qiang, Computers and Geotechnics, **64**, 20-31 (2015)
14. K.B. Min, L. Jing, Int J Rock Mech Min Sci., **40(6)**, 795–816 (2003)
15. X. Wang, Y. Zhao, X. Lin, Journal of Rock Mechanics and Geotechnical Engineering, **3**, 398-406 (2011)
16. Y. Zhang, D. Stead, D. Elmo. Computers and Geotechnics, **65**, 56-72 (2015)
17. N. Barton, Journal of Rock Mechanics and Geotechnical Engineering, **5(4)**, 249-261 (2013)
18. A.N. Stavrogin, B.G. Tarasov, *Experimental physics and rock mechanics (results of laboratory studies)* (A.A. Balkema Publishers, 2001)
19. A.N. Stavrogin, A.G. Protosenya, *Mehanika deformirovaniya i razrusheniya gornih porod* (Nedra, Moscow, 1992)
20. X. Yang, P.H.S.W. Kulatilake, H. Jing, S. Yang, Tunneling and Underground Space Technology, **50**, 129-142 (2015)
21. T. Yang, P. Wang, T. Xu, Q. Yu, P. Zhang, W. Shi, G. Hu, Tunneling and Underground Space Technology, **48**, 129-139 (2015)

22. W. Wittke, *Rock Mechanics Based on and Anisotropic Jointed Rock Model (AJRM)* (Wilhelm Ernst & Sohn, Berlin, 2014)
23. R. Usmanov, M. Rakočević, V. Murgul, N. Vatin, Applied Mechanics and Materials, **633-634**, 927-931 (2014)
24. R. Usmanov, V. Murgul, N. Vatin, Applied Mechanics and Materials, **633-634**, 1082-1085 (2014)
25. R. Usmanov, I. Mrdak, N. Vatin, V. Murgul, Applied Mechanics and Materials, **633-634**, 932-935 (2014)
26. I. Mrdak, M. Rakočević, L. Žugić, R. Usmanov, V. Murgul, N. Vatin, Applied Mechanics and Materials, **633-634**, 1069-1076 (2014)

Fire Resistant Panels for the Tunnel Linings

Marina Gravit[1], *Sergey* Antonov[2], *Oleg* Nedryshkin[1,*], *Ekaterina* Nedviga[1] and *Valerii* Pershakov[3]

[1]Peter the Great St. Petersburg Polytechnic University, 195251, Politekhnicheskaya st., 29, Russia
[2]CEO LLC Prozasq, 107076, 42, Krasnobogatyrskaya st., Russia
[3]National Aviation University of Ukraine, 03058, Kyiv st., 1, Ukraine

Abstract. Presents the results of studies of innovative materials in the field of experimental and theoretical research fire resistance fireproof panels Pyro-Safe Aestuver T. Owing to the assembly simplicity, materials cheapness, high ecological standard, recycling, reuse potential, are benefit. Research work is running to improve the knowledge about fireproof panels Pyro-Safe Aestuver T for tunnel lining, its basic performance, its long term behavior and in particular also its fire proof for example when used for the lining of road tunnels.

1 Introduction

Tunneling and transport tunnel construction has gained increasing importance both in Europe and internationally during the past 15 to 20 years. The vector of developments will continue in the future. The reason is provided by the results of a recently completed study by the Viennese Academy of Sciences.

This academy carried out an extensive survey on behalf of the Austrian Automobile and Touring Club on how tourist and goods traffic has developed in Europe. It reveals that private motoring in Europe will rise by around 40 % by the year 2030 compared with 1997. Europe's highways will come ever closer to reaching their capacity in the near future thus introducing more stop than go for traffic than ever. Other surveys are no less pessimistic. The OECD (Organisation for Economic Co-operation and Development) for instance, reckons that by the year 2020, there will rise by around 50 % compared with 1997.

Nowadays around 10 % of the urban population live in so-called mega cities with more than 15 million inhabitants. 26 mega cities distributed all over the globe, with no less than 18 of them located in Asia. The use of underground space will be required to a high degree to make sure that such mega cities and expansive urban regions in general are capable of functioning so that the quality of life is assured.

This applies to transport tunnels as well as to supply and disposal facilities and the storage of goods. In view of such developments, it is imperative that new logistics concepts, new ways to secure mobility and also new supplementary measures to improve the infrastructure are found. One solution is certainly to be found in exploiting the chances afforded by tunnelling pertaining to the overall securing of mobility, speeding up transportation processes and essentially for protecting the environment and countryside as

* Corresponding author: nedryshkin@gmail.com

well [1].

The current and future major tunnel projects undoubtedly represent particular challenges in fire safety. During 2012 - 2014 in Russia (Sochi, North Ossetia) facing work in the tunnels have been carried out in a volume of 140 000 m^2 with fireproof panels Pyro-Safe Aestuver T [2].

Pyro-Safe Aestuver T – fireproof panels of glass fiber lightweight concrete to protect metal and concrete constructions, including for use in corrosive atmospheric conditions (eg, fireproof construction of refineries and chemical plants) and for underground traffic facilities (tunnels). Pyro-Safe Aestuver T – it is non-combustible, purely mineral, hydraulically-cemented, fireproof panels of glass fiber lightweight concrete.

Pyro-Safe Aestuver T has a dense and evenly distributed fiber structure through the entire cross-section of the panel. This method allows the production of thin panels with high ultimate tensile strength and compression in bending resistance and a high abrasion. Smooth, homogeneous and stable surface joint allow different possibilities of operation and handling.

Data on the characteristics of the boards listed in this table are not official information, and can be required from the Supplier [2].

In the paper [3] were investigated the strength and strain of high-strength concrete elements with confinement and steel fiber reinforcement including the conditions of the effect of elevated temperatures. In the paper [4] were investigated physical-mechanical properties of the modified fine-grained concrete subjected to thermal effects up to 200°C.

The authors [5] conducted numerical analysis of thin steel panels loaded in shear at non-uniform elevated temperatures.

2 Methods

Testing experimental samples solid concrete plates with fireproof panels Pyro-Safe Aestuver T 15 mm and a density of 730 kg/m ± 10%, were conducted to determine the fire resistance of the experimental samples submitted in accordance with [6]. The corresponding normative documents use for each type of material construction in the European Union [7 – 15].

3 Experiment

Identification of experimental samples. In the research were test two solid concrete panels with fireproof panels Pyro-Safe Aestuver T 15 mm and a density of 730 kg/m^2 ± 10%. Experimental samples sizes 4200x2000x140 mm plates were made of heavy concrete class B 22.5, an average density of 2400 kg/m^3.

The thickness of the protective layer of concrete to the center of gravity of the working reinforcement from the bottom (heated) side was 20 mm. Installation of fire-retardant panels Pyro-Safe Aestuver T 15 mm and a density of 730 kg/m ± 10% was carried out on the bottom (heated) surface experimental samples concrete solid plates using self-tapping screws for concrete in accordance with the technical requirements.

Test conditions. Date of first experience – 09/17/2015, 24/09/2015; second – 24/09/2015. Ambient temperature during the first experience – 18 C°; second – 20 C°. Relative air humidity of first experience – 48 %; second – 50 %.

Air flow rate of first and second experience is not more than 0.5 m/s. Figure 1 shows prepared for testing experimental sample No 1 appended to the uniformly distributed load.

Fig. 1. Experimental sample number 1 appended to the uniformly distributed load

The experimental samples were installed on the experimental set-up and subjected to subject thermal effects on a standard temperature conditions.

The test of fireproof panels was conducted under a constant uniformly distributed load equal to 7.9 kPa (800 kg/m^2), without taking into account its own weight of the experimental sample. The magnitude of the load determined in accordance with the technical requirements of the customer. Pressing experimental samples carried iron weights weighing 25 kg and 330 kgs, which are placed evenly on the unexposed surface of the concrete plates. Experienced concrete plates experimental sample according to customer's specifications were 2-sided bearing on the hinged point and hinged point - movable support. The distance from the panels to the ends of the rocker bearings was 100 mm.

Thus, a working experimental sample span reinforced concrete floor plates is 4000 mm. Deflections of experimental samples in the middle span, during loading and during the test was measured by MP-3 deflectometer. Deflection panels after loading was at 1-st experimental sample of 30.0 mm in the 2-nd – 27.8 mm.

The temperature in the furnace fire chamber furnace was measured by thermocouples, uniformly distributed along the length of the experimental sample in six locations, and test experimental samples temperature was measured by thermocouples type TCA established in the amount of 5 pieces at the non-heated surface of the experimental sample. Figure 2 shows arrangement of fireproof panels Pyro-Safe Aestuver T to the concrete construction.

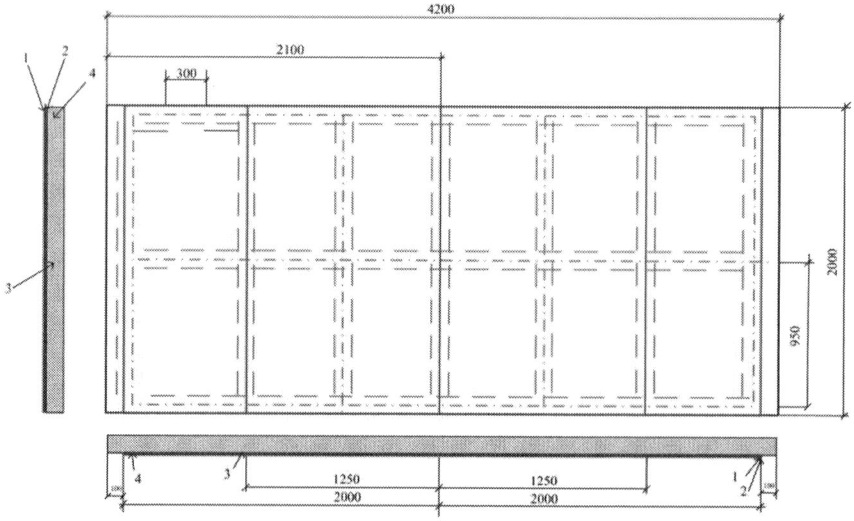

Fig. 2. Arrangement of fireproof panels Pyro-Safe Aestuver T to the concrete construction; 1 – fireproof panels Pyro-Safe Aestuver T; 2 – band of panels Pyro-Safe Aestuver T; 3 – concrete screw TSM-80 5x40 mm; 4 – concrete screw TSM-80 6x60 mm.

4 Results

The curves of temperatures and deflections experienced plates experimental samples with fireproof panels Pyro-Safe Aestuver T are shown in figure 3.

While testing experimental plate samples with fireproof panels Pyro-Safe Aestuver T, the following characteristics of samples behavior were observed:

a. 90 - 95 minutes – the appearance of surface cracks in the fireproof panels, and their gradual disclosure;
b. 110 - 115 minutes – there is a partial collapse of fireproof panels.

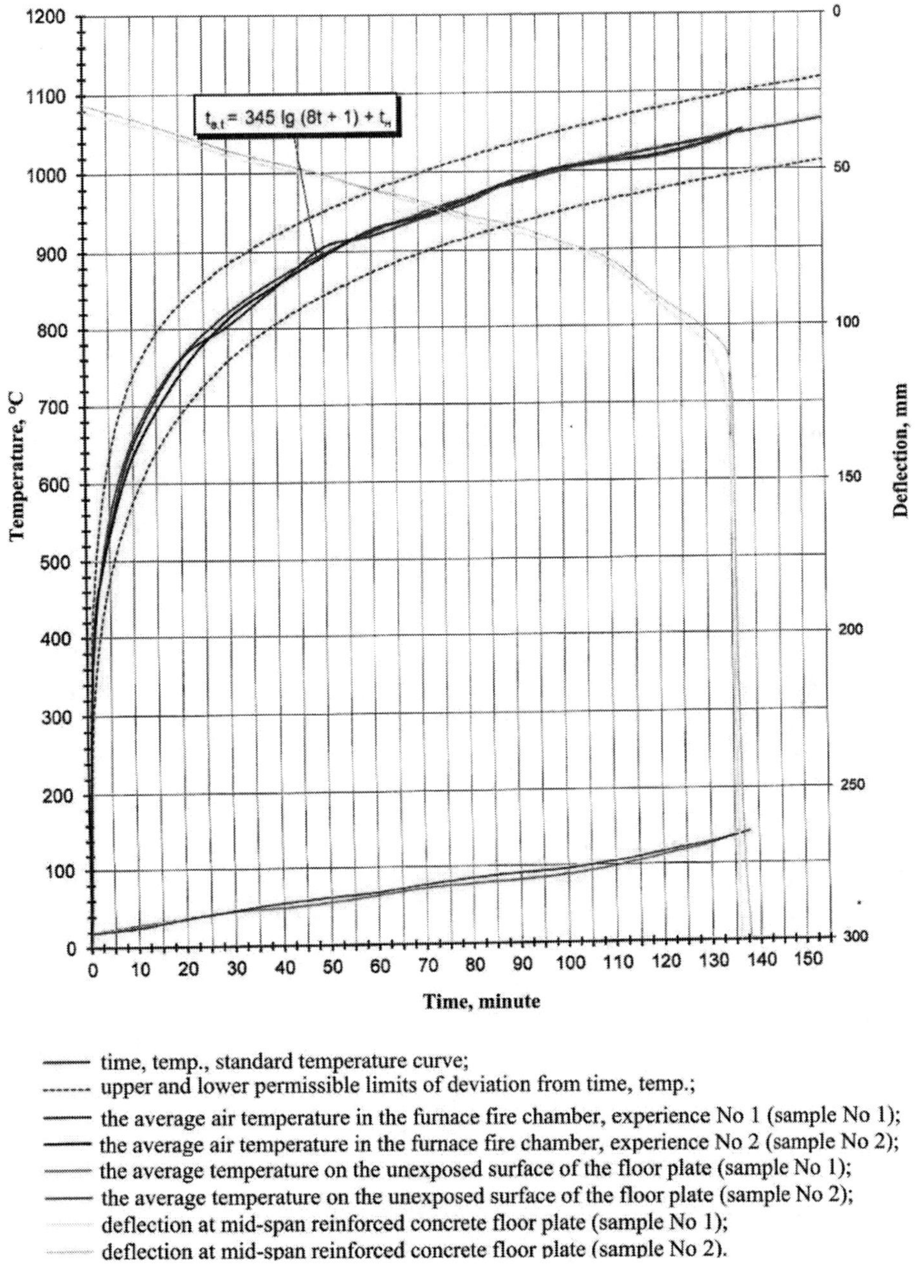

Fig. 3. Curves of temperature changes and growth troughs, experimental samples.

5 Discussion

At the 136th minute of the first test and at the 138th minute of the second test, experimental samples cross over in limiting condition characterized by a rapid increase in the bend and subsequent collapse of the experimental samples (refer with: Fig. 4).

As a result of the progressive achievement of the strain growth rate of fire resistance experimental samples loss of load-bearing structure ability (R) was detected at the 136th and at 138th minutes of the test (1st and 2nd experimental samples, respectively) more than 1.19 cm/min, the limit deflection of more than 200 mm, and the subsequent collapse of the experimental samples. At the time of achievement of experimental samples, a limiting condition for the loss of bearing capacity of the structure (R), the average temperature on the unexposed surface was 143 and 145 °C, for 1st and 2nd of the experimental samples, respectively. Temperature rise of the unexposed surface of the experimental samples overlap in one of the monitored points in comparison with the temperature before the test by more than 180 °C at the time of the collapse of the samples have been recorded. At the time of the collapse of experimental samples, through the formation of cracks or holes through which to penetrate the surface of the unexposed products of combustion or flames, not fixed.

In the paper [18] provides a review of the criteria that are normally employed in assessing the risk and consequences of fires in tunnels, and the selection of suitable time-temperature curves that describe the evolution of worst-case fires. The advantages and drawbacks of alternative fire protection methods, including sprayed-on fire protection materials, cementitious linings and the addition of polypropylene fibres, are briefly outlined. Effective passive fire protection can be provided by proprietary fireproof panels. The panels have good insulation properties and can withstand temperatures exceeding 1350 °C. They can be designed such that the temperatures at the face of the structural lining do not exceed 350 °C, which will prevent spalling and any changes in the mechanical properties of the concrete. The required panel thicknesses vary between 20 and 30mm depending on the temperature development over time.

The paper [19] author reports that the majority of existing tunnels worldwide rely wholly on active systems to ensure tunnel fire safety.

Passive fire protection is designed to be installed as a shield to protect the structure from fire at any time. They are not reliant on any initiation system as with active systems, and they always work. They maintain ventilation systems that are separated from the traffic by internal concrete structures, and also protect against catastrophic damage to third party property and life by preventing tunnels from collapsing.

Currently in Europe, both active and passive systems employed together are seen to be necessary for new tunnels in the future.

Pre-fabricated fire protection panels offer a clear advantage for box shaped tunnels where there are no curved tunnel walls or complex geometries. However, they are not well suited to curved profile tunnels and are generally 1.5 to 2 times more expensive than sprayed systems, which can prove cost prohibitive. The samples used in our work in the tunnels with a curved profile and have a concurrency price that makes fireproof panels Pyro-Safe Aestuver T the right solution in the field of fire safety [18 – 27].

Fig. 4. The experimental sample No 2 after the test.

6 Conclusion

The tragic fire accidents during the last years led to strong discussions in professional circles all over Europe and all over the world as well [20].

In the ongoing test concrete plates with fireproof panels Pyro-Safe Aestuver T 15 mm and a density of 730 kg/m^3 ± 10%, tested under constant uniformly distributed load equal 7.9 kPa (800 kg/m^2), without taking into account its own weight plate at 137 minutes, which corresponds to the classification of REI 120 [6].

Achieved the results of studies of innovative materials in the field of experimental and theoretical research fire resistance fireproof panels Pyro-Safe Aestuver T confirm the high quality of the present samples. The provision of passive fire protection to tunnels is not usually required, if the fire loads are low and the safety provisions are specified and maintained to a high standard. However, the existence of imported fire risks such as vehicle shuttles and fuel tank may imply significantly higher fire loads. In addition, there may be cases where the collapse of a tunnel may imply unacceptably high social costs. In such cases, a cost/benefit analysis, informed by a quantitative risk assessment, may demonstrate a strong case for additional fire protection measures, including passive fire protection and/or fire suppression systems as fireproof panels.

References

1 Haack, Innovations in Underground Construction and Tunnelling; special issue Tunnel, (2004)
2 Information on http://www.stalprotect.ru/pyro-safe_aestuver_t_tonnelnaya1
3 V. Korsun, N. Vatin, A. Franchi, A. Korsun, P. Crespi, S. Mashtaler, Procedia Engineering, **117(1)**, 970–979 (2015)
4 V. Korsun, N. Vatin, A. Korsun, D. Nemova, AMM, **633-634**, 1013–1017 (2014)
5 M. Salminen, M. Heinisuo, Journal of Constructional Steel Research, **97**, 105–113 (2014)
6 ISO 834-1:1999 Fire-resistance tests - Elements of building construction – Part 1: General requirements.
7 T. Eremina, M. Gravit, Y. Dmitrieva, Architecture and Construction of Russia, **8**, 24–29 (2012)
8 BS EN 13501-2:2007+A1:2009. Fire classification of construction products and building elements. Part 2: Classification using data from fire resistance tests, excluding ventilation services.
9 Information on http://www.eota.eu/en-GB/content/etags-used-as-ead/26/

10 ENV 13381-3:2002. Test methods for determining the contribution to the fire resistance of structural members – Part 3: Applied protection to concrete members.
11 BS EN 13381-4:2013. Test methods for determining the contribution to the fire resistance of structural members – Part 4: Applied protection to steel members.
12 ENV 13381-5:2014. Test methods for determining the contribution to the fire resistance of structural members – Part 5: Applied protection to concrete/profiled sheet steel composite members.
13 EN 13381-6:2008. Test methods for determining the contribution to the fire resistance of structural members. Part 6. Applied protection to concrete filled hollow steel columns.
14 ENV 13381-7:2002 Test methods for determining the contribution to the fire resistance of structural members – Part 7: Applied protection to timber members.
15 EN 13381-8:2013 Test methods for determining the contribution to the fire resistance of structural members – Part 8: Applied protection.
16 F. Tarada, M. King, Structural fire protection of railway tunnels.
17 F. Clement, Fire protection options for concrete tunnel linings, Fourth International Symposium on Tunnel Safety and Security, Frankfurt am Main, Germany, March 17-19, 2010.
18 A. Haack, Tunnelling and Underground Space Technology, **17**, 117–127 (2002)
19 Ahmad, S. Hassan, A. Ripin, M. Ali, S. Haron, Fire Safety Journal, **58**, 160–169 (2013)
20 S. Chang, S. Choi, J. Lee, Original Research Article Tunneling and Underground Space Technology, **54**, 1–12 (2016)
21 N. Bezgin, Construction and Building Materials, **95**, 279–286 (2015)
22 H. Ingason, Y. Li, A. Lonnermark, Tunnelling and Underground Space Technology, **43**, 327–335 (2014)
23 H.Lai, S.Wang, Y. Xie, Tunnelling and Underground Space Technology, **35**, 122–134 (2013)
24 S. Choi, J. Lee, S. Chang, Procedia Engineering, **62**, 907–915 (2013)
25 Z. Yuan, B. Lei, A. Kashef, Construction and Building Materials, **24**, 1984–1994 (2010)
26 J. Ćetković, M. Knežević, M., Žarković, V. Murgul, N. Vatin, AMM, **638-640,** 2465-2470 (2014)
27 M. Lazarevska, M. Cvetkovska, M, Knezevic, V. Murgul, N. Vatin, AMM, **627**, 276-282 (2014)

MATEC Web of Conferences **73**, 04008 (2016) DOI: 10.1051/matecconf/20167304008
TPACEE-2016

Evaluation of Interaction Between Flat Car and Container at Dynamic Coupling of Flat Cars

Andrew Nikitchenko[1], *Viktor* Artiukh[2,*], *Denis* Shevchenko[1] and *Raghu* Prakash[3]

[1]All-Union Research and Development Centre for Transportation Technology, 23 line Vasilievsky Island, 2, Saint-Petersburg, 199106, Russia
[2]Peter the Great St. Petersburg Polytechnic University, Polytechnicheskaya, 29, Saint-Petersburg, 195251, Russia
[3]Indian Institute of Technology Madras, C1-6-7, Second Loop Road, Madras Campus, Chennai, 600036, India

Abstract. Innovative materials and structures are analyzed in this paper. To calculate the strength of the collision of flat wagons for the transport of large containers, there is no clear methodology for determining effort interaction between the container and the platform. At high longitudinal acceleration of the container, it is set in motion, and the consideration of this problem in a static setting impossible. The relevance of this work is to develop a methodology that is based on the equations of motion and considers dynamic interaction between container and platform.

1 Introduction

There are different approaches to strength analysis of the flat cars intended for transportation of large-capacity containers, in particular, in what concerns evaluation of interaction between container and platform. The matter is that at large longitudinal accelerations the container may start moving and, respectively, the static scheme of analysis can not by applied [1]. Some specifications, e.g. [2], recommend to consider the dynamic effects of interaction by applying a vertical component of inertia force approximated by 50% of the longitudinal inertia of container which in principle, is incorrect [3-5].

2 Materials and Methods

The present analysis is focused on development of a technique which is based on formulation of relevant motion equations and considers dynamic interaction between container and platform [6-8].

The vertical component of the longitudinal inertia force acting on the container according to the scheme of Fig. 1 is defined as

* Corresponding author: artiukh@mail.ru

$$R = m_k a_x \frac{h}{2L}. \tag{1}$$

The total dynamic interaction of container and flat-car, respectively, is defined as:
- the dynamic force applied to fitting in direction of the flat-car loading:

$$RY_1 = \frac{m_K g}{2} + \frac{m_K a_x}{2} \cdot \frac{h}{L} = \frac{m_K}{2}\left(g + a_x \frac{h}{L}\right); \tag{2}$$

- and from the opposite side,

$$RY_2 = \frac{m_K}{2}\left(g - a_x \frac{h}{L}\right), \tag{3}$$

The scheme of inertia loads on a container (Fig.1)

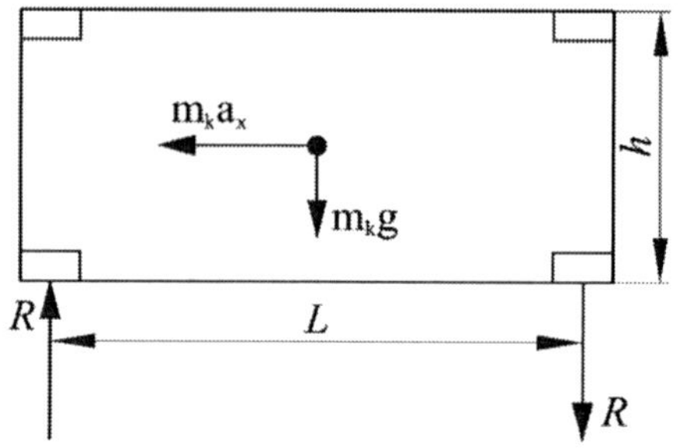

Fig. 1. Scheme of inertia loads on a container.

where a_x is the longitudinal acceleration of the container; when an only one container is installed on the flat car, the longitudinal acceleration is find by

$$a_x = \frac{P}{m_{\Pi} + m_K}, \tag{4}$$

where P is the longitudinal load transferred through the automatic coupling, m_{Π} – the mass of the flat car.

Substitution of (Eq. 4) into (Eq. 2) allows defining the vertical component of interaction between fitting of the container and the flat car:

$$RY_1 = \frac{m_K}{2}\left(g + \frac{P}{m_{\Pi} + m_K} \cdot \frac{h}{L}\right). \tag{5}$$

As follows from (Eq. 3) at longitudinal acceleration $a_x > gh/L$ the inertia load between fitting of the container and the flat car from outside, opposite to the direction of dynamic force caused by coupling of cars, becomes negative. If the container were fixed on the flat car the vertical direction, then formulae (Eq. 2) and (Eq. 3) would be true. However, since the fittings do not prevent vertical displacement of the container it is readily feasible that under dynamic coupling of cars the container would rotate around the fitting as shown in the scheme Fig. 2. At the same time, the interaction of the container and the flat car has to be obtained, specifically since it is necessary for the stress analysis [9-12] of the car structure.

To solve the problem of behavior of the system of rigid bodies loaded due to the dynamic automatic coupling of cars, Fig. 2, is considered.

The assumed system has two degrees of freedom: movement along the axis x and rotation, φ. Equations of motion of the system:

$$\frac{d}{dt}\left(\frac{\partial T}{\partial \dot{q}_i}\right)-\frac{\partial T}{\partial q_i}=Q_{qi}, \tag{6}$$

where T is the kinetic energy of system, q_i are generalized coordinates, Q_{qi} is the generalized external forces corresponding to q_i coordinate.

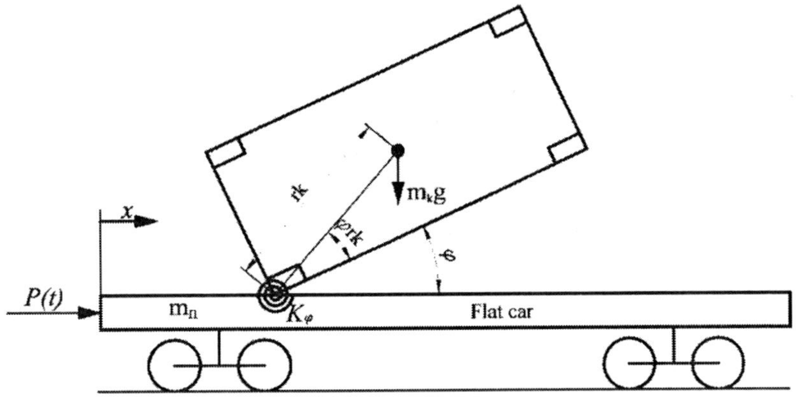

Fig. 2. The scheme of the mechanical system consisting of flat car and container

Kinetic energy T of the system is equal to the sum of kinetic energies of the two bodies, the flat car and the container:

$$T=T_{\Pi}+T_{K}, \tag{7}$$

where T_{Π} is the kinetic energy of the flat car, T_K is the kinetic energy of the container, which are, respectively:

$$T_{\Pi}=\frac{m_{\Pi}\dot{x}^2}{2}; \tag{8}$$

$$T_{K}=\frac{m_{K}\left(\dot{x}_K^2+\dot{y}_K^2\right)}{2}+\frac{I_K\dot{\varphi}^2}{2}. \tag{9}$$

The displacements of the center of gravity of the container are:

$$x_k=x+r_k\cos\left(\varphi_{rk}+\varphi\right);$$

$$y_k = r_k \sin(\varphi_{rk} + \varphi).$$

The respective velocities are:

$$\dot{x}_k = \dot{x} - \dot{\varphi} r_k \sin(\varphi_{rk} + \varphi);$$
$$\dot{y}_k = \dot{\varphi} r_k \cos(\varphi_{rk} + \varphi).$$

Substitution of these expressions into (Eq. 5) results in

$$T_K = \frac{m_K}{2}\left(\dot{x}^2 - 2\dot{x}\dot{\varphi} r_k \sin(\varphi_{rk} + \varphi) + \dot{\varphi}^2 r_k^2\right) + \frac{I_k}{2}\dot{\varphi}^2.$$

The total kinetic energy of the system is

$$T = \frac{\dot{x}}{2}(m_\Pi + m_K) + \frac{\dot{\varphi}^2}{2}(m_K r_k^2 + I_k) - m_K r_k \dot{x}\dot{\varphi}\sin(\varphi_{rk} + \varphi).$$

The differentiating of the kinetic energy by generalized coordinates x, φ and generalized velocities $\dot{x}$, $\dot{\varphi}$ results in:

$$\frac{\partial T}{\partial x} = 0;$$
$$\frac{\partial T}{\partial \varphi} = -m_K r_k \dot{x}\dot{\varphi}\cos(\varphi_{rk} + \varphi);$$
$$\frac{\partial T}{\partial \dot{x}} = \dot{x}(m_\Pi + m_K) - m_K r_k \dot{\varphi}\sin(\varphi_{rk} + \varphi);$$
$$\frac{\partial T}{\partial \dot{\varphi}} = \dot{\varphi}(m_K r_k^2 + I_k) - m_K r_k \dot{x}\sin(\varphi_{rk} + \varphi).$$

The time derivatives:

$$\frac{d}{dt}\left(\frac{\partial T}{\partial \dot{x}}\right) = \ddot{x}(m_\Pi + m_K) - \ddot{\varphi} m_K r_k \sin(\varphi_{rk} + \varphi) - \dot{\varphi}^2 m_K r_k \cos(\varphi_{rk} + \varphi);$$
$$\frac{d}{dt}\left(\frac{\partial T}{\partial \dot{\varphi}}\right) = \ddot{\varphi}(m_K r_k^2 + I_k) - \ddot{x} m_K r_k \sin(\varphi_{rk} + \varphi) - \dot{x}\dot{\varphi} m_K r_k \cos(\varphi_{rk} + \varphi).$$

The generalized forces along the coordinates x and φ:

$$Q_x = P(t);$$
$$Q_\varphi = m_K g r_k \cos(\varphi_{rk} + \varphi).$$

Substituting of these relationships into (Eq. 2) results in equations:

$$\begin{aligned} &\ddot{x}(m_П + m_K) - \ddot{\varphi} m_K r_k \sin(\varphi_{rk} + \varphi) - \dot{\varphi}^2 m_K r_k \cos(\varphi_{rk} + \varphi) = P(t); \\ &\ddot{x} m_K r_k \sin(\varphi_{rk} + \varphi) - \ddot{\varphi}(m_K r_k^2 + I_k) = m_K r_k g \cos(\varphi_{rk} + \varphi). \end{aligned} \tag{10}$$

It may be noted that equations (Eq. 6) characterize the dynamic equilibrium, the first is equilibrium equation of forces in the longitudinal direction, the second - equilibrium of the moments caused by rotation of the container.

Since the angle of the container rotation cannot be negative, the following restriction may be suggested that fittings are supported at the car structure by the rigid springs. With this suggestion the right-hand part of second equation in (Eq. 10) may be written as:

$$M = \begin{cases} m_K r_K g \cos(\varphi_{rk} - \varphi) \; npu \; \varphi > 0; \\ 0 \; npu \; \varphi = 0; \\ m_K r_K g \cos(\varphi_{rk} - \varphi) - K_\varphi \varphi \; npu \; \varphi < 0. \end{cases} \tag{11}$$

Any of the known numerical methods can be applied to solve the system (Eq. 10). Solving (Eq. 10) relatively $\ddot{x}$ and $\ddot{\varphi}$, by taking into account (Eq. 11), results in following equations:

$$\begin{aligned} \ddot{x} &= \frac{\left(P(t) + \dot{\varphi}^2 m_K r_k \cos(\varphi_{rk} + \varphi)\right)\left(m_K r_k^2 + I_k\right) - M m_K r_k \sin(\varphi_{rk} + \varphi)}{(m_П + m_K)(m_K r_k^2 + I_k) - m_K^2 r_k^2 \sin^2(\varphi_{rk} + \varphi)}; \\ \ddot{\varphi} &= \frac{M(m_П + m_K) - \left(P(t) + \dot{\varphi}^2 m_K r_k \cos(\varphi_{rk} + \varphi)\right) m_K r_k \sin(\varphi_{rk} + \varphi)}{(m_П + m_K)(m_K r_k^2 + I_k) - m_K^2 r_k^2 \sin^2(\varphi_{rk} + \varphi)}. \end{aligned} \tag{12}$$

The three versions of the longitudinal loading on the flat car may be considered:

1) permanent load $P(t) = 3.5$ MN;
2) linearly increasing loading;
3) the loading caused by collision of the cars equipped with the absorbing devices of the T1-class [13-15].

Firstly, the following example is analyzed:

- the mass of flat car is taken as $m_П = 22.0$ t;
- the mass of 20-foot container is $m_K = 24.0$ t;
- the distance from the fitting to the container center of gravity is $r_K = 3.2$ m;
- the container turning angle is $\varphi_{rK} = 24°$.

Fig. 3 shows displacement and velocity of the flat car, and displacement of the center of gravity of the system flat car - container installed on it. Fig. 4 illustrates the time-dependent turning angle and angular velocity of the container.

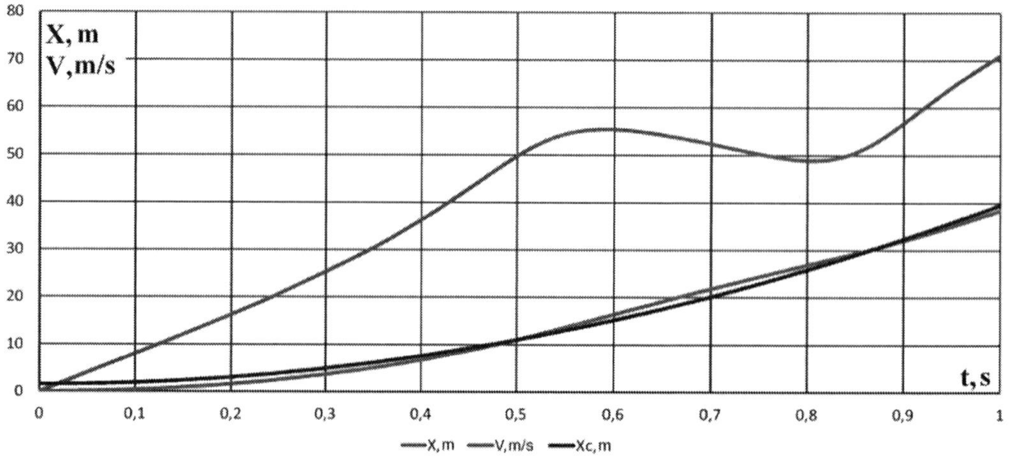

Fig. 3. Time-dependent displacement of the center of mass and of velocity of the flat car, and displacement of the center of mass of the system flat car-container.

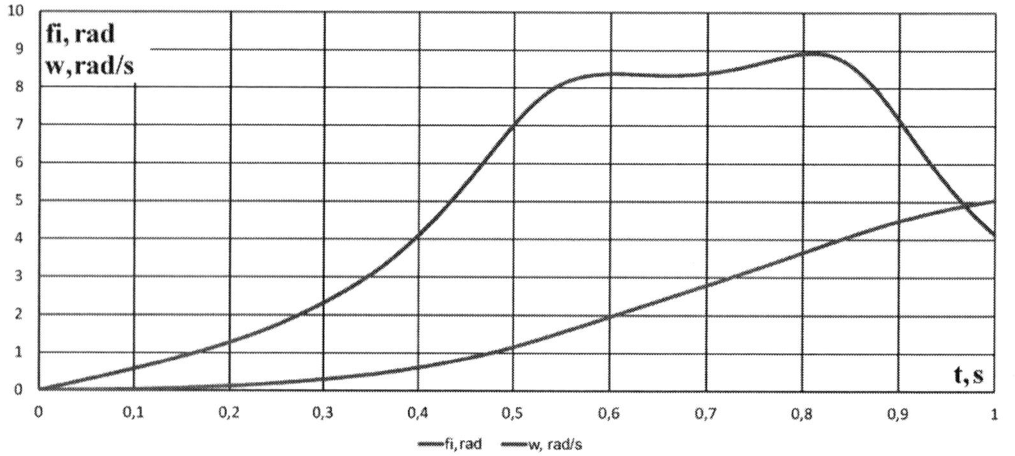

Fig. 4. Dependence in time of the rotation angle and angular velocity of the container with respect to the fitting from the side of impact.

As seen from Fig. 4, incase of application of the permanent load equal to 3.5 MN the container will inevitably overturn already through 0.5 sec.

From practice it is known that at impacts of cars it takes no more than 0.2 second that the longitudinal loading would reach the peak value [16, 17]. Therefore the linearly increasing loading case is considered.

In the second loading case the force applied to the car frame can be defined by the following relationships:

$$P(t)=\begin{cases}\dfrac{P_{max}}{t_1}t \; at \; 0\le t\le t_1;\\ P_{max}-\dfrac{P_{max}}{t_2-t_1}(t-t_1) \; at \; t_1<t\le t_2;\\ 0 \; at \; t>t_2,\end{cases} \quad (13)$$

where t_1 is the time through which the loading reaches a peak value; it is assumed that $t_1 = 0.2$ s; t_2 is the time for which loading from a maximum value drops down to zero; it is take as $t_2 = 0.05$ s.

Fig. 5 shows the time dependence of the displacement and velocity of the flat car, and also displacement of the center of gravity of the flat car - container system.

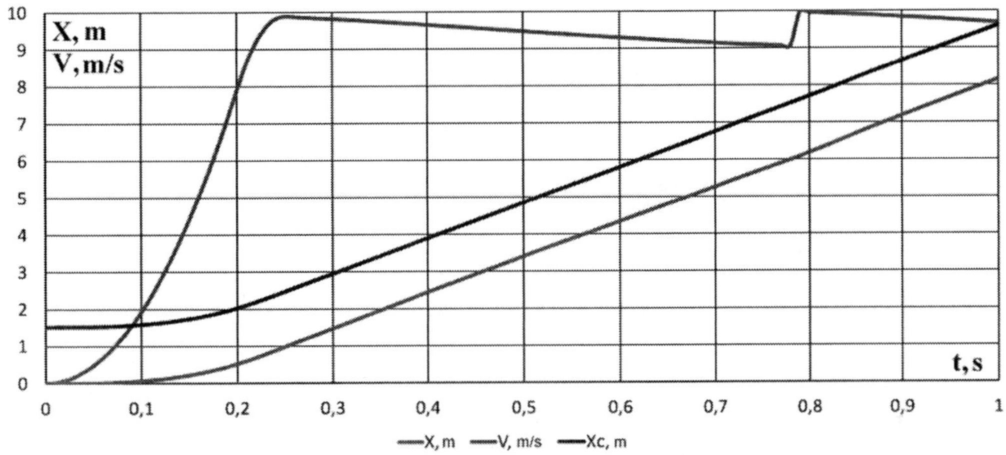

Fig. 5. The time dependence of the displacement of the center of gravity and velocity of the flat car, and of the displacement of the center of mass of the system flat car-container.

In Fig. 6 the time dependence of the rotation angle and angular velocity of the container are shown. Fig. 6 shows that the maximum angle of rotation of the container reaches $\varphi_{max} = 0.10190$ rad = 5.8386°. Maximum height which would reach the container fitting from the side opposite to the impact would be around $h_{max} = \sin(\varphi_{max})\cdot L = \sin(0.013881)\cdot 6 = 0.610$ m. So far, at the normative impact loads transferred through the automatic coupling the container by no means can overturn.

A more realistic result may be found if a case of collision of two cars equipped with the elastic-frictional absorbing class T1 devices would be considered [18, 19]. The mathematical model of the device is described in [20]. The scheme of the process of collision of cars is given in Fig. 7.

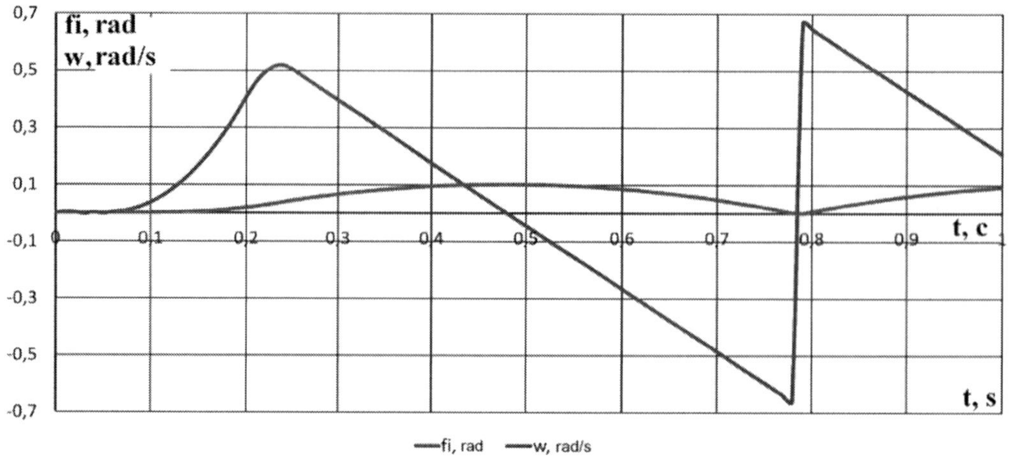

Fig. 6. The time dependence of the rotation angle and angular velocity of the container relative to fitting from the side of impact.

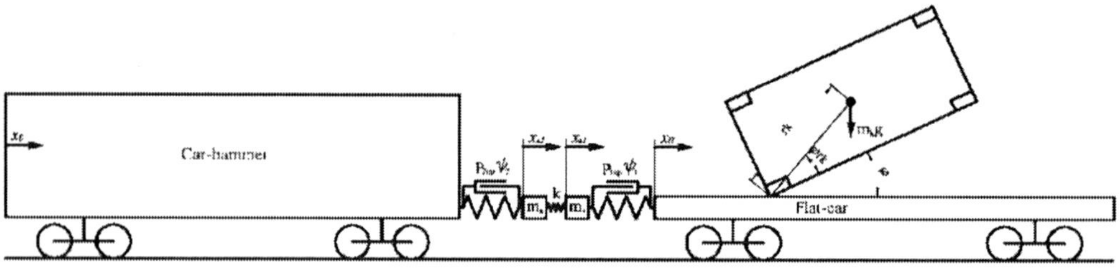

Fig. 7. Scheme of dynamic coupling of cars.

Equations of displacements of this system are as follows:

$$\begin{aligned}
&\ddot{x}_{П}\left(m_{П}+m_{K}\right)-\ddot{\varphi} m_{K} r_{k} \sin\left(\varphi_{rk}+\varphi\right)-\dot{\varphi}^{2} m_{K} r_{k} \cos\left(\varphi_{rk}+\varphi\right)-\psi_{1} P_{np1} sign\left(x_{a1}-x_{П}\right)=0; \\
&\ddot{x}_{П} m_{K} r_{k} \sin\left(\varphi_{rk}+\varphi\right)-\ddot{\varphi}\left(m_{K} r_{k}^{2}+I_{k}\right)-M=0; \\
&m_{a} \ddot{x}_{a1}+\psi_{1} P_{np1} sign\left(x_{a1}-x_{П}\right)-k\left(x_{a2}-x_{a1}\right)=0; \\
&m_{a} \ddot{x}_{a2}+k\left(x_{a2}-x_{a1}\right)-\psi_{2} P_{np2} sign\left(x_{Б}-x_{a2}\right)=0; \\
&m_{Б} \ddot{x}_{Б}+\psi_{2} P_{np2} sign\left(x_{Б}-x_{a2}\right)=0,
\end{aligned} \tag{14}$$

where ψ_1 is the transfer coefficient of the wedge system of the damping device of the flat car; P_{np1} is the compression force of the supporting block of the damping device of the flat car; ψ_2 is the transfer coefficient of the wedge system of the damping device of the car-hammer; P_{np2} is the compression force of the supporting block of the damping device of the car-hammer; m_a is the mass of the movable parts of automatic coupler; k is the longitudinal stiffness of the connected bodies of automatic couplers; $m_Б$ is the mass of the car-hammer.

The directions of forces applied by the container to the flat car fitting from the side of impact are shown in Fig. 8.

$$\vec{R}=\vec{R}_{x}+\vec{R}_{y}+\vec{R}_{r}+\vec{R}_{\tau} \tag{15}$$

The values of these forces can be obtained by formulae:

$$R_x = m_K\ddot{x};\quad R_y = m_K g;\quad R_r = m_K\dot{\phi}^2 r_k;\quad R_\tau = m_K\ddot{\phi} r_k. \tag{16}$$

The projections of the total force R on horizontal and vertical axes:

$$\begin{cases} \sum R_x = m_K\ddot{x} - m_K\dot{\varphi}^2 r_k \cos(\varphi_{rk} + \varphi) - m_K\ddot{\varphi} r_k \sin(\varphi_{rk} + \varphi); \\ \sum R_y = m_K g - m_K\dot{\varphi}^2 r_k \sin(\varphi_{rk} + \varphi) + m_K\ddot{\varphi} r_k \cos(\varphi_{rk} + \varphi). \end{cases} \tag{17}$$

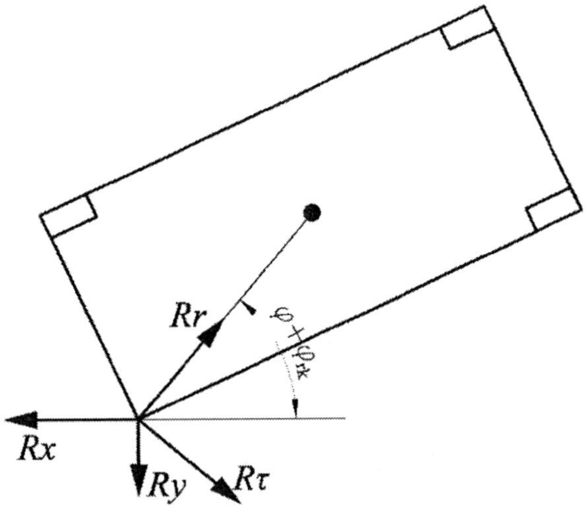

Fig. 8. The forces of interaction between the flat car and the container.

The variation in time of the vertical component of the interaction force between the fitting of flat car and container from the side of impact is presented in Fig. 9.

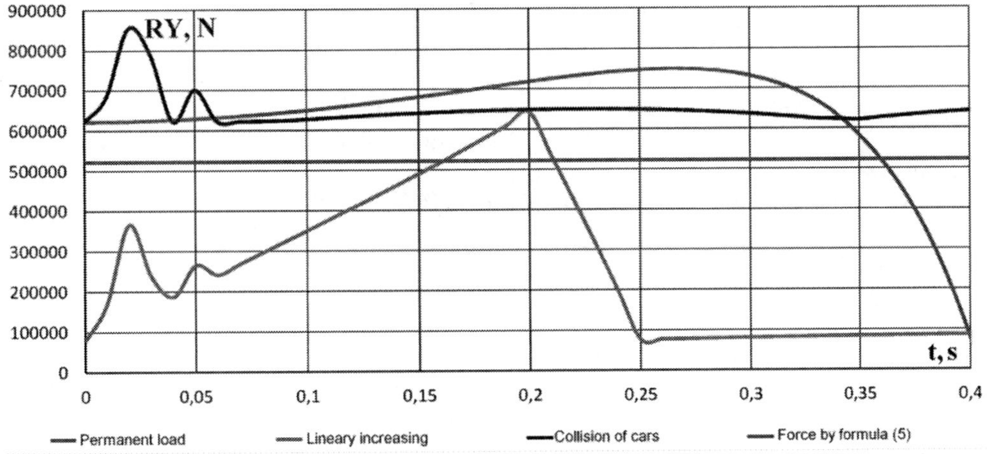

Fig. 9. Change in time of vertical force of interaction between the fitting of flat car and container from the side of impact.

3 Conclusions

It is seen from Fig. 9 that the load determined by (Eq. 5) is not overestimated and it can be applied for static strength analysis.

For the more detailed solution of the flat car-container system behavior it is necessary to develop model in which flexibility of spring suspension of a flat car would be considered. Also, the full-scale tests should be carried out with registration of forces on fittings and displacements of the container with respect to the flat car; the results of tests and of the analysis further should be compared.

The reported study was funded by RFBR according to the research project №16-08-00845a «Verification and development of models of inelastic deformation at the passive loading». The authors declare that there is no conflict of interest regarding the publication of this paper.

References

1. B. Melnikov, A. Semenov, Applied Mechanics and Materials, **617**, 187–192 (2014)
2. *Normy dlya rascheta i projektirovaniya vagonov zheleznyh dorog MPS kolei 1520 mm (nesamohodnyh)* (GosNIIV, VNIIZHT, 1996)
3. Nabeel S. Gharaibeh, Mohammed I. Matarneh, V.G. Artyukh, Research Journal of Applied Sciences, Engineering and Technology, **8(12)**, 1461–1464 (2014)
4. V.G. Artiukh, *Nagruzki i peregruzki v metallurgicheskih mashinah* (PSTU, Mariupol, 2008)
5. V.G. Artiukh, *Basis of Protection of Metallurgical Machines from Breakdown* (Publishing Group "University", Mariupol, 2015)
6. V. Mazur, V. Artyukh, G. Artyukh, M. Takadzhi, Engineering Designer, **37(1)**, 26–29 (2012)
7. Mohammed I. Matarneh, Nabeel S. Gharaibeh, V. G. Artyukh, International Journal of Engineering Science and Innovative Technology, **4(2)**, 1–7 (2015)
8. V.G. Artiukh, *Tochnost predohranitelei dlia metallurgicheskih mashin* (PSTU, Mariupol, 2000)
9. A.A. Alyamovskiy, *SolidWorks/COSMOSWorks. Engineering FEM analysis* (DMK Press, 2004)
10. V. Sevastyanov, A. Volvach, Computer design and technical flow of documents, **2**, 28–31 (2007)
11. A.B. Kaplun, E.M. Morozov, M.A. Olferieva, *ANSYS in Engineer's Hands. Practical issue* (Editorial URSS, 2003)
12. *SIMULIA Abacus (CAE User's manual)*, V.**6.7** (2008)
13. G.V. Artiukh, V.G. Artiukh, V.A. Korchagin, T.V. Korchagina, Ukraine Patent **82,142** (2007)
14. G.V. Artiukh, V.G. Artiukh, V.A. Korchagin, T.V. Korchagina, Ukraine Patent **85,962** (2009)
15. G.V. Artiukh, V.G. Artiukh, V.A. Korchagin, T.V. Korchagina, Russia Patent **2,373,092** (2009)
16. G.V. Artiukh, V.G. Artiukh, E.I. Ivanov, Protection of Metallurgical Machines from Breakdowns, **15**, 150-155 (2013)
17. Firas M.F. Al-Quran, M.E. Matarneh, V.G. Artukh, Research Journal of Applied Sciences, Engineering and Technology, **4(11)**, 1585–1589 (2012)
18. B.E. Melnikov, L.F. Khazieva, Advanced Materials Research, **915-915**, 68-71 (2014)
19. V.G. Artiukh, S.Yu. Karlushin, E.N. Sorochan, Procedia Engineering, **117**, 938–944 (2015)
20. S.V. Myamlin, N.E. Naumenko, A.A. Nikitchenko, Vestnik DIIT, **24**, 25-33 (2008)

Mathematical Model of Plates and Strips Rolling for Calculation of Energy Power Parameters and Dynamic Loads

Elena Sorochan[1], *Viktor* Artiukh[2,*], *Boris* Melnikov[2], and *Talzhan* Raimberdiyev[3]

[1]Pryazovskyi State Technical University, Universytets'ka, 7, Mariupol, 87500, Ukraine
[2]Peter the Great St. Petersburg Polytechnic University, Polytechnicheskaya, 29, Saint-Petersburg, 195251, Russia
[3]Hoja Akhmet Yasawi International Kazakh-Turkish University, Sattarkhanova, 29, Turkestan, 161200, Kazakhstan

Abstract. Presents the results of studies of innovative materials in the field of metallurgy. The article describes mathematical models for calculating energy-parameters of hot rolling plates and strips using a more correct evaluation of external friction, which is necessary for precisely determining the values of the dynamic loads in the drive's line of rolling mills stands. The results can be used for designing rolling mills drive lines' mechanical facilities.

1 Introduction

Iron and steel machines have to work under heavy dynamic modes, so requirements for design strength and reliability are high. The quality of these machines depends largely on how precisely their main parameters were calculated. At the same time, a machine must be considered as it actually is, i.e. with due regard to the material properties (elasticity), the actual character of application of external loads, taking into consideration the motor's characteristics and gaps in the drive lines, as well as the system's dissipative properties. Rolling machines typically have the heaviest dynamic modes; the character of their loads application is actually close to that of impact character. Thus, if dynamic excitation is neglected, it could lead to great errors in calculations for iron and steel machines. Dynamic models of various complexity levels are used for investigating the phenomena occurring in the main line of a rolling mill, when rolled metal is gripped by the rollers. The existing dynamic models have a drawback: a simplified mathematical model of the rolling process [1]. In some cases, the same simplified models of linear dependence of rolling momentum and force in time are used [2]. These simplifications lead to errors in calculations. Therefore, it is vital that we adapt the existing models of the process to dynamic calculations of the main line of a rolling mill.

* Corresponding author: artiukh@mail.ru

2 Materials and Methods

As a rule, time is a variable in dynamic models, so the problem is restricted to determining geometrical, kinematic and energy-power parameters of the rolling process at any time instant of metal filling the deformation area [3]. At strict target setting, the solution is reduced to the solution of a third degree equation. However, for determining kinematic and geometrical characteristics of the area of deformation, we can use the following dependence [4]:

$$V_x = V_0 + \left(\frac{\tau_x}{\tau_0}\right)^2 (V_1 - V_0), \tag{1}$$

where V_0, V_x – is the velocity of rolling at the entrance to the area at a current instant τ_x, respectively; τ_0 – time of metal gripping by the rolls, equal to

$$\tau_0 = 2\sqrt{R(h_0 - h_1)}/(V_0 + V_1); \tag{2}$$

R – is the rolls radius; V_1 – the velocity of rolling at the exit from deformation area; h_0, h_1 – are the initial and final thickness of the rolled stock.

The values of the rolled stock thicknesses at the current instant can be found from the law of constancy of the seconds volumes [5]:

$$h_x = h_0 V_0 / V_x . \tag{3}$$

Further calculations are made in accordance with one of the following methods of calculation, depending upon the value of the relation L / h_{cp}.

Now, let us consider the engineering methods of determining energy parameters for hot rolling relatively thin sheets and strips, i.e. at $L / h_{cp} > 2,5$ [6].

Having made in accordance with the methods applied in [7, 8] division of the integral area of deformation (Fig. 1, a) into the area of tardiness and the area of advance, with the lengths $L_{от}$ and $L_{оп}$, respectively, we will consider the condition of static equilibrium, detached within the tardiness area of elementary metal volume (Fig. 1, b) at projecting of all exerted forces at x-horizontal axis x:

$$\sigma_x h_x - (\sigma_x + d\sigma_x)(h_x + dh_x) + 2p_{xот} dx \sin\alpha_x / \cos\alpha_x - 2\tau_x dx \cos\alpha_x / \cos\alpha_x = 0, \tag{4}$$

where p_x, σ_x, τ_x – the current values of normal contact axial and tangent contact stresses, compression stresses are taken for positive values σ_x and τ_x – stresses with vector orientation with transfer of the rolled metal; h_x, α_x are the current values of the thickness of the rolled strip and contact angles, the numerical values of which at approximation of contact arcs with chords are equal $h_x = h_1 + \Delta hx / L$ and $\alpha_x = \alpha = arctg(0,5\Delta h / L)$.

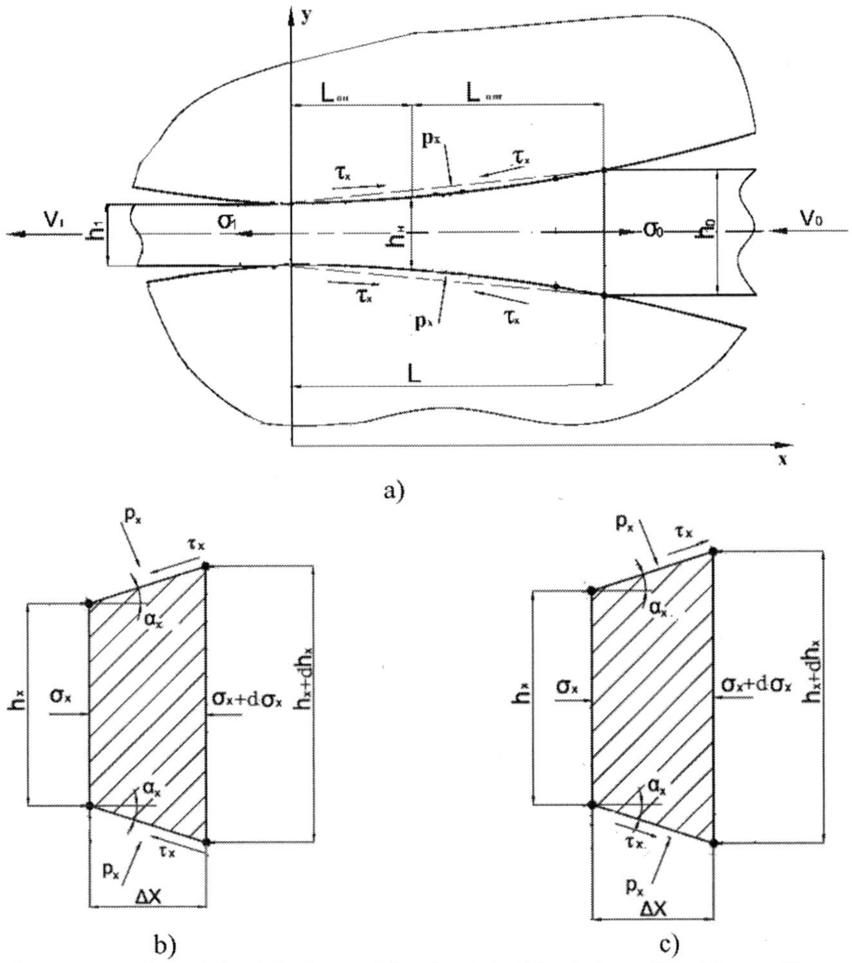

Fig. 1. Computational models of the integral focal point of the deformation (a) as well as elementary volumes of metal distinguished in the slower zone (b) and the faster zone (c), as applied to the engineering mathematical simulation of energy parameters of the process of hot symmetrical rolling of relatively thin strips.

Taking into consideration the law of external contact friction $\tau_x = 2K_c\mu$ (Siebel's law), recommended by the authors [9] $d\sigma_x dh_x \approx 0$ and disregarding infinitely small values of the second order along with taking into account the obvious purely geometrical relations $2dx\sin\alpha_x / \cos\alpha_x = 2dxtq\alpha = dh_x$ и $dx = dh_x /(2tq\alpha)$ the differential equation (4) can be represented as:

$$-\sigma_x dh_x - d\sigma_x h_x + \mathrm{p}_{\text{хот}} dh_x - 2K_c 2\mu dh_x /(2tq\alpha) = 0\,, \tag{5}$$

where μ are coefficients of plastic friction, the values of which are supposed to be constant along the entire length of deformation area.

From the complete formula of description of the conditions of plasticity [9]:

$$(p_x - \sigma_x)^2 + 4\tau_{xy}^2 = 4K_c^2\,, \tag{6}$$

and taking into account the results of two-dimensional analysis of metal stressed state [10, 11], according to which tangential components of stresses deviator τ_{xy} in the height of the deformation area sections are changed close to the linear, in compliance with the law, their average integral evaluation corresponds to the following equation:

$$\tau_{xyc} = \tau_x / 2 = 2K_c \mu / 2. \tag{7}$$

In the long run we will get the following for the values of normal contact stresses $p_{xот}$, occurring along the length of the area tardiness:

$$p_{xот} = 2K_c(\delta_\mu - a_k)\ln(h_0 / h_x) + 2K_c a_k - \sigma_0, \tag{8}$$

where $a_k = \sqrt{1-\mu^2}$ is an auxiliary variable used for a simplified form of description; $\delta_\mu = 2\mu L / \Delta h$ is an auxiliary variable characterizing the influence of boundary conditions within the deformation area at the process of hot rolling.

Likewise, for the area of advance of deformation area (see Fig. 1, a, c) we get:

$$\sigma_x h_x - (\sigma_x + d\sigma_x)(h_x + dh_x) + 2p_{xоп} dx tq\alpha_x + 2\tau_x dx = 0. \tag{9}$$

As a final view, with regard to the area of advance we will get:

$$p_{xоп} = 2K_c(\delta_\mu + a_k)\ln(h_x / h_1) + 2K_c a_k - \sigma_1. \tag{10}$$

Based on the condition of equality of normal contact stresses for the area of tardiness and the area of advance in the neutral section of deformation area $p_{xот}\big|_{h_x=h_н} = p_{xоп}\big|_{h_x=h_н}$ we will write the following:

$$2K_c(\delta_\mu - a_k)\ln(h_0 / h_н) + 2K_c a_k - \sigma_0 = 2K_c(\delta_\mu + a_k)\ln(h_н / h_1) + 2K_c a_k - \sigma_1, \tag{11}$$

from which the thickness of a strip in the $h_н$ section and the length of the area of advance L_a (see Fig. 1, a) could be defined as:

$$\begin{aligned} h_н &= \exp\{[\delta_\mu \ln(h_0 h_н) - a_k \ln(h_0 / h_1) + \sigma_1 / 2K_c - \sigma_0 / 2K_c]/(2\delta_\mu)\} = \\ &= \sqrt{h_0 h_1}\exp\{a_k \ln(h_1 / h_0) + \sigma_1 / 2K_c - \sigma_0 / 2K_c)/(2\delta_\mu)\} \end{aligned}; \tag{12}$$

$$L_{оп} = (h_н - h_1)L / \Delta h. \tag{13}$$

As the result of integrating of the calculated in accordance with (8) and (10) distributions of normal contact stresses $p_{xот}$, $p_{xоп}$ along the entire area of deformation:

$$n_\sigma = (1/L)[\int_0^{L_{оп}} p_{xоп} dx + \int_{L_{оп}}^{L} p_{xот} dx]. \tag{14}$$

An average integral value of the coefficient of stressed metal state could be determined: $n_\sigma = p_c / 2K_c$, which may be necessary for evaluation of an average integral value of normal contact stresses p_c and the value of the rolling force $P = 2K_c n_\sigma BL = p_c BL$

$$n_\sigma = \frac{1}{L}\{\int_0^{L_{on}} [(\delta_\mu + a_k)\ln(\frac{h_1 + \Delta hx / L}{h_1})]dx + \int_0^{L_{on}} (a_k - \frac{\sigma_1}{2K_c})dx + \\ + \int_{L_{on}}^{L} [(\delta_\mu - a_k)\ln(\frac{h_1 + \Delta hx / L}{h_0})]dx + \int_{L_{on}}^{L} (a_k - \frac{\sigma_0}{2K_c})dx\}. \quad (15)$$

From this equation, with due regard to substitution of variables of integrating limits $(h_1 + \Delta hx / L) / h_1 = U_1$, $(h_1 + \Delta hx / L) / h_0 = U_2$, and integrating itself, with subsequent mathematical transformation we will finally have:

$$n_\sigma = \frac{1}{L}\{\int_1^{h_н / h_1} \frac{h_1 L}{\Delta h}[(a_k + \delta_\mu)\ln U_1]dU_1 + (a_k - \frac{\sigma_1}{2K_c})L_{on} + \\ + \int_{h_н / h_0}^{1} \frac{h_0 L}{\Delta h}[(a_k - \delta_\mu)\ln U_2]dU_2 + (a_k - \frac{\sigma_0}{2K_c})(L - L_{on}) = \\ = (\delta_\mu + a_k)[\frac{h_н}{\Delta h}\ln\frac{h_н}{h_1} - (\frac{h_н - h_1}{\Delta h})] + (a_k - \frac{\sigma_1}{2K_c})(\frac{h_н - h_1}{\Delta h}) + \\ +(\delta_\mu - a_k)[\frac{h_0 - h_н}{\Delta h} - \frac{h_н}{\Delta h}\ln\frac{h_0}{h_н}] + (a_k - \frac{\sigma_0}{2K_c})(\frac{h_0 - h_н}{\Delta h}). \quad (16)$$

The value of the aggregate rolling momentum, taking into consideration the opposite direction of the tangential contact stresses τ_x in the areas of tardiness and advance (see Fig. 1), in this case corresponds to:

$$M_\Sigma = 2 \times 2K_c \mu RB(L - 2L_{on}) = 2 \times 2K_c \mu RBL[1 - 2(h_н - h_1) / \Delta h], \quad (17)$$

where R and B are the radii of the operating rolls and the width of the rolled strip.

On the whole, the presented dependences, combined with calculation of the average integral value of the doubled resistance to shift 2Kgf, with regard to elastic compression of the operating rolls and the presence of the elastic recovery [10] area, made up a complete algorithm for mathematical engineering modelling of energy parameters of the process of hot rolling of relatively thin strips, with due regard to more precise character of distribution of the boundary conditions of deformation area.

Correction of the original prerequisites (concerning the conditions of plasticity and analytical description of the coefficient of external contact friction) allowed us to develop engineering methods of calculation of energy- parameters of the process of hot rolling of relatively thin strips, with the degree of their non-correspondence, as compared to more strict two-dimensional analysis methods in relative measuring, not having exceeded 5%.

Let us now discuss the methods for determining the energy- parameters, in case of hot rolling of relatively medium and heavy plates and strips. i.e. at $L / h_{cp} < 2,5$.

The current level of computer technologies make it possible to apply sophisticated computer programs, based on the Finite Elements Method (FEM). ABAQUS/CAE [12] shell being one of them, allows us to take into consideration external friction coefficient, apart from more precise calculations, unlike the existing engineering methods.

This mathematical model was realized with Abaqus 6.5 FEM modelling using ABAQUS\Standard [12] programme package. For calculations, the operating rolls were simulated by means of not-deformed analytical surfaces, in accordance with the established assumptions whereas metal was simulated by 4-nodal , two-dimensional elements, equipped with CPE4R destruction control. We ensured quick formation of the original data for calculations by creating a parameterized model. The values of the original strip's thickness, the strip's thickness after rolling, radii of the upper and the lower rolls were varied without rearrangement of the model, as well as external friction coefficients, angular velocities of the operating rolls rotation and also the number of fragmentations in the strip's length and thickness for automatic generation of the finite elements network. Mechanical properties of the band's material were introduced by the corresponding values of resistance to deformation, depending upon the degree and velocity of deformation and metal temperature, determined by Andreyuk's method [13].

Mathematical simulation of the process itself was carried out in two steps. First, the yield of the strip by the operating rolls being simulated. Second; rotation of the operating rolls with specified angular speed was evaluated and hence, the band's transfer along the operating line, i.e. the direct realization of the process of rolling.

With the help of ABAQUS\CAE package, the coefficient of the stressed state was calculated, being expressed from the formula of rolling strength:

$$P = 2K_c LBn_\sigma, \qquad (18)$$

where $2K_c$ is a doubled average integral value of resistance to frequent shift , determined in accordance with the methods, employed in L.V.Andreyuk's works [13], or M.Y.Browman's works [14]; B is the width of the rolled strip; $L = \sqrt{R\Delta h - \left(\Delta h^2 / 4\right)}$ – general length of the area of plastic forming; h_0, h_1, $\Delta h = h_0 - h_1$ being the thickness prior and after rolling and the value of absolute reduction; R is the radius of the operating rolls of the rolling mill.

At the specified value of the rolling force and varied values, the specified distributions of the stressed state coefficient L / h_{cp} were determined, presented in Fig. 2. Here, the obtained dependences were compared to the conventional methods [15, 16].

The reliability of the results obtained by FEM was confirmed by comparing it to the methods used in the works of V.F.Potapkin [17] and M.Y.Browman [14], with regard to determining arm's coefficient, exerting influence upon the value of the rolling momentum:

$$M = 2 \cdot 2K_c n_\sigma L^2 B\psi. \qquad (19)$$

The values of arm's coefficient, expressed by the formula (19), for a known value of the rolling momentum M, calculated by means of FEM [18, 19] and also based upon the aforesaid methods, depending on L / h_{cp} is pictured in Fig. 3.

Proceeding from the specified distributions, given in Fig. 2 and Fig. 3, we may make conclusions regarding the reliability of the results obtained by FEM. This provides one of the most precise methods for the design of equipment and technologies of hot rolling mills. The generalized results for stressed state coefficient and arm's coefficient, determining the rolling momentum are represented in Fig. 4.

From the analysis of the represented specified distributions it becomes obvious that unlike the methods used in [13, 14, 17], the influence of the coefficient of external friction μ is essential enough even in $L/h_{cp} > 1$ range. Analytically, the results of the present investigation could be presented by the dependence with regard to coefficient of metal stressed n_σ (see Fig. 4, a) state as follows:

$$n_\sigma = 1 + 0,65\mu\left(L/h_{cp} - 1\right). \tag{20}$$

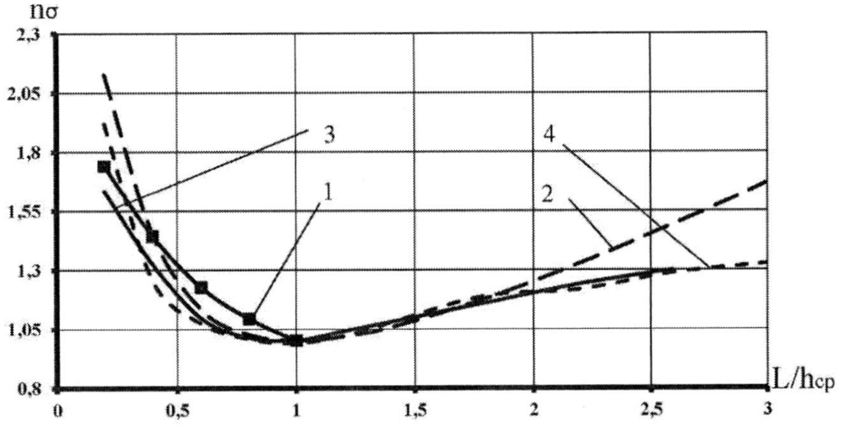

1 – Tselikov-Smirnov's methods [15]; 2 – M.Y.Browman's methods [14];
3 – V.Potapkin's methods [17]; 4 – numerical values, obtained by means of FEM.

Fig. 2. Specified distribution of coefficients of stressed state n_σ depending on the value L/h_{cp}, obtained on the basis of different methods.

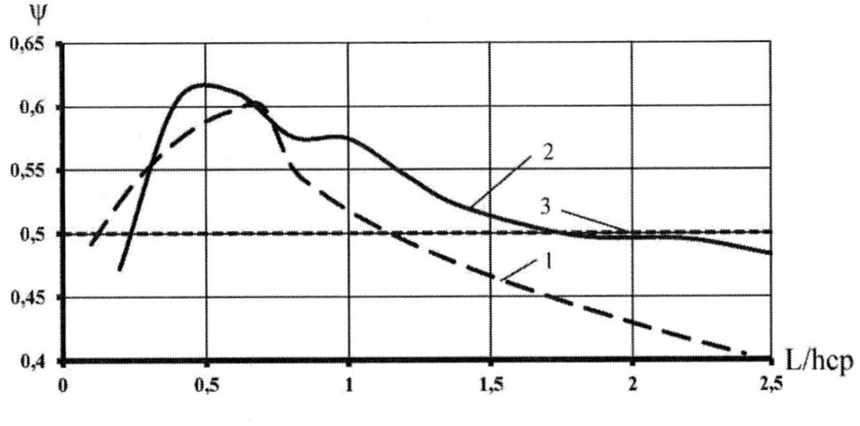

1 – V.F.Potapkin's methods [17]; 2 – numerical values, obtained by means of the method of finite elements; 3 – M.Y.Browman's methods [14].

Fig. 3. Estimated distribution of arm's coefficient ψ depending on the value L/h_{cp}, obtained with application of different methods.

DOI: 10.1051/matecconf/20167304009

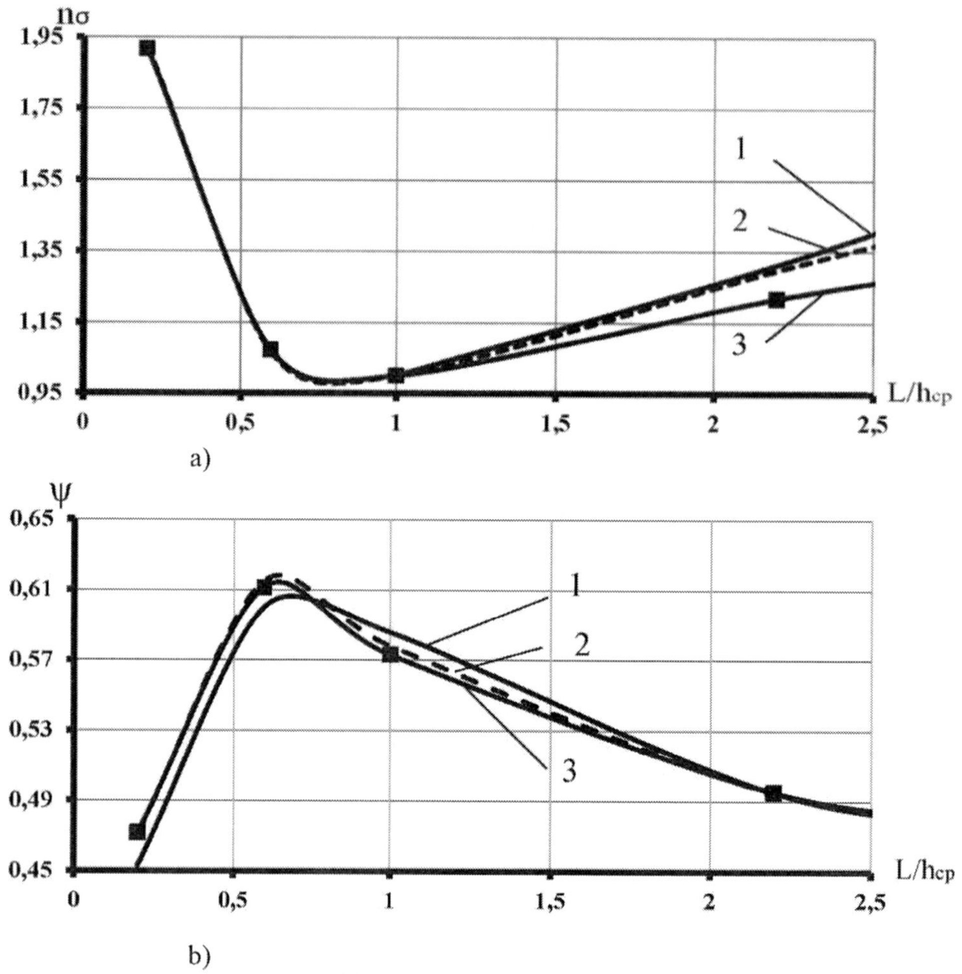

$\mu = 0,45$ – (1), $\mu = 0,35$ – (2), $\mu = 0,25$ – (3)

Fig. 4. Estimated distributions of the coefficient of stressed state n_σ (a) and arm's coefficient ψ (b) depending upon the relation between the contact arc and the average thickness of the ingot within the area of deformation L/h_{cp}, obtained at different values of friction coefficient.

The influence of the forces of external friction ψ, upon arm's coefficient determining the rolling momentum is not so essential (see Fig. 4, b) and can be described by a single analytical dependence like:

$$\psi = 0,62 - 0,109\left(L/h_{cp} - 0,6\right) + 0,023\left(L/h_{cp} - 0,6\right)^2. \tag{21}$$

Comparison of FEM to the existing methods of determination of the coefficient of stressed state helped to validate this method. At $L/h_{cp} < 1$, the reliable interval lies within $0,996 < X_i < 1,004$ limits; the reliable interval at $1 \leq L/h_{cp} \leq 2,5$ lies within $0,95 < X_i < 1,03$ which fully confirms the reliability of the results obtained by FEM, namely by ABAQUS\CAE package.

The developed engineering mathematical model was used for calculations of external loads at rolling in the corresponding dynamic model.

The dynamic model was based on compilation of the system of differential equations of the motion of inertia masses of a mechanical system of an operating stand of a rolling mill. The stage, when metal is gripped by the rolls and the stage of metal rolling after gripping, was analyzed separately. In accordance with recommendations, in [16] a dependence for determination of elastic momentum in *i*- phase of gripping was used:

$$\begin{aligned} M_{12} &= n_2 M_a \left(1-\cos\ pt\right) + M_{np}\left(1+\frac{\sin\ pt_0}{pt_0}\right)+ \\ &+n_2 M'_{np}\left(\frac{\lambda^2}{1+\lambda^2}e^{-t/T_M}+\frac{\lambda \sin\ pt-\lambda^2\cos\ pt}{1+\lambda^2}\right) \end{aligned}, \quad (22)$$

where $M_a = I_0\omega_n/\mathrm{t_p} - M_{np}\mathrm{T_M}/t_0$; $M'_{np} = M_{np}\mathrm{T_M}/t_0$; $n_2 = \dfrac{I_2}{I_1+I_2}$; $\lambda = 2\pi T_M/T_0$; t_o is the time of metal gripping by the rolls; T_o – the period of the system's own oscillations; t_p – time of the drive's take-off run; I_1, I_2 – momentums of electrical motor's rotor and rolls' inertia, respectively; $\omega_н$ – nominal angular velocity of drive's rotation; $\mathrm{T_M} = I_0 R / c_e c_м$ – electrical and mechanical constant of drive's time; I_o – general inertia momentum of the mill's drive; R – resistance in Ohms within anchor's chain of the electric motor; c_e, $c_м$ – parameters of the electric motor.

After metal gripping by the rolls, elastic momentum in the drive's line is determined according to the formula [8]:

$$\begin{aligned} M_{12} &= M_{npi} - n_2 M_{np}\left(\lambda^2 \mathrm{e}^{-t/T_M} + \lambda \sin\ pt_1 - \lambda^2\cos\ pt_1\right)/\left(1+\lambda^2\right)+ \\ &+A\cos\ pt_1 + B\sin\ pt_1 \end{aligned}, \quad (23)$$

here $M_{npi} = n_1 M_{np} + n_2\left(M_{np} + \mathrm{I}_0\omega_n/t_p\right)$; $t_1 = t - t_0$; $n_1 = I_1/(I_1+I_2)$;

$$\begin{aligned} A &= n_2 M_a\left(1-\cos\ pt_0\right) - M_{npi} + M_{np}\left(1\text{-sin pt}_0\right)/pt_0 + \\ &+n_2 M'_{np}\left(\lambda^2 e^{-t_0/T_M} + \lambda\sin\ pt_0 - \lambda^2\cos\ pt_0\right)/\left(1+\lambda^2\right); \\ B &= n_2 M_{np}\sin\ pt_0 + M_{np}\left(1-\cos\ pt_0\right)/pt_0 + \\ &+n_2 M'_{np}\left(-\lambda e^{-t_0/T_M} + \lambda\cos\ pt_0 + \lambda^2\sin\ pt_0\right)/\left(1+\lambda^2\right). \end{aligned}$$

As an example of numerical realization of the designed program means, a rolling process used for a narrow slab ($h_0 \times B_0 = 150\times1000$ mm, weight 15 tons) was investigated, the parameters of the slab being:

$$I_1 = 98\ t\cdot m^2;\ I_2 = 5\ t\cdot m^2;\ \mathrm{T_M} = 0{,}06\ s;\ n_1 = 0{,}95;\ n_2 = 0{,}05;\ p = 122\ s^{-1};$$
$$T = 0{,}051\ s.$$

In Fig. 5 and 6 the specified distribution of the rolling force and momentum were presented, here, the load peaks at the time of gripping and metal exit from the rolls were pointed out.

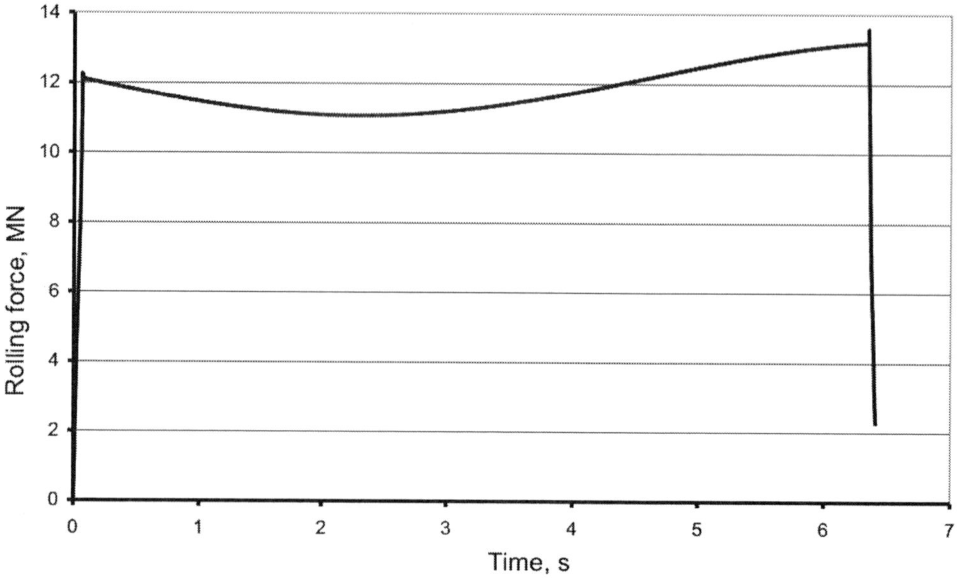

Fig. 5. Distribution of the rolling force.

In Fig. 7 the specified distribution of momentums inside elastic sections of the operating stand were pictured.

It can be seen from the dependences that the coefficient of dynamics at metal gripping was of 1.5 order, and at rolling it was equal to 1.2…1.35 order. The change of the momentum here has a complicated character. Attention should be given to this during the design of the elements of the line of stand's drive.

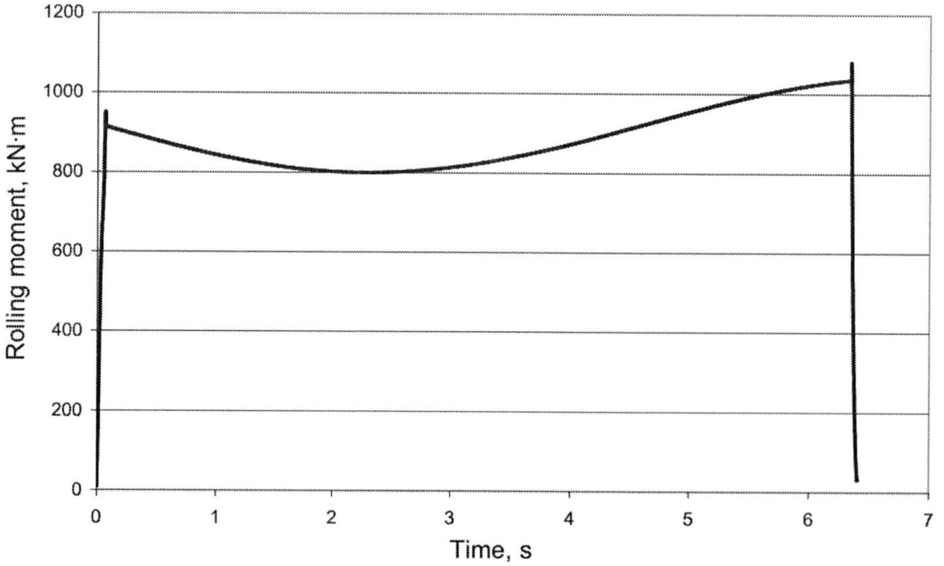

Fig. 6. Distribution of the rolling momentum.

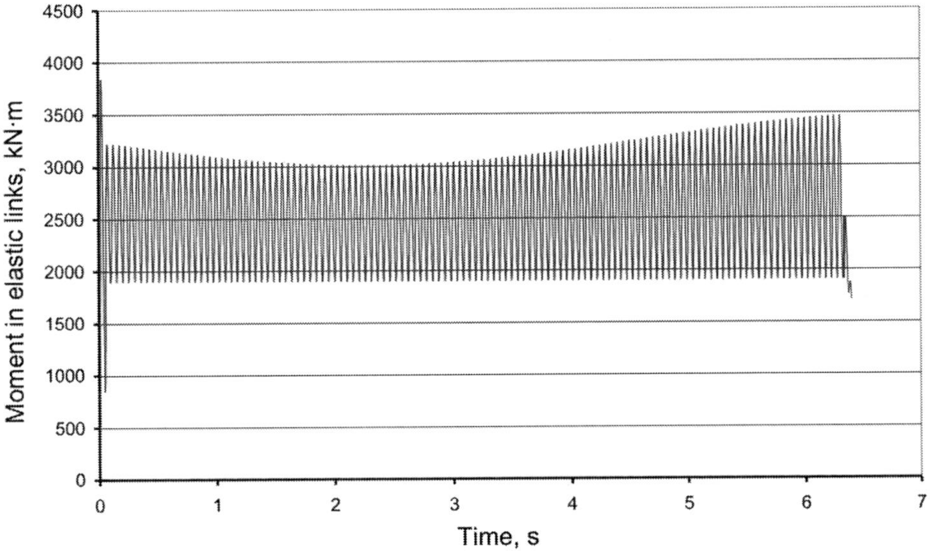

Fig. 7. Distribution of the momentums in elastic sections.

3 Results and Discussion

One drawback of the existing dynamic models is the necessity of a simplified dependence of the force and momentum of rolling in time. Correction of the original prerequisites in plasticity condition and analytical description of the coefficient of external contact time allowed us to develop engineering methods of calculating the energy parameters of hot rolling of relatively thin strips, with the relative degree of their error (as compared to more strict two-dimensional methods of analysis) less than 5%. It is obvious from the analysis of the specified estimate distributions that unlike the methods used in the cited works [20, 21], the influence of the external friction coefficient μ is essential enough even within $L/h_{cp} > 1$ range. As an example of the use of the prepared mathematical models for determination of energy parameters of the rolling practice, the results of calculation of changes of dynamic loads in the line of a drive of a blooming mill and also changes in momentum in drive's elastic lines were presented.

4 Conclusions

On the basis of two-dimensional FEM approach and numerical recurrent solutions of finite-differential forms of the condition of static equilibrium of detached elementary volumes, obtained by means of fragmentation of the strip along the rolling line and the analysis of the area of plastic deformation, mathematical models were developed for stressed-deformed metal state at hot rolling of sheets, plates and strips, which made it possible to improve dynamic models of drive's lines of the operating stands, by means of taking into account the real character of loads changes and geometrical parameters, as well as physical and mechanical properties of strip's material at rolling.

The reported study was funded by RFBR according to the research project №16-08-00845a «Verification and development of models of inelastic deformation at the passive loading». The authors acknowledge the assistance given by Ralph Williams who thoroughly edited the final version of the paper in English, making it acceptable for publication. The authors declare that there is no conflict of interest regarding the publication of this paper.

References

1. R.S. Adamia, *Optimization of dynamic loads of rolling mills* (Metallurgy, Moscow, 1978)
2. V. Mazur, V. Artyukh, G. Artyukh, M. Takadzhi, Engineering Designer. **37(1)**, 26–29 (2012)
3. V.V. Verenev, V.I. Bolshakov, A.M. Yunakov, Fundamental and applied problems of ferrous metallurgy: Works, Dnepropetrovsk. **19**, 346–358 (2009)
4. R.A. Yakovlev, Dynamic calculation of rolling mills (Bauman MSTU, Moscow, 1964)
5. V.V. Verenev, V.I. Bolshakov, N.I. Podobedov, Protection of iron and steel machines from failure. **3**, 35–39 (1998)
6. F.K. Ivanchenko, V.M. Gribennik, V.I. Shiryaev, *Calculation of machines and mechanisms of rolling shops* (Higher scool publishers, Kiev, 1995)
7. V.A. Fedorinov, A.V. Satonin, E.P. Gribkov, *Mathematical modelling of stresses, deformations and main indices of quality at rolling of relatively wide sheets and strips* (DSMA, Kramatorsk, 2010)
8. A.V. Satonin, Improvement of the processes and equipment for plastic working of metals in metallurgy and machine building: Volume of works. 298–302 (2000)
9. A.M. Laptev, Y.Y. Tkachenko, V.I. Zhabin, Plastic working of metals. **3(28)**, 129–135 (2011)
10. Nabeel S. Gharaibeh, Mohammed I. Matarneh, V.G. Artyukh, Research Journal of Applied Sciences, Engineering and Technology. **8(12)**, 1461–1464 (2014)
11. V.F. Potapkin, *Method of fields of sliding lines in the theory of rolling of wide strips* (DSMA, Kramatorsk, 2005)
12. *SIMULIA Abacus (CAE User's manual)*, V.**6.7** (2008)
13. L.V. Andreyuk, G.G. Tylenev, Steel. **9**, 545–547 (1972)
14. M.Y. Browman, *Application of theory of plasticity in rolling practice* (Metallurgy publishers, Moscow, 1991)
15. A.I. Tselikov, G.S. Nikitin, S.E. Rokotyan, *Theory of longitudinal rolling* (Metallurgy publishers, Moscow, 1980)
16. Mohammed I. Matarneh, Nabeel S. Gharaibeh, V. G. Artyukh, International Journal of Engineering Science and Innovative Technology. V.**4**, Issue **2**, 1–7 (2015)
17. V.F. Potapkin, *Method of the fields of sliding lines fields in the theory of rolling of wide strips* (DSMA, Kramatorsk, 2005)
18. E.A. Nekliudova, A.S. Semenov, S.G. Semenov, B.E. Melnikov, Applied Mechanics and Materials. **725–726**, 648–653 (2015)
19. I. Maniak, B. Melnikov, A. Semenov, S. Saikin, Applied Mechanics and Materials. **725–726**, 943–948 (2015)
20. Firas M.F. Al-Quran, M.E. Matarneh, V.G. Artukh, Research Journal of Applied Sciences, Engineering and Technology. **4(11)**, 1585–1589 (2012)
21. V.G. Artiukh, S.Yu. Karlushin, E.N. Sorochan, Procedia Engineering. **117**, 938–944 (2015)

Simulation of Pulling the Reinforcing Bar from Concrete Block with Account of Friction and Concrete Damage

Mikhail Beliaev[1,*], *Artem* Semenov[1] , *Sergey* Semenov[1], and *Andrey* Benin[2]

[1]Peter the Great Saint-Petersburg Polytechnic University, Polytechnicheskaya st. 29, St. Petersburg, 195251, Russia
[2]Petersburg State Transport University, Moskovsky pr. 9, St. Petersburg, 190031, Russia

Abstract. A direct finite element modeling of the pulling process of the steel reinforcing bar out from the concrete block with taking into account the cohesive behavior of steel-concrete bond and using elastic-plastic-damage constitutive equations for the concrete is considered. The comparison of obtained results of finite element simulations with experimental data is presented and discussed for the problem of pulling the reinforcing bar from the concrete block.

1 Introduction

One of the main problems in reinforced concrete elements is the contact between concrete body and steel rebars [1]. It allows them operates as a single monolithic body is a reinforced concrete bond. Reducing coupling leads to excessive crack opening, to the decreasing stiffness and to the reduced carrying capacity of the structure [2,3].

Pulling the rebar out of the concrete block shows the process of destroying reinforced-concrete bond. It is a complex multistage process, accompanied by the inelastic deformation, the cracking and crushing, the rupture of adhesive links, the presence of tribological phenomena and so on [4,5].

The objective in this paper is to compare the various numerical approaches to modeling of pulling the steel rebar out of the concrete block with experimental results.

2 Materials and Methods

2.1 Experiment

Pulling the reinforcing bar from the concrete block in accordance with RILEM/CEB/FIB [6] requirements is considered. Loading conditions and geometric parameters are shown in Fig. 1, 2.

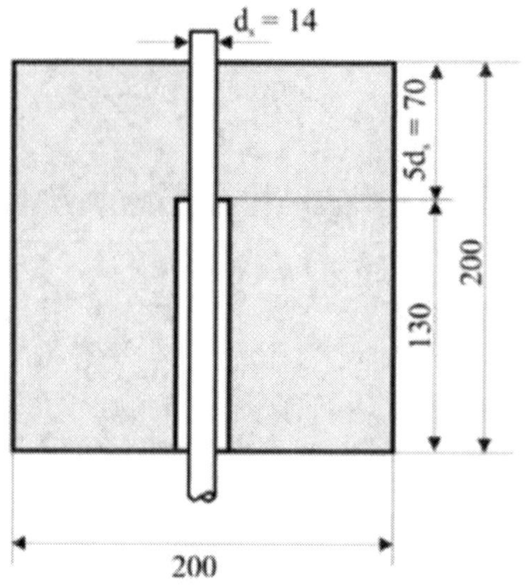

Fig. 1. Geometric parameters of concrete prism and steel rebar;

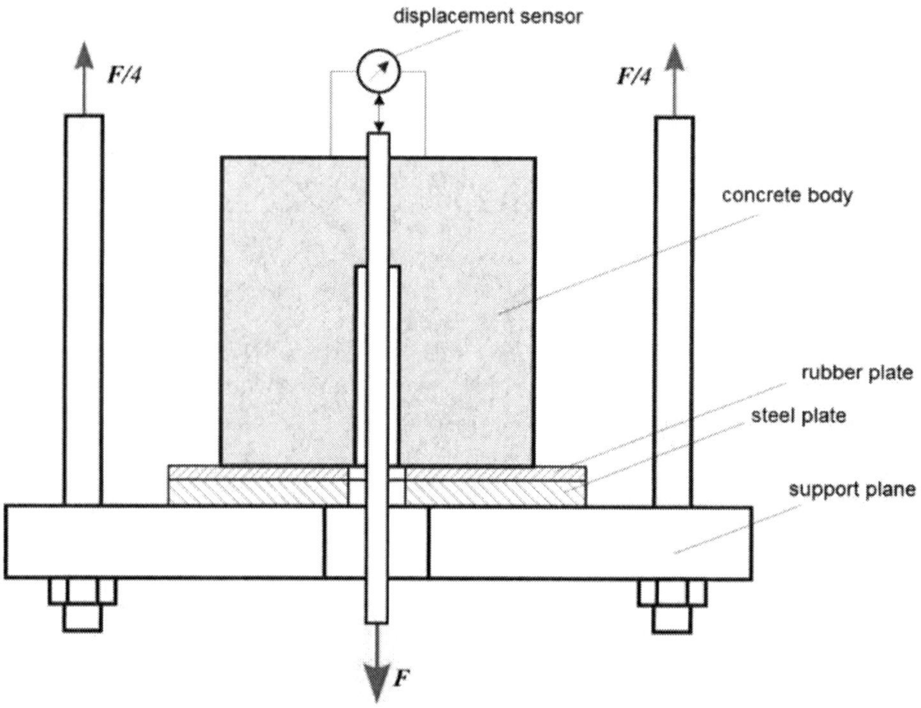

Fig. 2. Loading conditions.

Pull-out test experiment is shown in Fig. 3. It was conducted on tensile tester Shimadzu AG-300kN [7, 8]. Steel reinforcement Ø14 AIII and concrete B25 were used in experiments.

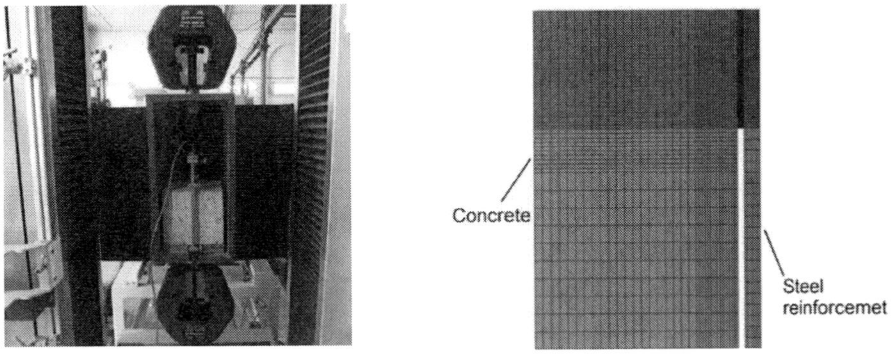

a) b)

Fig. 3. (a) Tensile tester Shimadzu AG-300kN; (b) 2D FE model.

2.2 Concrete model

The elastic, elastic-damage and elastic-plastic damage models are used in the present analysis [5,9-11]. In order to capture the degradation of the elastic stiffness of the concrete material as well as its irreversible deformations upon mechanical loading, the combined use of elastic-plastic constitutive equations along with continuum damage mechanics became vital to better describe the mechanical behavior of concrete. The coupled damage to concrete plasticity by defining damage growth as a function of accumulated plastic strains [9] is considered.

The extension of microcracks plays a critical role in the inelasticity of concrete, as it results in the degradation of the elastic stiffness. In the models it is taken into account by introducing of damage variables. In the simples case a single scalar damage variable is used. It ranging from 0 for the undamaged material to 1 for completely damaged material. The constitutive equation of elastic-plastic material with scalar isotropic damage (introduced by Kachanov [12] and further developed by Rabotnov [13] and others) takes the following form:

$$\boldsymbol{\sigma} = (1-D)\cdot{}^4\mathbf{C}_0^e \cdot\cdot (\boldsymbol{\varepsilon}-\boldsymbol{\varepsilon}^p) = {}^4\mathbf{C}^e \cdot\cdot (\boldsymbol{\varepsilon}-\boldsymbol{\varepsilon}^p) \tag{1}$$

where $\boldsymbol{\sigma}$ is Cauchy stress tensor, D is the scalar stiffness degradation variable, $\boldsymbol{\varepsilon}$ is the strain tensor, $\boldsymbol{\varepsilon}^p$ is the plastic strain tensor, ${}^4\mathbf{C}_0^e$ the initial (undamaged) elastic stiffness of the material, while ${}^4\mathbf{C}^e = (1-D)\cdot{}^4\mathbf{C}_0^e$ is the degraded elastic stiffness tensor. The effective stress tensor is defined by the relation $\overline{\boldsymbol{\sigma}} = {}^4\mathbf{C}_0^e \cdot\cdot (\boldsymbol{\varepsilon}-\boldsymbol{\varepsilon}^p) = \boldsymbol{\sigma}/(1-D)$.

The evolution of the scalar degradation variable is defined by function $D = D(\overline{\boldsymbol{\sigma}}, \widetilde{\varepsilon}^p)$, governed by a set of the effective stress tensor $\overline{\boldsymbol{\sigma}}$ and hardening (softening) variables $\widetilde{\varepsilon}^p$. In the used in further Lubliner model [9], the stiffness degradation is initially isotropic and defined by degradation variable D_c in a compression zone and variable D_t in a tension zone.

The cracking (under tension) and the crushing (under compression) in concrete are represented by increasing values of the hardening (softening) variables. These variables control the evolution of the yield surface and the degradation of the elastic stiffness. The yield function represents a surface in effective stress space $F(\bar{\boldsymbol{\sigma}},\tilde{\varepsilon}^{p}) \leq 0$ which determines the states of failure or damage.

Plastic flow is governed by a flow potential function $G(\bar{\boldsymbol{\sigma}})$ according to non-associative flow rule:

$$\dot{\boldsymbol{\varepsilon}}^{p} = \dot{\lambda}\frac{\partial G(\bar{\boldsymbol{\sigma}})}{\partial \bar{\boldsymbol{\sigma}}} \tag{2}$$

The flow potential G and yield function F are accepted in the forms:

$$G=\sqrt{(f_c - m\cdot f_t\cdot\tan\beta)^2 + 3\bar{J}_2} + \tfrac{1}{3}\bar{I}_1\cdot\tan\beta \tag{3}$$

$$F=\frac{1}{1-\alpha}\left(\sqrt{3\bar{J}_2}+\alpha\cdot\bar{I}_1+\theta(\tilde{\boldsymbol{\varepsilon}}^{p})\langle\bar{\sigma}_{\max}\rangle-\gamma\langle-\bar{\sigma}_{\max}\rangle\right)-\bar{\sigma}_c(\tilde{\varepsilon}_c^{p}) \tag{4}$$

where f_t and f_c are the uniaxial tensile and compressive strengths of concrete, respectively, β is the dilation angle measured in the $\frac{1}{3}\bar{I}_1-\sqrt{3\bar{J}_2}$ plane at high confining pressure, while m is an eccentricity of the plastic potential surface, $\bar{I}_1 = \mathbf{1}\cdot\cdot\bar{\boldsymbol{\sigma}}$ and $\bar{J}_2=\frac{1}{2}dev\bar{\boldsymbol{\sigma}}\cdot\cdot dev\bar{\boldsymbol{\sigma}}$.

Degradation of elastic modulus of concrete was shown in many experiments: tensile test [14-16], compressive test [17-21] of concrete.

2.3 Interface constitutive model

Cohesive behavior was used to describe destroying of reinforced-concrete bond. The available traction-separation model assumes initially linear elastic behavior followed by the initiation and evolution of the bond damage. The elastic behavior is written in terms of an elastic constitutive matrix that relates the nominal stresses to the nominal strains across the interface.

The nominal traction stress vector consists of three components: t_n, t_s and t_t, which represent the normal and the two shear tractions respectively. The corresponding separations are denoted by δ_n, δ_s and δ_t.

The elastic behavior can then be written as

$$\begin{Bmatrix} t_n \\ t_s \\ t_t \end{Bmatrix} = \begin{bmatrix} K_{nn} & K_{ns} & K_{nt} \\ K_{ns} & K_{ss} & K_{st} \\ K_{nt} & K_{st} & K_{tt} \end{bmatrix} \begin{Bmatrix} \delta_n \\ \delta_s \\ \delta_t \end{Bmatrix}. \tag{5}$$

The bond damage modeling allows to simulate the degradation and eventual failure of the bond between concrete body and steel rebar. The failure mechanism consists of two ingredients: a damage initiation criterion and a damage evolution law.

The bond damage initiation refers to the beginning of degradation of the cohesive response at a contact point. The process of degradation begins when the contact stresses and/or contact separations satisfy certain damage initiation criteria.

Maximum stress criterion: the bond damage is assumed to initiate when the maximum contact stress ratio (as defined in the expression below) reaches a value of one. This criterion can be represented as

$$\max\left\{\frac{\langle t_n\rangle}{t_n^0}, \frac{t_s}{t_s^0}, \frac{t_t}{t_t^0}\right\}=1. \tag{6}$$

Maximum separation criterion: the bond damage is assumed to initiate when the maximum separation ratio (as defined in the expression below) reaches a value of one:

$$\max\left\{\frac{\langle \delta_n\rangle}{\delta_n^0}, \frac{\delta_s}{\delta_s^0}, \frac{\delta_t}{\delta_t^0}\right\}=1. \tag{7}$$

The bond damage evolution law describes the rate at which the cohesive stiffness is degraded once the corresponding initiation criterion is reached

A scalar bond damage variable, D_b, represents the overall damage at the contact point. It initially has a value of 0. If damage evolution is modeled, D_b monotonically increases from 0 to 1 upon further loading after the initiation of damage. The contact stress components are affected by the damage according to

$$t_n=\begin{cases}(1-D_b)\bar{t}_n, & t_n>0\\ \bar{t}_n, & t_n\le 0\end{cases}, \quad t_s=(1-D_b)\bar{t}_s, \quad t_t=(1-D_b)\bar{t}_t, \tag{8}$$

where $\bar{t}_n$, $\bar{t}_s$ and $\bar{t}_t$ are the contact stress components predicted by the elastic traction-separation behavior for the current separations without damage. To describe the evolution of bond damage under a combination of normal and shear separations across the interface, it is useful to introduce an effective separation [22] defined as

$$\delta_m=\sqrt{\langle\delta_n\rangle^2+\delta_s^2+\delta_t^2}\,. \tag{9}$$

An evolution of the bond damage variable D_b was used for exponential softening, that reduces (in the case of damage evolution under a constant mode mix, temperature, and field variables) to

$$D_b=1-\left\{\frac{\delta_m^0}{\delta_m^{\max}}\right\}\left\{1-\frac{1-\exp\left(-\alpha\left(\frac{\delta_m^{\max}-\delta_m^0}{\delta_m^f-\delta_m^0}\right)\right)}{1-\exp(-\alpha)}\right\}. \tag{10}$$

In the expression above α is a non-dimensional parameter that defines the rate of damage evolution.

3 Results and Discussion

The results of simulation of the pulling the reinforcing bar from the concrete block (see Fig. 1) are obtained using axisymmetric finite element (FE) model (FE mesh is shown in Fig. 2b). The numerical model consists of 4963 8-node quadrilateral elements (CAX8R). The computations were performed using the FE software package ABAQUS [11].

In order to determine the optimal strategy for solving the problem the six different models representing combinations of the bond behavior model and concrete model are considered and compared:

1. Ideal bond (bond model without discontinuity and cohesive behavior) and elastic model for concrete;
2. Bond model with cohesive behavior and elastic model for concrete;
3. Ideal bond and elastic-plastic model for concrete;
4. Bond model with cohesive behavior and elastic-plastic model for concrete;
5. Ideal bond and elastic-plastic-damage model for concrete;
6. Bond model with cohesive behavior and elastic-plastic-damage model for concrete.

Material parameters used in simulations are: concrete compression strength $R_b = 18.5$ MPa, tensile strength is $R_{bt} = 1.55$ MPa, Young modulus is $E = 30000$ MPa, Poisson's ratio is $\nu = 0.2$. To describe linear elastic traction-separation behavior the following parameters were used: K_{nn}= 1500 MPa/mm, K_{ss}= 360 MPa/mm, and K_{tt}= 360 MPa/mm. Maximum stresses, which were used for damage initiation are: t_n^0=110 MPa, t_s^0=9 MPa, and t_t^0=9 MPa. The elasticplastic model with isotropic hardening [23] with $\sigma_{02} = 235$ MPa is used for steel rebar in FE computations.

Comparison of experimental data with simulation results of the pulling the reinforcing rebar from the concrete block for six considered models are shown in Fig. 4.

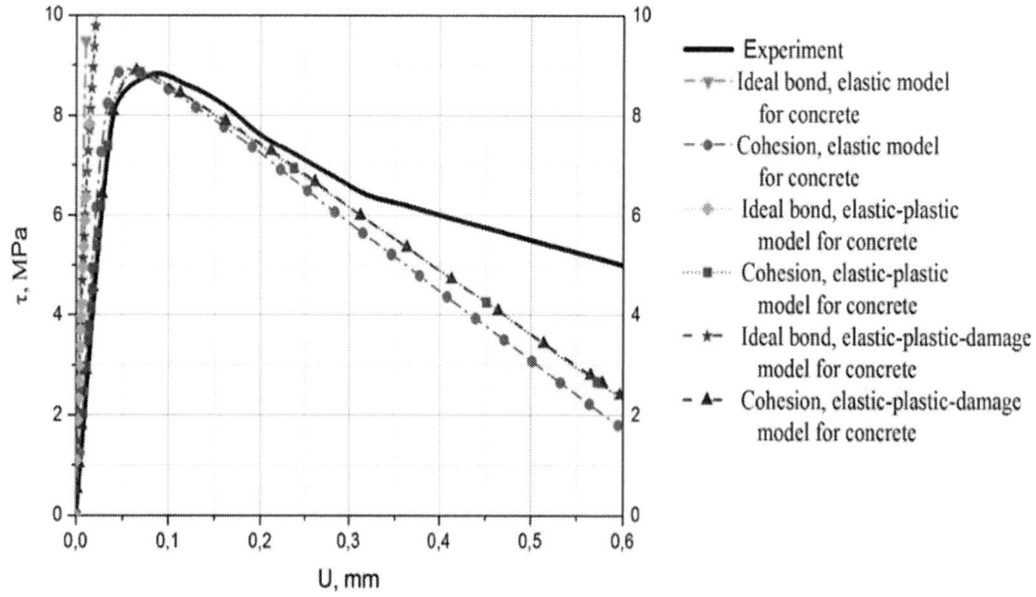

Fig. 4 Dependences of the shear stress in bond on the displacement of reinforcement.

All three models with ideal bond are unable to describe the appearance of a peak in dependence of the shear stress in bond on the displacement of reinforcement. Despite the concrete model, using the model of ideal bond without taking into account the effects of bond damage doesn't allow to describe correctly the process of pulling out of rebar from the concrete block. However using bond model [24] with cohesive behavior allows to describe the peak of shear stresses and rising branch on the diagram. Also it allows to simulate the degradation and eventual failure of the bond even with elastic model for concrete. As well as application of elastic model for concrete using the elastic-plastic model for concrete and the model of ideal bond doesn't allow to describe the pulling process. Using bond model with cohesive behavior gives near results for both elastic and elastoplastic models for concrete. There is no difference in results with using elastic-plastic and elastic-plastic-damage models for concrete. The obtained results point out on the dominant influence of accounting cohesion in bond model on pulling process description and weak influence of accounting model of concrete model. The proposed approach, which is based on direct FE modeling of reinforced concrete elements with account of nonlinear material model of concrete and cohesive behavior for interface between reinforcement and concrete, allows to describe correctly reinforced-concrete bond at pulling the rod out of the concrete body.

References

1. M. Ahmed Diab, E.Hafez Elyamany, A. Mostafa Hussein, M. Hazem Al Ashy, AEJ, **53(2)**, 355-371 (2014)
2. Y. Zhao, H. Lin, K. Wu, W. Jin, Constr. Build. Mater., **48**, 348-359 (2013)
3. A. Michou, A. Hilaire, F. Benboudjema, G. Nahas, P. Wyniecki, Y. Berthaud, Engineering Structures, **101**, 570-582 (2015)
4. A.V. Benin, A.S. Semenov, S.G. Semenov, Adv. Mat. Res., **831**, 264-369 (2013)
5. A.V. Benin, A.S. Semenov, S.G. Semenov, *Advances in Civil Engineering and Building Materials* (CRC Press, Moscow, 2012)
6. *RILEM/CEB/FIB. Recommendation on reinforcement steel for reinforced concrete. RC6. Bond test for reinforcement steel. 2. Pull-out tests* (1983)
7. A.V. Benin, A.S. Semenov, S.G. Semenov, B.E. Melnikov, MCE, **5**, 86-144 (2013)
8. A.B. Benin, A.S. Semenov, S.G. Semenov, B.E. Melnikov, MCE, **1(45)**, 23-40 (2014)
9. J. Lubliner, J. Oliver, S. Oller, E. Onate, Int. J. Solids Struct., **25(3)**, 299-326 (1989)
10. T. Jankowiak, T. Lodygowski, FCEE, **6**, 53-69 (2005)
11. *Abaqus/CAE User's Guide* (2013)
12. L.M. Kachanov, Proc. Academy of Sci. SSSR., **8**, 26-31 (1958)
13. Y.N. Rabotnov, *Creep problems in structural members* (NHPC, Amsterdam, 1969)
14. R.H. Evans, M.S. Marathe, Materials and Structures, **1(1)**, 61-64 (1968)
15. V.S. Gopalaratnam, P. Surendra Shah, ACI Journal proceedings, **82(3)**, 310-323 (1985)
16. H.W. Reinhardt, HERON, **29(2)**, 3-41 (1984)
17. I.D. Karsan, J.O. Jirsa, Proceeding of the American Society of Civil Engineers, **95(12)**, 2543-2563 (1969)
18. Mazars, Y. Berthaud, S. Ramtani, The unilateral Behaviour of Damaged Concrete, Eng . Fract. Mech., **35(4)**, 629-635 (1990)
19. S. Okamoto, S. Shiomi, K. Yamabe, Proceedings of Annual Architectural Instititute of Japan (AIJ) Convention, 1251–1252 (1976)
20. B.P. Sinha, K.H. Gerstle., L.G. Tulin, ACI J Proc., **61(2)**, 195-211 (1964)
21. Y.-H. Lee, K. Willam, ACI Materials Journal, **94/6**, 457-471 (1997)
22. P.P. Camanho, C.G. Davila, NASA/TM, **211737**, 1–37 (2002)
23. B.E. Melnikov, A.S. Semenov, S.G. Semenov, Proceedings of the A.N. Krylov Shipbuilding Research Institute, **53**, 85-92 (2010)
24. T.S. Phan, P. Rossi, J.-L. Tailhan, Numerical modelling of the concrete/rebar bond, Cement and Concrete Composites, **59**, 1-9 (2015)

MATEC Web of Conferences **73**, 04011 (2016) DOI: 10.1051/matecconf/20167304011
TPACEE-2016

Buckling Behavior of Cold-Formed Studs with Thermal Perforations

Marsel Garifullin[1,*], *Alexey* Sinelnikov[1], *Maria* Bronzova[1], *Boštjan* Kovacic[2], and *Rok* Kamnik[2]

[1]Peter the Great Saint-Petersburg Polytechnic University, Institute of Civil Engineering, 195251 St.Petersburg, Russia
[2]University of Maribor, Faculty of Civil Engineering, 2000 Maribor, Slovenia

Abstract. Studies have shown that the optimal structural scheme for low-rise buildings that meets all regulatory requirements is a frame system. In this connection, thin-walled cold-formed steel (CFS) profiles seem to be the best material for constructing light steel framed (LSF) walls. The framework of LSF walls is usually constructed from CFS C-shaped profiles. To increase the thermal effectiveness of a wall, CFS profiles usually have thermal perforations and thus are called thermoprofiles. However, these openings have a negative impact on bearing capacity of profiles and require accurate evaluation. In this article a relatively new reticular-stretched thermoprofile with diamond-shaped openings is considered. The article deals with the buckling analysis of perforated CFS C-sections subjected to compression.

1 Introduction

Thin-walled cold-formed steel (CFS) sections are widely used in various fields of industrial and civil engineering, bridges, storage racks, car bodies, railway coaches, transmission towers and poles, various types of equipment [1–6].

Over the last decade CFS structures have been actively used in the construction of LSF systems in residential housing [7, 8]. This caused a growing interest to the issue of their mechanical properties [9–11]. Rybakov [12, 13] presented four types of finite elements to analyze cold-formed members with various boundary conditions on the ends. Belyj [14] introduced a new method for the approximate estimation of CFS structures service life in buildings. Al Ali [15–17] investigated thin-walled cold-formed steel members with closed cross-sections. Tusnin [18–22] introduced thin-walled finite elements to analyze spatial cold-formed structures with open cross sections. Tusnina [23–25] presented a finite element analysis of cold-formed Z-purlins supported by sandwich panels. Nackenhorst [26] investigated the problem of initial geometric imperfections on buckling behavior of CFS C-shaped compressed columns. Prokic [27–29] analyzed the behavior of thin-walled open section I-beams under torsion and bending. Björk [30–32] studied the influence of residual stresses on the fatigue strength of CFS rectangular hollow sections. Ungermann [33, 34]

* Corresponding author: 273marcel@gmail.com

dealt with experimental and numerical investigations on plain channels subjected to coupled instabilities. Post-buckling behavior of cold-formed steel columns was observed by Schafer [35].

To increase the thermal effectiveness of exterior LSF walls in framed residential buildings CFS profiles are usually perforated on webs and thus are called thermoprofiles (Fig. 1). Traditionally, thermoprofiles with longitudinal perforations are mostly used in construction. In this article a new type of perforated thermoprofile is considered – reticular-stretched thermoprofile (hereinafter RST-profile). The key features of this profile are completely different (diamond-shaped) geometry of perforations and the presence of the intermediate longitudinal stiffener.

Due to the fact that RST-profile has been designed recently, nowadays very little research has been devoted to the investigation of its mechanical properties. Sinelnikov [36–38] and Nazmeeva [39, 40] investigated the buckling behavior of compressed cold-formed columns made of RST-profiles. Trubina [41–44] analyzed the problem of local and global buckling of RST-profiles in bending.

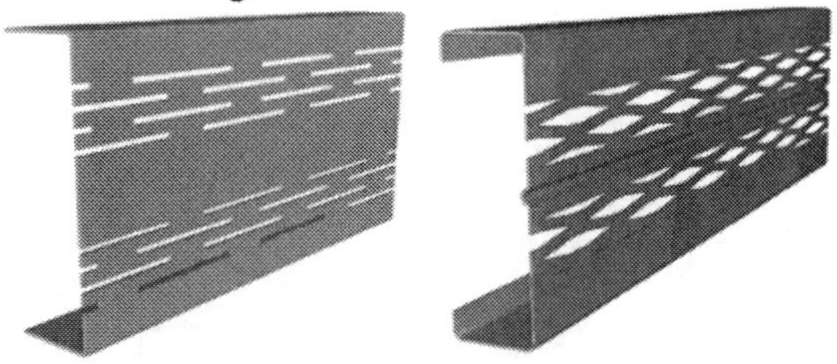

Fig. 1. Thermoprofiles: traditional (left) and reticular-stretched (right).

Despite the growing interest to RST-profiles, the influence of perforations on the resistance of profiles remains an open issue. Buckling behavior of these sections is also investigated very poorly.

This article deals with buckling performance of cold-formed C-sections made of RST-profile. Firstly, the experimental results of the compressed columns are provided. The second part contains the nonlinear FE analysis of these columns.

2 Experimental investigations

Laboratory tests were carried out for the two types of specimens:
- RST-profiled columns without nodal connections at the ends, Sn (n –sample number);
- RST-profiled columns as part of the nodal connection, Cn (n – sample number).

Experimental study of the profile was conducted in the laboratory of HAMK University, InnoSteel Sheet Metal Development Centre, Finland. The test specimens were produced by OOO Arsenal ST, Russia.

The steel used for the production of the profile had the following characteristics:
- Design strength: 350 N/mm^2;
- Zinc coating: 350 g/m^2;
- Zinc coating thickness: 25 microns.

The characteristics of the samples are shown in Table 1.

Table 1. Characteristics of the samples tested.

Type	S1, S2, S3	C1, C2, C3
Profile	TCc 200-45-2.0	TCc 200-45-2.0
Length, mm	1000	350
Boundary conditions	Top – pinned, bottom – pinned	Top – pinned, bottom – fixed

The ends of Sn specimens (Fig. 2a) were cut using a circular metal saw (not machined). Support wooden blocks (thickness 40 mm; edge is positioned 3 mm from the end of the profile) were placed inside the ends of the columns. The lower ends of the specimen were placed on a hinged steel support. The load of a hydraulic cylinder was applied through a thick steel plate to the upper ends of the specimens. All the specimens were loaded using the displacement control until the failure. The loading rate was 3 mm/min.

Every Cn specimen (Fig. 2b) was constructed from a vertical column (hereinafter – stud) and a horizontal element (hereinafter – rack). The 350 mm long studs were made of the same profile as Sn specimens. The racks represented U-shaped cold-formed profile without any perforations on webs. The flanges of every stud were fixed with 4+4 self-drilling screws on the flanges of the racks. The studs were fixed to the head of the hydraulic cylinder. The racks were fixed rigidly to the test frame. The specimens were loaded using the displacement control until the failure. The loading rate was 1.5 mm/min.

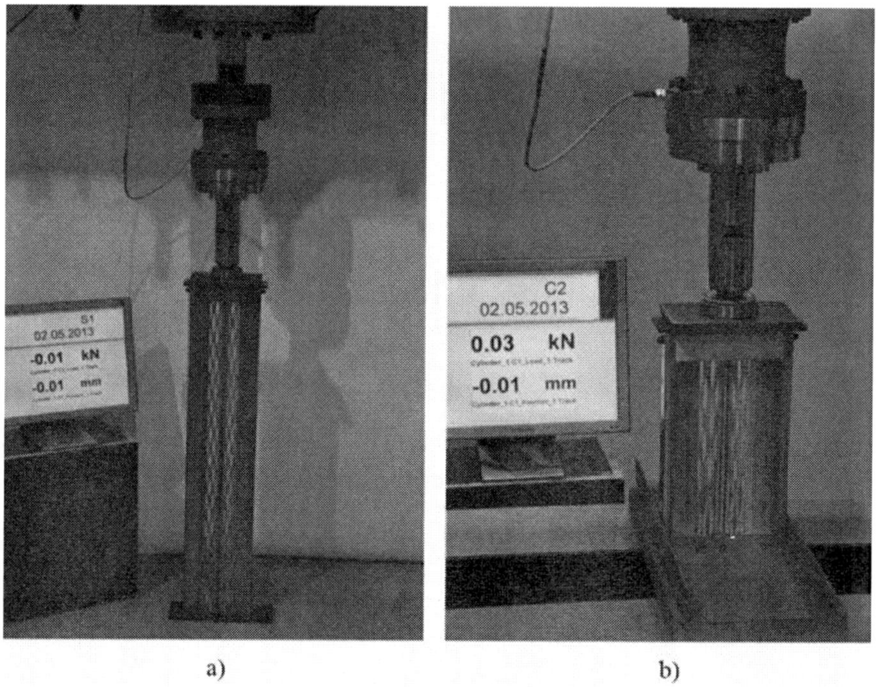

Fig. 2. Overall view of test setup: (a) S1; (b) C2.

During the experiments every tested specimen experienced local buckling with subsequent global (flexural) buckling. A summary of the buckling test results is given in Table 2.

Table 2. Characteristics of the samples tested.

Specimen	S1	S2	S3	mean	C1	C2	C3	mean
Length, mm	1000	1000	1000	1000	350	350	350	350
Force, kN	84.4	68.9	83.4	78.9	90.8	91	99.9	93.9

The observed failure mechanisms of the Sn and Cn columns are shown in Figs. 3 and 5 respectively. Figs. 4 and 6 illustrate the load-displacement curves for the tested columns.

Fig. 3. Observed failure mechanisms for tests on S2: (a) local buckling; (b) global (flexural) buckling.

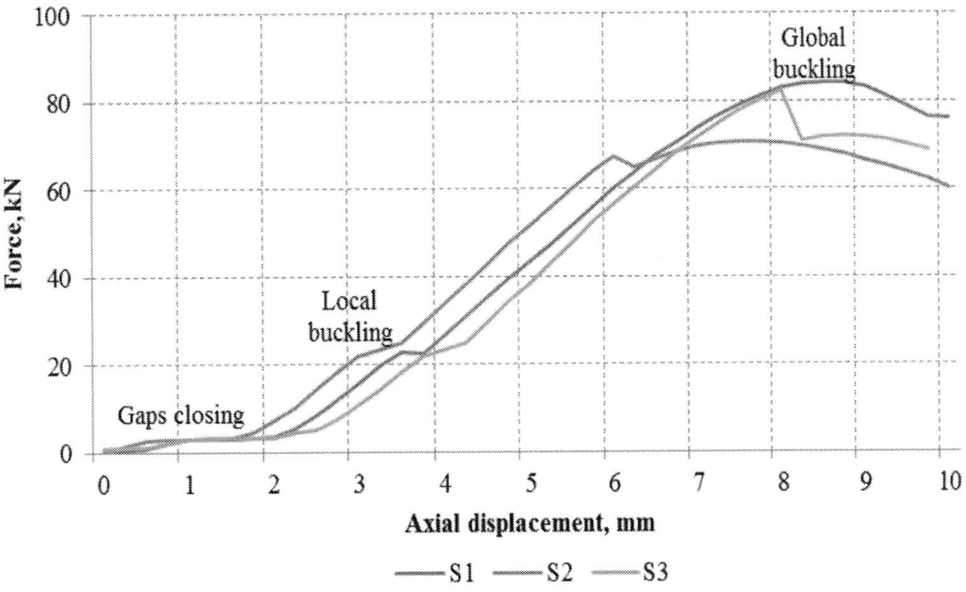

Fig. 4. Force-displacement response for Sn specimens.

a) b)

Fig. 5. Observed failure mechanisms for tests on C1: (a) local buckling; (b) global (flexural) buckling.

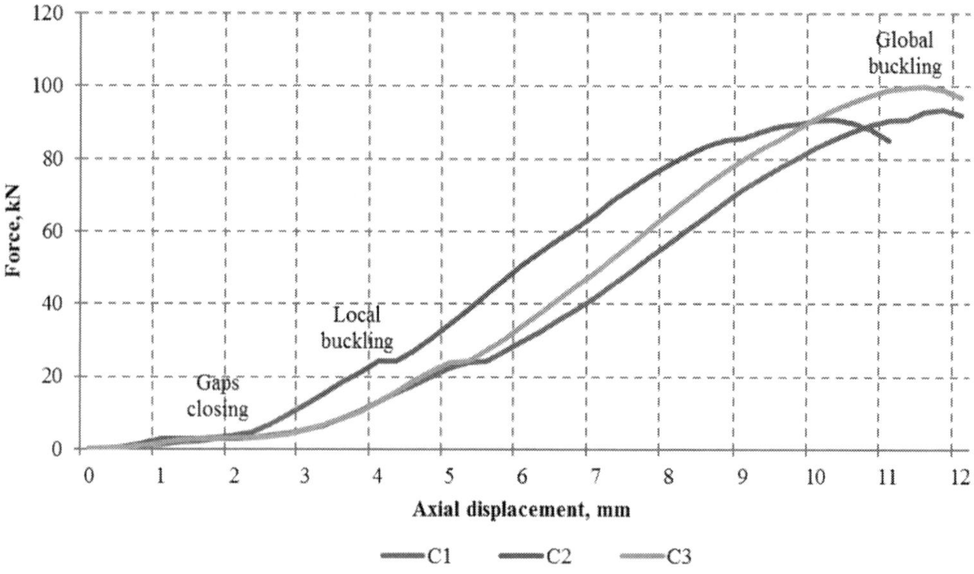

Fig. 6. Force-displacement response for Cn specimens.

Since the fact that the ends of all specimens were cut manually using a circular metal saw, on both load-displacement curves a short plateau shortly after the beginning of loading process was observed. On that step the gaps arisen as the result of cutting were closed making the force applied to the whole section of the profile. After the gaps were closed the curves behaved linearly with a very short plateau marking local buckling accompanied by a typical cotton. Then both curves continued growing gradually up to global buckling. One can see that for S1 specimen after the global buckling started the curve declined smoothly whereas for S2 and S3 specimens the noticeable drops were observed probably arisen as a results of initial geometric imperfections. At the same time, short specimens (Cn) did not have such drops.

3 Finite element analysis

To validate the experimental data FE buckling analysis in Abaqus was conducted. The simulation was arranged in such a way to repeat the tests: 1000 mm and 350 mm columns made of RST-profile were loaded by compressive loads. Geometrical properties of the RST-profile used for tests and modelling are presented in Fig. 7.

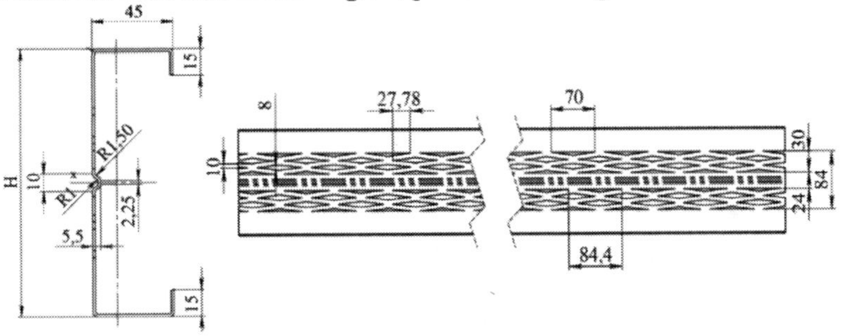

Fig. 7. Geometrical dimensions of RST-profile.

The mesh of the models was created using 4-node, quadrilateral, shell elements (S4R). The boundary conditions are shown in Fig. 8. All the models were loaded with concentrated compressive forces acting in Z direction. To simulate the wooden plates the rigid bodies at both ends of each specimen were created.

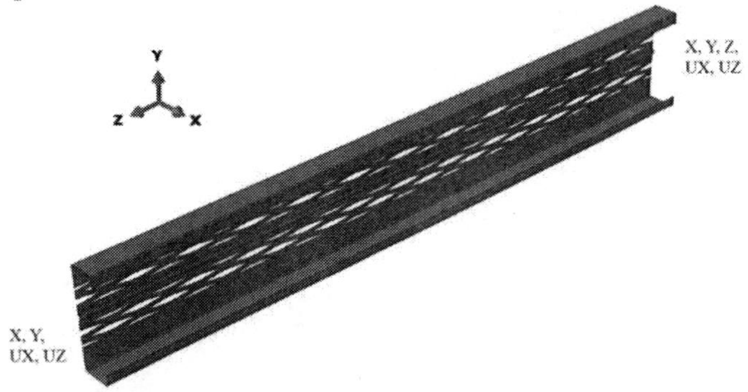

Fig. 8. Boundary conditions of the members tested.

For each column a buckling analysis with physical and geometrical nonlinearity ("Static, Riks") was performed. The results of the FE analysis are presented in Table 3. The table includes the buckling forces determined experimentally, by modelling and analytically by EN 1993-1-3-2009. Fig. 9 illustrates the deformed shapes and the load-displacement curves for the specimen tested.

Table 3. The results of FE analysis.

Specimen	Length [mm]	Force [kN]		
		Experiments	FEM	EN 1993-1-3-2009
S1	1000	84.4	84.8	49.7
S2	1000	68.9		
S3	1000	83.4		
mean		78.9		
C1	350	90.8	92.1	55.9
C2	350	91.0		
C3	350	99.9		
mean		93.9		

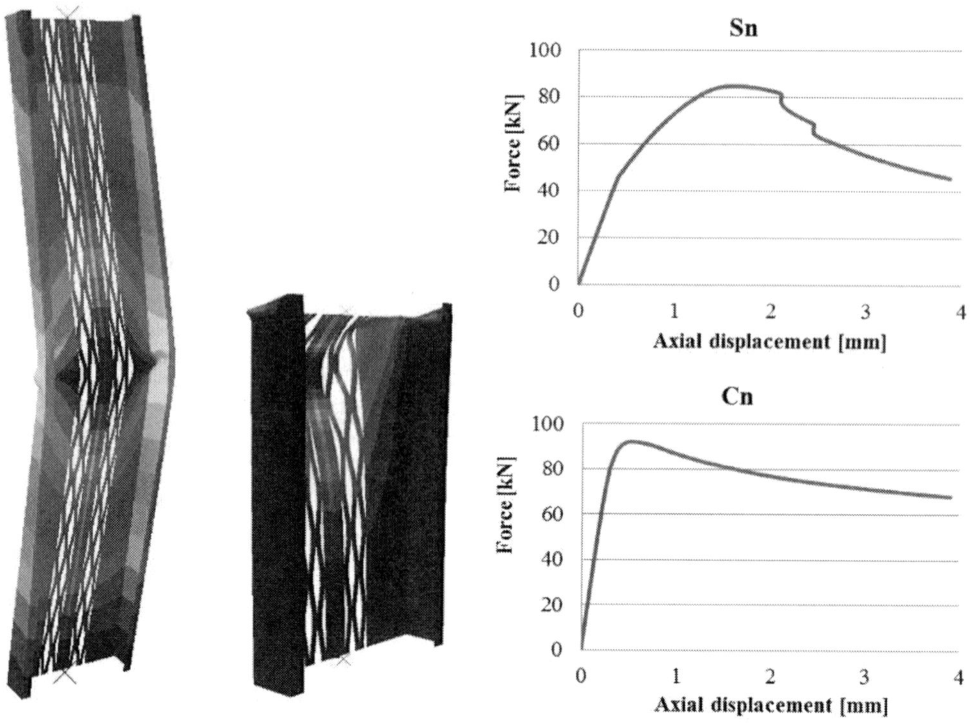

Sn Cn load-displacement curves

Fig. 9. Deformed shapes and load-displacement curves of simulation tests.

It can be seen that in case of modelling the load-displacement curves had the similar behavior as for experimental data but without closing gaps plateaus. Local buckling segment for Sn was reduced down to a certain point making the curve bilinear while for Cn it was not observed at all. For both cases global buckling was determined at points close to those for experimental tests.

FE analysis was also conducted for a solid-web profile (without any perforations on web) and a traditional thermoprofile to compare their bearing capacity to RST-profile. The results are presented in Table 4. For convenience, buckling forces of all profiles are divided to a buckling force of a solid-web profile (F_{SW}).

Table 4. Comparison of RST-profile with solid-web CFS profile and traditional thermoprofile.

Specimen	Length [mm]	F / F_{SW}		
		Profile with solid web	Conventional thermoprofile	RST-profile
Sn	1000	1	0.69	0.96
Cn	350	1	0.92	0.97

It can be seen that due to the presence on intermediate longitudinal stiffener RST-profile has bearing capacity considerably higher than conventional thermoprofile with longitudinal perforations (particularly for long elements) and almost as high as solid-web CFS profile.

4 Conclusions

Nonlinear FE models were developed and verified against previously conducted compression tests on cold-formed steel C-section columns of RST-profile. Due to rational geometry RST-profile has very high bearing capacity under compression, having relatively similar mechanical properties as CFS profile without thermal openings.

Computer simulations showed good agreement with the experimental data. Discrepancy between the results was less than 10%. Despite the close results, the values of buckling forces for the tests and simulations corresponded to the different values of displacements. This might be the result of closing the gaps, which, in their turn, occurred due to the initial geometric imperfections of the columns.

Simulation tests for short columns (350 mm) showed very close results, however, the deformed shape obtained from modelling was completely opposite to the one from the experimental tests. This demonstrates the need for a more detailed modelling with the particular attention paid to nodal connections and wooden plates in the top of the models.

Analytical calculations made in accordance with EN 1993-1-3-2009 showed considerably lower results. This probably means that the building codes provide more reliable values of buckling forces for this type of profile.

Anyway, RST-profile remains poorly explored and requires more careful and detailed research. It also seems reasonable to provide computer modelling taking into account initial geometrical imperfections of the elements, since the fact that CFS structures with such perforations are very sensitive to any mechanical deformations.

The experimental work was conducted with the help of Dr. Arto Ranta-Eskola, director of research, Rautaruukki Oyj (Finland). The authors also gratefully acknowledge the helpful comments of Prof. Milan Veljkovic (Luleå University of Technology).

References

1. M. Veljkovic, B. Johansson, J. Constr. Steel Res., **64,** 7–8 (2008) 816–821
2. M. Heinisuo, J. Lahdenmaa, T. Jokinen, *Proceedings of the 12th International Conference on Steel, Space and Composite Structures*, (2014) 85–97
3. M. Pavlović, C. Heistermann, M. Veljković, D. Pak, M. Feldmann, C. Rebelo, L.S. da Silva, J. Constr. Steel Res., **115,** (2015) 458–466.
4. M. Pavlović, C. Heistermann, M. Veljković, D. Pak, M. Feldmann, C. Rebelo, L.S. da Silva, J. Constr. Steel Res., **115,** (2015) 445–457
5. V.K. Vostrov, A.A. Vasilkin, Montaz. i Spetsial'nye Rab. v Stroit., **11,** (2005) 37–39
6. G. Kashevarova, P. Kosykh, Appl. Mech. Mater., **670–671,** (2014) 466–473
7. S. Kesawan, M. Mahendran, Thin-Walled Struct., **98,** (2016) 111–126
8. S. Kesawan, M. Mahendran, J. Constr. Steel Res., **115,** (2015) 191–205
9. V. Ungureanu, M. KoteŁko, R.J. Mania, D. Dubina, Thin-Walled Struct., **48**, 10–11 (2010) 818–826
10. C.D. Moen, B.W. Schafer, Eng. Struct., **31(12)**, (2009) 2812–2824
11. M. Garifullin, N. Vatin, Constr. Unique Build. Struct., **6,** (2014) 32–57
12. V. Lalin, V. Rybakov, A. Sergey, Appl. Mech. Mater., **578–579,** (2014) 858–863
13. V. Rybakov, A. Sergey, Appl. Mech. Mater., **725–726,** (2015) 746–751
14. G. Belyj, E. Serov, World Appl. Sci. J., **23(13)**, (2013) 160–164
15. M. Al Ali, Adv. Mater. Res., **969,** (2014) 93–96
16. M. Al Ali, M. Tomko, I. Demjan, V. Kvočák, Procedia Eng., **40,** (2012) 8–13
17. M. Al Ali, M. Tomko, I. Demjan, V. Kvocak, Komunikacie, **14(4),** (2012) 83–87
18. A.R. Tusnin, Montaz. i Spetsial'nye Rab. v Stroit., **2,** (2003) 4–7

19. A.R. Tusnin, Montaz. i Spetsial'nye Rab. v Stroit., **4,** (2003) 12–14
20. A.R. Tusnin, Montaz. i Spetsial'nye Rab. v Stroit., **9,** (2003) 2–4
21. A.R. Tusnin, Promyshlennoe i Grazhdanskoe Stroit., **6,** (2003) 59–6.
22. A.R. Tusnin, Montaz. i Spetsial'nye Rab. v Stroit., **3,** (2003) 2–4
23. O. Tusnina, Appl. Mech. Mater., **467,** (2014) 398–403
24. O. Tusnina, Appl. Mech. Mater., **475–476,** (2014) 1483–1486
25. A. Danilov, O. Tusnina, J. Appl. Eng. Sci., **12(2),** (2014) 153–158
26. M. Garifullin, U. Nackenhorst, Procedia Eng., **117,** (2015) 1078–1084
27. A.R. Tusnin, M. Prokic, J. Appl. Eng. Sci., **12(3)**, (2014) 179–186
28. A.R. Tusnin, M. Prokic, Procedia Eng., **111,** (2015) 789–796
29. A.R. Tusnin, M. Prokic, Mag. Civ. Eng., **5(49)**, (2014) 44–53
30. T. Björk, H. Saastamoinen, *Tubular Structures XIV - Proceedings of the 14th International Symposium on Tubular Structures, ISTS 2012* (2012) 167–176
31. S. Heinilä, T. Björk, G. Marquis, *ASTM Spec. Tech. Publ. 1508 STP* (2009) 200–215
32. T. Nykänen, T. Björk, H. Mettänen, A. V. Ilyin, M. Koskimäki, Fatigue Fract. Eng. Mater. Struct., **37**(3), (2014) 325–334
33. D. Ungermann, S. Lübke, B. Brune, Thin-Walled Struct., **81,** (2014) 108–120
34. D. Ungermann, B. Brune, S. Lübke, Steel Constr., **5(4)**, (2012) 205–211
35. Y. Shifferaw, B.W. Schafer, Thin-Walled Struct., **80,** (2014) 142–152
36. N. Vatin, J. Havula, L. Martikainen, A. Sinelnikov, A. Orlova, S. Salamakhin, Tests and FEM-Modelling, Adv. Mater. Res., **945–949,** (2014) 1211–1215
37. N. Vatin, J. Havula, L. Martikainen, A. Sinelnikov, L.L. Shurovkina, Appl. Mech. Mater., **725–726,** (2015) 722–727
38. N. Vatin, J. Havula, L. Martikainen, A. Sinelnikov, A. Orlova, S. Salamakhin, *Proceedings of the International Conference „Innovative Materials, Structures and Technologies"* (2014) 187–192
39. N. Vatin, T. Nazmeeva, R. Guslinscky, Adv. Mater. Res., **941–944,** (2014) 1871–1875
40. T.V. Nazmeeva, Mag. Civ. Eng., **5,** (2013) 44–51
41. D. Trubina, D. Abdulaev, E. Pichugin, M. Garifullin, Appl. Mech. Mater., **633–634,** (2014) 1052–1057
42. N. Vatin, A. Sinelnikov, M. Garifullin, D. Trubina, Appl. Mech. Mater., **633–634,** (2014) 1037–1041
43. D. Trubina, D. Abdulaev, E. Pichugin, V. Rybakov, M. Garifullin, O. Sokolova, Appl. Mech. Mater., **725–726,** (2015) 752–757
44. M. Garifullin, D. Trubina, N. Vatin, Appl. Mech. Mater., **725–726,** (2015) 697–70

MATEC Web of Conferences **73**, 04012 (2016) DOI: 10.1051/matecconf/20167304012
TPACEE-2016

Multi-objective Truss Optimization Using Different Types of the BB-BC Algorithm

Aleksandar Milajić [1,*] and *Dejan* Beljaković [1]

[1]University Union Nikola Tesla, 11000 Cara Dušana 62-64, Belgrade, Serbia

Abstract. Optimum design of truss structures is considered as a benchmark problem in the field of the structural optimization. In order to solve this hard combinatorial problem, it is necessary to implement adequate optimization tool that would provide sufficiently wide range of possible solutions within a reasonable time as well as to obtain good exploration and exploitation of search space. The aim of presented study was to compare efficiency of different multi-objective algorithms in solving this task.

1 Introduction

Size optimization of truss structures involves determining optimum values for member cross-sectional areas that would minimize the weight of a given truss structure while satisfying a number of inequality constraints that limit design variable sizes and structural responses. However, minimum weight is often not the only goal, because limited deflections and deformations are often more important demand in order to achieve usability demands in exploitation during the structure's life-cycle. These cases belong to the field of multi-objective optimization.

The main task of every optimization tool is to perform effective and efficient exploration of a given search space. This search for an optimal solution should be versatile enough to both intensively explore promising areas of the search space around high quality solutions, and to reach its unexplored areas. Two basic concepts for reaching these goals are usually called intensification and diversification. These terms originate from the tabu search field, but same or related concepts can also be found in other methods, such as evolutionary algorithms, denoted as exploitation (related to intensification) and exploration (related to diversification). The main difference between intensification and diversification is that intensification feature of a given search tool focuses on examining neighbors of elite solutions, while the diversification encourages the search process to examine unvisited regions and to generate solutions that differ in various significant ways from those seen before [1]. A meta-heuristic will be successful on a given optimization problem if it can provide a good balance between the exploitation of the accumulated search experience and the exploration of the search space to identify regions with high quality solutions in a problem specific, near optimal way [2].

* Corresponding author: aleksandar.milajic@gmail.com

In the last decades, different nature inspired evolutionary algorithms have been developed and employed for structural optimization. These algorithms do not require for a given function to be derivable and an explicit relationship between the objective function and constraints is not needed. Big Bang – Big Crunch (BB-BC) algorithm, introduced by Erol and Eksin [3], is relatively new optimization method that relies on one of the theories of the evolution of the universe namely, the Big Bang and Big Crunch theory and has a low computational time and high convergence speed. The aim of presented research was to explore applicability of different variants of basic and hybrid multi-objective BB-BC in order to find an efficient and effective optimization tool for optimum design of a given truss structure that would be able to deal with numerous variables and to provide good solutions. Obtained results confirmed that the hybrid versions of the BB-BC algorithm outperform the basic one in exploration and exploitation of a search space for remarkably less time, as well as in producing much better Pareto fronts of possible solutions. Analysis of the results shows that hybridizing basic BB-BC with other meta-heuristic or some of their features can further improve its performance in solving hard combinatorial problems such as optimum structural design of truss structures.

2 Problem formulation

Mathematically, the optimal design of a truss can be formulated as finding set of variables $[A_1, A_2, \ldots, A_n]$, $A_i \in D$, where A_i is the cross-sectional area of a member i, n is the number of members in a givent russ structure, and D denotes the allowable set of values for the design variable A_i, in order to minimize both the nodal displacements (1) and the total weight of structure (2):

$$\delta_j = \delta_{j,\min}, j = 1,2,\ldots,n_n \tag{1}$$

$$\mathrm{W(A)} = \sum_{i=1}^{n} \gamma_i A_i L_i \tag{2}$$

where n_n is nuber of nodes; $W(A)$ is weight of the structure; n is the number of members of the structure; γ_i represents the material density of member i and L_i is the length of member i, subject to constraints:

$$g_j(A) \leq 0, j = 1,2,\ldots,n \tag{3}$$

Since the presented problem has two objective functions, an appropriate solving method is the multi-objective tool that would be able to locate multiple Pareto optimal solutions in a single run. A solution is said to be Pareto optimal if and only if it is not dominated by any other solution in the performance space. If one solution dominates another, it implies that the first one is non-inferior to the second one for all the considered performance criteria but it is better than it for at least one criterion. All Pareto solutions form a Pareto front in the performance space.

Analysis of the whole Pareto front provides useful information on trade-off relationship between the fitness functions and enables a decision maker to consider different alternatives and make a choice that would represent acceptable compromise for conflicting objectives. In hard combinatorial problems such as this one it is impossible to conduct thorough search for Pareto solutions within the whole search space without appropriate optimization tool.

3 Optimization Method

Problem defined by the fitness functions (1) and (2) and set of constrains (3) is a multi-objective optimization problem with conflicting criteria. Consequently, there is no one optimal solution but a whole set of acceptable solutions called Pareto solutions or Pareto front. A solution is Pareto optimal if and only if it is not dominated by any other solution in the search space. Analysis of the whole Pareto front provides useful information on trade-off relationship between the fitness functions and enables a decision maker to consider different alternatives and make a choice that would represent acceptable compromise for conflicting objectives. In hard combinatorial problems such as this one it is impossible to conduct thorough search for Pareto solutions within the whole search space without appropriate optimization tool.

The BB-BC algorithm is relatively new evolution algorithm introduced in 2006 by Erol and Eksin [3]. The BB–BC method is inspired by a theory of the evolution of the universe, namely the Big Bang and Big Crunch theory. Every step of the algorithm consists of two phases: a Big Bang phase, and a Big Crunch phase. Similar to other evolutionary algorithms, initial solutions in the first Big Bang are created randomly and spread all over the search space in a uniform manner. The Big Crunch is a convergence operator that has many inputs but only one output, named as the *centre of mass*, where the term *mass* refers to the inverse of the fitness function value. The point representing the center of mass (X_C) and can be calculated according to:

$$X_C = \frac{\sum_{i=1}^{N} \frac{1}{f^i} X_i}{\sum_{i=1}^{N} \frac{1}{f^i}} \tag{4}$$

where X_i is a point within an n-dimensional search space, f_i is a fitness function value of this point, and N is the population size. The other possibility is to select the fittest individual as the center of mass [4]. After calculating the center of mass, the algorithm generates new candidate solutions for the next Big Bang phase using a normal distribution around the previous center of mass according to:

$$X_i^{k+1} = X_C^k + r_j \alpha \frac{X_{\max} - X_{\min}}{k+1}, i = 1,\dots,N \tag{5}$$

where k is number of iteration, r_j is a random number from standard normal distribution which changes for every candidate, α is a parameter for limiting the size of search space and X_{max} and X_{min} are vectors consisting of maximal and minimal acceptable values for all variables. After the new population is generated, algorithm moves onto the next Big Crunch phase by calculating new center of mass. This sequence of explosion and contraction is repeatedly carried out until a stopping criterion has been met, whether it is reaching maximum number of iterations or obtaining a convergence.

Although the BB-BC method has been shown to outperform classical Genetic Algorithm for numerous benchmark functions [3] and obtains good results in the fine search around a local optimum (exploitation), it has some drawbacks in exploration (covering the whole search space). In order to overcome these drawbacks, we have implemented hybridization of BB-BC and some characteristics of Particle Swarm Optimization (PSO) proposed in [5] because this approach has been proven to be successful in solving different design problems [5, 6]. PSO is optimization algorithm inspired by birds flocking or fish schooling in search for food [7, 8], where every individual (particle) adjusts its movements according to both its own experience and the population experience. This feature has been used for modifying the BB-BC into hybrid BB-BC (HBB-BC) by using not only the center of mass from the previous generation (X_c^k) but also the best position of a given individual solution up to iteration k ($X^{lbest(k)}$) and the best position in the whole population found up to the iteration k ($X^{gbest(k)}$) [5]:

$$X_i^{k+1} = \beta X_C^k + (1-\beta)\left[\gamma X^{gbest(k)} + (1-\gamma)X^{lbest(k)}\right] + r_j\alpha\frac{X_{\max} - X_{\min}}{k+1}, i = 1,\ldots,N \tag{6}$$

where β and γ are adjustable coefficients for controlling the influence of the global and local best results $X^{gbest(k)}$ and $X^{lbest(k)}$, respectively.

In order to examine applicability of different BB-BC algorithms for optimum design of truss structures, we have analysed four variants of the algorithm. The first two algorithms, namely BB-BC1 and BB-BC2, are two variants of the classic BB-BC algorithm, where the latter one do not use center of mass calculated using Equation (4) but takes the fittest individual from the last iteration as the new center of mass. This was done in order to examine how this change would affect the calculation time and quality of solutions. The other two variants, namely HBB-BC1 and HBB-BC2, are hybrid algorithms where the Big Bang phase is conducted using Equation (6) instead of Equation (5). In the former one, the center of mass is calculated using Equation (4), while in the latter one the center of mass is the fittest individual from the previous iteration.

4 Comparison Criteria

Basically, every multi-objective optimizer aims at three goals: a) to find out the true Pareto front or to converge as close as possible to it; b) to cover as wide as possible span of solutions, and c) to discover solutions as diverse as possible along the obtained Pareto front. In order to provide a quantitative performance assessment for different multi-objective optimizing algorithms, it is necessary to establish exact criteria for measuring and comparing their effectiveness. A variety of such metrics and methods have been proposed in literature [9–11]. Method used in this study is based on a set of three metrics, namely: Coverage, Spacing and Maximum Spread [11, 12].

Coverage (C-metric) provides comparison between two Pareto fronts. If A and B be are two approximations to the Pareto front, then $C(A, B)$ is the percentage of the solutions in B that are dominated by at least one solution in A:

$$C_{(A,B)} = \frac{\left|\{u \in B \mid \exists v \in A : v \ dom \ u\}\right|}{|B|} \tag{7}$$

Spacing (S-metric) indicates how evenly the solutions are distributed along the discovered Pareto-front:

$$S = \left[\frac{1}{n_{pf}}\sum_{i=1}^{n_{pf}}(d_i - \bar{d})^2\right]^{\frac{1}{2}}, \bar{d} = \frac{1}{n_{pf}}\sum_{i=1}^{n_{pf}} d_i \tag{8}$$

where n_{pf} is the number of members in Pareto front and d_i is the Euclidean distance (in the objective space) between the member i in Pareto front and its nearest member. A smaller value of S implies a more uniform distribution of solutions in Pareto front.

Maximum Spread (MS-metric) measures how "well" obtained Pareto front covers the true Pareto front:

$$MS = \left[\frac{1}{m} \sum_{i=1}^{m} \left[\frac{\min(f_i^{\max}, F_i^{\max}) - \max(f_i^{\min} - F_i^{\min})}{F_i^{\max} - F_i^{\min}} \right]^2 \right]^{\frac{1}{2}} \tag{9}$$

where m is the number of objectives, $f_{i\max}$ and $f_{i\min}$ are the maximum and minimum of the ith objective in obtained Pareto front, respectively, and $F_{i\max}$ and $F_{i\min}$ are the maximum and minimum of the ith objective in true Pareto front, respectively. A larger value of MS inicates a better spread of solutions. Since the true Pareto front in this study is not known, $F_{i\max}$ and $F_{i\min}$ are considered as the maximum and minimum of the ith objective in all obtained Pareto fronts by various algorithms.

5 Numerical example

Proposed algorithms were tested on a well-known example of 10-bars plane truss shown in Figure 1. The objectives are to minimize the weight of the structure and the displacement in y-direction at node 5 simultaneously using the cross-sectional areas of the 10 truss numbers as design variables. The cross-sectional area of every member is selected from a set of 32 discrete values predefined as {1.05, 1.16, 1.54, 1.69, 1.86, 1.99, 2.02, 2.18, 2.34, 2.48, 2.50, 2.70, 2.90, 3.10, 3.21, 3.30, 3.70, 4.66, 5.14, 7.42, 8.71, 8.97, 9.16, 10.00, 10.32, 12.13, 12.84, 14.19, 14.77, 17.10, 19.35, and 21.61}×10−3 m^2. External load is $P = 444.5$ kN, and material properties are modulus of elasticity E = 68.9 GPa and weight density ρ = 2712.6 kg/m3. Stress σ_j in each element has to be below the maximum allowable stress of σ_a = ±172 MPa.

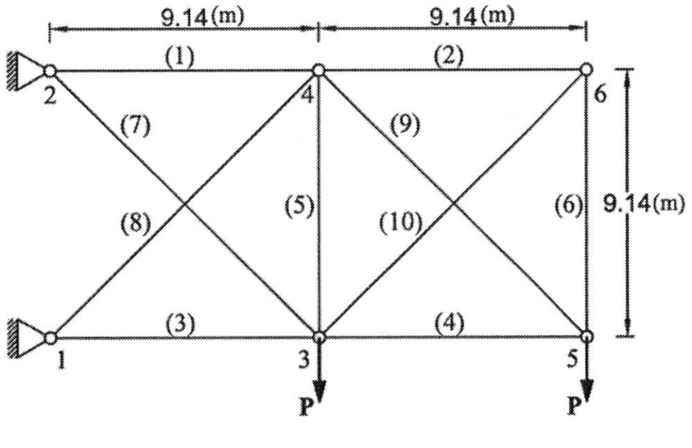

Fig.1. Geometry and loads for the 10-bars truss example.

6 Results and Discussion

In order to evaluate performance of described algorithms, they were run five times. Extreme solutions, required computational times for solving the problem example and number of solutions in each Pareto front are presented in Table 1 and the performance metrics values are presented in Table 2. Comparison of results for BB-BBC1 and BB-BC2, and for HBB-BC1 and HBB-BC2, respectively, indicates that using the fittest individual as the new center of mass instead of the one calculated using Equation (5) had no significant impact on the final solutions except the little improvement considering the calculation time.

On the other hand, comparison of results obtained by the algorithms BB-BC1 and HB-BBC1, and BB-BC2 and HB-BBC2, respectively, indicates that hybrid versions of the algorithm obtained better extreme results for the both fitness functions.

Table 1. Best results after five runs.

Algorithm	Best extreme solution 1 (kg, mm)	Best extreme solution 2 (kg, mm)
BB-BC1	(832.677; 135.98)	(5655.621, 29.91)
BB-BC2	(831.078; 140.85)	(5658.870, 29.89)
HBB-BC1	(825.251; 160.78)	(5698.841, 29.65)
HBB-BC2	(825.847; 159.24)	(5675.142, 29.68)

Differences in quality of obtained Pareto fronts can be further discussed according to the results of three considered performance metrics given in Table 4. Comparison of results obtained by BB-BC1 and BB-BC2, and HBB-BC1 and HBB-BC2, respectively, indicates that using the fittest individual as the new centre of mass had no significant impact on quality of results. However, hybridized versions of HB-BC outperformed the basic versions concerning all three metrics, so it can be concluded that hybridization is a good way of improving the performance of classic BB-BC algorithm.

Comparison of results obtained by BB-BC1 and BB-BC2, and HBB-BC1 and HBB-BC2, respectively, indicates that using the fittest individual as the new center of mass had no significant impact on quality of results. However, hybridized versions of HB-BC outperformed the basic versions concerning all three metrics, so it can be concluded that hybridization is a good way of improving the performance of classic BB-BC algorithm.

Table 2. Performance metrics values.

C-metric					
Algorithm	**Run 1**	**Run 2**	**Run 3**	**Run 4**	**Run 5**
BB-BC1 vs. BB-BC2	0.23	0.24	0.17	0.22	0.27
BB-BC2 vs. BB-BC1	0.19	0.26	0.12	0.20	0.20
BB-BC1 vs. HBB-BC1	0.17	0.23	0.11	0.16	0.18
HBB-BC1 vs. BB-BC1	0.27	0.27	0.38	0.32	0.36
BB-BC1 vs. HBB-BC2	0.15	0.20	0.23	0.16	0.20
HBB-BC2 vs. BB-BC1	0.30	0.38	0.29	0.36	0.34
BB-BC2 vs. HBB-BC1	0.21	0.20	0.17	0.14	0.15
HBB-BC1 vs. BB-BC2	0.25	0.30	0.32	0.36	0.31
BB-BC2 vs. HBB-BC2	0.16	0.21	0.18	0.14	0.16
HBB-BC2 vs. BB-BC2	0.36	0.32	0.38	0.34	0.33
HBB-BC1 vs. HBB-BC2	0.21	0.25	0.22	0.26	0.24
HBB-BC2 vs. HBB-BC1	0.27	0.26	0.23	0.22	0.22
S-metric					
Algorithm	**Run 1**	**Run 2**	**Run 3**	**Run 4**	**Run 5**
BB-BC1	28.042	29.982	28.674	27.054	29.002
BB-BC2	27.423	28.360	26.931	27.794	28.097
HBB-BC1	16.533	16.811	17.911	18.292	19.024
HBB-BC2	15.580	17.012	16.345	17.676	15.468
MS-metric					
Algorithm	**Run 1**	**Run 2**	**Run 3**	**Run 4**	**Run 5**
BB-BC1	0.921	0.959	0.977	0.945	0.953
BB-BC2	0.9544	0.9638	0.970	0.9687	0.935
HBB-BC1	0.995	0.989	0.9848	0.991	0.990
HBB-BC2	1.000	0.998	0.995	0.989	1.0000

7 Conclusion

Appropriate choice of optimization techniques and tools is extremely important for overcoming insufficiencies and inadequacies of conventional trial-and-error approach in optimum structural design, especially for complex problems such as exploring numerous different possibilities of optimum truss design. In this paper a heuristic population-based search inspired by the Big Bang and Big Crunch theory (BB–BC) of the evolution of the universe is implemented for solving the classical truss optimization problem. Comparison of obtained numerical results for the benchmark problem indicates that proposed method is efficient and sufficiently reliable for solving complex problems in the field of optimum structural design and that hiridization of the multi-objective Big Bang-Big Crunch algorithm with Particle Swarm Optimizer outperforms basic versions of the algorithm both in calculation time and the quality of solutions.

References

1. F. Glover, M. Laguna, *Tabu Search* (Kulwer Academic Publishers, 1997)
2. T. Stutzle, *Local Search Algorithms for Combinatorial Problems - Analysis, Algorithms and New Applications* (DISKI, Sankt Augustin, 1999)
3. O. K. Erol, I. Eksin, Advances in Engineering Software, **37**, 106–111 (2006)
4. H. M. Genc, I. Eksin, O. K. Erol, Turkish Journal of Electrical Engineering & Computer Sciences, **21**, 1359–1375 (2013)
5. A. Kaveh, S. Talatahari, Computers and Structures, **87(17–18)**, 1129–1140 (2009)
6. M. Sedighizadeh, M. Esmaili, Energy, **76**, 920–930 (2014)
7. J. Kennedy, R. Eberhart, Y. Shi, *Swarm intelligence* (Morgan Kaufmann Publishers, 2001)
8. J. Aghaei, K. M. Muttaqi, A. Azizivahed, M. Gitizadeh, Energy, **65**, 398–411 (2014)
9. K. Deb, *Multi-Objective Optimization using Evolutionary Algorithms* (John Wiley & Sons, Chichester, 2001)
10. K. C. Tan, T. H. Lee, E. F. Khor, Artificial Intelligence Review, **17(4)**, 253–290 (2002)
11. E. Zitzler, L. Thiele, M. Laumanns, C. M. Fonseca, V. G. Fonseca, IEEE Transactions on Evolutionary Computation, **7(2)**, 117–132 (2003)
12. A. Kaveh, K. Laknejadi, Acta Mechanica, **224**, 343–364 (2013)

MATEC Web of Conferences 73, 04013 (2016) DOI: 10.1051/matecconf/20167304013
TPACEE-2016

Examination of Quality and Operational Properties of Vibropressed Paving Elements

Kseniia Usanova[1], *Vladimir* Rybakov[1,*], *Viktoriia* Udalova[1], and *Anton* Kovylkov[1]

[1]St. Petersburg State Polytechnical University, 195251, Politekhnicheskaya, 29, Saint-Petersburg, Russia
[2]St. Petersburg State University of Trade and Economics, 194021, 50, Novorossiyskaya st. , St. Petersburg, Russian Federation

Abstract. This article presents research findings on examination of quality of paving elements. Were investigated and analyzed performance properties of paving slabs on following: accuracy of geometric dimensions, appearance, compressive strength, frost resistance, water absorption and abrasion. Based on these results recommendations improving the current regulatory and technical documentation are provided.

1 Introduction

An artificial paving stone is a concrete element with section height (thickness) of not less than 60 mm, bearing surface area of not more than 0.05 sq. m and maximum length of 28 cm and 30 cm, for rectangular and shaped artificial stone, respectively. Paving stones are produced with the thickness of 60, 70, 80 or 100mm. Requirements for artificial paving stones are indicated in the following documents: GOST 17608-91 " Concrete Sidewalk Plates. Specifications ", TS 5746-001-33157194-97" Artificial Stones of Road Topping, "TS 5746-018-03984296-2004" Concrete Sidewalk Products "and TS 5746-020-03984296-2005 "Concrete Sidewalk Plates with Textured Surface".

Concrete paving slabs are made of heavy and fine-grained (sandy) concrete, and are used for prefabricated pavements, pedestrian and park paths, pedestrian areas and public transport loading sites.

Therefore high requirements are imposed to the quality of the finished goods.of the artificial stones producers.

According to ISO 9000:2005 (Quality Management Systems — Fundamentals and Vocabulary), quality is a "degree to which a set of inherent characteristics fulfils requirements". There are various classes of characteristic, such as the following:

— physical (e.g. mechanical, electrical, chemical or biological characteristics);
— sensory (e.g. related to smell, touch, taste, sight, hearing);
— behavioral (e.g. courtesy, honesty, veracity);
— temporal (e.g. punctuality, reliability, availability);
— ergonomic (e.g. physiological characteristic, or related to human safety);
— functional.

* Corresponding author: fishermanoff@mail.ru

As a rule the physical, the sensory, the temporal and the functional ones are exposed to examination among the listed characteristics above during quality assessment.

In [4] scientist have developed a mathematical model and calculation method, which takes into account the connection between the paving stones in surface coating, their shapes and sizes. The influence of shapes, sizes, artificial stone paving and joints in the coating on the strength of the pavement are experimentally and theoretically investigated.

Developed practical recommendations on the calculation, design, construction and evaluation of the strength of road pavements coated with artificial stone paving. A significant influence of friction forces wedges between the stones has been revealed. A general formula for determining the stiffness of joints has been suggested.

Also A.V. Kochetkov, M.L. Ermakov, A.A. Shestopalov, E.I. Denikin, J.G. Barabanshikov, S.G. Nikolskiy, S. Belyaeva, A. Korsun, V. Korsun, V.D Staroverov [8-21] study the quality assessment elements paving and other building materials.

The aim of this work is to improve the methods and the regulatory guideline of quality control road surface paving elements.

For this control was carried out examination of the quality of paving slabs, consisting of two stages: quality control of products and an experimental analysis of the properties of finished products.

Experimental studies were carried out on the basis of research laboratory of the Department "Examination of Consumer Goods" VPO "SPbGTEU"

2 Monitoring and Examination of Quality of Paving Slabs

To carry out random check of paving slabs were taken 3 lots of paving slabs produced by Experimental Plant JSC:

• Lot No. 1- paving slabs of 1-P-8 "Brick" gray color; the geometrical sizes are 206x102x80 mm (Fig. 1 (a)) ,

• Lot No. 2 - 1-F-8 "Volna", gray colour. геометрические размеры 225x112x80 mm (Fig. 1 (b)),

•Lot No. 3- paving slabs 5-F-6 "Klassika-Rusto", gray colour геометрические размеры 119x119x60 mm (Fig. 1 (c)).

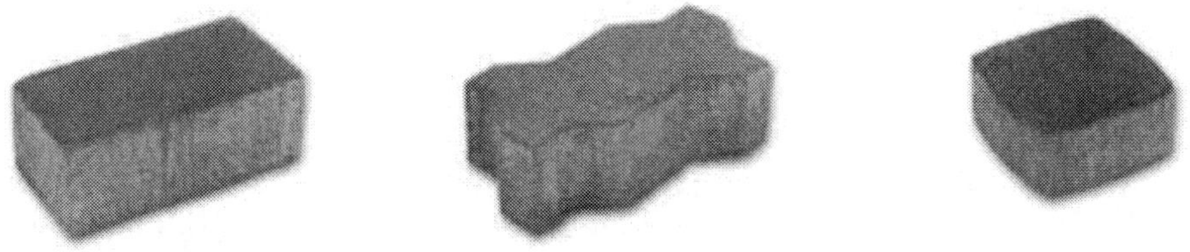

(a) 1-P-8 «Brick» (b) 1-F-8 «Volna» (c) 5-F-6 « Klassika-Rusto »

Fig. 1. The models of research

All the researched lots have the declared concrete class for compressive strength B35 and for frost resistance F200

Quality control was performed in accordance with GOST 17608-91 and TS 5746-003-23078401-08.

The size of paving slab lot was 80 units. Sampling was conducted in accordance with the requirements of GOST 17608-91. Five paving slab samples shall be taken from a lot of up to 100 units.

During the quality control inspection we checked the following: if there are transport documents, unhurt package, correct marking; if the products comply with the application form; if there are any defects in appearance; as well as the category of front concrete surface, dimensions, upper layer thickness, straightness of front surface profile, front surface flatness, perpendicularity of front and adjacent faces.

While inspecting transport documentation we checked the following: the certificate of compliance with GOST 17608-91 and TS 5746-003-23078401-08, installation instructions, warranty conditions, a shipping list.

During package checking we inspected package and its condition. Slabs are usually put on reusable pallets. Pallet with the products are wrapped with stretch film to prevent spillage of products during transportation, and fixed with metal, polypropylene, or any other tape, ensuring the safety of stones.

During monitoring marking accuracy we checked its compliance with the requirements of GOST 13015 - 2003 Reinforced Concrete and Concrete Construction Products. General Technical Requirements. Rules of Acceptance,

Marking, Transportation and Storage. Markings and labels are specified on the label attached to the pallet with the products. Marking label shall contain the following marks:

- A trademark of the manufacturer or its short title;
- Product designation;
- Indication of standard;
- A Stamp of QC department;
- Date of product manufacture;
- A number of products on a pallet.

Marking method shall ensure its safety before packing products; it shall be indicated in the technical documentation. In accordance with these requirements the following data in the marking of paving slabs shall be verified: marking shall be affixed indelibly on the end face of at least 10% of plates from a lot. Plates are marked in accordance with GOST 23009 Designs and Products, Concrete and Reinforced Concrete Prefabricated. Sign Designation (Marks).

Marks of square, rectangular, hexagonal, and bordering plates consist of alphanumeric groups which mean the following: the first digit - serial number size; a letter – plate type; a number after the letter - slab thickness in centimeters; which depends on foundation type. Marks of curly plates consist of alphanumeric groups separated by dots, indicating: the first digit - serial number of this configuration; the letter F - type plate; a number after the letter F - serial number plate; the last digit - slab thickness in centimeters.

During appearance control of paving slabs we visually checked if there are any of the following defects: cracks, chips, shells, sagging, uniformity and color intensity.

The category of front concrete surface is A6, the category of non-front surface - A7 according to GOST 13015 -2003 Reinforced Concrete and Concrete Construction Products. General Technical Requirements. Rules of Acceptance, Marking, Transportation and Storage.

Inspection of dimensions for compliance with GOST 17608-91 was performed using measuring tools, in particular calipers.

The thickness of the upper layer of paving slabs was tested with the help of calipers for compliance with the requirements of TU 5746-003-23078401-08

Deviations in straightness of front surface profile was checked with plumbing metal squares.

Deviations in front surface plane and in perpendicularity of end and adjacent faces were also checked with plumbing metal squares.

3 The Results of Quality Control Sampling of Paving Slabs

During quality control sampling we reveled the following:

1. Each lot has a compliance certificate, instructions for paving slab installation, warranty conditions, transport list; which meets the requirements;
2. Package of each paving slab lot is not damaged;
3. All samples of paving slabs are marked in accordance with GOST 17608-91, the product brand is also indicated.
4. While evaluating appearance of paving slab samples of the brands 1-P-8, 1-F-8 and 5-F-6 it was revealed that the samples of all three lots have no defects in appearance; which meets the requirements;
5. The dimensions of paving slab samples of 5-F-6 brand have no deviations in length, width and thickness.
6. The dimensions of paving slab samples of 1-F-8 brand have length deviation of +1mm in one sample; width deviation of + 2mm and + 1mm; thickness deviation of + 1mm and -1mm.
7. Upper layer thickness of all samples in three lots complies with documentation requirements equaling to more than 10 mm.
8. There is no deviation in straightness of front surface profile in all samples of three lot.
9. Deviation in the plane of the front surface of one sample of 1-P-8 brand is 1 mm, the same deviation occurs in one sample of 5-F-6 brand. There are no deviations in samples from 1-F-8 brand.
10. There is no deviation in perpendicularity of end and adjacent faces in all the samples from three lots.
11. Samples from all three lots comply with a standard of reference.
12. The category of the front surface of samples from all three lots complies with A6.
13. All samples have no surface cracks on stones from all three lots.

Based on the abovementioned facts, we can conclude that the paving slabs produced at Experimental Plant JSC of the brands 1-P-8, 1-F-8 and 5-F-6 have only minor deviations in analyzed parameters and fully comply with all the requirements of GOST 17608-91.

4 The Examination of Operational Properties of Paving Slabs. Examination Subjects and Methods

For examination of paving slabs produced at Experimental Plant JSC, fourteen brands of paving slabs were taken: 1-P-8, 1-F-8, 1-F-6, 2-F-8, 3-F-8, 1-P-6, 4-P-6, 7-P-8, 9-F-8, 5-F-8, 6-F-6, 7-F-6, 8-F-6, 2-P-6. These brands are the most common and produced in the largest amount in Experimental Plant JSC.

During tests, up to 0.1% of the total amount of paving slabs is subject to inspection. In order to do this, the cubes with an edge length of 70 mm are cut. Deviation in straightness of bearing surfaces must not exceed 0.1 mm. It is necessary to test compressive strength, flexural strength, abrasion resistance, frost resistance, water absorption, water permeability.

Determination of concrete strength consists of measuring the minimal effort, destroying specially prepared test concrete samples during static loading at a constant rate of load growth and the subsequent calculation of stresses in these efforts, assuming elastic behavior of the material.

The methods and test procedures are described in GOST 10180-90 Concretes. Methods of Determining Strength by Control Samples, GOST 13087-81Concrete. Methods of Determination of Attrition, GOST 10060.0-95 Concrete. Methods of Frost Resistance Determination.

During testing of compressive strength the control samples are put on a support plate of a hydraulic press and gradually loaded with the upper baseplate until fracture. A study of three test samples is presented.

During tensile flexural test the samples are put on two supports. The force is transmitted by a round steel bar of 10 mm in radius, installed in the middle of a slab parallel to the supports.

In order to determine the friability the control samples or parts of slabs obtained after their bending test are used. Tests are carried out using a special abrasion abrasive wheel with a rotational speed of 30 rev / min. It makes 110 turnovers. After this it is necessary to calculate amount (mass) loss of the sample with the abraded surface of 50 cm2.

Determination of frost resistance of concrete is performed on three control samples with the help of the second reference method. Frost resistance is the ability of concrete to maintain the physical and mechanical properties after repeated freezing and thawing. All samples were previously saturated with 5% aqueous sodium chloride solution and undergo 30 cycles of freezing (at -18 ° C) and thawing (at +18 ° C); there shall be no cracks and flaking external surfaces.

Water absorption is determined by the difference in product weight. A product shall be completely dried within several hours at a high temperature. The dry product weight is measured; then it is soaked for several hours. Then it is necessary to calculate the weight difference between wet and dry products. After than one shall calculate the percentage of water in the product.

5 The Results of Examination of Operational Properties of Paving Slabs

Analysis of the experimental data for paving slabs (1-P-8, 1-F-8, 1-F-6, 2-F-8, 3-F-8, 1-P-6, 4-P-6, 7-P-8, 9-F-8, 5-F-8, 6-F-6, 7-F-6, 8-F-6, 2-P-6 brands.) revealed that:

1. The strength limit of concrete for compressive strength varies from an average of 50.01 MPa to 51.9 MPa. All samples were analyzed by compressive strength exceed the requirements of GOST.

2. The value of handling concrete strength varies from 49.5 MPa to 50.8 MPa, 1-F-8 brand has the best results. For this indicator, paving slabs, manufactured at Experimental Plant also JSC exceeds the requirements of GOST.

3. The strength limit of concrete for tensile strength in bending all kinds of paving slabs is 6.8 MPa, which is higher than the requirements of GOST.

Thus after analyzing combined indicators of concrete strength for paving slabs produced at Experimental Plant JSC we can conclude that these products are of high quality in terms of strength and in all cases the deviation from GOST requirements is positive.

- Frost resistance analysis of experimental data for fourteen types of paving slabs shows the exact compliance with the requirements of GOST 10060.0-95 and equals to F200.
- As for water absorption all brands of paving slabs show results that are better than requirements of GOST 12730.3-78. Concrete. Methods of Water Absorption Determination.
- The average value of abrasion performance for all samples of analyzed paving slabs is 0.39 g/cm2. In terms of abrasion the paving slabs of Experimental Plant JSC have the best performance characteristics.
- As for service life of paving slabs produced at Experimental Plant JSC , the average repair of road pavements, pedestrian areas and areas of artificial roads paving stones manufactured at Experimental Plant JSC, which are not designed for the constant traffic of vehicles, shall be implemented after 9 years, major repairs after 24 years. This conclusion is based on analysis of repair periods of cement concrete and small sized pavements.

In order to confirm these repair terms we can mention the fact that some sections of the road surface of the pedestrian zone in Malaya Sadovaya Street (date of commissioning is 1998) made of artificial paving slabs produced at Experimental Plant JSC were repairs in 2007.

Thus, analyzing all the performance characteristics and service life of paving elements produced in at Experimental Plant JSC, we can conclude that the products meet the requirements of GOST 17608-91 and allow to create a strong road surface which maintains its continuity and smoothness under the influence of vehicle traffic and weather and climate factors. Evenness of pavement is a qualitative condition of the coating characterized by the absence of defects: cracks, crevices, ledges, shifts, waves, high joints

6 The Comparative Examination of Adhesion Properties of Paving Slabs with the Different Surface Textures

We conducted a comparative examination of adhesion properties of paving slabs of the type "brick", the brand 1-P-8 manufactured at Experimental Plant JSC with a smooth and relief textured front surface.

The front surface of paving slabs can be smooth or textured (relief). In the first option the face layer is made of only a mixture of sand and cement, and in the second - with inclusion of a filler. Due to adding various solid natural aggregates to the textured front layer of paving stones they have better mechanical strength and decorative properties if compared with stones with the smooth upper surface. There are requirements to face category for the quality of the smooth front surface that corresponds to the limiting values of the diameter or size of the shells, the height of the local influx (projection) or the depth of the depression, a depth of about ribs. The quality of textured surface shall comply with a standard of reference (in the form of a surface of a product fragment) which is approved by the company (GOST 13015-2003 Reinforced Concrete and Concrete Construction Products). The front surface of the stone paving provides a comfortable pedestrian traffic on the road surface, that is, has some traction with the soles of shoes pedestrians. In autumn and winter the front surface of the stone under the influence of moisture and rainfall becomes more slippery and traumatic.

7 The Results of the Comparative Examination

Five samples of the paving slabs of the brand 1-P-8 with textured and smooth surfaces were selected for the comparative examination of adhesion properties of paving slabs

To estimate the coupling properties of artificial stone paving can serve as friction coefficient.

The friction coefficient characterizes the ability of coverage to resist to sliding of car tires and represents the relation of maximum reactive force resisting to longitudinal displacement of the stopped wheel to vertical force (coupling weight) acting through the wheel on the coverage and applying perpendicular to its surface

The friction coefficient of front surface paving stones with shoes of pedestrians was determined with the device DM-T intending for measuring of the friction coefficient of wheel tire with the material of road surface.

Thus the road surface material is the part of the measuring device. Therefore the device allows to measure the friction coefficient in static conditions

The results of comparative examination of paving slabs of the type "brick" of the brand 1-P-8 with different surface textures, produced at Experimental Plant JSC are represented in Table 1.

Table 1. The Results of Determination of Adhesion Coefficient

№	Product Brand	Indicators to be Measured	Requirements STD	Actual values				
				Sample 1	Sample 2	Sample 3	Sample 4	Sam ple 5
1	1-P-8 with a smooth surface	Adhesion coefficient	no	0.51	0.48	0.49	0.49	0.50
2	1-P-8 with a textured surface	Adhesion coefficient	no	0.56	0.57	0.57	0.57	0.56

Analysis of the experimental data leads to the conclusion that the adhesion coefficient of the slabs with textured surface on the average by 0.07 higher. Thus, we can conclude that the performance characteristics of paving slabs with textured surface are the best in the autumn-winter season when it is necessary to provide the most adhere to the road with a shoe pedestrian.

Existing regulations do not reflect the requirements of the operational characteristics, in spite of its importance for the consumer. Therefore, it seems appropriate, the inclusion of the friction coefficient as a measure of quality regulations.

8 Summary

The paving slabs of Experimental Plant JSC are made with the equipment of the last generation of the German company MASA with the method of moist vibropressing. The high level of product quality is ensured by high-tech equipment, high-quality raw materials and professional staff.

On the basis of conducted research and data quality examination of paving slabs we can make the following conclusion:

- The paving slabs of all fourteen brands produced at Experimental Plant JSC for basic operational significantly exceeds the requirements of GOST 17608-91, resulting in a reliable, durable and practical material for durable, aesthetic, easily repaired pavement.
- Comparative examination of the “Brick” type paving slabs of the brand 1-P-8 with textured and smooth surfaces show that the paving stones have the highest performance with relief (textured) surface. It is preferably to use them for pedestrian areas with heavy traffic on the road sections and areas with a significant slope.
- It seems reasonable to include the adhesion coefficient as a quality indicator to regulation documents.

References

1. S.A. Anokhin, V.D. Kuznetsov, Stroitel'nye materialy, **12**, 16-18 (2000)
2. Y.S. Burov, *Technology of building materials and products* (Graduate School, Moscow, 1972)
3. G.A. Vasilyev, L.V. Osipov, *Improvement of marketing services industries: Overview* (TSNIITEIMS, Moscow, 2002)
4. J.B. Kostikov, *Vliyanie parametrov pokrytiya iz iskusstvennykh kamney moshcheniya na prochnost' dorozhnoy odezhdy, Dis. cand. tehn. sciences* (St.-Petersburg, 2005)
5. K.I. Lvovich, Populyarnoe betonovedenie, **1(15)**, 180-181 (2007)

6. A.M. Simanovskiy, *Scientific and Technical Report on "Development of recommendations regarding the construction of the pavement with a coating of artificial stone paving"* (St.-Petersburg, 2006)
7. M.V. Romanenko, *Sovremennye napravleniya tekhnologii stroitel'nogo proizvodstva: Stat'i i tezisy dokladov* (S-PbVITU, St. Petersburg, 2004)
8. O.N. Pertseva, S.G. Nikolskiy, Magazine of Civil Engineering, **3(47)**, 71-76 (2014)
9. S.G. Nikolskiy, M.M. Kazderko, Applied Mechanics and Materials, **633-634**, 1114-1116 (2014)
10. O.N. Pertseva, S.G. Nikolskiy, Advanced Materials Research, **941-944**, 1537-1543 (2014)
11. J.G. Barabanshikov, Mekhannika kompozitsionnykh materialov i konstruktsiyy, **10**, 211-223 (2004)
12. J.G. Barabanshikov, V.A. Gatitullin, A. Mhanna, Construction of Unique Buildings and Structures, **7(22)**, 166-175 (2014)
13. E.A. Epstein, V.A. Rybakov, Magazine of Civil Engineering, **4(4)**, 32-37 (2009)
14. J.G. Barabanshikov, S.V. Belyaeva, S.G. Nikolskiy, *Materials Science and Technology of construction materials. Assessment of the quality of building materials* (Publishing House of STU, St. Petersburg, 2008)
15. V. Yu. Ogvozdin, *Management of quality. The basics of theory and practice* (Moscow, 2009)
16. C. Gehlen, F. Duran, D. Lowke, Betonwerk und Fertigteil-Technik, **80(2)**, 40-41 (2014)
17. C. Thiel, T. Stengel, C. Gehlen, *Life cycle assessment (LCA) of road pavement materials Eco-Efficient Construction and Building Materials: Life Cycle Assessment (LCA), Eco-Labelling and Case Studies,* 368-403 (2013)
18. V. Korsun, N. Vatin, A. Korsun, D. Nemova, Applied Mechanics and Materials, **633-634**, 1013-1017 (2014)
19. V. Korsun, N. Vatin, A. Korsun, D. Nemova, Applied Mechanics and Materials, **723**, 445-450 (2015)
20. M.A. Akhmatov, S.R. Likhov, Bulletin VolSABU, **20(39)**, 70-75 (2010)
21. V.D. Staroverov, Tekhnologii betonov, **4(81)**, 32-35 (2013)

MATEC Web of Conferences 73, 04014 (2016) DOI: 10.1051/matecconf/20167304014
TPACEE-2016

Influence of Kinetics Concrete Hardening on Strength of Constructions

Mikhail Berlinov[1,*], and *Marina* Berlinova[1]

[1]Moscow State University of Civil Engineering, 129337 Yaroslavskoe sh. 26, Moscow, Russia

Annotation. In this article the question of the influence of the kinetics of hardening concrete on its strength in building structures through the stress function. Determined a direct correlation between the optimal structure of concrete and a guarantee of high strength and durability of building structures that ensured a coincidence given the amounts and extreme properties. A method for predicting the elastic properties of concrete based on known theory of strength of concrete model. It is shown by way of example, that in certain specific strength characteristics of concrete, properties of its components (cement stone and aggregates), the maximum value of the short-term strain, creep deformation and modulus of elasticity of concrete depends on its strength at the time of loading. The proposed method is the most convenient for practical approximation to obtain the equations of mechanical state to determine the total deformation of concrete.In this article the approach to determine the strength of concrete in building structures through the stress function under load. It is shown by way of example, that the results obtained by experiment and the proposed article approach to the assessment of the strength to coincide with high accuracy.

1 Introduction

Deformation properties of concrete depends on its structural characteristics, resulting dosage and quality components, but the intensity of deformations in concrete constructions can not be reliably predicted depends only on the strength of the composite, while retaining all other conditions equal to [4, 6]. In practice, this implies that the findings on the relationship between deformation and concrete strength to do is useless, if not provided in advance some of the boundary conditions and additional factors characterizing features of the concrete structure, which vary (or remain constant) simultaneously with the change in strength. This fact is proved by the authors [1, 2] under different operating conditions and loadings [3, 5].

Under the long-term strength (durability) understand the stress that causes the destruction of the material in a given time. Obtaining high quality concrete in advanced technology conglomerate materials today becomes a major factor both in science and in practice, the production of artificial materials and products. It is known that the favorable construction and performance of the concrete consist of: the largest - the strength and

* Corresponding author: berlinov2010@mail.ru

elasticity; least - creep, porosity and defects in the structure. According to the law "alignment" I.A.Rybev, which reflects a high quality performance products, provided that the specific values of the properties are at specified values, you must use the following principle: a comparison of materials in the research and practical developments are not as often made - equal to or under the same conditions (which can result in large errors) but only when the respective conditions, i.e. under optimal - similar to each other, structures.

2 Method

This assumption establishes a direct connection between the properties and the optimal structure parameters. In other words, the guarantee of high (if not higher) quality products to accepted technology of its production is given a coincidence and extreme properties of the amounts, which is possible only when the optimal structure.

In particular, to improve the reliability of the valuation of deformation characteristics of concrete, along with the strength necessary for the long-term deformation (creep and shrinkage) to take into account the specific (by volume) amount of water in the mixing of concrete, which is closely correlated with the initial moisture content of the concrete; for short-term deformation (initial elastic modulus and ultimate strains) to take into account the specific (by weight) of cement paste content in the concrete mix, which determines the ratio of cement paste and aggregate in the hardened concrete.

Prediction of the elastic properties of concrete can be made using methods known in the theory of strength of concrete models:

linear -

$$\frac{1}{E_b(\tau)} = b_0 + b_1 P_z + b_2 \frac{P_z}{R_b(\tau)} + b_3 \frac{1-P_z}{E_a} \tag{1}$$

and nonlinear –

$$\frac{1}{E_b(\tau)} = b_0 + b_1 P_z + b_2 \frac{P_z^{\varphi}}{[R_b(\tau)]^{\psi}} + b_3 \frac{1-P_z}{E_a} \tag{2}$$

Where $R_b(\tau)$- the strength of concrete loaded at a time τ , $E_b(\tau)$ - initial modulus of elasticity of concrete loaded on the age τ ; P_z - Specific (by weight) content of cement paste in compacted concrete mix; E_a - modulus of elasticity of the filler; b_0, b_1, b_2 and b_3 - deformability parameters taken from experimental data.

3 Results

Summarizing, we can write:

$$\frac{1}{E_b(\tau)} = a + \frac{b}{R_b(\tau)^m} = \frac{1}{R_b(\tau)^m}\left[aR_b(\tau)^m + b\right] \tag{3}$$

where $a = a(P_z, E_a); b = b(P_z)$; $m = 1$ and or 0.75.

Maximum short-term strain on the concrete « $\sigma - \varepsilon$ » curve:

$$\varepsilon_R = \frac{R_b(\tau)}{E_b(\tau)} + 274\times10^{-8} P_z \frac{0.3+1.8\bar{e}_0}{0.3+\bar{e}_0} \frac{E_b(\tau)}{R_b(\tau)} \quad (4)$$

where $\bar{e}_0 = e/h$ - relative eccentricity of load application.

Limit specific concrete creep deformation, constant load loaded at a time determined by the formulas:

$$C(\infty,\tau) = k_c \frac{W}{R_b(\tau)} \quad (5)$$

or

$$C(\infty,\tau) = k_c \frac{W}{R_m + \Delta} \quad (6)$$

here R_m - the strength characteristics of cement paste at the start of loading, W - the quantitative value of the aqueous component in the concrete mix, k_c - individual constant coefficient of concrete.

When the strength of concrete in the $\tau = 28 days.$ concrete creep age defined by the formula (5), where $k_c W = C(\infty,28) R_b(28)$.

Assume the notation

$$a_n = C(\infty,28) R_b(28) \quad (7)$$

and substituting (7) the expression (6), we obtain $C(\infty,\tau) = \frac{a_n}{R_b(\tau)}$, and taking into account the duration of the stress full expression to creep measures will be:

$$C(t,\tau) = \frac{1}{R_b(\tau)} a_n f(t-\tau) \quad (8)$$

Equation (6) differs from (8) by the presence of the parameter $\Delta \neq 0$, and is used instead of the matrix R_m strength concrete strength in the record $R_b(\tau)$ that is most convenient for practice approximations.

If we assume that the initial relative level of stress

$$\eta(\tau) = \sigma(\tau) / R_b(\tau) \quad (9)$$

and specific to them creep

$$\bar{S}_r(\tau) = \eta(\tau)\left[1 + V_r \eta(\tau)^m\right] \quad (10)$$

it can be written for each particular strain stress function:

$$S_r(\tau) = R_b(\tau)\bar{S}_r(\tau) \tag{11}$$

If $T(\tau) = 20^o C$, $W(\tau) = 50\%$ and $S(\tau) = const$, taking into account the record (3) and $m = 1$ obtain the record:

$$\varepsilon^{эт}(t,t_0) = \bar{S}_k[aR_b(t_0) + b] + \bar{S}_c(t_0)a_n f(t - t_0) \tag{12}$$

or

$$\varepsilon^{эт}(t,t_0) = \frac{R_b(t_0)}{R_b(t)}\bar{S}_k(t_0)[aR_b(t_0) + b] + \\ + \bar{S}_c(t_0)\left[a_n f(t - t_0) + b\left(1 - \frac{R_b(t_0)}{R_b(t)}\right)\right] \tag{13}$$

$$d\varepsilon^{эт}(t,\tau) = \delta_k(\tau)\frac{\partial}{\partial\tau}\bar{S}_k(\tau)d\tau + a_n f(t - \tau)\frac{\partial}{\partial\tau}\bar{S}_c(\tau)d\tau \tag{14}$$

where,

$$\delta_k(\tau) = aR_b(\tau) + b \tag{15}$$

a and b take the same meanings as in the formula (3), as a_n calculated from (7).

If the overall record of the equation of the mechanical condition of the concrete substitute (12 and 14), we obtain a record of private strain

$$\varepsilon(t,t_0) = \varepsilon^{эт}(t,t_0) + \int_{t_0}^{t} d\varepsilon^{эт}(t,\tau) \tag{16}$$

spending is not complicated iteration write

$$\varepsilon(t,t_0) = \delta_k(t)\bar{S}_k(t) + a_n f(0)\bar{S}_c(t) - \int_{t_o}^{t}\bar{S}_k(\tau)\frac{\partial}{\partial\tau}\delta_k(\tau)d\tau - \\ - \int_{t_0}^{\tau} a_n\bar{S}_c(\tau)\frac{\partial}{\partial\tau}f(t - \tau)d\tau. \tag{17}$$

We use the expression (13) for recording the total relative deformations

$$d\varepsilon^{эт}(t,\tau) = \frac{1}{E_b(\tau)}\frac{\partial}{\partial\tau}\sigma_k(\tau)d\tau + C^*(t,\tau)\frac{\partial}{\partial\tau}\sigma_c(\tau)d\tau \tag{18}$$

Further, by rewriting the expression (18) together with (13) we obtain

$$d\varepsilon^{\text{эт}}(t,\tau)=\delta_k(t,\tau)\frac{\partial}{\partial\tau}\bar{S}_k(\tau)d\tau+\delta_c(t,\tau)\frac{\partial}{\partial\tau}\bar{S}_c(\tau)d\tau, \tag{19}$$

where

$$\delta_k(t,\tau)=R_b(\tau)\left[a+{}^{b}\!/\!{}_{R_b(t)}\right] \tag{20}$$

and

$$\delta_k(t,\tau)=a_n f(t-\tau)+b\left[1-{}^{R_b(\tau)}\!/\!{}_{R_b(t)}\right] \tag{21}$$

Here,

$$\delta_k(t,t)=\delta_k(t)=aR_b(t)+b \tag{22}$$

and

$$\delta_k(t,t)=a_n f(0)=const. \tag{23}$$

After further changes, substituting (13 and 19) to (16), we obtain the equation for determining the mechanical condition of the total strain

$$\varepsilon(t,t_0)=\delta_k(t)\bar{S}_k(t)+a_n f(0)\bar{S}_c(t)-\int_{t_o}^{t}\bar{S}_k(\tau)\frac{\partial}{\partial\tau}\delta_k(t,\tau)d\tau-$$
$$-\int_{t_0}^{t}\bar{S}_c(\tau)\frac{\partial}{\partial\tau}\delta_c(t,\tau)d\tau. \tag{24}$$

From the foregoing, it is evident that under certain specific strength characteristics of concrete, properties of its components (cement stone and aggregates), the maximum value of the short-term strain, creep deformation and modulus of elasticity of concrete depends on its strength at the time of loading.

It is known that concrete creep deformation are separate relative to the initial level of stress in the structure $S(t,\tau)=\dfrac{\varepsilon_c(t,\tau)}{\beta(\tau)}$, where $\beta(\tau)=\dfrac{\sigma(\tau)}{R_b(\tau)}$, where t - time of the beginning of loading, τ - the current time of observation.

Creep practically do not depend on the age of the concrete and are the property of invariance with respect to the loading.

The relationship between stress $\sigma(\tau)$ and specific creep, ascribed to stresses $S(t,\tau)$, is given.

$$S(t,\tau)=C(t,\tau)R_b(\tau) \tag{25}$$

If creep measures in the form of Mack Henry and S.V.Aleksandrovskiy can accept an expression
$C(t,\tau)=C(\infty,28)\Omega(\tau)f(t-\tau)$, (25) can be rewritten as

$$S(t,\tau)=C(\infty,28)\Omega(\tau)R_b(\tau)f(t-\tau) \quad (26)$$

Concrete creep deformation $S(t,\tau)$ depends on the duration of loading and $(t-\tau)$ is not dependent on age concrete τ . In this case, the condition must be satisfied

$$\Omega(\tau)R_b(\tau)=k \quad (27)$$

where k - is a constant of the material.

This expression is determined by a function of aging concrete in building structures on the creep properties

$$\Omega(\tau)=\frac{k}{R_b(\tau)} \quad (28)$$

or prism strength of concrete.

$$R_b(\tau)=\frac{k}{\Omega(\tau)} \quad (29)$$

Then take concrete function of aging at the age of 28 days: $\Omega(\tau)=\Omega(28)=1$,
further in accordance with the condition (27) at obtain $k=R_b(28)$

$$R_b(\tau)=R_b(28)\Omega(\tau)^{-1} \quad (30)$$

Concrete strength depends on its age: $R_b(t)=0.7R_b(28)\lg t$,
aging function $\Omega(t)=0.5+0.7e^{-2\gamma t}$ for $\gamma=0.006$.

When loading at a time $t=100days$, depending on the concrete age, account for the strength of the prism will have the form $R_b(100)=0.7R_b(28)\lg 100=1.4R_b(28)$:
and a function of aging $\Omega(100)=0.5+0.7e^{-2\times0,006\times100}=0.71$.

According to the formula (30)

$$R_b(100)=R_b(28)\Omega(100)^{-1}=R_b(28)\frac{1}{0,71}=1.4R_b(28).$$

For concrete example, the age t = 1000days:

$$R_b(1000)=0.7R_b(28)\lg 1000=2.1R_b(28),$$

$$\Omega(1000)=0.5+0.7e^{-2\times0,006\times1000}\approx 0.5$$

and formula (30) $R_b(1000)=R_b(28)\frac{1}{0,5}=2R_b(28)$.

As seen from the above examples, the results obtained by experiment and by the same formula (30) with high accuracy.

Continuing iterations, multiply the right and left sides in (30) a setting deformation nonlinearity $\frac{1+V_k}{E_b(\tau)}$, we obtain $\frac{R_b(\tau)(1+V_k)}{E_b(\tau)} = \frac{R_b(28)(1+V_k)}{E_b(\tau)}\Omega(\tau)^{-1}$.

Limit the relative shortening of the instantaneous static sense uploading reflects left-hand side of this expression.

Then we get from where $\varepsilon_{k,R}(\tau) = \frac{R_b(28)(1+V_k)}{E_b(\tau)}\Omega(\tau)^{-1}$,

$$E_b(\tau) = \frac{R_b(28)(1+V_k)}{E_{k,R}(\tau)}\Omega(\tau)^{-1} \tag{31}$$

at $\tau = 28$ $\Omega(\tau) = 1$ and

$$E_b(28) = \frac{R_b(28)(1+V_k)}{E_{k,R}(28)} \tag{32}$$

Further, substituting (32) (31), we get

$$E_b(28) = \frac{R_b(28)(1+V_k)}{E_{k,R}(28)} \tag{33}$$

If we neglect the change in the limit of deformability of concrete under short-time loading, depending on the age, we get.

$$E(\tau) = E_b(28)\Omega(\tau)^{-1} \tag{34}$$

Experiments known scientists show that the age of the concrete limit deformability depends weakly, and its variations are within the experimental error. However, current regulations limit deformability take concrete constant.

4 Discussion

Our results give reason to assume that in this case, three mechanical characteristics of concrete: strength measure $\frac{1}{R_b(\tau)}$, elastic-instantaneous deformation $\frac{1}{E_b(\tau)}$ and creep C (t,τ) are proportional to the value of one common aging function expressing $\Omega(\tau)$, the changes considered indicators for any concrete age in relation to the reference age $\tau_1 = 28$ days.

Unique, besides a strong link between deformability and strength of concrete properties exists for the particular concrete with a fixed number of components of the composition and other indicators.

This fact is consistent with the "physical" theory O.Ya. Berg strength, as the process of changing these values is the result of a single cause: the hardening of concrete. From this it follows that to improve the reliability of deformation characteristics of concrete standardization must be considered, along with the strength of the influence of material characteristics, namely: the standardization of long deformations (creep and shrinkage) - specific (by volume) amount of mixing water in the concrete mix, which is closely correlated with an initial moisture content of the concrete; when standardization of short-term deformation (initial elastic modulus and ultimate strains) - specific (by weight) of cement paste content in the concrete mix, which determines the ratio of cement paste and aggregate in the hardened concrete.

References

1. M.V. Berlinov, E.A. Makarenkov, Industrial and civil construction, **11**, 46-49 (2013)
2. M.V. Berlinov, E.A. Makarenkov, Bulletin MGSU, **11**, 26-33 (2013)
3. M.N. Berlinova, A.V. Tvorogov, Natural and engineering sciences, **6(84)**, 530-532 (2013)
4. E.M. Zveryaev, M.N. Berlinova, Natural and Technical Sciences, **9-10(77)**, 358-360 (2013)
5. E.M. Zveryaev, M.V. Berlinov and M.N. Berlinova, International Journal of Applied Engineering Research, **8**, 5811-5816 (2016)
6. A.V. Tvorogov. Proceedings of the Eighteenth International interuniversity scientific-practical conference of students, undergraduates, graduate students and young scientists. Moscow, 1005-1007 (2015)

MATEC Web of Conferences 73, 04015 (2016) DOI: 10.1051/matecconf/20167304015
TPACEE-2016

On the definition deflections of monolithic slabs with the mixed reinforcing at the stage of limit equilibrium

Vitaliy Kuznetsov[1], and *Yulia* Shaposhnikova[1,*]

[1]Moscow State University of Civil Engineering, 129337 Yaroslavskoe sh. 26, Moscow, Russia

Abstract. The article considers innovative materials and structures for flat monolithic slabs. Overlapping mixed diagonal reinforcement where high strength is used as prestressing reinforcement in a flexible sheath type "Monostrend". The article discusses the theoretical solution to determine the deflection of a monolithic beamless slab without capitals in the process of limiting equilibrium. The formula based on the value of the limiting deflection cell, reinforcement class and plate thickness. Shows a plot of marginal values of deflections of different thickness slabs with reinforcement A500, for three cells to span 6, 7 and 9 m. These data set the initial stress in high-strength reinforcement and stress as a result acquired deflection in the slab floors with mixed diagonal reinforcement.

1 Introduction

In recent years, more attention is paid to monolithic slabs. Frame-monolithic is now one of the most popular types of construction [1].

To extend the range of span and loads for buildings of this type must be a more complete study of the work of monolithic slabs.

We present, developed by the authors, the theoretical solution to determine the deflection of a monolithic slab beamless floors without capitals in the stage of limit equilibrium [2, 3, 4, 5, 6].

Knowing the values of deflections beamless floors in the stage of limit equilibrium necessary in the calculations of overlap strength while using ordinary and high-strength reinforcement [7] to prevent constriction reinforcement and maximize its strength characteristics [8, 9, 10, 11].

Particular importance is the purpose of prestressing armature levels in the case of high-strength armature, such as "monostrend", without adhesion to the concrete, as damage or break a wire rope is not localized in the concrete, and leads to overstress the rest of the wires, a jumplike increase of deflections, excessive disclosure of cracks and destruction of all overlap [13, 14, 15, 16].

* Corresponding author: yuliatalyzova@yandex.ru

2 Experimental section

As the object of study, the authors selected monolithic reinforced concrete slab beamless floors in an extreme condition. The dimensions of the cell «a» and «b», plate with prestressed reinforcement "Monostrend" without adhesion to concrete.

Traditional stress distribution scheme based on the compatibility of deformations of concrete and reinforcement, in this case, does not work, because the adhesion between the high-strength reinforcement and concrete missing.

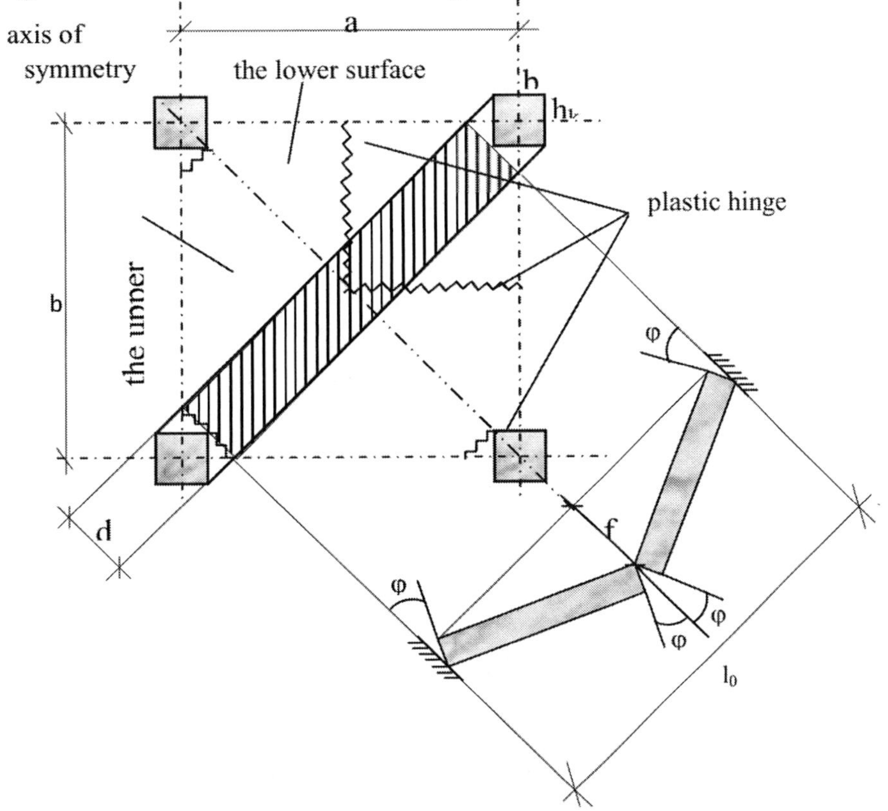

Fig.1. Scheme of the floor slab in a stage of limit equilibrium.

To display the calculated dependencies following assumptions adopted. Plate is in limiting equilibrium and consists of separate units, connected by a line hinges (Figure 1).

The diagram of stresses in the concrete compression zone is rectangular, the maximum stresses equal to the rated stresses - Rb. The maximum height of the compressed zone, which ensures the development of plastic deformations in the concrete and free of tension reinforcement $\xi \leq 0{,}3$ according to [5]. Free of tension stresses in the reinforcement equal to the rated voltage Rs, relevant to physical limit of elasticity (Figure 2). The initial stresses in high-strength reinforcement (excluding losses) is constitute 0.8 accounted for 80 percent of the conventional limit of elasticity Rsn.

So, under these conditions, the maximum positive moment M+ (plate per unit width) in the center of the cell is determined only by the amount of the longitudinal lower reinforcement As in a predetermined section and plate height.

$$M^{+}=R_sA_sz_b. \quad (1)$$

There zb - is shoulder of internal force couple.

$$z_b = h_0\text{-}x/2. \quad (2)$$

When the boundary value of x = 0.3h0 shoulder couple of forces is

$$z_b = h_0\text{-}0.3h_0/2 = h_0(1\text{-}0.3/2) = 0.85h_0. \quad (3)$$

The value of the positive moment in the midspan is

$$M^+\text{=}R_sA_sz_b = 0.85R_sA_sh_0. \quad (4)$$

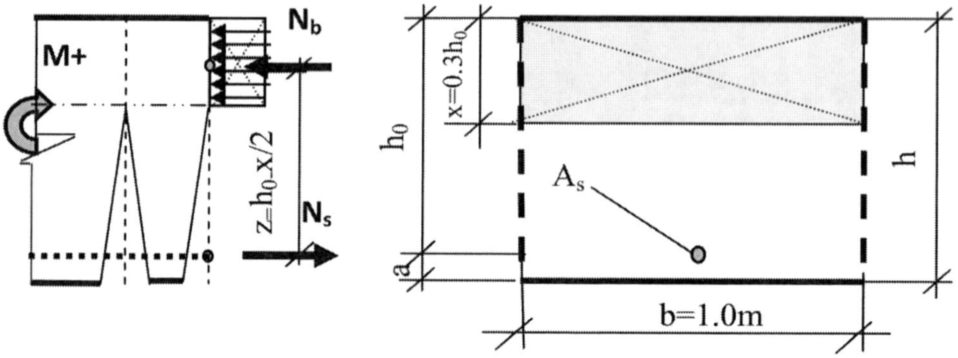

Fig. 2. The settlement scheme of the plate section in the span.

Similarly, the maximum negative moment on the supports-columns is determined by the amount of upper reinforcement A's, crossing the column contour (Figure 1).

M- = RsA'zb has a cross-section per unit width.

The amount of negative moment at the supports is

$$M^-\text{=}R_s\,A'_sz_b = 0.85R_sA'_sh_0. \quad (5)$$

To determine the deflection method is used when the cut between the columns notional diagonal strip width equal to the diagonal of a square or rectangular column. The strip is seen as hingedly supported beam, with the calculated span equal to the distance between the supporting plastic hinges (Fig. 1).

In finding the deflections must first calculate the curvature of the conventional lane plate $1/\rho$

$$\frac{1}{\rho} = \frac{M}{D}, \quad (6)$$

here:

D - bending stiffness of the strip,

M - acting at the section moment. Stiffness concrete element section with cracks may be determined according to [3] by formula

$$D\text{=}E_{s,red}\,A_sz(h_0\text{-}x_m), \quad (7)$$

where:

xm- height of the compressed zone, taken for the element after the formation of plastic hinges equal 0.3h0, shoulder internal pair of forces z = 0.85h0. Taking z = 0.8h0, we obtain an expression for the stiffness

$$D=E_{s,red}\,A_s z(h_0-x_m)\ \approx 0.7E_{s,red}\,A_s\,h_0^2 \tag{8}$$

Using the expressions (2) and (3), and taking Es,red = Es get expression of curvature

$$\frac{1}{\rho}=\frac{M}{D}=\frac{0.85R_sA_sh_0}{0.7E_s\,A_s\,h_0^2}=1.21\frac{Rs}{E_s\,h_0}\,. \tag{9}$$

Given that, as free of tension reinforcement recommended armature A500 classes (Rs = 435 MPa) or A400 (Rs = 350 MPa) and the steel elastic modul is Es = 200000MPa, we obtain the following formulas.

When using reinforcement class A400

$$\frac{1}{\rho}=1.21\frac{Rs}{E_s\,h_0}=1.21\frac{350}{200000h_0}=\frac{0.00212}{h_0}\,. \tag{10}$$

When using reinforcement class A500

$$\frac{1}{\rho}=1.21\frac{Rs}{E_s\,h_0}=1.21\frac{435}{200000h_0}=\frac{0.00263}{h_0}\,. \tag{11}$$

Obviously, the curvature of element 1/ρ for limiting equilibrium depends on two variable parameters: the value of the physical limit of elasticity of reinforcement Rs and the section height h0.

Taking into account that for a square cell 6×6m with columns 400×400mm calculated span is equal to the distance between the inner corners of columns l0 = 7.92 m, the deflection at the center of the cell (conditional beams) can be determined by well-known formula

$$f=S\frac{1}{\rho}l_0^{\,2}=\frac{5}{48}\frac{0.00263}{h_0}7.92^2=\frac{0.0172}{h_0}\,. \tag{12}$$

Then, at a thickness of plate 0.2m, the reinforcement class A500 and the section column 400×400mm, the deflection at the center of the cell is 10.12sm (1/78), and if reinforcement class A400 deflection f = 12.5cm (1/63).

When the reinforcement class is A500 and cell 7×7m, formula becomes

$$f=S\frac{1}{\rho}l_0^{\,2}=\frac{5}{48}\frac{0.00263}{h_0}9.48^2=\frac{0.0246}{h_0}\,. \tag{13}$$

When the reinforcement class is A500 and cell 9×9m, formula becomes

$$f=S\frac{1}{\rho}l_0^{\,2}=\frac{5}{48}\frac{0.00263}{h_0}12.16^2=\frac{0.0405}{h_0}\,. \tag{14}$$

3 Results section

Data limit deflections obtained from Eqs 12-14 for different sizes of cells and thickness of plates using background reinforcement A500 are summarized in Table 1.

Table 1. Deflection limits for different thicknesses of plates and cell sizes.

Cell size, [м]	Plate thickness, [м]				
	0.2	0.22	0,25	0.28	0.3
6x6	0.105	0.094	0.081	0.071	0.066
7x7	0.145	0.129	0.112	0.098	0.091
9x9	0.244	0.218	0.189	0.166	0.154
12x12	0.441	0.395	0.341	0.300	0.278

Fig. 3 shows graphs of limit values for deflections of the four cells with sizes 6, 7, 9 and 12 m, with different plate thicknesses overlap and the background reinforcement A500.

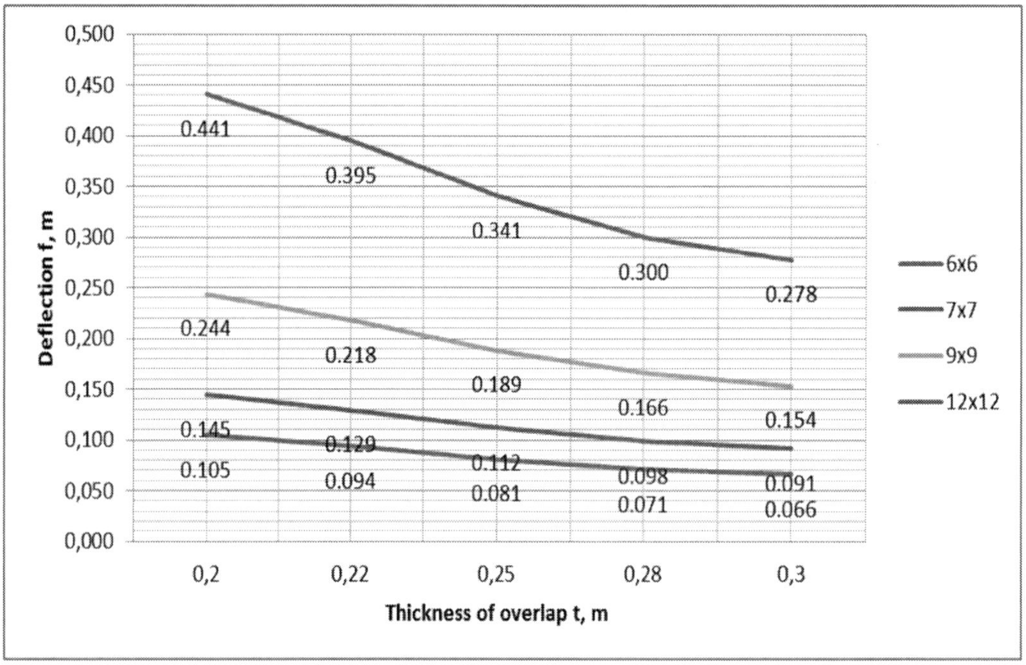

Fig. 3. Graphics of limit deflections to different cells.

Obviously and logically, that by decreasing plate thickness from 0.3 m to 0.2 m deflections increase significantly - for all the above cell size by 59%. Also, with the increase of cell size from 6×6m to 7×7m deflections increase - by 38%, with the increase of the cell to 9×9m deflections increase by 133%, to 12×12 - increased by 320%.

The graphs show that in the limiting state at least the thickness of the slab 0.2 m and the largest span of the cell plate 12×12m greatest deflections 0.441m.

Whereas at the lowest span 6×6m plate with a thickness of 0.2 m deflection is 0.105m in the limit state of the cell plate.

4 Conclusions

1. The proposed engineering method and presented graphics assess deflections slab beamless monolithic slab without capitals in the limiting state at different spans, slab thickness and classes free of tension reinforcement.
2. The proposed engineering method of calculation allows you to set the initial stresses in the high-strength reinforcement in the slab mixed diagonal reinforcement, and stresses as a result of the acquired deflection.

References

1. Yu.A. Bardysheva, V.S. Kuznecov, Yu.A. Talyzova, Vestnik MGSU, **6**, 44-51 (2014)
2. *Nauchno-issledovatel'skij institut betona i zhelezobetona Gosstroja SSSR (NIIZhB), Instrukcija po raschetu staticheski neopredelimyh zhele-zobetonnyh konstrukcij s uchetom pereraspredelenija usilij* (Strojizdat, Moscow, 1990)
3. V.S. Kuznecov, Yu.A. Shaposhnikova, Vestnik MGSU, **7**, 51-58 (2013)
4. S.L. Citnikov, *Patent na izobretenie №2427686 «Sposob izgotovlenija predvaritel'no naprjazhennyh zhelezobetonnyh konstrukcij i monostrend»* (Moskva, 2011)
5. *La norme NBN EN 1992-1-1 (Eurocode 2)* (1998)
6. P. Seinturiere, *Etat Limite de service, IUT,* (Génie Civil de Grenoble, 2006)
7. G.M. Paille, *Calcul des structures en beton arme* (AFNOR, Paris, 2013)
8. F. Freyssiner, *Naissance et perspectives du beton precontraint* (Montevideo, 1994)
9. I.O. Pogrebnoy, V.D. Kuznetsov, Inzhenerno-stroitel'nyy zhurnal, **3**, (2010),
10. N.I. Karpenko, *Obshhie modeli mehaniki zhelezobetona* (Strojizdat, Moskva, 1996)
11. A.S. Vol'mir, *Gibkie plastinki i obolochki* (Gosudarstvennoe izdatel'stvo tehniko-teoreticheskoj literatury, Moskva, 1996)
12. D.V. Portaev, *Raschet i konstruirovanie monolitnyh prednaprjazhennyh konstrukcij grazhdanskih zdanij* (ASV, Moskva, 2011)
13. V.S. Kuznecov, Yu.A. Shaposhnikova, PGS, **3**, 50-53 (2015)
14. V.A. Kremnev, V.S. Kuznecov, Yu.A. Talyzova, Vestnik MGSU, **9**, 48-53 (2014)
15. S.P. Timoshenko, S. Voynovskiy-Kriger, *Plastinki i obolochki, edition 2* (Nauka, Moskva, 1999)
16. A.E. Dorfman, L.N. Levontin, *Proektirovanie bezbalochnykh beskapitel'nykh perekrytiy* (Stroyizdat, Moskow, 1995)

MATEC Web of Conferences 73, 04016 (2016) DOI: 10.1051/matecconf/20167304016
TPACEE-2016

Assessment Stability and Informative of Diagnostic Matrix in Analysis States of Reinforced Concrete Slab

Vladimir Sokolov[1], *Tatiana* Musorina[1,*], *Ekaterina* Starshinova[1] and *Ivanna* Popovych[2]

[1]Peter the Great St. Petersburg Polytechnic University, 195251 Polytechnicheskaya 29, St.Petersburg, Russia
[1]National Aviation University, 03058 Kosmonavta Komarova, 1, Kyiv, Ukraine

Abstract. In this article was examined the issues of sustainability and informative of diagnostic matrix, which are used when it is necessary to determinate the technical condition of constructions. Determination based on the mathematical instrument of technical diagnostics. One of the most widespread methods of technical diagnostics – a statistical method of Bayes is used.

1 Introduction

To solve tasks of technical condition definition of building constructive elements, offered an approach, based on the use of probabilistic mathematical instrument of technical diagnostics. Diagnosis is performed using probabilistic methods states recognition of complex technical systems. Calculations of probabilistic parameters are executed with use of the statistical method of Bayes, most widespread in technical diagnostics [3].

Bayes's method provides collecting statistical information on diagnostic results over last years. These results have formed the content of diagnostic matrices. One of the most important tasks is to assess the stability of these matrices, that is make influence on changes of statistical information results of calculation by Bayes's formula[1].

In article [2], the solution of this problem is based on theoretical instrument of information theory. Estimated weight of diagnostic features and diagnostic value as a whole. It was shown that the exclusion of little information statistics from diagnostic matrix provides quantitative change distribution of state probabilities, however have not provide any qualitative changes of the results. Moreover, the result has been updated in the sense of being more certainty (entropy - degree of uncertainty - has decreased by 1.5 times).

Nowadays, in normative base not defined quantitative reduction bearing abilities of building structures. This situation is typical when using classification of technical state of building objects, both, their defects and damage, and using the classification of the technical conditions[4]. Quality of construction object state eventually decreases in comparison with the technical state put at design and during the construction. In [8, 9] rightly pointed out that the safe operation of buildings requires a constant examination of its

* Corresponding author: flamingo-93@mail.ru

technical condition, and science-based appointment of the relevant categories for its construction elements.

In this paper, the research of sustainability and informative diagnostic matrix is based on multivariate numerical experiment. Considered diagnostic matrix composed for concrete beams of reinforced concrete slab. The matrix consists of 9 signs, 5 of which are simple (answer the question "yes" or "no"), and 4 more complex (with bits and intervals).

The issues involved in assessing the technical condition of a number of authors. In these papers the assessment of technical conditions determined on the basis of: the theory of risk and structural safety [16], the logical-probabilistic approach [17], based on a combination of probabilistic and possibility approaches [18], based on the application of the theory of fuzzy sets [19, 20] and etc.

In [2] these paper the technical condition of the main and secondary beams and of plate sections and columns. The degree of certainty of the overall system is determined by taking into account the impact of each component, so that the technical condition of the system elements can either improve or become worse.

In [3] these paper calculating posterior probabilities of states for each element, carried out an analysis of the results obtained using the apparatus of probability theory and using the concept of information entropy. On this basis, constructed a multi-level model of diagnosis. This approach to the assessment of the technical condition of the building is offered for the first time.

In [4] on the basis of great theoretical and practical experience in the construction of a special purpose the authors suggest the possibility of monitoring the bearing elements of the building objects when exposed to static and dynamic loads. It is proposed to use an integrated method of diagnosing the state of construction projects under dynamic loading. The method is based on the search for diagnostic features reduce the carrying capacity of structure sand sub grade in the transfer function of the object model of the famous "black box."

These paper [5] is devoted to ensuring safe operation of buildings, which is determined by a complex of factors that depend on: the design decisions and their implementation during construction, the residual resources and the technical state of the object, the degree of change of object and environmental standards for the operation and the quality of their compliance during the lifetime of the object, and so on. Safety of operation it is not only technical, but also an economic problem, since the totality of the costs of reproductive events during the lifetime of the object is comparable to the cost of new construction, and often exceed them.

Operation phase is the most prolonged period of operation of the facility, so planning costs, and site restoration should be based on learning and forecasting of the technical condition of the building and its components, taking place over time, to determine the composition and scope of work.

Purpose of the research: Determine the information stability of the diagnostic matrix. Assess the degree of change between "input" priori statistical information and "output", which was estimated. For this it is necessary to solve the *following tasks*:

1. Analyze various options of signs realization by their exception one by one.
2. Assess how thus exceptions change the probabilistic parameters of states, which considered element of the system – the main beams, will take.

2 Materials and Methods

As noted above, this article study sustainability and informative of diagnostic matrix based on multivariate quantitative experiment. Quantitative experiment consisted of performing a large number of calculations, which was taken by using software PATC (Probabilistic

Analysis of Technical Condition) [2]. Software PATC based on the most widespread statistical method of Bayes. The method consists in the fact that if there is a state Si and simple sign kj, occurring in this state, then the probability of co-occurrence of events (the availability of the system state Si and feature kj) in accordance with the concepts of probability theory, will be obtained by the following Bayes's formula:

$$P(S_i / k_j) = P(S_i)\frac{P(k_j / S_i)}{P(k_j)}. \quad (1)$$

For a complex of signs:

$$P(S_i / K^*) = \frac{P(S_i)P(K^* / S_i)}{\sum_{s=1}^{n} P(S_s)P(K^* / S_s)}. \quad (2)$$

where $P\ (S_i / K^*)$ – probability of the diagnosis of S_i after results of diagnostics with a complex of signs were obtained.

The analysis carried out for reinforced concrete floor beams, diagnostic matrix has the form shown in Table 1. Thus, the table is based on preparatory statistical data, which means it includes priori probabilities of the states and features. The decision depends on rule that an object with a complex of features K^* refers to a state of S_i, if the posterior probability of this state $P\ (S_i / K^*)$ is the largest of all.

Table 1. Diagnostic matrix for the main beams of the monolithic ferroconcrete overlap.

№	Diagnostic features	kij	Categories of features	p (kij)	State S1	State S2	State S3	State S4	State S5
					P(S1)	P(S2)	P(S3)	P(S4)	P(S5)
					0.18	0.29	0.35	0.13	0.05
1	Damages of external surface	k11	yes	p(k11)	0.11	0.17	0.29	0.54	0.80
		k12	no	p(k12)	0.89	0.83	0.71	0.46	0.20
2	Longitudinal cracks in protective layer	k21	yes	p(k21)	0.06	0.14	0.23	0.77	0.60
		k22	no	p(k22)	0.94	0.86	0.77	0.23	0.40
3	Normal cracks(width of opening)	k31	< 0.4 mm	p(k31)	0.76	0.59	0.40	0.15	0.20
		k32	до 1.0 mm	p(k32)	0.18	0.34	0.43	0.62	0.40
		k33	≥ 1.0 mm	p(k33)	0.06	0.07	0.17	0.23	0.40
4	Inclined cracks	k41	yes	p(k41)	0.06	0.17	0.34	0.31	0.20
		k42	no	p(k42)	0.94	0.83	0.66	0.69	0.80
		k51	design strength	p(k51)	0.72	0.48	0.49	0.15	0.20

5	Strength of concrete	k52	≤ 30%	p(k52)	0.22	0.31	0.20	0.39	0.20
		k53	> 30%	p(k53)	0.06	0.21	0.31	0.46	0.60
6	Corrosion of armature	k61	< 5%	p(k61)	0.76	0.59	0.29	0.15	0.20
		k62	5 – 20	p(k62)	0.18	0.24	0.37	0.39	0.20
		k63	> 20%	p(k63)	0.06	0.17	0.34	0.46	0.60
7	Deflection	k71	limit	p(k71)	0.83	0.48	0.51	0.31	0.20
		k72	≤ 30%	p(k72)	0.11	0.38	0.23	0.46	0.20
		k73	> 30%	p(k73)	0.06	0.14	0.26	0.23	0.60
8	Strength condition on normal cross-section	k81	satisfied	p(k81)	0.94	0.90	0.86	0.69	0.20
		k82	no	p(k82)	0.06	0.10	0.14	0.31	0.80
9	Strength condition on inclined sections	k91	satisfied	p(k91)	0.94	0.83	0.60	0.61	0.20
		k92	no	p(k92)	0.06	0.17	0.40	0.39	0.80

Realizations of signs for G_1 (for example) beam (as an example they are highlighted in Table 1 by yellow color):

1. there are damages of external surface;
2. there are longitudinal cracks in protective layer;
3. there are normal cracks with width of opening more than 1 mm;
4. there are inclined cracks;
5. strength of concrete is lower than the design strength by 30%;
6. corrosion of naked armature is more than 20%;
7. deflection exceeds limit by more than 30%;
8. strength condition according the calculation by normal cross-section is not satisfied;
9. strength condition according the calculation by inclined sections is not satisfied.

When using data of table 1, a technique offered above and the specified conditions following values were received:

$$P\,(S_1)P(K^*/S_1)=0.18(0.11\cdot 0.06\cdot 0.06\cdot 0.06\cdot 0.06\cdot 0.06\cdot 0.06\cdot 0.06\cdot 0.06\cdot 0.06)=0;$$

$$P\,(S_2)P(K^*/S_2)=0.29(0.17\cdot 0.14\cdot 0.07\cdot 0.17\cdot 0.21\cdot 0.17\cdot 0.14\cdot 0.10\cdot 0.17)= 0;$$

$$P\,(S_3)P(K^*/S_3)=0.35(0.29\cdot 0.23\cdot 0.17\cdot 0.34\cdot 0.31\cdot 0.34\cdot 0.26\cdot 0.14\cdot 0.40)=0.02\cdot 10^{-5};$$

$$P\,(S_4)P(K^*/S_4)=0.13(0.54\cdot 0.77\cdot 0.23\cdot 0.31\cdot 0.46\cdot 0.46\cdot 0.23\cdot 0.31\cdot 0.39)=0.23\cdot 10^{-5};$$

$P(S_5)P(K^*/S_5)$=0.05(0.80·0.60·0.40·0.20·0.60·0.60·0.60·0.80·0.80)= $2.65\cdot10^{-5}$.

As noted above, the denominator of the generalized Bayes' formula is the same for all states, and will be obtained by next formula:

$$\sum P(S)P(K^*/S) = (0 + 0 + 0.02 + 0.23 + 2.65)\cdot 10^{-5} = 2.9\cdot 10^{-5}.$$

Then the a posteriori state probabilities for beams G_1:

$P(G_{11}/K_1^*)$= 0/2.9=0; $P(G_{12}/K_1^*)$= 0/2.9=0;

$P(G_{13}/K_1^*)$= 0.02/2.9=0.0068; $P(G_{14}/K_1^*)$= 0.23/2.9=0.079;

$P(G_{15}/K_1^*)$= 2.65/2.9=0.913.

Given decision rule above, the diagnosed beam $\boldsymbol{G_1}$ throughout the combination of technical state.

3 Results and Discussion

In this article were discussed two variants of state related to considered elements of building:

1. In first option, calculation results didn't lead to high-quality change distribution picture of states probabilities. It was noticedinbothcase of the most negative manifestation of signs, and in case most favorable.

2. In second option, results of calculation,in some cases,shows high-quality transition to new state compared with initial state, this transition was noticedinbothcases, when settlement picture of probabilitiesdistribution was deteriorated, and when its improvement takes place.

Condition of any building element can be estimated by various diagnostic signs [9] The authors have analyzed diagnostic matrix for the fifteen options of beams state in monolithic concrete reinforced overlap (G_1, G_2, …, G_{15}). As calculation results were outcome the values of posteriori (estimated) probabilities for five technical conditions, the largest of which says about assignment to appropriate state. Technical states are presented by five categories [9]: 1 - serviceable, 2 - usable, 3 - limited-usable, 4 - unacceptable, 5 - emergency.

Obtained posterior probabilities for the five states of the beam by different implementations of diagnostic features. You could see results of calculation in the tables below. In these tables, the first column is the initial state of the beam, which is obtained by calculation, without exception signs. The next columns show the results obtained by the sequence of their exclusion.

In most cases, the qualitative picture almost remained unchanged when comparison was from initial state, when the diagnosis includes all the features, to state obtained the exclusion of a particular sign.

The article presents the most vivid realization, by which one can assess the stability of the matrix. In other realization signs picture practically didn't change therefore author think it is unnecessary to give results for all considered cases. Examples are given below in paragraphs 1.1, 1.2, 2.1 and 2.2

1.1. In this example considered realization of signs in which all positive categories are excluded from consideration. Results demonstrate that the beam will accept the worst state

– *emergency* (S_5).It should be noted that this category says that exploitation of overlap is impossible. Therefore, the identification of this category is essential for research practices. This also applies to the previous category – *unacceptable* (S_4).

Table 2. Posteriori probabilities of beam G1.

Initial state	1st feature removed	2nd feature removed	3rd feature removed	4th feature removed	5th feature removed	6th feature removed	7th feature removed	8th feature removed	9th feature removed	Technical condition
0.000	0.000	0.000	0.000	0.000	0.000	0.000	0.000	0.000	0.000	S1
0.005	0.001	0.002	0.021	0.004	0.011	0.014	0.015	0.020	0.018	S2
0.147	0.048	0.079	0.274	0.172	0.245	0.228	0.261	0.463	0.240	S3
0.057	0.029	0.103	0.079	0.064	0.064	0.066	0.115	0.082	0.096	S4
0.791	0.921	0.816	0.626	0.760	0.679	0.693	0.608	0.435	0.646	S5

1.2. The realization of diagnostic features, when the element receives a steady third *limited-usable* (S_3) technical condition. The highest value of probability of these states indicates the appointment of categories state. Evidently, by results of considered signs realization, in all cases the third state is obtained, despite little changes of numerical values of probabilities, despite an exception of signs. Based on this, we can conclude that the qualitative picture is saved and diagnostic matrix demonstrates itself as a stable probabilistic model of beam state.

1. Realizations of signs for G2 beam:
2. there are not damages of external surface;
3. there are longitudinal cracks in protective layer;
4. there are normal cracks with width of opening less than 0,4 mm;
5. there are inclined cracks;
6. strength of concrete is the same as design strength;
7. corrosion of naked armature is fewer than 5%;
8. deflection exceeds limit by more than 30%;
9. strength condition according the calculation by normal cross-section is not satisfied;
10. strength condition according the calculation by inclined sections is satisfied.

Table 3. Posteriori probabilities of beamG2.

0.021	0.016	0.067	0.011	0.084	0.012	0.009	0.069	0.044	0.014
0.285	0.233	0.393	0.191	0.406	0.252	0.155	0.405	0.365	0.214
0.617	0.591	0.519	0.610	0.440	0.535	0.684	0.472	0.565	0.643
0.054	0.079	0.013	0.141	0.042	0.152	0.115	0.046	0.022	0.055
0.024	0.080	0.008	0.047	0.029	0.050	0.038	0.008	0.004	0.074

2.1. The probability distribution shows that the condition is within the range from 3 to 4 states. This is due to the fact that the highest probability is distributed between two states.

When excluding the most significant positive traits (4, 7, and 8) beam falls in a bad state, as it is required to prove in this article. The probability distribution shows that the condition is getting worse, when positive categories of signs did not taken into account. Also obtained that the entropy increases and the system becomes less certain.

Realizations of signs for G_3 beam:

1. there are not damages of external surface;
2. there are not longitudinal cracks in protective layer;
3. there are normal cracks with width of opening less than 0.4 mm;
4. there are not inclined cracks;
5. strength of concrete is the same as design strength;
6. corrosion of naked armature is fewer than 5%;
7. deflection does not exceeds limit;
8. strength condition according the calculation by normal cross-section is satisfied;
9. strength condition according the calculation by inclined sections is not satisfied.

Table 4. Posteriori probabilities of beam G3.

0.012	0.033	0.053	0.004	0.055	0.014	0.023	0.006	0.010	0.008
0.140	0.253	0.268	0.060	0.227	0.118	0.201	0.119	0.122	0.106
0.468	0.497	0.546	0.296	0.380	0.612	0.437	0.374	0.428	0.490
0.378	0.215	0.132	0.637	0.336	0.253	0.335	0.497	0.431	0.389
0.002	0.001	0.001	0.003	0.003	0.003	0.004	0.005	0.009	0.007

2.2. Compares two cases, first, where all the signs according to the initial state, the second where one of the features excluded. In case of exception of the fourth sign (presence of inclined cracks) and the seventh (deflection size), there is a qualitative change to more favorable technical state. The highest values of probability occur in the first three states.

Realizations of signs for G_4 beam:

10. there are not damage so external surface;
11. there are not longitudinal cracks in protective layer;
12. there are normal cracks with width of opening less than 0.4 mm;
13. there are inclined cracks;
14. strength of concrete is the same as design strength;
15. corrosion of naked armature is fewer than 5%;
16. deflection does not exceeds limit;
17. strength condition according the calculation by normal cross-section is satisfied;
18. strength condition according the calculation by inclined sections is not satisfied.

Table 5. Posteriori probabilities of beam G4.

0.140	0.121	0.122	0.089	0.397	0.099	0.072	0.089	0.131	0.413
0.343	0.318	0.327	0.282	0.344	0.363	0.228	0.377	0.337	0.358
0.516	0.559	0.548	0.625	0.258	0.535	0.697	0.533	0.530	0.229
0.001	0.002	0.003	0.003	0.001	0.003	0.002	0.002	0.001	0.000
0.000	0.001	0.000	0.000	0.000	0.000	0.000	0.000	0.001	0.000

4 Summary

In most considered cases it wasn't noted high-quality transition from one state to another in comparison with initial realization. Where was observed only quantitative changes between posteriori probabilities of the states. It shows the stability and informative of statistical data, which form the basis for creating diagnostic matrices, as well as the basis for constructing a multi-level procedure of diagnosing the state of overlap and buildings in general.

References

1. V.A. Sokolov, Civil Engineering magazine, **4(14)**, 49-58 (2010)
2. V.A. Sokolov, Journal of Civil Engineering, **7(25)**, 45-51 (2011)
3. S.H. Shakov, F.E. Kovalev, V.M. Vaskevich, V.Y. Ryzhikov, Civil Engineering Journal, **7(33)**, 76-88 (2012)
4. O.N. Popova, T.L. Simankina, Civil Engineering Journal, 7(42),40-50 (2013)
5. T.N. Soldatenko, Civil Engineering Journal, **7**, 52-61 (2011)
6. *GOST R 53778-2010 Rule of inspection and condition monitoring*
7. *GOST 31937-2011. Buildings and constructions. Rule of inspection and condition monitoring*
8. V.A. Sokolov, Civil Engineering Journal, **5(15)**, 31-37 (2010)
9. V.A. Sokolov, Civil Engineering Journal, **3(13)**, 27-31 (2010)
10. V.A. Sokolov, Civil Engineering Journal, **2(12)**, 3 (2010)
11. I.A. Birger, *Technical Diagnostics Engineering* (1978)
12. *SP 13-102-2003. Terms surveys bearing structures of buildings and structures. Code of practice for design and construction* (2003).
13. A.S. Zolotov, Civil Engineering Journal, **1(27)**, 14-21 (2012)
14. S.N. Savin, S.V. Demishin, I.V. Sitnikov, Civil Engineering Journal, **7(25)**, 33-39 (2011)
15. G.L. Koziniec, Civil Engineering Journal, **5(23)**, 43-48 (2011)
16. A.V. Perelmuter, S.F Pichugin, Magazine of Civil Engineering, **5(49)**, 5–14 (2014)
17. I.S. Ptuhina, M.E. Viatkin, T.A. Musorina, Construction of Unique Buildings and Structures, **9(14)**, 46-56 (2013)
18. J. Tamošaitienėa, E. Gaudutisb, M. Kračkaa, Procedia Engineering, 1151–1155 (2013)
19. B. Kovačič, R.Kamnik, M.Premrov, Survey Review, **230**, 150-161 (2011)
20. B.Kovačič, R. Kamnik, Geodetski List, **63(4)**, 315-327 (2009)
21. O.N. Popova, T.L. Simankina, Magazine of Civil Engineering, **7(42)**, 40-50 (2013)
22. B. Kovačič, Z. Kapović, M. Premrov, A. Breznikar, International Journal for Engineering Modelling, **13(3-4)**, 85-91 (2000)

MATEC Web of Conferences **73**, 04017 (2016) DOI: 10.1051/matecconf/20167304017
TPACEE-2016

Local Buckling of Axially Compressed Rectangular Concrete-Filled Steel Tubes

Ruslan Kanishchev[1,*] and *Vincent* Kvocak[1]

[1]Technical University of Kosice, Faculty of Civil Engineering, 04200 Vysokoskolska 4, Kosice, Slovakia

Abstract. The article deals with the theoretical analysis of the local stability of contemporary structures such as rectangular concrete-filled steel tubes (CFST)'s without imperfections. The work presents a numerical method for calculating the coefficient of critical stress by differential equation for slender walls of hollow and concrete filled closed right-angled profiles. The results of the method were compared with the modelled results of the construction elements in ABAQUS software. Based on this theoretical analysis, the direction of future research has been determined, which will focus on the impact of local stability on the cross-section of rectangular CFSTs in terms of their resistance.

1 Introduction

The construction of civil engineering and transport works increasingly use composite structures formed by joining two effective materials such as steel and concrete. One of the most common examples is concrete filled right-angled (square or rectangular) tubes. These elements have distinct advantages over hollow steel tubes as described in [1-8], but one of the main structural advantages is its significant resistance to loss of local and global stability (which allows designers to reduce the cross-section of the element).

Standards described in [9,10] already exist for the design of the above mentioned structures. However, these standards have limitations in practise such as mechanical properties of materials or the slenderness of steel elements of composite cross-sections etc. Therefore, research should be directed to the enhancement of validity of these standards. Authors such as Mouli and Khelafi [11], Uy [12] investigated the ultimate strength, stability and ductility characteristics of rectangular concrete-filled steel tubes (CFST)'s subjected to axial compression using high performance steels and lightweight concrete aggregates. Experimental tests, conducted by Lee [13], on high-strength concrete-infilled steel tube columns subjected to eccentric loads, showed the influence of width-to-thickness ratio, buckling length-sectional width ratio and eccentricity ratio on the behaviour of these structures. Patel *et al.* [14] proposed a multiscale numerical model for simulating the interaction of local and global buckling behaviour of eccentrically loaded high strength rectangular CFSTs with large depth-to-thickness ratios. Yang and Han [15] presented research aimed at experimentally investigating the behaviour of rectangular CFSTs loaded

* Corresponding author: ruslan.kanishchev@gmail.com

axially on a partially stressed cross-sectional area. On the grounds of above mentioned, we noted, that nowadays the direction of research on the development of world standards for the design of composite columns with concrete-filled hollow steel tubes is an actual topic.

2 Local stability of axially compressed rectangular CFSTs

Currently, the European Union uses Eurocode 4 to design rectangular CFSTs [16]. The basic disadvantage of the standard is its limitations regarding the slenderness of the wall of rectangular cross-section. The design of more efficient or economical composite structures is driving the research on class 4 hollow steel cross-sections [17] filled with concrete, which already lie beyond the validity of Eurocode 4.

The basic principles of designing class 4 cross-sections were stipulated by Bryan [18], which offers a critical analysis of the elastic stress σ_{cr} for local buckling of long right-angled wall elements with simple (hinge) supports on all the edges acting under the influence of a uniform pressure load. The term stress includes various boundary conditions and distribution of forces over the element with the aid of the coefficient of critical stress k_σ:

$$\sigma_{cr} = k_\sigma \frac{\pi^2 E}{12(1-\nu^2)}\left(\frac{t}{b}\right)^2 \tag{1}$$

The minimum values of the coefficient of critical stress k_σ are stipulated in Eurocode 3 part 1-5 [19]. In the case of internally compressed cross-sections (rectangular tube wall), this coefficient is changed from $k_\sigma = 4$ for pure pressure until $k_\sigma = 23.9$ for a combination of compression and bending. The coefficient can be used for hollow rectangular tubes. When the tube is filled with concrete, the standard does not provide a k_σ value.

The issue regarding the impact of local buckling resistance on thin concrete filled tubes was addressed in the work of [20], that assumed a coefficient of critical stress of $k_\sigma = 10.3$, which was discovered by Uy and Bradford [21-23] using the finite strip method. However, research work by Timoshenko and Gere [24,25] for changing boundary conditions from simply supported to clamped supports calculated a minimum value of $k_\sigma = 7$.

2.1 Stability of simply supported rectangular walls subjected to compression

The work of [25] povides a differential Germain's equation for a slender wall with a length a, and width b (Fig. 1a), which is simply supported around its perimeter:

$$C\left(\frac{\partial^4 w}{\partial x^4} + 2\frac{\partial^4 w}{\partial x^2 \partial y^2} + \frac{\partial^4 w}{\partial y^4}\right) + P\frac{\partial^2 w}{\partial x^2} = 0 \tag{2}$$

where w is the deflection of slender walls; P is the compression force.

The cylindrical wall stiffness C:

$$C = \frac{Et^3}{12\left(1-\nu^2\right)} \tag{3}$$

where E is the modulus of elasticity of the steel; t is the wall thickness; ν is the Poisson's ratio.

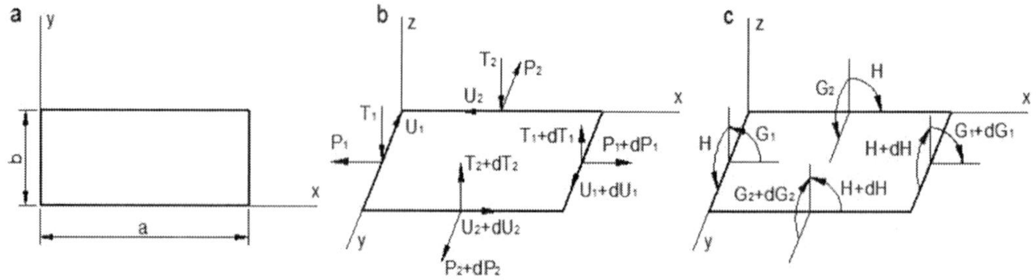

Fig. 1. Mathematical model of slender wall: (a) location of walls in the coordinate axes; (b) components force in a unit element of the wall; (c) components of the bending moment in a unit element of the wall.

A particular solution of the differential equation (2) represents the balance of forces in one of the possible buckling forms of its walls:

$$w = A\sin\frac{m\pi x}{a}\sin\frac{n\pi y}{b} \tag{4}$$

The given solution satisfies the boundary conditions (Fig. 1b,c): for $x = 0$ and $x = a \rightarrow w = 0$ and $G_1 = 0$; for $y = 0$ and $y = b \rightarrow w = 0$ and $G_2 = 0$. In addition, the following conditions are fulfilled for stress: $P_1 = -P$; $U_1 = U_2 = P_2 = 0$. Compressive force P can be determined by substituting expression (4) to equation (2):

$$P = C\pi^2 \frac{\left(\frac{m^2}{a^2} + \frac{n^2}{b^2}\right)^2}{\frac{m^2}{a^2}} \tag{5}$$

where m and n are half-waves along the length (in direction axis X) and across the width (in direction axis Y) of the wall (Fig. 1a).

Of all the possible buckling forms of the balance of force in the cross-section of the wall, we need to find one where the value of the force P will be minimal, so that the right side of equation (5), reflects the minimum. This will be achieved if $n = 1$, naturally, and the value m according to [25] is expressed by equation (6). From the above relationship it follows that the limit value for length a, at which the wall reaches its first buckling form, comprises the m half-waves.

$$a = b\sqrt{m(m+1)} \tag{6}$$

If we consider that the $P / t = \sigma$, where σ is the value of compressive stress, i.e. formula (5) adjustments for $n = 1$ and the cylinder wall stiffness are taken as:

$$\sigma = \left(m\frac{b}{a} + \frac{1}{m}\frac{a}{b}\right)^2 \frac{\pi^2 E t^2}{12(1-\nu^2)b^2} \tag{7}$$

According to Eurocode 3 part 1-5 [19] expression (7) is the value of elastic critical stress of the wall $\sigma_{cr} = k_\sigma \sigma_E$. Thus, the coefficient of critical stress k_σ is expressed as shown in (Table 1):

$$k_\sigma = \left(m\frac{b}{a} + \frac{1}{m}\frac{a}{b} \right)^2 \tag{8}$$

2.2 Stability of rectangular walls with two unloaded clamped sides and two loaded simply supported sides

Changing the boundary conditions of walls on the unloaded edges on clamped, requires a different solution of differential equation (2) which takes the shape defined by Timoshenko [25]:

$$w = A \sin \frac{m\pi x}{a} f(y) \tag{9}$$

The given solution satisfies the boundary conditions on the transverse edges of the wall (Fig. 1) at $x = 0$ and $x = a$. However, an unknown function $f(y)$ must be selected so that it satisfies the boundary condition at the longitudinal edges where $y = 0$ and $y = b \rightarrow w = 0$ and $\partial w / \partial y = 0$:

$$f(y) = C_1 e^{\alpha y} + C_2 e^{-\alpha y} + C_3 \cos \beta y + C_4 \sin \beta y \tag{10}$$

After considering the above boundary conditions we obtain a transcendental equation (11), the solution of which is used to determine the coefficient of critical stress for a variety of aspect ratios of slender walls (Table 1):

$$2(1 - \cos \beta b \cosh \alpha b) = \left(\frac{\beta}{\alpha} - \frac{\alpha}{\beta} \right) \sin \beta b \sinh \alpha b \tag{11}$$

3 The modeling of composite cross sections in ABAQUS software

The tube walls of hollow and concrete filled rectangular cross-sections which are uniformly compressed at the end edges meet the conditions stipulated in previous two chapters (Fig. 2). Therefore, the simulation of 10 samples of empty rectangular steel tubes and 10 samples of concrete-filled rectangular steel tubes was generated in ABAQUS 6.13-4 (Fig.3) to obtain a theoretical comparison of the coefficient of critical stress defined by Timoshenko [25] and identified after modelling values of the elastic critical stress of the walls according to Eurocode 3 part 1-5 [19]. An analysis was carried out on the rectangular cross section of tubes *RHS 200x100x3* (Class 4 according to [17]) without imperfections, the length of the column varied between 94 mm and 263 mm.

When modelling the steel cross-sections the material characteristics defined by Eurocode 3 [17] steel class *S235*, elastic modulus $E = 210\ GPa$, Poisson's ratio in elastic state $v = 0.3$ were used. Behaviour of material is modeled as elastoplastic with linear hardening according to EN 1993-1-5.

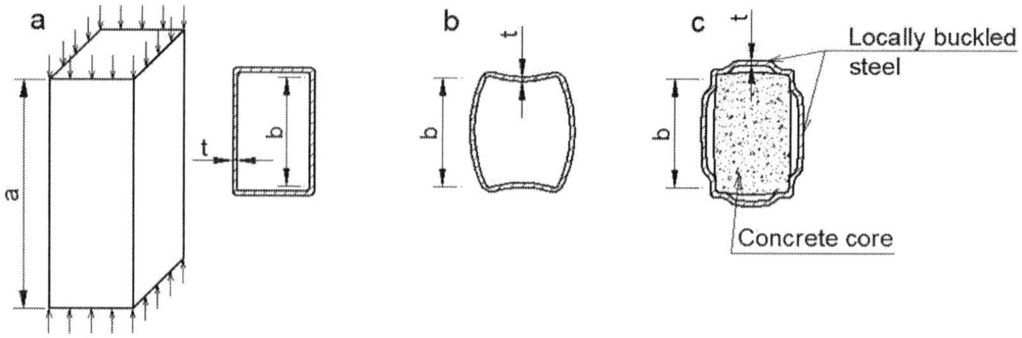

Fig. 2. (a) model of loaded rectangular column; (b) local buckling deformation of the cross-section of the hollow specimen at midpoint of length; (c) local buckling deformation of the cross section of the concrete filled specimen at midpoint of length.

The core of the hollow section was modelled as concrete C20/25 according to Eurocode 2 [26]. A library of ABAQUS elements was used for the modelling: the steel section was made from "shell-type" S4R elements and the concrete core from "solid-type" C3D8 elements. Boundary conditions were simulated so as to correspond to a real experiment. The interaction of the steel and concrete elements was modelled by means of two components: the "normal" one as compression of concrete on the steel section and the "tangential" one as shear resistance at the steel-concrete interface. Loading of columns was modelled as short-term, rising steadily at a constant rate, introduced through the plate of a hydraulic press.

4 Results and discussion

The elastic critical stress walls were investigated in the middle of a broader wall of the cross-section upon reaching first buckling form, corresponding to one half-wave (Fig.3-4). The results of comparison are shown in Tab. 1 and Fig. 5.

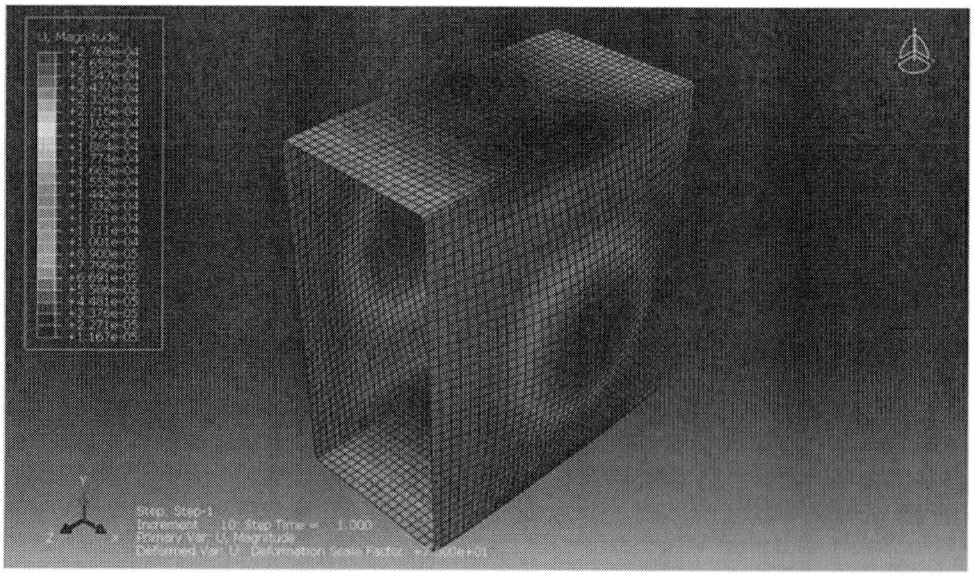

Fig. 3. 3D deformation of the hollow specimen, with length L = 188 mm in ABAQUS.

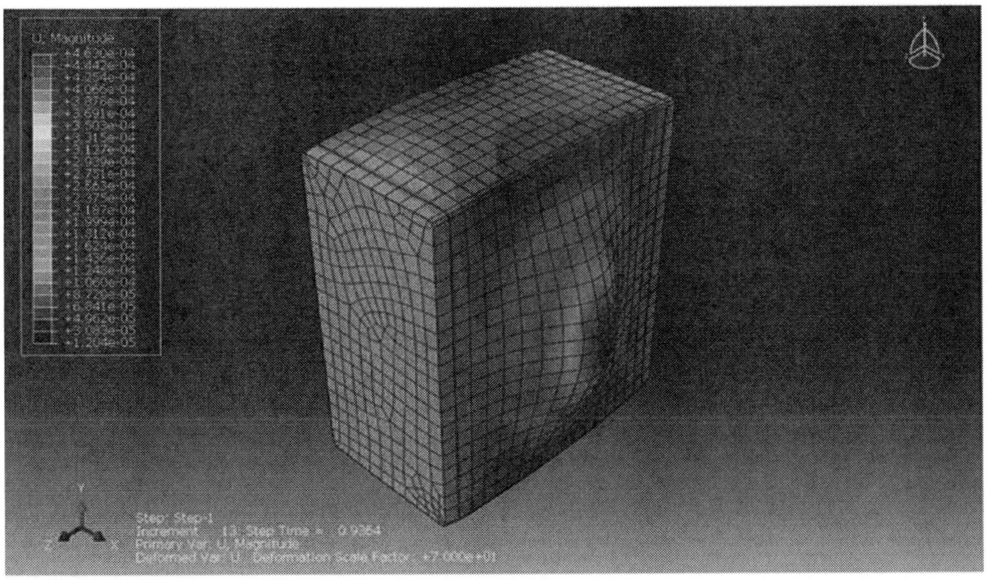

Fig. 4. 3D deformation of the specimen filled with concrete with length L = 188 mm in ABAQUS.

The minimum theoretical value for the coefficient of critical stress $k_{\sigma,min}$ calculated in case of the simply supported walls equals *4* for aspect ratio *a/b = 1* (Fig.4). This value is given in Eurocode 3 part 1-5 [19]. The modelling of hollow rectangular tubes in ABAQUS obtained a minimum value of $k_{\sigma,min}$ = *4.080* for aspect ratio *a/b = 0.8*, which exceeds the theoretical value of 2%. Therefore, the difference in aspect ratio shows that the modelling considered a rectangular cross section, where, due to various wall stiffness's, stress-redistribution was attained.

Table 1. Results of the comparison of coefficients of critical stress.

a [mm]	b [mm]	a/b	Numerical calculation (simply supported walls)		Hollow tubes in ABAQUS		Numerical calculation (walls with two unloaded clamped sides and two loaded simply supported sides)		CFSTs in ABAQUS	
			σ_{cr}, [MPa]	k_σ	σ_{cr}, [MPa]	k_σ	σ_{cr}, [MPa]	k_σ	σ_{cr}, [MPa]	k_σ
94		0.5	270.163	6.250	221.333	5.120	332.465	7.691	220.390	5.099
113		0.6	222.086	5.138	194.133	4.491	305.236	7.061	218.834	5.063
132		0.7	195.85	4.531	179.605	4.155	302.682	7.000	218.631	5.058
150		0.8	181.658	4.203	176.372	4.080	315.515	7.299	218.828	5.062
169	188	0.9	174.831	4.045	180.315	4.171	340.212	7.871	221.521	5.125
188		1.0	172.904	4.000	185.889	4.300	371.930	8.604	224.415	5.192
207		1.1	174.480	4.036	197.307	4.565	410.406	9.494	224.490	5.193
226		1.2	178.716	4.134	206.016	4.766	457.309	10.579	225.178	5.209
244		1.3	185.082	4.282	223.575	5.172	507.474	11.740	225.947	5.227
263		1.4	193.230	4.470	231.203	5.349	565.755	13.088	227.399	5.261

In the case of CFSTs, a numerical model gives a substantially higher coefficient of critical stress ($k_{\sigma,min}$ = *7.000*) compared to the results obtained by modelling of this elements in ABAQUS ($k_{\sigma,min}$ = *5.058*). Higher values of this coefficient do not permit the use of the numerical model to study the local stability of rectangular CFST cross sections without experimental verification.

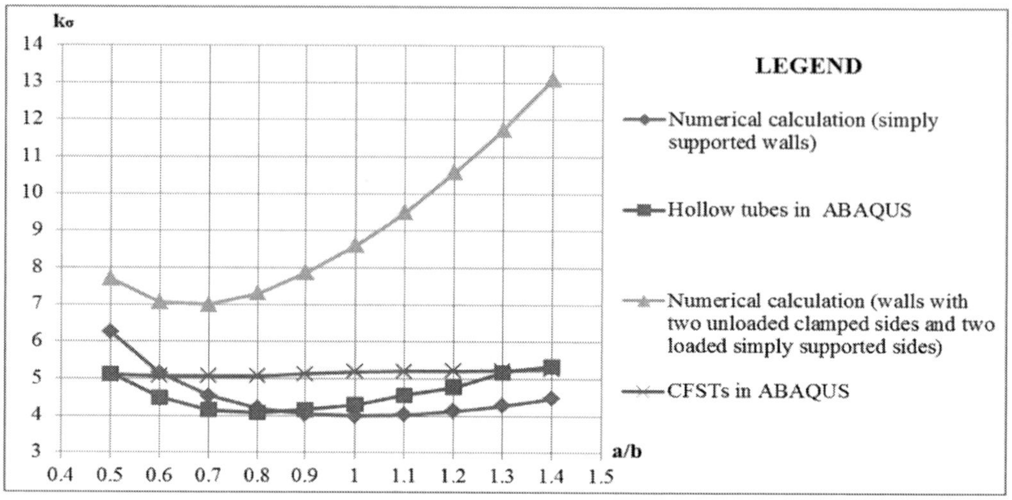

Fig. 5. Comparison of the numerical model to calculate the coefficient of critical stress and results in ABAQUS.

5 Conclusion

Based on the results of this work as general conclusions of the research are the following:

1. One of the basic types of composite sections in the form of a rectangular steel tube with a profile of RHS 200x100x3 (EN 10219), filled with concrete, which lie beyond the validity of Eurocode 4, was investigated.
2. Based on modelling of the behavior of the specimens under axial compression in ABAQUS was obtained the theoretical values of the coefficient of critical stress. These coefficients were compared with values, which was obtained by numerical calculation. This comparison shown, that development of the local buckling of the steel part of composite cross-section occurs at a higher value of the coefficient of elastic critical stress in comparison with normative data EN 1993-1-5 for tubes, not filled with concrete.
3. Based on the above theoretical analysis the direction for further research was chosen aimed at the development of numerical methods for calculating the coefficient of critical stress by differential equations of slender wall rectangular CFSTs.

Acknowledgements

The paper presented was supported by the projects: VEGA 1/0188/16 "Static and Fatigue Resistance of Joints and Members of Steel and Composite Structures" of the Scientific Grant Agency of the Ministry of Education, science, research and sport of the Slovak Republic and the Slovak Academy of Sciences.

References

1. R. Kanishchev, V. Kvočák, Vedecko-výskumná činnosť UIS 2013, 107-116, (2014)
2. V. Kvočák, G. Varga, R. Vargova, Proc. Eng. **40**, 469-474 (2012)
3. A. Ďuricová, M. Rovňák, *Navrhovanie oceľovo - betónových konštrukcií* (VEDA, Bratislava, 2008)
4. J. Kozák, Š. Gramblička, J. Lapos, *Spriahnuté a kombinované oceľobetónové konštrukcie pozemných stavieb* (Jaga, Bratislava, 2000)
5. A. Kikin, R. Sanžarovskiy, V. Trul, *Konstruksii iz stalnykh trub, zapolnennykh betonom* [Construction of steel tubes filled with concrete] (Stroyizdat, Moscow, 1974)
6. L. Storoženko, P. Plachotny, A. Černiy, *Raschet trubobetonnykh konstruktsiy* (Budivelnik, Kiev, 1991)
7. O. Oniščenko, S. Pičugin, V. Oniščenko, O. Semko, L. Storoženko, I. Emelanova, O. Landar, *Vysokoefektivni tekhnologii ta kompleksnye konstruktsii v budivnitstvi* (Formika, Poltava, 2009)
8. I. Garanzha, N. Vatin, J. Appl. Mech. and Mat. **633-634,** 965-971 (2014)
9. R. Kanishchev, V. Kvočák, IABSE Elegance in Str. Nara, 1-8 (2015)
10. W.-H. Kang, B. Uy, Z. Tao, S. Hicks, J. Advanced Steel Constr., **11 (2),** 165-184 (2015)
11. M. Mouli, H. Khelafi, J. Eng. Str., **29**, 1791-1797 (2007)
12. B. Uy, J. of Constr. Steel res., **64**, 748-754 (2008)
13. S. Lee, J. Steel and Comp. Str., **7(2)**, 135-160 (2007)
14. V. I. Patel, Q. Q. Liang, M. Hadi, J. Inter. and M. M., **5(2)**, 91-104 (2012)
15. Y.Yang, L. Han, J. of Constr.Steel Res., **65**, 1617-1630 (2009)
16. *EN 1994-1-1, Eurocode 4: Design of composite steel and concrete structures. Part 1-1: General rules and rules for buildings* (CEN, Brussels, 2004)
17. *EN 1993-1-1, Eurocode 3: Design of steel structures. Part 1-1: General rules and rules for buildings* (CEN, Brussels, 2005)
18. G. H. Bryan, Proc. of the London Mathem. Soc., **22**, 54-67 (1891)
19. *EN 1993-1-5, Eurocode 3: Design of steel structures. Part 1-5: Plated structural elements* (CEN, Brussels, 2008)
20. D.A. Nethercot, *Composite construction* (Spon press, New Yourk, 2004)
21. B. Uy, J. of Constr. Steel Res., **47**, 47-72 (1998)
22. B. Uy, M.A. Bradford, Fifth intern. Conf. on steel struct. Jakarta, 313-322 (1994)
23. M.A. Bradford, D.J. Oelhers, *Composite steel and concrete structural members, Fundamental behaviour* (Pergamon, London, 1995)
24. S.P. Timoshenko, J.M. Gere, *Theory of elastic stability* (McGraw- Hill, New York, 1961)
25. S.P. Timoshenko, *Ustoychivosť sterzhney, plastin i obolochek* (Nauka, Moscow, 1971)
26. *1992-1-1, Eurocode 2: Design of concrete structures. Part 1-1: General rules and rules for buildings* (CEN, Brussels, 2005)

MATEC Web of Conferences **73**, 04018 (2016) DOI: 10.1051/matecconf/20167304018
TPACEE-2016

Comparative analysis of strength and crack resistance of normal sections of bent elements of T-sections, made of rubber concrete, kauton reinforcement and concrete

Yuriy Potapov[1], *Alexey* Polikutin[1*], *Dmitry* Panfilov[1], and *Mikhail* Okunev [1]

[1]Voronezh State University of Architecture and Civil Engineering. 20-letiya Oktayabrya Str., 84, 394006, Voronezh, Russia,

Abstract. One of the urgent tasks of development of the construction associated with the development of new building designs, the use of which provides increased strength, crack resistance, reducing the flow of construction materials, reducing the labor intensity, energy consumption and cost. Ensuring effective functioning of structures during their operation in the harsh environments associated not only with the task of developing materials of higher strength and corrosion resistance, but also composites of increased strength and crack resistance, as structural materials crack resistance is largely determined by the ability of the structure to prevent the formation and growth of cracks. For structures operating in conditions of chemical action, the question of crack resistance is paramount, as the disclosure of cracks in aggressive environment, penetrating deep into the section and causing corrosion of the reinforcement will significantly impair the ability of the load bearing capacity of the element as a whole. The results of experiments of polymer concrete beams of the cross T-profile and comparison of the results with those obtained in an experiment similar to concrete and kauton reinforced elements are given. The observations of the stress-strain state of polymer concrete of flexural members, strength and crack resistance of these elements are presented.

1 Introduction

Creation on the basis of polybutadiene oligomer belonging to the class of liquid polymers, composite materials of new generation, reinforced longitudinal rod reinforcement – kauton reinforcement with high chemical resistance, crack resistance, strength and other favorable conditions is a promising direction in the solution of this issue. In practice, wide application of polymer concrete based on furan, epoxy, polyester, urea-formaldehyde and other resins

* Corresponding author: apl-sc@mail.ru

is seen in construction. However, industrial production of these resins in Russia in recent years has declined sharply, and was completely outside the territory of the country (furan resin), therefore bringing their cost to have risen sharply and close to world prices. In this situation, issues related to the protection of building structures from the aggressive effects of the environment are possible with the use of alternative types of commercially available polymers, for example, diene oligomers, belonging to the class of liquid polymers. Obtained on the basis of liquid polymers produced at the Voronezh industry is Synthetic Rubber (SC-2), the material - polymer concrete, has in addition to the favorable physical and mechanical properties and high chemical resistance and can, together with effective species of polymer concrete take its place at the decision for protection against corrosion problems of various products and structures operating in aggressive environments. In addition, the introduction of a mixture of polymer concrete of large technogenic waste allows to solve the environmental problem associated with their disposal. Creating a safe and efficient working flexural member is impossible without studying the stress-strain state, which occurs when exposed to different kinds of forces.

In the course of studies of the properties of rubber concrete and building structures based on it, was worked by: Borisov Yu.M. [1,2,3,4,5], Potapov Yu.B [6,7], Polikutin A.E. [8], Figovsky O.V. [9], Nguyen Phan Duy [10], Pinaev S.A. [11], Chmykhov V.A. [12] etc., the effectiveness of this material and designs based on it has been proven.

2 Experimental studies

The work is devoted to the experimental study of strength and crack resistance of normal sections of polymer concrete of flexural members of T-section with a shelf made of concrete and the wall made of polymer concrete at cross bending under the influence of short-term loads, and the comparison of their work with that of similar kauton reinforcement and reinforced-concrete flexural members of T-section.

To solve the problems we need to produce a series of samples of polymer concrete of flexural members of T-cross-section (Fig. 1).

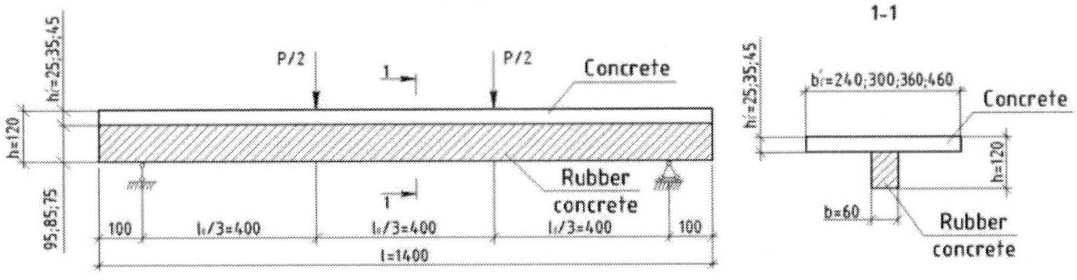

Fig. 1. General view and the diagram of the experimental load beams

The main variable parameters are proposed as follows: the height of the concrete flange hf/, flange width bf/ and percentage of longitudinal reinforcement μ. For comparison of work of polymer concrete of flexural members of T cross-section with the work of single-layered reinforced concrete flexural members and kauton reinforcement of cross T profile, we need to produce a similar series of concrete beams and kauton reinforcement with dimensions similar to the dimensions in (Fig. 1), and with the same variable parameters. All samples were tested for transverse beams two bending symmetrically arranged acting forces in the thirds of the span (Fig. 1). All field tests of the control samples and experimental beams were carried out in the center of collective use named after Professor Yu.M. Borisov at the Voronezh GASU.

Fig. 2. Experimental beam before and after the experiment

To get a clear picture of the stress-strain state of normal sections of double-layered polymer concrete of flexural member,the longitudinal deformation section height was measured by the means of strain gauges glued on the height of the section at mid-span (Fig. 3).

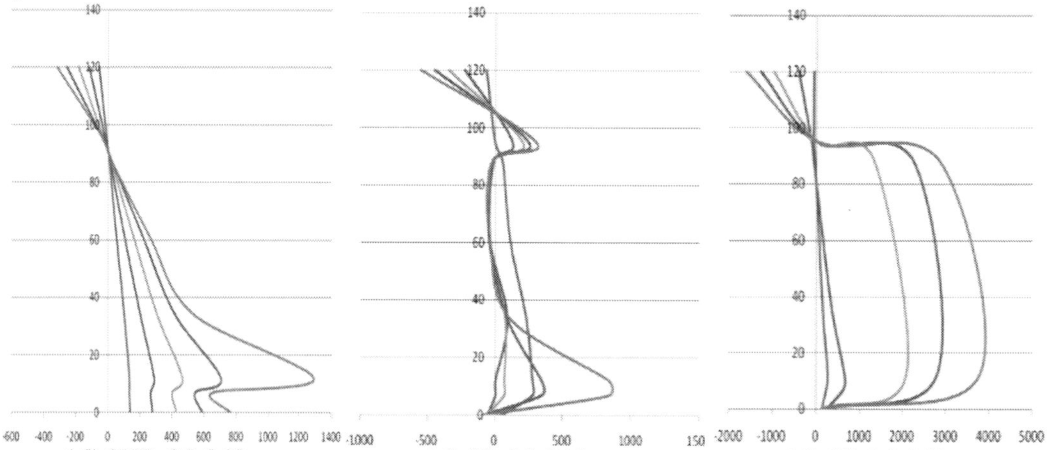

Fig. 3. Schedule the distribution of deformation height of the normal section beams

In order to move from strain to the stress, the relationship "$\sigma_b - \varepsilon_b$"is used in the concrete to determine compression and tension, resulting in a experiment of concrete test samples. For polymer concrete, the relationship "$\sigma_k - \varepsilon_k$" is used on the compression and tension, as obtained in the test control samples of polymer concrete.

Analysis of test results allowed to establish the nature of the distribution of the longitudinal strain at the height of the cross section in the zone of pure bending. Changes in strain and stress adjustment section under the influence of an external load are shown in Fig. 4.

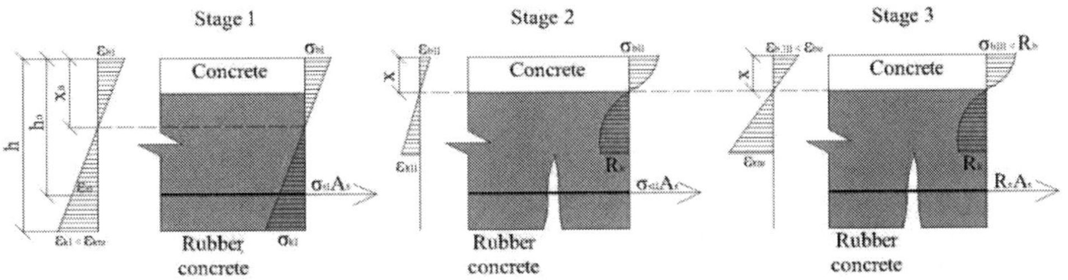

Fig. 4. Stages of stress-strain state of the normal section of double-layered rubber concrete of bent elements of T-sections

$\varepsilon_{b,I}$, $\varepsilon_{b,II}$, $\varepsilon_{b,III}$ и $\sigma_{b,I}$, $\sigma_{b,II}$, $\sigma_{b,III}$ – respective strains and stresses in the extreme fibers of compressed concrete in stages 1, 2, 3;

R_b – the compressive strength of concrete;

$\sigma_{s,I}$, $\sigma_{s,II}$ – respective fitting of the tensile stress in stages 1, 2;

R_{kt} – strength in tension of rubber concrete;

R_s – tensile strength reinforcement in tension;

A_s – the cross-sectional area of the valve.

Analyzing the dependence of the stresses on the sectional height of the load, it can be concluded that the normal section at mid-span of double-layered polymer concrete of flexural members undergoes 3 stages of stress-strain state.

Stage 1 - before the appearance of cracks to the normal of polymer concrete. Until the moment of the formation of cracks to the normal of polymer concrete, the deformation of compression and tensile strain is distributed linearly. In this stage, the stress in the compression zone of the concrete, polymer concrete and fittings is small, deformations are predominantly elastic in nature and the strains and stresses have a linear relationship. Prior to the formation of cracks in polymer concrete the deformation values of the stretched fiber of polymer concrete reach the limit. With further increase in load comes a new qualitative state of structural work.

Stage 2 - after the appearance of normal cracks in the polymer concrete. At this stage, there is a significant redistribution of stresses in normal sections. In these tensile areas where cracks are formed, tensile forces undergo reinforcement. In the intervals between the cracks in the tension zone the contact of the reinforcement with the polymer concrete is preserved and the distance from the edges of the cracks, the tensile forces in polymer concrete increases and decreases in the reinforcement. With further increase, the loads in the concrete of the compression zone increases inelastic deformation, stress diagram in a compressed zone is curved. Upper compression deformation fibers increases, but still have not reached the limit values. At this stage part of the concrete at the tensile zone (above the top of the crack) continues to resist external load and accept tensile forces. Here, there is formation of new and increase in the existing cracks. The height of the compressed zone is significantly reduced (Table 2).

Stage 3 - just before the destruction of the element. At this stage there are inelastic deformations in the concrete at the compressed zone, compressive stresses become curvilinear. The height of the compressed zone is significantly reduced. With a weak reinforcement, stresses in reinforcement reach the yield limit, stresses in the concrete at the compressed zone under the influence of the deflection element and reduction in the height of the compressed zone reach the values of temporary resistance to compression. Destruction of polymer concrete element starts from the reinforcement at tensile zone.

In this case, the work of normal sections of double-layered polymer concrete of flexural members is similar to the work of normal sections of single-layered reinforced concrete of flexural members.

As shown by the studies, the effective factor in the strength of the normal sections of double-layered polymer concrete of bending elements is the percentage of longitudinal reinforcement.

The graphs of relative deformations of the loads of some beams are shown in Fig. 5.

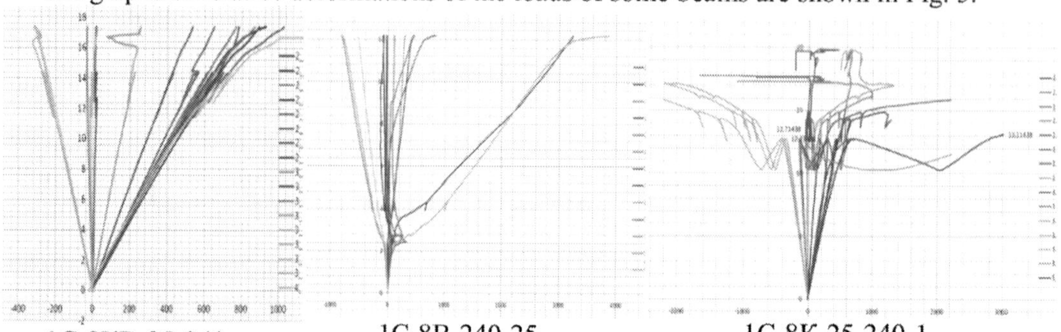

1C-8KB-25-240 1C-8B-240-25 1C-8K-25-240-1

Fig. 5. The graphs of relative deformations of the external loads for experimental beams

After analysis, the realized values of the bending moments in the destruction of elements (time flow valve) are summarized in Table 1.

Table 1. - Bending Test Results

Reference designation	The percent of longitudinal reinforcement μ, [%]	The bending moment of the destruction of the sample (at fluxion of reinforcement), $M_{p,}$ [кН*м]
KB_8	0.84	3.506
KB_12	1.89	6.540
KB_16	3.35	9.811
K_8	0.84	2.798
K_12	1.89	6.996
K_16	3.35	10.376
B_8	0.84	2.436
B_12	1.89	6.116
B_16	3.35	9.376

The test results of two-layered rubber concrete, single-layered kauton reinforcement and reinforced-concrete of bent elements of T-sections are presented in Table 2.

Table 2. Test results of bent elements of T-sections, made of rubber concrete, kauton reinforcement and concrete.

The beam codes	μ, [%]	x, [mm]	M_u, [кN·m]	$\varepsilon_{kt}\cdot10^3$	$\varepsilon_{ktk}\cdot10^3$	R_b, [MPa]	R_k, [MPa]	R_{kt}, [MPa]
1(2,3)C-8KB-25-240-1.	0,84	23.7	3.51	0.60	0.59	33.10	85.50	10.90
1(2,3)C-10KB-25-240-1.	1.31	24.4	5.13	0.64	0.62	32.45	82.90	12.40
1(2,3)C-12KB-25-240-1.	1.89	25.4	6.54	0.67	0.65	31.80	80.30	13.90
1(2,3)C-14KB-25-240-1.	2.57	25.4	8.46	0.71	0.69	32.15	80.80	13.15
1(2,3)C-16KB-25-240-1.	3.35	26.2	9.81	0.74	0.73	32.50	81.30	12.40
1(2,3)C-8B-25-240-1.	0,84	18.1	2.43	-	-	32.65	-	-
1(2,3)C-10B-25-240-1.	1.31	19.4	4.28	-	-	31.68	-	-
1(2,3)C-12B-25-240-1.	1.89	20.4	6.12	-	-	30.70	-	-
1(2,3)C-14B-25-240-1.	2.57	20.5	7.96	-	-	31.75	-	-
1(2,3)C-16B-25-240-1.	3.35	20.9	9.38	-	-	32.80	-	-
1(2,3)C-8K-25-240-1.	0,84	21.4	2.80	0.78	0.75	32.50	84.00	9.20
1(2,3)C-10K-25-240-1.	1.31	22.6	5.47	0.69	0.66	32.25	84.45	11.75
1(2,3)C-12K-25-240-1.	1.89	25.0	6.99	0.59	0.57	32.00	84.90	14.30
1(2,3)C-14K-25-240-1.	2.57	25.3	9.29	0.66	0.60	31.65	84.20	12.85
1(2,3)C-16K-25-240-1.	3.35	25.5	10.38	0.72	0.63	31.30	83.50	11.40

μ - the percentage of longitudinal reinforcement;
x - the height of the compressed zone before destruction;
M_u – breaking bending moment;
ε_{kt} – rubber concrete deformation in the test samples on tension;
ε_{ktk} - extreme fiber deformation of rubber concrete at the stretched zone during the priod of crack formation;
R_b - prism strength of concrete;
R_k - prism strength of rubber concrete;
R_{kt} – rubber concrete strength in tension.

The dependence in strength of the bending kauton reinforced, polymer concrete and concrete elements on the percentage of longitudinal reinforcement is shown in Fig. 6.

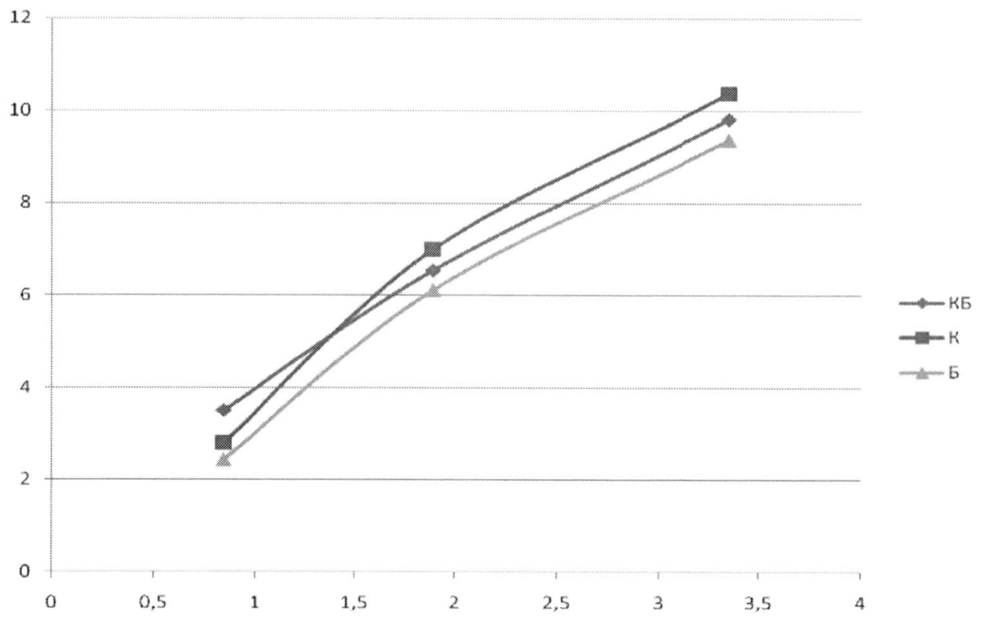

Fig. 6. The graph of dependence of the strength of the flexural members of kauton-concrete, kauton reinforcement, concrete on the percentage of longitudinal reinforcement

An analysis of the graph in Fig. 6 shows that the strength of the flexural members of kauton reinforcement, polymer concrete and concrete increases with increasing percentage of the longitudinal reinforcement. By increasing the percentage of longitudinal reinforcement from 0.84% to 3.35% the strength of normal sections of polymer concrete beams increases almost linearly.

The load bearing capacity of structures in the tension zone of polymer concrete is higher than those of reinforced concrete structures and this is due to the fact that the high grip of reinforcement and rubber concrete prevents rapid development of plastic deformation in the reinforcement, as well as part of polymer concrete around the tension zone between the cracks involved in construction as a whole.

It is pertinent to note that for a small percentage of reinforcement, polymer concrete beam collapses at a higher load than polymer reinforcement, but with an increase in interest of load bearing capacity kauton reinforcement of reinforcing elements becomes greater than polymer concrete by 5.4-6.5%.

On these assumptions at the time of cracking an abrupt increase in strain is noted. The value of the bending moment corresponding to the point of the hop deformation, is the moment of the formation of cracks in the polymer concrete.

Cracking moments of normal sections of double-layered polymer concrete of bent flexural members of T-sections, depending on the percentage of longitudinal reinforcement and comparison with the cracking moments of normal sections of reinforced concrete and kauton reinforcement of flexural members are shown in Fig. 7.

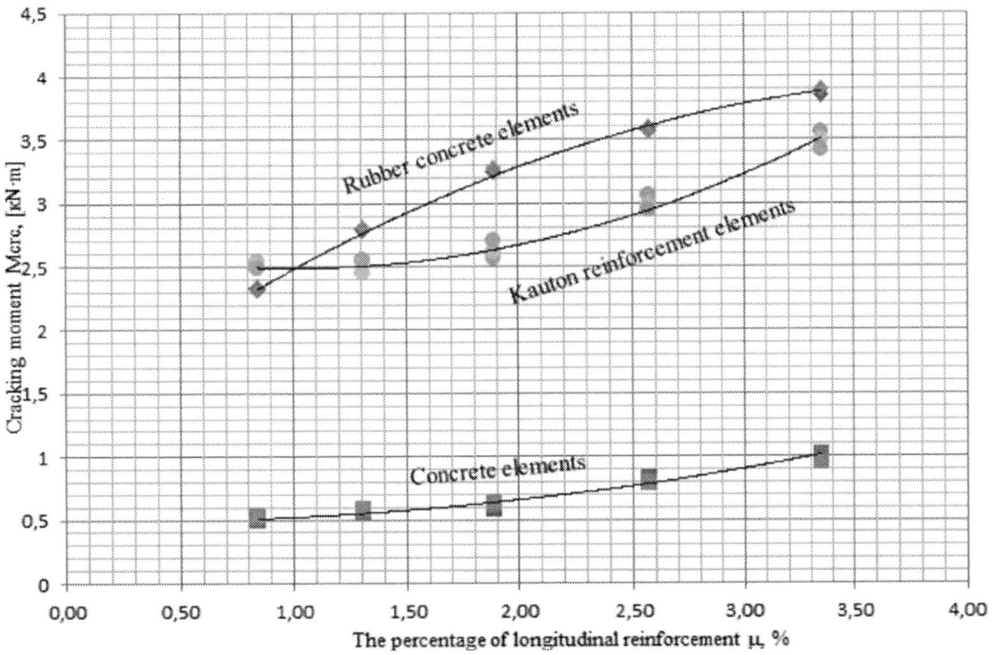

Fig. 7. Cracking moments of experimental beams, in relation to the percentage of longitudinal reinforcement

The testing results of two-layer rubber concrete, single-walled reinforced concrete bent elements and armorubber concrete and analysis revealed that fracture rubber concrete samples is significantly higher than the concrete (average 78.05%). Crack resistance rubber concrete samples higher than armorubber concrete (average 12.78%), but a small percentage of longitudinal reinforcement (μ=0,84%) samples from armorubber concrete showed the best evidence of fracture toughness than rubber concrete (at 9.96%).

The tests allowed to identify three stages of the formation of cracks in the normal bilayer rubber concrete bent elements of T-sections.

First stage. Disrupted strain distribution patterns in rubber concrete - strain in those sections where further cracks, increase the intensity compared to neighboring.

Second stage. The formation of cracks in these sections. This deformation in the reinforcement and the extreme fiber tensile verge abruptly increased. The height of the compressed zone is significantly reduced.

Third stage. When the load growth cracks develop on the sectional height. This continues the cracking in adjacent sections until the destruction of the element.

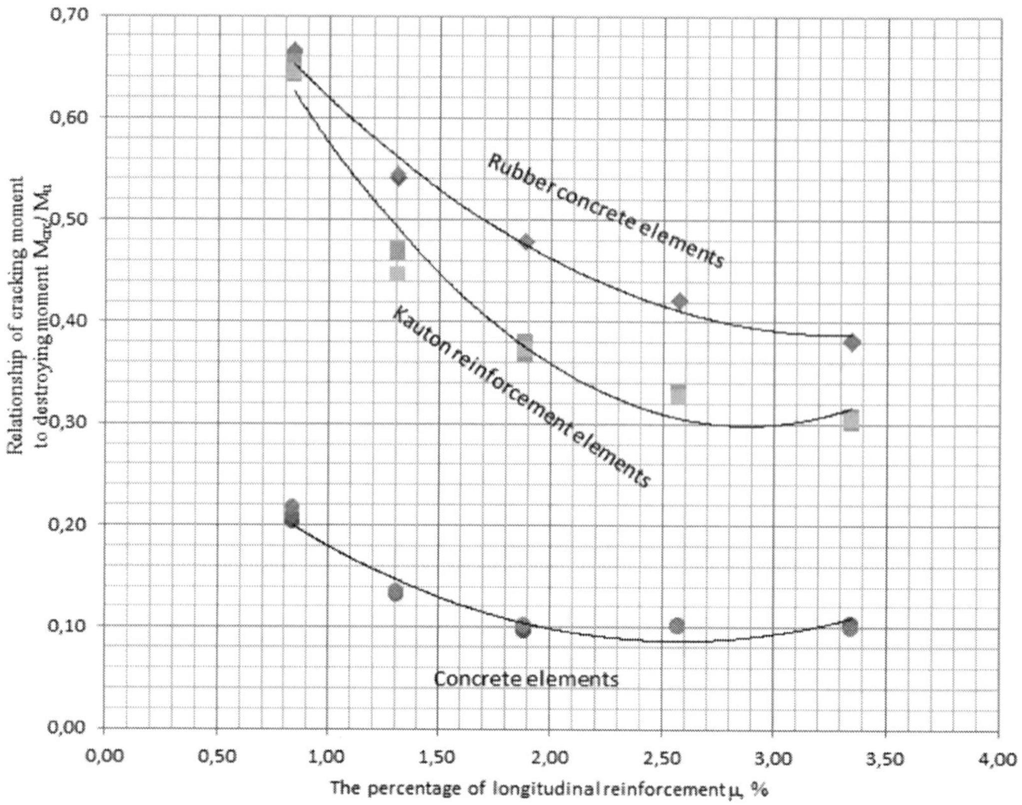

Fig. 8. Dependence of the relationship of cracking moments to breaking moments of the experimental elements of T-sections on the percentage of longitudinal reinforcement

The experimental results show that with the increase in the percentage of longitudinal reinforcement, the effect of raising the crack resistance decreases (Fig. 8). This is due to the fact that in a high percentage of reinforcement, significant part of force at the time of cracking experienced reinforcement and only a small proportion of the load acts on polymer concrete.

The results for the determination of the width of cracks to the normal to the longitudinal axis of the elements, depending on the ratio of the acting cracks to the breaking of the bending moments are shown in Fig. 9-13.

 DOI: 10.1051/matecconf/20167304018

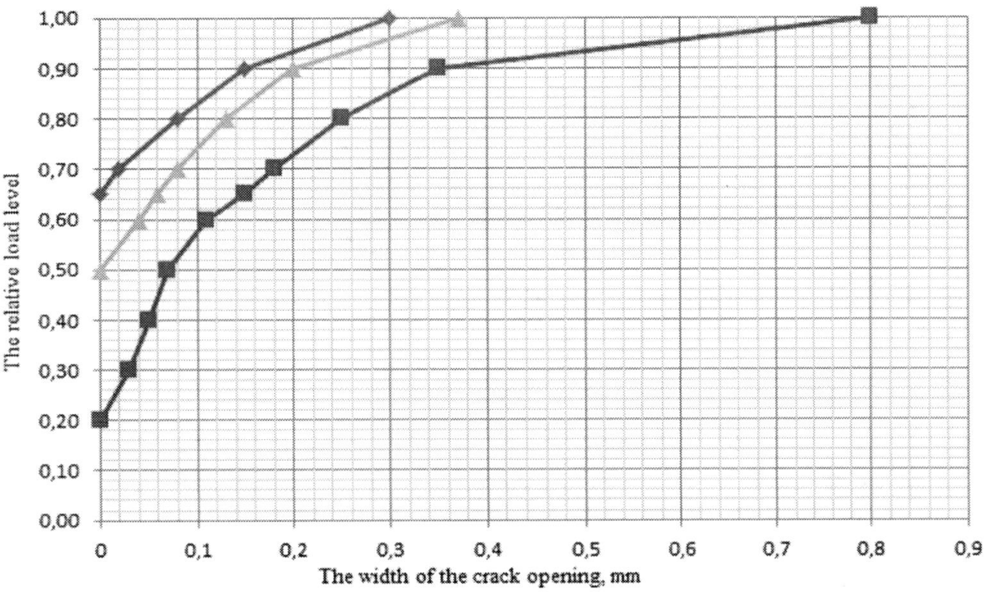

Fig. 9. The width of cracks in beams with the percentage of longitudinal reinforcement μ = 0,84%

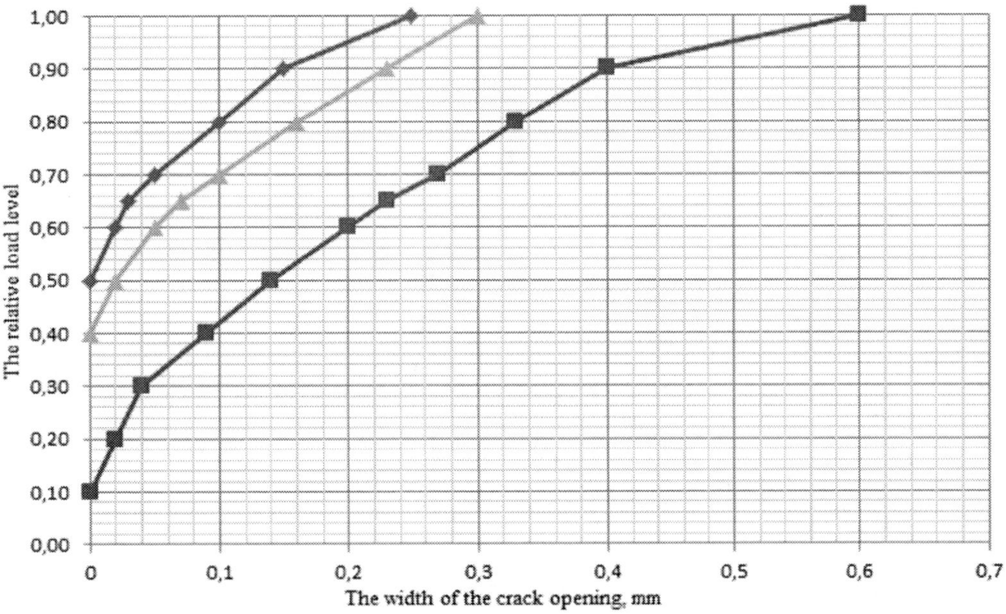

Fig. 10. The width of cracks in beams with the percentage of longitudinal reinforcement μ = 1,31%

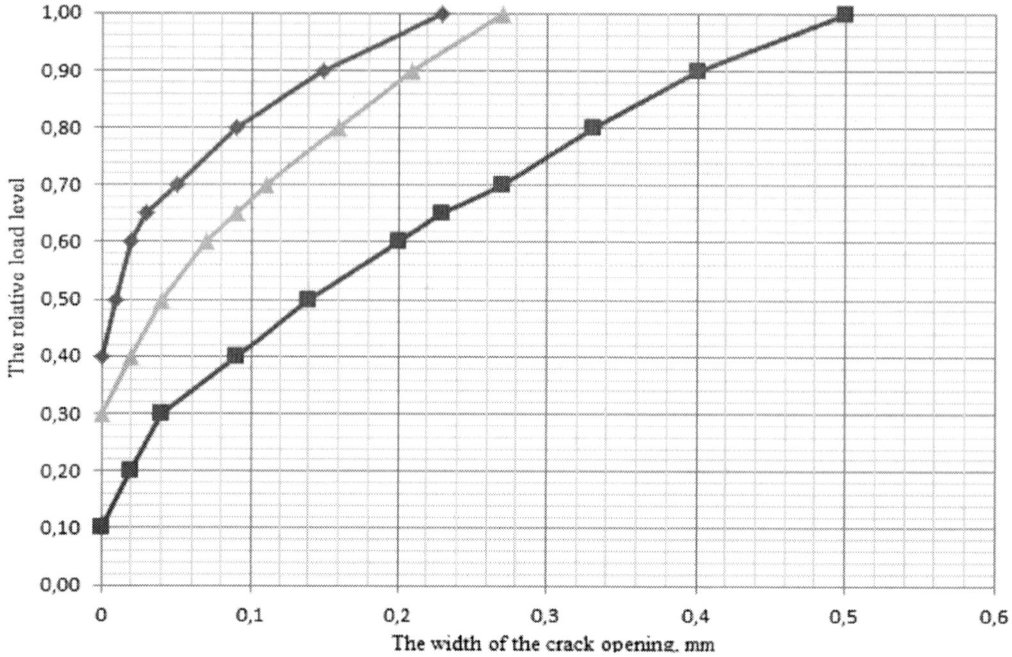

Fig. 11. The width of cracks in beams with the percentage of longitudinal reinforcement μ = 1,89%

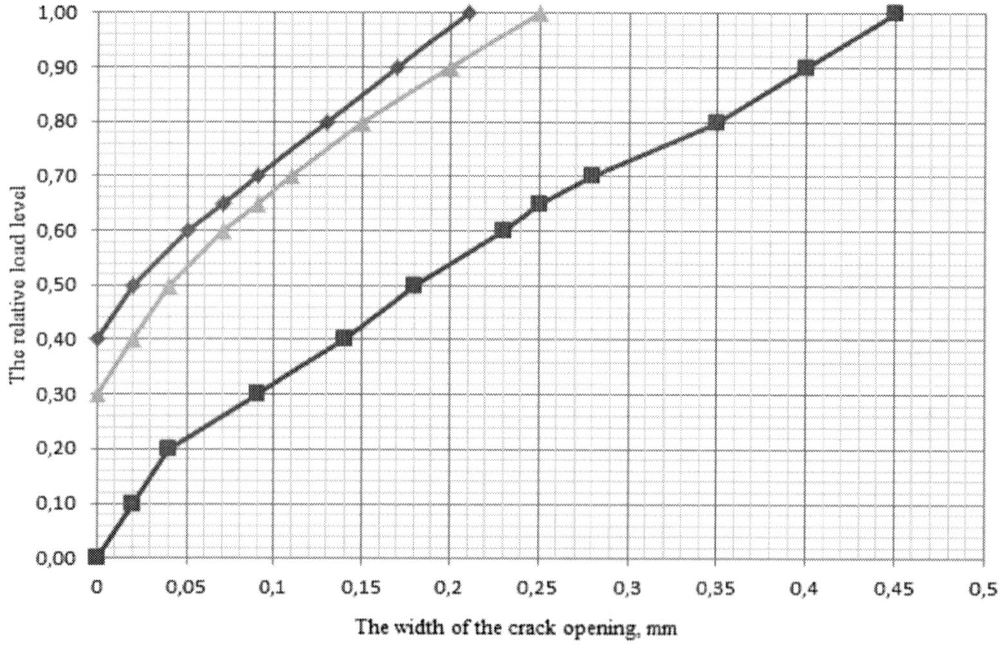

Fig. 12. The width of cracks in beams with the percentage of longitudinal reinforcement μ = 2,57%

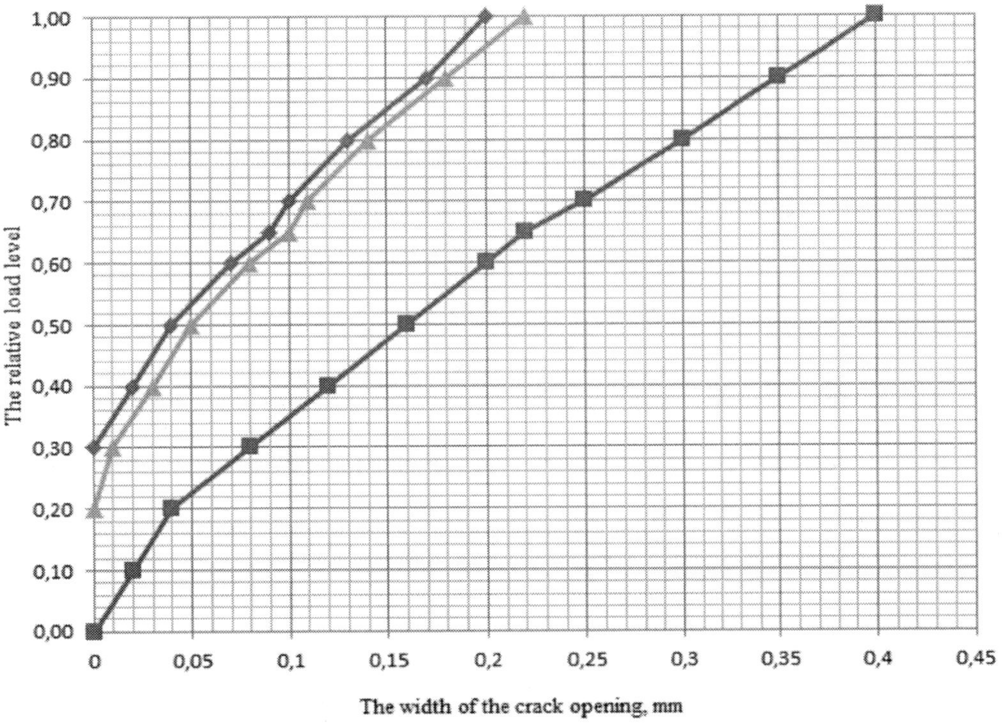

Fig. 13. The width of cracks in beams with the percentage of longitudinal reinforcement μ = 3,35%

On Fig. 9-13 blue line shows the crack opening width with respect to the relative load level for polymer concrete beams, red line - for concrete beams, green - for kauton reinforcement beams.

The width of cracks increased almost in proportion to the bending moments within the elastic reinforcement work. For beams, collapsing on the tension zone, the dependence of the width of the cracks from bending moments is close to linear.

Analysis of a graph of the width of the opening of normal cracks tested beams on the ratio of the relative load level also shows that the width of the opening of normal cracks of double-layered beams is largely dependent on the percentage of longitudinal reinforcement μ. Normal opening width of cracks decreases with increasing percentage of the longitudinal reinforcement.

3 Conclusions

1. The normal cross section of two-layered polymer concrete of flexural members in the middle of the span undergoes three stages of stress-strain state.
2. The strength of the flexural members kauton reinforcement, polymer concrete and concrete increases with the percentage of longitudinal reinforcement. By increasing the percentage of longitudinal reinforcement from 0.84% to 3.35% the strength of normal sections of polymer concrete beams increases almost linearly.
3. The load bearing capacity of structures in the tension zone of polymer concrete is higher than that of reinforced concrete structures. High grip reinforcement and polymer concrete prevents rapid development of plastic deformation in the reinforcement, as well as part of polymer concrete on tension zone between the cracks involved in construction as a whole.
4. In a small percentage of reinforcement, polymer concrete beam collapse under the

action of a higher load than in kauton reinforcement, but with an increased reinforcement, load bearing capacity of kauton reinforced elements becomes greater than polymer concrete by 5.4-6.5%.

5. Crack resistance of polymer concrete samples is significantly higher than that of concrete (average 78.05%). Crack resistance of polymer concrete samples is higher than kauton reinforcement (average 12.78%), but a small percentage of longitudinal reinforcement (μ = 0,84%) samples from kauton reinforcement showed the best evidence of crack resistance than polymer concrete (by 9.96%).
6. The tests allowed the identification of three stages of the formation of cracks in the normal bilayer polymer concrete of flexural members of the T-sections.
7. With increasing percentage of longitudinal reinforcement, effect of raising the crack resistance decreases. This is due to the fact that in a high percentage of reinforcement, significant part of the force at the period of crack experienced reinforcement and only a small proportion of the load acted on polymer concrete.
8. The difference in the deformation of polymer concrete on tensile zone of two-layered beams and extensibility limit of corresponding polymer concrete control samples, averagely is 10%. This is because high adhesion of polymer concrete with reinforcement increases the tensile limit of polymer concrete in constructions.

References

1. Yu.M. Borisov, *Bulletin of the Central regional branch RAASN: collection of scientific articles. - Voronezh: VGASU*, 133 – 137 (2010)
2. Yu.M. Borisov, *Scientific Bulletin VGASU "Construction and architecture". - Voronezh: VGASU*, 18-24 (2010)
3. Yu.M. Borisov, *Building mechanics and construction. - Voronezh: VGASU*, 76 (2012)
4. Yu.M. Borisov, Scientific Israel - Technological Advantages, **14 (2)**, 11 (2012)
5. Yu.M. Borisov, Industrial and civil construction, **7**, 47-49 (2010)
6. Yu.B. Potapov, *Proceedings of the international scientific conference (IV Academic Reading RAASN) "Actual problems of building materials": Sat. scientific. Article. - Penza*, 16-17 (1998)
7. Yu.B. Potapov, Building Materials of the XXI century, **9**, 9-10 (2000)
8. A.E. Polikutin, *The strength and crack resistance of the inclined sections of the bent structural elements of kauton reinforcement, dis. ... Cand. tehn. Sciences: 05.23.01* (2002)
9. O.L. Figovsky, Eastern European Journal of advanced technologies, 6(11), 14 – 20 (2014)
10. Nguyen Phan Duy *Double-layered rubber concrete of bent elements of building structures, diss. ... Cand. tehn. Sciences: 05.23.01* (2010)
11. Pinaev S.A. Short compressed elements of building constructions of an effective composite based on butadiene polymer [Text]: diss. ... Cand. tehn. Sciences: 05.23.01 / Pinaev Sergey. - Voronezh, 2001. - 191 p.
12. V.A. Chmykhov, *The resistance of the rubber concrete to aggressive media, diss. ... Cand. tehn. Sciences: 05.23.05* (2002)

Ecological-Economic and Technical Advantages of Reinforced Concrete Girders with Combined Reinforcement

Dmitriy Mailyan[1,*], *Anatoliy* Kubasov[1], and *Liya* Mailyan[1]

[1]Don State Technical University, 344022 Sotsialisticheskaya str. 162, Rostov-on-Don, Russian Federation

Abstract. Changing of usual reinforcement, having, as a rule, welding joints with the pre-stressed joint-free steel has a significant ecological meaning, reducing harmful emissions. Decrease of the concrete consumption is also solving this important ecological issue, since the volume of concrete and concrete dust is reduced. In the compressed elements extreme compressing forces in the reinforcement upon concrete destruction reach 400-500 MPa due to limited concrete compression, which is why use of extra-strong reinforcement in such elements is not reasonable. With the purpose of increase of extreme compressing forces in the reinforcement of compressed elements it is suggested to pre-compress it. Increase of limits of stress in the reinforcement allows to efficiently using extra-strong reinforcement in the compressed zone which will lead to substantial reduction of its consumption. Effect from the pre-compression of the reinforcement is the higher, the bigger is the reinforcement percent. Degree of increase of strength of the elements upon pre-compression of reinforcement with decrease of concrete strength increases.

1 Introduction

As it is known, reinforced concrete girders are the most common types of building constructions and spans of bridges. In such constructions axial forces are mostly operating, which allows to most efficiently use materials. In comparison with steel reinforced concrete girders in the spans of up to 30 meters provide for economy of metal of up to 70%. They are more durable and are distinguished with high fire-resistance and require less operational expenses. Reinforced concrete girders are almost twice lighter than reinforced concrete beams with solid wall.

Taking into consideration the indicated advantages of reinforced concrete girders the task of their improvement, the main trend of which is decrease of consumption of deficit steel and increase of technical, economic and ecological indices, gains special importance[1,2,3,4].

* Corresponding author: mailyan@rgsu.ru

It should be noted that changing of usual reinforcement, having, as a rule, welding joints with the pre-stressed joint-free steel has a significant ecological meaning, reducing harmful emissions. Decrease of the concrete consumption is also solving this important ecological issue, since the volume of concrete and concrete dust is reduced.

In the known girders the stretched lower chord, and sometimes the utmost rising diagonals are reinforced with extra-strong pre-stretched reinforcement to reduce steel consumption. Other elements of reinforced concrete girder containing non-extra-strong reinforcement are manufactured without pre-stressing.

2 Literature review

Ecological-Economic and Technical Advantages of Reinforced Concrete Girders with Combined Reinforcement is a very actually problem of modern building and constructions. A lot of scientists devote their scientific works to this problem. We can see the parts of this theme in the works of Ch. Alk. Apostolopoulos, M.P. Papadopoulos, Sp.G. Pantelakis; C.A. Apostolopoulos, M.P. Papadopoulos; V. Kumar, U. Kumar Sharma, B. Singh, P. Bhargava; E. Cadoni, M. Dotta, D.Forni, N. Tesio, C.Albertini; P. Dybeł, K. Furtak; P.Desnerck, Janet M. Lees, Chris T. Morley; O. Kelestemur, M. Halidun Kelestemur, S. Yildiz; E. Cadoni, L.Fenu, D.Forni, G. Riganti, E. Cadoni; X. Song, Y. Wu, X. Gu, Ch. Chen; P.Kopas, L. Jakubovičová, M.Vaško, M. Handrik; T. Xu, A. Castel ; J. Toribio, B. González, J.C. Matos, F.J. Ayaso; Michael D. Sangid, Garrett J. Pataky, H. Sehitoglu, Reginald F. Hamilton, Hans J. Maier; J. Yin, W. Wang, Z. Man, S. Khoo; R.R. Bhargava, S. Hasan; C. Xue, An He, H. Yong, Y. Zhou; M.V. Menshykova, O.V. Menshykov, Vita A. Mikucka, I. A. Guz; R.K. Joki, F. Grytten, B. Hayman, B.F. Sorensen; Di-Hua Tong, Xue-Ren Wu; M. El-Zeghayar, T.H. Topper, K.A. Soudki; Doo-Ho Cho, An-Dong Shin, Nam-Su Huh, Hyun-IkJeon; C. Fischer, C. Schweizer, T. Seifert; C. Mang, L. Jason, L.Davenne; M. Bruggi; M.Nematzadeh, A. Salari, J. Ghadami, M. Naghipour; Y. Xing, Q. Han, J. Xu, Q. Guo, Y. Wang ; P. Li, Yu-Fei Wu at their articles : Tensile behavior of corroded reinforcing steel bars BSt500s [1]; Stress concentration and fatigue of profiled reinforcing steels [2]; Tensile and low cycle fatigue behavior of corroded reinforcing steel bars S400 [3]; Effect of temperature on mechanical properties of pre-damaged steel reinforcing bars [4]; Mechanical behaviour of quenched and self-tempered reinforcing steel in tension under high strain rate [5]; The effect of ribbed reinforcing bars location on their bond with high-performance concrete [6]; Bond behaviour of reinforcing bars in cracked concrete; Improvement of Mechanical Properties of Reinforcing Steel Used in the Reinforced Concrete Structures [7]; Strain rate behaviour in tension of austenitic stainless steel used for reinforcing bars; Numerical simulation of the high strain-rate behavior of quenched and self-tempered reinforcing steel in tension [8]; Bond behaviour of reinforcing steel bars in early age concrete [9]; Fatigue Resistance of Reinforcing Steel Bars [10]; Modeling the dynamic stiffness of cracked reinforced concrete beams under low-amplitude vibration loads [11]; Effect of sudden load decrease on the fatigue crack growth in cold drawn prestressing steel [12]; High resolution analysis of opening and sliding in fatigue crack growth [13]; Modeling and analysis of gear tooth crack growth under variable-amplitude loading [14]; Crack-tip-opening displacement for four symmetrically situated cracks with coalesced interior yield zones [15]; Crack tip opening displacement in a linear strain hardening material; Interface cracks with initial opening under harmonic loading [16]; Determination of a cohesive law for delamination modelling – Accounting for variation in crack opening and stress state across the test specimen width [17]; Analysis of crack opening stresses for center- and edge-crack tension specimens [18]; A model of crack opening stresses in variable amplitude loading using smooth specimen fatigue test data for three steels [19]; Engineering estimates of crack opening displacement for non-idealized circumferential through-wall cracks in pipe [20]; Assessing steel strains on reinforced concrete members from surface cracking patterns [21]; Numerical limit analysis of steel-reinforced concrete walls and slabs [22]; A crack opening stress equation for in-phase and out-of-phase thermomechanical fatigue loading [23]; Crack opening estimate in reinforced concrete walls using a steel–concrete bond model [24]; A numerical method to generate optimal load paths in plain and reinforced concrete structures [25]; Stress-strain behavior of freshly compressed concrete under axial compression with a practical equation [26]; Experimental and numerical study on static behavior of elastic concrete-steel composite beams [27]; Stress–strain behavior of actively and passively confined concrete under cyclic axial load [28].

3 Ecological-Economic and Technical Advantages of Reinforced Concrete Girders with Combined Reinforcement

In the compressed elements extreme compressing forces in the reinforcement upon concrete destruction reach 400…500 MPa due to limited concrete compression, which is why use of extra-strong reinforcement in such elements is not reasonable. With the purpose of increase of extreme compressing forces in the reinforcement of compressed elements it is suggested to pre-compress it. This will lead to increase of compressing forces in the reinforcement upon concrete destruction by the value of pre-compressing forces. As the resultlimits of stress will be equal$R_{sc} + \sigma'_{sp}$, where$R_{sc}$is design resistance of reinforcement to compression, and$\sigma'_{sp}$is pre-compressive stress. Increase of limits of stress in the reinforcement allows to efficiently use extra-strong reinforcement in the compressed zone which will lead to substantial reduction of its consumption.

Decrease of steel consumption is explained by increase of limits of compression stress in the reinforcement of the compressed chord by the value of pre-compression of reinforcement. In extra-strong reinforcement of compressed chord of the girder limit compression stress equals to$R_{sc} + \sigma'_{sp}$,which will lead to decrease of steel consumption compared to elements without pre-stressing by$\left(R_{sc} + \sigma'_{sp}\right)/R_{sc}$times.

Besides, pre-compression of compressed chord reinforcement upon drawback of pre-stressing leads to girder bending which is substantially bigger than pre-stretch of lower chord reinforcement. It enables also increase of cracking resistance of the lower stretched chord [1,2].

Pre-stretched reinforcement could be produced from any extra-strong steel (rod, wire, cable, wire-rope), and reinforcement frame of the upper chord is manufactured from longitudinal extra-strong working rods and closed stirrups.

It should be noted that pre-compression of the extra-strong reinforcement in compressed girder elements is reasonable not only in girders with parallel chords, but in other types of girders. For example, in composite girders of trapezoidal and rectangular profile pre-stretching of lower chords and pre-compression of upper ones is carried out separately for each half, which before assembly are joined by welding of embedded and mounting steel elements.

In eaves girder it is convenient to pre-compress reinforcement of compressed chord and outmost vertical posts[3,4].

Elements of reinforced concrete girders containing pre-compressed and pre-stressed reinforcement, as a rule, have non-stressed reinforcement, as well, availability of which in the elements increases resistance of the latter upon transfer of strength of pre-stressing from the reinforcement to the concrete, allowing to increase the pre-stressing force.

For quantitative evaluation of relative content in the cross-section of pre-stressed reinforcement coefficients of stretching pre-stress$K_p$and compressing pre-stressK_c are introduced, numerical values of which are determined by formulas:

$$K_p = \frac{\sigma_{0.2}^{sp} A_{sp}}{\sigma_{0.2}^{sp} A_{sp} + \sigma_{0.2}^{sc} A_{sc} + \sigma_{0.2}^{s} A_s}, \tag{1}$$

$$K_c = \frac{\sigma_{0.2}^{sc} A_{sc}}{\sigma_{0.2}^{sc} A_{sc} + \sigma_{0.2}^{sp} A_{sp} + \sigma_{0.2}^{s} A_s}, \tag{2}$$

where$\sigma_{0.2}^{sp}$; $\sigma_{0.2}^{sc}$; $\sigma_{0.2}^{s}$; are conventional liquid limits of pre-stretched, pre-compressed and non-stressed reinforcement;

A_{sp}; $A_{sc}$and$A_s$are areas of cross-sections of pre-stretched reinforcement located in the stretched (or less compressed) zone of the element.

Similarly coefficientsK'_pandK'_c are determined, characterizing relative content of pre-stretched, pre-compressed and non-stressed reinforcement located in the compressed zone of the element.

Values of pre-stressing coefficient lay within the limits from 0 to 1.

For evaluation of efficiency of centrally compressed elements during pre-compression of the reinforcement let's review the element having symmetrically located pre-compressed and non-stressed reinforcement.

Resultant of pre-stress in the reinforcement will be equal to:

$$P_c = \sigma_{spc} A_{sc} + \sigma'_{spc} A'_{sc} + \sigma_{so} A_s + \sigma'_{so} A'_s, \tag{3}$$

σ_{spc}andσ'_{spc} – are pre-compressing stresses;

$\sigma_{so} = \sigma_8 - \sigma_{6,9}$; $\sigma'_{so} = \sigma'_8 - \sigma'_{6,9}$are stresses in non-stressed reinforcement, numerically equal to algebraic sum of losses from concrete shrinkageσ_8 (leading to compressing stress in the non-stressed reinforcement) and concrete creep (quickly creeping or long)$\sigma_{6,9}$and$\sigma'_{6,9}$. Stressesσ_{so}andσ'_{so}, as a rule, are compressing, since losses of pre-stress from shrinkage are more than from creep upon stretching by 2…4 times.

Before application of external load concrete section under the influence of forceP_c will turn out to be stretched. Destructing external compressing force Nupon central force application ($e_{ON} = e_{OP} = 0$ will be equal to:

$$N = R_B A + \varepsilon_{BR} E_s (A_{sc} + A'_{sc} + A_s + A'_s) + \sigma_{spc} A_{sc} + \sigma'_{src} A'_{sc} + \sigma_{sc}(A_s + A_{sc}) + \sigma_{sc}(A'_s + A'_{sc}), \tag{4}$$

and in case of absence of pre-stressing –

$$N_1 = R_B A + \varepsilon_{BR} E_s (A_{sc} + A'_{sc} + A_s + A'_s) + \sigma_8 (A_s + A_{sc} + A'_s + A'_{sc}), \tag{5}$$

Thus, element strength could be as a sum

$$N = N_1 + N_2, \tag{6}$$

where$N_1$is a force taken up by the same element, but without pre-stress and determined by formula (5) with consideration to compressing forces$\sigma_8$from concrete shrinkage; $N_2$is the additional force nulling the pre-stressing stretching force in the concrete and being strength growth of the element in pre-compression of the reinforcement, it equals:

$$N_2 = \sigma_{spc}A_{sc} + \sigma'_{spc}A'_{sc} - \sigma_{6.8}(A_s + A_{sc}) - \sigma'_{6.8}(A'_s + A'_{sc}), \quad (7)$$

It should be noted that if steel of low classes with physical liquid limit is used for non-stressed reinforcement, there might be a case when deformation in the reinforcement by the liquidity beginning will turn out to be less than limit compression of concrete $(\varepsilon_y < \varepsilon_{BR})$. In this case, rarely met in practice, in the formulas (4) and (5) stresses in the non-stressed reinforcement$A_s$andA'_sshould be taken as equal to liquid limitσ_y (instead of$\varepsilon_{BR}E_s$).

It should also be noted that formula (4) is true under the condition that summarized compressing stresses in the reinforcement upon element's destruction do not increase the conditional limit of reinforcement tension, i.e. subject to the condition$(\sigma_{spc} + \varepsilon_{BR}E_s) \leq \beta\sigma_{0.2}$, where$\beta = \sigma_{el}/\sigma_{0.2}$. Otherwise, which is rarely met in the practice of compressed elements designing, non-linear dependence should be used between stresses and deformations in the reinforcement.

Let us show on the example of centrally compressed short (non-flexible) element the efficiency of pre-compressing of reinforcement. Ratio of forces taken up by the longitudinal reinforcement and concrete upon depletion of strength of elements equals to:

$$\frac{N_s}{N} = \frac{\mu(R_{sc}+\sigma_{spc})}{R_B+\mu(R_{sc}+\sigma_{spc})}; \quad (8)$$

$$\frac{N_B}{N} = \frac{R_B}{R_B+\mu(R_{sc}+\sigma_{spc})}, \quad (9)$$

where$N$, $N_s$and$N_B$are stresses taken up by the reinforced concrete element, longitudinal reinforcement and concrete, correspondingly; $\sigma_{spc}$are value of pre-compression of the reinforcement.

With concrete of the classB30 and reinforcement with design compression resistance$R_{sc} = 400\ MPa$,If the reinforcement is not pre-compressed $(\sigma_{spc} = 0)$, dependence of the indicated stresses from the percentage of reinforcement is shown with slightly curved lines. By these lines it is seen that even with$\mu = 2\%$only 30% of total stress is taken by the non-stressed reinforcement. If this reinforcement is pre-compressed to reach resistance $\sigma_{spc} = 400\ MPa$, the stress taken up by it rises to 50%, and bearing capacity of the element increases by more than 30%.

Let us analyze the influence of various factors on the degree of increase of strength of reinforced concrete elements with pre-compressed reinforcement. In central compression of elements with symmetrical reinforcement strength of the element containing pre-compressed reinforcement, will be equal to:

$$N_c = R_B A + R_{sc}(A_s + A_{sc}) + \sigma_{spc}A_{sc}, \quad (10)$$

where the first two summands of the first part represent the strength of element without pre-stress N. Ratio

$$\frac{N_c}{N} = 1 + \frac{\sigma_{spc}\mu_{sc}}{R_B+\mu R_{sc}}, \quad (11)$$

depends on the prism strength of concreteR_B, value of pre-compression of the reinforcement$\sigma_{spc}$and reinforcement coefficient – summarized$\mu$and of pre-compressed reinforcement μ_{sc}.

4 Conclusions

With increase of values of pre-stress coefficients K_c, i.e. relative content of pre-compressed reinforcement, from 0.25 to 1 with $\mu = 2.55\%$ relative strength of elements rises by 20...25%. Effect from the pre-compression of the reinforcement is the higher, the bigger is the reinforcement percent – with its increase from 0.65% to 2.55% relative strength of the element rises by 17...22%. Degree of increase of strength of the elements upon pre-compression of reinforcement with decrease of concrete strength increases. Thus, with $K_c = 1$ and $\mu = 2,55\%$ decrease of concrete strength R_B from 40 to 20 MPa leads to growth of relative strength of the element by 13%, and with $\mu = 1,13\%$ - by 6%.

Thus, use of reinforced concrete girders of the suggested construction allows increasing technical, ecological and economical indices of such constructions.

References

1. Ch. Alk. Apostolopoulos, M.P. Papadopoulos, Sp.G. Pantelakis, Construction and Building Materials, **20**, 782-789 (2006)
2. C.A. Apostolopoulos, M.P. Papadopoulos, Construction and Building Materials, **21**, 855-864 (2007)
3. V. Kumar, U. Kumar Sharma, B. Singh, P. Bhargava, Construction and Building Materials, **46**, 19-27 (2013)
4. E. Cadoni, M. Dotta, D.Forni, N. Tesio, C.Albertini, Materials & Design, **49**, 657-666 (2013)
5. P. Dybeł, K. Furtak, Archives of Civil and Mechanical Engineering, **15**, 1070-1077 (2015)
6. P.Desnerck, Janet M. Lees, Chris T. Morley, Construction and Building Materials, **94**, 126-136 (2015)
7. O. Kelestemur, M.HalidunKelestemur, S. Yildiz, Journal of Iron and Steel Research, International, **16**, 55-63 (2009)
8. E. Cadoni, L.Fenu, D.Forni, Construction and Building Materials, **35**, 399-407 (2012)
9. G. Riganti, E. Cadoni, Materials & Design, **57**, 156-167 (2014)
10. X. Song, Y. Wu, X. Gu, Ch. Chen, Construction and Building Materials, **94**, 209-217 (2015)
11. P.Kopas, L. Jakubovičová, M.Vaško, M. Handrik, Procedia Engineering, **136**, 193-197 (2016)
12. T. Xu, A. Castel, Journal of Sound and Vibration, **368**, 135-147 (2016)
13. J. Toribio, B. González, J.C. Matos, F.J. Ayaso, International Journal of Fatigue, **76**, 53-59 (2015)
14. M. D. Sangid, G. J. Pataky, H. Sehitoglu, R. F. Hamilton, H. J. Maier, International Journal of Fatigue, **37**, 134-145 (2012)
15. J. Yin, W. Wang, Z. Man, S. Khoo, Mechanical Systems and Signal Processing, **40**, 105-113 (2013)
16. R.R. Bhargava, S. Hasan, Applied Mathematical Modelling, **36**, 5741-5749 (2012)
17. C. Xue, An He, H. Yong, Y. Zhou, Mechanics of Materials, **66**, 21-34 (2013)
18. M.V. Menshykova, O.V. Menshykov, Vita A. Mikucka, I. A. Guz, Composites Science and Technology, **72**, 1057-1063 (2012)

19. R.K. Joki, F. Grytten, B. Hayman, B.F. Sorensen, Composites Science and Technology, **128**, 49-57 (2016)
20. Di-Hua Tong, Xue-Ren Wu, Chinese Journal of Aeronautics, **27**, 291-298 (2014)
21. M. El-Zeghayar, T.H. Topper, K.A. Soudki, International Journal of Fatigue, **33**, 1337-1350 (2011)
22. Doo-Ho Cho, An-Dong Shin, Nam-Su Huh, Hyun-IkJeon, Engineering Fracture Mechanics, **142**, 236-254 (2015)
23. C. Fischer, C. Schweizer, T. Seifert, International Journal of Fatigue, **88**, 178-184 (2016)
24. C. Mang, L. Jason, L.Davenne, Archives of Civil and Mechanical Engineering, **16**, 422-436 (2016)
25. M. Bruggi, Computers & Structures, **170**, 26-36 (2016)
26. M.Nematzadeh, A. Salari, J. Ghadami, M. Naghipour, Construction and Building Materials, **115**, 402-423 (2016)
27. Y. Xing, Q. Han, J. Xu, Q. Guo, Y. Wang, Journal of Constructional Steel Research, **123**, 79-92 (2016)
28. P. Li, Yu-Fei Wu, Composite Structures, **149**, 369-384 (2016)

MATEC Web of Conferences 73, 04020 (2016) DOI: 10.1051/matecconf/20167304020
TPACEE-2016

Ecologically safe and techno economically efficient reinforced concrete constructions of equal resistance

Dmitriy Mailyan[1,*], and *Liya* Mailyan[1]

[1]Don State Technical University, 344022, Sotsialisticheskaya, 162, Rostov-on-Don, Russian Federation

Abstract. Creation of reinforced concrete constructions of ideally equal resistance is practically impossible, however maximum approximation to such solution should become the main purpose of constructions designing, since it provides for minimal materials consumption.
Changing of total cross section of pre-stressed reinforcement along the span with the purpose of approximation to elements of equal resistance is possible during creation of stepped section of the element.
In particular, the suggested solution allows to closely approximate the ideal beam of equal resistance, at which with decrease of external moment the height of element's cross-section and the area of cross-section of extra-strong pre-stressed reinforcement are also decreased.
In accordance with the existing reinforced concrete construction with the same class of concrete, reinforcement steel and load the suggested one due to more advanced profile is the most economic - the consumption of steel, concrete and weight of the beam are lower in it. This increases the ecological safety of constructions, as well.

1 Introduction

Taking into consideration that reinforced concrete constructions today and in the nearest future are being the basis of construction the issues of increase of their ecologic safety and techno economical attractiveness are rather pressing [1-5].

It is known that choice of sections of reinforced concrete constructions is carried out by maximum applied load. At that the major part of constructions has unjustified safety margin. It leads to excessive consumption of materials and increase of environmental pollution. Creation of reinforced concrete constructions of ideally equal resistance is practically impossible, however maximum approximation to such solution should become the main purpose of constructions designing, since it provides for minimal materials consumption.

* Corresponding author: mailyan@rgsu.ru

One of the known ways of constrictions creation approximating the elements of equal resistance is changing of dimensions of cross sections across the element in accordance with the moment diagram from external loads. For single-span beams, as well as consoles from metal or wood, in this way almost ideal elements of equal resistance could be achieved. In reinforced concrete it is more difficult, however creation of variable section of elements in this case also leads to significant effect [1,2,3].

2 Literature review

Ecologically safe and technoeconomically efficient reinforced concrete constructions of equal resistance is a very actually problem of modern building and constructions. A lot of scientists devote their scientific works to this problem. We can see the parts of this theme in the works of F. Bencardino, A. Condello; P. O. Awoyera, J.O. Akinmusuru, J.M. Ndambuki; Ch.K. Kankam, M. Adom-Asamoah; Sung-Won Yoo, Jinkyo F. Choo; Doo-Yeol Yoo, Young-Soo Yoon, N. Banthia; K. Celik, M.D. Jackson, M. Mancio, C. Meral, A.-H. Emwas, P.K. Mehta, P.J.M. Monteiro; J. Wei, C. Meyer; E. Syroka-Korol, J. Tejchman; Gritsada Sua-Iam, N. Makul; Khoa T. Nguyen, N. Ahn, Tuan A. Le, K. Lee; Jerry M. Paris, Justin G. Roessler, Christopher C; M. Aslam, P. Shafigh, Mohd ZJumaat, M.Lachemi; Hanaa K. Alwan Al-Bayati, Prabir Kr. Das, Susan L. Tighe, H.Baaj; Y. Geng, Y. Wang, J. Chen; P. Zhu, X. Qi Mao, W. Qu, Z. Li, Zhongguo J. Ma; T. Proske, S.Hainer, MoienRezvani, Carl-Alexander Graubne; J. Hu, K. Wang, James A. Gaunt; Se J. Choi, Soo-Hwan Kim, Su-Jin Lee, R. Won, Jong-Pil Won; I. Kovacic, V. Zoller; E. Moretti, E.Belloni, F. Agosti; F. Fedorik, M. Malaska, R. Hannila; A. Haapala; M. Maria Sesana, M.Grecchi, G. Salvalai, C. Rasica; C. Thomas, J. Setién, J.A. Polanco; M. Nematzadeh, A. Salari, J. Ghadami, M. Naghipour; Y. Xing, Q. Han, J. Xu, Qi Guo, Y. Wang, P. Li, Yu-Fei Wu; J. Yin, W. Wang, Z. Man, S. Khoo; R.R. Bhargava, S. Hasan; C. Xue, An He, H. Yong, Y. Zhou; M.V. Menshykova, O.V. Menshykov, Vita A. Mikucka, Igor A. Guz; R.K. Joki, F. Grytten, B. Hayman, B.F. Sorensen; Di-Hua Tong, Xue-Ren Wu; M. El-Zeghayar, T.H. Topper, K.A. Soudki; Doo-Ho Cho, An-Dong Shin, Nam-Su Huh, Hyun-Ik Jeon; C. Fischer, C. Schweizer, T. Seifert at their articles : Eco-friendly external strengthening system for existing reinforced concrete beams [1]; Green concrete production with ceramic wastes and laterite [2]; Strength and ductility characteristics of reinforcing steel bars milled from scrap metals [3]; Evaluation of the flexural behavior of composite beam with inverted-T steel girder and steel fiber reinforced ultra high performance concrete slab [4]; Predicting the post-cracking behavior of normal- and high-strength steel-fiber-reinforced concrete beams [5]; High-volume natural volcanic pozzolan and limestone powder as partial replacements for portland cement in self-compacting and sustainable concrete [6]; Improving degradation resistance of sisal fiber in concrete through fiber surface treatment [7]; Experimental investigations of size effect in reinforced concrete beams failing by shear [8]; Utilization of coal- and biomass-fired ash in the production of self-consolidating concrete [9]; Theoretical and experimental study on mechanical properties and flexural strength of fly ash-geopolymer concrete [10]; A review of waste products utilized as supplements to Portland cement in concrete [11]; Benefits of using blended waste coarse lightweight aggregates in structural lightweight aggregate concrete [12]; Evaluation of various treatment methods for enhancing the physical and morphological properties of coarse recycled concrete aggregate [13]; Time-dependent behaviour of steel tubular columns filled with recycled coarse aggregate concrete [14]; Investigation of using recycled powder from waste of clay bricks and cement solids in reactive powder concrete [15]; Eco-friendly concretes with reduced water and cement contents — Mix design principles and laboratory tests [16]; Recycling lead-based paint contaminated deconstructed masonry materials as aggregate for Portland cement concrete—

A cost effective and environmental friendly approach [17]; Mix proportion of eco-friendly fireproof high-strength concrete [18]; Building life cycle optimization tools for early design phases [19]; Innovative mineral fiber insulation panels for buildings [20]; Thermal and acoustic characterization Original [21]; Improving the thermal performance of concrete-sandwich envelopes in relation to the moisture behaviour of building structures in boreal conditions [22]; Methodology of energy efficient building refurbishment: Application on two university campus-building case studies in Italy with engineering students [23]; Structural recycled aggregate concrete made with precast wastes [24]; Experimental and numerical study on static behavior of elastic concrete-steel composite beams [25]; High resolution analysis of opening and sliding in fatigue crack growth[26]; Stress-strain behavior of freshly compressed concrete under axial compression with a practical equation [27]; Experimental and numerical study on static behavior of elastic concrete-steel composite beams [28]; Stress–strain behavior of actively and passively confined concrete under cyclic axial load [29]; Modeling and analysis of gear tooth crack growth under variable-amplitude loading [30]; Crack-tip-opening displacement for four symmetrically situated cracks with coalesced interior yield zones [31]; Crack tip opening displacement in a linear strain hardening material [32]; Interface cracks with initial opening under harmonic loading [33]; Determination of a cohesive law for delamination modelling – Accounting for variation in crack opening and stress state across the test specimen width; Analysis of crack opening stresses for center- and edge-crack tension specimens; A model of crack opening stresses in variable amplitude loading using smooth specimen fatigue test data for three steels; Engineering estimates of crack opening displacement for non-idealized circumferential through-wall cracks in pipe [34]; A crack opening stress equation for in-phase and out-of-phase thermomechanical fatigue loading [35].

3 Ecologically safe and techno economically efficient reinforced concrete constructions of equal resistance

In this case the most probable solutions include reinforced concrete beams with inclined upper or lower edges, double-wedge section, with curvilinear edges, oval profile, with stepped edges, etc. In reinforced concrete elements changing of the bearing capacity of various sections could be achieved also by way of changing of longitudinal reinforcement which is carried out by creation of interruptions of parts of fixtures or changing on some parts of the class and (or) diameter of one fixture to another with connection of them by jam welding.

Changing of bearing capacity of cross sections of reinforced concrete beams across their span with mixed reinforcing when preliminary tensioning is applied not to all extra-strong fixture but to only part of it is achieved by interruptions of non-stressed parts of reinforcement.

In the beams where all longitudinal reinforcement is pre-stressed, interruptions of its part in the span in the existing solutions are not provided for, which leads to excess margin of safety and cracking resistance on weakly loaded parts of single span bending elements [4,5].

Changing of total cross section of pre-stressed reinforcement along the span with the purpose of approximation to elements of equal resistance is possible during creation of stepped section of the element.

With the purpose of significant decrease of extra-strong steel consumption, as well as concrete, a new design solution is suggested, in which the beam has a stepped profile allowing to interrupt stressed reinforcement in the span in accordance with the materials' curve. Such a solution allows to maximally approximate the beam of equal resistance.

In the solutions suggested, unlike in typical ones, combined pre-stress is taken, in which extra-strong reinforcement of the stretched zone is pre-stretched, and compressed zone is pre-compressed. It allows not only to increase cracking resistance of the stretched zone, but also increase tension in the reinforcement of the compressed zone before deformation of compressed concrete to values $R_{sc} + d\sigma_{sp}^{'}$.

Reinforced concrete beam has concrete steps creating stepped profile. Interrupted pre-stressed reinforcement is anchored in concrete steps of the beam with stepped profile. Pretension effort from the cuddy to the reinforcement is translated via inventory pull rods, and initial force is translated the same way, but through rigid inserts. Anchoring of pre-stressed reinforcement is carried out by collet closing mechanisms.

The beam is manufactured as usual – assembly of reinforcement cage, its installation and fastening in the form – frame, beam casting, transfer of pre-stress from the reinforcement to the concrete to reach the transfer strength by the concrete, beam stripping.

To receive certain technical and economic data of the proposed solution and compare with typical beams the latter have been re-designed in accordance with the indicated proposal. For comparison a typical pre-stressed reinforced concrete I-pile beam with parallel chords - serie 1.462.1 - 1/88. Longitudinal reinforcement of typical beam along the whole length. Main working pre-stressed reinforcement - 4Ø20, A1000 heavy-weight concrete class B30.

Evaluation of efficiency of reinforcement use could be carried out with the help of determination of coefficient, equal to ratio of the area of moments' curve from external forces to the area of the moments' curve from internal efforts (materials' curve) for the suggested beam and the typical one, equal toη= Am/Abn. In the ideal beam of equal resistance this ratio equals 1. In the actual beams it is less than 1. With its decrease the efficiency of beam materials use, in particular, of the reinforcement, is decreased.

In the typical reinforced concrete beam, as well as in other reinforced concrete beams with parallel edges and pre-stressed reinforcement, the cross-section of which along the beam is constant, the coefficient of the reinforcement use is minimal, in the reviewed beam it isn=0.58. Its value could be significantly increased with the help of interruptions of pre-stressed reinforcement in the span of the beams with stepped profile. The suggested beam of stepped profile with combined pre-stress with full satisfaction of all technical requirements at the same time has the best economic values.

In the beams with interruptions of longitudinal pre-stressed reinforcement in the stretched and compressed zones the coefficient of reinforcement use efficiency isn=0.8.

Consumption of longitudinal extra-strong reinforcement in the stretched and compressed zones in comparison to typical solution is decreased by 12.5%. Total consumption of longitudinal reinforcement decreases by 18 ... 21%.

Results of calculations of the typical and suggested beams with the span of 12 m, reviewed above, show that ratio of the moment accepted by the cross-section to the moment from the external load *Ms/Mon* drawing nearer from the average cross-section of the single span beam to the support in the suggested beam increases in the substantially lower degree, than in the typical one. It is related also to the ratio of cracks formation moment *Mckc* to the external moment *M.*

In particular, the suggested solution allows to closely approximate the ideal beam of equal resistance, at which with decrease of external moment the height of element's cross-section and the area of cross-section of extra-strong pre-stressed reinforcement are also decreased. In the modern construction ribbed reinforced concrete slab roofs are rather widely used. These are mainly slabs with cross-section of flat roof with webs on axial edges in which with rather big spans, as a rule, extra-strong pre-tensioned reinforcement is located. Another type of reinforced concrete slab roofs is two-console slabs of the double-T type. In the latter axial edges are located ¼ of the width of the slab from the edges of the slab. At that console corbels of flat flanges are lessening the load of span cross-sections which makes the reinforcement of the flange easier.

Main working longitudinal reinforcement located in the longitudinal edges at its pre-stressing is stretched along the whole span- from one edge to the other. Reinforcement calculated by the highest bending moment in the middle of the span, at the near-support parts of single span slabs turns out to be more necessary, and pre-stress of these parts could lead to destruction of ends and opening of cracks on the upper backs. At that excess safety margin and cracking resistance are created on these parts.

For creation of more economic slabs approximating the elements of equal resistance it is suggested to use the same principle as in pre-stressed beams. With the purpose of decreasing of consumption of the expensive deficit reinforcement it is recommended to create stepped profile of longitudinal edges of slabs, which allows to interrupt pre-stressed reinforcement in the span in accordance with the curve of internal efforts (materials' curve) and anchor it in the steps of stepped profile.

For quantitative evaluation of efficiency of the suggested solutions of reinforced concrete pre-stressed slabs a full calculation of slabs of the suggested construction on both groups of limit states taking into consideration peculiarities related to stepped profile of elements and variable pre-stress gain along the span has been carried out.

It should be noted that bowing of certain elements across the span are changed not only because of changes of values of bending moments, but also in relation to changes in the height of element's cross-section, area of reinforcement cross-section and clamping force. Sags, cracks formation moments and width of cracks opening were determined with consideration to actual characteristic of each cross-section.

With practically equal technical characteristic in the suggested solutions the economic values are higher than in the typical slab. Consumption of extra-strong reinforcement classA1000 is by8 ... 9% less, and slab weight is by 120...125 kg less than in the typical slab. Re-designing of pre-stressed two-console reinforced concrete double-T slabs with the span of 12 m with wire-rope reinforcement based on the suggested principle allowed to obtain even bigger technical-economic effect compared with traditional solution.

In the reviewed slab the existing traditional solution is that main longitudinal working reinforcement in each of two edges is pre-tensioned. In the suggested solution reinforcing and pre-stressing are assumed as the same, however, pre-stressed reinforcement of the stretched zone, unlike the existing solutions, is interrupted in the span of two cross-sections in accordance with the moments' curve. Both types of two console slabs are designed to evenly distributed load.

4 Conclusions

Comparison of technical-economic indices of double-console pre-stressed reinforced concrete traditional and suggested slabs has shown the absolute advantage of the latter. As well as in other previously reviewed types of constructions, creation of conditions approximating the elements to the construction of equal resistance substantially increases their economic efficiency. In this case the economy of extra-strong reinforcement by 14% is achieved and substantial increase of coefficient of use of working reinforcement.

Positive factor in the suggested solution is also spreading of efforts of concrete clamping transferred to the edges of the construction. Unlike traditional solutions in the suggested one the clamping force is transferred not on the one end section, but on the ends of every part of stepped profile of the construction.

In accordance with the existing reinforced concrete construction with the same class of concrete, reinforcement steel and load the suggested one due to more advanced profile is the most economic- the consumption of steel, concrete and weight of the beam are lower in it. This increases the ecological safety of constructions, as well.

References

1. F. Bencardino, A. Condello, Composites Part B: Engineering, **93**, 163-173 (2016)
2. P. O. Awoyera, J.O. Akinmusuru, J. M. Ndambuki, Construction and Building Materials, **117**, 29-36 (2016)
3. Ch. K. Kankam, M. Adom-Asamoah , Materials & Design, **23**, 537-545 (2002)
4. Sung-Won Yoo, Jinkyo F. Choo, Engineering Structures, **118**, 1-15 (2016)
5. Doo-Yeol Yoo, Young-Soo Yoon, N. Banthia, Construction and Building Materials, **93**, 477-485 (2015)
6. K. Celik, M.D. Jackson, M. Mancio, C. Meral, A.-H. Emwas, P.K. Mehta, P.J.M. Monteiro, Cement and Concrete Composites, **45**, 136-147 (2014)
7. J. Wei, C. Meyer, Applied Surface Science, **289**, 511-523 (2014)
8. E. Syroka-Korol, J. Tejchman, Engineering Structures, **58**, 63-78 (2014)
9. Gritsada Sua-Iam, N. Makul, Journal of Cleaner Production, **100**, 59-76 (2015)
10. Khoa T. Nguyen, N. Ahn, Tuan A. Le, K. Lee, Construction and Building Materials, **106**, 65-77 (2016)
11. Jerry M. Paris, Justin G. Roessler, Christopher C., Journal of Cleaner Production, **121**, 1-18 (2016)
12. M. Aslam, P. Shafigh, Mohdz Jumaat, M. Lachemi, Journal of Cleaner Production, **119**, 108-117 (2016)
13. Hanaa K. Alwan Al-Bayati, Prabir Kr. Das, Susan L. Tighe, H. Baaj, Construction and Building Materials, **112**, 284-298 (2016)
14. Y. Geng, Y. Wang, J. Chen, Journal of Constructional Steel Research, **122**, 455-468 (2016)
15. P. Zhu, X. Qi Mao, W. Qu, Z. Li, Zhongguo J. Ma, Construction and Building Materials, **113**, 246-254 (2016)
16. T. Proske, S. Hainer, M. Rezvani, Carl-Alexander Graubner, Cement and Concrete Research, **51**, 38-46 (2013)
17. J. Hu, K. Wang, James A. Gaunt, Conservation and Recycling, **54**, 1453-1460 (2010)
18. Se J. Choi, Soo-Hwan Kim, Su-Jin Lee, R. Won, Jong-Pil Won, Construction and Building Materials, **38**, 181-187 (2013)
19. I. Kovacic, V. Zoller, Energy, **92**, 409-419 (2015)
20. E. Moretti, E. Belloni, F. Applied Energy, **169**, 421-432 (2016)
21. F. Fedorik, M. Malaska, R. Hannila, A. Haapala, Energy and Buildings, **107**, 226-233 (2015)
22. M. Maria Sesana, M. Grecchi, G. Salvalai, C. Rasica, Journal of Building Engineering, **6**, 54-64 (2016)
23. C. Thomas, J. Setién, J.A. Polanco, Construction and Building Materials, **114**, 536-546 (2016)
24. M. Nematzadeh, A. Salari, J. Ghadami, M. Naghipour, Construction and Building Materials, **115**,402-423 (2016)
25. Y. Xing, Q. Han, J. Xu, Qi Guo, Y. Wang, Journal of Constructional Steel Research, **123**, 79-92 (2016)
26. P. Li, Yu-Fei Wu, Composite Structures, **149**, 369-384 (2016)
27. J. Yin, W. Wang, Z. Man, S. Khoo, Mechanical Systems and Signal Processing, **40**, 105-113 (2013)
28. R.R. Bhargava, S. Hasan, Applied Mathematical Modelling, **36**, 5741-5749 (2012)
29. C. Xue, An He, H. Yong, Y. Zhou, Mechanics of Materials, **66**, 21-34 (2013)
30. M.V. Menshykova, O.V. Menshykov, Vita A. Mikucka, Igor A. Guz, Composites Science and Technology, **72**, 1057-1063 (2012)

31. R.K. Joki, F. Grytten, B. Hayman, B.F. Sorensen, Composites Science and Technology, **128**, 49-57 (2016)
32. Di-Hua Tong, Xue-Ren Wu, Chinese Journal of Aeronautics, **27**, 291-298 (2014)
33. M. El-Zeghayar, T.H. Topper, K.A. Soudki, International Journal of Fatigue, **33**, 1337-1350 (2011)
34. Doo-Ho Cho, An-Dong Shin, Nam-Su Huh, Hyun-Ik Jeon, Engineering Fracture Mechanics, **142**, 236-254 (2015)
35. C. Fischer, C. Schweizer, T. Seifert, International Journal of Fatigue, **88**, 178-184 (2016)

MATEC Web of Conferences **73**, 04021 (2016) DOI: 10.1051/matecconf/20167304021
TPACEE-2016

The Effectiveness of Materials Different with Regard to Increasing the Durability

Vladimir Erofeev[1,*], *Anna* Dergunova[1], *Anna* Piksaikina[1], *Andrey* Bogatov[1], *Evgeny* Kablov[2], *Oleg* Startsev[2], and *Aleksandr* Matvievskiy[3]

[1]Ogarev Mordovia State University, 430005 Saransk, Russian Federation
[2]All-Russian scientific research institute of aviation materials, 105005 Moscow, Russian Federation
[3]Maxmir' Open Joint-Stock Company, 121099 Moscow, Russian Federation

Abstract. The article considers contemporary materials and structures for construction of buildings. The article conducts an economic study of the problem of durability. It addresses the issue of increasing longevity, affecting the term of service to building structures and the efficiency of their operation. Revealed the main factors affecting the durability. It identifies measures its realisation. The method of calculation of economic efficiency of improving the durability of building constructions.

1 Introduction

Recently, in connection with the acceleration of introduction of scientific and technical progress, an increasing attention is paid to the introduction of new aircraft, industrial, electrical and construction materials and increasing of durability of machines, units, buildings and structures, as the most important indicators of the quality of industrial products.

Assessment of the level material allows to accurately set the relationship between the quantity produced in a particular product and indirect costs in material resources, as well as to evaluate the rationality of the latter. The indicator reflects the impact of increased material costs associated with the modernization of the production process, the production and the efficiency of economic activity in general.

The reduction of the material production is an important direction of increasing of economic efficiency of production as the economical use of energy and material resources ensures the continuous growth of output and reduction of production costs. Conservation of resources has become a crucial source to meet the growing needs of the country in raw materials, materials, fuel, energy.

2 Materials and Methods

In many sectors of the economy, reduction in material costs is achieved by the development and use of more economical types of products, materials and constructions, machines and

* Corresponding author: erofeev.vt@yandex.ru

mechanisms. Reduction of weight characteristics of machinery and equipment without compromising quality an indicator of their high technical level, deepening the intensification of production.

One of the ways to reduce material intensity is the use of progressive types of material resources: composite materials, metal powders, polymers, plastics and etc. Their application makes it possible not only to reduce the cost of raw materials, but also reduce the complexity of products, increase the loading of equipment. The important reserve of reduction of the material production is the organization of process stages, for example, rational cutting of metal and wood, as well as reduction of scraps and waste. This facilitates the introduction of advanced technological processes, increasing technological discipline, improving the organization of production and labor [1].

The annual economic effect of reducing the material intensity by two options of production, with a planned amount of output is determined by the formula:

$$E_g = ((S_b + E_n \cdot K_b) - (C_2 + E_n \cdot K_2)) \cdot Q, \tag{1}$$

where Sb – unit cost of the product when using technology in the base case; En – norm of efficiency; Kb – the specific capital investments using basic equipment option; C2 – unit cost of the product when using the technique in the second embodiment; K2 – specific capital costs when using technology of the second option; Q – volume of the planned products.

Durability is a property of a system to remain operational until the occurrence of the limit state with the necessary interruptions for maintenance and repairs. The limiting condition is determined by technical impossibility or inexpediency of further operation of the system and is caused by safety requirements or a permanent reduction in the efficiency [1].

Overall durability of industrial products is determined by economically optimal lifespan, representing a calendar time, during which their maintenance is economically justified and safe.

Durability has a huge impact on the efficiency of operation of devices, apparatuses, components and structures. For its estimation the most commonly used technical and economic indicators. Durability affects the level of national economic costs both in manufacturing and in the field of consumer products. Increase the durability of any items pushes the beginning of repairs by the consumer in their recovery and may extend the economic life of manufactured products. In the production of devices, apparatus, components and structures measures to improve the durability, usually require an additional one-off and ongoing costs at the same time by improving the durability may be reduced demand for its products, which in General leads to savings in capital investments for their creation. The impact of longevity on the scale of need for them is manifested through the change of life. The higher the economically optimum service life of the devices, apparatus, components and structures, the less the need for them.

Durability, having a major impact on the efficiency of the production and use of industrial products, in turn, depends on many factors acting at all stages: design, manufacturing and operation.

At the design stage the main factors influencing durability include:

- constructive decisions;
- selection of materials;
- production technology.

At the manufacturing stage there are the following factors, influenced on durability:

- level of technology, technology and culture of production;
- technological discipline;

• the quality of the materials and components used in the manufacture of apparatus, devices, products and constructions and their elements;

• the current output and quality control.

At the stage of operation the level of durability laid in the previous stages is realized, and it depends on the following factors:

• adherence to prescribed operation modes;

• level of qualification and discipline of staff;

• the quality of preventive maintenance;

• natural aging of the component parts of apparatus, instruments, products and structures;

• the state of the external environment in which products operate and design.

In domestic and world practice great attention is paid to improving the durability of the materials, industrial, aviation, electrical and construction purposes, operating in conditions of biological aggression. For further consideration of a large range of industrial materials select building materials and building structures on their basis.

Biological damage of building structures is one of the basics external factors that determines the rate of depreciation of buildings and other engineering structures (along with exposure to aggressive gases, acid rain, freezing, weathering, etc.) [2-16].

In the design and construction of industrial buildings and structures exposed to the biodestructors, you must strive to ensure that by appropriate measures for the protection of building structures from biological damage to ensure the smooth operation of basic funds of enterprises within the regulatory period of service [3]. This can be achieved by applying the following measures to protect against destroying structures [17-24]:

• creation and implementation of building materials and structures having improved resistance to biological damage;

• application of protective coatings or special additives;

• develop constructive measures of protection against biological damage;

• design and implementation of measures to protect against biological damage, carried out in the manufacture of building products and components intended for use in aggressive environment;

• design and implementation of projects constructed of materials resistant to the effects of biodestructors.

Activities to increase longevity are very diverse, and cost of holding are formed from the funds allocated for these purposes at all stages of the life cycle of structures. The cost required for increasing the input current and of quality control; the purchase of more expensive, but more reliable components and materials; the use of new constructive and technological solutions; stimulate training of employees; use of new equipment and production technologies; maintaining an economically optimal amount of redundant and spare components (blocks and units), etc [25-32].

Activities to improve the durability, from the point of view of economic efficiency, should be divided into two groups, each of which you can specify some features of the calculation of national economic benefits. The first group will be assigned those that are reduced to enhance durability of building structures and will not change the found for the building or facility the economically optimal lifespan. The commitment of this group of activities is quite legitimate, despite the fact that theoretically the durability (resource) of any element involved in the formation of optimal service life of buildings and structures. The fact is that not every increase of one resource element has such a significant effect on technical and economic parameters to "offset" the optimal lifetime. You must keep in mind the fact that the definition of economically optimal timing of service is based on anticipation of future changes in techno-economic parameters, which lead to its moral depreciation and impairment losses. Therefore, all qualitative changes of the elements are already accounted for.

The second group of measures for increasing longevity are those whose implementation leads to increased lifetime of buildings and structures. Included among the activities for the components and subsystems of complex structural systems, durability which limits their lifespan.

When the events of the first group, the economic effect can be obtained only when the service life of buildings and structures reduce the number and duration of works to restore and replacement of those elements that have promoted the resource. Savings resulting from this should exceed the growth in the value of elements that are part of buildings and constructions and used when conducting planned repairs (replacements).

3 Results and Discussion

In determining the boundaries of economically feasible improvement of durability should start from the optimal system repair, which under the influence of an increase in resource groups of elements can be changed so that there will be the task of the next optimization. Therefore, the problem of evaluating economic efficiency of increasing the longevity of one element cannot be viewed in isolation from other elements and the corresponding system of repair of the building or structure [25-29].

Clearly ineffective interventions include those that do not reduce through a number of planned replacements of elements. Denoting the number of planned replacements to increase longevity as ns, we have:

$$n_s = T_s / t_0 \tag{2}$$

where T_s – the service life, years; t_0 – the average life of components to enhance durability.

Here it is clearly inefficient for the building or structure to measures to increase longevity are those for which

where t_1 – average of service life after improve the durability; n_p - the number of planned replacements of elements after increasing durability.

If the number of planned replacements of elements reduces the result of an increase in resource, and the total number of scheduled repairs remains unchanged, i.e. under conditions

$$n_p < n_s,\ N_n = N_s, \tag{3}$$

where N_s and N_n is the number of scheduled repairs (replacements) of the elements of the design of structures for the service life before and after improve the durability of one of the groups of elements, in the sphere of consumption must increase prices because of increased longevity be compared with possible savings by reducing the number of planned replacements. Subject to the terms of the data saving from the use of more durable elements, in essence, boils down to cost savings on their purchase. This should include not only the cost of planned replacements (repairs) elements, but also their value, which is equal to the product of the number of items and their wholesale prices. Hence, taking into account the time factor, approximately the expenses of the consumers on the elements before and after to enhance their durability are defined by the formulas:

$$C_s = (n_s + 1)\, P_s\, \beta_e + n_s\, (C_r + C_v), \tag{4}$$

$$C_n = (n_n + 1)\, P_n\, \beta_e + n_n\,(C_r + C_v), \qquad (5)$$

where P_s and P_n – wholesale price of one of the element respectively before and after the increase its durability; β_e – the number of elements contained in the technical system; C_r – expenses for registration and issuance of one purchase order and receive the items for replacement from the supplier; C_v – expenses for the planned replacement of elements without regard to their value.

The difference between costs before and after increasing their durability represents the magnitude of the economic effect of increasing durability:

$$E_i = C_s - C_n. \qquad (6)$$

In accordance with this economically all options for increasing durability of elements justified are:

$$C_s - C_n > 0. \qquad (7)$$

The savings of the consumer is formed by reducing the number of planned replacements and repairs.

For the general case, this savings is defined as the sum of cost savings by the consumer as a result of reducing the number of planned replacements and repairs of items, i.e.

$$E_n = (n_s - n_n)(P_s\beta_e + C_r + C_v) = \Delta n C_s, \qquad (8)$$

where $\Delta n = n_s - n_n$.

Improve the durability, usually accompanied by rising costs in production, as a result of increased price leading to increased costs. Additional costs to the consumer in this regard, in the general case will be:

$$C_p = (n_n + 1)(C_n + C_s)\, \beta_e, \qquad (9)$$

It follows that the magnitude of the economic effect of the improvement of durability can also be determined in the form:

$$E_i = E_n - C_p. \qquad (10)$$

$$C_s - C_n = E_n - C_p.$$

The factor of time influences the magnitude of the economic effect of the time difference and cost savings from better resource. It is more correct to speak about the time difference just cost, since they are primary, and the economy and the secondary effect.

The factor of time affects the dynamics of the economic effect thus:

– leads to narrow the range of cost-effective solutions when increasing durability of elements;

– supports absolute and relative decrease in the economic effect.

When carrying out the activities to improve the durability of the economic effect can be achieved by increasing of service life of building structures. The formula for calculating the economic benefit of increased service life is as follows:

$$E_n = (C_s K_c K_s + \Delta E_m - C_n) V_n, \qquad (11)$$

where K_c – coefficient accounting for changes in performance as a result of increasing longevity; Ks – the factor of life; ΔE_m – economic effect, which is formed by the lower costs of resources in the process of operation, more durable building structures; V_n – annual production of new building structures in natural units.

When the same depreciation rates factor change of life is determined by the formula:

$$K_s = (N_o + Z)/(N_n + Z), \tag{12}$$

where N_o and N_n – norm of depreciation charges on full restore set respectively the old and new designs in fractions of a unit; Z – the coefficient of efficiency of capital investments equal to the annual interest rate.

4 Summary

Studies have shown that the evaluation of the effectiveness of measures to improve structural durability could be guided not only on the change in costs of production, but also from the standpoint of the individual consumer. And in different cases the measurement of effectiveness involves a comparison of costs before and after improvement of durability.

References

1. N. P. Makarkin, *Economics of reliability engineering* (Publishing Economics, Moscow, 2001)
2. V. T. Erofeev, *Microbiological degradation of materials* (Moscow, 2008)
3. V. T. Erofeev, *Protection of buildings and structures from microbiological damage of biocidal preparations on the basis of guanidine. Monography* (St. Petersburg, 2009)
4. V. I. Solomatov, V. T. Erofeev, V. F. Smirnov, A. S. Semicheva, E. A. Morozov, *Bio logical mechanics of materials* (Saransk, 2001)
5. V. T. Erofeev, *Frame building composites. The dissertation on competition of a scientific degree of the doctor of technical Sciences* (Moscow, 1993)
6. V. F. Smirnov, D. A. Svetlov, S. V. Kaznacheev, A. D. Bogatov, E. A. Morozov, O. D. Vasiliev, Y. M. Makarevich, V. A. Spirin, N. A. Patsyuk, *Protection of buildings and structures from microbiological damage of biocidal preparations on the basis of guanidine* (St. Petersburg, 2009)
7. V. T. Erofeev, V. F. Smirnov, *Microbiological degradation of materials* (Moscow, 2008)
8. D. A. Svetlov, V. A. Spirin, S. V. Kaznacheev, A. D. Bogatov, A. S. Boriskin, V. T. Erofeev, Transport construction, **2**, 21-23 (2008)
9. V. T. Erofeev, A. I. Rodin, A. D. Bogatov, S. V. Kaznacheev, V. F. Smirnov, E. N. Suraeva, M. A. Rodina, International Journal for Computational Civil and Structural Engineering, **3**, 81-92 (2012)
10. V. T. Erofeev, V. F. Smirnov, I. E. Kondakova, S. V. Kaznacheev, A. D. Bogatov, Izvestiya of the Tula state University. Technical Sciences, **7-2**, 310-325 (2013)
11. V. T. Erofeev, V. F. Smirnov, S. N. Bogatova, A. A. Bogatov, S. V. Kaznacheev, Bulletin of Volgograd state University of architecture and construction. Series: Construction and architecture, **16**, 122-126 (2009)
12. V. T. Erofeev, A. I. Rodin, A. D. Bogatov, S. V. Kaznacheev, V. F. Smirnov, D. A. Svetlov, News of the Tula state University. Technical Sciences, **7-2**, 292-310 (2013)

13. V. T. Erofeev, S. V. Kaznacheev, A. D. Bogatov, V. A. Spirin, D. A. Svetlov, S. N. Bogatova, Bulletin of the Dagestan state technical University. Technical Sciences, 26, 108-113 (2012)
14. V. T. Erofeev, A. D. Bogatov, S. N. Bogatova, V. F. Smirnov, News of Kazan state University of architecture and construction, **2(14)**, 21-217 (2010)
15. V. T. Erofeev, A. D. Bogatov, V. F. Smirnov, S. N. Bogatova, S. V. Kaznacheev, A. I. Rodin, *Proceedings of the International scientific-technical conference named after Leonardo da Vinci*, **1**, 83-97 (2013)
16. V. T. Erofeev, E. N. Suraeva, , A. A. Bogatov, S. V. Kaznacheev, V. F. Smirnov, A. I. Rodin, International Journal for Computational Civil and Structural Engineering, **8**, 93-100 (2012)
17. M. M. Kacimkina, D. A. Svetlov, S. V. Kaznacheev, A. D. Bogatov, A. S. Boriskin, V. T. Erofeev, Transport construction, **2**, 29-30 (2009)
18. M. E. Bazhanova, V. T. Yerofeev, A. N. Bobryshev, News of southwest state University, **5-2(38),** 415-421 (2011)
19. E. V. Zavalishin, V. T. Erofeev, V. F. Smirnov, E. A. Morozov, *Materials of the International scientific-technical conference*, 156-159 (2004)
20. V. T. Erofeev, A. D. Bogatov, S. N. Bogatova, V. F. Smirnov, E. A. Zakharova, Vestnik of Volgograd state University of architecture and construction. Series: Construction and architecture, **22**, 73-78 (2011)
21. M. E. Bazhanova, V. T. Erofeev, Belgorod state technological University, **1**, 31-33 (2012)
22. S. V. Houtorskoy, V. T. Erofeev, V. F. Smirnov, News of Kazan state University of architecture and construction, **2(24)**, 281-286 (2013)
23. V. T. Erofeev, A.V. Dergunova, Building materials, **2**, 88-89 (2008)
24. A. V. Dergunova, V. T. Erofeev, V. F. Smirnov, *Biodeterioration of biological corrosion and in the construction materials of the Third International scientific-technical conference. Russian Academy of architecture and construction Sciences,* 285-288 (2009)
25. A.V. Dergunova, A. D. Bogatov, V. T. Erofeev, *Biodeterioration biological corrosion and in the construction materials of the Third International scientific-technical conference. Russian Academy of architecture and construction Sciences,* 281-284 (2009)
26. V. T. Erofeev, V. F. Smirnov, N. V. Cherushova, E. A. Mitina, O. N. Smirnova, News of higher educational institutions. Construction, **2**, 36-41 (2003)
27. A. V. Dergunova, A. D. Bogatov, V. T. Erofeev, *Biodeterioration biological corrosion and in construction: proceedings of the third international scientific-technical conference,* 277-281 (2009)
28. V. T. Erofeev, A.V. Dergunova, Building materials, **2**, 88-89 (2008)
29. A.V. Dergunova, *Actual problems of architecture and construction: proceedings of the eleventh international scientific.-tech. Conf.,* 352-355 (2009)
30. R. Usmanov, V. Murgul, N. Vatin, Applied Mechanics and Materials, **633-634**, 1082-1085 (2014)
31. I. Mrdak, M. Rakočević, L. Žugić, R. Usmanov, V. Murgul, N. Vatin, Applied Mechanics and Materials, **633-634**, 1069-1076 (2014)
32. R. Usmanov, M. Rakočević, V. Murgul, N. Vatin, Applied Mechanics and Materials, **633-634**, 927-931 (2014)

Using Crumb Rubber from Waste Tires to Reduce Road Traffic Noise

Olga Frolova[1,*], *Brigita* Salaiová[1], *Tomáš* Olexa[1], and *Ján* Mandula[1]

[1]Technical University of Kosice, 04200 Vysokoskolska 4, Kosice, Slovakia

Abstract. The results of measurements of sound pressure levels on an experimental road section (road pavement with addition of crumb rubber from waste tires) in Mala Ida village (Kosice region, Slovakia) and on a road with stone mastic asphalt in Haniska village (Presov region, Slovakia) are presented in the paper. The results obtained using statistical pass-by methods were compared and they confirmed that road surface with addition of crumb rubber from waste tires produced quieter pavement than stone mastic asphalt road pavement alone.

1 Introduction

In general noise has a negative influence on life quality in urban areas located in noisy environments. People who live in these areas suffer from various problems [1-5]. However, it is not so easy to evaluate the effects of noise on people [6-9]. The largest contribution to road traffic noise is the tire/road noise at speeds above 50 km/h [7].

The following experimental findings were obtained concerning the relationship between the noise and road surface [7]:

1. Surface texture with characteristic particle length greater than 20 mm tends to increase noise.
2. Surface texture with characteristic particle length less than 10 mm tends to reduce noise.
3. Porosity of the pavement tends to reduce aerodynamic noise above 1500 Hz.
4. An elastic pavement surface reduces noise and other mechanical noise sources.
5. Better results tend to be obtained for negative texture (below the surface) than for positive texture (sticking out of the surface).

Nowadays in road construction porous materials absorbing noise (for example, PA – Porous asphalt pavement) are used, but this road cover type is difficult from the maintenance and cleaning points of view, especially in winter when the pores are clogged with dirt and snow and the surface does not serve as acoustic cover anymore [10]. For this reason asphalt mixtures with the addition of crumb rubber from waste tires with better porosity attract much attention. These mixtures have good slip resistance and acoustic properties [11].

* Corresponding author: olga.frolova@tuke.sk

For the correct assessment of the impact of road noise it is necessary to measure the noise correctly [12]. The most frequently used ways of measuring tire/road noise are the Close Proximity (CPX) method and the Statistical Pass-By (SPB) method [7, 13-16]. Both methods give accurate results.

The SPB method is used for the measurement of sound pressure levels from common traffic at a speed that is approximately constant. The speed and type of the vehicles passing the roadside microphone are recorded. Vehicles are classified in one of three categories (cars (category 1), dual-axle heavy vehicles (category 2a) and multi-axle heavy vehicles (category 2b) and one of three reference speeds is chosen according to the average operating speed of the road. A minimum of 100 vehicles in the category "cars" and 80 vehicles in the categories "dual-axle heavy vehicles" and "multi-axle heavy vehicles" have to be included in the measurement. Statistics are used to determine the noise levels for the road at different speeds and vehicle categories [7, 13, 17-19].

The Close Proximity Method (CPX) is based on a test in which the tire rolls over the road or the test track surface with measuring microphones located close to the tire surface. In the CPX method, the average A-weighted sound pressure levels emitted by two or four specified reference tires [20] are measured over an arbitrary or a specified road distance, together with the vehicle testing speed. The data are collected by at least two microphones, located close to the tires. For this purpose, a special test vehicle is constructed. The CPX method technology and results of measurements road traffic noise using CPX method are reported in [7, 11, 13, 16, 21-23]. The measurements using the CPX method are performed in accordance with [24].

Noise level measurements on a road section with crumb rubber additive presented in this paper were carried out using the SPB (Statistical Pass-By) method [17] because it can eliminate the variability of noise emissions from different vehicles within each type and noise measurements near the receptors along a roadway and it is more representative of actual traffic conditions..

2 Experimental

2.1 Road section with asphalt-rubber mixture.

Asphalt-rubber mixture is a solution to many problems with waste tires. The results of laboratory tests prove that mixtures with rubber have comparable or better properties than traditional asphalt mixes. The design of the mixture has a major influence on its properties. According to test results presented in [25], the optimal amount of rubber particles is 1.2 % of aggregate weight. The rubber content should be chosen taking into account the aggregate and binder content. The studied asphalt mixture was developed within the project "NFP 26220220051 Development of progressive technologies for utilization of selected waste materials in road construction engineering, supported by the European Union structural funds" at the Department of Geotechnics and Traffic Engineering, Technical University of Kosice. When the asphalt mixture went through laboratory testing an experimental road section was constructed. The original surface layer of this road was replaced with asphalt mixture with addition of crumb rubber. Additive Addibit L300 was used in this pavement because it improves the affinity between aggregate and bitumen. Crumb rubber from waste tires was supplied by the company V.O.D.S. a.s., and this product fulfils all current environmental and health standards in the EU. The experimental section of road is 200 m long and the surface layer is 40 mm thick. The weather during the laying process was acceptable with air temperatures from 15 °C to 25 °C. The temperature of the mixture after output from the plant was 160 °C. The laying and compaction were done one hour after mixing and the temperature of the pavement was over 135 °C. These processes were the same as for ordinary mixtures. Views on road section with addition of crumb rubber are shown in Fig. 1.

Fig. 1 Road section with addition of crumb rubber a) view of the mixture; b) view of the experimental road section.

2.2 Noise measurements

The measurements were carried out on the experimental road section consisting of asphalt concrete with added crumb rubber (Road section 1), and reference road section with stone mastic asphalt (SMA) (Road section 2).

Road section 1: Kosice region, Mala Ida village, road II/548, with width 7.5 m, reference speed 90 km/h, average value of road speed 71.14 km/h, and according to [17] the road has "a medium speed" category. Road surface material - AC 11 (Asphalt Concrete) with 1.2% crumb rubber (CR) from waste tires.

Road section 2: Presov region, Haniska village, road E50 connecting to highway D1. Reference speed 80 km/h, average value of road speed 74 km/h, and according to [17] the road has „a medium speed" category. Road surface material - SMA 11.

During the measurements the following experimental equipment was used:

Traffic Detection Device Sierzega SR4 to measure the instantaneous velocity, time of passage and the length of the vehicle. The device was located 1 m from the road surface at an angle of 30° from the direction parallel with the axis of the road. The measurement uncertainty for speed is ±3%, for the length of vehicle ±20%, and for time ±0.2s. The speed should be measured at the time instant when the vehicle passes the microphone [17].

Sound level meter type 2250 with frequency analysis module BZ 7223 was used. This is a sound analyzer with class 1 precision according to EN 61672 with a 120 dB dynamic range without switching. Prepolarized 1/2" free-field microphone of type 4189 was used.

3 Results and discussion

To evaluate the noise reduction effect of a specific pavement, it is necessary to carry out reliable acoustical measurements. For our measurements the SPB (Statistical Pass-By) method was chosen. The SPB method takes into account all aspects of traffic noise at the sideline of the highway, including engine, exhaust and aerodynamic noise. The method also detects variations in noise from vehicles of the same type. In the SPB method random sampling of vehicles is used. The details of the SPB method are specified in ISO 11819-1 [17].

During the measurements the maximum sound pressure level was recorded when the vehicle passed closest to the microphone, and the speed of each vehicle and its length were measured as well. The next step of data processing was data normalization. The values of maximum sound level of the vehicle and the decimal logarithm of the vehicle speed were obtained. Linear regression analysis was then used for data processing. The least-squares method was used for the parameters of regression line calculation for each category of vehicle. The coordinate of the sound pressure level on the regression line for each vehicle category at the equivalent reference speed is the sound pressure level of the vehicle L_{veh}. For this reason three values of L_{veh} were obtained at the specific location: for passenger cars, for two-axle heavy vehicles and for multi-axle heavy vehicles [17]. The average A-weighting sound pressure levels are determined separately for each category (see in Fig. 2). By using weighting factors that can be adapted to the traffic flow composition, an SPBI (Statistical Pass-By Index) ranking of surfaces suitable to the specific application can be generated (SPBI see in Fig. 3).

SPBI (Statistical Pass-By Index) is not an equivalent sound pressure level, but it can be used to describe the relative impact of the road surface of these levels. SPBI was calculated in accordance with [17].

$$SPBI = 10\lg\left[W_1 \times 10^{L_1/10} + W_{2a}(v_1/v_{2a}) \times 10^{L_{2a}/10} + W_{2b}(v_1/v_{2b}) \times 10^{L_{2b}/10}\right] dB \quad (1)$$

Where *SPBI* is Statistical Pass-By Index for normalized composition light and heavy vehicles, dB; L_1, L_{2a}, L_{2b} are acoustic pressure levels for vehicle category 1, 2a and 2b according to [17]; W_1, W_{2a}, W_{2b} are weighting factors which correspond to the probably proportion of vehicles categories in automobile traffic; v_1, v_{2a}, v_{2b} are reference speeds of single vehicles category, according to [17].

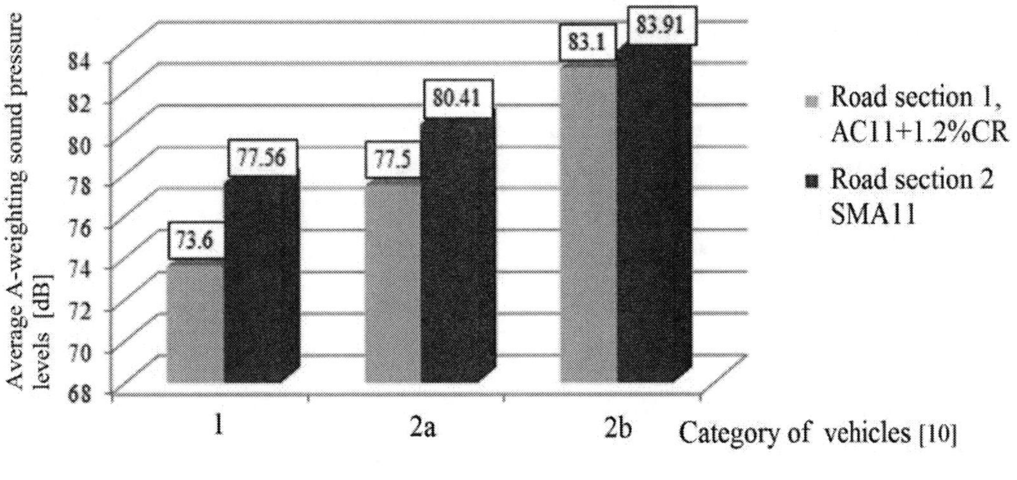

Fig. 2. Average A-weighting sound pressure levels for 2 road sections.

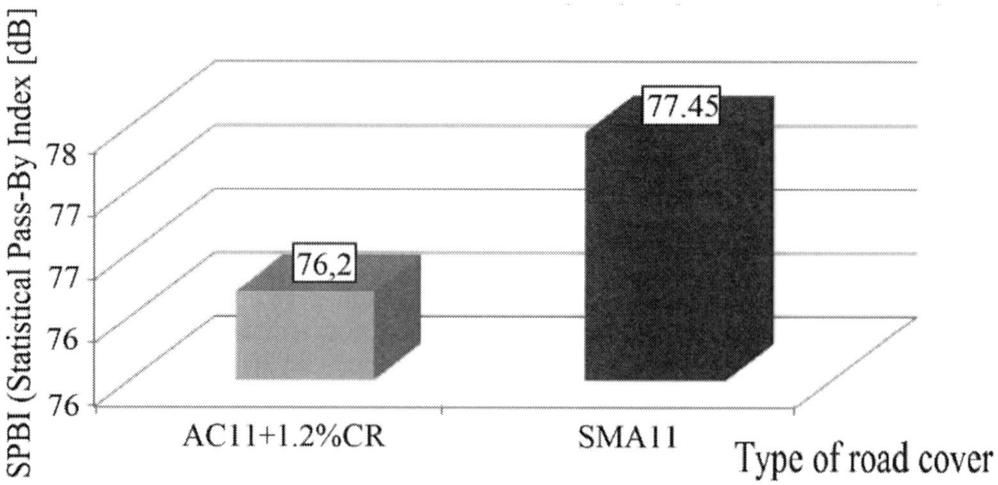

Fig. 3. Comparison of SPBI for 2 types of road cover, [dB].

Better acoustic properties for all three categories of vehicles (passenger cars, two-axle heavy vehicles and multi-axle heavy vehicles) were confirmed for the AC 11 surface with addition of 1.2% crumb rubber from waste tires, as can be deduced from the comparison of the average sound pressure levels (Fig. 3). The SPBI takes into account the impact of the vehicle category, reference speed of the road and weighting factors. The SPBI values also show that AC 11+1.2% CR cover has better acoustic properties (SPBI was 76.2 dB) versus SMA 11 cover (SPBI was 77.45 dB).

4 Conclusions

Detection of the acoustic situation is an important factor for the assessment of environment quality. The results presented show that it is possible to produce a quieter road pavement that is safe, durable and economical using current technology. Our measurements of acoustic properties of road surfaces carried out using the SPB method on two road sections in the Kosice and Presov regions confirmed that sound pressure levels for the road surface with addition of crumb rubber from waste tires revealed that it was a quieter surface than SMA road pavement. The comparison of sound pressure levels was carried out using the average sound pressure level and SPBI values.

5 Acknowledgements

The contribution was incurred within Centre of the cooperation, supported by the Agency for research and development under contract no. SUSPP-0013-09 by businesses subjects Inžinierske stavby and EUROVIA SK.

The research has been carried out within the project NFP 26220220051 Development of progressive technologies for utilization of selected waste materials in road construction engineering, supported by the European Union Structural funds.

References

1. G. H. Pandya, International Journal of Environmental Studies, **60(6)**, 595-602 (2003)

2. N. Riedel, , H. Köckler, , J. Scheiner, K.Berger, Journal of Environmental Planning and Management (2013)
3. Racioppi, F., Eriksson. L., Tingvall, C., Villaveces, A. Preventing road traffic injury: a public health perspective for Europe (2004) World health organization, 97 p.
4. V. P. Podolsky, *Vozdejstvie shuma, vibracij i elektromagnitnogo izluchenija na okruzajushchuju sredu v zone vlijanija avtodorog* (Voronezh State University of Architecture and Civil Engineering, Voronezh, 1996)
5. V.P. Podolsky, , V. G. Artyukov, V. S. Turbin, , A. N. Kanishchev, *Avtotransportnoje zagrjaznenije pridorozhnich territorij* (Voronezh University Press, Voronezh, 1999)
6. C. Smetana, *Hluk a vibrace. Měření a hodnocení* (Sdělovací technika 1999)
7. R. Bernhard, R. L. WaysonFinal Research Report Number: SQDH 2005-1. Purdue University, (2005)
8. J. Alberola, I. H. Flindell, A. J. Bullmore, Applied Acoustic, 66, 1180-1195 (2005)
9. B. Salaiová, J. Mandula, M. Kovaľaková, *Hluk z dopravy* (Technical university of Kosice, Civil Engineering faculty, 2010)
10. I. Gschwendt, *Konštrukcie a ich dimenzovanie* (Jaga Group, 1999)
11. J. Kudrna, , O. Dašek, Vysoké učení technické v Brně, (2009)
12. M. Hassall, M Zaveri, *Acoustic noise measurement: Fifth edition* (1988)
13. I. D.Hanson, S. R. James, C. NeSmith, *Tire pavement noise study* (Auburn University0 Auburn, 2004)
14. S. Pugh, D. Claughton, F. Kamst, Acoustics (2008)
15. V. Křivánek, K. Effenberger, Transactions on transport sciences, **3**, 109-116 (2014)
16. L. Nekula, V. Křivánek, Silnice mosty, **3**, 19-23 (2015)
17. *STN EN ISO 11819-1:1997. Acoustics. Measurement of the influence of road surfaces on traffic noise. Part 1: Statistical Pass-By method.*
18. M. Haider, G. Descornet, *Noise classification methods for urban road surfaces Classification Methodology. Silence. Integrated project – contract N. 516288*(2006)
19. J. M. Brennan, A. M. Kavanahgt, N. J. Sheahan, International Journal of Pavement Engineering, **2(2)**, 121-134 (2007)
20. P. Morgan, U. Sandberg, G. Blokland, INTER-NOISE 2009, **1**, 462-470 (2009)
21. J. Kachtík, O. Dašek, P. Coufalík, Silnice železnice, **4**, 74-79 (2013)
22. V. Křivánek, R. Cholava, IV Czech-Slovak Scientific Conference „Transport, Health and Environment", 213-239 (2010)
23. J. Smutný, L. Pazdera, J. Gottvaldová, Silnice-Železnice, **5**, (2011)
24. *ISO/CD 11819-2 Acoustics – Measurement of the influence of road surfaces on traffic noise – Part 2: The close-proximity method* (2000)
25. T. Olexa, P. Orolin, J. Mandula, B. Salaiová, Situation in ecologically polluted regions of Slovakia and Central Europe: 23 scientific symposium with international participation, 91-94 (2014)

MATEC Web of Conferences 73, 04023 (2016) DOI: 10.1051/matecconf/20167304023
TPACEE-2016

Studying Changes in Physical and Mechanical Properties of Polyethylene Pipes in Bending

Daniil Serebrennikov[1,*], *Vladimi* Berg[1], *Vadim* Ivanov[1], and *Natalia* Silnitskaya[1]

[1]Industrial University of Tyumen, 625001 Volodarskogo str. 38, Tyumen, Russia

Abstract. In recent years, polyethylene pipes are used more and more often in the gas industry. In the construction of pipeline, especially for trenchless laying and repairing, they are subject to considerable bending stresses. The magnitude of the stress in the pipe wall can briefly approach the yield point. The article is devoted to the study of mechanical characteristics of polyethylene pipes at short bending. The results can be used for engineering calculations at trenchless laying and repairing of pipelines.

1 Introduction

It is known that the peak of the construction of trunk pipelines in the Russian Federation took place in 60-80th years of the last century and today most of them are in need of overhaul including replacement of pipes [1- 9, 10, 11]. If we consider the pipeline repair in terms of manufacturing operations, it is as well as the laying includes:

- delivery of pipes manufacturers on pipe-welding bases;
- welding of individual tubes into stalks;
- development of the trench;
- welding a string whip into of the pipeline;
- laying into the trench;
- backfilling of the trench.

It should be noted that in the construction and repair of pipelines, most of the economic and labor costs (about 75%) are spent on earthworks. Therefore more attention is paid to trenchless pipeline repair methods recently that reduce the amount of required process operations including excavation, and hence increase the repair speed and reduce the cost [7, 8, 15]. Patent search [1, 2] conducted by the author showed that there is a growth of inventions, related to trenchless pipeline renovation in recent years. Published patents are directed both to modernization of existing units and aggregates of equipment for trenchless laying and repair, and to implementation of new methods. Patented methods are different on the principle of implementation, but all of them can be divided into several areas, namely:

- pulling existing pipeline into the adhesive shells;

* Corresponding author: daniilsereb@yandex.ru

- destruction of the old pipe and laying the new one in its place;
- protective coating of the inner surface of the pipeline;
- pulling the polymer pipe inside an existing metal pipe.

It is known that polymer materials are the most promising ones today because of the diversity and uniqueness of their properties, as well as the universality of their obtaining. This is due to the fact that except for the polymers, fillers, stabilizers, reinforcement and other additives, are widely used in their structure, through which the materials obtained can be used in almost all industries. The polymers can even exceed the commonly used materials such as metals by many properties due to lower specific weight, high corrosion resistance and thermal insulation properties, low manufacturing costs, etc. [1-24]

Based on the abovementioned, it can be argued that the method of pipeline carrying capacity recovery with the use of polymeric inserts is quite promising. As a consequence, in the development of such methods, it should be considered that during the repair work, polymer pipe may be subjected to a short-term bending on a large enough angle, and have considerable stretching influence on the traction at installation. Because of this, it is necessary to know if its strength characteristics are remaining [2, 3, 4].

2 Experimental part

To study the behavior of polyethylene pipe in bending, an experimental unit was made. The unit (Figure1) looks like a frame made of U-sections. In the framework guides of different radii are set. Fixing of the pipes was carried out with the quick-detachable support, which is attached to the frame. The bended surface of guides performed with radii, size of which is within the range of possible use under real conditions [6, 7]. Choosing bending radii guides is justified as follows. According to the requirements to ensure defined physical and mechanical properties at transportation and storage, polyethylene pipes must be wound on the bay with a diameter of at least 20 diameters of the pipe itself. Considering that that the thrust transmitting power from the traverse on investigated samples of pipes can endure a limited load, experiments were carried out with the pipes with a diameter up to 40 mm [5, 6]. The task of the study was to consider the possibility of single short pipe bend radius at less than what is presented above requirements while maintaining the strength of the pipe characteristics. Based on these conditions, the brackets radii were carried out as follows: 80 mm, 157.5 mm and 315 mm because these values are less than the radius at which providing the strength characteristics is guaranteed. Fig. 1.

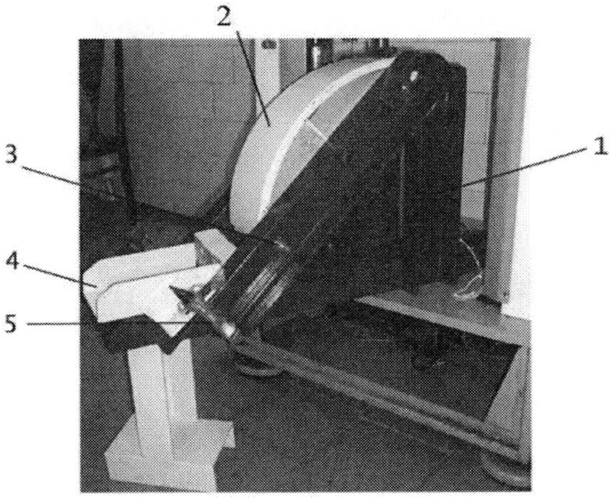

Fig. 1. Device for testing polyethylene pipes bending.; 1 – frame, 2 - guide; 3 – bolt joint;4 – quick-detachable support; 5 – clamp.

For the static tests on the deformation of pipe under bending, the sample was placed on the bracket and was fixed at one end. To the other end of the tube thrust was attached, in its turn, connected to the movable beam and dynamometer. During the experiment, a sample of the pipe was pressed to a bracket all over its bending surface. After a dense pressing, a mechanism switched off, measurements of changing the diameter of pipe samples at intervals of 0, 2, 5 and 10 min were carried out. (Fig. 2)

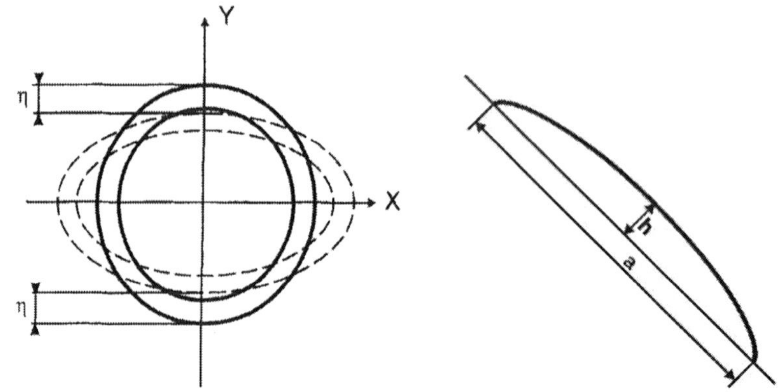

Fig. 2. Schemes of measuring deformation of the pipe.
a - the length of the chord; h - the height of the deflection; η- change in radius.

After 10 minutes, the load on the sample was stopped, it was taken from mechanism (Figure 3) and trough height measurements in relation to the chord were carried out (Figure 4). Further samples of pipes were kept for one day in vivo. 24 hours later the measurements of diameters and the height of bending in relation to the chord were carried out again in order to evaluate elastic properties of the pipeline over time.

The value of the height of depression and its changes over time, do not by themselves make it possible to evaluate the elastic properties and, in what range (acceptable or unacceptable) is made pipe bend. The criterion for estimation can be stated by a requirement: the ratio between the diameter of the pipe and diameter of the bay, on which the pipe is wound, should not be more than 5% (d pipe / D bay ≤ 5%). The height of deflection was estimated from this point. It can be argued that the pipe will not change physical and mechanical characteristics after application of bending load, if the following conditions are met:

1. The ratio of pipe diameter to the diameter of its bending after load application not exceed 5%;
2. The return of the deformable pipe diameter to the value of the original size is ensured.

During the test, pipe samples were pressed to a bracket with a known radius (Rcr, the center of the circle - O1). After removal of the load they straightened to a new circle (center of the circle - O2) with an unknown radius (R) and a central angle (ψ).

The calculation task included the definition of R in terms of the measured deflection height (h), followed by assessment of the ratio of d pipe / D bending and conclusions about the elastic properties and the level of deformation of pipes [13, 14, 19]. Fig. 3

 DOI: 10.1051/matecconf/20167304023

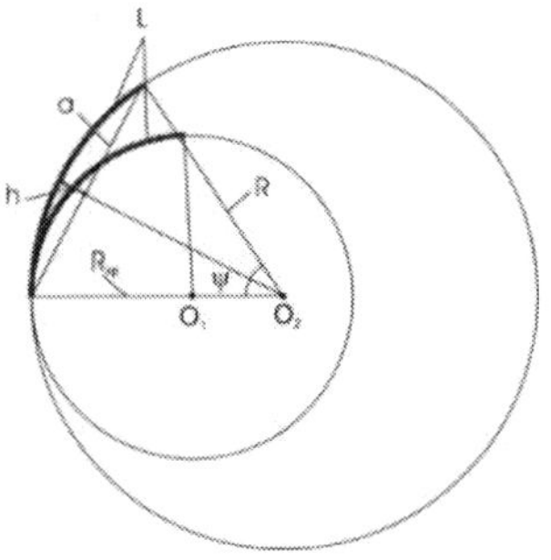

Fig. 3. Scheme to the determination of the bending radius of pipe samples at unloading.

The radius of the pipes samples after removal of the load can be determined from the ratio:

$$R = h + R \cdot \cos\psi / 2 \tag{1}$$

Central angle ψ after straightening of pipes samples can be determined, if the arc length of the bended part of sample pipes, which before and after the application of the bending force does not change and is calculated from equation:

$$L = R_{br} \cdot \frac{\pi}{2} = \psi \cdot R \tag{2}$$

From (1), taking into account (2) we can obtain:

$$\psi = \frac{\pi R_{br}}{2h}\left(1 - \cos\frac{\psi}{2}\right) \tag{3}$$

or

$$\frac{\pi R_{br}}{2h}\left(1 - \cos\frac{\psi}{2}\right) - \psi = 0 \tag{4}$$

After the calculation of angle ψ at removal of the load from the sample pipes, one can determine the radius of their bending R from relation $R = \frac{L}{\psi}$. The solution of the nonlinear transcendental equation (4) reduces to finding the value of ψ, in which the condition f (ψ) = 0 is satisfied. In a general view, a function f (ψ) has no analytical formulas for finding the roots. Therefore, in this case, one of the iterative methods for solving nonlinear equations - "bisection method" – is used. To find the root of the equation f(ψ)=0, change interval of angle ψ was divided in halves $\left[\frac{\pi}{180}, \frac{\pi}{90}\right]$, i.e. the initial approximation equal $\psi_0 = \left(\frac{\pi}{180} + \frac{\pi}{90}\right)/2$ was chosen. If a (ψ) = 0, than ψ0 is the root of the equation.

Otherwise that of the segments $\left[\frac{\pi}{180}, \psi_0\right]$ or $\left[\psi_0, \frac{\pi}{90}\right]$ was chosen, for which the value of the function f (ψ) have opposite signs at interval ends. This interval was divided in halves again, and selects that half, at the ends of which the function has opposite signs, etc. As a result, the calculation accuracy increased with each iteration [5, 7, 8, 9].

To substantiate the reliability of analytical calculations of stress states in the developed mathematical model, it was necessary to compare them to the results of experimental studies. For this purpose, the dynamic tests were carried out with sample pipes of different diameters. During the experiment, the samples were stretched with a walking traverse along the brackets with different radii. Measured efforts also allowed to estimate indirectly the voltages arising in a pipe, which then were calculated by the equation:

$$\sigma_s = \frac{Mz}{KJ}\left[1 - \beta\left(\frac{z}{r}\right)^2\right], \tag{5}$$

where M - the bending moment; z - coordinate of the height of the pipe cross section (-dn / 2 <z <dh / 2); J - axial moment of inertia; β - flattening ratio; K - reduction ratio sectional of rigidity [16, 18, 20].

Coefficient of reducing the cross section of rigidity (K) takes into account the effect of flattening the thin-walled section in bending on relation between the bending moment and the change in curvature and is defined by from the equation:

$$K = \frac{1 + 12\lambda^2}{10 + 12\lambda^2}, \tag{6}$$

where λ - geometric pipe coefficient.

Coefficient of flattening (β) considers the fact that the coordinate (z) on height of the cross section pipe required to calculate voltage value at flattening, changes its value regarding the neutral axis and is defined by the ratio:

$$\beta = \frac{6}{5 + 6\lambda^2}; \tag{7}$$

In its turn, pipe geometry coefficient (λ) is calculated by formula:

$$\lambda = \frac{\delta R}{r^2}; \tag{8}$$

where R - radius of curvature of the bracket; r - the outer radius of the test pipe sample [6, 7].

During the experimental studies, the amount of force required to forced delivery of polyethylene pipe with a different diameter [5, 7, 8, 9] was determined. The magnitude of this effort allows determining the magnitude of the bending moment:

$$M = N_S\left(R + \frac{d}{2}\right) + Q\left(R + \frac{d}{2}\right) = N\left(R + \frac{d}{2}\right)\cdot(\cos\varphi + \sin\varphi); \qquad (9)$$

The measurement results were processed using statistical methods. All parameters were determined: the arithmetic average value of the measured parameter, individual measurement error, root mean square deviation [12, 17] (Fig. 4).

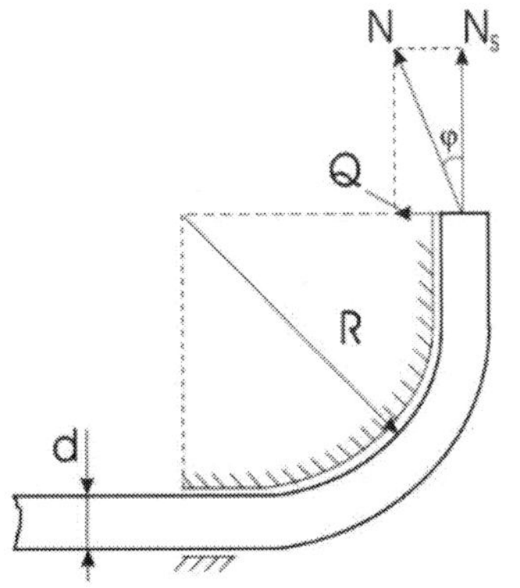

Fig. 4. The scheme to determine bending moment

It is known that the deformation experienced by a pipe in bending should take place within the elastic zone. Analysis of such changes is made on the basis of changes in pipe sample diameters, and their deflection values, and is presented in Table 1 [7, 8].

Table 1. The comparative fetch of results of experimental studies

R = 80 [mm]		R = 157.5 [mm]		R = 315 [mm]	
t = 0	t = 24 hours	t = 0	t = 24 hours	t = 0	t = 24 hours
d= 16 mm					
16.13±0,11	15.91	16.15±0.11	16.08	16.14±0,10	16.06
d= 25 mm					
25.07±0,09	24.51	25.15±0.16	25.04	25.14±0,08	25.07
d= 32 mm					
32.19±0.09	28.81	32.17±0.10	31.99	32.15±0,10	32.10
d= 40 mm					
40.18±0,07	37.17	40.16±0.14	39.94	40.20±0,09	40.13

3 Conclusion

After the experiment, the following conclusions can be made:

1. 24 hours after the removal of bending stress by using the bracket with radius of 315 mm, the average value of the diameters of all pipes samples is in the area of confidence ranges of measurements before load applications.
2. At the bracket radius with a radius of 157.5 mm, the same ranges are characterized only by samples with diameters of 16 and 25 mm. For pipes with diameters of 32 and 40 mm one day after the bending removal the load dimensions are not restored.
3. When using the bracket with a radius of 80 mm, for all pipes samples, diameters complete return to its original dimensions does not occur.

Change of pipe samples along the height deflection is shown in Figure 5.

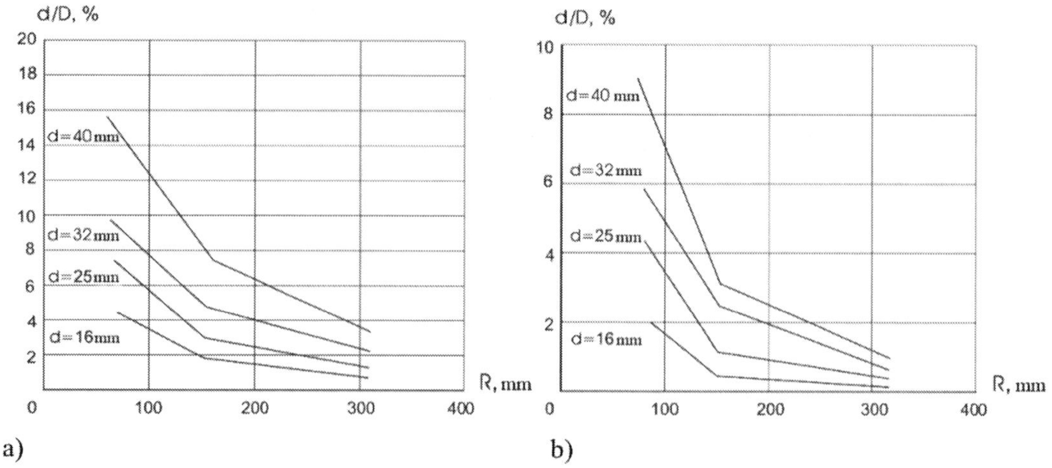

Fig. 5. The ratio of the diameter of the pipe samples to bend diameter.
a) after unloading, b) 24 hours after unloading

Taking into account that the indirect confirmation of the maintenance of physical and mechanical properties of the pipe is the return of its diameter to its original size and the ratio of pipe diameter to bend diameter of not more than 5%, it could be argued that the use of the radii of less than 80 mm is impossible.

Also, it can be argued that at bending of pipes of studied diameters, on the bracket with a radius of 157.5 mm, in the elastic area, only pipes with diameters of 16 mm and 25 mm work.

Thus, the results presented in the paper can be used for engineering calculations of mechanisms used in trenchless pipeline repairing with use of polyethylene inserts.

References

1. V. Ivanov, V. Kochurova, D. Serebrennikov, A device for trenchless laying of polyethylene pipelines. A patent for an invention RUS 2237783, 2003.
2. I. Lavrov, D. Serebrennikov, *Problems of transport systems functioning: Proceedings of All-Russian scientific-practical conference*, 247-251 (2012)
3. I. Lavrov, D. Serebrennikov, *Problems of operation of transport systems: Proceedings of All-Russian scientific-practical conference*, 178-181 (2008)
4. I. Lavrov, D. Serebrennikov, *Land transport and technological systems and tools: Proceedings of International scientific-technical conference*, 177-181 (2015)
5. D. Serebrennikov, I. Lavrov, Z. Khakimov, *Problems of transport systems functioning: Proceedings of All-Russian scientific-practical conference of students, graduate students and young scientists*, 344-348 (2012)
6. D. Serebrennikov, I. Lavrov, *Materials of International scientific-technical conference*, 130-134 (2005)
7. D. Serebrennikov, *Justification of machine operating parameters for trenchless laying of polyethylene pipelines, thesis of candidate of technical sciences* (TSOGU, Tyumen, 2004)
8. D. Serebrennikov, *Justification of machine operating parameters for trenchless laying of polyethylene pipelines, abstract of thesis of candidate of technical sciences* (TSOGU, Tyumen, 2004)
9. Yu. Zemenkov, V. Shalay, M. Zemenkova, Procedia Engineering, **113**, 254-258 (2015)

10. V. Kurushina, Yu. Zemenkov, WIT Transactions on Ecology and the Environment, **190(2)**, 881-888 (2014)
11. E. Kurushina, V. Life Science Journal, **11(11)**, 517-521 (2014)
12. A. Serebrennikov, S. Kurochkin, *Operation and service of transport technological machines: Proceedings of the conference*, 174-176 (2003)
13. A. Serebrennikov, I. Lavrov, Oil and Gas, **7**, 42-45 (2007)
14. A. Serebrennikov, Z. Khakimov, Scientific and technical bulletin of the Volga region, **6**, 382-386 (2012)
15. K. Voronin, Yu. Zemenkov, Oil and Gas, **3**, 70-72 (2012)
16. A. Serebrennikov, I. Lavrov, D. Serebrennikov, *Proceedings of the X international conference on permafrostology*, 467-469 (2012)
17. A. Serebrennikov, I. Lavrov, Z. Khakimov, *Science, Equipment, Innovations: Proceedings of the international scientific and technical conference,* 68-72 (2014)
18. A. Serebrennikov, I. Lavrov, V. Konev, Engineering bulletin of Don, **4**, 3271 (2015)
19. D. Serebrennikov, A. Davydov, *Innovations in modern science: Proceedings of VIII International spring symposium*, 130-135 (2015)
20. D. Serebrennikov, S. Yakubovskaya, Yu. Sysoyev, M. Chekardovsky, Modern problems of science and education, **2-3**, 45 (2015)
21. V. Vijayan, H. Bae, T. Marti, S. Choi, Polymer Testing, **53**, 338-346 (2016)
22. N. Sun, M. Wenzel, A. Adams, Polymer, **55(16)**, 3792-3800 (2015)
23. L.E. Pimentel Real, Polymer Testing, **49**, 88-93 (2016)
24. A. Leuteritz, K.-D. Döring, T. Lampke, I. Kuehnert, Polymer Testing, **51**, 142-147 (2016)

MATEC Web of Conferences 73, 04024 (2016) DOI: 10.1051/matecconf/20167304024
TPACEE-2016

Tank Structural Fatigue Fracture in Hydrocarbon Transportation

Alexey Pimnev [1,*], *Ruslan* Aginey[2] , *Svetlana* Yakubovskaya[1], and *Mikhail* Alexandrov[1]

[1]Industrial University of Tyumen, 625001 Volodarskogo str. 38, Tyumen, Russia
[2]JSC "Giprogazcentr", 603005 st. Alexeevskaya 26, Nizhniy Novgorod, Tyumen, Russia

Abstract. The article touches upon the reasons of cracks formation in a vertical steel tank wall used in oil transportation. Here are given the comparative research results of chemical composition and mechanical properties of steel, used in manufacture of the flat product: tensile strength, yield strength, impact strength and hardness.

1 Introduction

Modern studies have found that fatigue fracture occurs in constructions with a large number of technological and operational defects, which include quench cracks, weld defects, nonmetallic inclusions, pores, corrosion pits, nicks, cuts, scratches, "pittings", the traces of mechanical processing, etc [1,12,15,19-21]. Cyclic loading of vertical steel tanks (VST) leads to accumulation of damages starting with local plastic deformation, resulting in the occurrence of shear bands and micro-cracks, one and in some cases several of which, turns into a main crack leading to ultimate limit state — the destruction of the tank.

Fatigue durability or VST resource means the number of cycles N which a construction can endure before the occurrence of the limit state. The process of construction cyclic durability consists of two stages; and it can be represented by the following formula:

$$N = N_{OM} + N_{DC}, \quad (1)$$

where N_{OM} is the number of cycles before the occurrence of micro-cracks - the first stage; this value is determined by the Coffin-Manson Law; a crack of this size typically does not cause section destruction;

N_{DC} is the number of cycles before the development of cracks having a critical size – the second stage; the crack development can be described by the Paris Law.

The design stresses in the VST structural parts reach 0.6...0.8 of yield stress. Under such conditions, the occurrence of local defects, which are stress concentrators, lead to a formation of separate zones, where plastic deformation is so high that micro-cracks occur at a very early stage of loading. Different authors have found that the stage of micro-crack initiation is approximately 10% of the total number of cycles N.

There are a lot of cases when constructions, e.g. gas holders, having passed the acceptance tests with the load exceeding the estimated one 1.3... 1.4 times, have been

* Corresponding author: pimnev@mail.ru

destroyed after 3...5 years of operation; or during their examination cracks absent earlier were found.

Pressure vessels are very close to tanks by their stress-strain state which is confirmed by studies published by V. L. Berezin, A. G. Gumerov, R. I. Verevkin, etc. Thus we can conclude that qualitatively the picture of fatigue fracture of pressure vessels and tanks should be the same.

Currently, the average "age" of tanks used in Russia is over 30 years, i.e. the lifetime of the main part of the tanks has exceeded the standard. Therefore, we can assume that fatigue cracks in constructive elements of these tanks have already been initiated.

2 The Object of Study

For experimental research a vertical steel tank VST-20000, constructed in 1988 at a midstream operations pump station, was selected. A brief description of the tank is shown in table 1.

Table 1. Tank characteristics according to operational documentation.

Tank Parameters	Value
Tank type	Vertical Welded Tank VST-20000
Volume, m3	19400
Height, mm	11940
Tank apron ring diameter, mm	45600
Maximum loading height, mm	8700
Type of stored product	Oil
Metal information	09G2C
Design	Standard Design № 704-1-60, Melnikov NIIPSK, Moscow
Manufacturer	Novosibirsk Metal Tanks Plant
Installation contractor	Bashnefteprovodstroy Mobile Mechanical Division
Installation starting day	January 1988
Installation ending day	June 1988

The choice of this tank is based on the fact that in operation the tank leaked through a crack formed on the sheet of the second wall ring. The crack layout is shown in fig. 1. A version of mismatch of the wall material with the designed one was taken as a hypothesis about possible causes of cracks of the tank second ring. To check this assumption, the cut pieces have been studied for determination of chemical composition and mechanical properties of the metal.

3 Methods

During the study the chemical composition and mechanical properties of steel such as tensile strength, yield strength, impact strength and hardness were evaluated.

In cutting fragments from the tank wall the following aspects were considered. The internal stresses of the first ring metal are greater than those of the second one, because the heights of the tank wall and the fluid affecting the first ring are bigger than the heights of the same parameters affecting the second ring. However, despite this, the crack formed in the second ring, not in the first one. It was therefore decided to compare the metal properties of the first and second rings.

For metallographic studies two fragments of the size of 750×750 mm. were taken from the tank wall. One fragment with a crack was taken from the second ring; the other, without cracks, was taken from the first ring.

To determine the chemical composition of steel of the first and second rings fragments of the tank wall the contents of the main elements was obtained.

Mechanical properties (tensile strength, yield strength and elongation) were determined on flat samples of type XV according to State Standard 6996-84. Tests were carried out on a tensile testing machine R-10 in accordance with State Standard 1497-84 and State Standard 7564-84.

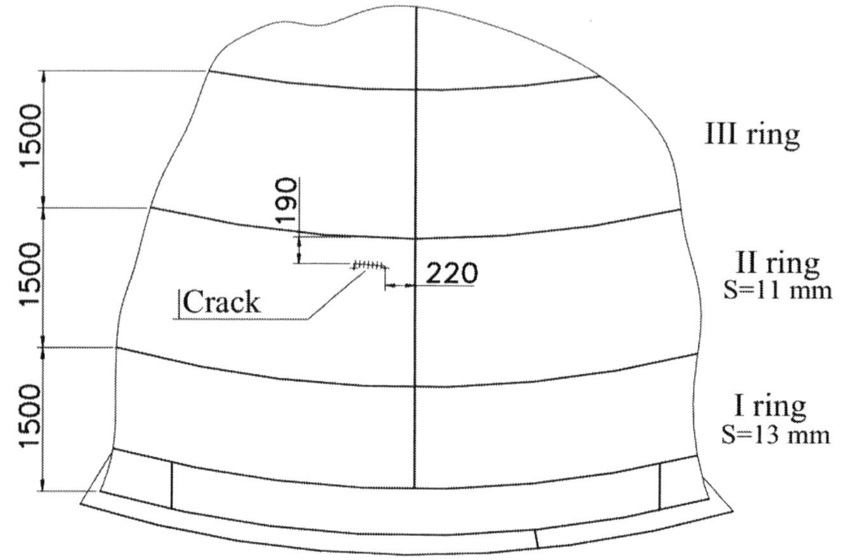

Fig.1 Crack layout on a tank wall.

The impact strength was determined on standard samples with size of 10×10×55 mm and U-shaped (sample VI) and V-shaped (model IX type) cuts according to State Standard 9454-84.

The type VI samples (Mesnager specimens) were tested at temperatures of +20°C and -40°C. The test purpose is to identify the compliance of the studied steels with State Standard 19282-84. The type IX samples (Charpy specimens) were tested at temperatures of +20°C, -5°C and -40°C. The temperatures of +20°C and -5°C were taken to determine the conformity of the studied steels to the requirements of Construction Rules and Regulations 2.05.06-85. The temperature of -40°C was adopted with regard to the area of the tank construction. Tests were carried out on the Charpy Machine. Negative temperature was obtained by means of liquid nitrogen. Aviation kerosene was used as a cooling medium. The temperature was measured with a low-temperature thermometer with a scale division of 0.1 °C. The tests were carried out in accordance with State Standard 9454-84.

The hardness measurement was carried out by the device TSH-2M with the diameter of the ball equal to10 mm and the load P=30000 kN. The tests were carried out in accordance with State Standard 9012-84.

4 The Results and Discussion

The chemical analysis showed that the first and second rings had been made of steel 09G2S. The chemical compositions of the steels of the first and second rings are shown in table 2. From the figure we can see that the carbon content in the metal of the second ring is significantly higher than the carbon content in the first ring. Moreover, the content of carbon in the second ring exceeds the values specified in State Standard 19281-89. However, the excess of the standard values are within the tolerance – ±0.02 %.

Table 2. Chemical composition of the studied steels.

Fragment	Element content, %					
	C	Si	Mn	Cr	Cu	Ni
Ring I	0.11	0.56	1.46	0.05	0.13	0.24
Ring II	0.14	0.50	1.57	0.15	0.28	0.24
Steel 09G2C State Standart 19281-89	< 0.12	0.5 – 0.8	1.3 – 1.7	≤ 0.3	≤ 0.3	≤ 0.3

Fig. 2 shows a tensile diagram of the transverse specimens cut from the first and second rings. It can be seen that the sample cut from the first ring has a distinct yield plateau. The sample cut from the second ring practically misses a yield plateau.

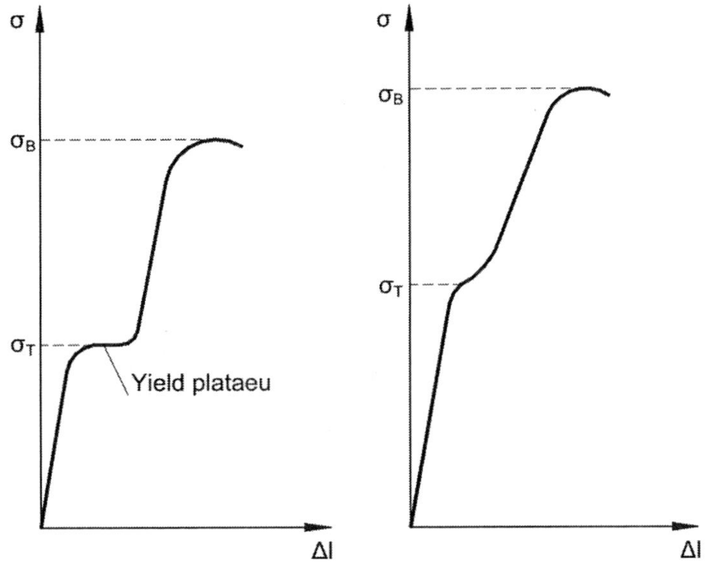

a) First Ring b) Second Ring

Fig.2 Qualitative tensile diagrams.

The results of the machine tensile diagrams processing are shown in Table 3; it also shows the values of the mechanical properties regulated by State Standard 19282-89. As shown in the table, the mechanical properties of the studied steels exceed the values regulated by State Standard 19282-89.

However, it should be noted that for the second ring the excess of the mechanical properties is significant. Such excess cannot be only stipulated by the fact that carbon content in the second ring is higher than in the first one.

Table 3. Mechanical properties of the studied steels.

Fragment	Ultimate strength σv, MPas	Yield strength σт, MPas	Elongation δ, %
Ring I	480	330	24
Ring II	540	353	18
Steel 09G2C State Standard 19281-89	450	325	21

Table 4 shows values of impact hardness for the examined steels at various temperatures. It also gives the values of impact hardness regulated by State Standard 19282-89. State Standard 19282-89 only provides data for samples with notch radius of 1 mm (KCU), because in metallurgy only these values are specified.

Table 4. Impact hardness of studied steels.

Fragment	KCU [J/cm2]		KCV [J/cm2]		
	+ 20°C	-40°C	+20°C	-5°C	-40°C
Ring I	100 – 105	65 – 74	83 – 86	—	35 – 41
Ring II	57 – 60	32 – 35	46 – 51	27 – 31	14 – 18
Steel 09G2C State Standard 19281-89	> 59	> 39	—	—	—
Steel 09G2C SNIP 2.05.06-85	—	—	—	—	≥ 29.4

KCV tests are tougher, as these samples have notch radius of 0.25 mm. Therefore, these tests make it possible to more accurately identify the tendency of steels to brittle fracture. In this regard, builders and operators are more interested in the KCV values. Table 4 shows the KCV values regulated by SNIP 2.05.06-85.

As shown in Table 4, the KCU and KCV values for the first ring exceed the values regulated by normative documents. For the second ring the situation is different. At room temperature KCU can be considered as corresponding to State Standard. However, at temperature of -40 ° C the impact hardness is below the normative values. KCU values differ slightly and KCV values differ significantly, almost twice.

Thus, it appears that the rings produced from the same grade of steel, have different properties. To find out what caused the difference in the properties, the hardness was measured. The results of hardness measurement are shown in Table 5.

Table 5. Steel hardness measurement results.

Fragment	Hardness value, HB, at measuring point			
	base metal	in metal bending area	in heat-affected base metal area	welded channel weld
Ring I	148	168	155	—
Ring II	166	275	176	170

As can be seen from the data shown in Table 5, the hardness value of the second ring is higher than the first. And the hardness value depends on the measuring point.

Thus, the deviation of the second ring top edge from the vertical may have contributed to the formation and further development of cracks. To prove this, an analysis of changes in the PBC-20000 tank wall radial movement caused by the operational load at different levels of loading was carried out [1]. It has been determined that in the area of the vertical site weld at the second ring level radial movements are maximal. They are caused by the initial imperfections of geometric shape of the tank wall in this area. This is where a crack formed, which together with the material properties data, gives evidence of fatigue fracture. To prevent the occurrence of similar failures it makes sense to implement a tank technical condition monitoring system [2-11].

5 Conclusion

The VST defects analysis showed that there is a change in metal properties [13,14,16-18]: strength, impact hardness and ductility, in particular. Thus, changing metal mechanical properties can be referred to as operational defects of tanks, used in hydrocarbon resource transportation.

The site of crack occurrence in the tank under test coincides with major wall deformation caused by operational load, which gives evidence of a more intensive development of fatigue processes in the tanks with imperfect geometry.

References

1. A. Pimnev, *The development of methodology for assessment of imperfections in tanks geometry in technical diagnosis, Thesis of Candidate of Technical Sciences* (TSOGU, Tyumen, 2006)
2. V. Kurushina, Yu. Zemenkov, WIT Transactions on Ecology and the Environment, 190, **2**, 881-888 (2014)
3. V. Antip'ev, A. Nevolin, Yu. Zemenkov, Neftyanoe Khozyaistvo - Oil Industry, **10**, 46-48 (1981)
4. Yu. Zemenkov, V. Shalay, M. Zemenkova, Procedia Engineering, **113**, 254-258 (2015)
5. Yu. Zemenkov, V. Shalay, M. Zemenkova, Procedia Engineering, **113**, 312-315 (2015)
6. E. Kurushina, V. Kurushina, Life Science Journal, **11(11)**, 517-521 (2014)
7. R. Mamadaliev, V. Kuskov, Yu. Zemenkov, A. Popova, Applied Mechanics and Materials, **770**, 19-22 (2015)
8. A. Shabarov, A. Ibragimov, M. Zemenkova, S. Podorozhnikov, Gas industry, **12 (715)**, 50-54 (2014)
9. A. Pimnev, M. Zemenkova, Pipeline transport: theory and practice, **5 (51)**, 43-47 (2015)
10. M. Zemenkova, Yu. Zemenkov, A. Pimnev, V. Petryakov, Business magazine Neftegaz.RU, **11-12**, 64-7 (2015).
11. A. Pimnev, M. Zemenkova, Fundamental researches, **11-2**, 292-296 (2015)
12. A. Tarasenko, *Deflected mode of tanks during repair works* (Nedra, Moscow, 1999)
13. G. Hopersky, *Research of the intense deformed condition of a wall of the tank at uneven rainfall of the basis, Thesis of Candidate of Technical Sciences* (Tyumen, TSOGU, 1998)
14. M. Zemenkova, *System monitoring of indicators of reliability of objects of pipeline transport, Dissertation of Candidate of Technical Sciences* (TSOGU, Tyumen, 2007)
15. A. Votinov, *Estimation of structural parameters of steel and resource characteristics of storage tanks in a continuous operation, Abstract of thesis of Candidate of Technical Science* (KGTU, Krasnodar, 2006)
16. S. Stepanov, N., Cheremisin, S., Sokolov, A., Altunin, A., Shabarov, *Computational method for relative permeability curves based on the Bernoulli generalized porous media equations, ECMOR 2012 - 13th European Conference on the Mathematics of Oil Recovery* (2012)
17. R. Levitin, Yu. Zemenkov, Neftyanoe khozyaystvo - Oil Industry, **1**, 110-114 (2016)
18. Ya. Gorelik, A. Shabarov, Yu. Sysoyev, Earth's Cryosphere, **12(1)**, 59 (2008)
19. I. Rozenshtein, Diagnostika materialov, **74(7)**, 48-51 (2008)
20. I. Kiryanov, S. Toropov, V. Toropov, Neftyanoe Khozyaystvo - Oil Industry, **8**, 48 (2015)
21. I. Rozenshtein, Oil and Gas Territory, **10**, 110-114 (2015)

Developing a Method of Calculating the Operational Flow of Methanol to Prevent the Formation of Crystalline Hydrates in the Operation of Underground Gas Storage Facilities

Anton Shipovalov[1], *Alexey* Gladenko[2], *Ivan* Seroshtanov[3], and *Anton* Ryabkov[4,*]

[1,3]Gazprom Transgaz Yugorsk, st. Mira 15, Yugorsk, Russia
[2]Omsk State Technical University, st. Mira 11, Omsk, Russia
[4]Industrial University of Tyumen, 625001 Volodarskogo str. 38, Tyumen, Russia

Abstract. When operating underground storage (UGS) of gas hydrates liquidation formed untimely could lead to serious consequences - a complete shut-in and elimination of its process. With a small fund operating wells with high daily output storage operation would entail a violation of technological regime, the failure of gas sales plans, increased hours of downtime operational fund. Therefore, ensuring the smooth and reliable operation of underground gas storage wells fund is an urgent task. The authors of the article developed a methodology for calculating the operational flow of methanol to prevent the formation of gas hydrates in UGS operation. On the basis of the developed technique using industrial operating data Punginskoye UGS made the study of technological modes of its work and recommendations to prevent hydrate formation in the underground gas storage wells.

1 Introduction

In production systems, processing and transportation of natural gas the formation of crystalline hydrates is a serious problem associated with the violation of technological processes working gas production equipment and pipelines. Typical locations of gas hydrate formation in field conditions are: a near-wellbore zone of wells, cables and infield collectors.

To restore the normal operation of the well after the occurrence of operational complications serious measures to thawing of long hydrate blockage are required. For the realization of these measures considerable forces and means are spent, but one can't stop hydrate formation in gas wells completely. Unlike gas deposits the characteristic feature of the operation of underground gas storage facilities (UGS) is the cyclical nature of the UGS work , when there is a periodic change in the direction of gas flow - from the formation (in

* Corresponding author: an100980@rambler.ru

season selection) into the reservoir (in the injection season). Under these operating conditions, the formation of hydrates, most often occurs in the gas sampling period from the formation at negative outdoor air temperature and higher sampling rates. Season selection is usually carried out from October to March, sometimes to the April.

2 The object of research

In field conditions the most common way to deal with hydrates is the use of volatile hydrate inhibitor - methanol. Methanol has a high degree of temperature reduction of hydrate formation, the ability to rapidly decompose already formed tube and mix with water in any ratio. It has low viscosity and low freezing point [3, 17]. The advantage of using methanol as the non-hydrate reactant is that this technology provides not only the prevention of hydrate formation, but under certain conditions, is an effective means to remove already formed deposits of hydrate.

Using the unique physical and chemical properties of methanol, in particular its ability to be mixed in any concentration and pass into the vapor phase without losing its initial characteristics, and the cyclic operation of the underground storage mode (injection of gas into the reservoir - the selection of the gas from the formation [19]) the technology of methanol supply in injection period in the bottom zone of the formation of individual wells was developed. Its technological mode of operation is accompanied with the formation of hydrates in the selection period. In practice, this warning method of providing of continued non-hydrate well operation is much more effective than the elimination methods of the already arisen problem.

The technology provides a method of methanol injection into the formation before the end of the season of filling gas storage [14, 16, 18]. Submission of hydrate inhibitor occurs in the flow of gas wells, technological mode of operation which is characterized as a hydrate. At the stage of technology development it is found out that it is necessary to deliver methanol to the bottom hole of wells for one - two months before the end of the gas injection season at UGS.

3 Methods

Currently there are various methods for calculating the parameters of technology using methanol [1, 2, 5, 6, 7, 8]. The works of the authors V.A.Istomin, V.G.Kvon, B.V.Degtyarev, Y.F.Makogon, E.B.Buhgalter, G.V.Ponomarev and others are well-known.

There is a method for calculating VNIIGAZ methanol consumption, which in accordance with the WFD 39-1.13-010-2000 involves calculation of hydrate formation temperature and indirectly reflects the influence of the gas composition [10, 20]. This technique is very effective for the calculation of indices in a continuous supply of methanol.

When the periodic injection into the reservoir with the change of the operating conditions it is necessary to create flexible methodologies for incorporating changes in the conditions, frequency of injection, methanol concentration and the relationship between process parameters [4].

To develop a new method which would allow to remove the hydrate problem securely and have the least possible negative consequences and measures to ensure the smooth operation of the gas transportation system, industrial researches in different technological modes UGS were carried out.

Initial data for calculation are adopted by dispatching data Punginskoye UGS operation. The composition of gases corresponds to average over a five year period and is identified as Gas №1 (Cenomanian), Gas №2, Gas №3 (Valanginian).

Underground storage of gas hydrates can be formed directly in the formation, if injection is conducted in the cooled aquifer [4]. Hydrates are accumulated directly in the bottomhole formation zone or at a considerable distance from the well bottom, if the high-temperature gas is injected under pressure, greatly exceeding the initial hydrostatic pressure [15].

The estimated dependence for determining the specific consumption of methanol injected into the gas stream to prevent hydrate formation in the "protected" area, is [2]:

$$G = (\Delta W \cdot C_2)/(C_1 - C_2) + [(100 - C_2)](q_{g1} - q_{g2} + q_{c1} + q_{c2}), \quad (1)$$

where ΔW - the amount contained in the gas (or condensate) liquid water, kg/1000m3;

C2 - minimum necessary concentration of methanol in the aqueous phase required to prevent hydrate formation in the protected point, % wt.;

C1 - methanol concentration in the injected gas (90 - 95% wt.);

qg1 - amount of methanol contained in the feed gas, kg/1000 m3;

qg2 - methanol, dissolving in the gas phase at a concentration in the aqueous solution C2, kg/1000m3;

qc1 – the amount of methanol contained in the feed gas hydrocarbon condensate, kg / 1000m3;

qc2 – the amount of methanol, dissolved in the hydrocarbon condensate at a concentration of water-methanol solution C2, kg / 1000m3.

The resulting image, depending on the flow rate of methanol pressure and temperature, as shown in [1], with a high correlation coefficient can be described by an equation of the form (for 60 to 75 kgf / cm2 pressure and gas temperature of 0 to 20 ° C):

$$G = N \cdot \ln p - 0{,}064 \cdot t - M \quad (2)$$

where p - pressure, kgf / cm2; t - temperature, ° C; N and M - factors that depend on the composition of the gas.

This dependence can be used to calculate the daily consumption of methanol by gas pressure for various compositions.

The technique of methanol consumption calculation, suggested by IUT, is based on the obtained inhibitor, depending on the pressure and consumption of gas composition at its constant flow rate [9]:

$$Q_{24} = N \cdot \ln p - M \quad (3)$$

where N and M - coefficients which depend on gas composition.

This dependence can be used to calculate the daily consumption of methanol by pressure directly to the different gas compositions. Example of dependence obtained for different gases is shown in Figure 1.

Obviously, the pressure increases [13] and, consequently, the temperature of hydrate formation, increasing the daily intake of methanol. On consumption of methanol, based on the received graphic dependence, the gas composition affects. With increase in the proportion of heavy hydrocarbons in the gas, methanol consumption curve depending on the pressure becomes steeper, it indicates a sharp, compared with methane, methanol consumption increases with increasing pressure [11, 12].

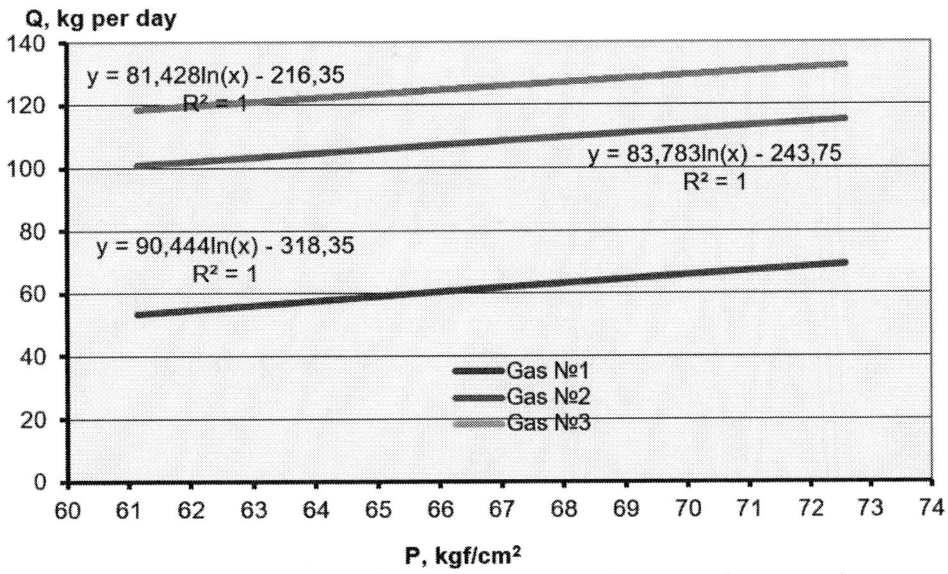

Fig. 1. The dependence of the methanol flow from the pressure and gas composition.

The dependence of the daily consumption of methanol from the gas composition can be expressed in terms of the coefficient B, which depends on the specific gravity. Changing the value of the coefficient B in the original model of hydrate, dependence of the coefficients N from B and M from B is obtained.

The authors obtained dependence according to the definition:

- The ratio N from B:
  N=63,83e0,017B,
- The ratio M from B:
  M=0,215B2+9,755B+37,27.

To determine the daily consumption of methanol at a constant gas flow rate of the proposed method, it is necessary to know the pressure and determine the coefficient B, depending on the specific gravity γ'_Γ , the concept, suggested by G.V.Ponomarev.

The concept of the specific gravity is the following one. The equilibrium conditions depend on the existence of hydrates in gas composition, which can be roughly characterized by a molecular or specific gravity. If we calculate the sum of the partial specific weights hydrate forming components $\sum Y_i' \cdot \gamma_i'$ which are a part of the gas, and divide it by the amount of mole hydrate forming components $\sum Y_i$, the resulting value γ'_Γ will characterize gas hydrate-forming ability more strictly:

$$\gamma'_g = \frac{\sum Y_i' \cdot \gamma_i'}{\sum Y'_i}.$$

Analytical dependence for determining the coefficient B (0.555 <γ'≤1) is based on the analysis of data obtained in UIT:

$$B = \frac{1{,}73}{(\gamma_g' - 0{,}5)^{0{,}8}} + 7{,}8 \tag{4}$$

where γ'g - reduced specific gravity.

The proposed relationship considerably simplifies finding the coefficient method in the calculation of the daily consumption of methanol.

Table 1 shows the values of coefficients for the gas №1 ... №3.

Table1. Value ratios N and M.

	B	N	M
Gas №1	24.25	0.552	1.744
Gas №2	15.7	0.471	0.860
Gas №3	14	0.458	0.706

To determine the amount of methanol, required for the injection before selection period, the following dependence is suggested:

$$Q_I = \frac{Q_{24} T_I}{C K_E} \tag{5}$$

where QI – the amount of methanol for well injection;

C – methanol concentration;

TI – methanol supply frequency;

KE – the coefficient, characterizing the effectiveness of the inhibitor for the period between the methanol supply.

4 Results and discussion

The resulting methanol dependence on the daily consumption of gas flow is shown in Figure 2-4.

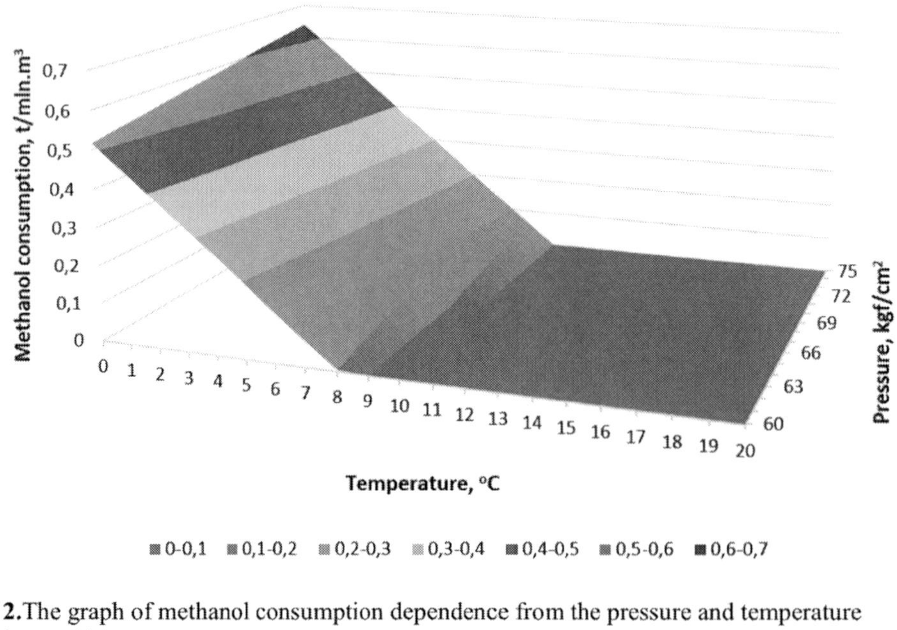

Fig. 2. The graph of methanol consumption dependence from the pressure and temperature for the gas №1.

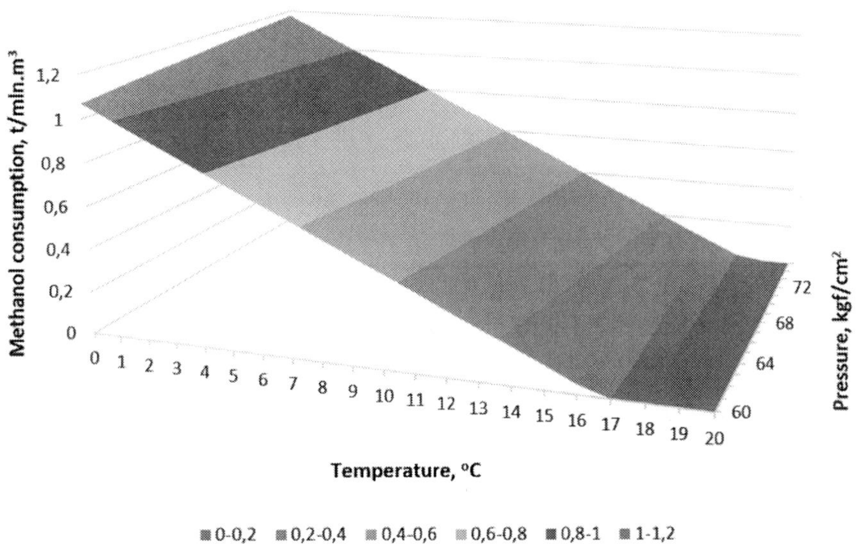

Fig. 3. The graph of methanol consumption dependence from the pressure and temperature for the gas №2.

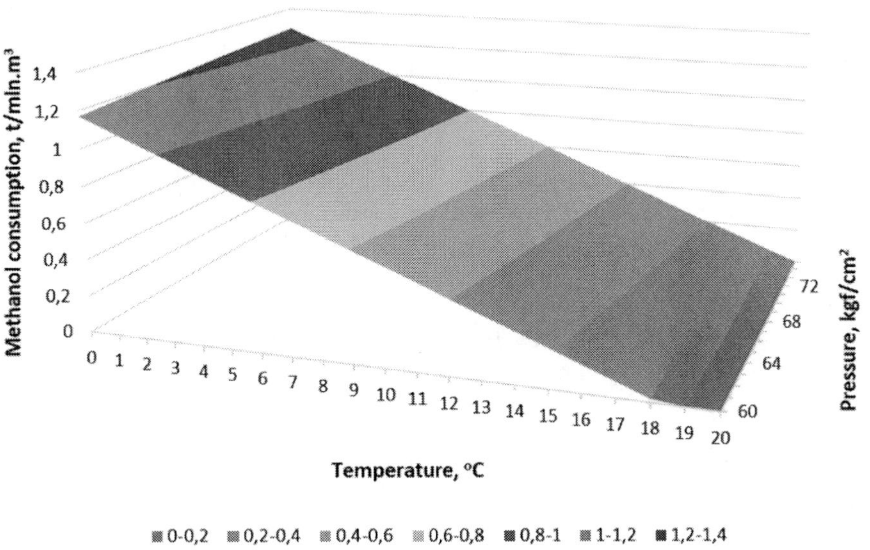

Fig. 4. The graph of methanol consumption dependence from the pressure and temperature for the gas №3.

We see that when increasing the proportion of heavy hydrocarbons in the gas, the dependence of methanol consumption from the gas flow increases due to the higher temperature of hydrate formation and, consequently, the greater minimum necessary concentration of methanol in the water phase which is required to prevent hydrate formation in the protected point.

Thus, according to the analysis of the developed technique for modeling of hydrate formation process in the conditions of UGS Punginskoye, heavy gas is more dangerous. A large amount of methanol is necessary to prevent its hydrating.

5 Conclusion

1. A method of preventing hydrate formation has been developed. It has been successfully tested in operation mode UGS of "pumping gas", by supplying methanol to the culling of problem wells, the operation of which has been complicated by hydrate formation in wellhead and ground communications.
2. It has been found that reliable non-hydrated modes of wells and pipelines on UGS are implemented with sufficient content of methanol in the whole volume of gas withdrawn from the storage. Methodology for calculation of the specific consumption of methanol to each non-hydrated well operation provides a certain margin of methanol feed, since the temperature drop at the end of the loop, the appearance of the produced water in the well production affect its consumption.
3. It has been found that for the calculation of the time and period of the methanol injection into the reservoir it is necessary to know exactly the start of the upcoming season selection. An ideal variant is methanol feed during the entire period of the injection season in problem wells. It requires additional feasibility study and may lead to a large reagent costs. It has been found that the positive effect of the methanol injection is achieved in 1-2 months before the start of the withdrawal season, i.e., start of processing bottom-hole zones - in August - September.

References

1. A. Shipovalov, Yu. Zemenkov, S. Toropov, M. Zemenkova, S. Podorozhnikov, I. Tyrylgin, V. Pavlov, *Aspects of technological reliability and economic efficiency of operation of underground storages of natural gas of Western Siberia* (TSOGU, Tyumen, 2012)
2. WFD 39-1.13-010-2000 Instruction on calculation of methanol consumption standards for the use in the calculation of the maximum permissible or temporarily agreed discharges of methanol for the facilities of "Gazprom"
3. S. Dudin, Yu. Zemenkov, A. Shipovalov, S. Podorozhnikov, Oil and gas, **2**, 65-69 (2011)
4. S. Dudin, V. Nekrasov, Yu. Zemenkov, Mining informational and analytical bulletin (scientific and technical journal), **3**, 53-61 (2013)
5. V. Istomin, V. Kwon, *Prevention and elimination of gas hydrates in gas production systems* (Ltd "Gazprom RPI", Moscow, 2004)
6. V. Istomin, V. Yakushev, V. Kwon, S. Dolgaev, *Hydrate technologies* (VNIIGAZ, Moscow, 2009)
7. Y. Makogon, A. Malyshev, A. Sedykh, Y. Shalyaho, K. Topchev, K. Unarokov, *Temporary instruction about the prevention and elimination of hydrates in gas production and transportation systems* (VNIIGAZ, Moscow, 1983)
8. Yu. Zemenkov, *Backup energy for the reliability of the gas supply system* (TSOGU, Tyumen, 2006)
9. A. Shipovalov, Yu. Zemenkov, S. Podorozhnikov, M. Zemenkova, Pipeline transportation: Theory and Practice, **6**, 25-28 (2010)
10. A. Shipovalov, Yu. Zemenkov, A. Smirnov, S. Podorozhnikov, Pipeline transportation: Theory and Practice, **1**, 27-30 (2011)
11. K. Voronin, Yu. Zemenkov, Oil and gas, **3**, 70-72 (2012)
12. K. Voronin, V. Pavlov, Yu. Zemenkov, Oil and Gas Terminal: Proceedings of the International scientific conference, 169-173 (2015)
13. K. Voronin, Natural and Technical Sciences, **5**, 208-211 (2011)

14. V. Nekrasov, R. Levitin, Fundamental and applied research: Problems and results, **13**, 223-228 (2014)
15. V. Nekrasov, R. Levitin, A. Maier, Oil and gas, **5**, 63-65 (2014)
16. V. Nekrasov, R. Levitin, Yu. Zemenkov, Equipment and technologies for oil and gas complex, **3**, 43-46 (2014)
17. Yu. Zemenkov, V. Shpilevoy, S. Podorozhnikov, A. Zakirzakov, *Energotechnological complexes at the design and operation of transport and storage facilities of hydrocarbons* (TSOGU, Tyumen, 2014)
18. K.Voronin, S. Dudin, M. Zemenkova, N. Zakirov, A. Gladenko, Modern problems of science and education, **2-3**, 32 (2015)
19. K. Voronin, A. Gladenko, S. Sokolov, M. Chekardovsky, Modern problems of science and education, **2-3**, 31 (2015)
20. M. Zemenkova, T. Puzina, S. Maslakov, I. Seroshtanov, V. Nikiforov, Mining informational and analytical bulletin (scientific and technical journal), **4**, 174-183 (2014)

Development of a Complex of Self-Cooling Bearing Systems as a Fundamentally New Direction in Construction on Permafrost

Vadim Passek[1,*], and *Aleksandr* Tsernant[2]

[1,2]Industrial University of Tyumen, 625001 Volodarskogo str. 38, Tyumen, Russia

Abstract. The article describes a new direction in constructing on permafrost (the new constructions are suggested) – a complex of self-cooling bearing systems, which is based on the winter transfer of cold from the atmosphere directly into the deeper layers of the soil by natural air convection. It presumes the spontaneous activating of cooling process in winter period and deactivating of it during summer. The complex is designed as a system of cavities and channels disposed at different angles to the vertical, which are partially located in the base soil and partially contact the open air environment. They may be closed and open for the access of the outside air, besides, all the elements of the system or some of them are the bearing structures of the construction object.

1 Introduction

Over 60% of Russian territory is situated in the zone of permafrost. These are the territories where the major natural resources are located. Their development determines the viability of the national economy, therefore, these regions require development, including the construction of residential and industrial facilities, transport systems. The relevance of this, in particular, is determined by the Presidential Decree № 537 of May 12, 2009 "Russian National Security Strategy until 2020", which involves the development and implementation of innovative scenarios for development of the natural, and especially the energy resources, of the Arctic zone of Russia, that covers 140 degrees of longitude in northern latitudes of the Eurasian continent.

Permafrost is a good base for buildings and transport systems, but only if the frost persists during operating period. If soils start to thaw, they represent a flowing mass, totally unsuitable for use as bases.

Permafrost soil may consist of layers of buried ice, soil ice (80% of which is ice), saline soils that freeze only at low negative temperatures. And the frozen ground conditions are getting worse northward. To compensate these negative impacts, it is necessary that after the construction works, the desired low temperature of permafrost is provided during the entire operating period. Although building regulations allow to build on permafrost in

* Corresponding author: tgasu.passek.vadim@mail.ru

accordance with two principles: preserving permafrost (principle 1) and letting it thaw (principle 2). The latter can be regarded only as an exception. It is applicable only to soils that do not lose strength properties during thawing –both directly under the structure, and in the adjacent soil masses. It is important to emphasize that thawing of a single soil body can cause the degradation of permafrost in neighboring areas, which can lead to major deformation processes not only of the structure itself, but of the whole built-up area. Furthermore, if a pile foundation zone starts thawing, the underlying layers of permafrost will thaw to a plastic-frozen (deformable) state which will gradually lead to the deformations of the entire soil body along with the piles. Thawing of the soil upper layers dramatically reduces the soil resistance to horizontal loads. Therefore, when building on permafrost, one should always try to achieve at least the preservation of permafrost, although lowering its temperature is better.

2 Review

In natural conditions the permafrost has formed in regions with a negative annual average temperature of air. At the same air temperature, the permafrost temperature and even presence depend on the surface layers in winter and summer. Construction of roads, cities, various facilities causes the sharp changes of conditions on the soil surface: there occur the areas with increased thickness of snow cover (which prevents the entry of the cold in winter), the vegetation cover is disturbed (which increases the heat input to the soil in summer), there occur the facilities with year round positive temperature, the regime of surface and ground water changes, etc. As a result, the frost starts either thawing (which reduces the load bearing capacity of soils), or even degrading.

Construction science and practice have been evolving for thousands of years beyond the permafrost zone. And no wonder that builders tried to use the past experience in the zone of permafrost when they faced it only at the turn of the XIX and XX centuries. And the endless deformations, fractures, accidents began to happen.

On the permafrost, the layout of a structure should be different! So the research and development work were started.

To compensate the defrosting influence of the facilities in Russia, Canada and the United States, a range of different measures has been developed [1, 2].

The efforts of the developers of these measures primarily were directed towards the formation of the best conditions for the heat transfer at the surface. One of the main measures – the ventilated underfloor space. In addition, different forms of insulation, riprap and other were suggested. Thus, the range of surface cooling measures had been formed, which in their nature were similar to factors, acting in a natural environment. This class of measures (generally very large and subdivided) can be considered to be formed according to the coverage of many different areas of practical applicability. We call it the first step of the theory and practice of constructing on permafrost.

But these measures turned out to be not enough in a large number of cases. Firstly, if the frost is formed over the centuries in natural conditions, these terms are not acceptable for conditions of building the real facilities (much shorter terms are required). Besides, some internal sources of heat appeared (e.g. heat pipelines). Under such circumstances, there occurred the necessity of rapid introduction of cold directly into the deep soil layers. In other words, there occurred a need to develop a class of deep cooling measures. This class of measures forms the second step of the theory and practice of constructing on permafrost, which expands the opportunities for construction by many times.

A range of this class measures has been developed: the liquid cooling systems of S.I. Gapeev [3], V.I. Makarov [4] or the vapor-liquid systems of Long [5], which were transformed into a range of variants and new versions in Russia [6]. All of them appear to be vertical or inclined pipes filled with different refrigerants and partially buried in the soil, partially protruding outward above the ground. The transfer of cold into the soil occurs only in winter. Each of these systems has its advantages, disadvantages and scope of rational application. However, all of them are located in the zone of engineering structure, but not inside of the structure itself, which means they tend to be some foreign body to it, the "medication". The mentioned installations can be very effectively used for technological purposes: for single initial freezing of the soil bodies, or during the operating period – for example, for eliminating the expected or unexpected thawing of soils [7-12].

Without belittling the mentioned projects, it can be assumed that the problem cannot be considered fully solved if the facility is "sick" and requires "medications", which are environmentally unsafe (as they contain kerosene, freon, etc.), non-durable, unreliable. Another thing is - if the facility is "healthy" and itself contributes to the desired regulation of the temperature regime of the bases soil.

3 Description of the research and developed design

The designed Complex of self-cooling bearing systems is based on the winter transfer of cold from the atmosphere directly into the deeper layers of the soil by natural air convection. It presumes the spontaneous activating of cooling process in winter period and deactivating of it during summer. The complex is designed as a system of cavities and channels disposed at different angles to the vertical, which are partially located in the base soil and partially contact the open air environment. They may be closed and open for the access of the outside air, besides, all the elements of the system or some of them are the bearing structures of the construction object.

The simplest and most developed part of the complex is a CSRIC thermal-control bearing. Figure 1 presents the exterior of the bridge with thermal-control bearings, and the principle of their operation is shown in Figure 2.

Fig. 1. Exterior of the bridge with thermal-control bearings.

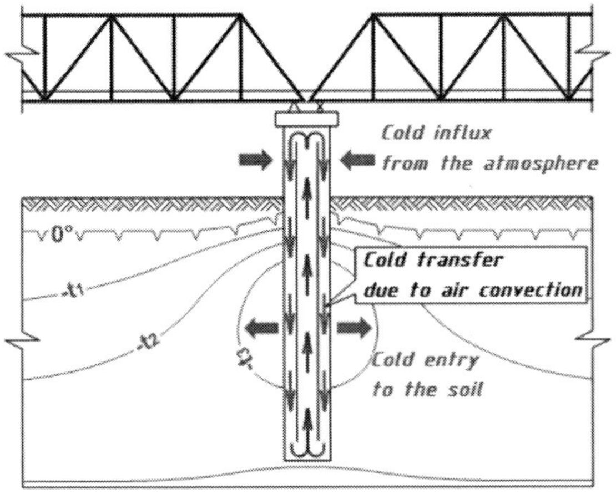

Fig. 2. The operation principle of thermal-control bearing of the bridge.

Fig. 3 is a schematic diagram of a cooling system in a building constructed on a box-shaped reinforced concrete foundation with intercommunicating air cavities. The upper chamber is located in the zone of the outer air. Warm air goes up the right pipe, then goes down the left pipe and enters the lower chamber being already cold.

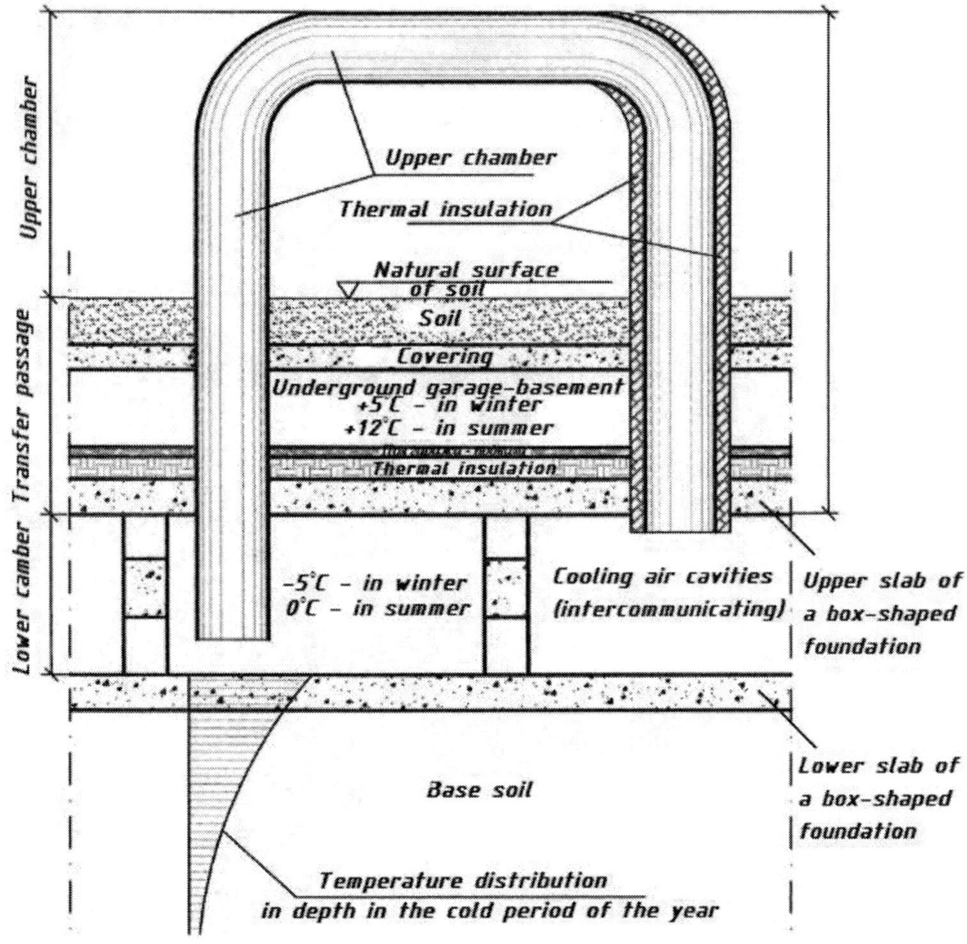

Fig. 3. Scheme of a cooling system in the building on a slab foundation.

Fig. 4 shows a cooling system based on a combination of vertical and horizontal heat fluxes. In winter, the cold air causes convection in the underlying layer of rock by passing through the through-cavity.

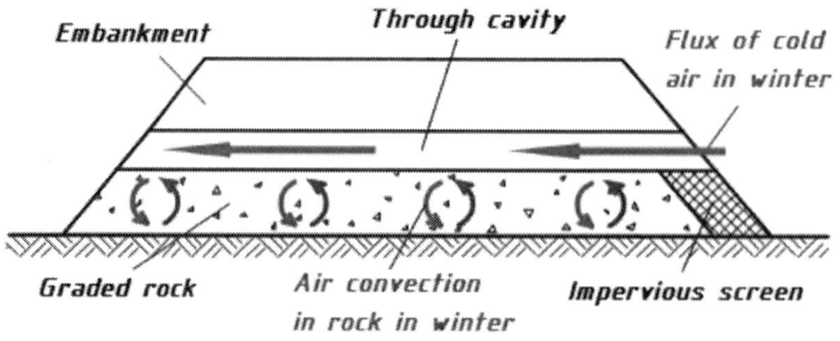

Fig. 4. The cooling system for the embankment.

Fig. 5 illustrates a cross-section of the oil pipeline transiting the steep bank of the river with the bedding of the plastic frozen soils. The air cooling system is also used here. The pipeline is located directly in the air cavity.

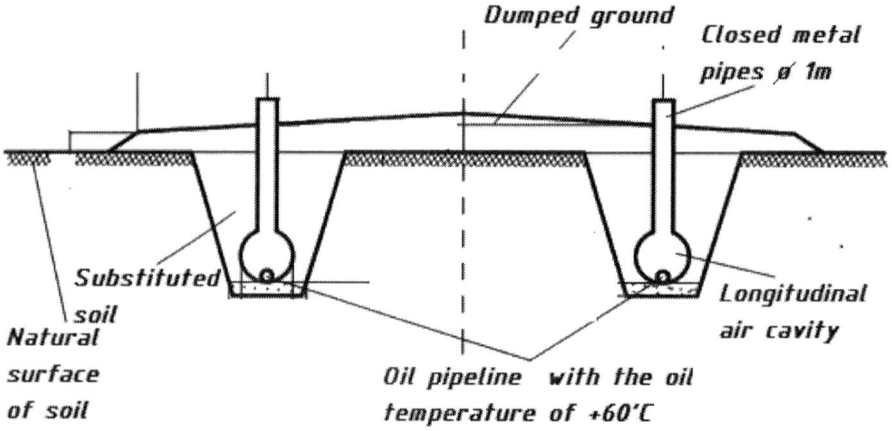

Fig. 5. The cross-section of the pipeline cooling system.

To determine the effectiveness of deep cooling in comparison with the surface cooling, a series of calculations were conducted, which involved comparing the speed of freezing the defrosted soil for a number of settlements in different climatic zones. The results were presented in the form of a diagram shown in Fig. 6. It can be seen that with the annual average reduced temperature of air (further AARTA) ta = -10°C, it takes the same amount of time to freeze the thawing soil for both surface cooling and deep cooling. So to ease the comparison, this value was taken as the time unit. When the AARTA increases, the rate of two curves changes. The curve of the surface cooling shoots up dramatically at AARTA above -4°C, indicating the fall of its effectiveness, while the deep cooling curve begins to acquire a slope only when the AARTA reaches positive temperatures. This suggests that short-term or global climate warming does not have a significant effect on the effectiveness of the deep cooling. The diagram is divided into zones in accordance with the borders of usage of the two cooling types.

Over the past forty years, dozens of different systems have been developed. The most notable designs are reflected in works [13-20].

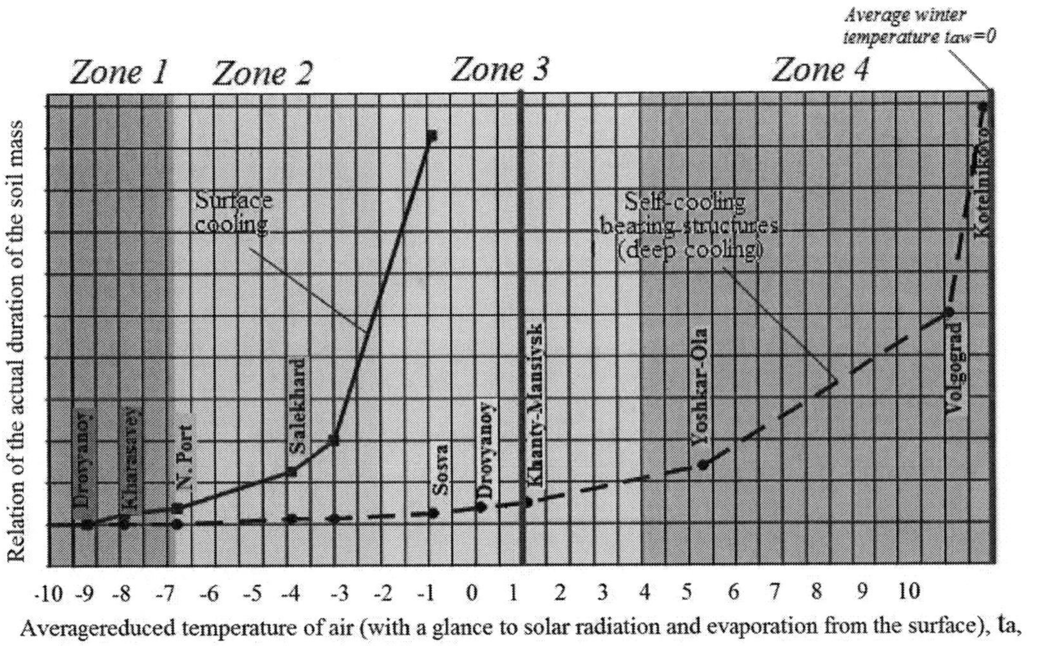

Fig. 6. Dependency diagram duration of the soil mass freezing with respect to the annual average reduced air temperature ta for the surface and deep coolings.

zone 1 - area of the equal effectiveness of the surface and deep coolings, zone 2 - area of the preferential use of the deep cooling, zone 3 - area of the deep cooling use only, zone 4 - area of the theoretically possible use of the deep cooling.

4 Conclusion

When designing the complex, the following has been developed: the theory of the soils deep cooling by natural air convection (in particular, 2 doctoral theses and 4 candidate's theses were defended, which were entirely or partially devoted to the analyzing issue); a set of specific structural and technological solutions; 27 patents were received; 23 scientific articles and a monograph were published; long-term (25 years) observations of the actual bridges built with the developed technical solutions, were conducted; 8 regulatory and guidance documentations were developed that allow to make calculations and designs.

The complex of the self-cooling bearing systems for the first time solves the problem of deep cooling by the fact that the constructed facility does it itself during the whole period of its operation. The complex is resistant to the global warming and offers new possibilities for construction technology and monitoring. It sums up the 45 year-period of the corporate authors' work, and marks the establishment of the theory of this issue (the theory of deep cooling by natural air convection), as well as the establishment of a large number of specific structural and technological solutions, which open the potential for new types of structures in relation to the permafrost conditions (Fig. 7).

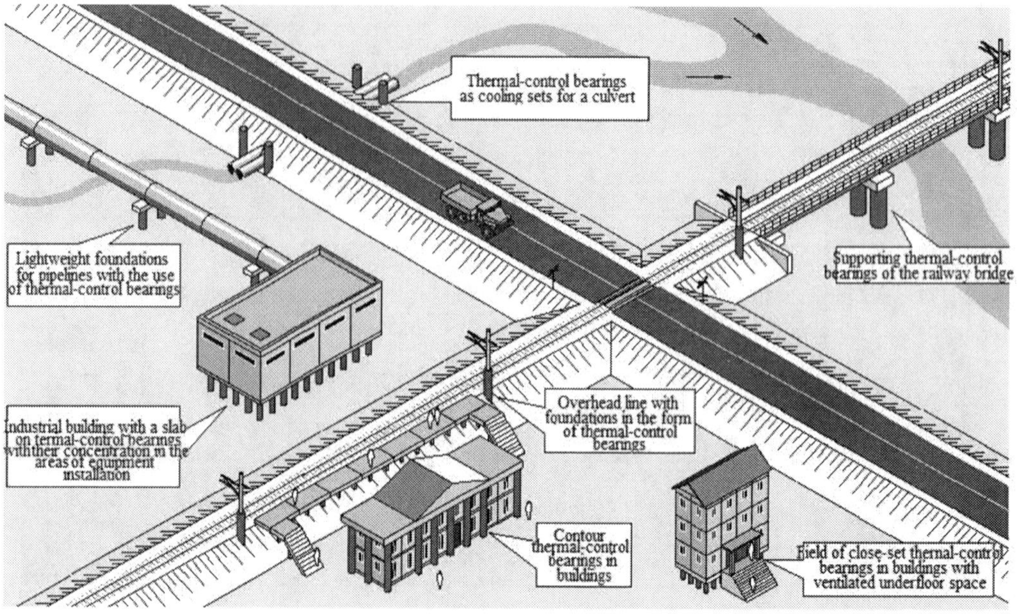

Fig. 7. Possible application areas of the self-cooling bearing systems complex in construction.

One can assume that this project pushed a significant leap forward and formed the second stage of the theory and practice of constructing on permafrost, which is defined by class of the deep cooling systems.

References

1. V. Passek, Sb. nauchnykh trudov JSC TSNIIS, **213** (2002)
2. A. Tsernant, Vestnik TyumGASU, **4**, 10 – 15 (2015)
3. S. Gapeev, *Ukreplenie merzlykh osnovaniy okhlazhdeniem* (Stroyizdat, 1999)
4. V. Makarov, *Termosifony v severnom stroitel'stve* (Nauka, Novosibirsk, 1999)
5. E. Long, Means for Maintaining Permafrost Foundations. Pat. USA. № 3, 217, 791, 165-45 (1964)
6. R. Bayasan, *Termostabilizatsiya grunta v osnovanii ob"ektov gazotransportnykh system «Teplomassoobmen – MMF. Mezhdunarodnyy forum»*, 34 – 37 (1988)
7. V. Passek, A. Nabokov, M. Baev, T. Maltseva, V. Bai, I. Palavoshev, Vestnik TyumGASU, **4**, 33 – 42 (2015)
8. V. Passek, M. Baev, A. Nabokov, I. Palavoshev,, A. Babukh, V. Bai, Vestnik TyumGASU, **4**, 43 (2015)
9. V. Passek, V. Gerasimov, V. Rudenko, V. Dubinin, V. Passek, I. Kozlov, A. Odrinskiy, R.F. Patent 91723 (2009)
10. V. Passek, V. Gerasimov, V. Velichko, V. Rudenko, I. Kozlov, A. Odrinskiy, R.F. Patent 103542 (2010)
11. V. Passek, V. Gerasimov, I. Kozlov, B. Drobyshevskiy, V. Velichko, R.F. Patent 110758 (2011)
12. A. Che, Z. Wu, P. Wang, Soils and Foundations, **54(4)** 639-647 (2014)
13. Y. Wang, H. Jin, G. Li, Cold Regions Science and Technology, **126**, 10-21 (2016)
14. M. Wei, C. Guodong, W. Qingbai, Cold Regions Science and Technology, **59(1)**, 3-11 (2009)
15. V. Passek, I. Palavoshev, V. Velichko, G. Poz, M. Paskov, R.F. Patent 136446 (2013)
16. V. Passek, I. Palavoshev, M. Paskov, R.F. Patent 147446 (2014)

17. V. Passek, I. Palavoshev, A. Nabokov, A. Babukh, M. Baev, V. Bai, R.F. Patent. (2015)
18. V. Passek, M. Baev, A. Nabokov, A. Babukh, V. Velichko, I. Palavoshev, G. Poz, V. Bai, R.F. Patent 156932 (2015)
19. V. Passek, A. Nabokov, I. Palavoshev, T. Maltseva, A. Kraev, A. Babukh, V. Andreev, R.F. Patent 156217 (2015)
20. T. Maltseva, A. Nabokov, A. Chernykh, Procedia Engineering, **117,** 239-245 (2015)

Strengthening Thawed Permafrost Base Railway Embankments Cutting Berms

Sergey Kudriavtcev[1], *Vladimir* Paramonov[2], and *Igor* Sakharov[3,*]

[1]Far Eastern State Transport University, 680021 Seryshev str. 47, Khabarovsk, Russia
[2]St. Petersburg State Transport University, 190031 Moskovsky pr. 9, Saint-Petersburg, Russia
[3]Saint-Petersburg State University of Architecture and Civil Engineering, 190005, 2-nd Krasnoarmeiskaya st. 4, Saint-Petersburg , Russia

Abstract. The paper presents the results of a long-term survey of freeze and thaw action conducted at the railroad sections under reconstruction on the permafrost area of the second line of the Baikal-Amur Mainline. The thermophysical research of freeze and thaw action was carried out as well as the calculation of stress-strain distributions and deformation ranges of the roadbed and the subgrade support. The calculations were made by the numerical modeling method with the «FEMmodels» complex. The modeling results help choose the roadbed rational designs that decrease the volumes of deformations to the acceptable ones.

1 Introduction

When developing new construction projects and reconstructing the existed ones, the whole of known methods to stabilized the subgrade support on permafrost soils is used. There is no reason to believe these methods are innovative in many ways because most of them are known from the times of construction the Baikal-Amur Mainline and its first period of exploitation. However, they have not been much spread and used either at that time or now.

It is considered, for example, that the methods of keeping the subgrade support frozen are not acceptable in the eastern part of the Baikal-Amur Mainline. As there would be no alternative to Design and Construction Principle II, no preventive measures providing stability of the subgrade support were put forward. When implementing Principle II, the only method of compensating the subsidence was to make some width and height reserves that allow the followed corrections of deformations with upgrades on ballast and side filling ups. Besides, the plastic yielding of heave deformations caused by an annual freeze-thaw cycle in wet soils was not taken into account.

A vivid example of the above said is a carried out project of a railway detour around the Bureya Hydropower Station in 2007. The mistakes made when constructing the Baikal-Amur Mainline such as wide berms, ill-provided drain, underestimated volume of post-construction subsidence, have discredited the application of Principle II. It can be reasonably used when the shift component in the subgrade support subsidence is excluded. Unfortunately, currently there is no shared opinion on this issue among the permafrost

* Corresponding author: tgasu.sakharov.igor@mail.ru

construction professionals. The problem of eliminating the shift deformations in the subgrade support during construction and especially in the long-last period of roadbed exploitation have not been essentially solved in the given region.

It is reasonable to use the mortised berms of stones wrapped in geosynthetic nets that keep the structures solid in thawed toes to decrease or avoid shift deformations.

2 Numeric modeling processes of freezing, frost heaving and thawing

A long-term observation and survey of the roadbed of the Baikal-Amur Mainline have been resulted in the soil survey reports, the analysis of gauge stability in warm and cold seasons according to a monitoring of the track measurement car and the aerographics observation data. They confirm that the rock cooling structures of berms and graded rock-fill on the roadbed slopes are reasonable as a project. The technical characteristics of the structures composed of graded rocks were developed and tested by the Tyndinsky Permafrost Station in the given cryological area and have shown their effective work for more than 20 years.

The numerical modeling was carried out in the programming complex «FEM-models» developed by the geotechnical engineers under the guidance of Prof. V. M. Ulitsky. The «Termoground» program is a component part of the complex which allows to research dimensionally by the finite element method the freeze, heave and thaw processes in a year cycle. The numerical problem solution of freeze, heave and thaw processes is modeled in two stages. The first stage, thermotechnical, is resulted in obtaining data on the temperature and wet fields for every period of time. The second stage is resulted in obtaining data on stress and strain conditions of soils that appear in the subgrade support because of the heave and thaw processes.

The model of the thermophysical processes in the «Termoground» program is based on the model of freeze, heave and thaw processes of N. A. Tsytovich, Ya. A. Kronik and V. F. Kiselev. The program calculation supposes that the area with the structure is divided into the finite elements with the same temperature and soil characteristics. Actual on the upper bound the time-variant heat flow rate is relevant to the air temperature and the coefficient of heat transfer.

The initial and limit conditions are assigned into the simulation model, thus the thermotechnical problem is solved to obtain the temperature and wet fields for every period of time. The initial conditions are the soil temperatures in depth; the limit conditions are the soil surface temperatures in time.

The numeric modeling was done on the FEM-models program complex developed by geotechnical engineers in St.Petersburg, Russia (Kudriavtcev S. et al., 2003, Kudriavtcev S. et al., 2012, Kudriavtcev S. et al., 2013). A component of the FEM-models is the "Termoground" program which helps study process of freezing, frost heaving and thawing in a year cycle in spatial programing of numerical modeling by the method of final elements. The core of a mathematical modeling of thermophysical processes in "Termoground" program is the model of high ice, thawed and frozen soils offered by N.A. Tsytovich, Y.A. Kronik and V.F. Kiselev. The task of numeric modeling of freezing, frost heaving and thawing is solved in two stages. Firstly, the thermotechnical part is implemented where moisture and temperature fields are determined for every period of time. Secondly, the stress strain behavior of soils in foundation is determined in the processes of frost heaving and thawing (Kudriavtcev S. et al., 2003).

General equation describing the freezing and thawing processes for a transient thermal regime in a three dimensional soil space can be expressed as following:

$$C_{th(f)}\rho\frac{\partial T}{\partial t}=\lambda_{th(f)}\left(\frac{\partial^2 T}{\partial x^2}+\frac{\partial^2 T}{\partial y^2}+\frac{\partial^2 T}{\partial z^2}\right)+q_V \quad (1)$$

where $C_{th(f)}$ - specific heat of soils (frozen or thawed), J/kgK; ρ-soil consistency, kg/m³; T – temperature, K; t – time, c; $\lambda_{th(f)}$ - thermal conductivity of soil (frozen and thawed), W/mK; x, y, z – coordinates, m; qv- internal heat source capacity, W/m³.

Under steady conditions the flow, inbound and outbound, in/from the elementary unit of soil is the same atany time. Thus, as the lefthand side of the equation is eliminated, the equationis reduced to:

$$\lambda\left(\frac{\partial^2 T}{\partial x^2}+\frac{\partial^2 T}{\partial y^2}+\frac{\partial^2 T}{\partial z^2}\right)+q_V=0 \quad (2)$$

$$C_{th(f)}=C_{th(f)}+L_0\frac{\partial W_W}{\partial T} \quad (3)$$

Where C $_{th(f)}$ - volume specific heat of thawed or frozen soil; L_0= 335 x 106J/m³ = 335 x 103kJ/m³ = 8975 Btu/ft3 = 79760kkal/m³ - water-ice latent heat of transition; Ww – moisture content of unfrozen water.

The volume specific heat C $_{th(f)}$ can be presented graphically on the heat exchange curve in thawed and frozen areas as shown in Figure 1.

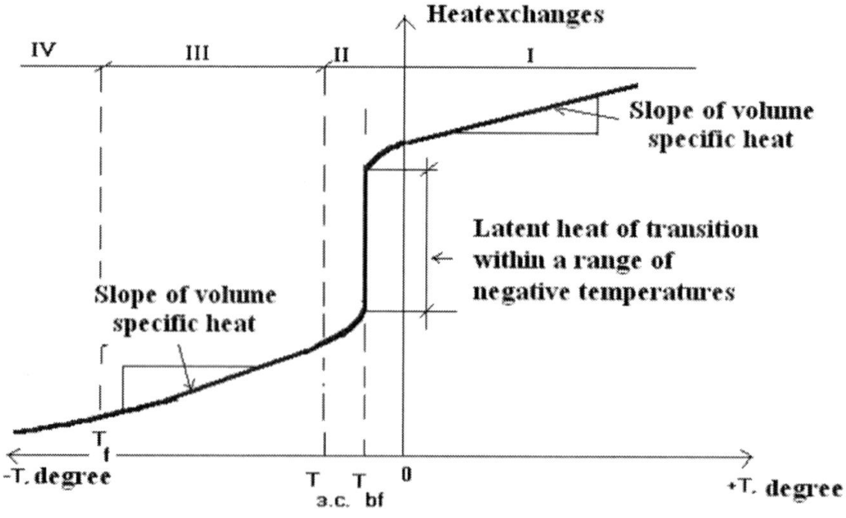

Fig. 1. Curve of heat exchanges in soil (freezing-thawing).

The term $L_0\frac{\partial W_W}{\partial T}$ in Fig.1 represents changes in latent heat of transition within a range of negative temperatures absorbed or yielded by soil due to changes in ground water phase.

Average soil moisture within a migration layer (amount of migration moisture is considered) is determined from the formula:

$$w_{wf} = \frac{Q_{wf}}{\gamma_d} \quad (4)$$

Where Q_{wf} – amount of migration moisture; γd – specific weight of dry soil. The mass of migration soil is determined from the formula:

$$Q_{wf} = q_{wf} \cdot A \cdot t \quad (5)$$

Where q_{wf} - migration flow intensity of moisture; A– cross-sectional area of migration flow; t – impact time.

As the migration flow is measured in the unit volume, the specific weight of dry soil γdequals to the mass of dry soil Q_d. Thus, the delta of average migration moisture per time period equals:

$$\Delta W_{wf} = \frac{Q_{wf}}{Q_d} \quad (6)$$

Having analyzed all existed ratios of the moisture increment $\Delta w1$(fr. unit) versus the rate of freezing V_f(m/time) in different types of soil, relevant approximating functions were selected . Average values of the approximating functions of ratios of moisture increment $\Delta w1$ versus rate of freezing V_f on the freezing front for different soils are expressed by a common formula as follows:

$$\Delta W_{wf} = b \cdot \exp^{c \cdot v} \quad (7)$$

where V- rate of soil freezing, b, c – empirical coefficients.

On the basis of abundant surveys in frost soil territories of Russia there are two maximums of ground water level fluctuations a year: spring and autumn.

The spring maximum is characterized by the highest level in a year.

The specialists from the Far Eastern State Transport University have a great experience in this field; their concept statements on the mortised berms design are included in "The technical standards for stability of deforming roadbeds on thaw subgrade supports in permafrost" (1993).

The cross-section of 1400 m long (Fig. 2) is projected on the basis of a geotechnical conditions analysis, and it solves the problem of soil stability in the subgrade support without restoration of its frozen condition.

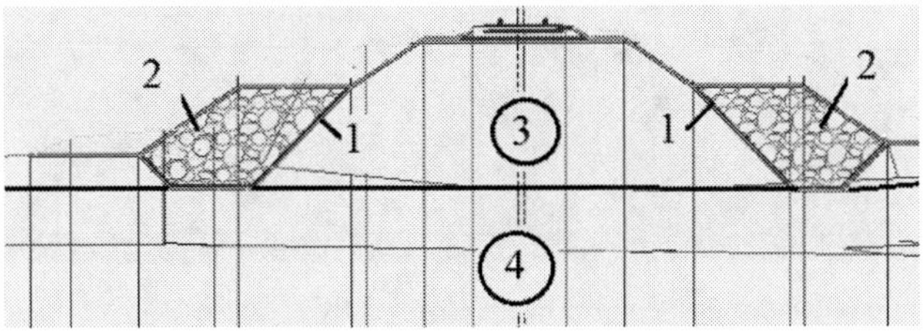

Fig. 2. Cross-section profile of railway embankment: 1- geogrid; 2 - stone; 3 - thaw soils; 4 - permafrost soils.

The results of thermophysical processes calculations without and with increased permafrost base berms are shown in Figure

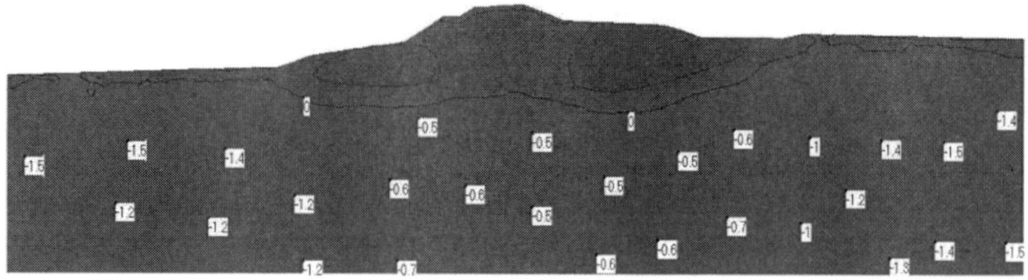

Fig. 3. Temperature distribution for 10 year of operation without increased

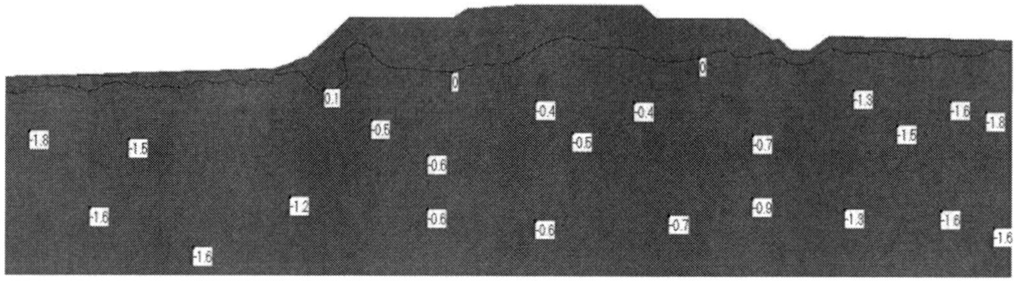

Fig. 4. Temperature distribution for 10 year of operation with increased.

This solution is considered optimal due to several reasons:

– as it is impossible to remove a weak thaw soil from under the subgrade support, some conditions preventing fill wash-out appear. The thaw soil being weak because of annual deep freeze is heaved. The near-toe area is heaved when a multiple rise and settlement of the slope lead to soothing of its gradient;

– a mortising part of the berms prevents heaving and creates the conditions for so called soil consolidation of the subgrade support under the main area. Besides, it provides a service lane during all the works on the site while the traffic keeps on active; moreover, the moss layer is not disturbed in the right-of-way;

– the berm's superstructure of graded stone creates the gravitational stability by its weight, tolerates uneven upper bound of permafrost and prevents further degradation of permafrost. In case the water drain works well, the cooling part of the berm would create a grout curtain of a frozen area around the berm's centerline in 3 – 5 years.

Fig. 5 shows the roadbed cross-section with the mortised berm design.

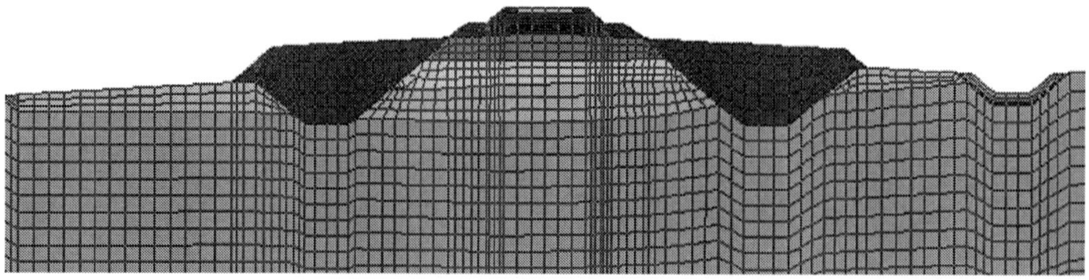

Fig. 5. Roadbed design with subgrade support reinforced by mortised berm.

Fig. 6 shows isolines of vertical displacements in the body and the subgrade support caused by railway traffic. The volume of vertical deformations has decreased by more than three times and is equal to 9 cm.

DOI: 10.1051/matecconf/20167305002

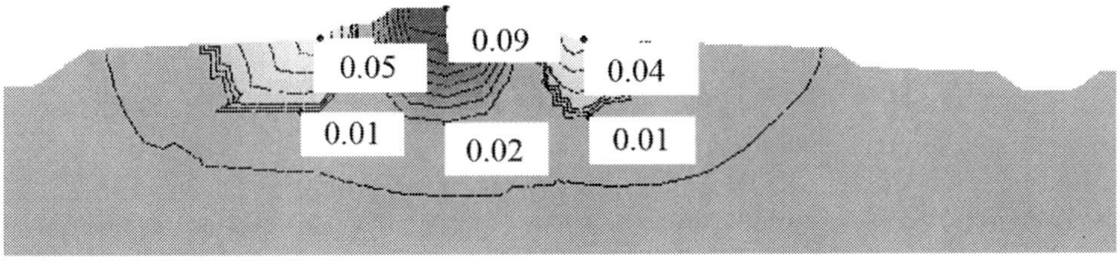

Fig. 6. Isolines of vertical displacements.

Fig. 7. shows the isolines of horizontal displacements in the body and the subgrade support caused by railway traffic. The volume of horizontal deformations has decreased by more than three times and is equal to 6 cm.

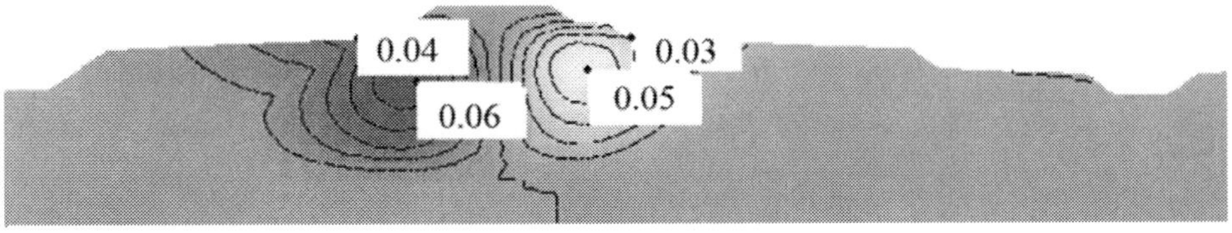

Fig. 7. Isolines of horizontal displacements.

3 Conclusion

The results of thermophysical calculations showed that the permafrost temperature under the body of railroad rises from 0.6 °C up to 1.0 °C forming a thawing basin.

The depth of soil thawing in field conditions beyond the embankment equals to 1.1-1.8 m.

The results of the calculations are suitable for a water drain system being in a working order while the defrost influence of atmospheric water infiltration as well as creep of water through the body and the embankment base are not taken into account.

The calculations were done for the applied cooling measures in a form of rock fill on the slopes and the mortised berms on the both sides of the embankment. These cooling and stabilizing measures keep the permafrost frozen and raise its upper bound from 3.0 m to 4.1 m. The combined calculations accounting the traffic loads on the embankment design were carried out to assess the efficiency effect of the mortised berms and the rock fills on the stress and strain condition of the embankment.

The results of the calculations for stress and strain state have proved the efficiency of rock fill on the slopes and the berms mortised into the toe soils.

The vertical and horizontal deformations of the roadbed of the railway embankment do not outreach 5 cm for the period of thaw maximum.

References

1. S.A. Kudryavtcev, Soil Mechanics and Foundation Engineering, **41(5)**, 177-184 (2004)
2. S.A. Kudryavtcev, U.B. Berestyanyy, T.U. Valtseva, N.V. Barsukova, 1st International conference on new developments in geoenvironmental and geotechnical engineering, 423-427 (2006)
3. S.A. Kudryavtcev, V.N. Paramonov, M.B. Lisuk, Asian Conference on Permafrost. Abstracts. Lanzhou (2006)
4. S.A. Kudryavtcev, U.B. Berestyanyy, T.U. Valtseva, L.A. Arshinskaya, A.Z. Zussupbekov, International workshop on scrap tire derived geomaterials. «Scrap Tire

Derived Geomaterials. Opportunities and challenges». Taylor & Francis Group/Balkema (2008)
5. S.A. Kudryavtcev, U.B. Berestyanyy, T.U. Valtseva, The Proceeding of the International Geotechnical Symposium "Geotechnical engineering for disaster prevention & reduction", 348-351 (2007)
6. S.A. Kudryavtcev, V.N. Paramonov, The Proceeding of the International Geotechnical Symposium "Geotechnical engineering for disaster prevention & reduction", 467-470 (2007)
7. S.A. Kudryavtcev, V.N. Paramonov, A.L. Nevzorov, E.V. Kriger, The Proceeding of the International Geotechnical Symposium "Geotechnical engineering for disaster prevention & reduction", 600-605 (2007)
8. S.A. Kudryavtcev, Y.B. Berestyanyy, T.U. Valtseva, V.N. Chylichkov, D.G. Tsvigunov, 8th International Symposium on Cold Region Development, 37-38 (2007)
9. S.A. Kudryavtcev, Y.B. Berestyanyy, T.U. Valtseva, Ground Improvement and Geosynthetics for Human Security and Environmental Preservation, 493-500 (2007)
10. S.A. Kudryavtcev, D.G. Tsvigunov, Extended Abstracts. Ninth International Conference on Permafrost. Institute of Northern Engineering. University of Alaska, 155-156 (2008)
11. S.A. Kudryavtcev, Y.B. Berestyanyy, T.U. Valtseva, E.D. Goncharova, Ninth International Conference on Permafrost. Institute of Northern Engineering. University of Alaska, 323-324 (2008)
12. S.A. Kudryavtcev, Y.B. Berestyanyy, T.U. Valtseva, Fourth International Conference on Scour and Erosion, 651-654 (2008)
13. S.A. Kudryavtcev, Y.B. Berestyanii, E.V. Fedorenko, T.U. Valtseva, R.G. Mikhailin, 3rd International Geotechnical Symposium on Geotechnical Engineering for Disaster Prevention and Reductio, 18-25 (2009)
14. S.A. Kudryavtcev, Y.B. Berestyanii, T.U. Valtseva, R.G. Mikhailin, E.D. Goncharova, 14th Conference on cold regions engineering, 45-48 (2009)
15. S.A. Kudryavtcev, V.N. Paramonov, I.I. Sakharov, The freezing and thawing of ground (practical examples and finite-element calculations) (Saint-Petersburg, 2014)
16. S.A. Kudryavtsev, Y.B. Berestyanyy, E.D. Goncharova, The 23rd International Offshore (Ocean) and Polar Engineering Conference Anchorage, 562-566 (2013)
17. S.A. Kudryavtsev, Y.B. Berestianyi, T.Y. Valtseva, A.V. Kazharskyi, E.D. Goncharova, The 1st International Symposium on Transportation Soil Engineering in Cold Regions. Sciences in cold and arid region, 404-407 (2013)
18. S.A. Kudryavtsev, A.V. Kazharskyi, The 5th International Geotechnical Symposium-Incheon, Geotechnical engineering for disaster preventional & reduction. Incheon, (2013)
19. S.A. Kudryavtsev, Y.B. Berestianyi, T.U. Valtseva, R.G. Mikhailin, E.D. Goncharova, The 24rd International Offshore (Ocean) and Polar Engineering Conference (2014)
20. S.A. Kudryavtsev, Y.B. Berestianyi, A.V. Kazharskyi, E.D. Goncharova, *The 10th* International Symposium on Permafrost Engineering in Cold Regions. Sciences in cold and arid regions, 474-478 (2014)
21. S.A. Kudryavtsev, T.U. Valtseva, R.G. Mikhailin, E.D. Goncharova, Using designs of variable rigidity on weak soils roads in the Russian Far East 6th International geotechnical symposium on Disasters Mitigation in Special Geoenvironmental Conditions (2015)
22. S.A. Kudryavtcev, V.M. Ulitskii, V.N. Paramonov, I.I. Sakharov, Soil mechanics and foundation engineering **52(5)**, 240-246 (2015)

The Use of Sheds for Thermal Management of the Permafrost Bases

Vadim Passek[1,*], and *Aleksandr* Tsernant[2]

[1,2]Industrial University of Tyumen, 625001, Volodarskogo str., Russia

Abstract. The article analyzes the use of various types of sheds as an effective measure for thermal management of the permafrost soils. It also analyzes the possible structural designs of sheds for railway and highway roadbed and the effectiveness of their thermal influence on frozen conditions.

1 Introduction

Over half of Russian territory is located in the permafrost zone. However, this area is not uniform in its natural environment. It is very diverse in its climatic, permafrost soil, landscape and other conditions. Type of construction and building structures is different in each zone, where absolutely different measures contributing to the temperature conditions stabilization are rational for implementation. Therefore it is necessary to develop new measures, to improve the old ones and to find areas of their rational application.

[1, 2] present the systems of approach to the classification of various measures. Different types of sheds may be one of the effective measures for thermal management of the permafrost soils. In summer, sheds allow to protect soils from precipitation and solar radiation and thus reduce heat influx, and in the winter - from snow deposits and thereby to increase the cold flow (to increase the efflux of heat) into the base soils. Despite the self-evident's seemingly of appropriateness of sheds, in fact the practical realization of this event is not self-evident and it's connected with a number of practical difficulties. Shed's operating efficiency depends on following factors:

- design of engineering structure. Efficiency of sheds for railway's and road's embankments depends on the steepness of slopes, berms presence, the width of the main site and etc.;

- geographical location of region. The structure of sheds depends on presence of wind and therefore movement of snow masses (for example, the north of Western Siberia) or its absence (eg, Central Yakutia).

- constructive solution of shed. Shed must have durability equal to the durability of the structure itself, but when you do so, complements shed cost can increase significantly.

In connection with the above, the relevance of the development of such a topic (development sheds for different conditions) is determined, on the one hand, the efficiency

* Corresponding author: tgasu.passek.vadim@mail.ru

sheds, and with another - lack of practical developments. This article is an attempt to somewhat fill this gap.

2 Review

The idea of shed using in order to generate the required heat and humidity regime appeared not only at the dawn of mankind, but generally at the beginning of all life. The most visible effect of using sheds in terms of permafrost soils was obviously reached by applying the ventilated underground spaces, the details of which are presented in [3] (usage within the Trans-Baikal railway in 1907) and which have found a wide use by now.

The use of sheds on permafrost for other building structures also took place in the middle of the last century. N.N. Romanovsky suggested using sheds as anti-icing measures. [4] Sheds are also known to be applied for dams [5].

In order to find the efficient ways of thermal management, theoretical studies led by V.S. Lukyanov were widely deployed in the 50 - 60s by CSRIC. The studies were devoted to thermal regime of the railways and roads in the permafrost. A variety of measures, including sheds were analyzed. Effectiveness of the sheds use is covered in [6]. Afterwards, the possibility of sheds using was analyzed in [7]. At the same time, other researchers were as well engaged into sheds application, for example, V. G. Kondratyev [8, 9, 10]. Sheds were also used abroad: in China [11, 12], in Alaska [13].

Thus, thermal effect that is obtained when using sheds is undeniable.

3 Description of the research and developed designs

The main problem lies in the embodiment of the sheds. Shed is an integral part of the engineering structure, which should function throughout the entire service life of the structure. Therefore, it should be constructed as thoroughly as a bridge or a culvert. At that point it becomes expensive and often already uncompetitive in comparison with other structural measures. That is why there is a problem of finding the way to simplify the sheds.

Let us examine the peculiarity of the sheds usage in relation to the conditions of a railways roadbed. One of the effective ways lies in the replacement of overall sheds with the local ones. Fig.1. shows various sheds designs.

Fig. 1 presents a shed in accordance with patent [8]. It contains the load carrying elements 1, completely enveloping the embankment 2 and obstruction clearance 3. The carrying elements 1 are mounted on the bearings 4. Continuous coating 5 is laid on the carrying elements 1. Snow deposits 6 in winter, solar radiation and rainfall in summer are held up by the shed over the entire surface of the embankment, which allows to dramatically reduce the temperature of the permafrost soils.

The advantage of this technical solution is that it achieves the maximum protection against the harmful effects of: snow drifts, solar radiation and rainfall.

The disadvantage of it is the high cost of the structure. The carrying elements 1 have a rather large span (about 25 - 30 m), and are located above the rolling stock. Consequently, the structure should be sufficiently robust (like a bridge), able to withstand the wind and snow load, dead load, etc. The design can be simplified by several times (Fig. 1b) [14]. The principal difference from the previous design is that the load carrying elements 1 are located below the transport passage. Firstly, it dramatically reduces the level of design responsibility, and secondly, the span decreases (by 2 - 3 times), which consequently reduces the material consumption by 4 - 9 times. As for effectiveness, it is practically the same as that of the first design (see Fig. 1a).

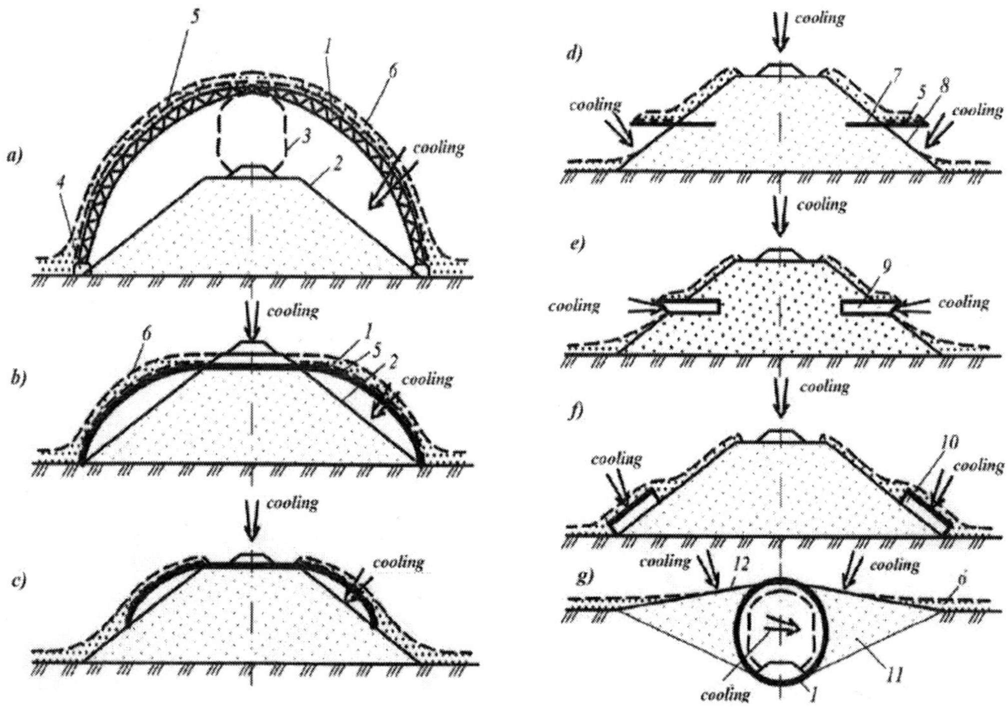

Fig. 1. Various schemes of sheds: a, g – overall; b – zonal; c, d, e, f – local

This is because within the primary platform of the embankment there is practically no snow in any climatic zones (both in regions with a strong transport of snow, and in regions with no transport of snow). And snow is the major factor that influences the temperature regime. The peculiarity of the Fig. 1b scheme is that instead of the bearings 4 (see Fig. 1a) the use of a cantilever rigid-fixing is implied: load carrying elements 1 are introduced into the upper part the embankment body 2.

But the maximum possible actuating quantity is not always needed. Often it is enough to partially improve the situation in order to provide the required temperature regime. Thus, local sheds become expedient. The first scheme of local shelter is shown in Fig. 1c. It mostly repeats the scheme of Fig. 1b, so it does not require explanation.

Fig. 1d [15] is a diagram of the local cantilevered shed. Supporting cantilever elements 7 are stacked during the process of soil layers placing. Deck 5 is placed on the supporting elements 7 on the outer side of the embankment.

This type of sheds is suitable for regions with a lack of snow drift (e.g. Central Yakutia). In this case, the part of the slope surface 8 is called bare (with no snow there). This surface is enough to provide the desired temperature regime. Fig. 1e presents a very effective scheme of the local shelter. [16]

When stacking the layers, the pipes with penetration of about 4-5m are placed in the embankment body. The pipes diameter is about 1m, the distance along the embankment - about 10m. On the degree of effectiveness, it is close to the scheme of Fig. 1d, but is simpler to install and is reliable in operation. Fig. 1f, [18] gives a scheme of the shed in the form of a sill hollow covering 10. In regions with a strong snow transit excavation it is very difficult to construct cavities: they are buried with snow, and passage is impossible. In this case, the use of a tunnel type cavity is possible.

We return to the scheme of an overall shed (see. Fig. 1a), but the supporting structure 1 envelopes only the obstruction clearance, so the cost is immeasurably lower.

The design in Fig.1g [19, 20] comprises a supporting structure 1 made of, for example, the corrugated steel elements.

Backfill soil 11 bounds the structure on the sides. The upper surface 12 is raised above the snow deposits in natural conditions. In regions with a strong transport of snow, the surface 12 becomes bare of snow and the cooling process passes through it. The cooling is also carried out through the tunnel passage cavity (assuming that the length of a tunnel makes the cold air aeration possible).

We can distinguish the following groups of sheds in relation to the railway and highway roadbed:

– on the degree of coverage of the cross section of the roadbed: overall (which cover the entire cross-section of the embankment or cavity), zonal (covering one or more elements of the cross-section, but not the entire cross-section), local (located in separate parts of the cross-section);

– on the degree of continuity: continuous and non-continuous (i.e. constructed with intervals);

– on the nature of the location in relation to the outer surface of the roadbed cross-section: external sheds (located on the outer surface side) and internal ones. The internal sheds can include the pipes inside the sloping part, as well as, for example, the horizontal layers of the thermal insulation from the side of a slope;

– on the scheme of bearing: closed (for example, in the form of a corrugated structure of the tunnel, laid instead of a cavity), pillar-type (arch or frame semi-closed profile with upright pillars) sill-type (e.g. box-shaped section structures, laid on a slope), cantilever-type (sheds on a horizontally stacked elements in the sloping part of the embankments or cavities);

– on the nature of air circulation under the shed: with the augmented thrust or with natural convection.

To test the effectiveness of local sheds (as in Fig. 1d), a series of thermal and physical calculations was run.

Fig. 2 shows the obtained dependence of the temperature tc in the center of the embankment foot (on its axis) on the relation b/ hem (b - width of the primary embankment platform, hem - height of the embankment) for the sheds of various width. The calculation results show that effectiveness of the shed is quite significant: even in the worst case, the soil temperature becomes low enough (-0,75°C).

The next effect turned out to be very interesting. When applying several local sheds installed with some spacing, the effectiveness increases in comparison with the case where sheds are arranged directly adjacent to each other.

Curves 3 and 4 in the Fig. 2 show the effectiveness of using two sheds of 2m in width arranged side by side (curve 3) and installed with the spacing of 2m (curve 4). In the latter case, the effectiveness is higher by 30%.

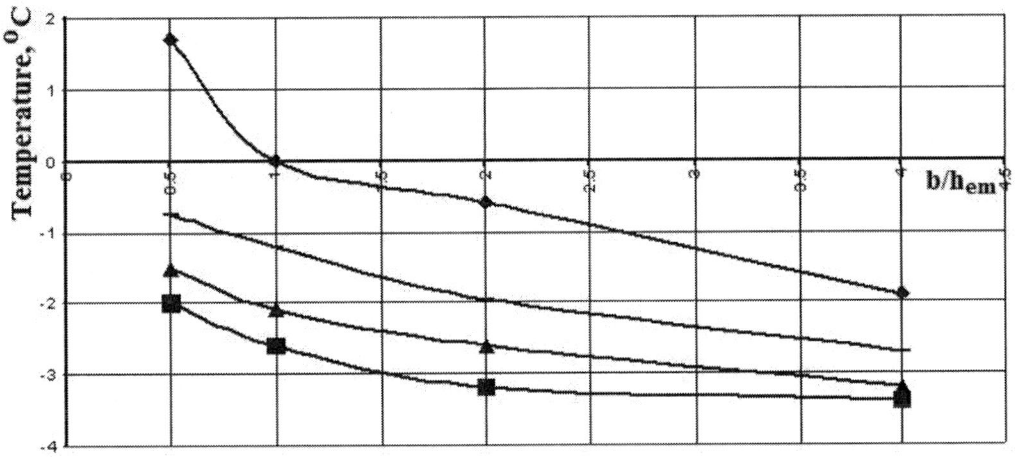

Fig. 2. Dependence of the temperature tc on the level of the embankment foot (on its axis) on the relation b/hem: 1 – with no sheds; 2, 3 – with the shed of 2,4m respectively; 4 – with two sheds of 2m, constructed with the 2m spacing

4 Conclusion

The study of different types of sheds from the perspective of thermal influence effectiveness showed the following:

- overall closed continuous external sheds are more appropriate only for cavities in regions with strong transport of snow (when the cavity is covered with snow). Embankments have a primary platform with dramatically decreased snow thickness in all the regions, so the use of the shed for embankments in this region is ineffective;
- zonal continuous external sheds within the slopes are appropriate for embankments in the regions with strong transport of snow;
- zonal or local (as calculated) internal and external non-continuous sheds constructed within the sloping part are suitable for Central Yakutia – both for cavities and embankments;
- apart from the artificial sheds, it is also appropriate to use the natural sheds in the formation of calculated temperature conditions in the soil bases. Those can be, for example, the span of the bridge, the culvert or ventilated underfloor spaces in buildings (especially in regions with low transport of snow).

References

1. Tsernant, Vestnik TyumGASU, **3**, 10 – 15 (2015)
2. V. Passek, Sb. nauchnykh trudov JSC TSNIIS, **213**, 28 – 46 (2002)
3. V. Statsepko, *Chasti zdaniy* (SPb, 1912)
4. N. Romanovsky,*Merzlotnye issledovaniya* (MGU,1963)
5. G. Biyanov, *Plotiny na vechnoy merzlote* (Energiya ,1975)
6. N. Tsukanov, V. Pataleev, *Sb. trudov KHabIIZHTa*, **41** (1970)
7. V. Passek, N. Tsukanov, A. Tsernant, *Issledovanie konstruktsiy tonnel'nogo tipa dlya peresecheniya vyemok zheleznoy dorogi v usloviyakh poluostrovaYamal* (JSC TSNIIS, 2005)

8. V. Kondrat'ev, A. Korolev, M. Karlinskiy, V. Pozin, A. Rozanov, Zheleznodorozhnyy put' na sil'nol'distykh vechnomerzlykh gruntakh. Patent na izobretenie № 1740555, (1992)
9. V. Kondrat'ev, Nauchno-tekhnicheskiy otchet po teme «Obosnovanie primeneniya solntseosadkozashchitnykh navesov dlya predotvrashcheniya degradatsii sil'nol'distykh mnogoletnemerzlykh gruntov osnovaniya zemlyanogo polotna zheleznodorozhnoy linii Tommot – Kerdem», TransIGEM (2007)
10. V. Kondrat'ev, i Lyu Tszyankun'. Stroitel'stvo. Proektirovanie, **1** (1996)
11. Liu Jiankun, Li Dongqing, Ma Wei, Zhang Luxin, *Proceedings of the Fifth International Symposium on Permafrost Engineering,* **2** (2002)
12. Z. Luxin, L. Jiankun, *International Symposium on Geocryological Problems of Construction in Eastern Russia and Northern China,* 2 (1998)
13. J. Zaling, A. Breley, Thaw stabilization of roadway embankments constructed over permafrost. Report NO FHWA-AK-RD-81-20 (1986)
14. V. Passek, N. Tsukanov, A. Tsernant, A. Datskovskiy, V. Gerasimov, B. Drobyshevskiy, V. Passek, Patent RF 72492 (2007)
15. V. Passek, N. Tsukanov, V. Pozin, N. Verbukh, A. Tsernant, A. Maleev, G. Pereselenkov, G. Orlov, N. Lukin, V. Volodin, N. Dedova, N. Banova, Patent RF 69883 (2007)
16. V. Passek, N. Verbukh, N. Tsukanov, A. Tsernant, B. Drobyshevskiy, V. Passek, V. Pozin, V. Sholin, G. Orlov, G. Pereselenkov , Patent RF 70267 (2007)
17. N. Verbukh, Patent RF 68534 (2007)
18. N. Tsukanov, V. Passek, V. Zakovenko, P. Dydyshko, R. Evstigneev, Patent RF 1139176 (1993)
19. V. Passek, A. Tsernant, A. Datskovskiy, V. Linnik, Patent RF 2275471 (2006)
20. Patent RF 2275472 na izobretenie «Tonnel' melkogo zalozheniya na vechnoy merzlote». A. Tsernant, V. Passek, A. Datskovskiy, V. Linnik, Opubl (2006)
21. T. Maltseva, A. Nabokov, A. Chernykh, Procedia Engineering, **117**, 239-245 (2015)

MATEC Web of Conferences 73, 05004 (2016) DOI: 10.1051/matecconf/20167305004
TPACEE-2016

Stabilization Techniques for Road Lower Structure and Roadbed Constructed on Permafrost Soil

Vyacheslav Vorontsov[1,*], *Aleksey* Kraev[1], and *Victor* Mironov[1]

[1]Industrial University of Tyumen, 625001 Volodarskogo str. 38, Tyumen, Russia

Abstract. The perspective development of Yamal-Nenets Autonomous District, which is rich in mineral resources is impossible without creation of the road network for communication between settlements, oil and gas fields and transport hubs. Yamal-Nenets Autonomous District is characterized by complex engineering-geological conditions where different geological and geocryological phenomena and processes are being developed causing specific approaches to design and construction of engineering structures to be used. In this regard, an urgent task is to develop constructional solutions making it possible to stabilize the road lower structure and the roadbed, prolong periods between repairs on individual sections and improve operational reliability of roads in general. The paper describes the proposed and implemented constructional and technological solutions to stabilize the road lower structure and the roadbed constructed on permafrost soils. The results of two-year geotechnical monitoring of the road sections with implemented solutions are given.

1 Introduction

Maintenance of roads built on permafrost soils is accompanied by a large number of repairs which are caused by cyclic deformations of the pavement structure and the roadbed. There are lots of causes of such deformations. The main reasons include errors in designing a road being located in severe climatic conditions and changing engineering-geocryological and hydrological conditions during operation. As a result, needed are the additional measures aimed at regulating water-and-temperature conditions of the roadbed and the road lower structure allowing traffic current to be available on the repair area. One such measure is the implementation of various schemes for reinforcing the roadbed and the road lower structure aimed at possible operation of the road during winter and in the period of maximum thawing of the roadbed and the lower structure of the road. Generalized results of research in this area are listed in domestic and foreign literature (Citovich N.A., Stefan I, Krylov M.M., Saltykov N.I., Hakimov X.R., Dalmatov B.I., Kudryavcev V.A., Melamed V.T., Melnikov V.P., Chzhan R V., Aldrich H.P., Paynter H.M., Ershov E.D., Ivanov N.S., Orlov

* Corresponding author: tgasu.voroncov.vyacheslav@mail.ru

V.O., Porhaev G.V., Pchelincev A.M., Tyutyunov I.A., Fedorov V.I., Feldman G.M., Cheverev V.G., Corte A.E., Xu X. and etc.)

2 Literature review

In 2011 a group of researchers from Tyumen State University of Architecture and Civil Engineering (TSUACE) carried out a survey of the road section (PK 593+00 – PK 600+50) "Surgut - Salekhard, Novy Urengoy - Nadym the 1st commissioning and start-up complex: Pangody (km 870) – Pravohettinsky (km 936); the survey was carried out in parallel with the engineering investigations. Photographic images of engineering-geodetic and geological surveys of the previous years were used. The following defects were identified after the analysis of the results obtained (Fig. 1.):

Roadbed

Deformation of the roadbed –vertical movements of the road border

Minor destructions of the roadbed near the head walls of the culverts.

Fig. 1. Defects of the road pavement and the roadbed.

Pavement

Development of longitudinal and transverse cracks could be seen on the surveyed section of the road pavement. Sections with a strong network of cracks were observed in the areas where the roadbed and road borders were deformed.

Cracks generated in the center of the pavement along the axis of the road with 3 cm of growth indicated that vertical movements of the road pavement were accompanied with the horizontal ones toward the slopes. Development of longitudinal cracks up to 23 m in length indicated the possibility of deformation of the roadbed and the road lower structure.

Development of transverse cracks on the surveyed sections of the road pavement indicated vertical movements both along the entire width of the roadbed and pavement and also along the local sections. This type of defect called a hollow spot was accompanied by development of perpendicular cracks thickened on both sides to a maximum vertical movement.

Culverts

Two culverts were constructed on the road section PK 593+00 – PK 600+50. Visible minor deformations of the roadbed slope near the entrant head wall occurred on PK 599+16. One could observe long-standing areas of water in ditches on the right and the left sides of the road indicating the inefficiency of the culverts. Destruction of slopes was not observed. Visual inspection of the defects did not make it possible to identify the causes, but it was possible to determine the direction and nature of roadbed and pavement deforming.

After the results obtained the road was divided into four sections:

Section №1 - PK 594+00 – PK 594+90 - destroyed was the right side of the pavement, the maximum deviation from the design marks was 28 cm. Vertical movements of the left and right sides of the road pavement were observed on PK 594+26 – PK 594+32. The mean value of vertical movements was 20 cm.

Section №2 - PK 596+40 – PK 597+00 - destroyed was the left side of the pavement incorporating two local zones of fracture. The maximum vertical movement on the left side of the pavement was 19cm.

Section №3 - PK 597+20 – PK 597+45 - destroyed was the left side of the pavement and the border. The maximum vertical movement on the left side of the pavement was 23cm.

Section №4 - PK 598+80 – PK 599+90 - identified were the local zones of fracture both on the left side of the pavement and hollow spot edgewise the road structure. On PK 599+46 – PK 599+60, PK 599+20 – PK 599+36 and PK 598+90 – PK 598+80 destroyed were the left sides of the pavements. The maximum deviation of the elevation points from the design marks was 39 cm. On PK 599+18 – PK 598+92 developed were the vertical movements edgewise the road pavement with the maximum value of 30cm.

Engineering-geological surveys resulted in making transverse profiles of the road with engineering-geological cross section view of the specified sections.

On the right side of the road the transverse engineering-geological cross section of section №1 was presented by the medium-sized water-saturated sands with water capacity of up to 2 m in the upper layer, and in the bottom layer - fine water-saturated sands. Permafrost was absent. On the left side of the road the cross section was presented by wet peat with moderate degree of decomposition of up to 1.8 m and icy plastic-frozen peat. The boundary of permafrost was at around 41.90 - 40.80 m, the distance from the daylight surface at 2.35 - 4.50 m. Cracks developed in the right lane of the given section. The absence of permafrost on the right side of the road bottom and the presence of permafrost border on the left side indicated that the road structure rested upon heterogeneous base. The longitudinal cracks along the road axis indicated the boundary separating the lower structure from permafrost soils and the lack of them. Deformations were associated with settlements occurring at the bottom of the right side resulting in horizontal movements of the roadbed structure toward the right slope.

On the right side of the road the engineering-geological cross section of the base on the road section №2 (PK 596+40 – PK 597+00) was presented by fine sands with moderate water saturation capacity of up to 2.1 m and medium-sized sands saturated with water. Permafrost was absent. On the left side of the road the engineering-geological cross section consisted of wet peat with moderate degree of decomposition and saturation capacity of up to 4m, plastic-frozen peat with the saturation capacity of up to 3m and water-saturated sand. The boundary of permafrost was at around 41.60 - 40.00 m, the distance from the daylight surface at 1.35 -. 2.50 m. The pavement had defects in the left lane of the carriageway. Deformations of the roadbed structure and covering of the roadway occurred due to the deformations in the left portion of the roadbed structure resulting from thawing of frozen peat layer. Thawing of the upper layers was worsened by long-standing surface water at the left slope of the road. Engineering-geological cross section of the road lower structure on the road section №3 (PK 597+20 – PK 597+45) was presented by the intermediate layer of medium-sized water-saturated sand with an average degree of water capacity varying from 0.5 to 2.3 m, widely spread throughout the whole structure. From the right side of the road the underlying layers of the engineering-geological cross section were presented by medium-sized water-saturated sand with water capacity of up to 1.2 m, the layer of medium-sized hard-frozen sand with water capacity of up to 2 m and medium-sized water-saturated sand. From the left side of the road the engineering-geological cross section was presented by peat with moderate degree of decomposition saturated with water up to 1.9m of capacity, plastic-frozen peat with moderate degree of decomposition saturated with water up to 2m and the intermediate layer of sandy loams with the capacity of up to 1 m. The bottom layer was presented by water-saturated medium-sized sand. Deformations developed in the left part of the roadbed structure and covering of the roadway. Defects were caused by the presence of weak soil under the left side of the road, i.e. peat with moderate degree of decomposition saturated with water, plastic-frozen peat with the total capacity of 3.5m. On the right side of the road the level of the steady groundwater was 43.7 m at the given section, and on the left side -. 42.3 m. This difference was capable of generating a hydraulic pressure gradient in the soil under the road structure, which often leads to mechanical suffusion and, as a consequence, the destruction of the roadbed structure. Engineering-geological cross section of the road lower structure on the road section №4 (PK 598+80 – PK 599+90) was presented by fine and medium-sized sands with moderate water saturation. Permafrost was absent in the lower structure.

The analysis of the geological-engineering surveys resulted in understanding the possible causes of defects occurring annually on the road pavement and the roadbed:

- Section №1 (PK 594+00 – PK 594+90) - the absence of permafrost on the right side in the lower structure of the road and the presence of permafrost border on the left side indicated that the road structure rested upon the heterogeneous base. Deformations occurred due to settlements in the lower structure from the right side of the road, thus resulting in displacement of the roadbed structure toward the slope;
- Section №2 (PK 596+40 – PK 597+00) – failures in the pavement were due to the deformations in the left side of the roadbed as a result of thawing of the frozen peat layer. The process of thawing in the upper layers was worsened by long-standing surface water at the left slope of the road;
- Section №3 (PK 597+20 – PK 597+45) – failures were caused by the presence of weak soil layers in the lower structure, i.e. peat with moderate degree of decomposition saturated with water and plastic-frozen peat with moderate degree of decomposition. Hydraulic pressure gradient in the soil under the road structure led to mechanical suffusion in the lower structure and the roadbed body.
- Section №4 (PK 598+80 – PK 599+9) – failures were caused by the pressure from the weight of the roadbed and covering of the roadway transmitted onto the weak soil resulting in settlements both in the road lower structure and the roadbed. The slope of the terrain and structure of the road (virage) in the direction of the deformable lane worsened the situation. These factors led to an increase of pressure on the weak soil under the left side of the road and an increase of stresses in the roadbed causing vertical and horizontal deformations of the lower structure and destruction of the left lane pavement of the carriageway.

It was necessary to work out measures to stabilize the road structure and preserve the covering of the roadway on the section "Surgut - Salekhard, Novy Urengoy – Nadym - the 1st commissioning and start-up complex: Pangody (km 870) - Pravohettinsky (km 936)" PK 593+00 – PK 60+50, since the errors were made in designing the road when choosing the type of the transverse cross-section profile of the roadbed, and the engineering-geological features of the road lower structure were not considered [1].

3 Task description

To eliminate the defects developing annually in the roadbed structure and on the pavement it is needed to develop options for constructional and technological solutions to stabilize the road lower structure and the roadbed allowing traffic current to be available on the repair area. The effectiveness of the proposed constructional and technological solutions to stabilize the road lower structure and the roadbed should be assessed by geotechnical monitoring of the section with the realized schemes during the period of three years. Geotechnical monitoring results will allow to evaluate the performance of the structures or develop additional measures aimed at improving their effectiveness, as well as help form the experimental database to develop the technique (forecasting) to calculate temperature, moisture content and the stress-strain state of the object under study. The most effective and efficient measures to stabilize the roadbed and the lower structure of the road are to be used in future on similar problematic sections arising on newly built roads and the existing ones. Preservation of the existing roadbed, covering of the roadway and availability of road traffic during the period of construction works and repairs is the advantage of the developed measures. These measures must agree with the specific features of the engineering-geocryological structure of the roadbed bottom and hydrological conditions [2, 3]. Since the section is located on permafrost, the proposed measures should take into account the peculiarities of construction and design of the roadbed on given sections [4].

Vertical reinforcement of the right side of the roadbed being deformed vertically and horizontally (Fig. 2.) is suggested on section №1 (PK 594+00 – PK 594+90). Reinforcement is performed by vertical placement of the reinforcing material - to a depth of H = 2 m into the roadbed bottom with a hard soil roller l2=2 m in width and h2=0.75 m in height. Its size h1 is chosen so that the roller tilt agrees with the slope of the given transverse profile of the roadbed. The roller is constructed in the sloping part of the roadbed at a distance of l1=8 m from the axis [5]. The soil roller located in the sloping part of the roadbed will block the deformations to be developed in the direction of the slope. To prevent the roller from horizontal movements it is reinforced with the material which is fixed in the bottom with the polyethylene tube 25 mm in diameter. This reinforcement allows horizontal and vertical deformations to be reduced when they occur in the roadbed body and the road lower structure during the period of maximum active layer capacity and prevent further strain on the given section. Geotextile "Geospan TH 80" is used as reinforcement.

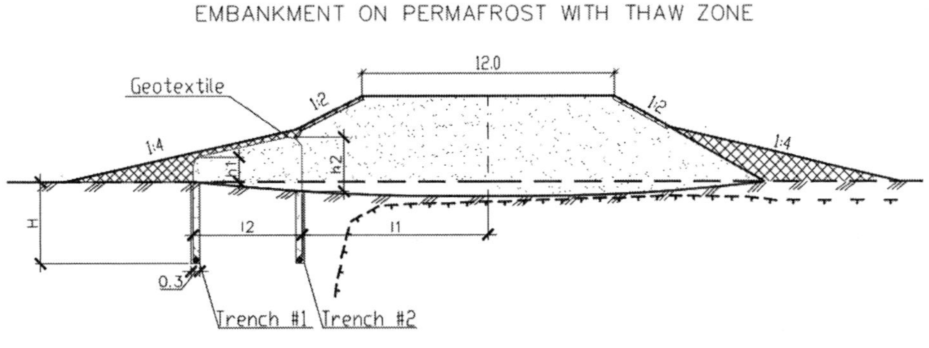

Fig. 2. Constructional and technological solution for roadbed stabilization on road section №1 (PK 594+00 – PK 594+90).

It has been proposed to reinforce the roadbed and the lower structure vertically in the left side of the road on road section №2 (PK 596+40 – PK 597+00) forming a hard soil roller wrapped with the reinforcing material around the outer perimeter (Fig. 3.). Vertical reinforcement is performed to a depth of H = 2 m into the roadbed bottom with the hard soil roller l2=2 m in width and h2=0.75 m in height. The roller size h1 is chosen so that the roller tilt agrees with the slope of the given transverse profile of the roadbed. The roller is placed in the sloping part of the roadbed at a distance of l1=8 m from the axis. The soil roller will reduce the deformations to be developed in the roadbed and the lower structure and prevent penetration of surface water into the body of the roadbed. Lowering of moisture content is achieved by reducing the filtration rate due to the reinforcing material placed vertically which prevents penetration of water into the body of the roadbed and the lower structure of the road. To prevent thawing of weak soil, heat stabilizers TK 32/10 10m in length and 32mm in diameter are added. Heat stabilizers are placed 2m from the outer sides of the reinforced roller. Heat stabilizers will make it possible to raise permafrost level to the daylight surface and reduce the capacity of easily deformable thawed soil layer. Heat stabilizers will allow a cylindrical volume of frozen soil of 1.3 m in radius to be formed for 2 years of operation. Step of 2 meters between the stabilizers allows the vertical frozen soil massif to be formed along the reinforcing elements; that will give additional stability to the road structure [3, 5].

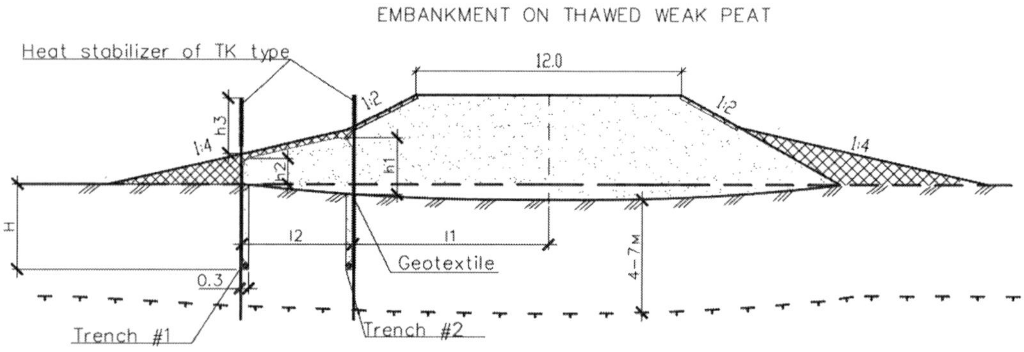

Fig. 3. Constructional and technological solution for roadbed stabilization on road section №2 (PK 596+40 – PK 597+00).

Stabilization technique for the roadbed on section №3 (PK 597+20 – PK 597+45) is identical to that of the section №1 [5]. Unlike section №1 stabilization here is performed on the left side of the road (Fig. 4). The same type of solution used here is connected with performance of vertical reinforcing elements installed at a depth of 2 m into the bottom of the road. The reinforcement wrapping the roller is vertically immersed into the soil below the permafrost making it possible to fix the reinforcement ends rigidly in the permafrost. Besides, lowering of the filtration rate due to vertical placement of the reinforcing material allows moisture-temperature conditions under the roller to be changed and raise the level of permafrost daylight surface [6]. Stabilization of the roadbed structure is carried out due to rigid binding of the reinforcement ends in the permafrost making it possible to reduce deformations in the given section toward the slope.

To ensure stability of the road structure and the roadbed slope on section №4 (PK 598+80 – PK 599+90), the constructional and technological solution has been developed in order to eliminate deformations occurring in the roadbed and the lower structure of the road (Figure 4.).

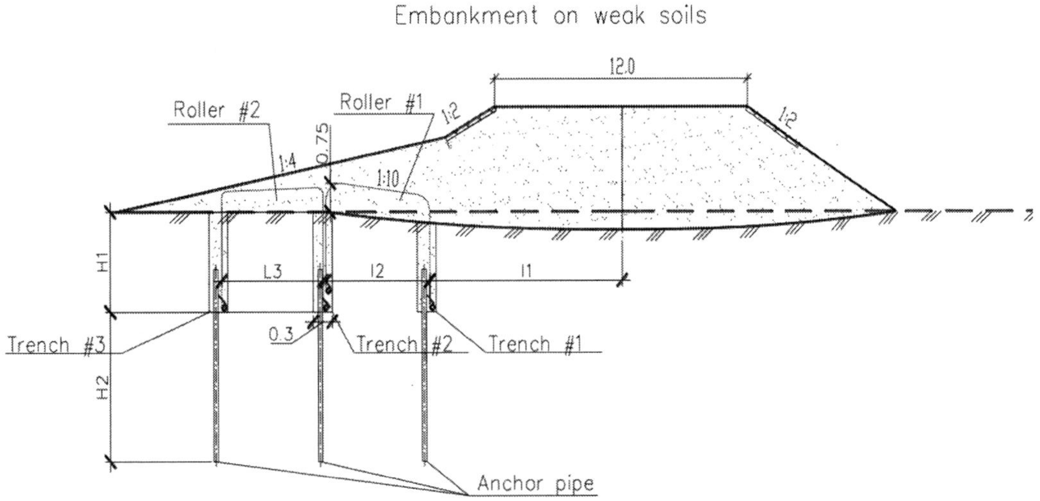

Fig. 4. Constructional and technological solution for roadbed stabilization on road section №4 (PK 598+80 – PK 599+90).

It is suggested to perform vertical reinforcement of the roadbed and the weak lower structure to a depth of H1= 2m with reinforcement to be anchored to steel tubes immersed in the underlying soils to H2=3m. Reinforcement is done with two rollers wrapped with the reinforcing material along their outer surface. Vertical reinforcement is placed in 3 rows with the distance from the axis of l1=6 m, l2=2 m and l3=2 m. A hard soil roller l2,3=2 m in width and h2=0,75 m in height is placed in the sloping part of the roadbed. These vertical rows of reinforcement will prevent soil movement toward the slope. Anchor tubes immersed in the weak subgrade soil to a depth of 3m allow the reinforcing element to be fixed firmly in the design position, and thus increase the stability of the slope and the roadbed structure.

In 2012 the proposed stabilization techniques for the road lower structure and the roadbed constructed on permafrost soils were implemented in the section (PK 593+00 – PK 600+50) - "Surgut - Salekhard, Novy Urengoy - Nadym the 1st commissioning and start-up complex: Pangody (km 870) – Pravohettinsky (km 936) (Fig. 5)..

The effectiveness of the constructional and technological solution to stabilize the roadbed is to be evaluated after the results of geotechnical monitoring of the sections with the implemented solutions during the period of 3 years. Geotechnical monitoring aims at selection and justification of the most rational roadbed on the given section - Novy Urengoy – Nadym - by analyzing the observation results of the stress-strain state and temperature conditions of the road lower structure soils and the roadbed as well as preparing the initial messages for design of road structures using the proposed roadbed.

Fig. 5. Reinforcement of the roadbed and the road lower structure: a) – Section №1 (PK 594+00 – P K 594+90); b) –Section №2 (PK 596+40 – PK 597+00); c) –Section №3 (PK 597+20 – PK 597+45); d) –Section №4(PK 598+80 – PK 599+90).

4 Description of research

Thus, to achieve the purpose when performing geotechnical monitoring and solve the related problems it was necessary to understand the stress-strain state and temperature conditions of the road lower structure soils developing in time [7, 8 and 9].
It was needed to:
- Measure stresses on the vertical platforms in the distinguished points;
- Measure stresses on the horizontal platforms in the distinguished points;
- Measure the vertical movements of soil particles in the road lower structure and its body;
- Measure the vertical and horizontal deformations of the road structure body;
- Measure soil temperature in the embankment base;
- Perform geodetic surveying of the roadbed body.

To monitor temperature conditions of the subgrade soil and the embankment body of the road, 56 thermometric tubes were immersed to a depth of 10 m. Vertical movements of soil layers in the road subgrade and the embankment were measured with 480 deep screw marks. Vertical movements of the marks were recorded by deflectometers made after a dial indicator ИЧ-1. To save measurements and protect instrumentation from adverse climatic factors, protective metal casings were installed above each motion sensor.

Measurement of total stresses and pore pressures were carried out by 480 membrane load cells made after the technique suggested by A.V. Golli. To record sensor readings the following devices were used: automatic meter of deformations ITC-01 and NationalInstruments. General view of the road section (PK 593+00 – PK 600+50) with the installed measuring equipment is shown in Figure 6.

Fig. 6. General view of the road section (PK 593+00 – PK 600+50) with the installed measuring equipment

5 Results

After the results of two-year monitoring of the road sections with the implemented solutions the following conclusions were drawn:

- The first year of geotechnical monitoring showed that such defects as cracks, subsidence and erosion of slopes appeared on the pavement and road borders of each section where stabilization works were carried on the roadbed and the road lower structure. The development of deformation was due to construction and installation works during winter. In maximum thawing of the roadbed and the lower structure of the road in spring and autumn, compaction of subgrade and roadbed soils was carried out using reinforcement (tension). In summer, during the first year of geotechnical monitoring, the pavement, road borders and road slopes were repaired; this was taken as the initial state of the road.
- In the second year of geotechnical monitoring a net of cracks appeared on the section №1 along the whole of the road pavement. Intensive development of deformations occurred due to a permanent flooded area providing water filtration into the body of the roadbed and the lower structure in the period of maximum thawing of the active layer. The level of roadbed flooding in spring was above that of the hard soil roller. In order to prevent water filtration into the body of the roadbed on section №1 it was recommended to develop the project and perform construction and installation works on overburden water removal berms made of non-drain soil on both sides of the road. Subsidence of pavement appeared on the reinforced side of the carriageway indicating that the soil must be further compacted on this section of the road. Constant flooding of the embankment bottom led to intensive development of defects on the given section. Rise of moisture content in the roadbed and the lower structure led to further development of deformations in the form of cracks and subsidence due to the flooding area. The designed drainage systems could not guarantee preservation of the natural hydrological regime of surface waters, and as a result, the frost-thaw flooding area appeared between the railway and the road. This resulted in deterioration of the roadbed frozen state and thawing of subgrade soils.
- In the second year of geotechnical monitoring three transverse cracks were observed on section №2. After measuring temperature on section №2, the negative temperatures were observed in the roadbed body and the lower structure on the left side of the road in 3 - 7 m

from the daylight surface. Reduction of temperature of the roadbed soils and the road lower structure as well as its expansion in the underlying soils made it possible to draw a conclusion about the effective operation of the heat stabilizers installed on the given section.

- In the second year of geotechnical monitoring not any defects were revealed on the road pavement and borders on section №3 indicating roadbed and lower structure stabilization.

- In the second year of geotechnical monitoring not any defects were revealed on the road pavement and borders on section №4 indicating roadbed and lower structure stabilization.

Snow deposits up to 1.6-2.3 m cumulating on the slopes of the road and at the foot of the roadbed due to the intense snow ridging and transport by the railway located nearby the sections had a negative effect on the inspected object. Snow deposits cumulating on the slopes of the road and at the foot of the roadbed did not allow the second principle of designing to be realized, since they blocked expansion of the negative temperatures in the roadbed body and the lower structure of the road. In spring, due to the accumulated snow transport, additional moisturizing of the roadbed and the road lower structure occurred.

6 Conclusion

As a result of the research three methods of stabilizing the roadbed and road base on permafrost soil, have been developed, implemented and tested. Application of our research's results will reduce the annual emerging defects in the roadbed and road surfaces.

References

1. Avtomobil'naja doroga na mnogoletnemerzlyh gruntah, R.F. Patent 2522268, MPK7 E 01S03/00 (2015)
2. V.V. Voroncov, A.N.Kraev, M.E.Igoshin, *Nauchno-tehnicheskij vestnik Povolzh'ja*, **2**, 119-123 (2014)
3. V.V. Voroncov, A.N. Kraev, M.E., *Vestnik Sibirskoj gosudarstvennoj avtomobil'no-dorozhnoj akademii,* **6(40)**, 67-72, (2014)
4. A.N. Kraev, T.V. Permitina, V.A. Zav'jalov, *Aktual'nye problemy stroitel'stva, jekologii i jenergosberezhenija v uslovijah Zapadnoj Sibiri*, 32-40 (2014)
5. M.E. Igoshin, V.V. Voroncov, A.N. Kraev, A.E. Ushakov, *Aktual'nye problemy stroitel'stva, jekologii i jenergosberezhenija v uslovijah Zapadnoj*, 21-28 (2014)
6. V.V. Passek, A.N. Kraev, M.V. Paskov, V.S. Andreev, *Vestnik Tjumenskogo gosudarstvennogo arhitekturno-stroitel'nogo universiteta*, **4**, 89-93 (2015)
7. A.N. Kraev, PhD Thesis (2009)
8. V.F. Baj, A.V. Nabokov, V.V. Voroncov, A.N. Kraev, Vestnik grazhdanskih inzhenerov, **2**, 74-76 (2009)
9. V.F. Baj, L.E. Mal'cev, T.V. Mal'ceva, et al. Jeksperimental'naja ustanovka dlja ispytanija grunta metodom odnoosnogo szhatija, Patent RUS 2213952 (2002)
10. L.E. Mal'cev, T.V. Mal'ceva, V.A. Demin, Neft' i gaz, **2**, 96-102 (2001)
11. A.V. Nabokov, PhD Thesis (2003)
12. V.F. Baj, A.N. Kraev, A.V. Nabokov, Ju.A. Novikov, *Nauchno-tehnicheskij vestnik Povolzh'ja*, **2**, 30-33 (2011)
13. V.F. Baj, T.V. Mal'ceva, A.N. Kraev, *Nauchno-tehnicheskij vestnik Povolzh'ja*, **5**, 108-111 (2014)
14. T. Maltseva, A. Nabokov, A. Chernykh, *Procedia Engineering*, **117**, 239-245 (2015)

Effect of Enclosed Space Configuration and Freezing Medium Volume on Refreezing Pressure

Vasiliy Ovchinnikov[1,*], *Dmitry* Gerasimov[1], *Lev* Maksimov[1], and *Pavel* Ovchinnikov[2]

[1]Industrial University of Tyumen, 625001 Volodarskogo str. 38, Tyumen, Russia
[2]Sergo Ordzhonikidze Russian State Geological Exploration University, Moscow, Russia (23, Miklukho-Maklya Str., 117997)

Abstract. In building underground structures (bases and foundations of residential and industrial buildings, boreholes for hydrocarbon extraction, etc.) in cryolithic zones, cavities of various shapes and sizes saturated with liquid media often develop. When they refreeze, the refreezing pressure occurs; this may result in failure of the structural member continuity, and further failure of the structure as it is. The paper describes the calculation techniques for the refreezing pressure depending on cavity shapes and sizes.

1 Introduction

After field-geophysical research it has been stated that various enclosed spaces can develop in perennial frozen rock in-situ in structures built and being built in the Far North. As a rule, the cavities develop due to icy sands and sandy loams. The following types of cavities occur [1 - 5]:

- cylindrical cavities are the most common form of cavities. The diameter of these cavities usually reaches 1m, the height – over 15 m;
- spherical cavities are enclosed cavities developing during freezing of the supports and bases of the idle structures in the cold season. Spherical pits and cavities not exceeding 1-1.5m in diameter develop. However, in long-term operation their sizes may increase;
- slatted cavities are formed in subsidence of thawed rocks after long-term operation of structures. Freezing occurs more rapidly in the axial direction [5].

All types of cavities usually are water-saturated, therefore the phase transformation are causes the occurrence of overpressure through the ice formation, which causes the destruction of built construction.

As a result, ability of their current forecast is very relevant for substantiation of a technology that promotes the prevention of occurrence of the incident.

* Corresponding author: tgasu.ovchinnikov.vasiliy@mail.ru

2 Analysis of calculation techniques

After the authors [6 - 10], the water-ice volume balance in freezing of the enclosed volume V0 by the value dV is described as follows:

$$\alpha \cdot mdV = [\beta_T \cdot (V_0 - V) + \beta_M \cdot V + \beta_0 \cdot V_0]dp, \tag{1}$$

where: α = (ρв – ρl) / ρв - coefficient of volume expansion during the phase transition of water into ice; ρв and ρl - density of water and ice; m - water content; βm, βM, β0 - coefficients of volumetric compression of the thawed zone, refrozen rocks and frozen rocks; V - the current value of the solid phase volume; dp – pressure increase in freezing of the refrozen media by the value *dV*.

Integrating the equation (1), and taking into account that at V0 the pressure in the system is equal to P0, we obtain finite pressure value as of freezing:

$$P = P_0 + \frac{\alpha \cdot m}{\beta_M - \beta_m} \cdot \ln\left[1 + \frac{(\beta_M - \beta_m)}{\beta_m + \beta_0} \cdot \frac{V}{V_0}\right] \tag{2}$$

According to [11 ÷ 15] the current radius of the freezing cavities can be given for:
- cylindrical cavities

$$\lambda \frac{T_{fn} - T_n}{r \cdot \ln \frac{R}{r}} = \sigma \frac{dr}{d\tau}, \tag{3}$$

- spherical cavities

$$\lambda \frac{T_{fn} - T_n}{(R - r)^{-1}} = \sigma \cdot r^2 \frac{dr}{d\tau}, \tag{4}$$

- slatted cavities

$$\lambda \frac{T_{fn} - T_n}{r} = \sigma \frac{dr}{d\tau}, \tag{5}$$

where: λ – thermal conductivity of the frozen rocks; σ – heat of phase transition; r – current radius of the cavity; RK – initial radius of the cavity; τ – time; Tfn and Tn – temperatures of phase transition and frozen rocks correspondingly. Transforming the equation (2) and integrating the equations (3) ÷ (5), we obtain the formulas for determination of the pressure during freezing for the following cavities:
- cylindrical

$$P_i = P_0 + \frac{\alpha \cdot m}{\beta_M - \beta_T} \cdot \ln\left[1 + \left(\frac{\beta_M - \beta_T}{\beta_0 + \beta_T}\right) \cdot \left(\frac{R_K^2}{R_K^2 - R^2}\right) \cdot \left(1 - \frac{r_i}{R_K}\right)^2\right], \tag{6}$$

$$\tau = \tau_{i-1} + \frac{m \cdot \sigma \cdot R_K^2}{4\lambda} \cdot \frac{\left\{ \left(\frac{r_i}{R_K}\right)^2 \cdot \left[\ln\left(\frac{r_i}{R_K}\right)^2 - 1 \right] - \left(\frac{r_{i-1}}{R_K}\right)^2 \cdot \left[\ln\left(\frac{r_{i-1}}{R_K}\right)^2 - 1 \right] \right\}}{T_{fn} - T - \left(\frac{P_i + P_{i-1}}{P_{max}}\right)}; \quad (7)$$

- spherical

$$P_i = P_0 + \frac{\alpha \cdot m}{\beta_M - \beta_T} \cdot \ln\left\{ 1 + \left(\frac{\beta_M - \beta_T}{\beta_0 + \beta_T}\right) \cdot \left(\frac{R_K^2}{R_K^2 - R^2}\right)^{\frac{3}{2}} \cdot \left[\left(1 - \frac{r_i}{R_K}\right)^2 \right]^{\frac{3}{2}} \right\}, \quad (8)$$

$$\tau = \tau_{i-1} + \frac{m \cdot \sigma \cdot R_K^2}{\lambda} \cdot \frac{\frac{1}{3} \cdot \left[\left(\frac{r_i}{R_K}\right)^3 - \left(\frac{r_{i-1}}{R_K}\right)^3 \right] - \frac{1}{2} \cdot \left[\left(\frac{r_i}{R_K}\right)^2 - \left(\frac{r_{i-1}}{R_K}\right)^2 \right]}{T_{fn} - T - \left(\frac{P_i + P_{i-1}}{P_{max}}\right)}; \quad (9)$$

- slatted (with one-sided freezing)

$$P_i = P_0 + \frac{\alpha \cdot m}{\beta_M - \beta_T} \cdot \ln\left[1 + \left(\frac{\beta_M - \beta_T}{\beta_0 + \beta_T}\right) \cdot \frac{h_i}{H} \right], \quad (10)$$

$$\tau = \tau_{i-1} + \frac{m \cdot \sigma \cdot H^2}{2\lambda} \cdot \frac{\left(\frac{h_i}{H}\right)^2 - \left(\frac{h_{i-1}}{H}\right)^2}{T_{fn} - T - \left(\frac{P_i + P_{i-1}}{P_{max}}\right)}, \quad (11)$$

where: R — outer diameter of the outer surface; Pmax - maximum possible freezing pressure at the given temperature.

3 Results

The calculation was done step by step [16 ÷ 20] in the following way:

- the value ri/RK was given, P1,τ1(τ0 = 0, τi-1/R = 1) was calculated in (P1 + P0)/ Pmax. Then the ratio r2/RK was given, P2 and τ2 were calculated using the values of τ1 and (P1 – P0)/Pmax, etc. At each step of volume V freezing to the value Vi, the pressure Pi, was taken as constant. It was calculated as the mean pressure of the pressure obtained in the previous step Pi-1, freezing V to the value Vi-1, and the pressure at the given step Pi.

Fig. 1 shows the graphs of pressure changes in freezing of cavities of various geometric shapes.

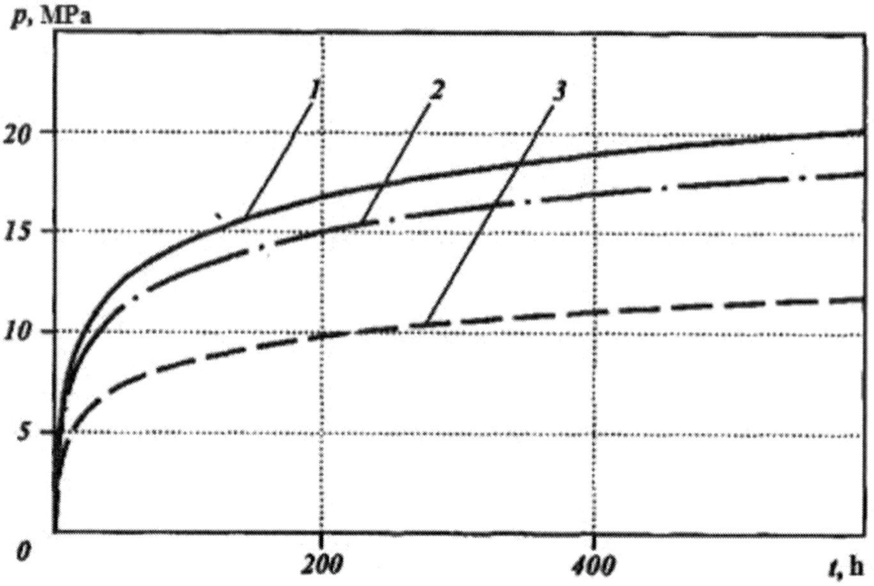

Fig. 1. Dynamics of pressure changes in freezing of enclosed cavities of various shapes: 1, 2, 3 — slatted, cylindrical and spherical correspondingly (t = minus 2 °C, $R_к$ = 0.5 м, β_0 = 0.001 MPa^{-1}, β_M = 0.005 MPa^{-1}).

The graphs show that the shape of the cavities affects only the rate of pressure increase. Thus, freezing of the cylindrical cavities develops faster than that of the slatted ones and slower than the spherical ones.

Depending on the volumetric compression of freezing media and the size of the cavities, the pressures may be lower than the maximum possible values of freezing pressure (Tab. 1).

Table 1. Pressures in cavities depending on temperature and compressibility of freezing media.

№ of experiment	T, °C	Compressibility of medium, 10^{-2} MPa^{-1}		Cavity radius , R_K, m	$P_{calc.}$ (P_{max}), MPa	Percent of freezing, (%)
		β_0	β_M			
1	-2	0.144	0.144	0.3	17.0(26.8)	63.4
2	-2	0.06	0.36	0.3	26.8(26.8)	100
3	-2	0.58	5.86	0.5	15.0(26.8)	56
4	-2	0.144	0.144	0.5	26.8(26.8)	100
5	-5	1.80	3.60	0.3	26.0(67.0)	38.8
6	-5	1.80	3.60	0.5	30.0(67.0)	44.8
7	-5	0.06	3.60	0.5	67.0(67.0)	100

This is due to the deformation of the medium inclosing wall occurring under the influence of the refreezing pressure (Pc) [20 - 22]. As a result, the volume of cavities increases and the pressure releases to the value (Pp). In this case the thermodynamic equilibrium of the system is violated: the unfrozen water is in the state of compressive suffusion. As a result, additional icing leads to pressure increase in the cavities which becomes equal to (Pc) again, and thermodynamic equilibrium returns. Significant movement of the inclosing walls can result in a situation when the whole of free water becomes icy, and the pressure never reaches its maximum possible value.

In experiments 1, 2, 4, and 7 (Table 1) the cavities freeze completely and the maximum pressure values are 63 - 100% from the maximum possible calculated ones. The cavities do not freeze completely in the lower values of freezing media compressibility and the maximum pressures at the given temperatures are obtained (tests 3, 5 and 6). Thus, the refreezing pressure value depends on the ratio of the ice volume and unfrozen liquid in the cavities at the given temperature.

The results obtained make it possible to assume that if the cavity is a cylinder filled with water, with an infinitely large wall thickness and a height substantially greater than its diameter, the cavity wall (support) undergoes only the elastic deformations; according to the field data it is valid, since the heights of the cavities generated in the cryolithic zone are ten times greater than their diameter, and compression of low-temperature ice-sandy rocks is negligible.

The modulus of elasticity E (Young's modulus) and the transverse elasticity coefficient μ (Poisson ratio) are the main indicators of the elastic properties of frozen rocks. After N.A. Tsytovich [15], the Poisson ratio of frozen rocks has weak temperature-moisture dependence and can be taken as constant in calculation for rocks of various degrees of dispersion. Thus, it can be taken in the range of 0.2-0.22 for frozen sand, and for clays its mean value is approximately 0.37.

Young's modulus depends on several factors: the composition of the frozen rocks, their iciness, values of negative temperatures and the external pressure [5]. Frozen sand obtains the highest modulus of elasticity - from 8..2 MPa (at minus 0.2 °C) up to 2.25 • 104 MPa (at minus 10.2 ° C); frozen clays obtain the lowest modulus of elasticity - from 6.8 • 102 MPa (at minus 1.2°C) up to 2.78 • 103 MPa(at minus 8.4 °C). The modulus of elasticity of silty loams and sandy loams is of the intermediate value.

Let us point out the cavity volume VK with two planes perpendicular to the longitudinal axis and spaced apart by a length unit:

$$V_K = \pi \cdot R_K^2 \cdot K, \tag{12}$$

where: $K = 1 - \left(\frac{R}{R_K}\right)^2$ — coefficient of proportionality.

Increase of the freezing cavity volume is compensated by the change of its original volume by the value ΔVK caused by the deformation of the frozen rocks, compression of ice ΔVL and the unfrozen water ΔVB. Let us take a final value of movement Uc at the inner radius of the cavity RK from the constant internal pressure Pc which can be evaluated by the formula:

$$U_C = \frac{1+\mu}{E} P_C R_K, \tag{13}$$

where: E, μ — elastic constants of frozen rocks.

Transforming the equation (1), considering $\beta_0 = 2\frac{1+\mu}{E}$ we obtain the formula:

$$P_P - P_0 = \frac{\varepsilon \cdot \chi}{\beta_L \cdot m + \beta_B \cdot (1-m) + A}, \tag{14}$$

where: $\chi = \frac{V_A}{V_K + \Delta V_K}$; $A = 2 \cdot \frac{1+\mu}{E + 2(P_C - P_0) \cdot (1+\mu)}$.

Then the dependence of the pressure that occurs when water freezes in the cavity (Pp) from the amount of generated ice is as follows:

$$P_P - P_0 = \frac{\varepsilon \cdot (\chi + \chi')}{\beta_A \cdot (\chi + m') + \beta_B \cdot (1 - \chi - \chi') + A}, \qquad (15)$$

where: $\chi' = \frac{V'_A}{V_K + \Delta V_K}$.

The volume of ice V'_L is evaluated by the formula:

$$V'_L = 0.0136 \cdot \Delta t \cdot V_B, \qquad (16)$$

where: VB — volume of the suffusion water; Δt— temperature of suffusion of water.

Having taken in the equation (4) $\chi + \chi' = 1$ and solving it with regard to K, we obtain the formula showing complete freezing of the liquid phase:

$$\frac{r_H}{R_K} \geq \sqrt{1 - \frac{2(P_C - P_0)\cdot(1+\mu)\cdot[1 - \varepsilon + (P_C - P_0)\cdot \beta_l]}{E[\varepsilon - (P_C - P_0)\cdot \beta_l]}} \qquad (17)$$

Thus, in the temperature range of frozen rocks from minus 2 to minus 5 ° C, corresponding to the conditions of the Far North of Western Siberia and the taken values: 1.0·103 ≤ E ≤ 2.5·103 MPa; μ = 0.22; ε = 0.083; βl = 1.1·10-4 1/MPa, the following formulas have been obtained:

-for temperature minus 2 °C: $\frac{R}{R_K} \geq 0.83$ (18)

-for temperature minus 5 °C: $\frac{R}{R_K} \geq 0.52$ (19)

To assess the adequacy of the obtained ratios the experiment was carried out at temperatures of minus 5o and minus 2 ° C in accordance with the technique described in the books [4, 18]. The results are shown in Table. 2.

Table 2. Pressure in freezing water in the enclosed volume .

Factors		Pressure, MPa			
		Results of research in experiments 1, 2, 3			Mean value,
t, °C	$\frac{R}{R_K}$	p_1	p_2	p_3	$\bar{p}$
-.2.0	0.83	0	0	0	0
-2.0	0.50	22.1	22.5	23.0	22.5
-5.0	0.52	17.5	19.8	18.2	18.5
-5.0	0.25	58.4	56.9	57.6	57.6

The results showed that the pressure at minus 2°C and the ratio $R/R_K = 0{,}83$ did not increase, and at minus 5°C and the ratio $R/R_K = 0{,}52$ it did not exceed 20 MPa.

While at lower pressure values R/R_k were close to their maximum possible ones evaluated by the formula (1) - 22.9 and 57.4 MPa correspondingly; this indicated the complete phase transition of water into ice in the system simulating the cavity with the boundary conditions (18).

4 Conclusion

Since the actual radius of the cavities generated in the range of occurrence of the frozen rocks is significantly greater, then after refreezing some unfrozen water will always occur in the cavities; in here, the pressure is below the maximum possible value.

References

1. R. Medvedskij, *Stroitel'stvo i jekspluatacija skvazhin na neft' i gaz v vechnomerzlyh porodah* (Nedra, Moscow,1987)
2. V. Kuznecov, PhD Thesis (Tjumen', 1991)
3. N. Weiss, D. Blok, B. Elberling, G. Hugelius, C. J. Jørgensen, M. B. Siewert, P. Kuhry, Sedimentary Geology, **340**, 38-48 (2016)
4. V. Ovchinnikov, D. Gerasimov, N. Aksenov, L. Kamenskij, Burenie i neft', **1**, 42-43 (2012)
5. V. Ovchinnikov, A. Nabokov, O. Danilov, D. Gerasimov, T. Haritonova, *Osnovnye razdely mehaniki sploshnoj sredy i ih prakticheskoe primenenie pri burenii i razrabotke neftjanyh i gazovyh mestorozhdenij. Uchebnoe posobie Tjumen'* (FGBO UVPO», Tjumen' TjumGNGU, 2015)
6. Q. Yu, Z. Zhang, G. Wang, L. Guo, X. Wang, P. Wang, Z. Bao, Cold Regions Science and Technology, **121**, 205-213 (2016)
7. M. Mel'cer, O velichine davlenija pri zamerzanii vody v ogranichennom obeme. (SibNIINP, Tjumen, 2010)
8. Z. Wen, Q. Yu, M. Zhang, K. Xue, L. Chen, D. Li, Cold Regions Science and Technology, **121**, 214-225 (2016)
9. M. Dubina, ZapSibNIGNI, **60,** 90-94 (1984)
10. S. Vjalov, *Reologicheskie osnovy mehaniki gruntov* (Vysshaja shkola, Moscow)
11. L. Guo, Y. Xie, Q. Yu, Y. You, X. Wang, X. Li, Cold Regions Science and Technology, **121**, 187-195 (2016)
12. T. Trofimov, Ju. Badu, G. Dubikov, *Kriogennoe stroenie i l'distost' MMP Zapadno-Sibirskoj plity* (Izd-vo MGU, Moscow. 1980)
13. A. Sel'vashhuk, Neftjanoe hozjajstvo, **5**, 15-19 (1999)
14. Y. Wang, H. Jin, G. Li, Cold Regions Science and Technology, **126**, 10-21 (2016)
15. C. Kneisel, A. Emmert, P. Polich, B. Zollinger, M. Egli, CATENA, **133**, 107-118 (2015)
16. V. Kuznecov, S. Jakubovskaja, D. Gerasimov, Izvestija VUZov «Neft' i gaz», **6**, 31-34 (2002)
17. S. Vjalov, Ju. Zareckij, S. Gorodeckij, *Raschety na prochnost' i polzuchest' pri iskusstvennom zamorazhivanii gruntov* (Strojizdat, Leningrad, 1981)
18. V. Kuznecov, V. Kucherjuk, Izvestija VUZov «Neft' i gaz», **6**, 36-42 (2001)
19. S. Vjalov, Ju. Zareckij, S. Gorodeckij, *Raschety na prochnost' i polzuchest' pri iskusstvennom zamorazhivanii gruntov* (Strojizdat, Leningrad, 1981)
20. M. Dubina, Prognoz davlenij obratnogo promerzanija II In-ta ZapSibNIGNI, **60**, 35-37 (1984)
21. B. Pobedrja, *Mehanika kompozicionnyh materialov* (Izd_VO MGU, Moscow, 1984)
22. A. Vasil'chenko, Fiziko-himicheskij mehanizm razvitija sverh geologicheskih davlenij smjatija obsadnyh kolonn, **981**, 23 (1987)

MATEC Web of Conferences **73**, 05006 (2016) DOI: 10.1051/matecconf/20167305006
TPACEE-2016

Solution of the Moisture Phase Transition Problem in the Insulation and Soils near Pipelines

Boris Aksenov [1,*] and *Svetlana* Kryaknina[1]

[1]Industrial University of Tyumen, 625001 Volodarskogo str. 38, Tyumen, Russia

Abstract. The analysis of various approaches to the heat conductivity modeling in soils with phase transition based on Stefan problem is provided. Herein the analytical solution for Stefan problem is received only for special cases.The approximate methods of mathematical modeling of heat exchange in soils and construction materials based on the Stefan problem without initial data are offered. The found analytical solution can be a basis for approximate solutions of a wide range of heat exchange problems at harmonious oscillation of temperature on one of the borders. The mathematical apparatus for estimating problems solution of freezing and melting in areas with plane-parallel and axial symmetry is considered. Solutions for applied problems of constructions building and operation in the permafrost conditions are provided.

1 Introduction

The prediction for freezing and melting boundary evolution is necessary to enable the successful constructions operation and working out measures to prevent the permafrost degradation. Stefan problem is a traditional mathematical model of heat conductivity in soil with phase transition. For one-dimensional semi-infinite domain the problem is defined as follows (soil thawing process is considered).

$$\frac{\partial t_1}{\partial \tau} = a_1^2 \frac{\partial^2 t_1}{\partial x^2}, \quad 0 < x < \xi, \tag{1}$$

$$\frac{\partial t_2}{\partial \tau} = a_2^2 \frac{\partial^2 t_2}{\partial x^2}, \quad \xi < x < \infty, \tag{2}$$

$$t_1(0,\tau) = f(\tau), \quad t_2(x,0) = \varphi(x), \quad t_1(\xi,\tau) = t_2(\xi,\tau) = t_*, \tag{3}$$

$$|t_2(x,\tau)| < M, \quad M = Const, \tag{4}$$

* Corresponding author: tgasu.aksenov.boris@mail.ru

$$\lambda_1 \frac{\partial t_1}{\partial x}\Big|_{x=\xi} - \lambda_2 \frac{\partial t_2}{\partial x}\Big|_{x=\xi} = \kappa w \rho \frac{d\xi}{d\tau}, \qquad (5)$$

where τ is the time; x is spatial coordinate; ξ is the interphase boundary coordinate; t_1 is the temperature in the thawed soil zone; t_2 is the temperature in the frozen soil zone; t_* is the phase transition temperature; $\lambda_1, a_1^2, \lambda_2, a_2^2$ are thermal conductivity coefficient and thermal diffusivity coefficient in thawed and frozen zones, respectively; ρ is the specific gravity of the soil skeleton; k is the latent heat of water melting; w is specific soil humidity.

The limitation condition (4) is necessary to enable the uniqueness of the problem solution (1) – (4).

The problem (1) – (5) is solved to determine the thawing boundary; it is also included in the mathematical models describing the various cryogenic phenomena such as frost heaving.

Analytical solution of this problem was obtained only for some special cases. Method of successive approximations in can be used to solve a nonlinear problem class of thermal conductivity, including the Stefan problem. There are several numerical solution schemes for the Stefan problem. In the scientific literature there are works constantly offering new approaches to solving this problem.

The variety of used methods indirectly indicates that this problem is not yet solved. The singularities of the type δ-function when $x = \xi$ complicates numerical solution. These singularities approximation introduces a poorly controlled error. This means there is no single numerical method suitable for all coefficients and boundary conditions, even for one-dimensional problems.

Approximate solutions based on the simplification of the problem (1) – (5) are used in engineering practice.

Assuming that $\varphi(x) = t_*$, we get the so-called one-phase Stefan problem. It means that the solution should be sought only in the thawed zone, since when $x > \xi$ at any value of τ temperature is known as $t_2 = t_*$.

Convenient analytical solutions are obtained for a single-phase problem when $f(\tau) = t_c = Const$ and for a two-phase problem when $\varphi(x) = Const$.

Further simplification is possible at a quasistationary approach. The boundary $x = \xi$ is believed to move slowly enough so at any instant of time τ temperature field in the domain $0 < x < \xi$ can be considered stationary.

Using quasistationary solution of a one-dimensional single-phase problem when $f(\tau) = t_c$ we receive the well-known Stefan formula. If $f(\tau) \neq Const$, then various modifications of the first and second L.S. Leybenzon methods are used. Those methods are also based on quasistationary approach.

From the heat balance standpoint quasistationary assumption is equivalent to the assumption of zero heat capacity in the "upper" zone ($0 < x < \xi$). The temperature in this zone varies but the thermal energy change due to the temperature change is not taken into account in the solution. Under certain conditions, this results in uncontrolled errors.

Without dwelling on the various attempts to consider the heat capacity of the "upper" zone which are widely represented in the literature, the authors should note that quasistationary approach is popular as it offers the simplest form of the boundary value problem solution in the domain $0 < x < \xi$. This same problem solution for the equation (1) has the form of a Fourier series, and is very inconvenient at the calculations.

However, if $f(\tau)$ is a periodic function (daily, annual temperature variation of the ground surface), then the closed form solution of heat conduction problem without initial data can be used in the upper zone. [1-13]

2 Simulation using heat conduction problem without initial values

Instead of problem (1) - (5) the problem (6) – (9) is solved:

$$\frac{\partial t_1}{\partial \tau} = a_1^{\ 2} \frac{\partial^2 t_1}{\partial x^2}, \quad 0 < x < \xi, \tag{6}$$

$$t_2(x,\tau) = t_*, \quad \xi < x < \infty, \tag{7}$$

$$t_1(0,\tau) = A + B \cdot \sin(\omega\tau + \varepsilon), \tag{8}$$

$$t_1(\xi,\tau) = t_2(\xi,\tau) = t_*, \tag{9}$$

where $A, B, \omega, \varepsilon$ are constants, .

When solving the problem (6) – (9) position of phase border $x = \xi$ is considered fixed when finding expression for $t_1(x,\tau)$. The law of the border movement $\xi(\tau)$ is found from the Cauchy problem solution for the ordinary differential equation.

$$\lambda_1 \frac{\partial t_1}{\partial x}\Big|_{x=\xi} = -\kappa w \rho \frac{d\xi}{d\tau}, \quad \xi(0) = 0 \tag{10}$$

Taking into account analytical solutions (6) - (9) to find $\xi(\tau)$ we receive a problem:

$$\frac{d\xi}{d\tau} = -\frac{\lambda}{\kappa w \rho}\left[\frac{\sqrt{2}Bp}{\sin^2 p\xi + sh^2 p\xi}\left(shp\xi \cos p\xi \cos\left(\omega\tau + \frac{\pi}{4}\right) - chp\xi \sin p\xi \sin\left(\omega\tau + \frac{\pi}{4}\right)\right) - \frac{A}{\xi}\right],$$

$$\xi(0) = 0. \tag{11}$$

where $p = \sqrt{\omega / (2a^2)}$.

Heat loss through the wall of a building is usually calculated on average daily temperatures. Approximate methods based on quasistationary approach take into account the daily heat loss dynamics.

The above method allows to consider the influence of both non-stationary and nonlinear heat transfer in the wall on the building heat loss. All construction materials contain moisture, so the correct thermal calculation for winter conditions should take into account that daily variations in ambient air temperature lead to variations of interphase boundaries ξ.

The Joule heat evolved or absorbed has a damping effect, resulting in a temperature variation phase displacement of the inner and outer building walls surfaces. This displacement is significant and should be taken into consideration when setting daily heating mode.

Variations in an external layer $0 < x < \xi$ are steady in nature so they can be described with a problem without initial data (6) - (9).

The swing value of ξ is small enough, so the problem of the phase boundary moving is solved in the single-phase formulation.

In a layer $\xi < x < H$ (where H is wall thickness) the influence of external air variations is considerably weakened, first of all by the phase transition boundary, so it is advisable to take the quasistationary approximation, in it is considered that the thermal flow at $x = \xi$ is equal to a thermal flow at $x = H$.

According to the calculations the daily air temperature variations, under the above assumptions do not change the total heat loss of the building. However, these variations significantly influence upon the phase displacement in temperature when $x = 0$ and $x = H$. Thus, this phase displacement is equal to the temperature phase displacement on the outer and inner wall surfaces. The phase displacement $\xi(\tau)$ referred to $t(\tau)$, which is mainly determined by the specific heat and humidity of the material, can be considered as material properties to develop economical heat supply schedule.

A problem without initial data is applied to modeling the temperature field in the pipes insulation when laying external heat supply networks, because the heat loss during winter is determined both by the average daily air temperature, and by the daily temperature variations.

The temperature field calculation method in the insulation layer is similar to the processes modeling methods in the frozen soil and in a flat wall, but inhere the heat transfer occurs in the domain with axial symmetry. Therefore, we used the procedure of reduction of axisymmetric problems to the problems with plane-parallel symmetry.

For external insulation layer we have a problem without initial data in cylindrical coordinates:

$$\frac{\partial t}{\partial \tau} = \frac{a^2}{x}\frac{\partial}{\partial x}\left(x\frac{\partial t}{\partial x}\right), \quad \xi < x < R, \tag{12}$$

$$t(x,\tau) = t_*, \tag{13}$$

$$t(R,\tau) = A + B\cdot\sin(\omega\tau + \varepsilon) = f(\tau), \tag{14}$$

where t is insulation temperature, R is preinsulated tube radius, A is average daily temperature, B, ω, ε are the constants indicating daily temperature variations.

The auxiliary problem (15) - (17) is taken to solve the problems (12) - (14):

$$\frac{\partial u}{\partial \tau} = a^2\frac{\partial^2 u}{\partial y^2}, \quad 0 < y < \xi_1, \tag{15}$$

$$u(\xi_1,\tau) \;=\; t_*, \tag{16}$$

$$u(0,\tau) \;=\; A + B\cdot\sin(\omega\tau + \varepsilon) = f(\tau). \tag{17}$$

where $\xi_1 = R - \xi$.

when substituting

$$x = R - y. \tag{18}$$

Coincidence of boundary conditions in problems (12) – (14) and (15) – (17) takes place. Following, the corresponding stationary problems solutions are considered. Connection between stationary problems solution in the field with plane-parallel symmetry and the problem solution in axisymmetric field is established.

As the phase boundary $x = \xi$ moves slowly, temperatures field in the area $\xi < x < R$ at each moment of time τ is considered stationary, we get the approximate linking formula between $u(x,\tau)$ and $t(x,\tau)$:

$$t(x,\tau) \approx t_* - \left(t_* - f(\tau)\right) \frac{Ln\left(1 + (\rho - 1)\cdot \dfrac{t_* - u(x,\tau)}{t_* - f(\tau)}\right)}{Ln\rho}, \quad \rho = \frac{R}{\xi}. \tag{19}$$

Stefan's condition on phase transition border is set in single-phase statement. So the law of the border movement is set by the equation:

$$\lambda \frac{\partial t}{\partial x}\Big|_{x=\xi} = -\kappa\gamma w \frac{d\xi}{d\tau}, \tag{20}$$

where λ is heat conductivity coefficient, γ is the specific mass of material, w is relative humidity.

In view of (19), equation (20) takes the form

$$\frac{d\xi}{d\tau} = \frac{\lambda}{k\gamma w} \cdot \frac{1-\rho}{Ln\rho} \cdot \frac{\partial u}{\partial x}\Big|_{x=\xi}. \tag{21}$$

In (20) – (21) the interphase surface change at changes ξ isn't considered because variations ξ are small.

This procedure is suitable for the areas with big daily air temperature variations, such as the North of Western Siberia.

Thus, the analytical solution of heat conductivity problem without initial data can be the base for approximate solutions of a wide class of heat transfer problems when the temperature variations on one of the boundaries is stationary harmonic in nature.

3 Estimates for problems solution of freezing and thawing in areas with plane-parallel and axial symmetry

The problem of definition of phase transition front can be solved by creating the narrowing estimates system.

If convergence of recurrent procedure of estimating is proved, then we get the specific structure solution with properties similar to properties of converging alternate series.

Namely, on each step it is possible to receive maximum value of absolute error with the error sign known which is important for applied problems.

I llustrates method of constructing a narrowing estimating system for the freezing and thawing problems in the area with plane-parallel symmetry.

Next the mathematical apparatus will be considered which allows to apply this method to heat exchange problems in the area with axial symmetry. The heat transfer problem with phase transition in moist fine disperse soil in an infinite one-dimensional axisymmetric domain is given by:

$$\frac{\partial t}{\partial \tau} = a^2 \frac{\partial^2 t}{\partial r^2} + \frac{a^2}{r} \frac{\partial t}{\partial r} - \frac{\chi}{c} \frac{\partial w(t)}{\partial \tau} \tag{22}$$

$$t(r_0,\tau) = F(\tau),\;\; t(r,0) = 0,\;\; \tau > 0,\;\; r > r_0,\;\; |t(r,\tau)| < M,\;\; M = const > 0,\;\; F(0) = 0, \tag{23}$$

where $a^2 = \frac{\lambda}{c}$, $t, c, \lambda, \tau, w, \chi$ are temperature, heat capacity, thermal conductivity, time, spatial coordinate, the content of unfrozen moisture, the latent heat of the phase transition of water accordingly.

For simplicity, we assume that $F(\tau)$ is monotonic function with limited variation differentiable at $\tau > 0$. The first two addends on the right side of (22) form the product of the coefficient on the one-dimensional Laplace operator in cylindrical coordinates, $\frac{\chi}{c} \cdot \frac{\partial w}{\partial \tau}$ is the source function. Here we have $\frac{\partial w(t)}{\partial \tau} = \frac{dw(t)}{dt} \cdot \frac{\partial t}{\partial \tau}$, where $\frac{dw}{dt} > 0$ is a function established for the given type of soil ("curve of unfrozen water").

For the integral representation of (22) - (23) we need the Green's function, and it has a simple form for only a semi-infinite flat domain.

It means that equation (22) can be formally regarded as an equation for a semi-infinite flat domain $r > r_0$.

Then the integral representation of the problem (22) - (23) has the form:

$$t(r,\tau) = t_0(r,\tau) - \frac{\chi}{c} \int_{r_0}^{\infty} \int_0^{\tau} G(r,\xi,\tau - y) \cdot \frac{\partial w}{\partial y} d\xi\, dy + a^2 \int_{r_0}^{\infty} \int_0^{\tau} \frac{\partial t}{\partial \xi} G(r,\xi,\tau - y) \frac{1}{\xi} d\xi\, dy, \tag{24}$$

where $G(r,\xi,\tau - y)$ is Green's function, $t_0(r,\tau)$ is the homogeneous heat equation solution with the conditions (23).

Applying the formula of integration by parts allows to eliminate the partial derivatives under the integral sign. Thus, we get:

$$t(r,\tau) = V(r,\tau) + \frac{\chi}{c} \int_{r_0}^{\infty} \int_0^{\tau} \psi(r,\xi,\tau - y) w(t(\xi,y)) d\xi dy - a^2 \int_{r_0}^{\infty} \int_0^{\tau} K(r,\xi,\tau - y) t(\xi,y) d\xi\, dy - \\ - \frac{a^2}{r_0} \int_0^{\tau} G(r,r_0,\tau - y) t(r_0,y) dy, \tag{25}$$

where $\psi(r,\xi,\tau - y)$ and $K(r,\xi,\tau - y)$ are function expressing the partial derivatives of the Green's function.

The integrated equation (25) has the form suitable for receiving the narrowing system of estimates of the problem solution (22) - (23). Suppose there are estimates t_1, t_2, which lead to $t_1 \le t \le t_2$. Having defined ψ_1, ψ_2 and K_1, K_2 similar to replacement in work and using estimates t_1, t_2 for $t(r,\tau)$ in equality (25), we receive the improved borders t_3, t_4 from the following derivations:

$$t_3(r,\tau) = V(r,\tau) + \frac{\chi}{c}\int_{r_0}^{\infty}\int_0^{\tau}\{\psi_1(r,\xi,\tau-y)w(t_1(\xi,y)) - \psi_2(r,\xi,\tau-y)w(t_2(\xi,y))\}\,d\xi\,dy -$$
$$-a^2\int_{r_0}^{\infty}\int_0^{\tau}\{K_1(r,\xi,\tau-y)t_1(\xi,\tau) - K_2(r,\xi,\tau-y)t_2(\xi,y)\}d\xi\,dy - \tag{26}$$
$$-\frac{a^2}{r_0}\int_0^{\tau}G(r,r_0,\tau-y)\,t(r_0,y)dy.$$

$$t_4(r,\tau) = V(r,\tau) + \frac{\chi}{c}\int_{r_0}^{\infty}\int_0^{\tau}\{\psi_1(r,\xi,\tau-y)w(t_2(\xi,y)) - \psi_2(r,\xi,\tau-y)w(t_1(\xi,y))\}\,d\xi\,dy -$$
$$-a^2\int_{r_0}^{\infty}\int_0^{\tau}\{K_1(r,\xi,\tau-y)t_2(\xi,\tau) - K_2(r,\xi,\tau-y)t_1(\xi,y)\}d\xi\,dy - \tag{27}$$
$$-\frac{a^2}{r_0}\int_0^{\tau}G(r,r_0,\tau-y)\,t(r_0,y)dy.$$

It's obvious that $t_1 \le t_3 \le t \le t_4 \le t_2$. Using t_3, t_4 under formulas similar to (26) - (27), we obtain the estimates t_5, t_6, and then the procedure is repeated recursively.

4 Results

Daily variation of external air temperature should be taken into account when calculating thermal fields in soil, construction materials and pipe line insulation. The control of daily temperature dynamics allows to establish phase displacement of temperature variations in and out of the premises resulting in correct daily mode of heating arrangement.

The algorithm of the freezing and melting problem solution can be used in engineering calculations, and also for testing various approximate methods. It should be noted that the similar solution can be constructed both for one-dimensional region with the central symmetry. Such a solution is especially valuable at high-quality research of freezing and melting processes in spherical shape regions.

5 Conclusion

At practical calculations of thermal fields in soil, construction materials and thermal*/ insulation of pipes it is expedient to consider daily fluctuation of temperature of external air. The accounting of daily dynamics of temperature allows to establish shift of phases of fluctuations of temperature in and out of the room that allows to organize the daily mode of heating correctly.

The scheme of the solution of a problem of frost penetration and thawing can be used by method of assessment directly in engineering calculations, and also for establishment of accuracy of various approximate methods. It should be noted that the similar solution can be constructed both for one-dimensional area with the central symmetry. Such a solution is especially valuable at high-quality research of processes of frost penetration and thawing in areas of spherical shape.

References

1. A. Ebrahimi-Moghadam, M. Farzaneh-Gord, M. Deymi-Dashtebayaz, Journal of Natural Gas Science and Engineering, **34**, 185-196 (2016)
2. V.V. Remizov, A.F. Shapoval, B.V. Moiseev, B.G. Aksenov, *Osobennosti stroitel'stva ob'ektov v neftegazodobyvajushhih rajonah Zapadnoj Sibiri* (Nedra, Moscow, 1996).
3. E. Rossi di Schio, S. Lazzari, A. Abbati, Applied Thermal Engineering, **102**, 227-233 (2014)
4. V.S. Sastri, Underground Pipeline Corrosion, 127-165 (2014)
5. J. Serrano-Arellano, J.M. Riesco-Ávila, J.M. Belman-Flores, K. Aguilar-Castro, E.V. Macías-Melo, International Journal of Heat and Mass Transfer, **102**, 26-35 (2016)
6. H. Mahgerefteh, P. Zhang, S. Brown, International Journal of Greenhouse Gas Control, **46**, 39-47 (2016)
7. S. R.Allahkaram, M. Isakhani-Zakaria, M. Derakhshani, M. Samadian, H. Sharifi-Rasaey, A. Razmjoo, Journal of Natural Gas Science and Engineering, **26**, 453-460 (2015)
8. M. A. Ersöz, A. Yildiz, International Journal of Refrigeration, **64**, 51-56 (2016)
9. R F.V.V. de Sousa, R.O. da Mota, J.P. Quintela, M.M. Vieira, I.C.P. Margarit, O.R. Mattos, Electrochimica Acta, **52(27)**, 7780-7785 (2007)
10. B. G., S. Song, A. Ghalambor, T. R. Lin, Offshore Pipelines, **2**, 113-124 (2014)
11. Y. Başoğul, A. Keçebaş, Energy, **36(10)**, 6156-6164 (2011)
12. Ju. S. Danielyan, D.K. Kalieva, P.Yu. Tret'jakov, Neftjanoe hozjajstvo - Neftedobyvajushhaja promyshlennost', 40-42 (2009)
13. Ju.A. Sigunov, RIO Surgutskogo gosudarstvennogo universiteta (2009)

MATEC Web of Conferences 73, 05007 (2016) DOI: 10.1051/matecconf/20167305007
TPACEE-2016

Forecast the Processes of Thawing of Permafrost Soils under the Building with the Large Heat Emission

Vladimir Paramonov [1], *Igor* Sakharov [2,*] , and *Sergey* Kudriavtcev[3]

[1]Saint-Petersburg State Transport University, 190031 Moskovsky pr. 9, Saint-Petersburg, Russia
[2]Saint-Petersburg State University of Architecture and Civil Engineering, 190005 2-nd Krasnoarmeiskaya St. 4, Saint-Petersburg , Russia
[3]Far Eastern State Transport University, 680021 Seryshev Str 47, Khabarovsk, Russia

Abstract. The paper describes a mathematical model of the soil experiencing freezing and thawing. The mathematical model is realized by the authors with the use of finite element method in the program «Termoground». The results of calculation the processes of the thawing of permafrost soil under the building with the large heat emission with various constructive measures for the conservation of frozen soil structure are given. Taking into account high costs of such measures it is recommended a combination of I and II principles of the design of foundations allowing partial thawing of soil, which requires the ability to predict the deformation of thawing soil.

1 Introduction

The prevalence of the processes of freezing and thawing of soils on the Earth is completely great. These processes are extended especially in the north latitudes on the Eurasian and American continents and also in the countries which are adjacent to the South Pole. Low temperature in winter periods is typical for mountainous areas, for example in Central Europe. Finally, the enormous Siberian spaces of the Russian Federation from the coast of the Arctic Ocean to the border with China composed of permafrost soils. Designing of foundations of buildings and structures on such soils has its special features which are differed significantly from the designing of bases and foundations of buildings on the non-frozen soils.

Besides the regions with the cold climate the problem of freezing and thawing of soils can also be found in the territories with high plus average annual temperatures during the underground works using freezing method. This method for example is widely used in the construction of tunnels especially inclined subway tunnels during the intersection by them the soft water-saturated soils of great thicknesses. The disadvantage of the method of freezing is significant settlement of territory after the degradation of ice-ground cylinders which causes deformations of buildings and structures located in the zone of influence of soil thawing.

* Corresponding author: tgasu.sakharov.igor@mail.ru

The volumetric strains characteristic for the soils in the process of their freezing and thawing can be up to ten percent or more. In this regard during the solution of this type of problem it is necessary to be able to predict not only temperature fields but also stress-strain state of system "structure – freezing (thawing) soil".

The practice of the construction and operation of buildings and structures when the influence of the processes of freezing and thawing of the soil was ignored has a huge number of negative examples. According to many years of experience of the authors we can mention the formation of lenses of frozen soils up to 6 meters under the buildings of Petersburg refrigerators; freezing of soils under buildings in the winter time, freezing of soils along the contour of pits in winter period, etc. In all these cases with the freezing or subsequent thawing of soil the through cracks were formed in the frameworks of buildings (walls or columns), and the sheet pile walls were deformed with the break of anchors. In the case of occurrence of permafrost soils in the base of foundations their thawing in the process of operation led to the settlements amounts to tens of centimeters which made further functioning of objects impossible [1-12].

2 Solution the problems of freezing and thawing of soils

Thus, the problem of the calculation of freezing and thawing of the soils – of base or medium, which contains construction, are extremely urgent. However, in these cases it is required to take into account the interaction between foundations and structures. Moreover the calculation of such interaction in connection with the development of computer calculations is possible only with the use of numerical methods which makes it possible to solve problems, also, for the three-dimensional conditions.

The calculation of the influence of freezing and thawing of soils assumes two-step solution of problems – first calculation of temperature fields and then the determination of the stress-strain state of the soil and the structure. Let us note that the Stephen solution for the solution of temperature problem cannot directly be used because in the cohesive soils the transition of water into ice and back occurs not at a zero temperature but in the range of negative temperatures. The freezing of the cohesive soils in most cases is accompanied by the migration of pore water to the front of freezing. In connection with this, the correct calculation of temperature fields requires taking into account the influence of the additionally entering water, which is essential for the balance of the entire heat exchange. Therefore the first step of calculations is calculation of temperature fields as a result of solution of hygrothermal problem.

The question about the lawfulness of using one or other expression for the thermal conductivity arises in solving the problem of hydrothermal problem.

The well-known equation for conductive heat transfer for the media with phase transitions is written as follows [1, 2]:

$$C_{th(f)}\rho_d \frac{\partial T}{\partial t} = \sum_{i=1}^{3} \frac{\partial}{\partial x_i}(\lambda_{th(f)} \frac{\partial T}{\partial x_i}) + q_v \tag{1}$$

where $C_{th(f)}$ – specific heat of soils (frozen or thawed); ρd – soil skeleton density; T – temperature, t – time; $\lambda_{th(f)}$ – thermal conductivity of soil (frozen and thawed); x, y, z – coordinates; q_v – internal heat source capacity.

The complete differential equation of thermal conductivity takes the form:

$$\rho_d (C_{th(f)} + L_0 \frac{\partial W_w}{\partial T}) \frac{\partial T}{\partial t} = \sum_{i=1}^{3} \frac{\partial}{\partial x_i}(\lambda_{th(f)} \frac{\partial T}{\partial x_i}) + q_v \tag{2}$$

where L_0 – water-ice latent heat of transition; W_w – water content of unfrozen water.

The calculation of temperature fields with the use of the expression (2) provides for the calculation of the heat transfer only by the mechanism of conduction. However as already mentioned the certain quantity water migrates to the front and into the frozen zone. In this regard the question appears what amount of heat the migrating water brings and whether it is necessary to take this heat delivered to the front due to convection into account in the heat equation.

The question about the relative influence of conductive and convective heat transfer in freezing soils studied R.L.Harlan and J.F.Nixon [3]. They have shown that the convective component should be considered only at high rates of mass transfer. However in the case of the migration of water in the relatively low-permeable cohesive soils the contribution of convective component of heat flux can be disregarded.

The similar conclusion can be approached on the other hand comparing the rate of freezing and water migration. In most cases the rate of the water migration is not more than ten percent of the rate of freezing, which makes it possible not to take into account the contribution of the heat delivered into the frozen zone by the migrating water. Thus, the use of expressions (1) and (2) for calculating the temperature fields in the freezing soils of low permeability is justifiable.

The use of rigorous theoretical solutions is thus far impossible for the forecast of an increase of the water content in the freezing zone due to the water migration. The authors of the article developed the «Termoground» program based on the well-known computer complex FEM models [8, 9] where the amount of inflowing water in the frozen zone is established on the basis of processing of experimental works of L.V.Chistotinov [4] and G.M.Feldman [5]. This amount of inflowing water depends on the type of soil, rate of its freezing as well as the distance between the front of the freezing and the groundwater level.

For the calculation the stress-strain state of the freezing soil it is necessary to assign the law of the soil deformation during the freezing depending on different factors. The calculated temperature fields can come out as this factor; however, similar solutions are applicable only for the closed systems [6]. More correct are established during the first step of calculations the fields of the water content of soil, an increase in volume of which during transition of water into ice will make it possible to estimate increase in the volume of soil with the freezing. An additional factor in increasing the soil volume during freezing is the growth of frost cracks. Finally, according to our experience in the majority of cases in the calculations of frost heave such the influence of factors as stresses in soil, etc. can be neglected.

For the solution of the problems of thawing there are different ways – from taking into account only the physical characteristics of frozen soil to the direct use "thermal" and "load" component of the standard characteristics of thawing placed in the Russian standards.

3 Numerical modeling of degradation of permafrost soil under the building with large heat emission

In this paper we consider a specific example of building with large heat emission on permafrost soil – the boiler house. Construction of the building is assumed according to the first principle, i.e., with the preservation of the frozen state of the soil. Foundations under the external walls of boiler house and under the boilers piled with piles length of 10 m. The cross section of building is shown in Fig. 1. The three-dimensional numerical model which includes soil, foundations and building is given in Fig. 2.

The temperature on the outer contour of the boiler is +55°C. It is assumed that the work of the boiler in the annual cycle is carried out from the middle of October before the middle of July. For a number of design considerations aired underground under the building is impossible.

First of all let us consider the solution of the problem without the use of additional measures for the conservation of frozen soil structure. This problem is solved for an example, since it is obvious that the construction of building with the large heat emission without such measures even on the little icy soil does not have a sense.

Fig. 3 shows the area of soil thaw in cross-section after 50 years of operation of a building. Estimated depth of thawing is about 16 m.

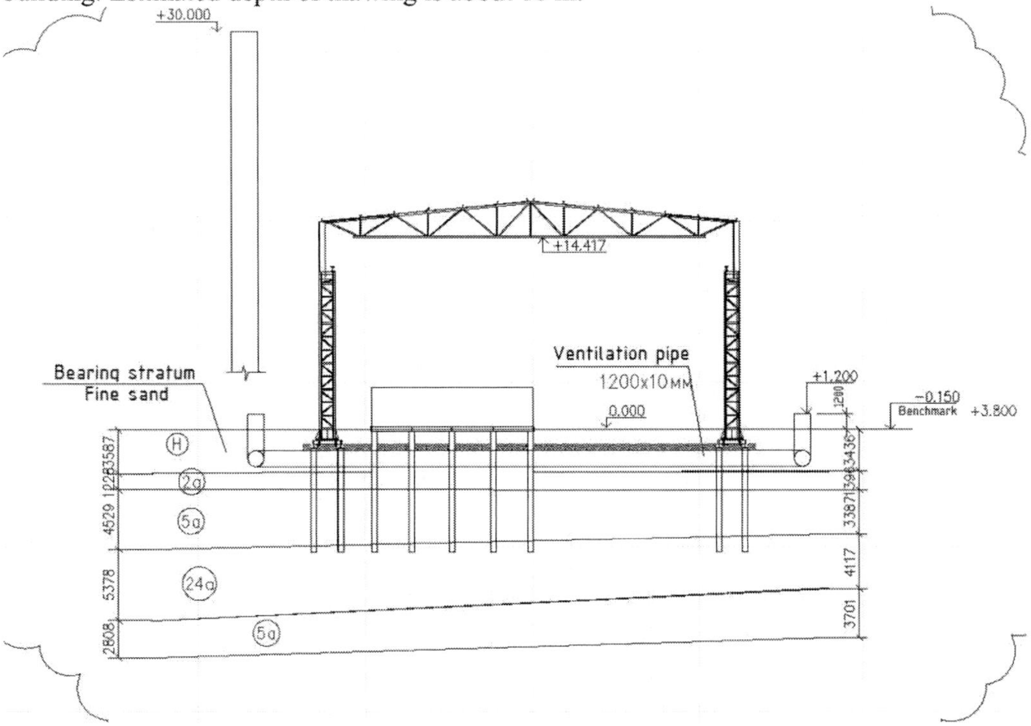

Fig. 1. Cross section of boiler house.

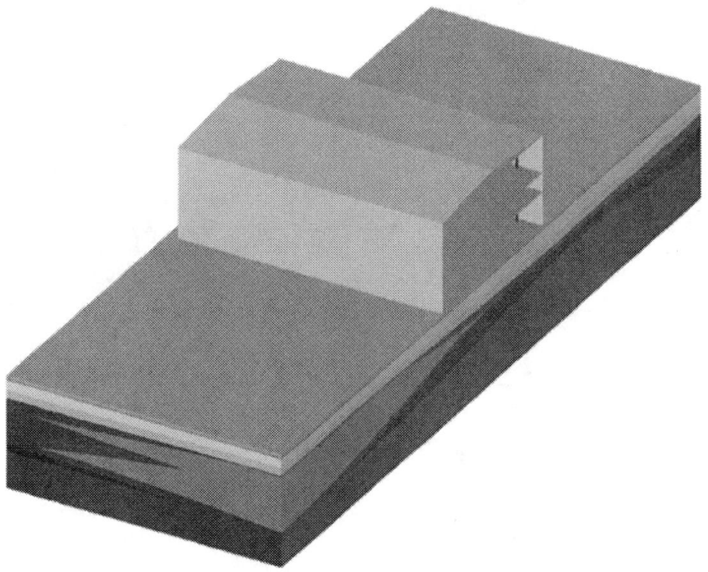

Fig. 2. Numerical models of building and soil.

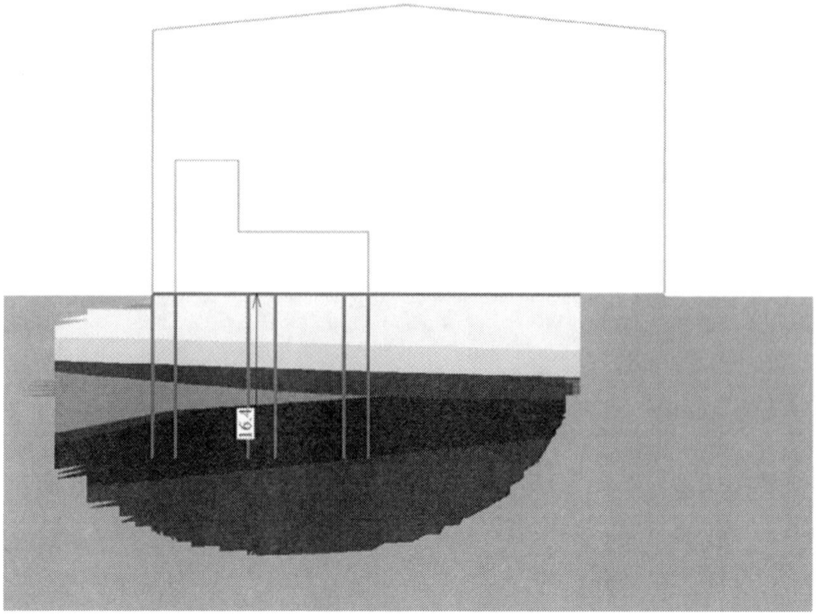

Fig. 3. The area of ground thaw under the building in the absence of measures for the conservation of frozen soil.

It is obvious that with this large depth of the soil thawing the development of catastrophic settlements should be expected. Fig. 4 shows a diagram of settlements of the building after 50 years of its operation in the construction of buildings on the shallow foundations.

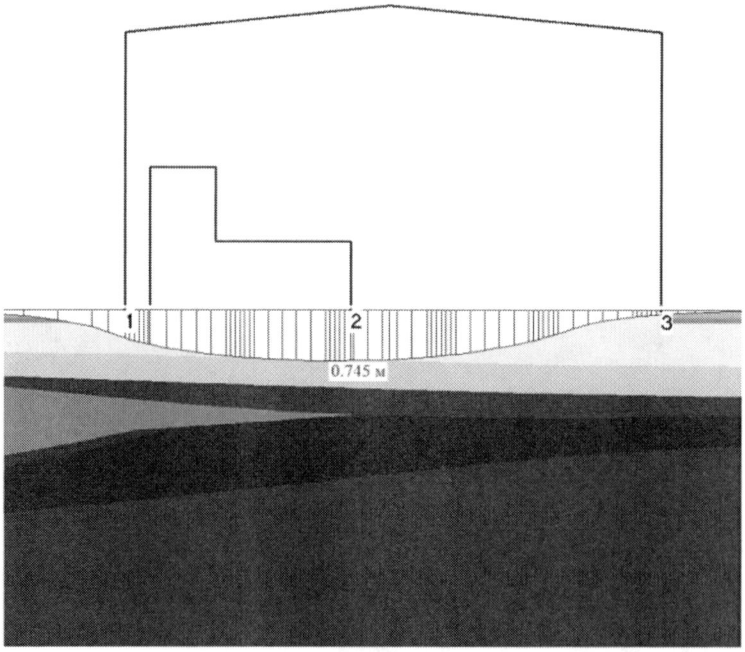

Fig. 4. Plot of ground settlements under the building in the absence of measures for the conservation of frozen soil structure and the construction of buildings on the shallow foundations

Diagram does not include the settlements of building due to its own loads. After 50 years the estimated settlement of the building is 75 cm. Fig. 5 shows the graphs of the development of settlements of building in the time for three characteristic points – two points (№1 also №3) along the edges of building, one point (№2) – under the center of building.

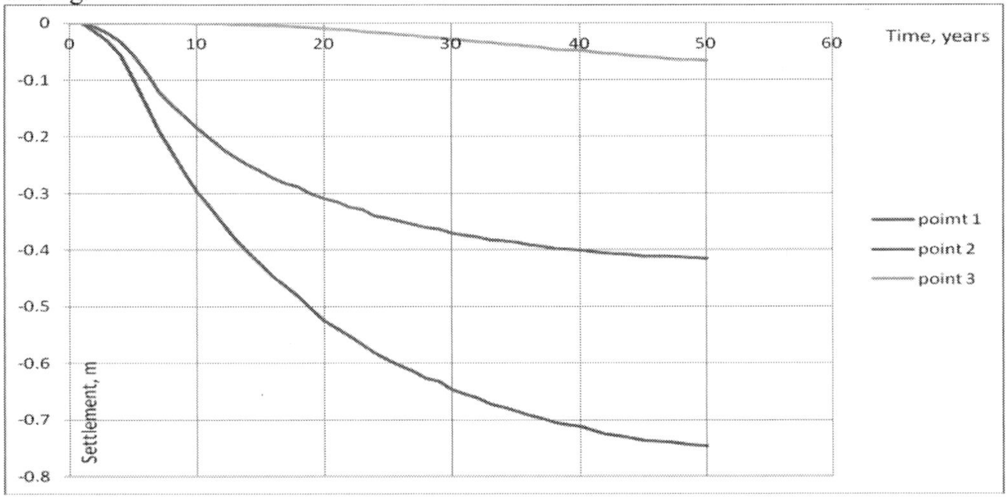

Fig. 5. Graph of settlements of building in the time.

The position of the points is shown in Fig. 4. As it can be seen from the graphs without any measures to preserve the frozen soil structure during the entire 50-year of operation of the building it will undergo settlements. Thus, it is necessary to find the solution to reduce the sizes of the area of thawing of the soil, since the thaw trough seizes foundations under the walls (left part of the section in Fig. 3).

Construction of buildings on the designed 10-meter length piles reduces the estimated settlements of the building almost in 2 time (Fig. 6), especially in the locations of the piles (to 47 cm in 50 years of operation of the building). However, in the span areas the settlements reduce insignificantly (to 62 cm).

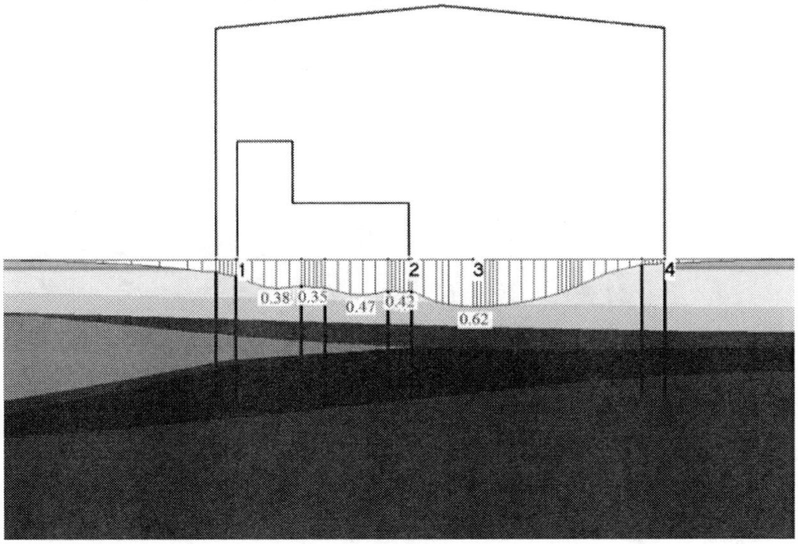

Fig.6. Plot of ground settlements under the building in the absence of measures for the conservation of frozen soil structure and the construction of buildings on the 10-meter length piles.

Graphs of the development of settlements in the time in Fig. 7 are shown for the same points under the building as on the graphs Fig. 5. Graph for the intermediate point (№3) with the greatest absolute value of settlement is additionally shown. The graphs show that the deformation of the soil appears almost immediately at the beginning of the thawing of the soil which is associated with the development of negative skin friction on the piles with the subsidence of soil thawing.

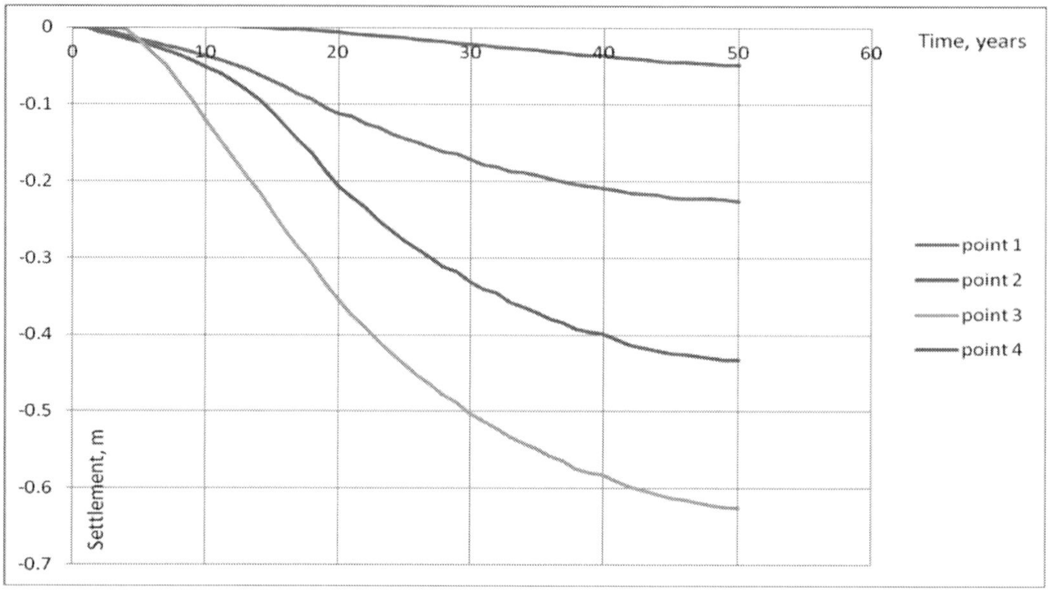

Fig. 7. Graph of settlements of building on piles in the time.

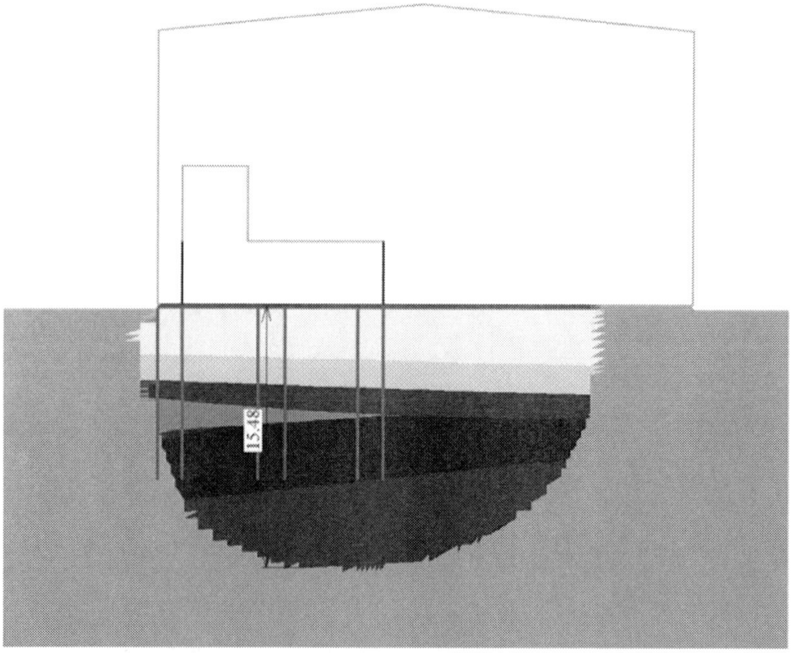

Fig. 8. The area of ground thaw under the building with the vertical thermal stabilizers on the outer contour of the building.

Let us estimate the effectiveness of the work of the vertical seasonal-cooling devices (SCD) for the case in question. Since the building does not have the aired underground, the installation of thermal stabilizers is possible only near the outer side of the building. The depth of the SCD was taken equal to the depth of piles. The work of SCD is accepted to evaluate by the effectiveness ratio of cooling. AS one of the most effective the thermal stabilizers of the type TMD-5 is considered with the anticorrosive aluminum alloy with the ratio of cooling efficiency to 0.75 developed be "Inter Heat Pipe" company. In our calculations we assumed the most common and cheap SCD made of steel with a diameter 54 of mm irrespectively of the type of thermal stabilizer with the effectiveness ratio of the cooling 0.5.

With the simulation of work of the SCD near the foundations of the building of boiler house the thermal stabilizers were arranged next to each pile of external rows. It is evident from Fig. 8 that the SCD eliminate the heat propagation in the ground outside of the building dimensions, actually ensures the work of the external piles, near which the SCD are installed, however the internal piles are located in the zone of thawing. These heat stabilizers do not affect the depth of thawing of the soil under the boilers.

One of the most effective for reducing of the soil thaw depth is often assumed the insulation between the ground floors of the building and the ground. Let us consider the case of 30-centimeters heat insulation made of polyfoam under the ground floor of the building in combination with the vertical SCD installed at the outer contour of the building.

It is evident from Fig. 9 that under these conditions the depth of thawing is reduced to 11.3 m. The internal piles are located completely in the zone of the thawing of soil. According to the experience of our calculations, the thermal properties of the 10-centimeter layer of polyfoam are equivalent to the 1-meter layer of soil. Thus to exclude the thawing of soil it will be required the 1-1.5-meter layer of thermal insulation.

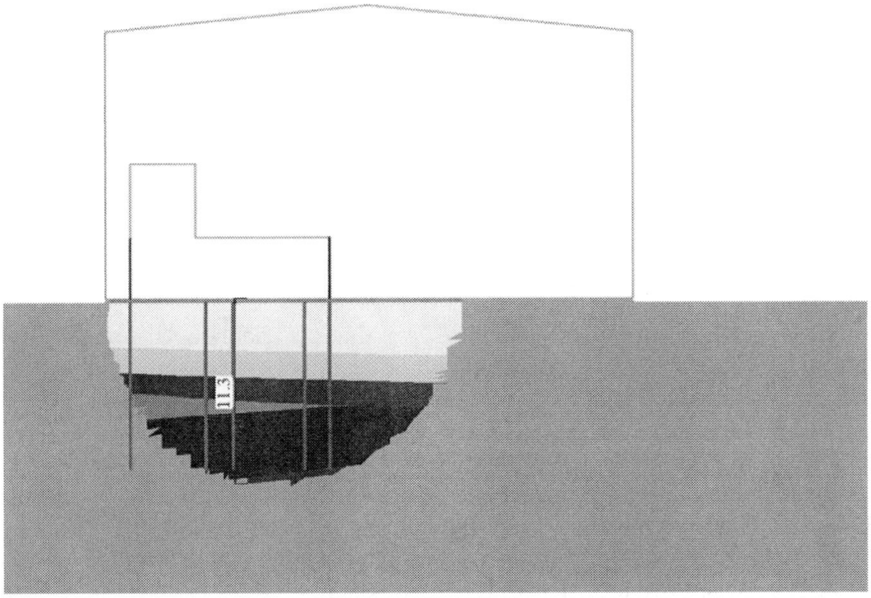

Fig. 9. The area of ground thaw under the building with the vertical thermal stabilizers on the outer contour of the building and 30-centimeters layer of heat insulation.

The horizontal ventilated channels under the building in combination with the 30-centimeter layer of heat insulation prove to be most effective for the building with the large heat emission. Under the conditions in question this version makes it possible to practically completely exclude the thawing of soil under the building. According to the calculations the maximum depth of thawing is fallen for the warm period, when the ventilated channels are turned off while the boiler is not working. This depth is 1.1 m (Fig. 10) that is almost equal to the depth of seasonal freezing and thawing of the soil.

Thus, for a satisfactory solution of the problem of minimization of soil thawing the ventilated channels are effective for a satisfactory solution of the problem of minimization of soil thawing.

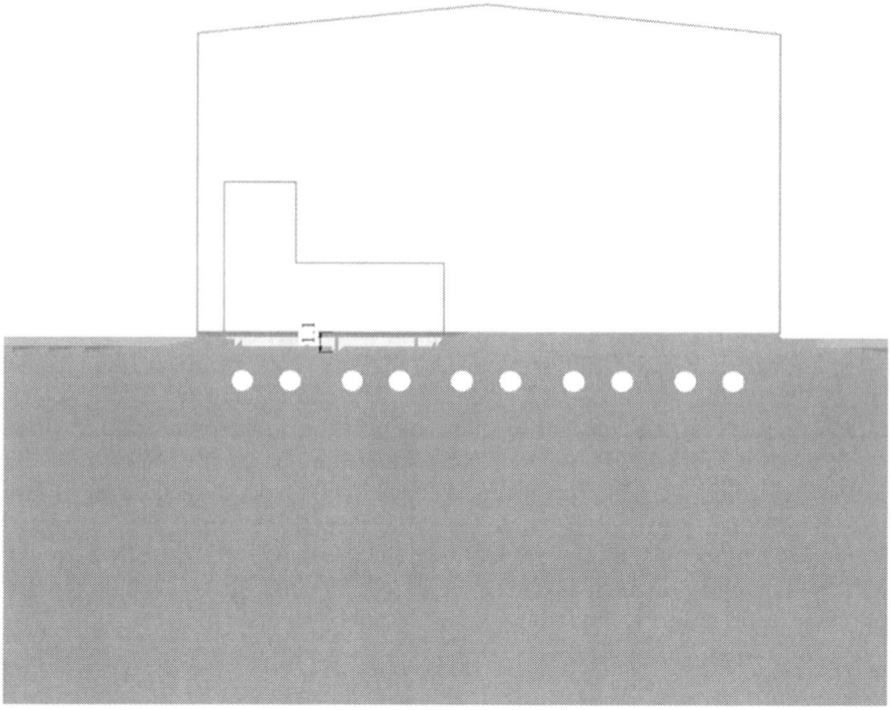

Fig. 10. The area of ground thaw under the building with the horizontal ventilated channels and 30-centimeters layer of heat insulation.

Conclusion

During construction of buildings with a large heat emission on the permafrost soils the dimensions of the area of soil thawing in both the horizontal and vertical directions can be tens of meters.

According to the results of calculations, ventilation channels under the building in combination with the horizontal heat insulation of relatively small thickness are one of the most effective measures.

The measures for conservation of frozen soil structure are quite expensive and energy intensive. In such conditions the combination of I and II principles of the design of foundations allowing partial thawing of soil seems appropriate. However, the possibility of a combination of these principles is not enough just only on the base of prediction of the temperature fields in soils. The correct solution of problems should include an assessment of the evolution of temperature fields and stress-strain state of the system "construction - freeze (thawed) soil" for the entire period of operation of the building.

References

1. Del S Guidice, G. Comini, R. Lewis, Int.J.Num.Anal.Meth.Geomech, **2**, 223-235 (1978)
2. A. Fadeev, *Metod konechnyh ehlementov v geomekhanike* (Nedra, Moscow, 1987)
3. R. Harlan, J. Nixon, *Geotekhnicheskie voprosy osvoeniya Severa: Teplovoy rezhim gruntov* (Nedra, Moscow, 1983)
4. L. Chistotinov, *Modelirovanie teplo-massoperenosa v promerzayushchih puchinistyh gruntah. Problemy geokrologii* (SO RAN, Yakutsk, 1998)

5. G. Feldman, *Peredvizhenie vlagi v talyh i promerzayushchih gruntah* (Nauka, Novosibirsk, 1988)
6. A. Fadeev, I. Saharov, P. Repina, OFMG, **5**, 6-9 (1994)
7. V. Borisov, Procedia Computer Science, **66**, 112-121 (2015)
8. S. Kudryavtsev, I. Saharov, V. Paramonov, *Promerzanie i ottaivanie gruntov prakticheskie primery i konechnoehlementnye raschet* (Gruppa kompaniy Georekonstruktsiya, SPb, 2014)
9. V. Ulitskiy, I. Saharov, V. Paramonov, Raschet sistemy osnovanie sooruzhenie pri promerzanii i ottaivanii gruntov s pomoshchyu programmy, "Termoground". OFM, **5**, 3-7 (2014)
10. P. Kotov, L. Roman, I. Saharov, V. Paramonov, M. Paramonov, OFMG, **5**, 8-13 (2014)
11. Y. Wang, H. Jin, G. Li. Cold Regions Science and Technology, **126**, 10-21 (2016)
12. G. Li, Q. Yu, W. Ma, Z. Chen, Y. Mu, L. Guo, F. Wang, Cold Regions Science and Technology, **121**, 258-274 (2016)

The Ecological Situation in the Russian Arctic Permafrost Zone

Sergei Petrov[1,*], *Natali* Mamaeva[2], *Maxim* Gabdullin[1] and *Alexey* Kraev[3]

[1]Tyumen Scientific Center, SB RAS, Tyumen, 625026, Russia
[2]Tyumen Scientific Center, SB RAS, IUT, Tyumen, 625026, Russia
[3]Industrial University of Tyumen, 625001 Volodarskogo str. 38, Tyumen, Russia

Abstract. The paper describes innovative approaches to ensure environmental safety in the production of hydrocarbon material in a permafrost zone. Studies the anthropogenic environmental factors, climatic and geographical and geological conditions of Purovskiy district of Yamalo-Nenets Autonomous Area (YaNAO). We consider the chemical characteristics of wastewater discharged into surface water objects, polluting emissions into the atmosphere. The conclusions of the environmental situation in Purovskiy and Ustpurovsk-Tazovskiy permafrost areas. Calculate the concentration of pollutants in the control section of the water object and the maximum ground-level concentrations of pollutants in the atmospheric air. The conclusions about the exceeding the maximum permissible concentration (MPC) in the atmospheric air for solids, carbon monoxide, nitrogen dioxide. Was examined the climatic conditions of the Far North. Correlational analysis was performed between human factors and temperature conditions of the northern territories, as well as between the climate and natural features cryological and disturbed permafrost soils.

1 Introduction

Russian Arctic - the most important and many-sided area, which is characterized by a combination of extreme cold temperatures originality helio geomagnetic background, unusual light conditions with periodic shortage of solar radiation, shifts biorhythmological processes and has huge reserves of hydrocarbons [1, 2]. Data on the hydrocarbon reserves of the Arctic published periodically in different countries of the world. However, their estimates vary considerably. According to UN statistics, the Arctic oil proven reserves amount to 100 billion tons, gas - 50 trln.m3. According to the calculations of a number of industry analysts, potential geological reserves of hydrocarbons in the Arctic are estimated at 200 billion tons of oil and 400 trln.m3 gas [3].

US Geological Survey Experts (USGS) believe that the Arctic is a fifth of the unexplored recoverable reserves of oil and natural gas. Potential reserves of oil in the region - 90 billion barrels, gas - 47.3 trln.m3, gas condensate - 44 billion barrels [4, 5].

* Corresponding author: tumiki@mail.ru

The largest total reserves of the Arctic, according to the USGS, concentrated in the West Siberian Basin - 3.6 billion barrels. of oil, 18.4 trln.m3 of gas and 20 billion. barrels. of gas condensate. This is followed by the Arctic shelf of Alaska (29 billion barrels of oil, gas - 6.1 trln.m3 and 5 billion barrels of gas condensate), The third place - the eastern part of the Barents Sea shelf (7.4 billion barrels of oil, 8.97 trln.m3 of gas and 1.4 billion. barrels. of gas condensate) [6, 7, 8]. Oil industry at risk of exposure to the environment ranked third among the 130 branches of modern production. Mining of hydrocarbons accompanied by a significant damage to the environment. Practically none of the existing oil and gas fields in Russia, does not apply to so-called "waste-free production". This is due to the imperfection of production technology, its disturbances, poor quality or deterioration of the technological equipment, natural features of oil and gas regions of the Arctic zone. Arctic landscapes are characterized by low resistance to anthropogenic influences. The more intense fluids removal is carried out, the more active is the formation of technogenic streams of various substances in the environment that have a negative impact on the atmosphere, the hydrosphere, soil, subsoil [9, 10]. Currently, environmental studies in the Russian Arctic, in particular Purovskiy region of Yamalo-Nenets Autonomous Area (YaNAO), are relevant.

2 Materials and methods

Studied anthropogenic factors, climatic geographical and geological environmental conditions of Purovskiy district YaNAO (settlement Sumburgh and Tarko-Sale, which according to the zoning of the permafrost zone belong to the North-Kharasavey Novy Urengoy subzone Ustpurovsk-Tazovskiy region and Central region Igarka-Numtinskoy subzone Purovskiy area) [11]. Sumburgh is located in the Urengoy oil and gas area with a high density of initial potential resources of hydrocarbons. Near the village is located Samburgskoye field gas condensate and 11 oil deposits (6 condensate, 2 oil and 3 oil with gas condensate) [12]. City Tarko-Sale is in the Pur-Tazovsky oil and gas area, which is characterized by high hydrocarbon resources. Near the city is a large of the East Tarkosalinskoye field gas condensate and oil with 17 deposits of hydrocarbons (1 gas, 8 gas condensate, 8 oil with gas condensate) [12]. To study the anthropogenic factors analyzed the quality of surface water objects: chemical characteristics of wastewater discharged in superficial water objects for different pollutants. Considered air pollution data in permafrost areas: emissions of pollutants into the atmosphere [13, 14, 15]. Been calculated concentration of pollutants in the water object control section (Cc.s. mg/l) and maximum ground concentration of harmful substances in the air (Cmax, mg/m3) [16, 17]. The calculation of the dispersion of surface concentrations of pollutants in the atmospheric air was made taking into account the physical and geographical and meteorological environmental conditions with OND-86 «Method of calculating the concentration in the atmospheric air of harmful substances in emissions of enterprises» [18] and software system unified program for calculating pollution of atmosphere (UPCPA) "Eco Centre» [19]. To study the climatic conditions are analyzed temperature characteristics: annual air temperature, average air temperature in the 13 p.m., the average minimum, absolute minimum temperature, maximum temperature and the amplitude of temperature vibrations of air per year. Climatic data were obtained from the Climate directories [20], as well as www.meteo.infospase.ru site [21]. In the study cryological conditions (power and humidity seasonally thawed layer) used data from the Institute of Earth Cryosphere of SB RAS [22].

Qualitative and quantitative traits subjected to statistical analysis using the «SPSS 11,5 for Windows» program. Been studied distribution pattern of the studied traits. When a Gaussian distribution used the following statistical parameters: the average (arithmetic mean, median, mode), variance and its derivative. Were compared significant difference (similarity) between the statistical characteristics by the criterion of the Student. The significance of differences between mean values calculated by standard error of the arithmetic mean value. In the absence of normal distribution using non-parametric methods of comparing two samples with the calculation of the pair Wilcoxona criterion, Spearman rank correlation coefficient.

3 Ecological analysis of Cryolithozone of Purovskiy district of Yamalo-Nenets Autonomous Area

3.1 Anthropogenic pollution of the permafrost zone of Purovskiy district of Yamalo-Nenets Autonomous Area

Technogenic impact on cryolithozone become particularly dangerous in recent decades. The intensity of air pollution emissions by pollutants from stationary and mobile sources, the quality of water objects are the main environmental factors. They affect the physical well-being of the population [9, 23]. Table 1 examined chemical characteristics of the waste into the air and water emissions Ustpurovsk-Tazovskiy and Purovsky permafrost areas. It was found that in the organized discharge of industrial wastewater in superficial water objects in the Purovskiy geocryological area compared to with Ustpurovsk-Tazovskiy geocryological region greater concentration of pollutants: iron (1.57 ± 0.17 tons and 0.37 ± 0.11 tonnes respectively at $p < 0.001$), chloride (0.02 ± 0.004 ths. tons and 0.01 ± 0.002 ths. tons, respectively, at $p < 0.05$), ammonia nitrogen (10.89 ± 1.66 tons and 1.01 ± 0.17 tonnes respectively at $p < 0.001$). Significant differences in other pollutants discharged into superficial water objects were not found.

Table 1. Chemical characterization wastewater and air emissions.

№	Substance name	Ustpurovsk-Tazovskiy	Purovskiy
1.	Wastewater		
1.1.	Synthetic surfactants, [tons]	0.04 ± 0.01	0.07 ± 0.02
1.2.	Total phosphorus, [tons]	0.25 ± 0.05	0.37 ± 0.09
.3.	Iron, [tons]	0.37 ± 0.11	1.57 ± 0.17**
1.4.	Sulfates, [ths. tons]	0.01 ± 0.001	0.02 ± 0.01
1.5.	Chloride, [ths. tons]	0.01 ± 0.002	0.02 ± 0.004*
1.6.	Nitrogen ammonium, [tons]	1.01 ± 0.17	10.89 ± 1.66**
1.7.	Nitrates, [tons]	7.55 ± 2.13	5.22 ± 2.55
1.8.	Nitrites, [tons]	0.11 ± 0.02	0.77 ± 0.45
2.	Emissions to air		
2.1.	Total of them, [tons]:	8609.35 ± 1574.71	395978.10 ± 34298.67**
2.1.1.	Solid, [tons]	104.88 ± 15.14	14210.65 ± 676.63**
2.1.2.	Gaseous and liquid, including, [tons]:	8504.53 ± 1576.76	381767.50 ±31927.93**
2.1.2.1.	Sulphur dioxide, [tons]	82.71 ± 11.54	929.47 ± 39.84**
2.1.2.2.	Carbon monoxide, [tons]	4023.41 ± 693.40	215444.20 ±24814.86**
2.1.2.3.	Nitrogen oxides, [tons]	1334.06 ± 235.74	23789.00 ± 1161.49**
2.1.2.4.	Hydrocarbons (no VOCs), [tons]	2972.47 ± 692.52	136530.20 ±14237.04**
2.1.2.5.	Volatile organic compounds (VOCs), [tons]	124.45 ± 43.55	5292.12 ± 732.57**

* - reliability of difference (* - $p<0.05$; ** - $p<0.001$).

Emissions of solid pollutants in the atmosphere longer (14210.65 ± 2676.63 tonnes from 104.88 ± 15.14 tons at $p < 0.001$), gaseous and liquid (381767.50 ± 31927.93 tonnes compared to 8504. 53 ± 1576.76 tons at $p < 0.001$) in Purovskiy area compared to Ustpurovsk -Tazovskiy geocryological region.

Been calculated concentration of pollutants in the water object control section (Cc.s., mg/l) and maximum ground concentration of harmful substances in the air (Cmax, mg/m3) (Table 2).

The concentration of pollutants in the water control zone was significantly greater in Purovskiy geocryological area compared to Ustpurovsk-Tazovskiy-over for the following substances: synthetic surfactants ($p < 0.05$), phosphorus ($p < 0.05$), iron ($p < 0.001$), chlorides ($p < 0.01$), ammonia nitrogen ($p < 0.001$). Concentrations of pollutants in the water control zone in Ustpurovsk-Tazovskiy and Purovskiy geocryological area does not exceed the maximum permissible concentration.

The maximum surface concentration of harmful substances in the atmospheric air in Ust Purovsk-Tazovskoye geocryological region significantly more for the following pollutants: nitrogen dioxide ($p < 0.01$), nitrogen oxide ($p < 0.05$), sulfur dioxide ($p < 0.001$). The maximum surface concentration of solids (soot) ($p < 0.001$) and carbon monoxide ($p < 0.001$) was significantly lower in Purovskiy geocryological area. Is investigated the maximum surface concentration summation of substances (sulfur oxide and nitrogen dioxide), which is significantly higher in Ustpurovsk-Tazovskiy geocryological region (0.540 ± 0.08 compared to 0.216 ± 0.01 for $p < 0.001$).

It was found exceeding the maximum surface concentration of pollutants in the atmosphere is 1.5 times MPCm.o. solids and 1.2 times MPCm.o. carbon monoxide Purovskiy geocryological area. (Table 2). This may cause the greenhouse effect in the Arctic zone. Revealed the excess of nitrogen dioxide Purovskiy and Ustpurovsk-Tazovskiy permafrost areas (1,2 and 3,0 times MPCm.o.). [24- 27]

Table 2. The concentration of pollutants in the water control zone, maximum surface concentration of harmful substances in atmospheric air and pollution standards.

Substance name	Ustpurovsk-Tazovskiy	Purovskiy	The criteria used to calculate	The value of the criterion	Hazard Class
The concentration of pollutants in the water of control zone:					
Synthetic surface-active agents (surfactants), [mg/l]	0.000001409 ± 0.00000035	0.000003963 ± 0.0000011*	MPCd.d.	0.5	3
Total phosphorus, [mg/l]	0.000008808 ± 0.0000018	0.000020951 ± 0.0000051*	MPCd.d.	0.0001	1
Iron, [mg/l]	0.000013036 ± 0.0000039	0.0000889 ± 0.0000096***	MPCd.d.	0.3	3
Sulfates, [mg/l]	0.000352331 ± 0.000035	0.001132492 ± 0.000566	MPCd.d.	500	4
Chloride, [mg/l]	0.00035 ± 0,00007	0.00113 ± 0.00023**	MPCd.d.	350	4
Nitrogen ammonium, [mg/l]	0.000035585 ± 0.000006	0.00061664 ± 0.000094***	MPCd.d.	2.0	3
Nitrates, [mg/l]	0.000266009 ± 0.000075	0.00029558 ± 0.000144	MPCd.d.	45	3
Nitrite, [mg/l]	0.000003875 ± 0.0000007	0.0000436 ± 0.0000254	MPCd.d.	3.3	2

The maximum surface concentration of harmful substances in atmospheric air:					
Nitrogen dioxide, [mg/m^3]	0.600 ± 0,1	0.240 ± 0.01**	MPCm.o.	0.20	3
Nitrogen oxide, [mg/m^3]	0.052 ± 0,01	0.031 ± 0.001*	MPCm.o.	0.40	3
Carbon (soot), [mg/m^3]	0.049 ± 0,007	0.225 ± 0.04***	MPCm.o.	0.15	3
Sulfur dioxide, [mg/m^3]	0.025 ± 0,003	0.009 ± 0.0004***	MPCm.o.	0.50	3
Carbon monoxide, [mg/m^3]	1.216 ± 0,21	6.00 ± 0.69***	MPCm.o.	5.00	4
Methane, [mg/m^3]	0.899 ± 0,21	1.379 ± 0.14	OEL	50	-
Groups of substances that have the combined harmful effects:					
Nitrogen dioxide, sulfur dioxide	0.540 ± 0,08	0.216 ± 0.01***	Cce	1.6	-

MPCd.d. - maximum permissible concentration of domestic and drinking water; MPCm.o. - maximum permissible concentration of maximum one-time; OEL - occupational exposure limits; Cce - the combined effect of the coefficient; * - reliability of differences (* - $p < 0.05$; ** - $p < 0.01$; *** - $p < 0.01$).

The effect of incomplete summation possess nitrogen dioxide, sulfur dioxide. Exceeding them in Ustpurovsk-Tazovskiy geocryological region was 2.7 MPCm.o. in Purovskiy – 1.08 MPCm.o. Majorants all the considered pollutants and summation group in Ustpurovsk-Tazovskiy geocryological region was 3 MPCm.o. in Purovskiy – 1.5 MPCm.o.

3.2 The climatic conditions of the permafrost zone Purovskiy district of Yamalo-Nenets Autonomous region

One of the most important characteristics is the climate temperature. It has a direct impact on humans, animals, plants, the work of various mechanisms and choice of materials, special technologies of production processes. From the temperature dependent freezing and/or thawing permafrost [22, 28]. Comparative analysis of temperature conditions Ustpurovsk-Tazovskiy and Purovskiy permafrost areas is presented in Table 3. It was found that the average temperature in the 13 p.m., the average minimum temperature, maximum air temperature authentically higher in Purovskiy geocryological area. It was calculated the difference between the maximum and the absolute minimum temperature for the year as a whole in the permafrost areas of data. It was found that in Purovskiy geocryological area for the year as a whole (76,64 ± 0,95 compared to 72,56 ± 0.65 for $p < 0.01$) air temperature variation bigger than in Ustpurovsk-Tazovskiy geocryological region. Table 3.

Table 3. Temperature characteristics of geological areas.

Substance name	Ustpurovsk-Tazovskiy	Purovskiy
The annual air temperature, [°C]	-7.05 ± 1.40	-5.51 ± 0.40
The average temperature at the 13.00 p.m., [°C]	-7.03 ± 0.44	-2.75 ± 0.75**
The average minimum temperature, [°C]	-12.75 ± 0.37	-10.19 ± 0.40**
The absolute minimum air temperature, [°C]	-45.98 ± 0.54	-47.74 ± 0.93
The maximum air temperature, [°C]	26.57 ± 0.30	29.09 ± 0.49**
The amplitude of the vibrations of air temperature over the year as a whole, [°C]	72.56 ± 0.65	76.64 ± 0.95*

* - reliability of difference (* - $p<0.01$; ** - $p<0.001$).

Correlation analysis revealed connection between the content of iron, chloride, nitrate, nitrite in the wastewater and maximum air temperature ($r = 0.800$ for $p < 0.01$; $r = 0.780$ for $p < 0.01$; $r = 0.660$ for $p < 0.05$; $r = 0.730$ for $p < 0.05$ respectively). The accumulation of iron, chloride, nitrate, nitrite in wastewater contributes to waterlogging and eutrophication and accumulation of heat. Overgrowing water bodies and soil the permafrost zone causes climate warming and enhances the environmental risks [25].

3.3 Correlation relationship between climate and cryological features of natural and disturbed soils of the Central geocryological zone

In the process of technogenesis being destroyed natural environment. Occurs degradation of permafrost soils. It was found that the correlation of mean monthly air temperature with humidity weighting the seasonally thawing layer (STL) of permafrost soils is dependent on their status. Correlation relationships of average monthly air temperature with humidity weight STL of natural (undisturbed) soil visible in the winter season to the soil depth of 0.6 m. ($r > 0.7$). In autumn and spring seasons - to the soil depth of 0.7 m. At the soil depth 0.85 m correlations disappear. The loss of correlations at the soil depth 0.85 m. is associated with the process of thawing soils. Permafrost soils are confining layer and prevent vertical water filtration, drainage area. Formed suprapermafrost the surface water.

Due to the active development of oil and gas fields of the northern territories appear disturbed soil. In this case, correlations between the of average monthly air temperature and humidity weight STL of soil in the winter season were in direct proportion to the depth of 0.85 m. In the spring and autumn seasons - to a depth of 1.1 m. The presence of correlations at a depth of 0.85 m. in disturbed soils indicates that thawing permafrost occurs to a greater depth than in natural soils. This shows the influence of anthropogenic factors on the biosphere of the Far North, in particular on the soil.

4 Conclusions

In the coming years the Russian sector of the Arctic will continue to play a leading role in the extraction of gas and oil. Main gas base of the country is remaining YaNAO [4]. Environmental problems of the Russian Arctic have an impact on the ecological situation in the region and the world at large.

In considering the quality of water bodies it has been revealed that the concentrations of pollutants in the water control zone in Ustpurovsk-Tazovskiy and Purovskiy permafrost areas do not exceed the maximum permissible concentration. Excess of solids in the atmosphere (carbon black) by 1,5 times MPCm.o. and by 1,2 times MPCm.o carbon monoxide in Purovskiy geocryological area. The content of nitrogen dioxide exceeded in Purovskiy geocryological area by 1, 2 times MPCm.o. in Ustpurovsk-Tazovskiy permafrost areas by 3,0 times MPCm.o.

The pollution of the atmosphere and water bodies by pollutants is accompanied by an increase in ambient temperature. There is a degradation of permafrost. The nature of the degradation of permafrost is also dependent on technogenic activities in the region. Climatic conditions associated with cryogenesis of the natural and disturbed soils of the Central geocryological zone. There are pronounced manifestations of thermal degradation of the permafrost zone in some northern areas of the Russian Arctic.

Comprehensive study of anthropogenic factors, climatic and geocryological conditions of the Far North insufficient attention is given. Scientific evidence scattered. The main tasks in ensuring the environmental safety of the Russian Arctic are:

- monitoring of the environment in the Arctic region;
- assessment of the environmental impact of economic activities;
- development and introduction of new techniques, technologies and materials adapted to the climatic conditions of the Arctic;
- reducing emissions of pollutants into the air, waste water discharge into water bodies through the introduction of resource-saving technologies in enterprises;
- compliance with the rules environmental protection, industrial safety and labor protection;
- rational usage of natural resources and energy, the constant optimization of their consumption;
- introduction of non-waste and low-waste technologies.

Acknowledgements

This work was supported by the grant of the RAS Presidium "Search basic research in support of development the Arctic zone of the Russian Federation" and the international grant RFFR and HIARC «Anthropogenic heat island in the Arctic - a window into the future regional climate conditions, ecosystems and society» (№ 15-55 -71004).

References

1. N.L. Mamaeva, S.A. Petrov, Oil & Gas, **5**, 148–152 (2015)
2. N.L. Mamaeva, S.I. Kvashnina, Oil & Gas, **5**, 123 – 125 (2011)
3. V.P. Melnikov, D.S. Drozdov, *Arctic Subarctic: mosaic, contrast, variation Cryosphere: proceedings of the international conference*, 243-246 (2015)
4. *"About the environmental situation in the Yamal-Nenets Autonomous District in 2012."* (Salekhard, 2013)
5. *Arctic Monitoring and Assessment Programme (AMAP): Black cardon and ozone as Arctic climate forcers* (Oslo, 2015)
6. N.L. Mamaeva, S.I. Kvashnina, S.A. Petrov, Oil & Gas, **4**, 119 - 122 (2011)
7. N.L. Mamaeva, S.A. Petrov, Oil & Gas, 99-104 (2015)
8. V.P. Melnikov, D.S. Drozdov, G.V. Malkova, Geology and Exploration, **6**, 75-82 (2009)
9. N.L. Mamaeva, S.A. Petrov, Tyumen state university herald, **6**, 122 - 127 (2013)
10. *Arctic Monitoring and Assessment Programme (AMAP): Snow, Water, Ice and Permafrost in the Arctic (SWIPA): Climate Change and the Cryosphere* (Oslo, 2011)
11. E.D. Ershov, *"Geocryology USSR. Western Siberia"* (Nedra, Moscow, 1989)
12. *Arctic Monitoring and Assessment Programme (AMAP): Human Health in the Arctic* (Oslo, 2009)
13. S.I. Larin, *"Atlas of the Yamalo-Nenets Autonomous District, Yamalo-Nenets District Administration"* (Tyumen, 2004)
14. A.N. Agryzin, *"Statistical Yearbook (1990 - 2003)", Part VII (II)* (Tyumen, 2004)
15. A.N. Agryzin, *"Environmental Protection in the Tyumen Region (1998 - 2003)"* (Tyumen, 2004)
16. A.N. Agryzin, *"Environmental Protection in the Tyumen Region (2002 - 2006)"* (Tyumen, 2007)
17. *"Method of calculation of maximum permissible discharges (MPD) substances in water bodies with wastewater,"* (VNIIVO, Kharkiv, 1990)

18. *"Guidelines on the calculation, regulation and control of polluting emissions of substances in atmospheric air (ext. And Revised)"* (JSC "SRI Atmosphere», St. Petersburg, 2012)
19. *OND - 86. "Methods for calculating concentrations in the air of harmful substances contained in industrial emissions"* (Gidrometeoizdat, Leningrad, 1987)
20. Information on: http://eco-c.ru/products/emission.
21. A.A. Shumanova, *"Climatological Handbook of the USSR", Issue 17, meteorological data for some years, Part I: Temperature, Part II: Precipitation, Part III: Snow cov*er, (Leningrad Hydrometeorological Publishing House, Leningrad, 1956)
22. Information on: http://meteo.infospace.ru
23. A.V. Pavlov, *"Monitoring of permafrost zone"*, (Acad. Publishing house "Geo", Novosibirsk, 2008)
24. *Arctic Monitoring and Assessment Programme (AMAP): Human Health in the Arctic* (Oslo, 2009)
25. *GN 2.1.5.1315-03 GN 2.1.6.2309-07*
26. Information on: http://base.consultant.ru/cons/cgi/online.cgi?req=doc;base=LAW;n=98704
27. A.I. Sysina *"List and codes of materials, polluting atmospheric air," Research Institute for Atmospheric Air Protection (SRI Atmosphere), Research Institute of human and environmental ecology* (St. Petersburg, SRI Atmosphere of 2012)
28. A.V. Pavlov, Publishing House of the University Book, **2**, 274-281 (2011)

MATEC Web of Conferences 73, 06001 (2016) DOI: 10.1051/matecconf/20167306001
TPACEE-2016

Location Criteria Relevant for Sustainability of Social Housing Model

Nataša Petković-Grozdanović[1,*], *Branislava* Stoiljković[1] *and Mikhail* Shubenkov[2]

[1]University of Niš, Univerzitetski trg 2, Niš, 18000, Serbia
[2]Moscow Institute of Architecture (State Academy), Rozhdestvenka, 11, 107031, Moscow, Russia

Abstract. Social housing models, which had began to develop during the last century, for their only objective had a need to overcome the housing problems of socially vulnerable categories. However, numerous studies have shown that these social categories, because of their low social status, are highly susceptible to various psychological and sociological problems. On the other hand a low level of quality, which was common for social housing dwellings, has further aggravated these problems by initiating trouble behaviours among tenants, affecting social exclusion and segregation. Contemporary social housing models are therefore conceptualized in a way to provide a positive psycho-sociological impact on their tenants. Therefore the planning approach in social housing should be such to: support important functions in daily life routines; promote tolerance and cooperation; influence on a sense of social order and belonging; affect the socialization of the tenant and their integration into the wider community; and improve social cohesion. Analysis of the influential location parameters of immediate and wider social housing environment strive to define the ones relevant to the life quality of social housing tenants and therefore influence on the sustainability of social housing model.

1 Introduction

The long-standing building tradition of the social housing complex, conceived on the idea of mass industrial construction, had resulted in spatial and social segregation, which further generated the various problems within these complexes. Numerous studies, some of which date back at the end of the last century, launched the question of the influence that complexes' of social housing and theirs immediate surrounding has on the level of social exclusion within their tenants (Newman 1970, Jacobs, 1961; Wilson, 1996; Munch, 2009). Some authors believe that precise urban and architectural approach in the construction of social housing complex had originated high level of different behaviour problems within these complexes (Jacobs, 1961; Newman, 1970; Petrović, 2004; Bolt, 2009), such as: early school abandoning, alcohol and drugs abuse, bullying and etc. Such housing environment further affected the population of these complexes, which resulted with a complete inefficiency of these housing models.

*Corresponding author : natasapetkovic83@gmail.com

These researches have shown that social housing cannot be reduced only to the provision of housing as a plane shelter. In the contemporary social housing models, the manner of their spatial positioning and organization; arrangement of the specific housing structures within residential area, and the quality of facilities intended for social housing therefore is increasingly associated, not just with urban or architectural planning, but also with social planning. In addition to achieve adequate housing and living standard, new social housing models tend to provide favourable sociological climate to their tenants as well. Socialization of the tenants by initiating social interactions within tenants of the social housing complex and its wider surrounding influences on the sense of belonging and safety of the tenants and encourage their integration within the neighbourhood.

Regarding this, it's not surprising that the basis of contemporary social housing models is relied on the manner how the urban planning of social housing complexes is managed, both from the aspects of positioning the social housing structures within the urban core and the belonging residential areas, and terms of planning and organizing such complex within their micro environment. However in urban planning approach of the new social housing models the emphasis must not be placed only on the quality of housing within those dwellings; on contrary the overall housing quality of social housing models has to be viewed as well through the sociological impact that these complexes have on the individual wellbeing and on society in general.

2 Socio-spatial short comings evidented in the existing social housing estates

The analysis of the examples of international practice in the construction of social housing estates had shown some common problems aroused during their exploitation time, which led to the conclusion of the certain common shortcomings:

• Locations of the social housing estates often were not understand in the context of existing urban matrix and its relationship with the surrounding urban fabric. As a consequence, large social housing estates were built in remote locations. Conceived like isolated urban islands, the position of these complexes had caused the creation of social segregation, which continued to generate a numerous problems within these residential areas (Jacobs, 1961; Newman 1970).

• The majority of social housing complexes build in the twentieth century was conceived and designed in a way to provide a cheap dormitory, all based on the idea of mass-industrial production (Jacobs, 1961; Newman 1970). The emphasis was on the quantity – building as much housing units under most economically favourable conditions. In such circumstances a very small amount of attention were paid on the housing quality and the individual needs of future tenants.

• Architectural uniformity arising from the mass-industrial construction was one of the main features of social housing complex (Jacobs, 1961; Newman 1970).

• High degree of social segregation - Different forms of problematic behaviour within these neighbourhoods (Jacobs, 1961; Newman 1970; Milić, 2006; Munch, 2009), have influenced on the sociological isolation of tenants of these areas.

Social housing models with detected shortcomings mentioned above, in the practice had given extremely negative results. Conceived on such principles, designed entirely for the use of social housing, became extremely unpopular places to live. As the residents of these areas have not been able to meet their basic housing needs, the high percentage of the apartments in these complexes was abandoned by their tenants. A large percentage of unoccupied units had influenced not only the facilities within social housing complexes but also had resulted in deterioration of both near and wide environmental surrounding.

3 Criteria for evaluation socio-spatial quality

Identified shortcoming have allowed more detailed assessment of the needs that social housing models should meet and also the necessary characteristics of the area within social housing complexes are planning to be located. In terms of immediate living environment in order to ensure higher housing and living quality, better integration of the social housing tenants within the housing areas; and the achievement of higher spatial and social cohesion is necessary in order to overcome the problems and shortcoming detected within existing social housing complexes.

The aim of this study, on the basis of observed shortcoming, is to define relevant criteria for the quality of social housing and the sustainability of social housing models, through the analysis of a large number of influential factors prevailing in the immediate and wider environment.

The supposed measurable aspects that largely influence the quality of housing and living environment and thus the value and the sustainability of social housing models are:

- Position of the social housing complex to the urban city core;
- Supporting facilities within residential area;
- Common spaces intended for use by the tenants of the social housing complexes;
- Heterogeneity of the social housing complex and its wider environmental surrounding.

As the problems, detected on the existing social housing complexes, show a close correlation between physical (position, manner of the planning and design of the complex social housing) and sociological (problematic forms of behaviour, levels of social exclusion, the degree of social segregation) aspects, this study analysis each criterion in comparison to both aspects, and strive to determinate the link between them.

4 Influential factors

The paper further analyses and evaluates the qualitative characteristics of alternative social housing solutions for individual spatial criteria.

4.1 Position of the social housing complex to the urban city core

The proximity to the urban city core, the distance from business, commercial, cultural and educational facilities, certainly affects the attractiveness of some location. However residential satisfaction surveys carried on the numerous residential areas show that this is not a crucial factor (Kauko, 2006; Geraedts & Voordt, 2007; radovi sa SCI). When it comes to residential satisfaction with environmental surrounding it mainly depends on the subjective attitude of potential users.

Conducted researches have shown that the proximity of residential area to the central city core is not the determining factor when it comes to the quality of the housing complex location (Kauko, 2006; Geraedts & Voordt, 2007). The research carried on in Finland (Kauko, 2006) had shown that within residential areas that were well infrastructural equipped and without difficult access, the residents have far more valued the quality of the living environment and the social climate.

However, while the above research shows that high frequency and the availability of the public transport are to a lesser extent when it comes to the choice of the housing environment, and that it is not of a crucial importance when it comes to successful of the housing complex realization, the research conducted on mono-centrically residential areas (Kauko, 2006) had show that the proximity to the urban city core, the availability of main traffic ruts and the frequency of the public transportation, were registered as a very important parameters, almost as important as the quality of living space.

Residential areas which are remote from the urban city core, however, do not necessarily suffer from the lack of housing quality. The example of the social housing complex Columbia Point in Boston, USA, shows that even spatially isolated and cut off from the rest of the urban core, have initiated even a greater sense of belonging among their tenants (Thebaud et al., 2008). Tenants have been faced to one another, which resulted in groups formed for cleaning, maintaining and interior decorating of the buildings, as well as their immediate environment. The positive affects was also noticed among parents who helped each other in babysitting.

Selection of the location directly dictates the initial cost of construction. As the economic sustainability is a basic prerequisite for the development of social housing models, the position of the complex intended for social housing is not possible to be at the "attractive" city locations. Primarily, for the location of such complex should be chosen those sites that could meet the basic tenants' needs; to reach the necessary facilities within a reasonable period of time, having in mind that among the tenants of social housing complexes very likely a lot of them who don't own a car. Therefore it is highly recommended that social housing complexes should be planned within the areas of central economic value (Fishman, 2004; Milic, 2006; NRIF, 2015). If this approach is not possible and requires the construction of social housing complex in a remote location, it is necessary to be positioned in the part with the higher service standard (Milić, 2006). This mainly refers to the position of social housing complex in suburban locations, which is very often the case in the current social housing models in Serbia. In such circumstances it is desirable to locate these complexes in central part of the area, or close to the amenities and service facilities.

4.2 Supporting facilities within residential area

Analysis of the previous criterion showed that most favourable areas are the ones with the higher quality of living environment and within walking distance to the service facilities, shops, workplace or social contents. But this factor precisely is the crucial in terms of economic viability of the social housing models. From the economic aspect, planning the social housing complex within a highly equipped residential area is unsustainable. In general, for housing to be affordable, it cannot count on high functional location standard, but rather the minimum; concerning the supply, education, child and health care and public transportation.

Mixed-use areas, which in their settings include some production or storage facilities, certain types of light industry or office spaces, are more suitable location in comparison to undeveloped suburbs (Milić, 2006; DSCI, 2015). Social housing can be easily integrated into those areas, because there are already a large variety of spatial and physical structures. In addition, introduction of the housing in these areas has the influence on their live cycle, because the hour cycles are complementary. Thus, the area maintains a certain level of necessary activities for 24 hours. Higher frequency during the day, supported from the business services, reduces the feeling of isolation among tenants and affects the sense of security; while the housing gives a human dimension to the area and provides an additional frequency after the working hours (DSCI, 2015). However it is necessary to combine such activities that are compatible with housing.

The potential working place for the unemployed and their relative proximity to the place of residence, reduced transportation costs, higher social inclusion of the residents by using the location support amenities are only some of the aspects that might have positive effect on the sociological component of social housing models.

In addition, the planning of social housing within mixed-use areas can provide a completely new dimension of the model sustainability, as it allows the use of some innovative approaches, such as: extensions of non-housing structures with additional housing units or conversion of industrial or warehouse buildings. Beside the economic and environmental sustainability, this approach has one additional advantage; it keeps the "identity" of space which has a positive influence on the sense of belonging.

4.3 Common spaces intended to be used by tenants of the social housing complexes

Due to economic circumstances the social housing units are very reduces in their size, therefore the existence of common spaces which could be used by the tenant, affects the housing quality as provides additional space that can be used by the tenants for recreation, interaction, playing and etc. Barring in mind obvious problems of social exclusion among the tenants of social housing complexes, in addition to functions mentioned above, the common areas should be organized in a way to encourage social interaction both between the tenants of the social housing complex, and with the residents of surrounding housing areas. By initiating social activities within the spaces intended for common use by the tenants, it is possible to encourage socialization and facilitate theirs' social inclusion, both in the immediate residential surrounding and in society in general (Unger & Wandersman, 1985; Forrest & Kearns, 2001; Huang, 2006; Cattell et al., 2008; Dai, 2011). The existence of playground, for example, besides influencing on the interaction among children which can encourage them to make friendship with one other, can also influencing the creation and development of the contacts among adults who are monitoring them (Huang, 2006; Faridan, 2013). Those adults already have a common topic - the children they are supervising, as well as numerous similar experiences that are sharing. Furthermore, the same parents are likely to meet at the same playgrounds often, and thus have the opportunity to develop real relationships. (Figure 1)

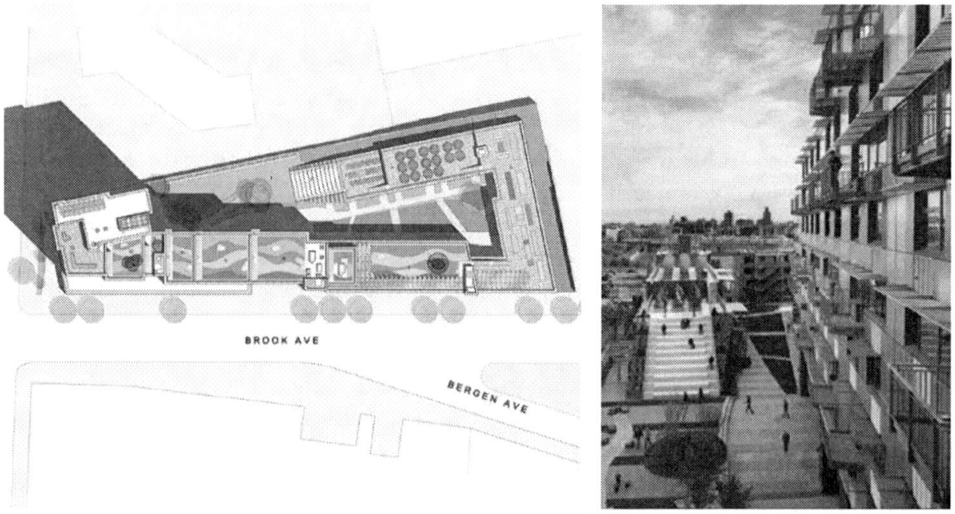

Fig. 1. Social housing complex Via Vrede Bronx, New York, USA. A dynamic garden serves as the organizing element for the community.

Secondly, for social housing complexes, due to the limited mobility of their inhabitants (elderly, people with poor health, people with disabilities, families with children and households with very low incomes) location and proximity of the common areas to the housing complex is a prerequisite for the development of social activities and social networks among the tenants (Dai, 2011; Ćetković, 2014).

The majority of free time of the beneficiaries of social housing complexes is related to the local contents. Therefore it is necessary that the facilities intended for the joint use by the tenants find within the complex of social housing, and if this is not possible, in their vicinity (Figure 1). If in the immediate vicinity of the site planned for the construction of social housing such facilities do not exist it is necessary to choose such location that is close to the spaces that can potentially be organized as common areas, with a minimum of intervention in terms of rehabilitation and reuse.

However, the physical existence and proximity to such areas are often not sufficient to conduct social interaction between the tenants. Researches show that the role of common open areas, to initiate social contacts among residents, largely depends on the physical features of the interactional areas.

The design of these areas and the existence of street furniture have been identified as very important elements for the quality of common open areas in order to achieve social interaction. Adequately planned and equipped common areas are attractive to users and affect the longer staying, and thus increase the possibility to develop social activities among tenants (Gehl, 1986; Carr et al., 1992). The provision of greenery in common open areas increases the opportunity for social activities (Kweon et al., 1998; Sullivan et al., 2004), and enhance social bonding among residents (Coley et al., 1997; Skjaeveland and Garling, 1997; Huang 2006). Interactional spaces should be welcoming and accessible to everyone. That means that it has to be in a place they can easily get to, and that they feel welcome when they get there. The characteristics of good spaces for interaction are that they provide reasons for people to go there; reasons to linger once they're there; safety and security; and a welcome and accessibility to all. All of these are, to a greater or lesser extent, dependent on design.

4.4 Heterogeneity of the social housing complex and its wider environmental surrounding

Although it is recommended that social housing complexes should be low or maximum medium density housing, construction of homogeneous complexes of social housing, both in structural and shaping sense, is abandoning. The recommendations concerning urban planning and design of the social housing complexes highlight the need for mixing different structure types, using diversity of housing concepts and including various physical housing structures (Jacobs, 1961; Newman, 1970; Milić, 2006).

The variety of housing structures, with different density area mixture, allow the development of a different housing forms and shapes, which results in a dynamic of the residential area. This approach aims to avoid the monotony that is considered to be one of the main shortcomings of the existing social housing complex.

Whenever it's possible it should strive toward the formation of individual housing units that are directly connected with the surrounding environment. Semi-detached houses, row houses or housing units in the form of duplexes with direct connection to the ground are a good solution, as they provide the necessary individuality to the households; and on the other hand locate the residents to be in direct correlation with the surroundings.

The individual housing units should be designed to have the small front yard, while the back yard should be larger in size and used as an open area that in summer should take a part of the everyday activities (Newman, 1970).

If this approach is not possible, it is desirable to combine family and low-rise and low-density multifamily housing structures (Jacobs, 1961; Newman, 1970; Milić, 2006; Radovic, 2014). Higher dwellings and higher density areas can reduce the positive aspects of this approach, which main aim is to prevent social exclusion of the tenants. The low-rise housing is preferably in terms of the better tenant's integration in the surrounding environment and the higher degree of social cohesion. The closeness to the surroundings affects the residents of these facilities to feel less isolated.

The possibility for parents to monitor their children, who are playing outside, strait from their homes, gives a greater sense of security. Relating older persons to the events in the immediate surroundings of their living space, on the street or in the neighbourhood, reduces their isolation.

All this have a positive effect on the tenants' satisfaction, encourage a sense of community within the residential area and contributes to better social climate in the area. In addition to the heterogeneity of the social housing complex itself, a certain level of heterogeneity within the wider housing areas is also mandatory (Figure 2).

Groups of multiple dwellings, intended for social housing, located within the complexes only intended for this type of housing, has proven to be extremely inefficient. In order to prevent spatial segregation the very important part of the planning is the proper positioning or rather the integrating of social housing complex in the existing urban matrix. Fragmental integration of various singular locations and combination of small group of social housing with other types of housing within residential areas of slightly higher standards are more favourable in terms of space and social segregation (Milić, 2006). (Figure 2)

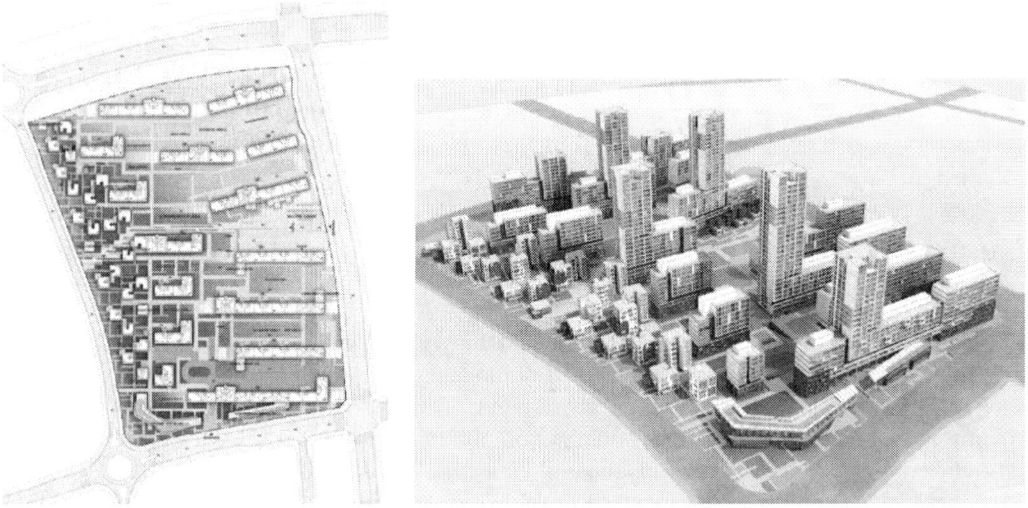

Fig. 2. Senjong 2-2 M2 housing complex Bronx, New York, USA. A multi-dimensional housing scheme with open organization of various types of building units that celebrates the diversity of residents.

Conclusions

The social housing models, in addition to their main demand to provide decent housing conditions, should also have positive effects on the sense of belonging among their tenants and social inclusion, through the stimulation of social competence and participation. These models should also provide support for the important everyday functions. By maintaining the tolerance, cooperation, sense of social order and belonging, these models should influence positively of the social cohesion in the social housing complexes and wider residential surrounding.

As tenants of these complexes are various socially vulnerable groups, which are more likely to spent their free time within the local residential area, the main focus of the social housing models' should be to find adequate spatial housing solutions that would have the potential to contribute to the social inclusion of the tenants and the level of social cohesion in the housing community.

In order to prevent the stigmatization of those areas, future complexes of social housing should meet the following (whether if it is a new dwelling construction or the revitalization of old or abandoned complex). 1) They need to be adequately positioned within the existing urban matrix and in the immediate vicinity of the necessary service facilities. If this is not the case, during the process of social housing complex planning, it is essential to anticipate the introduction of new service facilities, either by integrating them into social housing dwellings, or adapting the surrounding space for that purpose (removal of some existing buildings or its parts of in order to create additional service facilities space). 2) The wider housing area of social housing complexes should be heterogeneous in structure and form. This can be achieved trough: fragmental integrating of social housing dwellings in existing residential areas or areas of mixed-use; or revitalization of existing social housing complex by introducing different form and structure in a way to affect the homogeneity of the existing residential areas.

Better living conditions in social housing complexes achieved with the increase of quality of the immediate housing surrounding contribute to the better integration of their tenants and the increase of social cohesion, which have a direct impact on the quality of the area but also brings a lot of additional benefits to the society as a whole.

References

1. J. Jacobs, *The Death and Life of Great American Cities* (Random House, New York, 1961)
2. T. Kauko, Journal of Housing and Built Environment, **21**, 159–176 (2006)
3. O. Newman, *Defensible space* (MacMillan, New York, 1970)
4. W. Wilson, S. Fainstein, S. Cambell, *"The truly Disadvantaged: The hidden agenda". Reading the urban Theory* (Blackwell Publishier, Oxford, 1996)
5. G. Bolt, Journal of Housing and the Built Environ, **24**, 397–405 (2009)
6. K. Cabernite, *Community service through architecture: Social housing with identity* (Cigagna University of South Florida, 2009)
7. P. R. Geraedts, T. Voordt, Proceedings of International conference Sustainable urban areas, Roterdam, **4**, 23-31 (2007)
8. F. Thebaud, J. Haffner, E. Guerra, *Privately-Funded Public Housing Redevelopment: A Study of the Transformation of Columbia Point* (Institute for International Urban Development Cambridge, Masachusets, USA, 2008)
9. V. Milić, *Urbanistički aspekti socijalnog stanovanja*, (Arhitektonski fakultet univerziteta u Beogradu, Beograd, 2006)
10. S.Munch, Journal of the Housing and the Built Environ, **24**, 441–455 (2009)
11. G. Radovic, V. Murgul, N.I. Vatin, Applied Mechanics and Materials, **584-586**, 564-569 (2014)
12. G. Radovic, V. Murgul, N. Vatin, Applied Mechanics and Materials, **641-642**, 634-638 (2014)
13. J. Ćetković, M. Knežević, M. Žarković, V. Murgul, N. Vatin, Applied Mechanics and Materials, **638-640**, 2465-2470 (2014)

MATEC Web of Conferences 73, 06002 (2016) DOI: 10.1051/matecconf/20167306002
TPACEE-2016

Energy Efficiency in Architecture – a Technical Obstacle or a Chance to Find new Forms?

Nikolai Shchepetkov[1,*]

[1]Moscow Architectural Institute (State Academy), 107031 Rozhdestvenka st, 11, Moscow, Russian Federation

Abstract. The article exposes innovative materials and structures for improving the energy efficiency of lightning in architecture. The problem of architectural energy efficiency and its part – energy saving illumination of rooms and urban environment is very actual in developing countries. This problem can be successfully solved only by a complex of architectural, constructional, electrical and technological measures which will lead to new forms in architecture and lightning design. This article briefly reviews the benefits of light pipes technologies to enhance natural illumination of deep plan buildings.

1 Introduction

The point of architecture as a "foremother" of all kinds of art is to create an art image environment, if its objects could be ranked as expressive artworks, not only as a result of construction. In this case we are talking about architectural forms which are created by our visual impressions (there are also other art forms: music, literature, painting, scenic).

And the vision is impossible without light. Any light, natural or artificial, has its absolute autonomy, because it is the existence form of matter which consists of electromagnetic waves. Light is primary in the environment and is necessary for human to visually perceive objects and spaces, including architecture, which becomes the secondary source of light in this process, reflecting the light from optic primary sources like sun, sky, fire, electric lights.

We can state that exactly the light turns architecture into a figurative art if the dimensional form itself meets the requirements of harmony with that particularly light.

This objective fact is almost lost in the theory of architecture, architectural forms are usually analyzed without considering the lightning conditions, which could illusory change them beyond recognition compared with a certain archetype ("genetically" – with an average daylight).

Its obviousness is exposed in the rapid development of artificial lighting and light created by the expanding kaleidoscope of light forms in the exterior and interior (when the physical parameters of the architectural form are constant), which differ from the traditional day images in its novelty and emotion enhanced by the principles of visual perception in conditions of adaptation to dark [1, 2].

* Corresponding author : n_shchepetkov@inbox.ru

2 The interconnection between the energy-saving architectural form and the light

Created by the light of primary sources, visual images of architecture during the day are free, because global and inexhaustible natural light in the first half of day was given to us by nature. A man-made artificial light, consuming amount of which is increasing through time and space, costs money. And we often start to calculate those money without considering, for example, the connection between natural and electric light and possibilities for its joint use during the day in an increasing amount of windowless, underground and dimly lit rooms in modern architecture. This is a fundamental problem of methodology of architecture design, the interconnection between the architectural form and the light, hard perceptible formative (literally, not in a figurative sense) impact of the light, primarily natural, to the architecture. To this we must add the cost of not less important and closely related issue of heat-resistance of buildings, the wider the building, the more it is resistant to climate impact, but indoor space requires artificial lighting during the day, which is not rational; the more glass area in the enclosing surfaces of the room to ensure daylight, the more troubles and costs for heating and air conditioning ("glass mania" is still in fashion [3, 4]), etc.

Modern technologies of a "smart house" architecture are oriented mainly on expensive technical "stuffing" with guided engineering systems and effective heat and noise protective materials of walls, and also autonomous systems of alternative, environmental-friendly energy supply. Often the problem of choosing energy efficient architectural form is neglected, with the required minimum of façade and roof glazing, also considering its orientation to cardinals and the wind rose, with a minimum number of façade surface angles, with an efficient ratio of external walling area to the heated volume enclosed in it, etc.

This is the essence of a profession. Historical experience of our ancestors is forgotten, they used to build compact and beautiful buildings in hard climate conditions and they were not really involved into searching for a fashionable, extravagant form. These peculiarities are most obviously seen in the student projects and then they stick in the minds and practical work. Composite searching of expressive forms in studying is certainly priority, but in the case of their analytical support on the above issues, for example, in the essays and the explanatory notes, which is clearly not enough.

3 Natural, artificial and mixed lighting in the building

Thus, energy efficiency in architecture is a reasonable, calculated synthesis of architectural form, building materials, engineering and engineering systems. This triad is also related to the lighting of buildings - natural, artificial and mixed.

It is unreasonable to consider them separately. For example, today in the information field and in practice there is a forcing idea of indispensability of LED lighting as an energy-saving panacea. In fact it is only a part, promising though, of the general energy-saving strategy for artificial lighting, which shouldn't be dwelled on so much. The problem is much more complex, multifaceted, and, in the case of complex solutions of certain actions, promises great economic benefits.

Where and how can we save energy on lighting without breaking the requirements? According to data published at the beginning of the 21st century, developed countries spent 35% of consumed electricity for artificial interior lighting, and only about 1.5% for outdoor lightning. Perhaps, this ratio is different today due to the rapid development of architectural lighting in the cities, but it is unlikely to have changed fundamentally.

Thus, lightning design of interior in the first place allows us to save money, and the work in this direction is in progress, the release of new and various LED lamps and luminaires is growing rapidly, and the most common and widely produced devices with fluorescent last generation lamps (T 5) for industrial and public buildings have sufficiently high performance.

Only the integration of modern lighting control systems significantly falls behind, which are also an important tool for energy savings. For example, hundreds of thousands of fluorescent lamps illuminate staircases of multi-family housing around the clock, when there is no one there and no light required, and only a small automatic motion sensor is needed. Commercial developers in 2015 launched a fairly simple and affordable lighting control device of the new generation – Celian Smart Switch with traditional switch functions and IR motion sensor. Built-in sensor automatically switches the light off 10 minutes after people leave the room. Widespread replacement of old switches with new ones in all the buildings at once would give a tangible effect.

4 Light pipes

One of the most promising ways of energy saving in the daytime, especially in regions with a predominance of sunny weather, is the use of tubular hollow light pipes in the big wide rooms, in the underground and other areas with a lack or absence of natural light (parking lots, warehouses, transport platforms, underpasses etc.). Light pipes are physical structures used for transporting or distributing natural light for the purpose of illumination.

Current light transport technologies include the following (Figure 1) [5]:

1. Lenses,
2. Hollow prismatic pipes,
3. Light rods,
4. Mirrored light pipes
5. Fibre optics

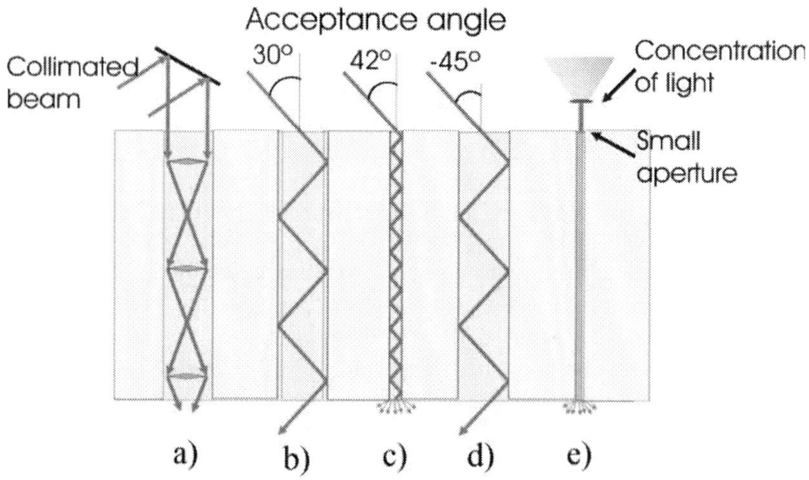

Fig. 1. Different light pipe technologies. a) Lenses; b) Hollow prismatic pipes; c) light rods; d) Mirrored light pipes and e) Fibre optics.

These pipes (Figure 2), widely spread in Italy, England, USA, Canada, Germany, Austria and other developed countries, have a number of light-, sound- and heat engineering and operational advantages over traditional systems of natural illumination of buildings [6] .

Fig.2. Integrated light pipe «Heliobus» in the school hall in the town of Schiers (Switzerland).

Captured by the receiving optical device outside the building (on the roof, the facade), natural light is transferred by means of mirrored inside pipes to different dark rooms almost lossless for tens of meters vertically and horizontally. Along the way, light can be separated if necessary, concentrated, dissipated, it can change direction and spectrum due to various technical devices. This light in the room is natural, it shows all the changes taking place with an open-air light, which is especially interesting for the light atmosphere in the interior in a cloudy-bright day. The design of the hollow light pipe can contain the installation of artificial light, which eliminates the lack of natural light on a cloudy winter's day or any day at dusk and provide lighting at night (so-called integrated system). If the crowning part of such pipe rises, for example, above the ground in the exterior pedestrian area, in the evening it looks like a light sculpture, locally illuminating the adjacent land area (the area of Potsdamer Platz in Berlin). Figure 3.

Fig.3. Light pipes located in Potsdam Platz Square in Berlin [8].

Light pipes located in Potsdamer Platz Square in Berlin, which are partly illuminating the platform of underground hall in the afternoon, and used as light sculptures in the evening.

In Russian projects this ecologically perfect system, unfortunately, almost does not occur, because everybody relies on solar panels which convert the energy of daytime radiation into electricity, and then in the light, for example. The efficiency of such double-conversion systems is low. It is better to sacrifice a minimum of internal volume for the light pipes, as we do it for ventilation systems, in order to deliver free natural light to the room in the original. Of course, the integration of this method will have some influence on the space-planning design of buildings.

5 Energy efficient light design

In some cases with a large built-up area with one-, two- and three-storey buildings, properly designed light well can be energy efficient as a medium element of space-planning design between traditional courtyards and atriums.

In any case, the main purpose of these designs is energy savings on lighting of rooms in the daytime, provided by architectural form. These designs can be original in forms too.

For illumination of underground spaces, method of translucent exploited coatings in the ground level is also effective. Sidewalk, crossing the square in front of the railway station in the city of Tours in France, at night lights with "free" light: it is made of glass bricks for daylight of underground parking and parking artificial lighting at night creates a spectacular light alley on the station area for free (Figure 4).

Fig. 4. Glowing at night pavement of glass blocks located above the underground parking at the railway station square in Tours, France.

An urban environment has its reserves and energy saving methods. The main trick is to "address" the electric lighting, which creates in the surrounding darkness optically finite light environment of the city as a system of discrete heterogeneous lightened spaces (within luminaire photometric bodies), which is fundamentally different from the homogeneous global day lightened space (from horizon to horizon through the zenith).

Targeted lighting of objects and spaces in exterior environment is the main reserve of energy savings in towns and suburban areas: the light in proper amounts needed only in the "habitable" pedestrian and vehicle spaces (on road surfaces, the faces and bodies of the pedestrians) for functional purposes and certain amount is needed for aesthetic needs (on the facades of objects).

These requirements are partially reflected in existing regulations, partially known to science. They are executed by the right choice and setting of lighting devices with limited light distribution. Luminous balls and similar devices should be excluded from the range of products for outdoor lighting as they are both environmentally unfriendly and energy wasteful.

The most promising way to solve the energy efficiency problem is a complex design of lighting systems integrated in architectural forms and material objects of the urban environment that will make them the light-carrying objects and allows you to use also the "secondary", essentially free light, for example, light of windows and interiors, passing through light apertures, and light informational systems like media facades for lighting urban pedestrian spaces (Figure 5, 6).

Fig. 5. Pedestrian street in Cologne, Germany, illuminated by the light of shop windows and lighting installations and advertising.

Fig. 6. Times Square in New York, where "free" static and dynamic light creates high lighting levels (up to 600 lux on the pedestrian's face).

6 Conclusion

Address lighting is largely (but not entirely, because it is possible only at full refusal of artificial light, i.e. return to the Middle Ages and beyond) takes the edge off a rather speculative environmental problem of "light pollution of sky", at least during a snow-free period [7]. Energy-saving LED technologies which are increasingly used in various groups of fixed lighting installations, in the utilitarian street, facade and landscape, architectural and information and advertising lighting, not to mention the temporary (holidays, seasonal, fair, exhibition) lighting systems, also making an increasingly important contribution to solve this urgent problem.

Thus, energy efficiency in architecture is not limited only to energy-saving lighting systems, or even to electrical measures at all. The answer is presented in the table below: energy-efficient lighting design. Energy-efficient lighting design begins and develops in the architectural environment and is organically linked with both its dimensional and functional structure and its imaginative qualities, as well as with other activities, tools and technologies (Table 1).

Table 1. Energy efficient light design.

<table>
<tr><td colspan="3">ENERGY EFFICIENT LIGHT DESIGN</td></tr>
<tr><td colspan="3">▼
▼</td></tr>
<tr><td>INDOOR
(interior lightning)</td><td></td><td>OUTDOOR
(exterior lightning)</td></tr>
<tr><td>• Rational dimensional problem solution
Maximum use (free access) of daylight in rooms, also in conditions of compact planning.
•Innovative technologies if natural light
(hollow fiber, light wells, light-refracting glasses, glasses regulated by SZU, quality, high-quality mixed lightening and etc).
• Improving standardization, methods of calculation and design of natural and artificial lighting
(scientific researches and experimental designs)</td><td>◀▶</td><td>• Integrated design of lighting systems and elements of the urban environment (general plan of the city illumination and its fragments with the predicted technical and economic parameters of the lighting).
• Providing and considering complex action of stationary lighting devices in the urban space
(outdoor light + architecture light + highlighted information)
• Improving standardization, methods of calculation and design of artificial lighting
(scientific researches and experimental designs)</td></tr>
<tr><td colspan="3">• Innovative tools and techniques of artificial lighting of spaces and objects (LEDs, etc.)
• Effective control system (multi-mode, programmable, interactive)
• High-quality operation of lighting installations (timely replacement of failed elements of lighting installations, cleaning and adjustment of lighting equipment, compliance with energy efficient lighting modes, etc.)</td></tr>
</table>

References

1. N. Zainordin, S. M. Abdullah, Z. A. Baharum, Procedia - Social and Behavioral Sciences, **36**, 51-60 (2012)

2. S. Paunović-Žarić, V. Radulović, E. Alihodžić-Jašarović, V. Murgul, Journal of Applied Engineering Science, **14(1)**, 140 – 147 (2016)
3. E. Kharkness, M. Mekhta, *Regulirovanie solnechnoy radiatsii v zdaniyakh* (Moscow, Stroyizdat. 1984)
4. N.V. Obolenskiy, *Arkhitektura i solntse* (Moscow, Stroyizdat. 1988)
5. V. Garcia Hansen, I. Edmonds, ISES Solar World Congress, **9**, 14-19 (2003)
6. Yu. B. Ayzenberga, *Spravochnaya kniga po svetotekhnike* (Znak, Moscow, 2006)
7. N. I. Shchepetkov, Svetotekhnika, **5-6**, 43-46 (1993)
8. Information on http://www.heliobus.com

MATEC Web of Conferences 73, 06003 (2016) DOI: 10.1051/matecconf/20167306003
TPACEE-2016

Tourist Quarter "Chinese-Baroque" of Dao Way District in Harbin City: experience, problems and perspectives of renovation.

Svetlana Levoshko[1,*], and *Igor* Kirichkov[2]

[1]The Research Institute of Architecture and Urban Planning Theory and History affiliate of Central Scientific Research and Desighn Institute of Ministry of Housing and Building in Russia, 105264 7-ya Parkovaya str. 21a, Moscow, Russia
[2]Harbin Institute of Technology. Architectural Design and Research Institute of HIT, China Post Index 150 001 Huanghe Road 73, Nangang District, Harbin City, China

Abstract. This article analyzes results of an unique experience of the Dao Wai historic district renovation project in Harbin of the 2010th. It includes an interpretation of the stylistic features of the Dao Wai building. Also, there was made a presumptive conclusion about the origins of the "Chinese Baroque", which is now famous Dao Wai, combining European order architecture and far Eastern decorative tradition. Presumptive conclusion was based on the construction area observing in 2011-2016 and on the Chinese sources. As a result of renovation, there was formed a new public space with high tourism potential. Social value and status of the Dao Wai has significantly grown. The significant cost increase of real estate and provided services is an essential consequence of the gentrification method. There are were noted increased problems of the native people forced to move from the center to the outskirts of the city. Also, this article analyzes the current stage of the second phase design of Dao Wai renovation project and the perspectives for its implementation.

1 Introduction

Today Dao Wai district (former name Fu Jia Dian) is one of the three largest touristic centers originally from the heritage Harbin city. It became this by his unique development and huge territory renovation project realization in 2010th. [1-3]

Memoirs and other nonscientific literature of early 1920th called Fu Jia Dian as a "Chinese city of the new formation", and on the plans of Harbin after 1917 Fu Jia Dian was called as a «city». However, Chinese researcher Mi Da Wei wrights, that it wasn`t separated city, but a district, which submitted to Chinese government, and this is the main specialty, that predetermined special and unique development of Fu Jia Dian [1]. Being

* Corresponding author: s_levoshko@mail.ru

independent from administrative system, it was actually the new phenomenon in Chinese urban planning organization, include functional and architectural design meaning.

The researching area is known as the "Chinese Baroque" [4-5]. It has a very limited dissemination in Harbin, no wonder that today the brand of Dao Wai exactly became this style, mainly focused on the Jing Yang Street .

The renovation development concept of Dao Wai district was made in the middle of 2000th. It was preceded by some experience of reconstruction area in the middle of 1990th, now estimated as negative. The big part of the heritage buildings on the North-West of Dao Wai was deconstructed, and on this place were created the new three-floor buildings, which don`t satisfy historical environment stylistic requirements and functional tradition.

"The Conservation and Renovation Project of Heritage Dao Wai District" was designed by Harbin Urban Design Institute group in 2007–2011 years. Project realization area estimated near to 20% of historical territory of Dao Wai. The Hao Min Ju Development Company , director of the company Mr. Wang Liren has been realizing this project for more than three years. This project was observed in professional community and won award as one of the best and capital-intensive project of 2013. The capital investments is more than 8.33 billion dollars, land area — 28.5 hectare, building area (with the floors) — 50 0000 m². [6]

2 History of Fu Jia Dian — Dao Wai district development

The situation schemes of Harbin city, founded by Russians, even from 1898 has a little spot of Chinese settlement that is called Fu Jia Dian. The translation of Fu Jia Dian from Chinese language is meaning "the trade house of the Fu`s family". Despite of this district has many other trade houses with different family clans, but this district exactly was named by this toponym. Originally it was the place for Chinese workers settlers, who build the railway bridge through the Sungari river (Song Hua Jiang). More than 10 000 of Chinese workers lived here on 1898, and then population became more than 25 000 on 1903 [1, 7]. As the rule, European cities in Asia were designed like urban systems that not include compact districts for the native people living. Harbin`s situation is the same. If the Russian part of the city has a pre-designed plan with regular and clear blocks of buildings, the planning system of Fu Jia Dian has a really spontaneous character. Even today we can see these huge contrasts between Fu Jia Dian urban planning and other districts of Harbin.

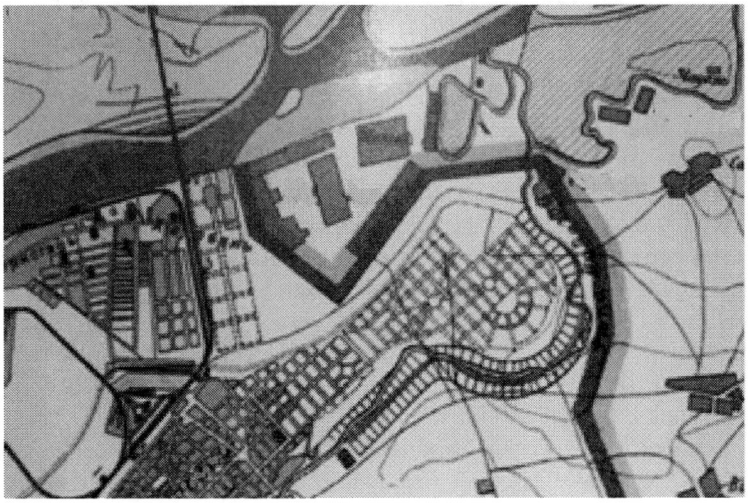

Fig. 1. Urban plan of Harbin on 1902 year.

At first sight chaotic view of Fu Jia Dian urban planning system, actually, had a traditional way of urban planning organization, where buildings and placed in them function (inside and outside) are closely related with each other. As a result, a lot of niches and recesses appear in the front of buildings intended for the trade and other services. And thin narrow lanes connected the main spaces of district area.

Fig.2. Historic view of Fu Jia Dian (historic photo).

This kind of urban planning design was not comfortable for the automobile transport, but, however, as everybody knows, the rickshaws represented the main transport of China in the early of XXth century, for which narrow and curved streets of Fu Jia Dian were not a barrier. Moreover, the streets being free from the traffic provided the close communication space for the residents: there were held a celebrations, children games, etc. The loss of the former purpose of the streets is the biggest problems of contemporary cities.

Gradually Chinese settlement became to one of the part of Harbin. Urban planning design schemes of 1946 fixed Fu Jia Dian dividing on two parts — West and East. Fu Jia Dian gradually were integrated to the Harbin city structure, that reflected to the new name — Dao Wai in 1956. Dao Wai from Chinese means the district "behind the railway", at the same time, other district Dao Li , former Russian name is Pristan, means "inside the railway".

According to the information of the memories desks in Dao Wai district facades, we can try to reestablish the function accessories of many buildings. Here, besides residential buildings were concentrated a huge number of the small and large factories, art galleries, public and commercial organizations, also including specific buildings with the "amoral" function. But the "amoral" behavior of Fu Jia Dian`s facilities didn`t prevent the work of Chinese theatres and cinematography, Chinese fashion industry was highly developed here. Russian writer Mart, W., who researched this district with ethnographic interests in early 1920th, called Fu Jia Dian as "a children of indigenous Harbin" [8]. It`s reasonable to note, that in spite of hundred years of district development, function of many buildings were saved, there are Christian Church, Synagogue, Cinematography, which work even at this day, pharmacy, banks, martial art club, some shops, etc.

Russian contemporaries noted contrasting neighborhood between Chinese "pathetic Fang Zi" and European order architecture, that contain the huge colorize significant of Fu Jia Dian architectural environment (perhaps, the word "Fang Zi" has Chinese etymology, which means 1-3 floor "house"). The photos of 1920th allow to us make sure in it.

Actually, "the new Chinese city", which located behind the East-Chinese railway, had completely exotic view, which polar to "Russian imagination of Harbin" with his factories, street advertisement, trade, rickshaws and native people. The unique cultural landscape was created exactly here. In architectural attitude, the most significant point of this landscape was a combination of European architecture and the East highly developed decorative tradition. According to Mi Da Wei historian researcher, Dao Li and Dao Wai districts like a “two different worlds”[1].

3 The "Chinese -Baroque"origins style and features

The new artistic phenomenon was born in Fu Jia Dian in 1920th, received the name in Chinese art history of 2000th "Chinese Baroque" — Local Interpretation of Europe Neoclassic. Chinese researches of architecture history called "Chinese Baroque" as a "successful combination of Chinese culturogical conception with house modeling and decorative tradition which include architectural systems and buildings technologies of the West" [9]. At this time, they note, that this style at first may be identify by saturated and sophisticated decoration of the front facades in National-Chinese color [5,10].

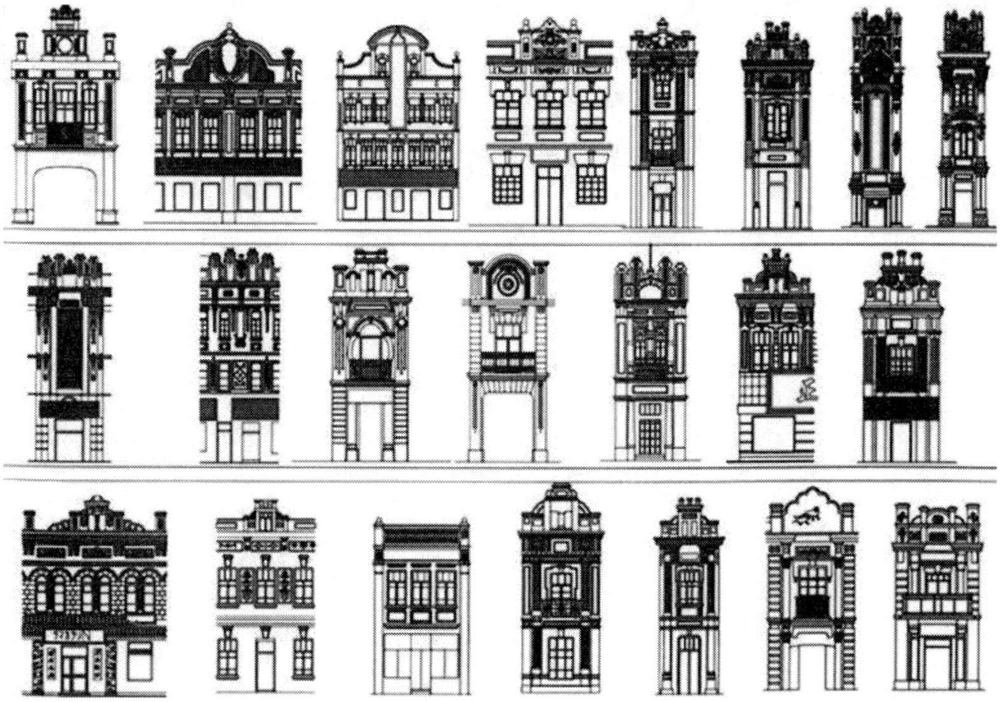

Fig.3. A combination of architectural styles - Modern, Baroque, Classic with Chinese Traditional style.

Authoritative Chinese art researcher Chang Huai Sheng (former professor of the Architectural School of Harbin Institute of Technology, dean of architectural environment design department) specifically refers this style to the Russian buildings elements (step attic, parapet balustrade, discrete fronton and other architectural elements) and to the baroque decor, which make a huge source for the different distinctive interpretations. It has a row of some buildings of Russian architects A. A. Miaskovsky and U. P. Jdanov in period of 1921 — 1933 years: The Store "Mazuura and Co", Glouberman Residence on Artilleriyskaya Street, Harbin South Manjouli Railway Association, Japan General

DOI: 10.1051/matecconf/20167306003

Consulate, Japan-Manzhouli Club, Post-Telegraph Department and some other buildings [7]. As a hypothesis, "Chinese Baroque" buildings also have an attitude to Japanese specialists, because in 1930th they have a very strong positions in Harbin architectural building market.

"Chinese Baroque" mainly focused on the Jing Yang Street, but as shown by research area, other place facilities in Harbin city were built in this style. For example, the three-span arches, which may be read as the prototype of Paifang (Far East Decorative Triumphal Arch). The arch style forming is symbolize of classical tradition and stylistic sample of Ancient Chinese Architecture.

Fig.4. An example of Chinese-Baroque style decoration (renovated part).

3.1 Dao Way urban environment . For the authorship of buildings and interferences

Buildings in Fu Jia Dian also were constructed by the European method, specifically by the method of forming quarterly front buildings abutting the firewalls and the facades to each other. Intradomestic confined spaces are typical not only for the income of European homes in XIX–XX centuries, but also for Chinese people tradition, for example, perimeter houses Si He Yuan, that have a garden in courtyard, often met in Ancient Beijing Architecture.

The most significant massive of contemporary Dao Wai with two or three floor buildings consists from the gray bricks buildings structure in rationalistic "brick style", that characterize laconic decoration and other elements of architectural ceramics.

T

Fig.5. The reconstructed quarter of Dao Wai on 2013.

Fig.6. The renovated facade. Rationalistic "brick style".

Remains some open questions, who were the authors of these building projects, these projects exist or not, from which place this experience came here. Chinese researches approved in idea, that the authors were "Chinese masters", "artels". But only one name of Chinese author is possible to know, design architect Zhuang Jun, he was the author of Harbin Communication Ministry Bank in period of 1928–1930 years. Perhaps, Dao Wai reused the Russian author's projects near 1917. There were used the typical project albums of the beginning of 1900th, which also could serve as a basis for self-construction activities. Chinese construction workers have got a huge practical experience on the building sites of Harbin — it is well known Chinese ability to fast "civilize", enterprise and assimilate European life style in all of his forms. The Russian architecture history of Harbin doesn`t have any facts about Russian architects, who works in Fu Jia Dian in 1920th, and even more previously. But since middle of 1930th, when The Buiding Department of Russian Immigrants was founded, regulated by civil engineers brother and sister Ivan Korablev and

Olga, the huge number of buildings for Fu Jia Dian were designed here (for example, Yi Tond Bank, Muslim mosque, three-floor apartment buildings and other). [11]

4 Problems of heritage preservation and tourism development

Almost for a century there was a significant degradation of Dao Wai. (ill. 8) Chinese researches noted: loss of the many valuable historical and architectural monuments, emergency condition of buildings, increased fire danger, irrational and unreasonable land use, high density of population, failure of the average living space per person, increasing share of non-working population by aging reason, inadequacy using of intradomestic space to the modern requirement, insufficient number of public buildings and open spaces for the rest, unsatisfactory transport infrastructure and functional planning structure of buildings for modern business. Therefore, district space organization and functional planning parameters of several buildings don`t satisfy to present requirements. As the result, there are urban environment destruction and economic recession. At this time situation in not renovated North-East part of Dao Wai remains the same – most part of buildings in emergency conditions, without roofs and fire hazard.

4.1 The renovation concept

The new concept of Dao Wai district renovation project means the transformation of it to one of the three biggest touristic centers of Harbin based on the districts, that were formed in the early period of city in 1900–1930. Two other touristic centers have got Chinese government attention more earlier, therefore they were reconstructed in the end of 1990th. Today this centers are "Architectural Arts Museum of Central Avenue" in Dao Li district (former Pristan) Zhong Yang Street (former Kitaiskaya Street), which opened in 2006 and center in Nan Gang district (The New Town) with dot reconstructed historical buildings, mainly located on Dong Da Zhi street (former Bolshoi Prospect) and Hong Jun street (former Vokzalniy Prospect).

Historical analysis of Dao Wai urban development shows, that many valuable qualities of the historic environment were saved. The authenticity of the buildings has not been lost. There are no high-rise buildings, which has become a real scourge in other areas of historical Harbin. As a consequence, the scale of Dao Wai buildings didn`t changed, also valuable views and perspectives, points and ways of perception. Many commercial brands were saved (famous old restaurants, stores , hotels, markets), folklore tradition (advertising, color, form of decoration, holding celebrations). All of these allow to say about integrity of Dao Wai historical environment, it is a key of the cultural landscape value.

4.2. Gentrification method and its problems

One of the modern renovation methods of dilapidated historical territory — the gentrification method ("refining"). What is new in it compared with other types of renovation, redevelopment of the dilapidated industrial and residential environment? The gentrification method has the new urban politics, cultural reassessment of inner city areas and their importance in the city structure. The gentrification method allows reusing urban sources in higher level [12]. A significant rise in cost of real estate and provided services is an inherent consequence of the implementation of this method.

The process of declined districts renovation by using gentrification method is focused on cultural landscape transformation, the real estate cost rising — in this regard on the first stage native people migrate to other areas by administrative way. This social moment is the most difficult (the conflict between the new and indigenous people, gradual substitution of the true face by another, oriented to the new needs) and subjected to public critics.

Dao Wai district is the same: the huge housing area of 500, 000 m2 becomes empty from his native people before the realization of renovation reconstruction works. The part of people — was moved to the North-West part of Dao Wai to relatively new seven-floor houses on Jing Yang street , other part — to the outlying north districts of Harbin city Jiang Bei behind the Sungari River. According to Hao Min Qu Development Company the realization of this project inhibits. People don`t want to leave their own houses, and as the result, we can see special created troubles with transport system, guards against native people, violent damages of houses and other. The cost of living necessarily will be increased, gentrification initially assumes a higher level of deferred income residents and focused on middle social class. But whether it was formed in Harbin — is the question. So, the social tension is growing up. How it showed by experience, this is negative specific of gentrification method in all of countries.

The concept of Dao Wai Renovation Project means except transformation to the touristic and business center of Harbin. At this day the renovated Dao Wai district remains low populated, many commercial buildings are empty — as a result, it wasn`t prepared to full life yet. Therefore it`s not the time to make conclusion, because the process of renovation is not full completed yet, and maybe, the social situation problem in Harbin will transform to other forms, than Europe or USA. Russian cities also have partial examples of gentrification method realization, mainly in the biggest cities like Moscow and Saint Petersburg, which also have own specifics.

5 The implementation of the project. Advantages and disadvantages

There were reconstructed the huge number of Dao Wai architectural monuments, approved many restoration technologies and the new "Baroque" decoration, renovated big massive of valuable buildings, complex decided public spaces improvement on required level [13-15]. The Dao Wai photographs of 1930s extensive flooding occurred in Harbin.

Fig.7. The courtyard as new public space.

There is also new "historicist", metaphorical, symbolic and figurative architecture. For example, main "entrance gate" to the housing area, which may be modeling on the associative level as a Paifang form, designed as sharply stylized three "buildings" facade planes. This modern architecture, intellectually and spiritually reached by the new meanings of historical forms, here is appropriate. There were applied covered passages, which connect parallel streets produce the necessary functional and figurative effect in this environmen.

Fig. 8. The old "new methods" of renovation.

There were realized some business facilities projects, strictly subordinated zoning and not exceeding the average height of the historical buildings. Structure of some new buildings is very ultra-modern. For example, Concert Hall on the main square of Dao Wai has unique glass covering structure with span near 30 meters. This building became "the visit card" of this district in line with other architectural monuments.

Fig.8. Concert Hall on the main square of Dao Wai.

The principal rejection of the high-rise construction in historical environment of the city is the new word in Harbin urban design politic. Maybe, the negative experience of the urban regulation non-availability in other historical districts is shown the consequences of "the point of no return", for example, Nan Gang district . However, according to observations by several years, the number of high-rise buildings in the historic center of Harbin (especially along the former railway lines) sharply increases. In spite of theatrical effect and "remakes of buildings", "Chinese Baroque" has already achieved originality of local architecture. And in the analyzed region was achieved identity preservation and local architecture personality by the restoration of more valuable buildings, reconstruction inner-house spaces, savings of composition specialties, scale and material. Dao Wai district is free from the transport, everything belongs to pedestrian, and this is very powerful factor of his attraction in overpopulated megalopolis.

Fig.9. Walking the streets of the renovated Tao Wai.

As shown by location observing area, the project realization in 2015 year has some engineer-technology and exploitation disadvantages: using of low-quality bricks, which have an efflorescence, unsatisfactory setting of rain tubes, leading to the masonry destruction, rented facilities don`t cope with his function. Probably, the manufacturing technologies and decoration fixing still not up to par, because in some cases even today we can see destruction. The green spaces are not enough. It seems that the metal sculptures in the renovated part of the Dao Wai are not always in perfect harmony with the urban environment . Also as the disadvantages may be referred the demolition of the large part of Dao Wai construction area, the reason for this demolition was the creation a central square with a fountain surrounded by signs of the zodiac. No one of Harbin architects can give an explanation of the link between historical Dao Wai and zodiac symbols. A significant number of buildings were demolished for the construction of large shopping facilities that surround central square.

Fig.10. Central square with a fountain in Dao Wai.

6 Perspectives of the renovation. New ideas

Preliminary design of the second phase of the Dao Wai renovation project started in March of 2016 by three architectural design organizations, including Architectural Design and Research Institute of HIT. The new concept of Dao Wai renovation project was based on bigger capital efficiency. The new project was focused on attraction of a larger number of people, by this reason, there is considered a construction of Dao Wai metro station entrance pavilions in the center or on the edge of district in future. The north-western part of Dao Wai is requested to do the most prestigious, with many boutiques, salons, premium restaurants. North-eastern part is requested to focus it on hotel business. The material level of modern residents of Harbin does not allow making a "souvenir streets" for the whole area, because art market in Harbin is not highly developed yet.

In the future, design of Dao Wai does not exclude the possibility of the development of Chinese Baroque. It is important to demonstrate that the Chinese Baroque is not "dead" style from the last century, and it has prospects of stylistic interpretation today and in the future.

Department of Urban Planning in Harbin solves the problem of mass building demolition, appeared in the 1970s-80s (5 - 8 storey residential buildings) surrounding all sides of Dao Wai and significantly reduces the historical and cultural significance of the historic environment. Today, historic district Dao Wai does not have concrete boundaries, just line of construction area. Most likely, the main street of Dao Wai will remain a traffic, and the side streets will be fully provided for pedestrians. Now huge market is situated in the north-western part of Dao Wai, it is possible to relocate this. As an effort to give a breathe for "new life" to the historic district is considered another one method, namely: using of deconstructivist forming for the creation of a new center, it will attract people by analogy with the large-scale renovation project in Shanghai, Shenzhen and Beijing (project of "Gallery Street 798"). According to Harbin architects opinion, Daniel Libeskind's deconstructionist projects demonstrate the effectiveness of this practice - the historic area becomes more interesting not only for architects, but also for ordinary people. Although, population of Harbin, according to recent data, exceeded 12 million people, Harbin continues to be referred to the third line of development. The first line includes Shanghai, Shenzhen and other cities. Development lines suggest the appropriate amount of public investment in urban infrastructure. Many modern experts emphasize the impossibility of the development of modern cities without relying on a rich historical past, which, in particular, has Harbin.

Fig.11. Renovated communication infrastructure.

7.Conclusion

This article does not consider the economic efficiency questions. However without doubt, business development, good furnished local zones systems for the rest, trade, entertainment — all of this will grow up touristic attraction of this district Dao Wai and give own urban, ecological and economical effect. From the unique architectural environment point, Dao Wai — is the worthy concurrent to other tourist centers in Harbin, making his stake on European architecture and unique culture from the Russian background. It seems, that "Chinese Baroque" district may take the focus not only for domestic tourists in future, but also may attract foreign people. The realization project, how it was remarked, on the finish stage, but even today this place makes an impressiveness by his complex approach, saturation of functional and aesthetic content of each environment fragment. There are increasing of social function and the status of Dao Wai as a part of urban space. The Dao Wai renovation project gives us the model, which may be analyzed and adapted to different countries.

References

1. M. D. Wei, D. B. L. Shi, *Hei - Ji - Liao Ji Dong Mang Tong Lan, 240* (Heilongjiang Publish Society Harbin, 2014).
2. W. Ting, R.Lifen, T. Wei, Hua Zhong Architecture, **3**, 128-132 (2011)
3. M. Huanwei, J.Li, Shanxi Architecture, **1**, 46-47 (2010)
4. A.P.Ivanova, H. P. Kradin, Vestnik TOGU, **4(35)**, 31-40 (2014)
5. W. Yang. L.Daping, L. Tong, Architectural History, **11**, 185-187 (2006)
6. Information on: http://news.my399.com/zt/content/2014-12/23/content_1400474.htm
7. C. H. Sheng, *The Art of Architecture in Harbin* (Harbin, 1989).
8. S.A. Zabiyako, A. P. Zabiyako, S. S. Levoshko, A. A. Hisamutdinow, *Russkiy Harbin: opit zhiznestroitel'stva v usloviykh dal'nevostochnogo frontira* (AmGU, Blagoveshchensk, 2015).
9. *Sino-Russian forum at the highest level on the theme of architectural styles and city design. Conference materials, 18* (Harbin, 2006).
10. C. X. Zhen, Guangdong province Architectural Design Institute, **4**, 153-156 (2010)
11. N.P. Kradin, *Harbin — Russkay Atlantida: ocherki* (Khabarovsk, 2010)
12. G. A. Ptichnikova, Sovremennay arkhitectira mira, **3**, 177-190 (2013)
13. L. X. Bo, Architectural Design and Research Institute of HIT, Harbin, **8**, 38-40 (2012)
14. L. C. Lai, Q. L.Yan, L.B. De, Sichuan Building Science, **8**, 111-113 (2011)
15. J. Hong, S.Yan, Shanxi Architecture, **4**, 12-15 (2010)

MATEC Web of Conferences 73, 06004 (2016) DOI: 10.1051/matecconf/20167306004
TPACEE-2016

The Formation of the Modern Urban Environment Using Landscape Design Methods

Liubov Shabatura [1,*], *Natalia* Bauer [1] and *Olga* Iatsevich [1]

[1]Industrial University of Tyumen, 625001 Volodarskogo str. 38, Tyumen, Russia

Abstract The authors discuss the challenges, which face the person in relation to the culture of creating a sustainable urban environment. Urban environment is seen as space of human existence, influencing his spiritual and physical health. All the specialists engaged in the urban environment, think about the culture of its formation, unfortunately, only from an institutional point of view. Differently organized human environment inevitably affects the behavior, psycho - emotional state of a person, his perception and understanding of space. The cultural heritage of society fits in the space of an artificial environment that man has created and determines human behavior. To solve the problems of the urban environment, it is necessary to apply a set of interrelated technical, social, environmental, aesthetic and other measures. In this capacity, landscape design can be seen as a set of methods to create a sustainable urban environment to ensure harmonious living. The given paper presents the results of studies of innovative materials in the field of the modern urban environment.

1 Introduction

The urban environment in contemporary anthropological versions of the ontological theories appears as a space of human existence. Being as a symbol of culture, the city is a mechanism of its transformation into real life forms, makes up a socio-psychological entity and the human personality.

The problems of interaction of man and nature increase in proportion to the acceleration of scientific and technological progress and transformations of the urban environment. During the 20th century cultural landscapes of Russian cities are rapidly degraded, representing a system in stable disequilibrium with society. This is indicated by the facts that characterize the deterioration of the domestic urban environment.

Increasing diversity of the phenomena of modern society profoundly changes the nature of the design, evealing its capabilities in shaping the environment and impacts the environment of a human being. Modern landscape design becomes the design of environment, human experience design, social context. The subject of the design (in our case the urban environment) extends to the engineering, style and lifestyle, for creation a synthesis of meanings and human values.

* Corresponding author: shabatura@tsogu.ru

The patterns and directions of development of the urban environment are interrelated with philosophy, religion, science, aesthetics, socio-economic relations, natural and climatic characteristics and culture of a particular country. The urban environment is being formed under the influence of processes taking place in society at each historical stage. At the same time, the urban environment has an impact on social well-being, worldview, attitudes, behavior and culture of the citizens [16]. Therefore, the urban environment must meet the functional, aesthetic, spiritual, cultural and human need, which contribute to the creation of favorable conditions for its vital activity.

The creation of habitat for human – creative, supportive, attractive, safe, harmonious, connecting the advantages of living in the city and the natural desire of citizens to nature, preservation of traditions and development, the introduction of new technologies – is one of the main challenges of a contemporary culture.

2 Methodological bases of formation of the urban environment

Questions of human perception of the city and the problems of mutual influence of the urban environment on a person have been considered by the thinkers long time ago. Thomas Moore in the XVI century described the ideal city in which domestic and spiritual needs of citizens are satisfied and social equality plays a dominating role. Next, Tommaso Campanella, Robert Owen, Saint-Simon, Charles Fure in their writings create an image of the "ideal" settlement. The problem of the "man in town" of the early urbanization affects Engels in "the condition of the working class in England". In his work "the City" M. Weber shows that cities in different countries and periods provide the basic needs of man in varying degrees. In the works of F. Braudel, "What is France?, L. Wirth, "Urbanization as a way of life", R. Park and E. Burgess, G. Simmel "Cities and spiritual life", F. Tennis "Community and society", J. Forrester "the Dynamics of city development", etc. the socio-cultural living environment in the city is widely considered.

At all stages of human development the question of interaction with the surrounding material and spiritual environment were discussed. The sole existence, isolated from people who are look like you, was impossible. Starting from the primitive communal to the present time man has created a variety of settlements, organizing the environment with industrial and socio-cultural relations [1].

In the work "Decline of Europe" Spengler talked about the fact that every culture, running internal creative capabilities is mortified and goes into the phase of civilization (civilization, according to Spengler, is the end of any culture, its crisis) associated with an aggressive outward expansion, materialism, and technicism, as well as urbanization [17]. Rationalistic civilization, according to Spengler, is the degradation of the higher spiritual values of culture, doomed to destruction. In this case, the self-development of culture is possible only in the context of awareness of its stakeholders in the importance of fixation procedures, measurements, and the formation of images of the external world, or the environment in our understanding. Spengler warns that the immorality of hostile mass culture remains relevant in our time.

With the name of I. Kant is the basis of the submissions about ethical and aesthetical values of the landscape is closely connected. He urged to preserve for our posterity the variety of landscapes as inexhaustible sources of emotional wealth. Thus, I. Kant considered the subject of aesthetics as "the beautiful in art," but aesthetics, due to Kant, examines not only lovely but also makes judgments about the beautiful and accordingly is the "critique of aesthetic able to judge" [9, p. 273].

G. Hegel first raised the problem of the aesthetic ideal as a theoretical problem of philosophy of culture and aesthetics. In his "Lectures on aesthetics", Hegel defined art as "the manifestation of the ideal". In the creative process, it is very important aesthetic ideal, since it forms the taste of the artist, the taste of the public [6]. In addition, the great importance of the creativity of the designer is in shaping of the urban environment, which forms the taste and culture of the citizens.

Reflecting their professionalism, both designers and architects, and all the other professionals involved in the urban environment, think about the culture of formation, unfortunately, only from an institutional point of view. Consideration of landscape design with the means of prevention of further degradation of the urban environment is associated with overcoming such negative qualities connected with functional and aesthetic disorder of the inexpressiveness. The mere filling of space, vegetation is `not enough to make the landscape with positive aesthetic qualities [3, p. 6]. . It is necessary to comprehend the possibilities of landscape design as one of the means of forming. It is worth saying, that in the "Lectures on the history of philosophy" the eminent philosopher argues,that philosophy is the basis of his age:" ...not only time comes when somebody begins to philosophize, but the folk develops a certain kind of philosophy, and this certainty, this view of thought, has the same certainty, which flows into all other historical sides of the national spirit; it is with them in the closest connection, with their Foundation". As G. Hegel says, this is "a certain image of philosophy, concurrent, therefore, with a certain way of the people among whom it acts, with their form of government, their morality, their social life, their healthy habits and comforts of life for their efforts and work in the field of art and science, with their religions ... in which he exercised his power this particular principle, and with the emergence and performance of new States, in which a higher principle finds its birth and development" [7, p. 111-112]. Mentioned "historic parts of the people's spirit" represent, in our opinion, not only the traditions of a culture; but they are important for research, according to Hegel, include public perception and "habits", and "convenience of life", and much more, i.e. a mentality.

German philosophers and theorists of culture A. Schopenhauer, F. Nietzsche, O. Spengler explained the existence of a self through a state of mind: intuition, human instincts, joy, fear, will, admiration, love, etc., at present the culture of the urban environment is being formed and does not account for essentially the informative presentation of human existence [18]. Culture of creating a sustainable urban environment is considered in dialectical unity of order and chaos: the gradual change of order and chaos.

The purpose and meaning of landscape design and of creating a sustainable urban environment is to connect the usual comforts of life and create a comfortable situation in the world corresponding to the spirit and intentions.

The attitude to the place of residence formed the basis of quality, satisfaction of higher (socio-cultural) and primary (basic) human needs and assess the degree of favorability (increasingly unfavourable) environment.

Based on the methodological principles of historicism and anthropology, urban environment should be as a continuum of coexistence, communication, interactions, relationships and needs of different nature and level, as a socio-cultural space of human life.

Life comfort, self-actualization, success or missuses, satisfaction or dissatisfaction with life depend on the environment the person creates by the deeds, actions, emotions and culture of those who organize the territory and lives in the area. The city is the way of the development and maintenance of sociality, reproduction and conservation values, the creation and distribution of intellectual and spiritual atmosphere, culture, attitudes, behavior. Residents coming in contact with the urban environment, give to it their own emotions, soul, consciousness, attitude, extracting from it what is interesting to them and valuable [12]. This environment, marked and transformed by the presence of a person is called the human environment, in relation to the city – urban environment.

Jean Baudrillard emphasized the interaction and mutual influence of a man and the surrounding space: "the Space given to him (man) as a distribution structure, and through control over the space he holds in his hands all options for the relationship..." [4, p. 31].

Philosopher A. Camus views the contact of man with nature, the surrounding world. He believes that man is helpless in connection with the nature, existing for millions of years ("I smell the grass and see the stars, but there is no knowledge on Earth which can give me the assurance that this is my world") [11, p. 160]. According to Camus, the meaning of life is deep in existence of man, and not in his external world (relationships, failures, successes). Therefore, the person is responsible for the preservation of natural landscapes and components of nature in the formation of the urban environment.

Philosophical search in the field of cultural theory, aesthetics of nature, landscape led to the formation of the subject-object system models in the understanding of beauty. Its origins are viewed in materialistic (Socrates, Aristotle, Diderot), and idealistic (Plato, Kant, Hegel, Soloviev) philosophical thought. The essence of this model is the recognition of objective foundations in aesthetic harmony of the surrounding world.

Tyumen philosopher L. N. Shabatura believes that in any activity, including project, it is necessary to preserve the "genetic code" that supports the traditions of the society. Saving the tradition of ancestors, we can pass to the next generation the nature of the cultural core of society [20]. True culture should be clear, accessible, rooted in the tradition.

Extending knowledge of the cultural values of the past contributes to the development of ecological, aesthetic, moral, historical and cultural education of future generations. From the standpoint of the structural-functional approach the urban environment can be seen as an open system (Park, square, street, promenade, Boulevard, yard, quarter, pedestrian street) and closed spaces (interiors of various functional areas). Differently organized human environment inevitably affects the behavior, psycho - emotional state of a person, his perception and understanding of space. The cultural heritage of society fits in the space of an artificial environment, created by a man.

Contemporary urban environment of the post-modernist period is not a harmonious space for better organization of work and rest of the population. The transformation of the urban environment in general is connected not only with economic aspects of production, and increasingly affects affects the ideological sphere, as it bringing up a new man, bringing certain aesthetic tastes, nature, beauty, harmonization to the surrounding space. Huge humanistic, socio-cultural potential of design was identified by prominent architects and designers in America and Europe. They are Alvar Aalto, Antoni Gaudi, Henry van de Velde, Walter Gropius, gio Ponti, George Nelson, Henry Dreyfuss, Charles Le Corbusier, Thomas Maldonado, Raymond Lowe [15]. But in the formation of the modern environment of Russian cities, this potential, in the majority, not used as one of the main factors of the culture of the urban environment.

Culture of creating a sustainable urban environment is a socio-cultural phenomenon that is formed and developed in the context of a specific social space, historical time. In the modern Russian society there are changes that we see as anti-cultural transformation. Such transformations destroy the historically established spiritual and moral foundations of society. The higher principles of the human being – kindness, justice, mercy, compassion, love of neighbour, have been replaced by the priority of the functional principles (efficiency, rationality, pragmatism).

3 The problems of the forming of the urban environment

As a result of anthropogenic loads a lot of valuable landscapes and suburbs were lost: deteriorating of water, air, chemical and mechanical composition of the soil. Highly urbanized environment affects health of people, causing physical, mental fatigue and, as a consequence, a disorder of the nervous system which cannot cope without medical intervention[13].

Contemporary urban environment contains a lot of unresolved social and cultural problems: the lack of General principles and laws governing the formation of the urban environment; growing environmental pollution; and ill-defined aesthetic; functional disorder; violation of architectural harmony with the needs of the patient; elimination of functional courtyard spaces; the destruction of the historic environment in the peripheral areas of the city; the erosion and disappearance of the spiritual filling of the urban space; the substitution of natural elements of the environment on artificially created; the impersonality of open spaces; the prevalence of utilitarianism and asceticism in shaping of the urban environment; irreversible changes in the chemical and mechanical composition of soils; replacement of courtyard spaces not authorized by the parking areas, etc.

Appeal to the problems of the urban environment allows one to realize the potential of landscape design to achieve a harmonious relationship of natural and artificial landscape components of the urban environment. The vector of design development should be focused on the transformation of the human environment, ensure compliance with the urban environment of new social relations. Considering design as a philosophical trend, one can conclude that the leading direction of a landscape designer is to create a sustainable urban environment should be the development of natural-environmental relations, the formation and maintenance of the environmental conditions, holistic systems, harmonizing the object - spatial environment and contributing to the preservation and development of culture.

The aim of landscape design consists of creating a sustainable urban environment as an organic fusion of architecture with natural components. The best works of architects (K. Tanghe, A. Aalto and F. Wright) based on the principle of holistic solutions for the building and its environment with a detailed design study of the elements of microlandscape. The landscape design provides the connection and organization of human environment, contributes to the synthesis of artificial and natural components of the environment. The necessary transition to the formation of the urban environment as a holistic space with a reassessment of its effects on the residents with the needs of the citizens in the required level of comfort and maximum preservation of natural components.

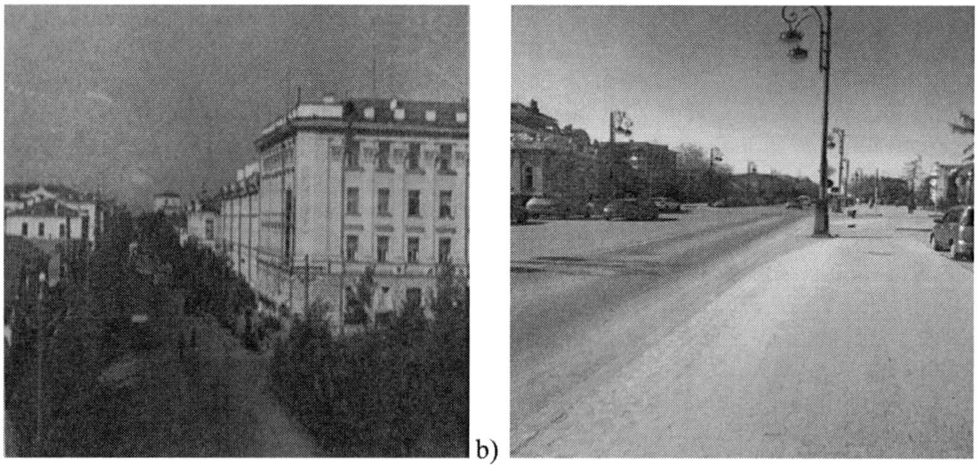

a) b)

Fig. 1. Tyumen, Respubliki Street. a) 1960; b) 2016

For human life is necessary not just a preserved nature in the urban environment but the landscape, aesthetically and structurally appropriate to the contemporary needs of society [13]. Thus, landscape design makes it possible to achieve functional order of urban spaces which is the most important quality of the environment associated with its sustainability.

To solve emerging problems and predict and prevent their rising should be taken implemented modern trends of landscape design in creating a sustainable urban environment: the transition to a holistic interpretation of urban space; efficient use of the landscape components of urban open spaces; consideration of human needs to ensure the necessary level of comfort; the achievement of functional orderliness, visual expression of landscape objects urban environment; improvement of the environment of urban spaces; creative transformation of nature; the formation of the awareness of urban space; the improvement of personal culture, the formation of a new worldview, a new co evolutionary interactions with the environment.

Landscape design projects a surrounding space as a full-fledged object of intellectual attention, the object with content and meaning. The appearance of the urban environment formed by the methods of landscape design is becoming more sensitive to the man, his relations with the outside world.

Operating with the space, its shape, color, light, bulges, depressions, rhythm, meter, contrast, nuance, landscape design affects a person, that is deeper than any word and meanings, what determines the sense of words and meanings, affects the core of being.

Landscape design, developing and enriching its own methodology, becomes the art of self-development. The task of landscape design is the individualization of the project results, the implementation in practice of modern design trends, methods of creating a sustainable urban environment.

The main task of design is not "beauty". It is an instrument, a consequence of deeper processes. The keyword for design is project. Landscape Design is not just about creating habitat. It is the construction of such a creation in principle, the making of a conscious, systematic, "deliberate", based on the rules, techniques, methods, which are realized by the Creator (designer). Landscape design – professional designing of a harmonious urban environment of self-worth in this activity. Methods of landscape design [19] is the theory in action. The main function of the methods of landscape design is the organization and management of practical transformation of the landscape object in a harmonious, aesthetically pleasing environment.

Implementing the features of the landscape design for the humanization of the environment, gives a person the chance to live, in the urban space with environmental sustainability, which implies the ability of nature to heal itself.

4 Ways and means of solving the problems of the urban environment

To solve the problems of creating a sustainable urban environment, it is necessary to apply a set of interrelated technical, social, environmental, ethical, aesthetic and other measures. At this point of view, we can use landscape design as a set of methods to create a sustainable urban environment for harmonious human life. Philosophical sense of the culture of creating a sustainable urban environment is to develop methods and principles that are commensurate with the severity of the real problems of society [3]:

The principle of environmental sustainability – the ability of nature to self-regulation through the use of "living" building materials, ensuring the conservation of nature, subject to the conditions of plant growth, the level of pollution of water, air, soil. The principle does not develop chaotic influence of the city on its natural environment.

The principle of aesthetic perfection – the light of artistic, architectural, compositional regularities of formation; organization of space with regard to their diversity, expressiveness, colors that creates visual comfort, eliminating uncomfortable faceless space through the rational use of natural ingredients.

The principle of universality is a fundamental principle of a sustainable urban environment. Creating an accessible environment for all categories of the population, taking into account the interests of all age groups of healthy people and the requirements of all categories of people with limited mobility.

The principle of economic efficiency implies an efficient use of labor and resources required to achieve a useful result.

The principle of humanization counts the needs of different socio-demographic groups of the population: it is necessary to include anthropometric requirements (height of steps and handrails of stairs, height and dimensions of tables and benches, tilting seat backs, dimensions of baby equipment); sanitary-hygienic requirements (regulate such parameters of the environment like temperature, light, noise, physical and chemical composition of the air); the emotional and psychological demands associated with the peculiarities of human perception of information (color, graphics).

The principle of continuity – provides the cultural sensitivity in the design of landscape objects in the urban environment.

The principle of multilevel vertical structure - provides multi-level landscaping with the traditional placement of green spaces in the ground level – vertical greening of roofs, terraces. To improve green planting in public spaces it is necessary to use all available plots: slopes, ramps, walls, engineering structures, terraces, etc.

The principle of diversity – takes into account the design of the sequence of exchangeable visual plans in the formation of green spaces in the city.

The principle of space integration, creating a single environment for living and recreation: the formation of private spaces (through the organization of apartment green areas – gardens); neighborhood spaces are created for residents of one yard; public spaces are formed for relaxation and communication between people regardless of place of residence.

The principle of the "flowing" interior and open space can be done using two approaches: use in interior environments of landscape inclusions (microlandscapes) becomes the condition of its high quality; design of stylobates (interpretation of the earth's surface surrounded by buildings).

The bionic principles (imitation of natural forms) – the principle of structural-functional analogies of nature and of a design object. The facilitator of the visual expression of design ideas are the objects of nature. Bionic principle in design is based on establishing the structural and functional unity of the formation process of artificial and natural systems, resulting in possible ergonomic and visual harmonization of objects and subjects of the universe.

The variety of the used principles allows to expand organizational opportunities of formation of a sustainable urban environment. Landscaping, being philosophically significant action, promotes design of an urban environment as the system, sociocultural, dynamically developing object of intellectual attention. Landscaping projects surrounding space as full-fledged object of intellectual attention, object with contents and sense [14]. The shape of an urban environment created by methods of landscaping becomes more sensitive to the person, his relations with world around. Operating with space, his form, color, light, cambers, hollows, a rhythm, meter, contrast, nuance, landscaping affects in the person that is deeper than any words and meanings, what defines with what there will be these words and meanings. Touches feeling of life. Landscaping, developing and enriching own methodology, becomes self-development art. A problem of landscaping is individualization of results of the project, introduction in practice of design of current trends, methods of formation of a steady urban environment.

5 Conclusion

The main task of design is not "beauty". It is an instrument, a consequence of deeper processes. The keyword for design – design. Landscape design is not just about creating habitat. It is the construction of such a creation in principle, the making of a conscious, systematic, "deliberate", based on the rules, techniques, methods, which are realized by the Creator (designer). Landscape design – professional designing of a harmonious urban environment of self-worth in this activity. Methods of landscape design is the theory in action. The main function of the methods of landscape design is the organization and management of practical transformation of the landscape object in a harmonious, aesthetically pleasing environment.

Implementing the features of the landscape design for the humanization of the environment, gives a person the chance to live in the urban space with environmental sustainability, which implies the ability of nature to heal itself

Spiritually rich, culturally shaped the urban environment has artistic taste, shapes the thinking, behavior, attitude, outlook and the whole culture of modern citizens. The preservation and development of culture prevents the extremity of civilization, to which both tends the world.

References

1. A.S. Akhiezer, L.B. Kogan, O.N.Yanitsky , Urbanizaciya, obshchestvo i nauchno-tekhnicheskaya revolyuciya, **2**, 44-52 (1969)
2. N.V. Bauer, L.N. Shabatura, Vestnik Ishimskogo GPI im.Ershova, **3**, 4-9 (2013)
3. N. V. Bauer, *Kul'tura formirovaniya ustojchivoj gorodskoj sredy metodami landshaftnogo dizajna: dis* (2015)
4. Z. H. Bodrijyar, *Sistema veshchej* (Rudomino, Moscow, 1999)
5. M. Veber, *Izbrannye proizvedeniya* (Progress, Moscow, 2012)
6. G.V.F. Gegel', *Lekcii po ehstetike* (Nauka, Sankt Peterburg, 1999)
7. G.V.F. Gegel', *Lekcii po istorii filosofii* (Nauka, Sankt Peterburg, 1993)
8. O.I. Genisaretskij, E.M. Bizunova, *Teoriya dizajna. Teoreticheskie i metodologicheskie issledovaniya v dizajne* (SHkoly Kul'turnyj Politiki, Moscow, 2004)
9. G. Zimmel', *Ponyatie i tragediya kul'tury, Izbrannoe* (Filosofiya kul'tury, Moscow, 1996)
10. I. Kant, *Kritika chistogo razuma* (Ehksmo, Moscow, 2007)
11. A. Kamyu, *Buntuyushchij chelovek* (Politizdat, Moscow, 1990)
12. V.A. Nefedov, *Gorodskoj landshaftnyj dizajn* (Lyubavich, Sankt Peterburg, 2012)
13. V.A. Nikolaev, *Landshaftovedenie: ehstetika i dizajn landshafta* (Aspekt Press, Moskva, 2003)
14. E. A. Orlova, *Problemno orientirovannoe sociokul'turnoe proektirovanie. Teoriya i metodologiya. Teoreticheskie osnovaniya kul'turnoj politiki* (RIK, Moscow, 1993)
15. D.O. Sajmonds, *Landshaftnaya arhitektura* (Strojizdat, Moskva, 1965)
16. L.G. Skul'movskaya, *Gorodskaya sreda kak substrat kul'tury goroda* (INFO-DA, Sankt Peterburg, 2004)
17. O.S. Hpengler, *Zakat Evropy* (Mysl', Moskva, 1998)
18. A. Schopenhauer, *Sobranie sochinenij: v 6 t.* (TERRA – Knizhnyj klub, Moscow, 2001)
19. N.-S. C. Genius Loci, *Paesaggio, Ambiente. Architettura*. (Electa, Milano, 1979)
20. L.N. Shabatura L.N., O.E. Yatsevich, Gumanitarnye i social'nye nauki, **2**, 156-162 (2014)

Evolvement of Public Open Spaces in Historical Environments of Yerevan City

Zaruhi Mamyan[1,*], *Artyom* Aloyan[1], and *Tatevik* Kartashyan[1]

[1]National University of Architecture and Construction of Armenia, 0009 Teryan str.105, Yerevan, Armenia

Abstract. In this article the role of public open spaces in Yerevan city as well as their correct and literate organisation from an architectural and planning perspective are studied and evaluated. In recent years a special attention has been paid to the development of Yerevan, the capital as well as other historical cities of Armenia in order to contribute to tourism development. This aspect resulted in the creation of new urban and architectural requirements. Among other issues the organisation of architectural environment of public open spaces - the role of urban design in our reality are also characterized. In these conditions urban and architectural organisation of public open spaces in historical cities are very important. Thus, the present study not only emphasizes the importance of studying the evolution of open spaces in Yerevan, but also stresses the significance of implementation of the international practice in this field in the city of Yerevan. The complex analyses, coordination and developing conditions of nowadays and historical requirements of historical cities are provided. The urban organisation and appropriate method selection for the current requirements of public open spaces are going to be analysed.

1 Introduction

This introduction will focus on the analysis of two aspects separately: public open spaces sand historical environment, at the same time it will concentrate on their relationship and the significant role each one has in sustainable development of urban environment in Yerevan city.

It should be noted that during the recent years several studies related to the urban development prospective of settlement and separate elements of urban design were done. These include Avet Miqayelyan "The prospective of Underground Urban Development in Armenia", 1997, Areg Kanayan "The Effect of urban Strategy in the Formation of Urban

* Corresponding author: zmamian@yahoo.com

Composition", 1997, Karen Berberyan "Modern Architecture and Urban Environment", 1997, Tigran Danielyan "Territorial Regulation Problems of Yerevan city in Current Economic Conditions", 1998.A special analysis was devoted to the study of separate sections of urban design as well as artificial lighting and green areas. These include Irina Aghavelyan "Artificial Lighting as a Factor of Organisation of Architectural Environments in an Example of Yerevan City Centre", 2009. Tzovinar Muradyan "The Use and Development Prospective of Green Roofs in Yerevan City", 2010.The formation of squares as a part of public open spaces are presented in Ara Harutyunyan "The issues of formation squares in Yerevan city", 2011. However, the deficiency of theoretical research in the field of contemporary problems of organising public open spaces in historical city environment leads to incomplete solutions to this sort of problems.

From the urban, architectural and visual point of view the historical character of every city in the world, as well as of the capital city of Armenia, Yerevan, is characterized by two aspects: on the one hand urban open spaces – walls and covers without fences, which are mostly open to the natural influences, such as streets, squares, residential courtiers, and, on the other hand, close areas, such as monumental or significant buildings and structures. Unfortunately, nowadays the architectural-planning structure that was created previously has partly changed, due to the development of the city life [3, 4, 5]. In these circumstances the modifications are also visible in organizations of public open spaces, the planning of which have to deal intensively with the question of how much of open space network has to be public and publicly available to allow a sustainable and community oriented urban development [6].

Therefore, the urban open space is the key concept, which will be analysed in this article. For that kind of understanding in landscape and natural engineering there is not an exact term, it's called "fragment of the urban environment", "Urban areas", "City landscape", "Urban interior", "Urban ensemble", etc.. Each of these terms define the phenomenon from their perspective, but three of them are most popular: "Open space", "Urban environment", "Urban interior", because these terms most precisely define the "environmental" essence of the object. Architectural open space is present in the city streets, alleys and parks, stadiums and squares, where human creativity sets borders to the "emptiness", so anything that visually limits the viewer, be it walls of buildings or rows of trees. All these things allow us to understand the essence of the concept "Urban Open Space" or "Urban Interior". The difference between urban interior from the usual one is the combination of elements that is created by the architect such as fragments of the landscape, horizon, street prospects, panorama, etc. It is clear that the main task of architect is to isolate, to limit the certain part of the space by material forms. The city's "Urban open spaces" in general are created by buildings and structures and especially when we deal with historical cities, these terms automatically become if not monumental buildings, the significant ones for the appropriate city environment [7].

2 Analysis of public open spaces in historical environment of Yerevan city during history

Public open spaces are critical components of life quality in Yerevan. Open space promotes the environmental sustainability of our city. Natural habitat provides sanctuary for wildlife species ranging from mammals, birds and insects to plants, trees and other types of vegetation provided in open space networks can reduce air pollution and wetlands can filter contaminants. The trails and streets of an open space network can also aid in reducing greenhouse gases, by providing alternative transportation routes and promoting cycling and walking.

Open spaces are critical components of every city including Yerevan and by necessity have to come in different forms and experiences. Open space provides the 'breathing' in a dense urban environment and promotes opportunities to engage in outdoor activities, access nature, enjoy scenic views, and experience the ecosystem of cities.

The Open Space Element in cities follows these guiding principles to ensure such holistic system: An integrated and multifunctional open space network would respond to a variety of needs for recreation and open space, better utilizing the existing resources. Yerevan is the center of ecological, economic, and cultural diversity. Open spaces should aim at building on our City's intrinsic qualities, both natural and cultural, and to reflect the values we place on cultural diversity. Open space and recreational programs should be equitably distributed. They should provide access for all residents, workers and visitors. Open spaces in historic core of our city are wholly connected. This aspect nowadays is very important. Open space should increase the City's capacity to be a safe and healthy place to live. Its design should promote social interaction, wellness, and a healthy lifestyle by providing opportunities for physical, cultural and social activities, and a connection to nature. With environmental sustainability as a driving theme, the quantity and quality of natural systems in the City should be preserved and expanded, by promoting aquatic and terrestrial biodiversity, by designing for watershed health, and by implementing environmental, ecological and conservation-minded strategies [8].

The public open spaces are evaluated during historical phases of Yerevan city. Our city as a locality has passed a long development, starting in 782 B.C., but as a city was founded already in 779. However, the term of "historical city" of the Yerevan is coming not only from the long term development in history, but also from the urban and artistic aspects, such us significant historical monuments, ensembles, municipal complexes, urban and natural landscape, cultural areas that also have archeological values [9, 10].

For further analysis of development of public open spaces in Yerevan city, within the frames of this article, the whole period of city's history can be conditionally divided into several phases.

Before 1924 (so called pre-Tamanian phase). A. Tamanyan was the author of master plan of Yerevan city. In this phase the city developed according to the plan designed by Mehrabov in1911.

With half-rounded configuration of flat part of the territory, an amphitheatre surrounding the central parts of northern and eastern highlands is formed. The main strong compositional and functional point of attraction forms the city centre with vibrant spaces of commercial and cathedral squares. From the centre departs a tree-beam system of ceremonial roads. The idea of city-garden is being formulated in the conception of saving in the city's widened green zone structure (98%), formed by estate gardens and public parks, flowing in the built-up neighbourhoods flanking the peripheral architectural areas.

Tamanyan's period 1924. The development of the second important project of planning Yerevan was initiated by academician A. Tamanyan in 1919. The project of Yerevan city reconstruction was finished in 1924. Tamanyan's project of Yerevan master plan retains the concept of the plan of 1911, with a corresponding development of urban areas and increasing the number of residence. According to this master plan the increase of the surface of public open spaces, parks, avenues, sidewalks, alleys, etc was provided. The project also provides brancing city streets: main transit movements, streets for residential areas, pedestrian ways. The main mass of the green is concentrated on the wide boulevard, with the ring covering the central part of the city.

After 1924. In 1971 the master plan of Yerevan (last master plan of soviet period) city was determined, one the hand, by Soviet Economy Development features, and the other hand by the city's area and landscape situation. In that period Yerevan city was the developed capital city of Armenia, unprecedented growth of population entailed increase of construction process, but on the whole public open spaces didn't "suffer". In 2005 the new project of master plan of Yerevan city was approved. City management and development of basic concept was developed in accordance with international cooperation of political, social and economic programs and high-tech investment [11, 12]. A survey showed that today together with an approved master plan a whole range of regulatory and urban planning documents function in the capital city of Yerevan. The below-mentioned list shows the essential normative and directory documents concerning the urban planning regulations and the preservation of the historical center of the city of Yerevan. The following urban planning documents can be found in the Republic of Armenia:

The decision of the RA Government on the zoning of Settlements, the development of projects, expertise, agreement, validating the verification and modification order /14.05.2001/;

The decision of the RA Government on the invalidity of decisions on RA communities (residential areas), development of master plans, expertise, agreement, verification and modification rule validating made on May 2, 2003;

Construction norms of urban and rural areas planning and building. /14.10.2014/;

RA Regulation on "the preservation and usage of historical and cultural monuments and historical environments". 1998;

RA Government. Decision on "Validating the norms of the realization of urban planning activities in urban planning areas\objects needing special regulation" 2009;

RA Government. Regulation on "the building of the core of the center of the city of Yerevan based on the urban planning regulation of the Republic of Armenia"/2013/ [13-19].

New open spaces, restored and consolidated historical public open spaces in a historical core of the city were designed and implemented. According to Table 1 during long term of Yerevan city's development, the situation of public open spaces was radically changed. Instead of open spaces, new buildings and structures were constructed. As a result, the presence of public open spaces decreased and the construction surface in a total land use scheme of Yerevan city increased [19, 20].

3 Discussion and results

Unfortunately due to the development of the city life the architectural-planning structure that was created in the previous period has partly changed. In these circumstances the modifications are also visible in organizations of public open spaces, the planning of which has to deal intensively with the question of how much of open space network has to be public. Contemporary tendencies of urban development also show that nowadays the recreation, restoration and reuse of public open spaces still continues. Tab. 1 shows how modern "quality and quantity" of public open spaces and their historical environment and image are changed.

Table 1. Modern "quality and quantity" of public open spaces and their historical environment and image are changed.

<table>
<tr><td>Phase I</td><td>Phase II</td><td>Phase III</td></tr>
<tr><td>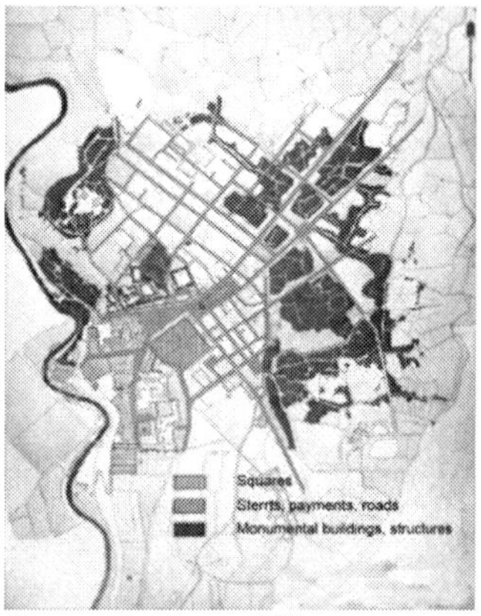

General plan of Yerevan city, 1911, by architect Mehrabov</td><td>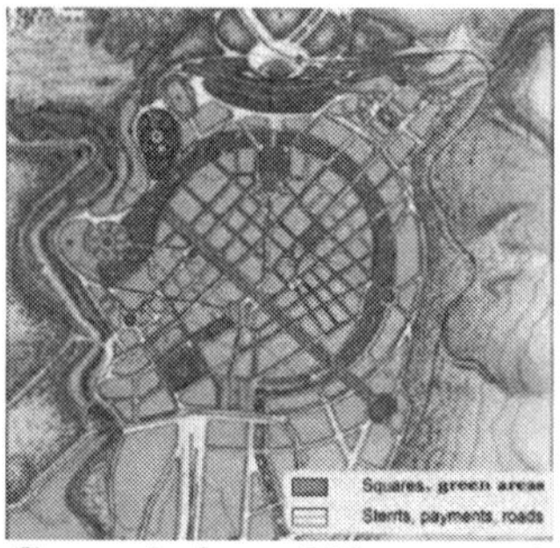

General plan of Yerevan city, 1924, by architect A. Tamanyan</td><td>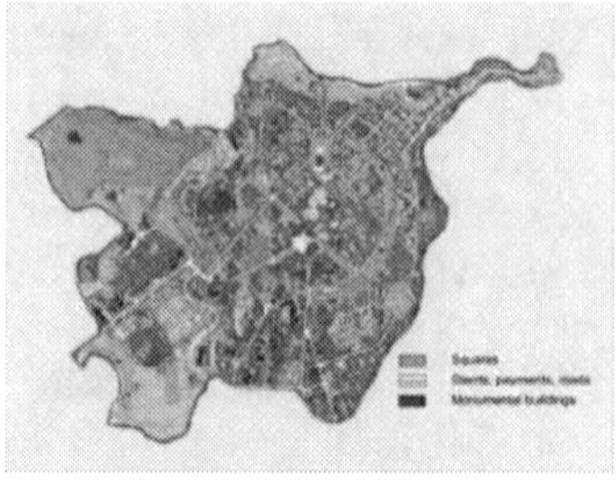
General plan of Yerevan city, 2005, "Yerevan Project"</td></tr>
<tr><td colspan="3">Amnoun of public open spaces in city</td></tr>
<tr><td>Phase I</td><td>Phase II</td><td>Phase III</td></tr>
<tr><td>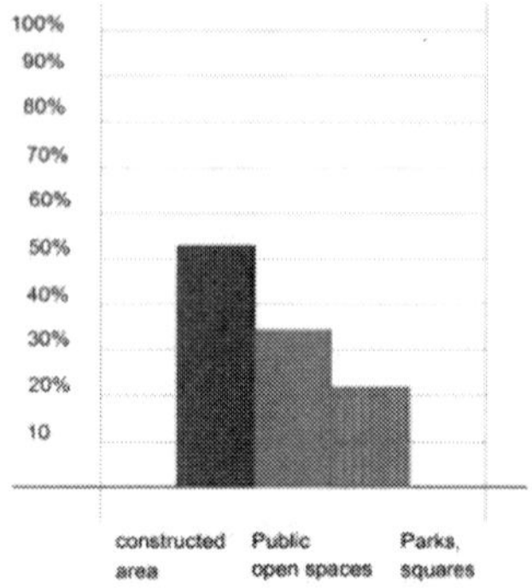
</td><td>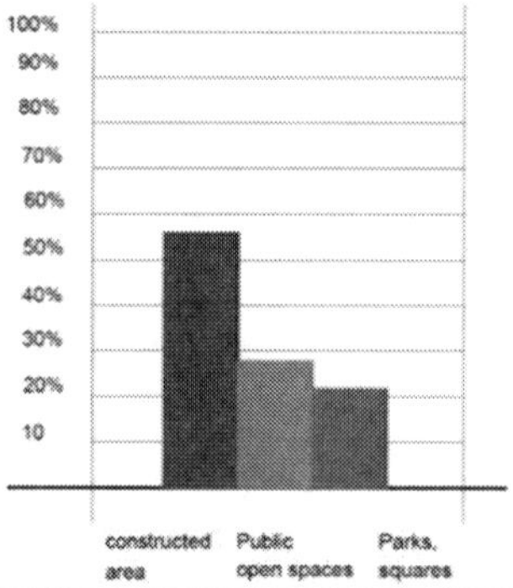
</td><td>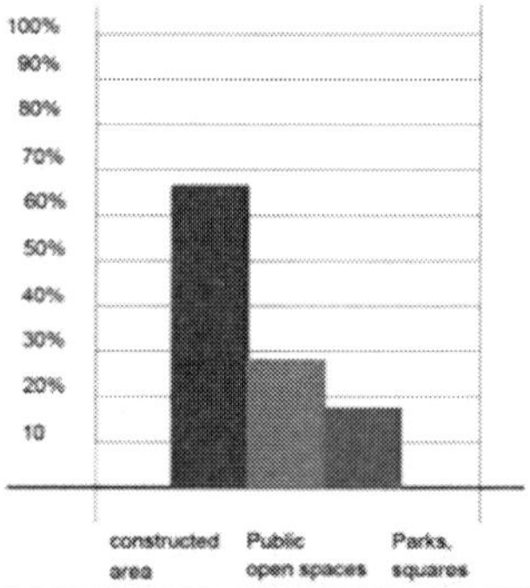
</td></tr>
</table>

Table 2 shows the present situation of public open spaces in Yerevan city. It is divided into three groups: public open spaces, in which the historical image of the environment is preserved, Public open spaces the historical image of which "is ruined" by new constructions, which change the scale of the existing environment, and new constructed public open spaces, in some cases in historical structure of Yerevan city, in some cases using the historical buildings and structures as fragments of newly created architecture.

Table 2. The present situation of public open spaces in Yerevan city.

N	Historical image of public open spaces	Combination of historical and contemporary image of public open spaces	Contemporary image of public open spaces
1	 Abovian street	 Abovian street	 Northern Avenue
2	 Abovian street	 Hanrapetutyan street	 Square of Moscow
3	 Hanrapetutyan street	 Hanrapetutyan street	 Aram-Byuzand streets
4	 Republic square	 Square of Aznavour	 Project of Old Yerevan

4 Conclusion

Today, when speaking about the historical urban planning of Yerevan we base our assumptions on single monuments of historical-cultural heritage, which are included in the list of preserved historical-architectural legacy. The recent transformations due to urban planning have changed the city's historical image, have altered a huge historical layer. We think that the historical monuments or ensembles should be given a significant role in environmental composition, they should be the starting points for reorganization of the space and they should determine the urban planning easements [20]. We also consider the possible preservation or renovation of 'complementary' environment for the monument, environment which is an open public space, environment which is not a monument in itself, but which 'tells about' the monument, 'leads' to it. It is this environment in its regenerated form that is seen as the main point for investment into city development [20, 21]. This in its turn can increase the anticipated potential profit from the cities as centers of touristic consumption. In general, the organizational actions of urban authorities, which are based on the orderly transformation, steady development and the preservation of historical and cultural environment within the architectural and planning organization of urban society spaces, must be realized together with the following essential steps:

- Formation of united planning connections of historical zones as well as involvment of objects which are being preserved or modernized into the planning structure.
- Improvement of the mechamisms realizing and controlling urban planning documents and projects.
- Provision of dominance of long-term projects over short-term projects by forming a regulating system and involving them into corresponding investors in order to have a complex realization of urban planning projects (engineering, transportation and the simultaneous realization of other substructures).
- The rising of the role of the civil society in the planning and realization of urban planning projects [16].

References

1. V. Harutyunyan, *Restoration of Architectural Monuments* (Dar, Yerevan, 2005)
2. *Rukovodstvo po rekonstrukcii gorodov* (Strojizdat, Moscow, 1979)
3. N.A Yastrebova, Vvestnik volgogr. gos. arhit. – stroit. un-ta. ser str-vo i arhit, **27(46),** 129-136 (2012)
4. M.V. Puchkov, Academia. Architecture and Construction, **3**, 57-60 (2009)
5. I.A. Dobricyna, Architecture and Construction of Moscow, **3**, 11-20 (2010)
6. R. Rogers, P. Gumuchdjian, *Cities for a small planet* (Faber and faber, London, 1997)
7. Tatev Kartashyan, *7th International Conference Contemporary Problems of Architecture and Construction*, Florence, 1-2 (2015)
8. "Recreations & open spaces", San Francisco Planning Department 1650 Mission Street, Suite 400 San Francisco, CA 94103-3114 www.sfplanning.org, pp 2-6,11.03.2016
9. Yu. Safaryan, M. Gasparian, A. Aloyan, *Proceeding of 3th International Conference on Contemporary Problems in Architecture and Construction*, 31-35 (2011)
10. A. Aloyan, Herald MSUCE, **7**, 7-14 (2013)
11. A. Aloyan, *7th international conference 2015 Florence International Conference on Contemporary problems of Architecture and Construction*
12. Yu. Safaryan, Collection of scientific works YSUAC, (32), 3-5 (2008)
13. *The decision of the RA Government on the zoning of Settlements areas, the development of projects, expertise, agreement, validating the verification and modification order.* /14.05.2001/ Based on order made on October 14, **263** (2014)
14. The decision of the RA Government on the invalidity of decisions on RA communities (residential areas), development of master plans, expertise, agreement, verification and modification rule validating made on May 2, 2003, No. 609 and March 4, 2010, No. 208
15. Urban planning. The Construction norms of urban and rural areas planning and building. /14.10.2014/
16. RA Government. Regulation on "the preservation and usage of historical and cultural monuments and historical environments" (1998)
17. RA Government. Decision on "Validating the norms of the realization of urban planning activities in urban planning areas\objects needing special regulation" (2009).

18. RA Government. Regulation on “the building of the core of the center of the city of Yerevan based on the urban planning regulation of the Republic of Armenia”.
19. S. Oganyan, *Penza State University of Architecture and Construction, "The problems of modern construction", a collection of scientific works of the international conference*, 76-78 (2009)
20. Information on: http://rcchd.icomos.org.ge/?l=E&m=4-4&JID=5&AID=38
21. Some issues of reconstruction of public centrs and conservation of historical cultural heritage in Yerevan, 6th International Conference on Contemporary Problems of Architecture and Construction, Technical University of Ostrava,2014, june, p.707
22. The Modern Area Planning Strategy of the Architectural Academy of the Republic of Armenia. Lraber, **12(1)**, 94-99 (2015)

MATEC Web of Conferences **73**, 06006 (2016) DOI: 10.1051/matecconf/20167306006
TPACEE-2016

Starting Points for Planning Business Zones in Slovenia in the Post-independence Period

Gregor Čok [1,*]

[1]University of Ljubljana, 1000 Zoisova 12, Ljubljana, Slovenia

Abstract. In Slovenia, during the period of transition practically all mechanisms for steering economic and spatial development changed. The planning of business zones is part of this framework; business zones are competitive business environments for the placement of economic activities in the physical space. Over the past 25 years, their design was influenced by various factors: introduction of free market principles, privatisation of social property, and interests of the young state to establish, using proper strategies and financial initiatives, an efficient network of zones as drivers of economic development. Despite the ambitious plans and adopted strategic spatial planning acts at the national level we nowadays find that the zones, in their current spatial and administrative form, are not an efficient enough development mechanism. The focus of the study was to define the starting points as the basis for planning business zones and shaping policies to improve the existing situation. To improve the business competitiveness of the existing zones and their role in the planning and design system, we will have to set up a more efficient data updating system and form a broader approach to their management, promotion, and marketing.

1 Introduction

After declaring independence in 1991, Slovenia entered a transitional stage, where its socio-political, economic, and spatial development concept changed completely. By introducing the sustainable development doctrine [1, 2], liberal principles of the free market, and the democratically founded local self-governance, the established practice of planning activities in space changed significantly as well.

The former planned socialist (social, economic, and spatial) development, which was in SFRY mostly realised within the constraint of the public finances [3], was replaced by the concept of selective investments following the principle of supply and demand.

Slovenia, as many other Eastern Bloc countries (Hungary, Poland, Czech Republic, Slovakia, etc.), formally opened out to western investments [4, 5], which, through establishment of enterprises under the "new democracies", mostly consumed cheaper labour force, more flexible environmental norms, and the "transitional" tax legislation

In the first years of independence, at the start of privatisation of social property, relatively low productivity, and the loss of established markets of the former SFRY

* Corresponding author: gregor.cok@fa.uni-lj.si

(Socialist Federal Republic of Yugoslavia) Slovenia's economy evidently stagnated. After a few years of consolidation, new development perspectives were formed. The state began to implement various measures to improve the business environment and, as a result, strengthen economic growth (tax policy, subsidies, attracting foreign investment, etc.) [6].

This includes the measure of building new and revitalise existing "business zones" to, above all, ease the establishment of new companies and secure their long-term existence in individual locations. Today, 25 years later, the spatial planning and economic professional community is trying to establish whether this measure was, in fact, effective.

The establishment of new zones resulted from various interests and motivations, which indeed followed the objectives of establishing a competitive business environment (strategic locations, cheap land, simple building conditions, etc.).

In the era of transition, the field of planning and design also changed considerably. Furthermore, spatial legislation, the financing method of public transport and municipal infrastructure (public-private partnerships), and the philosophy of real-estate investments itself changed [7]. Spatial capacities for site selection, placement, and organisation of economic activities in the physical space nowadays mostly take place as two spatial forms:

(a) old zones: former industrial, craft, storage, and logistical zones, as a consequence of extensive industrialisation of the territory of RS after WWII,

(b) new zones in new locations: established on the basis of various interests (needs of local entrepreneurship, national development strategies, investment motivations, etc.).

From the spatial planning perspective, the significance of these zones is two-fold. Economically, they are drivers of economic development while, spatially, these are extensive areas of urban fabric with various impacts on the environment. In this sense, "planning regulators" are important; through them we implement the goals set for all regulatory areas (settlement, demography, infrastructure, agriculture, etc.). This perspective is particularly important when preparing spatial development plans, where we have to coordinate various planning starting points and sectoral goals.

2 The problem and a hypothesis

In RS, the situation related to the "range of spatial development capacities" is nowadays the consequence of various factors from our recent past.

During the privatisation of former business zones after 1991, their ownership and spatial structure fragmented, which decreased the possibility of an efficient programmatic and structural regulation. A wide range of activities in the domain of small entrepreneurship (mostly trade and business activities) was introduced in these locations, while there was soon a lack of appropriate development areas for new investors, which resulted in the establishment of many new zones throughout Slovenia.

In the national spatial legislation and practice in 1995–2015 there was no efficient methodology to regulate the existing, and plan the new, zones, therefore regions and municipalities addressed this problem differently. Today, the problem is evident in an extensive network of various types of zones (approx. 400 zones); these zones have one thing in common: in the real-estate market and when implementing development programmes they are not recognised as an efficient tool for steering economic and spatial development (there is a lack of a common operator, of a single programmatic typology, of efficient promotion and marketing, of data on available internal areas, building conditions, etc.).

Under such conditions it is not possible to create an attractive business environment as are those in competitive locations in the wider region (Italy, Austria). This situation is confirmed by many expert guidelines and the recently conducted survey among foreign investors, by SPIRIT Slovenia – Public Agency for Entrepreneurship, Internationalization,

Foreign Investments and Technology. The survey showed that among the key obstacles for foreign investment in Slovenia there are problems related to the spatial perspective (high prices and/or availability of developed land, complex procedures to obtain various permits, lack of support activities, etc.). In this framework, the following research question was put forward:

What programmatic and spatial starting points provided the basis to plan business zones in the era of transition?

3 Hypothesis

The changed socio-economic conditions shaped the doctrine of supply and demand also in relation to investing into business zones. In this framework, despite considering, "in principle", strategic development starting points to strengthen national competitiveness, new, market-oriented implementing documents were produced. These mostly satisfied the current demand, following the principle of "anybody, anything, anywhere". The consequences of the non-systematic approach are reflected in the "regulatory inefficiency" of the business zone network as an instrument in the planning and design system.

4 Methodology

The focus of the study was to define all planning and design starting points, based on which business zones were planned over the past 25 years. This was done in two stages. In the first stage, a descriptive research method was applied to analyse 15 reference documents:

(1) strategic spatial development documents (3 documents at the national level),

(2) implementing spatial planning documents at the regional level (2 documents),

(3) implementing spatial planning documents at the local level, and an expert study for their preparation (5 + 5 documents).

From each document we took, separately for textual and graphical parts, all the mentioned starting points and justifications underlying the concept of developing the existing and new zones.

In the second stage a comparative research method was used, which was put in place in two steps. In the first step we reviewed the interim results and selected two groups of starting points: (a) global starting points: planning level, arising from the objectives of the Spatial Development Strategy (state, region, municipality) and (b) implementing starting points: project level, determining zone planning (programmatic design, size of zone, site characteristics). Then we categorised the starting points from each group according to their meaning into: (a) economic starting points and (b) spatial starting points. In the second step, conclusions were drawn up and the guidelines to remedy the existing situation were proposed.

5 Results

In the study we analysed three documents that in the period after 1991 strategically best defined the basic starting points for planning and designing business zones. They were as follows:

(1) Spatial Development Policy of the Republic of Slovenia (2002) [8] – defines the national vision of spatial development and is the umbrella document for coordination of sectoral policies in implementing sectoral measures and projects.

In this sense it does not provide detailed planning parameters for business zones, but rather it provides guidelines for their spatial integration (e.g. priority macro locations for economic development: border areas and regional centres; abandonment of energy-consuming activities; provision of developed land at the local level for development of local economy).

(2) Economic Development Strategy of Slovenia (SGRS) 2001–2006 [9] – its central goal is to achieve "complex competitiveness of the state" leading into shaping an efficient national economy for producing internationally competitive products and services. The spatial perspective of organising economic activities is integrated into various development measures affecting their range and territorial distribution.

This is mostly about revitalisation programmes for local development potentials and cooperation of the state in shaping a supporting environment for functioning of the existing and new companies (Table 1).

Table 1. Development measures and a "summary of starting points" for developing business zones when achieving complex competitiveness of the state.

A Regionally and spatially coherent development - Placement of new business zones is the basis to attract new investments and reduce regional disparities.	F Information and communication infrastructure - Its development will bring forward the surrounding locations, increase the possibility of home working, and viability of creating smaller business zones.
B Strengthening the role of spatial planning as a factor of economic development - In regional and local spatial documents it is necessary to provide for sufficient and economically attractive areas and define the various types of spatial integration of economic activities (point and continuous zones).	G Facilitating the development of small and medium-sized enterprises - By providing infrastructure, land and other elements, the state will participate in the shaping of an attractive support environment and thus directly impact the establishment and distribution of business zones.
C A supportive environment as a set of elements to facilitate technological development - Material infrastructure is an important part of the supportive environment.	H Foreign direct investment (FDI) - With various instruments, measures, and grants, the state will directly influence the location of business zones intended to attract FDI.
D Supporting the growth of private sector competitiveness - The enterprises whose expansion and development will grow beyond home working will move their businesses into the zone, where, in turn, the demand for new areas and locations will increase.	I Removal of administrative barriers (red tape reduction) - Along with simplification of administrative procedures we need to produce modern, more flexible spatial plans.
E Allocation of labour force - Distribution of business zones plays an important role in relation to labour commuting; a larger number of zones reduces the home–work–home flows.	J Economic infrastructure - Economic infrastructure is the central locational criterion in introducing new zones, its development will increase the number of locations and the quality of their services.

(3) Spatial Development Strategy of Slovenia (SPRS 2004) [10] – is the next key document that defines the national spatial and development goals and the institutional mechanisms to achieve them. It also defines the development measures with the starting points for developing business zones (Table 2).

Table 2. Development measures and a "summary of starting points" for developing business zones when achieving coordinated spatial development.

A Polycentric structure development - The concept of priority development will provide a range of jobs, housing, and public services in locations of interest. A strategy of developing new business zones will have an important role in this.	E Areas of production activities - Contemporary location criteria for optimum integration of economic activities in the physical space need to be defined. - In siting new zones, the viability of the development needs to be justified (the existing capacities checked).
B Border and disadvantaged areas - Establishment of small business zones for centres with a narrow impact area. - Zone typology needs to be defined which will comply with the specific development demands.	F Development of cities, towns, and settlements - Each economic activity contributes to global development. - Home working is encouraged along with collective organisation in the framework of business zones.
C Vitality and attractiveness of rural regions - Encouraging entrepreneurship to form a dispersed structure of individual economic structures. This trend is already evident and highlights the need for a specific spatial regulation.	G Regional integration - In spatial planning documents we need to define the criteria that will provide the basis for determining local attractiveness in the site selection of regional business zones.
D Development of settlement - In this context, different zones in relation to the specific types of urban settlements have to be designed. - It is necessary to provide the conditions for developing business zones with modern equipment and technology parks, particularly in Ljubljana, Maribor, and Koper, and in regional centres with a strong background in capital and knowledge in the fields of high technology and labour force.	H Transport terminals - Development of a transport logistics terminal will importantly affect the locations of business zones of regional and national significance.

All three documents provide distinctly "strategic starting points", whose implementation in lower-ranking documents and their actual impact is difficult to assess reliably. We found that many of these "starting points" (e.g. polycentric development, development of cities and infrastructure) were (in a somewhat different formulation) included already in the previous generation of the so-called "socialist" development strategies.

6 Implementing spatial planning documents at the regional level

In the previous decade, regional spatial documents were produced (the so-called Regional Concept of Spatial Development – RZPR) as partner documents between the state and municipalities.

This is a lower level of strategic and implementing documents that, on the one hand, summarised the starting points of national development strategies and, on the other hand, the municipal needs and development trends significant for both the individual municipalities and the region as a whole.

In the framework of this study we analysed 2 cases of RZPR. We found that there was a blending of "differently detailed" starting points for defining the concept of developing business zones. In one case (RZPR of SE Slovenia, 2006) [11] for the field of business zones a special expert study was produced (with detailed definition of location criteria for their placement), while in the second case (RZPR for the South Primorska Region) [12] the starting points were based on the content of the Regional Development Plan (RRP) [13]. In both cases, the spatial planning documents were not produced to the level of detail that would allow for definition of specific locations and the extent of individual zones. We also found that the regional scale was optimal for strategic planning of a zone network of various typologies, as it represents the appropriate framework for coordination of individual spatial interests.

7 Implementing spatial planning documents at the local level and the expert study for their elaboration

With the declaration of independence in 1991, all Slovenian municipalities had applicable spatial documents produced mostly in the period of 1980–90. They legitimised zoned land use with relatively large development reserves. All the amendments of these plans, or the production of new plans (the so-called OPNs) are largely based on the previous distribution of activities in space. In this study, we analysed 5 cases of spatial planning documents from various parts of Slovenia (3 OPN drafts: Žalec, Piran, Izola [14, 15, 16] and 2 cases of recently adopted OPNs: Ljubljana and Novo mesto, [17, 18]). To specify the starting points as the basis for the planning of the business zones in these documents we also analysed the relevant expert groundwork, [19, 20-23].

We found that the starting points are mostly covered in the chapter "Analysis of the spatial situation and trends" (AST). These analyses have different levels of detail and quality in terms of: (a) available financial frameworks and timeframes, and human capacity of their producers, (b) client cooperation, and (c) data availability.

By means of reviewing the individual content of spatial documents and their expert groundwork (Table 3) we found that when preparing the document the key problem of defining the planning and project starting points was the lack of reliable data: (a) on development potential of existing locations and (b) on actual (topical and envisaged) needs for establishing new zones.

By analysing the cartographic groundwork determining zoned land use we found that the traditional locations of the sites for industry, craft industry, storage, and logistics in all the cases analysed mostly preserve the existing scope.

The areas that give rise to concern from the environmental perspective, or were simply overgrown by cities/towns during their intensive growth, are cancelled.

These starting points are defined quite in detail in the strategic part of spatial plans, which also define other starting points, resulting from umbrella strategic documents at the state level.

The starting points regarding the data on the actual needs and prognosis of long-term development (quota of necessary and reserve areas, realisation timetable, programmatic design of individual zone types) are far more indeterminate. These starting points were designed with "varying quality" during the production of the expert study.

Table 3. Expert study for OPNs (a summary of findings): analysis parameters that the producers addressed as the starting points for producing the strategy for developing business zones.

Expert study for designing the strategy for developing business zones in producing OPNs, a summary of the 5 cases of expert studies		
AST activity (analysis of the spatial situation and trends)	**Analysis parameters demonstrating project starting points**	**Precision rate of data acquisition and processing (1 to 5)**
(a) analysis of existing capacities (zones according to the existing plan, active and non-active)	number and size of zones (m2)	5
	ownership of land and buildings (private, municipalities, other)	3
	reserved land	2 (difficult to define)
	level of municipal services	2 (difficult to define)
(b) analysis of potential capacities (new locations provided for – empty greenfield land)	Number and size of zones	4
	Ownership of land and buildings (private, municipalities, other)	4
	Reserved land	4
	Level of municipal services	4
(c) analysis of development initiatives provided during OPN production and in interviews	formal initiatives (number and scope)	4
	non-formal initiatives (positions of decision-makers)	3
	non-formal initiatives (positions of local entrepreneurs)	3
	plausibility of initiatives (assessment of actual needs)	2 (difficult to define)
(d) other sources of acquiring data on needs	situation in the market (supply and actual demand)	2 (difficult to define)
(e) consideration of guidelines of umbrella documents	sustainable development objectives as determined by strategic regional and national plans	3 (difficult to define)
(f) consideration of guidelines by those responsible for spatial development	spatial restrictions, sector conditionality, and protection and management schemes	5
Activity – producing the draft of a zone development strategy		
(g) definition of location-specific criteria for planning new zones and further development of the operation at existing locations	Actual needs (no reliable data)	2 (difficult to define)
	Spatial capacities	4
	Transport accessibility	5
	Environmental impacts	3
	Economic background	3
	Construction costs	2 (difficult to define)

Based on the interim results we selected a structural set of starting points. We found that, according to their "strategic status", we can distinguish between:

(1) General starting points (and guidelines), which mostly arise from the hierarchy of development documents (state, region, municipality). In the process of plan production they represent an "orientation" only, without specifically defined and quantified consequences that the zones would trigger in the economic and/or spatial sense.

(2) Specific starting points obtained during the production of the expert study for plan preparation. They provide the basis for planning the locations of zones, their dimensions, and the programmatic design.

According to their impact they were divided into two basic groups: (a) economic starting points (justifying the meaning of a zone as an instrument for encouraging economic development) and (b) spatial starting points (justifying the meaning of a zone as an instrument for regulating spatial activities) (Table 4).

Table 4. Economic and spatial starting points influencing the concept of business zone development in the analysed documents.

A – ECONOMIC STARTING POINTS justifying the meaning of a zone as an instrument for encouraging economic development A1 - global starting points, justifying the synergy between the zones and the national economy (e.g., zones in the function of an instrument for drawing on financial incentives used by the state to encourage economic development through the shaping of a competitive business environment, attracting FDI, etc.) A2 - specific local starting points (known needs of local entrepreneurs or creation of real spatial reserves for potential external entrepreneurs) quantifiable effect of the new zone (number of jobs, taxes, income tax, etc.) A3 - non-specific local starting points (establishing "reserve" zones, mostly real-estate interests, whose impact is difficult to quantify)
B – SPATIAL STARTING POINTS justifying the meaning of a zone as an instrument for regulating spatial activities B1 - global planning starting points, justifying the synergy between the zones and general spatial development (impact is difficult to quantify e.g. zones in the function of an instrument for regulating the environmental impacts, distribution of jobs, transport rationalisation, etc.) A2 - specific planning starting points (known needs regarding the number, size, typology of zones. These needs are: founded based on known actual or envisaged development initiatives, development potential of the existing zones, and the real-estate market situation) A2 - specific project starting points (known detailed demands for the planning of buildings within zones, known locational characteristics and restrictions, envisaged environmental impacts, known guidelines of those responsible for spatial planning)

Over the last 25 years, both general (desired global impacts of zones) and specific starting points (detailed parameters on the number, size, and typology of zones in a specific territory defined by a relevant document) were considered practically in all new plans; therefore it is incomprehensible that nowadays there is such a dichotomy between the large number of realised zones and, at the same time, their inefficiency both in the programmes for encouraging economic development as well as in the process of steering spatial development. This is the consequence of various conditions. Based on the analysis, the following conclusions were drawn:

(1) The problems of "efficiency of old zones" in the process of plan production and in their development promotion, particularly in the difficulty of defining their "actual development potential":

(a) spatial reserves within the existing zones are minimal and/or not available, (b) prevalence of complex ownership structure (result of non-systematic privatisation of social property), lack of data on the condition of infrastructure, interests regarding the shift of business and sale of real-estate, (c) unpredictable land prices of existing zones in the coastal zone and in the vicinity of urban centres, (d) non-digitised maps and obsolete regulations for urban and architectural design reflected in a diverse (or uncontrolled) image of the urban fabric (visually unattractive business environment).

(2) The problems of "efficiency of new zones" are mostly: (a) non-competitive land prices, (b) inefficient system of promotion and marketing, (c) inefficient system of joint management, (d) location; real-estate interests often affected the establishment of new zones (drawing on EU funds, single profit from the sale of developed land, etc.) rather than actual needs. Thus, locations were often selected that did not meet the criteria for zone construction – such zones will never thrive as efficient development zones.

8 Discussion

The planning and design of business zones in RS over the past 25 years was influenced by the following starting points: (a) strategic development starting points at national, regional, and local levels (to a smaller degree), (b) "implementing project" starting points (specific development needs at the local level – to a smaller degree), (c) real-estate motivations (to a larger degree), and (d) established spatial conditions as a legacy of former plans (the most). This study revealed that despite the ambitiously set development strategies, the zones in their current spatial, administrative, and market form are not an efficient enough instrument to steer spatial development. Based on this finding, we can fully confirm the hypothesis developed. In this framework, the results of the survey conducted among foreign investors (SPIRIT) become understandable, and can be interpreted as a suggestion for improving the existing situation. Based on these findings, we propose the following amendments:

(a) to elaborate a strategy for further development of zones at all levels of spatial documents, which would include priority locations, a uniform typology, and a timeline of their development (in the future, this is the basis for selectively guiding financial subsidies and other actions to boost competition), (b) form an efficient system for data updating (on free capacities, programmatic direction, public utilities, etc.), (c) in larger zones to set up a system of joint management, marketing, and promotion.

9 Conclusions

The planning and design of business zones, as a regulating instrument, is a complex social, professional, and administrative process. Despite the ambitiously set strategies for achieving global competitiveness, their impact is neutralised mostly at the implementing level. To achieve the desired efficiency in the field, a highly interdisciplinary approach and understanding of all parameters of the "current market economy" in the wider European space are needed.

References

1. *Agenda 21: Programme of Action of Sustainable Development* (United Nations Conference on Environment & Development, Rio de Janeiro, 1992)
2. *ESDP European Spatial Development Perspective* (Potsdam, 1999)
3. G. Balažič, *Relikt socializma v prostoru - primer mestne občine Koper, diplomska naloga, Fakulteta za humanistične študije* (Univerza na primorskem, Koper, 2006)
4. R. M. Agosin, R. Mayer, *Foreign Investment in Developing Countries. Does it Crowd in Domestic Investment? Discussion Paper No. 146* (UNCTAD, Ženeva, 2000)
5. P. A. Maksimenko, M. Rojec, *Foreign investment and privatization in Eastern Europe* (Palgrave Publishers, Hampshire, 2001)
6. J. H. Dunning, *Assessing the Costs and Benefits of Foreign Direct Investment: Some Theoretical Considerations, in: M. Patrick Artisien, M. Rojec (Eds.), Foreign Investment and Privatization in Eastern Europe* (Houndmills, Basingstoke, Palgrave. 2001)
7. K. Dimitrovska-Andrews, Local land & soil news, **6/II**, (2003) 25-26.
8. *Spatial Development Policy of the Republic of Slovenia* (MOP-UPP, Ljubljana, 2002)
9. *SGRS - Slovenija v novem desetletju: trajnost, konkurenčnost, članstvo v EU. Urad RS za makroekonomske analize in razvoj* (Ljubljana, 2001)
10. *Ordinance on Spatial Planning Strategy of Slovenia, Official Gazette of the RS* (2004)
11. *Regionalna zasnova prostorskega razvoja JV Slovenije* (Acer, Novo mesto, 2006)
12. *Regionalna zasnova prostorskega razvoja južne Primorske: Analiza stanja in razvojnih možnosti ter vizija prostorskega razvoja regije* (Acer, Novo mesto, 2006)
13. *Regionalni razvojni program za Južno Primorsko regijo 2007–2013* (RRC Koper, Koper, 2006)
14. *Municipal Spatial Plan of the Municipality of Žalec, Official Gazette of the RS, No. 64/2013* (2013)
15. *Municipal Spatial Plan of the Municipality of Piran (OPN) – strategic part, draft* (LUZ, Ljubljana, 2010)
16. *Municipal Spatial Plan of the Municipality of Izola (OPN) – strategic part, draft* (Locus, Domžale, 2013
17. Municipal Spatial Plan of the City Municipality of Novo mesto, Official Gazette of the RS, No. 101/2009.
18. *Municipal Spatial Plan of the City Municipality of Ljubljana, Official Gazette of the RS, No. 87/2010* (2010)
19. J. Kušar, G. Čok, U. Račečič Skrt, T. Berčič, *Strokovne podlage za gospodarska in industrijska območja: strokovne podlage za pripravo SPRO in PRO občine Žalec* (Univerza v Ljubljani, Fakulteta za arhitekturo, Uniarh, Ljubljana, 2006)
20. G. Čok, D. Kavaš, M. Košak, M. Koman, M. Čok, *Delo in turizem na domu: strokovne podlage za pripravo OPN Občine Piran* (U-M-A, Lokev, 2007)
21. M. Plazar Mlakar, A. Mlakar, L. Gosar, G. Čok, D. Kavaš, J. Hudoklin, M. Bidovec, I. Selak, T. Jerin, M. Kocjančič, J. Bagon, *Analiza stanja in teženj v prostoru za SPRO inPRO Občine* (Studio Mediterana, Izola, 2007)
22. R. Alihodzic, V. Murgul, N. Vatin, Applied Mechanics and Materials, **680**, 494-498 (2014)
23. G. Radovic, V. Murgul, N. Vatin, Applied Mechanics and Materials, **641-642**, 634-638 (2014)

MATEC Web of Conferences 73, 06007 (2016) DOI: 10.1051/matecconf/20167306007
TPACEE-2016

Design and analysis of information model hotel complex

Nikolai Garyaev[1,*]

[1]Moscow State University of Civil Engineering, 129337 Yaroslavskoe shosse 26, Moscow, Russia

Abstract. The article analyzes the innovation in 3D modeling and development of process design approaches based on visualization of information technology and computer-aided design systems. The problems arising in the modern design and the approach to address them.

1 Introduction

Modern computer technologies are being actively implemented in various spheres of human activity, their use becomes an essential condition for successful operation and competitiveness of today's market. In this regard, modern hotel - this is not just a place away from home, and the night, and a complex set, which includes a variety of functions, such as living space, food areas, recreational facilities, technical facilities and service infrastructure [1-14].

On this basis, the development of a hotel complex in front of designers and architects, a number of topical issues, such as:

• Wide range of realization of hotels, they create near airports, in remote from populated areas settlements where people could enjoy nature and get away from the bustle of the city (ekooteli) hotels short stays - Waiting plane or train, to negotiate and so on.

• Proper zoning and the creation of technological schemes of the hotel, in order to reduce the intersection of personnel and guests stream for a comfortable stay of the latter. The project must take into account the different nuances and parameters on which will continue to depend on the profitability and the successful functioning of the entire complex.

• The need for design and construction of building structures of complex geometries without seeing the final result - it's a big risk. In this connection there is need to prepare drawings of virtualization and the adjacent landscaped grounds in an easy, understandable and reflects the basic properties of the form.

• The original idea of the future of the hotel should be interesting and attractive for the guest.

* Corresponding author: garyaev@mail.ru

On the relevance of this research work has affected the company «VFM Leonardo Hotel E-Business». The key objective of the monitoring was to identify and analyze what hospitality professionals are doing and plan to do to increase online - the presence of the hotel (hotels), increasing the number of visitors (1 diag.). Research has shown hospitality tours (combination of 3d-panoramas, photo galleries, videos and animations): 66%, advertising and demonstration video: 62% 3d-panoramas (360 ° panoramas): 49%, search landing pages 36%, Images and Photos reflecting the life style: 34%, interactive brochures: 34%, therefore, the most effective is a combination of rich media tactics and tactics based on social media. In particular, the connection of the high-quality visual content like video and visual tours hotels (with circular 3d-panoramas, photographs, etc.) with user reviews, ratings and rankings.

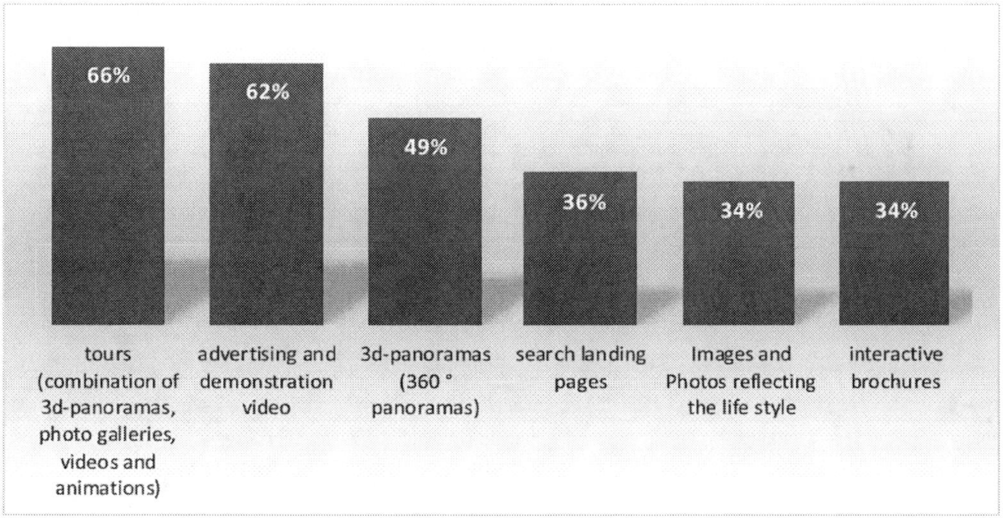

Fig. 1. The most effective multimedia according to «VFM Leonardo Hotel E-Business 2010".

2 Development

Having considered the above-mentioned problems, realizing the urgency of their solution, the problem arises - to analyze innovation in the 3D modeling and design using information visualization and computer-aided design technology based on information model property.

To solve this problem, you must:

1. Create a project exhaustive information architectural model of the hotel complex, with the aim of providing accessible information about the object, provided that the three-dimensional environment allows you to navigate through a virtual project in real time (it shows everything at once: the building materials used, equipment, interior fittings, including transport accessibility, landscaping level, significant specific points, d.) the interaction with the surrounding objects, and so on.

2. Development of recommendations on the use of new technologies for the design of the hotel complex.

It is well known: it is better to see once than hear a hundred times. This adage true for the modern construction industry, when to replace the traditional (2D) projection technology comes the 3D-modeling (see. Fig. 1) (a), (b), and then the building information modeling (the BIM). The project, created with the help of information modeling, not simply provides increased visibility of the results of and the most accurate reflection of the fruits of the architectural and technical imagination in a format close to reality, but creates a complete data base on BLM (Building Lifecycle Management) lifecycle management of buildings, and even longer. Data contained in it may be changed, supplemented, replaced by reproducing the current condition of the building.

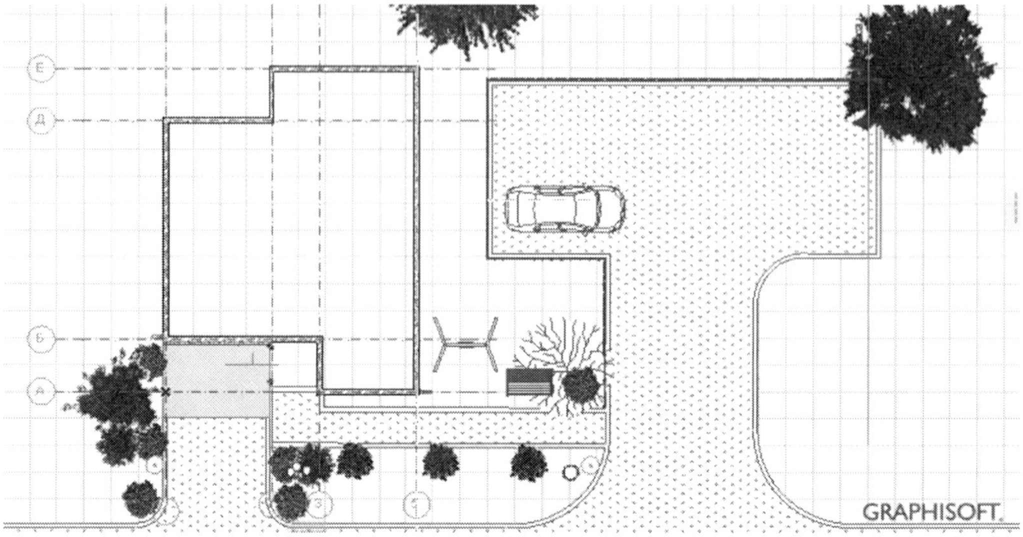

Fig. 2. (a) Detail of plan of the hotel complex in ArchiCAD.

Fig. 3. (b) a facility Virtualization.

So, how, work is underway to establish the information model.

First developed some blocks (family) - the primary elements of design that correspond to the construction products (windows, doors, floor slabs, etc.) and elements of equipment (heating and lighting, elevators, etc.) and much more that is directly related to the building, but produced outside the construction site and the construction of the object is not divided into parts.

The second stage - modeling that is created on the construction site. This foundation, walls, roofs, curtain walls, and more. This assumes the widespread use of pre-existing elements, such as mounting or framing components in the formation of curtain wall of the building.

This division into stages (first and second) when building information model is rather conditional - you can, for example, to insert a door in the simulated object, and then, again appeared reasons to change them, and the project will be assigned has already changed the door.

Built experts information model of projected object then becomes the foundation and actively used to create working documents of all types of development and production of building structures and components, facility equipment, economic calculations, the organization of the construction of the building itself. It also provides a solution to the technical error in the project, due to quality of coordination between departments and merge all the partitions in a single space model.

However, all the experts are also aware that, in spite of the high accuracy of the information models and the quality of computer software documentation for construction, all this will not make sense without taking into account the fact that a successful construction business, first and foremost, should be cost-effective. Namely, should always be money to build, and dating should always be sold in the future to function correctly.

To these obvious axioms become a reality, steps to attract investors and selling space in the new building begins long before the completion of his design. Ideally, they should already be completed before the completion of the design.

Thorough work with potential investors at the design stage involves the implementation of a minimum of three mandatory conditions, without which it is impossible to implement the funding:

1. The customer should be as clear picture of all the components of the building, its equipment, organization of construction, etc.

2. It should be possible promptly make changes to the project, taking into account the constantly emerging new customer requirements and without altering designated for all time.

3. At each adjustment of the project, each new version should be able to quickly retrieve its economic characteristics and all the necessary technical information.

The construction industry of developed countries, Investor has long been a high priority, and in this line of business building information modeling, and has established itself well, it was simply irreplaceable at the present rate and volume of construction.

As for individual housing construction, in particular hotels, here BIM efficiency is also very convincing. After all, the buyer of the building (or individual developers) usually takes care of the cost savings. This is organized through information modeling as optimization of the cost of building and construction of proper organization. This BIM also allows the well to predict the subsequent operating costs of the building, which in most cases can be crucial in the construction or operation of the construction site.

Maintenance and repair of buildings - one of the most important and highly costly activities of daily living, in which information modeling plays an important role. For example, the management company will be convenient to know in advance how much and where the bulbs should be replaced as paint or water pipes necessary for the overhaul of the building, how much it will cost replacement of equipment, furniture, construction and architectural features, where to find them at a better price and in what time You can carry out these works.

Thus, when designing the question as any man, from the architect and finishing manager of the hotel, I could see the information model in an understandable and clear as much as possible.

And realize this is possible, for example, with the help of such software as Graphisoft BIMx for ArchiCAD. This tool for architects, includes modern methods of communication and presentation projects. Any user of the program can now literally plunge into the project: a three-dimensional environment, without having special knowledge. 3D navigation area allows you to perform real-time 3D virtual tour of the architectural design. This can include stereootobrazheniya modes, gravity, and free flight, control layers, differentiate openings and solid objects. The program features the ability to save the passage of the building as an independent exe-file and its further demonstration without having to install additional software and a long tiring training.

Also, any element models can be filed data relating to him a passport for the equipment, material, dimensions, manufacturer and other information. There is no need to look for it all the folders on the disc, a bookshelf, closet, etc. A user can open the model and get from it all the necessary information in a structured way and the cost of the minimum time.

The program provides the user with information about the object in a wide range of species (plans, sections, presentation files, rendering and animation models, 3D models, and the like) technologically suitable for further processing by computer or other means.

3 Conclusion

Thus, we can conclude that the information model for today - is a kind of database that contains a 3D model of the passport facilities, library records and other information on the building project in a structured and interconnected form. Of course, the introduction of an information model implies the inclusion of expensive BIM-tool, but its use is quite reasonable and effective, as it allows you to perform and provide the following:

• Visualization, simulation and testing technology of labor-intensive construction and installation works;

• Eliminate time-space error due to visualization and analysis of construction and installation;

• Planning for the implementation of resources;

• Modeling and visualization of the processes of repair (assembly, disassembly, etc.);

• Creating a reference database with visual organization operative approach with a single access point provided in the executive form of the three-dimensional model, etc. This information model contains: all engineering and other data about the object: the information on the operation and maintenance, design documentation, results of installation and set-up data and the pre-launch tests and stuff;

• Building on the basis of them a single information space for coordination of interactions among the various units;

• Creating an information model based on 3D or panorama tour for visually spectacular and reasonable presentation of the institutions as partners and customers.

References

1. V.V. Garyaeva, N.A. Garyaev, *Computing in Civil and Building Engineering Proceedings 2014 International Conference,* 1336-1343 (2014)
2. P.N. Garyaev, *Computing in Civil and Building Engineering Proceedings 2014 International Conference,* 1618-1625 (2014)
3. N.A. Garyaev, *Computing in Civil and Building Engineering Proceedings 2014 International Conference,* 1578-1585 (2014)
4. V. Goryunov, I. Zayats, T. Konjkova, V. Murgul, Procedia Engineering, **117**, 825-831 (2015)
5. I. Yamshanov, V. Goryunov, V. Murgul, Procedia Engineering, **117**, 675-684 (2015)
6. V. Murgul, N. Vatin, I. Zayats, Procedia Engineering, **117**, 824-829 (2015)

7. G. Radovic, V. Murgul, N. Vatin, Applied Mechanics and Materials, **641-642**, 634-638 (2014)
8. R. Alihodzic, V. Murgul, N. Vatin, Applied Mechanics and Materials, **680**, 494-498 (2014)
9. B. Hardin, D. McCool, BIM and Construction Management: Proven Tools, Methods, and Workflows. 2nd Edition (Indianapolis, 2015)
10. M. Hergunsel Benefits of building information modeling for construction managers and BIM based scheduling (Master's Thesis, Worcester, 2011)
11. M. Day, AEC Magazine, 1 (47), 16–19 (2013)
12. L. Allen AutoCAD Professional Tips and Technique (San-Rafael, 2014)
13. A. V. Petelin 3D-modyelirovaniye v SketchUp 2015 in 2 volumes (St. Petersburg, 2015)
14. J. Vandezande, P. Read, E. Krygiel Autodesk Revit Architecture 2013: Autodesk Official Training Guide (San-Rafael, 2015)

MATEC Web of Conferences 73, 06008 (2016) DOI: 10.1051/matecconf/20167306008
TPACEE-2016

The Cluster Model of Development of the City

Aleksandr Panfilov [1,*]

[1]Industrial University of Tyumen, 625001 Volodarskogo str. 38, Tyumen, Russia

Abstract. The article deals with issues related to the current state and development of the concept of urban development and urban analysis, whose main task is to develop an evolutionary model settlements development. However, urban science covers many related areas of knowledge, and is becoming one of the areas of fundamental research on the theory of the development of society as a whole. It is in urban planning are reflected culture, sociology, economics, politics and much more, giving to Tectological characteristics. At the same mechanics of their interactions allows us to look at the city as a system matrix structures, interacting in time and space, and have the features of the cluster and fractal models. Cited in the article description of the author's concept of cluster model of development of the city raises fundamental questions about its formation, determines the key parameters and mechanisms for its operation.

1 Introduction

We are living in the cities for thousands of years, but so far failed to realize the essence of this phenomenon. What creates the city? What principles for their development? Why do some of them disappear, while others are becoming centuries-old legends? How formed the cloth of the city and how it functions?

What processes are invisible for us in centuries in moments of an urban life? All that remains unexplained mystery. But it is precisely on how we create our city ourselves depends our future! If you do not create a quality environment - rationally organized, comfortable accommodation, combining the needs of life-support and environmental protection - we do condemn their own little world of degradation and extinction. Equally, we can't clearly answer the question of its further evolution. There are many examples in history the rapid beginning of uncontrolled growth of cities and a sudden decline without objective reasons. The roots of all these processes are much deeper and it is with them and to understand.

People create the city, gradually formed its social order, laws and regulations - its society, which in turn develops, begins to re-convert the city for themselves. The circle has been closed! City - a paradoxical phenomenon, like the mythical centaur, organically combines the seemingly incongruous things. In the words of Daniel Danin, "Rider did not lash his horse, and the horse is not tries to throw the rider, as well as the rider can't get off the horse, and the horse can't get away from the rider" - these are two terms of a single equation. It is absolutely impossible to swap them. The city as a living mechanism of the

* Corresponding author: tgasu.panfilov.aleksandr@mail.ru

absolute level of difficulty, based on the six principles, the six "C" - five beneficial: compromise, consensus, consolidation, convergence, complementarity, and a depressing - confrontation. [8] Thus each six principles are being in harmony. City-Centaur does not involve struggle with the mandatory or desired victory for one of the diverse incarnations above the other - struggle with victory or defeat of one of the principles deprives the image of meaning.

The task of theory of urban planning are being formed in these conditions, the main aim of which is to define the principles, mechanisms and tools of the systemic organization of space, allowing to identify the conditions of the evolutionary development of the city.

2 The City as a problem

Speaking about the city, within the above mentioned positions of Centaurism, can be clearly divided our problem into several layers of perception. At first, the city can be regarded as a Tectological concept, as it is a single complex structural system, a kind of quasi-living organism. City, as well as the definition of an organized system Tectology based on the principle: "The whole is always greater than the sum of its parts." And it's true - the city is not a complex of buildings, roads, people, grass and trees. The city has always disproportionately - it is able to actively respond to internal or exogenous irritants, deformed, and adapting to the changed conditions, transforming them for suiting under the immediate needs. It is originally incorporated all the principles sufficient for its sustainable self-development, self-adaptation [19, 20], there is exists an invisible genetic relationship between all the elements allowing it to evolve here.

At the same time, the city as a "living organism" (and we have every right to percieve it that way) is equally constructed, and according to the laws of nature, one of the basic rules which is simplicity. Principle of Mopertyui (the principle of the least action) states: "Nature-lazy, she moves the body so as to make at least this effect", i.e., Nature does not like complexity, where is the possibility to make easy. [7]

However, sustainable development laws are constantly put before a choice: to anyone, even the most minimal change provides two ways of development: to create smith new or to improve existing. Both of these paths are equivalent justified and have their positive and negative sides, as well as its evolutionary boundaries.

This brings us to the second level of the problem: the city as a complex subordination structure consists of a system of such simple elements in a specific relationship and time-varying - city cluster. In essence, any city can be represented as a kind of mathematical matrix - a cluster consists of systems of equal or similar elements or opposite to each other elements, but it is the system. Working with each component of the system separately as a single element, we can more accurately track, process and send back on track its development. The gradual transition to more complex and large items, you can get some resultant vector and generalized indicators to work with the whole system and even combine them into larger super-system.

This provision opens up the third level concerns the city - the Fractal City. The Fractal City as a system possessing the self-similarity property, that is composed of several parts, each of which is similar to the whole figure as the whole, it brings us back to Centaurism basis of the concept of the city.

Fractal of the city, on the one hand, is a tool that allows us to work with smaller structures cluster model city, and on the other - allows you to combine clusters Tectological research model. It directly is this duality of the fractal city model: target and tool makes the possibility to further describing the principles of cluster model of development of the city.

3 The basic concept

The cluster model of development of the city.

There are many attempts to make sense of the city - this ancient Indian Vedic treatises [21] and socio-philosophical works of Thomas More [22] Tommaso Kompanello [23], Nicolo Machiavelli [24] and many others. However, in all these City (with a capital letter "C" as a philosophical term) is seen as an already existing in petrified form without its history and future development, given their model, but is not given its judgment. Exactly the same model, only much later advanced by Russian architects-constructivists: Ivan Leonidov, El Lissitzky, and others - they give the image of the City, transcending the scope of the reasonable to the highest limits. And of course, the artists of Russian "The Paper Architecture": Yuri Avvakumov, Mikhail Belov, Alexander Brodsky, Totan Kuzembaev, Ilya Utkin, Mikhail Filippov, Maxim Kharitonov and others.

However, the first person who begin to perceive the city as a complex spatial structure and try to translate it into a dynamically adaptable, this model became Makoto Sei Watanabe. In his project the INDUCTION CITY he raises questions of organization of dynamic spatial structure of the city, has built on a mathematically verified models [4, 5, 6]. In his works, the Japanese architect, shows that the city is able to become an integral part of the world. His INDUCTION CITY project is based on three complementary models:

SUNGOD CITY - city created in such way that each element receives the necessary its share of sunlight, heat and energy: "We cut tunnels in the structure so that the sunlight reaches the rear side of the device. We do the same for all the other units that do not have adequate access to sunlight. The result was the emergence of a huge "porous cube" with numerous cavities [6]. "

MOON GODDESS CITY - this component is the opposite of SUNGOD SITY. The main condition here is that the units should not be exposed to sunlight (it should not be heated). While exposure to sunlight is minimized for all items which are able for generating a cluster.

There is created a third model – it is based on these two directly the induction design, which is capable for generating a variety of forms, satisfying to both the previous requirements.

However, there is reviewed the only facet of extremely narrow understanding of the city as a multifaceted phenomenon - the relationship of the sun and the City.

Proposed to the reviewing basic concept of the cluster model of development of the city are designed to create a kind of universal evolutionary matrix allows for evolution and retrospective analysis, to identify the key points of development, and create the factorial-communication model and so on.

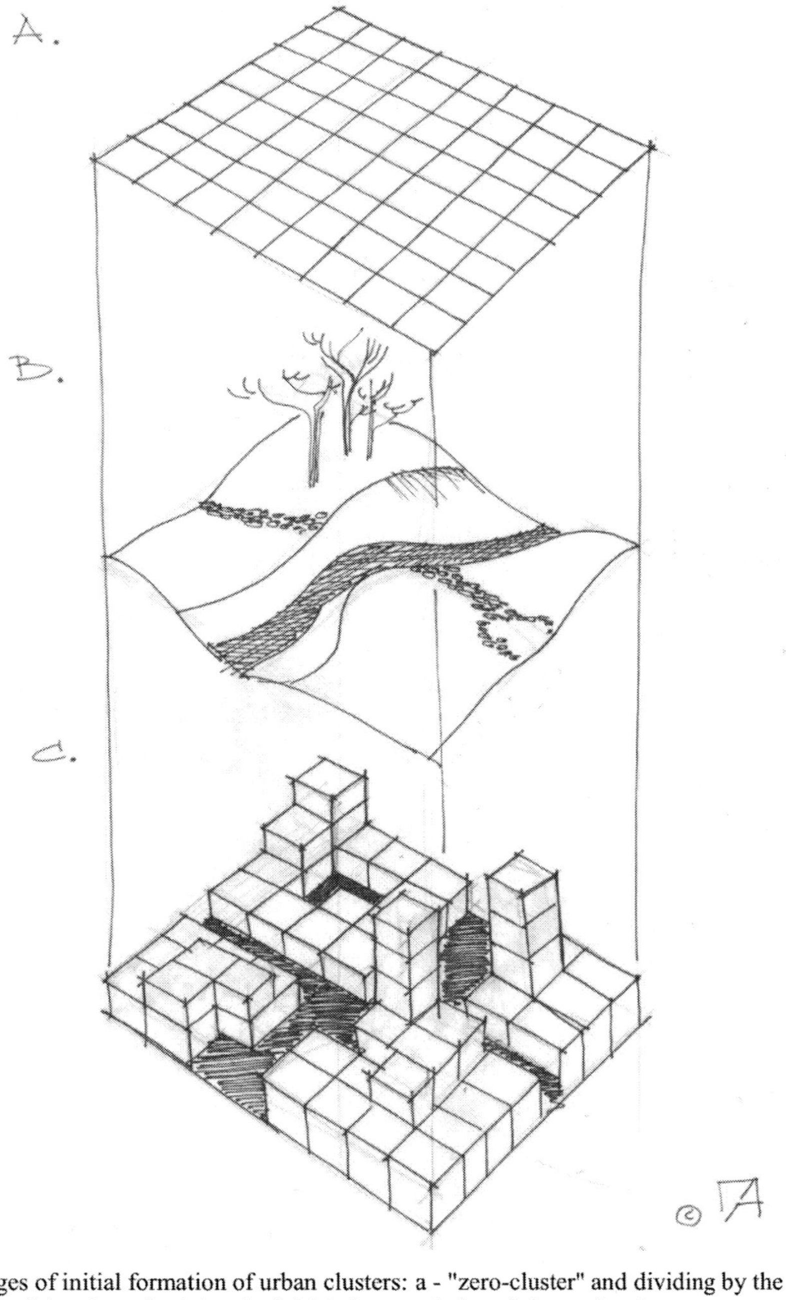

Fig. 1. The stages of initial formation of urban clusters: a - "zero-cluster" and dividing by the sub-clusters; b - "conditional area", giving the initial characteristics of cluster development; c - an example cluster reaction to given exogenous irritants.

Any of the space inhabitated and developing in time: the house, quarter, neighborhood, city, state, country or the entire world can be represented by a group of interrelated independent clusters and at the same time as a single cluster-live system. For the basic cluster (zero-cluster) adopted a common unit belongings space with the initial "zero" characteristics (see Fig. 1a): the absence of points of growth and degradation, lack of communication and factors of influences, the absence of binding to the terrain (even conditional), the lack of time factors, etc.

Placing a zero-cluster in terms of a real or synthesized space, set the initial conditions of development (see Fig. 1b.) Is determined by: the natural conditions, the network cluster communication, availability of attraction points and vectors of influence, presence or absence of adjacent clusters, etc. In this case, all of the factors. affecting the null cluster can have a progressive, so the regressive direction; play the role of active, passive or secondary factor; have different time duration of influence, changing in an infinite limit of instantaneous to a constant; be single or be a link in a chain of interrelated factors.

Develop models considered two types of zero-cluster reactions:

Reaction 1. General "mass" of the cluster remains the same, and it is only it's redistribution. This type of reaction is able to meet the needs of system analysis solely on the initial stage of the formation of the cluster model, as it shows only the first reaction unit or a null cluster introduced stimulus. However, this type of reaction can be used in the future as the effect of "instant camera" to get the data for certain types of reactions, bonds, irritants, dynamic progress points.

Reaction 2. His occurs deformation of elements of cluster system to the evolution of some elements, degradation and other third stopped (see. Fig. 1c). This type of reaction is a dynamic, developing in time and space of the process, which is characterized by constant reallocation of points active / passive / quiet within the cluster. Since the reaction of one object is an irritant to the other, the activity of the one element encourages the development of neighboring elements implicated him. At the same time an enhanced activation of the one plot of cluster may have a negative impact on its other portion or the adjacent portion of an adjacent cluster. It is worth noting that the result of such an impact can be even stronger as the revitalization of the dominant cluster, and the inverse process.

Another feature of the proposed cluster model of development of the city is being the fact that it and its constituent elements (zero-clusters) also possess characteristics of fractal systems, which implies the possibility of a division of a single cluster into sub-clusters nminus-order, and join them in larger ones nplus-order with identifying vectors of generalizing. The fractal cluster model allows us to differentiate the data and organize complex responses and influence fields for general evolutionary pattern. Concurrently, there is the possibility of identifying individual responses and irritants acting on the smallest sub-cluster members, which in turn opens the perspective analysis, differentiation and dot interference structure of the cluster.

In the course of its development, each element of the cluster (sub-cluster) can actively or passively interact with other elements of the cluster. This interaction can take place both directly to neighboring elements, and indirectly with all elements of the cluster by building a chain of direct interactions in response to perceived cluster and its elements of internal and / or exogenous irritants. It is worth noting that the interaction data can equally be seen as a cluster on the domestic level and at the level of inter-cluster cooperation.

It is possible to identify a number of typological characteristics which are responsible for the extent of the interaction. Firstly, each element (sub-cluster) can be actively or passively interoperate properly with the adjacent elements, as well as indirectly by all the elements of the cluster. The cluster in response to perceived an internal and/or exogenous irritants. The active position takes a cluster or a cluster member, received as a result of internal or exogenous irritants potential. That stimulus, for example, may be the construction or reconstruction of roads, the introduction into the fabric of the city a new attraction point (enterprise, office or shopping mall, residential education, etc.). However, it should be noted that this impact is almost always a complex consisting of a system of complementary factors, besides not always have the same direction of development.

Consider the following example. Building of the new company gives a positive vector in the development of the region and the surrounding regions, but in the conditions of absence of implement an advanced infrastructure model, the existing network (roads and utilities) can't cope with the efforts of a new point of attraction that will have a negative impact on development. At the same time the changed environmental conditions also reduces the development of the index in mind in the conditions of negative impact on the person. On the other hand do not creating a new center of gravity does not give an overall impulse in development of the territory, which also has a negative impact on the overall system. The result of all of the above is that there is need to create a system which capable for adequately balanced analysis of vectors of development for finding their optimal proportions and rational distribution of the all elements of urban infrastructure and its constituent objects and people.

Secondly, the type of influence can be characterized on several levels:

Level I.The direct contact - the influence exerted on the sub-cluster members directly adjacent to the study. There are contact, passing on the verge of sub-elements (more active) and through the corner points (equivalent of previous or less active).

Level II. Indirect contact - contact and interaction occur through a chain of close quarters. The greater the degree of influence, then longer the chain of k-contacts (k-contact - an indefinite number of gear cluster of interdependent elements which are in the chain of indirect touch)

Level III. Broadcast influence - the influence exerted in conditions of the absence of a chain close quarters, but it occurs when there is clearly directed vectors of interaction, mutual influence between the two separate units within a cluster and between clusters. Such communications may be the cause of a new transit corridor, entailing changes in the circuit adjacent cluster units of the structure. Chains borders of indirect contact sub-cluster system elements define the field of complementarity effect of each cluster unit, with it smaller the fields can be absorbed by the larger (more active) fields and in the enlarged analysis it does not take into account or give extra points to the dominant element. Conditions for the occurrence of III-level connections in contact or intersection of two or more fields appear with this the point of intersection or area association receives an additional stimulatory effect of development as a translator of communication.

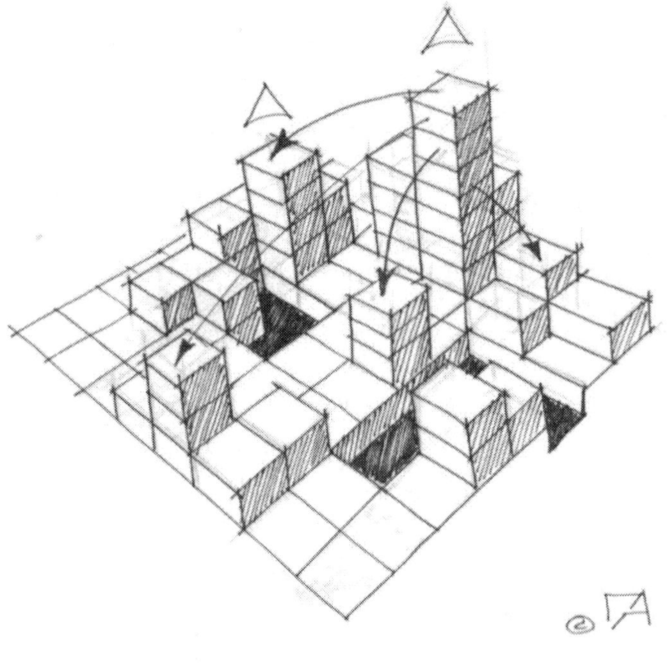

Fig. 2. An example of the further development of the urban cluster with a demonstration of the effect of different levels, the example of the dominant cluster unit.

There merged the forming a new cluster of clusters summarizing nplus-order with a sufficiently large number of connections between clusters III level and the appearance of the intersection zones of complementarity fields

Thus, there is getting a generalized model dynamically developing cluster model city. The model itself, which was originally described as a quasi-living system, which has a complex organizational structure, due to the constantly changing conditions and internal factors will continue to be in the process of evolution and making it more difficult and increasing its constituent sub-cluster elements.

At the same time it can be assumed that the process of growth of the spatial mass of sub-clusters can be in an infinite time, until it reaches the absolute dominance or balance the spatial structure of the cluster mass, with well-established old irritants and the absence of new exogenous irritants. However, none of the "tower", as well as any complex system, can't be infinitely large. At some moment of the time it begins to disintegrate under its own weight or over-weight in mind the internal conflict.

In this case, further evolutionary process can also go in two ways:

Way I. The process of the collapse of the dominant super-developed element generates one or more new zero-clusters, which are located in the proximity close to the main element.

Way II. The process of the collapse of the dominant super-developed element partially nullifies the results of previous evolutionary development of the cluster and by redistributing the sub-cluster members, in different levels and stages of development, as well as a system of relations and vectors of influence, re-launches the process of evolution in the numbers of changed conditions.

This time is the identifying the factors of influence and their boundary values, as in the influence of the factor of the plane, and the level of interaction between the factor; It is the identification system of intermediate values, enabling factors to differentiate between themselves and within themselves; It is the definition of models of complex influence factors on the development of sub-elements of the cluster.

It is worth noting the vast amounts of factors that have a potential impact on the evolutionary development of the cluster, which is an exemplary enlarged typology in differentiating their complex groups may include:

a)Natural: climatic, geological, territorial, etc. - this group of factors is responsible for the bulk of the initial characteristics and direction of the vector of zero-cluster (along with the human and economic, etc.).

b)Environmental/Ecological - group of factors measured up for the quality of the original characteristics of the "conditional areas" as well as for determining the quality of decisions in the form of a mechanism for determining the comfortable external conditions of Human existence.

c)Evolutional - this group of factors are responsible for the ability of the system development, and, as previously stated, the evolutionary means as positive (progressive) and as negative (of degradation) vector of development.

Chaotical: spontaneous, including the possibility of sudden violent changes - this group of factors is introduced, firstly, to the possibility of reverse evolutionary analysis of development of the city, when the placement of objects (territories) based is not on the comprehensive analysis, and the "pointing out" of solving old model of the planned economy; the possibilities of disasters and phenomena may also be included in this group of factors which wear a chaotic, but sufficiently predictable (in terms of an impact) character.

d)Anthropogenic, including ethnographic and anthroposophic - the most unpredictable group of factors proposed to operating this model, on the one hand, acting in terms of psycho-history, we can reverse the chaotic start of the Personality in the organizing structure of the City, on the other hand - the man is the overriding element of the system, because of it is the measure and the aim of all solutions.

e)Generalized City-Planing - this group includes the structural elements of the urban environment: facilities, infrastructure, territory and their functional features, as well as the degree of interference.

f) Economical - the group of factors in addition to political, natural and evolutionary characterize the ways and means of achieving the global goal: to create a harmonious, efficient and comfortable environment which are able to sustainable self-development.

g)Political – the group of factors are responsible, as previously marked, on the one hand for the introduction of a number of elements of the chaotic nature and destabilizing the system and for the will to detailed development and implementation of the proposed Tectological cluster model.

h)Medical, including biomedical and psycho-physical - this group of factors, along with other groups responsible for implementing the complex task of creating high-quality habitat for man, forming the qualitative characteristics of a comfortable life, because of it allows you to create a methodology and mechanism for monitoring vital signs as at the level of the person, and the level of the surrounding environment by combining with the environmental factors.

i) and many others.

As can be seen from the description of complex sets of factors, the reduction of the problem of sustainable urban development to any one (or even several) elements, as is happening at the moment, it does not seem not only possible, but, in the authors opinion, even unacceptable, because it leads to the dead end path of unilateralism. City the same way as a person, as mentioned earlier - is a multifaceted phenomenon, capable to the full development and self-realization only at the complex equilibrium development of all of its constituent elements.

The identity of the desired characteristics is possible as a result of a number of historical and analytical studies on the development of the most old with a long and ancient history, and the young, but rapidly growing cities - their master plans, building projects, etc. Determination of factors of all necessary for study will require a rethinking of the same degree of inclusiveness related to architecture and urban development of other branches of scientific knowledge. It is also planning a series of social studies and the establishment of clear mathematically verified models of factors impact.

On the basis of elaborated system of factors of development and the general model of cluster development of the city it seems possible to create a universal intellectual cybernetic system (virtual model) which is able to scroll mode or freeze time to assess the impact of decisions (both in town and on country or region) for searching optimal ways of development of human society.

In the result of analyze of the main provisions of the concept of cluster model of the city development, in addition to identifying a set of factors of influence, internal and exogenous irritants, orientation and characteristics reactions of the vectors and action of complementarity fields of influence, as well as identify of their qualitative and quantitative characteristics are need to consider the whole multitude of possible combinations to the terms of their relationship, complementarity and synergy, as well as the degree of influence of individual factors and their complexes in the development and expansion processes of the entire cluster system.

This is considered a range of issues relating to the generalization and development model of urban null cluster, and the entire cluster system:

- What indications of a positive development factor occurring leads to growth, transformation and evolution of the sub-cluster, the cluster and all the cluster system?

- What indications of a positive development factor is the division of the cluster to additional (internal) cluster subsystem happening?
- What conditions will terminate the active phase of the evolution of individual clusters and begins the process of unification and enlargement of the cluster units?
- How to determine the critical mass of cluster development in which there is its decomposition into independently developing clusters or is born (department) new cluster zero units?

4 Conclusion

The concept of creating a cluster model of development of the city proposed to the consider in some way can be compared with the science of psycho-history, applying mathematical methods for the study of the processes taking place in society. Thus it allows the determining or predicting the possible vector of development of a society with a high degree of accuracy. However, the excluding from the common system of such an unpredictable factor as a "man" we can with enough confidence to monitor and guide the urban development of the city in the right direction.

The proposed method of analysts by dividing urban space on the cluster system, having the properties of fractal systems, provides tools, both long-term planning of development of territories and conducting retrospective studies. In addition, this tool allows a high degree of accuracy to determine the position and characteristics of the main sites of the city or region to determine the system and model of economic development in terms of placement of the individual components in the observed area (country, region, city, county, etc.).

On the other hand the proposed analytical model of development, in the conditions of it's comprehensive study, testing and application could set a precedent formation of a new model of a planned economy, based on which the allocation of resources and production will be performed with the maximum rationality and efficiency for the overall development of the country, creating a comfortable and supportive environment for its citizens. This model because of it's maximum rationality contributes to the possibility of preventive evolutionary studies. Placing an object (building, plant, infrastructure element, and even the new city), we must understand that the object appeared not for a year or two, sometimes it comes to the centuries and even millennia, in which new challenges and crises can arise based on objective, measurable analysis, accounting, modeling and forecasting of cause-effect relationships. Therefore, the city, and perhaps, the whole country, constructed on the proposed model of cluster development, subject to the application of it as a well-established system of analysis / control / actions can be active a self-adaptive system which capable to the unlimited prosperity and sustainable development.

References

1. D. Haag, *Cluster specialization pattern and innovation styles* (1998)
2. V. Feldman, D. Audretsch, Specialization and Localized Competition-European Economic Review, **43**, 409-429 (1999)
3. S. Rosenfeld, European Planning Studies, **5**, 3-23 (1997)
4. Makoto Sei Watanabe. *Conceiving the City* (L`Arcaedizioni, 1998)
5. Makoto Sei Watanabe. *Theory+Work+Program: on the siten* (Maruzen-publishing, 2012)
6. Makoto Sei Watanabe. Information on: http://www.makoto-architect.com
7. A. Zakharov, Biophysics *(*1992)

8. D. Danin, Science and Life, 5-6 (1996)
9. D. Danin, *Centaurism. The course program for the humanities* (RSUH, Moscow, 1997)
10. S. Kostov, *Proceedings of the V All-Russian Internet Conference on econophysics and evolutionary economy*, 32-24 (2006)
11. A. Bogdanov, *Universal organizing science* (Tectology, Moscow, 1989)
12. V. Smolkov, International Journal of «Problems of Theory and practice of management», **3** (1997)
13. J. Hunt, Architectural Design, **11-12** (1998) Information on: http://www.i-home.ru/site.xp/049055048050124051056049055124.html
14. M. Shubenkov, M. Shubenkova, International electronic scientific and educational magazine «Architecture and Modern Information Technologies» (2015) Information on http://www.marhi.ru/AMIT/2015/special/shu /shub.pdf
15. N. Saprykina, Proceedings RAASN, **1**, 262-267(2012)
16. Ch. Landry, *Institute of cultural policy* (Moscow, 2005)
17. K. Lynch, *Image of the city* (Stroyizdat, Moscow, 1982)
18. T. Campanella, *Sun City. Q: Utopian novel XVII-XVIII centurie*, (Moscow, 1971)
19. A. Panfilov, *Proceedings of the international scientific-practical conference "Modern trends in theoretical and applied research*, **1(43)**, 134 (2013)
20. A. Panfilov, *Adaptive Architecture. The collection of materials II remote scientific conference with international participation on "Recent advances in science"*, 101-102 (2011)
21. *Vastu* (Philosophical Book, Moskva, 2001)
22. N. Machiavelli, *Selected Work*s (Fiction, Moskva, 1982)
23. T. Moore, *Utopia. Per., Joined. Art., comm. and the decree* (Publishing house Tbil. Univ., Tbilisi, 1983).
24. S. Khan-Magomedov, *The architecture of the Soviet avant-garde: Book 1: The problems of formation, Master and flow* (Stroyizdat, Moscow, 1996)

MATEC Web of Conferences **73**, 07001 (2016) DOI: 10.1051/matecconf/20167307001
TPACEE-2016

Improving the Efficiency of Administrative Decision-Making when Monitoring Reliability and Safety of Oil and Gas Equipment

Maria Zemenkova[1,*], *Viktor* Shalay[1], *Yuri* Zemenkov [2], and *Elena* Kurushina[1]

[1]Industrial University of Tyumen, 625001 Volodarskogo str. 38, Tyumen, Russia
[2]Omsk State Technical University, st. Mira 11, Omsk, Russia

Abstract. Methodology of rapid assessment of reliability index was developed based on system analysis of technological parameters. Within functioning of on-line monitoring system of reliability index of industrial facility this method allows to increase efficiency of making managerial decisions on technical and preventive maintenance. The technique is based on the analysis of technological parameters of operational modes of pipeline transport facilities registered by dispatcher controls. The created technique can be used by the operating, research, design institutes and oil and gas transport enterprises when declaring industrial safety. The received mathematical models allow federal services of supervision, the independent expert organizations to predict the development of reliability in the registered block of dispatching data either in real time mode, or taking into account the dynamics of service conditions of the object.

1 Introduction

The analysis of the existing legal and regulatory documentation, data bank of dispatch of oil and gas service companies suggests that a comprehensive assessment of the reliability of on-line mode requires a systematic model and universal mathematical models of the technical state of facilities with regard to the factors and parameters that change in long periods of operation and storage.

Modern database must contain all information to be registered by all monitoring tools for analysis, forecasting, decision-making in operation. The results of a retrospective analysis of objects indicate that the systems are not universal for oil and gas facilities, and require adaptation to a particular technical object, in accordance with the technological scheme, technical specifications and equipment models and other parameters that characterize a particular object [1-21].

Unfortunately, at present, the database for a comprehensive assessment of the reliability of on-line mode is not formed [6-10]. However, for the facility as a whole and its elements and subsystems at different levels certain operating parameters yi at fixed intervals are recorded. Diagnostic activities are performed in accordance with the schedule, the results of

* Corresponding author: muzemenkova@mail.ru

which determine the characteristics of the technical condition of the objects уід using experimental and computational methods. It should be noted that dispatching systems in oil-transport companies contain information about the actual operating parameters, which set common error standards defined by attorneys devices with a sufficiently small error (0.4%). Therefore, as the reference information database we use all the parameters available for analysis and recorded by various monitoring tools: performance instrumentation; diagnostics; the results of tests and examinations, etc. At the same time in order to gain proper results the database must be uniform and recorded at regular intervals.

2 Conditions of mathematical modeling

Accurate values of operational and admissible parameters y are set for dispatching parameters in accordance with technological scheme. The basic condition of safe system of operation is the condition of absence of parameter outcome out of critical limits $y_i \in (y_{\min cr}; y_{\max cr})$. At a deviation of parameters outside possible limits there is an exit of object out of operable state and inability to perform given functions, i.e. reliability. Loss of reliability can be associated with both factors depending on the operating conditions, and changing in the properties of the object in time, therefore it is necessary to evaluate conservation factors. Thus, failure can be seen as a physical and parametric. The problem of forecasting and reliability of the process y (t) is reduced to the prediction of non-stationary random process and the determination of the probability of not outcoming of the random process beyond the tolerance values of D for a certain period of time tpred. Comprehensive analysis of the reliability of the system in accordance with the developed scheme involves an application of individual modeling technology of operation of the facility and the creation of models to assess the reliability of their parameters.

3 Physical and mathematical model of reliability index

Studying objects and solving the given tasks are related to the development of physical and mathematical models of reliability, reflecting the peculiarities of the phenomena and processes characteristic of pipeline transportation facilities. Under the term mathematical model hereinafter, we will understand the general formulation of indicators and a system of mathematical equations, in which it is possible to study the dynamics of changes in the reliability, getting information about the parameters and processes taking place.

We will specify the developed model of technical conditions with following terms and provisions:

1. The technical condition of the object is characterized by its set of output parameters y0, yx, . . ., yn.
2. Characteristics of output signals (power, vibration, temperature, pressure, frequency, phase, duration of impulses etc.) or some system characteristics (productivity, efficiency etc.) are examined as output parameters.
3. Change of parameters is a random or deterministic process subjected to a certain distribution law.
4. Since the change of the parameters of technical objects (change in the vector $y = \{y_0, y_1 \ldots y_n\}$ of operational characteristics) is influenced by many factors, it is fair to consider the evolution of parameters as a random function of time:

$$y(\omega, \mathrm{t}), \mathrm{t} \in \mathrm{T}, \omega \in \Omega, \tag{1}$$

where Ω is the set of elementary events; T is the interval of operation of the facility.

The main feature of the operation of the facility is non-stationarity of the processes (due to the regime change and irreversible aging and wear), and General model of the variation of parameters is a random function on a probability space (Ω, F, P):

$$y(t) = \left\{ y_j(t) \right\}_{i=0}^{n} \tag{2}$$

where F is algebra of subsets of sets Ω; P - probability measure.

5. If the targeted intervention is carried out in the system to achieve certain values of operational parameters yi, the process during this period can be considered as stationary deterministic with a given distribution function values:

$$y(t) = F_y(t) \ . \tag{3}$$

6. The reliability index is determined by the function depending on two variables:

$$K_N = F(y,t) \ . \tag{4}$$

Thus, to build the mathematical model it is necessary to solve two problems:

1) determine the dependence $K_N = F(y)$ (5)

2)define the dependency $K_N = F(t)$ (6)

The problem of optimal operational planning assessment and forecasting is the management: a priori random process $y(t)$ for uncontrolled objects; a posteriori random process $y^{Ps}(t)$ with application of individual technical interventions for controlled objects.

In the control interval T_r predictable process can be observed in the presence of some noise (measurement error) ε (t). The overall task of the parameters drift forecasting can be formulated as a problem of prediction characteristics of a random process in the range of Tr according to the results of observations, distorted noise, which are obtained in the control range.

There are two components within creation of a model:

a) "observed" process:

$$z(t), t \in T \tag{7}$$

b) "unobservable" random process: $W(t), t \in T$ (statistically connected with $z(t), t \in T$).

Let in a time point

$$t \in T_p, T_p \subset T \tag{8}$$

where Tr - control interval (realization of "observed" process is known $z_{\omega}(t)$).

Changing of parameters is influenced by:

1. random factors;
2. deterministic processes.

Therefore, changes in the parameters are random or deterministic process, subject to a certain distribution law.

4 Formulation of the problem

Purpose of calculations:

1) evaluation of reliability index according to continuously changing technological parameter in real time;

2) definition of the function of parameter change over time;

3) forecasting of reliability index values in the prediction interval tprog.

Reliability assessment model in parameter is based on the following requirements of the established range of values $y_i \in (y_{\min cr}; y_{\max cr})$ (fig.1, 2):

$$1)\ \text{if}\ \ y = y_{\min \mathrm{Kp}},\ \ y = y_{\max \mathrm{Kp}},\ \ \mathrm{N}=0; \tag{9}$$

$$2)\ \text{if}\ \ y = y_{\min \mathrm{cr}} + \frac{y_{\min \mathrm{cr}} + y_{\max \mathrm{cr}}}{2},\ \ \mathrm{N}=1, \tag{10}$$

(in case the optimum technological parameter for object is not the middle of an interval $y_i \in (y_{\min cr}; y_{\max cr})$, we admit under $N=1$, $y=y_{opt}$).

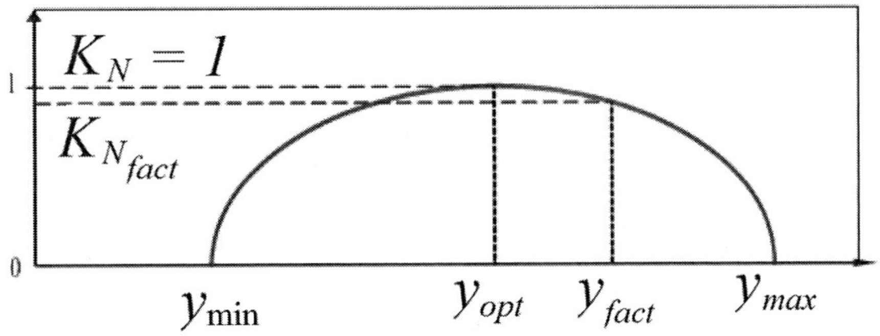

Fig. 1. Assessment scheme of reliability index in actual registered parameter.

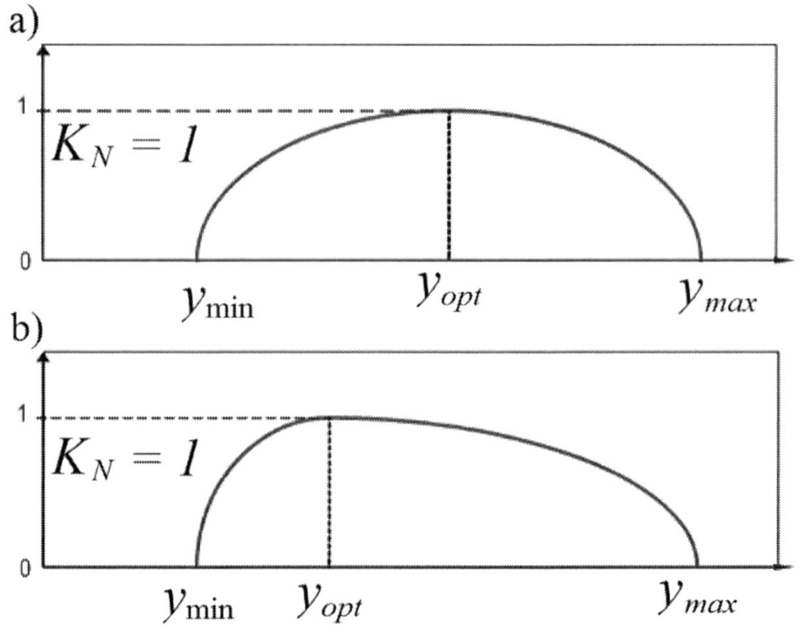

Fig.2. Dependence scheme in changing reliability index:
a) for symmetric and b) for asymmetric dependence of reliability function upon parameter change.

The advantage of the applied model is the continuous specification of distribution parameters in real time. Technique of realization of continuous monitoring of reliability is that at every moment of registration of the parameter calculation is performed on the reliability index of probability and statistical models.

Due to lack of a universal method it is necessary to develop the algorithm for determining distribution law of deviation of parameters, peculiarity of which is the ability to identify and clarify in real-time mode distribution parameters by statistical simulation modes.

Assume that for the solution of the given problem there is a number of operational registered parameters y of the volume n. Since the model works in real time, censored (incomplete) sampling is used for the calculations.

We will define the function of change of reliability depending on change of parameter of technical condition and the smallest volume of sampling for a distribution function assessment. According to classical methods of experiment planning at confidential probability 0.99 and the required accuracy up to 2% the minimum volume of selection makes (55 $\pm$ 19).

We will make statistical processing of the registered data by standard methods with the use of average arithmetic value.

As a result of data analysis we will also reveal the periods where function is determined or random.

Let the general function of distribution of parameter of technical condition in time be defined as the determined and casual constituent and has the following form (fig. 3):

$$F(x) = y_{\text{det}}(t) + y_{\text{def}}(\text{t}) + \varepsilon . \tag{11}$$

Then, in addition to the standard "releases", we will exclude from the sampling the periods, where the system parameter level is purposefully maintained at the same level y = const, i.e., factors determining the technical condition of the system are excluded.

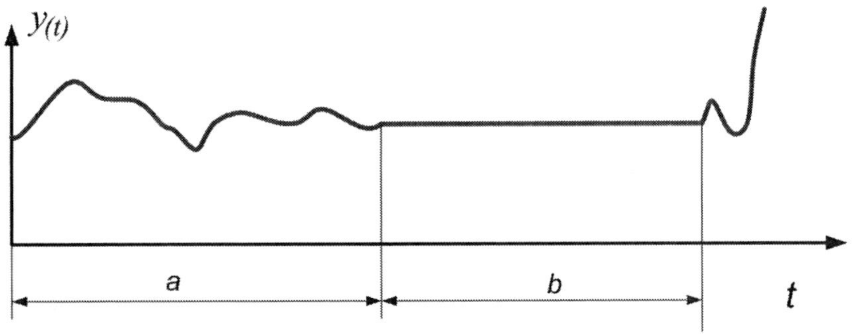

Fig. 3. Analysis of the function of the parameter over time.

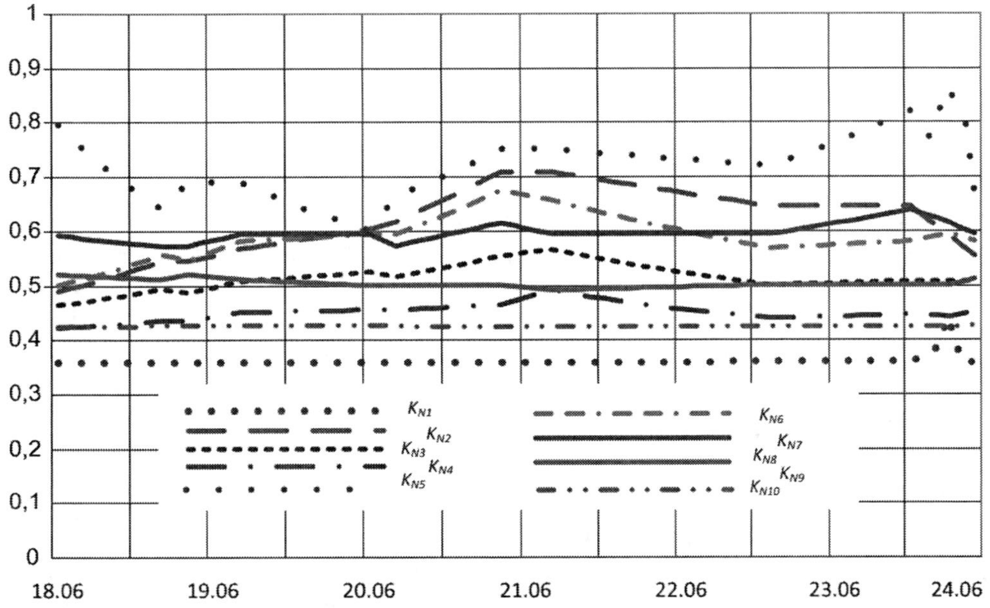

Fig.4. Monitoring results of reliability index of the pump of the retaining vertical pump at zero level.

As an example in fig. 4 it is shown that for one of object subsystems, for example, of the pump, various indicators of reliability can be received. The graph shows that in the analyzed period, the lowest rates are KN1, KN2, i.e. (according to interpretation) the lowest level of reliability have thrust bearing and the upper bearing of the pump due to high temperature and vibration and approximation of these parameters to the emergency values. Similar studies and calculations carried out by analytical and graphical dependencies at various levels of complexity of the system indicates that for the development of correct recommendations of elimination the causes of reduced reliability of pumping units the detailed and complex analysis of the modes and parameters during the entire period of operation of the equipment has to be carried out [3,5,11].

To obtain adequate models of reliability it is very important to choose the type of distribution of failures or events corresponding to the actual, which largely depends on the accuracy of the results. To determine the type of distribution we use four known methods based on the application of: engineering data; graphical evaluation methods; nonparametric methods and evaluation criteria.

The use of engineering data can prevent blunders at the stage of choosing a distribution. Here it is possible to estimate the kind of failure rate function and identify general information about the operated object. The use of graphical methods is justified on application of dispatching data or test data. Non-parametric methods are applied if it is not possible to evaluate satisfactorily the form of distribution, while the test time is equivalent to the required operating time of the equipment. Time as a factor, is possible to exclude and examine the signs of failure [15-21].

5 Conclusion

Thus, the developed technique allows carrying out an operational assessment of reliability index of an industrial facility and within monitoring increase efficiency of making administrative decisions on technical and preventive maintenance. The created technique can be used by research, design institutes and oil and gas transport enterprises when declaring industrial safety.

References

1. M. Zemenkova, *PhD Thesis* (TSOGU, Tyumen, 2007)
2. V. Kurushina, Yu. Zemenkov, WIT Transactions on Ecology and the Environment, **190(2)**, 881-888 (2014)
3. V. Antip'ev, A. Nevolin, Yu. Zemenkov, Neftyanoe Khozyaistvo - Oil Industry, **10**, 46-48 (1981)
4. A. Bacherikov, M. Pashkov, Yu. Bogatenkov, Yu. Zemenkov, V. Shalay, G. Vasiliev, A. Prokhorov, S. Sidelnikov, S. Podorozhnikov, V. Kehl, *Diagnosis in the maintenance of pipeline transportation* (Nedra, St.Peterburg, 2007)
5. S. Dudin, Yu. Zemenkov, A. Shabarov, N. Saranchyn, Oil and Gas, **6**, 63-68 (2010)
6. A. Shipovalov, Yu. Zemenkov, S. Toropov, et al., *Aspects of technological reliability and economic efficiency of operation of underground storages of natural gas of Western Siberia* (TSOGU, Tyumen, 2012)
7. A. Korolenok, *Technological forecasting of capital repairs of the main gas pipelines* (TSONIK GANG, Moscow, 1997)
8. Yu. Zemenkov, V. Shalay, M. Zemenkova, Procedia Engineering, **113**, 254-258 (2015)
9. Yu. Zemenkov, V. Shalay, M. Zemenkova, Procedia Engineering, **113**, 312-315 (2015)
10. M. Zemenkova, Yu. Zemenkov, A. Pimnev, V. Petryakov, Business magazine Neftegaz.RU, **11-12**, 64-70 (2015)
11. S. Stepanov, N., Cheremisin, S., Sokolov, A., Altunin, A., Shabarov, *ECMOR 2012 - 13th European Conference on the Mathematics of Oil Recovery* (2012) DOI: 10.399/2214-4609.20143270
12. S. Toropov, Yu. Zemenkov, S. Podorozhnikov, S.Yu. Gas industry, **720(720)**, 95-98 (2015)
13. M. Zemenkova, L. Fayzulina, I. Seroshtanov, Gas industry, **720(720)**, 59-64 (2015)
14. A. Pimnev, M. Zemenkova, Pipeline transport: theory and practice, **5(51)**, 43-47 (2015)
15. A. Shabarov, A. Ibragimov, M. Zemenkova, S. Podorozhnikov, Gas industry, **12(715)**, 50-54 (2014)
16. O. Ivantsov, *Reliability of building constructions of the main pipelines* (Subsoil, Moscow, 1985)
17. I. Ryabinin, *Reliability and safety of structural and difficult systems* (Polyequipment, SPb, 2000)

18. O. Abramov, A. Rosenbaum, *Forecasting of a condition of technical systems* (Science, Moscow, 1990)
19. V. Ostreykovsky, *Theory of reliability* (High school, Moscow, 2003)
20. E. Yasin, V. Berezin, K. Rashchepkin, *Reliability of the main pipelines* (Subsoil, Moscow, 1992)
21. I. Masur, O. Ivantsov, O. *Moldavanov, Constructive reliability and ecological safety of pipelines* (Subsoil, Moscow, 1990)

Traffic safety as a factor of improving logistics in construction sector

Slobodan Ognjenovic [1,*], *Tatiana* Sazonova[2], *Riste* Ristov[3], *Ivana* Nedevska [4] and *Anton* Kibkalo[5]

[1]University Ss. Cyril and Methodius in Skopje, 1000, Partizanski odredi, 24, Skopje, Republic of Macedonia
[2]St. Petersburg State University of Architecture and Civil Engineering, 2-Krasnoarmejskaja, 4, Saint-Petersburg, 190005, Russia
[3]Prostor DOO-Kumanovo, 1300 Mosa Pijade 2, Kumanovo, Republic of Macedonia
[4]Geing Krebs und Kiefer-Skopje, 1000, Boris Trajkovski str., Skopje, Republic of Macedonia
[5]Peter the Great St. Petersburg Polytechnic University, 195251 Polytechnicheskaya str. 29, St. Petersburg, Russia

Abstract. Security is a complex problem. It includes the driver's behavior, the features of the vehicle, those of the road and the driving conditions. Security can be approached from the aspect of the vehicle and pavement design, of driving habits, traffic regulation and application of laws. The good coordination of the driver, the vehicle and the road enables for better estimation of the traffic safety relations and interrelations giving the possibilities of improvement of the situation at present and in future. The driver has the main role in the determination of the success or failure of the road system. Over 90% of the accidents are due to improper behavior of people in traffic. Understanding of the human factors is the key factor in the optimal road design and traffic flow. Nevertheless, engineers need more understanding about driving behavior, as they are responsible for road design, for the traffic flow, and for the regulations intended to avoid errors wherever possible. The purpose of this paper is to inform about the current situation both in practice and in the researches of traffic safety, as well as to yield a broader review of the actual traffic safety situation on the roads in Republic of Macedonia.

1 Introduction

The terms of active and passive safety have been used in the field of transport since the 60-ties. Passive security can be improved much easier and for much lower price than active safety. Therefore, more attention is paid to improvement of passive safety.

Actice safety is becoming ever more essential in improvement of traffic safety. It includes the following:

- Dynamic sefety
- The behaviour of the driver

* Corresponding author: ognjenovic@gf.ukim.edu.mk

- Perception/visibility
- Ergonomy
- Local conditions in the vehicle (air conditioner etc.),

whereby the driver's behaviour is the key source in the improvement of active safety in view of prevention of traffic accidents [1-6].

The improper behaviour of the driver is the main reason of accidents besides the environment influences. It results from the lack of interaction betweent he driver and the vehicle and/or failure to understand the situations regarding the relation between the road and ghe environment [7-10].

In this reference, the most efficient methods of road safety improvement is the better interaction betweent he driver and the vehicle and designing roads on which the features relevant for driving are visible, understandable and well-designed, based on driving dynamism, and on the driver's behaviour from the psychological point of view [11-14].

2 Levels of action in driving

This basic relation can be seen on Figure 1.

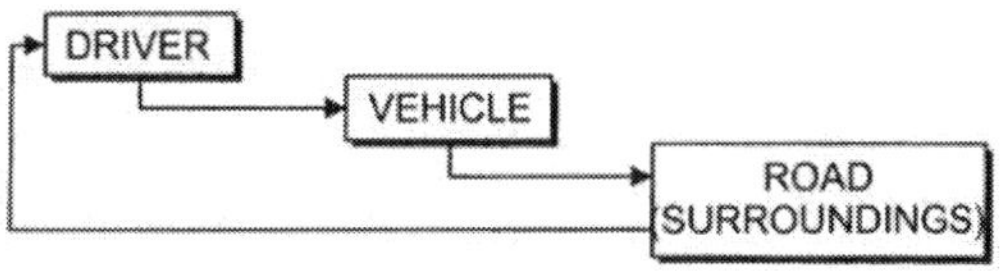

Fig. 1. Schematic presentation of the interaction driver - vehicle - road (surroundings).

Driving is mainly influenced by the applied and used information. Perceived information is compared to the information processed by the driver, which is followed by bringing a decision and taking action. Driving includes many discrete and related activities consisting of three levels of action:

-Control
-Guidance
-Navigation

Control: Control includes all activities related to interaction between the driver and the environment, as steering and velocity control, and the performance of the actions starting from the relatively simple (as for example, on the passenger vehicles) to more complex ones (as in the vehicles of several gears and clutches). Information is provided from the sense about the vehicle, from tis displays and from the pavement. Drivers perform accurate adjustments and and controls acccording tot he vehicle reactions. Once mastered, most control activities are performed automatically with little conscious effort.

Guidance: On this level the major activities of the driver are the maintenance of safe velocity and appropriate trajectory related toth e road and the traffic elements, as a response to the road and traffic conditions. Guidance activities are characterised by reasoning, estimation and foreseeing, in a dynemic and constrantly changing environment.

Navigation: The navigation level consists of a pre-journey phase, the one of planning the journey and choosing the route, and a journey phase that icludes the following of the previously made plan [14, 15].

All three levels are graphically presented on figure 2

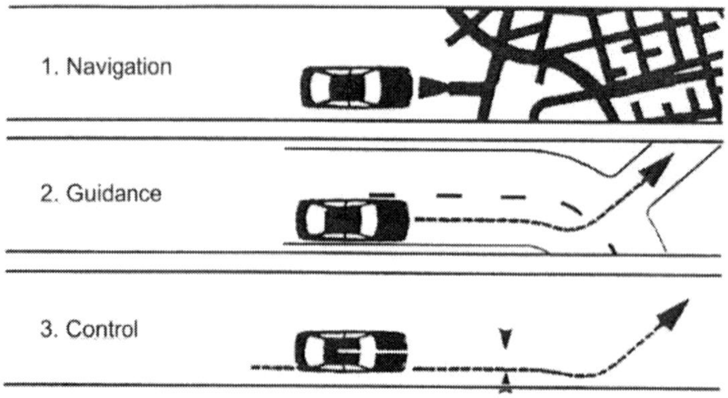

Fig. 2. Levels of action in driving.

3 Gathering of information

On eof the mosty important driving skills is th eperception of the lateral (side) force created upon change of the driving dirienction. The level to which the kinesthetic perception is relation to visual perception has no been clearly established. The capacity of save driving with high speeds can be largely influenced by the itneraction of these two senses. Drivers usually choose the velocity before taking a turning. If the curve is an unusual one, this choice can rely on visual perception. Therefore, when the driver selects the velocity, they envisage the lateral force that is to be felt in the curve and, refering to it, the velocity os going to be reversely proportional. The research result indicate a strong reverse relation between velocity and lateral acceleration for the velocities of over 32 km/h. The lateral force created in the curb is the first criterion in the chouce of velocity. The kinesthetic perception of the lateral force is the first option by which the driver receives information that will be used in driving. There is itneraction between this kinesthetic perception and the visual perception developped fast while the driving skill is being learnt, and can be degraded by reasons such as drugs, alcohol and ageing. [16]

The velocity that is too high for a certain situaiton is the main reason of accidents causing considerable material damage and injuries. Generally speaking it can not be supposed that the driver is consciously rushich towards danger if they choose high velocity. It could be said that the driver, as pursuant to their subjective assessment of the situation, estimates the danger and selects the velocity that they consider as appropriate. Ideally, the driver would be able to control the vehicle velocity by a speedometer, compare that velocity with the necessary one and coordinate both velocities.

In reality, however, the the value of the desired velocity, as based on the estimation of the driver's risk, is suspicious. For example, the necessary velocities depend on subjective factors, as the psychological state, the need to arrive sooner, drug abuse, etc. On the other hand, the actual velocity is controlled to esentially lower level by a speedometer. This means that the driver feels only by its senses, sight being the most important of them, while the role of the kinesthetic (movement) and the haptic (touch) is a supplemental, not a minor one.

4 Reaction time

The reaction time has an important role in the driver's mental capacity. Tghis is the time lapse from the receipt of information, its processing until the action taken. Reaction time varies from one driver to another, and is the function of readiness, complexity and expectance.

Complex decisions take longer reaction time compared to the simple ones. Reaction time and the driver's behaviour have similar effect. The longer the reaction time, the greater the possibility of an error while driving [17].

The relation between the reaction time in seconds and the copmplexity, expressed in bits, is given on figure 3 on an average 85% driver. A bit is the information necessary to solve the demur between two equal probabler eactions. The picture demonstrates that even the 0-bit decision (one alternative) should be developed in a certain time and that the reaction time increases exponentially. Only a few drivers can process 3-bit or 4-bit information (8 or 16 alternatives), as a transitory decision [18].

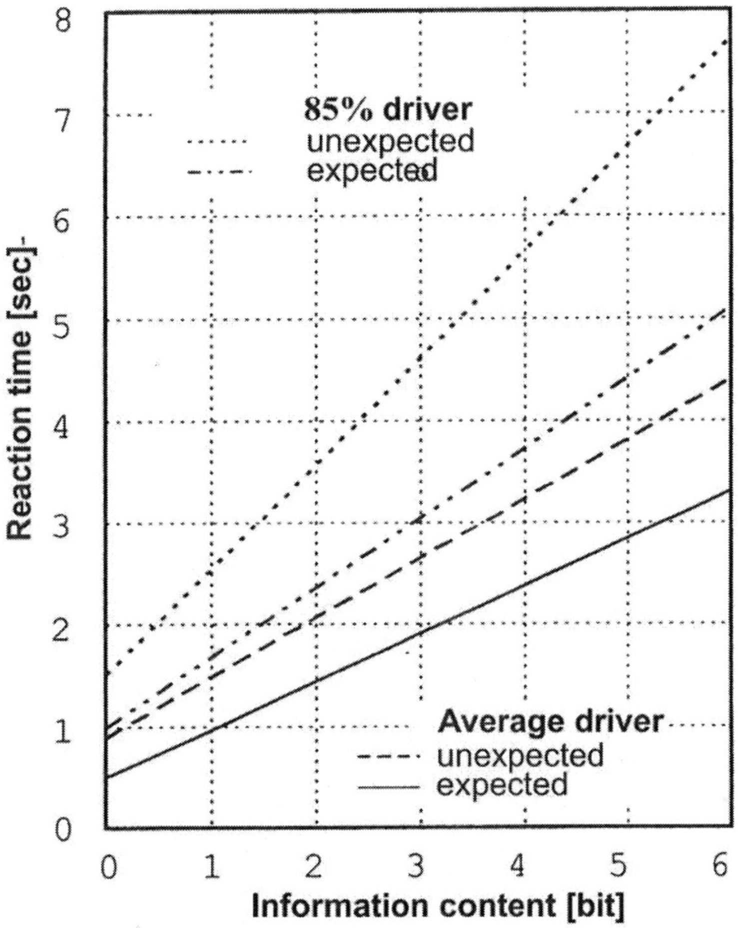

Fig.3. The reaction time of response between average and 85% of drivers.

Besides, the processing of a complex decision requires time and attention, apart from the other necessary sources of information. Thus, in situations requiring more detailed proecssing, it is usually more necessary or faster to bring several simple decisions, than one complex one. This means that the road design should be as simple as possible, without any complex sequences of elements and without many intersections. All important interactions referring tot he reaction time and the obvious distance should be processed in detail from the design point of view [19].

5 Traffic accidents coused by the driver's error

Many researches have been conducted in traffic psychology, their primary purpose being the establioshent of the psycological factors contributing to human error in driving. The principal feature of the locations with a large number of accidents is the fact that important or unusual requests are placed considering those locations regarding the processing of information on the drivers. The inefficient operaitons and the eccidents usually occur where yit is difficult to cope with the information. On locations where the road design is worse, the possibility of error and improper behaviour of the driver increases.

5.1 Error due to information ambiguity or to confusing information

The highest number of drivers' errors is due to false interpretation of the traffic situations and inattention. Such situations result from bad design solutions that can confuse the driver in the moment of decision-making. This phenomenon is especially critical on curves and in transitional curves and is called illusive curve.

That is a curve that does not look so sharp a sit is and the need to reduce speed can be completely disregarded. This is manifested by hard braking in the curve or, in extreme cases, by accidents. Thus, there is a divergence between the physical parameters of the curve and the psychological and physical reaction to the curve in the drivers. On subjective level, the driver understands that they enter the curve too fast. Therefore, it is supposed that an illusive curve creates a discrepancy between the physical parameters of the curve and its psycho-physical correlates. In such case, accidents result from the divergence between the physical and perceived curve, the lattrr influencing the velocity on entering the said curve.

Thus, regardless of the real curve, the partly conceald one looks "shallower" than the more visible curve with the same radius. It can be concluded that a curve with a large number of accidents is the less visible one, and not the curve similar to th eone with lower number of accidents.

This raises the question about the determinants of the driver's decision to reduce speed before a curve and the perceptive factors related to accidents. In reference therewith, the perceived curve sharpness seems to be the main determinant causing the drivers to reduce speed. The perceived width comes immediately after. Therefore, drivers should reduce speed in front of curves that look sharper, narrower and closer. The location of the driver related to the curve shuld also be taken into account. Obviouslym the driver does not possess all information necessary to estimate the real length and angle of the curve.

The following picture shows the edges and the central lines in a curve with a low number of accidents and in one where the number of accidents is high. The sharper the curve, the smaller the angle from the internal perspective.

The angle of the outer and of the inner perspective decreases with the increase of the real cruve angle (that is, the curve sharpening). However, for a given horisontal curve, the perspective angle is also under the influene of the vertical inclination and the level of elevation. Generally, the vertical curbe with a flexion increases the perspective angle, and reduces the hip-shaped curve making it look sharper.

5.2 Error due to site requirements

The improper motivation of the driver seems to be the main reason for accident. But, the potential road dangers, especially upon high speeds, indicate the necessity of care and readiness. If we do not suppose that people are very unreasonable, the failure of the motivational attitudes regarding the reduction of accidents indicates that the reasons for accidents can be such that even highly motivated people often fail to avoid errors. The basic conclusion reveals that there must be other cognitive mechanisms so strong, that they are able to surpass motivation as the main cause for errors.

5.3 Error due to reduced driver's capacity

Many erors are caused by inadequate capacities of of the drivers faced with an inadequate design or difficult situations. They contribute to accidents as their "secondary factors". For example, insufficient training often leads to incapacity to cope with the situation leading to slipping of the vehicle. Similarly, inadequate perceptory behaviour and taking risks can also lead to a driver's error.

Physical disabilities have limited influence on errors in driving. This is explained by the fact that the physical ability tests, for example the sight test within the procedure of acquiring a driver's license, eliminates an important part of possible accidents due to physical disabilities. On the other hand, drivers with such disabilities, as for example, elderly ones, can compensate them by driving carefully.

5.4 Characteristics of drivers by age groups

The highest accident risk is the one with the 18-24 years of age group of drivers. Insufficient experience, and the willingness of younger drivers to take higher risk are the main reasons thereof.

On the other hand, the number of accidents is lower in "more elderly" drivers (aged over 65 years). Their motorical capacities are more or less reduced, their eyesight, hearing capacity and memory are lower. They compensate these disadvantages by drivign slowlier, by avoiding stressful situations and relying on excperience. But the number of accidents is considerably high with this group of drivers as well, and they mostly include chain crashes and left-side turnings on intersections.

6 Conclusions

The purpose of this paper was to indicate the general mechanisms, from the theoretical point of view, regarding the adjustments of the drivers' behaviour related to sefety measures. A reasonable theory on the process of adaptation could, in special circumstances, be the basis of a homeostatic model of risk. The theory should not be defined as pursuant tot he accident. On the contrary it should foresee the accident level. From the theoretical point of view, it is possible to bring the following conclusions:

1. The mechanisms of adaptaiton are phenomena that are noticeable and have positive influence on security
2. The adapttion mechanisms can be observed as related tot he hierarchical stages of driving and a process to obtain results as a function of time should be developped
3. The notion of risk lies in the center of the theories explaining adaptation processing, and this notion has its objective and subjective component,
4. The subjective risk lies in the relation between the motivational components of the driver and the perception of the situaiton in which the driver is placed at a certain time.
5. The adaptation mechanisms are assisted by:

- The overestimation of the driver's own capacities, as in the perception function,
- The decision to accept high lisks,
- The incapacity of appropriate risk estimation.

References

1. R. Lamm, B. Psarianos, T. Mailaender, *Highway Design and Traffic Safety Engineering Handbook* (McGraw-Hill, New York, 1999)
2. N. Moray, Attention and Mental Models, Ergonomics, **33**, 365-372 (1990)
3. A. S. Cohen, Journal for Traffic Safety, **2**, 125-129 (1997)
4. *National Bureau of Statistics of the Republic of Macedonia* (Transport and other Communications, Skopje, 2013)
5. Z. Zafirovski, N. Vatin, Applied Mechanics and Materials, **725-726**, 349-354 (2015)
6. Z. Zafirovski, N.Vatin, Applied Mechanics and Materials, **725-726**, 214-219 (2015)
7. M. Jocovic, B. Melovic, N. Vatin, V. Murgul, Applied Mechanics and Materials, **678**, 644-647 (2014)
8. V.V Okrepilov, M.V. Leonidovich, Asian Social Science, **11 (7)**, 312-325 (2015)
9. V.V Okrepilov, V.L. Makarov, A.R. Bakhtizin, S.N Kuzmina, Economy of Region, **2**, 301-313 (2015)
10. V.L. Kvint, V.V Okrepilov, Herald of the Russian Academy of Sciences, **84 (3)**, 188-200 (2014)
11. V.V Okrepilov, V.N. Krutikov, G.I. El'kin, Measurement Techniques, **57 (2)**,109-116 (2014)
12. B. Kovačič, R. Kamnik, A. Štrukelj, N. Vatin, Procedia Engineering, **117** (1), 800-806 (2015)
13. B. Kovačič, R. Kamnik, M. Premrov, Survey Review, **43** (320), 150-161 (2011)
14. B. Kovačič, R. Kamnik, A. Štrukelj, Promet - Traffic - Traffico, **22** (1), 15-22 (2010)
15. V.V Okrepilov, A.Yu. Smirnov, Measurement Techniques, **56(1)**, 54-60 (2013)
16. V.V Okrepilov, Studies on Russian Economic Development, **24(1)**, 35-42 (2013)
17. V.V Okrepilov, Standarty i Kachestvo, **10**, 52-55 (2005)
18. V.V Okrepilov, Ivanova, G. Standarty i Kachestvo, **12**, 62-68. (2004)
19. V.V Okrepilov, Standarty i Kachestvo, **3**, 94-96. (2003)

MATEC Web of Conferences **73**, 07003 (2016) DOI: 10.1051/matecconf/20167307003
TPACEE-2016

Motivational Preferences of the Managers in Production and Promotion of Innovative Materials and Technologies

Nadezda Miloradova [1,*], and *Alexander* Ishkov [1]

[1]Moscow State University of Civil Engineering, 129337 Yaroslavskoye shosse 26, Moscow, Russia

Abstract. The article discusses the results of the comparative analysis of motivational profiles of managers, who work in the construction industry with innovative materials and technologies. In Russian construction industry the innovative building materials and technologies are being introduced slowly because of many reasons: the absence of a clear state policy in construction industry, dependence on the estate market. The purchasing possibility of the population is low. However, the innovation requires an appropriate motivational spirit of managers: "Interest and usefulness", "Creativity", "Achievement", "Variety and change".

1 Introduction

Productivity, efficiency and quality of labor activity depend not only on the qualifications of the staff, but also of their motivation. It is not surprising that employers prefer to highly motivated individuals whose motivational preferences are corresponded to the company organizational goals. Managers and leaders dream of working with highly motivated team, where its members have the same motivational orientation.

Now effective management is impossible without the knowledge about the origin of motivation, without the ability to recognize motivational preferences. The modern world provides wide opportunities to find work and people are actively using this opportunity. People want employers need to take into consideration their interests. Therefore, the heads tasks are to find motivated staff and keep the most valuable employees.

People always were interested in the origin of motivation and methods for the recognition of motivational preferences. The problems of motivation are actively discussed and studied in various social sciences: philosophy, sociology, history, management and others [1-3]. For the last century quite a lot of psychological theories of motivation were developed, many hundreds of empirical studies were carried out [4; 5].

In the motivational structure of human personality motives occupy different positions: some of them are major, defining human activities and behavior, the others are dominant, being the total emotional and motivational background, and another's are latent (hidden, implicit) [6].

* Corresponding author: milordos@gmail.com

Motivation is significantly associated with the development of society: the material and ideal objects, the person is ready to work for, change under the influence of society changers. The list of motivating factors is constantly expanding, changing. During the life a number of motivating factors person implements at the same time. His motivational sphere is fluid and converted [7].

Search of the motivating factors most often is conducted within the framework of hierarchical models of motivation by A. Maslow [8]. According to this theory, humans have five levels of needs: physiological, security, social, esteem and self-actualizing. In modern classifications of needs the first two levels are combined into one – the low (physiological). This level presents human needs of wage, financial rewards and comfortable working conditions. Today this level is considered just a condition of human survival, not motivation.

In the early XX-th century psychologists began to pay special attention to the social needs of human: public recognition, contacts with colleagues, etc. These motivating factors were considered important to a human regardless of the position which he occupies [9].

In the XXI-st century the values of the contemporary world community have had significant changes. The development of information technology leads to more individualization of work and as a result, the role of social motivation is reduced. The motivators of psychological (personal) level became the most important among the all motivators: commitment to diversity, changes or stability of structure; creativity and self-improvement [10].

The production and promotion of innovative materials and technologies is developing intensively [11; 12]. However, as the practice shows, innovations in the Russian's construction sphere are introduced in the last turn. As the innovative analytics, the main obstacle for the development and implementation of innovative technologies in the construction industry is the lack of clear state policy in the field of construction. One more reason of "innovative slowness" of construction industry is the fact that it is rigidly integrated in the real estate market, which does not seek to introduce a radical technological innovation. The purchasing power of the population is low, so cheap housing are constructed [13; 14].

In addition to the technical side of the development of innovative technologies, there is another side. This is the mentality of the people and their desire and ability to create something new and willingness to implement the new product. Russian people learn quickly, they are interested in new technologies and building materials, ready to accept innovation complex. They are aware of everything new that is being developed in other countries. But the question is whether they are ready to introduce new building materials and innovative technologies in the construction industry in their country. Sociological studies show that, on one hand, the residents are interested in high-quality innovative products. On the other hand, the Russian tendency of investor or the company's executive is to use the cheapest building material and technology. It should be noted that there is a "dictatorship" of builders in Russia, because the market is not saturated, the demand exceeds the supply a lot. The purchasing possibility of consumers is low [15].

The effectiveness of innovation venture involves the creation of a special organization culture, in which they have common motivational basis, focuses on the optimal combination of motivating factors [16; 17].

Researches of motivating factors of the managers have a particular meaning, because these people are forming the organization culture [18]. They are "trendsetters" in their organization: they implement their own ideas about the motivating factors that encourage some motivators and impede the implementation of others among subordinates.

2 Hypothesis

Development of innovative building materials and technologies, their promotion on the construction market and the widespread use in mass construction industry, the managers of these processes have such leading motivating factors as "Interest and usefulness", "Creativity", "Achievement", "Variety and change". This combination of motivators creates the unity of the motivational basis of man, leading him to the solution of creative tasks and the implementation of results of innovative searches.

The purposes of the research are:

1) description of motivational profile of Russian managers, who work in construction industry;
2) to make comparative analysis of motivational profiles of Russian and European managers.

3 Research methodology:

The test-questionnaire by Sheila Ritchie and Peter Martin has been selected to research a motivational profile [10]. Among the huge number of psychological tools to identify motivational preferences, the work of these authors has attracted our attention for several reasons.

Firstly, these authors have their own 12-factor model of motivational profile. In this model, traditional motivators (physiological and social) and also some new - psychological motivators are presented. Each motivator is described in detail.

Secondly, more than thousand managers from various European countries were tested by this test-questionnaire. The results of research by these British authors are presented in detail in their book. The results of the test-questionnaire may be presented in a graphical form as the motivational profile.

Russian version of the questionnaire adapted by A. D. Ishkov contains 33 statements, which have four possible answers each [19].

The study of motivational preferences has been hold in Moscow State University of Civil Engineering (MGSU) since 2004. During these years thousands of students from different faculties and more than 500 people, who attend (a) short training courses for managers of investment and construction sector and b) the two-year MBA program "Financial management", took part in this study.

This research was conducted in MGSU from 2014-2016.

The study involved:

- twenty seven top-managers aged from 34 to 46;
- fifty three middle managers at the age between 27 and 53.

4 Results of the research

Data obtained in the research were summarized and presented in graphical form (Fig.1) together with the results about Europeans managers received by W. Richie and P. Martin.

Fig. 1. Motivational profiles managers.

Motivational profile according to the method of W. Richie and P. Martin consists of twelve motivators, represented by corresponding numbers:

1. Money and tangible rewards.
2. Physical condition.
3. People Contact.
4. Relationship.
5. Recognition.
6. Achievement.
7. Power and Influence.
8. Interest and usefulness.
9. Structure.
10. Variety and change.
11. Creativity.
12. Self-development.

As we can see from the figure, the profiles of the Russian and Europeans managers vary considerably.

The results of research of Russian managers (top level) are: the leading motivator for them is only "Interest and usefulness". They are followed by motivators with the different degree of expressiveness: "Recognition", "Desire to achieve", "Self-improvement", "Commitment to Diversity and Change" (Figure 2). The motivators "Physical conditions of work" and the "Desire for structuring and procedure" located at the end of the motivational preferences list. These motivators, as shown in the research results, are initially insignificant. They said: "Constructions does not expect comfortable working environment and ideal order! Constructions is not the office, where people work with the papers!"

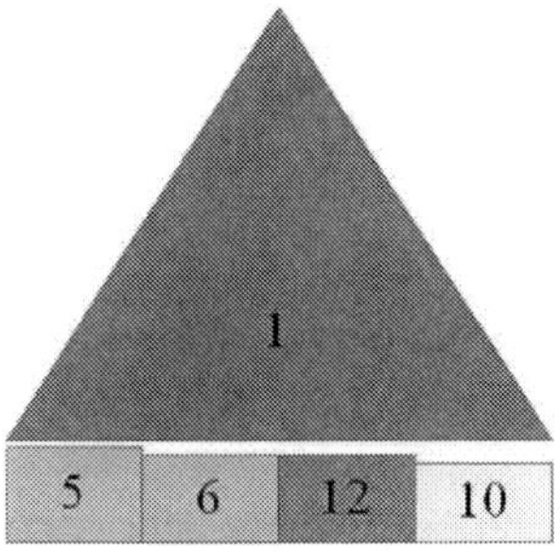

Fig. 2. The main motivators of Russian managers (top level).

The results of research of Russian managers (middle level) are: they have three leading motivators such as "Recognition", "Money and tangible rewards" and " Achievement." They are followed by the motivators with different degree of expressiveness: "Self-development", "Power and Influence", "Variety and change " (Fig. 3.) The same motivators "The physical conditions" and the "Structure", which top-managers are had at the end of the middle level managers motivational preferences list.

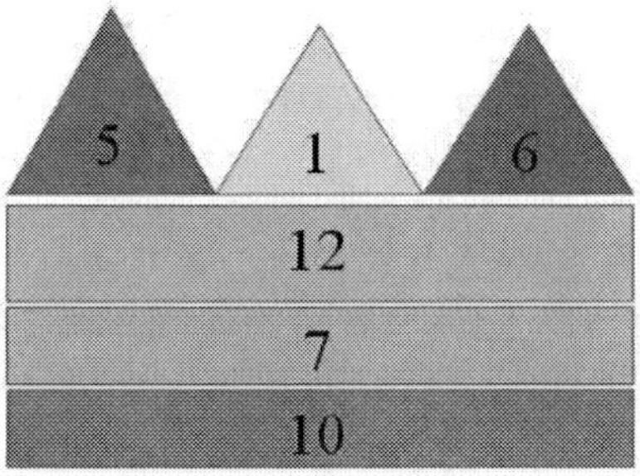

Fig. 3. The main motivators of Russian managers (middle level).

The results of research of Europeans managers by S. Richie and P. Martin are: the leading motivator for them is only "Interesting and useful work". They are followed by a motivators with almost the same degree of expressiveness: "Recognition", "Achievement", "Variety and change" and a little behind them is "Creativity" (Fig. 4). The motivators of physiological level "Money and tangible rewards", "The physical conditions " and "Relationship" are at the end of the motivational preferences list. As S. Richie and P. Martin noted, it is an indicate that these requirements have already been satisfied. These manages reached the financial level, which they considered appropriate for them. So these motivating factors don't determine their behavior any more.

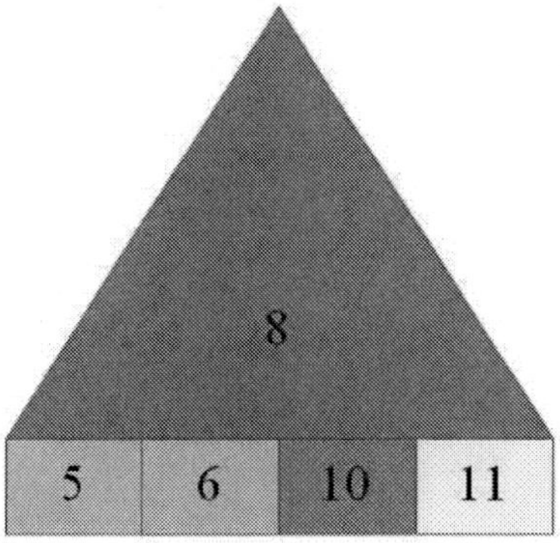

Fig. 4. The main motivators of Europeans managers.

The need of "Money and tangible rewards" is a leading motivating factors for the majority of Russian managers of all levels. The motivational profile of the university students in the construction direction is similar. This tendency has not changed during twelve years of studying motivational preferences. The attitude to such motivators as "Structure", which in the desire to work according to the rules, has not changed. As a "justification" Russian managers believe, that strict regulation prevents the creative process and don't inspire new ideas [20].

Thus we can state that the necessary conditions for development of innovations have at least two options: "Interesting and useful work' as a leading motivator and high level of such motivating factors as "Creativity", "Variety and change ", "Achievement". These motivating factors support the leading motivator. If these conditions are not observed, the innovative activity is really slow.

References

1. E.E. Lawler, *Motivation in work organizations* (Books/Cole Publishing Co, Monterey, Calif, 1973)
2. M.R. Ryan, E.L. Deci, Contemporary Educational Psychology, **25(19)**, 54-67 (2000)
3. C. Wiley, International Journal of Management, **18(3)**, 263-281 (1997)
4. G.A. Cole, *Management – Theory and Practice* (South-Western Cengage Learning, London, 2004)
5. F. Storseth, International Journal of human Resource Development Management, **4(3)**, 267-287 (2004)
6. D. Kun, *Osnovy psikhologii. Bol'shaya entsiklopediya. Vse tayny povedeniya cheloveka* (Praym–Evroznak, Sankt-Peterburg, 2008)
7. Z. Ivanova, I. Pichugin, J. Naimaviciene, Procedia Engineering, **117**, 154-161 (2015)
8. A. Maslow, *Motivation and personality* (Harper, New York, 1954)
9. E.V. Romanova, EHkonomika i predprinimatel'stvo, **6-3**, 793-796 (2015)
10. S. Ritchie, P. Martin, *Motivation Management* (Gower Pub Co, London, 1999)
11. A.D. Ishkov, A.V. Stepanov, S.V. Miloradov, I.V. Voronina, Applied Mechanics and Materials, **670-671**, 458-461 (2014)
12. A.D. Ishkov, D.A. Semernin, S.V. Miloradov, I.V. Voronina, Applied Mechanics and Materials, **741**, 500-503 (2015)
13. Information on https://imi.hse.ru/pr2013_1
14. E. Shnyrenkov, I. Pryadko, Procedia Engineering, **117**, 330-335 (2015)
15. Information on http://www.gks.ru/bgd/regl/B12_46
16. A.D. Ishkov, M.Y. Mishlanova, K.P. Grabovyi, International Journal of Applied Engineering Research, **11-3**, 1676-1679 (2016)
17. E.A. Lyubushina, E.V. Romanova, EHkonomika i predprinimatel'stvo, **6-3**, 925-928 (2015)
18. T.N. Magera, Vestnik MGSU, **6**, 527-532 (2011)
19. N.G. Miloradova, A.D. Ishkov, Ekonomika i predprinimatel'stvo, **5-2**, 891-896 (2014)
20. N.G. Miloradova, A.D. Ishkov, *Psikhologiya upravlencheskoy deyatel'nosti: izmenenie podkhodov* (Flinta, Moscow, 2014)

Project Management in the Field of Innovative Technologies of Environmental Management

Galina Potasheva[1,*]

[1]Moscow State University of Civil Engineering, 129337 Yaroslavskoe sh. 26, Moscow, Russia

Abstract. The article deals with the application of project management in the field of innovative technologies of environmental management. This article examines the sustainability of integrated development environment, and society in the "three ones" concept, which covers all major aspects of development and global change: ecological, social, economic-based planning. But the transition from the national level to the practice of management in the real institutional environment revealed the unresolved systemic issues such as: economic development is considered at the national level; the management of social processes is in the area of responsibility of regions; environmental issues are concentrated at the local, municipal level. To solve this problem, this article discusses the design aspects of environmental management. So in designing the required high degree of understanding of interdependencies and patterns of organizing itself independently of the system that explains the differences between technical and social design. Technical design relies on mechanistic management, while social planning is understood as a system or organic project management. Hence it is safe to state the fact that project management in the environmental management of the country is concentrated in specific design, planning, technical and management solutions, as well as in the creation interested in these effects in all spheres and strata of society.

1 Introduction

Addressing ensure comprehensive environmental sustainability of the environment and society directly depends on the state and development of investment and construction sector. The value of the investment and construction sector towards sustainable environmental urban development is closely related to environmental management. These areas were engaged in V.V. Asaul, U. Witt, I.P. Nuzhina, A.A. Gorbunov, S.N. Ivanov, A.A. Grebennikov, J.A. Schumpeter, L.E. Bygballe, M. Merklein, M. Anttonen, H. Minna, E. Houtbeckers [1, 2, 3, 4, 5, 8, 10, 12, 16]. Documents for sustainability integrated development environment and society were taken at the level of heads of States and governments for more than 20 years ago at the UN Conference on environment and development (June 1992, Rio de Janeiro). In this, sometimes called "tripod" covers all major aspects of development and global change: ecological, social, economic-based

* Corresponding author: potascheva@mail.ru

planning. However, the transition from the national level to the practice of management in the real institutional environment revealed the unresolved systemic issues such as: economic development is considered at the national level; the management of social processes is in the area of responsibility of regions; environmental issues are concentrated at the local, municipal level [7,15]. In such circumstances, the imbalance of interests of subjects of national, regional and municipal levels in addressing the "three ones" predefined questions. This is a contradiction of conventional development and the real control system with the existing institutions and interests can be solved from the standpoint of design at the scale of the territory or in the time scale [13, 15].

2 Theory

In designing the required high degree of understanding of interdependencies and patterns of organizing itself independently of the system that explains the differences between technical and social design. Technical design relies on mechanistic management, while social planning is understood as a system or organic project management. Element of project management is its ability to integrate, resulting in increased flexibility (flexibility), which performs set of projects to develop strategies for system - program of projects [6].

Note that in the framework of projects, conflicts might occur, as reflected in the matrix of SWOT - analysis.

Chance	Risk Uncertainty at the end	Potential High probability of success
The threat	Potential threats High chance of failure	Risk The uncertainty in the completion
	Weaknesses	Strengths

Fig.1. SWOT analysis in project management.

The task of project management is proper assessment of the project, contributing to the achievement of goals. Approach to the project can be regarded as successful if their implementation is taken into account by those who are directly or indirectly related to the project.

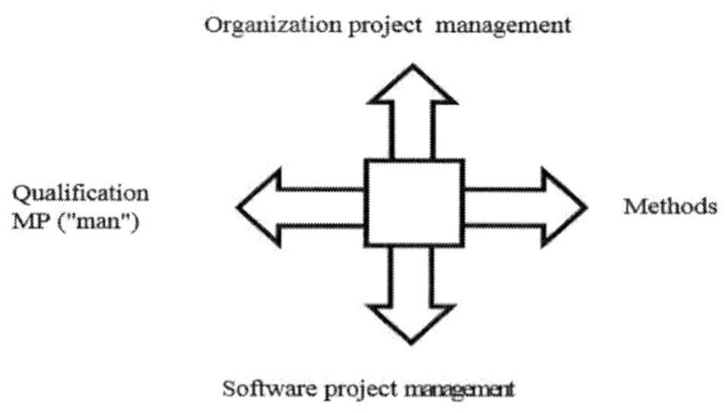

Fig.2.Factors of project management.

Therefore, in project management it is necessary to think not only in the direction of the customer, but primarily in the direction of the user, only then the system can be considered as optimal. Based on the networking of all components and functions of the system occurs systems engineering approach, which allows to achieve internal compatibility between system elements and external compatibility. When implementing a systems engineering approach is applied, the hierarchy of the design tasks: the level of social groups (design of municipal services in General); the level of production systems (systems engineering); the level of the subsystems (design of roads); the level of objects (design of specific buildings).

In design it is necessary to consider the classical form of the triple constraints, which reflects the ratio between cost, time and quality of the project [9]. This restriction forms the triangle of project management, where each side represents a constraint as it does not actually carry out the project with minimal time, minimal resources and maximum quality (improving one parameter degrades the other two). Hence the triangle rule States that for each project for a given technical task, there is a boundary triangle. (Fig. 3) The advantage of project management technology is the ability it approaches the boundary limit of the triangle [9].

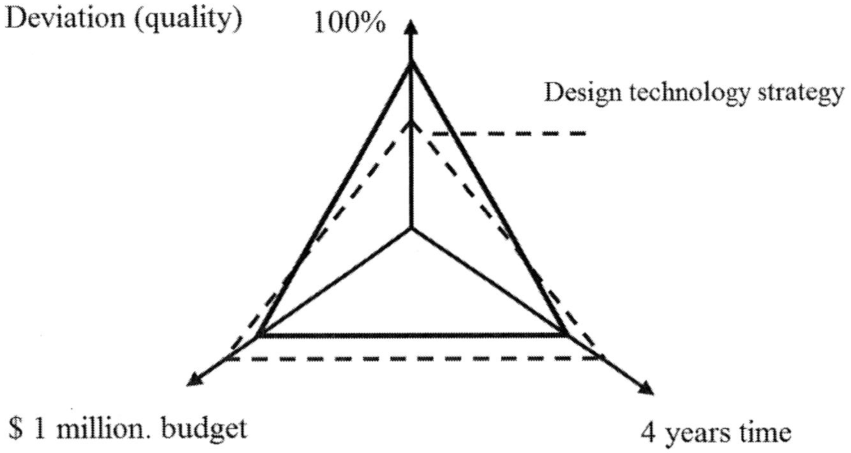

Fig.3. The triangle of project management

Classical Economics assumes that target aspects may behave as a substitution. However, modern project management requires that these aspects are correlated with each other partially, which would ensure the achievement of several aspects simultaneously. Target the aspects necessary to identify and achieve them within a predetermined corridor of acceptability within which it is possible to marking the upper and lower limits of the reached target aspects. By multiprojective possible simultaneously and in parallel manage multiple projects based on a matrix diagram that illustrates the involvement of the projects in relation to the provision of resources, staff, Finance, infrastructure and so on. The use of multiprojective suitable for those projects that have a very similar background to the implementation and, therefore, are in the nature of repetition.

Table 1. An example of the use of resources in case of implementation of several projects.

Projects / Resources	Project 1	Project 2	Project 3	Projects 1,2	Projects 2,3	Projects 2,3	Maximum power
Staff							
Finance							
Infrastructure							
Contract work							
The total number	Σ	Σ	Σ	Σ	Σ	Σ	

In the system of project management should be distinguished: a design system as existing parties; the product as the object on which the action is directed; the project itself as a system of action. Project management is a cybernetic process of planning, management and control (Fig. 4).

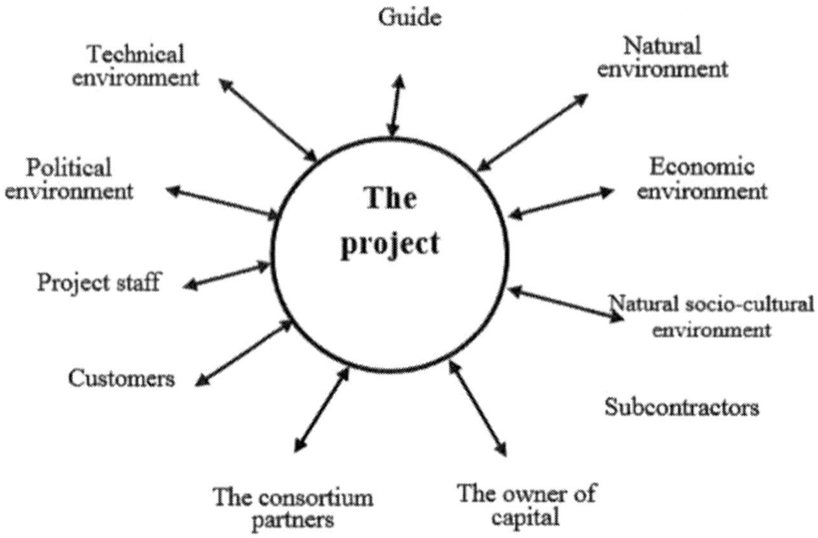

Fig.4. The system model of the relationship between the project system and its environment.

In the design of the following phase (phase - a period of time, which is characterized is characterized by achievement of a certain objective state of the project) - planning and implementation. In the construction industry distinguishes between the following phases: the formation of concepts, design, preparation and execution. The end of each phase is denoted by the so-called main landmark: in podpisac for differentiation can be used intermediate milestones. Phases can overlap each other in order to save preparation time or to provide timely transition into a new phase.

The relationship between project phases and arising costs according to phases is manifested in the fact that in the later phase of the project (production and supply, operation and maintenance) account for a large part of the costs.

In the project management applies the concept of extensive funding exists within the financial design that reflects accurate, specific to each project. Thus, financial engineering is a set of actions to implement financial consultations in the framework of financing the project.

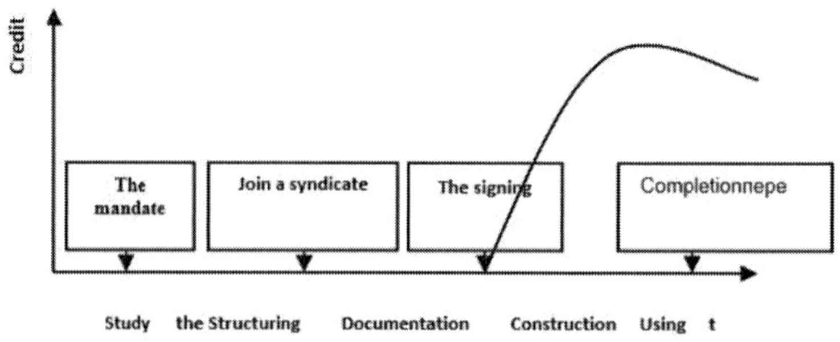

Fig.5. The life cycle of the project.

There are the following significant "project Phase" in the framework of their financing, which for better understanding in this respect the project duration is subdivided into three main stages:

At the preliminary stage of investment, within the framework of Finance explores the idea of the project and is supported by a market study, profitability study of the project and its technical feasibility.

At the stage of implementation of the investment and the Bank provides the financing concept, which includes a detailed financing structure.

At the stage of operation and use of the project is controlled by the return of funds.

As a special risk of financing the project are the following: the risk in the end; during dismantling risk; operational risk; market risk; political risk. It is obvious that it is necessary to limit the extent of possible risk to the minimum or to find a large number of parties willing to undertake the risks (Fig. 5) [6].

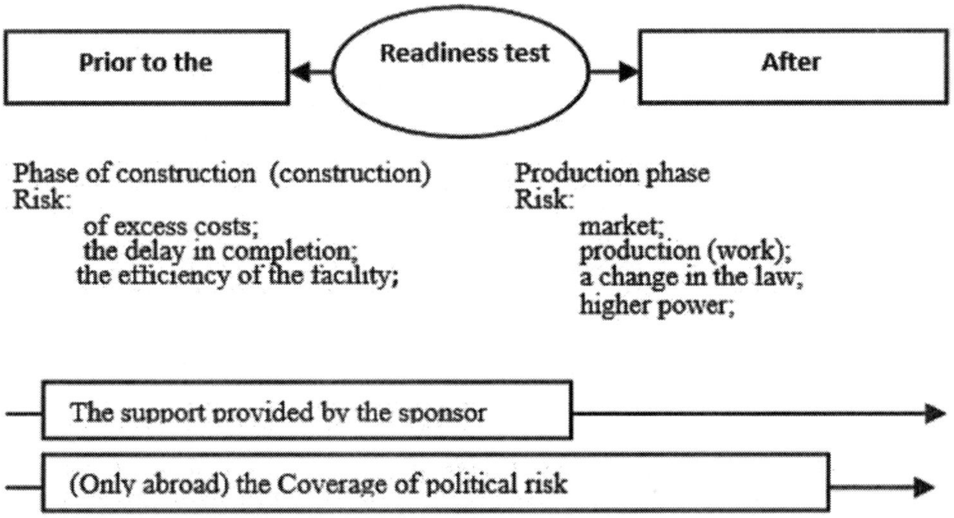

Fig.6. Financing structure of projects and the corresponding risk phases of completion.

The project cycle management - linked to project lifecycle and management objectives. As the main scheme of the model cooperative societies of project management can make the following (Fig. 7).

The project cycle, which underlies the model, is characterized by the following phases: identification; formation concepts; implementation. At these phases as milestones to achieve the next phase are the project idea, project objectives, project plan and its ending.

During the execution phases, decisions on the continuation of the project. Phase processes at certain intervals are characterized by uneven temporal development. The valuation of the project taking into account expenses incurred after the termination of the project, needs to be set against the benefits of its further carrying out. The decision on the termination of the project: KPD - KRA>D, where KPD is the cost of the proposed project; EDGE - costs on the termination of the project; RE - the benefits of the project. If the future benefits from the project is less than the cost of the proposed project net of expenses on termination of the project, the project should be terminated (Fig. 8) [6].

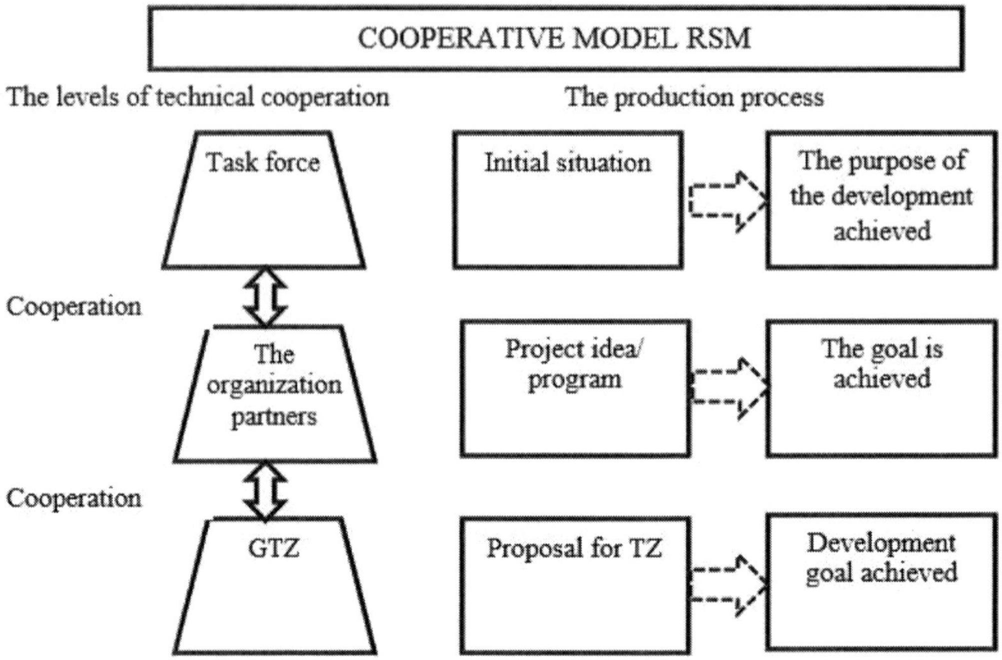

Fig.7. Cooperative management model project cycle.

GTZ – society for technical cooperation
TZ – technical cooperation

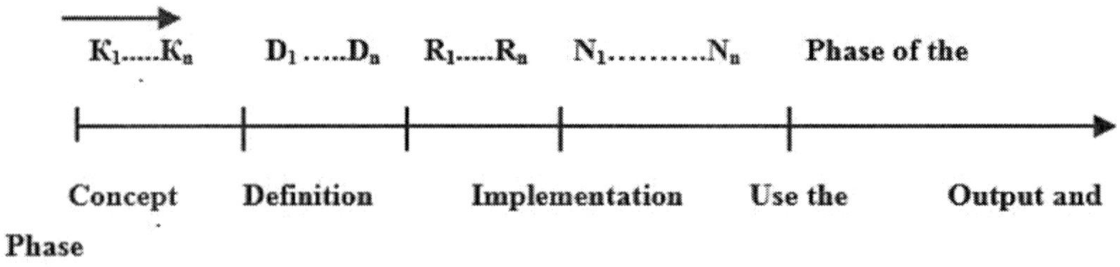

Fig.8. Possible milestones the termination of the project.

The planning aspect is Central in the field of project management as it is responsible for work plans, time and cost, for the commercial execution of orders, the preparation of financial statements and traced in the scheme of differentiation of the phases: definition; analysis; design; implementation; system test; acceptance; maintenance.

In the hierarchy of plan creation project management applied in a variety of positions. On the lower level of planned budgets (project expenditures), and the progress of the case at the secondary level. On a higher level through projects implemented strategies and implemented goals: Goals → Objectives → Strategy → Policy: main or auxiliary → Procedure → the Rules → Program: main or auxiliary and support → Budget: the program digital indicators.

The need for regulation arises from the deviations that appear as a result of changing the reference position on the basis of incorrect action, or new settings of the project during its implementation [1].

Variance analysis relates to the various hierarchical levels of the project, starting with work package, using the partial projects to the entire project. To prevent possible deviations, management needs to respond to the signals coming from the system. In this case, shorter reaction time in project management (Fig. 9).

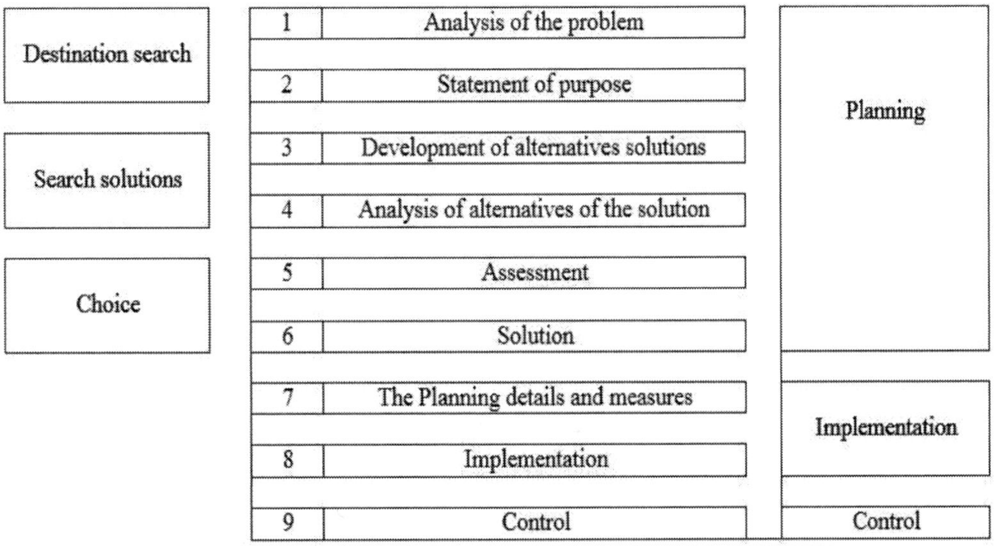

Fig.9. The elements of planning, implementation and control.

In the formation of projects as elements of system development, must adhere to the following principles: investing in research through projects, should seek advantages in price, quality or other characteristics; the results of research are always uncertain, so you need to take into account the probability of success and the amount of potential profit: if the probability of success is high, it makes sense to work on the project; projects with a high probability of success, but with small prospects of recoupment are worth implementing, if the costs are low; if the sales potential is great, and the probability of success of the project is small, this is based on luck or random good fortune.

3 Insights

In modern conditions it is safe to state the fact that project management in the environmental management of the country is concentrated in specific design, planning, technical and management solutions, as well as in the creation interested in these effects in all spheres and strata of society [13, 14, 15].

References

1. M. Anttonen, H. Minna, E. Houtbeckers, Journal of cleaner production, **45**, 89-103 (2013)
2. V.V. Asaul, *Scientific bases of the concept of economic innovation strategy building organizations on the basis of integration processes* (SPbGASU, SPb, 2005)
3. U. Witt, The Paperson Economics and Evolution, **0605**, (2006)
4. A.A. Gorbunov, S.N. Ivanov, A.N. Asaul, *Formation of a regional building complex in transitive economy* (Peter Fund, SPb, 1999)
5. A.A. Grebennikov, Regional Economics and Management, **6**, (2006)
6. G. Diethelm, *Project Management, T1. In 2: volumes* (Business – press, SPb, 2004)

7. The concept of long-term socio-economic development of the Russian Federation until 2020. Information on: www.youngscience.ru
8. M. Merklein, M. Geiger, Journal of materials processing technology, 532-536 (2002)
9. .S.A. Mishin. *Project activity: adapted model for Russia* (AST, Moscow, 2006)
10. I.P. Nuzhina, *Economy of Natural Resources and Conservation: a collection of articles XII International nauch.- Practical. Conf.*, 23-27 (2009)
11. A. Serpell, R. Alvarez, Procedia Engineering, **85**, 464-472, (2014)
12. Information on http://cheltoday.ru/articles/blogi/ustoychivoe-razvitie-i-ekologicheskaya-polupravda-95668/
13. N. Parthenopoulou, M. Malindretos, Original Research Article Materials Today: Proceedings, **3**, 898-912 (2016)
14. S. Bobylev, R. Flights, *Sustainable development in Russia* (2013)
15. Information on: http://cheltoday.ru/articles/blogi/ustoychivoe-razvitie-i-ekologicheskaya-polupravda-95668/
16. J.A. Schumpeter, *Essays on entrepreneurs, innovations, business cycles, and the evolution of capitalism, Clemence; with a new introduction by Richard Swedberg* (Transactions Publishers, London, 1983)

MATEC Web of Conferences 73, 07005 (2016) DOI: 10.1051/matecconf/20167307005
TPACEE-2016

Innovative Approaches in Distance Education in the Field of Environmental Management and Environmental Technologies

Mikhail Leontev[1,*]

[1]Moscow State University of Civil Engineering, 129337 Yaroslavskoe sh. 26, Moscow, Russia

Abstract. This article discusses the innovative structures and components of distance learning and education, discusses the results of application of approaches to teaching in the electronic environment based on the proposed andragogic and pedagogical models of teaching in cyberspace, for adult learners, bachelor graduates of "Management" for the training program "Introduction to environmental management systems". This program particularly addresses the role of environmental managers in a company activity, the implementation of ecologically clean technologies. The author proposed an innovative nonlinear andragogic model of learning. The model was mediated by the constructive approach and problem-oriented learning.

1 Introduction

Turning to adult education, which occurred in the past two decades, linked to labor market requirements, increasing demand for the competence and professionalism of staff [1, 2]. The need for adult education, including the training of environmental managers, due to the social, scientific and technological progress of society, changes in the content and nature of work, the creation and implementation of new technologies [3].

The current situation requires a finding of a balance between ecology and economics [4]. Find that balance is intended to environmental management [5], it is a new direction of the organizational management. These circumstances lead to an increasing demand for specialists in the field of environmental management, retraining, further training and acquire new skills, often through distance learning, as many experts are often uncompetitive [6, 7].

The role of environmental managers in the enterprise activities is determined by the extent of the impact of the enterprise on the environment and the level of their social responsibility. In addition to developing environmental management policies, the specialist is actively involved in the manufacturing process, oversees compliance with the applicable standards of production technology, developing the manuals on environmental safety of the company, instructs staff on environmental safety.

* Corresponding author: miillen@rambler.ru

In view of such a wide-scale problems, modern environmental managers should meet strict professional requirements, also requirements to personal qualities. Personal qualities of environmental managers are the basis for the formation and development of professional qualities. They allow to act within the law and comply with the basic principles of environmental audit. The environmental manager should be: 1) honest: truthful, sincere, restrained; 2) open to be able to accept alternative ideas or points of view; 3) diplomacy, able to interact tactfully with people; 4) supervisory, actively acquainted with the surroundings and activities; 5) insightful, intuitively assess the situations; 6) versatile, to be ready to different situations; 7) resistant, persistent, focused on achieving objectives; 8) able to make timely decisions based on logical reasons and analysis; 9) independent, be able to act and perform their functions independently, at the same time, effectively cooperate with colleagues [8].

Training of environmental managers often follow the way of discovery in Russian universities new master's degree programs based on undergraduate, using innovative approaches and methods, using problem-based learning and andragogic approach.

The program "Introduction to environmental management systems" involves the study of international experience in the application of environmental management systems in order ecologization of manufacturing processes and clean production output; organizing the preparation, planning and implementation of environmental management systems in the enterprises; assess the effectiveness, efficacy and impact of activities in the field of environmental management with environmental auditing procedure, the study of experience of application of environmentally friendly technologies as a means of sustainable development. The use of environmentally friendly technologies allows to ensure the protection of the environment, less pollution, reasonable use of resources, while recycling lends itself to a large part of the wastes and products, and all residual contaminants are recycled more acceptable manner to the environment compared with replaceable technologies. The development of the program involves the use of problem-based learning and andragogic approach.

2 Literature review

Problem-based learning is a pedagogical strategy, a special style of comprehending knowledge in which it becomes possible the complete mastery of the problem with deep, active, persistent, contextual learning in real-life contexts with maximum use of substantiated information resources. Thus, the role of the teacher in problem-based learning is to promote active discussion of a problem, providing a complementary work of learners, their active cooperation, usually in quantitatively small groups, providing the possibility of obtaining required information [9]. The teacher can play the role of an expert on the discussing problem, a head on the use of information sources and a consultant in performing group tasks, and the volume of direct imperative instructions is minimized [10].

Andragogic approach covers various areas of education, it was developed through continuing education as a new paradigm, which involves a rethinking of the concept of education, its aims, essence, principles, functions, modes of organization [11]. Modern teachers use the principles of the andragogic approach fully or partially in various conditions. Based on his own observations M. Knowles put forward five points that reflect his concept of andragogy. Adult learners: 1) are taking control of their learning; 2) bring a wealth of personal experience in educational institutions; 3) begin training, being motivated to it; 4) focused on solving problems in the learning process; 5) especially sensitive to factors of intrinsic motivation [12].

These provisions contrast sharply with the provisions of traditional pedagogy, where learners are dependent personalities who bring little experience (or not bring) in learning activities, motivation to learn is mainly external, for example, from parents, teachers, or connected with the threat of punishment. In the practice and theory of modern education there are studies showing different conditions of formation of educational motivation, including professional orientation of the personality of the future ecological managers [13]. Therefore, it is necessary to explain the essence of learning and to associate the logic of the academic discipline with personal needs of students [14].

The provisions of andragogics are applicable for e-learning environment. They should be used for conducting online courses by teachers, facilitation, to understand the needs of adult learners and to strengthen the collaborative distance learning environment. Today, teachers of online courses need to know your audience and understand its requirements, characteristics and expectations [15; 16]. Online courses must be carefully structured to provide flexibility of learning and understanding by learners of course objectives and tasks, to facilitate the exchange of experience between learners, for example, by using group projects and interactive discussions, thereby increasing the value of the course for participants [17; 18].

Teachers should provide time in order to the relationships between learners were developed based on modeling positive examples of joint activities. In addition, the course content must also have the opportunity to develop, in contrast to the rigidly prescribed regulations [19].

Distance learning involves interaction between the teacher and learners at a distance, while retaining all the inherent components of the educational process (objectives, content, organizational forms, methods, means of learning). Distance learning is being implemented specific innovative means of information technologies involving interactive communication.

Models of organization of educational process using distance educational technologies can also be classified on the basis of three parameters, proposed by V. E. Bochkov: 1) the degree of activity of pedagogical interaction of a teacher and learner in the implementation of the educational process; 2) the degree of integration of theoretical learning on a specific educational program with practical professional activity of a learner in relevant field; 3) the degree of organizational and methodological support and control of independent work of a learner from the educational institution [20].

Distance learning requires the application of learning strategies that differ from traditional teaching in a classroom, the teacher should be ready to provide support to learners for better assimilation rate. In addition to these widely recognized educational strategies, such as constructivist approach [21] and the model of problem-based education [22], andragogic approach to e-learning is highly effective among adult learners.

The basis of interaction between teachers and learners in distance environment is the assimilation of all participants of educational process the innovative educational actions as active and interactive learning technologies both in a classroom, and in distance learning have the same methods of organizing active process of learning. New learning technologies help the modeling new experience by the specialists, the formation of necessary knowledge, abilities and skills, professional attitudes and views.

The condition of training learners for collaboration in active educational environment is the ability of teachers who organize distance learning to create the environment. The cooperation of learners requires to create the safe environment in which they are not afraid to share ideas, experiences and to learn through conversations and information exchange.

It should be understood that every learner is included in the distance learning environment for the specific reason, whether personal desire to learn something new or the relevance of the course by the employer or institution [23]. If learners are unclear about specific goals and objectives of the course, the facilitator should take steps to help learners to identify their learning needs. Facilitator as the organizer of the training should be ready to provide assistance to learners. The learners determines learning needs through discussion with the facilitator the reasons at which the learner took the course, and by questioning (interviewing) learners about expectations from the study of this course. Learner's needs [24] are in the center of teaching, and a variety of learning resources should be available. Learners know about these resources in the course syllabus and the course lecturer directs the learners through the channels of Internet communications if necessary.

3 Research objectives

The aim of this study was to identify and describe the methods preferred by distance learning teachers from the point of view of adult learners, graduates at the bachelor's 38.03.02 "Management", which passed training under the program "Introduction to environmental management systems" in 2015. The development of the program and training course based on it was conducted on the basis of the system and activities approaches. The content of the curriculum meets the requirements of Federal state educational standard of the third generation.

4 Methods

The survey was used for data collection. It was used a modified technique of G. Conti [25], designed to detect the use of the principles of andragogics (personality centered learning) in all forms of teaching, including training adults in cyberspace. This technique is also designed to determine the key pedagogical approaches to education.

It was used a Likert scale from zero to five for assessment of responses, with 5 the highest score (support approach that matches the name of the factor) and 0 the minimal score (support the opposite approach). The reliability coefficient Cronbach alpha for the technique was 0,91 (N=78).

It was applied a factor analysis which allowed to identify seven factors, carrying to the teaching: 1) personally-oriented activities, 2) personal instruction to learners, 3) the use of experiment in teaching, 4) assessment of learners' needs, 5) the involvement of learners in the educational process, 6) flexibility in the personal development, 7) creation of a favorable psychological climate. The seven factors include main elements that describe the pedagogical activity in general. High scores for factors provide support for the concept implied in the title of factors. Low scores provide support for the opposite concept.

5 Results

The results of the study demonstrated that 78 interviewed adult learners, which took the distance learning program "Introduction to environmental management systems", believe that their teachers in the distance environment are mainly supported centered on a teacher (pedagogical) approach to teaching in cyberspace, although adult learners thought that teachers adhere to personal centered (andragogic) approach to teaching in some situations.

Table 1 presents the results of analysis for each of the seven factors of methods of the study. Given the average values for the sample across all seven factors, each was covered by several questions. The Table 1 also includes standard deviations.

Table 1. Standard deviations

Factors	M	SD
1. Activities centered on learner	2.54	1.12
2. Personalized approach to learning	2.49	0.93
3. The use of experiment in teaching	3.31	0.92
4. Assessment of learners' needs	2.97	0.98
5. The involvement of learners in the learning process	2.11	1.07
6. Flexibility in personality development	1.87	0.90
7. The creation of a favourable psychological climate	3.68	0.91

The data in the table show that the teachers in the distance environment have attributed low scores for items relating to four of the seven factors. Low scores are marked according to factors such as activities centered on learner, personalized approach to learning, the involvement of learners in the learning process, flexibility in personality development. High scores were marked on the following factors: the use of experiment in teaching, assessment of learners' needs, the creation of a favorable psychological climate.

These results show that although teaching in the online environment was implemented to some extent in providing andragogic manner, for example, took place the connection of knowledge with experience, assessing learners' needs and creating a favorable psychological climate, teaching methods are not centered on learners or to promote the activities focused on learners. The data for the factor 6 indicate that the teachers did not support the cooperation of learners in the learning process, considered themselves as providers of knowledge rather than facilitators. They never used such modern educational approaches, as the negotiations for curriculum and priority learning areas with learners or educational contract. First of all, teachers in distance environment has approved the rigidity and lack of sensitivity to the personal needs, valued memorization as the main teaching technique.

It was built the linear model according to the results of analysis of adult learners' perception of teaching strategies in distance environment. Adult learners faced the teaching strategies when they trained under the program "Introduction to environmental management systems". You can clearly understand that this pedagogical model is not conducive to adult learning in cyberspace.

Adult learners begin to learning remotely and face: 1) the dominant role of teachers; 2) a predominance of knowledge transmission; 3) teaching for the sake of passing tests, quizzes and examinations; 4) ignoring of the personal characteristics of adult learners; 5) teaching orientation on the achievement of formal indicators, unmanifested desire of teachers to develop capabilities and competences of learners.

In terms of pedagogical approach, teachers are inclined toward formal testing, relying more on standardized tests than on informal methods of assessment. First of all, teachers saw themselves as providers of knowledge, rather than facilitators. They emphasize the importance of mastering of knowledge and, as a rule, one basic practice teaching method. They believe that memorization can facilitate greater autonomy of thinking, are in favor of the lecture method and the dominance of a teacher in a classroom. Pedagogical approach does not encourage the search for examples and perform tasks related to real life situations, the desire to relate course content to everyday life. The teachers are not attracted to their adult learners to develop meaningful performance criteria to assess performance, planning activities, not negotiate the priorities of the curriculum with the learners or use of educational contracts. The teachers believe that well disciplined audience is the stimulus for learning.

6 Conclusion

According to the results it can be concluded that teachers in the distance environment did not consider adult learners as the adults, used mainly pedagogical methods that work well in education and training children. Such methods lead learners to achieve certain indicators and the low level of skills and competences.

At the same time, the author offers an innovative non-linear andragogic learning model. The model is mediated of constructive approach and problem-based learning.

Adult learners go to study, and in the learning process following factors take place: 1) negotiations on the curriculum with the learners; 2) educational contracts; 3) non-formal methods of assessment; 4) accounting and integration into learning process of the characteristics of adult learners; 5) the teaching centered on the development of abilities and personal transformation.

In particular, teachers, applying andragogic approach, plan learning activities, taking into account the previous experience of the learners, calling learners to link new knowledge with prior experience. They strive to ensure that each learner has found what he wanted, based on individual meetings and informal counseling. They also diagnose the gap between the learner's goals and current performance levels. Teachers establish a friendly and informal atmosphere, as a first step towards the andragogic model. There is the removal of barriers, it is used multiple competences as building blocks for educational objectives and the personal transformation [26]. Teachers urged to take risks and perceive mistakes like a natural part of the learning process, look at failures as the feedback channels.

Democratic approaches to teaching are based on the andragogic model. Andragogic methods take into account the interests and experiences of adult learners. Characteristics of mature person of adult learner are taken into account in the learning process, with adult learners are not treated like children. This model better meets the challenges of adult learning in distance environments than pedagogical one, in particular, the objectives of training under the program "Introduction to Environmental Management Systems". The final result of the andragogic model is the transformation and personal development. The choice of pedagogical or andragogic model by teachers depends on the specific social contexts, educational philosophy, the influence of culture and administrative preferences.

References

1. A.D. Ishkov, Procedia – Social and Behavioral Sciences, **142**, 54–58 (2014)
2. I.A. Kalinina, T.N. Magera, Ekonomika i predprinimatel'stvo, **3(2)**, 441-446 (2015)
3. E. Shnyrenkov, I. Pryadko, Procedia Engineering, **117**, 330–335 (2015)
4. I.P. Pryadko, International Journal of Applied Engineering Research, **10(21)**, 42147–42152 (2015)
5. N.G. Miloradova, A.D. Ishkov, Procedia Engineering, **117**, 246 (2015)
6. M.S. Knowles, *The Adult Learner: A Neglected Species (3rd Ed.)* (Gulf Publishing, Houston, 1984)

7. Z.I. Ivanova, O.V. Yudenkova, A.D. Ishkov, E.A. Shnyrenkov, International Education Studies, **8(5)**, 232–239 (2015)
8. A.V. Anisimov, *Ekologicheskiy menedzhment* (Feniks, Rostov-on-Don, 2009)
9. G. LloydJones, D. Margetson, J.G., Medical Education, **32**, 492–494 (1998)
10. D.T.A. Vernon, R.L. Blake, Academic Medicine, **68**, 550–563 (1993)
11. B.L. Delahaye, D.C. Limerick, G. Hearn, Adult Education Quarterly, **44**, 187–200 (1994)
12. M.S. Knowles, *Andragogy in action* (Jossey-Bass, San Francisco, 1984)
13. R.M. Conrad, J.A. Donaldson, *Engaging the online learner. Activities and resources for creative instruction* (Jossey-Bass, San Francisco, 2004)
14. M.S. Knowles, E.F. Holton III, R.A. Swanson, *The Adult learner: The definitive classic in adult education and human resource development* (6th ed. ELSEVIER, Butterworth Heinemann, 2005)
15. A.D. Ishkov, Mezhdunarodnyy zhurnal prikladnykh I fundamental'nykh issledovaniy, **10**, 122–123 (2011)
16. A.D. Ishkov, *Osobennosti realizatsii dopolnitel'nogo professional'nogo obrazovaniya v issledovatel'skykh universitetakh: monografiya* (MGSU, Moscow, 2011)
17. D.E. Hanna, M. Glowacki-Dudka, S. Conceicao-Runlee, *147 practical tips for teaching online groups. Essentials of web-based education* (Atwood Publishing, Madison, WI, 2000)
18. R. Palloff, K. Pratt, *Lessons from the cyberspace classroom: The realities of online teaching* (Jossey-Bass, San Francisco, 1999)
19. A.D. Ishkov, N.G. Miloradova, A.G. Chernyavskaya, *Psikhologo-pedagogicheskaya podgotovka prepodavateley v vysshey tekhnicheskoy shkole: realizatsiya kompetentnostnogo podkhoda: uchebnik* (Arkhitektura-S, Moscow, 2011)
20. V.E. Bochkov, G.A. Krasnova, V.M. Filippov, *Sostoyanie, tendentsii, problemy i rol' distantsionnogo obucheniya v transgranichnom obrazovanii: uchebnoe posobi* (RUDN, Moscow, 2008)
21. J. Brooks, M. Brooks, *In search of understanding: the case for constructivist classrooms* (Association for Supervision and Curriculum Development, Alexandria, 1993)
22. J.R. Savery, The Interdisciplinary Journal of Problem-Based Learning, **1**, 9–20 (2006)
23. E.A. Savina, Izvestiya Volgogradskogo gosudarstvennogo pedagogicheskogo universiteta, **10**(85), 32–36 (2013)
24. E.V. Romanova, Vestnik MGSU, **6**, 532–535 (2011)
25. G.J. Conti, Principles of adult learning scale (PALS). Information on: http://www.conti-creations.com/PALS-Articles_Packet.pdf
26. A. Ishkov, T. Magera, Procedia Engineering, **117**, 148–153 (2015) Information on: http://dx.doi.org/10.1016/j.proeng.2015.08.251

MATEC Web of Conferences **73**, 07006 (2016) DOI: 10.1051/matecconf/20167307006
TPACEE-2016

Economic Efficiency of Innovative Materials for Sectors of Economy

Tatyana Miroshnikova [1,*], and *Natalia* Taskaeva [2]

[1] Vladivostok State University of Economics and Service, 690014 Gogol str., 41, Vladivostok, Russia
[2] Moscow State University of Civil Engineering, 129337 Yaroslavskoe sh.26, Moscow, Russia

Abstract. The paper proposes an approach to the assessment of the economic efficiency of innovative anti-corrosion coatings for sectors of the national economy of the Russia on the basis of a synthesis of strategic sectoral and cost analysis. According to the authors, a comparative analysis of composite polymeric anticorrosion protecting coatings with similar products, estimating of direct and indirect economic effect and prognosis of implementation, forms a deeper understanding of the role of innovative technologies in the Russian state development of import substitution, the investment attractiveness of Russian industries in the new part technologies, applied research activities of private companies. Metal consumption sectors of the economy were chosen as an object of research, as they are characterized by the use of the following products: industrial construction and reconstruction, nuclear and thermal power, chemical, oil and gas, utilities, food processing, automotive, shipbuilding, aviation and rocket science, other industry. Basic modeling of implementation of anticorrosion protecting coatings in industrial enterprises was carried out on the basis of generating energy enterprises as one of the main end-users of anti-corrosive materials that also issue accurate statements.

1 Introduction

Analysis of the current state of innovative development of the Russian Federation and the authors' experience in the design of innovative development programs and models of basic industries (oil and gas industry, energy) shows the relevance of innovation when achieving a high innovation potential of the economy. Under this theme, we should note that commercial enterprises, leading their own innovations not only have effective practices in introducing them to the market, but also provide economic effect [13, 14]. We carried out a study of the impact of innovative development of the Russian company LLC"Antikor". The enterprise's development of corrosion protection of metals exposed to corrosive environments and thermal effects is used to protect metal surfaces and reinforced concrete structures, dead works and bottom pf vessels and hydraulic structures, steel structures in industrial atmosphere, gas transportation, production, storage and processing of oil, heat transfer equipment, drinking water supply facilities, food proceeding equipment, equipment exposed to vibration and vacuum, which are widely used in various industries, such as at

* Corresponding author: tanmir12@mail.ru

nuclear and thermal power plants, chemical and metallurgical plants, ships of various classes, utilities, equipment, water supply and food industry. High chemical resistance of coatings applied to the most common aggressive environment (acids, alkalis, salt solutions, and others.) [15]. Together with resistance to pressure changes and abrasion wear, exposure to vibration and heat, help to protect a variety of surfaces, and industrial machinery, providing it with additional lifetime from 10 to 20 years without implementing process-related maintenance and cleaning of a covered surface during a specified period.

2 Methodology

The purpose of this study is to evaluate the total economic impact of anti-corrosion coatings introduced by a group of companies, taking into account the prospects of its activities for sectors of the economy of the Russian Federation. The main problems to be solved in the work are: firstly, to confirm the hypothesis about the effectiveness of the protective properties of corrosion products for metals estimated by the end-users - representatives of industrial enterprises of the Russian Federation; secondly, to estimate economic effectiveness of developing aimed to protect corrosion prevention; thirdly, to evaluate the economic effect of industrial enterprises of all branches of the Russian Federation. Methodical approach to achieving this goal is to measure the direct and indirect economic impact of the group of companies "Antikor" for Russian industries. The direct economic effect refers to the effect received by the companies of national economy, that acquires unique covers of LLC"Antikor" to protect different surfaces and their production equipment. The direct economic effect is based on reducing the cost of protecting and extending the life of surfaces and equipment of the corporate consumers. It is calculated by determining the net present value of the economic effects of corrosion protection. In the process of calculation the average value of direct economic effect for anticorrosion coatings was determined on the basis of 7-8 corporate consumers per 1 sq. m. of the protected surface of equipment per year. In the course of these calculations average value effect equal to $136 year or about 4080 rubles [1].

The indirect economic effect refers to the following spillover effects.

First point. An effect that is achieved by eliminating the technological losses and downtime of production equipment caused by corrosion. Downtime due to equipment failure, is either completely absent or significantly reduced by using corrosion-resistant coatings. Time needed to produce core products is reduced. This effect is calculated as the cost of goods (volume of services), produced by corporate consumers for the time spent on downtime and loss.

Second point. The effect achieved by reducing the metal content in the industry. The calculation is based on the principle of comparing the effect of alternative methods of surface protection against corrosion of equipment, for example, by using corrosion-resistant metals (stainless steel, copper-nickel alloys, and other special alloys) instead of black metal corrode intensively. This effect is calculated as the value of the volume of non-corrosive metal, which would be sent to the corporate consumers of the anticorrosive coating to protect their surfaces and equipment in case they were not to protect their surfaces and equipment with innovative anti-corrosion coating [5].

Key assumptions and limitations taken into account are as follows: the economic impact is calculated using those industries in which there has been a practical use and application of these coatings.

The economic effect for the period of the actual existence of the enterprise

The economic effect was calculated for the period of the actual existence of the development enterprise (18y.) taking into account the prospects of application in the fields of enterprise development in the forecast period for the next 4 years. The economic effect was calculated for 4 years.

The economic effect for in post-forecast period

The economic effect in post-forecast period (fifth and subsequent years of the LLC"Antikor") was estimated on the Gordon model, which takes into account only the results of the company achieved in the 4th year of the forecast and the fixed rate of development. [2] This approach stems from the fact that it is difficult to predict the development and identify current trends in this field of science - the development of corrosion protection for the long term. Evaluation of the economic effect was based on cash flow forecasting formed LLC"Antikor" from the proceeds from the sale of protective coatings for a variety of surfaces and equipment.

Basic data for calculation

A basic set of values was set to carry out calculation: the total production of protective coatings of all kinds, the average cost of stainless steel, the average consumption of protective coatings, the annual increase in production of protective coatings, the discount rate, the average annual rate of inflation, the income of the enterprise from the sale of protective coatings. Evaluation of the effect of the forecast period is made in accordance with the methodology for determining net present value (NPV) based on the discount rate. Evaluation of the effect obtained after the discrete projection period made on Gordon model. When forecasting the economic impact for a long-range period, it was expected that the business of LLC"Antikor" would not deteriorate and production volume would remain at the level of production volumes of 2013. Cash flow from the sale of specialized equipment used in the protective coating was not included into the calculation of the effect estimation. Neither was the cash received in the form of royalty payments for research and development, for the received patents, copyrights, etc. As a base for the calculation, the following values were chosen: total production of all types of protective coatings in 2008 was 50 371 kg.; the average cost of stainless steel is 150 rubles. per 1 kg.; the average consumption of protective coatings was adopted in the amount of 1 kg per 1 square meter of treated surfaces; flow rate of 1 kg coatings was equated to about 7 kg of stainless steel; since the beginning of 2010 the forecast of 5% of annual increase in production of protective coatings has been laid; in the post-forecast period it is assumed that the growth rate of total production of protective coatings will reach 3% annually; throughout the forecast and post-forecast periods the discount rate is assumed to be 20%; throughout the forecast and post-forecast the average annual inflation rate is assumed to be 7%; the company will not have any other income other than from the sale of protective coatings; it is assumed that there will be no re-equipment, no substantial acquisition of fixed assets, and no increasing of its own production facilities.[1] In other words, the investment program is not planned. Any capital expenditures will be implemented by means of depreciation charges.

Evaluation of the effect in the forecast period. In accordance with the method used to determine net present value (NPV) of economic benefit, the results were given to the current value based on the discount rate by the formula [6]:

$$NPV = \Sigma Di / (1+K)^i, \quad (1)$$

where: NPV- present value; Di - the amount of economic benefit obtained in the i-th period of the discrete projection period; K - the discount rate; i - period. Evaluation of the effect in the post-forecast period. Residual economic benefit is the present value of economic benefit obtained after the discrete predicting period. The Gordon model was used to calculate this value:

$$V = D1 / (K-G), \quad (2)$$

where: V - the residual value in the fifth year; D1 - the amount of economic benefit obtained in the first year after the end of the discrete projection period; K - the discount rate; G - the expected growth rate of flow. This formula gives a rough estimate of the present value of the future economic benefits at the end of the discrete forecast period, based on a discount rate and a constant growth rate [7].

Stages of the research

The logic of the research involves the implementation of the following basic steps. In the first stage opinions and estimates of consumers from various sectors and regions of the Russian Federation were collected and analyzed. In the second stage an examination of existing economic calculations of the development enterprise, and manufacturer of the anti-corrosion technology was performed, taking into account the proposed assumptions and limitations. In the third stage the authors modeled demand and economic benefits for the other metal consumption sectors of the economy of the Russian Federation on the base of data concerning current target consuming industries.

Research methods

The source of information is the reporting of the development enterprise producing protective coatings and industrials statistics of Rosstat [3]. Primary information includes estimating of consumers - a group of companies "Antikor" collected through qualitative field research - in-depth interviews [4]. To conduct in-depth interviews it is necessary to obtain estimates of the enterprises of the usefulness of products to enhance the conservation of metal-working equipment. According to the interviews qualitative consumer product evaluation of "Antikor" and commenting statements of research participants were obtained. Questionnaire for in-depth interviews includes questions distributed on three thematic clusters:

*efficiency of production equipment for industrial enterprises;

*production efficiency for industry in general;

*innovative products for the industrial importance of the enterprise on the whole.

This estimation cannot be extrapolated to the entire population of the industrial enterprises of the country, but they give an idea of consumers on a number of problematic issues. The methodology of the studies included a series of in-depth interviews with industrial enterprises, purchasing products [8]. Task Force to make field research included directors of enterprises, experts in technology, production and supply [11, 12].

Use 170 x 250 mm paper size (W x H mm) and adjust the margins to those shown in the Table 1. The final printed area will be 130 x 210 mm. Do not add any page numbers.

Table 1. Sample field study.

The company name, of the responden	Industry	The city of the Respondent
1 ANTICOR	Construction, application of anticorpos. coatings	Perm
2 Beloyarsk NPP	Nuclear energy	Zarechny, Sverdlovsk region.
3 LLC "Gazstroytechnica"	Gazprom	Voronezh
4 LLC "Zaporizhzhya nuclear energy company	Nuclear energy	Zaporozhye
5 KuibyshevAzot	Chemical	Kuibyshev
6 Electrocontact	Manufacture and Assembly of printed circuit boards	Novosibirsk

7 JSC "Montajhimzashita".	special construction work.	Moscow
8 LLC "THERMO"	Nuclear energy	Yuzhnoukrainsk
9 LLC "chemical protection-2"	Fitter, chemistry	St. Petersburg
10 LTD. "Jenergozashhita"	Contractors, installation and construction Ref.	Rostov-on-don
11 Of separated subdivision "Yuzhnoukrainsk NPP"	Nuclear energy	Yuzhnoukrainsk
12 LLC "Ecomelectronics"	The power supply	St. Petersburg
13 LTD "Uralspetsavto"	Water treatment, wastewater	Chelyabinsk
14 "Speciments""	Equipment of power plants	Bor, Leningrad region
15 CYS-112	Custom enterprises	Moscow
16 Premigration	TPP	Togliatti
17 LLC "Visit system installation"	Heat transfer equipment	Omsk
18 JSC Atomenergoremont	nuclear power station	Desnogorsk, Smolensk region

Source: a Report on field research [9]

Summary evaluation of the effectiveness of production "Antikor" shows that the majority of the key criteria, representatives of the consuming enterprises give high positive ratings.

Table 2. Positive responses enterprises using corrosion products, in%.

Performance	**Answers**
1. The efficiency for the equipment of industrial enterprises.	
• Increases structural reliability of equipment	94
• Increases safety equipment	78
• Increases the life of the equipment	94
• Improves performance of equipment	44
2. Efficiency for industrial enterprises in General.	
• Increases system reliability	83
• Promotes a safe environment	50
• Affects the production efficiency of the enterprise	89
• Affects economic efficiency of enterprises	83
3. Innovative significance for industrial companies in General.	
• Meets production requirements of the enterprise	89
• Analogues Anticorrosive products on the Russian market - it is the best	60

Source: a Report on field research [9]

Almost all consumers point the rise of structural reliability of the equipment from the use of anti-corrosion products. Estimated consumer's structural reliability is increased by 50-70% on average, life of some tipes of equipment increases by 4-5 years. Almost three-quarters of consumers point the rise of production safety of equipment, which is to increase from 50% to 150%. The equipment works more reliably, requires fewer repairs. Almost all consumers point the increase of life of the equipment, since the service life increases to 50-80%. Less than half of consumers point the increased productivity of equipment and more than half believe that performance is maintained at the required level. Such factors as thin layer increase the degree of utilization of the working volume of the process equipment. Most users report about the increase of the system reliability of equipment, which increases by 50-80% on average. Equipment becomes more manageable, easier to identify the planned mode of production, requires less changeovers. Half of consumers point the improving safety of equipment for companies' environment up to 40%. The vast majority of consumers point the improving of the overall production performance of enterprises, as production increases at the expense of ensuring stability. The average increase in production efficiency can reach 10-15%. For example, it is possible to apply thicker layers of protective coatings; it is possible to increase the shelf life of the equipment and the complexity of such works. This technology also reduces labor costs, saves time for preparation equipment. Most consumers point the increasing of the economic efficiency of the enterprise to 5-10%. Extension of the service life of equipment reduces both the frequency and the cost of repair and commissioning. In nuclear power industry efficiency is the highest, because relatively inexpensive protective materials can be used, annual repair the equipment is not necessary, and the turnaround time can be increased to 5-10 years. According to most customers, this innovative technology meets the requirements of industrial enterprises, as the plating technology is very comfortable, even though it is specific. At this point - it is the only product in Russia, which provides increased corrosion protection. Nearly two-thirds of consumers point the absolute leadership in quality of this technology on the Russian market. Based on the above advantages of coatings and materials were evaluated economic effect of their application to various sectors of the economy of the Russian Federation was calculated.

3 Discussion

3.1. Evaluation of direct economic effect

The direct economic effect of the introduction of unique anti-corrosion coating for operating period of the enterprise (18y.) amounted to 1 380 130 000 rub.

The direct economic effect of the introduction of unique anti-corrosion coatings in the forecast period of the company is estimated to be 677,560,000 rub. The direct economic effect of the introduction of unique anti-corrosion coatings in post-forecast period of the company is estimated to be 672,116,000 rub.

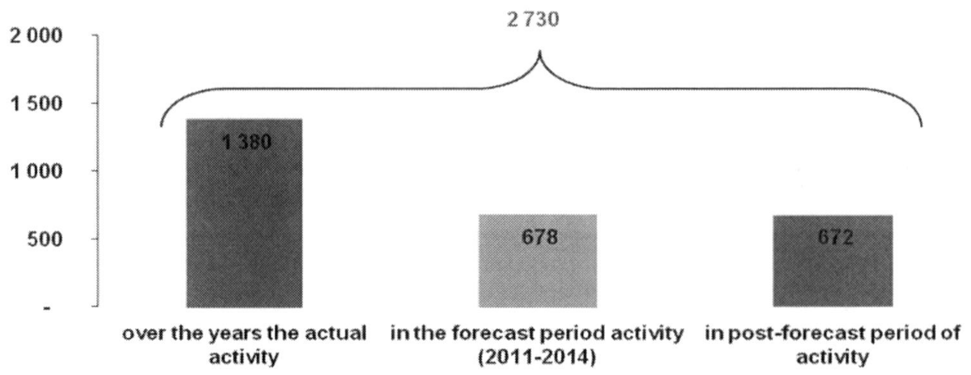

Fig. 1. The volume of direct economic effect from the use of products LLC"Antikor" in the Russian Federation, in mln. Rub.

Thus, the total direct economic effect of the introduction of unique anticorrosive coatings is estimated to be 2 729 800 000. Rub. Assessing the impact of direct economic effect from the introduction of protective coatings LLC"Antikor" was carried out in the following industries.

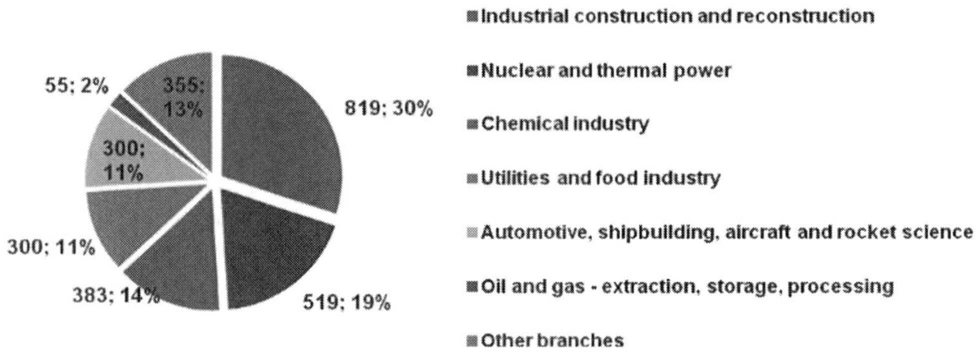

Fig. 2. Structure of the direct economic effect from the use of products "Antikor" by economic sector of the Russian Federation, in mln. Rub.

3.2 Evaluation of epy indirect economic effect

Indirect economic benefits by reducing metal consumption the total indirect economic benefits of the company by reducing the metal content following the introduction of the unique anti-corrosion coatings is estimated to be 859,380,000 rub. Evaluation of indirect economic effect by reducing metal content: from the introduction of anti-corrosion coatings for operating years of the company amounted to 598,210,000 rub.; in the forecast period the effect is estimated to be 219.8 million. rub.; in the post-forecast period the effect is estimated to be 41.37 million. Rubles

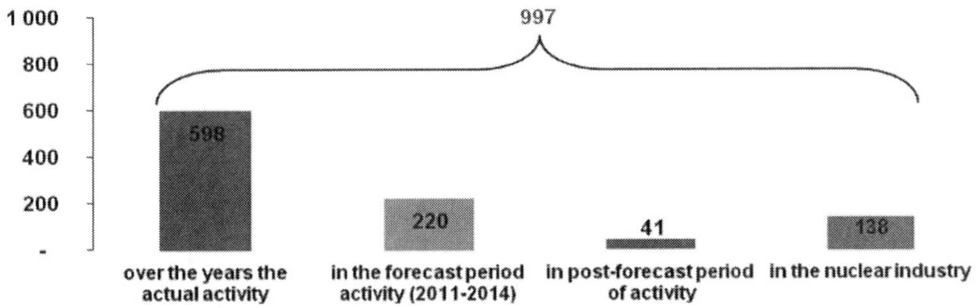

Fig. 3. Volume of indirect economic effect by reducing metal content from the use of products "Antikor" in the Russian Federation, in mln. Rub.

Reducing metal content by economic sectors can be described through the following industries-specific features: in nuclear power industry by increasing the life in 1.5-2 times the cost of copper, nickel and, titanium can be reduced by 30-50% in the manufacture of heat exchange equipment. For example, by applying a composite polymer anticorrosion coating at the Leningrad nuclear power plant, the saving of copper-nickel alloys MNZH-5-1 for a period of three years made 138 million. rub. The work for corrosion protection and recovery of heat exchange equipment is extremely effective. The use of new materials allows obtaining an indirect effect due to the exclusion of technological losses. [Compiled by the authors]

Table 3. The indirect economic impact due to the exclusion of process losses.

Industry	**direct economic impact, mln.rub**	**effect from the reduction of losses**		**the effect of the increase in electricity generation**	
Industrial construction and reconstruction	818.94				
Utilities and food industry	300.28				
Automotive, shipbuilding, aircraft and rocket science	300.28				
Chemical industry	382.17	4.2%	16.05		
Oil and gas - extraction, storage, processing	54.60	5.6%	3.06		
Nuclear and thermal power	518.66	3.1%	16.08	8%	41.49
Other branches	354.87				
Total	2729.80		35.19		41.49

Thus, the application of protective coatings can significantly reduce or eliminate the technological losses and downtime, allows to implement new advanced technologies and get the total economic impact of $ 16 245, 68 mln. Rub. You also need to take into account the effect of the introduction of new advanced products and technologies that make it possible to implement innovative projects, in particular - the submarine project Amur-1650 with modern habitable spaces of high purity; construction and equipment of factories for the destruction of chemical weapons; electrophoresis painting line newly built car factories in Russia foreign manufacturers. The total economic impact of the introduction of anti-corrosion coatings LLC"Antikor" during the existence of the enterprise, as well as the prospects for its activities (medium-term) is 18 975 480 000 rub.

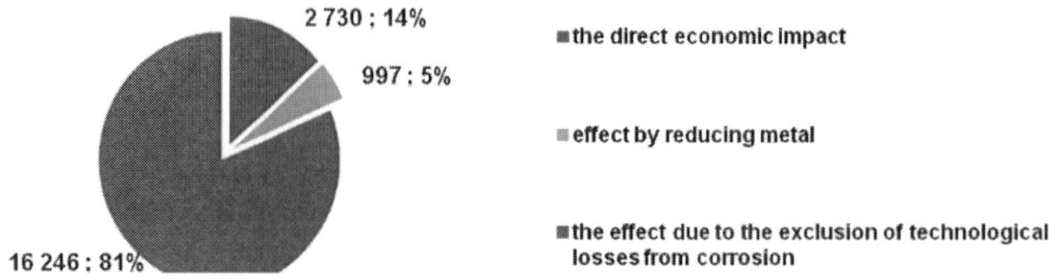

Fig. 4. Structure of the total economic impact of the introduction of anti-corrosion coatings LLC"Antikor" in the Russian Federation, mln. Rub.

According to estimation of the enterprises surveyed, the savings on running repairs in the case of using anti-corrosion materials can reach 5 - 10%. One of the main end-users of anti-corrosive materials that also issues accurate statements (as compared with the construction industry) is energy industry. To estimate the costs of current repairs of fixed assets, data on Leningrad NPP were taken. For example, in 2012 it was spent nearly 90 mln. Rub for repairing [2]. Then, these expenses are extrapolated for the volume of electricity to other nuclear power plants of Russia, as well as to heating plants. Thus, the average savings on current repairs on the nuclear power plant in 2012 was about 40 million Rub. While for thermal plants it was about 252 million Rub. Total saving of the energy complex made 292 million Rub. Then, based on the calculated value the economic effect of using anti-corrosion materials is designed through the indicator of production volumes for other branches of the Russian Federation, estimated in this paper.

The total economic effect from application of anti-corrosive materials by economic sectors in Russia during the analyzed period from 1992 to 2013 is estimated at 271 238 million Rub.

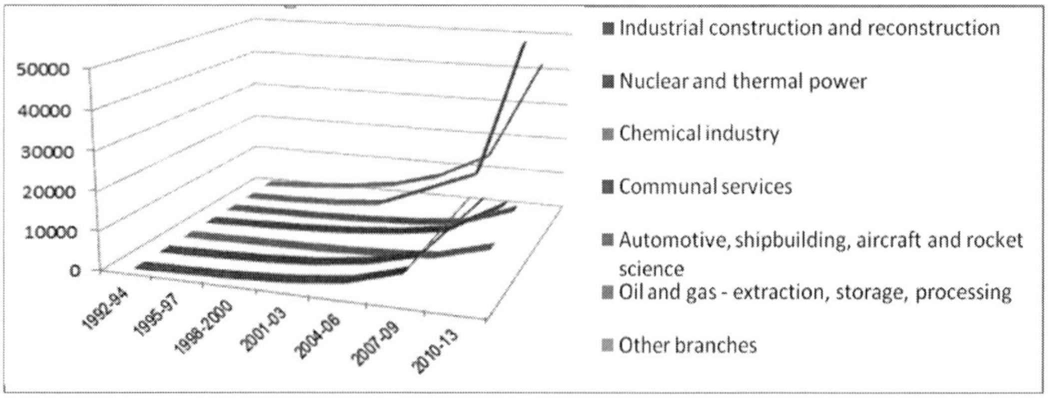

Fig. 5. Assessment of economic effect from application of anti-corrosive materials for various industry of the Russian economy, mil. Rub.

4 Conclusion

Results of the assessment of the economic efficiency of innovative anti-corrosion coatings for the economy of the Russian Federation was based on methodology for determining the present value (NPV), estimating the economic benefit results was based on the discount rate. Evaluation of the effect obtained after the discrete projection period was made on the Gordon model. First of all, it is clear that economic effect from application of anti-corrosive materials for various branches of the Russian economy includes direct and indirect economic effect. In this study, the forecasting of the economic efficiency of anti-corrosion technologies for Russian economy is based on allocating common features of increasing the efficiency of operations and maintenance activities of industries, forecasting development and ongoing assessment of consumer's product and calculating the direct and indirect economic effects.

References

1. *Internal reporting LLC «Antikor»* (2013)
2. *Marketing agency LLC «Business-Class», Report on a field research* (2013)
3. M. J. Gordon, *The Investment, Financing and Valuation of a Corporation* (Conn.: Greenwood, Westport, 1982)
4. Federal State Statistics Service (Rosstat). Information on http://www.gks.ru
5. Ministry of Chemical Industry of the Russia, *Method of determining the annual economic effect on the creation and implementation of new technology, inventions and innovations in the chemical industry* (1978)
6. A. Damodaran, *Investment valuation. Tools and methods for the evaluation of any assets(* Alpina Pablisher, Moscow, 2011)
7. Fisher, *The Theory of Interest* (Macmillan, New York, 1930)
8. N.K. Denzin, Y.S. Lincoln, *Handbook of Qualitative Research (*Sage Publications, London, 2000)
9. N.L. Kuchugin and other, *Marketing on the notes: a practical course on Russian examples (*Moscow, Market DS, 2008)
10. T.K. Miroshnikova, N.L. Kuchugin, Procedia Economics and Finance, **15**, 426-434 (2015)
11. K.M. Eisenhardt, J.A. Martin, Strategic Management Journal, **21(10-11)**, 1105-1121 (2000)
12. C.E. Helfat, M.A. Peteraf, Strategic, Management Journal, **24(10),** 997-1010 (2003)
13. R.M Grant, Strategic Management Journal, **17**, 109-122 (1997)
14. R. Gulati, Strategic Management Journal, **19(4)**, 293-317 (1998)
15. C. Freeman, L., *Soete, The Economics of Industrial Innovation* (London, Pinter, 1997)

Ecological Management and the Cosmogenic Mechanism of Earthquakes

Zhanna Mogiljuk[1,*] and *Michael* Hlystunov[1]

[1]Moscow State University of Civil Engineering, 129337 Yaroslavskoe sh. 26, Moscow, Russia

Abstract. Critical issue of ecological risk management in urban areas is to predict the evolution of the dangerous natural processes intensity. The special situation in the realization of these risks take the earthquake threat and the stresses emergency fluctuations in the geological environment of the buildings and structures bases. This article is devoted to one of the main problems of earthquake engineering - verification of the dominant mechanisms and causality of the earthquakes intensity dangerous evolution. In it discusses the comparative analysis results of the Earth gravitational interaction energy variations amplitudes with the Sun, with the Moon and the solar system planets. Also presented the comparative evaluations results of the Earth geospheres gravitational perturbations amplitudes with the Earth solar radiation energy with the energy of its own heat of the Earth. It is shown that the energy of his own heat and Sun exposure of the Earth much less energy to gravitational perturbations in the near-earth space. In the article presents the spectral analysis results of earthquakes global daily energy on the Earth before and after the Shoemaker-Levy comet explosion on Jupiter. It is shown that the seismic events number on Earth with magnitude greater than 2.5 on the Richter scale after the comet explosion increased in 10 times. In the earthquakes global daily energy spectrum shows the spectral manifestations of solar system planets gravitational resonances. In given article the researches results of natural disasters cosmogenic sources power allow us to argue that ecological risk effective management is impossible without the evolution forecast of the cosmogenic effects intensity on natural processes for sustainable urban development.

1 Introduction

The issue mainstreaming of the environmental risk management effectiveness has received high priority at a high international level [1, 2, 3]. In the US this problem has a half-century history of state regulation [4, 5, 6]. In Russia the environmental risks management is a relevant topic of public policy [7]. In a number of Russian universities textbooks widely presents the main methods of environmental management in the cities and enterprises of the country [8, 9].

* Corresponding author: dpsskmgsu@mail.ru

However in all these documents and textbooks the list of environmental control proposed methods are not considered modern problems of global climate change and the growing intensity of geophysical processes on Earth.

The ecological safety of industrial zones and cities territories largely depends on the fecal and storm sewers stability and buildings and structures of environmentally hazardous industries to natural processes extremal manifestations. Among these processes for most cities in the world include the the geological structures stability violation in the foundations of buildings and underground engineering structures.

Among them the most dangerous are the karsts collapse and landslides in seismoactive areas as a result of the geological structures stress state natural growth.

On the other hand, the stability of the foundations is reduced as a result of the increase in the heavy rainfall and floods intensity in many densely populated areas of Europe, Asia and America. These effects are another reason for the implementation of environmental emergencies.

In areas of high seismic activity are the most dangerous earthquakes and their hydrological consequences, such as tsunamis. It is a major tsunami caused an environmental disaster in the area of the Fukushima nuclear plant in Japan.

Environmental risk management in these areas is a major function of the current and perspective planning of urban development.

However, the effectiveness of this activity largely depends on the evolution forecast accuracy of the dangerous natural processes intensity and seismic activity primarily.

To date, however, the earthquakes reliable prediction is an unsolved problem of modern Geophysics. In this regard, definite prospects for the opening by professors V. P. Nikitsky and M. S. Khlystunov cosmogenic mechanisms of earthquake intensity global growth on Earth [10].

Currently construction in seismically active areas faced with a serious problem of powerful earthquakes risk reliable assessment at a urban planning depth (up to 100 years and more).

One of the key tasks of this problem is the need for verification of a poorly known causality of the earthquakes intensity dangerous evolution.

As a first step in solving this problem, the authors examine the analysis and identification of dominant power sources and mechanisms of seismic activity activation on Earth. Among these sources are undoubtedly gravitational perturbations in the near-earth space. To date, however, the world's scientific schools focused on endogenous causes and mechanisms of earthquakes.

This attitude to the problem by Geophysics explain the global capacity of its own heat of the Earth, the processes of heat and mass transfer and thermodeformative in the lithosphere.

However, the authors comparative assessment of energy these processes with energy cosmogenic gravitational perturbations of the Earth proved the fallacy of this approach.

First, from the modern science view point, the concept of the space (mainly solar activity) influence on the Earth's natural processes was formulated in the late 19th and early 20th century in the work of a scientists number from different countries [11,12,13] (Lehrbuch, 1903; Das Werden, 1907; First Report, 1926).

A comprehensive approach and interdisciplinary formulation of natural disasters and catastrophes cosmogenic concept were formulated by honorary member of the USA Sciences Academy, the Russian scientist Czyzewskiy A. L. [14].

Despite the scientists assumptions in the early twentieth century about the gravitational effects of the solar system planets movement to solar activity, and other evolutionary processes in all Earth spheres (atmosphere, hydrosphere, lithosphere, biosphere, society and other spheres of the Earth) this mechanism still remains one of the little-studied problems in the Cosmo-terrestrial relations field.

This problem stems from the fact that unexplained fluctuations in cycles of the climatic and geophysical processes intensity variations on Earth did not correspond to the stable orbital periods of gravitational perturbations in the solar system space.

The special interest of the Hlystunov's scientific school to this problem has arisen in connection with the unique result of this schools research in the orbital microgravity field and the theoretical opening in 1992, the radial gravidynamic resonances orbital space objects [10].

This discovery helped to explain the cause of gravitational perturbations instability cycles and the corresponding oscillation periods of the Geosphere processes intensity variations. Individual research results of the author's discoveries with colleagues on this topic have been published in a articles series dedicated to the total influence of orbital and resonant gravitational disturbances in near-earth space environment on the global meteorological and seismic processes evolution.

2 Radial Gravidynamic Resonance of Space Objects

After centuries of oblivion, the first step for the cosmogenic gravitational concept further development has resulted in the opening by the Russian scientists [professor Hlystunov M.S. and professor Nikitskiy V.P.[10] phenomena of orbital objects radial gravitational resonance and its impact on the natural disasters and natural catastrophes intensity evolution.

According to the formula of discovery when the excitation of the orbital object radial gravitational resonance on its orbital motion will be imposed radial resonant oscillations with an amplitude of A_{gr}.

Schematically this movement for a perfectly circular orbit (orbit radius R_{orb}), without considering the effect of the the angular momentum conservation law (assuming that the mass of the Central object M_g significantly less than the mass ms of the satellite $M_g >> m_s$, and the satellite radius rs is significantly less than the radius R_g of the Central object ($r_s << R_g << R_{orb}$) will appear, as shown in Figure 1. The resonant motion phase dependence from a time is equal $\varphi_{gr} = 2\pi t / T_{gr}$.

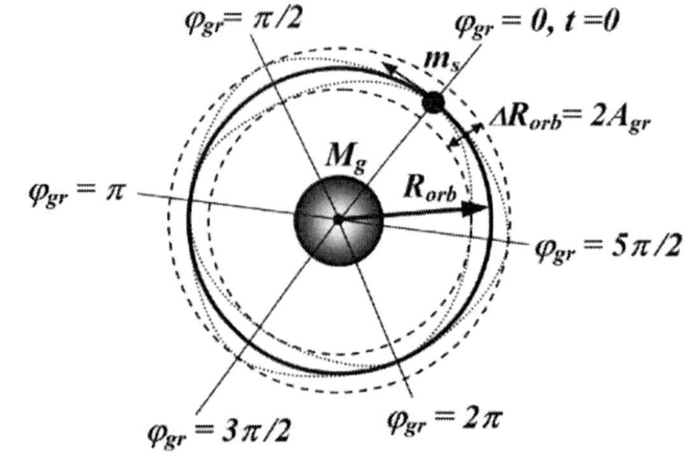

Fig. 1. Schematic representation of the resonant oscillations of the satellite (the planet) relative path in the form of the unperturbed circular orbit.

The opening formula of these resonances was first published by the authors in the International Aerospace Congress IAC'94 theses in August 1994 in the following terms: "Orbital object revolving around the center of gravitational attraction, has a radial gravidynamic resonance whose period is approximately equal to 2/3 of the period of his conversion" [10].

In the future, the hypothesis and the radial gravidynamic resonance theory have been verified by the authors of the discovery from data on the orbital flight trajectory height variations of the geodetic satellite «Topex-Poceidon» and satellite «GEO_IK».

Along with this was established the surprising fact that the radial gravidynamic resonance period, for example, at the sea level height should be 1 hour.

Attempts to find historical sources introduction in practice of such units (1 hour) have been unsuccessful. It should be noted that the satellites radial vibrations relative to the classical orbit was an imposing classical orbital motion, radial resonance frequency and this movement phase modulation result from the action of the angular momentum conservation law when changing the orbit radius by these resonant oscillations.

3 The Parameters of the Orbital and Resonant Gravitational Perturbation

The amplitude evaluation results of the planets and Sun orbital gravitational perturbations energy Eorb and the solar system planets radial gravidynamic resonances mid-frequency calculations (in accordance with the formula-opening) are shown in table 1.

Considering most powerful catastrophic Geosphere processes (endogenous and exogenous- cosmogenic), the authors drew attention to a very important fact. All major endogenous and exogenous processes of heat and mass transfer are realized due to the presence of the gravitational field or the direct action result of its perturbations.

Table 1. The spectral parameters of the orbital and resonance gravitational perturbations.

Sources of disturbance	The parameters of the perturbation *				
	F_{orb}, [Hz]	T_{orb}, [day]	F_{gr}, [Hz]	T_{gr}, [day]	E_{orb}, [J/kg]
Moon	4.25 E-7	27.32	6.01E-7	19.32	-
Mercury	1.32 E-7	87.97	1.87E-7	62.2	-
Venus	5.165 E-8	224.7	7.3E-8	158.9	6.1E33
Earth	3.177E-8	365.2	4.49E-8	258.3	5.55E33
Mars	1.69E-8	686.7	2.39E-8	485.6	-
Jupiter	2.68E-9	4332	3.79E-9	3063	3.3E35
Saturn	1.079E-9	10760	1.53E-9	7609	5.68E34
Uranium	3.782E-10	30689	5.35E-10	21700	-
Neptune	1.928E-10	60187	2.73E-10	42558	-

Sources of disturbance	The parameters of the perturbation *				
	F_{orb}, [Hz]	T_{orb}, [day]	F_{gr}, [Hz]	T_{gr}, [day]	E_{orb}, [J/kg]
Pluto	1.281E-10	90616	1.81E-10	64075	-

*Where F_{orb}, T_{orb}, F_{gr}, T_{gr}, E_{orb} – respectively, the frequency and period of the orbital motion, the frequency and period of the radial gravidynamic resonance, the specific orbital energy of the Sun gravitational perturbations

A comparison of the cosmogenic gravitational perturbations energy distribution average density in the Earth`s geospheres confirms their dominant role compared with the endogenous component (see table 2).

The gravitational interaction energy density of the Earth with the objects in the solar system and the Moon by several orders of magnitude higher than the solar radiation average energy density and its own heat of the Earth.

Table 2. The comparative energy parameters of the geospheres cosmogenic perturbations.

Sources of disturbance	The parameters of the perturbations*		
	E_g/E_{sr}, [r.u. / year]	E_g/E_{gt}, [r.u. / year]	Q_g/Q_{sr}, [r.u. / year]
Sun	1.0E+09	8.3E+12	1.81E+07
Moon	1.85E+05	1.52E+09	1.03E+04
Mercury	2.36E+02	1.96E+06	1.22E+02
Venus	9.02E+03	7.48E+07	7.56E+03
Mars	6.18E+02	5.13E+06	5.21E+02
Jupiter	2.27E+05	1.89E+09	7.38E+04
Saturn	3.27E+04	2.71E+08	6.41E+03
Uranium	2.36E+03	1.96E+07	2.39E+02
Neptune	1.78E+03	1.48E+07	1.09E+02
Pluto	1.31E-02	1.09E+02	6.41E-03

*Where E_g/E_{sr}, E_g/E_{gt}, Q_g/Q_{sr} – respectively, the amplitude variations ratio of the geospheres perturbation gravitational energy to the solar irradiation, the ratio of the geospheres perturbation gravitational energy amplitude variations to own energy dissipation of the Earth, the ratio of the geospheres gravitational energy density perturbations variations amplitude to the specific energy of the solar radiation (on a 1 kg mass of the Earth during the variations period).

The calculations for this table is made only taking into account the eccentricity of the solar system planets the around the Sun.

In this regard, particular interest to the authors of the opening was the planets resonant vibrations influence on the Geosphere processes under the gravitational perturbations action, for example, caused by the Shoemakers-Levy`s comet fall on Jupiter in July 1994.

According to preliminary estimates, the hydrogen bomb "father" Edward teller, when the comet collision with Jupiter the released energy, which is equivalent to energy blast 10 billion megatons of TNT, or a hundred million explosion energy of the Tunguska meteorite (2.77*1022 to 4.61*1025 J).

The resulting energy of the Jupiter gravitational perturbations can be much more solar irradiation of the Earth for a year.

The self-heat generation energy of the Earth for the year will be over five orders of magnitude lower than the Jupiter gravitational perturbations energy.

In this regard, it can be assumed that the comet explosion on Jupiter may have a significant impact on the intensity of seismic, volcanic and meteorological processes on Earth.

In this article, we will consider the global analysis results of the seismic activity evolution on Earth in the period before and after the collision of Shoemakers-Levy`s comet on Jupiter.

For example, use of data on global seismic activity on Earth in length 1024 days.

The original data were represented by the earthquakes magnitudes more than 2.5 points on the Richter scale. The array primary analysis showed that the number of such seismic events and their daily energy increased considerably after the fall of the Shoemakers-Levy`s comet on Jupiter.

Figure 2 shows the amplitude (a) and spectrum (b) variation of earthquakes daily energy on the Earth for a period of 2.8 years. Normalization of daily energy earthquakes curve was performed by the maximum value in the array.

4 Spectrum Analysis of Seismic Activity Daily Variations on Earth

As the array length is only 1024 days, and its spectrum is clearly observed only the gravitational perturbations manifestations of Mercury, Venus, Earth and Mars, as well as their combined harmonics. The background spectrum below 1% of the maximum amplitude in Figure 3b are not shown, since it had no resonance or cyclic manifestations and do not present opportunities for correct analytical conclusions. A comparative analysis of the spectrum and fundamental frequencies and combined harmonics of the gravitational perturbations in the near-earth space according to the data of table 1 reveals the following coincidence frequencies of the dominant resonant manifestations in the spectrum:

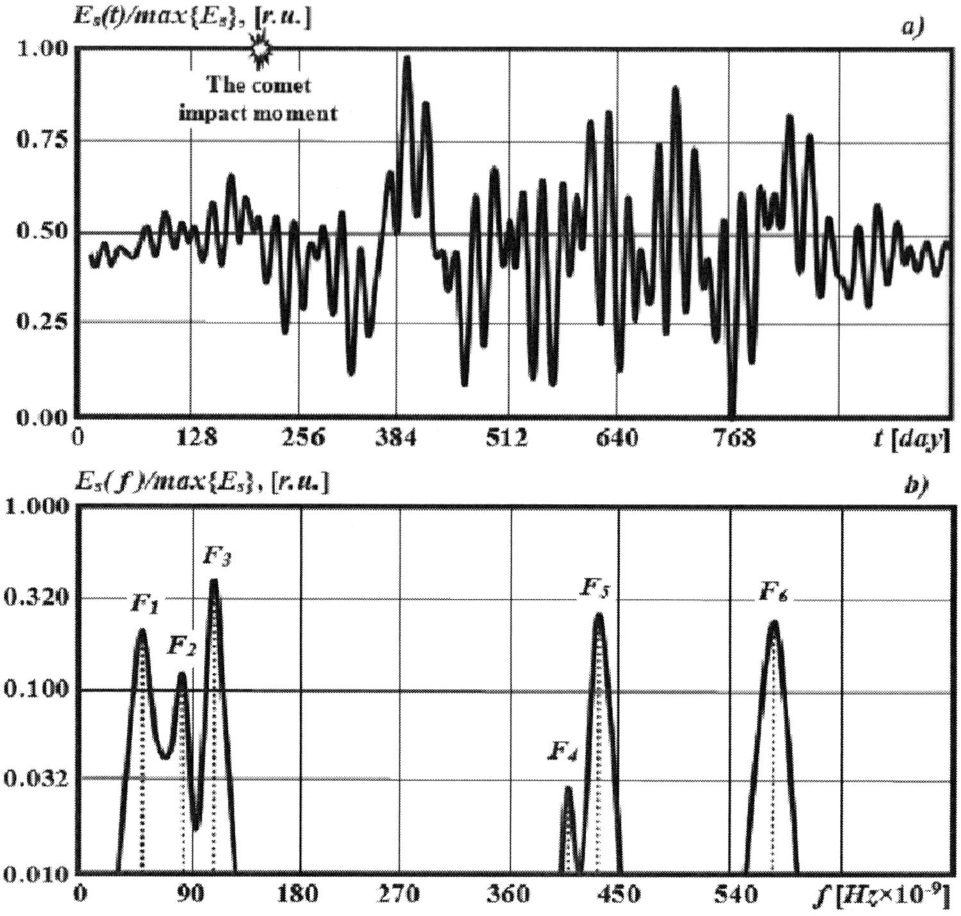

Fig. 2. The normalized amplitude and the spectrum of earthquakes daily energy variation in the period from 01.01.1994 on 19.09.1996:

- F1=5.16*10-8 equal to the frequency of Venus circulation around the Sun;
- F2=8.04*10-8 equal to the frequencies difference of Venus and Mercury circulations;
- F3=1.08*10-7 equal to the frequencies difference of the Mars radial gravidynamic resonance and the orbital circulation frequency of Mercury. In the frequency band of this peak is also the frequencies difference harmonic of radial gravidynamic resonances of Venus and Mercury, the frequencies sums harmonic of the Earth circulation and radial gravidynamic resonance of Venus, the second harmonic of the Venus orbital circulation around the Sun;
- F4=3.90*10-7 equal to the frequencies difference of radial gravidynamic resonance of the Earth and the Moon circulation around the Earth, but also in the frequency band peak of this resonance is the frequencies difference harmonic of the Moon circulation around the Earth and the Earth circulation around the Sun;
- F5=4.50*10-7 equal to the frequencies sum of the Mars radial gravidynamic resonance and the Moon circulation around the Earth, the speed of the layers of the Sun near the equator;
- F6=5.50*10-7 equal to the synodic period of the Moon circulation around the Earth, the frequencies difference of the Moon radial gravidynamic resonance and Venus circulation around the Sun.

5 Conclusion

The research considering results of natural disasters cosmogenic sources power allow us to argue that ecological risk effective management is impossible without the evolution forecast of the cosmogenic effects intensity on natural processes for the urban planning period. Critical issue of environmental risk management in urban areas is to predict the evolution of the dangerous natural processes intensity. The special situation in the realization of these risks take the earthquake threat and emergency fluctuations in the stresses in the geological environment of the buildings and structures bases. The studies results of the radial gravidynamic resonances and the solar system planets around the Sun on the evolution of the geophysical processes intensity on Earth allow to draw the following conclusions.

1. A comparative power analysis of endogenous and cosmogenic sources of geospheres excitation showed that the Earth cosmogenic gravitational perturbations power by several orders of magnitude higher than the total capacity of own heat and sun exposure of the Earth, and the technosphere power, including the military.

2. The variations analysis of the earthquakes daily energy on the Earth before and after the Shoemakers-Levy`s comet explosion on Jupiter showed that the number of seismic events on Earth with a magnitude of more than 2.5 points after the comet explosion increased in 10 times. This fact confirms the dominant gravitational influence presence of the Cosmo-terrestrial relations on the Earth seismic activity.

3. Reliable the intensity evolution prediction of natural disasters and natural catastrophes on Earth cannot be implemented without taking into account the effect, as the orbital gravitational perturbations and perturbations caused by the radial gravidynamic resonances excitation of the solar system planets, the Earth and the Moon.\

The results of this study can be useful for the natural risks assessment on the long term development of urbanized territories [15] in an era of global climate change [16] and growth activity of geophysical processes.

References

1. Rio Declaration on Environment and Development.-UN: A/CONF.151/26/Rev.1, **I**, 3–7 (1992)
2. *Ekologicheskiy uchet dlya predpriyatiy, Konferentsiya OON potorgovle i razvitiyu* (Finansyi statistika, Moscow, 1997)
3. *Directive 2004/35/CE of the European Parliament and of the Council of 21 April 2004 on environmental liability with regard to the prevention and remedying of environmental damage*
4. *Comprehensive environmental response, compensation and liability act (CERCLA) (*USA, 1980)
5. *National Environmental Policy Act (NEPA)* (Washington, 1970)
6. *Environmental Quality. The Twenty-Second Annual Report of the Council on Environmntal Quality together with The President's Message to Congress* (Washington, 1992)
7. *Gosudarstvennyy doklad «O sostoyanii okruzhayushchey sredy Rossiyskoy Federatsii v 2006 godu»* (Moscow, 2007)
8. R.S. Permyakov *Ekonomicheskiy mekhanizm ekologicheskogo menedzhmenta. Lektsii* (Izd-vo Rossiyskoy Akademii Gosudarstvennoy Sluzhby, Moscow 1998)
9. S.N. Bobylev *Ekologizatsiya ekonomicheskogo razvitiya: Uchebnoe posobie* (Izd-vo MGU im. M.V. Lomonosova, Moscow, 1993)
10. M.S. Khlystunov, V.P. Nikitskiy. *Inernational Aerospace Congress*, 225-226 (1994)

11. *Lehrbuch der kosmischen Physik* (Leipzig, 1903)
12. Das Werden der Welten. Ueb. von L. Bamberger L. (1907)
13. *First Report of the Commission appointed to further the study of solar and terrestrial relationships* (Paris, 1926)
14. A.L. Chizhevskiy. *Zemnoe eho solnechnyih bur* (Izdatelstvo «Myisl», 1976) ISBN 978-5-517-92874-0
15. *Managing the Risks of Extreme Events and Disasters to Advance Climate Change Adaptation Special Report of the Intergovernmental Panel on Climate Change* (UK: Cambridge University Press, 2012)
16. *Climate Change 2013. The Physical Science Basis/Working Group I Contribution to the Fifth Assessment Report of the Intergovernmental Panel on Climate Change* (UK: Cambridge University Press, 2013)

MATEC Web of Conferences **73**, 07008 (2016) DOI: 10.1051/matecconf/20167307008
TPACEE-2016

Biosphere Compatibility as a Principle of Sustainable Urban Development

Zinaida Ivanova [1,*], *Olga* Yudenkova [1], and *Jurga* Naimaviciene[2]

[1]Moscow State University of Civil Engineering, 129337 Yaroslavskoye sh. 26, Moscow, Russia
[2]Vilnius Gediminas Technical University, 10223 Saulėtekio al. 11, Vilnius, Lithuania

Abstract. The article covers safety and environmental issues, as well as the sustainable interaction between man and nature in the context of present-day cities. An important task, according to the authors, is the study of scientific approaches and practical achievements in this field, and then the detection of the most urgent critical issues. The coauthors provide an overview of numerous research projects on the treatment of nature, the anthropogenic load, the biosphere as a geological body, the noosphere, and ratios characterizing the condition of the biosphere. The coauthors believe that ecological management, aimed to reverse the treatment of nature, must be integrative. New legislation, effective control of its execution, new disciplines to be introduced at the level of secondary, vocational, and higher education, development of ecological schools of thought and advanced green standards and materials applicable to any construction operations may comprise an effective solution. The authors offer their vision of the challenges and outline the primary steps, including immediate implementation of urban planning solutions to be developed in accordance with the biosphere compatibility principles, with calculations of humanitarian balances of the bio-technosphere. The coauthors also consider the innovative approaches to the use of eco-friendly materials in construction projects.

1 Introduction

Safe and sustainable environment is the objective to be attained in the 21st century. Environmental problems are comprehensive; they arise from the improper treatment of nature [1]. These problems are especially grave in the areas exposed to strong anthropogenic impacts, including hazardous emissions, air pollution, soil and water contaminations, extinction of forests and parks. These problems cause social diseases, including hypodynamia, stresses, and irregular behaviour patterns that produce negative effects on human health (alcoholism, desolation, drug addiction, criminality, smoking, etc.) Russian cities (that accommodate about 74% of the country's population) are unsafe for their residents. According to the Global Liveable Cities Index (GLCI), published by the World Review of Science, Technology and Sustainable Development, Moscow has taken the 64th position in the rating of good-for-living cities. The Global Liveable Cities Index

* Corresponding author: ivanovazi@mail.ru

assesses urban culture, social environment, economy, ecological safety, and competitive strengths [2].

Ecological problems are particularly pressing in Siberia and the Arctic Region (Russia); these areas have a changeful landscape, and they are highly sensitive to any external influence. For example, the value of the scattering property of the atmosphere in Eastern Siberia is half the value of the atmosphere of the European territory of the Russian Federation. However, this factor was disregarded when industrial enterprises, cities and villages were constructed there. This is the reason why the majority of Siberian cities and all industrial areas of Siberia demonstrate excessive air and water pollution.

The Arctic Region is contaminated by oil, heavy metals, and radioactive substances. This is a unique territory, rich for islands, seas and archipelagos; it is frail and easy to exhaust. The consumption of it biological resources should be performed under strict control, with account for the regeneration patterns tailored to local organisms.

This is the reason why ecologists insist on (1) the sustainable development of Northern areas, (2) the control over any activities in offshore zones and in the areas of mineral/fuel resources recovery, (3) the prevention of any destruction of the natural environment exposed to anthropogenic impacts.

We are sure that man produced the initial negative impact on nature in the ancient times rather than in the 20ieth century. The culture of humans develops around the human ability to get accustomed to nature, on the one hand, and to change it, on the other hand. Nature has been exposed to numerous adverse impacts in the course of the human history, although the intensity of impacts was different in different centuries. Back in the ancient times, humans burnt grass away to get more space for their cows and sheep, and therefore, fires spread over extensive areas of forests and steppes. "The ecological effect of these fires caused succession of plants: moist forests gave way to shrubs and savannas, soil composition and structure changed, and the same about the climate" [3].

L.N. Gumilev, a Russian ethnologist and geographer, wrote that "the predatory treatment of nature by humans may occur in any social system; therefore, it can hardly be considered as a consequence of the social progress" [4]. However, the events of the 20th century cannot compare to the environmental changes that occurred in the course of dozens of millennia. Back then, the biosphere had a chance to restore, to generate new landscapes, to create and to multiply its flora and fauna.

2 Developments

Many scholars, who study the ecological and physical peculiarities of Earth and nature, believe that, unless urgent measures are taken now, environmental changes may turn irreversible. Researchers have done their best to surmount the crisis and to draw the public attention to sustainable development; however their efforts cannot convert into the public need, or into the collective consciousness of governments, municipal authorities, business communities, and urban residents. When setting limits to growth, D. Meadows, J. Randers, and D. Meadows warned: "Unfortunately, the environmental load imposed by humans keeps growing, despite any technological developments and efforts invested by public organizations. The situation turns more serious, as the humankind has already stepped beyond the limits into the instable domain. Nonetheless, the understanding of this problem is desperately poor worldwide. There is a need to reconsider personal and public values to reduce the impact produced on the environment and to roll back to the acceptable level, and a lot of time is needed to win the support from the politicians in this respect" [5].

The essence of sustainable urban development means the repletion of human needs without any damage to future generations. Russian scholar Vladimir Vernadsky generated this concept and developed the notion of the biosphere as a geological body, whose structure and functions were based on the features of Earth, Cosmos, living creatures, populations, species, or the living matter. He believed that any human activity was a natural step towards the development of the biosphere. "... in the recent centuries, the human society produced all the more pronounced influence on the environment encompassing the living matter. In the biosphere, or in the upper shell of our planet, this society turned into the sole agent of its kind whose strength grew as long as the time progressed"[6]. Vladimir Vernadsky employed the term "noosphere", initially introduced by Armand Marie Leroi, a French mathematician, to define the area of interaction between nature and society, or the future condition of the biosphere transformed by the collective mind and effort of the humans. Vladimir Vernadsky believed that the treatment of nature and its resources would turn more reasonable in the future and that it would be aimed at its preservation, renewal, and improvement.

Whenever we attempt to develop Vernadsky's research ideas, we realize that the noosphere is created by the man. The noosphere represents the future condition of the biosphere and the outcome of the scientific thought, capable of reengineering the biosphere into the noosphere as the realm of reason. Is it possible?

In his works, Vernadsky enlisted the conditions that set the stage for the positive development of the noosphere, with the conscientious attitude towards the consumption of natural resources being the top priority. He wrote: "The problems of food and production must be reconsidered. As a result, the upheaval of social principles, which control the public opinion, will follow" [6]. The main principles, developed by Vernadsky, include (1) the equality of all human races and confessions; (2) a higher impact produced by the grass-roots onto interior and international policies, (3) the freedom of scientific thought from any pressure produced by religious, philosophical, or political constructs, and setting the stage for free scientific research in public and social systems; (4) the growth of the public welfare. As for the biosphere, Vernadsky developed a sensible transformation methodology for the Earth's nature to satisfy all material, aesthetic, and spiritual needs of the ever-growing population.

Russian scientist Vyacheslav Ilyichev, member of the Russian Academy of Architecture and Construction Sciences, is developing the general ideas, initially proposed by Vernadsky, into specific actions, aimed at the improvement of the biosphere's development pattern. He authors a set of theoretical principles underlying the transformation of cities into biosphere compatible ones capable of developing humans. He also offers practical, technological and economic instruments to implement the transformation in question [7].

Ilyichev's major contribution is a pattern of calculations of humanitarian balances of the Bio-technosphere, designated for the identification of ratios between (f) the biosphere's life potential, population numbers, and the number of areas, where human needs are satisfied; (b) the needs for the biosphere's resources expressed by the humans and the technosphere, as well as the ability of the regions to satisfy them [8]. This collection of methods is a novel branch of urban research; it is in compliance with the concept of sustainable urban development. The main objectives of the research team, headed by Vladimir Ilyichev, are to assess the deterioration of the biosphere in different regions, to identify the extent of the biosphere compatibility in urban areas, and to calculate the ratios. The natural environment has already been assessed in several Russian cities, and the ratios of the bio-technosphere have been calculated there (the ratios have taken account of airflows, impacts produced by the vegetation, etc.). Moreover, Ilyichev has offered efficient power-saving technologies that transform urban pathologies into development resources. He also transforms architectural design projects by introducing natural ventilation and lighting elements.

The implementation of the V. Ilyichev's concept has been initiated. For example, the humanitarian balances of the bio-technosphere (including costs, expenses, maintenance, and restoration) were calculated in the course of development of the project for the redesign of the embankments of the Oka and Orlik rivers in Orel. The Ilyichev's team offered the numerical implementation for the methodology of calculation of the bio-technosphere in cities and villages and identified the feasibility values for specific urban functions. The calculations were made on the basis of the following two factors: the biosphere's resources available and the needs of residents. Moreover, the project developers took account of the town-planning regulations applicable to leisure zones located in the first and second belts of water protection areas. The methodologies at the disposal of the Ilyichev's team provide for the calculation of the humanitarian balances with account for the preservation of the atmosphere and development of comfortable urban environment safe for its residents. The average feasibility ratio for the functions of a biosphere compatible city reached 79.2% in terms of the project on the restructuring of the embankments of the Oka and Orlik rivers. [5, p. 98]

Present-day environmental research projects are developed with regard to the social constituents of biosphere compatible cities. Russian researchers, including V.A. Ilyichev, K.K. Kolin, A.D. Ursul, I.K. Liseev, study the social issues of urban planning [10-13]. Social and humanitarian problems, covered in the urban development doctrine, developed by the Ilyichev's team of researchers, are analyzed in numerous research articles [14-15].

The co-authors of the paper are involved in the research into socio-humanitarian features of biosphere compatible cities; however the co-authors admit that this problem does not have any technology-intensive solution. Therefore, humans need to change their mindset; the pattern of the human treatment of the biosphere needs improvements, and Earth should no longer be exploited. We must interact with Earth as if it were living matter. If we fail, no technology problems will be
solved even on paper. The coauthors also participate in the development of the fundamentals of ecological ethics; they think of advanced behavioral patterns originating from the basic ones typical for the archetypes of the collective unconscious, or ethnically mental and ethnically religious fundamentals of biosphere compatible villages, wide-scale migration, development of the positive ethnic climate in cities, and social aspects of the biosphere compatibility in architectural projects [16]. The coauthors and their colleagues solve problems of management competencies on the basis of the new regulations set by the Federal State Educational Standards in respect of the training of civil engineers, urban planners, and managers, who will be able to implement the principle of the biosphere compatibility in their work [17].

"The symbiosis with the biosphere is necessary and possible, if and when the man develops, if his philosophical, moral and ethical mindset focuses on the cooperation with the biosphere as a system that is superior to the human civilization," says V. Ilyichev [18] "... the formation of the post-industrial civilization shall involve not only a technological revolution, but also a spiritual reformation, criticism and reconsideration of several former basic values of the anthropogenic culture (its treatment of nature, its cult of power as the basis for transformation, its ideals of the consumer society based on the rising consumption of goods and energy, etc.)," concludes V.S. Stepin, a prominent philosopher and member of the Russian Academy of Sciences [19].

Researchers stand for the civilizational changeover; however this changeover is an extensive process that covers the life periods of several generations. What particular resources do we have at our disposal in order to transform the human mindset, to develop the need for a safe environment? For example, how strong is the need for the ecological housing, and how can this need be developed? The eco-sustainable architectural design and the green building philosophy represent the notions and concepts that are much spoken about, but so far they are relevant on paper, as all parties involved in the construction process – investors, architects, developers – fail to implement these concepts, although they realize the need to recognize the ecological requirements and to assure the natural comfort on construction sites and inside new buildings [20]. This state of affairs needs to be changed, but the psychological problem prevents any changes. "Any social changes, declared and adopted by the society, start working when they convert into a mental set that triggers and guides the individual behavior" [21]. For example, many Russian citizens cannot change their behavioral pattern when buying construction materials. Their price, rather than their ecological safety, is the purchase driver. The mission of the ecological management is to develop a conscious human need in safe construction and finishing materials.

Residents start making demands for ecological residential houses, although these demands are unconscious and weak [22]. According to A.D. Remizov, Chairman of the Board, Green Building Council, eco-sustainable architecture can only be boosted by the demand from the society, the business community, and state authorities. "… there is a need for a new mindset that does not oppose man to nature. A new attitude towards consumption must serve as a means for the implementation of new principles. Being a consumer, the man must take account of the concerns of other species and the planet as a whole: thus, the severe exploitation of Earth, the devastation of forests and wildlife habitats, the industrial and economic development, changing the climate on the planet, are inadmissible. And the man himself, being part of nature, needs a sustainable, healthy and society-oriented environment" [23].

According to A. Remizov, the mission of ecologically sustainable architecture and urban development is to develop and maintain:

- cultural values (originality of local areas, contextual and historic correlations, development of the sense of place);
- social values (social involvement, solidarity, good neighborliness, crime reduction, civil dignity development);
- economic values (rate of return, real estate cost growth, capital return);
- consumption values (job satisfaction, labour capacity improvement, health improvement, education availability, location satisfaction);
- environmental values (protection of species, health improvement, reduction of waste and CO_2 emissions, life fullness) [24].

The ideas, formulated by A. Remizov, comply with the concept, developed by Z. Gaevskaya and S. Mityagin, who believe that the capital construction management must be based on the economic capacity of the biosphere and on the interaction between natural and anthropogenic structures and processes [25].

The concepts proposed by the researchers, specializing in the study of the interaction between man and nature, are embodied in the effective legislative acts [26]. The provision of the Urban Planning Code of the Russian Federation that reads as: "Assurance of well-balanced consideration of ecological, economic, social and other factors in the course of performance of urban planning activities" serves as a main principle underlying architectural design and construction [27]. Besides, according to the Federal Law on Environmental Protection, when buildings, houses, structures and other facilities are designed, account shall be taken of the norms of the admissible human load on the environment, measures shall be arranged for preventing and eliminating environmental pollution and also for disposing of the industrial and consumption waste; resource-saving, low-waste and other best existing technologies shall be employed in environmental protection, restoration of natural environment, rational use and reproduction of resources [28].

Not only new construction should be "green", but also reconstruction and restoration of buildings and structures and all the real estate as a whole. «Renovation of the built and human environment must be considered instead and the holistic approach used. From this perspective, decision-making problems of renovation often involve a complex decision making process in which multiple requirements and conditions have to be taken into consideration simultaneously» [29, 30]. In the majority of the developed countries the amount of capital investments into housing hasn't greatly changed, but the structure of their use changed. The funds are devoted mainly not for new housing, but for renovation of the existing housing stock. The concept of the integrated analysis of built and human environment renovation developed by a Lithuanian researcher Prof. Arturas Kaklauskas from Vilnius Gediminas Technical University and his creative team allow controlling the whole lifecycle of buildings. The build-up environment is considered as a complicated constantly changing system, in which controlled and uncontrolled factors are operating and which depends on the definite number of variables on micro-, meso- and macrolevels. One of the most important factors is the exhaust of natural resources. "The construction activity should be organized in the way so that the negative impact on the environment and energy consumption would be minimal during the whole period of building existence (i.e. as of the moment of design, production of building materials, including recycling and recycled building materials and products, construction, operation, property management of buildings, and demolition). Towards this end, it is necessary to implement a system of actions" [31, 32]. Among them, there is the reduction of waste volume during building existence, increase of thermal insulation of buildings, hot water saving and heating, the use of renewable construction resources and energy.

3 Conclusions

On the basis of the above ideas, the coauthors propose several focus areas and urgent actions in the field of interaction between the man and the human society, on the one hand, and the biosphere, on the other hand.

№	Focus Area	Goals and Objectives	Urgent Actions
1	Legislative recognition of the values of nature. Tight control over the execution of laws.	Human health and safety protection of life. Preservation of human health and provision of safety. Enforcement of law and order for the sake of ecological safety.	1. Legislative strengthening of the humanitarian balance of bio-technosphere in regions (or phased transition). 3. Development of an efficient system of control over the execution of legislation by the organizations of the civil society.
2	Development of the ecological culture and ecological enlightenment, methods of ecological upbringing and education.	Development of the ecological conscience, ecological ethics and aesthetics, and deliberate personal needs for the harmonious coexistence with the natural world, a caring treatment of nature, and proactive behavior aimed at the attainment of the above objectives.	1. Expansion of ecological modules for programmes of general secondary, vocational secondary and higher levels of education. Continuity of the ecological education. 2. Development of integrated interdisciplinary educational modules; integrative education. 3. Assurance of higher ecological competencies of students majoring in areas different from ecology. 4. A stronger role of the social disciplines, designated to develop the new mindset.

3	Training of high quality specialists in ecology. Introduction of obligatory positions of environmental specialists.	Competent resolution of environmental problems. Ecological control over any economic activities to be performed by enterprises or organizations.	1. Introduction of a new speciality of an environmental specialist. 2. Development of a standard set of responsibilities to be performed by the future specialists.
4	Development of environmental sciences, environment friendly technologies and machinery.	Identification of trends in the development of the biosphere and projection of consequences of varied economic solutions. Development of theoretical models for the further development of the biosphere. Resolution of acute environmental problems through the employment of advanced scientific achievements.	1. Development of research schools of the environmental thought. 2. State support of research projects contemplating the development of innovative technologies. 3. Getting young and talented graduates involved in research activities; the training and maintenance of the high quality human capital. 4. Launching integrated research projects on the problems of the biosphere to be developed and implemented by the teams of researchers specializing in versatile areas of knowledge.
5	Urban planning solutions to be developed in furtherance of the biosphere compatibility principles, including the green building methodology and the ecologically sustainable architectural design. Development of a network of ecological villages.	A symbiotic coexistence of the city, the biosphere and human.	1. Integrated design projects are to be developed by teams of versatile specialists in architectural and structural design. 2. Scientific knowledge advancement in the fields of advanced construction materials and technologies, resources recycling, alternative energy sources, active and passive methods of resources saving.

Each of the actions above must be treated as the one of paramount importance and included into the list of Russia's strategic priorities. Initial actions should consist in the development of the state policy on the interaction between the society and the biosphere, as well as in the development of specific actions aimed at the implementation of this policy, provision of the state funding, and contributions of the business community into strategically important actions.

References

1. N. Vatin, D. Nemova, D. Nabiullina, A. Ovdienko, Applied Mechanics and Materials, **725-726**, 1434-1444 (2015)
2. T.K. Giap, W.W. Thye, G.A Aw. World Review of Science, Technology and Sustainable Development, **2(11)**, 176–196 (2014)
3. Information on http://spbhram.ru/2013/ecology/nehr/4698
4. Information on http://gumilevica.kulichki.net/EBE/ebe04.htm#ebe04chapter14yu

5. D. Meadows, J. Randers, D. Meadows. *Limits to Growth: The 30-Year Update* (Chelsea Green Publishing, 2004)
6. Information on: http://vernadsky.lib.ru/e texts/archive/Vernadsky_V.I.__Avtotrofnost_ Chelovechestva.html
7. V.A. Ilyichev. *Biosfernaya sovmestimost': Tekhnologii vnedreniya innovatsiy. Goroda, razvivayushchie cheloveka* (Librokom, Moscow, 2011)
8. V.A. Ilyichev, V.I. Kolchunov, S. G. Emelyanov, N. V. Bakaeva, Applied Mechanics and Materials, **725-726**, 1224-1230 (2015)
9. D. V. Danilevich, N. V. Bakayeva, I.V. Shishkina, Biosfernaya sovmestimost': chelovek, region, tekhnologii, **2(10)**, 88-100 (2015)
10. V.A. Ilyichev, A.M. Karimov, V.I. Kolchunov, V.V. Aleksashina, N.V. Bakaeva, S.A. Kobeleva, Zhilishchnoe stroitel'stvo, **1**, 2-10 (2012)
11. K.K. Kolin, Strategicheskie prioritety, **1**, 6-31 (2014)
12. A.D. Ursul, Strategicheskie prioritety, **1**, 31-41 (2014)
13. I.K. Liseev. Materialy Mezhdunarodnoy nauchno-prakticheskoy konferentsii. 12-13 marta 2014g. BGITA, Bryansk, 306-311 (2012)
14. I.P. Pryadko, K.V. Orlina, Kazanskaya nauka, **6**, 245-248 (2014)
15. S.N. Rumyantsev, A. L. Bachurin, Trudy 6-y Mezhdunarodnoy nauchno-prakticheskoy konferentsii "Innovatsionnye tekhnologicheskie resheniya dlya ekologicheski ustoychivogo stroitel'stva i gorodskogo kommunal'nogo khozyaystva". MGSU, Moskva, 127-135 (2014)
16. Z.I. Ivanova. Etnomental'nye, etnoreligioznye osnovy sozdaniya biosferosovmestimykh gorodov i poseleniy. Materialy Mezhdunarodnoy nauchno-prakticheskoy konferentsii «Biosfernosovmestimye goroda i poseleniya» 11-13 dekabrya 2012g. BGITA, Bryansk, 23-29 (2012)
17. A.D. Ishkov, N.G. Miloradova. Ekonomika i predprinimatel'stvo, **9**, 884-887 (2014)
18. V.A. Ilyichev.. Strategicheskie prioritety, **1**, 42-58 (2014)
19. Information on http://spkurdyumov.narod.ru/Stepin11.htm
20. O. Gamayunova, N. Vatin. Applied Mechanics and Materials, **725-726**, 1493-1498 (2015)
21. N.G. Miloradova, A.D. Ishkov. *Psikhologiya upravlencheskoy deyatel'nosti: izmenenie podkhodov* (MGSU, Moscow, 2014)
22. N.G. Miloradova, Psikhologicheskie problemy zhiteley bol'shikh Trudy pyatoy Vserossiyskoy i tret'yey Mezhdunarodnoy nauchno-prakticheskoy konferentsii «Gumanitarnye problemy sovremennosti». 17-16 dekabrya 2008 goda. MGSU, Moskva, 99-100 (2008)
23. Information on http://www.archpeter.ru/arkhiv/2011/07/ekoustojjchivaja-arkhitektura/
24. Information on http://i-stp.ru/archive/5040
25. Z.A. Gaevskaya, S.D. Mityagin, Applied Mechanics and Materials, **587-589**, 123–127 (2014)
26. S. Ershova, E. Smirnov, Applied Mechanics and Materials, **725-726**, 1002-1006 (2015)
27. Information on http://www.consultant.ru/document/cons_doc_LAW_172910.
28. Information on http://www.consultant.ru/popular/okrsred.
29. A. Mickaityte, E.K. Zavadskas, A. Kaklauskas, L. Tupenaite. International Journal of Strategic Property Management, **12**, 53-68 (2008)
30. L. Tupenaite, E.K. Zavadskas, A. Kaklauskas, Z. Turskis, M. Seniut. Journal of Civil Engineering and Management, **16**, 257-266 (2010)
31. A. Kaklauskas, E.K. Zavadskas, D. Bardauskienė, R. Dargis, *The harmonious development of real estate: a monograph* (Technika Publ., Vilnius, 2014)
32. N. Banaytiene, A. Banaytis, A. Kaklauskas, E.K. Zavadskas. Omega – International Journal for Management Science, **36(3)**, 429–441 (2008)

Management Methods by Development of Objects of Energy Supply Taking into Account the Combined Risks

Ariadna Kirillova[1,*], *Sergei* Beliakov[1] and *Anna* Krylova[1]

[1]Moscow State University of Civil Engineering, 129337 Yaroslavskoe sh. 26, Moscow, Russia

Abstract. In the article the methods of choice of optimal administrative decisions are examined on the development of objects energy supplies of housing and communal services. That would take into account nascent risks related to investment-building activity of energy supplying enterprises. It is shown that the basic condition of housing and utilities on an energy supply is the building of objects with a subsequent production, distribution and realization of electric energy, that allows at their reproduction to provide quality and reliability of energy supply to the consumers. Also the questions of decline of risks and exposure of factors, influencing on the processes of management development of objects of energy supply in the field of housing and utilities, estimation of organizationally-economic reliability and combined risks, estimation of possibility of the use of innovative energy technologies, alternative sources of energy supply, are investigated in the article. It is fixed, that risk indexes must be appraised not only on the investment stage of building object, but also on his operating stage (after commissioning) on the basis of account of the great number of risks characterizing the combined risk.

1 Introduction

In the conditions of steady tendency of accumulation of physical and moral deterioration of energy supply objects, and also increase of relative losses in the electric systems and decline of techno-scientific and building potential on a background of the permanent increase of rates on electric energy for consumers, including for a population, one of major strategic tasks of the state presently is a choice of optimal administrative decisions on further development of the systems energy supplies of housing and utilities, that would take into account arising risks related to investment activity of energy supplying. At the same time, it is necessary to promote the transfer of energy system objects in innovative technologies and use modern equipment and expand scientific and technical potential. [2]

Russia has huge stocks of traditional energy resources and significant reserves for the use of alternative sources. However, it is necessary to stimulate the process of construction of energy plants that use renewable sources of energy to ensure affordable, quality

* Corresponding author: serj-bel@yandex.ru

electricity to consumers, especially those that are outside of centralized energy systems. [5, 3]

Solution of problems of management of development facilities allowing to identify the impact of the cost of the investment - construction projects in the field of energy supply housing and utilities services tariffs on electricity businesses territorial entities is a timely and urgent task.

Despite the large number of studies on the implementation of the investment - construction projects, a number of issues related to the analysis of emerging risks and their impact on the efficiency of investment in the development of energy supply facilities of housing and utilities services remains poorly understood and requires further study. [3]

2 Methods and algorithm

The analysis of existing problems and the main directions of development of energy facilities in Russia showed that:

- A large part of the country with a population of about 20 million people do not have access to centralized energy grids and these areas generally have the highest fuel and energy prices;
- Part of areas lacking energy, that is in need of importation of fuel and energy supplies;
- Russia's gas, supply only 50 % of the city and about 35 % of rural settlements, the rest use oil and coal, which are sources of local environmental contamination;
- In a continuous rise in prices of energy and fuel, increasing the cost of connection to the networks of centralized energy and autonomous alternative energy in Russia are developing quite intensively: consumers try to provide themselves with their own sources of electricity and heat.

Under these circumstances, one of the main goals is a fundamental renewal of basic production assets of the enterprises of energy supply on the basis of domestic and international experience in the application of innovative technologies in the construction, increasing the reliability and security of energy supply provided slowdown in electricity tariffs. Achieving this objective in the construction of energy supply facilities required:

- Replacement of obsolete technologies and equipment companies' production, transmission and distribution for innovative;
- Development of new domestic innovative technologies in all areas of energy supply companies;
- Modernization of the equipment and separate units of energy plants and networks;
- Optimize the structure of generation capacity;
- Create an effective system of holistic management of functioning and development of the Unified Energy System of the country, contributing to minimize costs and reduce electricity tariffs.

Analysis of the causes and problems of the full restoration of energy supply facilities in the housing sector has allowed to classify the identified problems and to group them is presented in Table 1.

Table 1. Analysis of the causes of problems and the full restoration of energy supply facilities in the housing and utilities sector.

Management domain	Causes	Problems
Investment policy	- Lack of law framework. - Strong political influence. - The lack of sequence and systematics in making managerial decisions. - Inconsistency investment decisions at the federal and regional levels.	- Lack of funding. - The absence of effective mechanisms to attract private investment. - Lack of development and implementation of innovative technologies. - The lack of economic interest in increasing the efficiency of its operations, reducing losses and costs.
Tariff policy	- Inefficient existing system of tariff regulation. - Lack of regulatory framework. - Short-term regulation period tariffs for the population. - Politicized the process of setting tariffs for electricity.	- Sustained growth in electricity tariffs by region. - The deterioration of the socio - economic situation at the regional level.
Technical policy	-Lack of the system of strategic and technical policy. -Lack or imperfection of development programs.	-High degree of depreciation of fixed assets. -High accident on networks. -Low level of upgrade fixed assets. -Low reliability Infrastructure.
Energy efficiency and resource efficiency policy	-Weak motivation of producers and energy consumers in the use of energy and it saving.	-Wasteful use of electricity.
Personnel policy	-The loss of prestige of working professions. - Frequent personnel changes. - Lack of labour motivation.	- The low level of training. -Low Quality of the work performed. -Low level of wages.

The development of energy supply facilities of housing and utilities services of Russia territories depends on many factors. Questions of construction of energy plants that use alternative renewable energy sources are the most relevant for areas with decentralized energy supply systems, with a deficit of energy resources and where there are favorable natural - climatic conditions. The study of these factors showed that the interest of regional authorities and municipalities for the construction of such facilities is low, which can be explained from the point of view of the authors, the uncertainty associated with an increase in electricity prices due to the implementation in the field of alternative generation projects, the lack of teaching materials for the analysis of the profitability of investments and management of emerging risks. [6]

In assessing the profitability of investment in electricity facilities take into account the characteristics of their innovative development , attractiveness to investors and the choice of investment and construction projects in the territorial unit , as well as organizational and economic reliability of enterprise management mechanism , taking into account the overall risk level of attractiveness of the projects for the construction of energy supply facilities for the investors is invited to assess the according to formula (1) based on the model $Y(f)$.

$$Y(f) = \sum_{n=1}^{5} K(f)_n \times f_n, \text{ where } f_n = \sum_j K(f_{nj}) \times f_{nj} \qquad (1)$$

where $K(f_{nj})$ – weight factor fnj in the coefficients; $K(f)_n$ – weight factors of the group fn in the coefficients determined by expert assessments.

The level of attractiveness of the project is determined by the natural and climatic and geographical factors, socio-demographic, innovation and infrastructure, price, administrative and other factors. Proposed methods of management of development of energy supply facilities consist in choosing the organizational and economic decisions between the construction projects of energy plants that use alternative energy sources, close to the consumers, and the project of building the network management objects on the basis of the proposed instruments assess the profitability of investments in the development of energy supply facilities on the basis of risks analysis.

This approach is based on the justification of the maximum level of net present value (NPV) and risk analysis (R), associated with the implementation of each project individually. [10] Making decision (P_{dec}) in this case will be based on the following economic-mathematical model of selecting the direction of development of energy supply facilities of the territorial formation:

$$\begin{cases} P_{dec} = \max(NPV_{ce}, NPV_{al}) \\ NPV_{ce} = f(x_1, x_2, x_3, x_4, x_5, x_6, x_7, x_8, x_9, x_{10}, x_{11}, x_{12}, e) > \\ NPV_{al} = f(x_{13}, x_{14}, x_{15}, x_{16}, x_{17}, x_{18}, e) > 0 \\ (R_{ce}, R_{al}) \rightarrow min \end{cases} \qquad (2)$$

where NPV_{ce} – net present value of the project of building the network management objects based on their connection to the central energy grid, which allows for the construction of external conditions (x_1-x_6), characteristic of construction projects (x_7-x_9) and construction units (x_{10}-x_{12}): x_1 – the remoteness of the construction of the main production base ; x_2 – terrain complexity; x_3–climatic conditions of the territory; x_4 – the duration of the winter period; x_5 – the nature of the surface coating; x_6 – distance to the unloading station equipment and materials; x_7 – equipment and materials costs; x_8 – length of the heating main construction; x_9 – diverse nature of transmission lines; x_{10} – construction labor input; x_{11} – the degree of factory assembly equipment; x_{12} – kinds of used vehicles; NPV_{al} - net present value of the project of building energy plants using alternative energy sources (renewable energy), which depends on the parameters (x_{13}-x_{18}): x_{13} – factory equipment cost; x_{14} – the foundation costs; x_{15} – connection to net; x_{16} – electrical equipment and materials; x_{17} – resources of communication; x_{18} – labor costs; e – discount rate; R_{ce}, R_{al} – risk assessment for each project.

3 Discussion and outcomes

The construction of new energy facilities is expected to attract investment and, consequently, increase the price (tariff) for electricity. Therefore, at the stage of the investment decision is necessary to determine the investment component in the tariff, and to assess the level of electricity prices for end-users based on the project cost and the current outlay of electricity production on facility object, that is built and service provided.[9] Formula (3) offered for public utilities territorial entity that reflects the indicators related to the construction and subsequent operation of the new object of energy supply - for the calculation of energy prices for housing:

$$P_{crit} = \frac{A_{man} + A_p + A_{civ} + A_{com} + \sum_{t=1}^{T}(a_v \times A_f) \times q^{-t} + L \times q^{-T}}{\sum_{t=1}^{T} x_t \times q^{-t}} \tag{3}$$

where P_{crit} - critical energy unit selling price , energy production which refers to a time t; A_{man} – energy plant manufacturing costs; A_p – engineering costs; A_{civ} – the cost of civil and erection works; A_{com} – the cost of commissioning works; T – equipment life; a_v – payment per unit of electricity produced, depending on the volume of production; x_t – amount of energy (t = 1, 2, 3, 4, 5…); A_f – payments for the period of operation, not dependent on the amount of energy; q - sales volume; L - payment upon liquidation of equipment.

The sensitivity of the energy unit price to a change in the cost of building a new facility in the field of electricity allowed to disclose the risks associated with the implementation of the project. Risk indicators must be evaluated not only on the investment stage of construction of the object, but also its operational stage (after commissioning) by taking into account a number of risks that characterize the overall risk (R).

The processes of construction and further operation of energy are inextricably linked to the operation of enterprises engaged in both production and transport and distribution of electricity. Therefore, the study was considered a multi-purpose nature of the activity of the enterprises of energy territorial and their interaction with the environment through performance indicators (x_i), which include performance indicators, efficiency and profitability.

Accordingly, the proposed method of determining the reliability of the organizational-economic mechanism of management of the enterprises of energy supply modules on the basis of the definition of relative deviations (Δ_i) actually achieved the planned performance by the formula (4):

$$\Delta_i = \left| \frac{x_{i\,actual}}{x_{i\,plan}} - 1 \right| \tag{4}$$

where $x_{i\,actual}$ – the actual value of i- th performance index; $x_{i\,plan}$ – planned (or regulatory) value of the i-th index.

The average value of the relative deviation of actual achievements from the planned performance (or regulatory), defined by the formula (5) will characterize the organizational - economic reliability of management:

$$\bar{\Delta}_i = \sqrt[n]{\prod_{i=1}^{n} \Delta_i} \quad (5)$$

The indicator is relative and varies from 0 to 1, depending on its value, you can judge the degree of organizational - economic reliability of the multi-purpose oriented management mechanism of enterprises electricity territorial unit.

To this end, work is carried out with the participation of experts, which allowed to set the intervals for the mean relative deviation of actual performance against the planned (or regulatory).

The degree of mismatch can be taken as the criterion for inclusion the company of an area of risk according to the degree of organizational and economic reliability of mechanisms of management. In particular, if the calculated average deviation value falls within the range, characterized by:

0 - 5 % - low risk area;

6 - 10 % - permissible (moderate) risk area;

11- 15 % - high-risk area;

16 - 100 % - critical risk area.

Activities of enterprises in the field of public services of electricity includes uncertainties associated with the probability of occurrence of possible adverse events that give rise to production and technical, pricing, financial, investment and regulation risks are presented in Table 2.

Table 2. The set of combined risks in the sphere of management of development of energy supply facilities.

Events that have a negative impact on the organizational and economic reliability of management mechanisms	The emerging risks
- Lack of experience in implementation of investment projects	investment
- equipment failure; - overload electrical network equipment; - disruptions in fuel supply to energy plants; - natural disasters and force majeure; - the lack of wind, sun, etc. for renewable energy; - the human factor (erroneous actions of staff); - poor quality of management decisions; - lack of qualified personnel	production and technical
- prices encrese (fuel, etc.)	price
- reduction in solvency of demand	financial
- changes in legislation; - industry monopoly	regulatory

On the basis of the developed models, tools profitability of investment evaluation for selection of innovative projects for the construction of energy supply facilities and methodological approaches to the definition of the organizational and economic reliability of management mechanisms was developed the algorithm of the management of the development of energy supply facilities of housing and communal services area, which is shown in Figure 1.

In order to implement the proposed approach to the development of objects of small utilities generating proposes the creation of local investment companies for the development of facilities in the energy supply domain, which would carry out management of investment projects in construction and further exploitation of small generating facilities that use alternative energy sources that will help to solve a number of problems, which are of particular relevance to rural communities, small towns, areas remote from centralized energy supply system. Based on this is proposed organizational and economic model of management of innovative development of energy supply facilities of housing and communal services with the need to manage the construction and operation of energy plants that use alternative (renewable) energy sources, which is presented in Figure 2.

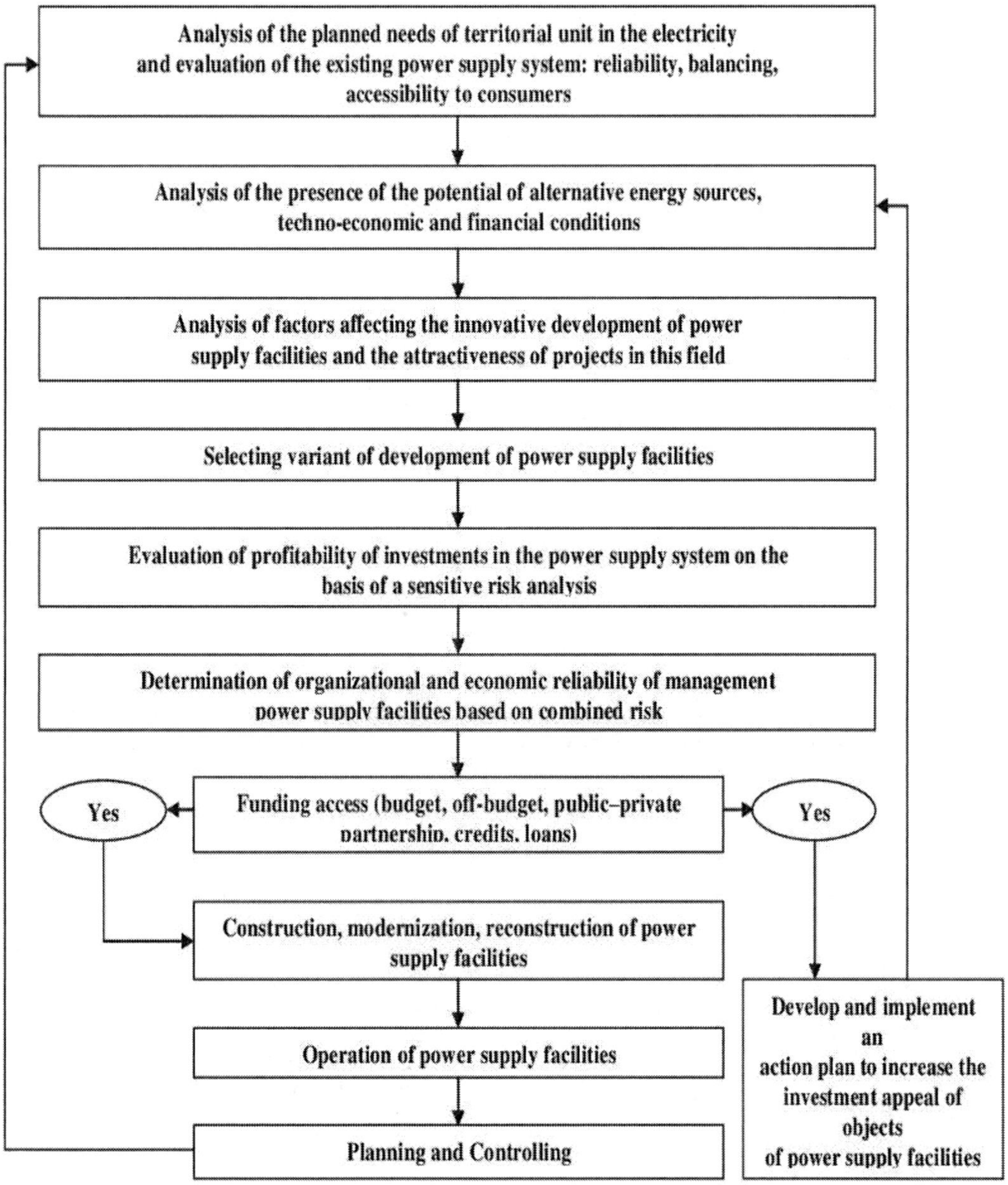

Fig. 1. Control algorithm of development of energy supply facilities housing and communal services, taking into account the combined risk.

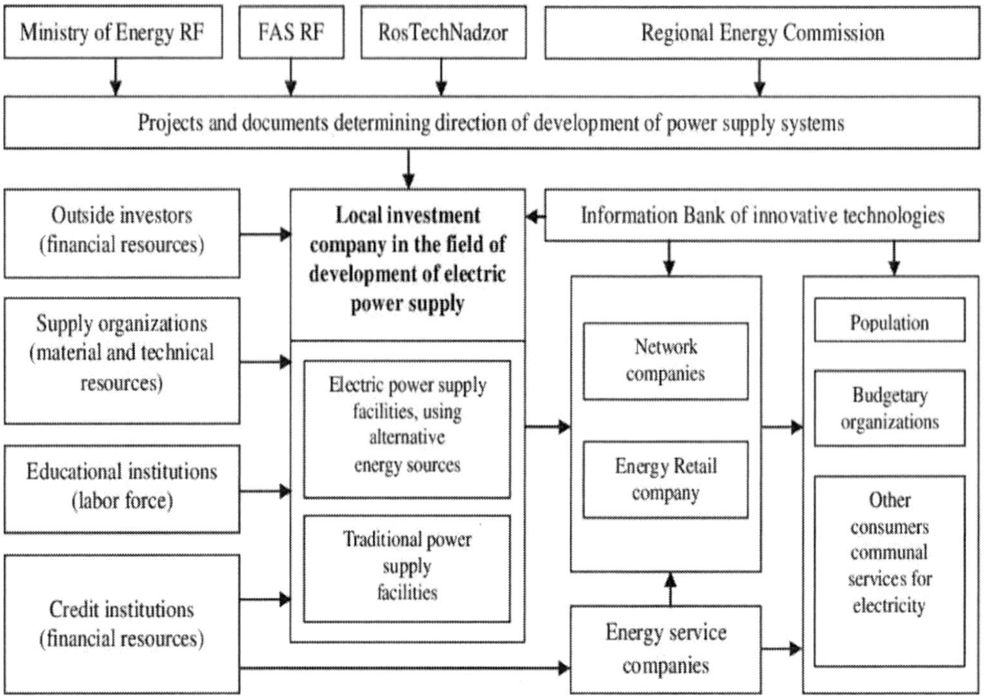

Fig. 2. Organizational and economic management model for innovative development of energy supply facilities of housing and communal territorial economy.

The study contains the calculation of the total cost of the project of construction of wind-energy installations (renewable resources), which amounted to 1 621 100 rubles, and the cost of the project to build two new transformer substations 10 / 0.4 kV, 35 kV networks and new transformer substation 35 / 10 kV worth a total of 2 700 000 rubles. Hereinafter, was an examination of the risks for each of the investment projects, taking into account regional factors on the basis of the proposed thesis methodical approaches, which showed that the risks of the project of construction of wind-energy installation is much higher in many positions than for the construction project grid facilities, as shown in Table 3.

Table 3. Expert assessment of the value of total risk in the provision of public electricity services (for example, the rural settlement of Lovozero, Murmansk Region).

№	Events that have a negative impact on the organizational and economic reliability	Expert assessment of the risk (renewable resources)	Expert assessment of the risk (for network objects)
1. Production and technical risks			
1.1	equipment malfunction	0.25	0.17
1.2	overload of electrical network equipment (insufficient capacity)	0.0	0.10
1.3	disruptions in fuel supply to energy plants	0.0	0.24
1.4	natural climatic events and other force majeure	0.75	0.12
1.5	the lack of wind, sun, etc. for energy plants using these types of energy	0.2	0.0

1.6	the human factor - the erroneous actions of the operating personnel at energy and dispatch management bodies	0.17	0.17
1.7	poor quality of management decisions	0.2	0.2
1.8	lack of qualified personnel	0.6	0.3
2. Price risks			
2.1	prices increase (fuel, etc.)	0.0	0.25
3. Financial risks			
3.1	slowdown of effective demand	0.27	0.27
4. Investment risks			
4.1	lack of experience in implementation of investment projects	0.6	0.15
5. Regulating risks			
5.1	legislative change	0.3	0.10
5.2	industry monopoly	0.2	0.10

The cost of wind-energy installation is lower than the cost of building the network objects. However, it should be noted the high risks for wind turbines associated with climatic phenomena, the lack of experience in implementing similar projects and a lack of qualified personnel, and the risks associated with network congestion and interruptions in fuel supply, that estimated at a minimum level for the territory of the Murmansk region. Less risky for the energy supply system of rural settlement Lovozero is the construction of grid facilities. For other territorial entities need to conduct similar assessments in accordance with the proposed approach.

4 Conclusion

The developed tools to assess the profitability of investments and the impact of the cost of building a critical selling price of energy for consumers on the basis of risk analysis allows to take into account the impact of the cost of production of the energy plant, engineering, construction, installation and commissioning, as well as ongoing operating costs of already existing object on the rise in the cost of electricity and return on invested capital. On the basis of the proposed instruments is possible to evaluate the profitability of innovative projects relating to the construction of generation facilities that use alternative renewable

The proposed methodological approach to the evaluation of risks associated with organizational and economic reliability of management mechanism of a multi-purpose activity of the enterprises in the field of territorial energy supply is based on a set of performance indicators, efficiency and profitability.

Methodical approaches to the estimation of the overall risk associated with the uncertainty in the area of energy supply, which allows to take into account the probability of occurrence of the events that have a negative impact on the organizational and economic reliability of management mechanisms, and emerging investment, pricing, financing, regulation and production and technical risks.

References

1. P. Lombardi, T. Sokolnikova, K. Suslov, N. Voropai, Z.A. Styczynski, Renewable Energy, **90**, 532-541 (2016)
2. V.A. Barinov, Analysis and forecasts, **3**, 322 (2010)
3. I.A. Bashmakov, Questions of Economics, **8**, 104-122 (2007)

4. T. W. Thurner, V. Roud, Journal of Cleaner Production, **112(4)**, 2851-2860 (2016)
5. S.Y. Esyakov, Housing and communal services, **4**, (2013)
6. E.E. Ermolaev, The economy of construction, **1(7)**, 24-29 (2011)
7. V. Vorotnikov, World Pumps, **3**, 30-33 (2014)
8. A. Korppoo, N. Korobova, Energy Policy, **42**, 213-220 (2012)
9. A.V. Svistunov, Modern economy: problems, trends, prospects, **5**, 13-15 (2011)
10. A.J. Vladova, J.R. Vladov, V. Kushnarenko, N.N. Bakhtadze, *IFAC Proceedings,* **45(6)**, 929-933 (2012)
11. S. Hunt, G. Shuttleworth, *Competetion and Choice in Electricity* (John Wiley, Chichester, 1996)
12. Y.N. Kondrashova, O.V. Gazizova, A.V. Malapheev, Procedia Engineering, **129**, 759-763 (2015)
13. T. Datsuk, G. Pozin, V. Ulyasheva, M. Kanev, Procedia Engineering, **117**, 874-882 (2015)
14. T. Kossova, M. Sheluntcova International Journal of Project Management, **34(3)**, 403-411 (2016)
15. S. Paiho, Å. Hedman, R. Abdurafikov, Ha Hoang, M. Sepponen, I. Kouhia, M. Meinander, Energy and Buildings, **66**, 706-713 (2013)

MATEC Web of Conferences 73, 07010 (2016) DOI: 10.1051/matecconf/20167307010
TPACEE-2016

Research of Ecological Motivation of the Heads and Specialists Getting an Education in the Field of Production and Application of Construction Materials

Elena Romanova[1,*]

[1]Moscow State University of Civil Engineering, 129337 Yaroslavskoe sh. 26, Moscow, Russia

Abstract. Production of construction materials causes serious damage to environment. On the other hand, this industry in big scales and with big effect uses waste of other industries that promotes improvement of an ecological situation. The advanced entities build the activities according to standards of the ISO 14000 and theEco-ManagementandAuditScheme (EMAS) series, implement the principles of ecological management. But process goes slowly. As objective obstacles the financial barrier (expenses), ignorance of effects to environment (unhealthy work conditions degree) and what operations need to be performed (to undertake) act. The main difficulty of the problem resolution of ecological production is the person, its values and motives, his consciousness and behavior. Results of research of ecological motivation of future specialists and heads of the entities on production and application of construction materials are given in article. The determining role of mentality and ecological consciousness at slow rates of implementation in production of eco-efficient technologies is proved.

1 Introduction

Objective threat of survival of mankind forces to build processes of interaction of society and the nature by the principles of mutual adaptation. Problems of ecological efficiency in production of construction materials take key places. Industry of building materials the largest consumer of natural resources. This industry annually gets and processes about 3 billion tons of raw materials: sand, clays, plaster, limestones, granites, basalts and other breeds. It causes serious damage to environment, pollutes the atmosphere and the Earth's surface. On pollution of the atmosphere dust the industry of construction materials wins first place among all industries. The second place is taken by motor transport; the third – thermal power [1].

On the other hand, this industry in big scales and with big effect uses waste of other industries (more than 300 million tons of various waste). For this reason consideration of environmental problems of the main productions and a possibility of their transition to a

* Corresponding author: romanova_e_v@mail.ru

wasteless (net) way of development begins with production of construction materials [2]. Inclusion of the entities on production of construction materials in wasteless territorial and production complexes and what and industrial parks allows to remove problems of waste recycling of other industries, reduces harmful effects on environment. Also it leads to essential economic effects due to economy and a saving of raw materials, materials, energy resources; decrease in losses; decrease in ecological payments and penalties and so forth [3].

Versatile ecological activities of the entities which allow not only to reduce the corresponding costs and expenses but also to get additional profit are called ecological efficiency.

Already today the western investors begin to demand from the Russian entities of confirmation not only economic, but also ecological solvency. Standards of the ISO 14000 and Eco-Management and Audit Scheme (EMAS) series regulate a production process, including the core and auxiliary business. They characterize functioning of system of ecological management and activities of a management for improvement of system/

The advanced western companies already at the end of the XX century have begun to use actively the principles of the ekologo-economic theory, have changed strategy and tactics of the development and have achieved considerable success [4]. From two hundred new net technologies implemented in developed countries, more than 50% at the same time reduce the level of pollution of the environment and energy consumption, 44% — reduce consumption of raw materials, 68% — give a prize in operating costs, and 11% — in volume of capital investments [5].

In modern conditions transition to ecologically safe production of construction materials it is in many respects caused by consumer behavior.

Buyers are ready to pay much higher price for environmentally friendly products and the producer or the supplier offer more advantageous conditions of sale of ecological materials or provision of services [6]. On the stock exchanges share value of ecologically solvent entities increases.

Thus, the relations of construction industry with society are more and more firmly governed by ecological reputation of the producer whom to support becomes economically.

However in Russia not all entities hurry to subordinate the activities to requirements of standards of the ISO 14000 series. The main obstacles the financial barrier (expenses), ignorance of effects to environment (unhealthy work conditions degree) and what operations need to be performed (to undertake) [7]. But as the most obvious obstacle in a way of increase of ecological efficiency of the country, providing its ecological safety and decrease in load of environment the consumer behavior of managers - power engineering specialists and officials acts [8].

Practice has shown that environmental problems cannot be solved at once. It is necessary to know what factors determine consecutive and steady development of production what the emergence of new quality of management turning it into ecological management depends on.

The main difficulty of the problem resolution of ecological production is the person, its values and motives, his consciousness and behavior. Emergence of eco-friendly equipment and ecological technologies also depends on it.

Interest of the person, an assessment of importance of problems, understanding of their role in further development of production – it is the determining factor of production management of construction materials. Therefore, for more widespread introduction in Russia of eco-efficient production of construction materials it is necessary to use the controlling mechanism behavior of accounting entities, to develop at them motivation on emission restriction, to stimulate use of "net" technologies.

It is possible only in case of prevalence of spiritual and moral values over material (aggressive consumer) motivation, following to the principles of eco-justice which are shown in understanding by a management of the entity of moral responsibility for negative impact on environment and irrational use of natural resources [9].

There is a large number of the approaches, theories and concepts explaining ecologically motivated behavior [10, 11].

One researchers believe that growth of attention to environment is connected with expression of post-materialistic values and self-expression as a result of the increasing wealth in developed countries. Others place emphasis on religious values.

The third claim that the ecological behavior is altruism manifestation. It is regulated by personal moral standards of the individuals convinced that private actions can pose threat for others (so-called understanding of adverse effects) and that the acts they can prevent these consequences (so-called attributing of responsibility to itself).

Guschina E. G., Badrak N. Y, referring to the website of the company by Energy Education, provide the list of motives which push people to ecological behavior. It is demonstration of care of society; environment protection and saving of financial resources; demonstration of social responsibility to people for whom you care; accomplishment of the necessary and important work.

Therefore, mechanisms and resources of impact on heads of the entities on production of construction materials, for the purpose of implementation of ecological management, should be looked for in consciousness of people. Important characteristics of consciousness are both interests, and values, and motives of activities. Goal achievement depends on their quality and system. Motivation of activities shall rely on the funds allocated for the solution of environmental problems. Administrative or especially organizational controls as practice shows are ineffective.

Besides, in case of implementation in production of ecological management it is necessary to follow the principle of commitment.

The purpose of implementation of eco-efficient technologies shall include components which reflect environmental problems and lash them in system of common problems of development of production.

The principle of professionalism is of also great importance. It consists in need of special training of managers, heads of the entities on production of the construction materials capable to operate with knowledge in the field of ecology.

Professional training gives the operating installations of management and allocation of priorities. The Russian heads lack it today catastrophically.

2 Tasks, methods and results of research

Studying of the motivational and valuable sphere of future heads of the entities of a construction industry, identification in it ecological orientations and comparison to the purchased McClelland's needs became a task of this research.

Research has been conducted in Moscow State University of Civil Engineering in 2015-2016. The group of respondents consisted of 112 students who are trained according to the "Production and Use of Construction Materials" program (Experts group) and 97 students who are trained according to the "Production Management" program (Leaders group). Research included three procedures: sociological poll, a technique "Valuable orientations" of Rokich and a technique of research of the acquired McClelland's needs.

Sociological poll included 17 statements about ecological aspects of activity of the person. Respondents had to express the attitude towards them. Results of research regarding the questions concerning future professional activity of respondents are given in table 1.

Table 1. Results of sociological poll.

Question	Replies of Experts group, %			Replies of Leaders group, %		
	No	Perhaps	Yes	No	Perhaps	Yes
People exaggerate value of negative impact of technical progress on the nature	89	4	7	85	6	9
Environment protection is so important that no restrictions for emission of pollution can be considered as too rigid	2	9	89	4	16	80
The environmental disasters happening on the planet concern me personally	57	33	10	62	26	12
It is necessary to improve habitat, without being considered with any expenses	5	31	64	23	32	45
"The person – the owner of the nature"	55	7	38	53	9	38
My future profession is connected with the relations "the Person – the Nature"	55	25	20	92	7	1

The given results show that in general nature of replies of two groups of respondents is similar. The greatest distinction is observed only in the answer to the last question. Students from Experts group are inclined to connect more the future professional activity with natural factors, than Leaders group. It partly confirms the statement pushed earlier about shortcomings of system of ecological attitude of heads of the entities on production of the construction materials which are a barrier to implementation of eco-efficient technologies. Other question reflecting distinction in system of the ecological-economic relations concerns expenses on improvement of habitat. As we see, the Leaders group gives less affirmative answers that testifies to its bigger economic orientation, than at Experts group.

Studying of valuable orientations of students was performed by means of Rokich's technique. The system valuable orientations substantially defines an orientation of the personality and constitutes a basis of its relations to world around, to other people, to itself, a basis of outlook and a kernel of motivation of activity, a basis of the vital concept and "life philosophies". Rokich allocates two groups of values: terminal – beliefs that the ultimate goal of individual existence is worth it that to aim at it; tool – beliefs that some conduct or property of the personality is preferable in any situation. This division corresponds to traditional division into values purposes and values means.

In this research only terminal values (values purposes) were considered. For the analysis they have been distributed on three equal groups: priority (1), significant (2), insignificant (3). In table 2 in percentage expression the frequency of hit of each value is provided to the specified groups.

Table 2. Rokich test results.

Terminal values	Experts group, %			Leaders group, %		
	1	2	3	1	2	3
active life (completeness and emotional saturation of life);	25	54	21	67	30	3
vital wisdom (the maturity of judgments and common sense reached by life experience);	3	17	80	5	43	52
health (physical and mental);	13	26	61	19	50	31
interesting work;	56	42	2	60	33	7
beauty of the nature and art (experience fine in the nature and in art);	0	4	96	1	9	90
love (spiritual and physical proximity with	48	48	4	42	50	8

darling);						
financially secure life (absence of material difficulties);	92	8	0	95	4	1
presence of good and loyal friends;	7	80	13	19	60	21
public calling (respect of people around, collective, fellow workers);	51	46	3	78	20	2
knowledge (possibility of expansion of the education, outlook, general culture, intellectual development);	40	33	27	39	30	31
productive life (most complete use of the opportunities, forces and capabilities);	47	40	13	51	40	9
development (work on, continuous physical and spiritual improvement);	28	20	52	39	30	31
entertainments (pleasant, easy pastime, lack of duties);	9	40	51	12	30	58
freedom (independence, independence in judgments and acts);	86	11	3	88	9	3
happy family life;	40	51	9	44	48	8
happiness of others (welfare, development and improvement of other people, all people, mankind in general);	1	13	86	4	22	74
creativity (possibility of creative activity);	10	18	72	15	18	67
self-confidence (internal harmony, freedom from internal contradictions, doubts).	75	24	1	78	10	12

From the table it is visible that in both groups have entered category of priority values: financially secure life, interesting work, public recognition, freedom, self-confidence. Difference consists only in one value: love and active life ("Experts" and "Heads" respectively). Distinctions and in the values concerning ecological consciousness, beauty of the nature and happiness of others are not revealed. Both of them have got to the third category – insignificant. This circumstance, adjusted for early age of respondents, can serve as an explanation of the difficulties arising in case of implementation of ecological management in activities of the Russian entities for production of construction materials.

Any significant communications between results of sociological poll and Rokich's technique it hasn't been established.

In the third part of research the technique of an assessment of the purchased requirements based on McClelland's theory has been used. It considers three types of requirements: power, achievement (success) and partnership (affiliation). McClelland has established that the motive of success allows the person (and even to the whole society) to achieve high economic results quicker. For success of the manager the developed requirement of dominion has the greatest value. The requirement of partnership is significant, generally for non-management employees of the entity.

McClelland has revealed, and S. E. Kock has experimentally proved that the greatest efficiency at the highest organizational levels is shown by heads at whom the high motive of the power (V) is combined with high motive of achievement (D) and low motive of affiliation (And). Such combination of motives (In + D-and) most promotes economic development of firm [12]. Research of features of motivation of future specialists and heads of the entities of construction materials I have shown availability of rank correlation of Spirmen between prevalence of motive of affiliation and an affirmative answer on the questions "Environment Protection Is So Important that No Restrictions for Emission of Pollution Can Be Considered as Too Tough" (0,71), "It is necessary to improve habitat, without being considered with any expenses" (0,73).

Thus, this research allows to assume that for the modern director on production of construction materials aimed at implementation and use in work of the principles of ecological management, the formula of an optimum combination of motives shall look as follows: In + D+ And. Such system of motives will provide systematic movement to the planned what efficiency increase purpose (motive of achievement), will allow to organize collective and to implement system of ecological management (motive of the power) in work. And the motive of affiliation inducing the head to correlate the decisions and actions to opinion and interests of society will act as the starting and supporting mechanism.

3 Conclusions

The conducted research of ecological motivation of future specialists and heads of the entities on production of construction materials has revealed low level ecological constituting in mentality of the Russian students. It finds further expression in slow rates of implementation in production of eco-efficient technologies, ecological management and standards of the ISO 14000 series.

Change of the existing behavior is possible through provision to people of thematic information and the appeal to personal values. The higher school shall provide development of the moral criteria of behavior limiting boundless material requirements [13, 14], putting in the forefront care about health and ecological safety of the nation and all mankind.

The central place in this process is taken by a factor of ecological education. It is backbone and can be considered in two foreshortenings – general ecological education and ecological education of the modern manager, his professional training in this area. The bigger number of class periods given to environmental problems in the Production and Use of Construction Materials program in comparison with the Production Management program is shown in more responsible attitude of students to environmental problems of a construction industry and in understanding of communication of results of the professional activity with environment [15].

Implementation of ecologically effective technologies and ecological management in the industry assumes obligatory involvement in conscious purposeful versatile ecological activities not only certain specialists, but also heads, the persons making decisions, factory personnel in general, and also interested in ecological activities of the entity of all other parties. The solution of a similar task is possible on the basis of essentially other motivation of activities which is possible for creating within general and professional education [16].

In the Russian state educational standards there are already several specialties of an ecological profile. The need for specialists in the field of ecological management already exists. Only specialists with complex ecological and managerial education will be able to satisfy it. Yet there are no such specialists in Russia as there is no excessive demand for them and. So far it is only a piece good. But it is necessary to train such specialists. Education shall be oriented not only to momentary requests of an industry, but also be able to expect future requirements which satisfaction will require specialists with new qualities. Training of such specialist takes any year.

References

1. Sh. I. Mamedova, *Construction Technologies and Architectural Aesthetics of the Information Society. The Third Planet From The Sun: Modern Theories And Research Practice In The Field Of Earth And Space Science* (2013)
2. A. D. Ishkov, D. A. Semernin, S. V. Miloradov, Applied Mechanics and Materials, **741**, 500-503. (2015).

3. S. F. Korenkova, Yu.V. Sidorenko, **10(20)**, 40908-40913 (2015)
4. B. G. Eddy, B. Hearn, J. E. Luther, R. Parsons, D. Piercey, G. Strickland, B.Wheeler, M. van Zyll de Jong, W. Bowers, Ecology and Society, **19**(3), 40 (2014)
5. S. V. Makarov, T. A. Ivanova, E. V. Aleksandrova,. EHkologiya i promyshlennost' Rossii, **8**, 25-30 (1998)
6. Z. Ivanovaa, I. Pichugina, Ju. Naimaviciene, Procedia Engineering, **117**, 154-161 (2015)
7. E. G. Guschina, N. Y. Badrak, Sovremennye problemy nauki i obrazovaniya, **5**, 121 (2012)
8. S. A. Mudrak, V. V. Glebov, EHkonomika i predprinimatel'stvo, **4-2(69-2)**, 457-462 (2016)
9. Z. I. Ivanova, O. V. Yudenkova, A. D. Ishkov, E. A. Shnyrenkov, International Education Studies, **8(5)**, 232-239. (2015)
10. I. A. Shmeleva, Psychology in Russia: State of the Art, **2**, 619-636 (2009)
11. P. C. Stern, Journal of Social Issues, **56 (3)**, 407- 424 (2000)
12. H. Hekhauzen, *Motivaciya i deyatel'nost'* (2003)
13. E. G. Gushchina, V. V. SHihovec,*Wykstalcenie i nauka bez granic `2005: mater. II Miedzynar. nauk.-prakt. konf. Psychologia i socjologia* (2005)
14. N. G. Miloradova, A. D. Ishkov, Procedia Engineering, **117**, 246-251 (2015)
15. E. Shnyrenkov, I. Pryadko, International Journal of Applied Engineering Research, **10(21)**, 330-335 (2015)
16. Z. I. Ivanova, O. V. Yudenkova, Asian Social Science, **11(1)**, 297-303 (2015)

Methods and techniques of research objects of forensic building-technical expertise (FBTE)

Svetlana Kolobova[1,*]

[1]Moscow State University of Civil Engineering, 129337 Yaroslavskoe sh. 26, Moscow, Russia

Abstract. This article discusses classical and modern methods and techniques of scientific and expert study objects of forensic building-technical expertise; dialectical peculiarities and logical methods of interdisciplinary methods, special scientific methods used in certain circumstances to establish causal linkages. Examples of their practical application. Is denoted by the prospect of the development of new methods and techniques of research to improve the quality of judicial construction and technical expertise.

1 Introduction

Forensic building-technical expertise (FBTE) remains very popular among different kind of expertise. The court considered the claims of multimillion-dollar construction companies and mostly from expert opinion depends on the court decision. Hence the importance of choosing the right experts, methodologies and methods of research facilities construction expertise.

Russian scientists R.S. Belkin, A.I. Winberg and N.T. Malahovskya, N.A. Selivanov, A.R. Shlyakhov, E.R. Rossinskaya and others [1, 2, 3, 4, 5] were at the beginnings of the creation of expert methodology. Scientists have developed a classification of the different methods used when conducting expert studies. R.S. Belkin, E.R. Rossinskaya have developed three-tier system proposed methods: universal dialectic method-shared (scientific) methods are special methods of science.

A.I. Winberg and A.R. Shlyakhov [6] substantiated scientific method system and expert study, taking into account their degree of commonality and subordination. This system can be represented as follows:

1. Universal dialectic method and methods of logic;
2. General (scientific) methods: observation, measurement, description, planning, experiment, simulation, etc., are used in many branches of science and practice;
3. Special methods, originally developed for a certain kind of (species) examination or borrowed from other areas of scientific or application activities.

The experts shall be guided by the following well known in their research methods.

* Corresponding author: KolobovaSV@mgsu. ru

2 The dialectical method

The methodological basis for all research involves the use of public and private research methods, idiosyncratic methods and techniques that depend on a particular task, the subject and object of research "[7]. the main logical methods include analysis, synthesis, abstraction, generalization of idealization, deduction, induction, analogy, extrapolation, modeling, hypothesis [8].

Determination methods and research of causality formulated in the early 17th century, f. Bacon, in the 19th century systematized and refined D.S. Mill. Hence their name: "the canons (principles) Bacon-Mill", in modern editions-"methods of Bacon-Mill", described in handbooks and instructional editions according to the logic of [9].

Similarity Method and the only difference method for singular method differences based on experiment, unlike the only similarity, which is based on observation.

For example, in the study of building collapse experienced by you can set the following:

-engineering survey errors and survey errors;

-errors of designers for loads on structures, strength

and the quality of the material;

-unfavorable reasons, reducing the quality and durability of materials and structures in the process of transportation, storage, installation, etc.;

-third-party effects (rain, wind, an abrupt change in temperature) and internal factors (breaking the rules) on the quality of the material.

For the conclusion about the cause of the events change, on the one hand, strength parameters of material, on the other hand, appearance, character and magnitude of load, carry out two experiments, one of which analyzed event occurs and the other not. These two cases are related in all circumstances, except one, which is present only in the first of them. The second case is a result of the Elimination of the circumstances. If such a removal will disappear and the phenomenon itself, the repaired fact can be considered a prerequisite for building structure collapse occurred.

While conducting experiments, including instrumental studies, it should be borne in mind that many construction materials are not homogeneous in content. That is evident in the construction sites, with long periods of operation, as well as subjected to intensive aggressive Wednesday. So, the characteristics of the same material, measured in different parts of the same design (for example, staying in the underground and above-ground parts) can vary widely in their numerical value.

The combined method of differences and similarities : If two or more cases when this phenomenon occurs, similar in only one circumstance, whereas two or more cases occur in the absence of this phenomenon and are distinguished only by the fact that this circumstance does not exist, then this circumstance, and there is probably a reason the phenomenon. This method applies when there is no possibility of an accurate accounting of the composition and nature of the circumstances leading up to the event being investigated, that there was a need in the separate use of similarities and differences.

Method of accompanying changes used when changing one changes and other phenomena, and other circumstances remain unchanged, i.e. between these phenomena a causal relationship exists. This method applies when modeling expert Builder process, consisting of a large number of components having different origins. For example, destruction of building structures may be linked to the violation of technology of building production.

Application of the method of residues when correctly identify clear disproportion causes investigated actions. If the activity in scope, scale or intensity does not match a known reason, the question of the existence of any other circumstances.

According to E.K. Voishvillo and M.G. Degtyarev: "If the cause of a phenomenon is not among some aggregate of previous circumstances, it must be among other prior circumstances or perhaps, among the circumstances that were in this case even unrecorded» [10].

Often the results can occur under the influence of more than one (particular) causes a number of reasons, acting independently (multiple reasons). For example, the causes of destruction of constructions buildings can be defects in steel structures, inaccurate installation (dock) details, process preparation of solutions.

"Because of the multiplicity of causes, says A.A. Eisenman, a" competition reasons "[11]. As a result of this "competition" "WINS" (expert) one or more of the number alleged.

The complexity of the method consists in choosing a single true cause of many. Causes may be the lack of special measuring instruments, testing base, inadequate and incomplete initial data, as from the date of the event that occurred, for example, collapse of building, pending a court decision on the FBTE time goes by. Then it is not one, but two or more possible causes that may have only an indicative value for the Court.

Unlike the multiplicity of causes mixing of consequence (action) occurs when the target event could occur only as a result of the joint impact of not one, but two or more factors. Each of them serves as a prerequisite for the coming result, but not a sufficient condition.

2 General (scientific) methods

These include sensuous-rational methods. Life phenomena must be understood not just as a set of separate, isolated elements, but as their set of systemized in a certain way [12]. Expert Builder studies most often uses such methods as surveillance, comparison or analogy relations and analogy items.

In FBTE is the most common method of graphic simulation. A graphical model is floor plans of buildings and constructions, land plans, case plans, etc. as a means of fixing the results of observations and measurements obtained by the expert. The need for building graphical models occurs in court cases about the legality of buildings, real estate section, share allocation, recognition of the right of ownership of building objects. In graphical models reflect the most significant for the proper dispute resolution characteristics of objects: orientation to the sides of the beam, the configuration and dimensions of the plot, the arrangement of residential and ancillary buildings and their location relative to the boundaries of the land, engineering networks and other communications, the availability of green spaces. Created by expert graphic models are required by an application expert elicitation and important evidential, supporting the conclusions.

Method of idealization is the establishment and exploration of ideal, mentally constructed objects. The value of this method to FBTE due to the specificity of its objects and the nature of the tasks performed by an expert. Real estate objects attached to land may not be presented directly to the expert. So their study is completed in two stages. In the first phase the expert fixes the basic characteristics of the objects relevant to the case. The second phase occurs mediated perception of the object when the material and symbolic information is converted to an ideal. Analyzing sketches, drawings, records, results measurement, expert "complements" the perfect image of investigated them in kind of object parts, which together with the already etched in the memory of moments create a system of signs, which gives an idea of the object's properties to solve other tasks in search of an answer to a question put by the Court. Using the method of formalization of concepts are replaced with special symbols and characters. The method is also used in the preparation of expert opinions. The examiner should give explanations in detention indicate that means one or another character (symbol), or it may be significantly more difficult. This method allows you to visually assess the results of the work carried out to all actors of the judicial process.

The second group of interdisciplinary methods-mathematical methods involving measurement, calculations and geometric building. When investigating a plot of land or a building complex configuration expert can produce about 100 measurements, allowing it later graphically reproduced in a certain scale positioning located on station buildings and structures or premises. Measurements shall be made on a number of occasions, either manually, by using a measuring tape and using the latest laser and optical measuring instruments, where it is technically feasible. Results are processed error and used to justify the withdrawal.

In any study of the applied computing. So calculated the cost and volume of completed and outstanding construction, construction and repair, commissioning, operation and durability of bearing and enclosing structures of buildings, constructions and buildings, etc.

The following General scientific periodical method-geometric build. Environment establishes the location of the object's sketchy plan. The plan gives a clear view on the mutual arrangement of items, various signs of investigated objects detected during inspection.

Now increasingly apply cybernetic methods that can help in many cases to meet the study objectives. Used linear programming methods. When determining the volume of excavation is used with success in computing, which helps determine the execution of mass haul volumes on a computing machine with vertical planning areas, for example, with a balance of zero.

Dynamic programming methods give possibility to choose the optimal ways of movement of construction machinery or best route construction machinery at the construction site, taking into account natural obstacles in the terrain, the shape of the construction site, wear of machines, their mobility and speed, fuel consumption and other indicators.

When construction of the facility shall develop schedules for different technological models that satisfy certain requirements, important of which is to match the managed object model, i.e. the criteria. Each of the models has the most appropriate scope. For example, the mission with the image stream in its development in time and space are most convenient when scheduling the construction of multistory and single-storey model buildings and structures.

Network models to accurately reflect the order of erection of a complex object, give reasoned decisions in the case of changing situations in the course of erection of the object. Especially efficient alternative network models with variable structure and enable you to display multiple variants perform the same process.

Successfully used in FBTE such as information systems (AIS) "House" AIS "Earth", developed by Members of the forensic expert institution of the Ministry of justice of Russia.

Modern information technology based on their use, let you search and automated processing of scientific, technical, legal and statistical information needed for the production of FBTE; produce in the course of research computation and mathematical modeling, as well as computer simulation in choosing, for example, a typical expert versions or reconstruction elements of the situation before, at the time and after the events, which became subject to investigation or prosecution.

3 Special scientific methods or special methods

Their application is limited to one or a few scientists or industries applied activities.

The most common research methods in FBTE are physical, chemical, physical-chemical for analysis of morphology, composition, structure, and properties of substances and materials used in the construction industry; methods for extrajudicial investigations. Using these methods, an expert Builder can determine the cause of the collapse of building structures, deformation etc.

In FBTE are used as proven methods, old and modern, using high-tech equipment.

Methods developed for extrajudicial investigations:

-establish the causes and conditions of the collapse of construction objects, determine their status and opportunities of reconstruction;

-verification of the characteristics and properties of products manufacture of building materials and products, and their compliance with these technical passports and certificates;

-determination of physical and technical characteristics of local building materials;

-selection of concretes, mortars, mastics, anti-corrosion and other construction formulations;

-sampling of soil, concrete and mortar;

-manufacturing of samples and testing (including weld joints);
-determination of concrete strength in structures by non-destructive methods and products.

A special place among the special methods of photographing method takes, as has a high degree of clarity, accuracy and completeness of the transmission conditions.

Research methods in FBTE can be divided into reproductive and heuristic. Reproductive, in turn, are subdivided into qualitative-descriptive and algorithmic, suggesting a certain set and sequence of operations expert.. Algorithmic methods are different from qualitative-descriptive in a strict sequence of actions. Under the Granovsky G.L. algorithm refers to exactly formulated a rule whose purpose is to be a guide for achieving the desired result [13].

Heuristic methods, in contrast to algorithmic and qualitative-descriptive compliance do not provide for a specific sequence of actions. Heuristic decision is considered to be found in the production process of examination by the expert himself [14] or when such use existing techniques, which selectively ignores a number of its provisions. Following the selected version, he "jumps" through the methodology stages (for example, verification of the opposite version), complements her new elements and study substantiates the results obtained. Heuristic methods are based on creative thinking and intuition expert, based on experience. The application of these methods has its limits. They can only be used at certain stages of the study, and the decision of the expert tasks cannot be based on heuristics.

Reasonable combination of classical and modern methods used by the Builder during the examination, such as dialectical and logical methods, scientific methods, including various modeling techniques: graphical simulation, cybernetic methods including mathematical and computer modeling will successfully meet the challenges FBTE and answer the questions of the Court, having worked on techniques and methods for material.

The inevitable technological advances in construction is happening rapidly, new building materials, innovative technologies and high-tech devices. Therefore, the term and the emergence of new methodological approaches to the study of objects FBTE.

Methodical recommendations on study of objects FBTE.

Depending on the nature of the task or tasks the experts choose certain methods, which are divided into generic (generic), generic and specific (private).

Generic (specific) expert technique is applied to a certain type of expertise. For FBTE it is more expedient to use the generic and not specific techniques. Each specific technique will aim to solve cluster close to its content. Among others, raises questions about the amount and cost of actually executed works, including the quality, on the quantity and market value of the construction works for the Elimination of deficiencies and defects of poorly executed works. This content is determined by the setting of trial, resolution of which requires the use of special expert knowledge-Builder, for example: the need for a decision by the Court of the question of the use of the building after the fire, whether the reconstruction and further exploitation, establish the cost of repair and reconstruction works in apartments (other areas), damaged the Bay (fire) [15].

The typical technique is popular or common tasks without any adaptation. The structure model of the methodology consists of the following elements:

-a description of the typical events that become the subject of legal proceedings;

-list of objects;

-methods and means of research;

-provisions concerning the sequence of conditions and procedures for the application of methods and tools;

-assumptions about the possible outcome of the use of methods and means; the characteristic of these results in terms of expert tasks.

Specific (private) technique is designed to address a particular (private) expert task. During the creation of specific individual elements of the methodology by the expert model methods are ignored and it is supplemented by new elements. As a result of expert based on the model technique and its knowledge develops methodology for each of its subtasks, as well as the program of study for this situation. The structure of a particular methodology is a way to study the combination of working methods or individual techniques and funds research.

Currently, a number of publications released [Borodina E.A. Butyrina A.Y. Davydenko O.V. Malysheva S.D. Popova A.N., etc.], which identified methodological approaches to research forensic examiner-Builder, such approaches are detailed, the choice of certain methods and means is justified. So far, however, has not yet created any official methodical system to tackle all challenges FBTE. This will be achieved on a particular, higher stage of development expertise of this kind. The absence of such a system, however, does not mean the impossibility of formalizing common tasks FBTE, develop algorithms to solve them. The need to update the Foundation and development of forensic FBTE methods in their writings of reputable scientists.

In the future it is necessary to consider that inevitable scientific progress in construction is happening rapidly, new building materials, innovative technologies and high-tech devices. Therefore it is inevitable and the emergence of new methodological approaches and methods of the study.

Reasonable combination of classical and modern techniques and methodologies used by the expert Builder during the examination, such as dialectical and logical methods, scientific methods, including various modeling techniques: graphical simulation, cybernetic methods including mathematical and computer modeling will successfully meet the challenges FBTE and answer the questions of the Court, having worked on techniques and methods for material.

The inevitable technological advances in construction is happening rapidly, new building materials, innovative technologies and high-tech devices. Therefore, in the long term the emergence of new methodological approaches and methods of the study.

So, you will need the following tools and methodological recommendations: nullify construction sites to the category of movable and immovable property; by definition, construction projects in the categories of capital or non-capital; establish opportunities and develop options for a real section of public and industrial buildings between their owners; to establish the functional purpose of basement and attic residential apartment buildings; to establish the causes, conditions, circumstances and destruction mechanism of the construction site [16].

With the development of modern construction technologies, both domestic and foreign, imported into our country, objects of FBTE have become complex multifunctional complexes, industrial and civil buildings and structures, as well as large-scale plots, functionally related. Among these objects could meet the tall buildings and unique constructions. In such cutting-edge objects space-planning and design as construction sites in General and their individual fragments, innovative building materials, designs, products, technical equipment, etc. All of the above was the development of the judicial building-technical expertise, and it required new methodological approaches and techniques. [16]

4 Conclusions

Formation of methodic base basic judicial construction and technological research can be conventionally passes through three stages. In the first phase, experts, receiving questions the Court will divide them into questions that can be answered by applying the technique of production expertise of a certain type and subspecies (private technique) [17] and on the remaining issues that need to be answered by applying a specific methodology study [18]. In the future a specific technique can be adapted to the new object of study. The first phase can last for a long time [19] to the need to address the qualitatively new expert tasks. Next, in the second phase, experts share experience in publications of articles of scientific and methodical nature of the reports at scientific conferences, in describing the emerging practice solutions expert tasks not covered previously in the literature. Conducted the peer review of the study of regularities expressed would trigger the development of new methodological approaches and techniques. In articles subjective expert opinion contrasted with emerging trends in practice, solving the tasks stated and disclosed their insolvency or difficulty, outlines the proper from the point of view of the authors, the system methods which provides science-based result [20].

The third stage of methodological framework includes preparation, approval, publication and implementation procedures (guidelines) forensic solutions to common challenges. In the system of forensic science institutions of adoption procedures (guidelines) is carried out by the scientific-methodical Council of experts under Ministry of justice of Russia. This three-tier structure of the process of formation of forensic of methodic base FBTE allows you to organize your modern methodical development of forensic experts-Builders according to the degree of development of their separate directions.

The past in October 2015 years Kazakhstan international legal forum "development of forensic expertise in the context of the Eurasian Economic Union» brought together participants of the modern court proceedings in an effort to transition to a qualitatively new level of forensic technology. Currently, the ways of upgrading expertise in Eurasia, such as standardization, accreditation, certification, validation and inter-laboratory proficiency testing. One of the priority directions for development of the Eurasian Economic Union is the coordination of scientific-methodical work, improving the provision of information, the development of new hi-tech investigation techniques [21].

Close regional and international cooperation will lead to the development of new effective methods and methodical recommendations of FBTE and hence improve the quality of forensic evidence.

References

1. R.S. Belkin, *Experimental method of obtaining samples for comparative studies* (Moscow, 1961)
2. A.I. Winberg, N.T. Malahovskaya, *Trial expertology (theoretical and methodological problems of forensic examinations)* (Volgograd, 1979)

3. N.A. Selivanov, *Controversial forensic issues* (Moscow, 1978)
4. A.R. Shlyakhov. *Actual problems of theory of forensic expertise*VNIISÈ, 33-49 (1984)
5. E.R. Rossinskaya, *Forensics in criminal, civil, arbitration process* (Norm, Moscow, 2006)
6. A.I. Winberg, A.R. Shlyakhov, *General doctrine of forensic methods*, 54-94 (1977)
7. A.I. Winberg, A.R. Shlyakho, General doctrine of forensic methods, 65 (1977)
8. N.I. Kondakov, *Logical* (Moscow, 1975)
9. E.K. Voishvillo, M.G. Degtyarev, Logic, 143-150, 273, 149-152, 293, 34 (2000)
10. E.K. Voishvillo, M.G. Degtyarev, Logic, (Moscow, 2000)
11. A.A. Eisenman, *Expert opinion (structure and scientific justification)* (Moscow, 1967)
12. *The foundations of scientific knowledge for lawyers: Stud. for universities* (Moscow, 1999)
13. G.L. Granovsky, Expert tasks and their solutions in the light of the STC, **42**, 27-44 (1980)
14. G.l. Granovsky, Mikrotrasologičeskih Research Methods (VNIISÈ, Moscow, 1983)
15. A.Yu. Butyrin, O.V. Lukovkina, *Methodology expert solution to the question of the cost of repair apartments. Damaged Gulf (fire),* 55-175 (2007)
16. E.B. Stativa,). Russian Federal Centre for forensic sciences under the Ministry of Justice of the Russian Federation, **4(36)**, 72-78 (2014)
17. Y.G. Koruhov, *Dictionary of basic terms of legal expertis*. (Moscow, 2007)
18. *The specific technique is the technique of manufacturing expertise in a particular case* (see ibid., p. 44)
19. *Expert production no. 823/19-3, no. 3456/19-2. Archive of experts under Ministry of justice of Russia,* (2013, 2014)
20. A.Yu. Butyrin, Principles divisibility criteria for real stuff and a section of the building sites of various functional purposes, Theory and practice of forensic examination, **1(29)**, 75-80 (2013)
21. A.I. Usov, Law and State, **1(70)**, 48-52 (2016)

Basic and Combined Economic needs Monitoring in the Industry Development Support

Dmitry Silka[1,*]

[1]Moscow State University of Civil Engineering, 129337 Yaroslavskoe sh. 26, Moscow, Russia

Abstract. The evolution of economic processes and phenomena occurs cyclically with occasional deceleration or acceleration. In general, this development is accompanied by all known phases of business activity. The article reveals a new evaluation mechanism of the achieved results of the business activity growth, based on the determination of the current economic organization structure generation, the criteria for economic needs and interests. This approach allows us to critically assess the growth limits, and to form the conditions for further progressive development.

1 Introduction

The formation of economic intercourse both in the whole country and in the sectoral format predetermines business activity and related processes of economic cycles. On the basis of the author's research it has been stated that the existing complex of economic intercourse that determines business activity factors and business cycles is reflected in the existing economic organization structure.

2 Basic and Combined Economic needs Monitoring

Economic organization structure predetermines business activity cycle in the investment and construction sector depending on the existing potential of economic needs and interests. For example, at a certain stage of Russia's recent history the government came to a conclusion that the society had the need in the shared financing of the construction. Such schemes had been implemented spontaneously in the contract form of "the construction at the stage of excavation". But the society had covert needs to improve the situation with transactions security. Subsequently, due to the state involvement, these schemes were documented in the regulated shared financing schemes (Federal Law - 214). In this case the state role in supporting business activity became obvious, though it was not active enough. There is another example. If we define a modern economic organization structure, where the advanced role of science and innovations is acknowledged, it is possible to identify economic needs, the potential of which is capable of helping to solve a lot of problems. As a variant, these needs can provide housing for workers in the sphere of education and

* Corresponding author: w220@yandex.ru

science. First and foremost, creating a mechanism to meet their interests will be the peculiarity of the economic organization structure. It will also be able to afford to create the conditions for business activity with the use of innovative ideas and solutions. This example expresses the potential of the economic structure needs more clearly, where the economic needs predetermine business cycle, etc. Hence, identifying of latent and explicit economic needs within the existing economic organization structure becomes absolutely necessary for business activity and business cycle control. In turn, business activity will have clear prerequisites for growth, if at the same time the existing economic structure meets the following criteria:

- latent and explicit economic needs of the state and society have been identified and are systematically identified;
- in each case, before we start the search for the ways to satisfy these needs it is admitted and taken into account that the identified needs are not finite, and in order to meet those the largest possible number of different economic needs related to each other should be identified;
- there is a clear ultimate economic need (for example, innovations in the industry), and another basic need (housing for science workers) has been identified, and taking them both into account we can generate a lot of different economic needs interconnected with each other;
- while identifying a finite need other economic requirements are being identified.

When comparing different economic structures, the structure will dominate (that is it will have the largest number of growth factors of business activity) if it has the highest level / criterion of combined economic needs included in the given basic economic need of the society (of the real sector of the economy, industry, sphere of activity, a group of entrepreneurs with consolidated interests etc.). All combined economic needs and the basic need have clear relationships and the scheme to meet the interests in sequence.

Indicator- "level /criterion of the combined economic needs included in the basic need" - appears in the form of the systematically important criterion if we do mean the concept of "the combined economic needs included in the basic need." If we consider the situation with the development of the standard technical documents regulating the development of the construction industry, where the basic need is the stimulation of innovations, the consideration of the needs will be the criterion. (The details are missed intentionally.):

1) the needs of the state and industry administrating authorities;
2) the needs of the scientists and the experts in the relevant sphere of knowledge;
3) the needs of the scientific and research organizations;
4) the needs of the enterprises of the construction sector designing and building up the completed construction products;
5) the needs of the construction products commercial users.

This example has been given for illustrative purposes only. To control business activities within the framework of the economic organization structure it is necessary to state that the criterion (C) in the development of the standard technical documents in the construction is a compulsory record keeping of the economic needs, the minimum quantity of which covers the above list of five items (taking into account decomposition $C = 5x$, where x is a fixed number of combined needs included in the basic need). In turn, the existing economic organization structure can be diagnosed on the compliance of the combined economic needs (included in the given basic economic need) with certain criteria. Next, we can define that the given economic structure refers to the first generation of the satisfying economic needs (as an option, according to the example above, criterion $C = 5x$). When we reach a new level of criterion ($C = 6x$), the economic structure will refer to the second generation, etc. But in general, the proposed approach aims at achieving the system of classification of the economic organization structures in the investment and construction field or in any other fields of the live economy sector. Economic organization structures are classified according to generations and dominant management entities. Generations depend on the criteria of the combined economic needs. The practical examples confirm that the satisfying of the economic needs growth is really a trend which scientists and economists are interested in worldwide. So, systematic measures are being developed for that. As an option, a complex of appropriate tools is included into the "Open Data" concept.

Both economic organization structures and economic entities are not isolated from each other. It was stated that for different economic structures transfer of knowledge and competencies is peculiar. Therefore, a variety of production intercourse, which represents the specific economic organization structures, may reflect growing interconnections that will also result in business activity. The increasing relationships between the economic structures allow an increase in the level / criterion of combined economic needs included in the basic need, and they also become a prerequisite for the growth of business activity. It is a good place to start the research of the generations of the existing economic structure in the investment and construction industry with the beginning of the 1990s period, when the economy had transformed into a market form. The study of the earlier period has a more scientific and educational purpose than a practical one.

The economic organization structure in the construction [1] has the characteristics, the content of its description both at the level of the managed systems and at the managing systems level. Managed systems characteristics of the economic organization structure are: 1) the types of the construction products which are being built, the types of works and services; 2) the types the physical resources, labour force, finances and other resources being used; 3) the construction technologies being used; 4) organizational mechanisms of construction activities; 5) the consumers of the completed construction products, works and services; 6) engineering systems and communications either centralized or local and the other. They have been reflected in the investment and construction activities projections, synthesized and shown above. We will consider and justify the methodology for determination of the managed systems characteristics in more details, giving an example of the description of the expectable economic organization structure.

So, for example, the economic organization structure has been identified, which is based on the real estate development business.

The implementation of the structure requires: 1) a complex of the capitalization of construction resources mechanisms for the complete project of the construction cycle, including the sales of the completed projects on the market; 2) the use of property, including land for the construction process; 3) the use of long-term investment resources and the relevant financial tools; 4) the use of both a wide range of physical resources and appropriate tools of inventory and logistics management; 5) a complex of certain categories of labour resources; 6) the level and the rate of the personnel employment on the site facilities; 7) the statutory regulation system of the types of works; 8) the state involvement in the strategic issues; 9) the resources taking into account the specifics of the construction process; 10) the application of innovations for enhancing the construction activity efficiency taking into account the duration of the construction process full cycle. Each of the ten points corresponds to the identified projections.

In the same way we can describe the economic organization structure, which is based on the cooperative activity, or the economic organization structure based on the share in the capital mechanisms of the households, etc. In fact, it confirms the possibility and adequacy of the description of the managed systems characteristics of the economic organization structure through the investment and construction activities consideration projections.

Due to the fact that the economic organization structure in construction is defined by the production relations, the managing systems characteristics are described right at the level of the subject relations. Describing the subjects of the economic organization structure we can distinguish: 1) the individual households that do not have large savings and investment resources; 2) individual households united by common interests and representing the potential in terms of possession of investment resources for regular organizing of the construction processes; 3) professional investment and construction institutions that do not have significant investment resources, but that are capable of attracting and developing the capital investments; 4) professional domestic investment companies, which are able to implement large-scale construction programs on the basis of the availability of their own adequate investment resources; 5) professional foreign investment companies, which are able to implement large-scale construction programs on the basis of the availability of their own adequate investment resources; 6) the state and specialized institutions implementing the investment and construction activity.

So, basing on the achieved results we will state that the category of the economic organization structure is of great interest for the research of business activity, its causes and performance. The criterion proposed above - "level /criterion of the combined economic needs included in the basic need" - belongs exactly to the economic organization structure, and as a result it specifies the business cycle parameters. Due to the dominance of the economic structure stability over the business cycle, identification of economic organization structures is the method to introduce and consider a higher level of abstraction in order to clarify the mechanism of the business cycle advent. As it is known, abstraction is a "distraction from non-essential sides, properties, connections of an object (thing or phenomenon) during the cognitive process in order to highlight their essential, inherent characteristics. The degree of abstraction of the discussed concept is called the level of abstraction. And depending on the goals and objectives we can speculate about the same object at different levels of abstraction. The need for abstraction is defined by the situation where the differences between the intellectual nature of the problem and the existence of the object in its concreteness become apparent " [2, 3].

The scope of business activity is determined by a complex of the construction types within the existing economic organization structures, which have common features. The economic interests in the framework of the economic organization structures should facilitate innovation activity in the construction sphere, where innovation activity is considered to be a combination of growth factors. Hence, business activity cycle will have positive trends, there will be prerequisites for the transition to a new phase with an efficiency increase of the investment and construction activities, the potential for investment and construction activities will be unlimited.

The manifestation of innovations is implemented improvements providing high-quality growth of efficiency of construction processes and construction products, works and services demanded by the market. A new or significantly improved product or a process of construction, a new sales method or a new organizational method in business practices, in workplace organization or in external relations all introduced in use are also innovations. Finally, not every improvement or a new development represents an innovation, but the only one that significantly (fundamentally) increases the efficiency of the current system. [4-8].

Thus, if it is obvious that business activity should be provided with innovations then the important questions arise: what should be a source of innovations, and what kind of relationships will facilitate innovation activity. Generations of the economic organization structures are of a key importance here. Some growth in the economy can occur within the economic structure of one generation without transiting to a new economic structure of a higher generation basing on activating of various potentialities, etc. We can even simulate the situation where economic growth continues while economic structure has transited to a lower generation level. We can assume that it is a transition from market instruments to a limited complex of administrative procedures. But we keep to the view that such situations do not occur simultaneously with the growth of innovations in economy. The transition both to the system of management and to the economic organization structure of a higher generation on the one hand and the growth of innovative economy on the other hand are compatible conditions. They also predetermine long-term growth factors at large.

The source of innovations is in the relations system of the economic organization structure. The development of innovations is a difficult process which requires both scientific research and theoretical substantiation and practical implementation. In this regard, if there are no innovations in the identified economic organization structures, in the determined content of the economic intercourse and in the achieved economic interests, it means that the business cycle is not accompanied by an innovative makeup. Thus, to facilitate a potential for growth in the business activity cycle there must be innovation processes within the economic organization structures of the investment and construction activities. Hence, identification and research of the innovative components of the economic structure are the most primary objectives during the analysis of economic organization structures. Since the development of innovations is a specific activity requiring considerable resources withdrawal from the production process we assume that primarily economic organization structures both of the state and private business without isolation from each other can generate the necessary innovations for the development of the construction industry.

It is obvious that the development of innovations plays an important part in business activity and hence the economic structure that provides the most favourable conditions for innovations and is characterized by a new generation of the economic structure is capable of becoming dominant in the economy. But in order to identify the economic structure facilitating the development of innovations we should analyse in detail the category of the economic organization structure itself. First of all, each structure requires self-identification. To do this we will reveal insights into common attributes and properties of the economic organization structure in the construction industry.

The survey showed that economic organization structures are a structured economic relations complex of economic entities, to which the processes of satisfying economic interests are inherent in accordance with the existing competencies. The presence of business activity in the relevant cycle phase means both the development of construction products in general and the presence of the expanded reproduction process in system. And in turn, special aspects of the expanded reproduction process may be revealed within the economic organization structure. The author revealed the insights into this issue in more details in the article [3].

Since the essence of economic relations and of satisfying economic interests processes is revealed by means of economic organization systems analysis hence through the understanding of the nature and the causes of the economic relations in the economic structure it is possible to identify the causes of the business cycle. Economic organization structures have certain characteristics and attributes in connection with the objectivity of the economic structures in the investment and construction sector.

After analyzing the main basic definition of the "attribute" [2, 3], the concept of the economic organization structure attribute can be expressed as follows.

Firstly, the characteristic of the economic organization structure is such a specific feature of this category by means of which we can determine that this is the economic organization structure exactly.

Secondly, we can also note that the attribute of the economic organization structure is a distinctive feature of the economic organization structure among other categories.

Thirdly, the attribute of the economic organization structure is the feature of the structure which we can recognize, determine or describe which represents the certain property or the designation of the economic organization structure.

According to the analysis of the established definitions properties differ from the attributes in that they are a complex of characteristics, inherences and peculiarities of an object or a phenomenon which allow us to clarify the essence of the identified object, phenomenon and other.

So firstly, the property of an economic organization structure is a characteristic or a complex of characteristics of the economic organization structure, which indicate a distinctive feature of the structure.

Secondly, the property of an economic organization structure is one of the quality aspects considering that the quality is permanent in the object (phenomenon, etc..) while properties may or may not come out.

Economic organization structures can have a great variety of properties. Generation of the economic structure is also its property which is a priori assigned and permanently inherent.

Thus, the attribute of the economic organization structure enables us to answer the question whether or not the economic organization structure is an economic structure, and thus to distinguish it from other categories. The attribute is based on the double opposite characteristics of the object, phenomenon and other. And only one characteristic can be true whereas the other one should be false.

The purpose and the final outcome of the investment and construction activity are to create the completed construction products. Therefore, during the analysis of any economic organization structure in the investment and construction industry it is necessary to make sure that, firstly, the economic structure is connected with the economic processes, in particular, with a certain type of mobilization, investment, development and capitalization of construction resources, and secondly, with the organizational processes (via the system of economic intercourse), in particular, with a certain type of building production. At the same time in favour of confirming mentioned above we will use the argument taking into consideration that the economic activity of the society expressed as a complex of relations in the system of production, distribution, exchange and consumption is the basic concept of the economy [2].

Conclusions

So, we can assume that it is feasible to define the economic organization structures attributes through a detailed analysis of the construction methods of production and the tools of mobilization, investment, development, and capitalization of resources.

Based on these statements and realizing that the growth of business activity depends on the transfer of the economic structures from one generation to a higher generation, we can note that the growing needs of the society are expressed in the automatic growth of the criteria of the combined economic needs included in the basic need. Therefore, in order to facilitate the progressive development of the society it is necessary not only to identify the spheres of economic efficiency increase of various processes and phenomena, but also to offer new needs, interests and the ways to satisfy those. The idea of the research to fix the criteria relying on the basic needs and to identify the generation of the economic structures and inherent attributes and properties gives a very objective and sustainable approach, and the system of knowledge to offer favourable conditions for economic and industry fields growth.

References

1. N.Y. Yaskova, Bulletin of Irkutsk State Technical University, **11(70)**, 278-280 (2012)

2. A.N. Azriliyan, *Large Economy Dictionary* (Institute of New Economy, Moscow, 2002)
3. N.Y. Yaskova, D.N. Silka, Bulletin of Irkutsk State Technical University,**11** (2012)
4. S.I. Agabeyov, D.I. Kokurin, K.N. Nazin, *Innovations in Russia: systematic institutional analysis. 2nd ed. Corr.* (TransLit, Moscow, 2011)
5. A.E. Varshavsky, *Problematic innovations: the risks for mankind. Economic, social and ethical aspects* (LENAND, Moscow, 2014)
6. J.V. Vertakova, E.A. Alpeeva, I.F. Ryabtsev, *Progress and innovations: analysis of systemic interdependence: a monograph* (INFRA-M, Moscow, 2014)
7. I. Miloslavskii, Izvestia, **99**, 4 (2009)
8. R.M. Nizhegorodtsev, *Information economy. Book 1. Information universe: Information bases of economic growth* (ICS RAS, Kostroma, 2002)

Components of functional information model of city environment reorganization in interactive mode

Olga Kuzina[1,*]

[1]Moscow State University of Civil Engineering, 129337 Yaroslavskoe shosse 26, Moscow, Russia

Abstract. The article contents the description of reorganization complementary information model provided for the lifecycle stage of facility management. The main components or assets of that model are functional, constructive, and organization-technological level of a declination from the project perfect model. There are also three-level model of building reorganization type definition. It represents an operation-automated algorithm for updating the technical specification of building reorganization.

1 Introduction

The main purpose of building reorganization (BR) is to bring the object to the state, to meet customer requirements by architectural, functional and planning transformation. That is important to understand that the nature of buildings and structures reconstruction is determined not only economic and functional expediency, historical, architectural and environmental restrictions, but also dependent on funds and customer preference [1]. The main tool for improving the efficiency of the organization processes of building reorganization is process of modelling at the planning stage of lifecycle.

Creating a functional model of BR, you can create operation description of system data and analyse the reality of upcoming improvements, changes, range of the necessary expenses of all types of resources, as well as the most well-coordinated work of the subsystems.

This approach will create basis for calculating the cost of reconstruction, creating a base for the project of construction organization (PCO), a technological project of works (TPW), schedules. Analysis shows the opportunity of decision-making on the feasibility of the project and determine its results. The customer has the choice of solutions not only upon the estimated cost, but also after evaluation of time and labour efforts [2].

2 Functional system of city reorganization

Functional system of city reorganization designed to provide:

* Corresponding author: kuzinaon@mgsu.ru

- eliminate physical deterioration or obsolescence,
- changes in the functional purpose,
- increase production capacity,
- equipment replacement,
- restoration of artistic monuments, etc.

Includes:

- designing, engineering,
- production of special materials, structures and equipment,
- development of special technologies, often in the continued operation of the facility [3].

Components of city environment reorganization functional system need to define type of reorganization for completing technical requirements and project of reconstruction organization. These components divided into three levels: functional need, structure damage, organizational and technological production process.

A key point of using building information modelling (BIM) during reorganization stage of life cycle is a principle of the collecting data and documents about all the resources and solutions used in the project. Following this methodology, the reconstruction process remains available for analysis and then for reasonable changes. Using BIM technology in general has important advantages: custom geometric modeling, administration and arrangement of layers, etc.

A significant number of objects in the system are project-oriented, with the possible use of the advantages of their parameterization, so that with small changes they can be reused in similar projects of reconstruction of historical objects. Currently, most libraries of objects with historical simulations are created for the purpose of the one project. It is necessary to expand the existing modelling technology to the modern level of technologies and an even higher level of realism [4-15].

3 Methodology of determining the type of organizational model

Functional modeling of BR help to identify the types of features of the facility and presenting them in the form of complementary information model that reflects the process of selecting the right kind of building reorganization. To define the functional operation for BR the manager should use three components of applied three-level model.

Each level represents a set of features: functional need for the reconstruction of buildings (system) or the individual element of the system, physical obsolescence of the specific system element, the elements of the organizational and technological production process as the third defining characteristic, which is determined depending on the structural and functional one. In addition, the third features is express in the organizational - technological solution for the production of works on BR site.

For each level designed the database with a set of possible properties. That database is open and expect to be enhance if it will have reasonable proofs.

Presented in Figure 1, a three-level model for determining the type of BR. The first level - the purpose of BR, a functional necessity. It consists of thirteen items regulated by existing regulatory documents.

The second level is a list of structural elements of buildings and structures of the system, consisting of 23 items. The third level - a list of organizational and technological production processes of BR depending on the functional requirements and structure elements of urban infrastructure system.

The third feature contains 93 items. It is important to note that these sets are open. After analyzing the three-level features, we can conclude what kind of reorganization is required on site.

Figure 1 also shows the correlation of BR structure with the standards of reorganizational types, regulated by the current regulations, and divided into separate species by researchers of higher education.

There are 8 standard information models (repair, reconstruction, restoration, renovation, recombination, reversation, retrivation, rehabilitation engineering (sanitation)). Each standard model has its information model, as a set of parameters on three grounds - the levels of the structure in Figure 1.The standard is a three-level model with a set of elements on three functional levels.

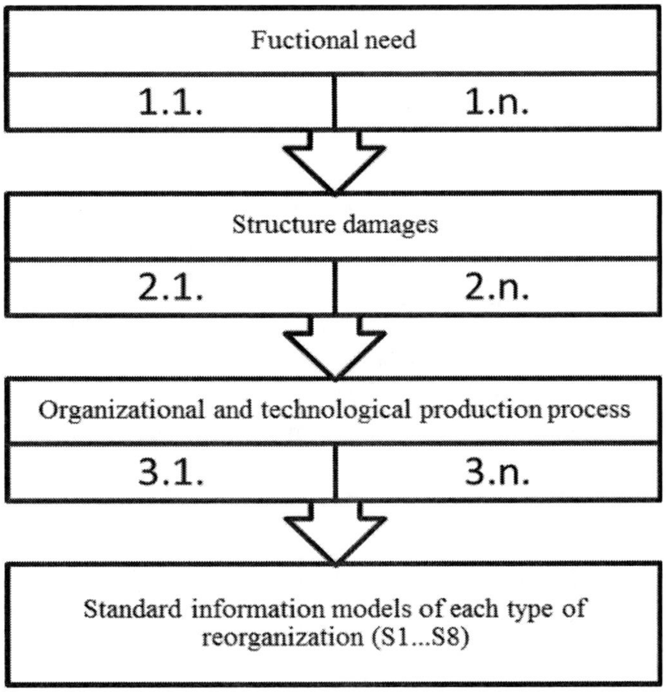

Fig. 1. Three-evel model for determining the type of BR.

4 Resulting modelling

Assessment of the BR of non-productive (city urban) sphere objects shows that there is a problem of varieties interpretations multiplicity of BR type, no uniform evaluation criteria, which can be classify as types of restructuring.

It reveals that modern business processes, innovative technologies in the production of construction materials and others define new types of BR, as they have special properties that allow them to identify as an independent kind of reorganization.

It is obvious that for calculating the value of certain types of work there is a need for the designation of operational sets, applying the current needs and innovative processes.

The basis of this choice is a method of dynamic programming, based on the principle of optimality, which can be formulated as follows: to get the best solution, it is necessary to follow the rule - whatever the way to achieve the test system of a state, subsequent decisions must belong to the optimal strategy for the remainder of the path, starting from this condition.

There are dynamic recurrence:

$$f_n(s) = \min\left[c_{sj} + f_{n-1}(f)\right], n = 1, 2, \ldots;$$

where $f_n(s)$ - cost, corresponding to the lowest cost strategy for the path from state s to the final state of the system, if the rest of the way consists of $\boldsymbol{n}$ steps;

$j_n(S)$ - solution to achieve $f_n(s)$;

c_{sj} - cost of translation system from the state s to the state j.

So, for the needs of BR, was complete the database including information of three basic levels: functional, organizational, and technological. The purpose of the perfect information model of building reorganization is to automate the processes of completing the documents for providing works of reconstruction or another type of reorganization (Figure 2). That approach would let to evaluate the expenses of money, labor, and time for reorganization processes at the stage of preoperating [3].

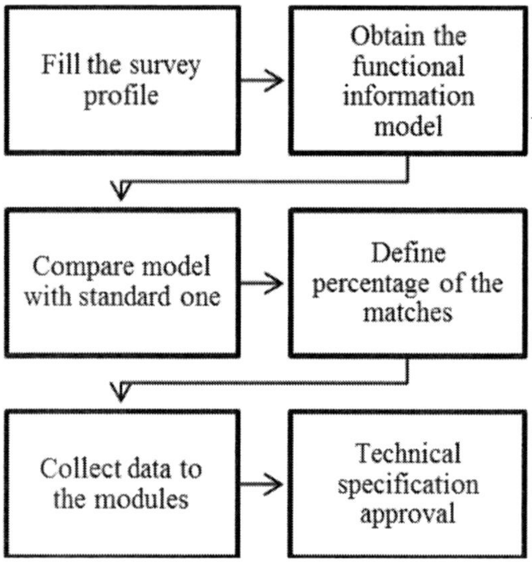

Fig. 2. Sequence of technical specification obtaining.

Besides varieties of BR types may also be isolated accompanying processes, which include the conservation and reconservation, recycling, industrialization, etc. [4].

Thus, the algorithm of updating the technical specification is the following:

1. The Contractor sends Customer the form of survey (profile).
2. The Customer, in conjunction with the Contractor completes the form, using the database for the three levels of features, defines the purpose of the work production (functional requirements), structural elements that require reorganization, organizational and technological production processes.
3. Then Contractor makes the comparison of standards model (S1-S8) with this obtained model of BR [5].
4. The Contractor determines the percentage of the matches the selected parameters to the results obtained through the three levels. Get the chain - modules for each type of BR that for further design recorded in the database of new information model for further processes [6].
5. Modules of the results (the recruitment process for each type of BR involved in the facility) should be agree with the Customer. [7]

References

1. A.A. Volkov, S.N. Shulghenko, A.I. Shirokov, E.N Kulikova, Vestnik MGSU, **6** (2011)
2. A.A. Volkov, Stroitelnie materiali, oborydovanie, tehnologii XXI veka, **7** (2001)
3. O.N.Kuzina, V.O. Chulkov, R.R.Kazaryan, PGS, **11** (2009)
4. O.N.Kuzina, V.O. Chulkov, *Sbornik trudov Stroitelstvo – formirovanie sredi jiznedeiatelnosti,* **23**, 67-72 (2009)
5. V.O. Chulkov, *Pereystroistvo. Organizacionno-antropotehnicheskaia nadejnost stroitelstva* (SVR-ARGUS, 2005)
6. K.A. Shreyber, *Variantnoe proektirovanie pri reconstrukcii jilih zdanii, Seria Zhilishche – 2000,* (Stroyizdat, Moscow, 1991)
7. O.N.Kuzina, V.O. Chulkov, PGS, **11** (2009)
8. 17. V.V Okrepilov, V.L. Makarov, A.R. Bakhtizin, S.N Kuzmina, Economy of Region, **2**, 301-313 (2015)
9. V.L. Kvint, V.V Okrepilov, Herald of the Russian Academy of Sciences, **84(3**), 188-200 (2014)
10. V.V Okrepilov, V.N. Krutikov, G.I. El'kin, Measurement Techniques, **57(2)**,109-116 (2014)
11. S. Rutešić, J. Ćetković, M. Žarcković, M. Knežević, N. Vatin, Procedia Engineering, **117 (1),** 905-915 (2015)
12. S. Rutešić, J. Ćetković, M. Knežević, M. Žarcković, N. Vatin, Procedia Engineering, **117(1)**, 642-650 (2015)
13. M. Jocovic, B. Melovic, N. Vatin, V. Murgul, Applied Mechanics and Materials, **678**, 644-647 (2014)
14. G. Radovic, V. Murgul, N. Vatin, Applied Mechanics and Materials, **641-642**, 634-638 (2014)
15. M. Penića, S. Golovina, V. Murgul, Procedia Engineering, **117**, 883-890 (2015)

Prerequisites of creation of the general branch informational system in construction

Irina Karakozova[1,*]

[1]Moscow State University of Civil Engineering, 129337 Yaroslavskoe shosse 26, Moscow, Russia

Abstract. The article is dedicated to the problem of the absence of any theoretical justification of the process of the creation of the general information space, which would make the effective use of the main types of resources possible during the processes management at all the stages of life cycle in construction. The problems of classification and the problem creation of the system of coding of the material resources considering their main properties and characteristics, as well as the need of creation of any requirements management unit as a part of the information system are separately considered.

1 Introduction

The relevance of the matter is determined by the need of the theoretical justification and the subsequent formation of any tools for their use within the creation of the general common information space for the purpose of reduction of terms, decrease in expenses and minimization of risks during all the life cycle of projects in construction. On the one hand there is an intensive updating and development of the construction branch and in this regard there is the urgent need to have an information systems allowing to make decisions in due time, qualitatively, openly and with all the responsibility. On the other hand the organizations in the construction branch are not fully ready to adopt such open format of communication, demanding considerable time and financial expenses. In this regard there is the need of justification of the creation of a uniform information space for the purpose of receiving and transfer of data at all the stages of the life cycle of a construction project from the conception of the idea of the construction and up to its elimination.

2 Prerequisites of creation of the general branch informational system in construction

It is possible to refer to relevance, unambiguity, orderliness and automation to the principles of development of various systems of classification and coding of various elements in construction. The developed systems are to reflect the existing condition of the market of modern resources, works and projects in construction and to provide

* Corresponding author: i.kar@inbox.ru

unambiguous perception of data during their description. As information technologies are widely used in construction now, the qualifiers being created and systems of coding have to represent the organizational structure of the interconnected elements, allowing obtaining information about the set element consistently for the purpose of the automated processing of the obtained information and as well for the search of the optimal solution of the task.

Nowadays various qualifiers of projects, works, structures, resources are used in construction. For example, the set of projects and complexes is presented in the All-Russian qualifier of fixed assets OKONF, which is acting since January 01, 1996. The types of works which are carried out in construction are listed in the All-Russian qualifier of types of economic activity. However the expanded list of types of the works performed in construction can be seen in the operating budget regulatory bases where budget costs of the performance of works are formed on the basis of element budget norms and single quotations. Classification of material resources is given in the All-Russian qualifier of production OKP (OK 005-93) which will be acting till January 01, 2017. However it will be replaced by the All-Russian qualifier of production according to the types of economic activity OK 034-2014 (KPES 2008), (Rosstandart) accepted and put into operation by the Order of Federal Agency for Technical Regulation and Metrology of January 31, 2014 No. 14-st. It is possible to name the following main shortcomings of the listed qualifiers:

• Excessive division into the making elements of separate groups or, on the contrary, their huge integration;

• Inexact assignment of the name to the structural units of classification that causes their double treatment;

• Orientation of the existing qualifiers to a certain circle of consumers or separate range of application (such as trade, taxation, etc.).

Now work on the creation of the state information system "Pricing in Construction" aimed at the solution of such tasks as the increase of reliability and transparency of the determination of cost of construction at all the stages of the investment-and-construction project, the improvement of competitive climate in the market of construction materials and the increase in the stability of the market of contract services [1] is being carried out at the moment. The functional subsystem of the formation of the qualifier and management of the catalog of construction resources which will store the information about the resources, used in construction, their actual prices, demands in the design-and-budget documents, etc. will be a part of this system.

This article presents the idea of the creation of the uniform concept of formation of the branch system of classification and codification of information which application due to timeliness, reliability and informational content will increase efficiency of the use of the resources necessary during planning, organization, management and control of works on the project in construction for the purpose of reduction of terms, expenses and risks. Works on the creation of such system must include:

• development of branch system of qualifiers for the purpose of creation of a unified information system which will be used during the information modeling of processes at all the stages of life cycle of a project;

• the direct development of all the qualifiers as a part of a unified information system for the use during the developing, justification and decision-making;

• ensuring the harmonization of the created branch system of classification and coding of information with the operating Russian and international qualifiers.

Databases including, various qualifiers, for example, qualifiers of objects, constructive decisions, works, resources (financial, material, labour, technical, the equipment, etc.), projects, participants (contractors), contracts (contracts), standards, expenses, budgets, risks, requirements, standard and legal documents, technical documents, etc. have to be the basis for the branch information system. Thus, the system being created will play a leading role in the accumulation of information, and also in its use during the development and optimization of the design decision as well as in control over its realization.

Creation of a full and reliable classification of material resources seems to be one of the labour-intensive processes in the offered system. First off all it is connected with the fact that it is necessary to consider all the properties and characteristics of resources which have to be united and identified within technologically uniform groups according to practice of their use in the investment-and-construction activity. Identification is meant to be an assignment to projects objects of the group code abbreviated names, their identification and definition of the place (identification) in the created classification system of material resources. At the same time the dictionary fund of the specified nomenclature of resources is based on the reflection of the main consumer properties of production as subjects of labour in the construction production.

The material resources used during the construction, reconstruction, repair of buildings and structures can be classified according to their appointment and origin, taking into account the main properties and characteristics. In this case the material resources will include materials, products, semi-finished products and structures used during construction works. Classification of material resources will include the following groups of resources formed originally on the basis of their appointment:

- construction materials for roadworks, foundations of buildings and structures;
- structural construction materials and products as a part of the bearing and protecting structures;
- functional construction materials for covering of structures;
- materials, products and equipment for technical systems;
- materials, products and equipment for wire systems;
- materials and products for improvement and green construction;
- materials, products and equipment of special and other purposes.

The structure of properties and parameters of the material resources used in classification have to include physical, mechanical, technological, chemical, operational, sanitary and hygienic and esthetic properties of material resources, and also their mass-and-dimensional characteristics. As a result when coding a material resource all the information about it can be recorded, from the application and origin and up to the key parameters influencing on the choice of resources according to the established production and engineering requirements of the design decision.

We will refer the set of the construction machines, mechanisms, equipment and tools participating in the process of creation of construction products as well as in the performance of different types of works to the technical resources used in construction.

The groups of technical resources created according to their appointment and entering classification of resources consist from:

- machines, mechanisms and equipment for preparatory works and works on gardening of the territory;
- machines and the equipment used for earthwork;
- machines and the equipment used for pile works;
- machines and the equipment used for concrete works;
- machines and the equipment used for road-building works;
- load-lifting machines, mechanisms and equipment for construction (repair-and-construction) works;
- machines, mechanisms and equipment for transportation of freights and loading and unloading works;
- machines, mechanisms and equipment for finishing works;
- machines, mechanisms and equipment for special works;
- machines, mechanisms and equipment for auxiliary works;

• mechanized tools.

Such parameters as capacity, productivity, loading capacity, useful height, synchronous frequency of fluctuations, capacity, etc. belong to the properties of technical resources acting as the for their classification.

Standard-and-technical and other documents validating the set parameters have to act as the basis for the creation of classification. Such documents can include, for example, state standards, industry standards, specifications, standards of large organizations (being customers of the specified products), etc.

When developing the classification the hierarchical method of classification, which characteristics give the possibility of the use of unlimited number of signs of classification and hierarchy of signs of the classification, is accepted. Namely, in this case the group formed according to one sign is divided into a set of classification groups according to a subordinate (subordinate) sign.

The created qualifier of material resources as a part of branch information system will allow to create the database, which structure besides the list of properties and technical characteristics of resources can include the following information:

• graphic materials of the resources used or intended to be used;

• factual information (operational data for providing operating conditions, energy consumption parameters, number and qualification of personnel, etc.);

• structure of suppliers and producers of material resources with a summary about them (location, requisites, the nomenclature of the delivered resources, the certificate of activity, etc.);

• price indicators of resources based on price lists and catalogs of suppliers and producers;

• statistical information on structure of resources, technologies, etc. depending on appointment and the size of object, a stage of its life cycle.

However the complexity of the creation of such a system is that in during the use it will demand continuous replenishment by new information and the updating which is already available as a part of the database.

It is also necessary to note the need of the use of this information system when developing resource-and-technological models in construction for the purpose of formation of budget indicators of various level of integration. For example, for the creation of a cost index the unified set of the labour, material and technical resources necessary for performance of separate types or complexes of works, creations of constructive elements or the building (structure) in general is formed. In this case the total cost of separate types of works as well as the cost of object in general depends on the structure of resources, their characteristics and price indicators. Completeness and reliability of the formed indicator can be obtained on the basis of the offered information system allowing to consider the diversity of decisions, to choose the most optimum and, if necessary to make changes in due time.

The unit of requirements management, representing also the hierarchical system where each branch has to reflect requirements to the designated group has to hold a specific place in information system. For example, the requirements to the structure can include the following groups [2]:

• requirements of government bodies (environmental protection, architectural expressiveness);

• requirements of the user during the operation of the structure depending on its appointment (durability, maintainability, productivity, etc.);

• requirements of the customer (economic efficiency, payback);

• requirements of supervision (observance of the existing normative documents);

• requirements of the contractor (technological effectiveness, labour protection, quality of resources, works, etc.).

Requirements to the project documents can include such groups as:

• requirements of government bodies (observance of acts and the existing standard and methodical documents);

• requirements of the customer (validity of the made architectural, technological and technical decisions, optimization of cost);

• requirements of supervision (observance of the existing standard-and-methodical documents, requirements to structure of documents, the requirement to completing of documents, validity of the used resources, etc.)

• requirements of the contractor (confirmation of the made decisions in the field of resource management, requirements to the structure of reporting documents, etc.).

The process of the formation and the use of such block consists of collecting, processing, analysis, formulation and approval of requirements with the obligatory subsequent tracking of their execution during design, construction, operation and elimination of the project. As for the customer he or she initially has to create the list of requirements for the purpose of the subsequent their transfer to the design and contract organizations and control for from performance. Owing to certain circumstances, such as discrepancy of the resource offered to application to the requirements of the set parameters, the requirements, established earlier can be modified or changed and that surely has to be confirmed by proving materials.

Thus, the information system has to comprise a large number of data which will get the necessary sense only in the certain conditions made, for example, by the design decision.

3 Conclusion

By now in the Russian Federation the certain experience in the field of the development of various systems of classification used in construction for the justification of the forthcoming expenses during the planning of production, design of structures, management of expenses has been saved up. Creation of a common information space will promote quick and flexible response to the happening changes that will allow to make managerial decisions taking into account all the influencing factors of the external and internal environment of the structure. Classification and coding of information used in construction are the most important prerequisites of the modern approach to the project management, as well as to logistics, development of resource and technological models, drawing up budget documents, information support of the life cycle. However despite a lot of work which is carried out in this direction, the search of the unified approach satisfying all the participants of construction is still proceeding.

References

1. Information on: www.nopriz.ru/upload/iblock/aa6/text_postanovlenie_pravitelstva_npa_39886.pdf
2. E.A. Gusakova, A.S. Pavlov, *Principles of organization and management in construction. In 2 parts, part 1* (Urayt, Moscow, 2016)
3. M. Jocovic, B. Melovic, N. Vatin, V. Murgul, Applied Mechanics and Materials, **678**, 644-647 (2014)
4. G. Radovic, V. Murgul, N. Vatin, Applied Mechanics and Materials, **641-642**, 634-638 (2014)

5. J. Ćetković, S. Rutešić, M. Zarković, M. Knežević, , N.Vatin, Procedia Engineering, 117 (1), 780-790 (2015)
6. S. Rutešić, J. Ćetković, M. Žarcković, M. Knežević, N. Vatin, Procedia Engineering, 117 (1), 905-915 (2015)
7. S. Rutešić, J. Ćetković, M. Knežević, M. Žarcković, N. Vatin, Procedia Engineering, 117(1), 642-650 (2015)
8. J. Cetkovic, S. Rutesic, T. Hanak, M. Knezevic, B. Melovic, Technics Technologies Education Management, 7(1), 285-293 (2012)
9. V.V Okrepilov, M.V. Leonidovich, Asian Social Science, 11(7), 312-325 (2015)
10. V.V Okrepilov, V.L. Makarov, A.R. Bakhtizin, S.N Kuzmina, Economy of Region, 2, 301-313 (2015)
11. V.L. Kvint, V.V Okrepilov, Herald of the Russian Academy of Sciences, 84(3), 188-200 (2014)
12. G. Radovic, V. Murgul, N. Vatin, E. Aronova, Applied Mechanics and Materials, **627**, 357-364 (2014)

The Factors Affecting Innovative Activities and Cost Management

Irina Karakozova[1], and *Yulia* Prokhorova[1,*]

[1]Moscow State University of Civil Engineering, 129337 Yaroslavskoe shosse 26, Moscow, Russia

Abstract. The article presents general and specific factors that influence the generation and management of facility construction cost. It is suggested to create a national information system for costs formation in construction that would correspond to the current requirements of a dynamic requirements. Efficiency of the proposed methods shall be ensured by increased accuracy of estimates at the phase of investment decision taking for any public capital investments, from direct budgetary expenditure to extra-budgetary funds resources, subsidies and grants.

1 Introduction

Within the past few years, along with the reputable practice-oriented approaches to construction cost formation, Russian economy has witnessed a gradual implementation of cost-oriented approach and cost engineering methods into management practices. In the broadest terms, their fundamental nature lies in determining and systematizing cost aspects of management activities within an enterprise or of project realization. However, academic and methodological literature on these issues, especially practically oriented one and applicable to the construction sector in the Russian Federation, is currently insufficient [1-4]. In the context of a politically and economically unstable situation, the importance of management for facility construction costs with support from federal budget is increased, because the issues of rational resource saving, efficient assets operation, and project realization in general become a priority at any level, starting from the state one to the level of separate companies and investment projects [5-6].

Construction sector is one of the key catalysts for economic growth in the Russian Federation. Public procurement for construction and installation activities accounts for 40 % of total volume, and it is expected that in the nearest years, the Russian Federation government shall invest several trillions of rubles into the upgrade and amplification of the country's infrastructure [1].

Construction sector plays an essential role in Russian economy, and the condition thereof is an indicator of economic and technological development of the state. However, there exists a series of external circumstances that "slow down" the income-generating sector of the country, and first of all, one should name rather low quality of facility engineering and construction as such. An important obstacle lies in the absence of a proper

* Corresponding author: eshly.06@mail.ru

basis for regulatory documents, updated and revised in accordance with the latest technological and technical amendments and decisions [3]. As far as the application of regulatory methodological documents to costs formation and costs estimate is being concerned, current situation is equally complicated.

The Minister of Construction Industry, Housing and Utilities Sector of the Russian Federation, M.A. Men, commissioned this issue to the Supervisory board of the federal autonomous organ called Federal center for cost formation in construction and construction materials industry, which approved strategic directions for its activities in 2016-2020 aimed at implementation of the Reform Philosophy for Cost Formation and Cost Estimate in Construction Industry. By 2020, the technological strategy of the reform implementation should result in elaboration of new national regulatory estimate basis, updated methodological guidelines, indications and manuals, new reference books of state estimate standards and quotations, and in development of a national information system (NIS). But the regulatory basis as such does not resolve the issue of interaction of the concerned parties during implementation of an investment construction project with regards to quality, terms, and costs, because engineering companies are normally directly dependent on the customer and are obliged to comply with his specific requirements, while the customer is often only driven by "optimization" of his own expenses [7-15].

2 Methodology

2.1 Complex cost management

There is a practical necessity to increase competitiveness of construction companies through expert assessment of civil construction costs in order to justify feasibility, volume and terms of capital investments. And this implies elaboration of a universal standard for managing facility construction cost, i.e., a basis that would allow a specific company to create its own costs managing system, while accounting for financing particularities and properties of the facility under construction[13].

The issues of complex cost management for investment projects at different life cycle stages are investigated in a number of studies by foreign authors, such as Larry R. Dysert, Williams Jr. R., Chilton C.H., Guthrie K.M., Remer D., Chai L., Vatavuk W.M., Hall Richard S., Lang H.J., Miller C.A., Hand W.E., Black J.H., Dysert L.R., and others, and their works served as the basis for the integrated approach to portfolios, programs, and projects management within the complex cost management.

However, the critical analysis of foreign studies revealed insufficient development of methodological issues in forming an innovative organizational and economic mechanism for managing facility construction costs on the basis of cost engineering methods, the reason for this being the absence of a universal approach to the construction cost management system, to the problem of analysis and synthesis for requirements and construction cost management indicators, to the issues of elaborating a conceptual model and estimating final efficiency.

In Russian practice, project cost management implies a sequence of planning processes applicable to resources that are required for project execution, estimation of their costs, formation of total costs for each cost item, and controlling the cost of the project throughout all stages of its execution.

Consequently, project financing is based on two documents: project estimate and project budget. Project estimate represents a structured list of expenses and income per cost item and per project sections, while project budget provides information on expenses and income items distribution with reference to time periods[7].

Russian experience of cost management is described in works by such authors as Reznichenko V.S., Lenintsev N.N., Dorozhkin V.R., Didkovskaya O.V., Kuzminsky A.G., and others [6].

Analysis of their studies reveals existing problems in cost estimates and cost formation related to issues of retrieving complete and true estimated expenses due to the lack of modern system for regulatory estimation standards used at different stages of an investment construction project, constant monitoring of resources prices, description of methodological approaches to estimated cost calculation in facility construction while preparing FEED, project, bid, and other documentation.

Besides, yet another problem is revealed: lack of an efficient marketing system for vendors selection based on a system for monitoring market prices in construction products (buildings, structures) [5].

If the complete life cycle of a project is considered from the point of view of expenses estimation, then at each stage we can see various types of assessment, from enlarged to more detailed ones.

This objective implies recurring to such methods of project cost estimation as parametric project cost estimation, estimation by equivalent facilities, estimation on the basis of bid proposals from potential contractors, “downward” estimation, “upward” estimation, and so on.

Depending on the project phase, on the established objective, on the evaluation method and on the resulting estimate document, Russian experience of project cost management comes up with error within the range of 0 ... 40%. At the same time, selected method depends on specific project, on contractor’s qualification, on availability and quality of a hierarchical structure of works, on degree of accuracy assumed in calculations, etc. [3].

It is well known that abroad, the majority of construction contracts, both public and corporate ones, are awarded through bids. In this case, the Customer prepares such documents, among others, as an Invitation for bid, a Request for proposal, a Request for quotation, but not the Contract price.

These documents usually contain a detailed list of activities and their scope of works, and they should be flexible enough to incorporate the contractors’ capabilities to improve the proposed contract conditions.

That is why totally different approaches to construction cost estimation are applicable for the investor and for the contractor: while the investor makes an approximate estimation of required financing for the construction in general, the contractor calculates with much more detail production costs and quotations for the specific range of works to be provided [14].

In contrast to the foreign experience, contemporary cost formation in Russian construction sector tries to apply the same method to different tasks, while using unit costs, which provokes reasonable criticism both from the investors and public customers, and from the contractors. The customer is unable to determine the contract price without preparing project, or better still, working documentation, and making detailed estimate calculations, while the contractor is unable to accurately calculate the quotation on the basis of the estimates that are inevitably averaged.

As construction contractors should meet specific requirements that differ from the conventional financial evaluation criteria, for example, from the investor’s (customer’s) point of view, a specific methodology is required.

This methodology should analyze and synthesize conventional financial and other factors applicable to construction companies’ activities, and would allow classifying contractors by their capability to perform within a certain scope of works within the construction process [10].

Thus, we might conclude that the organizational and economic mechanism for managing facility construction costs should be generated individually for each participant of the investment-construction process as a system where its objectives are being formed, and tasks aimed at achieving these objectives, and stimuli that allow revising its spiritual, environmental, social and material needs during the project execution, accounting for the final result oriented at the satisfaction of consumer's demand for this particular product.

2.2 Organizational and economic mechanism for managing facility construction costs

Organizational and economic mechanism might be structurally represented as an aggregate of three systems: target, information and functional ones.

A target system should comprise objectives and main performance indicators for investment and construction projects, as well as selection criteria and progress evaluation and results assessment criteria.

An information system is required in order to provide updated, complete and true information for all stages of operation within the mechanism being developed. It includes subsystems for providing legal, resource, regulatory technical, regulatory methodological and research information.

A functional system is oriented at consecutive resolution of tasks being set at the stages of planning, organizing, controlling, and creating/improving a database.

Efficient use of public investment requires creation of an updated regulatory legal and methodological basis for the domain of costs formation and costs estimates, that would ensure the balance of interests for all parties involved in the construction process, unity of approaches and methods for estimation and monitoring of construction products costs for all levels of public authorities, ministries and state organs, extended application of information systems, analytical software packages [2].

Number of subsystems within each system might vary, in accordance with the type and the scale of the facility under construction, with the number of participants in the respective investment and construction process, effect level of the external environment, and internal preparedness of the contractor [12].

Together with structural development of the organizational and economic mechanism for managing facility construction costs, we should define its principles of operation applicable when managing decisions in this domain are being taken and executed. Among the basic principles one might name the following: principle of innovation, principle of consistency, principle of adaptability, principle of business ethics, principle of reliability, principle of expenditure justification, etc.

The principle of innovative operation within the cost management system implies that all participants of an investment and construction process are oriented towards innovations and aspiring to discover new opportunities.

The principle of consistency implies an aggregate of all systems and subsystems within the certain area, accounting for the influence of external factors, in order to achieve the pre-established objective.

The principle of adaptability aims to provide timely reaction of all participants, accounting for current changes in external and internal environment. The principle of business ethics implies open and transparent relations between the process participants in their interactions with each other. The principle of reliability is intended to create conditions that would ensure uninterrupted system operation in the context of limited resources.

The principle of expenditure justification suggests selection of an efficient and transparent tool for costs calculation in order to obtain true and non-prejudicial cost for facility construction. Such principles allow integrating the organizational and economic mechanism for managing facility construction costs into the general management system applicable in the specific construction company [8].

Creation of the organizational and economic mechanism for managing facility construction costs shall include the following consecutive phases:

- analysis of development and of internal state of the company;
- external environment analysis;
- gathering information on previously constructed facilities (performed works) with the use of the database on contract (subcontract) construction companies, analysis of the obtained results, preparation of proposals for modifications of functional connections between elements of the costs management system, should any deficiencies be revealed within its structure;
- evaluation of a possibility to improve managing system for facility construction costs (certain scope of works cost) in the domain of development and further application of an efficient organizational and economic mechanism for costs management based on cost engineering methods (Figure 1)

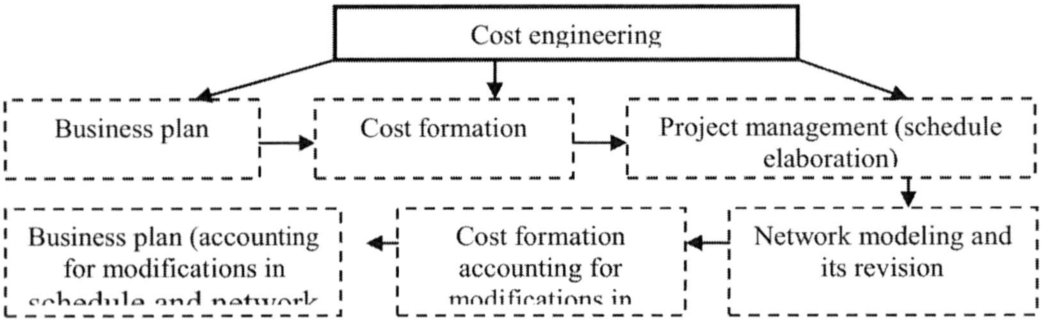

Fig. 1. Cost engineering.

- preparing a program of measures aimed at creation/improvement of the organizational and economic mechanism for managing facility construction costs, including a list of activities, distribution of required resources and responsibilities for measures implementation;
- implementation of measures aimed at creation / improvement of the organizational and economic mechanism for managing facility construction costs, evaluation of the obtained performance results, database update.

2.3 Cost formation for the facility construction

The organizational and economic mechanism for managing facility construction costs is based on the fundamental economic concepts, where the cost of a construction product is a calculated cost at the specific moment of time, expressed in monetary equivalent. At the same time, the costs for the contractor are usually higher than the costs calculated by economic methods and accounting for the customer's perspective.

It is therefore absolutely obvious that the facility construction costs shall be formed gradually at different stages of the investment process, accounting for interests of both the customer and the contractor (Table 1).

Which means that approaches to determining estimated facility costs shall be classified in accordance with stages of the investment and construction project and with accuracy level of the resulting estimates [17]

Table 1. Costs formation at different stages of the facility construction.

Contents	**Approach to cost formation**	**Methods of costs formation**	**Sources of information**	**Required basis**
Pre-investment phase				
Preliminary costs estimation	Income-based; Comparative	Direct capitalization; Discounting for future cash flows; Direct comparison; Comparison by analogy; Targeted corrections; Development of statistic models; Specific technical and economical indicators	Draft solutions	Nation wide automated information and reference system, allowing accumulating and integrating information on the cost, technical, economical and structural specifications of the constructed facility on the performed contract bidding as it pertains to contract price, and allowing convenient search of information ensuring its availability and transparency
Investment phase				
Stage 1 – preparation of cost estimate documentation (development of estimated cost of construction – investor's (customer's) price)	Expenditure-based	Base-index based; Aggregated cost estimation; Analogue-parametric	Project documentation, work scope data, customer's data base on the cost of completed projects	Modern system of cost estimation regulations with aggregation according to consolidation degree and application level, database on completed construction projects
Stage 2 – bidding (arrangement of offer – contractor's price)	Expenditure-based	Base-index based; Resource-based; Resource-index based; Aggregated cost estimation; Analogue-parametric	Cost estimate documentation, construction work scope information provided by customer; contractor's data base on the cost of resources required for a single unit of work	Modern system of cost estimation regulations with aggregation according to consolidation degree and application level, contractor's database on the cost of works considering adjustment factors, establishing of analytical institutions developing adjustment factors for contractors.
Stage 3 – preparation of contract price and finalization of agreement with successful bidder.	-	Fixed contract price with application of compensatory arrangements for adjustment of the finalized price, considering customer's and contractor's interests	Design and cost estimate documentation	Modern system of cost estimation regulations with aggregation according to consolidation degree and application level, predicted indices of construction cost escalation.
Stage 4 – settlements between customer and contractor within the agreed cost	-	Payments for the performed packages (stages) of works or completion percentage of the constructed facility.	Construction contract	Recommended reference procedure of designation of packages (stages) of works and procedure of determination of percentage of each stage within the total cost of the entire facility

This staged approach to construction cost formation allows facilitating state supervision of cost formation and utilization of federal funds. This raises the possibility of replacing fiscal punitive measures (fines, sanctions), with economical and preventive depending on:

- relevance and correctness of selection of initial (maximum) contract price;
- conformity of payments for construction works with the posted price;
- degree of completion of the paid stages (packages) of works;
- quality and consumer properties of completed works (raising responsibility of technical, state and designer's supervision).

Preparation of efficient economical and organizational scheme of construction management allows handling the problems of cost formation and planning of capital investments, cost management and formation of company's investment policy. To achieve this, a single information space must be created allowing flexible and timely response for changes and for management decisions with due consideration of external factors and internal condition of the company. Therefore, organizational and economical scheme of construction cost management provides a new format for interaction between the participants of construction project and for estimation of the costs of upcoming construction project.

3 Key findings

Academic aspect of challenges related to construction costs management addresses the issues of methodological approach development which would allow determining an algorithm for efficient expenditure accounting during facility construction based on budgetary allocations.

Performed studies and theoretical justifications in this direction lag behind the present day reality, and need further development. Important of the covered issue results from non-availability of innovative model allowing handling all challenges related with cost engineering of capital construction timely and accurately.

Performed analysis allows generating a model of the organizational and economic mechanism for managing facility construction costs.

True-to-life determination of the estimated costs for construction (reconstruction, extension, modernization, overhaul) of buildings and structures remains one of the main directions for costs estimates and costs formation system reform, that is currently being executed in the Russian Federation.

Due to specific particularities of the construction industry, when costs estimates should be approved long before the facility construction starts, construction professionals have to justify the future expenses with a large margin for error.

Besides, this process is additionally complicated by the imperfection of effective costs estimate regulations that are not completely suitable for strategic estimations and are more oriented at calculations for separate components on the basis of the working documentation. In order to perform a true and accurate estimation of a future facility construction costs, prior to engineering activities, a whole complex of enlarged regulations shall be applied, presenting costs for construction solutions distributed by type of facility or by scope of works [18]. (Figure 2)

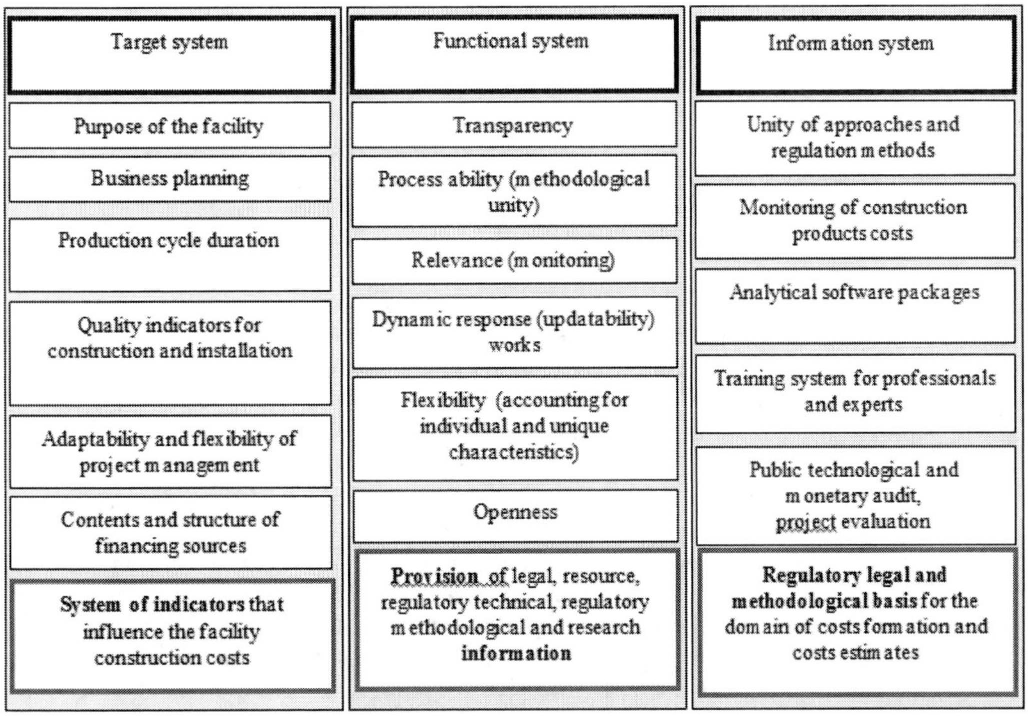

Fig. 2. Model of the organizational and economic mechanism for managing facility construction costs.

Development of methodological bases and detailed approaches to development of enlarged estimates regulations shall allow determining construction costs estimate according to existing circumstances of operations and to market economy relations in action. Under current market conditions, new concept of estimated costs formation shall be oriented at producing new methodological approaches to operation of enlarged costs estimate regulations for construction industry that would allow calculating limit estimated costs. At the same time, during the first improvement phase for the system of enlarged costs estimate regulations in construction, main methodological caveats of costs formation in construction sector shall be indicated, along with possible solutions [19].

4 Discussions

Application of the organizational and economic mechanism for managing facility construction costs should be accompanied by development and implementation of a requirement managing system which represents an uninterrupted information and documentation-based process to guarantee results that correspond to requirements and expectations of end users for construction products: customers, contractors, vendors, and others [11]. The requirement management system shall be created and implemented in order to reduce expenditures related to the construction (upgrade, modernization, etc.) of a facility, transparency and flexibility of the facility costs management, which in turn, shall increase the customer's confidence that the works performed at the facility are being executed in accordance with project, regulatory technological, and other documentation, with due quality, in full scope and within the time limits [16-19].

Requirement managing system shall be built in compliance with the hierarchical interrelations of the process participants, which shall allow structuring and better describing all the requirements and the degree of their implementation accounting for the needs of the participants, but, first and foremost, of the customer, who, as of today, is formulating the requirements to the final product during the initial phase of the project. Nowadays, in construction sector, the customer specifies his needs and prepares a technical assignment that represents a certain initial list of requirements for the facility under construction at the planning phase. Technical assignment serves as the basis for requirement formation, accounting for facility life cycle phases, including hierarchical presentation of the complete requirements list, their analysis, specification, and approval [20].

All types of requirements might be classified into three groups. Functional requirements correspond to the result of each and every decision that is being taken. Restricting requirements distinguish properties or separate parameters that should be present in the solutions being applied. Organizational requirements, known as limitations, are determined by the mission, objective and driving forces of the company. At the same time, all the above requirements should be measurable and verifiable [8]

Development and implementation of a requirement managing system implies automatic data gathering and processing for control, reporting, main indicators' monitoring, correction of the managing decisions that have been applied. Thus, it is a certain instrument for displaying and registering various types of data obtained at different life cycle phases of a facility, in order to evaluate performance, progress stage, variability of effective requirements, efficiency of applied solutions, etc. Measurable indicators for registering and correcting the requirements shall allow controlling arising deviations from the pre-established limit values, thus justifying the necessity to revise the requirements due to their biased reflection of current situation. While gathering and analyzing information on managing facility construction costs during the project execution, the customer might make up his own information data base that would allow considerably improving quality, completeness, and coherence of his requirements [9].

5 Conclusion

Management of facility construction cost is an art of managing and coordinating human and material resources along the life cycle of the project through application of advanced management methods and techniques in order to achieve certain results in contents and scope of performed works, costs, terms, quality and overall satisfaction of the project participants. Successful execution of a project implies achieving target indicators while respecting the pre-established limitations in: costs, duration of production cycle, quality of performed works, and specific requirements to the facility construction. Besides, final result should be approved and accepted by the customer.

As of today, management of facility construction costs is unavailable both at the federal level and in the regions. There are some single attempts to implement managing components: switch to a new regulatory basis for costs estimates, development of regional price lists and information circulation letters, introduction of regulatory methodological documents, costs optimization during bidding process and making contracting agreements, application of software complexes and creation of information databases with costs indicators. From the development perspective, it is suggested to create a national information system for costs formation in construction that would correspond to the current requirements of a dynamic requirements. Efficiency of the proposed methods shall be ensured by increased accuracy of estimates at the phase of investment decision taking for any public capital investments, from direct budgetary expenditure to extra-budgetary funds resources, subsidies and grants.

References

1. *The official website of the Federal Service of State Statistics of the Russian Federation* (2016), Information on: http://www.gks.ru/
2. *The official website of the Concept of the Russian Federation pricing reform* (2016), from: http://400days.ru/
3. I.I. Mazur, *Investment and construction engineering: Textbooks for high schools* (Elim, ZAO "Publishing House" Economics, Moscow, 2009)
4. I.V. Karakozova, Y.S. Prokhorova, Economy and Entrepreneurship, **3-2(56-2)**, 682-686 (2015)
5. V.R. Dorozhkin, *The pricing and cost management in construction* (Publisher by E.A.Bolhovitinova, Voronezh, 2003)
6. O.V. Didkovskaya, M.V. Ilina, O.A. Mamaeva, M.A. Konovalova, E.S. Spirina, *Methodical approaches to formation of value engineering system in the building: monograph* (Samara, 2013)
7. I.I. Mazur, V.D. Shapiro, N.G. Olderogge, A.Y. Zabrodin, *Investment and construction engineering: Textbook* (Elim, ZAO "Publishing house" Economy ", Moscow, 2009)
8. A.V. Tsvetkova, *Total coast management. Translated from English*, (Publishing House ZAO "PMSOFT", Moscow, 2013)
9. D. Tavassoli, *Requirements management. Ten steps on the path to perfection*. (2016), Information on: http://www.swd.ru/files/pdf/IBM_uspeh_edit1.pdf
10. *The control system requirements* (2016), Information on: http://www.diaprom.com/about/srm#prettyPhoto
11. *GOST R ISO / IEC 15288-2005 (2005) Information technology. Systems Engineering. The processes of the life cycle of systems. Approved and put into effect by order of the Federal Agency for Technical Regulation and Metrology of December 29, 2005 # 476-st.*
12. Y.N. Kulakov, T.N. Silanteva, J. "Bulletin SWSU", **2** (2012)
13. A. J. Makin, *Global Imbalances, Exchange Rates and Stabilization Policy* (Palgrave Macmillan, Basingstoke, 2009)
14. A. Novoseltseva, *Bulletin of the Volga. Episode 9: Studies of young scientists*, **8-2** (2010)
15. V.I. Resin, *The development of large cities in transition economies (systematic approach)* (Editorial URSS, Moscow, 2000)
16. *Russian Federation Government Decree dating from 02.02.2010. # 102-P "On the approval of the concept of the federal target program "Complex program of modernization and reform of housing and communal services for 2010-2020"//the legislation bulletin of the Russian Federation. 2010 # 7.*
17. Sir P. Abercrombie, *The Preservation of Rural England* (Hodder and Stoughton Ltd, London, 1926)
18. E.N. Smirnov, *Introduction to the world economy (economic geography of foreign countries)* (KnoRus, Moscow, 2008)
19. E.M. Stein, Economics and Entrepreneurship, **3(2)**, 258-258 (2015)
20. N.G. Verstina, N.N. Taskaeva, Economics and Entrepreneurship, **8(37)**, 244-249 (2013)

MATEC Web of Conferences **73**, 07016 (2016) DOI: 10.1051/matecconf/20167307016
TPACEE-2016

The Problems of project management software implementation in construction corporations

Pavel Kagan[1*], *Anna* Naumova[1], and *Yuriy* Vilman[1,]

[1]Moscow State University of Civil Engineering, 129337, Yaroslavskoe shosse, 26, Moscow, Russia

Abstract. The article is devoted to the main problems facing many companies, implementing software project management activities have been working construction corporation. The company transferred the task performing the activities for the implementation, as well as its purpose. Revealed rules that must be followed to achieve the best results in the implementation of software project management.

1 Introduction

Construction corporations are gradually coming out of the beginning stage of IT-development. Herewith, the main role in developing of Russian construction complex is acting by a construction companies, which are not only possessing of a significant investment potential for making a huge projects and programs of a branch development, but also interested in acceleration of a science-technical progress, qualitative changes of a branch complex and development of innovative technologies on different levels of a branch management. It is required the newest software for project management in order to automatize their activities. Implementing such a software in construction companies, it is necessary to take into consideration that it is possible to face many problems on any stage that can lead to failures from both of sides.

The Problems of project management software implementation

On practice, the project is a totality of an actions directed on the achievement of the goal to create some complicated system with specified characteristics of quality in limited deadlines and on limited quantity of resources.

It is impossible to make qualitative planning, realize efficient control and operative correction of a plan without the using of special software because of a big volume of planned actions, complexity of a distributions and plan optimization methods. The technology of network planning and management is in the basis of using software. The primary aspects of this technology were created in 70s.

* Corresponding author: kagan@mgsu.ru

Nowadays, the market of software products is full of different facilities that realize network-planning methods – from huge professional systems to the systems let you dispose of your working time, financials and resource allocations efficiently [4-7].

The main tasks of the company, which implements the project management methods, are:

- The automatization of business-processes in a corporation
- To set the software under the certain corporation
- The integration with using software
- Input of reference or analytical information
- The training of a corporation staff
- The main goal of a scientific organization of a labor is to increase the efficiency of work in organization. It means:
- More rational using of resources
- Increasing of product quality and quality of consumers service
- Reaching of higher labor producing; reduction of costs

The main problems of a software implementation:

- Unclear formulation of the task from customer

The common problem that is occurring in implementation of software is to find out all the tasks and to realize the solution. Such a problem can occur because of misunderstanding of work processes in corporation and all accompanying structures.

- The necessity of particular reorganization of structure and organization activities.

Before beginning to implement the automatization systems into the corporation, there is necessary to make particular reorganization of its structure and business management technology. So, one of the sages of implementation is full and reliable examination of all working aspects in corporation.

- The need to change the working technology with information and business-management principles.

Efficiently built informational system cannot deal without some changes of eventual planning technology.

- The resistance of employees.

With the implementation of new software, it can be resistance of employees in most of cases. It is a serious barrier for consultants and it can cause real delays of implementation. This problem is occurring because of several human factors: usual fear of innovations, conservatism, apprehension of losing the workplace or your indispensability, fear of responsibility [1].

- Temporary increase of employees load.

Workers have to learn new knowledges and technologies. During the experience exploitation and with transition to industrial exploitation of the system, it is necessary to make business as in new system as by traditional ways during some time.

After the implementation of new software into a longtime team, their earning capacity is decreasing sharply. It is over 60% of the original level of working team earning capacity. After the recovery period, the period of original level excessing is starting, it is over 10-20% more than original level. Each stage can go for different time in different corporations [2].

- The formation of skilled team of an implementation and system accompaniment, head of a team.

Some little working group (3-6 person) is starting to be formatted. The group is starting to learn full-max education of working with the system, than a huge part of implementation and accompaniment work is lying on this group. Especially important problem is choosing of the head of such group and administrator of the system. The head, excepting the knowledges of base computer technologies, must possess of the knowledges in the sphere of business making and project management [3]. (Figure 1, 2).

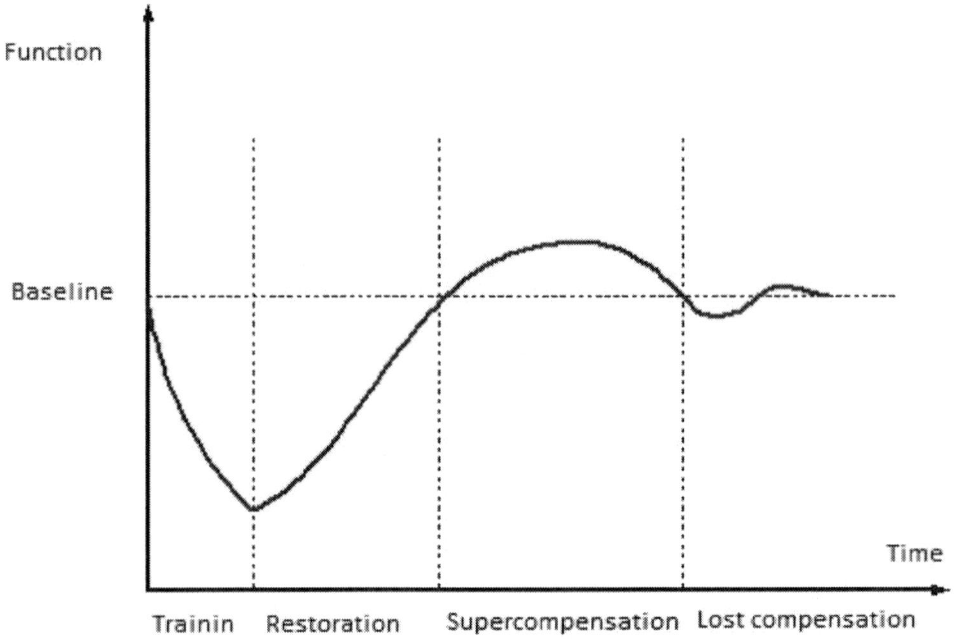

Fig. 1. Schedule changes in the load of employees during the implementation phase.

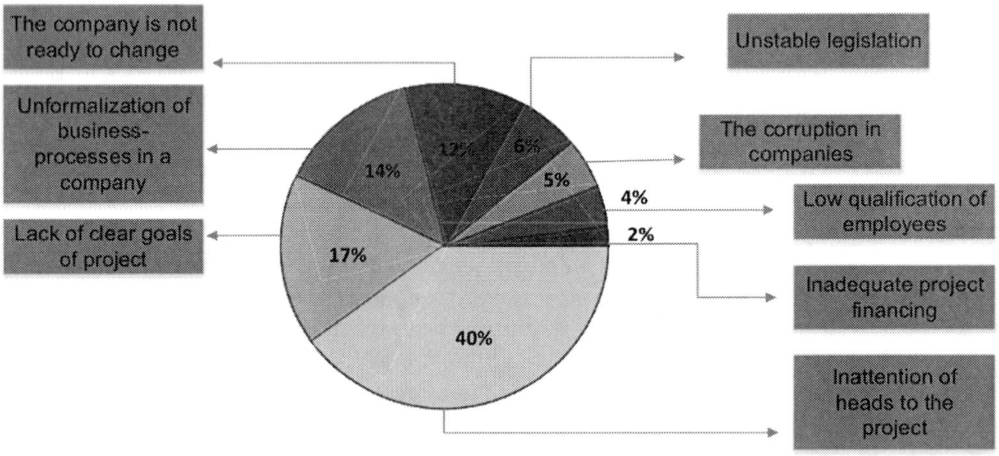

Fig. 2. Diagram reasons for unsuccessful implementation.

DOI: 10.1051/matecconf/20167307016

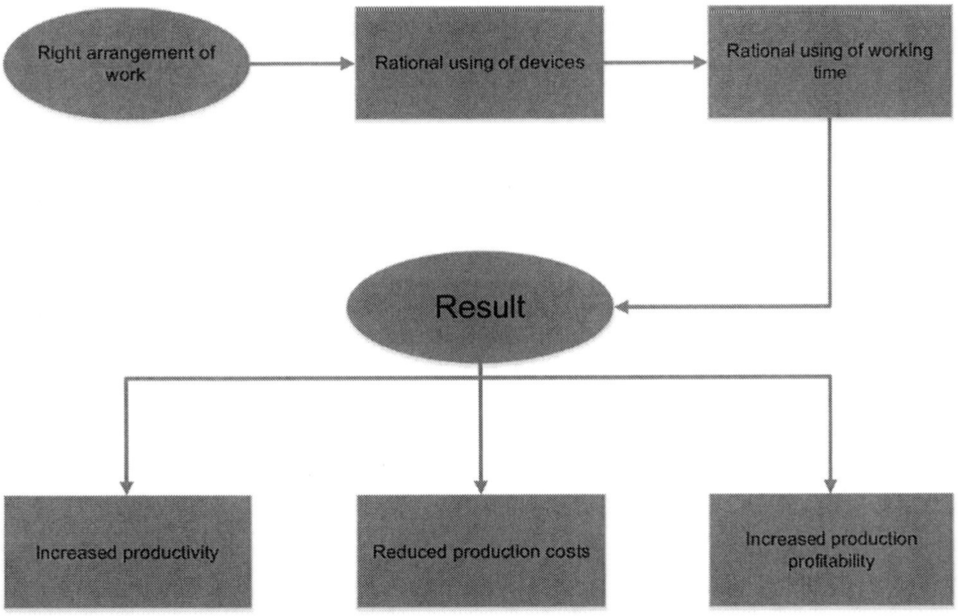

Fig. 3. The scheme rules software implementation in the organization.

Conclusions

The reasons that can cause unsuccessful implementation:

1. Inadequate project financing
2. Low qualification of employees
3. The corruption in companies
4. Unstable legislation
5. The company is not ready to change
6. Unformalization of business-processes in a company
7. Lack of clear goals of project
8. Inattention of heads to the project.

The rules to follow in order to reach the best result of a project management software implementation:

1. Right arrangement of work
2. Rational using of devices
3. Rational using of working time

References

1. M.L. Once, *Project management. Basics of project management: the textbook,* (KNORUS, Moscow, 2006)
2. D.V. Bogdanov, O.V. Kazantsev, Information technology, **2(22)** (1997)
3. D.A. Novikov, *Introduction to the institutional arrangements of project management,* (PMSOFT, 2007)
4. B.K. Pavel, V.G. Kulikov, Applied Mechanics and Materials, 951-954 (2013)
5. V.G. Kulikov, P.B. Kagan, L.V. Sukneva, Applied Mechanics and Materials, 405-408 (2013)
6. P.B. Kagan, T.A. Barabanova, Computing in Civil and Building Engineering, 2232-2237 (2014)

7. K.E. Anatolevna , K.P. Borisovich, International Journal of Applied Engineering Research, **10(23)**, 43456-43460 (2015)
8. M. Mazlum, A. F. Güneri, Procedia - Social and Behavioral Sciences, **210(2),** 348-357 (2015)
9. F. Backlund, D. Chronéer, E. Sundqvist, Procedia - Social and Behavioral Sciences, **119(19)**, 837-846 (2014)
10. M. Lazarevska, M., Cvetkovska, M., Knezevic, (...), V. Murgul, N. Vatin, Applied Mechanics and Materials, **627**, 276-282 (2014)
11. Z. Radovanović, R. Sinđić Grebović, S. Dimovska, N. Serdar, N. Vatin, V. Murgul, Procedia Engineering, **117**, 870-878 (2015)
12. J. Ćetković, S. Rutešić, M. Zarković, M. Knežević, , N.Vatin, Procedia Engineering, **117 (1)**, 780-790 (2015)
13. S. Rutešić, J. Ćetković, M. Žarcković, M. Knežević, N. Vatin, Procedia Engineering, **117 (1)**, 905-915 (2015)
14. A. Serpell, X. Ferrada, L. Rubio, S. Arauzo, Procedia - Social and Behavioral Sciences, **194 (2)**, 201-210 (2015)

International Standards in Vocational Education and Training for Financial Services Sector to Ensure Sustainable Development and Environmental Safety

Maria Sigova[1,*], *Inna* Kruglova[1] , and *Antonina* Ignatova[1]

[1]ANO HE International banking institute, 191023 Nevsky pr. 60, Saint-Petersburg, Russia,

Abstract. The article presents an overview of the Triple E Project developed by EBTN for the purpose to work out qualifications standards for finance services sector in the focus of sustainable development and environmental safety. The Triple E is based on the three instruments – European qualification framework and concept of learning outcomes (EQF), credit system (ECVET) and quality assurance (EQAVET) for vocational education and training. These three systems are the building blocks of European policy in vocational education and qualifications system with the task to strengthen the integration processes in the European educational area. They also take into account the concept of UNESCO in the field of education and sustainable development. Besides, the article traces the history of relations between educational providers, employers and public authorities in the Russian Federation in the context of setting requirements for learning outcomes and qualifications. The concepts of qualification, velocity of changes in educational standards, the establishment of a national system of qualifications of the Russian Federation, the emergence of professional standards are carefully analyzed. Current and potential problems in the assessment of qualifications are referred to. It is stressed that the volume of open data in the area of professional development, education and labor should be increased and the quality thereof significantly improved.

Introduction

The key idea of the contemporary economy, according to international and Russian strategies of development is a steady increase in the knowledge based economy. Scientific theories and best practice are the parent elements of this economy. A steady increase in the economy, based on the knowledge, can be provided only due to the high level quality of education. The contemporary economy has integrative nature therefore the present-day problem is the creation of international standards for higher and vocational education.

For several decades the entire system of higher education and vocational training in the Russian Federation has been undergoing the permanent and painful process of numerous reforms and adjustments. The educational system of the Former Soviet Union was closely

* Corresponding author: sigovamv@ibispb.ru

connected with its planned economy and the enterprises used to order a certain kind of specialists and the existing obligatory graduate placement scheme provided close connection between the country's educationalinstitutions and employers. After the FSU disintegration and the transition from planned to market- oriented economy, the above connection has disappeared. In the middle of 1990s Russia initiated the educational reform to meet requirements of the International Monetary Fund and immediately started to work out the Educational Standards. In 2003 Russia became part of the Bologna process with the aim to be also a part of European Higher Education Area (EHEA) thus facilitating international cooperation and academic exchange. To reach this goal Russian legislation was changed: instead of the former three level system "specialist degree - post-graduate degree - doctoral degree" the new system based on qualifications framework "bachelor's degree - master's degree - post-graduate degree - doctoral degree" was introduced, thus changing the award on specialty to award on qualification for each education level.

After two decades of reforms integration of the RF into EHEA a certain extent was implemented: the scope of academic exchange grew and international cooperation between educational providers accelerated. But a gap between educational content and employer's needs was rising because the majority of the Russian enterprises did not have either a power or an opportunity to change the content of educational programs. Therefore, they preferred on-site training for their employees by their own programs according to their own needs. To reduce the above gap, the RF introduced two new concepts: public professional accreditation (2011) and professional standard (2012).

One of the reason of educational reform was Russia's aspiration to join the World Trade Organization which was achieved in 2012, after nine years of being a part of the Bologna process. In addition to educational reform, the labor system was also changed, namely the National Qualifications Framework of the RF was implemented based on the principles of the European Qualifications Framework (EQF).

International Standards in Vocational Education and Training

The EQF [26] developed in the course of Bologna process, which is a process of integration of education systems of the European Union and European economic zone. EQF was developed as translation tool for communication and comparison between qualification systems of member-states and one of a number of tools for the education expanding and economy strengthening in EU.

The qualifications framework was created with purpose to compare qualifications awarded in different countries, to make correlative international educational programs, to include in qualifications elements both non-formal and informal learning. The EQF allows all education's stakeholders (students, education providers and employers) to compare the conferred qualifications in various countries and on different levels of educational system. Thanks to the Framework's explicitly formulated hierarchy of education and its stipulated system of levels.

In EQF adopted in 2008, the entire spectrum of existing qualifications is proposed to rank into eight levels. Every level defined by a set of descriptors indicating the learning outcomes relevant to qualifications at that level in any system of qualifications. Content of qualification is described in terms of the learning outcomes: knowledge, skills and competences and establishes the hierarchy of the qualification levels. Thus, for achieving the first level of qualification it is sufficient to have a basic general knowledge and skills allowing to work under direct supervision in a structured context. The EQF develops idea of three levels in higher education (bachelor, master, doctor), formulated within Dublin descriptors [7] and developed within Bologna process in the EHEA. Internationally the EQF is correlates with the

International Standard Classification of Education (ISCED) first revealed by the UNESCO Institute of Statistics in 1997 and last revised in 2011[16].

The EQF was approved by the European Parliament in April 23, 2013. It is assumed that each newly created qualification will be designed taking into account the above EQF levels. But this is not a case yet. The EQF is implemented on national level. Until it will fully implemented by member-states, there will be no reference new qualifications with EQF. There are only 13 NQFs (three of which were developed in the United Kingdom) operating at present [9]. Twenty countries have referencing reports of implementation agreed with the EU Commission and 36 countries are cooperating within EQF.

The implication is that every European member-state is supposed to create a national qualifications framework, showing a certain degree of similarity of the existing system of education and qualifications with that of EU. This framework is an instrument which allows any national qualifications framework effectively enter in European system of vocational education and labor market. Presently not all European countries agree with the above which is caused either by weak systematic policy and information base or because of the absence of adequate consideration and financing from their respective governments.

Furthermore, some states are strongly biased towards their own systems of education, which confirmed their effectiveness for many years and therefore they do not hurry to reject them and introduce EQF levels which can seem to some of them not quite reasonable to accept. At the same time the number of states carry out sufficiently strong policy in this direction, in particular France, Malta, Ireland and Great Britain. It was in Ireland and Britain where the list of qualifications (The register of regulated qualifications) [29] was first developed, realizing EQF concept.

Let us turn to the concept of qualification. There are two contexts of understanding this concept in European educational policy. As job requirements "qualification" means "knowledge, aptitudes and skills required to perform specific tasks attached to a particular work position". Another meaning of the term "qualification" as a formal qualification: "the formal outcome (certificate, diploma or title) of an assessment process which is obtained when a competent body determines that an individual has achieved learning outcomes to given standards and/or possesses the necessary competence to do a job in a specific area of work". We point out that the concept "learning outcomes" has completely definite meaning as "set of knowledge, skills and/or competences an individual has acquired and/or is able to demonstrate after completion of a learning process, either formal, non-formal or informal".

There are two contexts of qualification concept in Russia too. According to Labour Code of the RF, qualification of employee is defined as "a level of knowledge, skills, professional attainment and work experience" [17]. According to the Federal Law "About Education at the Russian Federation", qualification is "a level of knowledge, skills, attainment and competence, characterizing the readiness to perform a certain type of professional activity"[27]. In our opinion, there is a problem of distinguishing between the terms "skills" and "attainment" in the Russian language [15]. Both terms are used in all editions of State Educational Standards of the RF, but in Professional Standards the term "skills" is only used.

Russia has also been designing its own NQF. In 2006 the Russian Union of Industrialists and Entrepreneurs (RUIE) set up the non-governmental organization "National agency of the qualifications development" (NAQD) to coordinate actions of organizations representing business community, to stimulate interaction between business and state bodies during formation of the NQF, to create the NQF, professional standards and the system of qualifications assessment and certification. In 2008 NAQD together with Federal Institute for Educational Development presented NQF of the RF, being almost identical to EQF, but having one level more. It is assumed, that NQF will be a part of National Qualification System of the RF, that will include sectoral qualifications framework, educational and professional standards, national assessment system of learning outcomes and certification [6]. In 2012 the

Agency for Strategic Initiatives supported an entrepreneurial initiative "Development a national system of competences and qualifications" [5]. The road map for this initiative was designed in 2013, according to which the full implementation of national system of competences and qualifications should take place in 2015-2018. There are no data or reports as to the information whether the key performance indicators on this project were fully worked out. Only few reports for Krasnoyarsk Krai (2013) were published and the last reference to this project was in February, 2014. In April, 2014 the National Council under the President of the Russian Federation on Professional Qualifications was created with President chairing it and by its initiative Sectoral Councils on Professional Qualifications were created on the basis of the relevant sectoral employers' associations.

The legal bases for the level qualifications in Russia were developed in April, 2013. The order issued by the Ministry of Labor and Social Protection approves qualifications levels to develop projects for professional standards. The basis of this order is NQF for the RF worked out five years ago. Supplemented by description of the paths of achieving each level of qualification, i.e. education levels from the instruction on the first level qualification to the post-graduate and doctorate degree on the ninth level qualification. By EQF the sixth, seventh and eighth levels qualification correspond to qualification of bachelor, master and Ph.D. Respectively, while by NQF for the RF sixth and seventh levels qualification correspond to qualification of bachelor, specialist and master, the eighth level corresponds to qualification of specialist and master, and the ninth level corresponds to qualification of Ph.D.

In Russia, the requirements for learning outcomes, needed to obtain a certain level of qualification, are formulated according to the State Educational Standards being developed since 1994 and over three editions of those standards have been developed for the last 20 years. As was mentioned above, in European education policy results of learning are described in terms of knowledge and skills, and the ability to implement them in practice as competence. In curricula it is formulated as details the learner will know and can or able to do. Some of the member-countries also have educational standards, for example, the educational standards in Ireland are designed in accordance with the National Qualifications Framework of Ireland, and they contain generic requirements for the level of learning outcomes.

In Russia, requirements for learning outcomes started be formulated in the second edition of State Educational Standards (2000) with the structure of "to know - to able to do - to possess". The third edition of State Educational Standards in the requirements for learning outcomes introduced the concept of competence (2010) divided into two groups - general culture competences and professional competences. This standard establishes connection between competences and number of certain disciplines. In the content of State Educational Standards "3+ edition" (2014-15) the third group of competences was added – the general professional competences but there is no established disciplines in the standard, and the structure of describing competence as "to know - to be able to do to possess" disappeared. Generally speaking, the quantity requirements to content of learning outcomes decreases with each new standard.

In spite of serious work being conducted by state and commercial institutes to establish the qualifications framework, and especially their gradation along eight levels, proposed by the EQF, the desired unity of understanding and its realization is lacking. There is a certain clarity of requirements for design a higher education qualification since these requirements are formulated in Federal State Educational Standards, but there is no such clarity for the vocational qualification. Since the requirements to correlate vocational qualification with the professional standard are not yet legislatively established in RF at the legislative level.

In the RF requirements for employees set out in the Qualification Catalogs and Wage Rate books for jobs and professions, approved in 1998 and later. They contain job descriptions, minimum necessary knowledge and qualification (level of education and experience). In December, 2012 the Labor Code of the RF was supplemented with a new article 195.1, introducing the concept of professional standard in the Russian law practice. In April, 2013 several orders were issued containing rules and methods for the development of professional standards and the model of professional standard and special computer software "professional standard" were designed.

As far as the Russian vocational education is concerned, work here is conducted at two levels. On the one hand, at the level of professional associations as NAQD. On the other hand, the Ministry of Labor and Social Protection of RF officially (at the state level) asserts the developed professional standards and introduces them into the system of qualifications, reference books and registers.

It's premature assert that the educational standards (approved by the Ministry of Science and Education of RF) and professional standards (approved by the Ministry of Labor and Social Protection of RF) have the common denominator even despite the fact that real developers of standards often come out of education providers or academic methodological associations.

If the state educational standards are designed by Academic Methodological Associations, the professional standards are designed by employers' associations and some interested enterprises sometimes with participation of academic institutions. In April, 2014 the Ministry of Labor approved the State Program "Promoting employment of the population" after the realization of which conditions should be created to improve quality of workforce and to raise its professional mobility by implementation of professional standards. Standards could be developed by initiative of employers at their own expense or at the expense of the federal budget focusing on the priority areas of economy. According to this Program during 2014-15 800 professional standards should be designed but by the September 2015 only 471 standards were approved. Currently, there are 40 organizations indicated in the registry of professional standards as developers thereof including 10 private companies, 16 state educational providers, one private educational provider, 17 employers' associations and a few government agencies and institutions. Despite the fact that the number of participants taking part in the standards development exceeds their number in the respective registry the absence of open data in machine-readable format makes it hard to analyze who is actually responsible for the content of the standards.

The structure of professional standard contains a number of labor functions related with a certain position, which are made clear by description of necessary knowledge and skills for carrying out specific labor actions. As opposed to educational standards, which use "knowledge - skills - attainment" structure for learning outcomes, professional standards refer to requirements for an employee to possess necessary knowledge and skills for carrying out specific labor actions. The chosen structure meets international requirements to learning outcomes and qualification. According to the plan of the Ministry of Labor, these professional standards should create a base for the contents and structure of professional education to comply with requirements of labor market. The Federal Law "About Education in the Russian Federation" in sub-section 7 of Art. 11 Federal State Educational Standard (sub-section is entered after 2009), demands to consider provisions of the corresponding professional standards when developing educational standards.

The Ministry of Labor has a detailed plan for development of professional standards with [6] a section in this plan about "implementation of the mechanism of a professional qualification assessment on the basis of the professional standards", which forms basic rules for adopting a bill of an independent assessment of professional level of employees' qualification as of the first quarter 2015. According to this plan, the Assessment and Certification of Qualifications Centers (ACQC) should be formed since the third quarter of 2014. In fact, those Centers were created by RUIE and NAQD initiative already in 2009. In the same year the Ministry of Education for the RF approved the Provisions on the Formation of an Independent Assessment of the Quality of Vocational Education and after that some

Centers were created. Unfortunately, there is no unified register of these Centers on the website of NAQD at the present moment. According to RF Government Decree №881-p from 14 May 2015, there is 36 ACQC and 3680 individuals were certified in 2015. There isn't published any concrete information about this certificates, that allow analyze present situation about qualifications and their assessment.

It is essential that ACQC exercise their authority on the basis of a contract with the State Council (from NAQD) for certain financial compensation [6].

According to our analysis of a number of professional standards it is possible to assert that their developers correlate qualifications and EQF levels (predominantly in the professional standards they use levels 5-8). However, in our opinion, it is early to speak about intelligible and transparent relationship of Russian professional standards, qualifications and educational programs with those of other states.

One should add that in spite of indicated problems Russia undertakes active efforts in the right direction. Recent interest to educational system among Russian enterprises has grown as well as possibility to influence it in accordance with the real needs of the labor market. It is difficult to say that this phenomenon has general and mass character. Number of enterprises prefer to create their own vocational education centers to training staff according they own principles and in this case the common concept field of continues professional development could be missed.

There are several problems concerning the interaction between government, education providers and labor market:

- insufficiency of open data on all areas of work;
- published data are not adapted for computer processing and analysis;
- there is no traced interrelation between plans and their realization;
- there is no traced interrelation with data sources and calculations of the quantitative and

- qualitative performance for indicators specified in road maps, programs, etc.
- For example, Research Institute of Labor and Social Insurance should publish not only a text of a road map or an events plan, but these documents have to contain references to calculations of the chosen indicators with references to sources of statistical data (for example, by number of workers in different economy sectors, or graduates on particular specialties). Furthermore, it is necessary to add references to reports on realization of the planned actions in the same documents. Another example, the Register of developers of professional standards mentioned earlier, may contain attributes as "type of property" or "region", and it should be available in a machine-readable format, the same concerns also the data on professional standards.
- There is a number of disputed issues about an idea of assessment of qualification, offered by RUIE and NAQD.
- One of the issue is a distinguish between training and certification. According to NAQD it is not enough for a graduate to be assessed by education provider to earn a qualification. One disputed issue is an awarding body. Verification of qualification should be independent from learning process. Idea of NAQD is to divide training from verification by distinguish this function between two independent body – one is an education provider, other is an awarding body (ACQC). But in Russian practice frequently education provider wins the contest and it occurs as an ACQC. That means there is no any difference between training and awarding bodies. As other option, Triple E standard offers a concept of "Chinese wall" between training and verification. Education provider must provide functional independency and separation persons responsible for training and examination. This separation must be granted by documented procedure and must be verifiable by external auditors. In this case there is no necessity for ACQC.

- Another issue is a type of certificate. After finishing 6 to 9 levels qualification the state diploma is conferred. In the case of certification by the respective ACQC, graduates should also be conferred the state diploma. For now, ACQC gives a certificate of voluntary certification. What should be the difference between diploma and a certificate, if the educational provider and the ACQC is the same organization?
- The third issue is the accreditation of the ACQC. At present accreditation of educational providers is done by Government. Relationship between NAQD and the ACQC are built on the contract base, and in this case the ACQC cannot be quite independent in its assessment.
- Another section of the comprehensive plan for development professional standards [6, 18] contains the clause "revision concerning information resource for professional development (information on staffing, careers, opportunities for education and employment on the basis of existing information resources"). According to the clause, this kind of information resource covers the period of 2015-16. For now, in Runet there are two information resources on the subject of professional standard - one is inactive standartprof.ru and another one is an active profstandart.rosmintrud.ru. The last one does not publish information on staffing, careers, opportunities for education and employment. In EU this kind of information is usually published in national qualification frameworks resources.
- The problem of lack of interaction between education providers and employers can be solved by creation of effective communication technologies, primarily public information databases and environments that will effect a transfer of concepts and requirements among stakeholders, as well as create a database of publicly available methods to assess the professional qualifications.
- During the European integration process there is appeared a large number of different projects for development of sustainable economy through education policy and the EQF is one of them. There are some distinct projects and tools which are important for vocational education. One of the is the European Credit system for Vocational Education and Training (ECVET). ECVET being developed with the support of the European Commission since 2008 till present [10, 11, 13]. ECVET as a recommendation is ready. It's full implementation in EU is still in process.
- ECVET is created to ensure the mechanism of accumulation and transfer of learning outcomes. The effective work for implementation of international vocational qualifications based on the principles similar to the European Credit Transfer and accumulation System (ECTS).

ECVET tools [11] have been developed in order to increase the geographical mobility of trainees, to create the conditions for obtaining the qualifications in the educational organizations and the practical instruction on the territory of different states and to ensure the procedure of conducting this instruction. ECVET determines the minimum of the documented information, necessary for organizing the mobility of trainees [2]:

- the learning outcomes: the data about education, skills and competences, accumulation and transfer of learning outcomes connected with the professional sphere and the learning profiles, the criteria of comparability, the establishment of the equivalence between units;
- the process of learning: synchronization or sequence of studying curricula, units in ECVET format, process of assessment and verification;
- the organization of financing the mobility.

Besides this, ECVET established the set of rules and principles for unit's accumulation, which was learned by formal, non-formal and informal learning, and their transfer for the purposes of the qualifications acknowledgement. In ECVET the qualifications are described in terms of learning outcomes with the attribution of ECVET points.

Credits in the ECVET are determined on the basis [2,3]:

- the object of learning (theory, practice, base skills, the general education);
- the learning outcomes (knowledge, skills, competence)
- the place of learning (in class, apprenticeship, on the job training, at home),
- the context of learning (formal, non-formal, informal);
- the duration of learning (years, months, semesters).

Every stakeholder, student, education provider or employers can use ECVET to understand verification procedure for professional qualifications, which was obtained by studying in different countries.

The system of transfer and accumulation units for higher education works for decades (projects Tempus, Erasmus and other forms of international academic mobility), but exchanges in vocational education is not wide spread in RF. As far as we know, there are no procedure by including learning outcomes of non-formal and/or informal in regarding qualifications in RF at the present moment. There is kind of preparation legislative basis for the launching the above procedure. However, it concerned only e-learning (technology of distant learning).

Recently Russian enterprises actively use training, role games, teambuilding and similar methods of non-formal learning as the policy of professional development. However, as in continuous professional development of employees according to curricula created by enterprise, it is hard to find any accumulation and transfer system for credits of learning outcomes.

The main point to establish good vocation education is a quality assurance. In EU were created a recommendation for quality assurance in vocation education and training. European Quality Assurance in Vocational Education & Training (EQAVET) being developed by the European Commission since May 2001 till present [14]. EQAVET is a very large union [12] of representatives of educational committees and ministries of the European countries, candidates- countries and countries of European economic zone, social partners - commercial associations and trade unions, and also scientists from the European Commission and the European Center for the Development of Vocational Training (Cedefop).

As quality assurance tool, EQAVET is based on the Standards and Guidelines for Quality Assurance in European Higher Education Area (ENQA) and Common quality assurance framework (CQAF). The EQAVET's aim is to find and implement the most effective procedures of quality assurance in VET.

There are two levels of development quality assurance procedures:

- for national level EQAVET develops Quality Assurance for VET-system [22],
- for VET providers EQAVET develops Quality Assurance for VET-provider [21].

There was developed the web-application, which allows stakeholders to use EQAVET as a management tool for quality assurance. As a basis PDCA cycle is used, which consists from four stages - planning, implementation, assessment and review. One can find for each stage tentative descriptors, making it possible to evaluate whether the above stages correspond to the requirement of, and if corresponds, then to what degree. All tentative descriptors are accompanied by examples of the best practice of educational policies and/or organizations. Furthermore, methodology EQAVET is presented in two versions - for those, who only plan to build system of quality assurance (complete cycle) and for those, who want to assess the already existing system of quality assurance (stages of assessment and review) with the stress on data collection and analysis.

The total purpose EQAVET is to increase the culture of quality assurance of the educational services in the European Union and beyond, and also to stimulate the support of national policies of the states for improvement of quality in vocational education.

In the Russian Federation the attention is paid to quality assurance and control of education, including the quality of vocational education. State institutions such as the Federal Service for Supervision in Education and Science (Rosobrnadzor) and the Federal State Budgetary Institution "National Accreditation Agency" deal with the issues of verification of education quality.

The role of the public accreditation of educational programs is constantly growing. In Russia there is a number of agencies for the public accreditation of the educational programs, for example, non-governmental organizations "The National Center for Public Accreditation" (Yoshkar-Ola) or "The Agency for Quality Assurance in Higher Education and Career Development (AKKORK)" (Moscow). The national agency AKKORK formed the State Council for Independent Quality Assessment of Vocational Education consisting of two commissions: the Commission for independent assessment and certification of qualifications and the Commission for professional accreditation of qualification in VET. Some educational programs undergo professional accreditation by the international accrediting agencies, but the practice of professional accreditation of educational programs for vocational education is not yet wide spread in RF.

There are a few sectoral projects, which implement principles of quality assurance, transfer and accumulation of credits, or translate descriptors of qualifications framework to certain field. There are a few such projects for financial services sector (FSS) in European Union. For quality assurance in VET for FSS there is a QUALES project (2014-2015), for translation descriptors of qualifications framework there is a FIRST project (2009-2011), synthesizing approach is presented in Triple E project (2013-2015). We examine the last one.

The Triple E project was implemented by the European Banking & Financial Services Training Association - EBTN) [31] in the period from October 1, 2013 till September 30, 2015 with the support of the European Commission and Leonardo da Vinci Programme.

An international not for profit association European Banking & Financial Services Training Association A.i.s.b.l. (initially European Banking Training Network) unites 26 full members – education providers and employees' associations in FSS at the European Union and beyond, including the Russian Federation. There are two Russian organizations is full EBTN members presently - Banking Institute of the Association of Russian Banks (Moscow) and International Banking Institute (St Petersburg). Furthermore, EBTN includes 17 associate members, which provide financial education and training worldwide. EBTN is registered and located in Brussels (Belgium).

The main objective of EBTN is to represent the interest of the leading European bank training organizations and provide condition for improving professional culture in FSS through training and education both in the European Union and beyond its limits. To achieve this objectives EBTN leads various projects, most of them initiated by European Commissions, with the purpose to implement education projects of EU taking into account the specific character and needs FSS.

Triple E is aimed "to gather the main tools and recommendations related to existing modern EU qualifications and turn them into one integrated standard which is both aspirational and feasible for wider application in FSS" [8]. Those tools are – the European Qualification Framework (EQF), the European Credit system for Vocational Education and Training (ECVET) and European Quality Assurance in Vocational Education & Training (EQAVET). They are the cornerstones of contemporary policy in the sphere of vocational education and system of the qualifications of the

European Union. Taking into account the interest to vocational education for FSS, EBTN initiated the project for adaptation and propagation of European Union policy among financial and banking educational providers and associations.

The primary task of the project is to develop the quality standard and guidelines for qualifications in the European financial services sector on the basis of the above tools and to use this standard in the field of financial services. The project aims to strongly contribute to development of sectoral qualifications that will offer more flexibility, mobility and lifelong learning opportunities for the employees and employers of the sector. Moreover, the project supports the quality improvement of training systems through the development of innovative European Guidelines for the definition of professional qualifications in terms of learning outcomes, with the attribution of ECVET points, EQF level and according to quality assurance (EQAVET) criteria.

Triple E strives [8]:

1. To ensure comprehensive analysis and assessment of the best practices to design, development and realization of qualifications system for FSS.
2. To develop guidelines for designing sectoral qualification system in terms of learning outcomes (EQF), taking into account transfer of ECVET units and in compliance with EQAVET recommendations.
3. To create the quality standard, including criteria and tools for assessment on the basis of European system of credit transfer for the qualifications of vocational education for FSS.
4. To test best manuals and guidelines for developing the Triple E-based qualifications and standards of quality for the application by Triple E.

During teamwork of educational providers, employer's associations and other stakeholders EBTN developed the Triple E guidelines, which set the requirements for the qualifications of Triple

Guidelines were introduced for all EBTN members and interested stakeholders 24, September 2015. As an example, EBTN applied this standard for qualification European Foundation Certificate in Banking (EFCB).

"Our ambition is that the "Triple E" qualification standard will set up the top quality benchmark for the sectoral qualifications of financial services in the years to come. The standard is to bring a new dimension of cooperation and trust in the financial services sector. We want the "Triple E Qualifications" to be relevant for the sector, transparent for all users, reliable to deliver what they promise, supported and applied by people and organizations" [31].

Within the project framework several stakeholders were offered two surveys, on basis of which the number of recommendations regarding the qualifications of Triple E was accomplished [8]:

- to inform stakeholders about the above three tools;
- to use quality management system based on the standards of a ISO 9000 series, the national qualifications framework and accreditation procedure of VET providers;
- to use ECVET procedure without any adjustments and amendments;
- to involve stakeholders in design and assessment of qualifications, considering the interests of labor market and employers;
- to provide access to information about qualifications, learning outcomes, criteria of assessment;
- qualifications in FSS can be obtained by assessment of non-formal and informal learning, and consistent criteria of their assessment should be indicated;
- each qualification should be revised and updated with due consideration of the latest achievements
- to inculcate elements of continuous professional development (CPD) in qualifications.

Triple E standard consists from three blocks [32]:

1. Infrastructure (EQAVET)
2. Values

3. Content (EQF, ECVET)

The Standard states that quality assurance of the qualification infrastructure is achieved by following four steps: first, it is necessary to carry out the accreditation of educational organization at both the regional level and with the attraction of the international accrediting agencies. ISO certification and sector accreditation can be referred to as positive examples. Second, special attention should be paid to the unlimited access to the information about the qualification for the public. All the necessary information about the qualification should be published - lists of the required documents, procedures and actions, financial information, zones of responsibility and duration of learning. Third, the sectoral qualification councils should be created, which will provide accreditation both qualifications and educational provides. And fourth, all existing qualifications should be correlated with the EQF.

The content of qualifications should be specified in terms of learning outcomes - knowledge, skills and competences, and the size of qualification should be at least 50 hours. However, the indicated time must not be spent only in auditoriums. Qualifications can include the wide spectrum forms of learning. The size of each unit should comprise at least 10 hours and unit must have appropriate educational methodical documentation. EBTN introduces the Triple E concept of points: 1 triple (1 VET of credit) equal 10 hours of training. The amount of hours of a constant professional development (CPD) for each qualification should comprise at least 10 hours per year and is validated once every three years. The learning should not contain only on the job training or only self-assessment. A half of assessment could be a written report with confirmed evidence, or if there is no one, it should be confirmed by a third party. To assess the remaining 50 percent of knowledge obtained the accepted methods of assessment should be worked out taking into account the flexible approach to the methodology of evaluation.

Triple E establishes 10 values, relating both to the content and to the infrastructure of qualification. Qualifications must be relevant, transparent, reliable, impartial, flexible, effective, and creative, they should be part and parcel of professional activity, support the idea of lifelong learning and the application of knowledge in practice.

Since the objective of the standard is promotion information about EQF, ECVET, EQAVET values, its application by local institutions, i.e. by educational providers and by the associations of financial organizations, will contribute introduction into practice principles of the international educational policy into daily activity of organizations.

Conclusions

Attention, paid by the EBTN to the designated topic impels us as the members of association, to realize given principles in practice, to analyze the demands of the interested parties, to design programs on the basis of the proposed standard principles, to study the opinions of trainees and employers and, on the basis of the obtained data, to give feedback to EBTN and to Triple E developers.

This way - the standard, approval, recommendations for improvement, correction - the cycle of a constant improvement will be accomplished, which ultimately leads to an increase in quality of educational services, qualifications, and, finally, to economic growth and sustainable development and to an increase in the social stability and environmental safety.

Following the precise stages, i.e. standards, their approval, recommendations for their improvement, amendments and adjustments will help to go through the cycle of constant refinement which in the final analysis will lead to a rise in the quality of educational services and professional qualifications and thus to the growth of the economy, sustainable development and increasing social stability.

References

1. Information on http://www.ebtn-triple-e.eu/corsi
2. N. M. Aksenova, O. N. Oleynikova,Kazan Pedagogical Journal, **3(104)**, 33-46 (2014)
3. N. M. Aksenova, E. V. Sartakova, Bulletin of the Tver State University. Series: Pedagogy and Psychology, **3**, 122-134 (2014)
4. Information on http://www.edu.ru/db/portal/spe/index.htm
5. Information on http://asi.ru/npi/nskk/
6. *Development of the National Qualifications Framework: a collection of pre-normative documents* (Publishing House "Pen", Moscow, 2014)
7. Information on http://www.tcd.ie/teaching-learning/academic-development/assets/pdf/dublin_descriptors.pdf.
8. Information on http://www.ebtn-triple-e.eu/corsi
9. *EC. Find and Compare Qualifications Frameworks*, information on https://ec.europa.eu/ploteus/en/compare
10. *ECVET projects, The Education, Audiovisual and Culture Executive Agency (EACEA), 2010-2014* information on http://www.ecvet-projects.eu
11. *ECVET toolkit, Nationale Agentur Bildung für Europa beim Bundesinstitut für Berufsbildung (NA beim BIBB)* information on http://www.ecvet-toolkit.eu/
12. *EQAVET European Quality Assurance in Vocational Eduction and Training. - EQAVET Secretariat - Quality and Qualifications Ireland (QQI)* information on www.eqavet.eu
13. *European ECVET Network* information on http://www.ecvet-team.eu/en
14. *From ENQA-VET to EQAVET German Reference Point for Quality Assuranse in VET* information on https://www.deqa-vet.de/en/From-ENQA-VET-to-EQAVET-2190.php
15. E.P. Ilyin, Questions of psychology, **5**, 1984
16. *International Standard Classification of Education. – UNESCO Institute for Statistics, 2011* information on http://www.uis.unesco.org/Education/Documents/isced-2011-ru.pdf
17. *Labor Code of the RF*
18. A.N. Leibowitz, I.A. Voloshin, A.S. Perevertajlo, O.D. Pryanishnikova *An independent evaluation and certification of qualifications: Collection of documents and materials* (Publishing House "Pero", Moscow, 2014)
19. N. Mospan, Continuing professional education: Theory and Practice, **1-2**, 38-41 2014
20. O. G. Nefedova, Accreditation in Education, **8**, 62 (2006)
21. *Quality Assurance for VET-provider - EQAVET* information on http://www.eqavet.eu/qc/gns/home.aspx
22. *Quality Assurance for VET-system – EQAVET* information on http://www.eqavet.eu/qa/gns/home.aspx
23. M. B. Sinyagovskaya, The world of science, culture and education, **5**, 53-55 (2013)
24. *Terminology of European education and training policy-a selection of 130 key terms* (CEDEFOP, Luxembourg, 2014)
25. *The European Banking & Financial Services Training Association* information on http://www.ebtn-association.eu
26. *The European Qualifications Framework, European Commission* information on http://ec.europa.eu/ploteus/search/site?f[0]=im_field_entity_type%3A97

MATEC Web of Conferences **73**, 07018 (2016) DOI: 10.1051/matecconf/20167307018
TPACEE-2016

The Functional Model Approach to the Consulting for Vertically - Integrated Construction Group

Anna Pimenova [1,*], *Svetlana* Kuzmina [2], *Natalia* Morozova[3] and *Angela* Mottaeva[4]

[1]Saint-Petersburg State University of Economics, 191023, Sadovaya str. 21, Saint-Petersburg, Russia
[2]Saint-Petersburg University of State Fire Service of the Ministry of the Russian Federation for Civil Defense, Emergencies and Elimination of Consequences of Natural Disasters, 190005, Moskovsky pr. 149, Saint-Petersburg, Russia
[3]Saint Petersburg State Forest Technical University under name of S.M.Kirov, 190000 Institutskiy per. 5, Saint-Petersburg, Russia
[4]Moscow State University of Civil Engineering, 129337 Yaroslavskoe sh. 26, Moscow, Russia

Abstract. Managerial decision making in the framework of functional modeling of the consulting process have a direct effect on other business - processes of vertically - integrated group of construction companies. As a result, the experience of consulting companies tends to be used for the making managerial solutions. Consultancy is known as one of the most complicated types of buisiness process. It requires a huge and deep examines and researches of targeting area, therefore need to be provided with special methodology, included internal standards of the consulting companies. Correct methodological support, planning process and implementation of managerial solutions should be based on the survey of the direct and inverse connections and interdependence of all group's business – processes. Functional - process modeling of the vertically - integrated construction group could be considered as an instrument of examination and analysis of the issue how the managerial solution impact on the business-process for the construction group functioning. The main result of the research is the formalized process-oriented model – prototype of the business - processes of vertically - integrated group of construction companies.

1 Introduction

For modelling business processes of company, including vertically – integrated group it is essential to distinguish managerial, basic and supporting processes. Lists of business processes and functions of vertically integrated institution could be used as an "entry" for the further functional modelling.

The business processes reflect in details all the procedures that have to be done in frames of basic, supporting and managerial systems of an enterprise [1-5].

* Corresponding author: anna_pimenova@bk.ru

2 Methodology

The totality of the vertically integrated companies' business processes could have diverse combination. These combinations depend on the level of companies' concentration and diversification. Sometimes some of the operation processes could be excluded from the common integrated manufacturing system [6]. However, to provide compound approach to the management of vertical integration companies, which is aimed to the legal merger of all kind of the construction business processes, we suggest that all complex of business processes could be represented in the form of the process-oriented prototype model in the Figure 1.

1Management of vertically integrated construction company	1.1Strategic management	1.2General horizontal coordination and unification	1.3Cascading the strategy for a horizontal level	1.4Process oriented management support	1.5Financial accounting and taxation managing		
2.Customer interaction	2.1Sourcing of investment projects and tenders	2.2Contract formation	2.3Monitoring of relationship with the investors and project managers throughout the execution of project	2.4Interaction with customers at the stages of completion and post-project support			
3Constructional designing and marketing	3.1Investment projects and tenders market monitoring	3.2Constructional designing	3.3Coordination of project execution terms	3.4Brand management	3.5Product promotion		
4.Material and technical management	4.1Procurement management	4.2Inventory management and resource allocation	4.3Technical support management				
5.Manufacturing	5.1Construction planning	5.2Material manufacturing	5.3Construction works	5.4Related and subcontracting works	5.5Transport support	5.6Horizontal cost control	5.7Horizontal quality control
6.Facility commissioning and product sales	6.1Planning and organizing the commissioning of project results	6.2Certification and standardization of project rezult	6.3Other sales management	6.4Storage logistics of finished products			
7.Supporting business processes	7.1Safety and human resources management	7.2Research and development and testing subprocesses	7.3Software and information support	7.4Financial management	7.5Assets management		

Fig. 1. Reference prototype model of the business - processes that executed by the group of vertically – integrated construction companies.

The first level business processes of the vertically integrated group include those kinds of activities that provide the group's basic aim – the execution of a capital investment project. We presume that seven basic business processes could be interpreted as the first level processes of the model. Respectively, they include:

Enterprise management process - it provides setting of companies' goals as well as organizing of companies' interaction;

Customer interaction process - it is aimed at establishing of relationships with the customers, ensures the promotion and sale of the product of integrated group;

Constructional designing and marketing are targeting at development of competitive product of integrated companies that could meet the demands of the market;

Material and supply management is aimed at material, technical and resource support for the basic operating function;

Manufacturing process organizes the interaction of investment resources for the implementation of project which have to be legally and contractually relevant;

Facility commissioning process provides sales of group's product;

Supporting business processes include the processes that are diverse in a content but similar in functional meaning, they provide the provision and assistance for basic operating processes of integrated group.

The details of main business processes are reflected in the model as well. These details have an influence on the way how to specify the separation of tasks between the working group staff of consulting company and could be considered as a basis for contract relationship between the customers and consulting companies including the identification of consulting company's contract stages [7-12].

In accordance of the model the certain items of works could be represented as follows. The features of diversified company strategic management are reflected in the structure of the sub-processes that are belonged to the management business process of vertically integrated construction company. In frames of the process we could identify some sub-processes: the sub-process of strategic management (it generates key long-term strategic targets for results of the group's functioning); the sub-process of general horizontal coordination and unification (it generates certain methodological approaches that focus on the standards and uniformity of the group's elements, group's results and group's technology); the sub-process of cascading the strategy for a horizontal level (it includes certain controlling mechanism for each horizontal level); sub-process of process oriented management support (it aimed at collecting the data for business analysis); sub-processes of financial accounting and taxation managing (provides the collecting accounting data, bookkeeping and preparation of the financial, statistical and tax returns).

Customer interaction process includes some sub-processes. Firstly, it is the sub-process of the sourcing of investment projects and tenders for their implementation. It is targeted on meeting the construction market demands. Secondly, contract formation, thirdly, monitoring for correct relationship with the investor or project manager throughout the implementation of certain types of work. It includes the determination of contract conditions, contract planning process, the conclusion of a contract and control of contract fulfillment. In addition, the last sub-process involves an interaction with customers and the investors at the stages of completion and the maintenance of the finished project (post-project support).

Constructional designing and marketing involve conducting of market monitoring, where company – client operates. The next sub-processes are the execution of constructional design stages and the coordination of project execution terms, the brand management and, finally, the marketing promotion of the group product (sub-marketing policies and corporate product advertising).

Process of material and technical management consists of some sub-processes, namely, procurement planning, inventory management and resource allocation and technical support management (including energy, processing equipment, water supply, availability of appropriate facilities and machines tools and etc.).

Manufacturing process can be divided into seven parts. It essential to distinguish such kinds of activities as setting up of production (construction planning), the actual manufacturing process (including materials manufacturing, construction works, related and subcontracting works, transport service and support), the horizontal cost control of group's product, the quality and certificate requirements control of group's product.

Sales process involves stages (sub-processes) of planning and organizing the commissioning process of construction project result, certification support of construction object, related products sales management and product storage logistics.

Supporting business process involves safety (including ecological, energy security, fire protection) and human resources management sub-process, research and development and testing sub-processes (including laboratory tests of construction materials), software and information support of construction group, financial management and assets management.

The indicated sub-processes provide auxiliary but essential kinds of activities for the execution of the basic processes, the sub-processes and the procedures (Fig. 2).

			Functional tasks		
			Manufacturing	Management	Supporting
Business processes	Vertically integrated group management	K1			
			R11	R12	R13
	Customers support	K2			
			R21	R22	R23
	Designing and marketing	K3			
			R31	R32	R33
	Material management	K4			
			R41	R42	R43
	Manufacturing	K5			
			R51	R52	R53
	Sales	K6			
			R61	R62	R63
	Supporting business sub-processes	K7			
			R71	R72	R73

Fig.2. The consolidated function - process model of consulting services for a group of vertically integrated construction companies.

In a balance, reference prototype model of the business - processes that executed by the group of vertically – integrated construction companies could be used for the detailed description of key stages consulting services. The model could be interpreted as one of the "entrances" for functional - process model for vertically integrated companies. The second "entrance" for the model is the list of certain functional tasks that have to be done by consulting company.

The authors of researches devoted to the functional - process modeling claim that such functional tasks should be divided for some categories. For instance, other author admits that it is advisable to distinguish at high level followed basic functions of any enterprises:

Basic function tasks that is directly connected with conversion of external resources to the final company's product or service;

Managerial function that are devoted to company management;

Auxiliary function which provides supporting for manufacturing, commercial and managerial activities and creates certain conditions for them [13-16].

The target of creating function-process model is to determine the extent of details, the list and the content of selected functional tasks. The combination of consolidated function-process model and the reference prototype model of the business – processes could be used as a basis for general function – process modeling of a company. This model, in its turn, we believe, would be the basis for the working-out a consulting procedures' program.

For elaboration the most general function - process model of consulting services for vertically-integrated construction group it is essential to take into consideration the particular features that are indicated in the proposed referent prototype model of the business – processes for integrated companies (Figure 1) as well as consider those three basis function that are proposed by economists and were mentioned above. We think that such function-process model of consulting services could be represented by a certain visual form (Figure 2).

This is the matrix, the rows of which describe business processes, the columns include functions. In general the number of rows and columns depend on the required extent of details. The details are determined by the requirements of management level [17].

At this version of the model we set limits for the number of rows and columns by business processes of first level and by basic functions of enterprise. At the crossing of rows and columns (processes and functions) the structural element of the system should be offered. It describes the execution of the certain function for the certain business process.

Speaking about modeling particular tasks it is important to evaluate those elements. Such assessment reflects the significance and depth of function execution for the certain business process. As the result we obtain the function-process model of a construction group. The methodological approach to the solution of managerial tasks through the function - process modeling includes the following sequence (stages). The solution concerning individual functional tasks distribution within business – processes could be made after the analyses of crossing the areas of responsibilities by the process and functional description. In mentioned above reference prototype model we additionally introduced weighting coefficients K and R. These coefficients indicate the significance of the certain business process (K) and certain function (R) [18-20].

This modeling approach could provide advantages for consulting experts. Particularly it is useful for rationalizing the distribution of resources across groups' products and functional tasks. To evaluate the resource consumption for certain function (including consumption for all business processes) we need to sum up the value of resource along the matrix rows. To evaluate the resources requirement for certain business process for all its function we need to sum up the value of resource along the matrix columns.

3 Conclusions

The function-process model that we described above could be applied for the distribution of a predetermined amount of resources between system elements. However, it could be useful for the opposite aims. For instance, for evaluation of resources requirements for the system functioning in the conditions when the resource requirements for certain business process are known in advance (the value of matrix cells are known). Also it is possible to meet the situation when only requirements of some elements are known in advance and the aim is to evaluate the resource consumption for other elements and for the system in general, sometimes in a condition when restrictions for each of the required parameters exist. We suggest that for the management of innovative research in a group of vertically integrated construction companies at the stage of planning of managerial solutions concerning innovative research (according with the model of management of innovative research), it could be useful to apply the proposed method of the resources distribution, including finance resources, between the certain research stages and works.

To sum up, at the offered modeling approach we tried to reflect the basic innovative ways of consultancy development. We believe that our results could be considered as a contribution to the development of new, more complex and informative types consultancy services.

References

1. V.V Okrepilov, M.V. Leonidovich, Asian Social Science, **11 (7)**, 312-325 (2015)
2. S. Kuzmina, Journal of Economic Regulation, 3 (2012).
3. V. Vaganova, Publishing House of Peter the Great St. Petersburg Polytechnic University, 98-104 (2012).
4. Information on http://naukovedenie.ru/PDF/20evn412.pdf
5. L. Vorona-Slivinskaya, Saint Petersburg State University of Economics and Finance (FINEC), (2007)
6. V. Murgul, V. Pukhkal, Procedia Engineering, **117**, 891-899 (2015)
7. E. Aronova, N. Vatin, V. Murgul, Procedia Engineering, **117**, 771-779 (2015)
8. J. Ćetković, S. Rutešić, M. Zarković, M. Knežević, N. Vatin, Procedia Engineering, **117 (1)**, 780-790 (2015)
9. V.V Okrepilov, V.L. Makarov, A.R. Bakhtizin, S.N Kuzmina, Economy of Region, **2**, 301-313 (2015)
10. V.L. Kvint, V.V Okrepilov, Herald of the Russian Academy of Sciences, **84 (3)**, 188-200 (2014)
11. V.V Okrepilov, V.N. Krutikov, G.I. El'kin, Measurement Techniques, **57 (2)**,109-116 (2014)
12. V.V Okrepilov, A.Yu. Smirnov, Measurement Techniques, **56(1)**, 54-60 (2013)
13. V.V Okrepilov, Studies on Russian Economic Development, **24(1)**, 35-42 (2013)
14. V.V Okrepilov, Standarty i Kachestvo, **10**, 52-55 (2005)
15. V.V Okrepilov, Ivanova, G. Standarty i Kachestvo, **12**, 62-68. (2004)
16. V.V Okrepilov, Standarty i Kachestvo, **3**, 94-96. (2003)
17. S. Rutešić, J. Ćetković, M. Zarković, M. Knežević, N. Vatin, Procedia Engineering, **117 (1)**, pp. 905-915 (2015)
18. S. Rutešić, J. Ćetković, M. Knežević, M. Žarcković, N. Vatin, Procedia Engineering, **117 (1)**, pp. 642-650 (2015)
19. Lazarevska M., Cvetkovska M., Knezevic M., Trombeva Gavriloska A., Milanovic M., Murgul V., Vatin N. Applied Mechanics and Materials, **627**, 276-282 (2014)
20. Ćetković J., Knežević M., Žarković M., Murgul V., Vatin N. Applied Mechanics and Materials, **638-640**, 2465-2470 (2014)

MATEC Web of Conferences 73, 07019 (2016) DOI: 10.1051/matecconf/20167307019
TPACEE-2016

Regional Indicators of the Safety Development of the Entrepreneurship Economy

Elena Ivleva[1,*], *Khe* Sun Pak[1], and *Ksenia* Nagornaya[1]

[1]Saint-Petersburg Academic University, 190103 Lermontovsky pr. 44, Saint-Petersburg, Russia

Abstract. The present article is dedicated to identification of opportunities for stating and solving research problems related to evaluation of social and - economic development safety and efficient use of the regional economic potential. The analytical base is the statistical information for the North - West Federal District (NWFD) of Russian Federation. The authors offer the system of indicators of development safety and effective use of the economic potential. Each of them can be calculated individually for the territory. In practical terms, these indicators can be used by the executive authorities and regional administration to monitor changes.

1 Introduction

The importance and suspense of some economic, social and environmental issues makes the study of the whole set of issues related to the regional development safety management vitally important. The development of the management mechanism involves the development of assessment tools to measure the territory development safety, quantitative and qualitative indicators that allow researchers to draw conclusions and make the practical person take administrative decisions, considering the territory potential. Thus, the problem is not only of the theoretical but also of the practical interest. The system of indicators is offered to assess the social and economic development safety and effective use of the economic potential. This system can be used by the executive bodies of the Russian Federation territorial units to conduct the monitoring researches. The system of indicators shall be supplemented by ecological development indicators of the territories, as an approach from the standpoint of sustainability implies the solution of problems of social, ecological and economic character [1-23].

2 Theory

Regional safety is the main dominant of the territory sustainable development. The possibility to achieve sustainable development including at least clean technologies, eco-friendly products and use of sustainable resource, seems to be fairly remote. [1] This can be proven by facts. Let us pay attention at least to the following three of them:

* Corresponding author: ivlevaes@mail.ru

1. Opposite trends of economic and environmental cycles.
2. Mismatch of factors that determine economic and ecological components of sustainable economic growth, including the regional one.
3. Economic and ecological subsystems being developed in different directions, destabilizing the entire economic system.

This is the reality in which management decisions that determine the quality of life tomorrow are taken today.

In matters of sustainable development everything is controversial: theoretical feasibility, methodological argumentation, methodological support. The authors consider that the economic system does not aspire to a stability condition. This point of view is supported by the cycle theory [2]. Interrelation of development and safety is confirmed in the research papers of many authors, including ones of A. D. Ursul and A.L. Romanovich [3] and American scientist George Kirschner [4].

3 Results

The system of indicators is determined to assess not only the approach of the individual territorial development path to the level of safety, but also the effectiveness of managing this process on the principles of sociological, ecological and economic balance. To assess the level of social and economic safety of territories the system of 14 indicators is used (see. Table 1). It is necessary to explain the indicator stating "the ratio of financial assistance to the region from the federal budget to the volume of federal budget revenues received from the region." The latter is represented by a different phrasing due to the lack of information: the ratio of financial assistance to the region from the federal budget to the volume of revenues of the regional budget. The overall list of indicators includes also the innovative development index, introduced by the Institute for Statistical Studies and Economics of Knowledge of "Higher School of Economics", that reflects the level of scientific potential, social and economic status, openness of the government, and other indicators of the territorial units of the Russian Federation. The average median index of groups characterized by average, maximum and minimum figures for the period under review represents the threshold. The system of indicators presented in the article by I.E. Denezhkina, D. Suzdaleva [6] was used when preparing indicators to measure the economic safety of the regional development. Figures presented in the Moscow Government Decree № 707-RP "About approval of the Concept of complex safety of the city of Moscow"" [11] as of April 16th, 2010 can be used as basic indicators to measure the social safety of the regional development. These indicators include the following ones: the ratio of expenditures on social programs to the gross regional product (GRP), the ratio of incomes between the richest 10% and the poorest 10% of population groups, the proportion of the population below the poverty threshold, the ratio of minimum and average salaries, the crime rate (number of crimes per 100 thousand of population), the population life expectancy, the level of the population distrust to the authorities. We consider it important to add the following indicator to the proposed ones: "The amount of damage caused to the society as a result of corrupt activities of state and municipal employees." According to Rosstat (the Federal State Statistics Service), within the last few years there were significant changes in the structure of the layer of the richest: the share of wealthy officials (that determine the redistribution of wealth) rose, forcing out business people and professionals.

It is widely known that the multi-criteria assessment of safety includes the assessment of the economic potential of the region and determination of potential use effectiveness. The economic potential determines the material base for safe development of the territory. The search for reserves by the main factors affecting the level of regional safety is carried out to determine the effectiveness of the territory economic potential use. The following parameters are used for assessment: return on investment (the ratio of GRP to investment in the fixed capital), return on taxes (the ratio of tax revenues from the territory to the volume of GRP), return on assets, profitability of fixed assets, productivity, capital-labor ratio, return on wages and return on amortization [13].

Selection of indicators for assessment of the social and economic safety of the regional development in the North-West Federal District in 2013 is represented in the Table 1. It was used for the first time in the article [14].

St.Petersburg can be highlighted by the highest GDP per capita in comparison with the world average of 41% (the highest level of this indicator in the world as of 2013 is in Luxembourg - 6,091,568 rubles). The lowest indicator is in the Pskov region (2.7%).

The highest capital consumption level is registered in the Komi Republic (46.2%). The lowest value for this indicator is, however, registered in the Leningrad Region (32.4%) as this region has been characterized by the high level of investment activity for many years. The capital consumption figure in all districts of this region is below the threshold value.

In terms of the ratio of investment to GRP the Komi Republic stands out from the other regions - 40.6%. The mining industry, wood processing industry, chemical industry and mechanical engineering have contributed to these results. As of 2013 by this indicator the Komi Republic outperformed the Leningrad Region that had been the leader in this field for years. The lowest rate of this indicator is determined in Saint - Petersburg (19%) and the Republic of Karelia (19.4%) [10]. These values are below the threshold.

Saint - Petersburg belongs to the group of regions with high innovation activity (0.5382). The lowest index is in the Pskov Region - 0.2324. The Pskov Region belongs to the group of regions with moderate innovative activity [5].

The lowest unemployment rate is observed in Saint - Petersburg - 1.5%, while the highest is in the Komi Republic -7.1. In recent years there has been a significant decrease of employment in the electricity, coal (closure and liquidation of unprofitable mines) and gas industries, the pulp and paper industry and other industries. The unemployment rate in all other regions is well below the threshold.

The volume of the federal budget financial assistance to Saint - Petersburg accounts for 0.07% of the regional budget income - the factor indicates financial independence of the city from the center. It is the lowest rate in the district, while the highest one (0,41%) is in the Pskov Region. Evidently this region has a high level of a debt burden – 72.4%. In our opinion, in parallel with the indicator "ratio of financial assistance from the federal budget to the region to the volume of the regional budget revenues» we need to use another indicator - the amount of the debt burden. The level of a debt burden of the district regions is the highest in the Republic of Karelia - 96,3% [8]. They are equal to own revenues and financial assistance from the federal budget accounts for 34%.

The ratio of expenditures on social programs to the GRP is the highest in the Arkhangelsk Region - 32.5%, exceeding the threshold value (20%). The share of social expenditures of the regional consolidated budget compounds 67% in the general amount of expenditures, the level of a debt burden in the region is 62%, the volume of the federal budget financial assistance is 23% of the regional budget income. This situation will necessarily lead to reduction not only in non-social expenditures, but also in expenditures on social programs. The value of this indicator for other regions of the district doesn't not exceed the threshold.

The ratio of incomes of 10% of the richest population groups and 10% of the poorest population groups in all regions of the district exceeds a threshold value (10: 1); the highest ratio is in St. - Petersburg (20.2: 1), in Russia (16.8: 1). "By the ratio of revenues to be distributed between the poorest and the richest, we have already passed the threshold of social safety", - says Jacob Dubenetsky, the Head of Investment Centre of the Institute of

Economic Forecasting. The share of the population living in poverty in all regions of the district exceeds the threshold value, with the exception of Saint - Petersburg (8.3%). A very high percentage of the poor is in the Pskov Region (16.9%). The region has a high level of a debt burden, the lowest index of innovative development and the volume of GRP per capita.

The ratio of minimum and average wages in all the territorial units of the Russian Federation exceeds the threshold value (1: 3), the highest proportion is observed in the Komi Republic and Arkhangelsk Region that have the same value (1: 6). According to the data, the average salary in the Republic covers 4.2 minimum levels of subsistence, in the region –2.9 minimum levels of subsistence.

The crime rate in all regions of the district is below the threshold (5-6 thousand per 100 thousand of inhabitants). The high crime rate is observed in the Komi Republic (2,377 thousand per 100 thousand of inhabitants). In the Republic there is the highest ratio of minimum and average wages. The level of capital consumption and the share of the population living below the poverty line exceeds the threshold of safety. The lowest rate is observed in Saint - Petersburg and the Leningrad Region - 1,100 and 1,105 thousand per 100 thousand inhabitants, respectively. In Russia this indicator is 1,476 thousand per 100 thousand inhabitants. The life expectancy rate of the population in all territorial units of the Russian Federation exceeds the lower limit of the threshold value. The lowest life expectancy rate is observed in the Pskov Region (67,2 years).

The level of public distrust to the authorities is very high in the Kaliningrad Region (79%), the Republic of Karelia (77%) and the Novgorod Region (82%). Only in three regions - Saint - Petersburg (17%), the Pskov Region (17%), and the Vologda Region (24%) this indicator is below the threshold value (20 - 25%). In other regions the level of public distrust to the authorities is high (from 46 to 66%). The reasons are as follows: the low level of activity and effectiveness of the regional policy, poor implementation of demographic, social and family programs, poor quality of medical services (new technologies and equipment are not introduced, no kind of activity is led to achieve an accessible and quality health care in remote areas), few measures are taken to meet the cultural needs of the population [7].

The results are as follows:

1. There is a drop of the industrial production level not only in weak regions, but in the most developed ones, for example, in St. Petersburg – 8.2%.
2. A high degree of capital consumption is confirmed.
3. The investment activity of the regions is being reduced.
4. The debt burden of the regions increases.
5. By the level of capital consumption and the ratio of minimum and average wages, all territorial units of the Russian Federation passed the threshold of social safety.
6. The level of public distrust to the authorities is still very high.

Let us determine the task as follows: search of reserves and assessment of use of the regional economies potential.

Enhancement potential of the social and economic safety level of the regional development is determined by evaluating the effectiveness of the economic potential use by the indicators presented in Table. 2.

The highest returns on investment, the ratio of GRP to investments in fixed assets, are observed in St. Petersburg. The region executes major investment projects for construction of the central section of the road Western High-Speed Diameter and integrated development of the subway in the amount of 73 billion rubles. The volume of investments into development of the pharmaceutical cluster is 21.7 billion rubles. [9]. The lowest return on assets is in the Arkhangelsk Region, where in recent years there has been decrease in total amount of investments.

The leader in the return on assets (1.05 rub.) is the Kaliningrad Region. Every year fixed assets are renewed in the region. The outsider by this indicator is the Murmansk Region (0.21 rub.), where the degree of capital consumption in 2013 increased by 26% in comparison with 2012.

By the profitability of fixed assets the most advance region tends to be St. Petersburg (8.7%) with the highest capital productivity. The same cannot be said about the Pskov Region (0.77%), where the return on assets is the lowest. In the region there is the lowest volume of investment into fixed assets per capita among the territorial units of the Russian Federation.

The high capital-labor ratio remarks the Komi Republic (4.35%), which is one of the three regions of the Russian Federation with the largest volume of investments. The low capital-labor ratio is in the Kaliningrad Region (0.55%).

The largest amount of tax revenues from each ruble of GRP (returns on tax) is registered in the Komi Republic (24.51 kopecks), the lowest - in the Arkhangelsk Region (8.69 kopecks). With almost the same volume of GRP the amount of tax revenues in the Komi Republic is 3 times more than in the Arkhangelsk Region.

The Komi Republic is the leader in the efficient use of the territory economic potential; it has the highest labor productivity - 1.1 mln. rubles per person employed in the economic sector, the lowest one is in the Pskov Region - 0.35 million. rub. The latter belongs to the group of regions with the lowest rating by the social and economic status.

By the return on wages the Komi Republic (2.44 rub.) lags slightly from the leader by this indicator – the Leningrad Region (2.62 rub.), the lowest rate is in the Pskov Region (1.48 rub.).

The return on amortization in the Kaliningrad Region is the highest among the regions of the district - 3.02 rubles. The region belongs to the group of regions with the lowest level of capital consumption. The return on amortization in the Murmansk Region (0.55 rub.)and in the Komi Republic (0.54 rub.) is the lowest among the territorial units of the Russian Federation. They belong to the group of "regions with a high degree of capital consumption".

In general, by the efficient use of the territory economic potential the regions of the district demonstrate satisfactory results, especially the Republic of Komi, the Leningrad Region, the Kaliningrad Region and Saint - Petersburg. In these and other regions there is sufficient potential to improve the level of social and economic safety of the territory. Even the Pskov Region has significant potential: high transport capacity, good prospects for development of the energy infrastructure with connection to the largest sources of energy, development of customs and logistics terminals at the international checkpoints of the state border of the Russian Federation [9].

4 Conclusions

1. Monitoring studies must include the indicators of the territory development safety. The existing system of indicators to measure the quality of life, the level of social and economic development and the efficiency of public administration does not consider the basic parameters of the social safety.

2. To assess the efficiency of the social and economic safety management of development of the regions it seems appropriate to use structural indicators. They demonstrate constraints for the innovative growth and contradictory dynamics that not always can be explained from the standpoint of production. The indicators help to identify the factors destabilizing economic development and denying principles of sustainable development.

3. To illustrate a sustainable outcome of the study it is necessary to introduce a number of environmental indicators, including structural ones, such as the dynamics of all types of pollution, dynamics of production and consumption waste as a whole and per capita, and others.

4. It should be stated that the existing system of indicators to measure the quality of life, the level of social and economic development, and the effectiveness of government management almost doesn't take into account innovative components. We used the index of innovative development.

Table 1. Level of social and economic safety of the region.

Indicators	Threshold	Actual value										
		The Komi republic	The Republic of Karelia	Saint-Petersburg	Leningrad Region	Arkhangelsk Region	Pskov Region	Vologda Region	Murmansk Region	Novgorod Region	Kaliningrad Region	RF, NWFD (average)
Volume of GRP per capita, % from world average	100	8.65	4.5	41	6.5	6.4	2.7	4.9	5.9	4.5	4.6	NWFD-8.97
Level of decline in industrial production,%;	30 - 40	100.7	100 .3	91.8	100. 1	108.3	94.0	105.6	97.8	109. 2	108.7	NWFD -97,6 RF - NWFD - 101.7
Level of capital consumption,%	60	46.2	43. 4	36.8	32.4	41	40.4	43.1	39.5	43.8	34.7	RF – 46.3 NWFD – 39.0
Ratio of investment to VRP,%	20	40.6	19. 4	19.0	36.6	30.7	31.1	22.1	23.0	29.8	24.8	NWFD - 25.3
Index of innovation development	0.21 85	0.3728	0.2 755	0.53 82	0.38 07	0.317 1	0.23 24	0.3124	0.35 58	0.31 63	0.287 1	RF -0.3227
Level of registered and quasi unemployment, %;	8-10	8.1	6	1.5	4.5	7.3	6.5	5.6	6.7	3.7	5.4	RF -5.2 NWFD - 4.1

 DOI: 10.1051/matecconf/20167307019

assistance to the region from the federal budget to the volume of	0.5	0.11	0.3 4	0.07	0.11	0.23	0.41	0.18	0.14	0.19	0.35	NWFD - 0.26
Ratio of expenditures on social programs to VRP,%	20	11	16. 2	10.5	9	32.5	18	11	16.8	11	11.9	NWFD - 14.8
Ratio of revenues of 10% richest and 10% poorest population groups	10:1	16.6:1	11. 3:1	20.2 : 1	13:1	12.4: 1	11.5: 1	12.1:1	14:1	14.1: 1	12.9: 1	RF – 16.8 :1 NWFD -13.8 : 1
Share of population below poverty threshold,%;	10	14	15	8.3	11	14.1	16.9	13.1	11	13	12.3	NWFD -12.9
Ratio of minimum and average wages	1:3	1: 6	1:5	1 :4.3	1 : 4.2	1:6	1:3.5	1:4.5	1:3.6	1:4.2	1:3.2	RF -1:5.5
Crime rate (amount of crimes per 100 thousand persons), thousands	5-6	2.377	1.9 95	1.10 0	1.10 5	1.655	1.35 2	2.050	1.67 1	2.06 0	1.617	RF – 1.508 NWFD -1.476
Population life expectancy, years	65- 70	68.9	68. 4	74.1	70.4	70.3	67.2	69.8	70.5	68.4	70.8	RF – 71 NWFD -71.3
Population life expectancy, years	65- 70	68.9	68. 4	74.1	70.4	70.3	67.2	69.8	70.5	68.4	70.8	RF – 71 NWFD -71.3

Effective use of the economic potential of the North-West Federal District regions it is presented in Table 2

Table 2. Effective use of the economic potential of the North-West Federal District regions.

Territorial units of the Russian federation	Return on investme nt, rub	Return on assets, rub	Profi tabili ty of capit al asset s,%	Capital-labor ratio, mln rub/per capita	Return on taxes, kop	Produc tivity, mln.ru b	Return on wages rub	Return on amorti zation, rub.
The Republic of Karelia	5.87	0.36	2.43	1.61	11.43	0.59	1.77	0.84
The Komi Republic	3.02	0.25	3.3	4.35	24.51	1.1	2.44	0.55
The Arkhangelsk Region	2.67	0.37	1.08	2.34	8.69	0.85	2.19	0.89
The Vologda Region	5.77	0.28	1.19	2.15	11.78	0.59	1.96	0.64
The Kaliningrad Region	4.28	1.05	4.8	0.55	30.3	0.58	1.93	3.02
The Leningrad Region	2.94	0.3	3.88	3.03	24.28	0.92	2.62	0.94
The Murmansk Region	4.98	0.21	2.71	3.42	16.2	0.73	1.52	0.54
The Novgorod Region	3.38	0.45	2.25	1.27	12.28	0.57	2.04	1.03
The Pskov Region	4.19	0.36	0.77	0.98	13.82	0.35	1.48	0.89
Saint-Petersburg	6.8	0.57	8.57	1.69	22.62	0.97	2.2	1.56

Source: the table is prepared on the basis of the data "Social and economic situation in the federal districts" - 2013 [12].

References

1. I.E. Denezhkina, D.A. Suzdaleva Effective crisis management, **10**, 1-8 (2012)
2. *Report on effectiveness complex assessment of the executive bodies of the territorial units of the Russian Federation by the end of 2013* (Ministry of regional development, Moscow, 2013)
3. E.S. Ivleva, Modernization of the Russian economy: forecasts and reality: Collection of papers of the international scientific and practical conference on 25th of March 2015, 96-100 (2015)
4. J. Fiksel, *Design for Environment: Creating eco-efficient products and processes* (McGraw - Hill Companies, 1996)
5. I. Kalinkova, Magazine "Expert North-West", **15(662)**, 3-6 (2014)
6. J. Kirshner, Review of International Political Economy, **5(1)**, 64 – 91 (1998)
7. Kh.S. Pakh, Journal of Vlast i upravlenie na Vostoke Rossii, **2**, 42 – 47 (2012)
8. Information on http:www.complexdoc.ru
9. Information on http://gov.spb.ru/static/writable/documents/2014/04/24/
10. Information on http://www.gks.ru/bgd/regl/b13_20/Main.htm (2014).
11. A.D. Ursul, Journal Bezopasnost' Evrazii, **1**, 443–456 (2001)
12. Information on http://gtmarket.ru/news/2014/03/13/6628 (march 13, 2014).
13. L.G. Vorona-Slivinskaia, Kh. S. Pak, Journal Problemi upravleniya riskami v tekhnosfere, **1**, 119 -124 (2015)
14. N.V. Zubarevich, E.A. Gordina Sozialnie rashody v Rossii: federalnii i regionalnie budzheti, Information on http//csils.hse.ru/2015.

15. A. Gorshkov, V. Murgul, O. Oliynyk, MATEC Web of Conferences, **53**, Article number 01045 (2016)
16. V. Murgul, Procedia Engineering, **117**, 808-818 (2015)
17. G. Radovic, V. Murgul, N. Vatin, Applied Mechanics and Materials, **641-642**, 634-638 (2014)
18. V.V Okrepilov, Ivanova, G. Standarty i Kachestvo, **12**, 62-68. (2004)
19. V.V Okrepilov, Standarty i Kachestvo, **3**, 94-96. (2003)
20. J. Ćetković, M. Knežević, M. Žarković, V. Murgul, N. Vatin, Applied Mechanics and Materials, **638-640**, 2465-2470 (2014)
21. V.V Okrepilov, A.Yu. Smirnov, Measurement Techniques, **56(1)**, 54-60 (2013)
22. V.V Okrepilov, Studies on Russian Economic Development, **24(1)**, 35-42 (2013)
23. V.V Okrepilov, Standarty i Kachestvo, **10**, 52-55 (2005)

Innovative Aspects of Ecological and Economic Management of Investment and Construction Activities for the Sustainable Development of the Region

Angela Mottaeva[1,*]

[1]Moscow State University of Civil Engineering, 129337 Yaroslavskoe sh. 26, Moscow, Russia

Abstract. In the presented study examines innovative approaches in the process of formation of investment - construction complex on the basis of observance of environmental - economic aspect. Considered the ecological and economic aspects of the management of investment and construction activities. Attempted to develop recommendations for the management of investment and construction activities in a sustainable direction of economic development. Developed control system of the environmentally oriented development of the regional investment - construction complex, with the application of the ecosystem approach. Defined: the location of the investment - construction complex in the regional ecological - economic system; the totality of the relations and interactions of the enterprises of investment - building complex with elements of economic and ecological subsystems. It is revealed that the main innovative characteristic of the regional ecological - economic system are the processes of relationship and interaction between economic, natural resource and environmental elements of the site, which form the economic and ecological subsystemsThe proposed innovative system of management of ecological-economic component of the investment-construction activity will allow to refuse from the implementation of investment projects, production models, which do not provide for environmentally friendly economic and social development of the territory, do not contribute to the creation of favorable and comfortable environment for human life.

1 Introduction

The solution to the problems of sustainable development of the region's economy directly depends on the state and development potential of the investment and construction sector. In investment and construction sphere are various enterprises and organizations whose activity is aimed at the formation of the human environment and the creation of material-technical base of enterprises of other sectors of the economy. The value of the investment

* Corresponding author: angela-1309.m@yandex.ru

and construction sector to ensure economic growth of the region due to the location of the investment-construction complex in the regional ecological-economic system [1-19].

It is not only a performing creative functions, but also is influenced by activities of enterprises investment –construction complex on the state of individual ecosystems and the environment in General. In investment and construction sector directly develops and implements investment projects, and this means that the level of greening investment planning depends on the state of the environment and its change, and, consequently, the conditions of life of present and future generations [3, 6]. One of the most important areas of achieving strategic goals of socio-economic development of society is the development of investment and construction activity, which generates the basic capital, converted to urban and architectural environment, changing the ecological conditions of human life, the overall quality of life [1]. Of great importance are the reliability and safety of buildings and structures, environmental quality and ecological safety of construction production, as well as the impact of the environment on human health and its socio-psychological behavior. In addition to the direct impact on the ecological systems and environment of enterprises investment-construction complex indirectly influence the conditions of human life through developing design solutions that are in the process of their implementation are translated into real construction projects.

Therefore, the need of this study, due to objective necessity of development of methodology of management of investment-construction activities, appropriate balanced ecologically - oriented model of economic development, development of practical recommendations for the regulation of investment and construction activities in a sustainable direction of economic development.

The study theoretical and methodological bases and innovative aspects of ecological-economic investment-building, economic problems, effective management of the development of investment-construction complex in modern conditions, research on the possible use of innovative ecological materials in construction investment activities, dedicated to the works of scientists works Asaul V. V[2], Nuzhina I.P. [3], Gorbunov A.A. [4], Ivanov S.N. [4], Grebennikov A.A. [5], Mottaeva A.B. [6], Schumpeter Josef Alois. [8], Witt U. [9], Lena E. Bygballe, Malena Ingemansson, [10], Anttonen Markku, Minna Halme, Eeva Houtbeckers, [11], Alfredo Serpell, Romina Alvarez [12], Nicole Parthenopoulou K., Malindretos Michael, [13].

2 Methodology

An essential element of the regional investment - construction complex is a building product, as an object of interaction of subjects of regional investment - construction complex in the implementation of investment and construction activities.

The ecosystem approach in the management of economic development means a transition in management from the economic system to ecological and economic. Place of investment - construction complex in the regional ecological - economic system is determined by the totality of the relations and interactions of the enterprises of investment - building complex with elements of economic and ecological subsystems (see in Fig. 1) [3], the role of the Converter of investment resources in the form of capital investment and value of investment - construction complex in the formation of living environment, the development of individual elements and regional ecological - economic system as a whole. Managing the development of regional eco - economic system through the regulation of economic behavior of economic entities. The relationship of investment - construction complex is regulated using the tools and mechanisms of the institutional environment, acting as the balancing factor for development [2].

Thus, the mutual influence of subsystems, investment and environmental policies that are implemented by the economic subjects, identify trends in socio-economic development of the region as the ecological - economic system, the higher the level of development which imposes new requirements for the condition and development of all elements, including investment - construction complex.

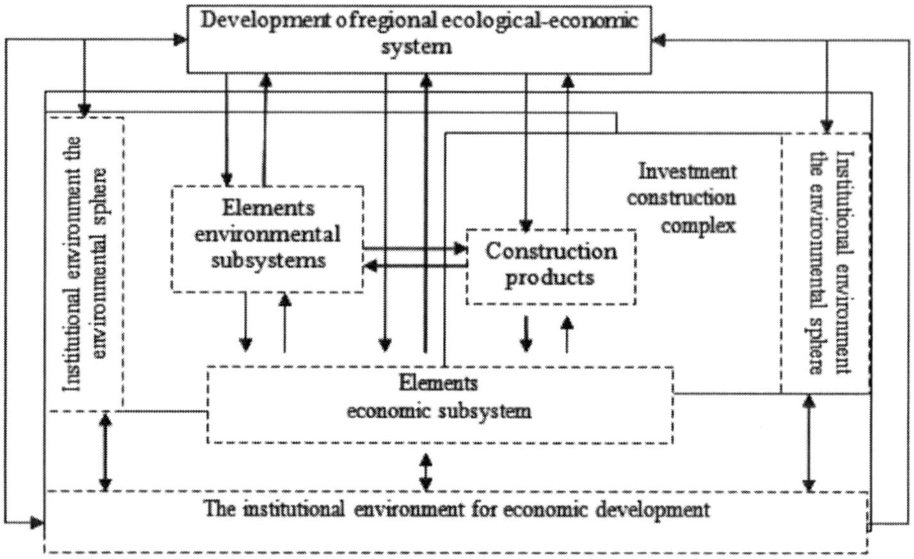

Fig. 1. Model of development and relations suit in the regional ecological-economic system.

The contents of the ecosystem approach is revealed in the formulation of strategic objectives and functions of ecologically - oriented management of investment and construction activities, on the basis of the relationship of investment processes, conditions of human life, the state of ecological systems and the environment in General. The main functions of ecologically-oriented management of investment and construction activities include:

1. The analysis of activity of the enterprises of investment - building complex, taking into account environmental factors in the development, which is based on qualitative and quantitative characteristics of ecological and economic interactions of investment - construction complex in the regional ecological - economic system.

2. The planning and design of investment projects in accordance with the requirements of greening the production processes of creation and exploitation of construction products. The level of greening investment planning determines operational indicators of a facility, including its resource-saving features.

3. Organization of ecological support of investment - construction activities, construction and installation works through the use of environmentally friendly technologies in the construction industry and environmental initiatives.

4. Control over execution of managerial decisions and regulation of investment and construction activities in accordance with objectives of ecologically oriented management.

We know that due to territorial tightness, the use of local raw materials for the production of basic construction materials and other features of construction production, investment and construction activity has a distinct regional character. The development of environmentally - oriented model of development of investment and construction activities, as an integral part of the overall strategic objectives of the formation of a model of socio-economic development of the region, predetermines the positioning of the region as a "regional ekologic-and-economic system" [4, 6]. The main system-forming sign of a regional ekologic-and-economic system are processes, due to the interrelation and interaction of economic, natural resource and environmental elements of the site, the totality of which forms an economic and ecological subsystems, respectively. The nature of the interactions of elements and subsystems defines a condition and dynamics of development of both separate elements, and the regional ecological - economic system as a whole. Accounting for this circumstance predetermines necessity of selection in the composition of regional eco - economic system of the subsystem of ekologic-and-economic interactions [5]. Thus, the regional ekologic-and-economic system should be regarded as an integrated territorial system of the meso-level, combining a set of interacting heterogeneous elements forming the environmental and economic subsystems, as well as a set of relationships between elements and processes under certain climatic and spatial-temporal parameters [2]. As the underlying determinants of environmental and economic aspects of investment and construction activities, highlight the natural, environmental, economic, social, regulatory (see Fig. 2) [3]. Ekologic-and-economic cooperation in investment - construction complex in the regional ekologic-and-economic system based on the use of natural and environmental resources, the impact on the environment and society.

3 Results

Using as methodological assumptions of the functions of the environment, the author proposed a classification of the ecological and economic interactions of investment and construction complex types, depending on their area of occurrence in the course of investment and construction activities, which includes: natural-resource interaction; technological collaboration; social interaction; institutional interaction; economic cooperation arising from economic expression of the totality of the processes of ecological-economic interaction of subsystems and elements using the scorecard.

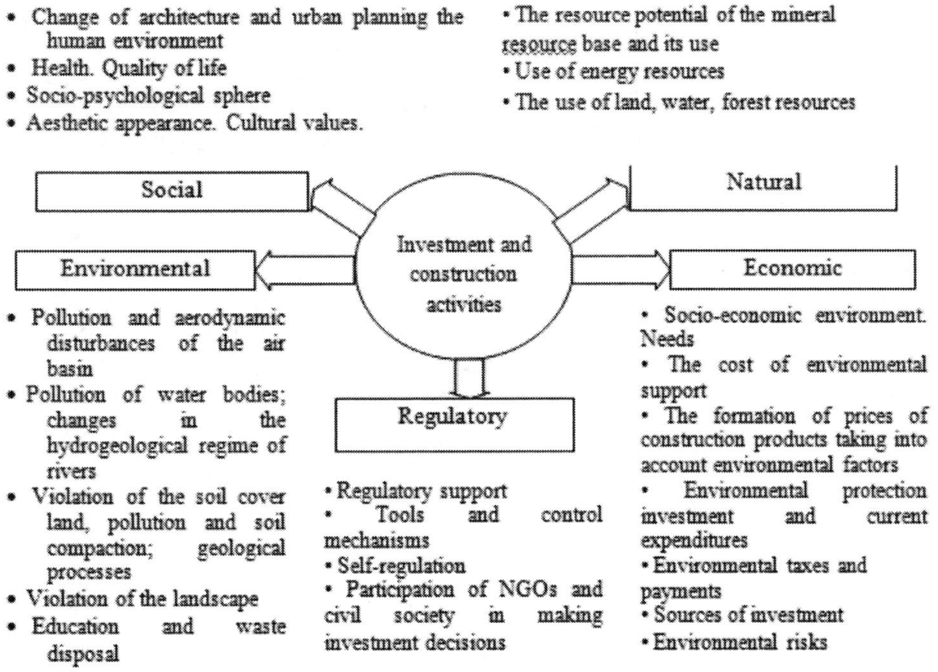

Fig. 2. Basic factors determining the ecological-economic aspects of investment-construction activities.

Considering the greening of investment and construction activities in the practical aspect, singled out the greening of production and greening of the mechanism of regulation of investment and construction activities. Precondition for the implementation of ecologically acceptable investing is not only greening the production processes, but the mechanism of regulation of investment and construction activities. In this case, greening means the mainstreaming of environmental orientation of the whole set of methods and tools used for the regulation of investment and construction activities, including legal, technical and economic regulation [3,5].

Theoretical and practical aspects of the greening of investment and construction activity are interrelated. System events the greening of investment and construction activities is the practical implementation of the ecosystem approach in the management of investment and construction activities. The implementation of the measures determined by the capabilities of the institutional environment of ecological and economic regulation of investment and construction activities, and the boundaries of their implementation – based on the life cycle of construction products. The capabilities and performance of ecological and economic regulation of investment and construction activity depend on effective forms of interaction between authorities, construction businesses and companies.

The proposed system of environmental-economic component of the investment-construction activity will allow refusing from the implementation of investment projects, production models, which do not provide for environmentally friendly economic and social development of the territory, do not contribute to the creation of favorable and comfortable environment for human life.

4 Discussions

Currently in Russia there are not enough comprehensive studies on ecological and economic regulation of investment and construction activities in the Modernization and actualization of the objectives of a green economy lead to the need to develop a balanced model of development of the national economy, individual economic sectors, environmentally competitive industries and activities, including investment and construction activities [7]. According to the state ecological monitoring, more than 54% of the urban population of the Russian Federation living in conditions of high and very high levels of air pollution. More than 40% of the population faces problems of water quality, on average, the Russians are losing up to one year of life due to air pollution, the magnitude of the economic damage caused by the impact on population health in connection with air and water pollution, is estimated by experts not less than 4-6% of GDP [8]. Therefore, the basic principle of strategic development, investment and construction sector in the twenty-first century should be economic parity, social interests and environmental needs of society.

In recent time gained significant scientific material and performed empirical research of economic problems of increase of efficiency capitalocratic investment, assessment of competitiveness of construction enterprises, improve the methods of management of investment projects, regional investment and construction complexes, etc. However, remain insufficiently explored aspects of ecological and economic regulation of investment and construction activities, the problems of economic estimation of investment projects taking into account environmental factors and investitsionno - building complexes. The transition to ecologically-oriented model of development necessitates the development of an integral concept of environmentally sound development of investment and construction sector, and in turn is the basis of mechanism and tools of ecological and economic regulation of investment and construction activities in the region [2,4,5].

According to the author, the greening of investment and construction activities should be considered in theoretical and practical aspects. In the theoretical aspect of the greening of investment and construction activities is the concept of environmentally sound development of investment and construction activities, the implementation of the main provisions of which will ensure a harmonious combination created a civilized architectural and town-planning environment of human life and the preservation of environmental quality, both for current and for future generations. The main directions of transition to environmentally sound development of investment and construction activities are:

– implementation of the ecosystem approach in managing the development of investment and construction activities. The formation of strategic programs of development of investment - building complex, taking into account priorities of regional environmental policies to increase energy efficiency and the reduction of nature intensity of socio-economic development. In the framework of the ecosystem approach the subjects of investment and construction activity taking management decisions with a view of the close relationship of investment processes, urban and architectural conditions of human life and environmental;

– the formation and development of the institutional environment of ecological and economic regulation of investment and construction activities in the region. Environmentally-oriented behavior of the subjects of investment and construction activities defines the institutional environment that shape organizations and institutions, legislative acts, formal and informal constraints, mechanism and tools of ecological and economic regulation of investment and construction activities;

– life cycle assessment of construction products, accountability of economic entities of investment and construction complex for environmental performance, which covers all points of contact with the environment, arising during and in connection with the establishment and operation of construction products;

– formation of effective system of interaction between authorities, business and society in the process of preparation, discussion, decision-making and implementation of investment projects. The main purpose of such cooperation is to harmonize the interests of the state, society and the construction business in the field of investment and construction activities in line with priorities of social and environmental policies. Achieving the goal of environmentally sound development of investment - building activity involves, along with the state and market regulation, active participation of civil society.

5 Conclusion

Thus, the conditions of ecologically acceptable development of regional investment - construction complex are: minimization of construction products are nature-based sustainable use of natural resources within the ecological life cycle; maximization of the generalized indicator of the ecological-economic efficiency of investment projects based on application of the best available technologies and environmental initiatives; modeling the equilibrium state, in which the increase of volume of performed works, manufactured products and the cost of ecological support of investment - construction activity does not violate the financial stability of economic entities.

The operation of construction companies as ecological - economic system necessitates evaluation of the effectiveness of activities based on ecological-economic analysis for structured components: an analysis of organizational aspects of environmental support productive activities; analysis of impact on components of the regional ecological-economic system; analysis of the ecological and nature intensity of work; analysis of the use of resources, providing ecological support of investment-construction activities; analysis of the impact on economic and financial performance of enterprises; investment analysis [2,3,5].

The practical significance of the study lies in the feasibility of using proposed scientific approaches, techniques in the practice of management, organization and planning of activity of the enterprises of investment-building complex, taking into account environmental aspects of development, in the activities of state administration bodies in the development of strategic plans and programs of development of investment and construction sector, relevant social and environmental priorities of the region development, the implementation of ecological-economic monitoring of construction industry.

The author believes that mainstreaming environmental knowledge in the regulation of investment and construction activities is possible through the expansion of the range of integration of ecoinstitut in the institutional environment of investment and construction sector and the institutional environment of ecological and economic regulation of investment and construction activities. The completeness of the integration processes is determined by the degree of greening of the totality of tools and methods of regulation of investment and construction activities.

References

1. Information on: www.youngscience.ru
2. V.V. Asaul, *Scientific bases of the concept of economic innovation strategy building organizations on the basis of integration processes* (SPbGASU, SPb, 2005)
3. I.P. Nuzhina, *Economy of Natural Resources and Conservation: a collection of articles XII International nauch.- Practical. Conf.*, 23-27 (2009)
4. A.A. Gorbunov, S.N. Ivanov, A.N. Asaul, Peter Fund, (1999)
5. A.A. Grebennikov, Regional Economics and Management, **6**, (2006)
6. A.B. Mottaeva, The methodology of the spatial distribution of business organizations in the region based on the development of transport infrastructure (2012)
7. A.B. Mottaeva, International Journal of Applied Engineering Research, **23**, 43446-43449 (2015)
8. On the way to sustainable development of Russia, **43**, 10 – 11 (2014)
9. J.A. Schumpeter, *Essays on entrepreneurs, innovations, business cycles, and the evolution of capitalism, Clemence* (1989)
10. U. Witt, The Paperson Economics and Evolution, **0605**, (2006).
11. L.E. Bygballe, M. Ingemansson, Industrial Marketing And Management, **43**, 512-524 (2014)
12. M. Anttonen, H. Minna, E. Houtbeckers, Journal of cleaner production, **45**, 89-103 (2013)
13. A. Serpell, R. Alvarez, Procedia Engineering, **85**, 464-472 (2014)
14. N. Parthenopoulou, M. Malindretos, Original Research Article Materials Today: Proceedings, **3**, 898-912 (2016)
15. M. Merklein, M. Geiger, Journal of materials processing technology, 532-536 (2002)
16. E. Aronova, N. Vatin, V. Murgul, Procedia Engineering, **117**, 771-779 (2015)
17. G. Radovic, V. Murgul, N. Vatin, Applied Mechanics and Materials, **641-642**, 634-638 (2014)
18. M. Jocovic, B. Melovic, N. Vatin, V. Murgul, Applied Mechanics and Materials, **678**, 644-647 (2014)
19. M. Lazarevska, M., Cvetkovska, M., Knezevic, (...), V. Murgul, N. Vatin, Applied Mechanics and Materials, **627**, 276-282 (2014)

The Psychoanalytic Interpretation of the Organizational Environment as a Management Tool for Sustainable Development

Elena Khripko[1,*]

[1]Moscow State University of Civil Engineering, 129337 Yaroslavskoe sh. 26, Moscow, Russia

Abstract. The article exposes contemporary materials and structures for sustainable development of organizational environment. Psychoanalytic modeling of organizational behavior makes it possible to identify out reflection, unconscious tendencies in individual, group and corporate behavior. This enables to significantly increase the effectiveness of measures for personnel management. Organizational Environment Researches base on psychoanalytic theory of object relations.

1 Introduction

Sustainable development of organization depends on the quality of application of managing function. And this issue in its turn depends on the functioning level and status of organizational environment.

The present research offers a new level of understanding of problematics of management decisions in organizations related to training, development, selection, motivation, as well as issues of staff efficiency as a whole. At present, systematic character have the problems which in general can be summarized as follows: the results of organizational activities in the field of human resources management aimed at increasing of activity efficiency do not match the initial requests.

Training and development staff programs, corresponded to organization standards, and performed with high quality, do not reach the target level of development of competencies.

In turn, the results of personnel evaluation may also be in conflict with the results of training activities, do not correlate with real image of organizational dynamics, do not provide complete information to answer the most important management problems. The revision of educational programs, organizational changes and personnel changes often do not achieve significant results. Without answering remains the question about what educational programs will bring the quality of organizational collaboration at the appropriate level?

Psychoanalytic approach allows us to put forward the hypothesis that there are hidden processes which affect the organizational environment, causing these contradictions, despite the taken measures. But the failures of personnel policy are not random at all, but

* Corresponding author: geleda@ya.ru

are due to underlying causes, and there is a cause-and-effect relation between the manifestations of organizational culture and individual psychological reactions of employees [1-29].

2 The psychoanalytic interpretation of the organizational environment

Consideration of organizational processes reveals another contradiction: personnel evaluation provides some information on the development of competencies, but on practice of the organizational behavior it is translated its another level. Often there is a situation where in spite of all taken measures appear a stable for a given organizational environment patterns of destructive behavior. The paradox of destructive stereotypes is that they are translated by employees that are different by their psychological characteristics. In such cases, any measures will be ineffective unless they are based on the idea of hidden content of organizational dynamics and do not include the tasks of identifying of the unconscious nature of destructive reactions.

Let me illustrate as noted above through the example of personnel evaluation. The results of personnel evaluation have revealed a certain level of necessary skills development, but in reality the same employees don't show the desired level of competence, on the contrary, their organizational behavior is deformed by unpredictable patterns of behavior and destructive influence on the surrounding activities. [2]

These facts go to prove that important aspects havent't been taken into consideration of personnel evaluation and were not displayed in its results. The lack of problems understanding of similar cases is that competency is the basis of a rational level among employees, its rational organizational behavior is modelled without notice of distortions caused by the unconscious tendencies.

The psychoanalytic interpretation of unconscious tendencies in the organization provides an opportunity to improve the quality of management decision making. Detection and interpretation of these non-reflexive stereotypes of organizational behavior significantly allows to improve the corporate governance as a whole.

The psychoanalytic research of social and group phenomena was proposed by Freud. The development of psychoanalytic theory and practice was as a basis of the modern psychoanalysis, where some fields of research were organizational processes. The psychoanalytic researches in organizational context are the most effective to me that are based on object relations theory [3, 4].

The psychoanalytic research of social and group phenomena is provided by means of met psychological models, reconstruction of the latent content and quality of object relations. The structure modeling from the perspective of the psychoanalytic approach presents the organizational content from the perspective of structures like the ego, the superego and the id. [5] The interaction between individuals and groups creates a specific range of defense mechanisms for all organization, and formal and informal organizational relationships get a certain quality. Analyzing communication skills, it becomes possible to create such a model where the organizational behavior is described as a result of the dynamic interaction between structures like the ego, the superego and the id.

Functioning of the organizational ego is simulated on the basis of information about the nature of communication, translation mechanisms of information, the interaction with the environment and the nature in organizational relationships.

Functioning of the organizational superego shows the organizational structure and the way it is related to the communication system. The nature of the superego functioning is found in the study of the incentive system and the system of penalties, methods and forms of control of assignments at different levels of management.

The id energy is manifested in traditions, communication affects. Conflict situations are the most informative in this context. Applying the structure modeling in the organizational context makes it possible to predict what organizational affects will be mobilized, what defense mechanisms will be involved, what conflict trends will be engaged into a given situation, and the most important thing is what hidden tendencies cause the distortion in the organizational context and what ways it is advisable to reduce their impact.

The system psychoanalytic study of organizational communications allows for the conclusion about the quality of object relations in the organizational context. Understanding the quality of object relations allows to predict the consequences of certain organizational measures to a high precision, to choose the optimal paths of implementing organizational changes.

The psychoanalytic position in the organizational context can attribute a symbiotic, separational or triad character of object relations. This view involves a range of assumptions. The reference of the organizational context to one of these three types is carried out by the fact what quality of object relations in organizational communications and what defense mechanisms are mainly interpreted.

Symbiotic cultures are characterized by the lack of borders differentiation in the the employee-organization system. The organizational context is marked by a number of defense mechanisms. First of all, it concerns imagination. At multiple communication levels you can view the prevalence of imagination (mostly about the future) than the interpretation of the organizational reality. In addition to the above there is a total control of information and activities.The discussion of innovations is accompanied by a primitive idealization and devaluation. Most of organizational situations are resolved through catharsis - rapid untested and unprepared appointments and dismissals, organizational reorganizations and other such actions. They show the desire to obtain instant results by radical and rapid problem resolution.

In the separation cultures you can trace diffuse boundaries in the employee-organization system, object relations are manifested with the lack of integration, the diffusion of responsibilities. It is clearly traced the perception of internal and external objects as absolutely "good" or absolutely "bad". Among the defense mechanisms the extrajection brings to the forefront. This defense mechanism is clearly appeared within the interaction with different units, whereby the organizational behavior of employees is characterized by instability and unpredictability.

The organizational context has the triode quality of object relations and shows mature defense mechanisms, and the organizational culture has an integrated identity. Managerial decisions are mainly thought out and widely discussed. The organizational conflicts of interests get a complex nature involving many participants, they can exist for a long time in a latent form.

In this article I want to fix on the most striking flashes, present a sort of sketch of object relations, as a detailed psychoanalytic description of the typology of corporate cultures requires the consideration of a considerable amount of facts that will be implemented in future publications on the subject.

In practice the organizational consulting of psychoanalytic research is carried out according to the request that the manager makes an inquiry.In some cases, the inquiry covers the organizational dynamics as a whole, and there is enough information about the psychoanalytic modeling in the organizational context.

In other inquiries the problem is specified, and the focus of the study is limited for this problem ofinformation content. The research should be carried out exclusively in the request scope, because the process of identification and interpretation of the hidden trends inevitably meets resistance from the organizational context, even in cases when the situation is complex and requires instant effective actions. At the moment the organizational context with a particular quality of object relations has a definite transfer of resources of certain hidden trends on a conscious level. The existence of an inquiry to change the organizational context does not indicate the existing availability to the realities of these changes. The realization of latent tendencies, the interpretation of a cause-and-effect relationship of different organizational phenomena is one of the most important working phases, but more difficult job is to change the current situation, its testing, the consideration and the implementation into the organizational context.

3 Conclusion

In practice of organizational consulting stick to the following steps:

- the discussion of the original inquiry, the detection of the scope of activities, the formulation of marking criteria results;
- the collection and analysis of information under the frame of the inquiry;
- the psychoanalytic modeling of an organizational context on the basis of provided information and the interpretation of the latent content;
- the discussion of options for possible changes;
- the evaluation of changes and the study on resistance in the organizational context;
- the implementation of coherent changes in the form of personnel decisions, development programs, and etc.;
- the feedback for the direction of this inquiry.

When applying the psychoanalytic approach, we can achieve the reduction of risks in management decision making, staff recruitment and key appointments. The detection of previously hidden cause-and-effect relationships makes it possible to clarify the worthwhileness of organizational effects and exercise an option of these effects to a qualitatively new level.

References

1. E.V. Romanova, Izvestiya VGPU, **89**, 115-123 (2014)
2. N.G. Miloradova, A.D. Ishkov, *Psychology of management activity: changing approach* (MGSU, Moscow, 2014)
3. O. Kernberg, *Ideology, Conflict, and Leadership in Groups and Organizations* (Nezavisimaya firma "Class", Moscow, 2015)
4. N. McWilliams, *Psychoanalytic Diagnosis: Understanding Personality Structure in the Clinic Process* (Nezavisimaya firma "Class", Moscow, 2006)
5. E. Berne, *Leadershipand group. Structure and Dynamics of Organizations and Groups* (Izdatel'stvo «E», Moscow, 2016)
6. J. Ćetković, S. Rutešić, M. Zarković, M. Knežević, , N.Vatin, Procedia Engineering, **117 (1)**, 780-790 (2015)
7. S. Rutešić, J. Ćetković, M. Žarcković, M. Knežević, N. Vatin, Procedia Engineering, **117 (1)**, 905-915 (2015)
8. S. Rutešić, J. Ćetković, M. Knežević, M. Žarcković, N. Vatin, Procedia Engineering, **117(1)**, 642-650 (2015)
9. M. Jevrić, M. Knežević, J. Kalezić, N. Kopitović-Vuković, I. Ćipranić, Tehnicki Vjesnik, **21(4)**, 873-879 (2014)
10. N. Pavličić, M. Perazić, D. Durić-Jocić, M. Knežević, Journal of Applied Engineering Science, **12(1)**, 11-18 (2014)
11. J. Ćetković, M. Knežević, M. Žarković, V. Murgul, N. Vatin, Applied Mechanics and Materials, **638-640**, 2465-2470 (2014)

12. M. Lazarevska, M. Milanović, M. Knežević, M. Cvetkovska, A. T. Gavriloska, T. Samadzioska, Journal of Applied Engineering Science, **12(1)**, 63-68 (2014)
13. J. Cetkovic, M. Knezevic, I. Vujovic, J. Cerovic, Technics Technologies Education Management, **7(4)**, 1646-1654 (2012)
14. M. Knežević, Journal of Applied Engineering Science, **10(1)**, (2012)
15. J. Cetkovic, S. Rutesic, T. Hanak, M. Knezevic, B. Melovic, Technics Technologies Education Management, **7(1)**, 285-293 (2012)
16. B. Kovačič, R. Kamnik, A. Štrukelj, Promet - Traffic - Traffico, **22(1)**, 15-22 (2010)
17. N. Gubeljak, J. Predan, D. Kozak, J. Tuma, B. Kovačič, P. Konjatić, J. Sertić, Strojarstvo, **51(4)**, 263-271 (2009)
18. B. Kovačič, R. Kamnik, M. Premrov, N. Gubeljak, J. Predan, Z. Tišma, Strojniski Vestnik/Journal of Mechanical Engineering, **54(5)**, 364-371 (2008)
19. B. Kovačič, R. Kamnik, International Journal for Engineering Modelling, **20(1-4)**, 77-84 (2007).
20. N. Lipovac, A.M., Jandriček, Prostor, **19(2)**, 443-455 (2011)
21. V.V Okrepilov, M.V. Leonidovich, Asian Social Science, **11(7)**, 312-325 (2015)
22. V.V Okrepilov, V.L. Makarov, A.R. Bakhtizin, S.N Kuzmina, Economy of Region, **2**, 301-313 (2015)
23. V.L. Kvint, V.V Okrepilov, Herald of the Russian Academy of Sciences, **84(3)**, 188-200 (2014)
24. V.V Okrepilov, V.N. Krutikov, G.I. El'kin, Measurement Techniques, **57(2)**,109-116 (2014)
25. V.V Okrepilov, A.Yu. Smirnov, Measurement Techniques, **56(1)**, 54-60 (2013)
26. V.V Okrepilov, Studies on Russian Economic Development, **24(1)**, 35-42 (2013)
27. V.V Okrepilov, Standarty i Kachestvo, **10**, 52-55 (2005)
28. V.V Okrepilov, G. Ivanova, Standarty i Kachestvo, **12**, 62-68 (2004)
29. V.V Okrepilov, Standarty i Kachestvo, **3**, 94-96 (2003)

Diagnostics of materials and structures of heating system as a component of ecological management in a modern city

Natalia Verstina[1,*]

[1]Moscow State University of Civil Engineering, 129337 Yaroslavskoe sh. 26, Moscow, Russia

Abstract. The maintenance of materials and structures of the underground space in modern cities which contains many engineering structures is considered in the article. It is claimed on the example of quantitative data that the underground structures exert a great influence on the ecological situation of the city. The matters of the research included the analysis of the main heating systems as the most responsible city structure; the conclusions obtained can be also used for smaller heat pipelines. The article includes the comparisons of modern technical methods of diagnostics of materials and heating systems structures according to the results of experiment, as well as the characteristic of norms of control of the materials condition at the stages of construction and maintenance applied in the developed countries. On this basis, the author has developed and presented a system of diagnostics taking into account the alternation of measures during the heating period as well as the season when heat supply is not required. For an effective ecological management in modern cities, it is suggested to create the automated data base of materials and structures of the heating systems on the basis of the developed scheme of their conditions and events, which needs to be used for forecasting of environmental risks and planning of recovery works with consideration of special features of the maintenance processes. The direction of the further research is defined: it is the expansion of the stated approaches on other types of underground communications in modern cities.

1 Introduction

The extensive discussion of the problem of changes of the ecological situation in the world to the deterioration which is being carried out in the expert community raises some questions of ensuring efficiency of functioning of structures in the city infrastructure in a new aspect and defines the new trends in the search of reserves for providing these processes. The transition of the economy from the resource-and-raw type to the resource-and-innovative type, which has begun according to the provisions of the "Concept of the Long-Term Social-and-Economic Development of the Russian Federation till 2020" is the

* Corresponding author: verstina@mail.ru

incentive to providing modern decisions on high characteristics of environmental friendliness of structures in the Russian Federation [14]. In order to solve the problem of reduction of the environmental risks in the ecological management complex it is important to consider that the underground space of modern cities has many engineering structures which have significant effect on the ecological condition in the city. From this point of view the heating systems, considered in the completed research, are the most dangerous. In the ecological aspect the main danger, produced by the heating systems for the urban development consists in the possibility of obtaining unorganized industrial emissions in the case of emergencies, when the heat carrier comes directly to soil in the form of streams of hot water (heat carrier), which is under pressure, without any cleaning. Scales of this problem in the Russian Federation are considerable and the experience of overcoming of these problems on a scientific basis can be useful to other countries which urban development has underground communications with the hot heat carrier of industrial or civil purpose. In the Russian Federation more than 17 thousand plants of heat supply work in the city settlements where heating systems have general extent of 172 019.3 km in two-pipe calculation (the direct giving pipe and the return one) [8]. While characterizing this problem more particularly it is necessary to provide the following data, defining its technical and economic aspects in the scales of the Russian Federation: according to Rosstat, in 2014 the quantity of the heating and steam systems, demanding repair, made up 49 738.1 km, that is more than one third of the general extent. At the same time during one year the considerable number of emergencies with the accompanying ecological damages has been noticed (6 782 cases), as the result 3 812 km of heating systems have been replaced in 2014. Taking into account that the estimated cost of replacement of 1 running meter of heating systems makes 10 thousand rubles, it is possible to estimate the total cost as 38 billion rubles. The similar quantitative assessment of "the generalized character" can be directly projected on any enterprise's activity in a modern Russian city in which the underground space has heating systems. At least one tenth of the heating systems property demands urgent repair and comprises some environmental risks; costs of their localization within the correct organization of the city management, according to the author, could be reduced by the size of about 30 percent. In the aspect of ecological management of the modern city existence of such risks supposes the promotion of measures of preventing character for diagnostics of heating systems which main elements are the diagnostics of technical condition of the structures of the heating systems under maintenance and the control of carrying out the measures for ensuring the efficiency of their functioning.

2 Methodology

For the definition of the directions and forms of the environmental risks related with the decrease in the properties of materials as well as with the failures of the heating systems structures, which need to be considered in the organization of the modern city management the conditions of the operating systems, methods of diagnostics of the technical parameters and the organization of their maintenance have been analyzed in the research. It is suggested to distinguish the conditions of the operating structures and the events, changing them:

- serviceable, at which the structure conforms to all the requirements of the specifications and technical documentation (STD);
- defective, at which the structure does not meet all the requirements of the STD;
- efficient, at which the values of all the parameters characterizing the ability to carry out the set functions; conform to requirements of the STD;.
- inefficient, at which the value at least one parameter characterizing the ability to carry out the set functions does not conform to requirements of the STD.

Full exhaustion of properties of materials and structures during the heating system maintenance is characterized by the achievement of its limit condition. The limit condition is the condition of a structure, when its further use according to its purpose is inadmissible or inexpedient, otherwise the restoration of its serviceable or working condition is impossible or inadmissible. The sign or the set of signs established by the STD act as the criteria of the limit condition of the heating system.

From the point of view of the assessment of materials and structures of the heating system some concepts were differentiated on the basis of observing the maintenance processes: the concepts "serviceability" and "defective" are wider than the concepts "efficiency" and "inefficiency". The maintenance of a heating system in the presence of dot corrosion on the metal surface of the pipeline or fittings which do not cause any considerable leakages of the heat carrier and not reducing the mode of its transportation, but which is already interfaced to damages of the ecological condition of the surrounding underground space can be an example. Defective condition of a heating system is seen, for example, in the violations of heat- and waterproofing layers of the pipeline, partial destruction of seams of reinforced concrete structures of channels and chambers.

But at the same time the heating supply system is efficient as it carries out its main function - which is the transfer of heat and no harmful ecological effects on the environment emerge. The transition of the system under maintenance from one condition into another results from some changes - damages and failures which are presented as environmental risks of different types. The events, consisting in the violation of the serviceable condition of the structure at the preservation of its efficiency, are considered to be damages. The events, breaking the efficiency of the structure, are considered to be failures.

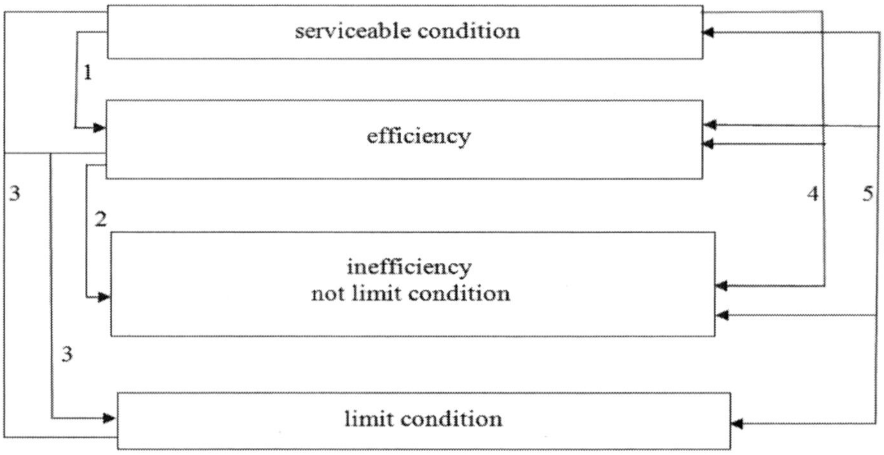

Fig. 1. Scheme of the main conditions and events of a heating system.

Deterioration in operational properties (event 1-3)

I - damage, 2 - failure, 3 - ineradicable violation in the heating supply maintenance (discrepancy to safety requirements, obsolescence, etc.);

Restoration (improvement) of the system properties:

4 - service and maintenance, 5 - repair.

The main difference among the aforesaid events is the extent of the influence on environmental of the underground space surrounding the structure and, respectively, the importance of the consequences for the managerial decisions making. It is considered, that the most serious consequences have the violations in the heating system maintenance, related with the metal corrosion of the pipeline - that makes up to 80 percent of the total number of failures in the practice of heating system maintenance in the Russian Federation.

Along with the corrosion processes the decrease in the maintenance properties of the heating systems happens because of the reasons of not corrosion character related with the deformation of elements, defects of metal, etc. The article includes the results of the author's research aimed on the introduction of the quantitative indicators, characterizing failures and damages of the heating system, which depend on the quality of their construction. At the same time the results of main heating systems survey of the city of Moscow, which due to the carried out functions cause the greatest environmental risks, were used. As a result of the completed analysis it was proved that the direct sites of the pipeline are the most common subject of failures. Fistulas in the material of the pipe and its gaps, which make up to 75-90% of the total number of failures of the heating systems, belong to the failures of corrosion character. They can be noticed on all the elements of the systems, which were produced from metal: pipelines, fittings, compensators, etc.

In order to find out the interrelation of the corrosion failures and the quality of construction the analysis of the reasons of emergence of failures was carried out. Seven groups of the reasons causing corrosion failures, four of them are the consequence of low-quality completion of construction works at the heating systems, were distinguished. The quality of such works as waterproofing of walls and seal of seams of channels and chambers (up to 62% of the total number of failures) influences most sufficiently. The second and the third places are held by the works on the passing drainage, sandy basis. The violation the requirement to the quality of welding works also influences emergence of corrosion failures, as the tightness is not provided [15].

Studying of allocation of these failures throughout the term of heating systems maintenance (25 years in the Russian Federation) has shown process of loss of opportunities of resistance of materials of metal elements of a thermal network of corrosion and, respectively, increase of risks of ecological damages. The first corrosion failures appear in 5-6 years from the beginning of the heating system maintenance. This results from the fact that the speed of corrosion processes averages 1 mm a year therefore at a thickness of a side of pipes ranging from 5 to 12 mm emergence of corrosion failures happens in the range from 5 till 12 years. It is noted that during from 5 to 16 years of value of an annual increment of a stream of corrosion failures constantly, i.e. the period of normal operation of a thermal network is observed. At this time corrosion failures are the consequence of poor quality of construction. Since 16 years from input of the heating system maintenance, a stream of failures increases in process of "aging" of materials of heating systems. From this age dominating in development of corrosion of materials of the heating systems there are maintenance factors - the period of wear and increase of environmental risks begins. Therefore owing to the nature of course of negative processes hidden from the direct survey, there is the following question, important for ecological management in the modern city – the organization of diagnostics of the condition of materials and structures of the heating systems.

3 Standards

In the methodical aspect the starting point of the diagnostics, which is presented by the standards, used for carrying out control measurements is important. The comparative analysis of the standard requirements imposed for this type of technical structures in the Russian Federation and abroad, which was carried out by the author and covered 322 standard documents, is interesting [10]. According to the results of the comparison of the STD on the nondestructive control of pipelines with those on the direct access to the pipelines (availability control of the pipeline is provided from all the directions), the essential difference was not revealed. Many Russian standards on the nondestructive control just are the foreign standards accepted for the use in Russia, since according to article 13 of the Federal Law "On Technical Regulation": "Documents in the field of standardization used in the territory of the Russian Federation include: the international standards, regional standards, regional sets of rules, standards of the foreign states and sets of rules of the foreign states registered in Federal Information Fund of Technical Regulation and Standards." [2, 5, 8] In the STD studied there are more than 40 types of nondestructive control of the main part of the heating systems, (i.e. the welded pipes material), carried out by the means of the instrumental methods and visual survey. That is a rather detailed study of possible options of control, use of them theoretically allows to exclude risks of emergencies during the heating systems maintenance and the accompanying environmental risks. At the same time, according to this list, the main part of nondestructive control is presented by the processes of pipes production (the automated technologies), and to a lesser extent by the check and acceptance of structures during the construction. Both foreign and Russian STD make the main accent on materials and that consists in the following fact: the pipes arriving on construction site are to already have some data on the completed control in their accompanying documentation. For example, according to the American standard. ASTM (ASTM A 106 Standard Specification for carbon Steel Pipe for High-Temperature Service) [3]. That means that both in foreign and domestic STD the priority of the safe maintenanse of the structures consists in meeting the strict requirements to the quality of structure. The only difference is that in the USA and EU the quality of installation is controlled by the independent expert (inspector) [3], and in the Russian Federation it is controlled by the commission of representatives of the operating enterprise, the contractor and Rostechnadzor. Formally the organization of control at this stage appears to be approximately equal, but the further maintenance of the structures of the heating systems proves that in the Russian practice more defects of structures are detected at this stage than in other countries. It was the very subject of the further analysis according the assumption that the revealed causes of failures will promote the development of more perfect materials, structures and methods of technical diagnostics.

In this regard the generalized characteristic of the pre-maintenance and maintenance processes concerning the equipment (including heating systems) in the foreign practice, minimizing the risks of the decrease in properties of materials, failures of structures and accompanying ecological damages, was prepared. According to the Directive PED, the equipment with the maximum admissible pressure of PS of more than 0.5 bars is the subject of obligatory certification. [15]. EU have no right to forbid or to limit commissioning of the equipment in the market of the EU on the basis of danger because of the existence of technological processes of work under pressure. At the same time, the interested parties are obliged to apply all the nesessary measures of control of allocation in the market of the products, which is meeting the requirements of the Directive and which has undergone the certification procedure with the right of drawing marking of SE. In the cases when the equipment, being the subject of action of the Directive, threatens the safety, the authorities of the EU are obliged to take the appropriate measures for the withdrawal of such products from the market of the European Union. The equipment has to be exposed to one of the procedures of control of compliance which can be chosen by the producer, depending on the category of the product.

As the result of the practical use of the rules, generalized above, the qualitatively constructed heating systems with polyurethane foam isolation, with the alarm system, warning about moistening of materials of structures, brought to the single automated control panel exist in the majority EU countries . Rigid control of the quality of the make-up water is also carried out and there are no heat exchangers which are "full of holes". All the arising cases of violation of properties of the materials of the pipes waterproofing are liquidated quickly and qualitatively, the pipelines serve on average 30 years. [8, 9] At the same time the temperature condition up to 130C is provided. In 30 years it is enough to open 2-3 sites and if the survey proves the materials and structures of the pipeline to be in good repair, that conclusion extends to all the heating systems of the same type. For the normal maintenance of the heating systems with the channel types of laying it is sufficient to provide the qualitative waterproofing of channels, the current maintenance of drainage systems and well-organized monitoring of the production processes of the transportation.

4 Results

The most exact information on the condition of materials and structures of the heating systems, their compliance with the considered STD can be obtained by the means of carrying out technical instrument diagnostics. [9, 13]. This type of diagnostics supposes the timely detection of places of thinning on the pipe side of the heating system because of external and internal corrosion, which makes up about 90 percent of the refusals of the system. The remote indirect methods became the basis for the analyzed technology of the technical diagnostics of the underground heating systems, which are inaccessible for any direct survey and measurements of parameters of maintenance. We should note that they do not reveal particular defects of the heating systems, but reveal certain abnormal zones, which are different from the standard values. Two methods which were considered in details are most perspective: they are the acoustic tomography (AT) and the method of contactless magnetic diagnostics (CMD) [3]. Their use allows to reveal the places of the heating system where there is increased metal tension on the metal side of the pipe. The method of thermovision infrared aerial photograph (TIRAP) is also interesting as a method, supplementing the aforesaid methods [3]. It shows the heat anomalies of the ground surface over the main heating system and that rather visually demonstrates the areas of an environmental risk on the city map.

For the formation of reliable data the check of the results received through the indirect methods is important for ecological management in the modern city. In this regard the analysis of the provided data on the basis of direct access to the separate sites of the heating system by the means of pitting. In general from 100% of the sites surveyed by the AT and BMD methods in 60 - 80% sites with anomalies have real defects of metal of pipes; the rest is the tension of metal without noticeable defects which can lead to loss of the heat carrier. Therefore such type of the result of these methods of diagnostics can be sufficient when for management requires to estimate the general reliability of a site, assuming a little extension of term of maintenance (till 1 year). As an additional condition of decision-making it is also necessary to consider the maintenance history, i.e. the previous repairs of small sites caused by the arisen damages.

In our opinion making such local decisions containing an environmental risk concerned possible leakages of the heat carrier cannot be applicable for all territory of the city. At the especially responsible territories of the city it is required to provide 100% reliability of maintenance, usually through traditional way of the full turn of a site instead of carrying out local repairs. Diagnostics of heating systems in this case can be carried out by the direct method on intra-pipe shells by the Norwegian firm, named “Breivoll”. [3] The measurement of thickness of a side of the heating system by the intra-pipe inspection device, which is in the water environment, is made by the means of technology of the acoustic resonance based on an ultrasonic method of nondestructive control.

Indirect methods of diagnostics are being developed and need an assessment of the results therefore for the comparison of one of the indirect methods of diagnostics with the IPD method the comparison of the results received through diagnostics of the same sites of the heating system was carried out. The specifications of application of the aforesaid methods are given in the Table 1.

Table 1. Specifications of application of IPD and AT-method.

Conditions of application of the method on the heating system	IPD - inspection shell, "Breivoll" firm, direct diagnostics	AT-equipment, "Rent Technologies", Ltd, indirect diagnostics
General description of the application conditions	The site of heating system without angles of rotation is chosen, taken out of service, cooled up to the admissible temperature. The laying is opened in one place before the access to pipes. In the top part of the surveyed pipe the window for loading of the device is cut out. The diagnostic shell is entered into the pipe filled with the cooled heat carrier. The shell with a cable of communication and a rope moves ahead until the end of a site and during the return pulling of the rope the diagnostic record is made.	The site of the heating system in working order in the maintenance mode is chosen. Acoustic sensors are fixed at the edges of the site in the access points (chambers, viewing hatches) at first on the giving pipe and then on the return pipe. The record is made separately for the giving pipe and for the return pipe. The record lasts 2 minutes for each pipe.
Data processing	By a highly qualified specialist using computer	By a highly qualified specialist using computer. In an operating mode is it usually done in 24 hours. In the emergency mode it can be processed on the site.
Removal after maintenance	Required	Not required
Temperature of the heat carrier	Not above +40°C at the stopped circulation	Up to +130C in the operation mode °
Length of the surveyed site for one work cycle	On direct sites on length only	40-300 metres irrespective of the scheme.
Nominal diameter of the pipe, mm.		50 - 1400
Data obtained	Side thickness on all the surveyed length with the exact binding on length	Abnormal zones with the increased metal tension with the binding on the site length on the ground surface

As the concrete example of comparison of methods the results of AT-diagnostics of the return pipeline with a nominal diameter of 600, 125 m long are presented in the form of the schedule (fig.1). The results of the detection of defects by the IPD method for this site are presented in the Table 2.

Table 2. Results of the diagnostics of a site of the heating system by IPD method.

№	Beginning of the interval, m	End of the interval, м	Length of the defective interval, m	Metal loss level, %	Result
1	5.1	6.5	1.4	29.5	External corrosion
2	69.7	71.6	1.9	48.2	External corrosion
3	71.6	72	0.4	39.7	External corrosion
4	87	90.8	3.8	27.2	External corrosion
5	99.5	100.3	0.8	27.5	External corrosion
6	109.2	111.5	2.3	39.6	External corrosion
7	117.0	117.7	0.7	39.6	External corrosion
8	59.6	60.2	0.6	Outgrowth up to 60 mm	Internal outgrowths on the structure

Similar diagnostics was also carried out by the AT-method, its results are presented graphically (Fig. 2). Axis "X" demonstrates the distance on a pipe from the initial point to final. Axis "U" demonstrates the energy of the signal of issue generated in the step-by-step analysis. The existence of defects is stated by signals above to the green line. The location of defect corresponds the position of "peak" (amplitude value) of the signal. According to the level of energy of issue, all the defects are divided into: subcritical (between green and red lines) and critical (above the red line).

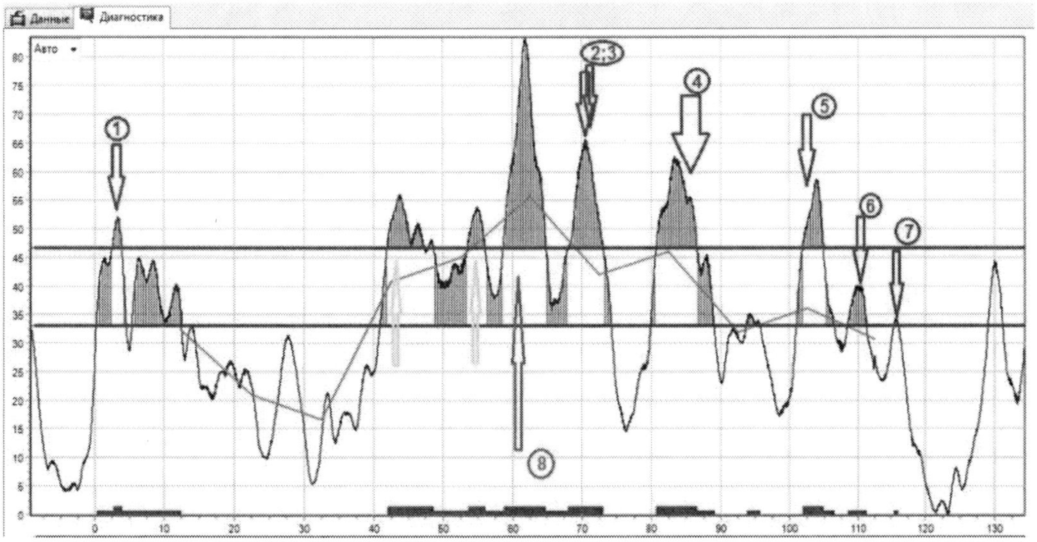

Fig. 2. Results of diagnostics of the site of the heating system by AT method.

The comparison of the results of diagnostics on this site shows that seven defects of external corrosion of the pipe metal, found by IPD were also found by AT-method. Besides, one interval with outgrowths inside the pipe was found by IPD (8 - green arrow in fig. 1). That indicates the existence of internal metal corrosion. This interval was found by AT-method too. In addition two more defective intervals of critical level were by AT-method revealed (yellow arrows in fig. 1). These defects are displayed in fig. 1 as arrows.

Total length of the heating system which is the subject of the comparison of results of the aforesaid types of diagnostics is 1958 running meter. 56 defective intervals were revealed by IPD method, 53 of them were revealed by AT-method. Besides AT-method indicates the existence of 15 more defective intervals which generally correspond to the heating system sites where the IPD method noted the existence of outgrowths inside the pipe, i.e. the internal corrosion. Thus, the coincidence of the results for IPD and AT methods makes up 82%.

5 Discussions

The developments, which basis is estimated for years of research activity in the field of management of the heating systems maintenance allowed to raise a question in such an aspect and to offer some methodical tools for ecological management in the modern city, concerning the activity of the heat supplying enterprises [10]. It is well-known that one of the main conditions of effective management of any process is the existence of necessary and sufficient data on the performance of the management subject at the moment of decision-making [4, 11] . Within the considered perspective the diagnostics of the performance of materials and structures of the heating systems on the basis of the standards and methods considered above in combination with the certain system of control measures makes up the basis for the formation of the automated data base at the operating enterprise. We will give the description of such data base in relation to the Russian practice which according to the official reports and the conducted research the performance of the heating systems is most problematic in the aspect of accident rate and has considerable environmental risks.

Considering the situation of making decisions on the existence of the environmental risk connected with the decrease in properties of materials and structures of the heating systems within the maintenance process it is possible to name the following parameters of obtaining of data: data have to be obtained within 10 minutes after the receipt of inquiry, and the relevance of data has to be no more, than 5 days after the date of inquiry. It appears very problematic to meet these requirements for the Russian heat supplying enterprises in the conditions of the modern city. During the construction, commissioning, repair works throughout rather long service life of the heating system data are fixed incidentally and irregularly. The main disadvantage of such accounting at the enterprises operating the heating systems is the significant amount of data which, despite the current possibility of automation of management, are still stored as papers or as the local databases separated on separate divisions (that is not less problematic). One more important aspect should be noted: in the case of the mixed "paper-and-electronic" data storage many doubts about the reliability of data exist, as it is impossible to compare them, in fact it is impossible to make any correct conclusions on the base of such data.

The most progressive enterprises operating the heating systems are guided by the use of the ready software which, is aimed generally on "drawing" of the heating system schemes on the electronic map of the city as well as on the hydraulic mode operation ("Zulu", "Stream", "Teploexpert") [3]. The existence of such data which are visually representing the heating systems in an electronic form is the step forward, however it is not exhaustive. Even in such circumstances in the particular situation of accident in the heating system decisions are made more intuitively, the wide range of options (from change of all the route to the local replacement of a separate site). In practice and as the extreme version the decision concludes the replacement of the damaged part of the heating system.

It is possible to correct this situation by starting permanent and consecutive work on the accumulation of the data acccording to the results of diagnostics using standard techniques of database formation [1, 7]. In various domestic STD there are general requirements about the need of fixing of these rounds, surveys, revealed defects, eliminated damages of heating systems. But in the case of the use of modern information technologies, it is necessary to develop some regular procedures of control and also some branch qualifiers of the performance of materials structures of the heating system on the basis of the received results [11, 12]. Domestic and foreign documents, despite the demands of management, do not include such qualifiers. It is suggested to use the performance and the maintenance events of the heating systems, applied earlier for the identification of damages and failures as the basis of such qualifiers.

We will present the procedures of control details. In the Russian practice of the heating systems maintenance (as compared with the heating systems in the EU) the technical performance of materials and structures of the heating systems vary much more. So, the heating systems, laid in channels according to the old technologies of construction has many unpredictable corrosion factors of failure. They include periodic flooding of the heating system, drops through the joints of overlapping plates, compensation violation, silting; that makes it impossible to define the performance of the heating system only by separate pitting as it has been described concerning the maintenance in EU. Thus the system of monitoring of the technical performance of the heating system is required to be more difficult. Rational allocation of diagnostic measures among the periods of the heating system maintenance (seasons) is important. The efficiency of survey by the indirect methods of AT and CMD is higher at full working loading of the heating systems, i.e. during the heating period, while control of metal in zones of the revealed anomalies (in pits or places of access) is better to be carried out in the summer period during shutdown of the site for pressure testing and repair. This obligatory requirement of the STD demands to carry out visually measuring and ultrasonic control only at the disconnected and emptied equipment [8]. The anomalies revealed by the indirect methods can be processed and verified by the means of thermovision aerial photograph, according to the production monitoring of the heating systems performance before the termination of a heating season. By the end of the heating season all the works within both AT and BMD methods have to be completed and the collection of data on the revealed anomalies has to be made for diagnostics by the methods of direct access – pitting and additional defect detection control on the pipe metal, purified of isolation. Intra pipe diagnostics is suitable in the summer, when it is technically simpler to switch-off the heating system for survey, the temperature of the heat carrier is lower and less time is taken to cool the water up to the acceptable temperature. All the control pit and additional additional defect detection control in them have to be completed before the following heating season begins. The actualized data on the performance of the heating system has to be prepared by then and the conclusion on the required measures for the localization of the detected environmental risks has to made on this basis.

6 Conclusion

Nowadays ecological management in the modern city has to be guided by objective data on the heating systems under maintenance, which materials and structures (as practice proves) are not always reliable and conclude risks of emergence of leakages of the heat carrier in the soil surrounding them. It is caused by the fact that they are operated in much more difficult technological environment than overland structures. The ground mass influences aggressively the heating system structures as compared with other engineering service lines due to the temperature conditions during their maintenance process [3]. Therefore for the minimization of this type of risks it is important to study the performance of the heating system during the maintenance period and to define which events, caused by the performance of materials and structures, lead to environmental risks. In the research the author defines the main type of failures of the heating system, related with corrosion of material of the pipeline, which causes the greatest number of the fixed heat carrier leakages. It is stated that violations of requirements of standard-and-technical documents both to quality of the materials used during the construction, and to the structure the heating system, when the standard maintenance conditions are not provided, are the reason of this type of defects. Therefore it is important to provide constant diagnostics of the performance of materials and structures of the heating systems under maintenance and to fix the obtained results in the data base which is specially created for this purpose. Obtaining objective data needs to be carried out on the basis of diagnostics which can be completed by the means of various indirect methods and methods of direct access to the materials and structures of the heating system, which are offered to be carried out according to a certain system taking into account alternation of the heating period and the season when the heat supply is not required. This approach to diagnostics will allow to save resources which are the important factor of management in modern conditions. In this aspect the options of diagnostics of materials and structures of heating systems related with the use of a certain type of technical means of diagnostics have been considered.

In our opinion, all the aforesaid results allow to reach systemic ecological management in the modern city concerning such material-intensive and complicated structure as heating system. The article proves that the importance of this matter in Russian and foreign countries differs. In the Russian Federation it is most acute, since in this country there are long-term traditions of the centralized heat supply in the cities on the basis of large plants, which were operating heating systems of considerable cumulative extent. However as the direction of the further research and practical measures the author considers the expansion of this approach on all the linearly - extended structures of the underground space in the modern city, to be very similar in the requirements to quality of materials, their use during the construction as well as similar in the character of the environmental risks related with unorganized industrial emissions.

References

1. A.U. Ahmed, C.F. Ahmed, M. Samiullah, N. Adnan, C.K.-S. Leung, Information Sciences, **354**, 60-85 (2016)
2. Information on: https://law.resource.org/pub/us/cfr/ibr/002/asme.b31g.1991.pdf
3. Information on: http://www.bhpipe.com/ASTM-A106-Seamless-Carbon-Steel-Steel-Pipe-for-High-Temperaure.htm
4. S.L. Beliakov, M.L. Beliakova, M.N. Savelieva, Management, control, diagnostics, **1**, 25-32 (2016)
5. Information on: https://law.resource.org/pub/us/cfr/ibr/002/asme.b31g.1991.pdf
6. Information on: http://www.icqc.eu/userfiles/File/PED_Directive.pdf
7. Dylewski Robert, Adamczyk Janusz, Energy and Buildings, **54**, 88–95 (2012)
8. Information on: http://www.breivoll.no/ date of the address 5/8/2016
9. Information on: http://www.consultant.ru/document/cons_doc_LAW_40241/
10. Federal Law of the Russian No 261-FZ, *«On energy saving and the increase in power efficiency and on the modification of separate acts of the Russian Federation"* (2009)
11. GOST R ISO 50001-2012. *Power management systems. Requirements and manual*
12. RD 03-606-03 *Instruction for the visual and measuring control. Approved by Ministry of Fuel and Energy of the Russian Federation; it is coordinated by Gosgortechnadzor of Russia* (Moscow, 2003)
13. E.Ya. Sokolov, Central heating and heating systems, the 7th stereotypic edition Moscow: MEI, 380-381 (2001)
14. *The concept of long-term social-and-economic development of the Russian Federation until 2020* Information on: www.youngscience.ru
15. Information on: http://www.slideshare.net/EbrahimKhayer/piping-inspection

MATEC Web of Conferences **73**, 07023 (2016) DOI: 10.1051/matecconf/20167307023
TPACEE-2016

The improving of the heat networks operating process under the conditions of the energy efficiency providing

Tatiana Blinova[1,*]

[1]Moscow State University of Civil Engineering, 129337 Yaroslavskoe sh. 26, Moscow, Russia

Abstract. Among the priorities it is important to highlight the modernization and improvement of energy efficiency of housing and communal services, as well as the transition to the principle of using the most efficient technologies used in reproduction (construction, creation) of objects of municipal infrastructure and housing modernization. The main hypothesis of this study lies in the fact that in modern conditions the realization of the most important priorities of the state policy in the sphere of housing and communal services, is possible in the conditions of use of the most effective control technologies for the reproduction of thermal networks. It is possible to raise the level of information security Heat Distribution Company, and other market participants by improving business processes through the development of organizational and economic mechanism in the conditions of complex monitoring of heat network operation processes

1 Introduction

The relevance of this study is due to the major directions of the state policy in the sphere of housing and communal services. Housing and communal services is a basic sector of the Russian economy, providing vital public services, and the industry provides the necessary infrastructure. The turnover of housing and communal services is 4.1 trillion rubles, which corresponds to 5.7% of the gross domestic product of Russia.

The main priorities, objectives and tasks of the state policy in the sphere of housing and communal services are defined by the Strategy of development of housing and utilities sector up to 2020, approved by the Federal Government on January 26, 2016 № 80-p. This strategy has been developed in accordance with the Presidential Decree of May 7, 2012 № 600 "On measures to ensure the citizens of the Russian Federation, affordable and comfortable housing and improving the quality of housing and communal services" and the Federal Law of the Russian Federation № 261. (2009, November 23). «On energy saving and energy efficiency and on Amendments to Certain Legislative Acts of the Russian Federation». The goal of the state policy in the sphere of housing and communal services is

* Corresponding author: tatianablinova@bk.ru

to improve the quality of life by improving the quality and reliability of utility services, as well as making them accessible to the public. [1] [2]

Among the priorities it is important to highlight the modernization and improvement of energy efficiency of housing and communal services, as well as the transition to the principle of using the most efficient technologies used in reproduction (construction, creation) of objects of municipal infrastructure and housing modernization. [3]

Quality and reliability of utility services contain the smooth delivery of services heating, hot and cold water supply, sewerage, electricity and gas, and other mandatory requirements for housing, established by the current legislation. The heating system is one of the largest and most high-tech elements of housing and communal services, on which effectiveness rates of return and profitability throughout the housing sector depend. Thus the current state in the housing heating system is characterized, in general, by low efficiency of thermal energy used by the consumer, that is caused by significant heat losses in transport systems and heat consumption plants, unplanned modes of heat, insufficient use of energy-saving equipment, low degree of automation of accounting and control of thermal energy and other factors. Inefficient heating leads to huge cost overruns of energy, material and financial resources. [4]

In order to achieve the priorities it is required to improve the housing and utilities management system in the field of legal regulation, as well as in the field of information security and to enhance transparency of the organizations in the sphere of housing and communal services for consumers, investors, regulatory authorities, for providing of which the state information system of housing and communal services (HCS GIS) has been created. Furthermore in order to ensure balanced development and reproduction of utility infrastructure systems, taking into account future needs mechanisms for spatial and investment planning must be coordinated.

The main hypothesis of this study lies in the fact that in modern conditions the realization of the most important priorities of the state policy in the sphere of housing and communal services, is possible in the conditions of use of the most effective control technologies for the reproduction of thermal networks. It is possible to raise the level of information security Heat Distribution Company, and other market participants by improving business processes through the development of organizational and economic mechanism in the conditions of complex monitoring of heat network operation processes.

The purpose of the research is the development of an organizational-economic mechanism of increase of efficiency of operating processes of thermal networks of the municipality in a complex monitoring and recording of the necessary financial and material resources for their reproduction.

The object of the research is heating networks as objects of heat-supply of real estate complex in the housing and communal services of the municipality.

The subject of research is the system of organizational and economic relations, conditions of formation, functioning and development to improve the efficiency of operation of heat networks process mechanism.

In the current macroeconomic situation, you must make maximum use of internal resources for the development of housing and communal services by improving the efficiency of infrastructure management, promotion of energy conservation and the transition to the use of the most efficient technologies used in modernization (reproduction) of utility infrastructure [1-14].

2 Methods

2.1 The process of reproduction of heating networks

The thermal network as property, have their own life cycle, from the three main stages of the life cycle: construction - operation - recovery and exploitation stage heating system to a greater extent is associated with a reduction in the quality of operation of the system, due to the accumulation of wear and tear.

The process of reproduction of heat networks - is primarily a process of restoration of the technical characteristics of thermal energy transportation and distribution system. Partial reproduction allows you to restore or improve the condition of some (not all) elements and structures of heating networks, and full reproduction is almost a complete renewal of the distribution system. Accordingly, the simple reproduction makes it possible to eliminate the physical wear and tear, and expanded reproduction involves the removal of functional deterioration or increase in heat networks by renovation and new construction.

In accordance with this definition, when the thermal networks are executed there are works carried out related to the dismantling of dilapidated utility networks and the construction of new, reconstruction, major overhaul and modernization of thermal networks, content control and communication facilities in the given design conditions, carrying out repairs and maintenance of heat networks in accordance with the existing legislation.

Features of operational processes of heating networks consist in the fact that the impact on the network and the construction is going on at the most prolonged period of time and has a decisive influence on the quality characteristics, which are largely determined by the engineering, technical and design decisions taken at the stages of design and construction.

The duration of the maintenance period of heat networks depends on a variety of factors and conditions, which include structural and technical characteristics of the object itself; provision of material, labor, financial resources to maintain the facility during the operational phase; quality and timeliness of maintenance and current repair and object; timeliness and quality of the overhaul facilities.

Management, maintenance and reproduction of the whole heating system should not be based on intuitive management decisions, but on economic calculations with the use of the system of scientific-methodical substantiation of a choice of primary objects of investments and sources of financing. The basis for this must be a constant monitoring of production, organizational, economic and legal processes in the municipality, analysis, forecasting and optimization of management decisions with use of modern technologies and methods of decision making.

In the conventional context, monitoring means the process of observing the object, assess its condition, monitoring the nature of the events, a warning of undesirable trends. Monitoring has targeted and action-oriented, that is applied to specific objects and processes for solving tasks. It is also an ongoing process, organizing a sufficiently long period of time, allowing you to assess trends of processes and predict the development of these trends. Monitoring is a continuous process of collecting, processing, evaluation and preparation of decisions aimed at achieving the goals and objectives of the organization.

2.2 Reproduction problems

The existing today information and methodological base of management decision making in the municipal administration, including the sphere of management of heating system does not provide a solution of reproduction problems in order to improve energy efficiency and sustainable development of the economy of the municipality. Among the problems of information support of the municipality the following can be distinguished:

- Orientation of public information systems to the needs of the federal government, while the specifics of the municipal and industrial development need to improve monitoring systems at the local level.

- Lack of adequate information on the development of heat supply system in the form of the creation of a single integrated system (monitoring). The uniform system of indicators of heat supply system development has not yet been formed due to lack of proper relationships, continuity and completeness of the different reporting forms and how one-time surveys in scope of certain populations, and on the composition of indicators.

- Existing indicators are enlarged, many developing processes can not be installed by them that derives primarily from the methodical information-gathering problems. Having existing today and constantly evolving computer technologies for the collecting, processing and reporting of information the solution of the problem is the use and storage of the database of the primary indicators.

- The most important task is providing of transparent technology acquisition, transmission, processing, storage and presentation of information, as well as the formation of analytical information on the basis of the integrated guidelines when key indicators of heat supply system development should be collected in a complex according to methodical uniformity of requirements.

- There is no possibility of not only automatic formation of a consolidated energy balance of the municipality, but even by hand it causes difficulties due to the fact that the methodology of its formation is outdated and does not correspond to international practice, there is no sufficient data for its formation, there are no specialists who know calculation methodology.

The municipality as the main owner of heating networks is a key element in the control system, heat supply companies, solve problems in the operation and safety of the heating system.

According to experts, in order to properly assess the condition of heating and other objects of heat it is necessary to carry out an objective inventory of all objects, as well as to coordinate the scheme of urban heating, allowing to remove the technical limitations of existing heat network infrastructures. According to Federal Law № 190-FZ of 27.07.2010 "On Heat Supply", it is necessary for each municipality to approve the heating circuit for the next 10-15 years, but unfortunately today, in many municipalities, the reconciliation process is not completed. [5]

Driving settlements heating is a document that contains pre-materials for substantiation of effective and safe operation of heating systems and directions of development in the long term (at least 15 years).

According to the non-commercial partnership "Energy Efficient City", in 83 regions of Russia there are 517 urban districts and 20544 urban and rural settlements, in total - 21 061 municipality in which according to the law "On Heat Supply" district heating schemes must be developed and approved. Adoption of these schemes is the responsibility of the Ministry of Energy Russia - for cities with a population of 500 thousand people or more, and local governments - for settlements with a population of less than 500 thousand people. Today, district heating schemes approval process continues. All background information can be divided into several main blocks:

1) spatial planning documents (general plan of settlement, land use planning scheme municipal districts), topographic base areas, etc .;

2) development plans and programs (the input of housing construction plans, investment programs of municipal utilities, software development systems for municipal infrastructure, energy efficiency program, and other plans and programs);

3) More information about the heat sources;

4) Details of heat networks. [6]

Often, 50% of the documents in the administration of municipalities do not have or are outdated and do not correspond to reality. Quality heating scheme will allow to make management decisions on the development of heat supply system of the municipality and to save not only the budget but also the consumers' money. As noted earlier, the improvement of energy efficiency in order to reduce public spending on public services is one of the priorities of socio-economic development of the Russian Federation. Currently, the total payment for municipal services is more than 10% of all expenses. This indicator is one of the highest in the world. [7]

The proposed approach to the formation of organizational and economic mechanism, based on monitoring as a tool for monitoring the heat network economic development parameters can be represented as a conceptual model, which includes a collection of objects of observation, the list of problem-oriented areas (indicators) and software-tools aggregation and reporting, as well as optimization and ground of management decisions, ensuring the highest possible degree of detail of the information provided for each indicator.

As a problem-oriented areas of process monitoring operation of heat networks are available:

-manufacture, trade and consumption of electricity;

-the indicator of the investment potential of the electric power companies;

-financial state;

-cost-structure and pricing;

-the indicator of the formation of the energy balance of the region;

-factors of institutional change.

2.3 Monitoring

In order to improve process control efficiency of operation of heat networks it is necessary to have a information provision of the organizational and economic mechanism, which allows to determine the level of the existing control system, its strengths and weaknesses. The basis of the organizational-economic mechanism is formed by the monitoring system not only processes the technical operation of real estate, but also its legal and economic framework, through the system of factors affecting them. Factors influencing the activity of the enterprises and the quality of heat supply taking into account the integrated approach of the analysis can be grouped into four main groups: resource and technological, organizational and economic mechanisms; political and legal.

The purpose of process monitoring operation of heat networks is to obtain information about the status of the test process and on this basis, setting targets to improve the existing control system with a target-oriented improvement of qualitative characteristics of thermal networks, reducing heat loss, ensuring compliance with regulatory requirements for the maintenance of facilities utility infrastructure, increase in investment activity in the sphere of reproduction, and so on.

The thermal network in operation should be under the systematic supervision and control of those responsible for the engineering and technical personnel. To this end, the operating organization creates a system of monitoring of thermal networks. Monitoring of the technical condition of heat network is a system of monitoring the technical condition, under which the collection, classification and information on their technical condition of thermal network analysis are made.

The basic principles for monitoring of the technical condition of heating networks:

-Validity of obtaining of information about the technical condition of heat networks;

- Continuous monitoring of the technical condition of heat networks;

- Openness access to the monitoring results.

Monitoring objectives:

- The prevention of emergency situations in the heat sector;

- Analysis of compliance activities for maintenance, repair and preparation of thermal networks for seasonal operation;
- Ensuring the efficiency of use of budgetary funds allocated for the maintenance, servicing and repair of heat networks;
- Determination of the service life of heating networks, as objects of monitoring and deadlines for carrying out major repairs of heating networks.
- Consideration of the effect on the service life of heating networks, as the quality of the monitoring facilities of work performed during the warranty period after the completion of construction, renovation and major repairs;
- Analysis of the course of the monitoring information; evaluation of the effectiveness and completeness of the implementation of laws and other normative legal acts in the field of maintenance and repair of heat networks;
- The integration of information flows generated by the different bodies involved in the monitoring. In addition to these tasks by inspection of buildings, the purpose of the technical inspection is to develop proposals to improve the technical operation of thermal networks, as well as the quality of all types of repairs.

Calendar terms of the general inspection of heating networks are established by the respective head of the unit for maintenance or the head of the operating organization. Calendar terms of the general inspection of heating networks are installed, depending on the climatic conditions. Monitoring ensures the completeness, comparability, timeliness and accuracy of information about the technical condition of thermal networks. And it is also aimed at realizing the following functions:

- analytic function, which assumes on the basis of continuous monitoring surveillance of operation of heat networks of the process, to analyze the state of the heating system, which includes the measurement and comparison of actual results with the parameters specified circuits of heating systems;
- diagnostic feature that allows you to get complete information on the status of heating networks in operation;
- predictive function, which consists in reasonable assumptions about possible changes in the state of thermal networks in operation;
- organizational and management function, which assumes control over the condition of heating networks and the preparation of recommendations for correction and proactive solutions.

The following can be taken as the main stages of the integrated monitoring of heat network operation processes:

- substantiation of methodical approaches to the organization and the monitoring of the processes of operation of heat networks, including the definition of principles, objectives and tasks of monitoring, a description of the object of monitoring;
- definition of key performance indicators (indicators);
- definition of the methods and frequency (frequency) data collection at different levels of the process control system operation of heat networks with reproduction, description of methods of analysis;
- Data collection, analysis, synthesis and interpretation of the description of the forecast of possible changes in the state of monitoring the facility;
- presentation of the monitoring results, including the conclusions and recommendations for correction and proactive management decisions.

In addition, comprehensive monitoring of operating processes of heating networks is a tool to make effective management decisions. Basing on the PDCA cycle (Plan -Do-Control-Activity), we can conclude that the monitoring operation of the processes of heat networks is the most popular in the implementation of management functions such as planning processes of reproduction and control over the implementation of these plans, because one of these two functions ends management cycle (control) and the other (planning), using monitoring results, starts the cycle again.

2.4 Indicators

Monitoring is supposed to be carried out using a system of indicators, which must comply with such requirements as

- Focus - indicators should be focused on the adoption of appropriate operational decisions,
- The timeliness and frequency - the frequency of incoming information to make timely decisions required to minimize the damage from adverse changes and maximizing the effect of the use of emerging opportunities,
- The dynamism, and the prospect of comparability,
- The accuracy, completeness or adequacy of indicators, taking into account the specifics of each monitored Enterprise,
- Efficiency - the effect of the use of identified opportunities in a timely manner should be greater than on monitoring costs.

A system to monitor the structure of indicators covering the main aspects of activity of the enterprises of housing and communal services is presented.

Monitoring indicators consist of five groups: 1) The operating activities of the company; 2) The costs of operating budget; 3) Applicability prices for contractors; 4) Applicability prices for businesses and organizations; 5) Analysis of financial condition (only for municipal enterprises).

3 Results

Among the most important results that make up the importance and novelty of the research, are the following:

proposed structural-logical model of monitoring management software energy efficiency operation of heat networks;

developed method of determining the parameters of complex monitoring system hierarchy process control entities operating heating networks;

proved management decisions in a system of hierarchy based on the information the subjects of management of complex processes for monitoring operation of heat networks;

proved practical use of information for monitoring operation of heat networks in the development of strategic and investment programs of heat network companies and other participants in the heat and power market.

Organizational-economic mechanism of energy efficiency in the operation of thermal networks based on the development (reproduction) of the property complex, through the use of integrated monitoring provides a synergistic effect in reducing the amount of energy losses and expenses for the development of heat distribution facilities;

4 Discussions

The management of the process of reproduction of heating networks involves many departments of the municipality administration (committees: housing and communal services; construction; maintenance of housing stock, the City Property Management, Labour and Social Protection, Energy and Engineering, Finance, Economy and Industry, and others.) . In this regard, in the management of the reproduction of thermal networks, there are certain difficulties of departmental nature that are aggravated by plenty of mutually independent participants in this process: banks, insurance companies, construction companies, asset management companies, real estate owners, etc. It is therefore necessary to take into account the optimal combination of volumes of expanded and simple reproduction of the housing fund, ensuring their compliance with economic opportunities and the needs of society.

In the set of measures that form a system of economic stabilization of heat supply system of the municipality, the crucial importance should belong to the pre-empt emerging problems. The emergence of these threats creates various investment component of the territory, its attractiveness for inward investment from a variety of sources, including but not limited public investment, funds of enterprises, credit institutions, private individuals and others.

In the practice of process control operation of heat networks there are various risks associated with inadequate resourcing of the main directions of its development. The ways to limit these risks are:

- Continuous monitoring of the implementation of program activities, taking into account the factors of the current state of heat networks;
- Determination of status indicators and evaluation indicators of heat networks;
- Allocation of financial resources in view of the possible risks and depending on the pace of achieving the objectives of the implementation of the process approach in the distribution of functions, powers and responsibilities of all participants in the heat supply market.

In addition, when the reproduction of heating networks generated socio-economic effect must be taken into account in its implementation. Thus, the successful development of the reproduction of thermal networks is provided by the development of tools for the identification, assessment and risk management, taking into account the factors that ensure the fuel and energy balance, improve of the quality of investment planning, ensuring effective interaction between all participants in the process of reproduction and is aimed at increasing of the competitive advantages of the territory as a whole.

Accordingly, the main objectives of ensuring of the development of heating network management reproduction and the whole housing sector, should be:

- Rational and efficient use of existing engineering infrastructure, including heating networks;
- Reduction of heat loss during transport;
- Economical consumption of material, labor resources in the reproduction of thermal networks;
- Improving the information management system; [8]

Analysis of the current state of the conceptual bases for the operation of real estate, as well as trends in the construction industry as a whole, proves the necessity and inevitability of the transformation of criteria and indicators of the effectiveness of the decisions. It is possible to optimize decisions on the preservation and reproduction of real estate based on energy-saving measures setting multi-task dependencies. Organizational and technological support for the implementation of the tasks of effective operation should be based on the improvement of planning methodologies and repair work, which will increase the energy efficiency of the heating system operation.

5 Conclusion

Comprehensive monitoring of the condition of heating networks in operation, as part of improving the management system forms the surveillance and control systems carried out regularly on a specific program for a comprehensive assessment of the status of objects heat network management, as well as to analyze the changes occurring in them and the timely identification of trends of these changes. Objective data obtained as a result of comprehensive monitoring and measurement, and based on them improvements are used in organization to continually improve the effectiveness of implemented thermal grid strategy.

A key landmark in the implementation of a strategy in the current conditions is to ensure the growth of the cost of the Heat Distribution Company. This issue is beyond the scope of this study and is a promising direction.

Since in the process of implementing of a specific strategy it is necessary to identify the most significant cost factors and include them as the comprehensive monitoring indicators to track the dynamics of promotion of Heat Distribution Company in the specified direction. The cost growth factors are always possible to determine its source. Correct definition of cost factors in the framework of a typical preferred strategy allows: in the planning - to establish their priority to ensure appropriate resources activities, and at the stage of execution – for managers and staff of the thermal grid companies to understand at what expense cost is maximized. Here introduce the paper, and put a nomenclature if necessary, in a box with the same font size as the rest of the paper. The paragraphs continue from here and are only separated by headings, subheadings, images and formulae. The section headings are arranged by numbers, bold and 10 pt. Here follows further instructions for authors.

References

1. *«Strategy of development of housing and utilities sector up to 2020»* (Federal Government on January 26, 2016)
2. Presidential Decree of May 7,2012 № 600 *«On measures to ensure the citizens of the Russian Federation, affordable and comfortable housing and improving the quality of housing and communal services»* (2012)
3. Federal Law of the Russian Federation # 261. *«On energy saving and energy efficiency and on Amendments to Certain Legislative Acts of the Russian Federation»* (2009)
4. Information on: http://www.cenef.ru/file/FINAL_EE_report_rus.pdf.
5. Federal Law № 190-FZ of 27.07.2010 *«On Heat Supply»*
6. Information on: http://www.energosovet.ru/bul_stat.php?idd=454
7. Information on: http://www.bp.com/en/global/corporate/about-bp/energy-economics/statistical-review-of-world-energy/2014-in-review.html (2014)
8. N. G. Verstina, T. S. Meshcheryakova, Biosci Biotechnol Res Asia, **12(2)** (2015)
9. J. Ćetković, S. Rutešić, M. Zarković, M. Knežević, , N.Vatin, Procedia Engineering, **117 (1)**, 780-790 (2015)
10. S. Rutešić, J. Ćetković, M. Žarcković, M. Knežević, N. Vatin, Procedia Engineering, **117 (1)**, 905-915 (2015)
11. J. Cetkovic, M. Knezevic, I. Vujovic, J. Cerovic, Technics Technologies Education Management, **7(4)**, 1646-1654 (2012)
12. M. Knežević, Journal of Applied Engineering Science, **10(1)**, (2012)
13. J. Cetkovic, S. Rutesic, T. Hanak, M. Knezevic, B. Melovic, Technics Technologies Education Management, **7(1)**, 285-293 (2012)
14. V.V Okrepilov, Studies on Russian Economic Development, **24(1)**, 35-42 (2013)

MATEC Web of Conferences 73, 07024 (2016) DOI: 10.1051/matecconf/20167307024
TPACEE-2016

Research into Behaviour Patterns Typical for Consumers of Construction Material as the Mission of Ecological Management

Zinaida Ivanova [1,*], and *Tatiana* Smetanina [2]

[1]Moscow State University of Civil Engineering, 129337 Yaroslavskoye shosse 26, Moscow, Russia
[2]Kozma Minin Nizhny Novgorod State Pedagogical University (Minin University), 603950 Ul'yanova str. 53, Nizhny Novgorod, Russia

Abstract. The objective of the co-authors is to study the motivation of purchasers of construction/finishing materials and the criteria that govern their selection. No systemic studies of consumer behavioral models and stereotypes in respect of residential housing have been performed so far. However, the environmentally determined management techniques, applicable to the production of building materials and construction of residential housing, are highly relevant both worldwide and in the Russian Federation. The co-authors have developed an original research methodology, drafted a questionnaire, and conducted a pilot survey. Its findings have proven that the price and quality of construction materials are the main factors that influence the decision making process in favor of particular items. Mere 14% of the respondents chose environmental friendliness as the decision making criterion. The findings of the focus group projects have also proven the trustworthiness of the stereotypes and behavioral models identified by the co-authors. The co-authors make a conclusion that further sociological surveys are needed to implement the patterns of environmentally determined management and to influence the value paradigms of the population.

1 Introduction

An urban resident spends about 90% of his or her spare time indoors, that is, at work, at home, or inside any public office. We believe that our houses or flats are 100% per cent safe, although the materials used to build or to renovate them may be dangerous for our health.

All construction and finishing materials may be broken down into ecological and non-ecological ones. The latter can do a lot of harm to the human health. Ecological materials include those that comply with the following requirements:

- they contain no irritant or toxic components ;
- they demonstrate minimal natural radioactivity;
- they are recyclable and safely reusable;

* Corresponding author: ivanovazi@mail.ru

• their production is based on the most advanced technologies that cause minimal damage to the environment [1].

In other words, an environmentally safe house must be made of the materials which do not produce any toxic impact on its residents, on the one hand, and the environment, on the other hand. Therefore, ecological materials include timber, building stone, clay, straw, paper, cotton, bamboo, glass, and aluminum, what are widely used, for example, in the construction of half-timbered buildings [2]. There is no doubt that 100% ecological materials no longer exist, and any advanced construction and finishing materials contain polymers or chemical components; however, this article covers the issue of their ability to comply with the environmental standards set by the law. Presently, those materials, which are considered conventionally safe, enjoy wider use. They are also made of particular natural resources, and they do no harm to the environment. These materials demonstrate high performance values. As a rule, they include bricks, finishing and roofing tile, aerated concrete.

Today a wide selection of construction materials, some of which fail to comply with the quality standards, is offered on the market. According to the developers of the Ecomaterial project, about fifty per centum of construction materials, offered on the unofficial market, do not comply with the All-Russian State Standards, and they may be a hazard to people and the environment. Their consumers do not have the slightest idea that the purchase of a new flat or the renovation of the old one can bring adverse consequences to their health.

Nowadays, environmentally safe materials are more often used in the process of construction of residential houses in Europe. Economically developed countries of the European Union spend substantial amounts on safe construction materials, used to build residential, office and industrial buildings, as the calculations have proven that these costs will pay off sooner or later. For example, The German builders, using "environmentally unfriendly" materials, have to pay taxes at higher rates than the builders using environmentally safe construction materials.

Having analyzed the German housing market, we have found out that customers tend to buy ecological houses for economic reasons there, as banks issue loans more willingly, if the purchased housing is ecologically safe. Besides, there is a special group of customers who take a particular care of their health, and their share reaches 30% of the market. The reason for buying green homes consists in the environmental protection and care for personal health. These customers collect detailed information about the flat, its construction and finishing materials, the extent of their harmlessness, and they choose the housing built in compliance with green standards.

German producers of green construction materials also offer their arguments to substantiate their viewpoint. Group one represents "ideologically motivated" producers, and they take care of the environment, while group two represents strong holding companies, which cannot operate in any other way due to their high profiles and market positions.

The EU constituent states practice environmental management, or environmentally determined management of the process of advanced production of construction materials; these countries do their best to make their policies as green as possible.

2 Literature review

Numerous recently published problem-oriented articles cover the environmental properties of construction materials. The first group of works deals with the research into the hazardous properties of construction materials. The Russian researchers, engaged in the analysis of impacts produced by construction materials on the environment, include A.S. Tsovrebov, E.G. Velichko, and E.A. Yaily. They study the chemical components of construction products and offer their recommendations in respect of the chemical composition, ecological, toxicological, and hygienic properties of construction materials, before the latter enter the market. They also develop their recommendations for consumers in respect of the prevention of any hazardous impacts [3, 4, 5]. The second group of researchers studies the greening of the process of production of construction materials. Several articles cover the greening of the cement industry, which is particularly harmful and dangerous [6, 7]. The findings of the research projects and new innovative materials are offered in the third group of works [8, 9, 10]. Vladimir Ilyichev, member of the Russian Academy of Architecture and Construction Sciences (RAACS), and his research team, draw a lot of attention to the production and application of environmentally friendly construction materials [11].

However, the overwhelming majority of articles, published in journals and newspapers, and distributed in the Internet, inform readers about the hazardous impact of particular construction materials or products that contain chemicals; thus, these publications act as the advertisements of other materials or products. This is the way producers of construction materials or construction companies compete for the market share for their products or services.

The importance of research into the motivation and behavioral patterns of consumers is acknowledged worldwide. The mission of entrepreneurs and market researchers is to change the behavioral patterns in favour of particular commodities. The business community is primarily governed by its willingness to generate cash; otherwise it will turn uncompetitive in the market. The social responsibility of producers and sellers is a complex issue, which is also influenced by the government policy, consumption preferences of the society, and the market position of producers [12].

We have not identified a single research into the consumer behavior in the market of construction materials and services in Russia. Some projects deal with the study of consumers of products manufactured by particular construction companies [13]. For example, in 2004, a thorough overview of the market of construction materials was made by the consulting department of the INEK Group; versatile groups of consumers and their preferences were the subject of research in this overview [14]. No major research projects of this kind have been implemented since then.

3 Research Objectives

What motivates the producers of dangerous construction materials? Why do these companies ignore environmental standards? What motivates the consumers of these commodities? What motivates the Russian buyers who purchase new housing or finishing materials for their flats? The objective of this article is to identify the motivation and the choice criteria.

The relevance of this research project is exceptionally high to the unavailability of any systemic studies of behavioral models and stereotypes in respect of consumers of construction materials. Its relevance is also driven by the unavailability of any methodology for the development of any conscious demand for green housing in Russia.

4 Methods

The main data collection methods, employed in the process of research into the motivation of producers and their preferences, include the analysis of secondary documents, polls conducted in the form of questionnaires, in-depth interviews, focus groups, and observations. Each of the above methods is equally valuable for the implementation of the process of environmental management.

The fundamental work, written by L. Blackwell, in which he analyses micro-cultures and their influence on the consumer behavior, can serve as the methodological basis for the research [15]. Research projects may also rely on the methodology applied by Boston Consulting Group, the international leader in the development of corporate marketing strategies. The title of their research reads as "The global green consumer survey". This poll was conducted in Canada, France, Italy, Japan, the USA, Germany, Spain, and the UK. Nine thousand consumers participated in the poll conducted over the Internet. Besides, consumers were interviewed directly in the seven largest cities of China. Mid-level and senior level employees of over twenty network retail companies were interviewed about their strategies and experience in dealing with green commodities [16].

The co-authors have developed an original methodology and drafted a questionnaire. The pilot sociological survey was conducted in Moscow in April 2016 over Internet.

Besides, the co-authors have conducted two group interviews (focus groups) with the first-year students, majoring in fire safety, and third year students, majoring in architecture at Moscow State University of Civil Engineering (MSUCE).

5 Results

The total number of respondents was 100. The age of the respondents, who participated in the pilot testing of the questionnaire, varied from 18 to 55; therefore, no analysis of the choice reasons was performed.

The project developers believe that their mission is to find out whether the respondents have their flats renovated and what arguments govern their choice of construction and finishing materials.

The respondents gave the following answers to the question "how many times did you have your flat renovated?": never – 19%, once – 23.8%, twice – 14.3%, and more than twice – 42.9%. This means that the respondents had a sufficient experience in renovating their flats and purchasing all the necessary materials. The next question read as follows: "what arguments drive your choice of finishing materials for your flat?" The majority of the respondents believed that the main criteria included the price, the quality, the visual appeal, and the style of their rooms. The environmental friendliness of the materials they bought took the interest of a small share of the respondents, or 14.3% (Fig. 1).

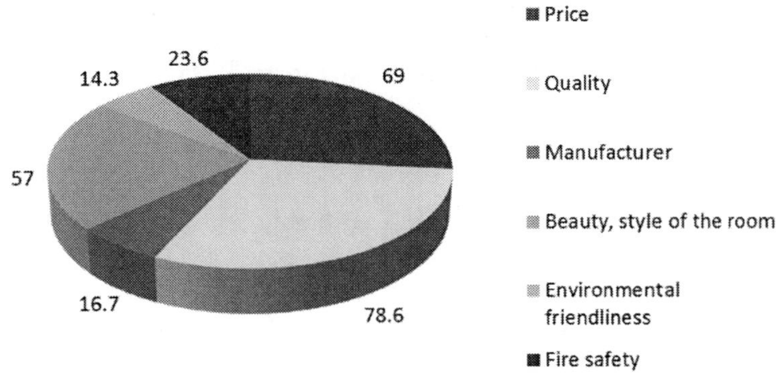

Fig. 1. The arguments of the choice of finishing materials.

One question dealt with the flammability of construction and finishing materials. The respondents turned out to be aware of the flammability properties. They gave the following answers to the question "do you know what materials are fire-proof?": yes – 50%, no – 50%. There was no doubt that the respondents had hardly completed an in-depth study of the fire safety of the materials they bought; however, nonetheless, they had some knowledge of the subject matter. For example, they knew that construction materials contained flammable chemicals, including polymers, lacquers and varnishes.

The project developers found out that the respondents asked for the advice of people, engaged in the provision of renovation services and/or in the distribution of construction or renovation materials. Our question read as: "do you ask anybody for advice when buying construction materials?"

The respondents answered that they had addressed their designer – 14.3%, a specialist in this area – 50%, a shop assistant – 42.9%, nobody – 14.3% (Fig. 2). The majority of the respondents addressed their questions to other people who know something about the renovation.

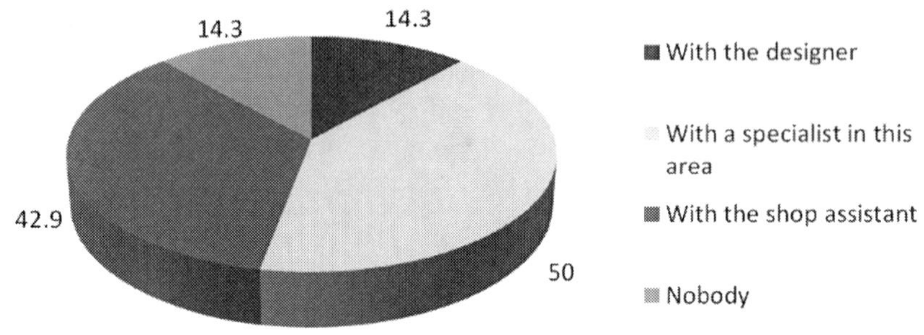

Fig. 2. The respondents asked for the advice of other people, when buying construction materials.

The respondents were asked to complete the following sentence: "The choice of finishing materials is a headache because…" Here are some examples:

1) "I need to find the most perfect material, which must represent a combination of quality, price, practicability, safety, and user friendliness";

2)"the wrong choice may threaten one's health and safety of residents";

3)"it is important to find the proper price to quality and user friendliness ratio";

4)"the choice is extensive, but the knowledge of the subject matter is poor."

Our group interviews identified that the students, majoring in the fire safety, were serious about the environmental safety of the construction or finishing materials they chose. Six out of ten students said that they had studied the environmental properties of materials before buying them, while four students say that their decision to make a purchase had been driven by the price. The future architects preferred the materials that could be used to easily implement their architectural and interior design projects.

6 Discussion

So far, the poll findings cannot be considered highly representative, because only a pilot pall has been launched. The objectives of our pilot poll are: 1) to learn more about the state of affairs in the industry, 2) to pilot test the questionnaire in respect of the adequate understanding by the respondents of the questionnaire questions.

Nonetheless, we believe that even this small sampling has helped us to prove our hypotheses:

1. The business community in Russian Federation is only interested in generating profits, while it does not care for the condition of the biosphere or the human health.
2. The purchase of green houses and materials is limited by the financial capacity of the customers, because green houses are very expensive.

Our hypothesis was proven by the answer to the question "what arguments drive your choice of finishing materials for your flat", because the majority of our respondents ignored the answer that read as "environmental friendliness" (14.3%), and believed that the price, the quality, the style were more important. Therefore, we have to admit that we choose cheap products and ignore our health and the health of our relatives. According to Leo Hollis, a British historian and urban specialist, we, the residents, are the main obstacle on the way to the improvement of the environmental friendliness of our cities [17]. However we shouldn't blame our respondents for their answers, as there is no malice in them. Their behavior can be easily explained by the: 1) insufficient environmental education, absence of any elementary knowledge about the environmental friendliness of structures, materials, instruments, and other items around us, the unwillingness of the business community to inform buyers about the true properties of the materials they produce, or the illiteracy of the business community [18] ; 2) the absence of the health culture or the unavailability of the practice of paying attention to the housing safety; 3) the overreliance upon producers; 4) no money to buy ecological finishing materials [19]. Any further development of the environmental sustainability in the construction industry needs to enjoy demand in the society, the business community, and public authorities. Consumers cannot pay extra for a mere "eco-" prefix, that's why they buy the housing which is harmful for their health at agreeable prices [20].

The breakdown of the answers to the question "do you address anybody for advice when buying construction materials?" may also be easily explained by the mentality of Russians. Few respondents address their designers because few of them have enough money to hire a designer. As a rule, the majority of Muscovites design their flats themselves, and their specialists in the field are the members of the team of workers invited to renovate their flats. The customers can hardly rely on their knowledge of the ecological properties of finishing materials; these specialists know more about their quality or durability. Assistants, employed with the stores of construction materials, have some knowledge of the subject matter; however, even if they have been working in this industry for a long time, they have no degrees in this field [21, 22]. Their mission is to sell as many commodities as possible to generate as much cash as possible.

The problem is that the government must set the stage for the investments to be made into the manufacturing of green products in the future, for the responsible executives to be encouraged to assure the compliance of their products with the green standards. There is a need to identify or to develop new methods and techniques for the distribution of construction products in the retail market through the target-oriented generation of awareness among the customers and the commodity positioning. Russia should adopt the example of Germany, where the government cuts the tax rates, applied to the producers of green products, and supports their consumers.

7 Conclusions

According to the theory, developed by J.N. Sheth, B.L. Newman, and B.L. Gross, consumer behavior is exposed to the influence of a wide range of values, including functional, conventional, social, emotional and epistemic ones [23, 24]. All these values shape up the choice process, as they help buyers to develop the need for the product due to its prestige. Cultural values also influence the behavioral model of Russian buyers; however, Russian buyers don't have environmental friendliness of construction/finishing materials on the list of their values, and this fact's been proven by the findings of the pilot survey and the secondary sources study implemented by the co-authors. We regret to conclude that the state, the producer and the buyer hardly feel the need for "green" construction processes, "green" construction materials or technologies. Only 94 construction facilities are certified according to international green standards BREEAM and LEED in Russia. For comparison, 4,977 construction facilities in Great Britain, 775 in

France, 684 in UAE, 240 in Germany, and 124 in the Czech Republic have green certificates issued [25]. Representative sampling and mass survey are needed to implement a deeper research into behavioral models in the market. The authors are willing to continue their research in order to develop the framework for the ecologically determined system of management in the construction industry.

References

1. Information on http://www.vashaibolit.ru/11557-opasnye-dlya-zdorovya-stroitelnye-materialy.html
2. D.S. Gavrikov, S.D. Mezentsev, Vestnik MGSU, **8**, 7-17 (2015)
3. E.S. Tskhovrebov, E.G. Velichko,Vop Stroitel'nye materialy, **5**, 99-103 (2014)
4. E.S. Tskhovrebov. Vestnik Kostromskogo gosudarstvennogo universiteta im. N.A. Nekrasova, **3**, 10–14 (2013)
5. E.S. Tskhovrebov, E.A. Yayli. *Obespechenie ekologi cheskoy bezopasnosti pri proektirovanii ob"ektov nedvizhimosti i provedenii stroitel'nykh rabot (*Izd-vo RGGMU, Sankt-Peterburg, 2013)
6. M.Ya. Bikbau, V.I. Zharko, Stroitel'nye materialy, oborudovanie, tekhnologii XXI veka, **7-8**, 22-25 (2015)
7. B. Kh. Sanzhapov, O.A. Cheremushkin, Biosfernaya sovmestimost': chelovek, region, tekhnologii, **2**, 89-95 (2013)
8. X. Liu, X. Z. Gong, Y. Liu, Z. G. Peng, Materials Science Forum, **814**, 430-434 (2015)
9. H. N. Abdelhamid, Materials Science Forum, **832**, 28-53(2015)
10. R. Spiegel, D. Meadows, *Green Building Materials: A Guide to Product Selection and Specification, , 3rd Ed* (John Wiley & Sons, Inc., NY, 2010)
11. V.A. Ilyichev. *Biosfernaya sovmestimost': Tekhnologii vnedreniya innovatsiy. Goroda, razvivayushchie cheloveka* (Librokom, Moscow, 2011)
12. N.G. Miloradova, A.D. Ishkov, Procedia Engineering, **117**, 246–251 (2015)
13. Information on http://stud24.ru/construction/analiz-potrebitelej-stroitelnoj-kompanii/430678-1545143-page3.html
14. Information on http://inec.ru/documents/stroymaterial-rus.pdf
15. R.D. Blackwell, P.W. Miniard, J.F. Engel, *Consumer Behavior. 10th Edition* (Thomson/South-Western College Pub, 2006)
16. J. Manget, C. Roche, F. Munnich, *Capturing the Green Advantage for Consumer Companies* (The Boston Consulting Group Year. 2009)
17. L. Hollis, *Cities are Good for You. The Genius of the* Metropolis (by Bloomsbury Press, New York, 2013)
18. E. Shnyrenkov, I. Pryadko, Procedia Engineering, **117**, 325-330 (2015)
19. I.P. Pryadko, International Journal of Applied Engineering Research, **10(21)**, 42147-42152 (2015)
20. Z.I. Ivanova, Strategicheskie prioritety, **3**, 28-39 (2014)
21. M. Arsenovic, Z. Radojević, Ž. Jakšić, L. Pezo, Ceramics International, **41(3)**, 4899-4905 (2015)
22. M. Arsenovic, Z. Radojević, Ž. Jakšić, L. Pezo, Ceramics International, **41(3)**, 4890-4898 (2015)
23. J.N. Sheth, B.L. Newman, B.L. Gross, Journal of Business Research, **22**, 159-170 (1991)
24. J.N. Sheth, The Journal of marketing, **37(4)**, 50-56 (1973)
25. Information from: http://reforum.ru/news/v-rossii-sertificirovano-94-zelenyih-zdaniya-10569.html

MATEC Web of Conferences **73**, 07025 (2016) DOI: 10.1051/matecconf/20167307025
TPACEE-2016

Development of modernization tools for construction complex through the mechanisms of enforcement

Natalya Yaskova[1], and *Tatyana* Alexeeva[1,*]

[1]Moscow State University of Civil Engineering, 129337 Yaroslavskoe shosse 26, Moscow, Russia

Abstract. The article focuses on increasing potential of leasing as part of institutional approach to national economy innovative development. The mechanisms of leasing relations enforcement are examined. The structure of leasing relations institutions in the context of the specification of leasing participants' rights are researched. The authors highlight the features of leasing agents interaction, and on this basis justify the need for change in leasing relations institutional structure. Engineering institute implementation in the institutional system of leasing is proposed in the article. It will contribute to harmonization of leasing relations agents' interests, prevention of their opportunistic behavior, leasing participants transaction costs reduction and improvement of leasing relations enforcement.

1 Introduction

One of the national economy priority tasks (including construction complex) is the movement towards innovation-based development. Elaboration and introduction of innovations in construction sphere will enhance the quality of construction products, improve its technical characteristics and consumer features, ensure environmental safety and energy saving. The problem of Russian economy movement towards a new path of development is reflected in the works of many scientists [1-14]. Experts say that engineering plays an important role as an effective mechanism of its modernization. It is a complex of research, productive services, including the preparation of investments analyses, recommendations on production management and development of innovation activity of organizations etc. These services are provided by engineering companies [1-5, 7].

Previous studies of the authors have shown that in times of crisis and sanctions it is worth using leasing technologies to ensure innovative development of construction complex [1-4, 7, 11-13].

In the Russian Federation leasing is mainly used as a necessary property acquisition mechanism aimed at the development of production and economic process and tax optimization. Current state of leasing relations development showed that leasing

* Corresponding author: w220@yandex.ru

mechanism needs improvement in order to fully realize its potential to intensify innovative development of construction complex and national economy as a whole.

For better understanding of construction complex innovative development characteristics, leasing relations specific features and engineering activity it is necessary to use institutional approach. A lot of scientists comment on this problem [1, 5, 6, 8-10, 12]. From the institutional approach viewpoint the concept of "institution" is defined as "rules of the game" in society, serving as certain restraining factors that organize and structure the relationship among economic agents. Institutions include formal regulations (laws, president decrees, codes, etc.), informal rules (traditions, social norms, stereotypes, etc.) and their enforcement mechanisms, which consist of the police, tribunals, public reprimand, etc. The complex of institutions forms institutional social environment [5]. L. Davis and D. North analyze it as macroeconomic level relations and as institutional framework that restricts contracting parties' terms of agreements. The complex of interrelated institutions represents institutional system [10]. In the system of institutions proper performance of established rules under condition of economic agents relations inconsistency can be provided by enforcement. This term means certain measure or condition directed to compulsion of parties to fulfill their contractual obligations. Enforcement through compulsion of economic agents to follow rules and perform obligations provides for their rights being fully realized, ensures respect of these rights, creates the incentive to act in good faith and thus stabilizes social and economic relations in Russia.

Development of modernization tools for construction complex

Different mechanisms of enforcement exist in practice. Among them are: state enforcement, group enforcement, self enforcement. State enforcement is realized by the establishment of legislative and administrative control on property rights regulations. Judiciary in its turn resolve disputes among economic agents.

Group enforcement is based on certain benefits from economic agents' participation in group relations and negative consequences in case of acting in bad faith. Self enforcement exists in relations where one partner is very interested in fulfillment of contractual obligations and the other partner realizes it. The possibility of loss of a trade partner in some cases can be sufficient for proper execution of obligations.

The framework of institutional theory and distinctive features of enforcement form the basis for research of the leasing relationships system in construction.

Federal legislation of the Russian Federation gives the definition of leasing: the complex of economic and legal relations arising from leasing contract including purchase of its subject (buildings, facilities, equipment, transportation, etc.) [19]. Among the advantages of leasing are the possibility of property purchase despite the lack of finance, taxes economy, decrease of innovation development risk, etc.

Leasing is the set of formal and informal rules and mechanisms aimed at enforcement of these rules (the institute of property, financial lease, crediting, insurance, etc), ensuring the regulation of economic and financial relations arising from the purchase of property and its further transfer for a certain term and fee [1].

The above mentioned rules and norms are necessary for stability and certainty of relations between leasing agents and defense of their rights and financial interests.

Formal institutions of leasing relations consist of legislative and legal regulatory acts in this sphere. Public relations traditions of different regions leasing markets serve as informal institutions. Production specialization of different regions influences the choice of the leasing subject and the tradition of public relations at the regional leasing markets.

The lessor, the lessee, the seller and other participants of leasing are the agents of leasing relations and their interaction is a process consisting of certain stages. At the first stage a future lessee chooses the seller of leased assets and the lessor. Then he draws an application for funding and sent it to the lessor. In case of positive decision after examination of all required documents on the financing of the leasing transaction lease agreement is entered into. In case of insufficient amount of lessor's funds to finance the deal he may apply to the bank for necessary credit resources. Lease contract is accompanied by sales contract concluded between the lessor and the seller of the leased assets. The next stage of leasing relations - is assignement of leased assets for use of the lessee. At the same time leasing object insurance is carried out to avoid various property risks. By mutual agreement of leasing transaction both parties (the lessor and the lessee) may act as the insured party. Thus leasing includes relations arising from finance leases, lending, sale and purchase of property, etc. Therefore formal leasing relations institutions include institutions of property, finance leases, credit relations and other institutions of market economy. |In the system of leasing relations institutions the performance of established formal and informal rules and regulations under the conditions of contradictory interests of economic agents of leasing is provided by enforcement which is also an essential element of specification of leasing participants' rights, largely determining the leasing contract parties' negotiating positions and other conditions accompanying leasing deal. Thus, the specification of leasing participants' rights should be implemented at the level of the institutional environment and contractual relations. The institutional environment determines exclusiveness regime, that is a set of rights of leasing relations participants, later secured by lease agreement and other agreements accompanying lease deal. It also defines a further opportunity to protect the interests and rights of each of the leasing relations participants. In practice, there could be the institutional environment in which the functioning of formal enforcement is hindired or, on the contrary, it can contribute to its proper functioning. Available studies show that institutional environment which was developed in our country is characterized by instability and backwardness of many of its elements [9].

Let us consider the institutional environment of leasing relations, defining substantial characteristics of leasing participants' rights specification. One of the formal institutions of leasing relations is the institution of property, based on property rights, which represents authorized relationships between economic agents regarding the use of limited resources [5]. Proper execution of the established rules by economic agents is proved by enforcement of property rights. In the leasing relationships the leasing assets property right pertains to the lessor, which gives him certain powers: the right to earn profits, the right to transfer authority to another person, and others. At the end of leasing term ownership rights can be modified when property after full depreciation is transferred to the lessee's ownership at depreciated zero value. Leasing allows determining the object of property rights, actors, scope of their powers and the mechanism ensuring its compliance - all contribute to the creation of a stable economic environment and stable profit from leasing transaction.

Formal leasing relations institutions also include the institution of financial lease: the interaction of the lessor as the property owner and the lessee as the property user. The lessor receives revenue by leasing his own property and the lessee – from the use of the property in production and business activities, and at the same time the latter has the chance to avoid large one-time costs for the acquisition of property.

If the lessor lacks own funds for the purchase of property to be transferred to lease, he may apply to the bank, and in this case, the interaction between the lessor and the bank will be regulated by formal rules governing the relations of economic agents in the sphere of crediting. The specification of the lessor's rights (who is acting as the borrower) within enforcement mechanisms, is determined by formal rules in the sphere of crediting.

In the framework of other institutions operating in the system of leasing relations related contracts necessary to implement the leasing transaction are usually entered into. Formal enforcement of such contracts is established by the relevant terms and regulations.

In business practice formal and informal institutions can complement each other, but contradictions can also arise. Consequently, for the development of leasing relations and their effective implementation in the context of innovation it is necessary to eliminate the inconsistency of institutions. Harmonization of these institutions can be achieved only by combination of leasing agents interests. The analysis has shown the lessee is interested in the possibility of reducing the initial costs of necessary equipment, machinery and other property purchase, receiving income from the leased assets operation, tax saving and other economic benefits. The lessor in his turn is interested in the receipt of income from leasing property under lease agreement, tax saving and other economic benefits. Interests of the seller of leased assets include benefit from products realization using leasing technologies, profit from ensuring technical maintenance of leased assets, and others. Leasing relations agents pursuing their economic interests is inclined to demonstrate opportunistic behavior. For example the lessor in course of transactions seeks to inflate the size of the commission, to hide the calculation mechanism of lease payments sums. The lessee strives to reduce the amount of lease payments by inclusion into leasing contract more favorable lease payment schedule, etc. The lack of information provided to leasing agents leads to certain difficulties in economic decision making. Participants of leasing relations obtaining more information enjoy certain advantages in the justification of leasing transaction. In the absence of sufficient information and qualification, potential lessees is unable to determine which equipment or machinery they need for the development of innovation activity and to assess the quality of property, which is supposed to be purchased, the amount of lease payments, to develop an effective leasing scheme, select a vendor, lessor, insurance company, etc. It is also difficult for the lessor to assess potencial risk of non-payment.

Taking into consideration the aforesaid, we propose to modify the structure of leasing relations institutions. The institute that will harmonize leasing relations, reduce risks and transaction costs, improve enforcement is really necessary. According to the research the institute of engineering is expected to perform all these functions. That is a set of rules, norms and mechanism of enforcement which regulate the engineering relationships aimed at ensuring stability, certainty and protection of participants' economic rights and interests [1]. Services of engineering companies include technical, technological audit of organizations (potential lessees), efficient and competent justification of equipment that is needed for innovation activities development, determination of the need for technological change and development of a project for modernization and innovation & technological re-equipment of construction complex organizations. As part of this project engineering companies will be able to develop an effective scheme of its financing. On mutually beneficial conditions these companies will ensure the selection of leasing assets provider, the lessor and insurance company, help in the process of leasing transaction support, etc.

Conclusions

In the Russian Federation as practice shows engineering companies often collaborate with particular leasing company, which provides services for projects financing, such as construction of small power plants (Mini-TES). The lessor traditionally carries out the process of development lease financing schemes and leasing management. We propose a new concept, according to which lease management is carried out by engineering company developing a project "turn-key".

In this concept the lessor preserves the functions of purchase and transfer of property to leasing, whereas the function of lease assets selection, its seller, insurance company and the lessor is transferred to engineering company. Therefore rights and obligations related to the implementation of these functions are conferred on engineering company. It also provides risk management services, advisory services on the conclusion of lease and related contracts. While operating leased assets engineering companies monitor lease payments, organization of technical maintenance of leased assets, training of lessee-company personnel, etc.

Modifications in the structure of leasing relations institutions and introduction of «engineering institute» allow forming one of the most effective management mechanisms of national economy modernization - "innovative leasing engineering" and create a new concept of leasing management [1]. This concept envisages leasing management services performed by engineering companies in practice, aimed at harmonization of leasing agents relationships, prevention of their opportunistic behavior, risk avoidance, leasing participants transaction costs reduction, improvement of leasing relations enforcement and innovative activities development. Practical application of the new concept proposed by the authors will help to prevent economic crises in the Russian Federation and ensure national economy movement toward innovative way of development.

References

1. T.P. Alekseeva, N.Y. Yaskova, Bulletin of Moscow State Construction University, **2**, 146-158 (2016)
2. M.I. Kamenetskii, N.Y. Yaskova, Studies on Russian Economic Development, **26(2)**, 124-131 (2015)
3. T.R. Alekseeva, Bulletin of Moscow State Construction University, **1**, 170-180 (2016)
4. T.R. Alekseeva, Economy and Business, **4-2(57-2)**, 583-590 (2015)
5. T.R. Alekseeva, Bulletin of Moscow State Construction University, **5**, 152 – 161 (2014)
6. A.A. Auzan, *Institutional Economics: A New Institutional Economic Theory* (Infra-M, Moscow, 2011)
7. V.I. Bazhenova, Problems of Economics and Law, **12**, 16-23 (2011)
8. E.E. Ermolaev, R.A. Durov, S.Y. Kopelchuk, D.N. Silka, *Engineering of investment and construction projects for industrial use. Monograph* (Stroyinformizdat, Moscow, 2014)
9. K.A. Krutchankova, T.I. Buchtiyarova, Basic Research, **6**, 1485 – 1492 (2013)
10. J. Ćetković, S. Rutešić, M. Zarković, M. Knežević, , N.Vatin, Procedia Engineering, 117 (1), 780-790 (2015)
11. S. Rutešić, J. Ćetković, M. Žarcković, M. Knežević, N. Vatin, Procedia Engineering, 117 (1), 905-915 (2015)
12. S. Rutešić, J. Ćetković, M. Knežević, M. Žarcković, N. Vatin, Procedia Engineering, 117(1), 642-650 (2015)
13. M. Knežević, Journal of Applied Engineering Science, 10(1), (2012)
14. M. Jocovic, B. Melovic, N. Vatin, V. Murgul, Applied Mechanics and Materials, **678**, 644-647 (2014)

MATEC Web of Conferences **73**, 07026 (2016) DOI: 10.1051/matecconf/20167307026
TPACEE-2016

Innovative Development of Building Materials Industry of the Region Based on the Cluster Approach

Asiiat Mottaeva[1,*] *Alexander* Zheltenkov[2], *Irina* Stukanova[2], *Sergei* Ryabichenko[2], and *Sophia* Zhuk[3]

[1]Moscow State University of Civil Engineering, 129337 Yaroslavskoe sh. 26, Moscow, Russia
[2]Moscow regional state University ,129125 Radio str. 10, Moscow, Russia
[3]Financial University under the government of the Russian Federation, 125993 Leningrad pr. 49, Moscow, Russia

Abstract. The article discusses issues of innovative development of building materials industry of the region based on the cluster approach. Determined the significance of regional cluster development of the industry of construction materials as the effective implementation of the innovative breakthrough of the region as an important part of strategies for strengthening innovation activities may be to support the formation and development of cluster structures. Analyses the current situation with innovation in the building materials industry of the region based on the cluster approach. In the course of the study revealed a direct correlation between involvement in innovative activities on a cluster basis, and the level of development of industry of construction materials. The conducted research allowed identifying the factors that determine the innovation process, systematization and classification which determine the sustainable functioning of the building materials industry in the period of active innovation. The proposed grouping of innovations for the construction industry taking into account industry-specific characteristics that reflect modern trends of scientific and technological progress in construction. Significance of the study lies in the fact that the proposals and practical recommendations can be used in the formation mechanism of innovative development of building materials industry and the overall regional construction complex of Russian regions by creating clusters of construction.

1 Introduction

The development of innovation and leadership in the global economy is the trend of development of science and technology policy, and this process does not depend on the

* Corresponding author: doptaganka@yandex.ru

type of state or of a political regime. However, the level of innovation in selected countries and regions is uneven.

As a rule, it depends on the willingness of a particular State to organize the innovation process, to manage this process and regulate the relations arising in the sphere of innovations. In this context, it becomes relevant to use strategic approaches to innovation management, allowing purposefully lead the process. This strategic management of innovative activity should be carried out at the level of the Executive Innovation (Company), and the meso and macro levels - regional level, the industry, and the state as a whole.

A key factor in the success of innovation in this situation, is the ability to link listed in the single-level strategy. In this paper, it is possible to implement it with the help of the new Russian economy forms of organization of economic subjects, science, infrastructure and state - clusters.

Building materials industry is a major area of the economy, including more than 15 branches and forming the basis of the material and technical base of construction, the cost of building materials is the main section of the building complex costs.

Meanwhile, the gap in the innovative development of the industry of construction materials leads to a stagnation of the entire building complex, both because of the use of obsolete materials and technologies, as well as due to the high cost of their purchase.

Theoretical and methodological basis of studies of scientific approaches to the formation of an innovation cluster in the building materials industry served as the scientific works of domestic and foreign authors. The basis of the innovation theory is easy foreign scientists: J. Schumpeter [3] B. Twiss [5], B. Santo [4]. Strategic management of innovative activity also engaged Drucker [2], I. Ansoff [1] R.

A significant influence on the development of the theory of clusters had M. Porter [8], C. Ketels [6] O. Solvell [7], Solène Goy, Donal Finn [9] Frank J. Kaltsonetti, Diane M. Miller, Neil Reed [10], Fundeanu Daniela Doina, Cosmin Sandu Badele [11] M. Khairullina [12]. An important role in the development of a building complex and building materials industries LN Asaul [13], AV Voronin [14], IM Greenberg, [15] M. Casas [16] S. Kolyada [17] and etc.

2 Methodology

At all times the status of the construction industry characterized the level of well-being and opportunities of development. In addition, the building complex, which includes the production of construction materials, plays a leading role in the economy of any region, and its condition is one of the indicators of economic and technological development.

Despite the fact that the construction industry is quite conservative, the new conditions of the market economy, focused on the use of new technologies and resursos-beregayuschih materials led her to certain changes in investment-making insulating complex with intensive use of innovative programs. The need for innovation is dictated primarily demand to reduce the cost of construction and operation of housing, as well - reducing construction time and improve the quality of living. [18].

In this regard, the main objectives for the construction materials industry enterprises are: the implementation of programs for the modernization of production, increase in the volume of production and sales, improving energy efficiency requirements and resources as the production processes and the products themselves.

The most popular have become the technology and construction materials related to energy saving, as now in Russia and around the world, energy efficiency is a key requirement of modern construction. One of the major obstacles standing in the way of implementation of many innovations is the problem of ensuring the effective interaction between stakeholders in the innovation process.

This problem is particularly acute in the early stages of innovation. It is at these stages of innovation requires investments at the maximum level of uncertainty.

The uncertainty arises from the fact that in the early stages there is no evidence of promising innovations, in addition, innovators can be difficult to outline the range of potentially interested in the subjects of innovation.

Difficulties in co-innovation participants in the early stages may also arise due to the fact that different stakeholders have different opinions innovation, considering them from a different perspective.

Thus, the scope of innovation experts note that the authors of innovations often overestimate their capacity, not taking into account the existing substitutes, inertia in consumer behavior as well as other psychological aspects of the adoption of new products by the market. When communicating with investors, who are interested primarily commercial potential of innovation, the authors often delve into the technical details, and the commercial component is characterized by superficial or overly optimistic.

At the same time, it is often the quality of the interaction between the participants of innovation depends on whether it will be implemented, in what circumstances and in what time frame. The effectiveness of innovative development of building materials industry is caused, linking the main factors:

- The potential of science, new ideas, discoveries and developments;
- Structure and production requirements;
- The resource base of innovative development;
- The availability of an effective management system that is adequate to the requirements of the time. [26]

These factors must interact so that the potential of science is constantly ahead of the needs of innovative enterprise development.

On this basis, by co-measuring with the resource potential in the planned period necessary to carry out selection of the most effective programs and activities of innovative development.

The main components that meet the specified requirements are available at the enterprise of scientific and technical base; innovative infrastructure of the region, contributing to the implementation of innovative development activities; degree of innovation susceptibility staff, its professional and educational level; opportunities and conditions for the implementation of programs and projects of innovation development, reflecting the interaction of the company in this aspect with other companies and organizations directly or indirectly interested in the development of the system [25].

Based on the foregoing appears to be the mechanism of formation of innovation potential of construction companies (Fig.1)

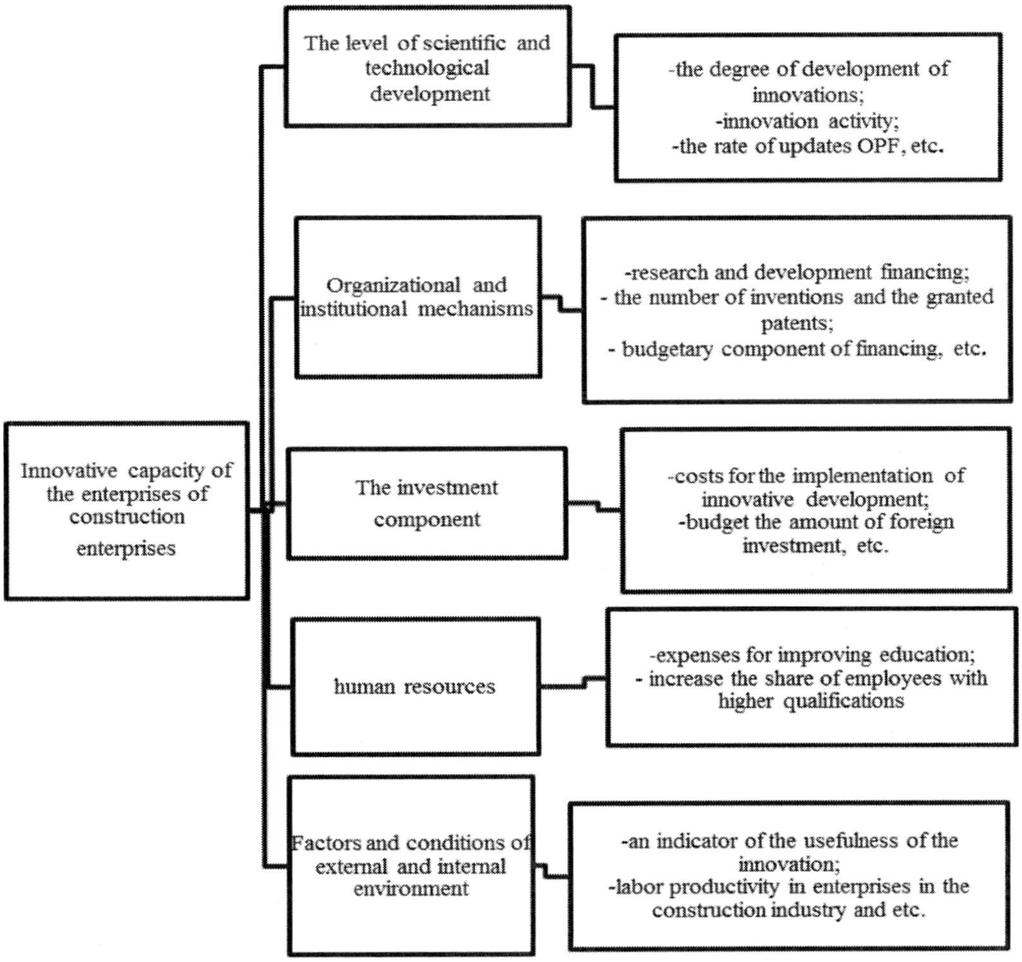

Fig. 1. The mechanism of formation of innovation potential of construction companies.

Based on the above, an important area of the innovation system in the industry production of building materials is to identify and encourage the development of regional cluster [24].

The role of the cluster in the development of innovation lies in the fact that within it formed stable relations between the participants of the production process, but also increases the proportion of competitive building materials produced by local construction companies, through the introduction of innovative methods in the organization of production (Fig. 2) [21,22].

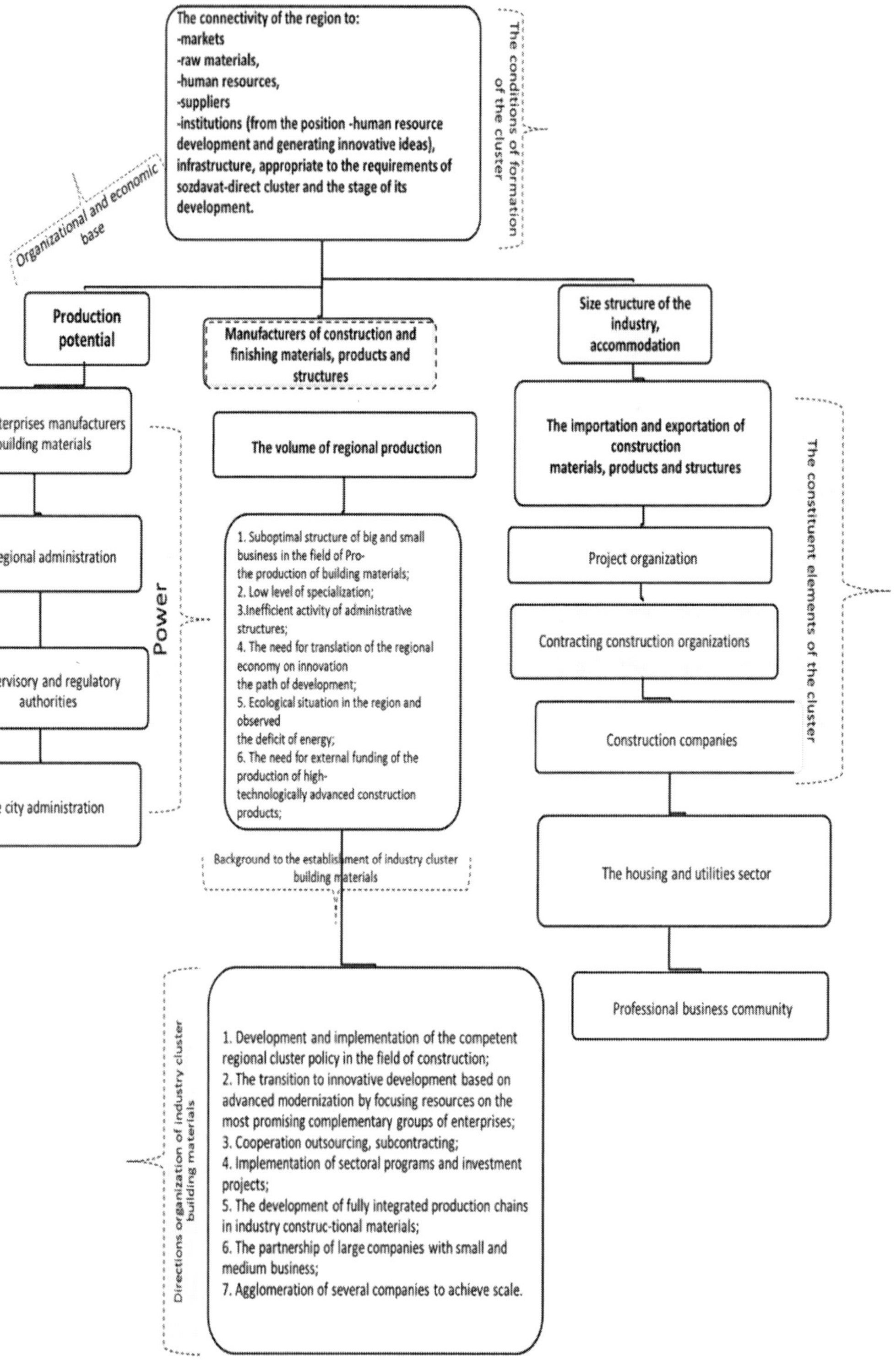

Fig. 2. The model of innovative development of building materials industry of the region based on the cluster approach.

The most important mechanism for the spread of innovation in the regional production clusters is a partnership of large companies with small and medium-sized businesses. The mechanisms of this partnership are outsourcing, subcontracting, implementation of joint programs, market research, training and other activities. Besides, the companies involved in the cluster, can benefit from the fact that businesses operate near the same sector or group of production, even if they do not interact with the competitors and in the development of new products. The benefit is to reduce transaction and production costs, lowering the barrier to exit the marketing and supply of raw materials markets. With the support of the government is gradually going evolutionary process of the consolidation of the manufacturers on the conditions of cooperation - to develop a common policy on procurement of raw materials for production, output at the federal regional product and foreign markets and the creation of a unified distribution structure [23].

From an analytical point of view, referring to the cluster as an already established phenomenon, they have to be considered outside the usual production activities.

On the one hand, an association of companies that is located at a relatively compact area, which in itself provides the appearance of a synergistic effect. Between them there is a functional dependence in the production and sale of goods and services, competition and cooperation. They are focused on the production of a specific group of goods and at the same time focus on the development and production of innovative products.

On the other hand, is a cluster of informal education, uniting on a voluntary basis, the major leading companies, medium-sized and small companies, research and educational organizations. The complexity of the implementation of the cluster policy is that it is located at the junction of the industrial policy, economic and regional, under the responsibility of different ministries and departments, as no arms control industry sectors and industries, as it was under the planned economy, no longer exists. Despite this, the government plans to use various measures to support small and medium-sized businesses to promote cluster initiatives in the regions. In turn, the cluster approach gives regional authorities the tools of effective interaction with business, a better understanding of its characteristic parameters and tactical tasks and enables a focused and motivated strategic planning.

3 Results

The proposed mechanism of the organization of innovative development of the industry of building materials in the region based on the cluster will create conditions for:

- Formation and realization of competent regional cluster policy in the field of construction;
- Transition to innovative development based on advanced modernization by concentrating resources on the most promising complementary groups of companies;
- Cooperation, outsourcing, subcontracting;
- The implementation of sectoral programs and investment projects;
- The development of fully integrated process chains in the field of building materials industry;
- Partnerships with large companies to small and medium-sized businesses;
- Agglomeration of several construction companies to achieve scale.

Besides:

- Given the high potential of the construction industry in many areas and regions of Russia, which have well-developed industrial base, modern manufacturing facility, significantly surpassing their own needs in the region (30% of the volume produced in the area of production is sold in other regions of the country), a rich mineral and raw material base, developed transport infrastructure - the proposed mechanism will generate the "points of growth" of innovative development of the industry production of building materials on the basis of the cluster;

- In terms of the number of research and training centers in the regions will create innovation cluster building materials industry with research and design centers, higher education institutions, as well as expand the number of participants in the cluster by the specialized innovative organizations - developers and technologist's production of innovative building materials. Also to introduce production technology and management innovations from other sectors of economic activity in the construction industry;

- Regional offices will be able to actively develop a new direction: the cooperation between small enterprises and major domestic manufacturers.

4 Discussions

Research shows that there is currently no uniform state policy on development of the industry of building materials, including innovation. Strategic approaches are presented only at the level of individual regions. In this regard, is it necessary to the formation of a strategic approach to the development of innovative building materials industry. Formation of a regional building cluster in a given situation will link industry with the development of the study of public policy priorities, as well as with those of other participants of the construction sector and with scientific basis.

However, analysis of the state of the industry in the context of the accession of the Russian Federation was held in the World Trade Organization has identified factors that may adversely affect the activities of enterprises and organizations of the industry of building materials production.

In the first place - is the lack of technical regulations and other regulatory documents governing the quality of products. In addition, there is no system of monitoring and state control of materials used in construction. And existing base of documents establishing the requirements to the characteristics of building materials and products requires a significant upgrade.

As a consequence, and also due to a lower level of regulatory requirements for quality construction materials produced in other countries, the regional construction market is projected delivery of large volumes of lower quality products, which will cause a decrease of quality of construction in the regions.

The arrival of the domestic construction market of imported technologies and materials will lead to a sharp decline in demand for products of Russian developers. The lack of market protection mechanisms and incomparably greater resources do not allow importing building materials, structures and locally produced products to compete with imported products.

Among sector problems include insufficient staff and management efficiency to ensure competitiveness, range of discrepancy, packaging, product certification requirements of world standards. And of course, the problem is the lack of available credit resources effective [19].

In addition, one of the reasons for lack of competitiveness of the industry is the lack of technological equipment with modern equipment such as industrial enterprises, as well as contractors. There are practically no production on the development and production of equipment for the creation of high-tech enterprises of the construction industry, modern means of transportation of materials [20, 24].

In this regard, today we need to be able to combine modern domestic design and research experience, and giving them official status, are clearly distinguished in separate directions in development. All this requires the regional authorities to carefully plan and carry out work on the development of innovative activity in the construction industry, which aims to create conditions for the transition to the current scenario of the entire regional economy. Since the need for a transition economy to an innovative way of development of the Russian Federation recognized as a major national task of the strategic nature of the Government, which will provide a solution to the translation of the economy to a higher technological way, and ultimately ensure its global competitiveness. For regions it is necessary to understand what industry, the formation of potential clusters and what they need to help them to become competitive on a global scale. This requires competent regional cluster policy.

5 Conclusion

It should be noted that in most regions of the development of the industry of building materials prevents the monopolization of regional construction markets, the organization of which has a local character. Competitive contracting companies provided their access to raw materials and production bases. As a rule, the main developers in the regions - a vertically integrated company, having in its structure a career, plants. A lack of competitive environment limits their motivation to optimize the cost of construction products.

For example, for the construction of 15-storey brick-monolithic houses in Russia requires a minimum of one and a half years, the complexity of the construction is not less than 500 thousand. Man-hr. This high cost. In China, there are examples where the building of 16 floors built in six days, time-consuming 27 thousand. Man-hours. Thus, virtually without using cement. This record delivered by advanced logistics, efficient organization of work at the site and the use of modern materials. The link between developers and manufacturers materials needs to be designers and architects, laying in the projects of modern materials and technology. To their use became widespread, we need projects that laid innovation [23].

The Russian construction market and construction materials industry is very unstable and very sensitive to external factors, market fluctuations. The main reason of instability - in the inefficient organization of the value chain, and generally ineffective structure of the cost of production and services. [23]

To the traditional building materials industry to become one of the links in the global supply chain and not to lose not only engineering, but also domestic production, it is necessary to create an alternative construction industry on the basis of modern building science. Joining together in a cluster, it can carry out not only the construction companies, but also metallurgical concerns, organizing metal service centers and production of the metal, and logistics companies, providing a regular supply of materials, and other industries. In addition, for the active development of the construction industry in the region, the modernization of old and creation of new enterprises, increase production capacity, enterprises snap high-tech equipment necessary to stimulate the sector, attracting investment into it by government policy, as well as the activities of professional societies. This entire means that the region will be able to provide not only the construction of any facilities, but also to produce diverse class housing.

References

1. I. Ansoff, *New corporate strategy: first* (Peter Combe, SPb, 1999)
2. P.F. Drucker, *Management Challenges in the XXI century* ("Williams" Publishing House, 2000)
3. J. Schumpeter, *The Theory of Economic Development* (Progress, Moscow, 1982)

4. B.Santo, B.V. Sazonov, *Innovation as a tool for economic development: first. with Hung* (Progress, Moscow, 1990)
5. B. Twiss, *Management of scientific and technical innovations: abbr. per. from English* (Economics, Moscow, 1989)
6. C. Ketels, G.Lindqvist, O. Solvell, Stockholm:Center for strategy competitiveness, **42**, (2006)
7. O. Solvell, *Cluster balancing evolutionary and constructive forces - Odeshog: Danagards Grafiska* (2009)
8. M. Porter, *Competition: Per. from English* ("Williams" Publishing House, M., 2000)
9. S. Goy, D.Finn, Energy Procedia, **78**, 3391-3396 (2015)
10. F. J. Kaltsonetti, D. M. Miller, N. Reed, Applied Geography, **34**, 265-273 (2012)
11. F. D. Doina, C. S. Badele, Social and Behavioral Sciences, **124**, 405-414 (2014)
12. M. Khairullina, Procedia, **16**, 88-94 (2014)
13. L.N. Asaul, Construction Economics, **6**, 16-25 (2004)
14. A.V. Voronin, Regional economy: theory and practice, **16**, 65-69 (2008)
15. I.M. Greenberg, Industrial and civil construction, **3**, 25-27 (2006)
16. M. Casas, *Industrial Economics of building materials and constructions* (Publishing Association building universities, Moscow, 2004)
17. S. Kolyada, Building materials, **7**, 4-8 (2008)
18. V.Y. Khmelev, Time, **1**, 22-26 (2011)
19. Mr. Bogachev, Growing power of the construction industry Industrial Construction Review, **145**, 49-52 (2012)
20. R. Mikhailov, Industrial and Construction Review, **138**, 3-5 (2011)
21. L.S. Belousova, *Sustainable development of a building complex in the region: monograph* (Publishing house SPSUEF, SPb, 2010)
22. M.I. School, *Organizational-economic mechanism of development of small pre-neurship in territorial investment-building complex: Abstract* (Cand. e. Sciences, Volgograd, 2008)
23. D. Sanatov, The era of traditional construction industry comes to an end, But the Industrial and Construction Review, **137**, 6-8 (2011)
24. A.B. Mottaeva, The methodology of the spatial distribution of business organizations in the region based on the development of transport infrastructure (2012)
25. S.V. Koshcheev, Herald of the Sochi State University for Tourism and Recreation, **1(15)**, 100-106 (2011)
26. I. Kuznetsov, Transport Russian business, **12(73)**, 62-67 (2009)

The Question of Construction Cluster Efficiency in Russian Regions

Aleksandr Voronin[1,*], *Aleksandr* Kuvshinov[1], and *Miroslava* Gusarova[2]

[1]Tyumen State University, 625003 Semakova str. 10, Tyumen, Russia
[2]Industrial University of Tyumen, 625001 Volodarskogo str. 38, Tyumen, Russia

Abstract. This article is devoted to the questions of building complex clasterization in Russian regions. The correlation analysis between the existence of the formed building cluster in each region and the volume of construction works fulfilled during the period under review is being conducted on the basis of data about the development of Russian regions generalized by Federal state statistics service. The authors examine the volumes of construction works in leading regions from the position of clusterization features, describing and analyzing key efficiency problems of construction. The authors draw a conclusion that declared cluster creation is not always based on factual building complex clusterization. As result, current situation in Russian economy seems to be problematic: regions without legally formed clusters are, in fact, more clusterized then those with announced formation.

1 Introduction

Regional building complex is a certain basis for social-economic changes at all levels of economic development. It should be noted, that the significance of building complex is great for both social-economic development and formation of positive environment for favorable living conditions. This makes it necessary to study regional building complex structure and economic aspects of investment and construction activities regulation [15].

Undoubtedly, the volume of construction works affects Gross Regional Product (Graph 1 – the summary volume of construction works in Russia), it is confirmed by the analysis of statistic data [12] about Russian regions development during the period of 2010-2013.

* Corresponding author: vrnnaea@mail.ru

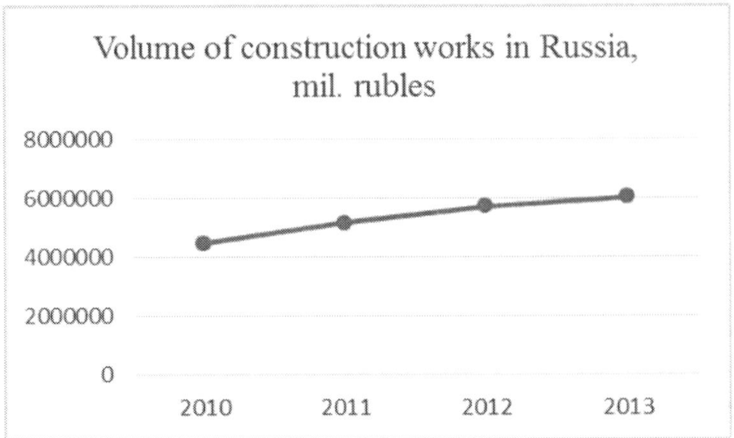

Fig. 1. The summary volume of construction works in Russia

It should be taken into consideration that the GRP dynamics` graph extensively follows the graph of construction works volume dynamics` and on this basis it is possible to make a conclusion about their correlation (Fig. 2).

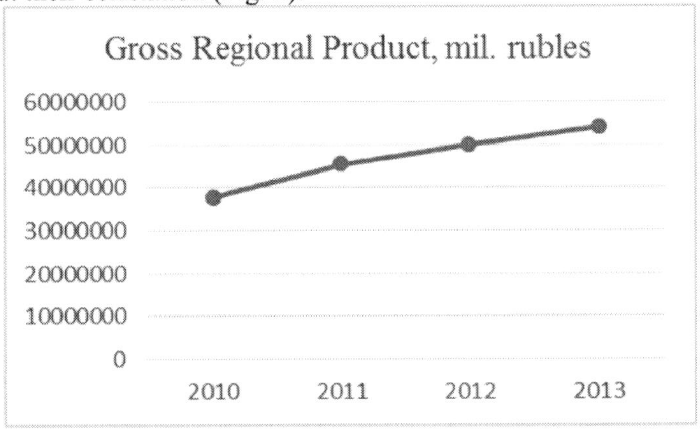

Fig. 2. Gross regional product

The graph of Gross Regional Product looks similarly and presents more precise statistic index (Fig. 3).

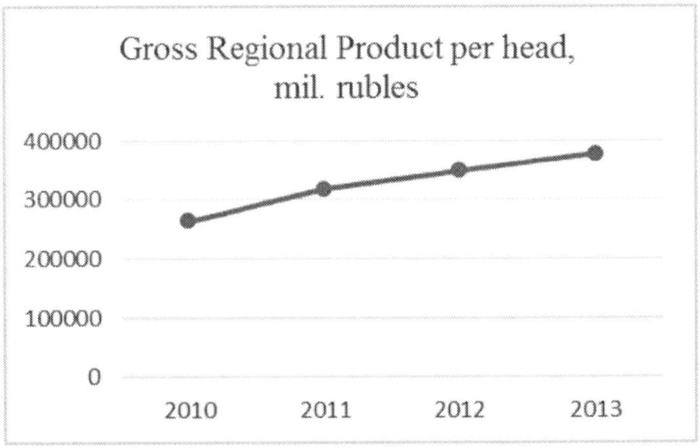

Fig. 3. Gross regional product per head

In that way, increase of building complex efficiency in Russian regions is especially important for basic assets extension, related industries development and general stimulation of economic growth.

As Garkina I.A. and Garkin I.N. noted [17], nowadays, most of regional departments and ministries in the construction industry have become no more than construction project requesters, which are , in turn, financed from the budget. Self-regulated unions of builders, designers and engineers work independently while construction industry, infrastructure, construction science and professional education have been left out of integration process. The lack of precise interaction between organizations of construction brunch becomes an obvious problem. So, it`s necessary to create construction clusters to form the foolproof collaboration mechanism. It should be mentioned, that cluster implies a concentrated group of interconnected organizations (commercial, educational, social etc.), which supplement each other and intensify competitive advantage of separate companies and cluster in general [12].

2 Theory or experimental methods

The object of study are the building complexes in Russian regions, analyzed from the position of cluster approach.

«In this case, interpretation of construction cluster regarding the content is close to the concept of "investment-building complex". Most theoretical and practical publications authors widely presented in economic literature, agreed, that investment-building complex is a number of organizations, not only practicing construction , reconstruction and enlargement of functioning industrial and non-industrial building and constructions, but also providing the fulfilling of these works with material and technical resources and transport services, carry out theoretical and practical works, designing development activities and preparing the staff» [19].

It should be noted, that economic cluster in Russian economy is being interpreted from positions of wide approach, because the category "cluster" is often used as a synonym for categories "sector", "complex" and "brunch" [5].

As Matveeva O.A. noted [8], the social effect of clusterization in construction cluster appears to be an increasing level of habitation availability. The statistic data shows that joining of companies within a cluster resolves this problem.

In regions, where construction cluster is being formed, residential housing volumes are considerably bigger. The presence of providing element (banks etc.) in cluster, legal state regulation and cluster system stability allow to increase this index by providing citizens with available hypothecary credits with guarantied job entry terms.

3 Results

The analysis shows that the following regions are the leaders in construction works volume in 2013: leading by ahead, Moscow with 674276.8 mil. rubles, Tyumenskaya oblast (506941,6 mil. rubles, where 229549.6 mil. rubles are provided by Khanty- Mansi Autonomous Okrug — Yugra), Krasnodarskiy kray (480663.6 mil. rubles), Saint-Petersburg (375803.6 mil. rubles), Tatarstan Republic(297459,9 mil. rubles) , Moskovskaya oblast (289351,4 mil. rubles). It should be noted, that every region from this list works on the basis of the cluster conception in construction or already has a developed construction cluster.

The Moscow innovation industrial-building cluster has been functioning since 2015 in Eastern administrative district, but specific weight of industrial brunch organizations was accumulated during a long period of time and we can now talk about the functioning of the cluster in potential form, consolidating 16 companies [9].

The statistics of 2013 shows that the reconstruction of 56 investment objects (with value over 100 mil. rubles) had been finished in Krasnodarskiy kray [11]. It`s extensively caused by a global preparation for Sochi Olympic Games 2014, within which the cluster approach in construction was widely used and consolidated investors, builders, contracting organizations, responsible for infrastructure objects' development, state authorities and banks etc. [16].

It is also possible to establish the natural (Tyumenskaya oblast, Moskovskaya oblast) or purposeful (Saint-Petersburg, Tatarstan) clusterization in other regions as well.

But the experience of regions, which aren`t leaders in construction works volume, but already have a formed construction cluster, is more interesting for regional economy in context of building complex clusterization efficiency evaluation. Nowadays, there are active construction clusters in Voronezhskaya oblast, Kaluzhskaya oblast, Novosibirskaya oblast, Sverdlovskaya oblast, Nizhegorodskaya oblast, Samarskaya oblast, Belgorodskaya oblast, Vologodskaya oblast and Permskiy kray.

These regions are included into the Top-25 of 83 Russian regions (2013, before constitution of Crimea Republic and federal city Sevastopol) regarding the volume of construction works except Vologodskaya oblast (35 place) and Kaluzhskaya oblast (39 place) in 2013. However, the comparison of Russian regions using other criteria is necessary for more detailed analysis.

The key factor, providing the construction works volume, is investment. The analysis of investment in building complex in 2013 clarifies the leadership of Moscow (575129.3 mil. rubles), Saint-Petersburg (261807.1mil. rubles), Moskovskaya oblast(347298.2 mil. rubles) and Tyumenskaya oblast(828802.3mil. rubles), Tatarstan (332261.4 mil. rubles) and Krasnodarskiy kray (727868.5 mil. rubles) are under the average level of investment (within Russia - 9560.3 mil. Rubles).

The investment in construction complex is comparatively lower in the regions with formed clusters: Sverdlovskaya oblast – 164811.8 mil. rubles, Nizhegorodskaya oblast - 144374,4 mil. rubles, Samarskaya oblast 139454 mil. rubles, Voronezhskayaoblast – 126935.1 mil. rubles, Permskiy krai – 103601.2 mil. rubles, Novosibirskaya oblast – 97150.9 mil. rubles, Belgorodskaya oblast – 82560.4 mil. rubles, Saratovskaya oblast – 60248.6 mil. rubles, Kaluzhskaya oblast – 59340.8 mil. rubles, Vologodskaya oblast – 50456.6 mil. rubles.

It should be mentioned that organizational factors, such as work organization level in construction, builders` qualification, business process mechanization, level of technical equipment, new building technologies integration, supply chain service etc. are very important. Provided that the other things are equal, these factors will guarantee the competitive advantage. The clusterization of building complex in Russian regions is aimed at development of these factors.

So, the regions with active construction clusters show more progressive results under comparable investment (see table 1).The remaining behind - Vologodskaya oblast 50456.6 mil. rubles of investment and construction works volume of 48862.7 mil. rubles) demonstrates a big separation from the following Tverskaya oblast (investment in construction – 50254.4 mil. rubles) with 16457.2 mil. rubles. In that way, it`s generally possible to speak about greater investment efficiency in regions with institutionalized construction clusters.

Table 1. Compare of construction works volume depending on investment in Russian regions with and without clusters [million rubles]

Region (with cluster)	Construction works volume in 2013	Investment building complex in 2013	Comparable region without cluster	Construction works volume in 2013	Investment building complex in 2013
Sverdlovskaya oblast	107171.5	164811.8	Tyumenskaya oblast (without Autonomous Okrugs)	135564.4	160502.2
Nizhegorodskaya oblast	142638/3	144374.4	Rostovskaya oblsast	126462.5	141268
Samarskaya oblast	139333	139454			
Voronezhskaya oblast	68720	126935.1	Sakha (Yakutia) Republic	65690.5	128974.8
Permskiykray	85927.1	103601.2	Irkutskaya oblast	94617	106233.5
Novosibirskay aoblast	69909.2	97150.9	Chelyabinskay aoblast	78704.3	94799.1
Saratosvskaya oblast	59730.6	60248.6	Volgogradskaya oblast	58597.4	76670.6
Belgorodskay aoblast	71282.4	82560.4	Primorskiy kray	62685.6	82450.9
Kaluzhskaya oblast	42436.3	59340.8	Lipetskaya oblast	34266.8	64396.2
Vologodskaya oblast	48862.7	50456.6	Tverskaya oblast	16457.2	50254.5

The analysis of introduced data clearly shows that some regions with an active construction cluster are more effective in implementation of investment in building complex. Comparison of these regions is interesting and purposed to discover cluster advantages and disadvantages in concerned regions of Russian Federation.

As U.G. Lavrikova and S.N. Kotlyarova [7] noted, that every cluster was created with different aims according to the region needs. The Ural construction cluster in Sverdlovskaya oblast declares an increase of regional building complex efficiency and competitiveness, saving of market positions and widening of market to be the general goal. The aim of Saratov interregional research-and-production cluster is formation of machine-building complex for innovation materials integration in construction in order to decrease the price of a habitation square metre, in other words, the increase of construction works volume wasn`t even presumed as a main goal. The Volgograd international cluster focuses on foreign progressive technologies attraction in order to develop the new brunch – timber house-building. The success of such an achievement directly influence construction works volume as well, so it is clear to see within the introduced statistic data.

Let`s separately examine problems of least successful regions in 2013 with active construction clusters.

The Ural construction cluster began its activity in 2013. As U.N. Chumerin [18], the director of noncommercial partnership "Ural construction cluster" noted, that the actual problem of Sverdlovskaya oblast is the fact, that the biggest building and infrastructural projects, such as 2018 FIFA World Cup, inevitably attract engineering companies from Moscow and Saint-Petersburg, which constantly win the tenders,so it`s very difficult for companies from Sverdlovskaya oblast to compete with them. Exactly this problem caused the necessity of cluster creation in order to consolidate the efforts of the local building organizations to fight for local projects.

Nowadays, the activity of the cluster pays off [10].The cluster consolidates 167 organizations, 15 mil. rubles are put into operations. In that way, this cluster provides more than 10% of the whole construction works volume of Sverdlovskaya oblast.

The problem of Novosibirsk construction cluster (common for many Russian clusters) is the fact, that it was institutionalized, but actually this structure doesn`t correspond to cluster concept by Michael Porter. The reason is insufficient degree of clusters understanding in Russia, connected to aspiration for their presence declaration as soon as possible. As A. Nizkovskikh [4], the director of State government institution of Novosibirskaya oblast "The center of regional development" noted, this center`s aim for the next few years is an annual formation of 2-3 clusters. As a result, the cluster creation happens for the government`s sake, but the degree of cluster maturity (accumulation of companies, their interdependence, active collaboration with scientific institutions etc.) isn`t taken into account.

There is the same problem in Permskiy kray . The building complex is concentrated in some degree, but it would be incorrect to speak about the presence of the formed cluster. V. Fedorovskiy, the minister of building , housing and communal services of Permskiy kray ,describes a cluster as 4873 building companies and institutions [2]. Accordingly, the regional government makes an attempt to clusterize the huge building complex at the same time instead of gradual attraction of companies to a big thematic organisation (or several of them). As a result, it`s rather difficult to organize a system of effective intercommunications between companies fast. However, the positive dynamics of cluster development is seen already today, especially in habitation building. Permskiy kray enters more then1 Mil. Square metres of habitation for the second year running. The annual temps of habitation building growth make up to 10%. It should be mentioned, that the region couldn`t achieve the index of 1 Mil. Square metres for more then years. Lately, Permskiy kray entered over 1 Mil. Square metres – 1 269 000 – in 1990 (former Permskaya oblast).

Fast growth of construction works has been showed since 2012 – 9,9%, from 76 933,6 up to 84 523,6 Mil. rubles in actual prices.

The cluster "Industry of building materials" in Voronezhskaya oblast is also insufficiently worked over, that was confirmed by its failure within the contest of Ministry of economic development getting some state support, where the independent expert commission sifted it out at the first stage, because one of the most important criteria was collegiate style. This cluster was examined as a project of "Domostroitelniy combinat" corporation with the calling of other companies.

It should be mentioned, that requests of at least several legal entities with government`s support and technical basis were supported within this competition. Although it was often clear under more precise studying, that companies doesn`t trust each other or universities – colleagues within the cluster. Their common projects were not worked over from the point of logistics and scientific solutions, and it`s almost impossible to handle the common deal without confidence and common priorities. In other words, participants have to clear the common interests and expectations, but not only a way to share possible subventions from the state exchequer [3].

In that way, the incorrect approach to cluster`s organization is the main problem of "weak" construction cluster in Russia. First, precise understanding of the cluster conception in the classic way, which was developed by M. Porter[13], is needed, while separate brunch of regional economy is being called "cluster" in many Russian regions. Moreover, cluster creation should not happen for political fashion`s sake, because a declaration about cluster creation doesn`t mean the organization of all conditions for its effective functioning. Actually, some clusters are just legal unions of separate companies without corresponding organizational content [1, 6].

4 Conclusion

The regions with formed clusters generally don`t issue the challenge of leadership in the country, but they examine clusters as a possibility to overcome their unsatisfactory position. The leader-regions are less motivated in cluster creation, because they have already satisfied their needs. On the other hand, the leader-regions anyway have collected specific weight of intercommunicating organizations of building complex and attendant economic sectors. In addition, on the contrary, there are some regions, which declare creation of a full cluster, but actually, these clusters can just be unions of organizations with poorly organized system of collaboration and intercommunication.

As result, factual clusterization degree in some regions without legal formed clusters is higher, then in regions, where cluster creation was declared. In that way, creation of conditions for cluster functioning and motivation of its potential participants must become the most important task for regional government. As international experience shows, the most powerful clusters appear to be most natural and the initiative of business, science or government is needed only for its formal fastening. Moreover, cluster creating "from above" is often connected with the absence of conditions for its effective functioning.

References

1. V. L. Abashkin, A. D. Bojarov, E. P. Kutsenko, Forsajt, **3**, 16-27 (2012)
2. Bol'she metrov, Ezhenedel'naja delovaja gazeta «Permskij obozrevatel'», **31**, 5 (2015)
3. Information on: http://36on.ru/magazine/province/5613-klastery-voronezhskoy-oblasti-provalili-konkurs-minekonomrazvitiya
4. Information on: http://infopro54.ru/news/4423/
5. O.V. Kostenko, Innovacionnaja nauka, **9**, 166 (2015)
6. E. P. Kutsenko, Forsajt, **3**, 6-15 (2012)
7. Ju.G. Lavrikova, P.N. Kotljarova, Jekonomicheskie i social'nye peremeny: fakty, tendencii, prognoz, **3(33)**, 173 (2014)
8. O.A. Matveeva, Serija «Jekonomika i jekologicheskij menedzhment», **1**, 3 (2014)
9. Information on: http://www.fondvao.ru/580.html
10. Information on http://www.spsi-sro.ru/?id=649
11. Information on: http://economy.krasnodar.ru/strategic-planning/the-program-of-development-of-krasnodar-region/files/otchet_ser_2013.pdf
12. M.E. Porter, Harvard Business Review November-December, 78-79 (1998)
13. M.E. Porter, Economic development quarterly, **14(1)**, 15-34 (2000)
14. Information on: http://www.gkp.ru/bgd/regl/b15_14p/Main.htm
15. Information on: http://uecp.ru/uecs-27-272010/item/359-2011-03-25-13-21-41
16. Information on: http://expert.ru/expert/2014/07/nauchilis-delat-megaproektyi/
17. Ju. V. Smirnova, Aktual'nye voprosy jekonomiki i upravlenija: materialy mezhdunar. nauch. Konf, 42–45 (2011)
18. Information on: http://cluster.hse.ru/news/141/
19. B.P. Zhiharevich, Opredelenie perspektivnoj specializacii hozjajstvennogo kompleksa goroda (ocenka profil'nosti), po materialam seminara «Osnovy territorial'nogo strategicheskogo planirovanija na principah shirokogo obshhestvennogo uchastija» – Sankt-Peterburg, 13 (2004)

MATEC Web of Conferences **73**, 07028 (2016) DOI: 10.1051/matecconf/20167307028
TPACEE-2016

Technology Audit: Assessment of Innovative Portfolio

Viktoria Kurushina[1] , *Elena* Kurushina [1,*] and *Alexander* Shabarov[2]

[1]Industrial University of Tyumen, 625001 Volodarskogo str. 38, Tyumen, Russia
[2]Tyumen State University, 625000 Semakova str. 10, Tyumen, Russia

Abstract. The article discusses the features of the technological audit performing in the companies of oil and gas sector of Russian economy. To measure the innovations quality level the scale was developed based on the Theory of Inventive Problem Solving and the theory of technological structures. Figures of the innovations quantity by levels, volume and quality of the innovative portfolio are offered for assessment the innovative portfolio quality. The method was tested on an example of oil and gas transporting enterprises. The results of the comparative analysis of innovative portfolio are shown.

1 Introduction

Modern trends in the world energy consumption growth [1], as well as cooperation developing between China and Russia in this area require accelerated introduction of innovative technologies in the oil and gas transportation industry. The innovation development pace depends not only on the duration of the innovation cycle, but also on the quality of implemented innovations. For successful innovative management in the world technology audit procedures are applied. Technological audit (hereinafter referred to as TA) in accordance with the procedure of the IRE (Innovative Regions of Europe) methodology is conducted in order to "evaluate the ability of the company / organization to integrate new technologies and to work with technology partners, as well as to create an understanding of what is necessary in order to integration or, on the contrary, the technology transfer were carried out most successful" [2, p. 9]. Following the procedure of OSEO anvar (National Innovation Agency, which coordinates the work of the French Innovation Relay Centres – IRC) during the technological audit process the company innovativeness as well as its ability to innovate is assessed. Also, analysis of a specific innovation project can be carried out [same, p. 10]. The data array formation and the choice of assessment criteria depend on the audit objectives.

In Russian practice technology audit is focused on the company competitiveness assessment in the area of innovation. As noted in the Methodological materials on the creating of innovative development programs (hereinafter referred to as IDP), TA is "comprehensive and documented analysis containing an adequate assessment of the existing technological level of the company in comparison with comparable companies in Russia and abroad" [3, p. 4].

* Corresponding author: kurushina.tsogu@yandex.ru

2 Research object (model, process, unit, synthesis, experimental procedure, etc.)

The research object is the oil and gas sector enterprises of Russian economy. The research tasks include:

1. Investigation of the features of technological audit procedures in the Russian oil and gas companies sector.
2. The development of scale for innovation quality assessment.
3. Assessment of the innovation quality level on example of enterprises of oil and gas transportation sector.
4. Formation of an innovative methodology for corporate portfolio assessment.
5. Comparative analysis of the innovative portfolio.

3 Research methods

As the main research method the content analysis of methodological materials and innovative development programs was used. Development of the innovations quality scale was made based on the analogies method, and assessment was carried out according to comparison method. Approbation of techniques was carried out using computational and analytical methods as well as comparative analysis methods.

4 Results and Discussion

Carrying out of TA in accordance with IDP of OJC "Gazprom" provides: 1) choice of foreign companies for comparison; 2) analysis of open information sources; 3) analysis of foreign companies figures that are suitable for assessment of the technological level; 4) the figures calculation at OJC "Gazprom" and those at foreign companies; 5) technological level comparison; 6) the development of KPI (key performance indicators) list for the technological level [4, p. 25]. In the ranking of technological level figures of OJC "Gazprom" take in comparison with foreign companies:

- 2nd place in terms of reliability and safety;
- 5th place in terms of energy efficiency and perspective technologies;
- 7th place in terms of ecological compatibility;
- 10th place in terms of investment in research and development.

The rating data make it possible to assess the technological leadership relative to its competitors in which quality the largest by market capitalization energy companies are used such as Exxon Mobil, BP, Shell, Statoil, PetroChina and other. There are total 19 companies. Accordingly, changes in introduction pace of innovations (e.g. their slowing) in the industry have an impact on the objectivity of innovation development assessments.

It should be noted that the competition level in the industry is constrained by a high degree of the capital concentration as well as the continuing mergers and acquisitions processes. Accordingly, the motivation of accelerated innovations introduction to strengthen the competitive position declines. The technology promise is estimated at OJC "Gazprom" not by the depth of innovative transformation but by the susceptibility and development degree of new technologies. The elaboration degree is assessed by experts.

The structure of key performance indicators (KPI), assessing the technological leadership, includes quantitative and qualitative indicators in Methodological materials on the development of IDP [3]. Programs of Russian oil and gas transporting corporations provide only quantitative figures. Following figures are included in the list KPI of technological level of OJSC "Gazprom": 1) the quantity of patents and licenses obtained for the calculation period and the last 2 years, 2) the quantity of patented technologies developed within the framework of the IDP for the calculation period and the last 2 years [4]. The figure of technological leadership of OJC "AK" Transneft "," is assessed only on basis of patents quantity assigned to the company's balance [5]. The recommended by Methodological materials quality portfolio estimation based on a balance between the breakthrough projects and improving projects in IDP in the survey, is not performed.

Taking into account the given above, in this article it is offered to assess additionally the individual innovation quality and innovative portfolio as a whole on the basis of objective evaluations to enhance the validity of the TA conclusions about promising technologies and projects. As the methodological basis for measurement of the quality innovation level modern theories can serve, such as the Theory of Inventive Problem Solving of G. Altshuller and the theory of technological structures of S. Glazyev. To assess the quality of certain innovations the appropriate scale was developed by the authors of the article, providing the levels differentiation according to the inventions characteristics [6, 7]. As seen in the Table 1, the criteria for the quality level estimation of innovation on this scale, are: 1) the inventions level (by G. Alshuller), 2) the degree in changes technology, 3) the depth of the system changes, 4), the origin and the innovations incidence.

Table 1. Scale of innovation quality assessment.

Criterion	**Innovations quality level**				
	1	**2**	**3**	**4**	**5**
Level of inventions	**Easiest invention**	**Easy invention**	**Average invention**	**Major invention**	**Largest inventions**
Changes degree in technology	Small improvemen t in technology	Moderate improvement in technology	Major step in the development of technology	Almost fundament al one	The fundament al one
Depth of the system changes	Change in the element	Change in the element	Partial change of other elements	System changing	Creation of essentially new system
Origin and the innovations incidence	In the present system	In related systems	Within the branch of science	Outside of the branch of science	The emergence of a new branch of technology

Inventions assessment of G. Altshuller allowed to obtain the data that 32% of inventions correspond to the first (lowest) level, 45% of those – to the second one, 19% – to the third one, less than 4% – to the fourth one and 0.3% – to the fifth inventions level [6, p. 68]. The inventions of the first three levels make 96%. Correlation of this scale with the theory of technological structures of Glazyev shows that the first 3 levels are improving (or evolutionary) rather than revolutionary innovations. Therefore, it is not enough to measure the technological leadership by the patents quantity. On the developed scale the quality level estimation of innovation was performed provided by IDP of Russian oil and gas transporting corporations. Since the IDP data sheet of OJC "AK" Transneft" does not specify exactly what the company understands under the future technologies, "nanotechnologies in oil pipeline transport" as well as "systems and technical means of oil pipelines monitoring," the assessment for quality level estimation of these technologies has not been made. The estimation results are shown in Table 2.

Of the fourteen being assessed technologies of OJC "AK" Transneft" 2 of them can be referred to the revolutionary, and 12 – to the evolutionary ones. Of the five main pipeline transport technologies represented in IDP of OJSC "Gazprom" 2 are assessed as revolutionary and 3 as evolutionary ones. The research had not taken into account the company innovations in the hydrocarbon production and storage areas etc.

Table 2. Quality level estimation of innovations provided by IDP in the oil and gas transportation sector of the economy.

Qualitative level of	Main oil transportation *	Main gas transportation **
	Evolutionary technologies	
1	Creation of systems for productivity increasing of pumping by drag hydraulic reduction in the main oil and oil product pipelines	-
2	- The development of highly reliable pump units with an increased efficiency factor. - Development of an integrated oil-heating system with nominal capacity of 50 MW as well as integrated heating and unloading systems of oil and fuel oil from railway tanks. - Technologies for production and use of non-contact bearings using permanent magnets or electromagnetic bearings for the electric motors of main and booster pump units of various capacities. - Development of monitoring system for technical condition of pipeline. - Development of technical solutions for the manufacture and installation of the pupping stations facilities elements as well as oil pipelines when operation in permafrost conditions. - Energy saving technologies.	- Technologies of highly efficient gas compression. - The technology of energy generation through its use.
3	- Development of a unified management system (UMS) for main oil pipeline. - Creation of highly reliable import-substituting equipment. - Creation of technology, equipment and facilities for the Polar region, offshore and marine zone. - Development of variable frequency electric motor for pumping units. - New composite materials for repair performing without cutting by method of composite-coupling technology.	Technologies of energy generating due to the use of low power generating units (1-10 kW), based on alternative and renewable energy sources for electric power supply of linear part consumers of main gas pipelines
	Revolutionary technologies	
4	- Development of the complex of the high-precision in-line diagnostic devices. - Development of the leak detection system and activity control for temperature and vibroacoustic operating principle.	- Gas transportation technologies in liquid and multiple-phase state - Construction and operation technologies of high-pressure pipelines.
5	-	-

* Compiled on the basis of the IDP data sheet [5]
** Compiled on the basis of the IDP data sheet [4]

For the most complete assessment of the corporation innovation portfolio it is advisable to perform calculation on the parameters system given below.

1. The innovations quantity in the portfolio (Nп), including by levels one (Ni).

The innovations quntity presented in the IDP data sheet of OJSC "AK" Transneft", makes 14 innovations, including that of the first level – 1 innovation; of the second one – 6; of the third – 5 and of the fourth – 2 innovations. In OJC "Gazprom" 2 of 5 innovations on the main transportation presented in the IDP belong to the second level; 1 of them – to the third one and 2 innovations – to the fourth level.

2. The volume of the innovative portfolio (Oip):

$$O_{ip} = \sum_{i=1}^{5} N_i \cdot I, \tag{1}$$

where I – innovation significance coefficient of the ith level; Ni – the innovations quantity of ith level.

The level number can be used as the significance factor. Distribution of "weights" will be, respectively, 0.067 (1st level), 0.133 (2nd level), 0.200 (3rd level), 0.267 (4th level), 0.333 (5th level). The innovations volume coefficient at OJSC "AK" Transneft", calculated according to Table. 2 data, is 2.4 c. i. u. (conventional innovations units), given to the first quality level. At OJSC "Gazprom" this figure is only 1 c. i. u. for gas transportation innovative portfolio.

3. The quality of innovative portfolio K_{un}:

$$K_{un} = \frac{N_r}{N_e} \cdot 100 \tag{2}$$

where Nr – quantity of revolutionary innovations (those of the 4th and 5th levels); Ne – quantity of evolutionary innovations (those of the 1st, 2nd and 3rd level).

According to the performed above calculations (2) the innovative portfolio quality of oil transportation corporation is equal to 16.7%, and that of gas transportation one – 66.7%. The results of the comparative analysis of innovation portfolio are shown in Table 3.

Table 3. Comparative analysis of innovative portfolio of IDP of Russian oil and gas transportation sector.

Index	Unit of measure	Main oil transportation *	Main gas transportation **
1. Total number of innovations including:	qty.	14	5
1st level	qty.	1	-
2nd level	qty.	6	2
3rd level	qty.	5	1
4th level	qty.	2	2
5th level	qty.	-	-
2. The volume of innovative portfolio	c. i. u.	2.4	1.0
3. The quality of innovative portfolio	%	16.7	66.7

The obtained estimations indicate a higher quality of innovative portfolio of OJC "Gazprom". Assessment of innovation development based on introduction of the revolutionary innovations allows the identifying the patterns development of technical systems. The results of the conducted researches on the example of Russian oil pipeline system containing qualitatively new conclusions were reflected in the works [8, 9, 10].

5 Conclusion

Assessment of the innovative portfolio quality can be used as an additional criterion for the selection of promising technologies in the creation of corporation innovation development programs. Taking into account the degree of technology innovation [11-21] will accelerate the innovation development.

References

1. Yu. Zemenkov, E. Kurushina, V. Kurushina, Oil and Gas Terminal, **3**, (2009) 32-34.
2. G. Pilnov, O. Tarasova, A. Yanovsky, *How to conduct the technological audit, Project EuropeAid «Science and Technology Commercialisation"* (2006)
3. Information on: http://www.docs.cntd.ru / document / 902 306 418 .
4. Information on: http://www.gazprom.ru/f/posts/97/653302/programma-razvitia.pdf.
5. Infromation on: http://www.dalmn.transneft.ru/u/section_file/7181/pasport.pdf .
6. G. Altshuller, *Find an Idea: Introduction to Theory of Inventive Problem Solving* (2013)
7. A. Kvashnin, *How to manage a technologies portfolio and intellectual property, Project EuropeAid «Science and Technology Commercialisation"* (2006)
8. Yu. Zemenkov, V. Kurushina, Mining informational and analytical bulletin (scientific and technical journal), **3**, 85 (2013)
9. V. Kurushina, Yu. Zemenkov, WIT Transactions on Ecology and the Environment, **190(2)**, 881-888 (2014)
10. E. Kurushina, V. Kurushina, Development of strategic corporation management and regional innovation policy: Proceedings of the conference, PSTU, Perm, 163-168 (2013)
11. *A technique by definition of damage to surrounding environment at accidents on the main oil pipelines. Ministry of Fuel and Energy* (Transpress, Moscow, 1996)
12. R. Mamadaliev, V. Kuskov, Yu. Zemenkov, A. Popova, Applied Mechanics and Materials, **770**, 19-22 (2015)
13. V. Antip'ev, A. Nevolin, Yu. Zemenkov, Neftyanoe Khozyaistvo - Oil Industry, **10**, 46-48 (1981)
14. A. Shipovalov, Yu. Zemenkov, S. Toropov, M. Zemenkova, S. Podorozhnikov, I. Tyrylgin, V. Pavlov, *Aspects of technological reliability and economic efficiency of operation of underground storages of natural gas of Western Siberia* (TSOGU, Tyumen, 2012)
15. Yu. Zemenkov, V. Shalay, M. Zemenkova, Procedia Engineering, **113**, 254-258 (2015)
16. Yu. Zemenkov, V. Shalay, M. Zemenkova, Procedia Engineering, **113**, 312-315 (2015)
17. E. Kurushina, V. Kurushina, Life Science Journal, **11(11)**, 517-521 (2014)
18. V. Kurushina, Yu. Zemenkov, E. Kurushina, Fundamental researches, **2-8**, 1632-1636 (2015)
19. E. Kurushina, Canadian Journal of Science, Education and Culture, **2(6)**, 378-384, (2014)
20. E. Kurushina, *Integration of power policy: management of energy efficiency and power safety, Problems of formation of a common economic space and social development in the CIS countries: Proceedings of the International scientific and practical conference*, (TSOGU, Tyumen, 2011)
21. A. Pimnev, M. Zemenkova, Pipeline transport: theory and practice, **5(51)**, 43 (2015)
22. M. Zemenkova, Yu. Zemenkov, A. Pimnev, V. Petryakov, Business magazine Neftegaz, **11-12**, 64-70 (2015)

AUTHOR INDEX

AUTHOR INDEX

AUTHOR INDEX

AUTHOR INDEX